WITHDRAWN
FROM
UNIVERSITY OF PLYMOUTH
LIBRARY SERVICES

Electrospray and MALDI Mass Spectrometry

Electrospray and MALDI Mass Spectrometry

Fundamentals, Instrumentation, Practicalities, and Biological Applications

Second Edition

Edited by
Richard B. Cole

A JOHN WILEY & SONS, INC., PUBLICATION

Published by John Wiley & Sons, Inc., Hoboken, New Jersey.
Published simultaneously in Canada.

For general information on our other products and services or for technical support, please contact our Custom Care Department within the United States at (800) 762-2974, outside the United States at (317) 572-3993 or fax (317) 572-4002.

Wiley also publishes its books in a variety of electronic formats. Some content that appears in print may not be available in electronic formats. For more information about Wiley products, visit our web site at www.wiley.com.

Library of Congress Cataloging-in-Publication Data:

Electrospray and MALDI mass spectrometry : fundamentals, instrumentation, practicalities, and biological applications / edited by Richard B. Cole. – 2nd ed.
p. cm.
Rev. ed. of: Electrospray ionization mass spectrometry. c1997.
Includes index.
ISBN 978-0-471-74107-7 (cloth)
1. Electrospray ionization mass spectrometry. 2. Biomolecules–Analysis. I. Cole, Richard B., 1956- II. Electrospray ionization mass spectrometry.
QP519.9.M3E44 2010
572′.36–dc22
2009035898

Printed in the United States of America

10 9 8 7 6 5 4 3 2 1

Contents

Foreword: Desorption Ionization and Spray Ionization: Connections and Progress

Mass spectrometry has the advantage of a broad definition. Only in a tangential way is it a form of spectroscopy. Rather, it is the field dealing with the study of a particular state of matter, the gaseous ionic state. This means that much chemistry and even physics and biology falls within its province. Practical applications come with this, driven by advances in instrumentation and techniques; but less obviously, so much fundamental science—kinetics and thermochemistry and reaction mechanisms—also emerges. Nowhere is this range more evident than in the methods and phenomena involved in the conversion of samples into gas-phase ions.

This Foreword starts with a consideration of the driving forces that have led to the elaboration of improved methods of ionization. A high point was reached with the introduction of the electrospray ionization (ESI)[1,2] and matrix-assisted laser desorption ionization (MALDI)[3,4] methods of the mid-1980s for the ionization of biological molecules. These methods are the prototypes of the two broad classes of ionization methods: spray ionization and desorption ionization. The first edition of this book[5] was concerned chiefly with ESI and its applications as developed in the decade after its introduction. The present volume deals with subsequent developments in ESI and also with MALDI. The forces driving improvements in ionization continue to operate, and emerging applications and variations on ESI and MALDI will be evident throughout this volume. In this Foreword, I cover the topics listed below. These are topics that I find fascinating and which appear to have potential to grow in importance in the future. Alternative examples could have been chosen and the astute reader will find many such in this volume.

1. Forces driving the development of ionization methods
2. Spray and desorption ionization
3. Ambient ionization
4. Molecular imaging
5. Enhanced analysis
6. Fundamental science
7. Opportunities and unsolved problems

FORCES DRIVING THE DEVELOPMENT OF IONIZATION METHODS

Over the past 50 years, mass spectrometry (MS) has been repeatedly transformed and its applications extended by new methods of ionization. These methods have had a cumulative effect as each has added to the capabilities available from previous work. The clear importance of the objectives involved, together with the use of striking names and acronyms, has given ionization methods great visibility, both amongst mass spectrometrists and among the wider scientific public. Earlier ionization methods that received significant attention include field desorption, plasma desorption, molecular ("static") SIMS, laser desorption, fast atom bombardment, and thermospray. Some of these methods are now little known, but all were significant in laying the groundwork for new capabilities.

It is useful to consider the several *desiderata* that drove these developments. The most prominent was the mandate to ionize and so record mass spectra of thermally unstable, nonvolatile compounds of biological origin and of increasing molecular weight. This task has served for several decades as a singularly powerful driving force in mass spectrometry, one that was effectively satisfied by the nearly simultaneous development of MALDI and ESI in the mid-1980s. However, developing mass spectrometric methods for high-molecular-weight compounds has not been the only driving force in the development of ionization methods. An early consideration was that ionization be accompanied by limited and simplified fragmentation. One of the great successes of chemical ionization[6] is the fact that it allows, through choice of the reagent ion, control over the thermochemistry of the ionization reaction and hence over the degree of dissociation seen in the mass spectrum. A third driving force has been the wish to examine aqueous solutions directly, especially after the introduction of aqueous-phase chromatography for biomolecules. The thermospray method[7] satisfied this requirement before being supplanted by the related but far more convenient electrospray method. There are additional sample characteristics and analytical questions that have driven the development of ionization methods. One was the desire to achieve atomic, then molecular, and then biochemical analysis of the surfaces of materials. This problem was first addressed in the 1960s by laser desorption and by secondary ion mass spectrometry (SIMS). Successes were hard won but came in the special cases of sub-monolayers of organics on metals using SIMS[8,9] and with the introduction of early non-UV-active matrices for both SIMS and LD.[10] These early matrices (typically inorganic or organic salts) had dramatic effects on the intensities and quality of mass spectra and they led to the fast atom bombardment, liquid matrix method.[11,12] All this work contributed to the energy-absorbing matrix work embodied in the MALDI method.[3,4] A different type of analytical task that contributed to ionization method development was the desire to achieve depth profiling in solids. The use of polyatomic projectile ions, most notably $C_{60}{}^{+}$, is the most recent contribution to resolving this problem.[13] Elemental and isotopic analyses present other analytical requirements that have driven ionization method development. There are special issues here too which place exceptional demands on ionization methods, including the avoidance of molecular ion formation in elemental analysis and issues of source stability and high ionization efficiency in isotopic analysis. Plasma ionization and glow discharge ionization are the common solutions to the problems of elemental analysis, while high-precision isotopic analysis often employs electron ionization (lower-mass elements) or thermal ionization (higher-mass elements).

The previous paragraph represents a brief survey of the forces that drove the development of ionization methods from around 1965 through the introduction of the commercial ESI and MALDI sources in the late 1980s. Subsequent efforts have been motivated by attempts to (i) increase the throughput of mass spectrometry by decreasing the time and effort taken in sample preparation and ionization, (ii) ionize biological compounds while preserving features of the mass spectra which are associated with higher-order structure and biological activity, not merely with overall mass and charge, (iii) reduce the complexity of ESI spectra by simplifying the charge-state distribution, either during ionization or by subsequent modification of the set of molecular ions that are observed, and (iv) ionize intact clusters, especially protein complexes while retaining the solution-phase cluster composition and structure. Dealing with each in turn: (i) Successful high-throughput methods in ESI involve miniaturization and automation of standard ESI nanospray ion sources[14,15]; in MALDI they involve automation of the matrix deposition,[16] that is, automated control of the laser position and of the sampling process to find "sweet" spots. (ii) Preservation of features characteristic of the native structures of biomolecules calls for gentle methods of ionization. This can be achieved using versions of ESI in which one starts with very small droplets and uses mild desolvation conditions. The use of an air amplifier[17] or of the high-velocity method of electrosonic spray[18–20] yields small initial droplets and proteins are ionized to give charge-state distributions dominated by a single charge state. These same methods achieve objective (iii) , simplification of the ESI mass spectra. Remarkable progress in the ionization of intact complexes has also been achieved.[21,22]

SPRAY IONIZATION AND DESORPTION IONIZATION

It is convenient to organize the ionization methods used in organic and biological MS into the broad groups shown in Table 1.

These methods are distinguished in various important ways. For example, only one type, electron ionization, completely avoids bimolecular chemical processes, and this is the reason for its intrinsically high reproducibility. EI normally gives radical cations (less commonly, radical anions), and these often fragment too readily and by too-complex routes, even in the small molecules to which EI is applicable. Chemical ionization solves this problem but still imposes the onerous requirement that the sample be vaporized before it can be ionized. By far the largest fraction of all studies on biological compounds is performed using ESI and MALDI. As noted, these are the exemplars of the modern desorption ionization (DI) and the spray ionization (SI) classes of methods. In spray ionization, the sample is introduced as a solution that is nebulized, with the resulting charged droplets being made to undergo desolvation and the resulting gas-phase ions being mass analyzed. In desorption ionization, by contrast, condensed phase samples are examined by the impact of energetic particles. A wide range of such energetic projectiles is available, leading to a rich variety of ionization methods of the DI type.

The desorption ionization event involves the input of sufficient energy to dislodge the analyte from the surface and also to ionize it, if it is not already charged. These requirements

Table 1 Classification of Ionization Methods for Organic Compounds

Method	Abbreviation	Example(s)	Analyte State	Characteristics
Electron ionization	EI	Electron impact; electron capture	Vapor phase	Highly reproducible spectra
Chemical ionization	CI	Chemical ionization	Vapor phase	Control of internal energy and fragmentation
Desorption ionization	DI	SIMS; MALDI	Solid phase	Matrix to protect analyte; singly charged ions
Spray ionization	SI	Thermospray ionization; electrospray ionization	Solution	Multiply charged ions; little fragmentation

call for significant energy input, and this in turn demands that the sample be protected from the direct effects of the energy. This is done using a matrix that is present in great excess (typically 1000-fold) and which transforms the incoming energy into a form suited to gentle desorption of analyte molecules from the surface. Coupling of the incoming energy (whether provided by a translationally excited atom, a UV-photon, or another source) into the internal modes of the locally and temporarily excited surface is a key step in the ablation process that results in desorption. Returning to the two steps involved in desorption ionization, namely desorption or phase change and ionization or charge-state change, one notes that it has long been known[23] that precharged species (e.g., organic cations when the organic compound is in the form of an organic salt) give exceptionally high yields in such DI experiments as LD and SIMS. If the analyte is neutral when it leaves the surface, subsequent gas-phase chemical reactions are needed to yield ions and yields drop correspondingly.

The spray ionization experiments do not require matrix because this is essentially provided by the solvent in which the sample is carried. However, the solvent must be removed, a demanding task if the analyte itself is to be preserved intact. Water, the solvent for biological molecules, is difficult to remove although the evaporation processes in ESI are remarkably successful.

The most striking difference between the desorption and spray ionization methods is the charge state in which the analyte is formed. In the DI experiments, the coulomb energy cost of removing multiply charged ions from surrounding counterions means that spectra are dominated by singly charged ions. In SI experiments, on the other hand, the evaporation of water (typically) from a microdroplet that has ionic constituents (buffer, a non-neutral pH) means for the same reason that the analyte tends to take a number of charges that is roughly in proportion to the size of the molecule. This dramatic difference in charge states has had a profound effect on the most appropriate type of mass analyzer for each type of ionization method. The DI methods demand an instrument with a large mass/charge range, whereas the SI methods yield ions for molecules of the same mass using mass analyzers with much smaller mass/charge ranges. It will be seen below (next section) that some newer DI methods yield highly charged ions.

In both the DI and the SI methods, the amount of internal energy acquired by a molecule during ionization affects the degree of fragmentation, as well as the level of molecular disruption that is not seen as fragmentation but which leads to loss of biological activity through changes in structure. The loss of these structural features is undesirable in the gas-phase ions because it precludes experiments such as H/D exchange to determine the number of active sites in the biomolecule that might have been responsible for biological activity. If biological ions can be generated in the gas phase with their original solution-phase structures, then molecular cross sections can be obtained by ion mobility measurements,[24] thereby providing valuable information on 3D structures and folding.

If one is successful in minimizing the input of internal energy during ionization, then spectra will be dominated by the ionized molecule, e.g. the singly protonated molecule in positive ion MALDI, or the multiply deprotonated molecule in negative ion ESI. The successful generation of a form of ionized molecule which retains the mass of the original condensed-phase molecule allows determination of its molecular weight. Structural characterization normally requires the formation of fragment ions and in both the DI and SI methods this is achieved by activation followed by dissociation. Since both the precursor and product masses are of interest, two stages of mass analysis (i.e., tandem mass spectrometry) are needed. This can be done using the same analyzer after a temporal delay (as in ion traps) or by using two separate mass analyzers. The activation methods available[25] include the common collisional activation in the electron-volt collision energy range (multiple collisions) or in the kilo-electron-volt range (single collisions) as well as single- or multiphoton activation, surface collisions (surface-induced dissociation), or electron capture or electron transfer activation and dissociation. Note that the electron capture[26] and electron transfer[27,28] processes change the charge state as well as cause internal energy uptake and are suited to structural studies on multiply charged ions. The type of MS/MS experiment[29] selected depends on the type of mass analyzer, and this choice is affected by the ionization method itself, as already noted.

AMBIENT IONIZATION

Recently, it was been shown that an electrospray emitter can be used to provide translationally excited projectiles (charged microdroplets) which serve as projectiles for desorption and ionization of condensed-phase analytes present on surfaces. This hybrid technique, DESI (desorption electrospray ionization),[30] is applicable to analysis of samples in the ambient environment. The practical advantage is that the sample can be examined directly, without any preparation; hence the experiment is extremely fast (typically <5 s) and can be conducted in a high-throughput fashion and tandem mass spectrometry can be used for identification of components of complex mixtures. These features provide the speed of analysis and high chemical specificity that are needed in applications such as public safety monitoring.

The DESI method is placed in the context of conventional and newer vacuum-based desorption ionization methods in Table 2. Consideration of the data on kinetic energy per nucleon makes clear just how gentle the method is. DESI was introduced simultaneously with another ambient ionization method, DART (direct analysis in real time),[31] in which the initial projectile is a metastable atom that arrives at the surface as a water ion. The relationship between these methods and the older method of atmospheric-pressure MALDI (AP-MALDI)[32] is summarized schematically in Figure 1. These new types of experiments illustrate the rapid developments still taking place in ionization methods in mass spectrometry. This reinforces the need for a thorough understanding of the essentials of ionization methodology, such as that provided in this book. Second, it illustrates a merging of the DI and SI methods into new forms that carry the advantages of both. The separate consideration of

Table 2 Velocities, Radii, and Kinetic Energies of Projectiles in Some Desorption Ionization Methods

Method	Projectile Velocity	Projectile Size	Kinetic Energy per Nucleon
DESI	0.1 km/s	~3 μm	0.6 meV/u
Massive cluster ionization (MCI)[a]	5 km/s	~80 μm	100 meV/u
Electrospray droplet impact (EDI)[b]	11 km/s	~20 μm	1 eV/u
Molecular SIMS with C_{60}[c]	10–100 km/s	~1 nm	80–300 eV/u

Source: Adapted from Venter, A.; Sojka, P. E.; Cooks, R. G. Droplet dynamics and ionization mechanisms in desorption electrospray ionization mass spectrometry. *Anal. Chem.* **2006**, *78*, 8549–8555.

[a]Mahoney, J. F.; Perel, J.; Lee, T.D.; Martino, P. A.; Williams, P. *J. Am. Soc. Mass Spectrom.* **1992**, *3*, 311–317

[b]Hiraoka, K.; Asakawa, D.; Fujimaki, S.; Takamizawa, A.; Mori, K. *Eur. Phys J. D*, **2006**, *38*, 225–229

[c]Winograd, N. *Anal. Chem.* **2005**, *77*, 143a–149a.

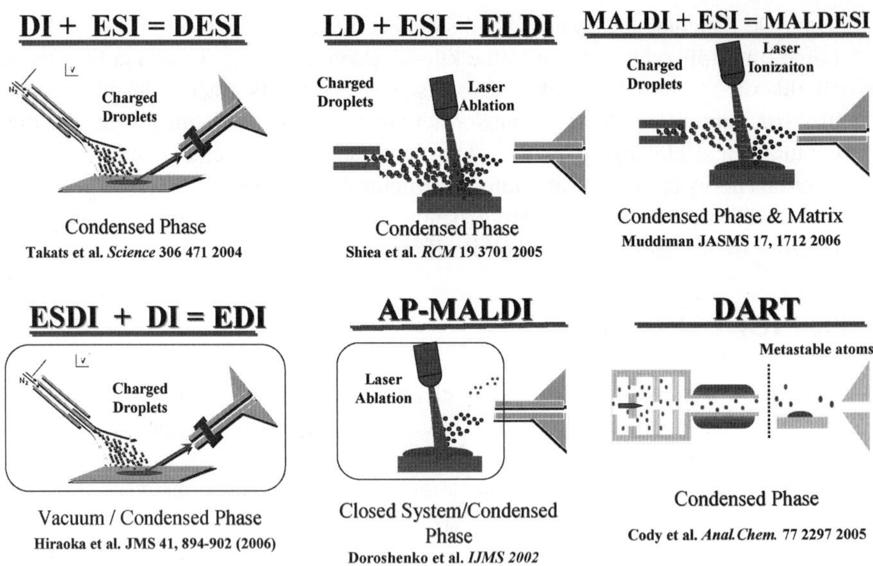

Figure 1. Ambient ionization methods for ionizing condensed phase samples on surfaces using droplets, atom beams, electrical discharges and laser beams. For comparison, some vacuum methods are also shown (See color insert).

these two main classes of ionization method is no longer appropriate, and this edition of Richard Cole's book provides a strong basis for understanding the current literature in both DI and ESI and for contributing to the development of the newer methods that will inevitably follow.

It is noteworthy that chemical reagents can be included in the spray solvent (or gas) in DESI (and related methods) to cause ionization of specific classes of compounds.

MOLECULAR IMAGING

Imaging of materials by mass spectrometry has been an important topic since at least the 1960s, when ion microscopes and ion microprobes were used for spatial elemental analysis. These experiments were based on sputtering in vacuum using energetic (kilo-electron volt) primary ions; that is, they were SIMS experiments. Using primary ion beams of different fluxes, it is possible to sputter without revisiting damage sites in order to perform 2D analysis. By alternating between dynamic and static SIMS, one can "mill" through a surface to perform 3D elemental or molecular analysis. With the introduction of TOF mass analyzers in place of the elegant sectors of the earlier ion microscopes, the spatial resolution of these devices has improved to a few tens of nanometers. At the same time, the use of SIMS to image molecules has improved with the introduction of cluster ion sources. As a result of these developments, TOF SIMS imaging[33] has acquired excellent capabilities in terms of sensitivity and resolution (sub-micron allowing subcellular structural studies), although the quality of the molecular information obtained is limited, in part because there is often significant fragmentation and in part because some instruments do not have MS/MS capabilities. More recently, imaging using laser beams has become a significant topic, especially the MALDI-based experiments.[34] These latter experiments have a spatial resolution that is in the 100 μm range, but they make up for this by providing information more rapidly and they give mass spectra that provide molecular-structure information (through fragmentation) on high-molecular-weight compounds.

The MS imaging methods are often contrasted favorably with optical methods because they are "label-free." However, in the SIMS method the samples are "fixed," and in MALDI they are treated with matrix; in both experiments the sample is introduced into the vacuum environment for imaging. These requirements are disadvantages that to some extent offset the label-free characteristics of the method. With the development of the ESI-based DESI method has come a new imaging procedure performed on unmodified biological materials in the ambient environment.[35] These advantages offset the still limited spatial resolution of the method (~200 μm). Applications to human tissue have been reported, and the location of cancer margins by the patterns of phospholipids is especially significant. The lack of sample preparation, surface specificity, and MS/MS capabilities of DESI imaging provide advantages. DESI imaging and MALDI imaging appear to be highly complementary methods, with the former ionizing phospholipids particularly effectively and the latter giving higher-quality protein data. The considerable promise of the methods in due course in clinical practice is illustrated by the human liver cancer data shown in Figure 2.

Looking at the current state of imaging MS, MALDI, and DESI, it can be argued that mass resolution is a lower priority than increased chemical information that can be derived through MS/MS experiments and improved molecular interpretation (structure–spectrum correlations). This latter undertaking requires further development of knowledge of ion chemistry, especially for classes of molecules like lipids that have not been thoroughly investigated by mass spectrometry. Complementing this analytical–organic chemistry task are the interlocked bioinformatics tasks of data collection, reduction, and manipulation that are needed to maximize the power of imaging MS. Another requirement in MS imaging is to develop more specific analytical methods—for example, methods in which

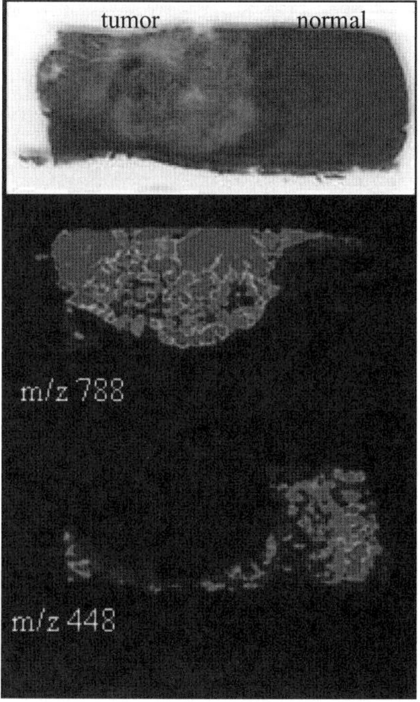

Figure 2. Human liver sample. (**a**) H&E (hematoxylin and eosin) stain. (**b**) Negative ion images of phosphatidylserine (ion of m/z 788). (**c**) Chenodeoxycholic acid (ion of m/z 448). (Unpublished DESI imaging data of Wiseman, Caprioli, Puolitaival, and Cooks; false color intensity; tissue shown is 17 mm by 8 mm by 12 μm.) (See color insert).

images are recorded for particular compounds on the basis of MS/MS signals (product ion or multiple reaction monitoring experiments). Such a front-end capability will reduce the quantity of raw data processed and increase the informing power of the questions asked during an imaging experiment as well as increasing the quality of the quantitative data obtained.

ENHANCED ANALYSIS

The range of applications of ESI and MALDI has been extended to a remarkable degree over the past several years. Extensions are to larger and more complex molecular systems, especially to biological problems of increasing complexity. This progress has been based on technological factors (as is so frequently the case in mass spectrometry), especially improvements in mass analyzers and in methods of ion activation. In regard to mass analyzers, notable developments are FT-ICR instruments of higher field strength and the introduction of the Orbitrap mass spectrometer with its unique nonmagnetic field high-performance mass analyzer. However, most ultra-high-mass work has used simpler systems, including hybrid time-of-flight instruments.

The development of activation methods for MS/MS has seen the extension of the electron capture dissociation (ECD) method from ICR to Paul ion trap instruments, providing users of modest-sized mass spectrometers the advantages of a fragmentation method that is complementary to collision-induced dissociation and, hence, access to high protein/peptide sequence coverage. Electron transfer dissociation (ETD), the more versatile chemical analog of ECD, has rapidly impacted biological analysis since its introduction in 2004.[27,28] This experiment is complementary to the more traditional collision-induced dissociation experiment and is rapidly making a large contribution to increasing the information on post-translational modifications in proteins. Also worthy of note are McLuckey's experiments on charge-changing collisions,[36] experiments that allow the charge state of a collection of ions to be altered to suit the measurement at hand—whether it is a high-accuracy mass measurement or collision-induced dissociation or an ion–molecule reaction. His remarkable "ion parking" experiment,[37,38] based on the strong velocity dependences of charge-changing reaction cross sections, means that trapped ions can be resonantly excited to give them high translational energies, thereby protecting them from reaction. This allows the reversal of the characteristic dispersion of molecules over a range of charge states in ESI so that quantitatively all molecules are "parked" in a chosen charge state. The ability to move at will between charge states—like scale transposition in a musical piece—is sure to have a large impact on mass spectrometry and the development has important implications for top-down proteomics.

Other noteworthy developments in the realm of analytical applications are:

(a) *High-mass Complexes.* Some of the most impressive applications of mass spectrometry over the past decade have been to protein complexes.[21,22] Recently, experiments in which micelles incorporate proteins have yielded to the ESI methodology.[21]

(b) *Inorganic Compounds.* Application of ESI to inorganic complexes and organometallic compounds is generally straightforward and a valuable way of characterizing these classes of compounds. Limitations are met in terms of the reactivity of many of these compounds with air and water, but nonaqueous solutions can be sprayed and MALDI is also useful.

(c) *Chiral Analysis*. The growing importance of chirally pure drug substances has led to great emphasis on enantiomeric control in synthesis and to increased demands for chiral analytical methods. This latter need has been satisfied in large part by chiral stationary phases for liquid chromatography. However, mass spectrometry and especially ESI can be used for chiral analysis.[39] One experiment involves generating a cluster ion involving a selected chiral reference (R), the analyte (A), and a metal salt (M). The resulting cluster ion (e.g., $[MR_2A - H]^+$ is mass selected and dissociated by competitive loss of the chiral reference or the analyte. The ratio of the two product channels depends on the chirality of the reference, and a plot of the abundance ratio versus % enantiomeric excess is linear. Unknown enantiomeric excess values can be quickly read from a calibration curve.

FUNDAMENTAL SCIENCE

Fundamental aspects of ESI and MALDI and the fundamental chemistry that they facilitate are large subjects. A recent book[25] has an extensive coverage of these topics, and there is also much on these subjects in the present volume. Some additional comments on particular topics are as follows:

(a) *Fundamentals of Field Ionization*. By levitating a charged droplet and applying a strong electric field, the Taylor cone phenomenon and associated ion emission processes have been studied with unprecedented precision.[40]

(b) *Binding Affinities*. ESI experiments are a powerful source of information on fundamental thermochemical quantities. Amongst these, none is more important in biochemistry than the binding affinity. The ESI experiment offers different methods to characterize this quantity: examination of ligand exchange ion/molecule reactions provides intrinsic (solvent free) values; alternatively, examination of solutions containing competitive ligands allows relative binding affinities to be established under appropriate conditions.[41]

(c) *Reaction Mechanisms of Organic Ions*. Somewhat lost sight of in the bioanalytical applications of ESI is its role in providing mechanistic connections between the vacuum (isolated phase) environment and the ordinary solution environment. Some reactions, like the Eberlin transacylation shown in Scheme 1, have been studied in the vacuum environment, the ESI interface, under ambient pressure conditions, and in solution.[42] The polar transacetalization (Scheme 1) parallels the condensed-phase transacetalization. The reactant acylium ion and the cyclic ionic acetal product mimic the free reactant carbonyl compound and the cyclic acetal product of the condensed-phase reaction, respectively.

(d) *Retention of Biological Activity: Native Protein Configuration*. The already remarkable reach of ESI mass spectrometry into the biological sciences would be enhanced if the ionization method were even more gentle. For some years there has been evidence of retention of native protein structures in the gas phase, starting with McLafferty's data on H/D exchange in proteins.[43] The most direct evidence of retention of biological activity comes from electrosonic spray ionization (ESSI) followed by mass selection and ion soft landing into an aqueous medium.[44] Demonstrations of enzymatic activity—for example, phosphorylation of typical substrates using protein kinase A—were successful.

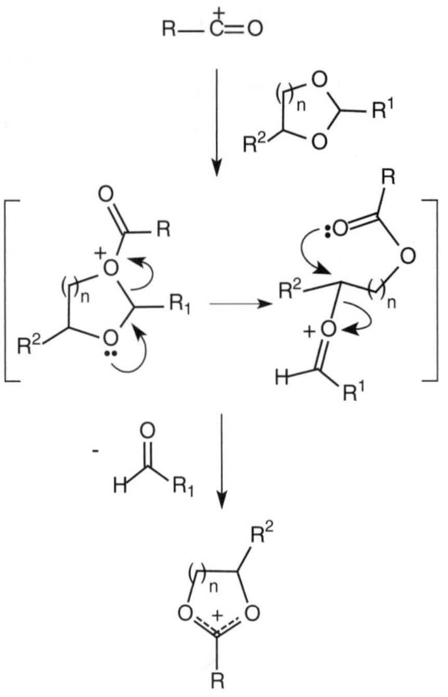

Scheme 1. Ionic transacetalization (Eberlin reaction) in the gas phase under high vacuum and at atmospheric pressure. (Adapted from Cooks, R. G.; Chen, H.; Eberlin, M. N.; Zheng, X.; Tao, W. A. Polar acetalization and transacetalization in the gas phase: The Eberlin reaction. *Chem. Rev.* **2006**, *106*, 188–211.)

(e) *Cluster Science*. The use of electrospray and to a more limited extent, MALDI, to generate noncovalently bound clusters illustrates the role of mass spectrometry in cluster science. The evaporation step that the electrosprayed droplets undergo imposes a Le Chatelier driving force in favor of clustering. The clustering of simple coding amino acids is a case in point with the clustering of proline to form a magic number dodecamer[45] being a feature of the ESI mass spectrum. The proposed structure of the proline 12-mer in its protonated form is shown in Figure 3.

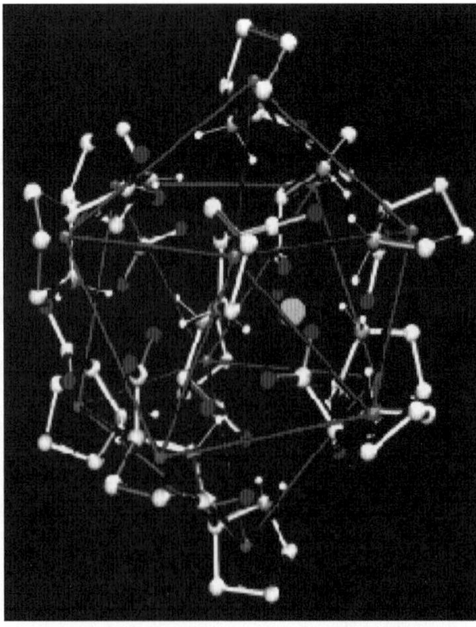

Figure 3. Calculated structure of the protonated form of the homochiral proline-12 icosahedral cluster. (From Clemmer, D. E., et al. *J. Am. Chem. Soc.*, **2006**, *128*, 15988–15989.)

OPPORTUNITIES AND UNSOLVED PROBLEMS

In spite of the spectacular progress in ESI and MALDI addressed throughout this volume, there remain major challenges and hence opportunities. Notable amongst them are:

(a) *Low Efficiency of Ionization.* Virtually all mass spectrometry ionization methods except for thermal ionization of low ionization energy elements have very low ionization efficiency (conversion of atoms/molecules of sample to gas-phase ions entering the vacuum system of the mass spectrometer). The best reported numbers are for nanospray where values are on the order of 0.1%. Obviously, there is much room for improvement. Optimization of the transfer of droplets and ions from the atmospheric-pressure region into the vacuum region is needed. Ion funnels[46] have had considerable success in the collection and transfer of free ions. The collection and transfer of droplets remains to be improved.

(b) *Higher Ion Currents.* Efficiency of ionization is important when sample sizes are limited. When this is not the case, higher-quality mass spectral data can be obtained by increasing ion currents. This cannot be done by simply increasing solution flow rates, although it might be achieved using multiple sprays.[47]

(c) *Hydrated Ions.* It is remarkably difficult to create micro-solvated ions by electrospraying solutions.[48] This is because solvents like water bind strongly to other water molecules and conditions which lead to desolvation tend to produce completely "dry" ions. This is just as well; otherwise the value of ESI in chemical analysis would be much reduced. Because of the enormous importance of solvation phenomena, along with attempts to understand solvated ion structures both spectroscopically[49,50] and through *ab initio* calculations,[51] new routes to the preparation of solvated ions would be of great value. One approach is based on the observation that electrospraying pure solvent in the presence of the vapors of an organic compound allows the uptake of that molecule into the charged microdroplets of solvent.[52]

(d) *Simulations at Higher Pressures.* The remarkable successes of ion simulation programs in assisting in the development of mass spectrometers has been based on the ability to calculate ion trajectories in vacuum, allowing for the effects of space charge and collisions. Both the widely used program SIMION and other more specialized programs have been used. However, at higher pressures than commonly encountered in mass analyzers, pneumatic effects on ion motion are not insignificant and appropriate computational procedures are still being developed. These pressures include those met in ESI and related ion sources, so clearly ion source development has proceeded without the aid of simulations. As such simulations become available, one can expect rapid progress in addressing issues like those mentioned in a) and b) above.

(e) *Multiplexed High-Throughput Analysis.* In spite of the value of ESI-based mass spectrometry, it is often coupled with slow chromatographic methods making the mass spectrometry experiment fast enough to be done in a serial fashion. Exceptions to this are the use of an array of micromachined electrospray sources[15] and the four-LC column MUX systems.[53] Multichannel mass spectrometers, in which multiple ESI ion sources are connected to multiple analyzers and multiple detectors, have been reported[54] but not yet commercialized.

(f) *Vapor Phase Analysis*. In an experiment patented by Fenn and Ffurstenau,[55] it was shown that ESI is applicable to these analytes by an extractive process in which the droplets of the electrosprayed solvent pick up vapor-phase compounds present in air. It is interesting to note that one of the variants on the DESI experiment, extractive electrospray ionization (EESI), uses a related principle to examine solutions for particular analytes.[56] In this experiment, a pure solvent spray, subjected to a high potential to give charged droplets, intersects the spray of a liquid. The advantages of this experiment, which appears to work by a solvent transfer process, is that the analyte is transferred from a solution that might be very complex, and hence difficult to handle by conventional ESI, into a pure solvent. The same concept has recently been applied by Zenobi and Chen to the analysis of nonvolatiles in breath[57] and to a high-throughput analysis of practical materials like foodstuffs for contamination and spoilage.[58] These compounds are presumably present in the form of an aerosol.

(g) *Detection of High-Mass Ions*. With the problem of creating high-mass ions well in hand, the limitations in high-mass MS lie increasingly in the detection step. Several innovative solutions are being developed: A cryo-detector in which the enthalpy of ion impact is measured has been developed and commercialized.[59,60] Another approach is to combine the detection and mass measurement step using laser elastic scattering from ions trapped in electromagnetic traps. The mass/charge ratio is measured from the ion frequency.[61] The limitations to this type of detector are that the size of the ion must not be too small. In addition, the experiment is effectively limited to the examination of small numbers of ions.[62]

(h) *Stand-Off (Remote) Detection*. The ability to move ions some meters, from the source of an ambient mass spectrometer to the mass spectrometer, has provided a modest stand-off capability that could be valuable in several areas of MS application. This is a convenience in many potential applications, not the least of which are those clinical applications in which the mass spectrometer would be intrusive but the DESI probe could be used to generate and then sample ions.

(i) *Miniature (Fieldable) Mass Spectrometers*. Given the importance of ESI for the detection and identification of biological compounds, there would be great advantages in several areas of application (environmental, clinical, food safety, public safety, etc.) if ESI or related methods like DESI could be adapted to small mass spectrometers. There has been progress in producing portable instruments based on laser desorption[63] and progress with DESI on a transportable instrument.[64]

The richness of the subject of ionization in mass spectrometry is the result of the not-always-easy coexistence of fundamental chemistry and physics with applied science; most certainly the role of instrumentation is an important factor in the continuous invention and re-invention that has characterized mass spectrometry in the past three decades. In searching for reasons for the vigor of mass spectrometry, one might argue that the scale of the instrumentation (in terms of physical size but even more in terms of complexity) has been appropriate: Mass spectrometers and their components are sufficiently complex that the instrumentation development is a difficult and rewarding scientific and technological undertaking, but it is not so complex that it can only be done with large budgets in a few places. (Parenthetically, this situation might change, and in the area of miniature instrumentation is already changing, to favor those with access to semiconductor and MEMS fabrication facilities.) Especially in the area of development of ion sources, the cottage industry aspect of instrumental mass spectrometry has been an engine for progress, as seen throughout the development of the desorption ionization and spray ionization methods. This book appears at the right time to help spur this progress; it is a work of importance, and the community owes Richard Cole its thanks as well as its congratulations.

REFERENCES

1. Yamashita M.; Fenn J. B. *J. Phys. Chem.* **1984**, *88*, 4451–4459.
2. Yamashita M.; Fenn J. B. *J. Phys. Chem.* **1984**, *88*, 4671–4675.
3. Hillenkamp F.; Karas M.; Holtkamp D.; Klusener P. *Int. J. Mass Spectrom. Ion Proc.* **1986**, *69*, 265–276.
4. Karas M.; Bachmann D.; Hillenkamp F. *Anal. Chem.* **1985**, *57*, 2935–2939.
5. Cole R. B., Ed. *Electrospray Ionization Mass Spectrometry*, John Wiley & Sons, New York, **1997**, 577 pages.
6. Munson M. S. B.; Field F. H. *J. Am. Chem. Soc.* **1966**, *88*, 2621–2626.
7. Blakeley C. R.; Vestal M. L. *Anal. Chem.* **1983**, *55*, 750–754.
8. Benninghoven A. *Ann. Phys.* **1965**, *15*, 113.
9. Benninghoven A. *Z. Phys.* **1969**, *20*, 159.
10. Liu L. K.; Busch K. L.; Cooks R. G. *Anal. Chem.* **1981**, *53*, 109–113.
11. Barber M.; Bordoli R. S.; Sedgwick R. D.; Tyler A. N. *Nature* **1981**, *293*, 270–275.
12. Morris H. R.; Panico M.; Barber M.; Bordoli R. S.; Sedgwick R. D.; Tyler A. *Biochem. Biophys. Res. Commun.* **1981**, *101*, 623–631.
13. Winograd N. *Anal. Chem.* **2005**, *77*, 142a–149a.
14. Wachs T.; Henion J. *Anal. Chem.* **2001**, *73*, 632–638.
15. ESI Chips™, www.advion.com.
16. Aerni H. R.; Cornett D. S.; Caprioli R. M. *Anal. Chem.* **2006**, *78*, 827–834.
17. Hawkridge A. M.; Zhou L.; Lee M. L.; Muddiman D. C. *Anal. Chem.* **2004**, *76*, 4118–4122.
18. Hirabayashi A.; Sakairi M.; Koizumi H. *Anal. Chem.* **1994**, *66*, 4557–4559.
19. Hirabayashi A.; Sakairi M.; Koizumi H. *Anal. Chem.* **1995**, *67*, 2878–2882.
20. Takats Z.; Wiseman J. M.; Gologan B.; Cooks R. G. *Anal. Chem.* **2004**, *76*, 4050–4058.
21. McKay A. R.; Ruotolo B. T.; Ilag L. L.; Robinson C. V. *J. Am. Chem. Soc.* **2006**, *128*, 11433–11442.
22. Demmers J. A. A.; van Dalen A.; de Kruijff B.; Heck A. J. R.; Killian J. A. *FEBS Lett.* **2003**, *541*, 28–32.
23. Cooks R. G.; Terwilliger D. T.; Ast T.; Beynon J. H.; Keough T. *J. Am. Chem. Soc.* **1975**, *97*, 1583–1585.
24. Clemmer D. E.; Jarrold M. F. *J. Mass Spectrom.* **1997**, *32*, 577–592.
25. Laskin J.; Lifshitz C. *Principles of Mass Spectrometry Applied to Biomolecules*, John Wiley & Son, Hoboken, NJ, 2006.
26. Zubarev R. A. *Mass Spectrom. Rev.* **2003**, *22*, 57–77.
27. Coon J. J.; Syka J. E. P.; Shabanowitz J.; Hunt D. F. *Int. J. Mass Spectrom.* **2004**, *236*, 33–42.
28. Syka J. E. P.; Coon J. J.; Schroeder M. J.; Shabanowitz J.; Hunt D. F. *Proc. Natl. Acad. Sci. USA* **2004**, *101*, 9528–9533.
29. Busch K. L.; Glish G. L.; McLuckey S. A. *Mass Spectrometry/Mass Spectrometry: Techniques and Applications of Tandem Mass Spectrometry*, VCH Publishers, New York, **1988**.
30. Takáts Z.; Wiseman J.; Gologan B.; Cooks R. G. *Science* **2004**, *306*, 471–473.
31. Cody R. B.; Laramee J. A.; Durst H. D. *Anal. Chem.* **2005**, *77*, 2297–2302.
32. Laiko V. V.; Baldwin M. A.; Burlingame A. L. *Anal. Chem.* **2000**, *72*, 652–657.
33. Cannon, D. M., Jr.; Pacholski, M. L.; Ewing, A. G.; Winograd, N. In Gillen, G., Lareau, R., Bennett, J., Stevie, F.,(Eds.), *Secondary Ion Mass Spectrometry (SIMS XI)*, John Wiley & Sons, New York, 1997; pp. 489–492.
34. Caprioli R. M.; Farmer T. B.; Gile J. *Anal. Chem.* **1997**, *69*, 4751–4760.
35. Wiseman J. M.; Puolitaival S. M.; Takats Z.; Cooks R. G.; Caprioli R. M. *Angew. Chem. Int. Ed.* **2005**, *44*, 7094–7097.
36. Pitteri S. J.; McLuckey S. A. *Mass Spectrom. Rev.* **2005**, *24*, 931–958.
37. Chrisman P. A.; Pitteri S. J.; McLuckey S. A. *Anal. Chem.* **2006**, *78*, 310–316.
38. McLuckey S. A.; Reid G. E.; Wells J. M. *Anal. Chem.* **2002**, *74*, 336–346.
39. Young, B. L.; Wu, L.; Cooks, R. G. Mass spectral methods of chiral analysis. In Busch, K. W.; Busch, M. A.,(Eds.), *Chiral Analysis*, Elsevier Science, Amsterdam, The Netherlands, 2006, pp. 595–659.
40. Grimm R. L.; Beauchamp J. L. *J. Phys. Chem. B* **2003**, *107*, 14161–14163.
41. Powell K. D.; Ghaemmaghami S.; Wang M. Z.; Ma L. Y.; Oas T. G.; Fitzgerald M. C. *J. Am. Chem. Soc.* **2002**, *124*, 10256–10257.
42. Cooks R. G.; Chen H.; Eberlin M. N.; Zheng X.; Tao W. A. *Chem. Rev.* **2006**, *106*, 188–211.
43. Wood T. D.; Chorush R. A.; Wampler F. M.; Little D. P.; Oconnor P. B.; McLafferty F. W. *Proc. Natl. Acad. Sci. USA* **1995**, *92*, 2451–2454.
44. Ouyang Z.; Takats Z.; Blake T. A.; Gologan B.; Guymon A. J.; Wiseman J. M.; Oliver J. C.; Davisson V. J.; Cooks R. G. *Science* **2003**, *301*, 1351–1354.

45. Myung S.; Lorton K. P.; Merenbloom S. I.; Fioroni M.; Koeniger S. L.; Julian R. R.; Baik M. -H.; Clemmer D. E. *J. Am. Chem. Soc.* **2006**, *128*, 15988–15989.
46. Belov M. E.; Gorshkov M. V.; Udseth H. R.; Anderson G. A.; Smith R. D. *Anal. Chem.* **2000**, *72*, 2271–2279.
47. Tang K. Q.; Lin Y. H.; Matson D. W.; Kim T.; Smith R. D. *Anal. Chem.* **2001**, *73*, 1658–1663.
48. Lemoff A. S.; Bush M. F.; Williams E. R. *J. Am. Chem. Soc.* **2003**, *125*, 13576–13584.
49. Robertson W. H.; Diken E. G.; Price E. A.; Shin J. W.; Johnson M. A. *Science* **2003**, *299*, 1367–1372.
50. Weber J. M.; Kelley J. A.; Nielsen S. B.; Ayotte P.; Johnson M. A. *Science* **2000**, *287*, 2461–2463.
51. Jorgensen W. L. *Acc. Chem. Res.* **1989**, *22*, 184–189.
52. Schlosser G.; Takats Z.; Vekey K. *J. Mass Spectrom.* **2003**, *38*, 1245–1251.
53. Yang L. Y.; Mann T. D.; Little D.; Wu N.; Clement R. P.; Rudewicz P. J. *Anal. Chem.* **2001**, *73*, 1740–1747.
54. Tabert A. M.; Goodwin M. P.; Duncan J. S.; Fico C. D.; Cooks R. G. *Anal. Chem.* **2006**, *78*, 4830–4838.
55. Fuerstenau, S. D.; Fenn, J. B.Method for detection and analysis of inorganic ions in aqueous solutions by electrospray mass spectrometry. U.S. Patent 5,523,566, 1996.
56. Chen H.; Venter A.; Cooks R. G. *Chem. Commun.* **2006**, 2042–2044.
57. Chen H. W.; Wortmann A.; Zhang W. H.; Zenobi R. *Angew. Chem. Int. Ed.* **2007**, *46*, 580–583.
58. Chen H. W.; Sun Y. P.; Wortmann A.; Gu H. W.; Zenobi R. *Anal. Chem.* **2007**, *79*, 1447–1455.
59. Frank M.; Labov S. E.; Westmacott G.; Benner W. H. *Mass Spectrom. Rev.* **1999**, *18*, 155–186.
60. Wenzel R. J.; Matter U.; Schultheis L.; Zenobi R. *Anal. Chem.* **2005**, *77*, 4329–4337.
61. Peng W. -P.; Cai Y.; Chang H.-C. *Mass Spectrom. Rev.* **2004**, *23*, 443–465.
62. Koppenaal D. W.; Barinaga C. J.; Denton M. B.; Sperline R. P.; Hieftje G. M.; Schilling G. D.; Andrade F. J.; Barnes J. H. *Anal. Chem.* **2005**, *77*, 418a–427a.
63. English R. D.; Cotter R. J. *J. Mass Spectrom.* **2003**, *38*, 296–304.
64. Gao L.; Song Q.; Patterson G. E.; Cooks R. G.; Ouyang Z. *Anal. Chem.* **2006**, *78*, 5994–6002.

Purdue University,
West Lafayette, Indiana

R. Graham Cooks

Foreword to the First Edition

When Richard Cole asked me to write a foreword for this volume, I was highly flattered, but hardly eager. That reticence was due in part to a pile of deferred chores that grows as relentlessly as the universal entropy. A more forbidding inhibition was that in a world now teeming with electrospray (ES) users, I really didn't know what to say that hadn't already been said many times in many ways. Even so, the deadline was then far away and distant commitments always loom small, so I said yes. Tomorrow is now here and none of the nine Muses have come to my rescue. In desperation I have succumbed to banality by presenting "The Inside Story" of what has been called the electrospray revolution or "Things About Electrospray You Won't Find in the Papers!" What follows, therefore, is a highly personal account of how I became involved in the developments that led to the basis for this book.

The prologue to the story was written in 1937 at Berea College in Kentucky, where I was a senior majoring in chemistry. Jobs of any kind were scarce, so I applied for graduate study at several universities and was lucky enough to be accepted, with financial support, at both Northwestern and Yale. At first, I leaned toward Northwestern because its chemistry department was supposed to be better and its pay definitely was. But for various reasons, including a door-to-door free ride to Connecticut with my gear, I decided that fate wanted me in New Haven rather than Evanston. As a result, my chance to meet Malcolm Dole, then an Assistant Professor at Northwestern, was delayed half a century. At the time, I knew only that he had written a textbook on electrochemistry, not that he was at Northwestern. When I arrived at Yale, oddly enough as it turned out, John Zeleny was still on the Physics Faculty. Thirty years earlier, he had led the definitive pioneering studies on electrostatic dispersion of liquids into charged droplets. I was oblivious to that work for another 35 years, when I learned about Dole's electrospray technique and found that its roots were in Zeleny's investigations.

Meanwhile, I finished graduate study in 1940 and spent the next dozen years in applied research and development at Monsanto Chemical Company in Anniston, Alabama, Sharples Chemicals in Wyandotte, Michigan, and, beginning in 1945, at Experiment, Inc., a contract research and development company in Richmond, Virginia. There I became involved in combustion research for Project Bumblebee, a Navy program to develop a ramjet-powered antiaircraft missile. The experience led, in 1952, to the directorship of Project SQUID, another Navy program, administered by Princeton, on pure and applied research in "those fields of science relating to jet propulsion": combustion, fluid flow, and heat transfer. In 1955, the Navy arranged for me to spend a year at the London Branch of the Office of Naval Research. While there I came across a 1954 paper by E. W. Becker and K. Bier describing the production of intense beams of hydrogen molecules from rarefied supersonic flows as proposed in a 1951 paper by A. Kantrowitz and J. Grey. I had been musing about using molecular beam scattering experiments to study combustion reactions, but was discouraged because the expected reaction cross sections were too small for measurement with beam intensities available from the effusive sources of Otto Stern, Nobel Laureate and father of molecular-beam research. Moreover, the activation energies for the most interesting reactions were above 1 eV and would require source temperatures above

5000 K, much too high to be feasible. The Becker–Bier results indicated that the intensity barrier could be scaled with source gas flows that were convective as opposed to effusive. It occurred to me that if heavier reactant molecules like oxygen or chlorine were seeded into flows of hydrogen or helium, they should be accelerated to the high flow velocities achievable with those light gases. Calculations showed that, depending on their mass, such seed molecules could reach translational energies of several electron volts or more with modest source temperatures.

Back from London, I found Princeton's Department of Mechanical Engineering willing to provide a home for a research project on this idea. A proposal to NSF was fortunate enough to win support for a program that started in the fall of 1959. To make a long story short, we found that supersonic free jets from very small orifices could indeed produce intense beams of molecules with energies as high as 10 eV or more. Alas, we also found that these high translational energies were not effective in promoting chemical reaction. Meanwhile, we had learned that supersonic free jets had many other valuable features that have, ever since, played a key role in my research. Another major byproduct of the beam project was my transition from Director of Project SQUID to Professor of Mechanical Engineering, an achievement undreamt of in the philosophy of a young chemistry major at Berea College, 23 years earlier.

In 1967 I decided to accept an invitation to join Yale's newly organized Department of Engineering and Applied Science. About that time, the Bendix Corporation, producers of time-of-flight (TOF) mass spectrometers, was looking for a way to make intact ions from large polymer molecules so they could sell their instruments to the burgeoning plastics industry. They interested Malcolm Dole in the problem and he came up with his now-famous idea of using Zeleny's technique to disperse dilute solutions of macromolecules as a fine spray of charged droplets into bath gas at atmospheric pressure. He reasoned that evaporation of solvent from such a droplet would increase its surface-charge density up to the Rayleigh limit, at which Coulomb repulsion overcomes surface tension and the droplet breaks up into offspring droplets. Each offspring droplet would repeat that sequence, ultimately producing droplets containing only one solute molecule, which would become a free ion by retaining some droplet charge after all solvent had evaporated. This charged-residue model (CRM) for ion formation is one of the two models most often invoked to explain the formation of ES ions. The other is the ion-evaporation model (IEM) proposed some years later by Iribarne and Thomson. As discussed in the chapter by Kebarle and Ho, neither model has achieved unanimous acceptance.

Under Dole's leadership, a group at Bendix assembled an apparatus to test his idea. Sample solution was injected through a small tube or needle into a flow of dry-bath gas (nitrogen) through a cylindrical glass chamber. The needle was maintained at several kilovolts, relative to a plate constituting the end wall of the chamber. The resulting high field at the needle tip dispersed the emerging liquid into small charged droplets that drifted down the potential gradient concurrently with the flow of bath gas toward the end plate. A small orifice admitted some of the resulting mixture of ions and gas into the vacuum system as a supersonic free jet. I find it interesting that even though the Bendix people were expert in TOF techniques, a retarding potential method was chosen for mass analysis. The underlying idea was that during free-jet expansion, the ions would be accelerated to the easily calculated velocity of the nitrogen bath gas. Well downstream of the orifice, where the jet-gas density is too low to have any further effect on ion velocity, the ions passed through a set of grids on their way to a Faraday-cup electrode monitored by a sensitive electrometer. A scan of the grids' potential produced a current–voltage curve in which a dip occurred when that potential became equal to the kinetic energy of some of the arriving ions, thus preventing them from contributing to the current at the Faraday cup. The mass of the excluded ions was readily obtained from this measured value of their energy along with a value of the velocity taken as equal to that calculated for the nitrogen in the jet. Promising results for polystyrene oligomers with molecular weights (M_rs) up to 500,000 were presented in the *Journal of Chemical Physics* in 1968, the year after I arrived at Yale.

Not an avid reader of the literature, I was unaware of Dole's paper, but it stirred the interest of Professor Seymour "Sandy" Lipsky in the Medical School. A long-time mass spectrometrist, he was excited by the possibility that Dole's technique might work with large biomolecules. Sandy showed the paper to Csaba Horvath, a colleague of mine in Chemical Engineering, who had been working closely with him on the development of HPLC, a now-invaluable methodology to which they made major contributions. Dole had kindly referenced some of our Princeton papers on the acceleration of heavy molecules by light carrier gases in free-jet expansions. When Csaba saw those references, he told Sandy that I was now at Yale, so Sandy tracked me down to show me the paper. Always on the lookout for new applications of free jets, I was very much intrigued and managed to interest a new graduate student, Mike Labowsky, in repeating Dole's experiments, which by then had been confirmed in a new apparatus at Northwestern and reported in a 1970 *Journal of Chemical Physics* paper. Our vacuum system was much bigger than Dole's and had much higher pumping speeds. Moreover, we had had more experience with free jet expansions. I realized that, in his experiment, the concurrent flow of bath gas and droplets meant that solvent vapor was present in the unheated mixture of gas and ions that entered the vacuum system. It was, therefore, highly likely that adiabatic cooling during free jet-expansion resolvated the ions to an appreciable extent, thus adding to their masses by an unknown amount. As a good chemical engineer, Mike knew that desolvation would be much more effective with bath gas flowing countercurrent to the drift of droplets and ions toward the end plate. Moreover, resolvation would be avoided, because only dry bath gas would then enter the free jet. With these changes, he obtained results somewhat different from and more reproducible than Dole's.

Both Dole and we had been persuaded by the work of R. Beuhler and L. Friedman that ions as large as we hoped we were producing would not generate secondary electrons at a dynode unless they were accelerated to about half a million volts. Therefore, we did not attempt to use ion-multiplier detectors, thus depriving ourselves of the mega-fold gain in sensitivity that mass spectrometrists take for granted. The currents of ions we could get into the vacuum system were very small and the vibrating-reed electrometer we used to measure them was very balky. Moreover, we knew from our earlier studies that there could be substantial slip effects during acceleration of very heavy molecules by a lighter carrier gas. Indeed, we estimated that the actual velocities of Dole's macroions were probably as much as 40% less than he had assumed. For these and other reasons, we abandoned further experiments. By that time, Dole had retired from Northwestern and moved to Baylor University, where he continued his experiments. However, instead of retarding potential measurements of mass, he was using mobility measurements to characterize the ions.

In 1981, our group was joined by Masamichi "Gado" Yamashita, a young scientist I had met during a stay at the University of Tokyo. During discussions about a possible project, I suggested that it might be interesting to take another look at Dole's ES ionization. Instead of macromolecules as analytes, we would use species with molecular weights less than 400 so they could be "weighed" with a small quadrupole mass analyzer we had in the laboratory.

Gado was a marvelous experimentalist and soon had converted a small "minibeam" apparatus into our first ES mass spectrometer. In an extensive set of exploratory experiments, with species small enough for our analyzer, he found that almost any solute organic molecule containing polar atoms such as O, S, N, and P would produce ions comprising the parent molecule with an anion or cation adduct. He also found that inorganic cations and anions could be produced, but generally in much lower abundances than he routinely obtained with organic solutes. As we later learned, the Aleksandrov group at the University of Leningrad (now St. Petersburg) had also begun to investigate ESI at about the same time as we did. From the fragmentary reports that we later obtained, it appeared that their emphasis was on ions from inorganic solutes.

One day in 1982 Sandy Lipsky, who had kept abreast of our work, brought VG's Brian Green to visit our lab. Brian was very interested and asked whether the technique would work with larger species. We pointed out that the upper mass limit for our analyzer was only 400 u so he arranged for VG to lend us a quadrupole with a mass range up to 1500 u. By that time, Gado had gone back to Japan and Craig Whitehouse, who had been working in Sandy Lipsky's lab, became a graduate student in Chemical Engineering. After joining my group, he designed and built a new system incorporating the VG analyzer and a modified version of Gado's electrospray source. That new system gave beautifully clean spectra for gramicidin S and cyclosporin, two cyclic decapeptides with almost the same M_r value. A provocative difference was that most cyclosporin ions had one charge, whereas most of those from gramicidin S had two. Moreover, for some slightly larger peptides, we obtained ions with three charges. Such multiple charging was most intriguing, so we decided to explore the phenomenon further with poly(ethylene glycol)s. They were attractive as test species because their oligomers were linear polymers whose composition and structure were essentially the same no matter what their size. (Moreover, samples over an M_r range from 200 to 20,000 were available at no charge from Union Carbide!) We found that the number of charges (Na^+) per oligomer increased with size, reaching at least 23 for an M_r of 20,000. Oddly enough, when we presented the data at the 1987 ASMS meeting in Denver, the only interest shown was by Charles McEwen, coauthor (with Barbara Larsen) of Chapter 5 and one of the first to recognize the virtues of ESI. He said right away that those PEG results were by far the most significant at the meeting. In contrast, a reviewer of our later paper dismissed the bands of overlapping peaks for large oligomers as spectra of dirt in the system, not of PEG oligomers!

To avoid the spectral complexity due to the wide distribution of molecular weights in PEG samples, we turned to nature to obtain samples comprising large molecules all of the same size, that is, pure proteins. We soon obtained spectra with the coherent sequences of peaks that have become the cachet of ES spectra for large molecules. The initial reaction of most mass spectrometrists to such peak multiplicity was one of horror. They were convinced that the resulting spectral complexity would make interpretation difficult, if not impossible. Moreover, the distribution of total charge among so many peaks would inevitably decrease signal/noise! They were wrong on both counts because they had reckoned without the powers of modern computers (as I still do!). One afternoon, I remarked to Matthias Mann, then a graduate student in my lab, that each of these multiple peaks really constituted an independent measure of parent ion mass. One should therefore be able to use signal-averaging methods to obtain a more accurate and reliable value of M_r. Two days later he came back with a deconvolution algorithm that allowed our little quadrupole, with a nominal mass limit of 1500 u, to determine M_r values up to 30,000 or more with an accuracy of 0.01%! At that time, M_r values from most other methods seldom had accuracies better than 5% to 10%. We presented these findings in San Francisco at the 1988 ASMS meeting;

the rest, as they say, is history. There were seven ES papers at that meeting: three from Henion's lab, two from Dick Smith's, and two from ours. Six years later, in 1994, at Chicago that number had climbed to over 300 where it remained in 1995 at Atlanta. I'm told that in the archival journals covered by the Citation Index, the term "electrospray" appeared in the title and/or abstract of over 300 papers in both 1994 and 1995. I'm sure Dole never dreamed that the seed he planted would bear so much fruit. Unfortunately, germination took so long that he didn't live to see the full magnitude of the stampede that began after our results with proteins became known.

My only face-to-face encounter with Malcolm Dole was in 1985 at the San Diego ASMS Meeting. At a session in which I gave a paper, I saw him in the audience and asked him to stand up and be recognized. Even though electrospray was not yet famous, the audience was generous in its applause, for which he (and I!) were grateful. My paper included some discussion of the importance of countercurrent drying gas to keep solvent vapor out of the free-jet expansion to avoid resolvation of the ions. At the end of the session, Dole came up to say he was delighted to learn about resolvation in the free jet, a possibility of which he had been unaware. He thought such resolvation might well account for an anomaly in his drift-tube experiments that had been puzzling him for a long time—the small differences between mobilities of ES ions from large and small molecules. (I now suspect that multiple charging of the larger ions, the extent of which didn't become clear until several years later, was as much or more to blame.) After our 1989 paper on ESMS of large biomolecules appeared in *Science*, Dole wrote me a nice note thanking me for the references to his papers and asking for a reprint. I sent it right away, along with one of our papers in *Mass Spectrometry Reviews* that included a more complete and complimentary account of his work. Three months later, I received a very apologetic note saying that he had taken the reprints along on a cruise and had somehow lost them. He would be most grateful if I could find another copy of each. He enclosed a twenty dollar bill to pay for whatever costs I might incur! I returned his money with the reprints and received a very gracious note of thanks, along with a copy of his privately printed autobiography. In it he made specific mention of my asking him to stand up and take a bow at the San Diego meeting, a gesture that was obviously very meaningful to him. He also quoted, clearly with great appreciation and satisfaction, my sometime remark to the effect that his electrospray idea was "extremely ingenious." Unfortunately, I never saw or heard from Dole again. Having not been invited, I did not attend the Electrospray Workshop in November 1991 at which he was present. I was delighted to learn that he had received much well-deserved attention, so that before he passed away some months later, he could begin to realize the abundance of the harvest from the seeds he had planted.

There is a presumption in many cultures that the longer one has lived in the past, the further can he or she see into the future. That may be one reason why Editor Cole suggested that my views on electrospray's future would be welcome. However, there have been so many surprises in its past that my crystal ball sees only great risks in any attempt to predict its future. Who would have dreamed 20 years ago that by now investigators would be able to examine the behavior of biopolymer ions in the gas phase, to determine their masses with accuracies approaching parts per billion, to study the kinetics and dynamics of their inter- and intramolecular reactions and processes, and to obtain detailed information on their composition, structure, and conformation that would provide insight on these properties in solution, *in vivo* as well as *in vitro*? Who could have anticipated that living organisms could be ionized, transferred into vacuum, recovered, and found to retain their viability? Yet this *tour de force* was recently accomplished with viruses by Siuzdak and his colleagues at Scripps Institute in La Jolla, California. With these so recently unbelievable achievements of the electrospray methodology already in hand, who would dare to imply any limit on its possibilities by undertaking to guess what the future might reveal? Instead, I would simply urge investigators to exorcise their inhibitions and exercise their imaginations. May the platform of past accomplishments, so ably described in this volume, serve as a launching pad for their flights of fancy into the future. Let them dare to take off for any star whose twinkle beckons. Let them hearken to the words of Robert Browning in "Andrea del Sarto": "Ah, but a man's reach should exceed his grasp, or what's a heaven for?"

Richmond, Virginia JOHN. B. FENN

Preface

One dozen years have passed since the book *Electrospray Ionization Mass Spectrometry: Fundamentals, Instrumentation, and Applications* was first published. During that time, the expansion of mass spectrometry has taken on global proportions. In addition to its habitual place in chemistry and physics laboratories, advanced mass spectrometry instrumentation has truly become a staple in biological and biochemical research facilities. In many ways, the biochemical revolution, sparked by past efforts such as the human genome project, has driven the development of mass spectrometry over the past two decades. But the benefits of this progress are indeed far-reaching, and they are being reaped in the traditional areas of chemistry and physics from which mass spectrometry arose, in addition to the more recently emerged frontier domains such as proteomics, metabolomics and materials science. The purpose of this new book is to update experts and inform novices about the most recent trends in modern mass spectrometry. The chapters are written in a style intended to serve as both critical reviews and tutorials. The content is designed to provide a thorough coverage of the major topics in current day mass spectrometry. The intent was to make this volume highly suited both as a wide-ranging reference tome and as a classroom tool to support a graduate-level course in Advanced Mass Spectrometry.

The development of electrospray ionization and matrix-assisted laser desorption/ionization as "soft" mass spectrometric ionization methods occurred virtually in parallel. The sudden growth of each began at almost the same time and was spurred primarily by the new possibilities that their advent provided for analyses of biopolymers, especially proteins. The impact of the two approaches on analytical chemistry—and in particular, biomolecule analysis—was both major and complementary. The overlapping histories of development of electrospray and MALDI over the past quarter century are like two sides of the same coin; for this reason, both have been included in this new volume.

Part I of the book is dedicated to explaining fundamental aspects of the electrospray process. First, the detailed fundamentals of electrospray are considered from a mechanistic viewpoint (Chapter 1). Special attention is then given to the root causes of the observed selectivity of ionization in electrospray (Chapter 2). Inherent to electrospray ionization sources are electrochemical processes that are explained in detail in Chapter 3. The ES fundamentals section is completed with a comparative inventory of source hardware (Chapter 4).

Part II turns to the MALDI side, and opens with an overview of ionization mechanisms in MALDI (Chapter 5). This is followed by a presentation of the progressive development of MALDI hardware up to the present (Chapter 6). As practitioners know, the art of MALDI lies in sample preparation, and an overview of matrices is given in Chapter 7. The special application of MALDI to obtaining two-dimensional images of the spatial distribution of compounds on surfaces is presented in Chapter 8.

Each of the remaining chapters considers both electrospray and MALDI in addressing the respective topic of interest. Part III examines the coupling of these ionization techniques to various mass analyzers. First up, quadrupoles, quadrupole ion traps, linear quadrupole ion traps, and the Orbitrap are discussed in Chapter 9. The development of MALDI and electrospray has also led to a renaissance in time-of-flight technology as detailed in Chapter 10. Ultra-high-resolution Fourier transform ion cyclotron resonance mass spectrometers as well as magnetic sector mass analyzers are examined in Chapter 11. Last comes Chapter 12's presentation on ion mobility spectrometry that is particularly oriented toward biological issues.

In Part IV of the book, analytical and practical issues of electrospray and MALDI are confronted. Chapter 13 presents an analytical comparison between electrospray and MALDI, with atmospheric-pressure chemical ionization thrown in for good measure. An investigation into the factors that influence charge state distributions in electrospray and MALDI appears in Chapter 14. Next comes a detailed overview of the utility of electrospray and MALDI for investigating noncovalent interactions that are preserved as gas-phase ions (Chapter 15). Attention then turns to discussion of the various ion activation techniques that are used to provoke dissociation of the typically stable intact precursors generated by electrospray and MALDI in preparation for tandem mass spectrometry experiments (Chapter 16). Chapter 17 presents a primer on interpretation of mass spectra specifically targeting decompositions of the even-electron ions produced by soft ionization techniques.

Part V targets applications that are organized according to specific compound classes of analytes. The role of mass spectrometry in peptide and protein characterization and in proteomics is the subject of Chapter 18. Next, the topic of carbohydrate analysis by ESI and MALDI is tackled in Chapter 19. This is followed by an examination of ESI and MALDI applications to lipid analysis (Chapter 20). Finally, the important subject of drug discovery is addressed in Chapter 21, including *in vitro* ADME (absorption, distribution, metabolism and excretion) profiling and pharmacokinetic screening.

I must thank all of the contributing authors who, in my humble opinion, rank among the finest mass spectrometrists in the world. Their dedication and perseverance throughout this project was truly remarkable. Out of necessity, the lives of scientists in the 21st century have become more multifaceted than ever, and the time and effort that is dedicated to a project like this book must be carved out of schedules that are fuller than the day has hours. Nevertheless, this group of consummate professionals delivered their contributions in extraordinary fashion. In this same vein, I extend my utmost gratitude to Professor R. Graham Cooks for agreeing to write the Foreword for this new edition. It is important to point out that each of the chapters was reviewed by independent experts in the field, whom I thank enormously for their time and effort. I must also thank my super-assistants Ms. Ashley A. Smith, Mr. Kevin M. McAvey and Ms. Priya Bariya for their vital help with the assembling and formatting of the final product. Finally, I thank Dr. Dominique M. Custos for *terra firma* support throughout the entire project.

New Orleans, Louisiana Richard B. Cole

Contributors

Dr. Carlos Afonso, Laboratoire de Chimie Structurale Organique et Biologique, Université Pierre & Marie Curie, Paris, France

Dr. I. Jonathan Amster, Department of Chemistry, University of Georgia, Athens, Georgia

Dr. Mark E. Bier, Center for Molecular Analysis, Department of Chemistry, Mellon College of Science, Carnegie Mellon University, Pittsburgh, Pennsylvania

Dr. Andries P. Bruins, Mass Spectrometry Facility, University of Groningen, The Netherlands

Dr. Alain Brunelle, Institut de Chimie des Substances Naturelles, CNRS, Gif-sur-Yvette cedex, France

Dr. Nadja B. Cech, Department of Chemistry and Biochemistry, University of North Carolina at Greensboro, Greensboro, North Carolina

Dr. Richard B. Cole, Department of Chemistry, University of New Orleans, New Orleans, Louisiana

Dr. R. Graham Cooks, Department of Chemistry, Purdue University, West Lafayette, Indiana

Dr. Jay. J. Corr, ABI/MDS-Sciex, Concord, Ontario, Canada

Dr. Robert J. Cotter, Middle Atlantic Mass Spectrometry Laboratory, Department of Pharmacology and Molecular Sciences, Johns Hopkins School of Medicine, Baltimore, Maryland

Dr. Thomas R. Covey, ABI/MDS-Sciex, Concord, Ontario, Canada

Dr. Jason Dunsmore, Department of Chemistry and Biochemistry, University of California—Los Angeles, Los Angeles, California

Dr. Christie G. Enke, Department of Chemistry Chemical Biology, University of New Mexico, Albuquerque, New Mexico

Dr. Kristina Håkansson, Department of Chemistry, University of Michigan, Ann Arbor, Michigan

Dr. Xialin Han, Division of Bioorganic Chemistry and Molecular Pharmacology, Department of Medicine, Washington University School of Medicine, St. Louis, Missouri

Dr. David J. Harvey, Glycobiology Institute, Department of Biochemistry, University of Oxford, Oxford, United Kingdom

Dr. Mahmud Hossain, Rieveschl Laboratories for Mass Spectrometry, Department of Chemistry, University of Cincinnati, Cincinnati, Ohio

Dr. Gordana Ivosev, ABI/MDS-Sciex, Concord, Ontario, Canada

Dr. Hassan Javaheri, ABI/MDS-Sciex, Concord, Ontario, Canada

Dr. Daniel B. Kassel, Takeda San Diego, Inc., San Diego, California

Dr. Paul Kebarle, Department of Chemistry, University of Alberta, Edmonton, Alberta, Canada

Dr. Vilmos Kertesz, Organic and Biological Mass Spectrometry Group, Chemical Sciences Division, Oak Ridge National Laboratory, Oak Ridge, Tennessee

Dr. John S. Klassen, Department of Chemistry, University of Alberta, Edmonton, Alberta, Canada

Dr. Richard Knochenmuss, Novartis Institute of Biomedical Research, Basel, Switzerland

Dr. Peter Kovarik, ABI/MDS-Sciex, Concord, Ontario, Canada

Dr. Olivier Laprévote, Institut de Chimie des Substances Naturelles, CNRS, Gif-sur-Yvette cedex, France

Dr. J.C. Yves LeBlanc, ABI/MDS-Sciex, Concord, Ontario, Canada

Dr. Yan Li, Protein Chemistry Technology Center, University of Texas Southwestern Medical Center, Dallas, Texas

Dr. Patrick A. Limbach, Rieveschl Laboratories for Mass Spectrometry, Department of Chemistry, University of Cincinnati, Cincinnati, Ohio

Dr. Shirley H. Lomeli, Department of Chemistry and Biochemistry, University of California—Los Angeles, Los Angeles, California

Dr. Joseph A. Loo, Department of Chemistry and Biochemistry, Department of Biological Chemistry, David Geffen School of Medicine, University of California—Los Angeles, Los Angeles, California

Dr. Sonal Mathur, Max Planck Institute for Chemical Ecology, Jena, Germany

Dr. John A. McLean, Department of Chemistry, Vanderbilt Institute of Chemical Biology, and Vanderbilt Institute of Integrative Biosystems Research and Education,Vanderbilt University, Nashville, Tennessee

Dr. Alexis Nazabal, CovalX AG, Zurich, Switzerland

Dr. Peter B. O'Connor, Department of Chemistry, University of Warwick, Coventry, United Kingdom

Dr. Rachel R. Ogorzalek Loo, Department of Biological Chemistry, David Geffen School of Medicine, University of California—Los Angeles, Los Angeles, California

Dr. Bradley B. Schneider, ABI/MDS-Sciex, Concord, Ontario, Canada

Dr. J. Albert Schultz, Ionwerks Inc., Houston, Texas

Dr. Jean-Claude Tabet, Laboratoire de Chimie Structurale Organique et Biologique, Université Pierre & Marie Curie, Paris, France

Dr. Mario Thevis, Department of Chemistry and Biochemistry, University of California—Los Angeles, Los Angeles, California

Dr. Gary J. Van Berkel, Organic and Biological Mass Spectrometry Group, Chemical Sciences Division, Oak Ridge National Laboratory, Oak Ridge, Tennessee

Dr. Udo H. Verkerk, Center for Research in Mass Spectrometry, York University, Toronto, Ontario, Canada

Dr. Jeremy J. Wolff, Department of Chemistry, University of Georgia, Athens, Georgia

Dr. Amina S. Woods, Cellular Neurobiology, Intramural Research Program, National Institute on Drug Abuse, National Institutes of Health, Baltimore, Maryland

Dr. Yongming Xie, Department of Chemistry and Biochemistry, University of California—Los Angeles, Los Angeles, California

Dr. A. Jimmy Ytterberg, Department of Chemistry and Biochemistry, University of California—Los Angeles, Los Angeles, California

Dr. Renato Zenobi, Department of Chemistry and Applied Biosciences, ETH Zurich, Zurich, Switzerland

Part I

Fundamentals of ES

Chapter 1

On the Mechanism of Electrospray Ionization Mass Spectrometry (ESIMS)

Paul Kebarle [*] ***and Udo H. Verkerk*** [†]

[*] *Department of Chemistry, University of Alberta, Edmonton, Alberta, Canada*
[†] *Center for Research in Mass Spectrometry, York University, Toronto, Ontario, Canada*

Electrospray and MALDI Mass Spectrometry: Fundamentals, Instrumentation, Practicalities, and Biological Applications, Second Edition, Edited by Richard B. Cole

1.1 INTRODUCTION

1.1.1 How It All Started

Electrospray ionization (ESI) is a method by which solutes present in a solution can be transferred into the gas phase as ions. The gas-phase ions can then be detected by mass spectrometric means (ESIMS). Remarkably, ESI can handle solutes such as polymers, nucleic acids, and proteins that have a very high molecular mass such as hundreds of megadaltons for proteins. The analytes present in the solution may be ions, such as the inorganic metal ions M^+ and M^{2+} protonated amines or negative ions such as the halide ions X^- or deprotonated carboxylic acids, sulfates SO_4^{2-}, and so on. They can be also compounds that are neutral in the solution that is sprayed. In that case, the analyte is charged by association with one or more ions present in the solution. This charging process is part of the electrospray mechanism. ESIMS is the ideal method for detection of analytes from high-performance liquid chromatography or capillary electrophoresis. ESIMS is particularly valuable to biochemical, biomedical, and pharmacological research. The significance of ESIMS was recognized by a Nobel Prize in 2002 to John Fenn, who was the major developer of the method.[1]

ESIMS is actually the brainchild of Malcolm Dole. In the 1960s, Malcolm Dole was very interested in the determination of the molecular mass of synthetic polymers and developing a method with which one could observe such macromolecules by mass spectrometry. It was clear to him that mass spectrometric analysis of the polymer molecules could answer many questions and solve many problems. But how could one get large polymers into the gas phase without decomposing them? He had the idea that if one uses a very dilute solution of the analyte and then nebulized the solution into extremely small droplets, one might obtain many droplets that contained only one analyte molecule. Evaporation of the droplets would then lead to a transfer of the analyte molecules to the gas phase. If the analyte was not charged, as was often the case for synthetic polymers, the presence of an electrolyte, such as Na^+ and Cl^- in the solution, could lead to charging of the polymer. Evaporation of a droplet that happens to contain one polymer molecule and one

Na^+ would lead to the desired charged analyte. However, for charging to occur, there has to be one or more functional groups on the analyte with which the ion can form a fairly strongly bonded complex in the absence of the solvent. For details of such complex formation see Section 1.2.13.1. In other droplets, there would be one Cl^- and one polymer, and that would lead to a negatively charged analyte that could be observed with the mass spectrometer in the negative ion mode. Such statistical charging was known to occur[2] and to be a rather inefficient source of ionized analytes.

Dole was preoccupied with thoughts on how to increase the efficiency, when a possible solution presented itself. While working as a consultant for a paint company,[3] he witnessed the electrospraying of paint on automobile bodies. The paint was sprayed on the cars very efficiently by very small charged paint droplets using a process known as electrospray. Applying electrospray to polystyrene solutions, Dole et al.[4] were able to develop an apparatus and demonstrate the production in the gas phase of polystyrene ions with molecular masses in the kilodalton range. While Dole's methods and results had some flaws and ambiguities, they clearly indicated that electrospray is a very promising soft ionization method for the mass spectrometry of macromolecules.

Dole et al.'s paper[4] caught the eye of Professor Seymour Lipsky at Yale Medical School. Lipsky, who was also involved with mass spectrometry, was very excited about the potential of electrospray for the mass spectrometric study of proteins. Dole et al.[4] had used a nozzle-skimmer system as the interface between the atmospheric pressure required for electrospray and the mass analysis region, and their paper contained a reference to work by John Fenn, who was a specialist in the field of molecular beams and their production by nozzle-skimmer systems. Through this reference, Lipsky got in touch with Fenn, who was also at Yale but at the Department of Mechanical Engineering. This contact inspired Fenn to start research on electrospray mass spectrometry. Since only a low-mass-range quadrupole was available in Fenn's laboratory, the first, pioneering papers[5] involved studies of small ions in the positive[5a] and negative ion[5b] mode. The ES ion source and interface to the mass analysis region were similar to those used by Dole, but included a few important changes such as the use of nitrogen gas counter flow at atmospheric pressure to remove solvent vapor caused by the droplet evaporation. This measure led to clean and relatively simple mass spectra that could be easily interpreted.

Subsequent work by Fenn and co-workers,[6] using a quadrupole mass analyzer with a high m/z range extending to m/z 1500 and a heated capillary as the interface between the spray at atmospheric pressure and the vacuum containing the mass analysis section, clearly demonstrated that ESIMS could be used very effectively for analysis of peptides and proteins with molecular mass m which could be much higher than $m = 1500$ daltons. This was possible because the use of ESI led to molecular ions that had multiple charges z so that the m/z value was lower than $m/z = 1500$. This work had a big impact and started the ESIMS revolution that is continuing to this day.

The development of ESIMS is clearly due to two people, Malcolm Dole and John Fenn. Dole's early death occurred before the full impact of ESIMS became evident. Fenn,[1] in his account on the development of the method, clearly acknowledges the seminal significance of Dole's work. Much of the information in this section is based on Fenn's account.[1]

1.1.2 Aims of This Chapter

This chapter is written for users of ESIMS. It presents an account of "how it all works." It addresses those just entering the field and more advanced users. Understanding of "how it all works" is desirable not only from a standpoint of intellectual curiosity, but also for practical reasons. The mass spectra that one observes depend on a large number of parameters. These start with (a) a choice of solvent and concentrations of the analyte, (b) a choice of additives to the solution that may be beneficial, and (c) their concentration, choice of the flow rates of the solution through the spray capillary, the electrical potentials applied to the spray capillary (also called "needle"), and the potentials on the electrodes leading to the mass analysis. The choice of these parameters requires not only an understanding of conventional mass spectrometry but also an understanding of the electrospray mechanism as well as some familiarity with chemistry in solution as well as ion-molecule reactions in the gas phase. In early work on ESIMS, many of the parameters were established experimentally by trial and error; but now when a better understanding of the mechanism is at hand, it is certainly more efficient and analytically rewarding to understand the reasons for the choices.

Unfortunately, not all the processes that occur in ESIMS are well understood, and this has led to some controversy. However, enough is known to make the study of the mechanisms in ESIMS worthwhile.

The present chapter presents only a limited account of electrospray. A much more extensive coverage is provided by a review by Smith et al.[7a] While Smith's review is quite old, most of the material is still very relevant.

1.1.3 Electrospray, Other than Mass Spectrometric Applications

ES existed long before its application to mass spectrometry. It is a method of considerable importance for the electrostatic dispersion of liquids and creation of aerosols. The interesting history and notable research advances in that field are very well described in Bayley's book[7b] *Electrostatic Spraying of Liquids*. Much of the theory concerning the mechanism of the charged droplet formation was developed by researchers in aerosol science. A compilation of articles in this area can be found in a special issue[7c] of the *Journal of Aerosol Science* devoted to electrospray.

1.2 PRODUCTION OF GAS-PHASE IONS BY ELECTROSPRAY AND ELECTROSPRAY IONIZATION MASS SPECTROMETRY

1.2.1 The Overall Process

There are three major steps in the production of gas-phase ions from electrolyte ions in solution. These are: (a) production of charged droplets at the ES capillary tip; (b) shrinkage of the charged droplets by solvent evaporation and repeated droplet disintegrations leading ultimately to very small highly charged droplets capable of producing gas-phase ions; and (c) the actual mechanism by which gas-phase ions are produced from the very small and highly charged droplets. Stages (a)–(c) occur in the atmospheric pressure region of the apparatus (see Figure 1.1).

Some of the ions resulting from the preceding stages (a)–(c) enter the vacuum region of the interface leading to the mass spectrometer either through a small orifice or through a sampling capillary (see Figure 1.2a,b). The ions may be clustered with solvent molecules and other additives and are subjected to (i) a thermal declustering "clean-up" in the heated capillary leading to the partial vacuum (pressure of a few torr) of the first chamber

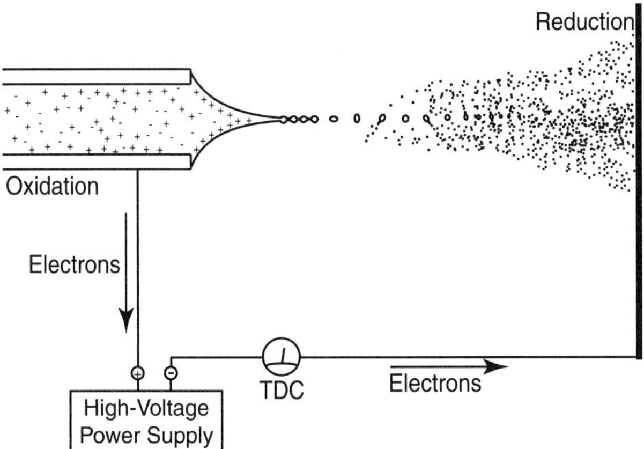

Figure 1.1. Schematic of major processes occurring in the atmospheric pressure region of electrospray. TDC stands for total droplet current (*I*). This figure illustrates major processes occurring in the atmospheric pressure region of ESI run in the positive ion mode. Penetration of the imposed electric field into the liquid leads to formation of an electric double layer at the meniscus. The double layer is due to the polarizabilty and dipole moments of the solvent molecules and an enrichment near the meniscus of positive ions present in the solution. These cause a destabilization of the meniscus and formation of a cone and a jet charged by an excess of positive ions. The jet splits into droplets charged with an excess of positive ions. Evaporation of the charged droplets brings the charges closer together. The increasing Coulombic repulsion destabilizes the droplets that emit a jet of smaller charged progeny droplets. Evaporation of progeny droplets leads to destabilization and emission of second-generation progeny droplets, and so on, until free gas-phase ions form at some point.

and (ii) collisional activation due to an electric potential difference imposed between the sampling capillary exit and the skimmer leading to the second, high-vacuum chamber that is the housing of the mass spectrometer.

1.2.2 Production of Charged Droplets at the ES Capillary Tip. The Electrophoretic Mechanism

As shown in the schematic representation of the charged droplet formation (Figure 1.1), a voltage V_c, of 2–3 kV, is applied to the conductive capillary, which is typically 1 mm o.d. and located 1–3 cm from the counterelectrode. The counterelectrode in ESMS may be a plate with an orifice leading to the mass spectrometric sampling system or a sampling capillary, mounted on the plate, which leads to the MS (see Figure 1.2a). Because the spray capillary tip is very thin, the electric field E_c at the capillary tip is very high ($E_c \approx 10^6$ V/m). The value of the field at the capillary tip opposite a large and planar counterelectrode can be evaluated with the approximate relationship[8]

$$E_c = 2V_c/[r_c \ln(4d/r_c)] \tag{1.1}$$

where V_c is the applied potential, r_c is the capillary outer radius, and d is the distance from capillary tip to the counterelectrode. For example, the combination $V_c = 2000$ V, $r_c = 5 \times 10^{-4}$ m, and $d = 0.02$ m leads to $E_c \approx 1.6 \times 10^6$ V/m. The field E_c is proportional to V_c, and the most important geometry parameter is r_c. E_c is essentially inversely proportional to r_c, while E_c decreases very slowly with the electrode separation d, due to the logarithmic dependence on d. For potentials required for electrospray, see Section 1.2.4.

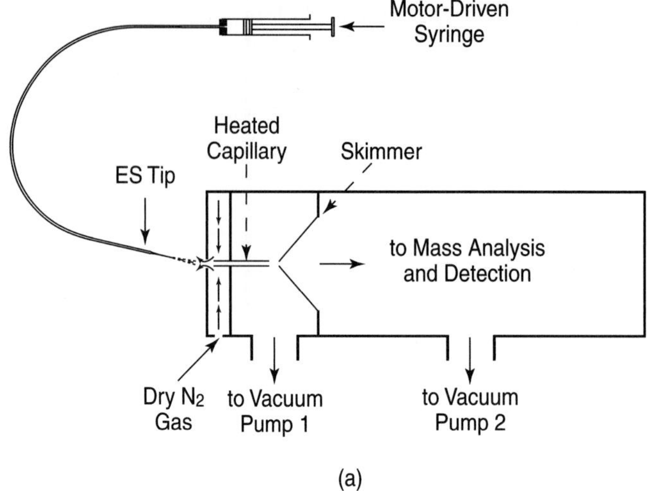

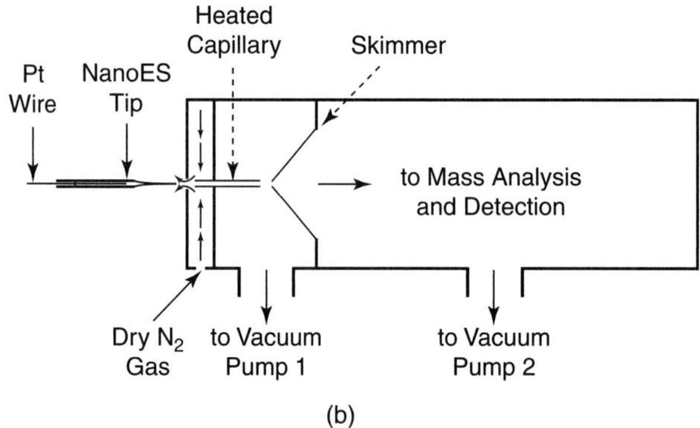

Figure 1.2. (**a**) Schematic of electrospray and interface to mass spectrometer. Solution containing analyte is supplied to the ES spray tip by a motor-driven syringe via flexible glass capillary tubing. A positive potential is applied to the spray tip (positive ion mode). The spray of positively charged droplets emerges from the spray capillary tip (see Figure 1.1). Solvent evaporation of the charged droplets leads to gas-phase ions. A mixture of ions, small charged droplets, and solvent vapor in the ambient gas enters the orifice leading to the nitrogen countercurrent chamber. The weak nitrogen countercurrent removes the solvent vapor; but the ions, driven by an electric potential and pressure difference, enter the heated capillary pathway into the low-pressure chamber. An electric field between this capillary and the skimmer cone accelerates the ions for a further collision activated "clean-up" of the ions. The potential difference over the cone orifice and downstream ion optical elements transports the ions into the high-vacuum region of the mass analysis chamber. (**b**) Same as Figure 1.2a but for nanoelectrospray. Large-diameter end of NanoES tip capillary is "loaded" with microliter amounts of solution. The electrical potential is supplied to the Nano tip either by a Pt wire or by a metal film coating the outside of the capillary. A spray of charged nano droplets results from the pull of the electric field on the polarized meniscus of the solution at the capillary tip.

A typical solution present in the capillary consists of a polar solvent in which the analyte is soluble. Because ESIMS is a very sensitive method, very low concentrations, 10^{-7}–10^{-3} moles/liter (M) of analyte need to be used. Methanol (or methanol–water) or acetonitrile (or acetonitrile–water) is often used as the solvent. For simplicity we will consider that the analyte is ionic and only the positive ion mode will be considered in the subsequent discussion.

When turned on, the field E_c will penetrate the solution near the capillary tip. This will cause a polarization of the solvent near the meniscus of the liquid. Assuming that the solvent is water, the field will align the permanent dipoles of H_2O so that on the average there will be many molecules oriented with the H atoms pointing downfield. Due to the polarizability of the H_2O molecule, induced dipoles with the same downfield orientation will also result. In the presence of even traces of an electrolyte, the solution will be sufficiently conducting and the positive and negative electrolyte ions in the solution will move under the influence of the field, This will lead to an enrichment of positive ions near the surface of the meniscus and negative ions away from the meniscus. The combined downfield forces due to these processes cause a distortion of the meniscus into a cone pointing downfield (see Figure 1.1). The increase of surface due to the cone formation is resisted by the surface tension of the liquid. The cone formed is called a Taylor cone (see Taylor[9] and Fernandez de la Mora[10]). If the applied field is sufficiently high, a fine jet emerges from the cone tip, whose surface is charged by an excess of positive ions. The jet breaks up into small charged droplets (see Figure 1.1; for a more accurate representation of the cone jet mode, see Figure 1.3a, due to Cloupeau[11a]).

It is apparent from Figure 1.3a that the size of the droplets formed from the cone jet is dependent on the jet diameter $2R_J$, and therefore all droplets produced could be expected to be approximately of the same size—that is, approximately monodisperse. This was proposed by Cloupeau[11a] and confirmed by studies of Tang and Gomez.[12] Also shown in Figure 1.3a is a much smaller "satellite" droplet. The satellite droplets are commonly observed, but their role in the ultimate formation of gas-phase ions out of the charged droplets is probably minor.

The droplets are positively charged due to an excess of positive electrolyte ions at the surface of the cone and the cone jet. Thus, if the major electrolyte present in the solution was ammonium acetate, the excess positive ions at the surface will be NH_4^+ ions. This mode of charging, which depends on the positive and negative ions drifting in opposite directions under the influence of the electric field, has been called the electrophoretic mechanism.[11b,c] The charged droplets produced by the cone jet drift downfield through the air toward the opposing electrode. Solvent evaporation at constant charge leads to droplet shrinkage and an increase of the electric field normal to the surface of the droplets. At a given radius, the increasing repulsion between the charges overcomes the surface tension at the droplet surface. This causes a coulomb fission of the droplet, also called a coulomb explosion. The droplet fission occurs, via formation of a cone and a cone jet that splits into a number of small progeny droplets. This process bears close resemblance to the cone jet formation at the capillary tip (see Fernandez de la Mora[10] and references therein). Further evaporation of the

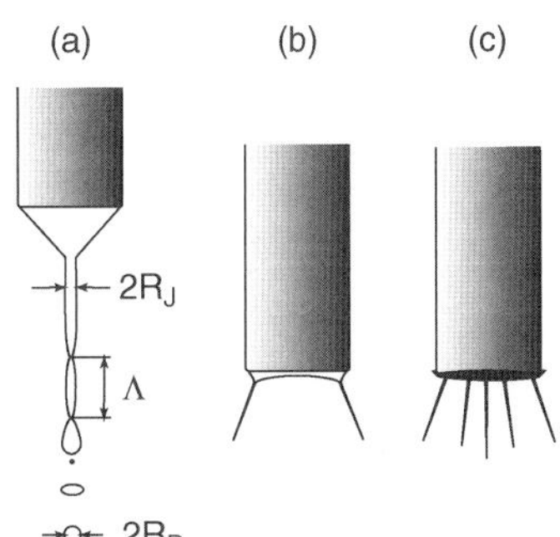

Figure 1.3. Different forms of electrospray at the tip of the spray capilary. (**a**) Cone jet mode. Relationship between radius of droplets and radius of jet: $R_D/R_J \approx 1.9$. Much smaller, satellite droplets can also be produced (see small droplet in this figure). These are more common at higher flow rates. In such cases, two types of monodisperse droplets are observed: the large progeny droplets and the small (satellite) progeny. (**b**), (**c**) Multijet modes result as the spray voltage is increased and the flow rate imposed by the syringe is high. (After Cloupeau.[11a])

parent droplet leads to repeated fissions. The progeny droplets also evaporate and fission. More details on these processes are given in Section 1.2.6. Very small charged droplets result that lead ultimately to gas-phase ions by processes that will be described in detail in subsequent sections.

The cone-jet mode at the spray capillary tip described and illustrated in Figure 1.1 and Figure 1.3a is only one of the possible ES modes. For a qualitative description of this and other modes, see Cloupeau[11a–c]. More recent studies by Vertes and co-workers[11d] using fast time-lapse imaging of the Taylor cone provide details on the evolution of the Taylor cone into a cone jet and pulsations of jet. The pulsations lead to spray current oscillations. The current oscillations are easy to determine with conventional equipment and can be used as a guide for finding conditions that stabilize the jet and improve signal-to-noise ratios of the mass spectra. The cone-jet mode is most often used in ESIMS. It is also the best-characterized mode in the electrospray literature[7,10–12] and references therein.

1.2.3 Electrospray as an Electrolytic Cell of a Special Kind

At a steady operation of the electrospray in the positive ion mode (see Figure 1.1), the positive droplet emission will continuously carry off positive charge. The requirement for charge balance in such a continuous electric current device, together with the fact that only electrons can flow through the metal wire supplying the electric potential to the electrodes (Figure 1.1), leads to the supposition that the ES process must include an electrochemical conversion of ions to electrons. In other words, the ES device can be viewed as a special type of electrolytic cell.[13] It is special because the ion transport does not occur through uninterrupted solution, as is normally the case in electrolysis. In the positive ion mode, part of the ion transport occurs through the gas phase where positively charged droplets and later positive gas-phase ions are the charge carriers. A conventional electrochemical oxidation reaction should be occurring at the positive electrode—that is, at the liquid–metal interface of the capillary (Figure 1.1). This reaction should be supplying positive ions to the solution. The nature of these ions depends on the experimental conditions. If the spray capillary is made out of metal, the neutral metal ions can become oxidized to the positive ionic state by releasing electrons to the metal electrode and entering the solution [see Eq. 1.2]. The other alternative is the removal of negative ions present in the solution by an oxidation reaction as illustrated below [Eq. 1.3] for aqueous solutions:

$$M(s) \rightarrow M^{2+}(aq) + 2e \qquad \text{(on metal surface)} \tag{1.2}$$

$$4OH^{-}(aq) \rightarrow O_2(g) + 2H_2O + 4e \qquad \text{(on metal surface)} \tag{1.3}$$

One expects that the reaction with the lowest oxidation potential will dominate, and that reaction will be depend on the material present in the metal electrode, the ions present in the solution, and the nature of the solvent. Proof for the occurrence of an electrochemical oxidation at the metal capillary was provided by Blades et al.[13] When a Zn capillary tip was used, release of Zn^{2+} to the solution could be detected. Furthermore, the amount of Zn^{2+} release to the solution per unit time when converted to coulombs/second was found to be equal to the measured electrospray current, I (Figure 1.1). Similar results were observed with stainless steel capillaries.[13] These were found to release Fe^{2+} to the solution. These quantitative results provided the strongest evidence for the electrolysis mechanism.

It should be noted that the oxidation reaction described in Eq. 1.2 adds ions that were not present previously in the solution. On the other hand, the oxidation, Eq. 1.3, provides the positive current by removing the negative counterions of positive ions that are already present in the solution. The excess unipolar ions provided to the solution when expressed as concentrations in the solution amount to very low concentrations. Taking the Zn capillary tip as example, a solution of 10^{-5} M NaCl in methanol at a flow rate $V_f = 20\,\mu$L/min was found to lead to an electrospray current of 1.6×10^{-7} A. The Zn^{2+} concentration produced by the Zn-tipped capillary evaluated from the current was $[Zn^{2+}] = 2.2 \times 10^{-6}$ M. Assuming that the Na^+ ion was the analyte ion, the concentration of the ions produced by the oxidation at the electrode is only ~1/5 of that of the analyte. It will be shown later that the electrospray current, I, increases very slowly with the total electrolyte concentration. Therefore, at higher total electrolyte concentrations due to analyte and additives, the ions produced by oxidation at the electrode may not be noticed in the mass spectrum because of the nature of the charges on the droplets that will ultimately lead to the detected ions. That is, one must consider all of the positive ions in solution and not only the positive ions produced at the positive electrode.

Van Berkel and co-workers, in a series of publications, have examined the consequences of the electrochemical processes to ESIMS. For a summary, see Chapter 3 by Van Berkel in this book. For example, they were able to demonstrate[14] that ions produced by the electrolysis process, such as hydrogen ions, can in some cases have important effects on the mass spectra obtained with pH-sensitive analytes such as nondenatured proteins.

Surprisingly, some skepticism on the significance to ESIMS of the electrochemicaly produced ions has been expressed in the literature and this led to a special issue of the *Journal of Mass Spectrometry*.[15] The consensus derived from this special issue is that for the typical ESIMS apparatus that provides a continuous unipolar charged droplet current, see Figure 1.1, an electrochemical process does occur at the spray electrode, in agreement with Blades et al.[13] and the work of van Berkel and coworkers. Most often, the ions created by the electrolytic process do not interfere with the analytical MS work. However, there are cases when one needs to consider the effect of the electrolytic ions.

1.2.4 Required Electrical Potentials for ES. Electrical Gas Discharges

D. P. H. Smith[16] was able to derive a very useful approximate equation for the required electric field, E_{on}, at the capillary tip, which leads to an onset of instability of a static Taylor cone and to the formation of a jet at the apex of the cone.

$$E_{on} \approx \left(\frac{2\gamma \cos\theta}{\varepsilon_0 r_c}\right)^{1/2} \tag{1.4}$$

This equation for the onset field, when combined with Eq. 1.1, leads to an equation for the potential, V_{on}, required for the onset of electrospray:

$$V_{on} \approx \left(\frac{r_c \gamma \cos\theta}{2\varepsilon_0}\right)^{1/2} \ln(4d/r_c) \tag{1.5}$$

where γ is the surface tension of the solvent, ε_0 is the permittivity of vacuum, r_c is the radius of the capillary, and θ is the half-angle for the Taylor cone. Substituting the values $\varepsilon_0 = 8.8 \times 10^{-12}\,J^{-1}\,C^2$ and $\theta = 49.3$ (see Taylor[9]), one obtains

$$V_{on} = 2 \times 10^5 (\gamma r_c)^{1/2} \ln(4d/r_c) \tag{1.6}$$

Table 1.1. Onset Voltages, V_{on}, for Solvents with Different Surface Tension, γ

Solvent	CH_3OH	CH_3CN	$(CH_3)_2SO$	H_2O
γ (N/m)	0.0226	0.030	0.043	0.073
V_{on} (kV)	2.2	2.5	3.0	4.0

where γ must be substituted in Newtons per meter and r_c in meters to obtain V_{on} in volts. Shown in Table 1.1 are the surface tension values for four solvents and the calculated electrospray onset potentials for $r_c = 0.1$ mm and $d = 40$ mm. The surface of the solvent with the highest surface tension (H_2O) is the most difficult to stretch into a cone and jet, and this leads to the highest value for the onset potential V_{on}.

Experimental verification of Eqs. (1.5)–(1.7) has been provided by Smith,[16] Ikonomou et al.,[17] and Wampler et al.[18] For stable ES operation, one needs to go a few hundred volts higher than V_{on}. Use of water as the solvent can lead to the initiation of an electric discharge from the spray capillary tip, particularly when the capillary is negative—that is, in the negative ion mode. The electrospray onset potential V_{on} is the same for both the positive and negative ion modes; however, the electric discharge onset is lower when the capillary electrode is negative[17,18] and metallic. This is probably due to emission of electrons from the negative capillary which initiate the discharge. Use of capillaries that are made out of glass where the electrical potential is applied via an internal metal wire that is embedded in the solution reduces the risk of electric discharge. For an illustration see Figure 1.2b. Neat water as solvent can be used with this arrangement and nanoelectrospray in the positive ion mode without the occurrence of electric discharge.

The occurrence of an electric discharge leads to an increase of the capillary current, I. Currents above 10^{-6} A are generally due to the presence of an electric discharge. A much more specific test is provided by the appearance of discharge-characteristic ions in the mass spectrum. Thus, in the positive ion mode the appearance of protonated solvent clusters such as $H_3O^+(H_2O)_n$ from water or $CH_3OH_2{}^+(CH_3OH)_n$ from methanol indicates the presence of a discharge.[17] The protonated solvent ions are produced at high abundance by ES in the absence of a discharge, only when the solvent has been acidified—that is, when H_3O^+ and $CH_3OH_2{}^+$ are present in the solution.

The presence of an electric discharge degrades the performance of ESMS, particularly so at high discharge currents. The electrospray ions are observed at much lower intensities than was the case prior to the discharge, and the discharge-generated ions appear with very high intensities.[17,18]

The high potentials required for electrospray show that air at atmospheric pressure is not only a convenient, but also a very suitable, ambient gas for ES, particularly when solvents with high surface tension, such as water, are to be electrosprayed. The oxygen molecules in air have a positive electron affinity and readily capture free electrons. Initiation of gas discharges occurs when free electrons present in the gas (due to cosmic ray or background radiation) are accelerated by the high electric field near the capillary to velocities where they can ionize the gas molecules. At near-atmospheric pressures, the collision frequency of the electrons with the gas molecules is very high and interferes with the electron acceleration process.

The presence of gases that capture electrons and convert them to atomic or molecular negative ions suppress the electrical breakdown. SF_6 and polychlorinated hydrocarbons also capture electrons and are more efficient discharge-suppressing gases than O_2. SF_6 has been used to advantage for the suppression of discharges in electrospray.[16–18] Use of these trace gas additives prevents gas discharges even when neat water is used as solvent.[17,18]

1.2.5 Electrical Current, *I*, due to the Charged Droplets. Charge and Radius of Droplets

The electrical current, *I*, due to the charged droplets leaving the ES capillary is easily measured (see Figure 1.1) and of interest because it provides a quantitative measure of the total number of the excess positive ionic charges that leave the capillary and could be converted to gas-phase ions.

Fernandez de la Mora and Locertales[19] have proposed the following approximate relationships, on the basis of experimental measurements of the current, *I*, and droplet sizes and charges, together with theoretical reasoning:

$$I = f\left(\frac{\varepsilon}{\varepsilon_0}\right)\left(\gamma K V_f \frac{\varepsilon}{\varepsilon_0}\right)^{1/2} \tag{1.7}$$

$$R \approx (V_f \varepsilon / K)^{1/3} \tag{1.8}$$

$$q \approx 0.7[8\pi(\varepsilon_0 \gamma R^3)^{1/2}] \tag{1.9}$$

where γ is surface tension of solvent, ε is permittivity of solvent, ε_0 is permittivity of vacuum (free space), $\varepsilon/\varepsilon_0$ is dielectric constant of solvent, K is conductivity of solution, E is applied electric field at capillary tip [see (Eq. 1.2)], R is radius of droplets produced at capillary tip, q is charge of droplets, and V_f is flow rate, volume/time. $f(\varepsilon/\varepsilon_0)$ is a numerical function tabulated by the authors[19]; the value of $f(\varepsilon/\varepsilon_0)$ is approximately 18 for liquids whose dielectric constant, $\varepsilon/\varepsilon_0$, is ≥ 40. This includes water ($\varepsilon/\varepsilon_0 = 78$) and water–methanol mixtures as well as acetonitrile and formamide. The relationships were obtained for solutions having conductivities K larger than 10^{-4} S m^{-1}. For polar solvents like water and methanol, as well as for electrolytes that dissociate essentially completely to ions, this requirement corresponds to solutions with concentrations higher than $\sim 10^{-5}$ mol/L—that is, a concentration range that is commonly present in ESMS. The flow rates used[19] were below 1 μL/min and are thus close to the flow rates used in conventional ESIMS. The equation is valid when the spray is operated in the cone jet mode.

More recently, a theoretical treatment by Cherney[20] has confirmed the deductions of Fernandez de la Mora and Locertales and provided a more detailed description of the conditions existing in the conejet. The treatment is also in agreement with recent experimental data by Chen and Pui.[21]

1.2.6 Solvent Evaporation from Charged Droplets Causes Droplet Shrinkage and Coulomb Fissions of Droplets

The charged droplets produced by the spray needle shrink due to solvent evaporation while the charge remains constant. The energy required for the evaporation is provided by the thermal energy of the ambient air. As the droplet gets smaller, the repulsion of the charges at the surface increases and at a certain droplet radius the repulsion of the charges overcomes the cohesive force of the surface tension. A Coulomb instability results and leads to a fission of the droplet that typically releases a jet of small charged progeny droplets. The condition for the Coulomb instability is given by the Rayleigh[22] equation:

$$Q_{\mathrm{Ry}} = 8\pi(\varepsilon_0 \gamma R^3)^{1/2} \tag{1.10}$$

where Q_{Ry} is the charge on the droplet, γ is the surface tension of the solvent, R is the radius of the droplet and ε_0 is the electrical permittivity. The shrinkage of the droplets at constant charge and the fission at or near the Rayleigh limit with the release of a jet of small, close to monodisperse, charged progeny droplets has been confirmed by a number of experiments. A single droplet from the spray was introduced at ambient pressure into an apparatus that allowed determination of the diameter and the charge of the droplet and the droplet observed until jet fission occurred.[23–28] Most studies have used some form of electrodynamic balance (EDB) that confines the charged droplet to a small region of space where it can be observed by optical methods. The EDB method is not well-suited for rapidly evaporation solvents such as methanol or water. Therefore, EDB has been applied mainly to relatively nonvolatile solvents. Davis et al.[23] were able to study water using a quadrupole ion trap but with the use of a surfactant that slowed down the evaporation.

A second method used by Gomez and Tang[24] is better suited for volatile solvents. It relies on phase Doppler interferometry (PDI). This method allows *in situ* measurements of the size and velocity of the electrosprayed droplets but not the charge, which must be inferred from other data. More recently, PDI was used by Beauchamp and co-workers,[25] who obtained the charge of the droplets from a comparison of the measured and calculated droplet mobilities. A concise summary of results by these and other authors (Taflin et al.,[26] Richardson et al.,[27] and Schweizer et al.[28a]) is given in Table 1.2. One can deduce from the

Table 1.2. Experimental Observations of Charged Droplets and Their Breakup

Authors	Solvent	Droplet diameter range (μm)	Onset of Instability (%) of Rayleigh limit	% of Mass Lost in Breakup	% of Charge Lost in Breakup
Beauchamp and co-workers[25a]	Water	10–40	90	N.D.[a]	20–40
	Methanol	10–40	110	N.D.	15–20
	Acetonitrile	10–40	100	N.D.	15–20
Grimm and Beauchamp[25b]	*n*-Heptane	35–45	100	N.D.	19
	n-Octane		87	N.D.	17
	p-Xylene		89	N.D.	17
Davis and Bridges[23]	Water with surfactant	4–20	90	1–2	15–25
Gomez and Tang[24]	Heptane	20–100	70	N.D.	N.D.
Taflin et al.[26]	Low-vapor-pressure oils	4–20	75–85	2	10–15
Richardson et al.[27]	Dioctyl Phthalate	Not reported	102–84	2.3	15–50
Schweizer and Hanson[28a]	*n*-Octanol	15–40	96–104	5	23
Duft et al.[28b]	Ethylene glycol	20–30	100	0.3	33

[a] N.D., not determined.

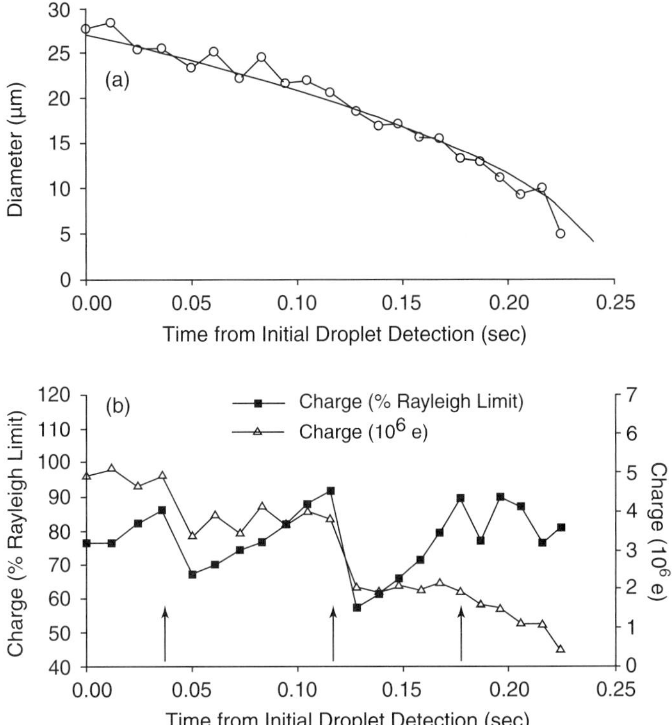

Figure 1.4. Evaporation and discharge of a positively charged water droplet in nitrogen gas at ambient pressure and 317 K and a weak (51 V/cm) electric field. (**a**) Variation of droplet diameter with time. Also plotted (smooth curve) is the predicted change of diameter due to evaporation of a neutral water droplet in a vapor-free N_2 gas at 317 K. (**b**) Variation of droplet charge with time, represented as number of elementary charges and as percent of the Rayleigh limit. Arrows indicate discharge events. Note that water droplets undergo a Coulomb fission at approximately 90% of the Rayleigh limit and are at approximately 65% of the limit after the Coulomb fission. (Figure from Grimm and Beauchamp,[30a] with permission from the American Chemical Society.)

table that the dependence on the type of solvent is relatively small. Thus, droplets from all solvents experience Coulomb fissions close to, or at, the Rayleigh limit. The loss of mass on fission is between 2% and 5% of the parent droplet, and the loss of charge is much larger—that is, some 15–25% of the charge of the parent droplet.

Beauchamp and co-workers[25] also provide information on the charge of the parent droplet immediately after the droplet fission. An example of such data is given in Figure 1.4, where the charge of the droplets before and after the fission is given as percent of the Rayleigh condition [Eq. 1.10]. These, along with results for the other solvents studied,[25] show that the evaporating charged droplets oscillate at all times between fairly narrow limits of the Rayleigh condition. This finding has bearing on the discussion of the mechanism by which non-denatured proteins enter the gas phase (see charged residue mechanism in Section 1.2.10). Notable also (see Figure 1.4a) is the observation that the diameter of the charged parent droplet undergoing evaporation *and* Coulomb fissions remains very close to the diameter of an uncharged droplet that loses mass only due to evaporation. [For equations used to evaluate the change of diameter of the uncharged droplet with time due to evaporation only, see Eq. (1.2)–(1.4) in Grimm and Beauchamp.[25b]] This result supports the observations of Davis and Bridges,[23] Taflin et al.,[26] and Richardson et al.[27] (see Table 1.2) that the total mass that goes into the charged progeny droplets is very small. Duft et al.,[28b] using ethylene glycol droplets with a radius of 25 μm and an arrangement that allowed high-speed microscopic images of the droplets to be obtained, report that 33% of the charge and only 0.3% of the mass goes into about 100 progeny droplets and that the progeny droplets, when formed, are below but very close to the Rayleigh limit.

When the sprayed solution contains a solute, such as a salt, the continuous evaporation of the droplets will lead to very high concentrations of the salt and finally to charged solid particles—the "skeletons" of the charged droplets. These skeletons can reveal some aspects of the droplet evolution. Fernandez de la Mora and co-workers[29] have used this approach to study charged droplet evolution. This work is of special relevance to the ion evaporation model and is discussed in Section 1.2.9.

Fission of charged droplets can be forced to occur also below the Rayleigh limit when the charged droplet is exposed to a strong electric field. In this case, the excess charges on the droplet are forced to move downfield and accumulate on the downfield side. The force exerted by this asymmetric charge distribution overcomes the surface tension of the droplet and causes a downfield distortion of the droplet and the formation of a cone and cone-jet that splits into charged progeny droplets. This field-induced droplet ionization (FIDI) has been studied in some detail by Grimm and Beauchamp,[30] who created the droplets with a vibrating orifice aerosol generator (VOAG). An example of this work is shown in Figure 1.5, which documents the gradual distortion with time of the droplet and the ultimate formation of the cone-jet and progeny droplets. Notable are the low charges of the droplet that still can lead to droplet fission when a high enough field is present. Thus a droplet with only 4% of the Rayleigh charge can be forced to emit a cone-jet by a field, $E = 2.15 \times 10^6$ V/m (see Figure 1.5).

This field is not far from the field needed to start conventional ES, with a spray capillary [see Eq. 1.1 in Section 1.2.2]. This indicates that near the spray capillary, Coulomb fissions of the charged droplets could be occurring before droplet evaporation has brought the droplet diameter down to the Rayleigh limit.

Grimm and Beauchamp[30b] have shown that the gas-phase ions ultimately resulting from FIDI can be detected with a mass spectrometer and that the method might have some analytical potential. These experiments bear some resemblance to earlier work by Hager et al.,[31] who also produced neutral droplets with a vibrating orifice and then exposed them to a nonhomogeneous electric field near a rod electrode and detected the ions with a mass spectrometer. These authors named the method "droplet electrospray." As could be expected, due to the small number of droplets, the sensitivity of the droplet method[31] was much lower than that obtained with ESIMS.

A variation of the FIDI method demonstrated by Grim and Beauchamp[30a] could have significant analytical utility. In this case, neutral droplets containing an electrolyte were exposed to a homogeneous high electric field. The field forced the negative ions in the solution to one side of the field and forced the positive to the other side. The induced instability of the droplet leads ultimately to a symmetric fission of the droplet, and negatively and positively charged jets of progeny droplets are emitted on the opposite sides. This experiment should allow the mass spectrometric observation of the positive and negative ions in the solution, which in certain cases could be of special analytical interest.

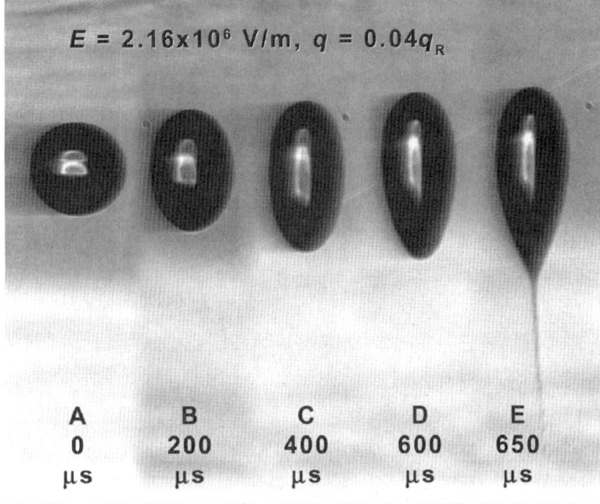

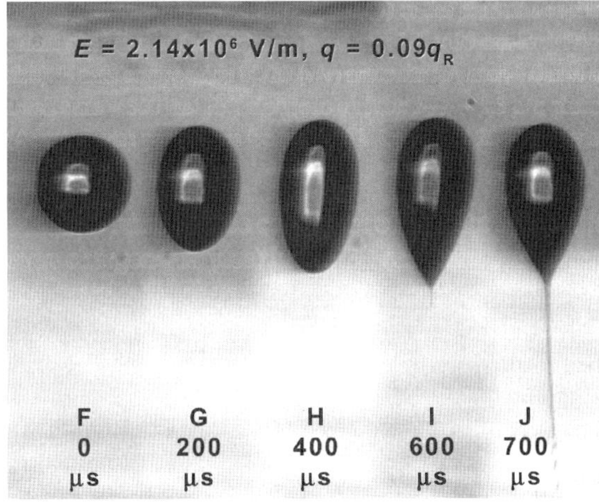

Figure 1.5. Sequences of 225-μm-diameter methanol droplets with a charge of $q = 0.04q_R$, where q_R is the charge when the droplet is at the Rayleigh limit. Droplets undergoing asymmetrical distortions at an applied electric field, $E = 2.16 \times 10^6$ V/m and in the second row droplets with $q = 0.09q_R$ and $E = 2.14 \times 10^6$ V/m. In both cases these fields represent the minimum for which field-induced droplet ionization (FIDI) is observed. The jets in frames E and J demonstrate capillary instability and the formation of progeny droplets of approximately 10-μm diameter. For the $q = 0.04q_R$, the time of 650 μs for jetting to begin is identical to that observed for a neutral droplet. (From Grimm and Beauchamp,[30b] with permission from the American Chemical Society.)

1.2.7 Evolution of Droplets by Evaporation and Coulomb Fissions Producing Smaller and Smaller Progeny Droplets that Lead Ultimately to Minute Charged Droplets that Produce Ions in the Gas Phase

It is clear that the process of repeated droplet fissions of parent droplets that lead to smaller parent droplets and progeny droplets, along with the evaporation of the progeny droplets that lead to second generation progeny, will ultimately lead to very small charged droplets that are the precursors of the gas-phase ions. The mechanisms by which the gas phase ions are produced from the very small "final" droplets is considered in Section 1.2.8. Here we examine some of the details of the evolution of the initial droplets, formed at the spray capillary, to droplets that are the precursors of the ions. The whole process is driven by the decrease of droplet volume by evaporation. The continuous evaporation is possible because the thermal energy required for the evaporation is provided by the ambient gas, air or pure nitrogen, at near atmospheric pressure. As will be shown below, a very large loss of solvent by evaporation occurs before the very small droplets that lead to ions are formed.

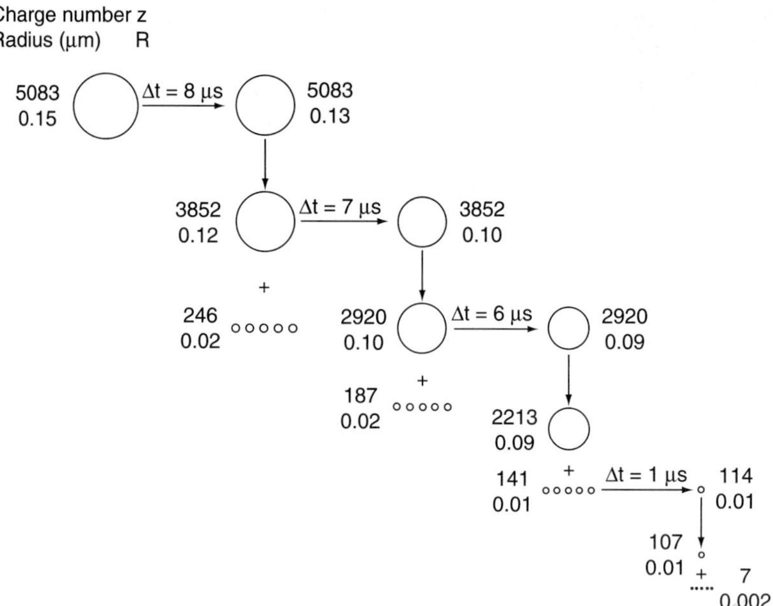

Figure 1.6. Droplet history of charged water droplets produced by nanospray. The first droplet is one of the droplets produced at spray needle. This parent droplet is followed for three evaporation and fission events. The first-generation progeny droplets are shown, along with the fission of one of the progeny droplets that leads to second-generation progeny droplets. *R* is the radius of the droplets and *Z* gives the number of charges on the droplet. *Z* corresponds to the number of excess singly charged ions near the surface of the droplet. The parents charge is $Z = 0.9Z_R$ just before the fission and $Z = 0.7Z_R$ just after the fission (as observed in Figure 1.4), while the progeny droplets have $Z = 0.7\ Z_R$ just after the fission of the parent. (Based on Figure 1.1 in Peschke, Verkerk, and Kebarle.[32])

It is desirable to be able to estimate the increase of solute concentration due to the volume loss, particularly so in certain applications of ESIMS. An example is the determination of association constants of protein–ligand complexes by the titration method (see Section 1.2.13.3).

A droplet evolution scheme is shown in Figure 1.6. It deals with droplets produced by nanoelectrospray. The nanospray droplets are better suited for the evaluation of droplet histories because the evolution is much shorter. The very much smaller initial droplets reach much faster the final stage of very small charged droplets that lead to ions.

The assumptions with which the scheme (Figure 1.6) for water as solvent was obtained are described in detail in the section entitled "Calculations and Experimental" in Peschke et al.[32] The stability limits of droplet fission at droplet charge $Z = 0.9Z_R$ (just before the droplet fission) and $Z = 0.7Z_R$ (just after the fission due to Beauchamp and co-workers[25a] for water) were used (see Figure 1.4 in the present work). These are for droplets of radii in the 13- to 3-μm range, while the nano-droplet evolution scheme (Figure 1.6) involves close to a hundred times smaller radii. It is not known to what extent the droplet fissions of such small droplets follow the same stability limits. Unfortunately, no measurements for such small droplets exist because these droplets evaporate and fission very fast, within several microseconds, while the large droplets[30a] fission within intervals of some 40 ms (see Figure 1.4).

Using the droplet radii, one can evaluate that approximately 40% of the volume is lost between each fission. A corresponding increase by 40% of the solute concentration must also occur. This means that after 10 successive fissions of the parent droplet, its volume will decrease 29-fold and the concentration of solutes in the droplet will also increase 29-fold.

1.2.8 Mechanisms for the Formation of Gas-Phase Ions from Very Small and Highly Charged Droplets: The Charged Residue Model (CRM) and the Ion Evaporation Model (IEM)

Two mechanisms have been proposed to account for the formation of gas-phase ions from very small and highly charged droplets. The first mechanism, proposed by Dole et al.,[4] depends on the formation of extremely small droplets that could contain one analyte molecule and some ionic charges. Solvent evaporation from such a droplet will lead to a gas-phase analyte ion whose charge originates from the charges at the surface of the vanished droplet. This assumption is now known as the *charged residue model* (CRM). Early support for the mechanism was provided by Rollgen and co-workers.[33]

Iribarne and Thomson[34] proposed a different mechanism for the production of gas-phase ions from the charged droplets. The Iribarne *Ion Evaporation Model* (IEM) is described in some detail in the next section. It predicts that, after the radii of the droplets decrease to a given size, direct ion emission from the droplets becomes possible. This process, which they called *ion evaporation*, becomes dominant over Coulomb fission for droplets with radii of $R \leq 10\,\text{nm}$ (*vide infra*).

Iribarne and Thomson[34] did not use electrospray to produce the small droplets. They were interested in the nature of the charged species produced from very small droplets obtained from pneumatic "atomization" of a liquid such as water containing a solute such as NaCl. Some of the droplets will be charged due to statistical imbalances between positive and negative electrolyte ions present in the droplets. The charged droplets of different polarities were separated by the application of a weak electric field. The charged droplets evaporate rapidly and lead to charged salt particles—that is, charged salt clusters. Iribarne and Thomson obtained ion mobility spectra of the charged particles and observed an isolated high-intensity peak at the high mobility end of the spectrum and a very broad and dense region of peaks at low mobilities. Changing the concentration of the salt did not change the position of the high-mobility peak. These results suggested that the high-mobility peak is due to single ions such as Na^+ while multiply charged, larger salt aggregates were producing the broad and dense region of peaks at low mobilities. Much earlier (1937), on the basis of extensive mobility studies of such charged species, Chapman,[2a] also using "atomization," had come to the same conclusion. In later work by Thomson and Iribarne,[34b] the ions were sampled with a mass spectrometer, and the high-mobility peak was identified as due to ions originating from the sprayed solution. Thus, $Na^+(H_2O)_n$ with $n = 3$–7 were observed when aqueous solutions of NaCl were sprayed. These early mobility studies[2,34] can be considered as the first good experimental evidence for IEM.

Iribarne and Thomson[34] came to the conclusion that the formation of abundant high-mobility gas-phase ions is possible only if a considerable fraction of the charges (i.e., ions) escaped from the droplets before the complete evaporation of the droplets. The theory for this escape process, ion evaporation, was developed in the same paper.[34a]

1.2.9 The Iribarne–Thomson Equation for Ion Evaporation from Small Charged Droplets and Subsequent Experimental and Theoretical Work Examining the Validity of IEM

Iribarne and Thomson derived an equation that provided detailed predictions for the rate of ion evaporation from the charged droplets.[34a] The treatment is based on transition state theory, used in chemical reaction kinetics. The rate constant k_I for emission of ions from the droplets is given by

$$k_I = \frac{k_B T}{h} e^{-\Delta G \neq /kT} \qquad (1.11)$$

where k_B is Boltzmann's constant, T is the temperature of the droplet, and h is Planck's constant. The free energy of activation, $\Delta G^{\neq}$, was evaluated on the basis of the model shown in Figure 1.7. The transition state selected by the authors resembles the products more than the initial state; that is, it is a "late" transition state. The advantage of this choice is that the energy of such a state can be expressed by a closed equation based on classical electrostatics and thermodynamics. However, the transition state could in reality be occurring earlier—for example, as the ion disrupts the droplet surface. The energy of such an early transition state would be much more difficult to evaluate. If a higher free energy barrier did occur at the earlier stage, the predictions of the Iribarne–Thomson model would be less sound.

The top of the barrier in the Iribarne–Thomson transition state is due to the opposing electrostatic forces: (a) the attraction between the escaping ion and the droplet, arising from the polarizability of the droplet, and (b) the Coulomb repulsion of the escaping ion by the remaining charges of the droplet. The ion-polarizability attraction is larger at short distances, but as the distance is increased, it falls off much faster than the Coulomb repulsion between the ion and droplet charges. The transition state (see Figure 1.7) occurs at a distance $x = x_A$, where the attractive and repulsive energies are equal.

The free energy due to these two electrostatic forces, ΔG(E), is not the only term that enters the $\Delta G^{\neq}$ expression. An ion desolvation term, ΔG^0_{dsol}, that is independent of x, is also involved. The desolvation free energy $-\Delta G^0_{\text{dsol}}$ corresponds to the free energy required to desolvate the ion—that is, transfer the ion from a neutral droplet of the same size to the gas phase. For relatively strongly solvated ions like the alkali ions, the most favorable path is for the ion to take several solvent molecules with it. For small, stronger solvating ions like Li^+, more solvent molecules are taken by the core ion. Thus, for water as solvent, the lowest ΔG^0_{dsol} occurs for $Li^+(H_2O)_7$, while for the larger Cs^+, the minimum occurs for $Cs^+(H_2O)_5$ (see Table III in Tang and Kebarle[35]). This reference provides also literature sources for the solvation energies of different ions. The individual properties of the ion enter into the ion evaporation rate constant mainly through the ΔG^0_{dsol} term.

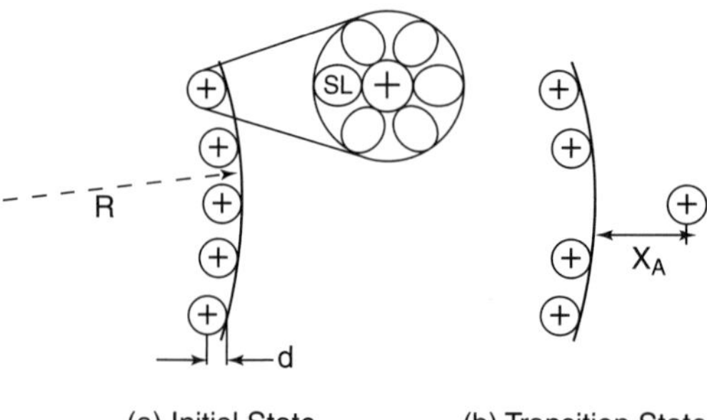

Figure 1.7. Model used in derivation of the Iribarne–Thomson equation. Part of surface of droplet with radius R is shown. (**a**) Excess of positive ions corresponding to the droplet charges located just below the surface. Each ion is solvated by several solvent molecules. (**b**) The transition state. Due to thermal activation, one of the ions has moved outside the droplet. (From Tang and Kebarle,[35] with permission from the American Chemical Society.)

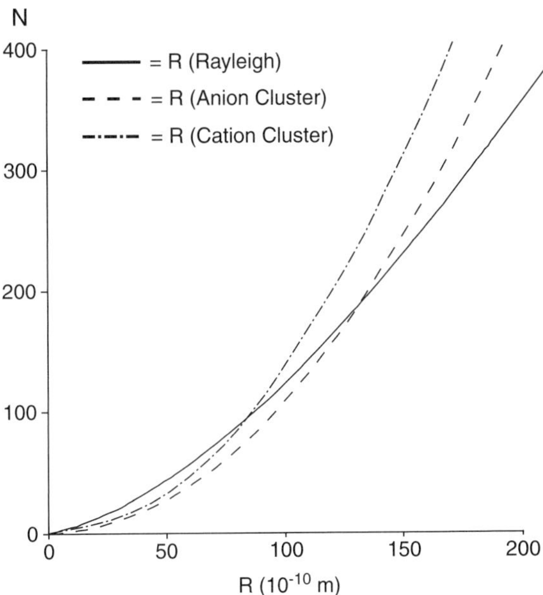

Figure 1.8. Predictions of the ion evaporation theory.[34] The Rayleigh curve provides the droplet radius R and the number of elementary charges N at which a charged water droplet will be at the Rayleigh limit. Moving at constant charge to a smaller radius R through solvent evaporation will cause a Coulomb fission. Similarly, the curves Cation Cluster and Anion Cluster show the threshold of ion evaporation at a given charge N and droplet radius R. For negatively charged droplets, moving at constant charge to a smaller radius R due to solvent evaporation will lead to negative ion evaporation when the radius $R = 140$ Å (1 nm = 10 Å) and for postively charged droplets at $R = 84$ Å, where the ion evaporation (Cation Cluster) and Rayleigh curves cross. Below this radius, ion evaporation replaces Coulomb fission. Thus taking a radius of $R \approx 100$ Å provides a useful benchmark for the region where ion evaporation takes over.

The curves shown in Figure 1.8, due to Iribarne and Thomson, summarize some of the predictions of the ion evaporation theory.[34] The initial charged droplets produced at the spray tip will lose volume by evaporation until they come close to the Rayleigh limit where the first Coulomb fission will occur. From then on, the droplet will experience repeated evaporation and fission events staying fairly close to the Rayleigh limit curve (as shown by experiments such as the results in Figure 1.4). Assuming that the droplets are positive, the + ion curve, crossing the Rayleigh curve, R(Rayleigh), at a radius $R = 84$ Å, (1 nm = 10 Å) indicates that ion evaporation will replace Coulomb fission at lower radii. As R decreases continuously by solvent evaporaton, for radii smaller than $R \approx 84$ Å, there will be continuous loss of charge by ion evaporation, such that R and N follow exactly the R (cation cluster) curve. Because the $\Delta G^0_{\mathrm{dsol}}$ values entering the calculations leading to Figure 1.8 were not exactly known, it was suggested[34] that the crossing point for positive ions could be in the region of $R = 70$–120 Å. Thus, taking $R \approx 100$ Å provides a useful benchmark for the region where ion evaporation takes over.

While the Iribarne–Thomson equation provides a good conceptual description of the ion evaporation process, it has not proven useful for predictions of the observed relative intensities of ions in the ESI mass spectra. Iribarne and Thomson were interested only in relatively simple ions like the alkali ions. Even for these there can be ambiguities in the choice of the values of the parameters entering the ion evaporation equation.

Attempts were made[35–37] to examine the validity of IEM by evaluation of the ion evaporation rate constant k_I [see Eq. 1.11, where the ΔG^0_{dsol} $(M^+(H_2O)_n)$ for ions like Li^+, Na^+, K^+, and so on, were obtained from experimental data in the literature. The rate constants $k_I(Li^+)$, $k_I(Na^+)$, and so on, were compared with experimentally determined ion intensities, $I(Li^+)$, $I(Na^+)$, and so on, when equal concentrations of the salts LiX, NaX, and so on, were present in the solution. The observed ion abundance ratios should correspond to the evaluated rate constant ratios if IEM holds. The data used for the evaluation of ΔG^0_{dsol} $(M^+(H_2O)_n)$ had error limits that led to some scatter of the evaluated ΔG^0_{dsol} $(M^+(H_2O)_n)$ (see Table 1.1 in Kebarle and Peschke[36a]). The results showed a trend of decreasing free energy of activation, $\Delta G^{\neq}$ (M^+) from Li^+ to Cs^+, which should lead to increasing ion evaporation rate constants from Li^+ to Cs^+.

Unfortunately, the experimentally observed ion abundances determined in different laboratories did not agree. Results from this laboratory[36a] showed similar intensities for the Li^+–Cs^+ series, while results from other laboratories such as Cole[36c] gave increasing intensities in the order Li^+ to Cs^+, as expected on the basis of the calculated ion evaporation rate constants. The triple quadrupole mass spectrometer used in this laboratory[36a] exhibited large decreases of ion transmission with increasing m/z, and the corrections for the transmission changes used by Kebarle and Peschke[36a] could have been unreliable.

Fernandez de la Mora and co-workers[29] have used a different approach to provide strong evidence for the qualitative validity of the ion evaporation mechanism. They circumvent the difficulty caused by the so far impossible direct observation of the evolution of the very small rapidly evaporating droplets and the determination of their radius and charge. Instead, using solutions that contained a dissolved electrolyte, they focused on the sizes and charges of the solid residues formed after evaporation of the solvent. Since the solid residues had been "charged droplets" just before the last of the solvent evaporated, their sizes and charges represent, to a fair approximation, the sizes and charges of evolving charged droplets that are now frozen in time and thus amenable to measurement.

A first effort using this approach[29a] provided results and interpretation that were in agreement with IEM. Using very low flow rates and high conductivity (high electrolyte concentration) solutions, very small initial droplets were produced by ES. These droplets were chosen because they are expected to reach the Iribarne ion emission radius (if IEM holds) before experiencing Rayleigh disintegration. The mobility of the charged solid residues formed from such droplets was determined. The mobility depends on both the radius and the charge of the residues. The radius could be determined independently by using a "hypersonic impactor" apparatus that provides the mass of the residues. The radius was obtained by assuming that the density of the residues was the same as that of the solid salt. The charge of the residues was then deduced from the known mobility and the now known radius of the solid residue. The charge q determined for a given solid residue with known radius was found to be considerably lower than the charge q_{Ry} required by the Rayleigh stability equation. This finding is in agreement with the prediction of the Iribarne Ion Evaporation Theory that when the droplets reach a given small radius, ion evaporation requires a smaller overall charge than does Coulomb fission (see Figure 1.8).

These findings could be considered as a strong experimental support for IEM. However, an obvious objection is the assumption that the density of the residues is the same as the density of the solid salt. The morphology of the solid residues formed by ES could be much more complex than that of solid crystals of the salt, and thus the density of the residues could be lower. Additional work[29b,c] using more advanced experimental methodology removed some of the problems of the previous work,[29a] such as the problem of the unknown density of the residues. For further details see Fernandez de la Mora and co-workers[29b,c] or a less technical, brief discussion in Kebarle.[36b]

Much greater problems in applying the Iribarne–Thomson equations are encountered with organic ions. Many of these can be surface-active, and even very weak surface activity is expected to have a very significant effect on the relative intensities observed with ESIMS. The surface activity will have a twofold effect on the ion evaporation rate constant: (a) The surface-active ions will be favored as charges at the surface of the droplets because they are surface-active. (b) The surface-active ions will have lower desolvation energies because the hydrophobic groups responsible for the surface activity are not well-solvated. Thus organic ions with higher surface activities will have higher ion evaporation rate constants. For a very effective use of surface activity of analytes, see the chapter by Cech and Enke (Chapter 2) in this volume.

Nohmi and Fenn[38a,b] have also provided experimental evidence that supports IEM.

Theoretical work involving simulations of ion evaporation from charged droplets can also provide valuable insights into IEM. A good example is the work by Vertes and co-workers[38c] on the evaporation of H_3O^+ ions from charged water droplets. The authors used classical molecular dynamics simulations to study droplets of 6.5-nm diameter. Checks were made that the parameters used led to predictions of properties such as the radial distribution function, the enthalpy of evaporation, and the self-diffusion coefficients that are in agreement with experimental values. Droplets of 6.5-nm diameter, which have some 4000 water molecules, are expected to lose charges by IEM when charged to the Rayleigh limit (see Figure 1.8). The droplet was charged with H_3O^+ ions, but ion pairs corresponding to a dissolved solute were not added. At equilibrium, not all H_3O^+ ions were located on the surface of the droplet as generally assumed for IEM. The fluctuations of water molecules at the droplet surface became much more accentuated as the H_3O^+ charges were added and some of these fluctuations developed into large protuberances that separated as hydrated H_3O^+ ions. Generally, the "solvation shell" of the departing H_3O^+ consisted of some 10 water molecules. Interested readers can observe the simulation of such ion evaporation at the website of Vertes: http://www.gwu.edu/~vertes/publicat.html.

In summary, the ion evaporation mechanism is experimentally well-supported for small inorganic and organic ions, but development of quantitative predictions based on this model remains a considerable challenge for the future.

1.2.10 Large Analyte Ions Such as Proteins and Dendrimers Are Most Probably Produced by the Charged Residue Model (CRM)

Denatured and non-denatured globular proteins and protein complexes are routinely produced by ESI. Native proteins will, in general, remain folded when sprayed in a water–methanol solution—that is, at a neutral pH of ~7. However, some proteins may be very sensitive to the conditions used and denature partially. Denaturing is easily recognized from the observed mass spectrum. Folded (non-denatured) proteins lead to mass spectra consisting of a compact series of peaks that correspond to the molecular mass of the protein charged (in the positive ion mode) by a number Z of H^+ ions. A small protein like lysozyme (Lys) is observed to lead to three peaks due to three different charge states with $Z = 8, 9,$ or 10. When the protein is denatured, the charge distribution is observed to be very broadened, covering a large number of peaks that extend to much higher charge states, Z. Obviously, it is of special interest to understand the mechanisms that lead to these observations.

An early study by R. D. Smith and co-workers[38d] provided good evidence that proteins are produced via CRM. If CRM holds, one would expect that when small charged droplets evaporate, there could be not only one protein, but also more than one protein, in the droplets. Therefore, one should observe in the mass spectra not only monomers, but also dimers, trimers, and higher multimers. The authors[38d] observed a preponderance of multiply charged monomers and much lower and rapidly decreasing intensities of dimers and trimers, where the charge to total mass ratio, z/m, value decreased with the degree of multimerization. Using a quadrupole mass spectrometer that had a very high mass range, the authors[38d] were able to observe even higher multimers and came to the conclusion that the results are consistent with CRM and a droplet evolution following a scheme of the type shown in Figure 1.6.

In later work, Smith and co-workers[39a] found an interesting empirical correlation between the molecular mass, M, and the average charge, Z_{av}, of starburst dendrimers [see Eq. 1.12]. These multibranched alkyl-amine polymers have relatively rigid structures that are close to spherical, with shapes resembling those of globular proteins. Results for non-denatured proteins were found also to fit Eq. 1.12.

$$Z_{obs} = aM^b \tag{1.12}$$

Z_{av} is the observed average charge state and M is the molecular mass of the ion while a and b are constants. The value $b = 0.53$ led to the best fit. Standing and co-workers[39b] observed an identical relationship for a large number of non denatured proteins, where the value of b was between 0.52 and 0.55. (The term "average charge state for proteins" was first introduced and defined by Wang and Cole.[39c])

Independently, Fernandez de la Mora[40] using the dendrimer data[39a], and including also additional data from the literature for non-denatured proteins, was able to show not only that the empirical relationship [Eq. 1.12] holds, but also that the relationship can be derived on the basis of the charged residue mechanism. The plot shown in Figure 1.9 is based on the data used by Fernandez de la Mora, but includes also the protein data of Ens and Standing.[39b]

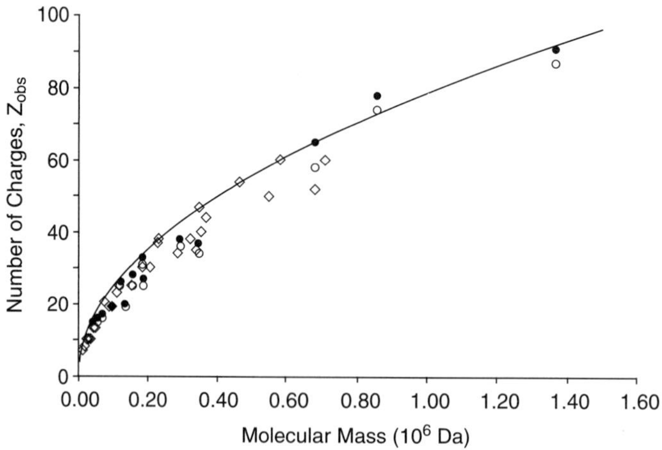

Figure 1.9. Reproduction of a plot used by Fernandez de la Mora[40] and extended to include also data by Ens and Standing.[39b] Z_{obs} is the number of charges observed on proteins produced by ESIMS under non-denaturing conditions. ((●) Highest charge; ((○)) lowest charge in mass spectrum (Fernandez de la Mora[40]); ((◇) average Z_{obs} (Ens and Standing[39b]). Solid curve corresponds to charge Z predicted by Eq. 1.1.2.

The derivation of Fernandez de la Mora[40] was based on the following arguments. There is theoretical[40] evidence that the evaporating charged droplets (which in the present context are assumed to contain one globular protein molecule) stay close to the Rayleigh limit. This is supported by more recent, experimental results[25] (see Figure 1.4b in Section 1.2.6) which involve charged evaporating water droplets of 35- to 5-μm diameter. These show that the charge is approximately 95% of the Rayleigh limit when the droplets experience a Coulomb fission and approximately 75% of the Rayleigh limit immediately after the Coulomb fission. Thus, the droplets stay at all times within the limits of 95–75% of the Rayleigh limit, and both of these values are close to the Rayleigh limit. Fernandez de la Mora[40] reasoned that when the charged water droplet, containing one protein molecule, evaporates completely, the charges on the droplet will be transferred to the protein. He assumed also that the protein will be neutral when all the water is gone so that the charges on the surface of the droplet become the charge of the protein observed in the ESI mass spectrum of the protein.

The radius of the protein can be evaluated with Eq. 1.13, where φ is the density of the protein.

$$(4/3\pi R^3\varphi)N_A = M \tag{1.13}$$

$N_A = 6 \times 10^{23}$ is the number of molecules per mole; R is the radius of the protein, and M is the molecular mass of the protein. Fernandez de la Mora assumed that the non-denatured proteins have the same density φ as water, $\varphi = 1\ \mathrm{g/cm^3}$. Evidence in support of that assumption based on mobility measurements by Jarrold and Clemmer[41a] is given in Section 2.2 of Fernandez de la Mora.[40] The charge of the protein is taken to be the same as the charge of a water droplet of the same radius, R, that just contains the protein and is at the Rayleigh limit. The charge can be obtained by evaluating R with Eq. 1.13 and substituting it in the Rayleigh equation [Eq. 1.10] and expressing the charge, $Q = Z \times e$, where Z is the number of elementary charges and e is the value of the elementary charge. The result is given in Eq. 1.14:

$$Z = 4(\pi\gamma\varepsilon_0/e^2N_A\varphi)^{1/2} \times M^{1/2} \tag{1.14}$$

$$Z = 0.078 \times M^{1/2} \tag{1.14a}$$

where Z is the number of charges of the protein, γ is the surface tension of water, ε_0 is the electrical permittivity, e is the electron charge, N_A is Avogadro's number, φ is the density of water, and M is the molecular mass of protein. The constant 0.078 in Eq. (1.14a) gives the number of charges on a protein of molecular mass M in mega daltons.

The solid line in Figure 1.9 gives the predicted charge based on Eq. (14a). Good agreement with the experimental results is observed. Notable also is the predicted exponent of M, which is 0.5, while the exponent deduced from the experimental data[39,40] is 0.53. It should be mentioned that Fernandez de la Mora's treatment[40] was preceded by a similar but much less detailed proposal by Smith and co-workers.[39a]

The agreement of Eqs. 1.14 and (14a) with the observed charges Z can be considered as strong evidence that globular proteins and protein complexes are produced by the charged residue mechanism.

Most of the data points in Figure 1.9 were obtained not with neat water as solvent, but from solutions of water and methanol. These solutions are easier to use in ESIMS because neat water, due to its high surface tension, can initiate electrical gas discharges at the spray needle (see Section 1.2.4). Nevertheless, the surface tension of water, $\gamma = 73 \times 10^{-3}$ (N/m) (see Table 1.1), was used by Fernandez de la Mora[40] in Eq. 1.14 for the surface tension of the droplets. The surface tension of methanol, $\gamma = 22.6 \times 10^{-3}$ (N/m), is much smaller. However, the assumption can be made that the evaporating droplets will lose methanol preferentially because methanol has a much higher vapor pressure, and therefore the final droplets that contain the protein will be very close to neat water droplets. This can be expected because there has been a very large loss of solvent by evaporation before the final very small droplets containing the protein are formed (see Figure 1.6 and associated discussion).

A recent compilation of data by Heck and van der Heuvel[41d] has shown that the square root dependence of the charge Z on M [see Eq. (14a)] holds also for protein complexes.

Fernandez de la Mora[40] did not consider the actual chemical reactions by which the charging of the protein occurs. These reactions will depend on what additives were present in the solution. Thus, in the presence of 1% of acetic acid in the solution, the charges at the surface of the droplets will be H_3O^+ ions. Charging of the protein will occur by proton transfer from H_3O^+ to functional groups on the surface of the protein that have a higher gas-phase basicity than H_2O. The gas-phase basicities are relevant because the solvent will essentially have disappeared. There are plenty of functional groups on the protein that have gas-phase basicities that are higher than that of H_2O. These could be basic residues or amide groups of the peptide backbone at the surface of the protein. Gas-phase basicities of several representative compounds are given in Table 1.4 in Section 1.2.12.

Recent results by Samalikova and Grandori [42a] have provided evidence that contradicts Fernandez de la Mora's CRM model. Solvents that have a lower surface tension than water should lead to lower charge states of the proteins, because the droplet charge at the Rayleigh limit [see Eq. 1.14] is proportional to the square root of the surface tension, γ. Use of solvents such as 1-propanol, with $\gamma = 23 \times 10^{-3}\,\mathrm{N\,m^{-1}}$ compared to water $\gamma = 73 \times 10^{-3}\,\mathrm{N\,m^{-1}}$, should lead to a decrease of the observed charge by a factor of $\sqrt{73/23} = 1.8$. However, essentially no charge change was observed for the proteins (ubiquitin, cytochrome C, and lysozyme). The nonaqueous solvents that were used, such as 1-propanol, were not neat propanol but instead a solution containing some 30% propanol and 70% water (by volume). Grandori et al. made the assumption that the solvent with the higher vapor pressure (H_2O in the present case) evaporates first, leaving the less volatile solvent (1-PrOH) behind.

Samalikova and Grandori[42] were not the first investigators to use solvent mixtures involving water and another less volatile solvent with lower surface tension. Earlier work by Iavarone and Williams[43] also examined the effect of solvents with different surface tension on the charge states of macromolecules, such as dendrimers. Of special interest are their results for dendrimers because these have stable, close-to-spherical structures that are not affected by nonaqueous solvents. One of the solvents used was 2-propanol, which bears some similarity to the 1-propanol used by Samalikova and Grandori. For the DAB 60 dendrimer, they found the average charge to be $Z_{av} = 2.63$ (propanol) and $Z_{av} = 4.72$, (water), which leads to $Z_{av}/Z_{Ry} = 1.09$ for propanol and $Z_{av}/Z_{Ry} = 0.98$ (water), in good agreement with the CRM[40] expected result. For the DAB 64 dendrimer, the $Z_{av} = 7.6$ (2-propanol) was considerably smaller than $Z_{av} = 9.9$ (water) as expected from CRM.[40] The ratios were $Z_{av}/Z_{Ry} = 1.27$ (propanol) versus $Z_{av}/Z_{Ry} = 0.9$ (water). Obviously the Z_{av} for 2-propanol is higher than it should be, but still acceptable as support for the CRM.

Iavarone and Williams[44] suggest that the lack of change of charge, Z_{av}, when a solvent of low surface tension is used, observed by Samalikova and Grandori, is caused by a conformational change of the proteins due to the presence of the nonaqueous organic solvent in the solution that was sprayed. Such a partial denaturing increases the size (radius R) of the protein and is thus expected to lead to a larger Z_{av}. The charge Z depends on $R^{3/2}$, while the dependence on the surface tension, γ, is much weaker, $\gamma^{1/2}$. Thus, it would take a relatively small change of R, due to unfolding, to compensate for the change of γ. Conformational changes of proteins caused by organic solvents have been well documented (see, for example, Karger and co-workers[45]). Evidence for partial unfolding of proteins, when alcohols were added to the solution used for ESIMS determinations, has been provided by Kaltashov and Mohimen.[46]

It is well known that denaturing of a protein introduces not only an increase of the charge state, but also a characteristic broadening of the charge state distribution. Examining the mass spectra in Figure 1.1 of Samalikova and Grandori,[42a] one finds significant broadening for cytochrome C and ubiquitin, but no broadening for lysozyme. Thus only lysozyme provides evidence that decrease of the surface tension does not lead to a decrease of Z_{av} as expected from Eq. 1.14 and the charged residue model. This result is significant and therefore, at this time, one cannot completely reject the Grandori results and interpretation. Clearly, additional work is desirable, such as (a) an examination of the reproducibility of the results[42] and (b) additional experiments with dendrimers in different solvents.

An extension of the charged residue model was provided recently by Kaltashov and Mohimen.[46] Determinations of ESI mass spectra and charge states of some 22 proteins extending in mass from 5 to 900 kDa were made, using the same experimental conditions in all measurements. In this manner, the authors[46] avoided having to use mass spectra from the literature that were obtained at a variety of conditions. Representative mass spectra for a small protein (insulin) and a large protein (human serum transferin–transferin receptor complex) are given in Figure 1.10, upper row. Both spectra show a narrow charge

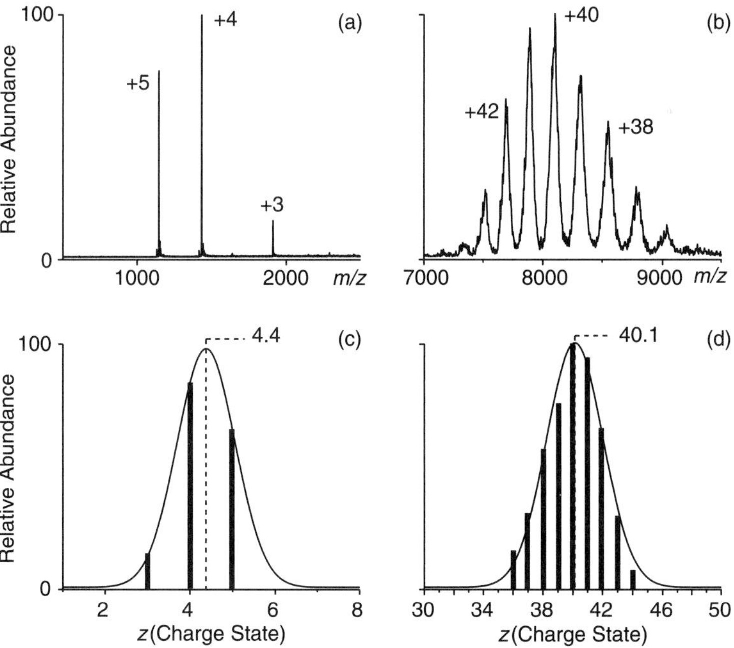

Figure 1.10. ESI mass spectra of insulin, A. Human serum transferrin-transferrin receptor complex, B. These spectra were used to calculate the average number of charges Z_{av} on the proteins, as illustrated with diagrams C and D. (Reprinted from Kaltashov and Mohimen,[46] with permission from the American Chemical Society.)

distribution indicating that the protein structures were not denatured. The average charge, Z_{av}, was determined by integration of the peaks as shown in Figure 1.10, lower row.

The authors[46] reasoned that the spherical approximation used by Fernandez de la Mora[40] is only seldom a realistic representation of the shape of native proteins. While the spherical shape provides a best fit for the spherical charged droplets, erroneous results are expected when the protein shape deviates from spherical. Therefore a better approach would be to use the protein surface area.

The correlation obtained with the surface area, S, of proteins evaluated from their known crystal structures and the charge states Z_{av}, obtained from the mass spectra of the 22 proteins, is shown in Figure 1.11. The fit of the data points [linear plot insert in Figure 1.11, corresponding to Eq. (15a)] is very much better than the fit in the Fernandez de la Mora[40] plot (Figure 1.9), which is based on the spherical protein approximation and Z_{av} data from different literature sources. The straight line obtained from the ln–ln plot (Figure 1.11) can be represented by Eq. (15b), where $\ln A$ is the intercept with the $\ln Z_{av}$ axis and a is the slope of the straight line. The value $a = 0.69$ is obtained from the plot.

$$Z_{av} = A \times S^a \tag{1.15a}$$

$$\ln Z_{av} = \ln A + a \ln S \tag{1.15b}$$

The authors[46] were able to show that the exponent a of the slope, S, could be derived if one assumes that the charging of the proteins occurred as shown in the schematic representation

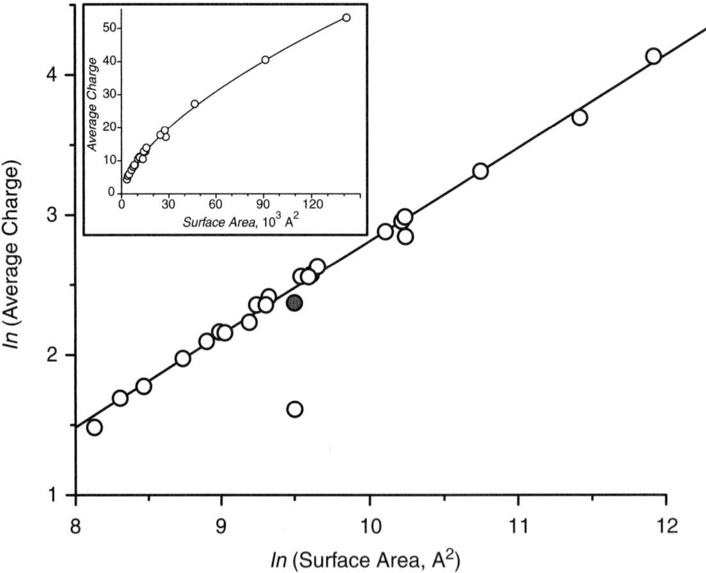

Figure 1.11. Correlation between average charge state of protein ions generated by ESI under near-native conditions (10 mM ammonium acetate, pH adjusted to 7) and their surface areas in solution whose calculation was based on their crystal structures. The data are plotted in ln (natural logarithmic) versus ln scale (a graph using linear scales is shown in the inset). A gray-shaded dot represents a data point for pepsin, and the open circle underneath represents the maximum charge expected for pepsin if the extent of multiple charging was limited by the number of basic residues within the pepsin molecule. (Figure and text reprinted from Kaltashov and Mohimen,[46] with permission from the American Chemical Society.)

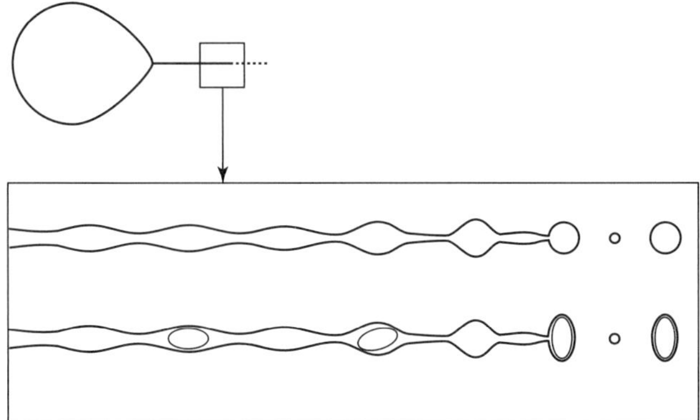

Figure 1.12. Schematic representation of a fission process for a droplet whose charge is equal to the Rayleigh limit. The straight line (jet) projecting form the pointed (cone) end of the droplet (top diagram) represents emission of a highly charged (at the surface) liquid, in the form of a jet. The axisymmetric disintegration of the jet is represented schematically in the boxed drawings. Capillary breakup in the absence of internal constraints produces two groups of homogeneous droplets (main and satellite). Presence of internal constraints in the stream (such as particles of a certain size) is likely to have a significant influence on the size and shape of the droplets. (Figure and text reprinted from Kaltashov and Mohimen,[46] with permission from the American Chemical Society.)

(Figure 1.12). A droplet at the Rayleigh limit emits a jet that separates into individual progeny droplets. The protein that escapes the parent droplet is shown to be fitting a progeny droplet. Under these conditions, the charging of the progeny droplets and the progeny that contains the protein will be determined by parameters associated with the formation of the jet.[46,47] These are different from the Rayleigh spherical droplet stability conditions. It is the former conditions[46] that lead to slope $= a$, which is in agreement with the experimental results (Figure 1.12).

It is interesting to see what type of fit to the Fernandez de la Mora equation [Eq. (14a)] is obtained with the proteins and their charges determined by Kaltashov and Mohimen.[46] A plot of the charge Z versus the molecular mass M is shown in Figure 1.13. The experimental points fit very well the curve obtained with Fernandez de la Mora's Eq. (14a), $Z = 0.078 \times M^{0.5}$. The fit is very much better than that observed in Figure 1.9 that used data obtained from several different laboratories. The best-fit curve to the data in Figure 1.13 leads to $Z = 0.041 \times M^{0.547}$. Considering the good visual fit in Figure 1.13, one could argue that the use of the surface area of the proteins is not an improvement. However, one experimental point[46]—that is, that for ferritin—provides clear evidence that the surface area counts. Ferritin in the Fernandez de la Mora plot (Figure 1.13) shows a large deviation to a higher charge, while in the surface area plot (Figure 1.11) there is no deviation. As pointed out by Kaltashov and Mohimen,[46] the apo-ferritin used is approximately spherical, but has a large cavity on one side, which increases the surface area substantially and causes the higher charge. For an informative discussion of Kaltashov and Mohimen's data, see Benesh and Robinson.[47]

In summary, the Charged Residue Mechanism has allowed quantitative predictions of the protein charge state in the gas phase and is well-supported for large proteins of widely varying mass. The rather simple assumptions at the basis of the mechanism have recently been examined and have led to the insight that proteins with large deviations from the spherical shape lead to better correlation of the charge with the surface of the protein.[46]

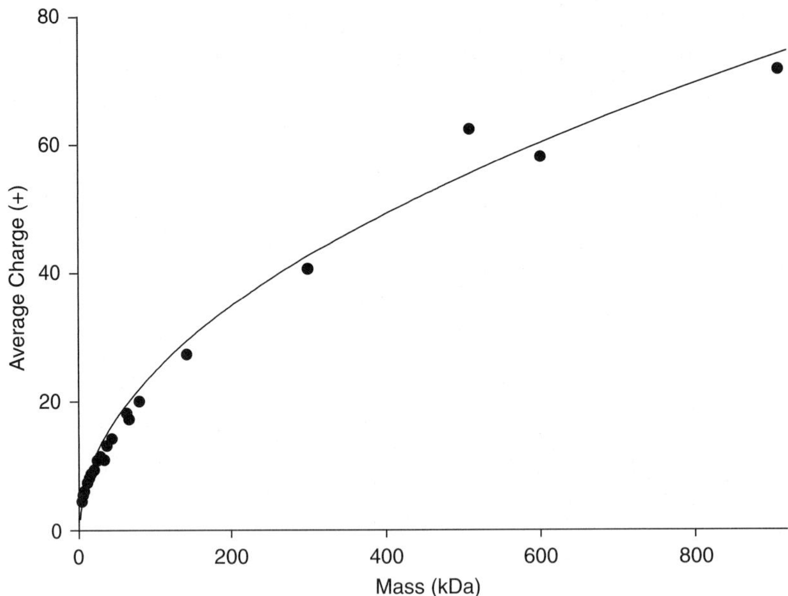

Figure 1.13. Plot of average charge of proteins observed by Kaltashov and Mohimen[46] versus molecular mass of protein. The solid line curve gives average charge predicted by the Fernandez de la Mora equation [Eq. 1.14]. A very good fit is observed except for one experimental point—that is, for ferritin (mass ≈ 510 kDa), which has a significantly higher charge Z. This protein is approximately spherical but has a cavity that increases its surface. This experimental point fits well in the charge versus protein surface plot (see Figure 1.11). (Figure 1.13 plot was graciously provided to the authors by Dr. Justin Benesh.)

The dependence on the surface of the protein and the model depicted in Figure 1.12 lend themselves to an extension. In solution, most of the ionized residues of native proteins are on the surface of the protein. When the protein approaches the surface of a positively charged larger droplet, several of the ionized basic residues of the protein may become part of the charge on the droplet. The specific shape of the protein and the closeness of the charges on the protein may introduce an instability of the droplet surface such that the droplet experiences a fission below the Rayleigh limit and the protein enters the first droplet of the cone jet caused by the fission. Because the surface-to-volume ratio increases as the droplet size decreases, this effect will become most important in the final stages of the droplet evolution when the droplets have become very small. Proposals that attachment of the protein to the surface of the droplet can introduce droplet instability and fission such that the protein leaves the droplet have been made before (see discussion of work by Karas and co-workers in Section 1.2.14). Some support for these models can be given also on the basis that proteins like pepsin that have only very few basic side chains are detected at very low sensitivities in the positive ion mode because the lack of positive sites reduces the ability of the protein to situate itself at the droplet surface.

It is notable that the generation of a charged protein in the gas phase by the above model falls somewhere between the CRM and the IEM. With CRM, the protein stays in the droplet until the solvent has evaporated and the charges of the droplet land on the protein. With IEM the protein leaves the droplet before it has evaporated, whereas in the model above, the protein stimulates the release of charged progeny droplets and leaves the parent drop in the first progeny drop that is of similar size to the protein.

1.2.11 Dependence of the Observed Ion Abundance of Analytes on the Nature of the Analyte, on Its Concentration, and on the Presence of Other Electrolytes in the Solution

The dependence of the sensitivities of different analytes observed with ESIMS on the nature of the analyte, on its concentration, and on the presence of other electrolytes in the solution is of interest both to the practicing experimental mass spectrometrist and to workers trying to relate the observations to the mechanism of ESI. The analytes considered in this section are not macromolecules like the proteins but are, instead, much smaller molecules that most likely enter the gas phase via the Ion Evaporation Model (IEM).

The dependence on various parameters of the total droplet current, I, produced at the spray capillary was given in Eq. 1.7. Relevant to the present discussion is the dependence of the current on the square root of the conductivity of the solution. At the low total electrolyte concentrations generally used in ESI, the conductivity is proportional to the concentration of the electrolyte. Thus, if a single electrolyte, E, was present, one would expect that the observed peak intensity, I_E, will increase with the square root of the concentration C_E [see Eq. 1.7]. Equation 1.7, is valid for the cone jet mode, which is used most often. At flow rates higher than the cone jet mode, the dependence on the concentration is lower than the 0.5 power.[35] Because ESIMS is a very sensitive method and the detection of electrolytes down to 10^{-8} M is easily feasible, one seldom works with a one-electrolyte system. In general, even with a single analyte ion, A^+, there will be most often also impurity electrolyte EX present, where the E^+ ions are generally Na^+ and NH_4^+ at levels below 10^{-5} M. Therefore, there are two concentration regimes for the analyte:

(a) C_A much higher than C_E. In that case, the I_A is expected to increase with the square root (or slower) of C_A.

(b) C_A much smaller than C_E. In that case, the I_A is expected to increase with the first power of C_A because now I_A will depend on a statistical competition between A^+ and E^+ for being charges on the droplets.

To cover both regions, Tang and Kebarle[35] proposed Eq. (16a) for a two-component system in the positive ion mode.

Two components: $$I_{A^+} = pf\frac{k_A C_A}{k_A C_A + k_E C_E} I \qquad (1.16a)$$

Three components: $$I_{A^+} = pf\frac{k_A C_A}{k_A C_A + k_B C_B + k_E C_E} I \qquad (1.16b)$$

Equation (16b) is for three components: A^+, B^+, and E^+. C_A, C_B, and C_E are the concentrations in the solution, I is the total electrospray current leaving the spray capillary, I can easily be measured (see Figure 1.1), and p and f are proportionally constants (see Tang and Kebarle[35]) while k_A, k_B, and k_E are the sensitivity coefficients for A^+, B^+, and E^+, which depend on the specific chemical ability of the respective ion species to become part of the charge on the surface of the droplet and, once there, to enter the gas phase.

In the regime where $C_A \ll C_E$, Eq. (16a) reduces to Eq. (16c).

$$I_A = \text{const} \times C_A, \quad \text{where const} = k_A I_E / k_E C_E \qquad (1.16c)$$

The experimental results[35] shown in Figure 1.14 give an example of a two-component system where the protonated morphine, $MorH^+$, is the analyte A, used at different

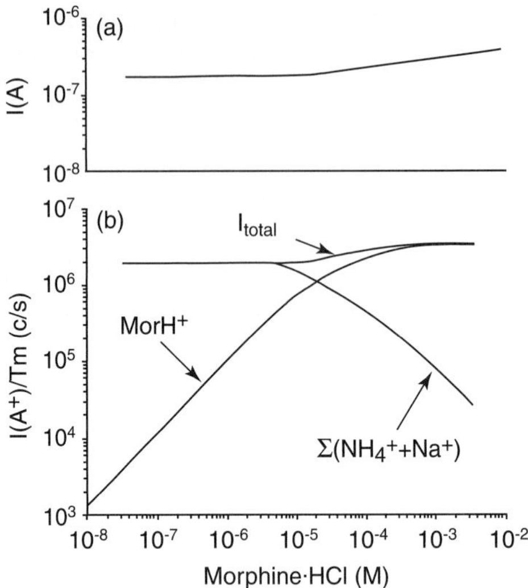

Figure 1.14. (**A**) Total electrospray current (Amp) with increasing concentration of analyte morphine · HCl. Due to presence of impurity ions (Na^+ and $NH_4{}^+$) at 10^{-5} M, I_{total} remains constant up to the point where the analyte reaches concentrations above 10^{-6} M. (**B**) Analyte $MorH^+$ ion intensity (corrected for mass-dependent ion transmission, Tm, of quadrupole mass spectrometer used) is proportional to concentration of morphineln · HCl up to the point where the morphine · HCl concentration approaches concentration of impurity ions. (Reprinted from Tang and Kebarle,[35] with permission from the American Chemical Society.)

concentrations, and the impurity ions $NH_4{}^+$ and Na^+, present at constant concentrations, are the electrolyte, E. The observed linear region of $MorH^+$ in the log–log plot used has a slope of unity at low concentrations, 10^{-8}–10^{-5} M, which means that the $MorH^+$ ion is proportional to the morphine concentration. This region is suitable for quantitative determinations of analytes. At about 10^{-5} M the increase of the $MorH^+$ intensity slows down because the $MorH^+$ concentration used comes close to that of the impurity electrolyte. Above that region where $MorH^+$ Cl^- becomes the major electrolyte, the peak intensity of $MorH^+$ can grow only with the square root, or even a lower power, of the electrolyte concentration.

Expanding the system to three components, two analytes and the impurity *E*, leads to an unexpected result (see Figure 1.15). In this experiment the concentrations of the two analytes A = tetrabutyl ammonium and B = cocaine (upper figure) or codeine (lower figure) are increased together such that $C_A = C_B$. The impurity C_E is constant. The experiment was made in order to determine the relative sensitivities, k_A and k_B, of A and B. In the log–log plot used, the difference $\log I_A - \log I_B$ equals the difference $\log k_A - \log k_B$ and one would expect that the difference will be constant for $C_A = C_B$ concentrations. However, this is not the case. The difference is constant only at high $C_A = C_B$ concentrations and becomes zero at low concentrations.

The tendency of k_A/k_E to approach unity at low C_A, C_B indicates[35] that there is a **depletion** of the ion that has the higher sensitivity *k*. This is the (A = tetrabutyl ammonium) ion in the present example. At $C_A = C_B \ll 10^{-5}$ M, the current *I* and the total charge *Q*, of the droplets and the number of charged droplets are maintained by the presence of the electrolyte, *E*, whose concentration is much higher. Under these conditions, ionic species like A^+ and B^+, when present at very low concentrations but having large coefficients k_A and k_B, find plenty of droplet surface to go to and ion evaporate rapidly. This results in a depletion of their concentration in the interior of the droplet. The ion A of higher sensitivity is depleted more than B, and this leads to an apparent value $k_A/k_B = 1$.

Experimental determination of the coefficient ratios k_A/k_B were performed[35] for a number of analytes in methanol by working at high concentrations C_A/C_B, where Eq. (16c) holds. The results in Figure 1.15 at high concentration represent two such determinations. It was found that the singly charged inorganic ions Na^+, K^+, Rb^+, Cs^+, NH_4^+ had low

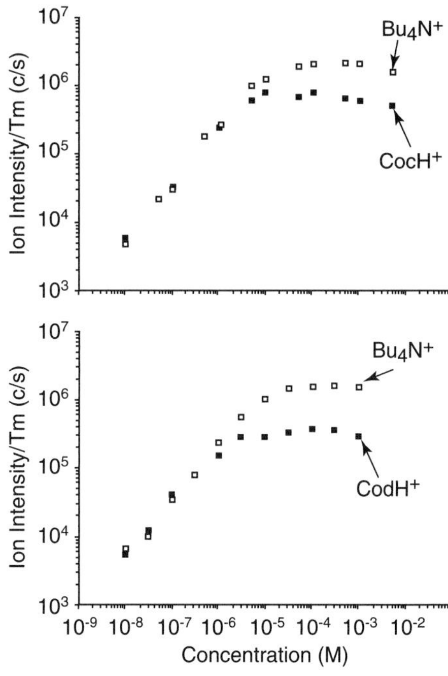

Figure 1.15. Ion intensities (corrected for mass-dependent ion transmission, Tm, of quadrupole mass spectrometer used) for pairs of analytes at equal concentration in solution. The different ESI sensitivities of the analytes are observable only at high analyte (above 10^{-5} M) concentrations. (Reprinted from Tang and Kebarle[35] with permission from the American Chemical Society.)

sensitivity coefficients, while analyte ions that were expected to be enriched on the droplet surface (i.e., which were surface-active) had high coefficients that increased with the surface activity of the ions. Thus, assuming that $k^{+}_{Cs} = 1$, the relative values k_A for the ions were: $Cs^+ \approx 1$; $Et_4N^+ = 3$; $Pr_4N^+ = 5$; $Bu_4N^+ = 9$; $Pen_4N^+ = 16$; $heptNH_3{}^+ = 8$, where Et, Pr, Bu, and so on, stands for ethyl, *n*-propyl, *n*-butyl, and so on (see Table I in Tang and Kebarle[35]). The tetraalkyl ammonium salts and alkylammonium salts, and especially those with long chain alkyl groups, are known surfactants.

Assuming that IEM holds, ions from the droplet surface will leave the droplets and become gas-phase ions. In this case, the gas-phase ion sensitivity coefficient, k_A, for ions A^+ will depend on the relative surface population of the droplet surface—that is, on the surface activity of ions A^+X^- given by a surface activity equilibrium constant K_{SA}—and on the rate constant for ion evaporation. The rate constant for ion evaporation is also expected to increase with the surface activity of the ion, because surface-active ions have low solvation energies (see Section 1.2.8). A third effect can be expected also. The very small droplets that lead to ion evaporation will, in general, be first- or second-generation progeny droplets (see Figure 1.6). Because the progeny droplets have higher surface-to-volume ratios relative to the parent droplets, an enrichment of the surface active ions is expected for the progeny droplets.

More recent work by Enke[48] made very significant advances starting from somewhat different premises. The role of the ion currents I, I_A, and so on, was represented by the molar concentrations of the ions. Thus the role of I was replaced with $[Q]$ moles of charge and the role of the currents I_A, I_B, ... was replaced by the molar concentrations of charges on the droplets due to the given ion species. Thus, $[A^+]_S$ is the molar concentration of charges on the surface of droplets due to A^+ species, and so on. The analyte A^+ was assumed to distribute itself between the interiors of the droplets with a concentration $[A^+]_I$ and as charge on the surface of the droplets $[A^+]_S$. An equilibrium between $[A^+]_S$ and $[A^+]_I$ was

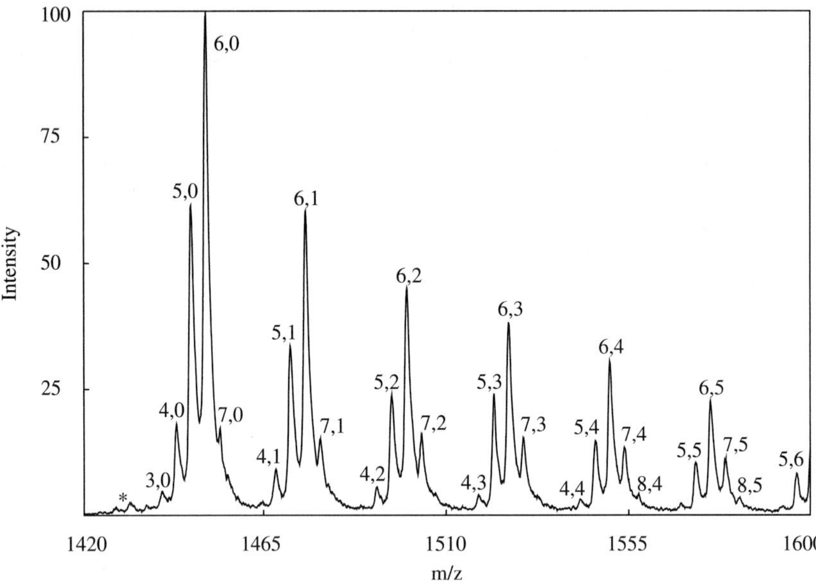

Figure 1.16. Mass spectrum of the major $Z = +6$, charge state of ubiquitin obtained from an aqueous solution of 25 μM ubiquitin containing 1 mM NaI. In the absence of NaI, essentially only one peak is observed corresponding to ubiquitin + $6H^+$. The observed first group of peaks corresponds to ubiquitin, where the H^+ charges are gradually replaced with Na^+. The largest peak in the first group, labeled 6,0, corresponds to ubiquitin + $6Na^+$. The second group has the same composition as the first group, but each ion contains also one Na^+ and one I^-. The next group contains two Na^+ and two I^-, and so on. The number *n* of Na^+ ions is indicated with the first number over the peak, while the number *m* of Na^+I^- is given by the second number. (From Verkerk and Kebarle,[58a] with permission from Elsevier.)

assumed. The other electrolytes were treated in the same way. Introduction of equations of charge balance and mass balance for each electrolyte led to an equation that predicts values for $[A^+]_S$, $[B^+]_S$, and so on, on the basis of the parameters [Q], which is known [see Eq. 1.3, the constants K_A, K_B, and so on, and the concentrations C_A, C_B, and so on. The assumption was made that $[A^=]_S$, $[B^+]_S$, and so on, will be converted to gas-phase ions and are therefore proportional with the same proportionality constant, *pf* [see Eq. (16)], to the ion currents I_A, I_B, and so on. The equation of $[A^+]_S$, $[B^+]_S$, and so on, is of the same form as Eq. (16) in the high concentration range but not in the low concentration range. By taking into account, via mass balance, the depletion of the concentration C_A and C_B of the analytes with high coefficients $k_A = K_A$, $k_B = K_B$, the equation of Enke provides an excellent fit of the ion abundance curves over the full concentration range, preserving a constant k_A/k_B ratio. Further development by Enke and co-workers has led to a most successful formalism (see Chapter 2, this volume).

1.2.12 Noncovalent and Ionic Interactions in Solution and in the Gas Phase and other Relevant Differences Between Gas Phase and Solution

In ESIMS the analytes experience both solution and gas-phase conditions, and therefore the mass spectrum of the analyte will reflect both gas- and solution-phase conditions. Many practitioners of ESIMS with a biochemical background are well-acquainted with non-covalent and ionic reactions in aqueous solution, but don't have a strong background in gas-phase interactions. The present section provides a brief comparison of the interaction in the liquid and gas phase.

Whereas in water, the separation of an ion pair such as NaCl is spontaneous, in the gas phase, the separation requires the very large energy input of 148 kcal/mol, a value much larger than the energy required to break a C–C or a C–H covalent bond.

$$\mathrm{NaCl}_{(g)} = \mathrm{Na}^+_{(g)} + \mathrm{Cl}^-_{(g)} \qquad \Delta E \approx 148\ \mathrm{kcal/mol}\ (\text{NIST database}) \tag{1.17}$$

The binding energy increases when the positive and negative ions are small such as LiF, and it decreases for large ions such as CsI. A salt like NaCl is very soluble in water because of the very strong solvation interactions of the Na^+ and Cl^- with the water molecules.

Very nonpolar compounds like the hydrocarbons are very weakly soluble in water, because the highly polar and hydrogen bonding water molecules interact very strongly with each other and thus exclude the hydrocarbon molecules.[50] The hydrocarbon molecules have weak van der Waals interactions with water molecules and somewhat stronger van der Waals interactions with each other. At very low concentrations of the hydrocarbon in water, this leads to formation of hydrocarbon islets in the water medium. Higher hydrocarbon concentrations lead to formation of two separate liquid phases.

Unfolded (i.e., denatured) proteins have regions that are hydrophobic and other regions, particularly where the ionized residues are located, which are strongly hydrophilic. The denatured protein will tend to fold spontaneously, such that the ionic residues face and interact with the water molecules while the hydrophobic sections interact with each other.

When charged analytes such as proteins are present in the evaporating droplets, ions such as Na^+, due to salt impurities, will begin to ion pair with the ionized acidic residues of the protein. This ion pairing will be driven by the very large loss of solvent by evaporation from the droplets. The bonding of these ion pairs is so strong that no amount of collision energy provided in the clean-up stages of the interface leading to the mass analysis region can remove these undesired Na^+ adducts. Examples of such ion pairing effects are given in Section 1.2.13.

The gas-phase positive ions such as the alkali ions Na^+ and K^+ form fairly strongly bonded adducts with polar as well as aromatic compounds B. The bond dissociation energies[51–55] corresponding to the reaction

$$\mathrm{M}^+\mathrm{B} = \mathrm{M}^+ + \mathrm{B} \tag{1.18}$$

for a number of such ion–ligand complexes are given in Table 1.3.

Of special interest are ligands that model functional groups of proteins. For example, *N*-methylacetamide, which models the amide groups $-CH_2CONH-CH_2-$ of the protein backbone, forms a bond with Na^+ that is equal to 36 kcal/mol. Comparing this with the value for acetone, which is 30 kcal/mol,[53] the stronger interaction with the methylacetamide must be due to a contribution of the $-NH_2-$ group. For the amino acids serine and proline, the binding energy (~45 kcal/mol) is even higher, probably due to participation of neighboring groups. These high values indicate that Na^+ adducts to the amide groups of the proteins also will not dissociate in the clean-up stages of the ESI mass spectrometer. Thus, the presence of Na^+ impurity in the solution is very undesirable because the mass shift due to the addition of Na^+ to ionized acidic residues and to the amide groups of the protein interferes with the interpretation of the mass spectrum.

Table 1.3. Bond Enthalpies, ΔH^0_B, for Reaction: $M^+B = M^+ + B$

M^+	H_2O	NH_3	Iso-PropOH	Acetamide	*N*-Methyl acetamide	Serine	Proline
Na^+	22.1[c]	25.6[c]	27.0[e]	35.6[c]	38.0[d]	45.0[c]	44.2[c]
K^+	16.9[b]	17.8[b]	—	—	—	—	—

[a]All values in kmol/mol at 298 K. ΔH^0_B value only weakly sensitive to temperature. For additional data on Na^+B complexes see Ref. 50, and for K^+B complexes see Refs. 50 and 52.

[b]Davidson and Kebarle.[51]

[c]Hoyau, et al.[53]

[d]Klassen et al.[54]

[e]Armentrout and Rodgers.[55]

The charging of an analyte such as a protein in the vanishing droplets will involve the excess charges at the surface of the droplets. When the solvent is a mixture of water and methanol, the methanol will have evaporated from the droplet; and if the droplet was at a somewhat acidic pH, due to the presence of millimolar acetic acid, the charging of the protein will involve proton transfer from H_3O^+ to basic residues and the amide groups of the protein.

The gas-phase basicities of the solvent molecules (H_2O in the present case) and of compounds modeling functional groups of the protein are expected to determine the outcome of the proton transfer reaction. Table 1.4 provides gas-phase basicity data for several representative compounds.[49] Gas-phase basicity data for thousands of compounds are available on the NIST database.[49] The reaction rate constants of the vast majority of ion molecule reactions in the gas phase such as proton transfer or ligand transfer can be predicted on the basis of the thermochemistry of the reactions.[56] Thus, for example, the proton transfer reaction $AH^+ + B = A + BH^+$ will proceed at collision rates, that is, without activation energy when the reaction is exoergic, that is, when the gas-phase basicity GB(B) is greater than GB(A). The absence of activation energy is due to the attraction between the charge of the ion and the polarizable molecule. This attraction leads to a "collision" where the ion and molecule spiral around their center of mass until they collide with each other. These spiraling collisions lead to rate constants that are very large and can be evaluated with an equation based on Langevin's work.[56] Furthermore, the energy released by the "collision" is sufficient to activate the reaction.[56]

Table 1.4. Gas-Phase Basicities of Bases B

$GB(B)^a = \Delta G^0_{298}$ for reaction: $BH^+ = B + H^+$			
H_2O	157.7	NH_3	195.7
$(H_2O)_2$	181.2	CH_3NH_2	206.6
CH_3OH	173.2	$C_2H_5NH_2$	210.0
$(CH_3OH)_2$	196.3	$(CH_3)_2NH$	214.3
C_2H_5OH	178.0	*n*-PropNH_2	211.5
CH_3CN	179.0	*N*-Methyl acetamide	205.0
$(CH_3)_2O$	182.7	Pyridine	214.8
$(C_2H_5)_2O$	191.0		

[a]GB(B) = Gas-phase basicity. Values in kcal/mol.

All values from NIST Databases.[50] Often used are also the proton affinities. They correspond to the ΔH^0 value for the gas-phase reaction $BH^+ = B + H^+$.

Proteins, in general, will be easily protonated because they have basic residues with relatively high GBs. Thus GB(lysine) modeled by GB(*n*-propyl amine) = 212 kcal/mol (Table 1.4) has a much higher GB than GB($(H_2O)_2$) = 181 kcal/mol, and a proton transfer to lysine is expected. Proton transfer to the backbone amide group modeled by GB(N-methylacetamide) = 205 kcal/mol is also expected. In the statements above, we have assumed that the proton donors in the proton transfer are $H_2OHOH_2^+$ species at the surface of the droplet. This is a plausible assumption, particularly if the charged residue model (CRM) holds.

On the other hand, protonation of a sugar-like glucose is unlikely because GB(glucose), roughly modeled by GB(ethanol) = 178 kcal/mol (Table 1.4), is too low. Experimentally, it was observed already quite early[57] that good ESI mass spectra of carbohydrates can't be obtained by protonation. However, spectra could be obtained by sodiation, achieved by the addition to the solution of mM concentrations of sodium acetate.[57] The sodium result is not surprising. Binding energies of sodium ions to sugars and even monosaccharides such as glucose do not seem to have been determined. However, high sodium affinities are expected because hydroxy groups on adjacent carbon atoms are present, which will lead to bidentate interactions with the sodium ion.

The gas-phase basicities and bond affinities such as in Tables 1.3 and 1.4 should be applied to charging of the analytes by the ion charges on the surface of the evaporating droplets with caution because of the complexity of the actual process that involves also a liquid surface. The application of the gas-phase data is much more straightforward for true gas-phase ion–molecule reactions that occur in the atmospheric region before the sampling capillary (or orifice) leading to the MS interface and in the following low-pressure stage before the skimmer (see Figure 1.2). Typically, these could involve charged solvent molecule clusters and analyte molecules, particularly when a countercurrent gas such as nitrogen was not used (see Figure 1.2). Such reactions are considered elsewhere (see Chapter 2, this volume).

1.2.13 Some Examples of Effects on Mass Spectra of Proteins Due to the ESI Process

1.2.13.1 Ion Pairing of Salt ions with Ionized Residues of Proteins

As discussed in the preceding sections and specifically Section 1.2.12, the formation of the gas-phase analyte ions occurs only after a very large loss of solvent from the charged droplets is produced by electrospray. This is the case for both mechanisms, CRM and IEM. In the presence of salts in the solution, the increase of concentration with solvent loss will lead to pairing of the positive with negative ions. This pairing will start long before all the solvent has evaporated. Pairing with salt ions will also involve ionized sites of the protein, that is, the ionized acidic and basic residues.

The pairing is most clearly observed in the mass spectrum when certain salts are used. One such salt is NaI. Shown in Figure 1.16 is the spectrum of folded ubiquitin obtained[58a] using nanospray of an aqueous solution containing 25 μM ubiquitin and 1 mM NaI. The $Z = +6$ charge state is shown. This is by far the major charge state observed for ubiquitin.[58a] A series of groups of peaks containing Na and NaI is observed, and the composition of these ions is given in the figure caption. The observed composition can be rationalized by the operation of two processes: (a) replacement of the H^+ charges with Na^+ charges where the Na^+ come from the surface of the droplets such that the ultimate charging of the protein involves exclusively Na^+ rather than H^+ charges and (b) ion pairing of Na^+ ions in the

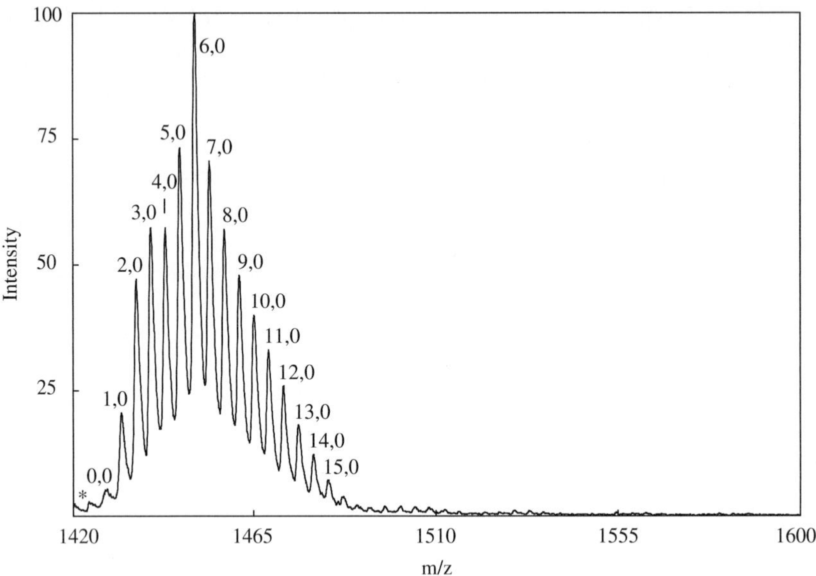

Figure 1.17. Mass spectrum of ubiquitin $Z = 6$ charge state under the same conditions as in Figure 1.16 but at high collisional activation (CAD). All the peaks that contained $Na^+ + I^-$ pairs have disappeared and are replaced by uncharged Na adducts. This process is consistent with loss of HI where the I^- ions paired to an ionized basic residue dissociated as HI. (From Verkerk and Kebarle,[58a] with permission from Elsevier.)

solution of the droplet with ionized acidic residues and I^- with ionized basic residues at the surface of the protein.

The spectrum (Figure 1.16) was obtained at low nozzle (spray capillary) to skimmer potential, so that there was little collisional activation of the protein. At a high potential, multiple loss of HI was observed (see Figure 1.17). All the peaks that contained $Na^+ + I^-$ pairs have disappeared and were replaced by peaks with unpaired Na adducts.

The observations based on Figures 1.16 and 1.17 can be rationalized as follows: (a) *Charging* reactions by Na^+. The sodium ion goes either on ionized acidic residues or on the amide groups of the protein back bone. The Na^+ bonding at both these sites is strong (see Table 1.3), and therefore no loss of Na is expected even at high CAD conditions. (b) Ion pairing reactions involving both Na^+, going to same sites as in (a), and I^-, going predominantly to the ionized basic residues. *No charging* occurs. At high CAD conditions, HI is formed by reaction 1.19, where the basic residue shown is lysine:

$$\text{Protein}-(\text{CH}_2)_4-\text{NH}_3^+-I^- = \text{Protein}-(\text{CH}_2)_4-\text{NH}_2+\text{HI} \qquad (1.19)$$

Experiments involving the salts NaCl and NaAc, where Ac stands for the acetate anion, were made and it was found[58] that the energy required for the dissociation decreased in the order HI, HCl, HAc. It was also shown,[58a] on the basis of theoretical data, that this order is expected and the energy required is well within the range provided by the CAD used.

Additional evidence that the basic side chains are involved in the dissociation reaction was obtained by experiments where much higher concentrations of sodium acetate were added so as to be certain that all the ionized acidic sites were paired with Na^+ and all basic sites with the acetate anions. The mass spectrum of the $Z = 6$ charge state for ubiquitin and sodium acetate at high CAD is shown in Figure 1.18. The sharp break of peak intensity past

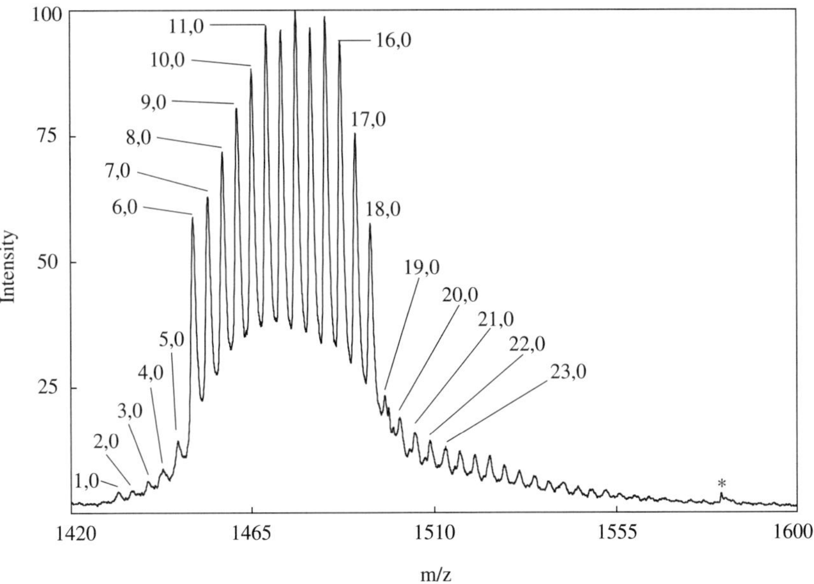

Figure 1.18. Mass spectrum showing $Z = 6$ charge state obtained with ubiquitin (25 μM) and a high (5 mM) concentration of sodium acetate at high CAD where all acetate anions have fallen off as acetic acid molecules. The sharp break of peak intensity past $n = 18$ indicates that the observed ubiquitin has 6 Na^+ charges and 12 ionized acidic sites on the protein which are sodiated. (From Verkerk and Kebarle,[58a] with permission from Elsevier.)

$n = 18$ indicates that the observed ubiquitin has 6 Na^+ charges and 12 ionized acidic sites paired to Na^+ ions, giving a total of $n = 18$. The acetate anions that had ion paired with the ionized basic sites have fallen off as acetic acid in a reaction analogous to Eq. 1.19. An examination of the number of acidic residues and the terminal acidic group of ubiquitin showed a total number of 12 acidic sites and all of these were near the surface of the protein, in agreement with the proposed ion pairing mechanism.[58a]

1.2.13.2 Why Is Ammonium Acetate Such a Popular Salt Additive to Solutions Used for ESIMS[58a]

Ammonium acetate at millimolar concentrations is very often used as an additive to solutions, particularly so when the analytes are proteins. It acts as buffer, albeit a weak one, and generally the assumption is made that the buffer action is the cause for its popularity. But ammonium acetate has another most useful property. It leads to very clean mass spectra solely due to the protonated protein. As a result of the "salting-out" precipitation procedure used in the isolation of proteins, sodium ions are a very common impurity in protein samples. Use of ammonium acetate prevents the formation of sodium adducts to the protein. The results described in the preceding part (a) provide the mechanism by which this occurs. The ion-pairing reactions in the presence of ammonium acetate are shown in Eq. 1.20.

$$\begin{aligned} &{}^{+}H_3N(CH_2)_2-\text{Prot}-(CH_2)_2COO^- + NH_4^+ + CH_3COO^- \\ &\quad = CH_3COO^{-}\,{}^{+}H_3N(CH_2)_2-\text{Prot}-(CH_2)_2COO^-NH_4^+ \end{aligned} \tag{1.20}$$

For simplicity, the equation shows only one ionized basic and acidic residue of the protein. The much larger concentration of the ammonium acetate relative to the sodium ion impurity leads to ion pairing that involves only acetate and ammonium ions. Subjecting the resulting protein to collisional activation in the gas phase leads to facile loss of acetic acid and ammonia as shown in Eq. 1.21.

$$\begin{aligned} CH_3COO^- \, {}^+H_3N(CH_2)_2-\text{Prot}-(CH_2)_2COO^- NH_4^+ &\rightarrow CH_3COOH \\ &+ H_2N(CH_2)_2-\text{Prot}-(CH_2)COOH + NH_3 \end{aligned} \tag{1.21}$$

The dissociation is facile because the bond energy for the dissociation is only15 kcal/mol for acetic acid and 11 kcal/mol for ammonia.[58a] The net effect of the complete process [Eqs. 1.20 and 1.21] is equivalent to a proton transfer from the ionized basic site to the ionized acidic site. Thus the molecular weight of the protein was not changed, but the positive and negative groups were neutralized. Thus a clean mass spectrum of the protein is obtained without a mass change.

Equations 1.20 and 1.21 show that any negative ion impurities due to anions of strong acids, such as the phosphate or trifluoroacetate anions, which would have led to strongly bonded adducts to the ionized basic residue [see Eq. 1.20], would also have been prevented by the relatively much higher concentration of the acetate ion from the ammonium acetate used. An important example is the use of trifluoroacetic acid mobile phase in liquid chromatography–mass spectrometry.[58b] The presence of the TFA anion in the effluent leads to a large loss of sensitivity in ESIMS and addition of ammonium acetate to the solution removes that loss.[58b]

1.2.13.3 Determinations by Electrospray of Equilibrium Constants of Association Reactions in Solution and Possible Sources of Error Due to the ESI Process

The determinations of equilibrium constants by ESIMS can be divided into two categories: (a) equilibria involving macro ions such as proteins and ligands which may be large organic molecules and (b) equilibria involving small reactant ions and ligands. Uncritical application of ESIMS can lead to erroneous results due to the complexity of the electrospray process. The determination of the association constants of protein–substrate complexes via NanoESI is well-developed and used, while that for small molecule complexes is not. Therefore, only the protein work will be considered here. Readers interested in small reactants equilibria can find useful literature information in the recent work by Zenobi and co-workers.[59] This work reports also new very interesting results based on measurements of laser-induced fluorescence and phase Doppler anemometry with which it was possible to measure the state of equilibrium inside individual charged and evaporating small droplets!

The formation of noncovalently bonded complexes of proteins with substrates is an extremely important class of reactions in biochemistry.[41d] It involves processes such as enzyme–substrate interactions, receptor–ligand binding, assembly of transcription complexes, and so on. The determination of these constants in solution has been an important component of biochemistry for many years. The established methods have various limitations, such as the requirement of large samples (X-ray crystallography), limited analyte molecular mass range (NMR), and poor mass resolution (gel electrophoresis). The major advantage of ESI is the very much smaller quantity of analyte required and the ability to identify also complexes involving multiple components. The first ESIMS study of noncovalent complexes involving proteins, by Ganem and Henion,[60a] was followed by a large number of studies which included quantitative determinations of the equilibrium constants.

References [60b–63] provide a sample of this work. Standing and co-workers[60b] demonstrated that reliable equilibrium constants for protein–substrate equilibria can be obtained with ESIMS.

Consider the general reaction [Eq. (22a)] where P is the protein and S is the substrate, and the reaction has reached equilibrium in the solution used. The equilibrium constant K_{AS} for the association reaction [Eq. (22a)] is given by Eq. (22b), where [P], [S], and [PS] are the concentrations at equilibrium.

$$P + S = PS \qquad (1.22a)$$

$$K_{AS} = [PS]/[P] \times [S] = I_{PS}/I_P \times I_S \qquad (1.22b)$$

Sampling the solution with ESI, the concentrations can be replaced with the ESIMS observed peak intensities, I_P, I_S, and I_{PS}. It should be noted that the charges on P, S, and PS leading to the ions I_P, I_S, and I_{PS} are, in general, not equal to the charge of these species in the solution, because multiple charges such as multiple H^+ ions are supplied from the surface of the very small droplets leading to the gas-phase analytes by either ion evaporation IEM or the charged residue CRM model (see Sections 1.2.9 and 1.2.10).

One can repeat the experiment at several, gradually increasing concentrations of S and examine if the association constant, evaluated with Eq. (22b), remains constant. This procedure is called the "Titration Method." ESIMS determinations of $I_{PS}/I_P \times I_S$ with this method have been often in agreement with the requirements of Eq. (22b) and have also provided K_{AS} values in agreement with data in the literature obtained by conventional methods.[60b–63] When the molecular mass of S is much smaller than those of P and PS, as is often the case, erroneous results may be obtained due to m and m/z factors including the transmission of the MS analyzer. Therefore, it is advantageous to use only the ratio of I_{PS}/I_P, because P and PS have a similar mass when S is smaller than P, which is often the case. Zenobi and co-workers[63] have provided an equation for the determination of K_{AS} with the titration method in which I_S is eliminated.

From the standpoint of the mechanism of ESI, the agreement of the K_{AS} values determined via ESIMS with values obtained with other methods may appear surprising. One could expect that the very large increase of the concentration of the solutes in the charged droplets, due to evaporation of the solvent from the droplets, before the analytes are converted to gas-phase ions, will lead to an apparent K_{AS} that is much too high. However, this equilibrium shift need not occur if rates of the forward and reverse reactions leading to the equilibrium are slow compared to the time of droplet evaporation.

Peschke et al.,[32] using experimental information[25a] on the evaporation of charged water droplets produced by ESI and their fissions, evaluated an approximate droplet history scheme for water droplets produced by nanoelectrospray. The early part of this scheme is shown in Figure 1.6 (Section 1.2.7). Because the initial droplets produced by nanospray are very small (>1 μm in diameter), their evaporation is very fast so that they reach the Rayleigh instability condition in just several microseconds (see Figure 1.6). It could be established[32] that the first-generation progeny droplets will be the major source of analyte ions. Assuming even the fastest possible reaction rates (i.e., the diffusion limit rates) in water for the forward reaction P + S = PS, it could be shown that the time can be too short for the equilibrium [Eq. (22)] to shift in response to the increasing concentration due to solvent evaporation. The rate constant at the diffusion limit decreases with an increase of the substrate S size. Substrates of medium size such as erythrohydroxy aspartate, adenosine diphosphate, and adenosine triphosphate, with diffusion-limited rate constants from $k = 10^6$–$10^7\,M^{-1}\,s^{-1}$, are too slow to cause an equilibrium shift that will lead to a significant error in the equilibrium constant (K_{AS}) determination via ESIMS. Thus, an equilibrium shift for substrates that are not too small is not expected at least with nanospray.

The droplet evolution history[32] was based also on determinations by Gomez and Tang[24] which indicated that a small number of progeny droplets is emitted at each Rayleigh fission and that the diameters of the progeny are relatively large. The recent results by Duft et al.[28b] (see Table 1.2) predict a very much larger number of very much smaller progeny droplets. Because the smaller progeny droplets will evaporate much faster, one might expect that substrates S of considerably smaller size will also not lead to an equilibrium shift.

Because proteins are most probably transferred to the gas phase via CRM, another question must be examined also. Assuming that close to equal concentrations of the protein and substrate, in the 10 μM range, were used, the evaluation[32] shows that in most cases there will be one protein and one substrate molecule in the average first-generation progeny droplet that has evaporated down to the size of the protein. In that case, the P and S will form a nonspecific complex because the substrate S makes a random encounter with the protein and has "no time" to find the site of specific strong bonding. Since the mass of the nonspecific complex is the same as that of the specific complex, the observed peak intensity, I_{SP}, will lead to an apparent K_{AS} that is too high. But this need not be the case. The weakly bonded nonspecific complexes are expected to fall apart easily in the clean-up stages of the interface to the mass analysis region, thereby minimizing the nonspecific contribution in the measured K_{AS}.[32]

The above *a priori* assumption about nonspecific complexes, while logical, may not predict the correct outcome in all cases. It neglects to consider the strong ion-neutral bonding in the gas phase discussed in Section 1.2.12 and Table 1.3 and the fact that the protein will be multiply charged as the charged droplet dries out. Strong hydrogen bonds can form between charged (protonated) sites of the protein and functional groups of the substrate such as -OH and -NH_2 groups. A recent study by Wang, Kitova, and Klassen[65] described exactly such effects for protein–carbohydrate protein–substrate complexes. Not only did they observe PS and SPS complexes in the gas phase, but the nonspecific bond was found to be stronger than the specific bond. This suggested that the nonspecific S–P bond was to a protonated site of the charged protein. The bond energies were determined with the Blackbody Infrared Radiative Dissociation (BIRD) technique with a modified Fourier Transfer Ion Cyclotron (FTICR) mass spectrometer.[65] Later work from this group[66] describes a method with which corrections can be made for the presence of nonspecific protein–substrate complexes.

1.2.14 Nanoelectrospray and Insights into Fundamentals of Electrospray—Nanospray

Nanospray was developed by Wilm and Mann,[67,68] whose primary interest was an electrospray with which much smaller quantities of analyte are required. Such a device would be particularly important to applications in biochemistry and particularly the analysis of proteins where, in general, very small samples are available. With ESI, most of the analyte is wasted. The large diameter of the spray tip produces large droplets whose evolution to small droplets requires the presence of a large distance between the spray tip and the sampling orifice (or sampling capillary) (see Figures 1.1 and 1.2a). As a result, only a very small fraction of the ultimate, very small droplets enters the sampling orifice (capillary). With nanospray, the spray tip has a much smaller diameter. Also, the flow is not a forced flow, due to a driven syringe, as used in ESI (see Figure 1.2a); instead, the entrance end of the spray capillary is left open. A "self-flow" results, which is due to the pull of the applied electric field on the solution at the capillary tip (see Section II.B). The self-flow is controlled by the diameter of the tip of the spray capillary.

In their first effort,[67] using what was essentially an electrospray source, Wilm and Mann developed an equation for the radius of the zone at the tip of the Taylor cone from which the charged droplets are ejected. This radius is related to the resulting droplets' radii. It was found that the radius depends on the 2/3 power of the flow rate. To minimize the radius of the zone, a modified electrospray ion source with a smaller orifice was developed which led to a "microspray" version of ESI. Further development[68] using capillary orifices as small as 1–2 μm in diameter led to nanospray. Such small orifices could be obtained[68] by pulling small-diameter borosilicate capillaries with a microcapillary puller. About 1 μL of solvent is loaded directly into the entrance end of the capillary. The droplets produced past the capillary tip had a volume that was close to 1000 times smaller than the volume of droplets obtained with conventional ESI. Such small droplets will evaporate very rapidly, so that the capillary tip can be placed very close to the sampling orifice that leads to the mass spectrometer, thereby minimizing sample loss and allowing efficient use of a large fraction of the solution subjected to MS analysis. Thus, even though the amount of analyte sample is 10–100 times smaller than used with ESI, the observed mass spectrum peak intensities are equal to, if not better than, those in conventional ESI.

Another advantage of nanospray was the observation that with nanospray one can use neat water as a solvent without causing electric gas discharges as is the case with electrospray (see Section 1.2.4). Water as solvent is more suitable for the analysis of proteins, some of which may denature at least partially in other solvents.

Nanoelectrospray has proven to be of enormous importance to the analysis of biochemical and biopharmaceutical samples. However, it is also important to research on the fundamentals of electrospray—nanospray. Karas and co-workers[69–71] have been major contributors to this research. The experimental finding that mass spectra of analytes such as proteins obtained with nanospray are much less affected by the presence of impurities in the solution such as sodium compared to electrospray spectra is an advantage of nanospray. The reasons for this were examined[69] and the following reason was given. Gas-phase ions are produced from charged droplets only when the droplets are very small. This holds for both IEM and CRM. Therefore, if one starts with relatively small initial droplets, as is the case with nanospray, much less solvent evaporation will be required to reach the small size droplets required. Therefore, in the presence of impurities such as sodium salts the concentration increase of the salt will be much smaller with nanospray.

Mass spectra were obtained[69] with nanospray and electrospray. First, large concentrations (10^{-2} mol/L NaCl) in a solution of H_2O:MeOH:HOAc (48:48:4 vol%) and no protein were used. The nanospray spectra showed distinct peaks $Na_nCl_{n-1}^+$, where n was between 9 and 32 and $Na_nCl_{n-2}^{2+}$ with roughly twice higher n. With electrospray the same peaks were observed, but also a forest of peaks spread over the whole mass range as would be expected from final droplets that had much higher solute concentrations. (*Note*: All mass spectra discussed were obtained in the m/z 200–2000 range.)

More interesting were the spectra obtained with nanospray and electrospray and the solutes, insulin at 10^{-5} mol/L and NaCl 10^{-2} mol/L in the same H_2O:MeOH:HOAc solvent (see Figure 1.2b,c in Ref. [69]). The electrospray spectrum was dominated by a forest of closely spaced peaks due to Na and Cl containing clusters so that the protein could be barely found. With nanospray, the protein peaks were very clearly visible and in spite of the 1000 times larger concentration of NaCl, the combined peak intensity of the insulin ions was close to equal, if not larger than, that of all the only Na- and Cl-containing ion clusters in the spectrum. This and other evidence made the authors consider that the charged insulin gas-phase ions are produced by a special process that leads to high peak intensity. Insulin is a rather small protein and can be considered also to be a peptide, but the implication of the discussion[65] is that the same process could be also valid for many, if not all, proteins.

It was proposed that the charged protein leaves the parent droplet. Thus, this is an IEM-type model but with special features. It was assumed that insulin is surface-active and therefore is present in enriched levels on the droplet surface. Some of the positively charged functional groups, such as protonated basic residues, reach the surface and become part of the charges at the surface of the droplet. If the basic residues were not protonated, they could become protonated by H_3O^+ charges on the surface of the droplet. Na^+ ions on the droplet surface can also form complexes with basic groups of the protein. Significantly, the basic sites on the surface of the protein will be closer together than the normal spacing of the charges on the droplet. The resulting uneven Coulomb repulsion between the charges on the droplet and the protein will force the protein charges outward, and this causes a bulge in the droplet (see Figure 1.5 in Ref. 65). The charged bulge will destabilize the droplet and lead to a droplet fission that expels the protein before the Rayleigh limit. This process may lead to an especially high yield of gas-phase protein ions.

The insulin spectra (see Figure 1.2b in Ref. 69) show groups of peaks for each charge state, $Z = 3$–5, with $Z = 5$ being the dominant group of peaks. The major peak of each charge state is due to all charges being H^+ ions. Peaks where one, two, and so on, H^+ charges are replaced by Na^+, with gradually decreasing intensities, follow. The charging by H^+ should in large part be due to the presence of HOAc in the solution. The pH was 2.5, which means a $[H^+] \approx 3 \times 10^{-3}$ mol/L. This is close to the 10^{-2} mol/L NaCl concentration used. Therefore, one can expect that the charges on the final droplets will not only be due to Na^+ ions, but will also be due to H^+ ions.

It was also observed[65] that the ratio of Na^+ to H^+ charges of the protein increases greatly with decreasing charge *Z* state of the protein (see Figure 1.2b in Ref. [69]). The authors[65] explain this observation on the basis of the same model. They assume that the lower charge ions result from progeny droplets originating from parent droplets that had experienced many Coulomb fissions and, due to the solvent loss required for each fission, contain a large concentration of electrolyte. This large concentration increases the probability of negative ions such as Cl^- or $CH_3CO_2^-$ to form ion pairs with ionized positively charged basic residues of the protein. These proteins then lose HCl or CH_3CO_2H, respectively, in the heated (200 °C) transfer capillary to the ion trap MS that was used. For evidence of such loss processes see Section 1.2.13 and Verkerk and Kebarle[58a]. The overall process reduces the charge of the observed protein ions. Due to the large $Na^+ X^-$ concentration in these parent droplets, the protein ions will also have incorporated more Na impurities as observed in the mass spectrum (Figure 1.2b in Ref. 69). (*Note*: The above discussion of the model proposed by Karas and co-workers[69] differs somewhat from that given in Ref. 69. This was done in order to include information based on more recent work[58a]).

In an extension and expansion of the above work, Schmidt et al.[70] studied the effect of different solution flow rates on the analyte signal. A series of analytes was chosen: detergents, peptides, and oligosaccharides, which have decreasing surface activities. The authors found that the ion abundance of the analyte with low surface activity was suppressed at higher flow rates, while at the very lowest flow rates the suppression disappeared. The high flow rates were close to the regime for conventional ESI, whereas the lowest flow rate was in the low nano region. This result was rationalized as follows. At high flow rates, the charged droplets emitted from the spray tip are much larger and the droplet history involving evaporation, droplet fission, progeny droplet evaporation, and fission until the very small final droplets are formed that lead to gas-phase ions is much longer. Surface-active analytes that will be enriched at the droplet surface will preferentially enter the progeny droplets and therefore will be enriched in the final generation progeny droplet, and this will lead to a high ion abundance for the surface active analyte. For the very small initial droplets obtained with lowest-flow nanospray, the evolution to the final droplet will be very short. In the extreme, there would be no such evolution, and this will lead to minimal discrimination against the non-surface-active analytes.

Another well-documented work by Chernushevich, Bahr, and Karas[71] deals with a disadvantage of nano relative to conventional electrospray. Using nanospray and repeated mass scans, some analytes were found to appear with delays of tens of minutes and a few were not detected at all, while no such suppression was found with ESI. The effect was found to be related to cation exchange chromatography on glass surfaces where the glass surface is negatively charged and retains positive ions. Peptides and proteins having a localization of positive charge where two or three ionized basic residues of the peptide or protein were in adjacent positions were most delayed or even completely suppressed. The very-small-diameter spray capillaries used in nano ESI lead to a high capillary surface-to-volume ratio and thus to a greatly increased exposure of negatively charged surface which delays or even traps the positive groups of the analytes. Replacing the glass with silica capillaries removed the analyte discrimination problem. Silica capillaries do not develop negative surfaces, whereas glass (sodium silicate) does. Presumably, with sodium silicate glass, some of the Na^+ gets washed out, leaving behind negative silicate sites on the surface facing the solvent, whereas with fused silica no such process is possible.

REFERENCES

1. Fenn, J. B. Electrospray wings for molecular elephants. Nobel Lecture available on web: http://nobelprize.org/nobel_prizes/chemistry/laureates/2002/fenn-lecture.pdf
2. (a) Chapman, S. Carrier mobility spectra of spray electrified liquids. *Phys. Rev.* **1937**, 52, 184–190. (b) Chapman, S. Carrier mobility spectra of liquids electrified by bubbling. *Phys. Rev.* **1938**, 54, 520–527. (c) Chapman, S. Interpretation of carrier mobility spectra of liquids electrified by bubbling and spraying. *Phys. Rev.* **1938**, 54, 528–533.
3. Dole, M. *My Life in the Golden Age of America*, Vantage Press, New York, **1989**, p. 169.
4. Dole, M.; Mack, L. L.; Hines, R. L.; Mobley, R. C.; Ferguson, L. D.; Alice, M. B. Molecular beams of macroions. *J. Chem. Phys.* **1968**, 49, 2240–2249.
5. (a) Yamashita, M.: Fenn, J. B. Electrospray ion source. Another variation of the free-jet theme. *J. Phys. Chem.* **1984**, 88, 4451–4459. (b) Yamashita, M.; Fenn, J. B. Negative ion production with the electrospray ion source. *J. Phys. Chem.* **1984**, 88, 4672–4675.
6. Whitehouse, C. M.; Dreyer, R. N.; Yamashita, M.; Fenn, J. B. Electrospray interface for liquid chromatographs and mass spectrometers. *Anal. Chem.* **1985**, 57, 675–679.
7. (a) Smith, R. D.; Loo, J. L.; Ogorzalek Loo, R. R.; Busman, M.; Udseth, H. R. Principles and practice of electrospray ionization mass spectrometry for large peptides and proteins. *Mass Spectrom. Reviews* **1991**, 10, 359–451.(b) Bayley, A. G. *Electrostatic Spraying of Liquids*. John Wiley & Sons, New York, **1988**. (c) Electrospray: Theory and applications. Special issue of *J. Aerosol Sci.* **1994**, 25, 1005–1252.
8. (a) Loeb, L.; Kip, A. F.; Hudson, G. G.; Bennet, W. H. Pulses in negative point-to-plane corona. *Phys. Rev.* **1941**, 60, 714–722. (b) Pfeifer, R. J.; Hendricks, C. D. Parameteric studies of electrohydrodynamic spraying. *AIAAJ* **1968**, 6, 496–502.
9. Taylor, G. I. The stability of horizontal fluid interface in a vertical electric field. *J. Fluid. Mech.* **1965**, 2, 1–15.
10. Fernandez de la Mora, J. The fluid dynamics of Taylor cones. *J. Annu. Rev. Fluid. Mech.* **2007**, 39, 217–243.
11. (a) Cloupeau, M.; Prunet-Foch, B. Electrohydrodynamic spraying functioning modes: A critical review. *J. Aerosol Sci.* **1994**, 25, 1021–1036. (b) Cloupeau, M. Recipes for use of EHD spraying in one-jet mode and notes on corona discharge. *J. Aerosol Sci.* **1994**, 25, 1143–1157. (c) Cloupeau, M.; Prunet-Foch, B. Recipes for use of EHD spraying of liquids in cone-jet mode. *J. Aerosol Sci.* **1989**, 22, 165–184. (d) Marginean, I.;

Parvin, L.; Heffernan, L.; Vertes, A. Flexing the electrified meniscus: The birth of a jet in electrosprays. *Anal. Chem.* **2004**, 76, 4202–4207.

12. (a) Tang, K.; Gomez, A. On the structure of an electrospray of monodisperse droplets. *Phys. Fluids* **1994**, 6, 2317–2322. (b) Tang, K.; Gomez, A. Generation by electrospray of monodisperse water droplets for targeted drug delivery by inhalation. *J. Aerosol Sci.* **1994**, 25, 1237–1249. (c) Tang, K.; Gomez, A. Generation of monodisperse water droplets from electrosprays in a corona-assisted cone-jet mode. *J. Colloid Sci.* **1995**, 175, 326–323. (d) Tang, K.; Gomez, A. Monodisperse electrosprays of low electric conductivity liquids in the cone-jet mode. *J. Colloid Sci.* **1996**, 184, 500–511.
13. Blades, A. T.; Ikonomou, M. G.; Kebarle, P. Mechanism of electrospray mass spectrometry. Electrospray as an electrolysis cell. *Anal. Chem.* **1991**, 63, 2109–2114.
14. Van Berkel, G. J.; Zhou, F.; Aronson, J. T. Changes in bulk solution pH caused by the inherent controlled-current electrolytic process of an electrospray ion source. *Intern. J. Mass Spectrom. Ion Process.* **1997**, 162, 55–62.
15. Fernandez de la Mora, J.; Van Berkel, G. J.; Enke, C. G.; Cole, R. B.; Martinez-Sanchez, M.; Fenn, J. B. Electrochemical processes in electrospray ionization mass spectrometry. *J. Mass Spectrom.* **2000**, 35, 939–952.
16. Smith, D. P. H. The electrohydrodynamic atomization of liquids. *IEEE Trans. Ind. Appl.* **1986**, 22, 527–535.
17. Ikonomou, M. G.; Blades, A. T.; Kebarle, P. Electrospray mass spectrometry of methanol and water solutions. Suppression of electric discharge with SF_6 gas. *J. Am. Soc. Mass Spectromt.* **1991**, 2, 497–505.
18. Wampler, F. W.; Blades, A. T.; Kebarle, P. Negative ion electrospray mass spectrometry of nucleotides: Ionization from water solution with SF_6 discharge suppression. *J. Am. Soc. Mass Spectrom.* **1993**, 4, 289–295.
19. (a) Fernandez de la Mora, J.; Locertales, I. G. The current emitted by highly conducting Taylor cones. *J. Fluid. Mech.* **1994**, 260, 155–184.
20. Cherney, L. T. Structure of the Taylor cone jets: Limit of low flow rates. *J. Fluid. Mech.* **1999**, 378, 167–196.
21. Chen, D. R.; Pui, D. Y. H. Experimental investigations of scaling laws for electrospray: Dielectric constant effect. *Aerosol Sci. Technology* **1997**, 27, 367–380.
22. Lord, Rayleigh; On the equilibrium of liquid conducting masses charged with electricity. *Philos. Mag. Ser. 5* **1882**, 14, 184–186.
23. Davis, E. J.; Bridges, M. A. The Rayleigh limit of charge revisited; light scattering from exploding droplets. *J. Aerosol Sci.* **1994**, 25, 1179–1191.
24. Gomez, A.; Tang, K. Charge and fission of droplets in electrostatic sprays. *Phys. Fluids* **1994**, 6, 404–414.
25. (a) Smith, J. N.; Flagan, R. C.; Beauchamp, J. L. Droplet evaporation and discharge dynamics in electrospray ionization. *J. Phys. Chem. A.* **2002**, 106, 9957–9967. (b) Grimm, R. L.; Beauchamp, J. L. Evaporation and discharge dynamics of highly charged droplets of heptane, octane and *p*-xylene generated by electrospray ionization. *Anal. Chem.* **2002**, 74, 6291–6297.
26. Taflin, D. C.; Ward, T. L.; Davis, E. J. Electrified droplet fission and the Rayleigh limit. *Langmuir* **1989**, 5, 376–384.
27. Richardson, C. B.; Pigg, A. L.; Hightower, R. L. On the stability limit of charged droplets. *Proc. Roy. Soc. A.* **1989**, 417–423.
28. (a) Schweitzer, J. W.; Hanson, D. N. Stability limit of charged drops. *J. Colloid Interface Sci.* **1971**, 35, 417–423. (b) Duft, D.; Achtzehn, T.; Muller, R.; Huber, B. A.; Leisner, T. Rayleigh jets from levitated microdroplets. *Nature* **2003**, 421, 128–128.
29. (a) Locertales, I. G.; Fernandez de la Mora, J. Experiments on the kinetics of field evaporation of small ions from droplets. *J Chem. Phys.* **1995**, 103, 5041–5060. (b) Gamero-Castano, M.; Fernandez de la Mora, J. Kinetics of small ion evaporation from the charge and mass distribution of multiply charged clusters in electrosprays. *J. Mass Spectrom.* **2000**, 35, 790–803. (c) Gamero-Castano, M.; Fernandez de la Mora, J. Direct measurement of ion evaporation kinetics from electrified liquid surfaces. *J. Chem. Phys.* **2000**, 113, 815–832.
30. (a) Grimm, R. L.; Beauchamp, J. L. Dynamics of field induced droplet ionization: Time resolved studies of distortion, jetting and progeny formation from charged and neutral methanol droplets exposed to strong electric fields. *J. Chem. Phys. B* **2005**, 109, 8244–8250. (b) Grimm, R. L.; Beauchamp, J. L. Field induced droplet ionization mass spectrometry. *J. Chem. Phys. B* **2003**, 107, 14161–14163.
31. Hager, D. B.; Dovichi, N. J.; Klassen, J. S.; Kebarle, P. Droplet electrospray mass spectrometry. *Anal. Chem.* **1994**, 66, 3944–3949.
32. Peschke, M.; Verkerk, U. H.; Kebarle, P. Features of the ESI mechanism that affect the observation of multiply charged noncovalent complexes and the determination of the association constant by the titration method. *J. Am. Soc. Mass Spectrom.* **2004**, 15, 1424–1434.
33. Schmelzeisen-Redeker, G.; Butfering, L.; Rollgen, F. W. Desolvation of ions and molecules in thermospray mass spectrometry. *Int. J. Mass Spectrom. Ion Proc.* **1989**, 90, 139–150.
34. (a) Iribarne, J. V.; Thomson, B. A. On the evaporation of small ions from charged droplets. *J. Chem. Phys.* **1976**, 64, 2287–2294. (b) Thomson, B. A.; Iribarne, J. V. Field induced ion evaporation from liquid surfaces at atmospheric pressure. *J. Phys. Chem.* **1979**, 71, 4451–4463.
35. Tang, L.; Kebarle, P. Dependence of the ion intensity in electrospray mass spectrometry on the concentration of the analytes in the electrosprayed solution. *Anal. Chem.* **1993**, 65, 3654–3668.

36. (a) KEBARLE, P.; PESCHKE, M. On the mechanism by which the charged droplets produced by electrospray Lead to gas phase ions. *Analyt. Chim. Acta* **2000**, 406, 11–35. (b) KEBARLE, P. A brief overview of the present status of the mechanisms involved in electrospray mass spectrometry. *J. Mass Spectrom.* **2000**, 35, 804–817. (c) COLE, R. B. Some tenets pertaining to electrospray ionization mass spectrometry. *J. Mass Spectrom.* **2000**, 35, 763–772.
37. (a) FENG, X.; BOGDAN, M. J.; AGNES, R. Coulomb fission event resolved progeny droplet production from isolated evaporating methanol droplets. *Anal. Chem.* **2001**, 73, 4499–4507. (b) ROMERO, S.; BOCANEGRA, R.; FERNANDEZ DE LA MORA, J. Source of heavy molecular ions based on Taylor cones of ionic liquids operating in the pure ion evaporation regime. *J. Appl. Physics* **2003**, 94, 3599–3605.
38. (a) NOHMI, T.; FENN, J. B. Electrospray mass spectrometery of polyethylene glycols with molecular weights up to five million. *J. Am. Chem. Soc.* **1992**, 114, 3241–3246. (b) NGUYEN, S.; FENN, J. B. Gas phase ions of solute species from charged droplets of solutions. *Proc. Nat. Acad. Sci.* **2007**, 104, 1111–1117. (c) ZNAMENSKIY, V.; MARGINEAN, I.; VERTES, A. Solvated ion evaporation from charged water nanodroplets. *J. Phys. Chem. A.* **2003**, 107, 7406–7412. (d) WINGER, B. A.; LIGHT-WAHL, K. J.; OGORZALEC LOO, R. R.; UDSETH, H. R.; SMITH, R. D. Observations and implications of high mass-to-charge ratio ions from electrospray ionization mass spectrometery. *J. Am. Soc. Mass. Spectrom.* **1993**, 4, 536–545.
39. (a) TOLIC, R. P.; ANDERSON, G. A.; SMITH, R. D.; BROTHERS, H. M.; SPINDLER, R.; TOMALIA, D. A. Electrospray ionization Fourier transform ion cyclotron resonance mass spectrometric characterization of high molecular mass starburst (TM) dendrimers. *Int. J. Mass Spectrom. Ion Proc.* **1997**, 165, 405.(b) CHERNUSCHEVICH, I. V.In Ens, W.; Standing, K. G.; Chernuschevich, I. V. (Eds.), *New Methods for the Study of Biomolecular Complexes*, Kluwer Academic Publishers, Dordrecht, **1998**, p. 101. (c) WANG, G.; COLE, R. B. Effect of solution ionic strength on analyte charge state distributions in positive and negative electrospray. *Anal. Chem.* **1994**, 66, 3702–3708.
40. FERNANDEZ DE LA MORA, J. Electrospray ionization of large multiply charged species proceeds via Dole's charged residue mechanism. *Anal. Chim. Acta* **2000**, 406, 93–104.
41. (a) VALENTINE, S. J.; ANDERSON, J. G.; ELLINGTON, D. E.; CLEMMER, D. E. Disulfide intact and reduced lysozyme in the gas phase: Conformations and pathways of folding and unfolding. *J. Phys. Chem B* **1997**, 101, 3891–3900. (b) HUDGINS, R. R. High resolution ion mobility measurements for gas phase conformations. *Int. J. Mass Spectrom.* **1997**, 165, 497–507. (c) SHELIMOV, K. B.; CLEMMER, D. E.; HUDGINS, R. R.; JARROLD, M. Protein structure *in vacuo*: Gas phase conformations of BPTI and cytochrome C. *J. Am. Chem. Soc.* **1997**, 119, 2240–2248. (d) HECK, A. J. R.; van der HEUVEL, R. H. H. Investigation of intact protein complexes by mass spectrometry. *Mass Spectrom. Rev.* **2004**, 23, 368–389.
42. (a) SAMALIKOVA, M.; GRANDORI, R. Protein charge state distributions in electrospray ionization mass spectrometry do not appear to be limited by the surface tension of the solvent. *J. Am. Chem. Soc.* **2003**, 125, 13362–133650. (b) SAMALIKOVA, M.; GRANDORI, R. Testing the role of surface tension in protein ionization by mass spectrometry. *J. Mass Spectrom* **2005**, 40, 503–510.
43. IAVARONE, A. T.; WILLIAMS, E. R. Mechanism of charging and supercharging molecules in electrospray. *J. Am. Chem. Soc.* **2003**, 125, 2319–2327.
44. IAVARONE, A. T.; WILLIAMS, E. R. Private communications to the authors.
45. VICAR, S.; MULKERIN, M. G.; BATHORY, G.; KHUNDKAR, L. H.; KARGER, B. L. Conformational changes in the reversed phase liquid chromatography of recombinant human growth hormone as a function of organic solvent: The molten globule state. *Anal. Chem.* **1994**, 66, 3908–3915.
46. KALTASHOV, I. A.; MOHIMEN, A. Estimates of protein areas in solution by electrospray ionization mass spectrometry. *Anal. Chem.* **2005**, 77, 5370–5379.
47. BENESCH, J. L. P.; ROBINSON, C. V. Mass spectrometry of macromolecular assemblies: Preservation and dissociation. *Current Opin. Struct. Biol.* **2006**, 16, 245–251.
48. ENKE, C. G. A predictive model for matrix and analyte effects in the electrospray ionization of singly-charged ionic analytes. *Anal. Chem.* **1997**, 69, 4885–4893.
49. NIST Database, http://webbook.nist.gov.
50. KAUZMANN, W. Some factors in the interpretation of protein denaturation. *Adv. Protein Chem.* **1959**, 14, 1–63.
51. DAVIDSON, W. R.; KEBARLE, P. Binding energies and stabilities of potassium ion complexes from studies of the gas phase ion equilibria: $K^+ + M = K^+M$. *J. Am. Chem. Soc.* **1979**, 98, 6133–6138.
52. SUNNER, J.; KEBARLE, P. Ion-solvent molecule interactions in the gas phase. The potassium ion and Me_2SO, DMA, DMF, and acetone. *J. Am. Chem. Soc.* **1984**, 106, 6135–6139.
53. (a) HOYAU, S.; NORRMAN, K.; MCMAHON, T. B.; OHANESSIAN, G. A. Quantitative basis scale of Na^+ affinities of organic and small biological molecules in the gas phase. *J. Am. Chem. Soc.* **1999**, 121, 8864–887. (b) MCMAHON, T. B.; OHANESSIAN, G. An experimental and *ab initio* study of the nature and binding of gas phase complexes of sodium ions. *Chem. Eur. J.* **2000**, 6, 2931–2935.
54. KLASSEN, J. S.; ANDERSON, S. G.; BLADES, A. T.; KEBARLE, P. Reaction enthalpies for $M + L = M + L$, where $M = Na^+, K^+$ and L = acetamide, *N*-methyl acetamide, *N,N*- dimethylacetamide, glycine, glycylglycine, from determinations of the

collision-induced thresholds. *J. Phys. Chem.* **1996**, 100, 14218–14227.

55. (a) Armentrout, P. B.; Rodgers, M. T. An absolute sodium cation affinity scale: Threshold Collision induced dissociation energies and *ab initio* theory. *J. Phys. Chem. A* **2000**, 104, 2238–2244. (b) Amicangelo, J. C.; Armentrout, P. B. Relative and absolute bond dissociation energies of sodium cation complexes determined using competative collision-induced dissociation experiments. *Int. J. Mass Spectrom.* **2001**, 212, 301–325.
56. (a) Su, T.; Bowers, M. T. Ion–polar molecule collisions: The effect of ion size on ion–polar molecule rate constants; the parametrization of the average-dipole-orientation theory. *Int. J. Mass Spectrom. Ion Phys.* **1973**, 12, 347. (b) Talrose, V. L.; Vinogradov, P. S.; Larin, I. K. On the rapidity of Ion–molecule Reactions. In *Gas Phase Ion Chemistry*, Vol. 1, Bowers, M. T. (Ed.), Academic Press, New York, **1979**, Chapter 8.
57. Reinhold, V. N.; Reinhold, B. B.; Castello, C. E. Carbohydrate molecular weight profiling, sequence, linkage and branching data. *Anal. Chem.* **1995**, 67, 1772–1784.
58. (a) Verkerk, U. H.; Kebarle, P. Ion–ion and ion-molecule reactions at the surface of proteins produced by nanospray. Information on the number of acidic residues and control of the number of ionized acidic and basic residues. *J. Am. Soc. Mass Spectrom.* **2005**, 16, 1325–1341. (b) Shou, W. Z.; Naidong, W. Simple means to alleviate sensitivity loss by trifluoroacetic acid (TFA) mobile phases in the hydrophilic interaction chromatography–electrospray tandem mass spectrometric bioanalysis of basic compounds. *J. Chromatogr. B.* **2005**, 825, 186–192.
59. Wortman, A.; Kistler-Momotova, A.; Zenobi, R.; Heine, M. C.; Wilhelm, D.; Pratsinis, S. E. Shrinking droplets in electrospray ionization and their influence on chemical equilibria. *J. Am. Soc. Mass Spectrom.* **2007**, 18, 385–393.
60. (a) Ganem, B.; Li, Y.; Henion, J. D. Detection of non-convalent acceptor–ligand complexes by mass spectrometry. *J. Am. Chem. Soc.* **1991**, 113, 6294–6296. (b) Ayed, A.; Krutchinsky, A. N.; Ens, W.; Standing, K. G.; Duckworth, H. W. Qualitative evaluation of protein–protein and ligand–protein equilibria of a large allosteric enzyme by electrospray ionization time of flight mass spectrometry. *Rapid Commun. Mass Spectrom.* **1998**, 12, 339–344.
61. Loo, J. Studying noncovalent protein complexes by electrospray ionization mass spectrometry. *Mass Spectrom. Rev.* **1997**, 16, 1–23.
62. Daniel, J. M.; Friess, S. D.; Rajagopalan, S.; Wend, S.; Zenobi, R. Quantitative determination of noncovalent binding interactions using soft ionization mass spectrometry. *Int. J. Mass Spectrom.* **2002**, 216, 1–27.
63. Daniel, J. M.; McCombie, G.; Wend, S.; Zenobi, R. Mass spectrometric determination of association constants of adenylate kinase with two noncovalent inhibitors. *J. Am. Soc. Mass Spectrom.* **2003**, 14, 442–448.
64. Peschke, M.; Verkerk, U. H.; Kebarle, P. Features of the ESI mechanism that affect the observation of multiply charged noncovalent complexes and the determination of the association constant by the titration method. *J. Am. Soc. Mass Spectrom.* **2004**, 15, 1424–1434.
65. Wang, W.; Kitova, E. N.; Klassen, J. S. Bioactive recognition sites may not be energetically preferred in protein–carbohydrate complexes in the gas phase. *J. Am. Chem. Soc.* **2003**, 125, 13630–13861.
66. Sun, J.; Kitova, E. N.; Wang, W.; Klassen, J. S. Method for distinguishing specific from nonspecific protein–ligand complexes in nanoelectrospray ionization mass spectrometry. *Anal. Chem.* **2006**, 78, 3010–3018.
67. Wilm, M.; Mann, M.; Electrospray Taylor–Cone theory, Dole's beam of macromolecules at last?. *Int. J. Mass Spectrom. Ion Proc.* **1994**, 136, 167–180.
68. Wilm, M.; Mann, M. Analytical properties of the nanoelectrospray ion source. *Anal. Chem.* **1996**, 68, 1–8.
69. Jurashek, R.; Dulks, T.; Karas, M. Nanoelectrospray—More than just a minimized-flow electrospray ion source. *J. Am. Soc. Mass Spectrom.* **1999**, 10, 300–308.
70. Schmidt, A.; Karas, M.; Dulks, T. Effect of different solution flow rates on analyte signals in nano-ESI-MS, or when does ESI Turn into nano-ESI. *J. Am. Soc. Mass Spectrom.* **2003**, 14, 492–500.
71. Chernushevich, I. V.; Bahr, U.; Karas, M. Nanospray taxation and how to avoid it. *Rapid Commun. Mass Spectrom.* **2004**, 18, 2479–2485.

Chapter 2

Selectivity in Electrospray Ionization Mass Spectrometry

Nadja B. Cech[*] *and Christie G. Enke*[†]

[*]*Department of Chemistry and Biochemistry, University of North Carolina at Greensboro, Greensboro, North Carolina*
[†]*Department of Chemistry and Chemical Biology, University of New Mexico, Albuquerque, New Mexico*

Electrospray and MALDI Mass Spectrometry: Fundamentals, Instrumentation, Practicalities, and Biological Applications, Second Edition, Edited by Richard B. Cole

2.1 INTRODUCTION

The versatility of electrospray ionization mass spectrometry (ESI-MS) as an analytical technique is truly phenomenal. It has been successfully employed to study proteins,[1] carbohydrates and glycoproteins[2,3] (Figure 2.1), lipids,[4] industrial polymers,[5] oligonucleotides,[6] metal clusters,[7] environmental pollutants,[8] drugs and drug metabolites,[9] and countless other important analytes. Given this widespread success, it is easy to be lulled into the perception of electrospray as the solution for any analytical problem. However, electrospray is a selective ionization technique, ideal for some analytes but poorly suited for others. While this selectivity can be advantageous in that it simplifies the analysis of complex solutions, it can be problematic if the analytes of interest are not responsive to the technique. In addition, competition among analytes often occurs in analyses with electrospray, and it is common for the signal of the analyte of interest to be suppressed, sometimes completely, by other components of the solution.[10] In this chapter, we explore the factors that contribute to selectivity and signal suppression in ESI-MS from both fundamental and practical perspectives. Ion formation, the nature of the droplet charge, its location, and its eventual vaporization are discussed, and we present a detailed model that helps to explain selectivity in electrospray response, the "equilibrium partitioning model." In addition, practical influences on selectivity, such as solvent choice and ionization mode, are addressed. An understanding of the topics presented here should help users to predict which analytes will be analyzable with electrospray ionization mass spectrometry and provide insight into the selection of appropriate conditions for the analysis of these species.

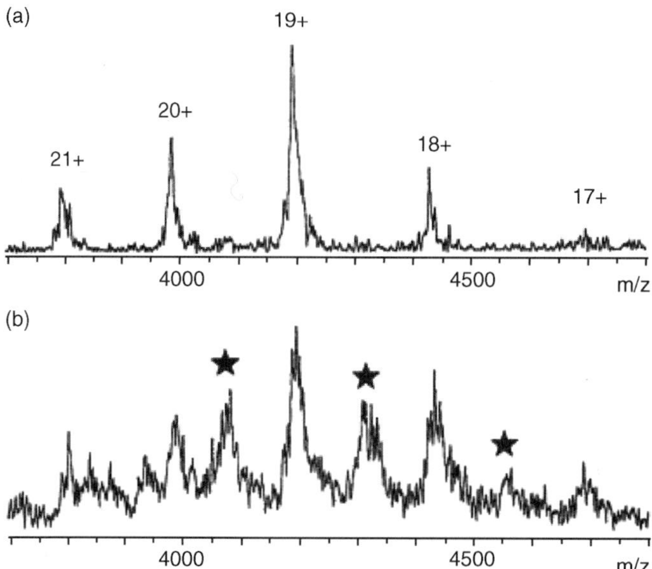

Figure 2.1. Electrospray ionization mass spectrum of the glycoprotein transferrin from (**a**) a normal patient and (**b**) a patient with a metabolic disorder known as a CGD (congenital disorder of glycosylation). The mass spectrum of the patient with abnormal transferrin displays peaks (marked by stars) that are shifted in mass because the protein is missing an oligosaccharide chain. Mass spectrometric analysis of transferrin currently plays a central role in the clinical screening of patients for CGD. (Reprinted from Refs. 2 and 3, with permission.)

2.2 MECHANISMS OF ION FORMATION

The first and foremost source of selectivity in electrospray ionization mass spectrometry is whether or not the molecule of interest can become ionized. Species that cannot become vapor-phase ions with electrospray, of which there are many, will not be detectable by the mass spectrometer. Electrospray is a soft ionization technique, meaning that it accomplishes ionization with minimal fragmentation. This is advantageous because it facilitates identification of the molecular ion of the species of interest and makes the analysis of intact biological and industrial polymers possible. However, the softness of the electrospray process also translates to selectivity in terms of which species are charged and how efficiently they are charged. The mechanisms by which charging of the analyte can occur in electrospray are discussed in the following sections.

2.2.1 Ionization by Protonation or Deprotonation and the Role of Analyte pK_a (or pK_b) and Solution pH

Electrospray ionization mass spectrometry is most often used for the analysis of species that already exist as ions in the bulk solution. In this case, the ions are not formed by the electrospray process, but rather electrospray serves to separate them from their counterions and to transfer them from solution into the gas phase where they can be sampled by the mass spectrometer. Analytes suitable for analysis in this way are species that are inherently charged, such as quaternary ammonium salts, and those that can be charged either in the bulk solution or in the electrospray droplets themselves by the gain or loss of one or more protons. This last category is perhaps the most common type of analyte that is studied by electrospray. A wide array of analytes can be observed in the positive ion mode as protonated species, including peptides and proteins, some surfactants and polymers, and small molecules with ammonium, phosphonium, or oxonium moieties. Those that can be observed as deprotonated species in the negative ion mode include oligonucleotides, some lipids and oligosaccharides, fatty acids, and small acidic molecules such as carboxylic acids and phenolic compounds.

When predicting whether an acidic or basic analyte will be analyzable with electrospray ionization mass spectrometry, it is useful to consider the analyte pK_a (for the case of acidic species) or pK_b (for basic species). In the simplest case, these values can give insight into whether or not a given analyte will be chargeable by protonation or deprotonation, informing the analyst about whether analysis in the positive or negative mode is more appropriate. In addition, studies by Kebarle and co-workers[11] and Ehrmann et al.[12] have demonstrated that for protonatable analytes, positive electrospray response is correlated with the bulk concentration of the protonated (charged) form of the species. In studies conducted by these groups, electrospray response was shown to be very low for weakly basic analytes (those with high pK_b values), for which the protonated form would not be expected to predominate in solution. Strongly basic analytes were observed to be more responsive to analysis with ESI-MS. Typically, pK_a or pK_b values are reported for aqueous solutions rather than the organic or aqueous/organic mixtures that are employed as solvents for analysis with electrospray ionization mass spectrometry; thus, the use of aqueous values to predict responsiveness is only approximate. Furthermore, other factors such as affinity for electrospray droplet surfaces and gas-phase proton transfer can also influence the relative responsiveness of charged analytes to analysis with ESI-MS, as discussed later on. However, consideration of analyte pK_a or pK_b is an excellent starting point for deciding whether a particular analyte will be well-suited for analysis with ESI-MS.

In the analysis of protonatable or deprotonatable analytes, it might be assumed that solution pH would have an important role in determining selectivity. While this is sometimes the case, the relationship between electrospray response and pH is far from simple.[13–16] For example, Mansoori et al.[13] reported only slight variations in response for both protonated (Figure 2.2) and deprotonated ions (analyzed in the positive-ion and negative ion modes, respectively) at pH values ranging from 3 to 11. Indeed, it is often the case that protonated species can be detected by electrospray analysis of solutions where the bulk pH is well above the analyte pK_a, and conversely that deprotonated species can be detected from solutions with very low pH.[14,16–18] A number of rationalizations have been proposed to explain this observation, including that the formation of ions is favored by the energetics of the electrospray process,[19] that the pH of electrospray droplets changes as solvent evaporates,[20,21] and that gas-phase proton transfer reactions contribute to charging of species that would not be charged in the bulk solution.[16]

Regardless of the mechanism by which it occurs, the fact that in many cases acidic or basic analytes can be ionized at a wide range of pH values is convenient for the sake of interfacing between ESI-MS and chromatographic separations. It is often possible to operate at the pH that gives optimal separation, and it is also to switch between positive- and

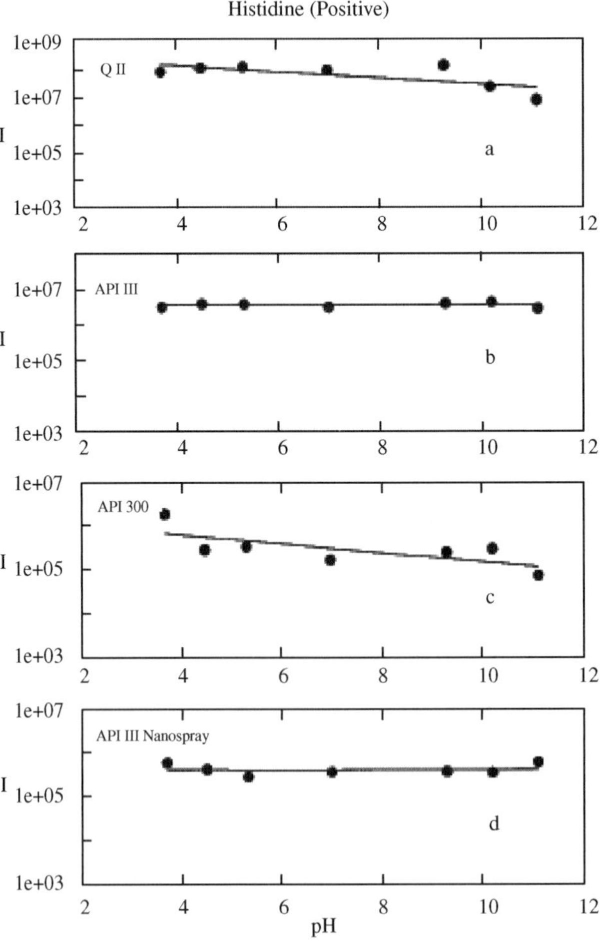

Figure 2.2. Demonstration of how electrospray response of histidine changes very little as solution pH is adjusted from 3 to 11. Each panel displays data collected on a different triple quadrupole mass spectrometer: (**a**) Quattro II (Micromass) with an electrospray source, (**b**) API 300 (Sciex) with a pneumatically assisted electrospray source, (**c**) API III (Sciex) with a pneumatically assisted electrospray source, (**d**) API III (Sciex) with a nanospray source. (Reprinted from Ref. 13, with permission.)

negative-ion mode analysis without changing the solvent system.[22] It is worth noting, however, that the use of nonvolatile HPLC or CE buffers to adjust pH is problematic when the separation will be interfaced to ESI-MS, and nonvolatile buffers must be used instead.

In cases where the pH range is not limited by the needs of the separation, it is common practice to utilize acidic solvents for the analysis of protonated species with positive ion ESI-MS. Indeed, a number of investigators have reported that the response of weakly basic species to analysis with ESI is optimal at low pH.[12,23,24] The most commonly used electrospray solvents consist of small amounts (0.1–0.5%) of volatile acids such as acetic or formic acid dissolved in methanol/water or acetonitrile/water. Nonvolatile acids such as hydrochloric acid are to be avoided because they cause signal suppression.[11] Signal suppression can also result when a volatile acid is added to the solution in too great an amount,[14,25] so the concentration added should be only enough to optimize protonation of the analyte and, in the case of HPLC/ESI-MS, to facilitate successful HPLC separation.

By the same arguments used to rationalize operation at low pH for positive-ion electrospray, it might be assumed that addition of a volatile weak base to the solution would improve analyte response for negative-ion mode analysis. While some authors recommend this approach,[26,27] others report optimal operation in the negative-ion mode at acidic pH.[15,25] Buffers containing volatile bases such as ammonium hydroxide or triethylamine are widely used in ESI-MS analyses conducted in the negative-ion mode, but often their purpose is to adjust pH to optimize chromatographic separations rather than to optimize electrospray response. The quirks and intricacies of negative ion electrospray (as far as they are understood) are presented in Section X.

2.2.2 Ionization by Adduct Formation

Analytes that are not inherently charged and cannot be protonated or deprotonated can still sometimes be ionized with electrospray. One mechanism by which this can occur is by the formation of adducts between polar (but neutral) organic species and either (a) cations such as sodium, lithium, and potassium or (b) ammonium or anions such as chloride or acetate. Salts present as trace impurities in the solvent may facilitate this adduct formation. For example, adduct ions detectable by ESI-MS can be formed between the analyte and sodium that leaches from glassware. Salts can also be intentionally added to the solution to facilitate formation of detectable salt adducts. An example is the use of lithium chloride for the analysis of oligosaccharides (Figure 2.3)[28] or glycosylated lipids.[29] With this method, the analytes can be detected as lithium adducts in the positive-ion mode or chloride adducts in the negative-ion mode. It is important to note that the presence of excess salt in the analyte solution can cause signal suppression due to the formation of salt clusters that decrease the analyte signal or compete for excess charge with the analyte ions (see Section V). To prevent this problem, salt concentrations below 1 mM (usually in the 100 μM range) are generally employed.

2.2.3 Ionization by Electrochemical Oxidation or Reduction

Redox reactions are a third mechanism by which analytes can be ionized in electrospray. In order for current to exist in the electrospray circuit, oxidation (for positive-ion mode analysis) or reduction (for analysis in the negative-ion mode) must occur at the contact between the high voltage supply and the electrospray solution. Depending on the particular

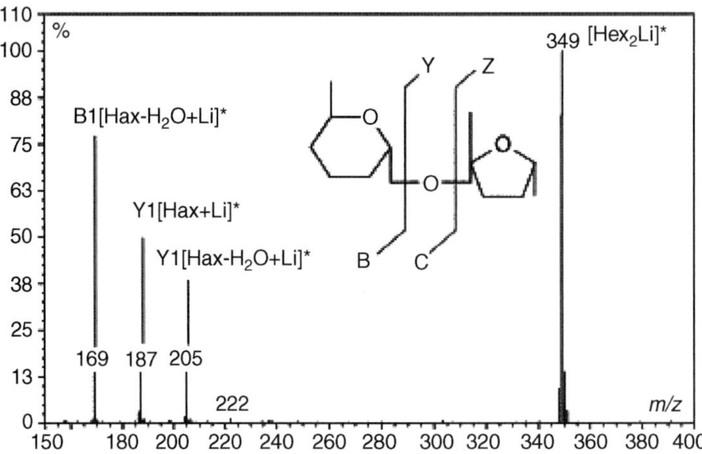

Figure 2.3. Electrospray mass spectrum of sucrose obtained with a single quadrupole mass spectrometer (Thermo Electron) operated in the positive-ion mode. Fragmentation was achieved by in-source collisionally induced dissociation. The post-column addition of submillimolar LiCl to the analyte solution facilitated the formation of lithium adducts. (Reprinted from Ref. 28, with permission.)

conditions of the analysis, the species that undergo oxidation or reduction can be the analyte, the solvent, additives present in the solution, or the metal contact to the solution itself.[30] Charging of the analyte by oxidation or reduction at the contact to the electrospray solution can facilitate detection with electrospray ionization. For example, polyaromatic hydrocarbons can be oxidized in the electrospray process to form detectable ions if the appropriate solvent system is used.[31] However, the oxidation or reduction reactions that are inherent to the electrospray process can also have a deleterious effect if they form ionizable products from the solvent or additives. Such species complicate mass spectra and can compete with the analytes for excess charge as described later on.

2.3 IONIC COMPOSITION OF ESI DROPLETS

In order to work at all, the solution to be electrosprayed must be conducting. The mobile charge carriers that make polar solutions conducting are dissolved ions. The concentration of ions present in the electrospray solution as well as the nature of those ions can have a significant effect on the performance of the electrospray ion source.

Typically, the analyte is only one of many other ionic species present in the solution being analyzed by electrospray ionization mass spectrometry. Hydronium and hydroxide ions come from the autoprotolysis of water, and other ions are often present as low-level contaminants even in “pure” solvents. For example, solvents stored in glass containers may have micromolar concentrations of sodium ion,[32] and reagent grade solvents contain trace levels of ammonium, calcium, chloride, fluoride, and other ions. In addition, ionic species may be intentionally added to the electrospray solution. Electrolytes in the form of volatile weak acids, weak bases, or salts of the same are often added to improve ESI operation and adjust solution pH, and salts may be added in support of adduct ion formation. Finally, ions are generally included among the products of the unavoidable electrolytic reaction that occurs at the contact between the high-voltage solution and the ESI power supply.

2.3.1 Ions Formed at the Electrolytic Contact with the Solution

The droplets leaving the electrospray have an excess of cations or anions, depending on the polarity of the applied voltage. It is the presence of these excess ions that give the droplets their charge. The "charging" of the solution is performed through the electrical contact between the ESI power supply and the sprayed solution. Since this is a contact between an electron conductor (the wire, tubing coupling, or metallic spray needle) and an ionic conductor (the solution), the conduction mechanism at the contact is necessarily an electrolytic reaction. The concentration of the excess charge $[Q]$ (in mole-charges per liter) can be calculated from the electrical current in the spray circuit i (in amperes) and the volume flow rate Γ (in liters per second) of the solution to the electrospray tip [Eq. (2.1)][33]. Here F is Faraday's constant (96,485 coulombs per mole of charge).

$$[Q] = \frac{i}{\Gamma F} \tag{2.1}$$

For example, if the circuit current is 1×10^{-7} A and the flow rate is 6 μL/min, the concentration of excess charge is approximately 1×10^{-5} mole-charges per liter.

From Eq. (2.1), it is apparent that the concentration of excess charge depends directly on the ratio of the circuit charge flow rate (the current, i) and the solution flow rate Γ. Significant reductions in solution flow rate are best accomplished in conjunction with the use of smaller-diameter spray capillaries. The circuit current is reduced somewhat,[34] but not in proportion to the flow rate. Therefore, microflow and nanoflow spray capillaries produce initial droplets with increasingly higher concentrations of excess charge. Conversely, the circuit current can be changed independently of the flow rate by altering the concentration of electrolyte in the sprayed solution. This is rarely done because the range of electrolyte concentrations over which the electrospray is analytically optimal is rather limited.

In positive electrospray ionization, the ions most commonly formed by the electrolysis current are ferrous ions from the oxidation of the steel tip. For more electrochemically inert contact materials, the oxidation products can be hydrogen ions and hydrogen peroxide or oxygen. In negative ESI, the reaction at the contact must be reduction and since the contact material cannot be further reduced, something in the solution must be reduced. This reduction can sometimes lead to a significant change in the composition of the sprayed solution.

2.3.2 Ions Formed by the Addition of Additives and Autoprotolysis of the Solvent

In virtually all practical applications of ESI, the concentration of excess charge is small compared to the total ionic concentration in the solution. The common addition of 0.1% formic acid would, in water, produce a pH of about 2.8 and a concentration of hydronium and formate ions of greater than 10^{-3} moles per liter—over 100 times greater than the concentration of excess charge. Many of the additives used in ESI-MS are weak electrolytes; that is, their dissolution into their component ions or their reaction with the solvent to form ions is not complete. The actual ion concentration is often in the millimolar range even when the concentration of additive is much greater. However, if the additive has reacted with an analyte to increase the latter's charge, the concentration of charge will increase by that mechanism as well. For instance, if a formic acid molecule reacts to protonate a peptide to a higher charge state, the ionic composition of the solution is increased by the more highly charged peptide and by the formate anion produced.

A much smaller contribution to the ionic composition of the sprayed solution than any of the above effects is from the autoprotolysis of the solvent. In practical analytical solutions, this effect is rarely greater than it is in water where the maximum concentration of ions formed is about 1×10^{-7} M and that only in a neutral solution.

2.4 EXCESS CHARGE AND ITS LOCATION

We now have a picture of the charged droplet at the time of its formation. It contains dissolved cations and anions with a slight excess of one or the other depending on the polarity of the applied voltage. The excess of cations or anions gives the droplet its charge. Since ions are the mobile charge carriers in this conducting solution, they are free to move. Principles of electricity dictate that the excess charge carriers will separate themselves from each other as far as possible. This means that they will be at or very near the surface of the droplet, spread out across its surface area. In fact, the distribution of the charge relative to the surface is the same as it is for the solution–electrode interface in electrochemistry. If the conductivity of the solution is high, the excess charge will be in the "compact" double layer at the interface. As the ionic concentration in the solution decreases, more of the charge goes into the "diffuse" double layer and the excess charge distribution extends further from the interface.[32]

The reason for the excess charge to be at or near the droplet surface is simply the repulsive force of like charges. As solvent evaporates from the droplet, the concentration of excess charge increases, the charge density near the droplet surface increases, and the repulsive force increases. There are two proposed mechanisms for the droplet to relieve itself of some of this excess charge. By the Iribarne–Thomson ion evaporation mechanism, ions more loosely bound to the droplet surface may be released from the droplet by the increasing repulsive force among the charges.[35] In the Coulomb fission mechanism,[36] the forces of charge repulsion continue to increase until they are, at the Rayleigh limit, equal to the forces of surface tension of the liquid. At this point, the droplet is no longer stable and it spawns a string of much smaller highly charged droplets. Some of its excess charge density has been relieved by the increase in overall surface area. In fact, the fraction of the original excess charge now in the spawned droplets may be significantly larger than the fraction of the volume accompanying the spawning. From this, it follows that those ions carrying the surface charge are now more highly concentrated in the spawned droplets. As the evaporation and spawning continues for both the spawned droplets and the original droplet, this transfer of the surface excess ions to the increasingly smaller droplets continues. Conversely, ions preferentially in the droplet interior become increasingly concentrated in the parent droplets. Because it is the small, highly charged offspring droplets that eventually lead to the most predominant gas phase ions, this process can lead to an enhancement in signal for the species that existed on the surface of the initial droplets.[37]

The two mechanisms for relief of excess charge during evaporation are not mutually exclusive. As a droplet approaches the Rayleigh limit through evaporation, some surface ions may be lost through ion evaporation while others are later transferred to spawned droplets. In either case, we see that there is a preference for surface excess charge to be transferred directly to the vapor phase or to the smaller droplets, which will themselves evaporate more quickly and reach ion evaporation and spawning more rapidly and perhaps more efficiently than will the parent droplets. Ions constituting the excess charge remaining in the parent droplets, however, may preferentially remain there until only charged salt remains. Ions formed in this way are said to be formed by the "charged residue" mechanism.[36] From this, we can see that surface and interior ions could follow different paths to vapor ion formation and that these paths could differ greatly in the efficiency of the ion vaporization process. This concept is explored from the standpoint of ion partitioning between the droplet surface and interior in the next section.

2.5 A MODEL FOR COMPETITION AND SATURATION IN SOLUTION

The postulates of the equilibrium partitioning model[33] are that each droplet has a limited amount of excess charge, that the species carrying the excess charge will be preferentially at or near the droplet surface, that there is competition among all the cations (in the case of positive ESI) to carry the excess charge, and that the surface ions carrying the excess charge are more likely to appear in the ESI mass spectrum. The basis for and consequences of each of these postulates is explored here.

The rate of excess charge deposition in the sprayed fluid is given by the current supplied by the power supply in the circuit. The rate of flow of the sprayed solution is either forced by an external pump or created by the electrospray process. In either case, the molar concentration of excess charge, $[Q]$, is the ratio of the charge and solution flows as given by Eq. (2.1). As mentioned before, a typical value of $[Q]$ for normal electrospray is $\sim 10^{-5}$ M. This is a much smaller concentration than the sum of the total ionic species in the solution, so it follows that some of the ions will be the carriers of the excess charge and others will not. Those that are not will be paired with an equivalent number of counterions (anions, in the case of positive ESI). An immediately obvious consequence of this postulate is that $[Q]$ is the maximum mole-charge concentration of ions in the solution that can be converted to vapor phase ions.

Because of charge repulsion, the ions carrying the excess charge are at or near the surface of the droplet. These ions are generally a small fraction of the total ionic composition of the droplet. These unpaired surface ions are free to exchange places with the ions in interior of the droplet. Since the environment of the surface ion and the interior ions are different, they can be considered to be in different phases of the droplet. The exchange of ion location between the interior and surface can be treated by standard partition equilibrium expressions for partition between two phases. For example, charged solution additives (E^+) can partition between the unpaired surface state and the paired (with X^-) state as shown in Eq. (2.2). Here E^+ represents cationic electrolytes (such as Na^+ or $NH_4{}^+$) or protonated solvent (such as $CH_3OH_2{}^+$ or H_3O^+). The latter species would likely predominate in solutions where the solvent is acidified, with added acetic or formic acid, and where the metallic salt concentration is at the trace level. Of course, true sample solutions will likely contain multiple different ionic additives and contaminants. However, one of these will often be the major carrier of excess charge, and the discussions here are simplified by considering that the analyte(s) compete with only a single charged additive E^+.

$$(E^+X^-)_i \leftrightarrow (E^+)_s + (X^-)_i, \qquad K_E = \frac{[E^+]_s[X^-]_i}{[E^+X^-]_i} \tag{2.2}$$

Analytes A^+ and B^+ can also partition between the interior and the surface, as shown in Figure 2.4. Equations (2.3) and (2.4) describe this partitioning. In this case, A^+ and B^+

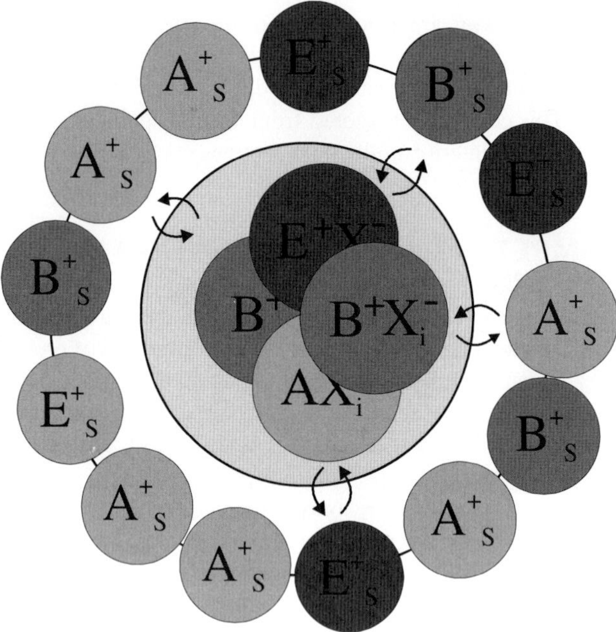

Figure 2.4. Schematic of a droplet with two separate phases, surface excess charge and internal electrically neutral. Analytes (A^+ and B^+) compete with electrolyte (E^+) for the surface excess charge phase of the droplet. A postulate of the equilibrium partitioning model is that the analytes that are part of the surface excess charge phase are most likely to become gas-phase (analyzable) ions. Those ions that reside in the droplet interior will be paired with counterions and consequently will not be detectable by the mass spectrometer. (Reprinted from Ref. 77, with permission.)

are analytes that have single positive charges, such as protonated amines or tetraalkylammonium ions.

$$(A^+X^-)_i \leftrightarrow (A^+)_s + (X^-)_i, \qquad K_A = \frac{[A^+]_s[X^-]_i}{[A^+X^-]_i} \tag{2.3}$$

$$(B^+X^-)_i \leftrightarrow (B^+)_s + (X^-)_i, \qquad K_B = \frac{[B^+]_s[X^-]_i}{[B^+X^-]_i} \tag{2.4}$$

For those species that would prefer the surface position to that in the interior, the equilibrium constant (K) will be high. Factors such as polarity, charge density, and basicity will determine the magnitude of K for a given ion. Analytes with very high K's will exist mostly on the surface of ESI droplets, so that at high analyte concentration, they will be capable of carrying a large fraction of the excess charge. Analytes with small K's will exist in the droplet interior and be matched by counterions, and thus they will not be in a position to carry the excess charge. Partitioning between the two phases in the droplet will occur very early in the droplet formation process, and the factors favoring an analyte's existence in the surface phase will increase as desolvation proceeds.

Now, using the equilibria shown in Eqs. (2.2)–(2.4), mass balance equations can be generated to solve for the concentrations of species between the two phases. For example, the mass balance equation for A^+ is shown in Eq. (2.5), where C_A is the analytical concentration of A in the droplet system. Similar equations will exist for C_B, C_E, and any other positively charged species in the solution.

$$C_A = [A^+]_s + [A^+X^-]_i \tag{2.5}$$

The last mass balance equation, Eq. (2.6), is that which sets the total mole-charge concentration of charged species in the surface phase of the droplet equal to $[Q]$, the concentration of excess charge.

$$[Q] = [E^+]_s + [A^+]_s + [B^+]_s \tag{2.6}$$

It is probable that the species that carry the excess charge on the droplet surface are the species that will end up as ionic species in the vapor phase. Thus, it is of interest to solve for the surface concentrations of these analytes.

In the case of a single analyte species, A^+, and a single electrolyte species, E^+,

$$[A^+]_s = C_A\left(\frac{K_A/K_E}{K_A/K_E - 1 + C_E/[Q]}\right), \qquad C_A \ll [Q] \tag{2.7}$$

From Eq. (2.7), it is apparent that the fraction of the analyte in the surface phase depends on the ratio of its partition coefficient to that of the electrolyte. Figure 2.5 is a plot of this equation for various values of K_A/K_E. In this figure, $[Q]$ is 10^{-5} M and C_E is 10^{-3} M. From this figure, we can see the anticipated plateau at a surface concentration of 10^{-5} M. It is also apparent that the surface concentration of the analyte decreases with decreasing K_A/K_E. The surface concentration also decreases with increasing concentration of electrolyte. This demonstrates that the analyte is always in competition with the electrolyte for the surface charge. Even though the K for the analyte may be much greater than that for the electrolyte, the much greater concentration of the electrolyte makes it a factor.

Solving for the surface concentrations of two analytes reveals the nature of the competition between analytes. The equation below is a slight modification of the one

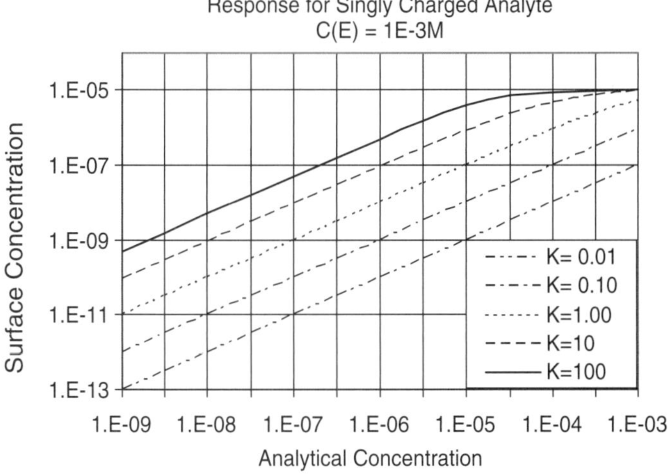

Figure 2.5. The surface concentrations of analytes are plotted versus their sample analytical concentrations for several values of K_A/K_E. The electrolyte concentration is 1 mM and $[Q]$ is 1×10^{-5} M. For the highest value of K_A/K_E, almost all the analyte is on the surface; for the lowest, only 0.01% of it is. This ratio is maintained almost to the point of saturation. For lower values of K_A/K_E the concentration of saturation is much greater than $[Q]$. (Reprinted from Ref. 33, with permission.)

previously published, which included several typographical errors.[33]

$$[\mathrm{A}^+]_S^3\left[K_B - K_A + K_E\left(1 - \frac{K_B}{K_A}\right)\right] +$$

$$[\mathrm{A}^+]_S^2\left[C_A(2K_A - K_B - K_E) + C_B\left(K_B - K_E\left(\frac{K_B}{K_A}\right)\right) +\right.$$

$$\left. C_E K_E\left(1 - \frac{K_B}{K_A}\right) + [Q]\left(K_A - K_B - K_E\left(1 - \frac{K_B}{K_A}\right)\right)\right] -$$

$$[\mathrm{A}^+]_S C_A[[Q](2K_A - K_B - K_E) + C_B K_B + C_A K_A + C_E K_E] + C_A^2[Q]K_A = 0 \quad (2.8)$$

A plot of the surface concentrations of analytes A and B as a function of their equal analytical concentrations is given in Figure 2.6. In this plot, the K_A/K_E for analyte A is 1 while K_B/K_E is 100. Two important points may be gleaned from this plot. One is that analyte B suppresses the response of analyte A at high concentrations (approaching [Q]). The second is that analyte B does not suppress the response of analyte A at low concentrations. This is because, in the lower concentration regime, both analytes are competing effectively with the electrolyte and not with each other. In the higher concentration regime, the analyte with the higher K/K_E (analyte B) is taking the bulk of the excess charge, thus suppressing the surface concentration of analyte A. In this plot, the concentration of electrolyte is very low, so both analytes respond similarly at low concentrations. If the electrolyte concentration were in the millimolar range, one would observe a lower response factor for the analyte with the lower K/K_E throughout its concentration range.

The final postulate of the equilibrium partition theory is that the surface concentration of analyte, $[\mathrm{A}^+]_s$, would be related to its appearance in the mass spectrum. This was demonstrated in the paper that originally introduced the equilibrium partitioning model,[33] in which experimental data of Kebarle and Tang using tetraalkyl ammonium salts was fit to Eqs. (2.7) and (2.8). Kebarle later reconciled the equations from the equilibrium partition

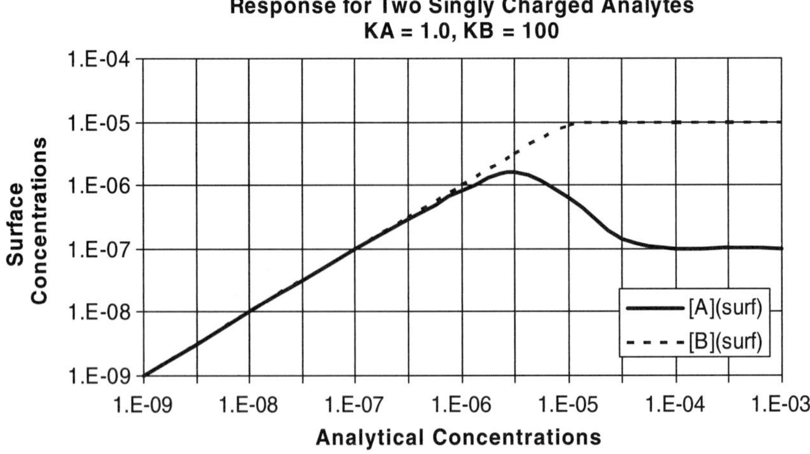

Figure 2.6. The surface concentration of two analytes, A and B, are plotted against their mutual concentrations in the sample solution. $C_E = [Q] = 10^{-5}$ M, $K_A/K_E = 1$, and $K_B/K_E = 100$. As the analyte concentrations approach $[Q]$, they begin to compete with each other for the charge, and the one with the greater surface activity suppresses the other's response. (Reprinted from Ref. 33, with permission.)

model with those he earlier derived using the ion evaporation model of Iribarne and Thomson.[38] Some analytes such as small cations and undenatured proteins may not be surface-active and yet can still be observed in the ESI mass spectra, though often at greatly reduced sensitivity. These ions are most likely vaporized by the charged residue mechanism, in which ions will compete for the charge left in the droplet after extensive fission and evaporation. Ions vaporized in this way will not follow the equilibrium partition model.[39]

It has been over a decade since the introduction of the equilibrium partition model. In that interval, many researchers have been involved in the verification and application of that theory. Haddrell and Agnes[40] have summarized a number of the verification studies in their fluorescence microscopy study of the location of Rhodamine 6G ions in electrosprayed droplets. SjÖberg et al.[41] have developed a variation of Eq. (2.7) which allows the determination of K_A/K_E for a given analyte from only two data points, enabling a study of the effects of spray position, spray potential, gas flow rates, ionic strength, and solvent composition on the value of K_A/K_E.[42] Other researchers have realized that the analytes are not the only species that could equilibrate in the droplet or on its surface and that these equilibria could affect, or be studied by examining, relative electrospray response. Host–guest complexation[43] and metal ion complexation[44,45] have both been explored. Several investigators have also attempted to correlate acid–base equilibria with electrospray response of protonatable species.[12,46] Although the relationship is not simple, proton transfer reactions in solution do appear to influence selectivity to analysis by ESI-MS.

2.6 SURFACE ACTIVITY AS A SOURCE OF SELECTIVITY

Electrospray ionization mass spectrometry is typically performed in polar solvents such as water, acetonitrile, methanol, or a combination of these. Thus, analytes with significant nonpolar regions should favor the air–solvent interface at the surface of electrospray droplets where these nonpolar regions can be desolvated. Such analytes are termed "surface-active." A relationship between response in atmospheric pressure ionization mass spectrometry and analyte surface activity was postulated as early as 1983 by Iribarne et al.,[47] the original authors of the ion evaporation theory. Such a relationship was also observed by Kebarle and co-workers,[48,49] who included a factor related to analyte surface activity in their models.

The equilibrium partition model also predicts quantitatively how the more surface-active analytes are likely to have higher response factors and out-compete less surface-active ions for the excess charge. This prediction has been verified in several ways. For example, Figure 2.7 shows a mass spectrum of an equimolar mixture of tripeptides that differ only by substitution of different amino acids on their C-terminal ends. There is a twofold difference in response between the least responsive (GGG) and most responsive (GGF) peptides. The peptides whose response is displayed in Figure 2.7 all have the same N-terminus. Thus, the part of the molecule that carries the charge is the same for all of the analytes, and differences in response are likely attributable to differences in the less polar region of the molecule. Indeed, the electrospray response of the peptides in Figure 2.7 is positively correlated with the nonpolar surface area of the C-terminal amino acids.[50] A large number of other studies have also demonstrated a positive correlation between analyte nonpolar character and electrospray response factor.[47,48,50–52]

Given that response in electrospray has been shown to correlate with the presence of nonpolar regions in the analyte structure, it might be expected that increased retention time

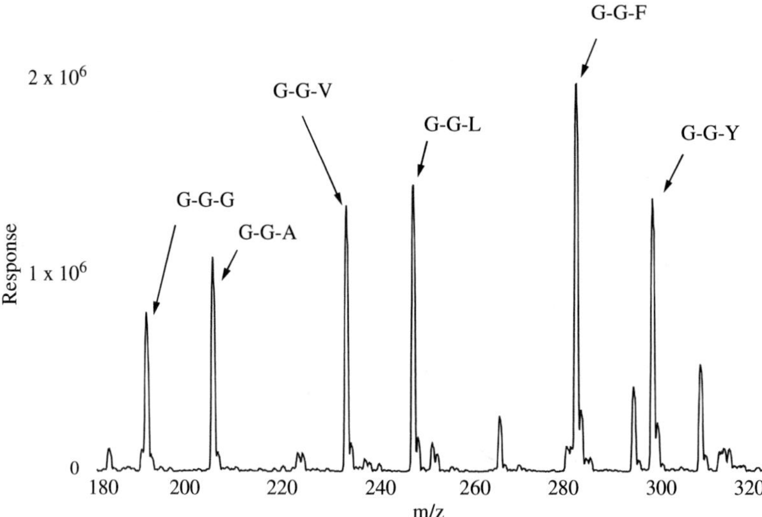

Figure 2.7. A mass spectrum comparing the responses of the six tripeptides whose structures differ only by what R-group is attached to the C-terminal residue. Although the peptides are prepared at equimolar concentration, their responses differ significantly. The additional small peaks in the mass spectrum represent peptide methanol clusters, whose response is insignificant with respect to that of the bare peptides. (Reprinted from Ref. 50, with permission.)

in reversed-phase HPLC would be correlated with increased electrospray response. This relationship has been verified for several simple series of analytes, tripeptides,[50] and tetralkylammonium ions.[53] Other studies have shown that log *P*, a measure of analyte polarity, is correlated with responsiveness of a diverse array of analytes in negative-ion electrospray ionization mass spectrometery[51,52] (Figure 2.8). These examples provide

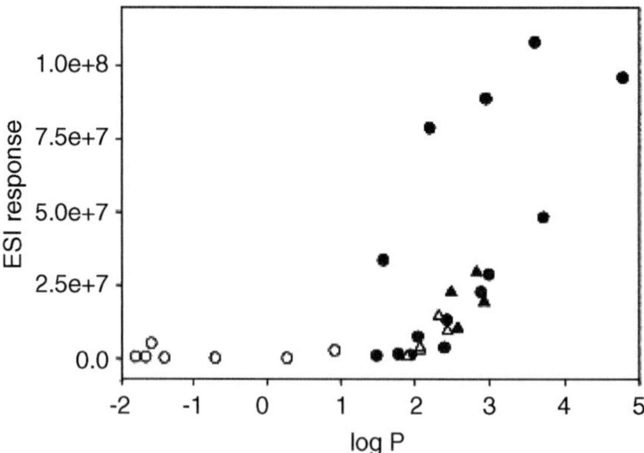

Figure 2.8. Electrospray response is positively correlated with log *P* for analytes having log *P* values greater than approximately 1.5. The analytes investigated were all dissolved in methanol and analyzed by flow injection. The peak area for the selected ion trace of the relevant deprotonated molecular ion (at 0.1 μM) was used as a measure of analyte response. The analytes studied were phenols (filled circles), benzoic acids (open triangles), phenoxyalkanoic acids (filled triangles), and triazines (open circles). The four analytes that deviated by a higher response are *p*-nitrophenol, DNOC, bromoxynil, and ioxynil (left to right). (Reprinted from Ref. 51, with permission.)

further evidence that charged analytes with significant nonpolar regions tend to be the most responsive species to ESI-MS analysis.

2.7 COMPETITION AND SIGNAL SUPPRESSION

When electrospray is performed with solutions containing high analyte concentration (concentrations $\geq$10 μM range), saturation in signal is observed. For complex solutions containing multiple analytes and/or additives, it is common to observe signal suppression of the desired analyte in this concentration range. This suppression has previously been explained on the basis of competition for space on droplet surfaces and competition for the total amount of charge present to charge the analytes,[33,49] both of which may be contributing factors. Limitation in space on droplet surfaces is addressed elsewhere in this book, and competition for droplet charge is discussed in the context of Kebarle's version of the ion evaporation theory in Chapter 1, as well as in the context of the equilibrium partitioning model here.

2.7.1 Signal Suppression Due to Competition Among Analytes

As long as the total concentration of all analytes in the solution is significantly lower than the concentration of excess charge, competition for excess charge will only occur between the analyte and the charged solvent or other solution additives. In solutions containing only singly charged analytes, competition among analytes for excess charge will begin to occur when the total concentration of all analyte charge approaches the concentration of excess charge. Thus, for multiply charged analytes, this competition will occur at even lower concentrations, significantly below the concentration of excess charge.

Charged analytes with higher surface activities will compete most effectively with other analytes, charged solvent molecules, or solvent additives for surface excess charge. This effect has been demonstrated by experiments in which the concentration of an analyte with low surface activity (a polar amino acid) is held constant while the concentration of a very surface-active surfactant (octadecylamine) is increased (Figure 2.9a). At low concentrations, the response of the amino acid is unaffected by the presence of the surfactant. However, in the saturation range (where analyte concentration approaches the concentration of excess charge), the response of the analyte is severely suppressed by the presence of the surfactant. When the experiment is performed in reverse, where the surfactant concentration is held constant and the concentration of the polar amino acid is increased (Figure 2.9b), the response of the surfactant is hardly affected by the presence of the amino acid. These observations can be rationalized by the proposed partitioning of surfactants in electrospray droplets shown in Figure 2.10; the surfactant is so effective at competing for the surface charge in electrospray droplets that it prevents the less surface-active amino acid from becoming charged. Thus, the polar amino acid is unable to displace the surfactant from the droplet surface even at high concentrations.

In light of the findings displayed in Figure 2.10, it is not at all surprising that surface-active contaminants or additives in the electrospray solution have deleterious effects on analyses with electrospray mass spectrometry. Such surface-active species include polymer contaminants that are often present in reagent grade solvents, soap residues from improperly rinsed glassware, or surfactants sometimes used in sample preparation or to facilitate HPLC or CE separations. All of these can seriously suppress the analyte signal, and they should be

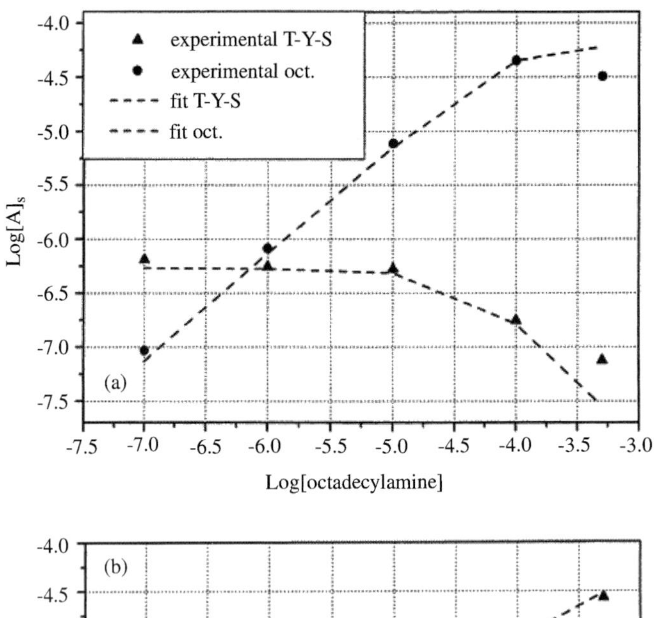

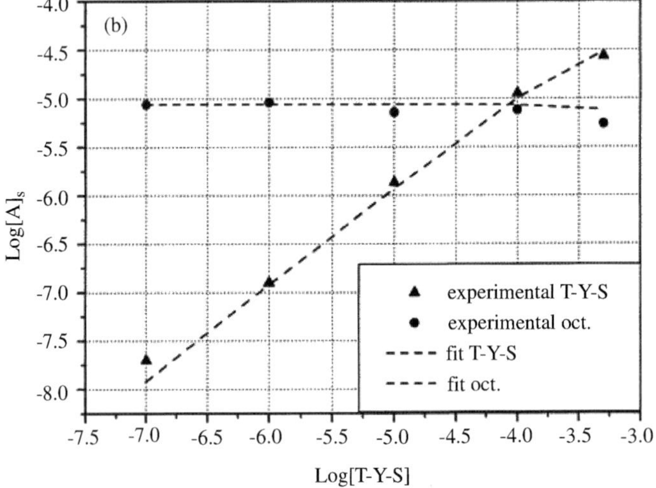

Figure 2.9. Competition in ESI-MS response between a surfactant and a polar analyte. (Reprinted from Ref. 50, with permission.) (**a**) Responses of a polar tripeptide (TYS) at constant concentration (10 μM) in solutions containing varying amounts of the surfactant octadecylamine (ODA). The TYS signal is unaffected by the presence of the surfactant octadeclyamine (ODA) at low concentrations of surfactant, but TYS response is significantly suppressed by the addition of high concentrations of surfactant. Response values were measured for solutions prepared in 50 : 50 methanol:water with 0.5% acetic acid and directly infused into a triple quadrupole mass spectrometer (Thermo Finnigan TSQ 7000). The experimental data were fit using the equilibrium partitioning model using the following values: $K_{TYS}/K_{SH}{}^{+} = 1.8$, $[Q] = 6.3 \times 10^{-5}$, $K_{ODA}/K_{TYS} = 48.6$, and $C_{SH}{}^{+} = 2 \times 10^{-3}$ M. (**b**) Response of the surfactant ocyadecylamine at constant concentration (10 μM), with addition of increasing concentrations of the polar peptide TYS. The addition of TYS concentrations ranging from 0.1 μM to 0.5 mM has little, if any, effect on the signal for the surfactant. Values for the fit were: $K_{TYS}/K_{SH}{}^{+} = 4.9$, $[Q] = 6.3 \times 10^{-5}$, $K_{ODA}/K_{TYS} = 48.6$, and $C_{SH}{}^{+} = 2 \times 10^{-3}$ M.

avoided in electrospray analyses. If signal suppression arises due to competition among analytes, the problem can sometimes be resolved by diluting the sample, but some clean-up (such as solid phase extraction) or separation (such as HPLC) prior to analysis is very often necessary for successful electrospray mass spectrometric analysis of complex solutions.

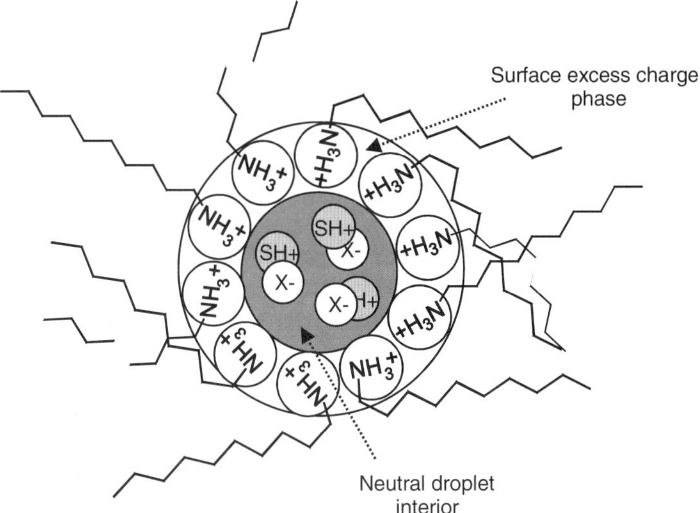

Figure 2.10. Schematic of an electrospray droplet showing partitioning of surface-active surfactants to the charged droplet surfaces (where SH^+ = protonated solvent and X^- = any negatively charged ion). The surfactants represent ideal analytes for ESI-MS because they are both surface active and chargeable. Species that do not have high affinity for droplet surfaces (represented in this case by SH^+) will reside in the electrically neutral droplet interior where they will be paired with counterions, and they will be lost as neutrals rather than be detected by the mass spectrometer. (Reprinted from Ref. 77, with permission.)

2.7.2 Suppression from Salt

It is important to note that although surface-active species cause the most severe signal suppression at the lowest concentrations, any charged solution additive present at high enough concentration can compete with the analyte for excess charge and can consequently cause signal suppression. For example, in a study conducted by Constantopoulos et al.,[32] it was observed that analyte signal was suppressed at very high concentrations (mM) of NaCl. Further evidence of high salt concentration causing suppressed electrospray response was provided by investigations conducted by Haddrell and Agnes.[40] They studied electrospray droplets formed from solutions containing 1.7 M NaCl and observed a surface layer of NaCl and an interior layer containing the analyte (a fluorescent dye). In these studies, MALDI-TOF mass spectrometry and fluorescence microscopy were used to profile the shape of single, dried electrospray droplets, providing for the first time direct evidence of the presence of two separate phases in electrospray droplets. On the basis of their results, Haddrell and Agnes argued that decreased response with high salt concentration occurs due to both (a) partitioning of the analyte from the charged droplet surface into the electrically neutral droplet interior and (b) decreased efficiency of droplet evaporation as a consequence of the surface salt layer.

The two previous examples demonstrate that high salt concentration has a deleterious effect on electrospray response. However, it is worth noting that the addition of smaller amounts of salt can actually improve electrospray response (even of analytes that are not detected as salt clusters) (Figure 2.11).[32] The mechanisms responsible for this effect have been postulated to be increased electrospray current with increasing salt concentration (that effectively increases $[Q]$) and/or decreased thickness of the electrical double layer at the surface of electrospray droplets.[32]

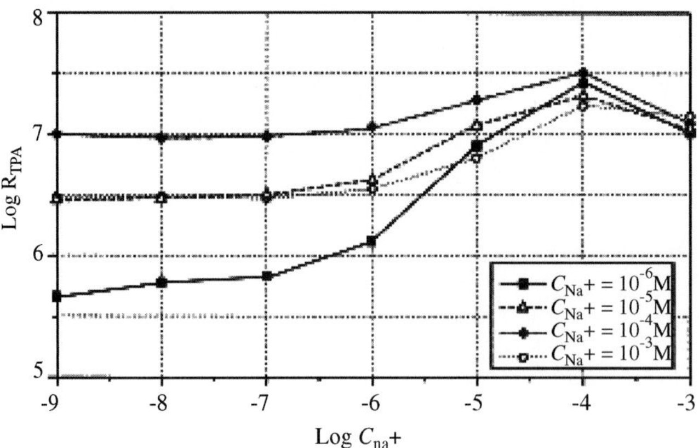

Figure 2.11. The influence of added salt on response curves for tetrapentylammonium (TPA) bromide in methanol. The relationship between response and added sodium concentration can be observed by comparing the TPA response (R_{TPA}) at a given concentration (i.e., 10^{-6} M) in solutions with various amounts of Na^+ (C_{Na^+}). The total response increases with increasing concentration of Na^+ between 10^{-6} and 10^{-4} M Na^+ added Na^+, but it decreases at a Na^+ concentration of 10^{-3} M. Thus, low salt concentrations improve the response of the analyte, but response is suppressed by the addition of high amounts of NaCl. (Reprinted from Ref. 22, with permission.)

2.8 GAS-PHASE PROTON TRANSFER AS A SOURCE OF SELECTIVITY

Prior to being sampled by the mass spectrometer, analytes freed from electrospray droplets enter a region at atmospheric pressure that contains a number of solvent molecules and other vapor-phase species. Proton transfer between the analytes of interest and these solvent molecules can sometimes affect the species detected with electrospray mass spectrometry. Gas-phase proton transfer has been shown to alter the charge states of proteins analyzed with electrospray mass spectrometry,[54,55] and in some specific cases it may be a source of selectivityin electrospray analysis of small molecules,[11,56]

Gas-phase proton affinity (PA) or gas-phase basicity (GB) are parameters typically used to predict whether a given analyte or solvent is likely to undergo proton transfer in the gas phase. Equation (2.9) shows a theoretical protonation reaction. Gas-phase proton affinity is defined as the negative change in enthalpy for this reaction at 298 K, while gas phase basicity is defined as the Gibbs free energy change for the same reaction.

$$M + H^+ \leftrightarrow MH^+ \tag{2.9}$$

A molecule with a high proton affinity can transfer a proton to a molecule with a lower proton affinity through a gas-phase proton transfer. For example, Eq. (2.10) shows the reaction between protonated water (PA = 691 kJ/mol) and methanol (PA = 754 kJ/mol).

$$H_3O^+(g) + CH_3OH(g) \leftrightarrow CH_3OH_2^+(g) + H_2O(g) \tag{2.10}$$

When considering whether gas-phase proton transfer will take place, it is important to note that relative basicities among molecules may differ in the gas phase and in solution. Water is more basic than methanol in solution, but in the gas phase it less basic.[57] Thus, electrospray ionization mass spectrometric analysis of solutions containing mixtures of water and methanol will show only methanol clusters.[56] This is the case because water, which exists in the protonated form in solution, transfers its protons to methanol in the gas phase.

If gas-phase proton transfer reactions occur in electrospray ionization mass spectrometry, they typically involve the solvent, because it is present in great excess. These reactions could cause signal suppression in the positive-ion mode if the protonated analyte were neutralized by proton transfer to the solvent. Conversely, signal suppression for acidic analytes in the negative-ion mode could result from the transfer of protons from the solvent to the negatively charged analyte. The former case has been documented experimentally. Amad et al.[56] showed that 2,2,2-trifluoroethanol can be detected when electrosprayed from aqueous solutions but not from methanolic solutions. These results were explained by that fact that the gas-phase proton affinity of this compound is 700.2 kJ/mol,[58] higher than that of water but lower than that of methanol. Fortunately, most of the species typically analyzed by positive-ion electrospray ionization mass spectrometry, including proteins and most small basic organic molecules, have significantly higher PA values than the usual electrospray solvents (water, methanol, acetonitrile). Thus, at least for positive-ion ESI-MS, ion suppression due to proton transfer only presents a problem under unusual circumstances.

If gas-phase proton transfer can occur in electrospray ionization mass spectrometry, it might be expected to be advantageous in some cases. Specifically, if neutral analytes could escape from electrospray droplets, they might become charged due to transfer of protons from the solvent or other solution additives. In this way, gas-phase proton transfer might facilitate the detection in positive-ion ESI-MS of species that are neutral in solution. Improvement in electrospray response due to such proton transfer have been predicted by a number of investigators.[11,59,60] However, in a recent investigation by Ehrmann et. al.,[12] positive-ion electrospray response trends were observed to correlate with basicity in solution rather than in the gas phase. In the Ehrmann study, all of the species investigated had much higher gas-phase basicities than did the solvent, but only those analytes that were strongly basic in solution were highly responsive in positive-ion ESI-MS. For the analytes investigated in the Ehrmann study, it appeared that some limiting factor (as of yet not identified) prevented ionization via gas-phase proton transfer. These results suggest that the contribution of gas-phase proton transfer to charging in ESI-MS is a topic worthy of further investigation.

2.9 EFFECTS OF NEBULIZATION AND FLOW RATE

Electrospray ionization mass spectrometry can be carried out over a wide range of flow rates. Technically, the term "electrospray" can be applied to any flow rate (it refers to the method of ionization), but additional terms have been coined to describe low-flow electrospray. "Microspray" describes electrospray conducted at intermediate flow rates (~0.1–10 μL/min), and "nanospray" refers to very low flow electrospray (typically < 100 nL/min).

One of the major attractions of ESI is its ability to serve as an interface between liquid chromatography and mass spectrometry. There are currently a number of low-flow HPLC systems on the market that are compatible with electrospray, microspray, and nanospray sources. Capillary HPLC systems are interfaced with electrospray conducted in the μL/min flow regime, while nanoflow systems can accommodate nL/min flow rates. When electrospray is coupled with conventional HPLC, it is necessary to accommodate a higher sample flow rate (~0.1–2 mL/min) than normal electrospray can tolerate. To facilitate operation at these higher flow rates, a technique called "pneumatically assisted electrospray" or "ion spray" is employed, in which sample nebulization by a flow of gas is used to stimulate a more rapid production of droplets from the electrospray tip.[61] The droplets produced using pneumatically assisted electrospray are initially further from the Rayleigh limit and require a greater evaporation time before vapor-phase ions are produced. To compensate for this, the nebulizer gas is often heated to temperatures as high as 700 °C.

A limitation with operation of ESI-MS at high flow rates is the requirement that the spray capillary be placed far back from the inlet aperture, which means that only a fraction of the spray is transferred into the mass spectrometer, and the detection limit of the technique suffers. This limitation can be overcome with the use of microspray or nanospray, provided that smaller inner-diameter capillaries are employed. The smaller inner-diameter capillaries and lower flow rates employed with nanospray and microspray result in the production of smaller droplets, which allows the capillary to be placed closer to the inlet aperture than is possible with high-flow operation. Consequently, a larger fraction of the analyte may enter the mass spectrometer, and the mass detection limit may improve. Another advantage of low-flow electrospray is that it often eliminates the need for nebulizing gas.

A number of investigators have observed that there are differences in performance of nanospray versus electrospray ionization.[62–64] The very different conditions (temperature, capillary diameter, use of nebulizer gas) typically employed for conventional and low-flow electrospray make absolute comparisons of sensitivity at widely different flow rates difficult. The current perception, however, is that very low flow rate systems have greater mass sensitivity than higher-flow-rate systems.[65–67]

Selectivity in electrospray is also affected by flow rate. This can happen in several ways. Tang and Smith[68] have shown that as the droplets undergo fission in the electrospray plume, the satellite droplets tend to concentrate on the outer edges of the plume cone. Since these droplets are enriched in the more surface active of the analytes present,[37] the relative response ratios of analytes would be affected by the needle position relative to the aperture.

Another factor affecting the relative response of analytes as the flow rate is decreased is the increasing value of $[Q]$. Based on the equilibrium partition model, one can see that an increase in $[Q]$ should increase the dynamic range and the concentration of interferent that would be required for suppression of an analyte's response. This advantage may be complicated by a corresponding increase in the electrolysis product, depending on its potential for causing interference. The increase in $[Q]$ as the flow rate is decreased to the nanoflow region, along with its effect on analyte matrix effects, has been studied by Schmidt, Karas, and Dülcks.[67] They recommend the lowest possible flow rates for the reduction of suppressive effects, but they warn that response factors will still favor compounds with some surface activity. A complementary study was reported by Smith's group,[34] and Liang et al.[69] have reported that even an isotopic internal standard could be involved in analyte signal suppression and contribute to nonlinearity in calibration curves for target drugs. Gangli et al.[70] developed a splitter such that flows as low as 0.1 μL/min could be delivered to a nanospray tip while input flow rates as high as 200 μL/min could be accommodated.

2.10 DIFFERENCES BETWEEN POSITIVE AND NEGATIVE ION ESI-MS

Successful analysis in the negative-ion mode cannot always be accomplished by simply reversing the polarity of the applied voltage. A number of factors complicate analyses in the negative-ion mode, and such effects can be a source of great consternation among electrospray users. Nonetheless, it is often necessary to operate in the negative-ion mode for the analysis of acidic species, and negative-ion mode operation can be advantageous because it sometimes generates spectra with less background chemical noise than is obtained with positive-ion mode operation. An understanding of how analyses in the positive and negative-ion mode differ is, therefore, important to the discussion of selectivity with ESI-MS.

The most important difference between positive- and negative-ion mode operation is the increased likelihood of corona discharge when operating with a negative applied voltage. Corona discharge is ionization of the atmosphere surrounding the tip of the spray capillary, which leads to the formation of an array of detectable products. These products can complicate mass spectra and obscure the signal for the species of interest. Other disadvantages of corona discharge in electrospray analyses include unstable (fluctuating) signals and irreproducible results.

To stabilize the spray current and prevent unwanted chemical noise, it is desirable to take steps to eliminate corona discharge. Corona discharge can be suppressed by the use of chlorinated solvents[71] or electron scavenging sheath gases such as O_2 or SF_6.[72,73] Another way to decrease the likelihood of occurrence of corona discharge is to use a pneumatically assisted electrospray, as described previously, in which a nebulizing gas is used to aid in droplet formation. Pneumatically assisted electrospray can be accomplished at lower applied voltages than are necessary to accomplish conventional electrospray, which makes the occurrence of corona discharge less likely. It is especially useful for aqueous solutions, which, due to their high surface tensions, require higher applied voltages than methanolic solutions when pneumatically assisted ionization is not employed.

A further difference between positive- and negative-ion electrospray ionization mass spectrometry is the fact that reduction, rather than oxidation, must occur at the high-voltage contact to the solution to enable flow of current in the electrospray circuit. For positive-ion electrospray, in a typical experimental design, the species oxidized is the metal at the high-voltage contact; and the products observed, metal cations, are poorly responsive to electrospray analysis and tend not to be a major source of chemical noise. However, as mentioned earlier, the species that undergoes reduction in negative ion electrospray cannot be the metal contact, because it cannot be further reduced. Therefore, some other species in the solution must be reduced. At high analyte concentrations, this species could be the analyte; but for the typical low analyte concentrations analyzed with ESI-MS, the predominant species that undergoes reduction will be the solvent or some high concentration solvent additive. The disadvantage of electrochemical reduction of the solvent is that it can produce reactive species that interfere with the chemical analysis of the analyte.

From the preceding discussion, it can be surmised that the ideal solvent system for analysis with negative ion ESI-MS would suppresses corona discharge, be easily reducible to produce noninterfering and nonreactive species, and facilitate ion formation through adduct formation with or deprotonation of the analyte. For analyses that require on-line separation (HPLC or CE), the solvent would also have to be compatible with separation conditions. Finally, the surface tension of the solvent would need to be appropriate for the generation of a stable electrospray, and it would ideally be nontoxic and easy to work with. No such perfect solvent system has been developed, but a wide array of recommendations of solvent systems for negative ion ESI-MS analysis has been published. For negative-ion mode analysis of low-polarity species that cannot be deprotonated, Cole and Zhu[74] showed success with 10 : 1 carbon tetrachloride:chloroform, which both suppresses corona discharge and forms chloride adducts with the analyte. For the analysis of deprotonated oligonucleotides, 1,1,1,3,3,3-hexafluoro-2-propanol (400 mM in methanol) adjusted to pH 7 with triethylamine is often employed. This solvent system was introduced by Apffel et al.[75] with the intention to promote ion paring for optimal chromatographic separations; but hexafluoro-2-propanol, which has a low reduction potential, may serve the additional advantage of facilitating stable reduction at the high-voltage contact with minimal generation of chemical noise.

For analysis of deprotonated analytes, it would be assumed that optimal response would be obtained at high pH. Indeed, Deguchi et al.[26] recommend the volatile weak base imidizole for negative-ion mode analysis of negatively charged oligonucleotides, and Huber and Krajete[27] recommend triethylammonium bromide (in both cases these additives were diluted in acetonitrile and added post-column). Surprisingly, however, several investigators have observed suppression of the negative ion ESI-MS response of deprotonated analytes when operating under basic conditions.[15,25] In fact, in a number of studies the response of deprotonated analyes has been observed to be optimal when operating under *acidic* conditions, typically achieved through the use of small amounts of acetic or formic acid.[15,25,76] These findings further highlight the complex (and still not fully understood) relationship between solution pH and analyte response in ESI-MS.

REFERENCES

1. Smith, R. D.; Shen, Y.; Tang, K. Ultrasensitive and quantitative analyses from combined separations-mass spectrometry for the characterization of proteomes. *Acc. Chem. Res.* **2004**, *37*(4), 269–278.
2. Wada, Y. Mass spectrometry for congenital disorders of glycosylation. *CDG. J. Chromatogr. B* **2006**, *838*(1), 3–8.
3. Wada, Y.; Nishikawa, A.; Okamoto, N.; Inui, K.; Tsukamoto, H.; Okada, S.; Taniguchi, N. Structure of serum transferrin in carbohydrate-deficient glycoprotein syndrome. *Biochem. Biophys. Res. Commun.* **1992**, *189*(2), 832–836.
4. Khaselev, N.; Murphy, R. C. Electrospray ionization mass spectrometry of lysoglycerophosphocholine lipid subclasses. *J. Am. Soc. Mass Spectrom.* **2000**, *11*(4), 283–291.
5. Ku Bon, K.; Fernandez de la Mora, J.; Saucy, D. A.; Alexander, J. N. Mass distribution measurement of water-insoluble polymers by charge-reduced electrospray mobility analysis. *Anal. Chem.* **2004**, *76* (3), 814–822.
6. Wu, J.; McLuckey, S. A. Gas-phase fragmentation of oligonucleotide ions. *Int. J. Mass Spectrom.* **2004**, *237* (2–3), 197–241.
7. Micciche, F.; van Straten, M. A.; Ming, W.; Oostveen, E.; van Haveren, J.; van der Linde, R.; Reedijk, J. Identification of mixed-valence metal clusters in drier solutions for alkyd-based paints by electrospray ionization mass spectrometry (ESI-MS). *Int. J. Mass Spectrom.* **2005**, *246*(1–3), 80–83.
8. Lin, Z.-P.; Ikonomou, M. G.; Jing, H.; Mackintosh, C.; Gobas, F. A. Determination of phthalate ester congeners and mixtures by LC/ESI-MS in sediments and biota of an urbanized marine inlet. *Environ. Sci. Technol.* **2003**, *37*(10), 2100–2108.
9. Musshoff, F.; Trafkowski, J.; Kuepper, U.; Madea, B. An automated and fully validated LC-MS/MS procedure for the simultaneous determination of 11 opioids used in palliative care, with 5 of their metabolites. *J. Mass Spectrom.* **2006**, *41*(5), 633–640.
10. Sun, W.; Wu, S.; Wang, X.; Zheng, D.; Gao, Y. An analysis of protein abundance suppression in data dependent liquid chromatography and tandem mass spectrometry with tryptic peptide mixtures of five known proteins. *Eur. J. Mass Spectrom.* **2005**, *11*(6), 575–580.
11. Ikonomou, M. G.; Blades, A. T.; Kebarle, P. Investigations of the electrospray interface for liquid chromatography/mass spectrometry. *Anal. Chem.* **1990**, *62*, 957–967.
12. Ehrmann, B. M.; Henriksen, T.; Cech, N. B. Relative importance of basicity in the gas phase and in solution for determining selectivity in electrospray ionization mass spectrometry. *J. Am. Soc. Mass Spectrom.* **2008**, *19*(5), 719–728.
13. Mansoori, B. A.; Volmer, D. A.; Boyd, R. K. Wrong-way-round electrospray ionization of amino acids. *Rapid Commun. Mass Spectrom.* **1997**, *11*, 1120–1130.
14. Wang, G.; Cole, R. B. Disparity between solution-phase equilibria and charge state distributions in positive-ion electrospray mass spectrometry. *Org. Mass Spectrom.* **1994**, *29*, 419–427.
15. Wu, Z.; Gao, W.; Phelps, M. A.; Wu, D.; Miller, D.; Dalton, J. T. Favorable effects of weak acids on negative-ion electrospray ionization mass spectrometry. *Anal. Chem.* **2004**, *76*, 839–847.
16. Zhou, S.; Cook, K. D. Protonation in electrospray mass spectrometry: Wrong-way-round or right-way-round? *J. Am. Soc. Mass Spectrom.* **2000**, *11*, 961–966.
17. Jemal, M.; Hawthorn, D. J. Effect of high performance liquid chromatography mobile phase (methanol versus acetonitrile) on the positive and negative ion electrospray response of a compound that contains both

an unsaturated lactone and a methyl sulfone group. *Rapid Commun. Mass Spectrom.* **1999**, *13*, 961–966.

18. Kelly, M. A.; Vestling, M. M.; Fenselau, C. C.; Smith, P. B. Electrospray analysis of proteins: A comparison of positive-ion and negative-ion mass spectra at high and low pH. *Org. Mass Spectrom.* **1992**, *27*, 1143–1147.
19. Cech, N. B.; Enke, C. G. Practical implications of some recent studies in electrospray ionization fundamentals. **2001**, *20*(6), 362–387.
20. Van Berkel, G. J.; Zhou, F.; Aronson, J. T. Changes in bulk solution pH caused by the inherent controlled-current electrolytic process of an electrospray source. *Int. J. Mass Spectrom. Ion Proc.* **1997**, *162*, 55–67.
21. Zhou, S.; Prebyl, B. S.; Cook, K. D. Profiling pH changes in the electrospray plume. *Anal. Chem.* **2002**, *74*, 4885–4888.
22. Cech, N. B.; Enke, C. G. Electrospray ionization mass spectrometry: How and when it works. In The Encyclopedia of Mass Spectrometry, Vol. *8*, Niessen, W. M. A. (Ed.), Elesevier, Oxford, **2006**, pp. 171–180.
23. Marwah, A.; Marwah, P.; Lardy, H. Analysis of ergosteroids VIII: Enhancement of signal response of neutral steroidal compounds in liquid chromatographic–electrospray ionization mass spectrometric analysis by mobile phase additives. *J. Chromatogr. A* **2002**, *964*(1–2), 137–151.
24. Straub, R. F.; Voyksner, R. D. Determination of penicillin G, ampicillin, amoxicillin, cloxacillin and cephapirin by high-performance liquid chromatography–electrospray mass spectrometry. *J. Chromatogr.* **1993**, *647*(1), 167–181.
25. Cuyckens, F.; Claeyes, M. Optimization of a liquid chromatography method based on simultaneous electrospray ionization mass spectrometric and ultraviolet photodiode array detection for analysis of flavonoid glycosides. *Rapid Commun. Mass Spectrom.* **2002**, *16*(24), 2341–2348.
26. Deguchi, K.; Ishikawa, M.; Yokokura, T.; Ogata, I.; Ito, S.; Mimura, T.; Ostrander, C. Enhanced mass detection of oligonucleotides using reverse-phase high performance liquid chromatography/electrospray ionization ion-trap mass spectrometry. *Rapid Commun. Mass Spectrom.* **2002**, *16*(22), 2133–2141.
27. Huber, C. G.; Krajete, A. Sheath liquid effects in capillary high-performance liquid chromatography–electrospray mass spectrometry of oligonucleotides. *J. Chromatogr. A* **2000**, *870*, 413–424.
28. Bruggink, C.; Maurer, R.; Herrmann, H.; Cavalli, S.; Hoefler, F. Analysis of carbohydrates by anion exchange chromatography and mass spectrometry. *J. Chromatogr. A* **2005**, *1085*(1), 104–109.
29. Han, X.; Cheng, H. Characterization and direct quantitation of cerebroside molecular species from lipid extracts by shotgun lipidomics. *J. Lipid Res.* **2005**, *46*, 163–175.
30. Van Berkel, G. J. Electrolytic corrosion of a stainless-steel electrospray emitter monitored using an electrospray-photodiode array system. *J. Anal. At. Spectrom.* **1998**, *13*(7), 63–67.
31. Van Berkel, G. J.; McLuckey, S. A.; Glish, G. L. Electrochemical origin of radical cations observed in electrospray ionization mass spectra. *Anal. Chem.* **1992**, *64*, 1586–1593.
32. Constantopoulos, T. L.; Jackson, G. S.; Enke, C. G. Effects of salt concentration on analyte response using electrospray ionization mass spectrometry. *J. Am. Soc. Mass Spectrom.* **1999**, *10*(7), 625–634.
33. Enke, C. G. A predictive model for matrix and analyte effects in electrospray ionization of singly-charged ionic analytes. *Anal. Chem.* **1997**, *69*(23), 4885–4893.
34. Tang, K.; Page, J. S.; Smith, R. D. Charge competition and the linear dynamic range of detection in electrospray ionization mass spectrometry. *J. Am. Soc. Mass Spectrom.* **2004**, *15*(10), 1416–1423.
35. Iribarne, J. V.; Thomson, B. A. On the evaporation of charged ions from small droplets. *J Chem. Phys* **1976**, *64*, 2287–2294.
36. Dole, M.; Mack, L. L.; Hines, R. L.; Mobley, R. C.; Ferguson, L. D.; Alice, M. B. Molecular beams of macroions. *J. Chem. Phys.* **1968**, *49*(5), 2240–2249.
37. Cech, N. B.; Enke, C. G. The effect of affinity for charged droplet surfaces on the fraction of analyte charged in the electrospray process. *Anal. Chem.* **2001**, *73*(19), 4632–4639.
38. Kebarle, P. A brief overview of the present status of the mechanisms involved in electrospray mass spectrometry. *J. Mass Spectrom.* **2000**, *35*, 804–817.
39. Pan, P.; McLuckey, S. A. Electrospray ionization of protein mixtures at low pH. *Anal. Chem.* **2003**, *75*, 1491–1499.
40. Haddrell, A. E.; Agnes, G. R. Organic cation distributions in the residues of levitated droplets with net charge: Validity of the partition theory for droplets produced by and electrospray. *Anal. Chem.* **2004**, *76*, 53–61.
41. Sjöberg, P. J. R.; Bockman, C. F.; Bylund, D.; Markides, K. E. A method for determination of ion distribution within electrospryayed droplets. *Anal. Chem.* **2001**, *73*, 23–28.
42. Sjöberg, P. J. R.; Bockman, C. F.; Bylund, D.; Markides, K. E. Factors influencing the determination of analyte ion surface partitioning coefficients in electrosprayed droplets. **2001**, *12*, 1002–1010.
43. Sherman, C. L.; Brodbelt, J. S. Partitioning model for competitive host–guest complexation in ESI-MS. *Anal. Chem.* **2005**, *77*(8), 2512–2523.
44. Mollah, S.; Pris, A. D.; Johnson, S. K.; GwizdalaIII, A. B.; Houk, R. S. Identification of metal cations, metal complexes, and anions by electrospray mass spectrometry in the negative ion mode. *Anal. Chem.* **2000**, *72*, 985–991.
45. Wang, H.; Agnes, G. R. Kinetically labile equilibrium shifts induced by the electrospray process. *Anal. Chem.* **1999**, *71*, 4166–4172.

46. Constantopoulos, T. L. Fundamental studies of electrospray ionizaton mass spectrometry. University of New Mexico, Albuquerque, **1999**.
47. Iribarne, J. V.; Dziedzic, P. J.; Thomson, B. A. Atmospheric pressure ion evaporation–mass spectrometry. *Int. J. Mass Spectrom. Ion Phys.* **1983**, *50*, 331–347.
48. Tang, L.; Kebarle, P. Dependence of ion intensity in electrospray mass spectrometry on the concentration of analytes in the electrosprayed solution. *Anal. Chem.* **1993**, *65*, 3654–3668.
49. Kebarle, P.; Tang, L. From ions in solution to ions in the gas phase. *Anal. Chem.* **1993**, *65*(22), 972A–985A.
50. Cech, N. B.; Enke, C. G. Relating electrospray ionization response to non-polar character of small peptides. *Anal. Chem.* **2000**, *72*(13), 2717–2723.
51. Henriksen, T.; Juhler, R. K.; Svensmark, B.; Cech, N. B. The relative influences of acidity and polarity on responsiveness of small organic molecules to analysis with negative ion electrospray ionization mass spectrometry (ESI-MS). *J. Am. Soc. Mass Spectrom.* **2005**, *16*(4), 446–455.
52. Schug, K.; McNair, H. M. Adduct formation in electrospray ionization mass spectrometry II: Benzoic acid derivatives. *J. Chrom. A* **2003**, *985*, 531–539.
53. Bokman, C. F.; Bylund, D.; Markides, K. E.; Sjöberg, P. J. R. Relating chromatographic retention and electrophoretic mobility to the ion distribution within electrosprayed droplets. *J. Am. Soc Mass Spectrom.* **2006**, *17*, 318–324.
54. Hautreux, M.; de Kerdaniel, A.; Zahir, A.; Malec, V.; Laprevote. Under non-denaturing solvent conditions, the mean charge state of a multiply charged protein ion formed by electrospray is linearly correlated with the macromolecular surface. *Int. J. Mass Spectrom.* **2004**, *231*(2-3), 131–137.
55. Ogorzalek Loo, R. R.; Smith, R. D. Proton transfer reactions of multiply charged peptide and protein cations and anions. *J. Mass Spectrom.* **1995**, *30*, 339–347.
56. Amad, M. H.; Cech, N. B.; Jackson, G. S.; Enke, C. G. Importance of gas phase proton affinities in determining the electrospray ionization response for analytes and solvents. *J. Mass Spectrom.* **2000**, *35*, 784–789.
57. Kebarle, P.; Haynes, R. M.; Collins, J. G. Competitive solvation of the hydrogen ion by water and methanol molecules studied in the gas phase. *J. Am. Chem. Soc.* **1967**, *89*(23), 5753–5757.
58. NIST, In http://webbooknistgov/chemistry, accessed April 18, 2008.
59. Ogorzalek Loo, R. R.; Smith, R. D. Investigations of the gas-phase structure of electrosprayed proteins using ion–molecule reactions. *J. Am. Soc. Mass Spectrom.* **1994**, *5*(4), 221–229.
60. Yen, T.-Y.; Charles, M. J.; Voyksner, R. D. Processes that affect electrospray ionization-mass spectrometry of nucleobases and nucleosides. *J. Am. Soc. Mass Spectrom.* **1996**, *7*, 1106–1108.
61. Bruins, A. P.; Henion, J. D.; Covey, T. R. Ion spray interface for combined liquid chromatography/atmospheric pressure ionization mass spectrometry. *Anal. Chem.* **1987**, *59*, 2642–2646.
62. Juraschek, R.; Dulcks, T.; Karas, M. Nanoelectrospray—More than just a minimized-flow electrospray ionization source. *J. Am. Soc. Mass Spectrom.* **1999**, *10*(4), 300–308.
63. Valaskovic, G. A.; Utley, L.; Lee, M. S.; Wu, J.-T. Ultra-low flow nanospray for the normalization of conventional liquid chromatography/mass spectrometry through equimolar response: Standard-free quantitative estimation of metabolite levels in drug discovery. *Rapid Commun. Mass Spectrom.* **2006**, *20*(7), 1087–1096.
64. Warriner, R. N.; Craze, A. S.; Games, D. E.; Lane, S. J. Capillary electrochromatography/mass spectrometry—A comparison of the sensitivity of nanospray and microspray ionization techniques. *Rapid Commun. Mass Spectrom.* **1998**, *12*(17), 1143–1149.
65. Asperger, A.; Efer, J. R.; Koal, T.; Engewald, W. On the signal response of various pesticides in electrospray and atmospheric pressure chemical ionization depending on the flow-rate of eluent applied in liquid chromatography–tandem mass spectrometry. *J. Chromatogr. A* **2001**, *937*, 65–72.
66. Page, J. S.; Kelly, R. T.; Tang, K.; Smith, R. D. Ionization and transmission efficiency in an electrospray ionization–mass spectrometry interface. *J. Am. Soc. Mass Spectrom.* **2007**, *18*, 1582–1590.
67. Schmidt, A.; Karas, M.; Dülcks, T. Effect of different solution flow rates on analyte ion signals in nano-esi ms, or: When does ESI turn into nano-ESI? *J. Am. Soc. Mass Spectrom.* **2003**, *14*, 492–500.
68. Tang, K.; Smith, R. D. Physical/chemical separations in the break-up of highly charged droplets from electrosprays. *J. Am. Chem. Soc.* **2001**, *12*, 343–347.
69. Liang, H. R.; Foltz, R. L.; Meng, M.; Bennett, P. Ionization enhancement in atmospheric pressure chemidcal ionization and suppression in electrospray ionization between target drugs and stable-isotope-labeled internal standards in quantitative liquid chromatography/tandem mass spectrometry. *Rapid Commun. Mass Spectrom.* **2003**, *17*, 2815–2821.
70. Gangli, E. T.; Annan, M.; Spooner, N.; Vouros, P. Reduction of signal suppression effects in ESI-MS using a nanosplitting device. *Anal. Chem.* **2001**, *73*, 5635–5644.
71. Cole, R. B.; Harrata, A. K. Charge-state distribution and electric-discharge suppression in negative-ion electrospray mass spectrometry using chlorinated solvents. *Rapid Commun. Mass Spectrom.* **1992**, *6*, 536–539.
72. Straub, R. F.; Voyksner, R. D. Negative ion formation in electrospray mass spectrometry. *J. Am. Soc. Mass Spectrom.* **1993**, *4*, 578–587.
73. Wampler, F. M.; Blades, A. T.; Kebarle, P. D. C. Negative ion electrospray mass spectrometry of nucleotides: Ionization from water solution with sulfur

hexafluoride discharge suppression. *J. Am. Soc. Mass Spectrom.* **1993**, *4*(4), 289–295.

74. Cole, R. B.; Zhu, J. H. Chloride ion attachement in negative ion electrospray ionization mass spectrometry. *Rapid Commun. Mass Spectrom.* **1999**, *13*(7), 607–611.

75. Apffel, A.; Chakel, J. A.; Fischer, S.; Lichtenwalter, K.; Hancock, W. S. Analysis of oligonucleotides by HPLC-electrospray ionization mass spectrometry. *Anal. Chem.* **1997**, *69*(7), 1320–1325.

76. Seto, C.; Bateman, K. P.; Gunter, B. Development of generic liquid chromatography-mass spectrometry methods using experimental design. *J. Am. Soc. Mass Spectrom.* **2002**, *13*(1), 2–9.

77. Cech, N. B. Understanding, predicting and improving the electrospray response of small biological molecules. University of New Mexico, Albuquerque, **2001**.

Chapter 3

Electrochemistry of the Electrospray Ion Source

Gary J. Van Berkel and Vilmos Kertesz

Organic and Biological Mass Spectrometry Group, Chemical Sciences Division, Oak Ridge National Laboratory, Oak Ridge, Tennessee

Electrospray and MALDI Mass Spectrometry: Fundamentals, Instrumentation, Practicalities, and Biological Applications, Second Edition, Edited by Richard B. Cole

3.1 INTRODUCTION

3.1.1 Electrospray Overview

Dispersion of a liquid into small charged droplets by an electrostatic field is a phenomenon whose observation dates back at least two centuries.[1] However, the first detailed experimental studies of this phenomenon, using an electrostatic sprayer similar to that used in today's electrospray mass spectrometry (ES-MS) experiments, were carried out by Zeleny in the early 1900s.[2] From the time of Zeleny's work until the present, an enormous amount of literature describing fundamental and applied studies of electrostatic spraying have appeared in the scientific literature.[1,3,4] Dole and co-workers[5–8] in the late 1960s and early 1970s are usually credited with the first attempts at using an atmospheric pressure electrostatic sprayer (i.e., an ES ion source) as a means to produce gas phase ions from macromolecules in a liquid solution for analysis by mass spectrometry. The first successful demonstration that macroions could be liberated from electrically charged droplets and detected using mass spectrometry may more appropriately be attributed to the late 1970s work of Iribarne and co-workers.[9,10] In any case, it was the successful, concurrent, and independent coupling of an ES ion source and a mass spectrometer by Fenn and co-workers[11–13] and Alexandrov et al.[14] in the mid-1980s that truly gave birth to the field of ES-MS. Within a couple of years of Fenn's and Alexandrov's first ES-MS publications, the groups of Smith[15–17] and Henion[18–21] were pushing the field forward quickly by demonstrating the applicability of ES-MS for the analysis of all types of nonvolatile, polar, or thermally labile compounds. Twenty years later the applications and users of ES-MS are too numerous to succinctly summarize. The interested reader is directed to other chapters in this book, which represent some of the most recent ES-MS application overviews.

Occurring in parallel with the development of ES-MS as an analytical tool (and certainly contributing to its rapid growth in utility and popularity) were numerous fundamental studies aimed at elucidating the individual steps in the ES process responsible for the liberation of gas-phase ions from the analytes in solution. These types of studies, which continue today, strive for a complete understanding of the relationship among the various instrumental components and operational parameters of the ES device, the physical and chemical nature of the solvents and the analytes in solution and charged droplets, the gas-phase ions generated from the charged ES droplets, and the ions ultimately observed by the mass spectrometer. The evolution in this understanding continues to lead to new and better analytical and fundamental applications of ES-MS. Readers are directed to recent reviews of the fundamental processes in ES-MS by Cech and Enke[22] and Cole[23] and to Chapters 1 and 2 in the present book.

Today there is generalized agreement that the ES process involves three main steps prior to mass analysis: the generation and charging of the ES droplets; droplet evaporation and the production of gas phase ions; and secondary processes that modify the gas-phase ions in the atmospheric and the subatmospheric pressure sampling regions of the mass spectrometer. The details of these individual steps and their associated analytical implications are a continued focus of study and debate. *Integral to the generation and charging of the ES droplets are electrochemical reactions that occur at the conductive contact/solution interface within or near the ES emitter to maintain the quasi-continuous production of charged droplets and ultimately gas phase ions.* The electrochemical reactions that take place at this ES emitter electrode may influence the gas-phase ions formed and ultimately analyzed by the mass spectrometer, because they change the composition of the sprayed solution. The composition and chemistry of the electrosprayed solution can affect not only the ions ultimately produced in the gas phase, but also the relative sensitivities with which they can be detected.[22] Thus, the nature and extent of the electrochemical reactions in the emitter, which alter solution composition, are important parameters to consider in ES ion formation. Moreover, the compositional changes caused by the electrochemical processes can include modification of the mass or charge state of the analytes under study.

In this chapter, the evolution in understanding of the fundamental electrochemical characteristics of the ES ion source is overviewed and the electrochemical fundamentals as they relate to the most common types of ES ion source configurations are detailed. These fundamental issues are presented so that the analytical implications of this inherent electrochemistry are apparent to the general ES-MS user, with particular emphasis on utilizing the chemical processes to avoid unwanted pitfalls in an analysis and to expand the scope of ES-MS studies. For further information on this topic, readers might also consult (a) the chapter by Van Berkel[24] in the previous version of this book and (b) the recent overview of electrochemistry and mass spectrometry by Diehl and Karst. [25]

3.1.2 Early Development of an Electrochemical Understanding of the Electrospray Ion Source

Figure 3.1 is a simple schematic diagram of a typical ES ion source and the nominal electrical circuit involved in the positive-ion mode.[26] In this scheme, the ion source is comprised of two electrodes, namely, the metal ES capillary (usually stainless steel) and the atmospheric sampling aperture plate of the mass spectrometer (also stainless steel) that are connected together (and ultimately to ground) via a high-voltage supply (up to about $\pm 6\,\mathrm{kV}$). Under typical ES-MS operating conditions, a solution containing the analyte of interest is pumped through the ES capillary (i.e., the working electrode) held at high voltage and sprayed toward the aperture plate (i.e., the counter electrode). Addition of any ionic species (i.e., electrolytes) to the solution other than the analyte (except for small amounts of ES-friendly acids, bases, or buffers to ionize the analyte) is usually avoided when possible, because their presence in solution tends to suppress the formation of gas phase ions from the analytes of interest.[27–30] However, some number of ions, either the analytes, contaminants, or deliberately added electrolytes, must be present in the solution or the ES device will not form charged droplets.[31] Under the influence of the applied electric field, ions of the same polarity as the voltage applied to the ES capillary migrate from the bulk liquid toward the liquid at the capillary tip. When the buildup of an excess of ions of one polarity at the surface of the liquid reaches the point that coulombic forces are sufficient to overcome the surface tension of the liquid, droplets enriched in one-ion polarity (positive ions in this case)

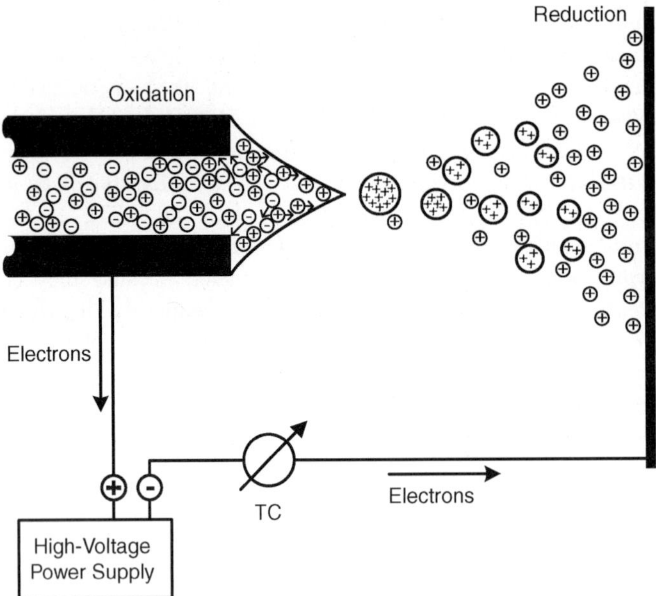

Figure 3.1. Schematic representation of the processes that occur in electrospray in positive ion mode. The imposed electric field between the emitter electrode and counter electrode leads to a partial separation of positive from negative ions present in solution at the meniscus of the solution at the metal capillary tip. This net charge is pulled downfield, expanding the meniscus into a cone that emits a fine mist of positively charged droplets. The droplets carry off an excess of positive ions. Solvent evaporation reduces the volume of the droplets at constant charge, leading to fission of the droplets. Continued production of charged droplets requires an electrochemical oxidation at the emitter electrode–solution interface—that is, a conversion of ions to electrons. Electrochemical reduction is required to be the dominate process in negative ion mode. (Adapted from the original figure in Ref. 26.)

are emitted from the capillary and travel toward the counter electrode. This results in a continuous steady-state current at the counter electrode.[32–35]

For this continuous production of an excess of positive ions in the charged droplets to be maintained, anions in solution must be neutralized or positive ions must be created. As we now understand it, this process involves electrochemical oxidation/reduction of the components of the metal ES capillary and/or one or more of the species in the solution, ultimately leading to electron flow to or from the high-voltage supply depending on the polarity applied to the capillary.[24–26,36–41] Specifically, oxidation reactions predominate in the ES capillary in positive-ion mode (electron flow to the voltage supply), while reduction reactions predominate in negative-ion mode (electron flow from the voltage supply). The current that can be measured at the working electrode owing to these reactions (i.e., the faradaic current, i_F) is equal in magnitude to the current measured at the counter electrode (i.e., the ES current, i_{ES}), but of the opposite polarity. Reduction/oxidation of some species at the counter electrode (i.e., the front aperture plate of the mass spectrometer) occurs to complete the electrical circuit (and accounts for the current we measure at this electrode).

Knowledge of the fact that the fundamental operation of electrostatic sprayers involves an electrochemical process pre-dates their use as an ion source in ES-MS. The circuit diagrams for electrostatic sprayers that are found in the literature from the early 1900s are drawn much the same as that shown in Figure 3.1. This circuit implies the occurrence of electrochemical reactions at both electrodes to complete the circuit. Therefore, some might consider these electrolytic reactions an obvious consequence of Kirchhoff's current law.[42] Yet, no mention of the details of this electrochemical process nor any implications (other than possibly charge-balance) for the particular application of the spray device are easily found in the literature before the 1970s. However, in 1974, Evans and co-workers[36] discussed the electrochemical process in connection with their incorporation of an electrostatic sprayer as an organic MS ion source in what is known as electrohydrodynamic mass spectrometry (EH-MS). EH-MS is a technique very related to ES-MS in that it uses an electrostatic sprayer to transfer ionic species present in solution into the gas phase for mass analysis. In ES-MS, the gas-phase ions are formed from analytes in charged droplets at atmospheric pressure, whereas with EH-MS ions are extracted directly from solution into vacuum. The circuitry of both sprayers is analogous, and therefore the characteristics of the charge-balancing processes that take place in the respective spray capillaries should be similar. In their initial organic EH-MS studies, Evans and co-workers[36] reported that their stainless steel spray capillary underwent electrochemical corrosion producing Fe^{2+} ions in solution as witnessed by the observation of a series of doubly charged ions in their mass spectra that could be attributed to Fe^{2+} attachment to glycerol clusters (G_5–G_8) and Fe^{2+} attachment to sucrose + glycerol (Suc + G_3, Suc + G_4, and Suc_2 + G_3). They later reported that this corrosion and appearance of solvated Fe^{2+} ions could be avoided by the use of a pure platinum capillary in the EH ion source.[37] In his review of EH-MS in 1984, Cook[38] discussed further the electrochemical effect in EH and included some speculation regarding redox reactions of solution species (e.g., Cl^- oxidation to form Cl_2 gas) that might cause the spray instability sometimes observed in EH-MS. He also noted in discussing ES-MS that "overall electroneutrality is presumably retained [in the ES ion source] by electrochemistry at the emitter surface, as in EH-MS."

While the literature shows that the electrochemical nature of electrostatic sprayers was no mystery prior to the development of ES-MS, it was the work of Kebarle and co-workers[26] in 1991 that first brought the electrochemical nature of the ES ion source to major attention in the mass spectrometry community. Like the earlier EH-MS work of Evans and co-workers,[36,37] the work of Kebarle's group served to alert the users of ES-MS that the redox reaction products formed in the metal ES capillary might be detected in the gas phase. More than that though, their work represented the first attempt to characterize the electrolytic nature of the ES ion source in formal electrochemical terms. On the basis of the electric circuit shown for the ES ion source in Figure 3.1, the electrophoretic charge separation mechanism for droplet charging and formation, and charge-balance considerations, Kebarle and co-workers[26] state succinctly the reason why redox reactions must occur in the metal capillary of the ES ion source:

> *Considering the requirements for charge balance in such a continuous electric current device and the fact that only electrons can flow through the metal wire supplying the electric potential to the electrodes, one comes to the conclusion that the electrophoretic charge separation mechanism [of droplet charging and formation] requires that the [positive-ion] electrospray process should involve an electrochemical conversion of ions to electrons [within the metal ES capillary].*

This means that charge cannot flow through these circuit junctions except via heterogeneous electron transfer chemistry. Heterogeneous electron transfer chemistry is electrochemistry. Thus, electrochemistry is inherent to the operation of the electrostatic sprayer used in ES-MS.

In their words, this meant that the ES ion source could be "viewed as an electrolytic cell of a somewhat special kind . . . insofar as part of the ion transport [between electrodes] does not occur through solution [as in a conventional electrolytic cell] but through the gas phase."[26] It was surmised that the redox reactions with the lowest redox potentials would predominate in this charge-balancing process, with the actual reactions occurring dependent on the particular solvent and solution composition. Furthermore, the reactions might be expected to involve neutral as well as ionic species, including the metal spray capillary (as Evans' group[36] had noted with EH-MS), solvents, additives, or contaminants in the solution. Given "wet methanol" containing a chloride salt as an electrolyte, they postulated that when the ES capillary was held at a high positive voltage (i.e., positive-ion mode), the excess positive charge would result from electrochemical oxidation reactions neutralizing the negative ions (e.g., oxidation of Cl^- or OH^- anions), reactions producing positive ions (e.g., oxidation of water or the metal capillary to produce protons or metal cations, respectively), or both types of reactions. When the capillary emitter electrode was held at a high negative voltage (i.e., negative-ion mode), the excess negative charge might be supplied by neutralizing the positive ions, by the production of negative ions from neutrals, or by both of these processes. No possible reactions were spelled out for this latter mode of operation. One should note that many of these specific reactions put forward by Kebarle's group are the same ones either observed or discussed by Evans and co-workers[36] and Cook[38] with regard to the electrochemical nature of the EH ion source.

Kebarle's group chose to demonstrate the electrochemical nature of the ES ion source by "forcing" the metal comprising the ES capillary to oxidize in the hope that the metal ions so produced would be observed in the ES mass spectrum. This task was accomplished by selecting zinc, which is very easy to oxidize ($E^0 = -0.76$ V vs. SHE), as the metal for the ES capillary tip. The Zn^{2+} ions released to the solution by oxidation of the zinc tip were detected in the gas phase with the amount of Zn^{2+} observed (as determined from calibration with zinc salts) corresponding to the amount required to maintain the ES current on the basis of Faraday's first law. In more precise electrochemical terms, the current at the ES emitter electrode (actually measured at the counter electrode), i_{ES}, was found to be equal in magnitude to the current due to the redox reactions occurring in the ES capillary—that is, the faradaic current at the working electrode, i_F (in this case due only to the oxidation of the zinc capillary tip). Kebarle's group presented no direct evidence for the occurrence of redox reactions other than those involving the metal ES capillary material. However, when using silver ($E^0 = 0.80$ V vs. SHE) as the capillary tip, the gas-phase signals observed for Ag^+ were less than that required to account for complete charge balance by this reaction alone. They speculated that because silver was relatively hard to oxidize compared to zinc, oxidation of some more easily oxidized trace level component in the methanol solvent probably provided the remainder of the current.

Direct evidence for the involvement of solution species in these redox reactions was reported at about this same time by our group. We found that under certain solution conditions, the molecular radical cations ($M^{+\bullet}$) of some divalent metal porphyrins (e.g., Ni^{II}octaethylporphyrin (NiOEP), ZnOEP, and VOOEP) formed by this electrochemical process could be observed in positive-ion ES mass spectra.[39,43] Certain other easy-to-oxidize species like polyaromatic hydrocarbons (PAHs), aromatic amines, and heteroaromatics were also oxidized at the emitter electrode and observed as cationic radicals. Molecular ions formed by loss of an electron had not been observed in ES mass spectra prior to those reports. Our work served to illustrate that analyte species, under the appropriate operational conditions, could be directly involved in the redox reactions in the metal spray capillary and that the products of their reactions could be observed in the gas phase.

Furthermore, this work showed that the electrochemical process could be exploited for analytical purposes as discussed in more detail in Section 3.3.

With continued studies, we were able to show that the ES ion source, as depicted in Figure 3.1, operates electrochemically in a fashion analogous to that of a conventional controlled-current electrochemistry (CCE) flow cell.[41] A conventional CCE flow cell houses a working electrode, a counter electrode, and a controlled-current source to set the magnitude of the current through the cell—that is, the cell current, i_C.[44–46] In our view, the charged droplet formation process is the controlled-current source with the "cell current"—that is, the ES current, i_{ES}—equal to the product of the rate at which charged droplets are formed and the average number of charges per droplet. This description is more akin to describing the ES ion source as a current-limited device (such as a phototube or flame-ionization detector[47,48]) as opposed to describing it as a somewhat special form of a conventional electrochemical cell.[26] Altering the output of this current source—that is, altering the magnitude of i_{ES}—requires that the rate of droplet production and/or the average number of charges per droplet be altered. This can be accomplished by changing one or more of several operational parameters including the electric field at the emitter tip, the solution flow rate, and the conductivity of the solution.[31,49,50]

In direct analogy to two-electrode, controlled-current electrolysis, one expects that the potential at the metal–solution interface in the ES capillary (i.e., the potential at the working electrode/solution interface, $E_{E/S}$), which ultimately determines which redox reactions can occur, will be that value for a given magnitude of i_{ES} that is necessary to oxidize/reduce sufficient species in the solution within the ES capillary to maintain that current (i.e., $i_{ES} = i_F$). The individual species oxidized/reduced to supply i_F will do so in order of their increasing redox potentials until the required current is supplied. Furthermore, the extent of any reaction involving solution species will be affected both by the rate at which the species flow through the capillary (the tubular working electrode) for a given magnitude of i_{ES} and by the rate of mass transfer of the species to the electrode surface. In summary, the particular electrochemical reactions that actually take place, and the extent to which they take place, are governed by the magnitude of i_{ES} (which is related to the nature of the solvents, solution conductivity, and electric field at the capillary tip), the respective concentrations and redox potentials of the various species in the system (including the metal(s) comprising the ES capillary), and the availability of a species for reaction at the metal–solution interface (which is determined by the rate of mass transport to the surface and therefore is related to, among other factors, capillary length, solution flow rate, and species concentration).

3.2 ELECTROCHEMICAL BASICS OF THE MOST COMMON ELECTROSPRAY SYSTEMS

To explain the electrochemistry of the ES ion source in more detail and to explain the advancements in the understanding of this process since our prior review,[24] it may be best to focus on the electrical circuits involved and on the three major principles central to electrochemistry in general, namely, current (faradaic) at the working electrode, interfacial electrode potential at the working electrode, and mass transport to the working electrode. The current ultimately determines the extent of the reaction possible on the basis of Faraday's first law, the potential determines what reactions are possible and the rate at which they take place, and the mass transport determines which materials and in what amounts are available for reaction. Of additional consideration is the working electrode material and the electrochemical characteristics of the particular solvents, electrolytes/additives, and analytes in the system.

3.2.1 Electrical Circuits

The two simplest electrical configurations used for ES ion sources are shown in Figures 3.2a and 3.2b. These can be referred to as the grounded emitter system (Figure 3.2a) and the floated emitter system (Figure 3.2b). The floated emitter system in Figure 3.2b is in fact analogous to the ES circuit shown in Figure 3.1. Examination of these ES configurations from an electrical perspective reveals that they comprise simple series circuits. That is, the current path is in one direction, and the current is of the same magnitude at each point in the circuit. A thorough evaluation of the equivalent circuit pertinent to these source configurations was presented by Jackson and Enke.[48] In these most common manifestations, the ES ion source is a two-electrode system. One of the cell electrodes is a metal capillary or other conductive material (usually stainless steel) in contact with the solution placed at or upstream of the point at which the charged ES droplet plume is generated (i.e., the ES emitter). In this system we define this as the working electrode, because the reactions of

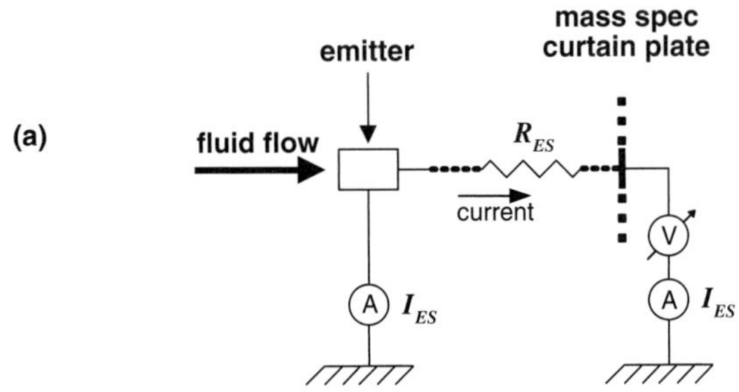

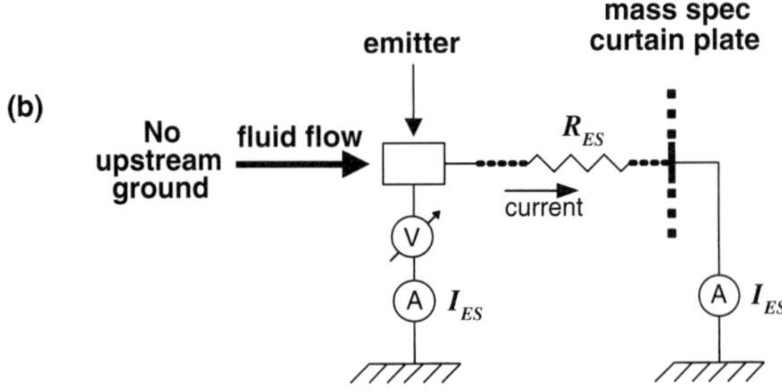

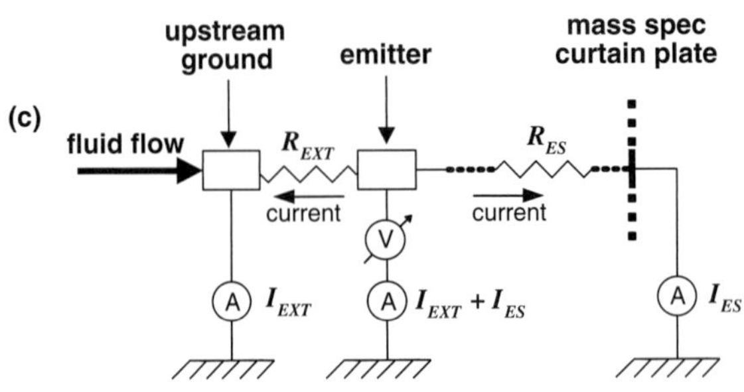

Figure 3.2. Diagrams of the electrical circuit for ES emitter systems of the (**a**) grounded and (**b**) floated emitter type. These geometries contain only the (downstream) ES circuit (with nominal resistance R_{ES} resulting in ES current I_{ES}) between the emitter and the mass spectrometer. (**c**) Electrical circuit of a floated emitter system with the typical commercial configuration incorporating an upstream grounded contact in the solution stream. In addition to the downstream ES circuit, this geometry includes an external upstream current loop (with resistance R_{EXT} resulting in current I_{EXT}) between the upstream grounding point and the emitter electrode. The total current at the emitter electrode (I_{TOT}, equivalent to the faradaic current at the emitter electrode) is the sum of I_{ES} and I_{EXT}.

analytical significance in this system occur at this electrode. The second electrode, or counter electrode, is the atmospheric sampling aperture plate or inlet capillary and the various lens elements and detector of the mass spectrometer. In the grounded emitter design (Figure 3.2a), the high voltage is placed at the counter electrode side of the circuit and the conductive contact to solution is grounded. A high negative voltage is used in positive-ion mode, and a high positive potential is used in negative-ion mode. With the floated emitter design (Figure 3.2b), the high voltage is applied at the conductive contact to the solution and the counter electrode is held near ground. A high positive voltage is used in positive-ion mode, and a high negative potential is used in negative-ion mode. In both situations, oxidation reactions predominate at the emitter electrode in positive-ion mode (reduction predominates at the counter electrode) and reduction reactions predominate at the emitter electrode in negative ion mode (oxidation predominates at the counter electrode). As mentioned before, the charged droplet formation process at the ES emitter can be viewed as a controlled-current source.[41,48] In practice, the magnitude of the ES current is altered most significantly and most readily (within the range of normal ES operational parameters) by altering the conductivity of the solution through the addition of an electrolyte and by changing the voltage drop between the ES emitter capillary and the counter electrode.

Complicating this relatively simple electrical description of the common ES ion source geometry is the fact that in floated emitter systems a grounded contact is often placed upstream of the emitter electrode (Figure 3.2c). This is a safety consideration on commercial instrumentation that protects operators and upstream equipment (e.g., HPLC equipment) from coming in contact with the high voltage. However, when this grounded contact (i.e., another electrode) is used, a second, upstream circuit is added to the system and electrochemical reactions occur at this electrode as well. For example, in positive-ion mode, oxidation reactions dominate at the emitter electrode and reduction reactions dominate at the upstream grounding point (and vice versa for negative-ion mode). Furthermore, the current in this circuit leg (i_{EXT}) can be an order of magnitude (or more) greater than that in the downstream circuit formed by the emitter electrode and the counter electrode of the mass spectrometer.[51,52] This extra current is additive at the emitter electrode, meaning that the total faradaic current, i_{TOT}, is far larger in the floated emitter configuration ($i_{TOT} = i_{ES} + i_{EXT}$) than with the grounded emitter ($i_{TOT} = i_{ES}$). This also means that the effect of the electrochemistry on the ES experiment and mass spectra is potentially far more significant with this source geometry (see below).

As described by Konermann et al.,[51] the magnitude of i_{EXT} depends on the conductance, $G = 1/R_{EXT}$, of the connection between the upstream ground point and the ES emitter electrode and the voltage applied to the ES capillary (i.e., $i_{EXT} = V_{ES}/R_{EXT}$). R_{EXT}, in turn, depends on the solution conductivity, $\lambda^0_m C_E$ (λ^0_m is the limiting molar conductivity of electrolyte, C_E is the concentration of electrolyte), and the length, L, and cross-sectional area, A, of the tubing connection between the electrodes (i.e., $i_{EXT} = V_{ES}A\,(\lambda^0_m C_E)/L$, where V_{ES} is the voltage applied to ES capillary emitter). Thus, i_{EXT} will be minimized by the use of long narrow tubes and weakly conductive solutions, whereas i_{EXT} will be maximized through the use of shorter, wider bore tubing and substantially conductive solutions.

3.2.2 Connection Between the Current Magnitude and the Extent of Electrochemical Reactions

The extent of the one or more electrode reactions that occur at the ES emitter electrode, and, in part, the resulting solution compositional change is determined by the magnitude of the faradaic current at the ES emitter electrode in the case of the ES configurations shown in Figures 3.2a and 3.2b. The current measured at the working electrode of an electrochemical cell is due to charging current (changing the interfacial potential of the electrode, that is, charging the electrode as a capacitor; $i_{\text{charging}} = C\frac{\Delta E_{\text{E/S}}}{\Delta t}$, where C and $E_{\text{E/S}}$ are, respectively, the capacity and the interfacial electrode potential and t is the time) and the faradaic current. In the stable cone jet mode of operation of an ES ion source, with a fixed solvent system composition, one might expect a constant emitter electrode potential ($i_{\text{charging}} = 0$) with all current measured at the emitter electrode being faradaic ($i_{\text{TOT}} = i_{\text{F}}$). This interpretation is supported by experiments in which the amount of electrochemical product measured correlates well with the measured value of i_{ES}.[26,53]

The concentration of excess charge in the charged ES droplets, [Q], in equivalents per liter, can be calculated from Faraday's current law, expressed as

$$[Q] = i_{\text{TOT}}/Fv_f = \sum_j n_j[EP]_j \tag{3.1}$$

where i_{TOT} is the faradaic current at the ES emitter, F is the Faraday constant ($9.648 \times 10^4\,\text{C mol}^{-1}$), and v_f is the volumetric solution flow rate through the ES capillary electrode. One can also derive from Eq. (3.1) the concentration of the individual species created, depleted, or altered electrochemically, where j is the number of different electrolysis reactions that occur, n_j is the molar equivalent of electrons involved in the production of 1 mol of electrolysis product in reaction j, and $[\text{EP}]_j$ is the concentration of electrolysis product j. This relationship is shown graphically by the solid diagonal lines in Figure 3.3 for a range of typical ES current values and flow rates used in ES-MS, assuming that only one redox reaction takes place ($j = 1$). The plot from this equation illustrates that the concentration of electrolysis products will be minimized at high flow rate, low ES

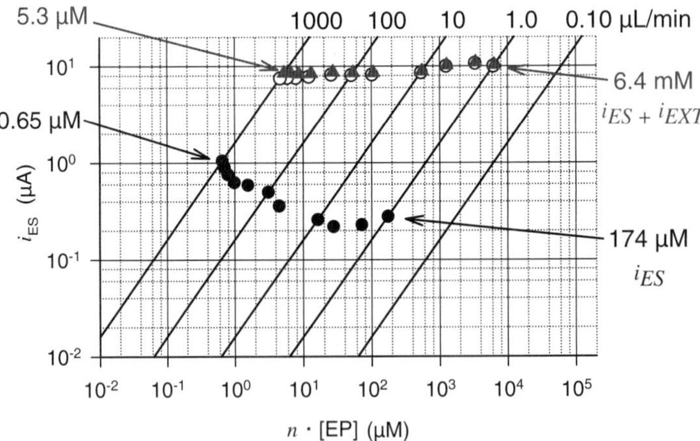

Figure 3.3. Theoretical solid line plots of the concentration of the electrochemical product, [EP], added to or removed from the solution sprayed via the electrochemical processes in the electrospray emitter as a function of flow rate through the emitter. Plots were calculated using Eq. (3.1), assuming that only one electrochemical reaction, j, occurred in which $n = 1$. Actual experimental currents measured at flow rates from 1.0 µL/min to 1000 mL/min using a pneumatically assisted floated ES source with and without an upstream ground are also plotted: black circles represent i_{ES}, white circles represent i_{EXT}, triangles represent $i_{\text{ES}} + i_{\text{EXT}}$. Solution composition was acetonitrile/water (1/1 v/v), 5 mM ammonium acetate, 0.75% by volume acetic acid (pH 4). Voltage drop between the ES emitter electrode and counter electrode of mass spectrometer is 4 kV.

current, and large n. The concentration of any one product may be decreased even further if more than one reaction takes place ($j > 1$). The product concentration reaches its maximum when only one reaction takes place ($j = 1$), the flow rate is low, and the available ES current is high with $n = 1$.

Actual measurements of i_{ES} and i_{EXT} were made in our lab with a typical ES solvent system at flow rates from 1.0 μL/min to 1.0 mL/min using a floated emitter system with and without an upstream ground point. These data are also plotted in Figure 3.3. On the basis of the measured i_{ES} alone, assuming $j = 1$ and $n = 1$, the maximum possible concentration of product created, depleted, or altered would range from about 0.65 μM at a flow rate of 1.0 mL/min up to 174 μM at a flow rate of rate of 1.0 μL/min. When i_{EXT} is factored in, the concentration of product ranges from about 5.3 μM to 6.4 mM in this same flow rate range. The significance of these high concentrations is apparent when one notes that only rarely would an actual upper limit for analyte concentration be more than 10–100 μM for an ES-MS experiment. Thus, it is possible that the concentration of a product created, depleted, or altered electrochemically will rival or at times far exceed the analyte concentration, or even the concentration of buffers added to solution.

In any case, the greatest impact of electrochemistry at the ES emitter electrode on solution composition, all other factors being constant, occurs in experiments using the lowest solvent flow rates. The impact is greater with a floated emitter system incorporating an upstream grounded emitter because of the increased faradaic current at the emitter electrode. For the typical ES solvent system we tested, the current at the emitter electrode with an upstream ground was about an order of magnitude greater than when there was no upstream ground. For this reason, the grounded emitter system and floated emitter with upstream ground are not equivalent electrochemically.

3.2.3 Interfacial Electrode Potential, Possible Electrochemical Reactions, and Influence of Electrode Material

The interfacial potential at the working electrode ultimately determines what electrochemical reactions in the system are possible as well as the rates at which they may occur. However, given the controlled-current electrochemical behavior of the ES ion source, the interfacial potential is not fixed during an experiment, but rather adjusts to a given level depending upon a number of interactive variables to provide the required current. The variables that are expected to affect the interfacial electrode potential are the current density at the electrode, the redox character and concentrations of all the species in the system, the solution flow rate, the electrode material and geometry, and any other parameters that affect the flux of reactive species to the electrode surface. Thus, knowing or even estimating the interfacial potential, and therefore knowing what electrochemical reactions can take place under a given set of ES conditions, is not a simple proposition.

These general operational characteristics are made more apparent by reference to the hypothetical i_{TOT} versus $E_{E/S}$ plots in Figure 3.4. The solid curve on the right represents a situation for positive ion mode ES in which three oxidizable species, A, B, and C, are in solution at equal concentration and each has the same diffusion coefficient. The standard redox potentials for oxidization of A, B, and C are E^0_A, E^0_B, and E^0_C, respectively. This illustration shows that as the magnitude of i_{TOT} increases, the value of $E_{E/S}$ increases so that a sufficient amount of these species available for reaction can be oxidized to supply the required current. The electroactive species react in order of increasing electrode potential (i.e., the easiest to oxidize species starts to be oxidized first, and so on). Thus, the total

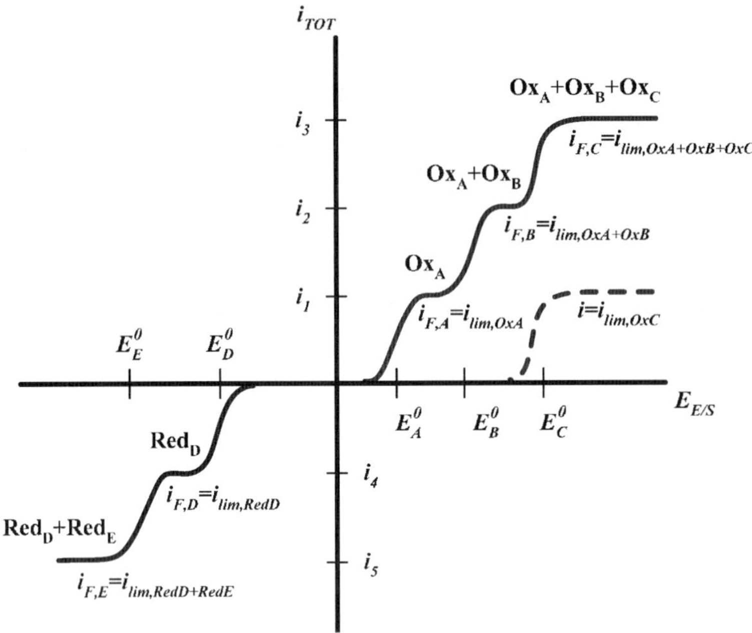

Figure 3.4. Schematic illustration showing the expected interdependence of the potential at the electrode–solution interface, $E_{E/S}$, in the ES capillary as a function of the faradaic current, i_F, and as a function of the composition of the electroactive species in the solution. The solid curve on the right represents a situation for positive-ion mode ES in which three oxidizable species, A, B, and C, are in solution at equal concentration and each has the same diffusion coefficient. The standard redox potentials for oxidization of A, B, and C are E^0_A, E^0_B, and E^0_C, respectively. The total faradaic current, i_F ($i_F = i_{TOT}$), is equal to the sum of the individual faradaic currents ($i_F = i_{F,A} + i_{F,B} + i_{F,C}$) resulting from oxidation of the respective individual species A, B, and C. The dashed curve represents a situation with only species C in solution. The solid curve on the left represents the situation in negative-ion mode in which two reducible species with the same diffusion coefficients, D and E (with E^0_D and E^0_E, respectively), are in solution at equal concentration.

faradaic current, i_F ($i_F = i_{TOT}$), is equal to the sum of the individual faradaic currents ($i_F = i_{F,A} + i_{F,B} + i_{F,C}$) resulting from oxidation of the respective individual species A, B, and C. This diagram also demonstrates that changing the composition of electroactive species in solution can alter the magnitude of $E_{E/S}$ for a given i_{TOT}. For example, with species A, B, and C present in the solution, and with an i_{TOT} corresponding to $i_{F,A}$, only A is oxidized and $E_{E/S}$ is around E^0_A. However, if only species C is present in the solution (or is the only species available for reaction), $E_{E/S}$ will be E^0_C since no species other than C can be more easily oxidized to maintain the required current (see the dashed curve in Figure 3.4). Analogous behavior is expected in negative-ion mode as illustrated in this same figure by the solid curve on the left representing a situation in which two reducible species, D and E, with similar diffusion coefficients are in solution at equal concentrations. The standard redox potentials for reduction of D and E are E^0_D and E^0_E, respectively.

In general then, one would expect that the occurrence of redox reactions of the solvent or any other species sufficient to supply all the required current will effectively maintain or "redox buffer"[45,54] the interfacial potential of the emitter electrode at or near the E^0 values for those particular reactions. The interfacial potential in the emitter is expected to be only that magnitude necessary to supply the required current for a given availability of material for reaction (see Section 3.2.4). Redox buffering exploiting emitter electrode corrosion or making use of discrete chemical redox buffers (e.g., hydroquinone) has been used effectively in ES-MS to control $E_{E/S}$.[55,56]

Computational simulations[57] have been used to estimate the interfacial potentials, and special ES emitter systems[58,59] have recently been used to actually measure the potential within a capillary emitter electrode in a simple floated or grounded emitter system. In general, the simulations and measurements are in agreement and show that the interfacial potential is highest at the spray tip of the capillary ($\leq$2.5 V vs. SHE) and decreases rapidly upstream into the capillary. The magnitude of the potential reached at all points along the electrode increases with increasing ES current and conductivity of the solution. That is, the interfacial potential is not uniform along the length of the electrode (nor is the current density). This nonuniform spatial distribution of the potential relates to the effective length (or area) of the emitter electrode, which is less than the actual physical length (or area) (see the Section 3.2.4).

Off-line chronopotentiometric measurements[60] have been used to estimate the interfacial potential under typical ES-MS conditions using stainless steel, gold, and platinum working electrodes. Different interfacial potentials were reached with the different electrodes, and these differences were most apparent at low current densities. The interfacial potential differences diminished as current density increased to the point where solvent oxidation/reduction supplied the majority of the current. There were also differences in the potential response for a given electrode as the pH of the solution was altered, but again these differences diminished as current density increased. For example, with the application of the low anodic current densities (equivalent to the current densities for tubular emitter electrode lengths of 1.0 cm and 1.0 mm), the interfacial potentials at the stainless steel electrode had negative values that ranged between −0.2 and −0.6 V due to corrosion of the stainless steel. When applying an intermediate current density of 0.32 mA/cm^2 (corresponding to the current density for an emitter electrode length of 100 μm), electrode corrosion alone could no longer support the current demand, and the electrode potential increased to a more positive value to oxidize other species in the system at a sufficient rate.

In lieu of modeling data or experimental measurements, one might simply postulate the interfacial potential of the capillary emitter electrode for a particular ES experiment by recourse to tabulated equilibrium potentials. Table 3.1 lists many of the reactions and the corresponding standard potentials for the components one finds in a typical ES solvent system including H_2O, CH_3OH, HCOOH, CH_3COOH, NH_4OH, or CH_3COONH_4 and for the most widely used capillary emitter material stainless steel and for platinum.[61] For example, one might anticipate that the magnitude of the interfacial electrode potential will be significantly limited in the cathodic direction if CH_3OH is the solvent rather than H_2O, because CH_3OH ($CH_3OH + 2H^+ + 2e^- = CH_4 + H_2O$, Table 3.1) appears much easier to reduce than H_2O ($2H_2O + 2e^- = H_2 + 2OH^-$, Table 3.1). However, this may or may not be the case. The standard potentials in Table 3.1 are the result of thermodynamic calculations; and the exact potentials for a significant amount of CH_3OH or H_2O reduction, or the potential necessary for any of the other reactions in Table 3.1 to actually occur, are influenced by the nature of the electrode (different reaction rates/exchange current densities on different materials), by the solvent system, and by solution additives (e.g., pH and electrolytes).

To understand the danger in proposing which reactions takes place to provide the required ES current on the basis of the standard electrode potentials alone, one must consider that several other factors influence the actual rate of a specific reaction. Assuming that mass transport is not the limiting factor, other processes such as the electron transfer rate at the electrode surface, chemical reactions preceding the electron transfer, or chemical

Table 3.1. Standard Potentials for Some of the Reactions Anticipated to Occur at a Stainless Steel or Platinum Capillary ES Emitter Electrode when Using Typical ES-MS Solvent Systems and Solution Additives

Oxidation (Positive-Ion Mode)	E^o (V) vs. SHE	Reduction (Negative-Ion Mode)	E^o (V) vs. SHE
Solvent System Reactions		*Solvent System Reactions*	
$HCOOH = CO_2 + 2H^+ + 2e^-$	−0.20	$O_2 + 4H^+ + 4e^- = 2H_2O$	1.23
$H_2O_2 + 2OH^- = O_2 + 2H_2O + 2e^-$	−0.13	$H_2O_2 + 2e^- = 2OH^-$	0.88
$CH_3OH + H_2O = HCOOH + 4H^+ + 4e^-$	0.10	$O_2 + 2H^+ + 2e^- = H_2O_2$	0.70
$2NH_3 + 2OH^- = N_2H_4 + 2H_2O + 2e^-$	0.10	$CH_3OH + 2H^+ + 2e^- = CH_4 + H_2O$	0.58
$OH^{*} + 2OH^- = HO_2^- + H_2O + e^-$	0.18	$2H_2O + O_2 + 4e^- = 4\ OH^-$	0.40
$HO_2^- + OH^- = O_2^- + H_2O + e^-$	0.20	$O_2^- + H_2O + e^- = HO_2^- + OH^-$	0.20
$CH_3OH = HCHO + 2H^+ + 2e^-$	0.23	$HO_2^- + H_2O + e^- = OH^{*} + 2OH^-$	0.18
$2HCOOH = H_2C_2O_4 + 2H^+ + 2e^-$	0.25	$HCOOH + 4H^+ + 4e^- = CH_3OH + H_2O$	0.10
$4OH^- = 2H_2O + O_2 + 4e^-$	0.40	$2H_2O + 2e^- = H_2 + 2OH^-$	0.07
$NH_3 + 2OH^- = NH_2OH + H_2O + 2e^-$	0.42	$NO_3^- + H_2O + 2e^- = NO_2^- + 2OH^-$	0.01
$ClO^- + 2OH^- = ClO_2^- + H_2O + 2e^-$	0.68	$2H^+ + 2e^- = H_2$	0.00
$H_2O_2 = O_2 + 2H^+ + 2e^-$	0.70	$O_2 + 2H_2O + 2e^- = H_2O_2 + 2OH^-$	−0.13
$2OH^- = H_2O_2 + 2e^-$	0.88	$CH_3COOH + 2H^+ + 2e^- = CH_3CHO + H_2O$	−0.13
$Cl^- + 2OH^- = ClO^- + H_2O + 2e^-$	0.89	$CH_3OH + H_2O + 2e^- = CH_4 + 2OH^-$	−0.25
$2H_2O = O_2 + 4H^+ + 4e^-$	1.23	$O_2 + e^- = O_2^-$	−0.33
$2NH_4^+ = N_2H_5^+ + 3H^+ + 2e^-$	1.28	$O_2 + H_2O + 2e^- = HO_2^- + OH^-$	−0.83
$NH_4^+ + H_2O = NH_3OH^+ + 2H^+ + 2e^-$	1.35		
$2Cl^- = Cl_2 + 2e^-$	1.36		
$H_2O + Cl^- = HClO + H^+ + 2e^-$	1.48		
$2H_2O = H_2O_2 + 2H^+ + 2e^-$	1.77		
$OH^- = OH^{*} + e^-$	1.89		
$H_2O = OH^{*} + H^+ + e^-$	2.72		
Emitter Electrode Reactions		*Emitter Electrode Reactions*	
$Fe + 2OH^- = Fe(OH)_2 + 2e^-$	−0.87	$Pt(OH)_2 + 2e^- = Pt + 2\ OH^-$	0.16
$Fe + 3OH^- = HFeO_2^- + H_2O + 2e^-$	−0.80	$Fe(OH)_3 + e^- = Fe(OH)_2 + OH^-$	−0.56
$Fe = Fe^{2+} + 2e^-$	−0.44	$Fe(OH)_2 + 2e^- = Fe + 2OH^-$	−0.87
$Fe = Fe^{3+} + 3e^-$	−0.03		
$Pt + 2OH^- = Pt(OH)_2 + 2e^-$	0.16		
$Pt + 2H_2O = Pt(OH)_2 + 2H^+ + 2e^-$	0.98		

reactions following the electron transfer [e.g., protonation, catalytic decomposition, or other reactions (e.g., adsorption/desorption)] influence the reaction rate observed at a given potential, at a particular electrode, in a particular solvent system.[45] Simple reactions involve only heterogeneous electron transfer of nonadsorbed species. More complicated reactions involve a series of electron transfers, protonations, electrode modifications, and so on. The rate of a particular reaction (the current for that reaction) at a given potential is ultimately limited then by the slowest of these different processes not simply by the E^0 value for the reaction.

To further illustrate the points above, hypothetical current density–potential curves for reactions with different reaction rates (exchange current densities) and/or standard electrode potentials were calculated and plotted (Figure 3.5). These curves were calculated based on the Erdey–Gruz–Volmer (Butler–Volmer) equation [Eq. (3.2)] describing the relation of current density and potential when the rate of reaction (current) is controlled solely by the rate of the electrochemical charge transfer process (activation kinetics) and the effect of mass transfer limitation on the current is not considered:

$$j = j_0(\exp(\alpha_a nF(RT)^{-1}\eta) - \exp(-\alpha_c nF(RT)^{-1}\eta)) \tag{3.2}$$

where j is the current density (current divided by the electrode area, i.e., normalized reaction rate), j_0 is the exchange current density, α_a and α_c are respectively the anodic and cathodic electron transfer coefficients, n is the number of electrons transferred between the reduced and oxidized forms, F is the Faraday constant, R is the universal gas constant (8.314 J K^{-1} mol^{-1}), T is the thermodynamic temperature, and $\eta = E - E_0$ is the overpotential—that is, the difference between the applied potential and the standard potential of the electrode reaction.

Figure 3.5a shows the effect of the exchange current density (charge transfer rate) on the current density–potential curve ($\alpha_c = \alpha_a = 0.5$). The interfacial electrode potential must be increased if the exchange current density is decreased to reach the same current density at the electrode (e.g., 5 $\mu A\,cm^{-2}$, dashed horizontal line in Figure 3.5a). This means that species with different reaction rates that happen to have the same E^0 value will provide the same current magnitude only at different interfacial potentials. Curves of the same type are observed for an individual reaction that has different exchange current rates on different electrode materials. For example, hydrogen generation on platinum is virtually overpotential-free (the reaction rate is very high at low overpotentials due to chemisorption of hydrogen prior to the electron transfer reaction), whereas the same reaction requires a very high overpotential on mercury electrodes.[62] Furthermore, if there are competing reactions that have different standard electrode potentials and exchange current densities, conditions may be such that the reaction with the higher E^0 value may be the dominant reaction. This phenomenon is illustrated in Figure 3.5b, where reaction 1 ($j_0 = 1$ $\mu A\,cm^{-2}$, $E^0 = 0.1$ V) is the dominant oxidation reaction instead of reaction 2 ($j_0 = 0.01$ $\mu A\,cm^{-2}$, $E^0 = 0$ V), which has lower standard potential. Furthermore, the actual products of a reaction can be different, depending on the electrode material and the electrolyte.[63]

It is also important to consider that the emitter electrode material may participate in electrochemical reactions (e.g., anodic corrosion of iron in a stainless steel emitter). Table 3.2 lists the standard electrode potentials for some of the possible oxidation processes of metals used as ES emitters. The potential at which oxidation of a metal may take place strongly depends on the solution composition and follow-up reactions (e.g., precipitation, complexation, etc.) as well. As an example, the standard electrode potential for the reversible Ag/Ag^+ couple is 0.80 V vs. SHE. However, at pH 14, Ag_2O forms during the

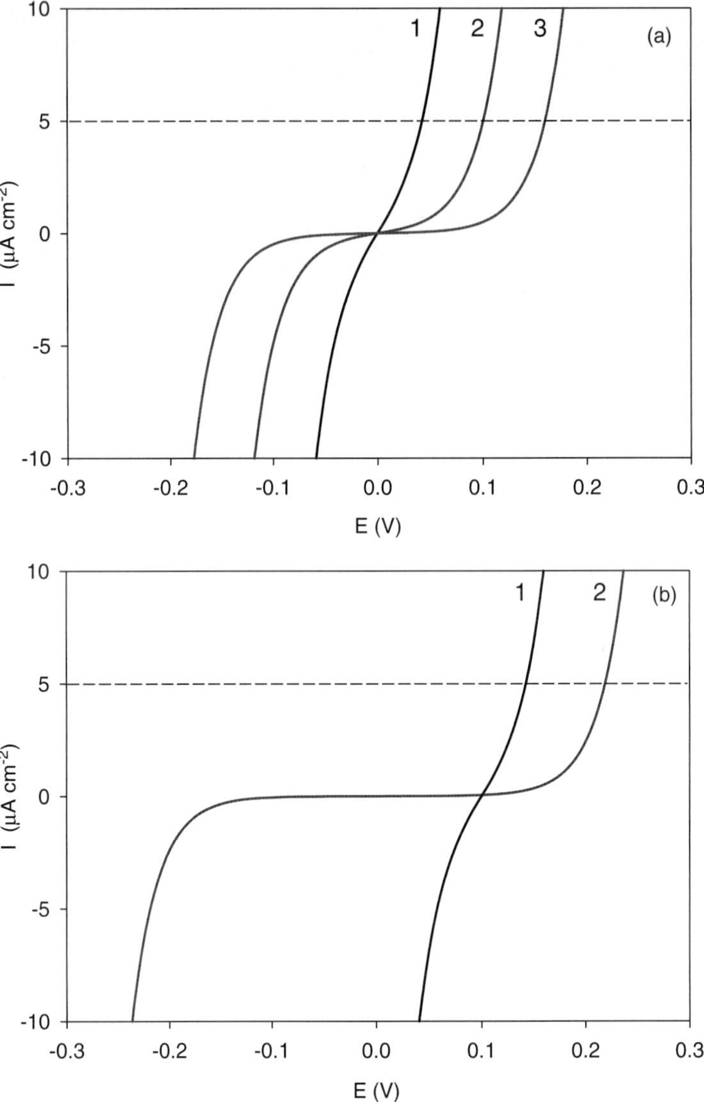

Figure 3.5. (**a**) Polarization curves of reversible systems with standard potential of $E^0 = 0$ V and exchange current densities of $j_0 = 1\ \mu A\ cm^{-2}$ (1), $j_0 = 0.1\ \mu A\ cm^{-2}$ (2) and $j_0 = 0.01\ \mu A\ cm^{-2}$ (3). (**b**). Polarization curves of reversible systems with $j_0 = 1\ \mu A\ cm^{-2}$ and $E^0 = 0.1$ V (1), and $j_0 = 0.01\ \mu A\ cm^{-2}$ and $E^0 = 0$ V (2). The horizontal dashed line indicates the potential where the current densities equals $j = 5\ \mu A\ cm^{-2}$.

oxidation of silver and the standard electrode potential of the reversible Ag/Ag_2O couple is 0.34 V vs. SHE, which is 0.46 V less than that of the Ag/Ag^+ couple. The standard potentials when ammonia or carbonate is present are 0.37 V and 0.47 V for the $Ag/Ag(NH_3)_2^+$ and Ag/Ag_2CO_3 redox couples, respectively. Thus, when considering possible reactions of the emitter electrode material, the solution composition is a very important factor.

Table 3.2. Standard Potentials for Some of the Possible Reactions Involving the ES Emitter Electrode Material

Oxidation	E^0 (V) vs. SHE	Oxidation	E^0 (V) vs. SHE
Zinc		***Silver***	
$Zn + 4OH^- = Zn(OH)_4^{2-} + 2e^-$	−1.29	$2Ag + 2OH^- = Ag_2O + H_2O + 2e^-$	0.34
$Zn + 2OH^- = Zn(OH)_2 + 2e^-$	−1.25	$Ag + 2NH_3 = Ag(NH_3)_2^+ + e^-$	0.37
$Zn + 4NH_3 = Zn(NH_3)_4^{2+} + 2e^-$	−1.04	$2Ag + CO_3^{2-} = Ag_2CO_3 + 2e^-$	0.47
$Zn = Zn^{2+} + 2e^-$	−0.76	$Ag_2O + 2OH^- = 2AgO + H_2O + 2e^-$	0.64
		$2AgO + 2OH^- = Ag_2O_3 + H_2O + 2e^-$	0.74
Iron		$Ag = Ag^+ + e^-$	0.80
$Fe + 2OH^- = Fe(OH)_2 + 2e^-$	−0.87		
$Fe + 3OH^- = HFeO_2^- + H_2O + 2e^-$	−0.80	***Platinum***	
$Fe(OH)_2 + OH^- = Fe(OH)_3 + e^-$	−0.56	$Pt + 2OH^- = Pt(OH)_2 + 2e^-$	0.16
$Fe = Fe^{2+} + 2e^-$	−0.44	$Pt + 2H_2O = Pt(OH)_2 + 2H^+ + 2e^-$	0.98
$Fe = Fe^{3+} + 3e^-$	−0.03		
Copper			
$Cu + 2OH^- = CuO + H_2O + 2e^-$	−0.29		
$Cu_2O + 2OH^- = 2CuO + H_2O + 2e^-$	−0.22		
$Cu + 2NH_3 = Cu(NH_3)_2^+ + e^-$	−0.10		
$Cu = Cu^{2+} + 2e^-$	0.34		

3.2.4 Mass Transport in Tubular Electrodes

Heterogeneous electron transfer chemistry cannot occur unless the material to react is transported to the electrode surface on passage through the emitter capillary electrode

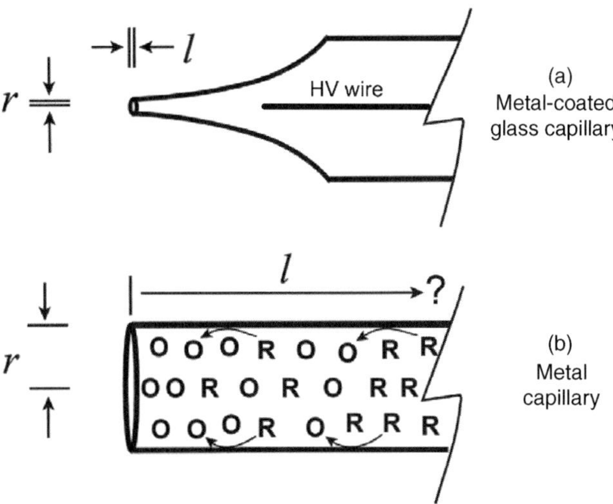

Figure 3.6. Schematic illustration of the effective electrode area in (**a**) a metal-coated, glass nano-ES emitter and (**b**) a metal capillary ES emitter. The necessary length of the electrode, l, sufficient for all of reduced species R to contact the electrode surface on passage through the electrode and react forming oxidized species O (complete electrolysis) is also illustrated in part b. A metal wire in solution to make electric contact is shown as an alternative electrode arrangement to a conductive contact at the tip part (part a). (Adapted with permission from Ref. 60. Copyright 2000, Elsevier.)

(Figure 3.6b, R $\rightarrow$ O). The fraction of a particular species that can contact the inner walls of a tubular electrode and react on passage through the tube can be calculated. Using the equation developed by Chen et al.,[64] the fraction of a particular species that can contact the inner walls of a tubular electrode and react on passage through the tube, f_A, can be estimated (assuming diffusive–convective flux to the inner wall of the tubular electrode under parabolic flow and total concentration polarization conditions). It is interesting to note that the value of f_A is independent of the tube radius. Making the tube more narrow does not improve mass transport because this results in a higher linear velocity through the tube (shorter residence time) that effectively counterbalances the shorter transport distances to the walls.

Van Berkel[60] calculated f_A as a function of tubular electrode length for volumetric flow rates of 0.01, 0.1, 1.0, 10, and 100 μL/min and diffusion coefficients of 5×10^{-6} and 2.4×10^{-5} cm^2/s. The former value of D is typical for an organic analyte in aqueous solution,[45] whereas the latter value is more typical of an organic analyte in a nonaqueous solvent (e.g., ferrocene in acetonitrile).[65] The calculations showed that for a typical ES emitter of 100-μm i.d. operated at 1.0 μL/min (the lowest operational flow rate for a capillary of this diameter), a 1.28-cm-long electrode is required for complete electrolysis given $D = 5 \times 10^{-6}$ cm^2/s, and only a 2.67-mm-long electrode is required for $D = 2.4 \times 10^{-5}$ cm^2/s. Complete electrolysis is defined here as $f_A = 0.99$. If the flow rate is 100 μL/min, electrodes of 128 cm and 26.7 cm in length would be needed for complete electrolysis of analytes with these two different diffusion coefficients, respectively. Thus, the length of the electrode needed for all of a particular species passing through the capillary electrode to contact the electrode surface increases linearly with flow rate.

Metal capillary ES emitter electrodes are normally at least several centimeters in length. Thus, the physical electrode length required for complete electrolysis at the low flow rate end of their operational range is readily available. This conclusion is supported by experimental studies such as that by Van Berkel and Zhou,[66] which showed complete electrolysis of ~10 μM NiOEP using a 6.5-cm-long platinum capillary at a flow rate of 1.5 μL/min in nonaqueous solvents. In contrast, the physical length of the ES emitter electrode required for complete electrolysis at 100 μL/min or more (tens of centimeters) is not usually available on commercial ES ion sources. Experimental studies corroborate this, showing that analyte electrolysis efficiency in tubular ES emitters dramatically decreases at high flow rates.[41,66,67]

An additional consideration beyond the simple physical length of the electrode is the length of the electrode over which electrolysis reactions actually take place. This "effective electrode length" (l in Figure 3.6) might be far less than the physical length, because of limited electric field penetration into the liquid at the spray tip or because of other limits to ionic transport in the solvent. Initial computational simulations of the ES process in a floated emitter (no upstream ground) indicated[57] that the majority of the current in a metal capillary emitter of these dimensions occurred only about 300 μm back into the capillary, or roughly a distance three times the diameter of the capillary. Refinements to that model, along with use of solutions of higher conductivity providing a more realistic ion current ($i_{ES} = 0.5$ versus 0.05 μA at 5 μL/min), predicted that a significant fraction of the total current was generated as far back as 1 cm into the capillary.[68] Spatial measurement of current within a grounded ES emitter capillary also found that a significant fraction of the total current occurred at distances approaching 1 cm into the capillary from the spray tip.[69] Thus, the effective electrode length under the most usual ES conditions will be about 1 cm even if the physical length of the capillary electrode is much longer.

In addition to a conventional tube, one might also consider the very short "tubular" electrode presented by a metal-coated, pulled-glass nano-ES glass capillary (e.g., 10-μm-i.d. tip)

(Figure 3.6a). Operated at its normal lowest flow of 10 nL/min, calculations reveal that an electrode length of 128 μm and 26.7 μm would be required for $f_A = 0.99$ given a diffusion coefficient of $D = 5 \times 10^{-6}$ cm^2/s and $D = 2.4 \times 10^{-5}$ cm^2/s, respectively. In a typical nano-ES capillary of this type, the maximum physical length of the electrode at the spray tip is no more than a few micrometers. Thus, precluding phenomena such as through-glass conduction and turbulent flow or eddy currents in the Taylor cone[70] that may affect mass transport to the electrode, one predicts that mass transport in a metal-coated nano-ES emitter (or other extremely short tubular electrode configuration) will be much less efficient (particularly in the aqueous solvent) than in a typical metal capillary ES emitter at the appropriately low flow rate for each.

3.3 ALTERNATIVE EMITTER ELECTRODE CONFIGURATIONS

As illustrated in the previous sections, the interfacial potential at the tubular working electrode, the magnitude of the ES current, the nature of the electrode surface, and mass transport to the electrode are all important parameters in determining which reactions can occur at the emitter electrode, their rates, and their extents. In this section, we describe alternatives to the most common tubular electrode emitter configurations and their utilities in various ES-MS experiments.

3.3.1 Comparison of Tubular, Planar Flow-By, and Porous Flow-Through Emitter Electrodes

The typical tubular electrode configuration of ES emitter systems presents limited means to alter the electrode material and mass transport to the electrode. In tubular electrode emitter designs, the emitter electrode and emitter spray tip are one. The inner and outer dimensions of the tube—and therefore, electrode geometry, area, and mass transport distances—are restricted to a small range that provides a stable spray. In addition, a limited number of conductive materials are commercially available in tubular form of the appropriate dimensions.

A common working electrode design in electrochemical flow cells is the planar flow-by electrode. This design consists of an electrode embedded smoothly in one wall of a rectangular duct through which electrolyte flows.[71] A schematic diagram is shown in Figure 3.7. With such a design, the rate of mass transport is controllable over a very wide range by varying the volumetric flow rate or by altering the channel height or width, and in particular the electrode length, from microns to millimeters.

Thin-channel, planar flow-by electrode ES emitter devices have been described in the literature.[72–74] The system developed by Van Berkel et al.[73] made use of a separate planar electrode upstream of a nonconductive tubular emitter spray tip (Figure 3.8). This design allowed electrode area, electrode material, electrode covering (such as a membrane

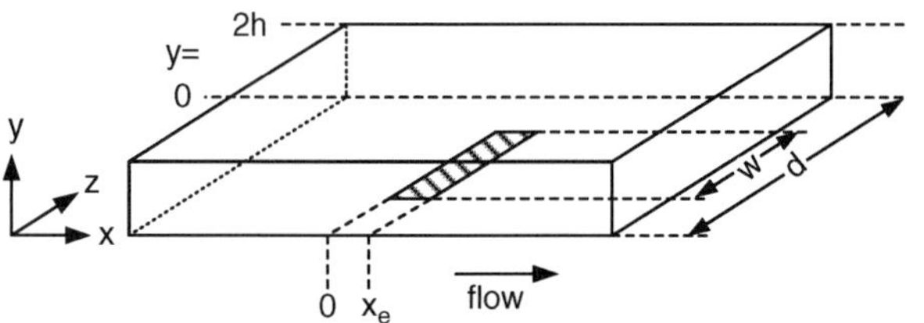

Figure 3.7. A typical channel flow cell, with the conventional *x*, *y*, and *z* directions marked. (Adapted with permission from Ref. [71]. Copyright 1998, Wiley-VCH Verlag GmbH.)

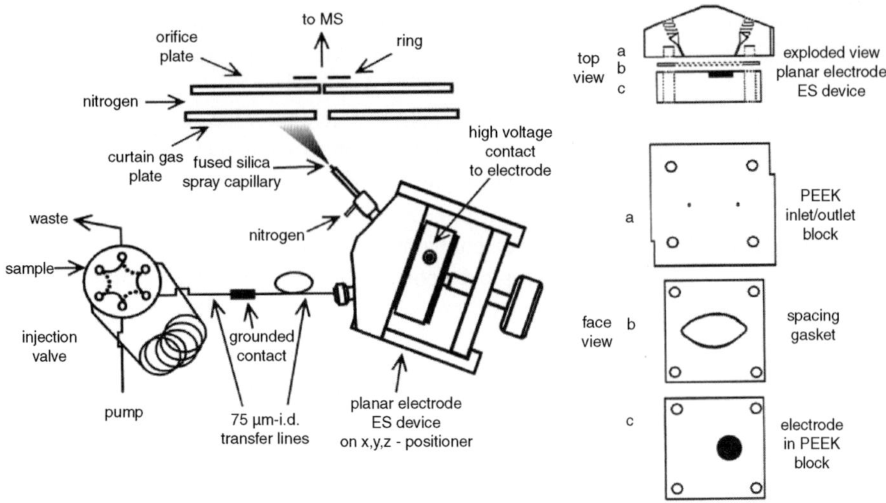

Figure 3.8. Schematic illustration of the planar electrode ES emitter setup. Exploded view is drawn to scale. (Reprinted with permission from Ref. 73. Copyright 2002, American Chemical Society.)

to inhibit mass transport to the electrode), and flow path channel height (or width) to be rapidly and conveniently changed. Thus, this design provided numerous means to control mass transport to the electrode surface while more fully exploiting the known role of electrode material in influencing electrochemical reactions, without negatively impacting spray stability. For nonelectroactive analytes, the analytical performance of this ES ion source was very similar to traditional ES ion sources that use tubular electrodes. For electroactive analytes, this device provided the means to easily alter the extent of analyte oxidation from an insignificant fraction to near 100% of the total amount of material flowing through the system at flow rates below 30 μL/min. Low analyte electrolysis efficiency was achieved at higher flow rates using relatively thick gaskets, small area electrodes, and/or using electrodes made from metals with low oxidation potentials (redox buffering with zinc or copper). High analyte oxidation rates were achieved by lowering the flow rate, using thin gaskets, using large area electrodes, and using electrodes made from materials with high oxidation potentials (platinum, glassy carbon).

When high electrochemical conversion efficiency of dilute solution species is required (e.g., electrochemical detection), a working electrode design common in coulometric flow cells is a porous flow-through electrode.[75] Recently, this type of emitter electrode was implemented into an ES emitter system (Figure 3.9).[52,76] This electrode design, because of the very small pore size, provided for efficient mass transport to the electrode surface even at flow rates of several hundred microliters per minute.

The difference in mass transport among the tubular, planar flow-by, and porous flow-through electrodes can be visualized by comparing the efficiency of electrolysis with each at different flow rates. The electrolysis efficiency (Eff) for each system was calculated by dividing the limiting current, i_{LIM} (maximum current possible based on the mass transport to the electrode from literature equations)[77,78] by the current equivalent of the analyte actually delivered by/through the electrode at every flow rate ($I_{\mathrm{MAX}} = nFCv_f$).

For tubular electrodes, Eff was calculated from Eq. (3.3):

$$\mathrm{Eff},\ \%_{\text{tubular electrode}} = 5.43 \times 10^4 D^{2/3} X^{2/3} v_f^{-2/3} \tag{3.3}$$

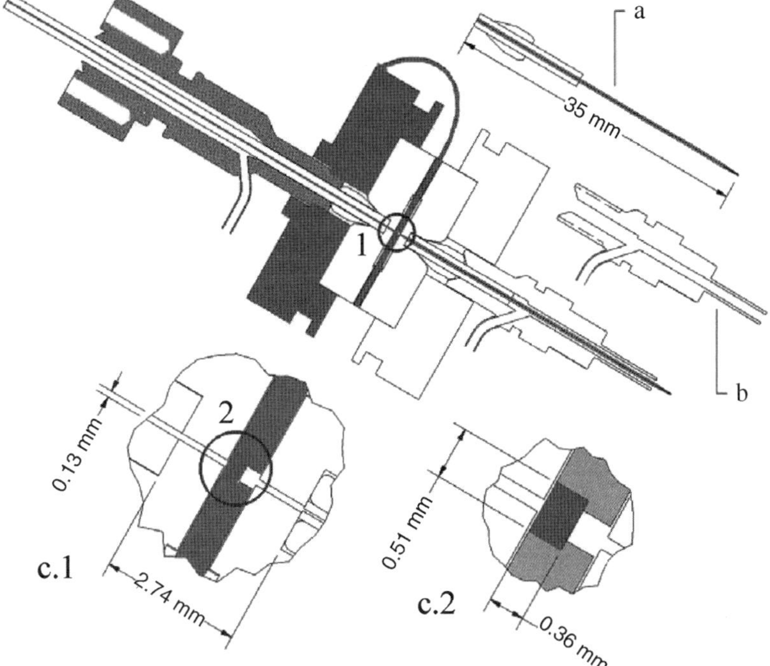

Figure 3.9. Illustrations of a porous flow through electrode emitter mounted in TurboIonSpray™ source: (**a**) Fused silica electrospray emitter capillary; (**b**) nebulizer capillary assembly modified with sidearm adapter; (**c.1**) expanded view of dimensioned fluid pathway through PFT electrode assembly; (**c.2**) expanded view of part c.1 showing PFT electrode details.

where D is the diffusion coefficient of the electroactive species in cm^2/s, X is the length of the electrode in cm, and v_f is the volumetric flow rate in μL/s. For the planar flow-by electrode, Eff was calculated from Eq. (3.4):

$$\text{Eff, \%}_{\text{thin layer}} = 2 \times 10^5 w D^{2/3} x_e^{2/3} v_f^{-2/3} h^{-2/3} d^{-1/3} \tag{3.4}$$

where w and x_e are, respectively, the width and the length of the electrode in mm, and h and d are the half-height and width of the channel in μm and mm, respectively.

To simplify the equation for calculating the efficiency for the porous flow-through electrode, it was assumed that all pores accessible to solution were linear, that the solvent flow was evenly split between each of the accessible pores, and that these pores were parallel with the direction of solution flow. With these simplifications, this electrode can be modeled as a series of microtubes where the total flow rate evenly distributes among the microtubes. For the porous flow-through emitter, Eff was calculated from Eq. (3.5):

$$\text{Eff, \%}_{\text{PFT}} = 5.43 \times 10^6 D^{2/3} t_e^{2/3} v_f^{-2/3} t_p^{2/3} (d/d_p)^{4/3} \tag{3.5}$$

where d and t_e are, respectively, the diameter and the thickness of the electrode in mm, t_p is the total porosity of the electrode in percentage, and d_p is the average pore diameter in μm.

The calculated electrolysis efficiency among the tubular, planar flow-by, and porous flow-through electrodes is illustrated by the plots in Figure 3.10. These plots represent

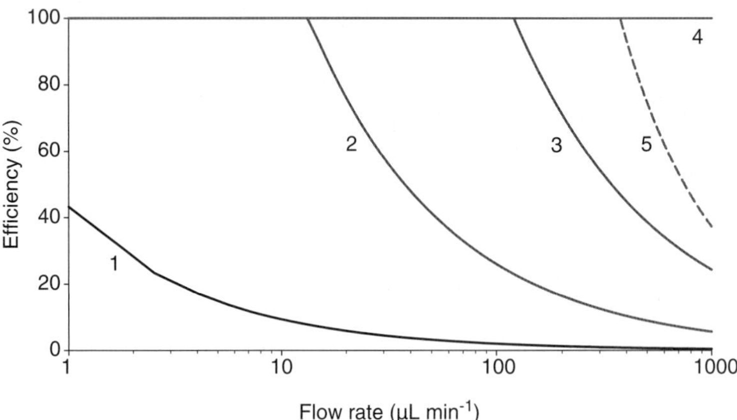

Figure 3.10. Calculated electrolysis efficiency curves for a (1) fused silica emitter with a 0.075-cm-long stainless-steel zero dead volume union; (2) a 3.5-cm-long platinum emitter; (3) a channel flow-by cell with 6-mm width and 6-mm length containing a 6-mm-wide electrode using a 16 μM gasket; (4) a porous flow-through electrode with 0.508-mm diameter, 0.36-mm width, 40% total porosity, and 800 nm average pore radius; and (5) the same cell as in (4), but the current is limited to 6 μA. $n = 2$, $D = 5 \times 10^{-6}\,cm^2\,s^{-1}$, and $c = 5\,\mu M$ were used for creating the curves.

the calculated maximum oxidation efficiency for two tubular emitter electrode lengths (3.5 and 0.075 cm) reported in the literature,[79] the approximate dimensions of the planar electrode emitter system described by Van Berkel et al.,[73] and the published parameters for a porous flow-through electrode emitter.[52,76] For this illustration, a 5 μM concentration of a redox active analyte that undergoes a two-electron transfer was assumed. Plots were made for both (a) the case when there was more than sufficient current for complete electrolysis of all the analyte passing by/through the electrode and (b) the case when the current was limited to 6 μA (approximate current magnitude achieved with floated emitter system employing an upstream ground and relatively conductive solution).[52]

In case of the very short tubular electrode (Figure 3.10, 1), equivalent to where high voltage makes contact with the solution through the bore-through in a metal zero dead volume union,[79] electrolysis efficiency only approaches 50% at flow rates as low 1 μL/min. In contrast, when a 3.5-cm-long tubular electrode is used (Figure 3.10, 2), calculations show that 100% electrolysis efficiency is possible up to flow rates of ~10 μL/min. However, because the effective length of the tubular electrode is probably less than the physical length of the electrode (~1 cm, see above) the actual electrolysis efficiency may be poorer than that plotted. The efficiency curve of the planar flow-by electrode (Figure 3.10, 3) shows that 100% electrolysis efficiency is possible up to ~100 μL/min using a 16-μm gasket. In practice, the observed efficiency was somewhat less[73] because of the current limitation of the actual experimental setup that did not use an upstream ground. With the porous flow-through electrode, electrolysis efficiency was not limited by mass transport at flow rates as high as 1 mL/min (Figure 3.10, 4). Only when the current was limited to that reasonably expected in the ES-MS experiment (6 μA) did the efficiency drop below 100%, and then only at flow rates greater than about 500 μL/min (Figure 3.10, 5). These calculated results are in agreement with the published experimental data.[76]

3.3.2 Miscellaneous Electrode Configurations

A number of other electrode arrangements have been used in ES ion sources. We limit the discussion here to those arrangements that are fundamentally different from those described in the previous section. These include those arrangements that remove the electrode from the flow stream and two different nano-ES arrangements for which the electrochemistry can profoundly affect the experimental results.

One general class of electrode arrangements in the ES emitter is that which removes the electrode from direct contact with the flow stream. This can be done, for example, through the use of membranes[73,80] or by placing the electrode in the split flow to waste in a nano-ES system.[81] With these types of electrode arrangements, the products of the electrode reactions are not produced in the flow stream that ends up producing gas-phase ions. These types of arrangements can be used to avoid the alteration of the analytes under study or to avoid the generation of gas in the flow stream that can lead to signal instability. However, charge balance requires ionic transport to and from the region where the electrode is located into the main flow stream. As a result, the composition of the electrosprayed solution is still altered.

Compared to the continuous flow nano-ES emitters that make use of a conductive tip, the electrochemical processes in two other nano-ES emitter electrode arrangements have more serious analytical implications. We have referred to the first of these nano-ES emitter electrodes designs as the wire-in-a-capillary bulk loaded nanospray emitter (Figure 3.6a).[82] Rather than making electrical contact to solution at the emitter tip, a metal wire at the ES high voltage is immersed in the solution in the emitter capillary. The size and shape of the wire and the solution flow rate will affect the current density along the electrode and the mass transport to the electrode. Each of these factors, in turn, influences the interfacial potential at the electrode and the nature and extent of the various electrochemical reactions that occur. Because the electrical contact to solution is made upstream of the emitter spray tip, there can be a significant delay time between the formation and the gas phase detection of the electrolysis products. Unlike a pumped ES device, the products of the electrochemical reactions are not necessarily swept directly from the capillary out into the spray. The whole volume of the solution within the capillary can be altered because of mass transport of the products formed at the spray tip end of the wire throughout the volume. Analytically, this translates into changes in the mass spectrum during the duration of an experiment. The choice of the wire electrode material also has substantial effect on what reactions occur and on what ions are observed in the gas phase.[82]

The second nano-ES emitter electrode design to consider is that developed by Schultz and co-workers[83,84] (Figure 3.11). The electrode in this system takes the shape of the interior walls of a conductive pipette tip that holds the sample solution to be sprayed. The basic spray device is a microfabricated array of ES nozzles on a planar surface of a silicon wafer or chip. In the system that makes use of the chip, one sample at a time is brought to the chip and a new nozzle and pipette tip is used for each sample, thus eliminating any possibility for cross-contamination among samples. The system might be viewed in simple terms as a robotic continuous-infusion nano-ES system that offers zero carryover. In practice, the device has been shown to be a reproducible, controllable, and robust means of producing a nano-ES of a liquid solution.

From an electrochemical perspective, this is a bulk-loaded tubular electrode system, like the prior system discussed. The large electrode area, limited volume, and very low flow rate provide a situation where a redox active analyte is particularly susceptible to

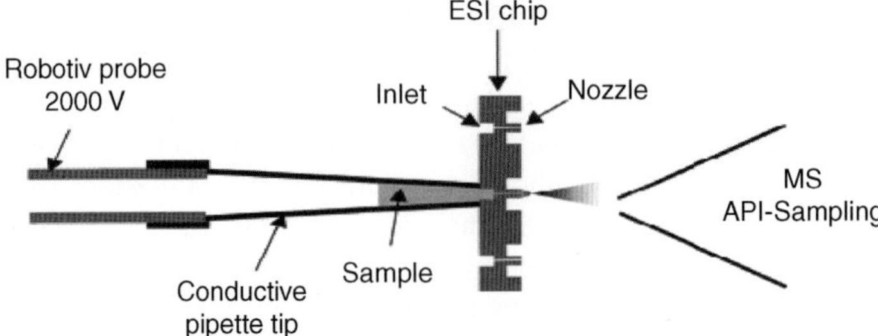

Figure 3.11. Illustration showing the interface between the pipette tip sample delivery system and the ESI chip. A robotic probe delivers sample (up to 10 μL) through a conductive pipette tip, which interfaces directly to the back plane of the ESI chip. Voltage required for nanoelectrospray along with a slight positive pressure (N_2) is delivered to the sample through the robotic probe. The ESI chip was positioned near the atmospheric pressure ionization (API) sampling orifice of a mass spectrometer. (Reprinted with permission from Ref. 84. Copyright 2003, American Chemical Society.)

electrolysis. With simple conductive carbon-impregnated tips as the electrode, analytes such as reserpine were observed to diminish rapidly in intensity during an experiment because of the electrochemical conversion to a species of another mass (Figure 3.12, Trace A).[85] This process was inhibited by the use of a chemical redox buffer to control the interfacial potential below that necessary for the oxidation of reserpine (Figure 3.12, Trace B). However, the use of a chemical redox buffer is not always possible or practical. Redox buffers to maintain the potential required may not be available, and the products of reaction of the buffers may not be innocuous (see Section 3.4.4). The current commercial device eliminates many of the issues with direct electrochemical reactions of analytes by coating the tips with a proprietary layer that allows ionic transport to and from the electrode, but inhibits mass transport of the analytes (Figure 3.12, Trace C). The electrochemical reactions that take place are altered because of the altered mass transport characteristics of the system. This prevents or slows down the direct contact of the analyte with the surface of the electrode, but it does eliminate the electrochemistry.

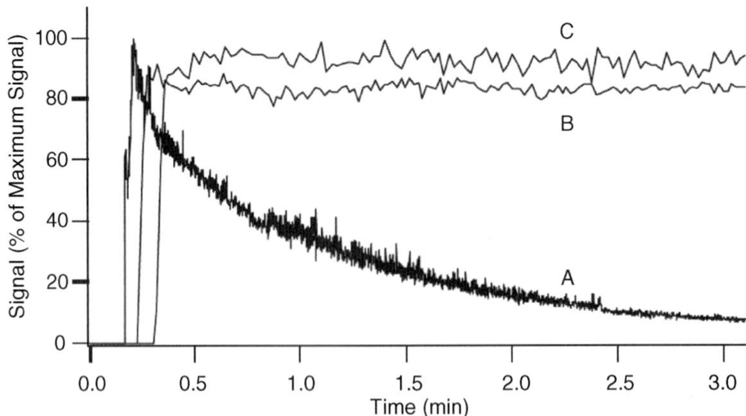

Figure 3.12. Signal from the molecular ion of a 100 pg/μL reserpine solution infused at 200 nL/min. Trace A, carbon-impregnated polypropylene tip; trace B, addition of hydroquinone to A; and trace C, modified tip without hydroquinone. (Reprinted with permission from Ref. 85. Copyright 2005, Elsevier.)

3.3.3 Multiple-Electrode Emitters

Electrospray emitter electrode systems that employ more than one electrode have particular utility to gain control over which reactions take place in the emitter.

3.3.3.1 Three-Electrode, Controlled-Potential Emitter Cells

As emphasized above, the interfacial potential at the working electrode ultimately determines what electrochemical reactions in the system are possible, as well as the rates at which they may occur. The inability to precisely control the potential at the emitter electrode in the typical ES ion source may have undesirable analytical consequences and limits the ability to exploit the electrochemistry for analytical purposes. One means devised to gain control over the electrochemical reactions at the emitter electrode is to incorporate the emitter electrode as the working electrode of a three-electrode, controlled-potential electrochemistry (CPE)–ES emitter cell.[86–89] This intertwining of a three-electrode cell into the ES source produces a circuit in which a potentiostat and the auxiliary, reference, and working electrodes of a three-electrode cell are parallel to the counter electrode of the ES ion source (Figure 3.13).

The combined circuits provide a means to control the emitter (working) electrode potential and, thus, determine which reactions are possible at this electrode. Incorporation of a three-electrode cell into the circuit also overcomes limitations in electrolysis efficiency caused by the relatively low current (<1 μA) supplied by a grounded emitter system or floated emitter system without an upstream ground. In the larger picture, three-electrode emitters provide the opportunity to perform electrochemical experiments on-line with ES-MS without interference from the inherent controlled-current electrochemical processes of the ES ion source. Thus, the three-electrode emitters provide new capabilities such as electrochemical reduction of an analyte in positive-ion mode and electrochemical oxidation of an analyte in negative-ion mode. Cole's group showed evidence, for example, of limited nitrobenzene reduction in both positive- and negative-ion modes using a CPE-ES emitter,

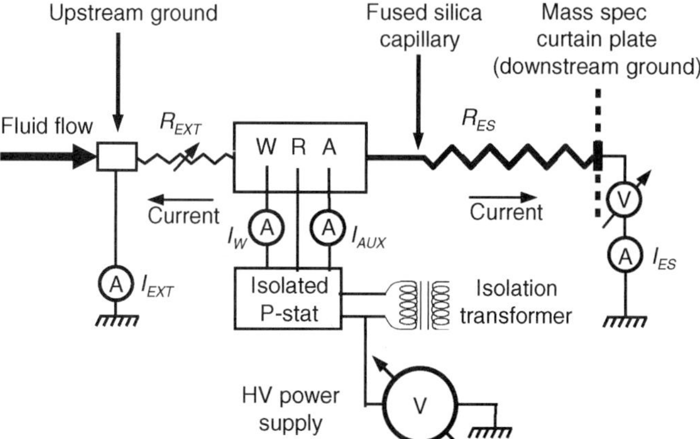

Figure 3.13. Diagram of the electrical circuit for a controlled-potential electrochemistry-ES emitter. Floated potentiostat common is that of the ES high-voltage supply. W, R, and A represent the working, reference, and auxiliary electrodes in the electrochemical cell, respectively. I_W, I_{AUX}, I_{ES}, I_{EXT} are the currents in the working electrode, auxiliary electrode, ES spray current, and upstream external current loop, respectively. In standby mode the electrodes remain connected to the ES high voltage, but the working electrode potential is not controlled with the potentiostat. (Adapted with permission from Ref. 89. Copyright 2005, American Chemical Society.)

$\rightleftharpoons$ $(MeOEP)^{\bullet+}$ + e^- $\rightleftharpoons$ $(MeOEP)^{2+}$ + e^-

$(6)^{\bullet+}$ = *m/z* 590

$(7)^{\bullet+}$ = *m/z* 591

$(6)^{2+}$ = *m/z* 295

$(7)^{2+}$ = *m/z* 295.5

6 (Me = Ni^{2+})

7 (Me = Co^{2+})

Figure 3.14. Nickel (6, Me = Ni(II)) and cobalt (7, Me = Co(II)) octaethylporphyrin structure, oxidation pathway, and ions observed. (Adapted with permission from Ref. 89. Copyright 2005, American Chemical Society.)

but they did not explore the effect of the potential on these processes in detail.[86,87] Our group demonstrated the oxidation of 3,4-dihydroxybenzoic acid in negative-ion mode, reduction of methylene blue in positive-ion mode and selective oxidation of different metalloporphyrins within a mixture.[89] In the latter case, this potential-selective oxidation/ionization enabled independent detection of the two porphyrins that would have otherwise been obscured by molecular ion isotopic overlaps (Figures 3.14–3.16).

One must note that in these three electrode systems, electrochemistry does take place at the auxiliary electrode. To minimize electrochemistry of the analytes or other dilute solution species at this electrode, either the auxiliary electrode must be removed from the flow stream or mass transport to this electrode must be limited by another means (e.g., small surface area). The same is true for the controlled-current system discussed below.

3.3.3.2 Two-Electrode, Controlled-Current Emitter Cells

The previous section described a means to control the emitter (working) electrode potential and, thus, control which reactions are possible at this electrode. This can also be accom-

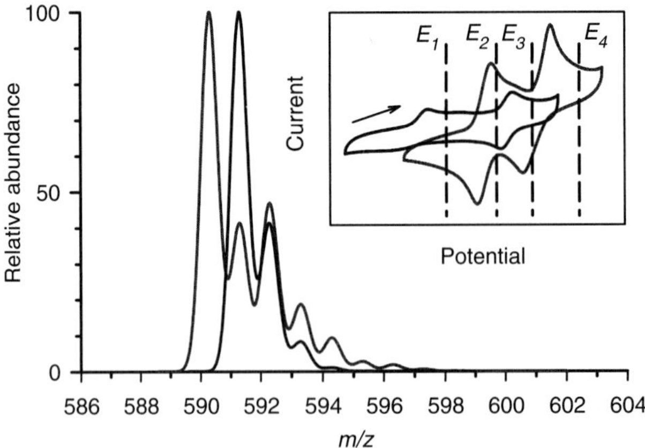

Figure 3.15. Theoretical isotope patterns for the single-charge cations of NiOEP (gray line) and CoOEP (black line). **Inset**: Off-line cyclic voltammograms of 0.5 mM NiOEP (gray line) and 0.5 mM CoOEP (black line) (50/50 (v/v) acetonitrile/methylene chloride containing 50 mM lithium triflate) at a scan rate of 50 mV/s. E_1–E_4 represent the potential conditions under which the data shown in Figure 3.16 were collected. (Adapted with permission from Ref. 89. Copyright 2005, American Chemical Society.) (See clor insert).

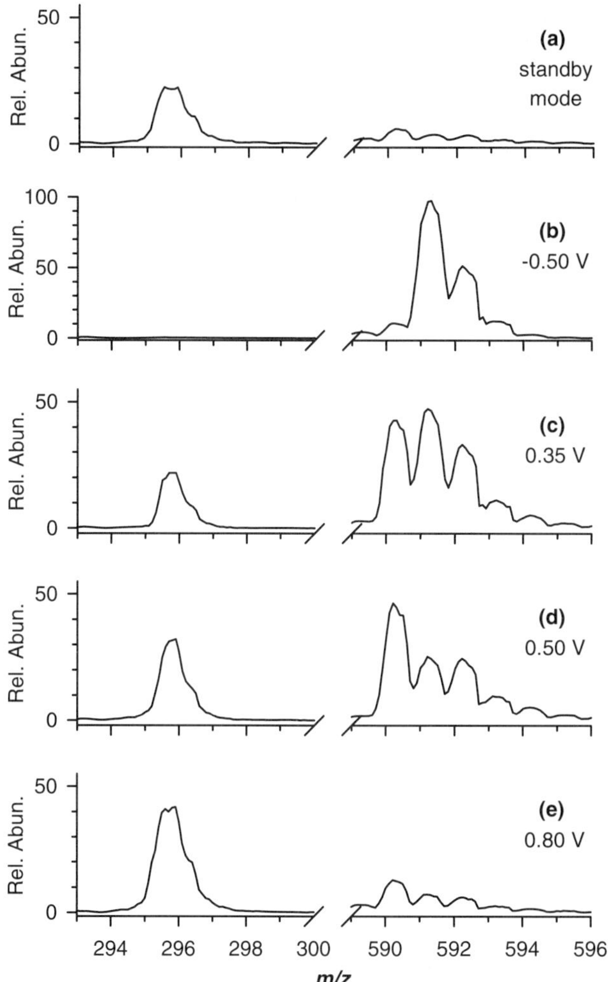

Figure 3.16. Positive-ion mode ES mass spectra showing the singly charged and doubly charged molecular ion regions that were obtained at the plateau region of a 50 μL injection of a mixture of 1.0 μM NiOEP and 10 μM CoOEP in 50/50 (v/v) acetonitrile/methylene chloride containing 1 mM lithium triflate acetonitrile at a flow rate of 50 μL/min under (**a**) open-circuit conditions and working electrode potentials of (**b**) −0.50 V (E_1 in Figure 3.15), (**c**) 0.35 V (E_2 in Figure 3.15), (**d**) 0.50 V (E_3 in Figure 3.15), and (**e**) 0.80 V (E_4 in Figure 3.15). (Adapted with permission from Ref. 89. Copyright 2005, American Chemical Society.)

plished to some degree with a single electrode emitter system by controlling the magnitude of the current at the emitter electrode through adjustable parameters such as solution conductivity, ES voltage drop, and/or upstream ground loop resistance.[41,51,52,66,67,79] Lower current magnitudes either limit the reaction or lower the current density at the electrode so that the potential at the electrode drops to a level lower than that required for the analyte reaction (but still sufficient for another reaction (e.g., solvent electrochemistry) to provide the required current).

Emitter working electrode potential control, intermediate to that of the single-electrode emitter and that of the three-electrode controlled-potential emitter system, can be obtained using two-electrode emitter systems with current control. In this case, one of the two electrodes is used as the working electrode, while the other acts simultaneously as the quasi-reference and the auxiliary electrode. This configuration results in somewhat limited control over the working electrode potential because the potential of the second, quasi-reference/auxiliary electrode is not well-defined. The potential of the latter is a function of many parameters including current density, solution conductivity, and solution composition. That being said, one statement is always true regarding a two-electrode cell:

On either electrode a cathodic current results in reduction and an anodic current results in oxidation.

We demonstrated a floated ES emitter system utilizing two tubular stainless electrodes in the mid-1990s,[90] and more recently Brajter-Toth and co-workers[91] discussed a very similar two-electrode tubular emitter cell. In each case, these simple battery-powered cells were used to either enhance the mass spectral signal level of the molecular ion of analytes that could be ionized (oxidized) by electron transfer or follow the reaction path of analytes that could be oxidized in the emitter cell. More recently, our group described a practical and economical two-electrode emitter cell using a porous flow-through working electrode.[92] The system is very similar to our previously described controlled-potential setup,[88,89] except that the auxiliary and reference electrodes were connected together and used as a sole auxiliary electrode in the controlled-current setup. To adjust the current on the working electrode in one version of the device, the system included a simple circuit built by using just one resistor and a toggle switch; it was powered by a small-size 3-V watch battery with all the components small enough to fit on the body of the emitter cell. By controlling the current magnitude on the large surface area working electrode, this two-electrode emitter system provided the ability to enhance/completely turn off the oxidation/reduction of analytes in positive/negative ion mode, respectively.

3.4 ANALYTICAL IMPLICATIONS AND BENEFITS

In the preceding sections we have discussed in detail the fact that the electrochemical nature of an ES ion source is a fundamental part of its operation as a generator of charged droplets and, ultimately, the gas-phase ions sampled and analyzed by the mass spectrometer in ES-MS. It is important to recognize that under some circumstances, the fact that electrochemistry is ongoing in the ES ion source can be very important analytically. For example, as discussed below, changes in the composition of the solution entering the capillary as a result of these electrochemical processes might be an important consideration in mechanistic interpretations of the ES process. The electrochemical reactions might also result in the generation of chemical noise or an unexpected modification of the analyte. It is also important to recognize that the electrochemical nature of an ES ion source positions one to use ES-MS as a means to perform on-line, electrochemistry–mass spectrometry. This combination may be used for the benefit of ionization and detection of specific analytes by ES-MS or, from an electrochemical perspective, as a means by which to study redox reactions or otherwise utilize the electrochemical reactions analytically.

3.4.1 Alteration of Solution Composition

On the basis of Eq. (3.1), and knowledgeable "guesses" as to the particular redox reactions that take place at the ES emitter electrode, one can examine the compositional change in the solution that takes place for a given current value and solution flow rate. As illustrated above in reference to the data in Figure 3.3, these changes can, under certain circumstances, significantly alter the chemical equilibria in the solution exiting the ES capillary from that in the initial solution. This possibility is especially important to consider when attempting to discern certain mechanistic aspects of ion production in ES and to rationalize the ions ultimately observed in an ES mass spectrum.

A simple and important illustration of solution composition change is the alteration of solution pH by the oxidation of water in positive ion mode ES (Table 3.1, $2H_2O =$

$O_2 + 4H^+ + 4e^-$). For illustration, assume oxidation of water to be the only redox reaction to occur when spraying a weakly basic analyte from an aqueous solution initially at pH 7.0 with no buffering capacity. The occurrence of this reaction is equivalent to adding strong acid to the solution (without the counterion). The concentration of protons produced—that is, $[H^+]_{elec}$—can be calculated from Eq. (3.6):

$$[H^+]_{elec} = i_{TOT}/nFv_f \quad (3.6)$$

where i_{TOT} is the total ES current at the emitter electrode, n is the molar equivalent of electrons involved in the production of one mole of H^+ (in this case $n = 1$), F is the Faraday constant), and v_f is the volumetric solution flow rate. The electrochemically produced protons will cause a pH change that is given by Eq. (3.7):

$$\Delta pH_{max} = -\log([H^+]_{initial} + [H^+]_{elec}) - pH_{initial} \quad (3.7)$$

Equations (3.6) and (3.7) show that the magnitude of the expected pH change is directly related to the magnitude of the current and inversely related to the flow rate. Assuming a constant value for i_{ES} of 0.1 μA, the pH of the solution exiting the capillary for flow rates of 10, 1.0, and 0.1 μL/min would be 5.2, 4.2, and 3.2, respectively. Van Berkel and co-workers[93] showed by optically monitoring the pH exiting the emitter capillary electrode that these pH changing reactions altered the pH of solution by up to 4 pH units under certain conditions. These changes in pH are quite dramatic and would be expected to alter the ES response of a weakly basic analyte through an increase in the degree of protonation in solution. They showed that this pH change could alter the magnitude of gas-phase ion signal from the protein cytochrome *c*. Data presented by Konermann et al.[51] indicated that the pH change caused by the electrochemical oxidation of water in the emitter electrode caused the protein cytochrome *c* to unfold, shifting the observed charge-state distribution to higher protonation states. This change in pH was magnified by the use of an upstream ground connection that increased the total current at the ES emitter electrode.

In a different study using pH indicators and time-lapse photography, Van Berkel et al.[82] studied the time-dependent change in pH in operating nano-ES emitters. As with the previously referenced study, they showed that the pH could be changed by one or more units over the course of a lengthy (>10 min) experiment. For analytes in which the equilibrium distribution of ions in solution is a function of pH (compounds with basic or acidic sites), or in situations where solution interactions that are to be preserved into the gas phase are a function of solution pH (e.g., noncovalent interactions), the electrochemically induced pH change in a nano-ES capillary could be expected to significantly affect the ions and/or ion abundances in a spectrum. Thus, one should be cautious of the electrochemical alteration of solution pH in lengthy experiments with a nano-ES emitter of the "wire-in-a-capillary" design (see Figure 3.6a).

McLuckey and co-workers,[94] using a wire-in-a-capillary nano-ES emitter, showed that reduction of water in negative-ion mode ES-MS increased solution pH ($2H_2O + 2e^- = H_2 + 2OH^-$, Table 3.1). This changing pH was exploited to alter the spectrum of a mixture of proteins of various pI, namely, ubiquitin (pI = 5.2), myoglobin (pI = 7), and cytochrome *c* (pI = 10.6). At an initial solution pH near 7, only ubiquitin was expected to carry a net negative charge in solution. Consistent with this, for the first 5 min of spraying, only ubiquitin ions were observed in the mass spectrum (−5 charge state most abundant). After 20 min, with the solution made more basic by the reduction of water, the signal for the ubiquitin ions increased in intensity and shifted to higher charge state (−6 charge state most abundant). Also, a more narrow distribution of charge states for myoglobin was observed.

After 40 min of spraying, the solution became even more basic, increasing further the intensity of the ubiquitin and myoglobin ions and shifting them to slightly higher charge states. Several peaks that could be assigned as multiple-charged cytochrome *c* ions were also observed.

Klassen and co-workers[95] measured the protein–carbohydrate binding affinities using a wire-in-a-capillary nano-ES setup. The binding affinities of protein–carbohydrate complexes and more general protein ligand complexes are often sensitive to the solution pH. They found that the measurements needed to be made within the first 10 min of the initial spraying. Beyond that time the decrease in pH in positive-ion mode brought about by the oxidation of water lowered the pH by what appeared to be at least 1 pH unit, thereby dramatically lowering the apparent association constant determined from the ions observed in the mass spectrum.

3.4.2 Chemical Noise

Evans and co-workers[36,37] viewed the Fe^{2+} adduct ions observed in their EH mass spectra (resulting from the electrolytic corrosion of the stainless-steel spray capillary) as unwanted chemical noise. To avoid this problem, they switched to a platinum capillary. As Kebarle's group demonstrated,[26] Fe^{2+} associated ions could be observed in ES-MS if ES conditions were altered significantly to enhance their gas-phase abundances. Under more typical ES-MS conditions, ions derived from electrolytically produced Fe^{2+} are not usually observed. This is due, at least in part, to the lower currents (and, therefore, lower concentrations of Fe^{2+} produced) in ES-MS than in EH-MS. However, Ijames et al.[96] reported observing ions at *m*/*z* 622 and 538 in ES mass spectra of acidified methanol–water solutions that they identified (on the basis of accurate mass measurements and tandem mass spectrometry) as the complex $Fe_3O(O_2CR)_6(L)_{0–3}$ where (O_2CR) is the acid added to the solution, and L is one of several ligands corresponding to the solvent or water. Apparently, the source of the iron ions is the electrolytic corrosion of the stainless-steel spray capillary. The ES current in these experiments (although not reported) was probably relatively high owing to the high acid content and corresponding high conductivity of the solution, which would account for a high iron concentration in the solution. These iron-associated ions are nominally electrochemically derived chemical noise in the spectra, the presence of which would be eliminated by switching to a platinum capillary. Whether a "cleaner" spectrum would result depends on the electrolysis reactions and products produced within the platinum capillary. These electrochemically produced metal ions can also be analytically useful when employed to form metal–analyte complexes as shown by Van Berkel's group[55,82] and by Girault's group.[97] Other analytically useful chemical follow-up reactions are discussed in Section 3.4.4.

One might expect that reactive species such as H_2O_2 or $OH^{\bullet}$ might be produced (see Table 3.1) under ES conditions where extremes of electrode potential are generated (e.g., small area electrode + high current = very high current density). These species might then react with other solution species to generate new species that might be observed in the mass spectrum. However, at this point we have found nothing definitive in the literature that clearly indicates that these types of reactions are an origin of chemical noise.

3.4.3 Direct Electrochemical Reaction of Analytes

With the common capillary electrode emitter setup, it may be considered fortunate, or a result of application bias, that the electrochemistry inherent to ES has been for the most part of little concern or consequence to the average user of ES-MS. When compared to conventional electrolysis studies, the analytes that are normally the subject of analysis in ES-MS (i.e., analyte ions in solution such as peptides, proteins, and oligonucleotides) are "spectator" electrolytes in the solvent system. That is, under typical ES conditions, these ionic molecules aid the formation of charged droplets and the transfer of charge between the electrodes in the cell, but they do not participate in the redox reactions at the working electrode. The electrochemical character of these analytes is such that many other common species in the system (e.g., the emitter capillary electrode material and the solvents) are easier to oxidize or reduce. Moreover, as illustrated above, the relatively poor mass transport characteristics of the common tubular emitter electrodes make high electrolysis efficiency for dilute solution species impossible except at the lowest-flow-rate end of their operational capability. Therefore, most analytes, under most common ES-MS operational conditions, are not usually directly affected by the electrochemical process.

Nonetheless, electrochemical reactions of the analytes under study in an ES-MS experiment can and do take place, and these reactions can substantially alter the analytes such that the ions observed in the gas phase have a different mass and/or charge than the species originally in the solution (Table 3.3). Many fundamental studies of the electrochemical processes in ES have made use of analytes that are relatively easy to oxidize/reduce and have both stable oxidized and reduced forms that differ in mass, each of which can be detected in the gas phase by ES-MS. Some compounds of this type that we have used include various metalloporphyrins, aniline dimers, the dyes thionine and methylene blue, and the drug reserpine.[52,55,67,73,79,88,92] The ability to observe both the starting material and the redox products with compounds of this type enhances the ability to study the electrochemical reactions that occur. These same molecules and those others with similar redox character are potentially those that can be altered unexpectedly in an ES-MS experiment.

Table 3.3. Analyte Electrolysis in a Tubular Emitter Electrode

Minimize Analyte Electrolysis	Maximize Analyte Electrolysis
Why?	**Why?**
• Avoid confusion in the analysis of unknowns—change in mass or charge	• Electrochemical ionization
• Preserve initial solution state of analyte	• Create novel ionic species
• Avoid dispersion of charge among different ionic species	• Study analyte redox chemistry
	• Perform electrosynthesis
How?	**How?**
• Electrode isolated from analyte flow path	• Electrode in analyte flow path
• Sacrificial electrode/redox buffer	• Inert electrode (platinum)
• High solution flow rate ($\geq$30 μL/min)	• Low solution flow rate ($\geq$2.5 μL/min)
• Short electrode ($\ll$1 mm)	• Long electrode (up to several cm)
• Low solution conductivity ($\leq 10^{-4}$ S/m)	• High solution conductivity ($\gg 10^{-4}$ S/m)
• Low electrospray voltage drop ($\leq$4 kV)	• High electrospray voltage drop (>4 kV)

The great expansion in use of ES-MS over the last decade means that it is being applied to more analytes that undergo electrochemistry readily (e.g., metals, metal–ligand complexes, organometallics, conductive polymers, and many low-mass organics and drugs). Coupling this with the rise in the use of low-flow-rate ES-MS, there are more reports of the electrochemical nature of ES and its influence on particular ES-MS analyses. A particular example from nano-ES is that already mentioned above for reserpine.[85] Reserpine is relatively easy to oxidize (~0.6 V vs. SHE) and under the right conditions can be partially or even totally converted to products of different mass. Corkery et al.[85] found in nano-ES that the signal for reserpine was fleeting because of this electrochemical transformation. In single-ion montoring or selected reaction monitoring, this electrochemical alteration translated into a very diminished signal even though the analyte was present in substantial amounts.

Another apparent, and somewhat extreme, example of electrochemical analyte alteration was reported by Hop et al.,[98] who examined two borane salts ($[(Me)_4N][B_3H_8]$ and $Cs[B_3H_8]$) by ESMS in positive-ion mode. Acetonitrile solutions provided the most informative spectra; nearly all signals observed were reported to correspond to cationic cluster ions of the general formula $\{[\mathrm{cation}^{m+}]_x[\mathrm{anion}^{n-}]_y\}^{(mx-ny)+}$. In contrast, methanol solutions of these salts produced only $B(OCH_3)_4^-$ cluster ions under otherwise identical conditions. Off-line ^{11}B NMR analyses showed that while the borane $B_3H_8^-$ anion was present in the methanol solution entering the ES capillary, the $B(OCH_3)_4^-$ anion exited the capillary. An electrolytic methanolysis reaction in the ES capillary involving the conversion of $B_3H_8^-$ to $B(OCH_3)_4^-$ was assumed to take place, but no reaction mechanism was put forward.

In contrast to the unwanted electrochemical alteration of the analyte just discussed, electrochemical ionization via oxidation or reduction of the analyte can be very advantageous. Using the ES ion source to "electrochemically ionize" an analyte provides the means to expand sensitive analysis/detection by ES-MS to include certain types of neutral analytes that otherwise are ES inactive, as well as expanding the overall universality of ES as an ionization source. Those analytes found to be most amenable to electrochemical ionization and detection in ES-MS form relatively stable ionic species upon electrochemical oxidation/reduction (Table 3.4). This is required, because the electrochemically

Table 3.4. Analytes reported to be electrochemically ionized in ES capillary during ES-MS

Compound Class	References
Oxidation	
Metal–organic complexes (including porphyrins, chlorins, metallocenes, etc.)	39, 40, 43, 89, 98–108, 206
Aromatic and highly conjugated systems (including PAHs, heteroaromatics, carotinoids, etc.)	39, 66, 67, 109–112
Ferrocene-based electrochemically ionizable derivatives	40, 52, 113–117, 124
Fullerenes and derivatives	118–120
Reduction	
Metal – Organic Complexes	121
Halogens	122
Quinones	124
Fullerenes and Derivatives	123–131

generated ions must survive in solution from the time of their formation in the ES capillary until they are sprayed and transported into the mass spectrometer as discrete gas phase ions. The compounds most amenable to electrochemical ionization via oxidation have relatively low potentials for oxidation and structural characteristics that aid in stabilization of the positive ion formed (often a radical cation). Compounds fitting this description are, for the most part, aromatic or highly conjugated systems that might also contain heteroatoms with lone-pair electrons and/or electron-donating groups such as -OH, -OCH_3, -$N(CH_3)_2$, and -CH_3.[132] Compounds most amenable to electrochemical ionization via reduction are relatively easy to reduce and have high electron affinities. Compounds of this type are usually comprised of aromatic or highly conjugated systems substituted with electron-withdrawing groups such as halides, -NO_2, or -CN.[133] Metal-containing organics of various types are often amenable to electrochemical ionization via either oxidation or reduction.[132,133]

The solvent system can be a very important factor in electrochemical ionization.[39,41,66] Typical solvent systems for ES-MS are comprised of various combinations of methanol, acetonitrile, and/or water along with a small amount of acidic or basic additives. Such solvent systems are chosen because of the solubility characteristics of the more common analytes, because they produce a stable spray, and because they allow for solution phase ionization of the compounds (typically ionization via salt dissolution or acid/base chemistry). In the case of electrochemical ionization, a more careful selection of the solvent may be required because particular ionic species produced may be consumed by several types of rapid reactions in solution, thus certain solvents/additives may have to be avoided. For example, particularly important with radical cations are reactions with nucleophilic solvents or solvent additives.

Cole and co-workers' study[40] of the electrochemical ionization of metallocenes in ES-MS cites several examples of the occurence of chemical follow-up reactions. For example, the electrochemical oxidation products of ruthenocene (Cp_2Ru^+) and osmocene (Cp_2Os^+) are known to lack solution-phase stability and to be very susceptible to nucleophilic addition reactions. The ES mass spectra of these metallocenes sprayed at a relatively slow flow rate (1.6 μL/min) from a methylene chloride/0.5–1.0% trifluoroacetic acid (TFA) solution include peaks corresponding to the chloride ion addition products ($[Cp_2RuCl]^+$ and $[Cp_2OsCl]^+$) and, in the case of osmocene, the trifluoroacetate addition product ($[Cp_2Os + \text{trifluoroacetate}]^+$).

Electrochemists have been able to avoid, or minimize chemical follow-up reactions, thereby "stabilizing" and extending the lifetime of radical cations in solution, through the judicious choice of solvents and solvent additives.[132] Protic solvents as well as nucleophilic solvents (e.g., water and methanol) and nucleophilic solvent additives (e.g., acetate anion) are typically avoided. Aprotic, non-nucleophilic solvents such as rigorously purified and dried acetonitrile and methylene chloride are commonly employed in the electrochemical generation and preservation of radical cations. These same solvent systems have been used very successfully in some of our own electrochemical ionization work.[39,41,66] Chemical follow-up reactions are also exacerbated as flow rates through the ES capillary electrode are decreased in order to increase electrolysis efficiency, because of the greater time the ion remains in solution following its formation. Therefore, a trade-off may have to be made when using a tubular electrode between a low flow rate that offers the highest electrolysis efficiency and a high flow rate that minimizes the possibility of the chemical follow-up reactions for certain "unstable" ionic species.

Four major requirements are key to utilizing electrochemical ionization for efficient and low-level, gas-phase detection in ES-MS in addition to favorable redox properties of the analyte and proper characteristics (reactivity) of the solvent species. First, for thermodynamic reasons, the value of $E_{E/S}$ must be appropriate to oxidize/reduce the analyte. Second, the magnitude of i_{TOT} must be sufficient for oxidation/reduction of all the analyte passing through the electrode plus all species available for reaction that are as easily or more easily oxidized/reduced than the targeted analyte. Therefore, it is prudent to eliminate from the solvent system all species that are easier to oxidize/reduce than the analyte, which might include certain solvents, contaminants, electrolytes, other analytes, and, particularly in positive-ion mode, the stainless-steel ES capillary. This also makes it most likely that the value of $E_{E/S}$ will be that necessary to oxidize/reduce the analyte. If necessary, the magnitude of i_{TOT} can be increased. In practice, a simple means to substantially increase the current over that obtained under optimized ES conditions (e.g., $V_{ES} \approx 4$–5 kV with a fixed solvent system, capillary size, and ES source geometry) is to increase solution conductivity by addition of an electrolyte to the solvent system. The electrolytes normally employed in electrochemical experiments [e.g., alkali metal and tetraalkylammonium nitrates, perchlorates, tetrafluoroborates, hexafluorophosphates, and trifluoromethanesulfonates (triflates)] might be most suitable for this purpose because they are difficult to oxidize/reduce and, therefore, do not contribute to the faradaic current.[45] However, in some cases they may cause signal suppression (see below).

The third requirement is that the analyte be available for reaction at the electrode solution interface in the ES emitter. This demands that the time for transport of the analyte to the electrode be shorter than the time the analyte spends in the typical capillary emitter electrode (i.e., the electrolysis time). The flow rate of the analyte through the capillary and the capillary length (effective electrode length) will have a major affect on this availability. In general, operating at slower flow rates enhances electrolysis efficiency by increasing the availability of the analyte for reaction as discussed above. However, alternative electrode geometries such as the planar flow-by electrode and the porous flow-through electrode emitter permit the use of significantly higher flow rates (compared to a traditional tubular emitter) while still providing efficient analyte electrolysis due to their efficient mass transport characteristics (Figure 3.10). In the case of the porous flow-through electrode, efficient electrolysis is possible at flow rates approaching 1 mL/min. The ability to operate at these high flow rates should also mitigate problems with chemical follow-up reactions given the shorter time the electrochemically created ions spend in solution before liberation into the gas phase.

The fourth requirement is to avoid operational conditions that inhibit the efficient formation of gas-phase ions from the electrochemically generated analyte ions in solution. A major obstacle in this regard is the sometimes necessary addition of an electrolyte to the analyte solution to increase the current magnitude to efficiently electrolyze the analyte. Tang and Kebarle[27,28] were among the first to show that that the mass spectrometrically detected ion current for an analyte of interest may be suppressed by the presence of other ions ("foreign electrolytes") in solution that have a greater propensity for formation of gas-phase ions. Therefore, to minimize problems with signal suppression, it is necessary to first select an electrolyte whose propensity for gas-phase ion formation is small relative to that of the analyte ion and then use it at the lowest concentration possible to provide the required faradaic current. In more nonpolar solvents, lithium triflate has been shown[66] to have a relatively low suppression effect, which enabled its use at concentrations up to a few millimolar to increase current without compromising the generation of gas-phase analyte ions. In aqueous solvents, the common modifier ammonium acetate at concentrations of 5–50 mM has been used as an electrolyte.[69,89]

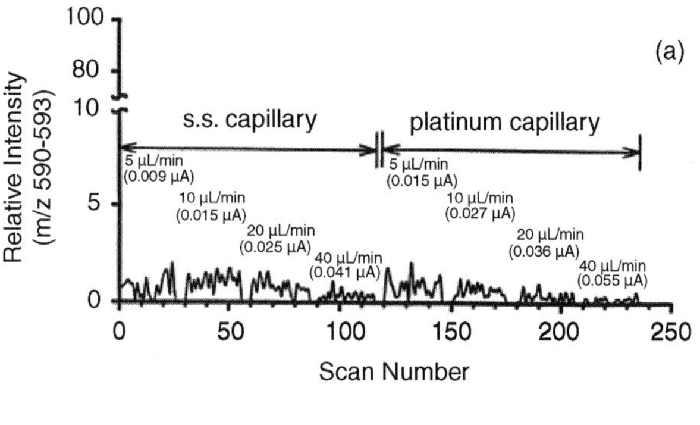

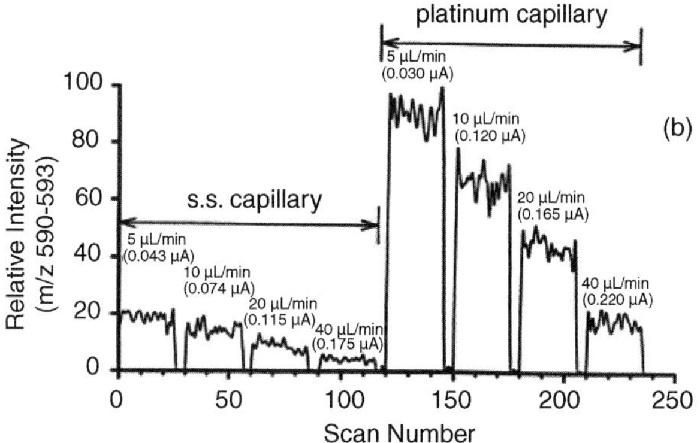

Figure 3.17. (**a**) Ion current intensities for the radical cation NiOEP (*m/z* 590–593) measured in continuous infusion experiments in which acetonitrile/methylene chloride (1/1 v/v) solutions of NiOEP (10 μM) containing (**a**) no added electrolyte and (**b**) 0.1 mM lithium triflate were sprayed from stainless steel and platinum capillaries at flow rates of 5, 10, 20, and 40 μL/min. The respective flow rates and measured values of i_{ES} are also shown in the figure. The signal levels in parts a and b are each normalized to the maximum signal recorded in part b. (Adapted with permission from Ref. 66. Copyright 1995, American Chemical Society.)

Data shown in Figure 3.17 illustrate how the composition of the metal ES capillary, either stainless-steel or platinum, the addition of an appropriate electrolyte to the solution to increase i_{ES}, and the solution flow rate through the capillaries affect the observed gas-phase ion abundances for an electrochemically generated analyte ion. When no electrolyte is added to the solution (Figure 3.17a), the value of i_{ES} and the gas-phase analyte ion signal for the radical cation of NiOEP are low for both capillaries at all flow rates. On the basis of Eq. (3.1), the magnitude of i_{ES} necessary for complete oxidation of the porphyrin (10 μM) at each of the respective flow rates, assuming that no other reactions supply i_F, are 0.08 μA (5 μL/min), 0.16 μA (10 μL/min), 0.32 μA (20 μL/min), and 0.64 μA (40 μL/min). The currents measured at each flow rate (assuming again only oxidation of the porphyrin) are sufficient to oxidize a maximum of only 5–20% of the total amount of porphyrin present, which explains the low gas-phase ion signals. With 0.1 mM of electrolyte added to the solution (Figure 3.17b), the magnitude of i_{ES} with both capillaries, at all flow rates, is substantially increased. As a result, the extent of analyte electrolysis/ionization and the gas-phase ion signal levels are also increased. Note that the NiOEP radical cation signal observed when using the platinum capillary is substantially greater than that observed when using the stainless-steel capillary owing to the supply of faradaic current by oxidation of the iron in the stainless steel. The occurrence of this oxidation reaction reduces the amount of i_F that might otherwise be supplied by oxidation of an analyte in solution. Using the ES capillary fabricated from platinum, which is much more difficult to oxidize than the iron in stainless steel, allows a greater fraction of i_F to be supplied by the oxidation of solution species, including the analyte. As such, more NiOEP ions are created in the solution within the platinum capillary compared to within the stainless-steel capillary, all other factors being equal. The data also show that regardless of the capillary material, the gas-phase signal for the NiOEP radical cation decreases as flow rate increases due to the increased rate of NiOEP transfer through the capillary without a sufficient increase in i_{ES} to enable the same extent of analyte oxidation. Another contributing factor to reduced signal levels as flow rate increases is the inefficiency of the mass transport.

In summary, when the key requirements for efficient electrolysis and sensitive gas-phase detection are met, it has been possible to ionize and detect neutral molecules such as the porphyrins discussed above, at levels that are comparable to the lowest levels achieved for preformed ionic compounds.[66] This is true even for species relatively difficult to oxidize (i.e., $E > 1.0$ V vs. SHE). Some demonstrative examples are presented below.

Shown in Figure 3.18a are extracted ion current profiles for the radical cation of perylene (*m/z* 252, $E_{1/2} = 1.29$ V vs. SHE) obtained from three replicate injections of a blank solution and increasing quantities of perylene into a flowing stream of acetonitrile/methylene chloride (1/1 v/v, 20 μL/min) containing 1.0 mM of electrolyte. The detection level between 130 and 270 fmol was comparable to those levels that were recorded on the same instrumentation for many preformed ionic compounds under similar flow-rate and solution conditions. Shown in Figure 3.18b is the averaged, background-subtracted mass spectrum obtained from the first 270 fmol injection recorded in Figure 3.18a. The signal level for the radical cation is several times higher than that of the background noise, providing a clear identification of the compound.

The ultimate detectability demonstrated so far in electrochemical ionization may be for ferrocene-based, electrochemically ionizable derivatives.[117] Figure 3.19a shows the selected reaction monitoring (SRM, *m/z* 613 → *m/z* 245) results for four replicate injections of a blank and derivative standards (ferrocenecarbamate esters of cholesterol) ranging

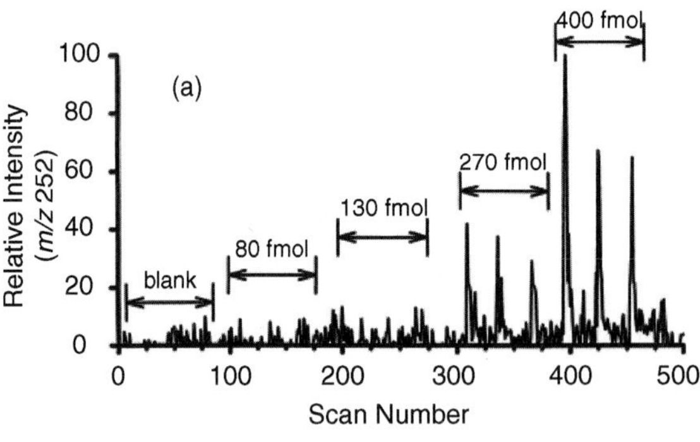

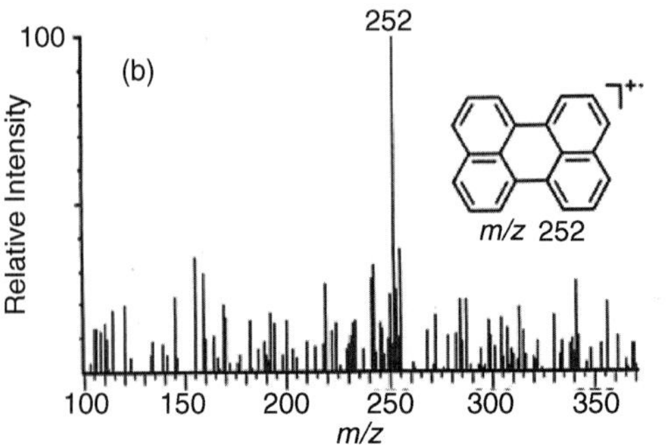

Figure 3.18. (**a**) Extracted ion current profile for the radical cation of perylene (*m/z* 252) obtained in a flow injection experiment in which three replicate injections (0.5 μL) of a blank solution and analyte solutions of increasing analyte concentration (concentration shown in figure) were made into a flowing solution (20 μL/min) comprised of acetonitrile/methylene chloride (1/1 v/v) containing 1.0 mM lithium triflate. The perylene standards were prepared in a solution of the same composition as the carrier solution. (**b**) The averaged, background-subtracted ES mass spectrum obtained from the first 270 fmol injection of perylene as recorded in part a. (Adapted with permission from Ref. 66. Copyright 1995, American Chemical Society.)

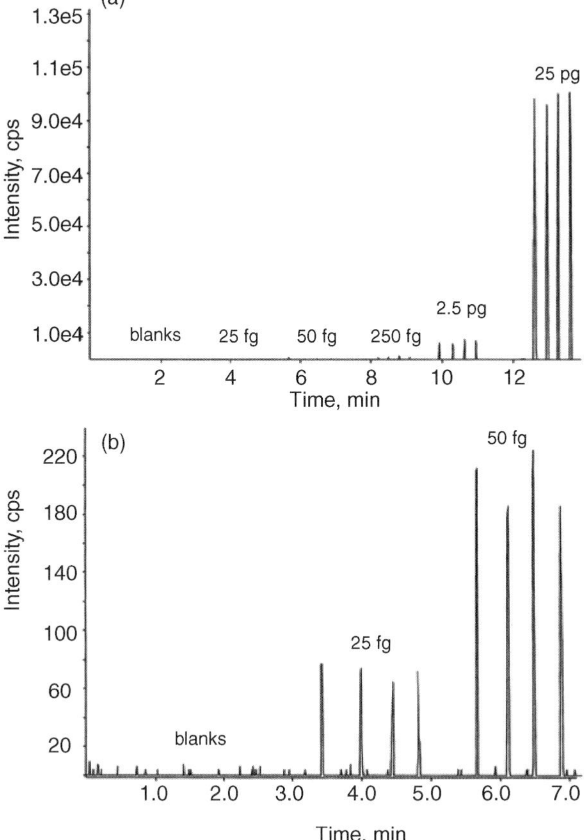

Figure 3.19. (**a, b**) Flow injection experiment performed using selected reaction monitoring (*m/z* 613 → *m/z* 245) to assess the detectability of the ferrocenecarbamate derivative of cholesterol. Standards were prepared by first dissolving the derivative in CH_2Cl_2 and then diluting appropriately with the same solvent used in the flowing stream to yield concentrations from 25 fg/μL to 25 pg/μL. Standards were injected (1.0 μL) into a flowing stream (50 μL/min) of CH_3CN/H_2O (9/1 v/v) containing 100 μM lithium triflate. Injected amounts appear in the figure. Data were obtained on a quadrupole-based instrument. $V_{ES} = 5.0$ kV. (Adapted with permission from Ref. 117. Copyright 1998, American Chemical Society.)

from 25 fg to 25 pg. Solvent conditions (CH_3CN/H_2O, 9/1 v/v with 100 μM lithium triflate) and flow rate (50 μL/min) were chosen to mimic what might be typical microbore reverse-phase HPLC separations conditions for derivatives of this type. The expanded view in Figure 3.19b illustrates the excellent signal-to-background level achieved, even for 25 fg (41 amol) injected. This level of detection is exceptionally good, particularly given operation at 50 μL/min, which might be expected to lower electrolysis efficiency.

As mentioned above, the alternative planar flow-by and porous flow-through electrodes dramatically enhance mass transport so that the flow rate limitations on high electrolysis efficiency are greatly relaxed. This makes the utility of electrochemical ionization more versatile and practical. For example, one can easily carry out electrochemical ionization with samples introduced via microbore or larger bore chromatography ($\geq$50 μL/min). As a specific example, consider the oxidation of the ferroceneboronate derivative of pinacol which transforms it into the ES active cation (Figure 3.20). To demonstrate the utility of this oxidation process in a real-world ES-MS analysis scenario, we performed on-line HPLC/ES-MS separation and detection of the derivative.[52] The mass chromatogram and ES mass spectrum in Figures 3.21a and 3.21b, respectively, illustrate the chromatographic peak profile and mass spectrum obtained for this derivative using a porous flow-through electrode emitter. The base peak in the spectrum corresponds to the molecular radical cation at *m/z* 312, and the observed isotope pattern is that expected on the basis of the molecular formula

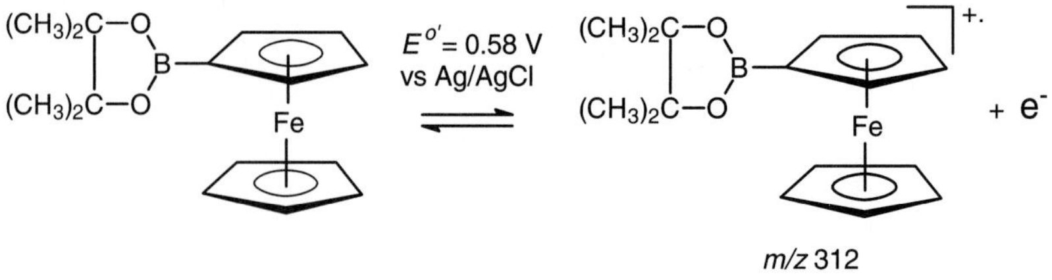

$(CH_3)_2C{-}O$, $(CH_3)_2C{-}O$, B, Fe $\underset{}{\overset{E^{o'} = 0.58\ \text{V vs Ag/AgCl}}{\rightleftharpoons}}$ $(CH_3)_2C{-}O$, $(CH_3)_2C{-}O$, B, Fe $]^{+\cdot}$ + e^-

m/z 312

Figure 3.20. Oxidation of ferroceneboronate derivative of pinacol. (Adapted with permission from Ref. 52. Copyright 2004, Elsevier.)

of this molecule. The sensitivity and detection levels for this derivative were compared in a separate set of on-line HPLC/ES-MS experiments (single ion monitoring) using a platinum capillary electrode emitter and porous flow-through electrode emitter system (Figure 3.22). Comparison of the plots of chromatographic peak areas versus derivative amounts injected indicates the superior sensitivity of the porous flow-through electrode emitter. The detection levels were as low as 8 fmol injected with the porous flow-through electrode electrode emitter compared to no less that 40 fmol with the platinum capillary electrode emitter. The detection levels in both cases would scale to considerably lower levels with selected reaction monitoring (SRM) detection as suggested on the basis of previous studies for related derivatives.[117]

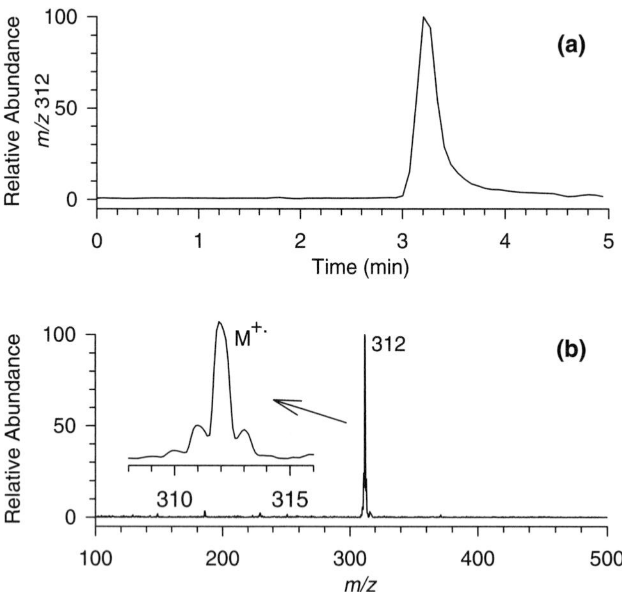

Figure 3.21. (**a**) Mass chromatogram obtained from the on-column injection of 4.5 pmol of the ferroceneboronate derivative of pinacol using the porous flow-through electrode emitter. (**b**) Background subtracted, averaged mass spectrum for the chromatographic peak. Inset shows an expansion of the molecular ion region of the spectrum. Derivative diluted in eluant and injected (1.0 μL) onto a 1.0-mm-i.d. × 15-cm-long PAH Hypersil column (5-μm particles, 120-Å pore size). Eluant composition 80/20 (v/v) acetonitrile/water, 200 μM ammonium acetate, flow rate = 50 μL/min. $I_{TOT} \sim 1.6\ \mu A$. (Adapted with permission from Ref. 52. Copyright 2004, Elsevier.)

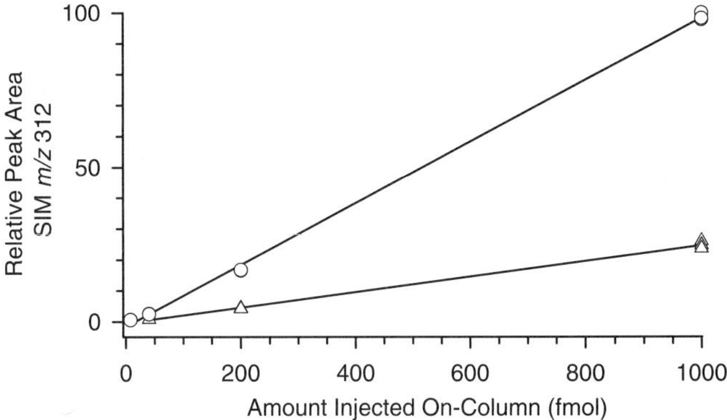

Figure 3.22. Plots of relative peak areas of the radical cation of the ferroceneboronate derivative of pinacol (*m/z* 312) as a function of the amount injected on-column (three replicate injections at each amount) using either the porous flow-through electrode emitter (○) or the platinum capillary electrode emitter (Δ). Derivative diluted in eluant and injected (1.0 μL) onto a 1.0-mm-i.d. × 15-cm-long PAH Hypersil column (5-μm particles, 120-Å pore size). Eluant composition 80/20 (v/v) acetonitrile/water, 1.0 mM ammonium acetate, flow rate = 50 μL/min, SIM dwell time = 500 ms. PFT electrode emitter: $I_{TOT} \sim 1.5\ \mu A$; platinum capillary electrode emitter: $I_{TOT} \sim 2.1\ \mu A$. (Adapted with permission from Ref. 52. Copyright 2004, Elsevier.)

One must also be cognizant of the fact that while electrolysis of neutral molecules (or even charged analytes) in the ES capillary can be the source of ionic products observed in the gas phase, chemical electron-transfer reactions in solution and or in the gas phase might actually be the source of the ions observed. In fact, several literature examples can be found in which the electrolytic mechanism for ion formation appears to have been incorrectly invoked.[24] Our experience using the electrochemical nature of ES for analyte ionization has indicated that understanding the electron-transfer chemistry of the analyte of interest, in the solvent system being used, is highly desirable to avoid incorrectly interpreting the origin of certain ions observed in the ES mass spectra. UV–visible spectra of the analyte solution acquired before and after addition of particular solvent additives may in many cases be used to confirm or rule out the role of solution-based chemical electron-transfer chemistry in ion formation. Cyclic voltammograms, from the literature or acquired in-house, are extremely useful in determining if the proposed electrolytic reaction leading to the observed ion is feasible (e.g., determining if the redox potential for electron transfer is within the potential window available) and whether the ionic products of electron transfer are stable in solution for a long enough period to be detected in the gas phase.

3.4.4 Utilizing Electrochemically Initiated Homogeneous Reactions

The chemical follow-up reactions or homogeneous reactions that can be detrimental to simple electrochemical ionization can be analytically useful in their own right. In such a case, the initial products of the electrolysis react with the analyte of interest and the product of this reaction provides an analytical advantage in the experiments. In some ways this mirrors the coulometric titrations used in classic electrochemistry.

Experiments that have taken advantage of the electrochemically initiated homogeneous reactions in the ES ion source can be distinguished into two basic groups. In the first group, the product of the heterogeneous reaction was used to improve the ionization efficiency of the analyte of interest. To improve the electrospray ionization efficiency of non-ES ionizable (under standard ES conditions) chromium and tungsten carbene complexes, Sierra and co-workers[134,135] have used known electron-donor compounds as hydroquinone and tetrathiofulvalene. They suggested that the electrochemical reduction of the electron-donor compounds at the emitter was followed by a charge-transfer reaction resulting in the radical anion of the carbene complex followed by elimination of a hydrogen radical resulting in the negatively charged carbene complex anion detected in the mass spectrum. The same fundamental principles were followed by Marjasvaara and co-workers,[136] who used a mediator (2,2′,6,6′-tetramethylpiperidine-*N*-oxyl radical, TEMPO) in order to oxidize a substituted benzyl alcohol to its corresponding carboxylic acid form.

In the second group of homogeneous reactions, the product of the heterogeneous reaction underwent a consecutive chemical reaction with the analyte of interest to provide structural information about the analyte. Girault and co-workers[74,78,137–142] have investigated electrochemically induced tagging of free cysteines in peptides/proteins by substituted hydroquinones using ES-MS with applications in protein analysis. The benzoquinone generated at the emitter electrode in positive-ion mode underwent a 1,4-Michael addition with the free cysteines on a peptide/protein in the same solution to tag the cysteines with quinone, causing a defined shift in the mass of the parent molecule. The mass shift was used to "count" the number of free cysteine residues on the peptide chain. They showed a higher tagging yield by using nano-ES instead of a conventional ESI interface,[142] demonstrating the importance of the time available for the diffusion of the electrochemically generated benzoquinone to participate in the homogeneous reaction with the protein. Based on their kinetic calculations and simulations,[74,141] proteins should have a residence time on the order of 10 s in their nano-ES channel cell emitter for an efficient quantitative tagging to occur. To further optimize tagging efficiency, they studied[139,140] substituted hydroquinones and concluded that carboxymethylhydroquinone is the optimal probe for electrochemical tagging of cysteine residues in proteins. They also explored[138] the effect of the number and relative location of cysteine moieties on the tagging efficiency and concluded effective tagging even in the case of two consecutive cysteines in the peptide sequence. Their numerical simulations[137] on multiple tagging of cysteinyl peptides established the range of optimum conditions for the determination of the number of cysteine groups in peptides. They also proved[138] that the determination of cysteine content in the tryptic peptides of a protein mixture provided powerful information in order to enhance identification scores as well as the discrimination against other protein candidates. As another example of the powerful use of the tagging reaction, it proved successful in probing the thiol reactivity/steric conditions around the free cysteine moiety of a single-thiol-containing protein.[78] In parallel work, Van Berkel and Kertesz[76] presented unprecedented complete tagging of a two-cysteine-containing peptide. They improved the tagging efficiency by using a porous flow-through electrode as both the upstream ground point and the emitter electrode (Figures 3.23 and 3.24).

An example of a different use of the homogeneous reactions is illustrated in the work of Duckworth and co-workers.[143] They studied the influence of water on the observed gas-phase population of negative ions for the undiluted ionic liquid 1,3-butyl-methyl-imidazolium hexafluorophosphate. They observed enhanced production of tetrafluorophosphate and difluorophosphate anions from hexafluorophosphate anion caused by the electrolytic reduction of water at the emitter. This unexpected reaction was used to determine the water content of the ionic liquid being sprayed.

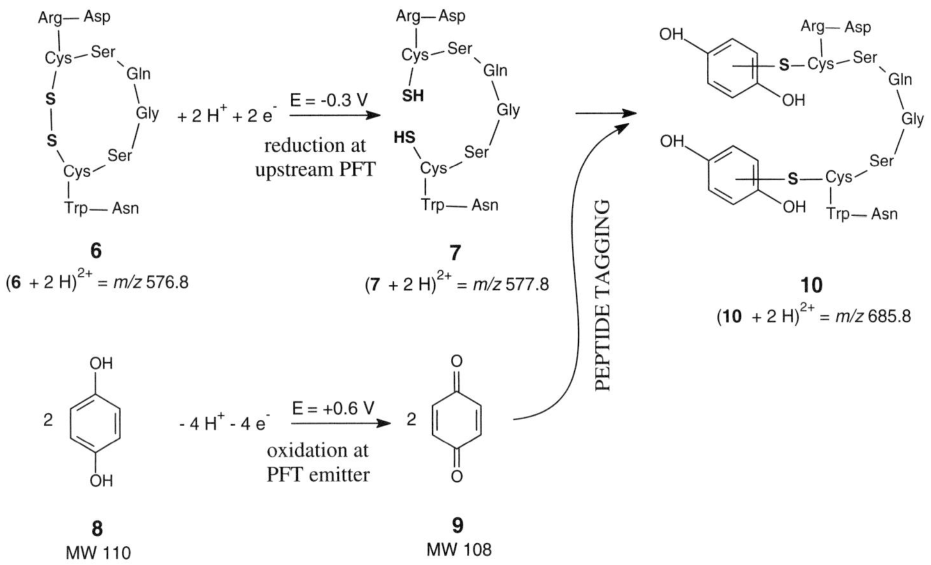

Arg— Asp
Cys—Ser
Gln
S
S
Gly
Ser
Cys
Trp— Asn

6

($\mathbf{6}$ + 2 H)$^{2+}$ = m/z 576.8

+ 2 H$^+$ + 2 e$^-$ E = -0.3 V →

reduction at upstream PFT

Arg— Asp
Cys—Ser
Gln
SH
HS
Gly
Ser
Cys
Trp— Asn

7

($\mathbf{7}$ + 2 H)$^{2+}$ = m/z 577.8

PEPTIDE TAGGING

OH
Arg— Asp
S—Cys—Ser
Gln
OH
Gly
OH
Ser
S—Cys
OH
Trp— Asn

10

($\mathbf{10}$ + 2 H)$^{2+}$ = m/z 685.8

OH
2
OH

8
MW 110

- 4 H$^+$ - 4 e$^-$ E = +0.6 V →

oxidation at PFT emitter

2
O
O

9
MW 108

Figure 3.23. Tagging pathway of oxidized Asp-Arg-Cys-Ser-Gln-Gly-Ser-Cys-Trp-Asn containing disulfide bridge 3–8 (6) with hydroquinone (8) and ions observed. (Adapted with permission from Ref. 76. Copyright 2005, American Chemical Society.)

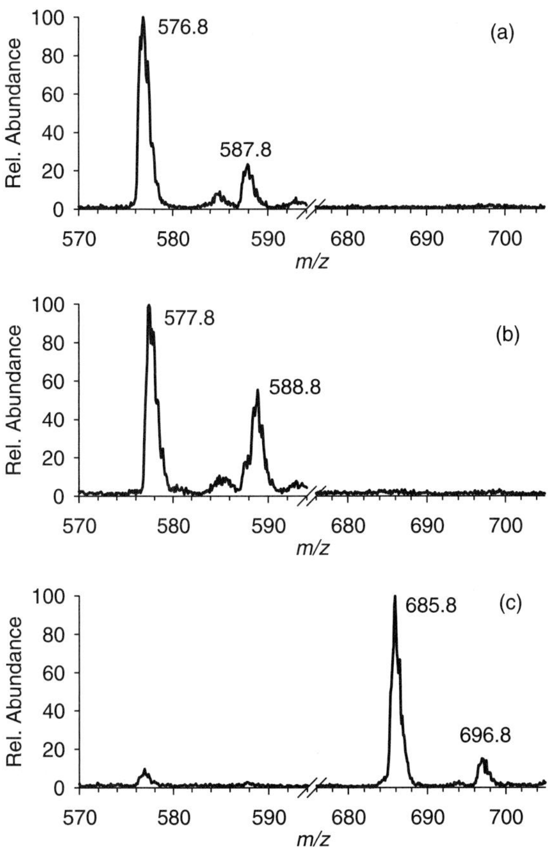

Figure 3.24. Positive-ion mode mass spectra obtained at the plateau region of a 100-μL injection of a 25 μM solution of oxidized peptide (6) in 50/50 (v/v) water/acetonitrile, 50 mM ammonium acetate, 7.5% by volume acetic acid, at a flow rate of 10 μL/min using (**a**) stainless-steel bore-through unions as both the upstream ground and emitter electrodes, (**b**) a PFT electrode assembly as the upstream ground, and (**c**) two PFT electrode assemblies, one as the upstream ground and the other as the emitter electrode with 100 mM 1,4-hydroquinone (8) in 50/50 (v/v) water/acetonitrile, 50 mM ammonium acetate, and 7.5% by volume acetic acid mixed in the main flow stream between the upstream ground and the emitter at a flow rate of 2.0 μL/min. (Adapted with permission from Ref. 76. Copyright 2005, American Chemical Society.)

3.4.5 Study of Redox Reactions

From an electrochemical point of view, ES-MS might be viewed as a means to monitor the products of a controlled-current electrolysis on-line with mass spectrometry. Carrying out electrochemistry/mass spectrometry (EC/MS) in this fashion requires only a very small amount of sample and provides the molecular weight, and potentially the structure (e.g., via tandem mass spectrometry experiments), for the ionic products of the reactions. Given an experiment for which controlled-current electrolysis is suitable,[54] utilizing the inherent electrochemistry of ES would be expected to be much simpler than combining discrete electrochemical cells on-line with ES-MS for the same purpose.[25]

Electrochemical ionization in ES-MS is in effect a use of ES-MS to study the products of exhaustive controlled-current electrolysis on-line with mass spectrometry. Used for this purpose, detection of the stable ionic products of oxidation/reduction in the ES capillary is the analytical goal. Of course, electrolytic reactions involving unstable intermediates and the products of multiple electron transfer might also be studied. The homogenous reactions used by Girault's group might also fall into this area. However, the typical single-electrode emitter or even the two-electrode emitters with current control are not the ideal cell systems for EC/MS studies due to the limited ability to precisely control the working electrode interfacial potential. Because of the ability to control the working electrode potential, the three-electrode, controlled-potential emitter cell provides a more powerful tool for the study of redox reactions with ES-MS.

Discrete controlled-potential cells of several types have been on-line coupled with ES-MS, upstream of the emitter cell, to study redox reactions.[25] With such setups, one can perform many of the types of electrochemical experiments that one might envision carrying out with a controlled-potential multiple electrode emitter cell. Nonetheless, there are advantages to the latter. As described in the Section 3.3.3, intertwining of a three-electrode cell into the ES source produces a circuit that provides the possibility to control the emitter (working) electrode potential. This design allows one to perform electrochemical experiments on-line with ES-MS without interference from the inherent electrochemical processes of the ES ion source compared to the addition of a discrete cell upstream of the emitter electrode. Also, the inherently larger residence time of the reaction intermediates and products in the solution using an upstream cell prohibits the detection of short-lived species.

A three-electrode controlled-potential emitter cell was first applied by Cole's group in the mid-1990s to investigate (a) the oxidation of PAHs[86] and (b) the intermediates and products of anodic oxidation of diphenyl sulfide, anodic pyridation of 9,10-diphenylanthracene, and reduction of nitrobenzene.[87] After these initial studies, this area of work was dormant until our group[88,89] implemented a commercially available three-electrode coulometric cell into the interface of a mass spectrometer, keeping it easily interchangeable with common ES emitters. This setup utilized a porous flow-through working electrode that was demonstrated[52] to allow flow rates of several hundred microliters per minute while preserving high electrolysis efficiency, thus enabling this technique to be successfully coupled to other analytically important techniques (e.g., HPLC). We believe that this newer controlled-potential emitter system has more fully established the approach to be an effective and powerful tool to study electrochemical reactions on-line with mass spectrometric analysis. This paves the way for more extensive use of this EC/ES-MS technique in solving electrochemistry-related problems in the future.

3.5 SUMMARY

Electrochemistry is an inherent aspect in the operation of an ES ion source as typically configured for operation with mass spectrometry, and this electrochemistry has implications for both (a) the fundamentals of gas-phase ion formation from the charged droplets produced by the device and (b) the analytical utility of the device. In this review, we have laid out the fundamental electrochemical aspects of thc ion sourcc in a way that the general user is better positioned to understand when and how this electrochemistry might be important analytically in terms of their particular ES-MS studies or applications. It is our hope that the general user also exploits this information to bring to light new possibilities for ES-MS that take advantage of this inherent electrochemistry. The use of alternative electrode configurations and multiple electrode emitters that allow one to control the electrochemical reactions that take place and the means to utilize electrochemically initiated homogeneous chemistries for analytical advantage are particular areas for growth. It seems clear that the understanding of the electrochemistry of the ES process has and will continue to contribute to the expanding utility of ES-MS.

ACKNOWLEDGMENTS

The work described herein performed in the authors' laboratory at ORNL was sponsored by the United States Department of Energy, Office of Basic Energy Sciences. ORNL is managed and operated by UT-Battelle, LLC, for the United States Department of Energy under contract DE-AC05-00OR22725.

REFERENCES

1. Baily, A. G. *Electrostatic Spraying of Liquids*, John Wiley & Sons, New York, **1988**.
2. Zeleny, J. Instability of electrified liquid surfaces. *Phys. Rev.* **1917**, *10*, 1–6.
3. Michelson, D. *Electrostatic Atomization*, Adam Hilger, New York, **1990**.
4. *J. Aerosol Sci.* **1994**, *25*. Special Issue: Electrosprays: Theory and Applications.
5. Dole, M.; Mach, L. L.; Hines, R. L.; Mobley, R. C.; Ferguson, L. P.; Alice, M. B. Molecular beams of macroions. *J. Chem. Phys.* **1968**, *49*, 2240–2249.
6. Mack, L. L.; Kralik, P.; Rheude, A.; Dole, M. Molecular beams of macroions. II. *J. Chem. Phys.* **1970**, *52*, 4977–4986.
7. Clegg, G. A.; Dole, M. Molecular beams of macroions. III. Zein and polyvinylpyrrolidone. *Biopolymers* **1971**, *10*, 821–826.
8. Teer, D.; Dole, M. Electrospray mass spectroscopy of macromolecule degradation in the electrospray. *J. Polymer Sci.* **1975**, *13*, 985–995.
9. Thomson, B. A.; Iribarne, J. V. Field induced ion evaporation from liquid surfaces at atmospheric pressure. *J. Chem. Phys.* **1979**, *71*, 4451–4463.
10. Iribarne, J. V.; Dziedzic, P. J.; Thomson, B. A. Atmospheric pressure ion evaporation–Mass Spectrometry. *Int. J. Mass Spectrom. Ion Phys.* **1983**, *50*, 331–347.
11. Yamashita, M.; Fenn, J. B. Electrospray Ion Source. Another variation on the free-jet theme. *J. Phys. Chem.* **1984**, *88*, 4451–4459.
12. Yamashita, M.; Fenn, J. B. Negative ion production with the electrospray ion source. *J. Phys. Chem.* **1984**, *88*, 4671–4675.
13. Whitehouse, C. M.; Dreyer, R. N.; Yamashita, M.; Fenn, J. B. Electrospray interface for liquid chromatographs and mass spectrometers. *Anal. Chem.* **1985**, *57*, 675–679.
14. Alexandrov, M. L.; Gall, L. N.; Krasnov, M. V.; Nikolaev, V. I.; Shkurov, V. A. Mass-spectrometric analysis of thermally unstable compounds of low volatility by the extraction of ions from solution at atmospheric pressure. *Zh. Anal. Khim.* **1985**, *40*, 1272–1236.
15. Olivares, J. A.; Nguyen, N. T.; Yonker, C. R.; Smith, R. D. On-line mass spectrometric detection for capillary zone electrophoresis. *Anal. Chem.* **1987**, *59*, 1230–1232.
16. Smith, R. D.; Olivares, J. A.; Nguyen, N. T.; Udseth, H. R. Capillary zone electrophoresis–mass spectrometry using an electrospray ionization interface. *Anal. Chem.* **1988**, *60*, 436–441.

17. Smith, R. D.; Barinaga, C. J.; Udseth, H. R. Improved electrospray ionization interface for capillary zone electrophoresis–mass spectrometry. *Anal. Chem.* **1988**, *60*, 1948–1952.
18. Bruins, A. P.; Covey, T. R.; Henion, J. D. Ion spray interface for combined liquid chromatography/ atmospheric pressure ionization mass spectrometry. *Anal. Chem.* **1987**, *59*, 2642–2646.
19. Covey, T. R.; Bruins, A. P.; Henion, J. D. Comparison of thermospray and ion spray mass spectrometry in an atmospheric pressure ion source. *Org. Mass Spectrom.* **1988**, *23*, 178–186.
20. Covey, T. R.; Bonner, R. F.; Shushan, B. I.; Henion, J. The determination of protein, oligonucleotide and peptide molecular weights by ion-spray mass spectrometry. *Rapid Commun. Mass Spectrom.* **1988**, *2*, 249–256.
21. Lee, E. D.; Henion, J. D.; Covey, T. R. Microbore high performance liquid chromatography-ion spray mass spectrometry for the determination of peptides. *J. Microcolumn Sep.* **1989**, *1*, 14–18.
22. Cech, N. B.; Enke, C. G. Practical implications of some recent studies in electrospray ionization fundamentals. *Mass Spectrom. Rev.* **2001**, *20*, 362–387.
23. Cole, R. B. Some tenets pertaining to electrospray ionization mass spectrometry. *J. Mass Spectrom.* **2000**, *35*, 763–772.
24. Van Berkel, G. J. The electrolytic nature of electrospray. In Cole, R. B. (Ed.), *Electrospray Ionization Mass Spectrometry*, John Wiley & Sons, New York, **1997**, Chapter 2, pp. 65–105.
25. Diehl, G.; Karst, U. On-line electrochemistry—MS and related techniques. *Anal. Bioanal. Chem.* **2002**, *373*, 390–398.
26. Blades, A. T.; Ikonomou, M. G.; Kebarle, P. Mechanism of electrospray mass spectrometry. Electrospray as an electrolysis cell. *Anal. Chem.* **1991**, *63*, 2109–2114.
27. Tang, L.; Kebarle, P. Effect of the conductivity of the electrosprayed solution on the electrospray current. Factors determining analyte sensitivity in electrospray mass spectrometry. *Anal. Chem.* **1991**, *63*, 2709–2715.
28. Tang, L.; Kebarle, P. Dependence of ion intensity in electrospray mass spectrometry on the concentration of the analytes in the electrosprayed solution. *Anal. Chem.* **1993**, *65*, 3654–3668.
29. Enke, C. G. A predictive model for matrix and analyte effects in electrospray ionization of singly-charged ionic analytes. *Anal. Chem.* **1997**, *69*, 4885–4893.
30. Constantopoulos, T. L.; Jackson, G. S.; Enke, C. G. Effects of salt concentration on analyte response using electrospray ionization mass spectrometry. *J. Am. Soc. Mass Spectrom.* **1999**, *10*, 625–634.
31. Ikonomou, M. G.; Blades, A. T.; Kebarle, P. Electrospray-ion spray: A comparison of mechanisms and performance. *Anal. Chem.* **1991**, *63*, 1989–1998.
32. Pfeifer, R. J.; Hendricks, C. D. Parametric studies of electrohydrodynamic spraying. *AIAA J.* **1968**, *6*, 496–502.
33. Smith, D. P. H. The electrohydrodynamic atomization of liquids. *IEEE Trans. Ind. Appl.* **1986**, *1A-22*, 527–535.
34. Hayati, I.; Bailey, A. I.; Tadros, T. F. Investigations into the mechanisms of electrohydrodynamic spraying of liquids. I. Effect of electric field and the environment on pendant drops and factors affecting the formation of stable jets and atomization. *J. Colloid Interface Sci.* **1987**, *117*, 205–221.
35. Hayati, I.; Bailey, A. I.; Tadros, T. F. Investigations into the mechanisms of electrohydrodynamic spraying of liquids. II. Mechanism of stable jet formation and electrical forces acting on a liquid cone. *J. Colloid Interface Sci.* **1987**, *117*, 222–230.
36. Simons, D. S.; Colby, B. N.; Evans, C. A. Jr. Electrohydrodynamic ionization mass spectrometry—The ionization of liquid glycerol and non-volatile organic solutes. *Int. J. Mass Spectrom. Ion Phys.* **1974**, *15*, 291–302.
37. Stimpson, B. P.; Evans, C. A., Jr. Electrohydrodynamic ionization mass spectrometry of biochemical materials. *Biomed. Mass Spectrom.* **1978**, *5*, 52–63.
38. Cook, K. D. Electrohydrodynamic mass spectrometry. *Mass Spectrom. Rev.* **1986**, *5*, 467–519.
39. Van Berkel, G. J.; McLuckey, S. A.; Glish, G. L. Electrochemical origin of radical cations observed in electrospray ionization mass spectra. *Anal Chem.* **1992**, *64*, 1586–1593.
40. Xu, X.; Nolan, S. P.; Cole, R. B. Electrochemical oxidation and nucleophilic addition reactions of metallocenes in electrospray mass spectra. *Anal. Chem.* **1994**, *66*, 119–125.
41. Van Berkel, G. J.; Zhou, F. Characterization of an electrospray ion source as a controlled-current electrolytic cell. *Anal. Chem.* **1995**, *67*, 2916–2923.
42. Diefenderfer, A. J. *Principles of Electronic Instrumentation*, Saunders College Publishing, Philadelphia, **1979**, p. 5.
43. Van Berkel, G. J.; McLuckey, S. A.; Glish, G. L. Electrospray ionization of porphyrins using a quadrupole ion trap for mass analysis. *Anal. Chem.* **1991**, *63*, 1098–1109.
44. Bockris, J.O'M.; Reddy, A. K. N. *Modern Electrochemistry*, Vol. 1, Plenum Press, New York, **1973**.
45. Bard, A. J.; Faulkner, L. R. *Electrochemical Methods*, John Wiley & Sons, New York, **1980**.
46. Štulik, K.; Pacáková, V. *Electroanalytical Measurements in Flowing Liquids*, Ellis Horwood, Chichester, West Sussex, England, **1987**.
47. Malmstadt, H. V.; Enke, C. G.; Crouch, S. R. *Electronics and Instrumentation for Scientists*, Benjamin/Cummings, Menlo Park, CA, **1981**.
48. Jackson, G. S.; Enke, C. G. Electrical equivalence of electrospray ionization with conducting and

nonconducting needles. *Anal. Chem.* **1999**, *71*, 3777–3784.
49. Kebarle, P.; Tang, L. From ions in solution to ions in the gas phase: The mechanism of electrospray mass spectrometry. *Anal. Chem.* **1993**, *65*, 972A–986A.
50. Juhasz, P.; Ikonomou, M. G.; Blades, A. T.; Kebarle, P. In Standing, K. G.; Ens, W.,(Eds.), *Methods and Mechanisms for Producing Ions from Large Molecules*, Plenum Press, New York, **1991**, pp. 171–184.
51. Konermann, L.; Silva, E. A.; Sogbein, O. F. Electrochemically induced pH changes resulting in protein unfolding in the ion source of an electrospray mass spectrometer. *Anal. Chem.* **2001**, *73*, 4836–4844.
52. Van Berkel, G. J.; Kertesz, V.; Granger, M. C.; Ford, M. J. Efficient analyte oxidation in an electrospray ion source using a porous flow through-electrode emitter. *J. Am. Soc. Mass Spectrom.* **2004**, *15*, 1755–1766.
53. Van Berkel, G. J. Electrolytic corrosion of a stainless-steel electrospray emitter monitored using an electrospray—Photodiode array system. *J. Anal. At. Spectrom.* **1998**, *13*, 603–607.
54. Lingane, J. L. *Electroanalytical Chemistry*, Interscience Publishers, New York, **1958**.
55. Van Berkel, G. J.; Kertesz, V. Redox buffering in an electrospray ion source using a copper capillary emitter. *J. Mass Spectrom.* **2001**, *36*, 1125–1132.
56. Moini, M.; Cao, P.; Bard, A. J. Hydroquinone as a buffer additive for suppression of bubbles formed by electrochemical oxidation of the CE buffer at the outlet electrode in capillary electrophoresis/electrospray ionization–mass spectrometry. *Anal. Chem.* **1999**, *71*, 1658–1661.
57. Van Berkel, G. J.; Giles, G. E.; Bullock, J. S., IV; Gray, L. J. Computational simulation of redox reactions within a metal electrospray emitter. *Anal. Chem.* **1999**, *71*, 5288–5296.
58. Li, Y.; Pozniak, B. P.; Cole, R. B. Mapping of potential gradients within the electrospray emitter. *Anal. Chem.* **2003**, *75*, 6987–6994.
59. Pozniak, B. P.; Cole, R. B. Negative ion mode evolution of potential buildup and mapping of potential gradients within the electrospray emitter. *J. Am. Soc. Mass Spectrom.* **2004**, *15*, 1737–1747.
60. Van Berkel, G. J. Insights into analyte electrolysis in an electrospray emitter from chronopoteniometry experiments and mass transport calculations. *J. Am. Soc. Mass Spectrom.* **2000**, *11*, 951–960.
61. Dobos, D. *Electrochemical Data*, Elsevier, Amsterdam, **1975**.
62. Wang, J. *Analytical Electrochemistry*, 2nd ed., Wiley-VCH, New York, **2000**.
63. Inzelt, G. *Az Elektrokemia Korszeru Elmelete es Modszerei (Modern Theory and Methods of Electrochemistry), Part II*, Nemzeti Tankonyvkiado, Budapest, **1999**, p. 142.
64. Chen, L.; Colyer, C. L.; Kamau, M.; Myland, J. C.; Oldham, K. B.; Synons, P. G. Microtube flowing coulometry. *Can. J. Chem.* **1994**, *72*, 836–849.
65. Kuwana, T.; Bublitz, D. E.; Hoh, G. Chronopotentiometric studies on the oxidation of ferrocene, ruthenocene, osmocene and some of their derivatives. *J. Am. Chem. Soc.* **1960**, *82*, 5811–5817.
66. Van Berkel, G. J.; Zhou, F. Electrospray as a controlled-current electrolytic cell: electrochemical ionization of neutral analytes for detection by electrospray–mass spectrometry. *Anal. Chem.* **1995**, *67*, 3958–3964.
67. Van Berkel, G. J.; Zhou, F. Observation of gas phase molecular dications formed from neutral organics in solution via the controlled-current electrolytic process inherent to electrospray. *J. Am. Soc. Mass Spectrom.* **1996**, *7*, 157–162.
68. Giles, G. E.; Van Berkel, G. J. Unpublished data.
69. Posniak, B. P.; Cole, R. B. Current measurements within the electrospray emitter. *J. Am. Soc. Mass Spectrom.*, in press.
70. Hayati, I. Eddies inside a liquid cone stressed by interfacial electrical shear. *Colliods Surf.* **1992**, *65*, 77–84.
71. Cooper, J. A.; Compton, R. G. Channel electrodes—A review. *Electroanalysis* **1998**, *10*, 141–155.
72. Rohner, T. C.; Rossier, J. S.; Girault, H. H. Polymer microspray with an integrated thick-film microelectrode. *Anal. Chem.* **2001**, *73*, 5353–5357.
73. Van Berkel, G. J.; Asano, K. G.; Kertesz, V. Enhanced study and control of analyte oxidation in electrospray using a thin-channel, planar electrode emitter. *Anal. Chem.* **2002**, *74*, 5047–5056.
74. Rohner, T. C.; Josserand, J.; Jensen, H.; Girault, H. H. Numerical investigation of an electrochemcially induced tagging in a nanospray for protein analysis. *Anal. Chem.* **2003**, *75*, 2065–2074.
75. Štulik, K.; Pacáková, V *Electroanalytical Measurements in Flowing Liquids*, Horwood, Chichester, **1987**.
76. Van Berkel, G. J.; Kertesz, V. Expanded electrochemical capabilities of the electrospray ion source using porous flow-through electrodes as the upstream ground and emitter high-voltage contact. *Anal. Chem.* **2005**, *77*, 8041–8049.
77. Blaedel, W. J.; Klatt, L. N. Reversible charge transfer at the tubular platinum electrode. *Anal. Chem.* **1966**, *38*, 879–883.
78. Dayon, L.; Roussel, C.; Girault, H. H. Probing cysteine reactivity in proteins by mass spectrometric EC-tagging. *J. Proteome Res.* **2006**, *5*, 793–800.
79. Kertesz, V.; Van Berkel, G. J. Minimizing analyte electrolysis in an electrospray emitter. *J. Mass Spectrom.* **2001**, *36*, 204–210.
80. Severs, J. C.; Smith, R. D. Characterization of the microdialysis junction interface for capillary electrophoresis/microelectrospray ionization mass spectrometry. *Anal. Chem.* **1997**, *69*, 2154–2158.
81. Martin, S. E.; Shabanowitz, J.; Hunt, D. F.; Marto, J. A. Subfemtomole MS and MS/MS peptide sequence analysis using nano-HPLC Micro-ESI Fourier

transform ion cyclotron resonance mass spectrometry. *Anal. Chem.* **2000**, *72*, 4266–4274.

82. Van Berkel, G. J.; Asano, K. G.; Schnier, P. D. Electrochemical processes in a wire-in-a-capillary bulk-loaded, nano-electrospray emitter. *J. Am. Soc. Mass Spectrom.* **2001**, *12*, 853–862.
83. Schultz, G. A.; Corso, T. N.; Prosser, S. J.; Zhang, S. A fully integrated monolithic microchip electrospray device for mass spectrometry. *Anal. Chem.* **2000**, *72*, 4058–4063.
84. Dethy, J.-M.; Ackerman, B. L.; Delatour, C.; Henion, J. D.; Schultz, G. A. Demonstration of direct bioanalysis of drugs in plasma using nanoelectrospray infusion from a silicon chip coupled with tandem mass spectrometry. *Anal. Chem.* **2003**, *75*, 805–811.
85. Corkery, L. J.; Pang, H.; Schneider, B. B.; Covey, T. R.; Siu, K. W. M. Automated nanospray using chip-based emitters for the quantitative analysis of pharmaceutical compounds. *J. Am. Soc. Mass Spectrom.* **2005**, *16*, 363–369.
86. Xu, X.; Lu, W.; Cole, R. B. On-line probe for fast electrochemistry/electrospray mass spectrometry investigations of polycyclic aromatic hydrocarbons. *Anal. Chem.* **1996**, *68*, 4244–4253.
87. Lu, W.; Xu, X.; Cole, R. B. On-line linear sweep voltammetry–electrospray mass spectrometry. *Anal. Chem.* **1997**, *69*, 2478–2484.
88. Van Berkel, G. J.; Asano, K. G.; Granger, M. C. Controlling analyte electrochemistry in an electrospray ion source with a three-electrode emitter cell. *Anal. Chem.* **2004**, *76*, 1493–1499.
89. Kertesz, V.; Van Berkel, G. J.; Granger, M. C. Study and application of a controlled-potential electrochemistry–electrospray emitter for electrospray mass spectrometry. *Anal. Chem.* **2005**, *77*, 4366–4373.
90. Zhou, F.; Van Berkel, G. J. Electrochemistry combined on-line with electrospray mass spectrometry. *Anal. Chem.* **1995**, *67*, 3643–3649.
91. Zhang, T.; Palii, S. P.; Eyler, J. R.; Brajter-Toth, A. Enhancement of ionization efficiency by electrochemical reaction products in on-line electrochemistry/electrospray ionization Fourier transform ion cyclotron resonance mass spectrometry. *Anal. Chem.* **2002**, *74*, 1097–1103.
92. Kertesz, V.; Van Berkel, G. J. Expanded use of a battery-powered two-electrode emitter cell for electrospray mass spectrometry. *J. Am. Soc. Mass Spectrom.* **2006**, *17*, 953–961.
93. Van Berkel, G. J.; Zhou, F.; Aronson, J. T. Changes in bulk solution pH caused by the inherent controlled-current electrolytic process of an electrospray ion source. *Int. J. Mass Spectrom. Ion Processes* **1997**, *162*, 55–67.
94. Pan, P.; Gunawardena, H. P.; Xia, Y.; McLuckey, S. A. Nanoelectrospray ionization of protein mixtures: Solution pH and protein p*I*. *Anal. Chem.* **2004**, *76*, 1165–1174.
95. Wang, W.; Kitova, E. N.; Klassen, J. S. Influence of solution and gas phase processes on protein–carbohydrate binding affinities determined by nanoelectrospray Fourier transform ion cyclotron resonance mass spectrometry. *Anal. Chem.* **2003**, *75*, 4945–4955.
96. Ijames, C. F.; Dutky, R. C.; Fales, H. M. Iron carboxylate oxygen-centered-triangle complexes detected during electrospray use of organic acid modifiers with a comment on the Finnigan TSQ-700 electrospray inlet system. *J. Am. Soc. Mass Spectrom.* **1995**, *6*, 1226–1231.
97. Rohner, T. C.; Girault, H. H. Study of peptide on-line complexation with transition-metal ions generated from sacrificial electrodes in thin-chip polymer microsprays. *Rapid Commun. Mass Spectrom.* **2005**, *19*, 1183–1190.
98. Hop, C. E. C. A.; Saulys, D. A.; Gaines, D. F. Electrospray mass spectrometry of borane salts: The electrospray needle as an electrochemical cell. *J. Am. Soc. Mass Spectrom.* **1995**, *6*, 860–865.
99. Paim, L. A.; Augusti, D. V.; Dalmázio, I.; de, A.; Alves, T. M.; Augusti, R.; Siebald, H. G. L. Electrospray ionization and tandem mass spectrometry characterization of novel heterotrimetallic Ru(η5-C5H5)(dppf)SnX3 complexes and their heterobimetallic Ru(η5-C5H5)(dppf)X Precursors. *Polyhedron* **2005**, *24*, 1153–1159.
100. Rondeau, D.; Rogalewicz, F.; Ohanessian, G.; Levillain, E.; Odobel, F.; Richomme, P. Electrolytic electrospray ionization mass spectrometry of quaterthiophene-bridged bisporphyrins: Beyond the identification tool. *J. Mass Spectrom.* **2005**, *40*, 628–635.
101. Schoener, D. F.; Olsen, M. A.; Cummings, P. G.; Basic, C. Electrospray ionization of neutral metal dithiocarbamate complexes using in-source oxidation. *J. Mass Spectrom.* **1999**, *34*, 1069–1078.
102. Henderson, W.; Olsen, G. M. Application of electrospray mass spectrometry to the characterization of sulfonated and ferrocenyl phosphines. *Polyhedron* **1998**, *17*, 577–588.
103. McCarley, T. D.; Lufaso, M. W.; Curtin, L. S.; McCarley, R. L. Multiply charged redox–active oligomers in the gas phase: Electrolytic electrospray ionization mass spectrometry of metallocenes. *J. Phys. Chem. B.* **1998**, *102*, 10078–10086.
104. Bond, A. M.; Colton, R.; Fiedler, D. A.; Field, L. D.; He, T.; Humphrey, P. A.; Lindall, C. M.; Marken, F.; Masters, A. F.; Schumann, H.; Sühring, K.; Tedesco, V. Coupled redox reactions, linkage isomerization, hydride formation, and acid–base relationships in the decaphenylferrocene system. *Organometallics* **1997**, *16*, 2787–2797.
105. Hop, C. E. C. A.; Brady, J. T.; Bakhtiar, R. Transformation of neutral rhenium compounds during electrospray ionization mass spectrometry. *J. Am. Soc. Mass Spectrom.* **1997**, *8*, 191–194.

106. Lau, R. L. C.; Jiang, J.; Ng, D. K. P.; Chan, T.-W. D. Fourier transform ion cyclotron resonance studies of lanthanide (III) porphyrin-phthalocyanine heteroleptic sandwich complexes by using electrospray ionization. *J. Am. Soc. Mass Spectrom.* **1997**, *8*, 161–169.
107. Kane-Maguire, L. A. P.; Kanitz, R.; Sheil, M. M. Electrospray mass spectrometry of neutral π-hydrocarbon organometallic complexes. *Inorg. Chim. Acta* **1996**, *245*, 209–214.
108. Van Berkel, G. J.; Quiñoñes, M. A.; Quirke, J. M. E. Geoporphyrin analysis using electrospray ionization–mass spectrometry. *Energy & Fuels* **1993**, *7*, 411–419.
109. Guaratini, T.; Vessecchi, R.; Pinto, E.; Colepicolo, P.; Lopes, N. P. Balance of xanthophylls molecular and protonated molecular ions in electrospray ionization. *J. Mass Spectrom.* **2005**, *40*, 963–968.
110. Li, H.; Tyndale, S. T.; Heath, D. D.; Letcher, R. J. Determination of carotenoids and all *trans*-retinol in fish eggs by liquid chromatography–electrospray ionization–tandem mass spectrometry. *J. Chromatogr.* **2005**, *816*, 49–56.
111. Guaratini, T.; Vessecchi, R. L.; Lavarda, F. C.; Campos, P. M. B. G. M.; Naal, Z.; Gates, P. J.; Lopes, N. P. New chemical evidence for the ability to generate radical molecular ions of polyenes from ESI and HR-MALDI mass spectrometry. *Analyst* **2004**, *129*, 1223–1226.
112. Maziarz, III, E. P.; Wood, T. D. Gas phase dimerization of dimethylaniline in an external electrospray Fourier transform mass spectrometer. *J. Mass Spectrom.* **1998**, *33*, 45–54.
113. Quirke, J. M. E.; Van Berkel, G. J. Electrospray tandem mass spectrometry study of ferrocene carbamate ester derivatives of saturated primary, secondary, and tertiary alcohols. *J. Mass Spectrom.* **2001**, *36*, 179–187.
114. Quirke, J. M. E.; Hsu, Y.-L.; Van Berkel, G. J. Ferrocene-based electroactive derivatizing reagents for the rapid selective screening of alcohols and phenols in natural product mixtures using electrospray tandem mass spectrometry. *J. Nat. Prod.* **2000**, *63*, 230–237.
115. Willams, D.; Young, M. K. Analysis of neutral isomeric low molecular weight carbohydrates using ferrocenyl boronate derivatization and tandem electrospray mass spectrometry. *Rapid Commun. Mass Spectrom.* **2000**, *14*, 2083–2091.
116. Young, M. K.; Dinh, N.; Williams, D. Analysis of *N*-acetylated hexosamine monosaccharides by ferrocenyl boronation and tandem electrospray ionization mass spectrometry. *Rapid Commun. Mass Spectrom.* **2000**, *14*, 1462–1467.
117. Van Berkel, G. J.; Quirke, J. M. E.; Dilley, A.; Tigani, R.; Covey, T. R. Derivatization for electrospray ionization mass spectrometry. 3. Electrochemically-ionizable derivatives. *Anal. Chem.* **1998**, *70*, 1544–1554.
118. Kozlovski, V.; Brusov, V.; Sulimenkov, I.; Pikhtelev, A.; Dodovov, A. Novel experimental arrangement developed for direct fullerene analysis by electrospray time-of-flight mass spectrometry. *Rapid Commun. Mass Spectrom.* **2004**, *18*, 780–786.
119. Rondeau, D.; Martineau, C.; Blanchard, P.; Roncali, J. Probing electrochemical properties of π–Conjugated thienylenevinylenes/fullerene C60 adducts by ESI-MS: Evidence for dimerized cation-radicals. *J. Mass Spectrom.* **2002**, *37*, 1081–1085.
120. Rondeau, D.; Kreher, D.; Cariou, M.; Hudhomme, P.; Gorgues, A.; Richomme, P. Electrolytic electrospray ionization mass spectrometry of C_{60}–TTF–C_{60} derivatives: High resolution mass measurement and molecular ion gas phase reactivity. *Rapid Commun. Mass Spectrom.* **2001**, *15*, 1708–1712.
121. Mayer, C. R.; Roch-Marchal, C.; Lavanant, H.; Thouvenot, R.; Sellier, N.; Blair, J.-C.; Secheresse, F. New organosilyl derivatives of the Dawson polyoxometalate $[\alpha_2\text{-}P_2W_{17}O_{61}(RSi)_2O]^{6-}$: Synthesis and mass spectrometric investigation. *Chem. Eur. J.* **2004**, *10*, 5517–5523.
122. Hiraoka, K.; Aizawa, K.; Murata, K.; Fujimaki, S. Electrochemical reduction of highly-sensitive analysis of iodine in electrospray mass spectrometry. *J. Mass Spectrom. Soc. Jpn.* **1995**, *43*, 77–83.
123. Barrow, M. P.; Feng, X.; Wallace, J. I.; Boltalina, O. V.; Taylor, R.; Derrick, P. J.; Drewello, T. Characterization of fullerenes and fullerene derivatives by nanospray. *Chem. Phys. Lett.* **2000**, *330*, 267–274.
124. Dupont, A.; Gisselbrecht, J.-P.; Leize, E.; Wagner, L.; Van Dorsselaer, A. Electrospray mass spectrometry of electrochemically ionized molecules: Application to the study of fullerenes. *Tetrahedron. Lett.* **1994**, *35*, 6083–6086.
125. Jinno, K.; Sato, Y.; Nagashima, H.; Itoh, K. Separation and identification of higher fullerenes by high-performance liquid chromatography coupled with electrospray ionization mass spectrometry. *J. Microcolumn Sep.* **1998**, *10*, 79–88.
126. Qi, L.; Zhang, C.; Wei, X.-W.; Wu, M.-F.; Xu, Z.; Zhou, K.-Y.; Cao, Y.-C. The direct negative-ion ESI-MS analysis of neutral fullerenes. *Acta Chim. Sinica* **1997**, *55*, 498–502.
127. Khairallah, G.; Peel, J. B. Identification of dianions of C_{84} and C_{90} by electrospray mass spectrometry. *Chem. Phys. Lett.* **1998**, *296*, 545–548.
128. Zhou, F.; Van Berkel, G. J.; Donovan, B. T. Electron-transfer reactions of $C_{60}F_{48}$. *J. Am. Chem. Soc.* **1994**, *116*, 5485–5486.
129. Drewello, T.; Franuendorf, H.; Herzschuh, R.; Goryunkov, A. A.; Strauss, S. H.; Boltalina, O. V. The formation of long-lived fluorofullerene dianions by direct electrospray ionization. *Chem. Phys. Lett.* **2005**, *405*, 93–96.
130. Khairallah, G.; Peel, J. B. Cyano adduct anions of C_{70}: Electrospray mass spectrometric studies. *J. Phys. Chem. A* **1997**, *101*, 6770–6774.

131. Felder, D.; Nierengarten, H.; Gisselbrecht, J. P.; Boudon, C.; Leize, E.; Nicoud, J. F.; Gross, M.; Van Dorsselaer, A.; Nierengarten, J. F. Fullerdendrons: synthesis, electrochemistry and reduction in the electrospray source for mass spectrometry analysis. *New J. Chem.* **2000**, *24*, 687–695.
132. Bard, A. J.; Ledwith, A.; Shine, H.;In Gold, V.; Bethell, D.,(Eds.), *Advances in Physical Organic Chemistry* Vol. *13*, Academic Press, New York, **1976**, pp. 15–278.
133. Foster, R. *Organic Charge Transfer Complexes*, Academic Press, New York, **1969**.
134. Sierra, M. A.; Gómez-Gallego, M.; Mancheño, M. J.; Martínez-Alvarez, R.; Ramírez-López, P.; Kayali, N.; González, A. Electrospray mass spectra of group 6 (Fischer) carbenes in the presence of electron donor compounds. *J. Mass Spectrom.* **2003**, *38*, 151–156.
135. Wulff, W. D.; Korthals, K. A.; Martínez-Álvarez, R.; Gómez-Gallego, M.; Fernández, I.; Sierra, M. A. Study of the ESI–mass spectrometry ionization mechanism of Fischer carbene complexes. *J. Org. Chem.* **2005**, *70*, 5269–5277.
136. Marjasvaara, A.; Torvinen, M.; Vainiotalo, P. Laccase-catalyzed mediated oxidation of benzyl alcohol: The role of TEMPO and formation of products including benzonitrile studied by nanoelectrospray ionization Fourier transform ion cyclotron resonance mass spectrometry. *J. Mass Spectrom.* **2004**, *39*, 1139–1146.
137. Dayon, L.; Josserand, J.; Girault, H. H. Electrochemical multi-tagging of cysteinyl peptides during microspray mass spectrometry: Numerical simulation of consecutive reactions in a microchannel. *Phys. Chem. Chem. Phys.* **2005**, *7*, 4054–4060.
138. Dayon, L.; Roussel, C.; Prudent, M.; Lion, N.; Girault, H. H. On-line counting of cysteine residues in peptides during electrospray ionization by electrogenerated tags and their application to protein identification. *Electrophoresis* **2005**, *26*, 238–247.
139. Roussel, C.; Dayon, L.; Jensen, H.; Girault, H. H. On-line cysteine modification for protein analysis: New probes for electrochemical tagging nanospray mass spectrometry. *J. Electroanal. Chem.* **2004**, *570*, 187–199.
140. Roussel, C.; Dayon, L.; Lion, N.; Rohner, T. C.; Josserand, J.; Rossier, J. S.; Jensen, H.; Girault, H. H. Generation of mass tags by the inherent electrochemistry of electrospray for protein mass spectrometry. *J. Am. Soc. Mass Spectrom.* **2004**, *15*, 1767–1779.
141. Roussel, C.; Rohner, T. C.; Jensen, H.; Girault, H. H. Mechanistic aspects of on-line electrochemical tagging of free L-cysteine residues during electrospray ionization for mass spectrometry in protein analysis. *ChemPhysChem 2003, 4*, 200–206.
142. Rohner, T. C.; Rossier, J. S.; Girault, H. H. On-line electrochemical tagging of cysteines in proteins during nanospray. *Electrochem. Commun.* **2002**, *4*, 695–700.
143. Lu, Y.; King, F. L.; Duckworth, D. C. Electrochemically-induced reactions of hexafluorophosphate anions with water in negative ion electrospray mass spectrometry of undiluted ionic liquids. *J. Am. Soc. Mass Spectrom.* **2006**, *17*, 939–944.

Chapter 4

ESI Source Design

Andries P. Bruins

Mass Spectrometry Facility, University of Groningen, Groningen, The Netherlands

Electrospray and MALDI Mass Spectrometry: Fundamentals, Instrumentation, Practicalities, and Biological Applications, Second Edition, Edited by Richard B. Cole

4.1 ELECTROSPRAY NEBULIZATION

The nebulization of liquids by electrical forces is carried out on a very small scale in the nanoelectrospray accessory for mass spectrometry, but it is performed on a grand scale in the electrostatic spray painting of automobiles and the electrostatic spray deposition of pesticides on crops.[1] Electrospray is the dispersion of a liquid into electrically charged droplets, and thus it combines two processes: droplet formation and droplet charging.[2–4] The formation of small, micron-sized droplets does not present a problem if the liquid flow rate, surface tension, and electrolyte concentration are low (1–10 mM). An increase of one or more of these variables makes it more difficult for the electric field to produce the desired charged aerosol for mass spectrometry. The electric field strength at the sprayer tip can be increased to try and overcome the adverse effects of the aforementioned three variables, but too high an electric field will give rise to an electrical discharge that accompanies the electrospray process. A discharge can be tolerated in some spray applications, but is detrimental in electrospray mass spectrometry. Electrical discharge is particularly troublesome in the formation of negatively charged droplets.

Modifications to the simple electrospray system as shown in Figure 4.1 are aimed at increasing the tolerance toward adverse effects of high liquid flow rate, surface tension, and electrolyte concentration. Dilution of an aqueous solution with an organic solvent reduces surface tension. Coaxial addition of a sheath flow of methanol, acetonitrile, ethanol, isopropanol, or 2-methoxyethanol to the sample solution at the tip of the spray capillary was first used for the combination of capillary electrophoresis with electrospray MS and later also used for sample infusion and LC coupling with electrospray MS.[5,6] In sheath-flow-assisted electrospray, it is still the electric field alone that has to disperse *and* charge the liquid in one operation. In industrial applications of electrosprays the input of mechanical energy is used for the dispersion of liquids at high flow rates. Droplet charging is done by exposing the mechanical sprayer to a high electric field. For example, a rotating disk combined with an electric field created by a 100-kV power supply may be used in spray painting.

The assistance of a high-velocity gas flow or ultrasonic vibration is used in electrospray mass spectrometry.[7–9] In a simple approximation, the pneumatic or ultrasonic nebulizer takes care of aerosol formation, while the electric field does the droplet charging. When

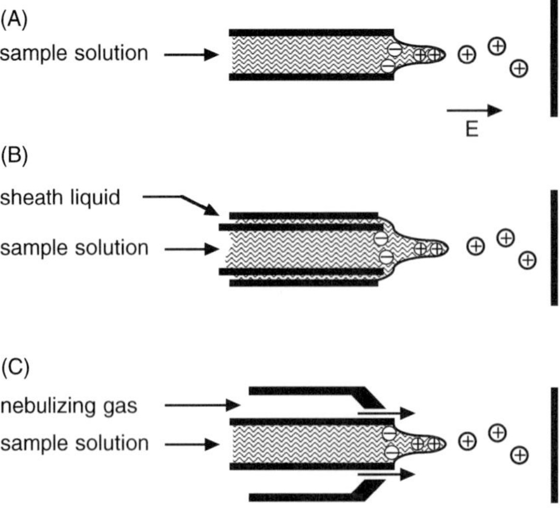

Figure 4.1. Aerosol formation by electrospray. (**a**) Simple electrospray. (**b**) electrospray with sheath flow. (**c**) Electrospray with pneumatic assistance.

compared with "pure" electrospray, pneumatically assisted electrospray can handle aqueous solutions and higher flow rates without the need for critical adjustment, and it can be operated at a lower field strength so that electrical discharge is eliminated. Ultrasonic assistance may offer the same advantages, but is not offered as a commercially available option by major mass spectrometer manufacturers. Trade names are IonSpray (SCIEX) and Ultraspray (Analytica of Branford).

Charged droplets generated by electrospray spread out in a wide angle. A moderately fast coaxial gas flow can be used to try and keep the droplets in a narrower beam. Such pneumatic focusing can be combined with a coaxial sheath flow[6,10] and with ultrasonic nebulization.[9] The focusing gas flow is not intended for pneumatic nebulization. Pneumatic focusing increases the number of ions transported into the vacuum envelope of the mass spectrometer. Pneumatically assisted electrospray produces a narrow beam of droplets by the very nature of the high-velocity gas flow used in the concentric nebulizer. High-flow electrospray makes use of pneumatic nebulization together with the supply of heat in order to assist the evaporation and shrinking of the primary aerosol droplets.

Pneumatically assisted electrospray can work at liquid flow rates up to several hundred microliters per minute. Pure, unassisted electrospray can be extended to very low flow rates. Microelectrospray or nanospray has the advantage of extremely efficient use of a sample solution: One microliter can produce ions for about 40 min, long enough for performing a number of MS/MS experiments. Very low flow rate nanospray is compatible with the electroosmotic flow in a capillary electrophoresis column, eliminating the need for a make-up flow.

A regular electrospray device is constructed from approximately 0.3-mm-o.d., 0.1-mm-i.d. stainless-steel tubing, or a coaxial arrangement of fused silica and stainless steel capillaries. Stability at low flow in nanospray requires a miniaturized version having narrower i.d. and o.d. A 10- to 50-μm-i.d. fused silica capillary is tapered on the outside by etching with hydrofluoric acid[11] or by mechanical abrasion.[12] Nanospray needles can be fabricated by pulling glass or fused silica to a tip of a few-micrometer internal diameter. Examples of electrical contact with the liquid stream are the application of silver paint,[11] or by deposition of gold.[12] Alternatively, the electrical contact can be made upstream from the tip by clamping the narrow-bore fused silica capillary in a metal union.[13,14] Caprioli and co-workers[14] have prepared an integrated microcolumn/microsprayer with the aim of on-line concentration of dilute samples.[14] Wilm and Mann[15] have developed nanoelectrospray for the analysis of small volumes of peptide solutions at a flow rate of about 25 nL/min. A 1-mm-o.d, 0.6-mm-i.d. glass capillary is pulled out to an i.d. of 1–3 μm by means of an electrode puller. The outside of the capillary is coated with gold. Approximately 1 μL of a sample solution is loaded into the capillary, and gas pressure is applied to push the liquid toward the tip. Nanoelectrospray emitters can be purchased from a number of suppliers.

4.2 ELECTROSPRAY CONSTRUCTION AND OPERATION

4.2.1 High-Voltage Connection

The tip of an electrospray capillary is exposed to a high electric field. In principle, the field can be generated by (a) connecting the sprayer to a high-voltage supply and the ion source to ground, (b) grounding the sprayer and connecting the source to high voltage, and (c) connecting both the sprayer and the source to separate power supplies set to different

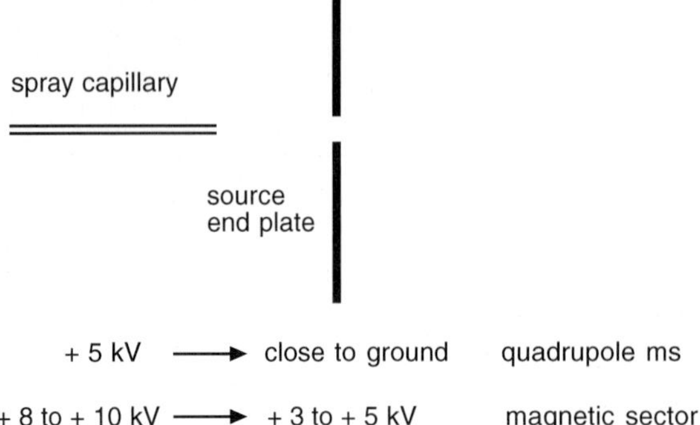

Figure 4.2. Voltage arrangements for electrospray capillary and source; no isolation between source end plate and ms accelerating voltage.

voltages. Selection of either of these options depends on the kind of mass analyzer and on the system chosen for the transportation of ions from the atmospheric pressure source region into the vacuum of the mass analyzer. In a quadrupole mass spectrometer the ion source can be at ground potential and the sprayer can be at up to ± 5 kV.[7,16] In most magnetic sector mass spectrometers the accelerating voltage is limited to 3–5 kV during electrospray operation, so that the voltage on the sprayer is 8–10 kV with respect to ground. See Figure 4.2.

Safety is not a major problem, since the source can be designed for protection of the operator by the use of appropriate insulation and safety interlocks.

If the sprayer is at high voltage, the connection to the HPLC or infusion pump can be made from fused silica capillary or PEEK tubing. One should keep in mind that a fused silica capillary filled with an electrolyte solution is not an insulator but a resistor, which may conduct enough to give the operator an itch if the needle or plunger of a syringe in an infusion pump is touched during electrospray operation. For safety reasons, one should not make or break a connection between syringe and transfer line when the electrospray high voltage is on. As a further safety measure, the needle of the syringe in an infusion pump can be connected to ground. A number of manufacturers place a grounded metal union upstream from the sprayer, and in this way the above-mentioned safety problem is eliminated.

Of more practical value is the question of whether or not the high voltage on the sprayer interferes with operation of a capillary electrophoresis system, and does the high voltage have an influence on the transportation of a sample through a fused silica or PEEK transfer line?

Often overlooked is the possibility of electromigration in the transfer line between a grounded union and the electrosprayer as shown in Figure 4.3. To conserve the sample, the liquid flow rate may be reduced to 1 μL/min, corresponding to a linear velocity of 50 cm/min in a 50-μm-i.d. transfer line or 12 cm/min in a 100-μm-i.d. line, or 4 cm/min in a 175-μm-i.d. line. If a positively charged sample molecule has a mobility of 10^{-3} $cm^2/V \cdot s$ and if a 50-cm-long transfer line is used with 5 kV on the electrospray capillary, the electrophoretic velocity of the sample is 6 cm/min in the direction from the sprayer back into the union. In this example, there should be no problem if a 50-μm-i.d. transfer line is used: The liquid velocity far exceeds the opposing electrophoretic velocity. A 100-μm-i.d. capillary (liquid velocity 12 cm/min at 1 μl/min) can still be used at 5 kV. The situation would be critical at 1 μL/min in a wider-bore transfer line (PEEK tubing), and also in the case of electrospray in a magnetic sector instrument, where the sprayer may be at $+8$ kV or higher. The narrower the internal diameter of the transfer line, the higher the liquid velocity and the smaller the chance that electrophoretic velocity exceeds the liquid velocity.

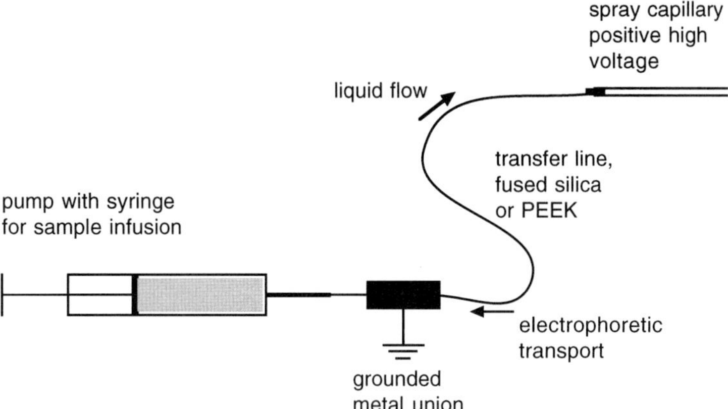

Figure 4.3. Possibility of electromigration of positively charged sample ions against liquid flow.

Operation of the electrospray capillary at ground potential is a practical advantage in CE/MS coupling, in an "on-chip LC" with integrated injector valve and ESI-tip, and in experiments with an electrochemical cell connected to an ESI ion source. The electric field necessary for the formation of positively charged droplets is generated by floating a counterelectrode inside the ion source at a negative high voltage. Ions are drawn from the electrspray capillary toward the opposing electrode. In order to accelerate ions into a mass analyzer inside the vacuum system, a more negative potential has to be applied, which would lead to a very unpractical arrangement of negative high voltages. An elegant system that eliminates the need for acceleration of positive ions toward negative high voltage is the glass capillary for the transportation of ions from the atmospheric pressure spray region into the vacuum of the mass spectrometer (see Figure 4.4).[17] Ions generated in the atmospheric pressure region of the source either by electrospray or by atmospheric pressure chemical ionization are drawn into the vacuum region via a glass tube that is metallized at both ends. The linear velocity of the gas flow in the tube is high enough to drag positive ions into the vacuum even if the inlet of the tube is at a negative high voltage and the outlet is at ground

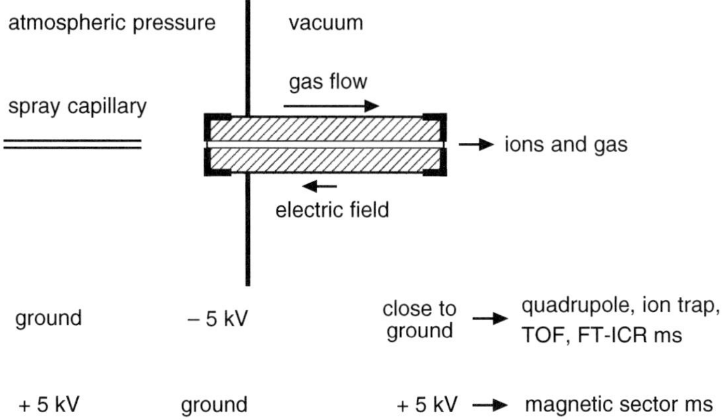

Figure 4.4. Voltage arrangements for electrospray capillary and source; isolation between source end plate and accelerating voltage by a glass capillary for ion transport.

potential. The grounded outlet now serves as a source of ions that can easily be guided into a quadrupole, ion trap, TOF, or FT-ICR mass analyzer. The atmospheric pressure source and ion optics plus mass analyzer are electrically isolated from one another, which is particularly attractive for magnetic sector instruments. The outlet end of the glass capillary is at the accelerating potential of the mass spectrometer, but there is no need to have the spray chamber and the spray capillary at increasingly high voltages. In the example given in Figure 4.4, a magnetic sector instrument has the spray chamber and capillary inlet at ground, while the electrospray capillary is held at $+5$ kV.[18,19] Magnetic sector mass spectrometers have become less relevant for LC-MS. At present, TOF, FT-Orbitrap, and FT-ICR mass spectrometers provide high-resolution mass spectra to LC-MS users.

4.2.2 Electrospray and Electrical Discharge

Corona discharge between the electrospray capillary and its counterelectrode drastically changes the appearance of the mass spectrum. Without a discharge, the spectrum represents ions present in the solution that is dispersed into charged droplets. When a discharge takes place, the spectrum represents the products of ion–molecule reactions, as is the case in atmospheric pressure chemical ionization. At the same time, many ions present in solution can no longer be observed.[16] A distinction must be made here between sample ions that exist only as ions in solution, such as quaternary ammonium ions, and ions that can be formed in solution, but can also be generated by ion–molecule reactions, such as amines and other reasonably volatile neutrals. Quaternary ammonium ions completely disappear when a discharge is established. Ionized neutrals may still be observed.

Electrical discharge takes place very readily if the electrospray needle is at a negative potential with respect to the spray chamber, since field emission of electrons from a sharp point does not require a very high field. Electrons are eliminated by flushing the space around the spray tip with an electron scavenging gas such as oxygen[20] or SF6, or by the addition of a halogenated solvent to the liquid stream.[21,22] In a discharge, both positive and negative charge carriers are formed, which will recombine with droplets having the opposite charge.[23] As a result, droplet charge is neutralized and the formation of sample ions from charged droplets can no longer take place. Electrical discharge is prevented to a large extent by reduction of the electric field at the sprayer tip. Pneumatically assisted electrospray can be operated at a lower electric field than pure electrospray. In routine operation of pneumatically assisted electrospray, the use of air as a nebulizing gas is preferred over nitrogen if a discharge has to be minimized.

A controlled discharge can be used to reduce the number of charges on multiply charged ions generated by electrospray.[24,25] Ion–ion reactions can be initiated with a radioactive α particle source[24] or with a corona discharge.[25]

4.2.3 Flow Rate and Sensitivity; Electrospray Versus Nanospray

Electrospray is a low-flow-rate technique. When the flow rate is increased for a given sample concentration, the analyte ion signal does not increase. So in terms of sample concentration, sensitivity remains constant, but in terms of mass flow the sensitivity drops when the flow rate of the sample solution is increased. This is an unusual situation in mass spectrometry because a mass spectrometer operating in the electron ionization or chemical ionization mode is a mass-flow-sensitive detector.

Table 4.1. Relationship between Some Key Variables and Liquid Flow Rate in Electrospray Ionization

Flow Rate (μL/min)	6	0.2	0.006
Droplet diameter (μm)	**1**	0.3	0.1
Spray current (nA)	**500**	90	16
Droplets/second	2.10^8	2.10^8	2.10^8
Sample molecules per droplet	940	30	1
Charges per droplet	15,000	2700	500
Charges: sample molecules	16:1	90:1	500:1

Example: sample 3 μM, acetic acid 1%.

Apparent concentration sensitivity of electrospray can be attributed to a decrease in droplet charging efficiency and a shift toward larger diameters in the droplet size distribution if the liquid flow rate is increased, according to the scaling laws in Eqs. (4.1) and (4.2):

$$\text{diameter} \propto [\text{flow}]^{1/3} \tag{4.1}$$

$$\text{current} \propto [\text{flow}]^{1/2} \tag{4.2}$$

Both effects lead to a lower ionization efficiency. In Table 4.1, a calculation is given for a few key figures at different flow rates. Reasonable assumptions are made for spray current and droplet diameter at the flow rate of 6 μL/min. The numbers for other flow rates are calculated using Eqs. (4.1) and (4.2). Numbers in Table 4.1 refer to the primary aerosol. The generation of offspring droplets is not taken into account. Furthermore, an extrapolation to a flow rate above 6 μL/min is not presented in this table because operation above 10 μL/min is always done with pneumatic assistance, and in this case the scaling equations no longer apply. Two conclusions can be drawn from the numbers presented in Table 4.1: First, the ratio of charges to molecules is most favorable at low flow rates; and second, at these low flow rates there is one sample molecule (or sample ion) per primary aerosol droplet. It is very reasonable to assume that the release of ions from droplets is most favorable if molecules face less competition for charge. When molecules have acquired a charge, they have to compete with other sample ions in order to be released from an offspring droplet. At low flow rate, this latter competition may already be absent in the primary aerosol, as exemplified in the case of 6 nL/min flow rate and 3 μM sample concentration. The reduced efficiency of the release of ions from droplets at higher flow rates is approximately compensated for by the introduction of more sample molecules per unit time into the interface so that the signal level at the detector remains constant. Schneider et al.[26] have demonstrated that apparent concentration sensitivity is largely due to insufficient evaporation of solvent from the electrospray aerosol and insufficient transportation of ions into the vacuum and mass analyzer. Passing the aerosol through a heated desolvation chamber, as shown in Figure 4.5, restores mass flow sensitivity in the nanospray flow rate regime.[26]

If the liquid flow rate is increased too much, the ion signals become lower and less stable. The practical upper limit to flow rate in pure electrospray is 10–20 μL/min, depending on the composition of the solvent and on the use of a coaxial sheath flow (see part 4.1). Pneumatically assisted electrospray has been used up to 1 mL/min. High-flow electrospray is always combined with the supply of heat to assist evaporation of solvents. In a commercial embodiment of high-flow ionspray (called Turbo-IonSpray) the spray plume

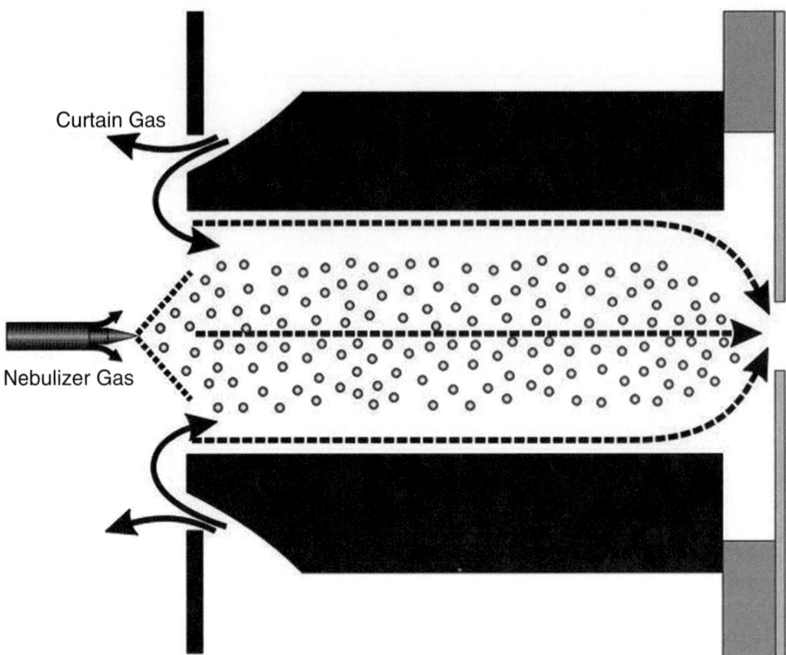

Figure 4.5. Desolvation chamber for use with nanospray. (Reproduced from Ref. 26.)

is mixed with a hot air flow from one or two heat guns in the region in front of the ion sampling orifice of the source.[27]

In ion sources based on the heated tube design by Chait and co-workers[28] as used in ThermoFinnigan mass spectrometers, the tube temperature is raised in the case of high-flow operation. Desolvation of charged droplets takes place mostly inside the atmospheric pressure source region, and ions are drawn into the heated tube together with solvent vapor and gas supplied to the source. Not only desolvated ions, but also charged aerosol, is drawn through the tube. Desolvation of the aerosol inside the tube generates sample ions that are transported into the mass analyzer, but nonvolatile material may be deposited inside the tube and eventually block the tube.

High-flow electrospray is compatible with 2-mm-and-wider bore columns in liquid chromatography. The need for eluent splitting is eliminated, but at the same time the load of nonvolatile contaminants on the ion source or transfer tube is much higher. It may be better to feed a low flow rate into the sprayer and use a splitter, unless the complexity of a splitter or sample losses to waste must be avoided.

A splitter can be set up using two capillary tubes of different length and diameter. At a given pressure drop the flow rate through a capillary is proportional to the fourth power of the diameter and inversely proportional to the length of the capillary. For example, a 1-m-long 50-μm-i.d. transfer line together with a 16-cm-long 100-μm side arm gives a split ratio of 1 : 100.

4.2.4 Position of the Sprayer Inside the Source

In the first publication of successful electrospray mass spectrometry[16] the spray capillary was positioned in the center of the source—that is, directly facing and concentric with the the ion sampling orifice of the ion source. When the experiments were repeated in Henion's group at Cornell University, the ion source of the SCIEX mass spectrometer was fully open, which allowed the sprayer to be moved around during full operation. By moving to an off-axis position, the ion signal observed by the mass spectrometer was more stable, and at least as high as in the on-axis position (Figure 4.6). It is desirable to take ions but no droplets into the mass analyzer. Charged droplets transported into the mass spectrometer may impinge on elements of the ion optics or mass analyzer and create bursts of ions that appear as spikes in the mass spectra. In the core of the aerosol generated by electrospray, the droplet diameter is larger than in the perimeter of the spray. Liberation of sample ions from small droplets should be more effective in the perimeter than in the core of the spray. There is also more chance of incomplete droplet desolvation in the core of the spray than in the perimeter.

Off–axis positioning is taken a step further by placing the sprayer in a diagonal or orthogonal position inside the source. The spray is not aimed at the sampling orifice but at a position 1 cm beyond, in order to reduce the chance of shooting droplets into the mass spectrometer. Diagonal or orthogonal positioning is most effective for pneumatically assisted electrospray. Stability is improved without loss of sensitivity. Pure

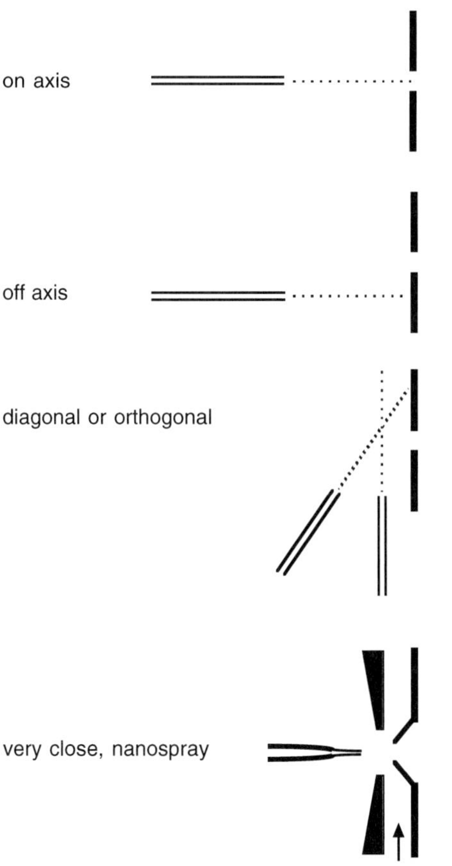

Figure 4.6. Arrangements for the position of the electrospray capillary in the ion source.

electrospray without pneumatic or ultrasonic assistance is more limited in freedom of positioning than the assisted sprays. The distance and voltage difference between the tip of the sprayer and the nearest part of the source determine the electric field that creates the spray and influences the performance of the spray. Optimization of sprayer position and spray voltage are interrelated. In the case of pneumatically assisted electrospray the generation and direction of the aerosol is determined by gas flow. The electric field has limited influence on the spray. Aerosol generation is independent of the position of the sprayer in the source. Optimization of sprayer position is sample-dependent. The best sprayer position for pneumatically assisted electrospray of proteins is closer to the sampling orifice and is in a narrower region than the best position for small molecules.

An important difference exists between regular electrospray and nanospray. In regular electrospray the distance between the tip of the sprayer and the ion sampling orifice of the mass spectrometer is around 1 cm. In pneumatically assisted electrospray, this distance is $\geq$2 cm, combined with a spray voltage of a few kilovolts; but in nanospray the capillary can be positioned exactly on axis, and the distance from the ion sampling orifice is reduced to a few millimeters, with concomitant reduction of the spray voltage[15,26] (Figure 4.6).

4.2.5 Electrical Aspects of the Combination of Capillary Electrophoresis with ESI

Capillary electrophoresis is usually conducted by applying approximately 30 kV between anode and cathode. In CE/MS with positive-ion operation, the CE anode is at 30 kV, while the CE cathode is at the high voltage of the electrospray interface–for example, at 5 kV. The voltage drop over the column is reduced to 25 kV, resulting in a longer analysis time.

There are additional electrical implications to CE/MS other than just the voltage drop over the CE column. For the sake of simplicity, the current passing through the CE column is assumed to be 100 μA, and the current carried by charged droplets in ESI (spray current) is assumed to be 1 μA in each section of Figure 4.7. In Figure 4.7A the simplest CE/MS system is drawn schematically. Current flowing from the CE power supply is carried away only to a minute extent as charge on sprayed droplets. Most of the current has to flow into the electrospray power supply, which has been designed to *supply* current, not to *sink* current. If the supply has to sink too much current, its internal feedback circuit that stabilizes the output voltage to a preset value cannot keep the power supply under control, resulting in an uncontrolled rise of output voltage and unstable electrospray. By connecting an additional load resistor as shown in Figure 4.7B, 125 μA has to flow through 40 MΩ at 5 kV (Ohm's law). Thus, the power supply is forced to deliver 125–99 = 26 μA, and output stability is maintained. In Figure 4.7C, the CE capillary is connected to a negative high-voltage power supply, while electrospray is operated in the positive mode. This situation exists, for example, if a coated CE capillary is used in which a reversed direction of the electroosmotic flow (EOF) takes place. Now, the electrospray power supply has to supply the sum of the load current through the resistor and supply the CE current through the CE column. Figures 4.7D through 4.7F depict the current flow in the case of negative-ion operation of the ESI source. In Figure 4.7 the convention of current is: positive charge flowing from plus toward minus.

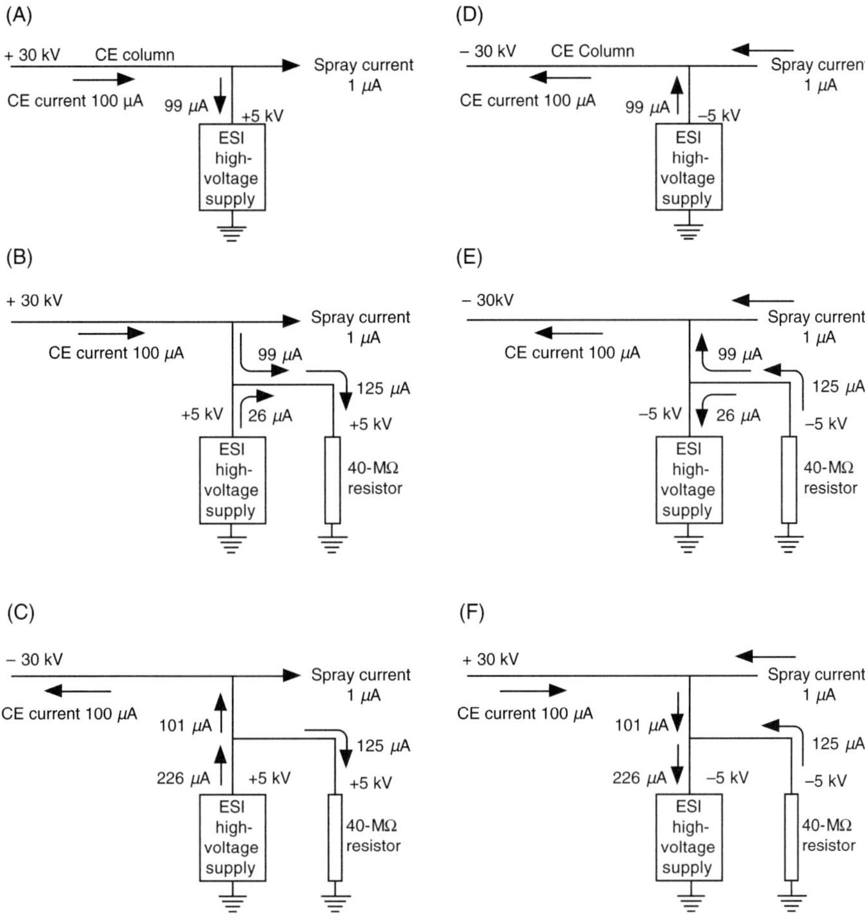

Figure 4.7. (**A**, **D**) Currents flowing in a simple CE/MS system using a CE high-voltage supply and an electrospray high-voltage supply; positive (**A**) and negative (**D**) ion operation, electrospray power supply has to sink 99 μA. (**B**, **E**) 40-MΩ load resistor connected between electrospray supply and ground carries 125 μA at 5 kV (Ohm's law); electrospray power supply has to deliver 26 μA to maintain current balance. (**C**, **F**) Currents flowing in reversed polarity CE connected with positive (**D**) and negative (**F**) ion electrospray with load resistor; electrospray power supply has to deliver 226 μA.

The user can make a reasonable estimate of the value of the load resistor by combining the expected values of ESI voltage and CE current in Ohm's law:

$$R_{\text{load}} \leq \frac{\text{ESI high } V}{\text{CE current}}$$

The user should also make sure that the ESI power supply can deliver the higher currents in Figures 4.7C and 4.7F. If the power supply cannot deliver sufficiently high currents, the load resistor should be left out in the case of reversed polarity of the CE high voltage (Figures 4.7C and 4.7F). The power dissipated in the load resistors in Figure 4.7 is $IV =$ 0.625 W, which cannot be handled by one small 1/8-W metal-fim resistor. It is recommended to use several resistors in series–for example, 10 resistors of 3.9 MΩ, 1/4 W each.

Of course these CE/MS current considerations do not apply if the spray capillary is at ground.

4.3 ATMOSPHERIC PRESSURE IONIZATION SOURCE CONSTRUCTION

4.3.1 Free-Jet Expansion into Vacuum

Electrospray ionization is one of the ionization techniques available for an ion source operating at atmospheric pressure.

Two problems are intimately related in atmospheric pressure ionization (API) and electrospray mass spectrometry: (a) the transport of ions from an atmospheric pressure region into the vacuum system of the mass spectrometer and (b) the strong cooling of a mixture of gas and ions when expanding into vacuum. The resulting condensation of polar neutrals (notably water and solvent vapor) on analyte ions produces cluster ions having a mass far beyond the range of most mass analyzers.

$$X^+ + nH_2O \rightarrow X^+(H_2O)_n$$

A simple scheme of the expansion of a gas into a low pressure region is given in Figure 4.8A. The principles of the generation of molecular beams[29,30] will be explained in qualitative terms. Inside the nozzle opening and behind the nozzle, the gas molecules gain a high velocity, and gas molecules with entrained ions follow straight stream lines originating approximately in the nozzle, the highest intensity of the gas flow being on the axis of the nozzle. Far away from the nozzle, the gas is pumped away, and gas molecules move at

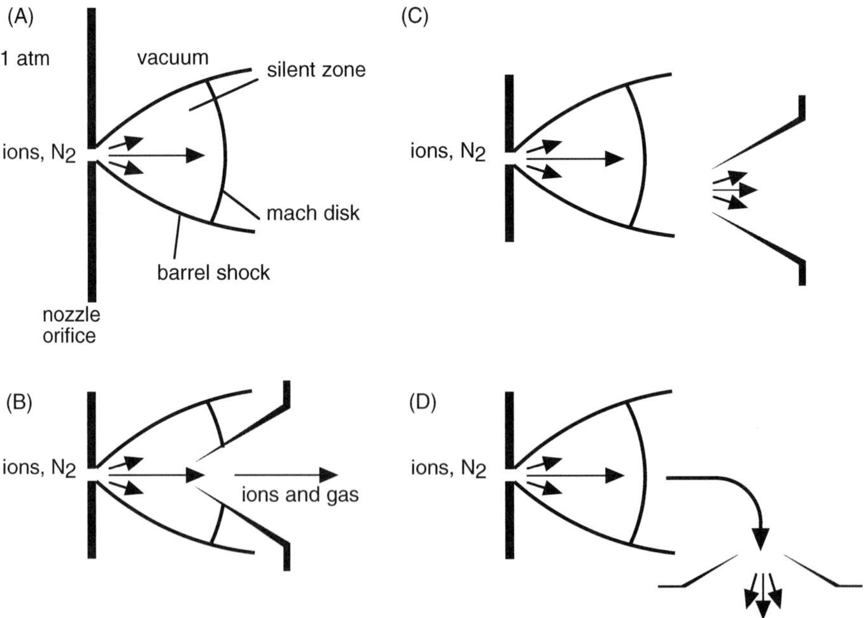

Figure 4.8. Free-jet expansion of gas and ions into vacuum. (**A**) Basic principle. (**B**) Arrangement with skimmer penetrating into the silent zone. (**C**) Skimmer located more distant than the mach disk. (**D**) Skimmer downstream and at right angle.

random. In the transition between directed motion and random motion, called shock waves, ions and gas molecules undergo many collisions, with scattering of the beam of ions and molecules as the result. The region inside the barrel shock wave and between the nozzle and the mach disk is called the silent zone, where ions and gas move at equal speeds in the same direction and undergo strong and rapid cooling.

The location of the mach disk is important for the design of an API mass spectrometer. The distance from the nozzle is given by

$$x_M = 0.67D_o\sqrt{\frac{P_0}{P_1}}$$

where D_o is the orifice or nozzle diameter, P_0 is the upstream (atmospheric) pressure, and P_1 is the downstream pressure in the vacuum chamber.[29] For a 0.1-mm nozzle diameter and 10^{-5}-torr vacuum generated by cryogenic pumping, the mach disk is located at 670 mm i.e. beyond the dimensions of a quadrupole mass spectrometer. In an expansion through a 0.3-mm-diameter orifice into a region pumped down to 1.2 torr by a 35-m^3/h rotary pump, the mach disk is located at 5 mm from the nozzle opening. x_M is fully determined by the pumping speed available in the area near the nozzle opening. A small pump or narrow pumping line, or other obstructions, will result in a small x_M. A 16-m^3/h rotary pump would give $x_M = 3$ mm. The gas flow through a 0.1-mm-i.d. (or bigger) nozzle is too large for the turbomolecular or oil diffusion pumps of the mass analyzer. Following the practice of molecular beam machines, the central portion of the beam can be sampled into the next vacuum stage by means of a skimmer,[29,30] as depicted in Figures 4.8B and 4.8C. In Figure 4.8B, the core of the beam is sampled from the silent zone, and the ions plus gas continue their movement in straight lines. Collection of ions from the gas flow through the skimmer should be efficient, due to the directional effect of the free-jet expansion. However, since the gas flowing in the molecular beam has strongly cooled upon expansion, ions will cluster with water molecules, if the latter are present. In Figure 4.8C, a sample is taken from behind the Mach disk. In Figure 4.8D, a further modification is depicted. The skimmer is located in an orthogonal position, and it is far downstream from the Mach disk. Ions and gas have undergone extensive scattering in Figures 4.8C and 4.8D, and the extraction and focusing of ions is more difficult. On the other hand, the gas temperature has risen through collisions in the Mach disk and during the turbulent or laminar flow toward the skimmer, with concomitant breaking of hydrogen bonds in cluster ions. The transmission of ions and gas through the skimmer is akin to the flow of ions and gas from a regular chemical ionization source.

The circumference of the skimmer opening in Figure 4.8B should be very sharp and free from burrs. The full angle of the skimmer cone is usually about 60°C in molecular beam apparatus[29,30] but can be wider in an API mass spectrometer. Fortunately, adequate transmission of ions (instead of a beam of neutrals) through the skimmer is subject to less stringent conditions of mechanical quality of the skimmer. A detailed description of the flow of gas and ions through a skimmer is presented by Jugroot et al.[31,32]

4.3.2 Cluster Ion Formation

In any design of an API mass spectrometer, the problem of formation of cluster ions is addressed. Polar molecules that tend to cluster with ions are water and solvent vapor present in air, or generated by the evaporation of the eluate from a liquid chromatograph or electrophoresis instrument connected to the API source. All practical designs of API instruments are aimed at either (a) prevention of clustering or (b) curing the problem by breaking the clusters.

4.3.2.1 Prevention

If ions are allowed but water vapor and other neutrals can be excluded from entrance into the vacuum system, formation of clusters is prevented. This can be achieved simply by forcing ions and neutrals in opposite directions by the opposing action of an electric field and a flow of dry gas, usually nitrogen, as is customary in ion mobility spectrometry. The flow of dry gas can be restricted to the area just in front of the ion sampling orifice, but can also be extended to the entire ion source. Figure 4.9A shows the essential part of the SCIEX API source. Ion sources by Agilent and Bruker work according to the same principle. Ions are pushed toward the curtain plate by the electric field from a corona discharge needle or electrospray capillary. The region between the interface plate and the sampling orifice plate is continuously flushed with dry nitrogen (SCIEX curtain gas: Agilent and Bruker drying gas). Part of the nitrogen flow goes to the left into the ion source, pushing water, neutral contaminants, and dust away from the sampling orifice. Ions that come close to the interface plate are driven to the right toward the sampling orifice by a 1000-V potential difference between the curtain plate and the orifice plate. The other part of the nitrogen flow goes through the sampling orifice and carries sample ions into the vacuum of the mass analyzer. Ions have thus passed through a so-called dry nitrogen gas curtain. Clusters of ions and nitrogen are not formed, in spite of strong cooling during the expansion into the vacuum system, since nitrogen lacks the ability to form hydrogen bridges.

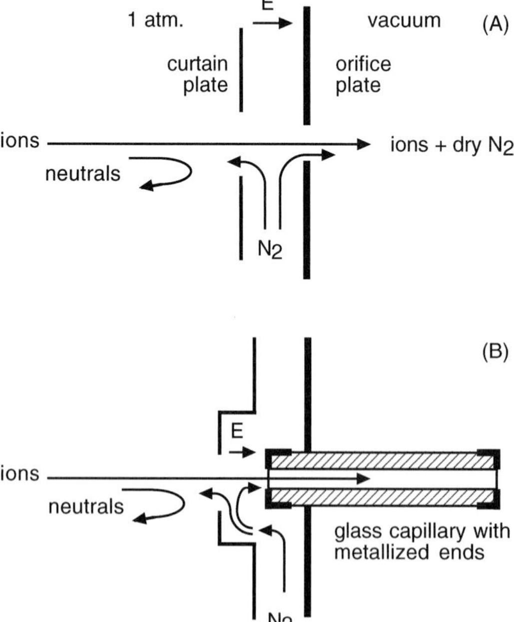

Figure 4.9. Prevention of cluster-ion formation and protection of sampling orifice or tube by means of: (**A**) gas curtain (SCIEX) and (**B**) countercurrent flow of bath gas (Analytica of Branford) or drying gas (Agilent and Bruker).

The countercurrent flow of nitrogen (bath gas or drying gas) through the electrospray ion source developed by Fenn and co-workers[17,20] and incorporated in Agilent and Bruker mass spectrometers is different in details but serves the same purpose (see Figure 4.9B).

A curtain gas or countercurrent flow is a very elegant solution for the clustering problem. As a further benefit, the gas curtain largely avoids contamination and blockage of the sampling orifice, making the API source more reliable. If the ion source is heated, the temperature of gas and ions remains high enough after the temperature drop resulting from the expansion, to prevent clustering. Another way to preheat the mixture of ions and neutrals prior to expansion is the heated transfer tube system developed by Chait and co-workers[28] and adopted by ThermoFinnigan. Ions and vapors pass through an approximately 20-cm-long, 0.5-mm-i.d. tube heated to 100–200°C. The heated tube may also be used to help in the desolvation of ions contained in droplets.

4.3.2.2 Curing

Ions pass through a region during the expansion into the vacuum, where the neutral gas density is falling rapidly. If ions moving in a low-density gas (pressure 10^{-3} to 1 torr) are accelerated by an electric field in this region, either between the nozzle orifice and the skimmer or between the skimmer and the first ion optics element, collisions take place that lead to the breaking of hydrogen bonds in cluster ions. Such declustering by collision-induced dissociation has already been described in an early study on corona discharges in air.[33] Ten years later, declustering by CID was adopted by Kambara and Kanomata.[34] Moderate acceleration of clusters is effective and widely used for declustering. The application of too much collision energy will not only strip off solvent molecules but also lead to fragmentation of the sample ion.

In an expansion system built according to Figures 4.8C and 4.8D, ions and gas pass through the mach disk where initially formed cluster ions are "heated" by collisions with randomly moving background gas. Water and solvent molecules will be partially removed from clusters. Downstream from the Mach disk, the ions and gas are in a viscous flow regime. By passing the viscous flow through a heated zone in Figures 4.8C and 4.8D, a final desolvation takes place and free sample ions can be delivered into the next vacuum stage. The Waters Z source and the Ionics HSID atmosphere–vacuum interface are built along these lines.

The temperature of heated zones can be tuned to accomodate the effects of liquid flow rate and the use of a high percentage of water in the eleuent of an HPLC.

4.3.3 Focusing of Ions at Atmospheric Pressure

The very short mean free path and high collision rate at atmospheric pressure prevent classical ion focusing inside an atmospheric pressure ion source. A correct combination of gas flow and electric field helps to guide the ions toward the orifice. Careful attempts to drive ions to the orifice were made by Eisele.[35] His aim was the sampling of very low concentrations of ions present in ambient air due to natural or other conditions. Ions were guided through a high-pressure ion optics region designed by computer simulation of electric fields. In spite of an extended static electric field converging toward the sampling orifice, no increase in ion signals was observed. The conclusion was drawn that the electric field immediately in front of the sampling orifice was most important. Calculations and experiments by Sunner and co-workers[36,37] have shown that electric fields in the atmospheric pressure ionization region are mainly dominated by space charge. Schneider et al. have demonstrated that an atmospheric pressure ion lens can improve signal stability and ion abundance, both in nanospray[38] and in nebulizer-assisted electrospray.[39]

4.3.4 Vacuum System Design

Electrospray API instruments can be divided into single-stage vacuum systems and multiple-stage vacuum systems. All present-day electrospray mass spectrometers make use of a muliple-stage vacuum system.

A single-stage instrument is shown schematically in Figure 4.10. Ions flow from the sampling orifice through ion optics into the mass analyzer. The ion optics stage can be either a stack of regular lenses or an RF-only quadrupole (hexapole, octopole). The pressure tolerated by the mass analyzer sets the upper limit to the vacuum in the mass spectrometer. This pressure, together with the speed of the vacuum pump, sets the upper limit to the throughput of gas and thus to the size of the sampling orifice. The higher the gas throughput, the more ions are transported through the sampling orifice. In older API instruments, a 10- to 25-μm-diameter orifice compatible with a 1000-L/s oil diffusion pump was used. The original single-stage atmospheric pressure ionization instrument built by SCIEX made use of a 120-μm orifice protected by a gas curtain, combined with pumping by cryogenic surfaces surrounding the ion optics and quadrupole mass analyzer, rated at 100,000 L/s.

High-speed cryogenic pumping is good from a theoretical point of view and has afforded very reliable performance in practice, but needs a big and expensive helium compressor. Manufacturers and researchers prefer a multiple-stage vacuum system that can be built with less expensive vacuum pumps (Figure 4.11). Ions are drawn through a sampling orifice into a first chamber, pumped to approximately 1 torr by a rotary pump. Part of the expanding beam of gas and ions is taken into the second chamber, pumped down to 10^{-2} torr or less. Most of the gas is pumped away in the second stage, and a suitable ion optics arrangement guides the ions into the mass analyzer region, which is pumped down to 10^{-5} torr. The second and third stages are usually the original source housing and analyzer housing of a standard differentially pumped mass spectrometer designed for chemical ionization. This simple scheme may be elaborated in different ways, to comply with particular geometrical or electrical constraints. In some implementations of the multiple-stage vacuum system, the first or the second stage has been split in two, which increases the total number of stages to four.

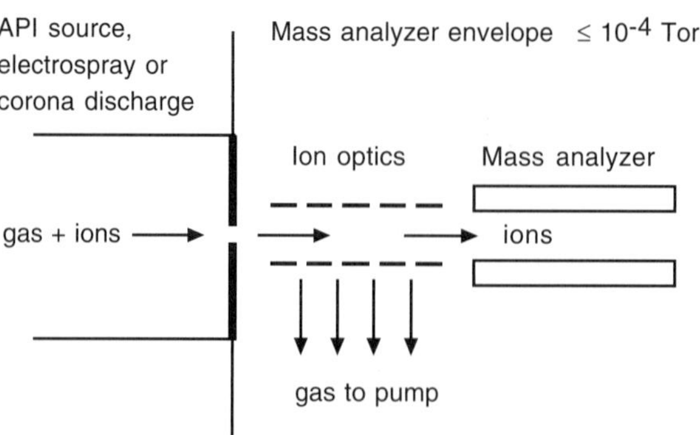

Figure 4.10. Electrospray (atmospheric pressure ionization) source with single-stage vacuum system for the mass analyzer.

Typical dimensions are 0.25–0.5 mm for the orifice and 0.6–4.0 mm for the skimmer. Such dimensions are strongly dependent on size and speed of turbomolecular and rotary pumps. In attempts to build electrospray API sources for magnetic sector instruments, the addition of a fourth pumping stage appeared advantageous. Problems with sector instruments are arcing due to the high voltage on the ion source, electrical discharges in pumping lines, and collisions between ions and background gas in the acceleration region. Manufacturers appear to have overcome such problems. Magnetic sector mass spectrometers are of minor importance for ESI; high-resolution ESI is achieved by time-of-flight, Fourier transform ion cyclotron resonance and Fourier transform Orbitrap mass analyzers.

4.3.5 Vacuum System and Sensitivity

Multiple-stage vacuum systems use larger orifices, which can take more gas and ions into the first vacuum stage of the mass spectrometer. Unfortunately, a large throughput through the sampling orifice and high sensitivity do not always go hand in hand. In a discussion about sensitivity, the first vacuum stage in Figure 4.11 will be called the molecular beam stage; the second, the ion optics stage; and the third, the analyzer stage.

The goal for the design of the system for ion transport from the electrospray source into the mass analyzer is to introduce as many ions, but as little gas, as possible. Ideally, ions should be separated from gas. In a simple approximation, the sensitivities of different API mass spectrometers can be judged from the throughput of gas in the relevant vacuum stage.[40]

For simplicity, we assume that no significant enrichment of heavy ions over light gas molecules takes place during passage through the first vacuum stage into the skimmer. We assume that the ion optics region is transparent to neutrals that are removed by the high vacuum pump connected to the second vacuum stage. We also assume that no ions are lost during transport through the ion optics region. In short, the ions–neutral-gas ratio is significantly increased in the ion optics stage.

The flow of ions into the mass analyzer is proportional to the gas flow (throughput). The throughput of gas is given by the pumping speed multiplied by the pressure in the vacuum system. For 10^{-5} torr in a single-stage system that is equipped with a 100,000-L/s cryopump, the throughput is 1 torr · L/s (SCIEX API 3).

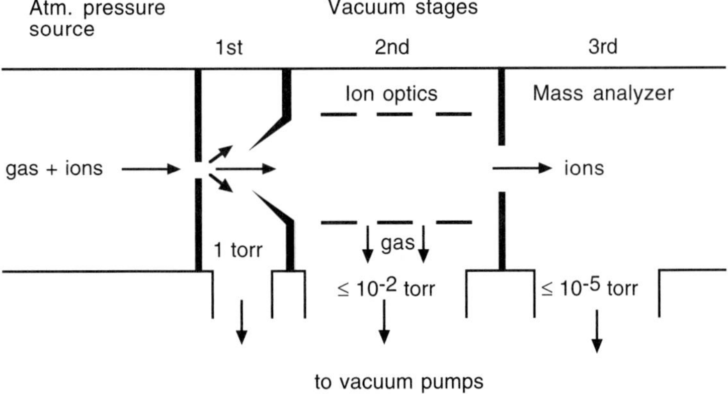

Figure 4.11. Vacuum system with multiple stages: first stage: free jet expansion, molecular beam stage; second stage: ion optics; third stage: mass analyzer.

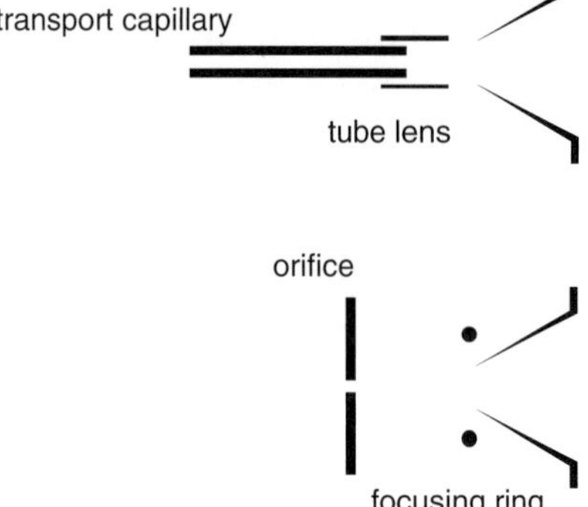

Figure 4.12. Tube lens (ThermoFinnigan) or ring (SCIEX) located in the molecular beam stage for forcing ions through the skimmer orifice.

In the discussion of multiple-stage vacuum systems, we assume for simplicity that an ion optics stage transmits all ions toward the mass analyzer, while gas is pumped away, thereby increasing the ions–neutral-gas ratio; a molecular beam stage is supposed to transfer a portion of the gas plus ion beam, without changing the ions–gas ratio. However, most versions of the molecular beam stage include a tube lens or focusing ring or focusing cone for increasing the ions–gas ratio as shown in Figure 4.12.[41] Positive ions that have spread out during the free-jet expansion are forced back toward the center through the skimmer by the application of a positive voltage on the tube, ring, or cone. A negative voltage is used for negative ions. Sensitivity is increased by a factor two to five, depending on the geometry.

The ion optics stage may be equipped with lenses—for example, an RF-only quadrupole (hexapole, octopole) or other elements such as a stack of ring electrodes connected to an RF power supply that generates a traveling wave of ions.[42]

In a three-stage system, the first stage does not dramatically increase the ions–gas ratio. The ion optics stage is crucial, and with a 500-L/s turbomolecular pump and 7×10^{-3} torr (approximate values for a Sciex API 3000 triple quadrupole) a throughput of 3.5 torr·L/s can be obtained. In all instruments built along these lines, the pumping and pressure in the ion optics stage determine the sensitivity. A compromise on pumping will compromise sensitivity. The efficiency of ion transport through an RF-only quadrupole (hexapole, octopole) increases if the pressure is raised to approximately 10^{-2} torr.[43] At this high pressure, it becomes difficult to select the right vacuum pump. Diffusion pumps cannot be used at an inlet pressure above 10^{-3} torr, and they suffer from considerable backstreaming of diffusion pump oil. Turbomolecular pumps have reached their limits at 10^{-2} torr. Continuous operation of a turbomolecular pump in the 10^{-3} to 10^{-2} torr range leads to a high temperature of the pump body and also leads to a reduced lifetime of the pump bearings of air-cooled pumps. Dual inlet turbomolecular pumps are equipped with (a) a high-pressure port for pumping the RF-only quadrupole and (b) a regular low-pressure inlet port for pumping the mass analyzer.

The prediction of sensitivity based on vacuum considerations alone is a first approximation. The effects of the quality of the molecular beam components, ion optics, and mass analyzer cannot be included in a simple generalization. An ion funnel made of a stack of rings can be incorporated in the first vacuum stage.[44] This technique has been implemented in the Bruker Apollo II source of their FT-ICR mass spectrometer. Guiding the ions in the first vacuum stage by means of a small quadrupole is another method to achieve a better transport of ions and higher sensitivity.[45]

The atmospheric part of the electrospray source on each vacuum system was assumed to be equally efficient in each case. A comparison between different instrument designs would

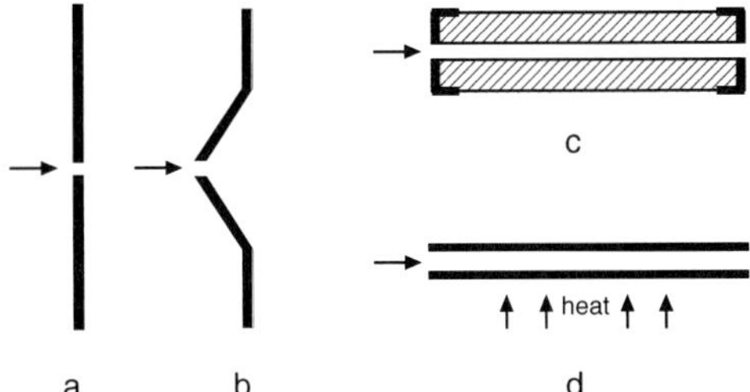

Figure 4.13. Sampling of ions and gas from the atmospheric pressure ionization source via different shapes of orifices and tubes: (**a**) Orifice in a flat disk (Sciex API 3000); (**b**) orifice in the top of a cone (Waters, Sciex API 4000/5000); (**c**) glass tube with metallized ends (Analytica of Branford, Agilent, Bruker); (**d**) heated metal tube (ThermoFinnigan).

require running the same sample in different instruments under well-defined and carefully controlled conditions.

4.3.6 Ion Sampling Orifice

Various arrangements of the sampling orifice are given in Figure 4.13. A hole in a disk (a) or in the top of a cone (b) can be used equally well. A 20-cm-long × 0.5-mm-i.d. glass tube (c) or heated metal tube (d) may also be used instead of an orifice.

The formation of molecular beams from nozzles of various shapes has been reported.[46] Campargue[30] recommends a short straight capillary with a ratio of length to diameter equal to or greater than 2. In our homemade ion source, a 0.3-mm-diameter hole in a flat 1.0-mm-thick stainless-steel disk protected by a gas curtain has proven to be very robust and reliable in experiments with corona discharge ionization and electrospray.

The operation of an API source at any desired voltage, independent from the mass analyzer, has advantages for magnetic sector instruments and for certain liquid sample inlet systems. The transport of ions through a glass capillary instead of a metal nozzle gives the desired insulation as discussed earlier in this chapter. The flat ends of the capillary are metal plated for efficient sampling of ions into the capillary and acceleration of ions at the exit. The gas flow drags ions through the capillary into the vacuum system, even if the electric field existing between the metallized ends opposes such a transport. Chait and co-workers[28] have used a long heated metal capillary for the desolvation of ions from charged microdroplets generated by electrospray and introduction of the core ions into a quadrupole mass spectrometer.

Another purpose served by the capillary is the transport of ions over a longer distance (e. g., 20 cm) between the API source and the ion optics of the mass analyzer, giving more freedom in mechanical construction.

Ion transport efficiency through capillary tubes has been studied in detail by Lin and Sunner[47] and by Guevremont et al.[48] Transmitted currents were very similar for glass and metal capillaries.[47] It was more difficult to obtain stable and reproducible transmitted currents with glass than with metal capillaries. Experiments were made with very long (0.6–15 m) capillaries, so that shortcomings observed are probably more pronounced than in "real life" in mass spectrometry, where the capillary length is 20 cm or shorter.

Contamination of the sampling orifice or tube can be troublesome. The buildup of charge on a layer of contaminants inside an orifice or tube can effectively stop the passage of ions, while the flow of neutrals (as read from vacuum gauges) is not affected. An electrospray source should be designed for ease of removal and cleaning of the orifice or tube.

Blockage of a tube or orifice is prevented by the use of a gas curtain (Sciex). The drying gas in an Agilent or Bruker source serves the same purpose. Foolproof operation cannot be

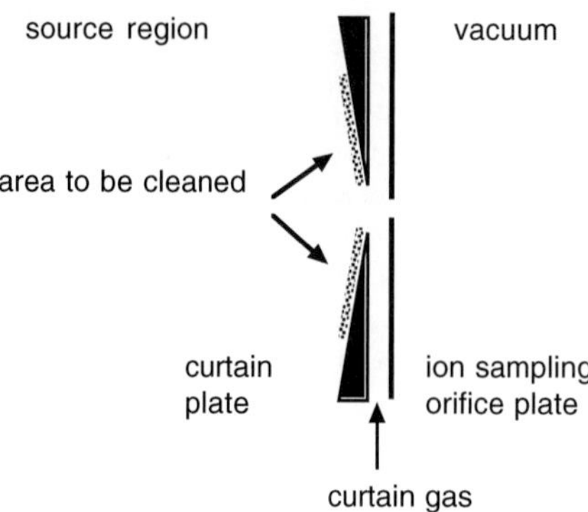

Figure 4.14. Area to be cleaned inside the atmospheric pressure source region.

guaranteed. One should be careful and avoid spraying straight at the orifice or tube since droplets can penetrate through the gas curtain. As discussed above in this chapter, diagonal or orthogonal positioning not only increases stability and abundance of ion signals, but also helps to prevent contamination of the sampling orifice or tube. If such precautions are taken together with protection by a gas curtain, it is possible to use nonvolatile buffer components in CE/MS, or use post-column addition of a dilute sodium chloride solution in LC/MS to promote formation of $[M + Na]^+$ ions of polar neutral samples such as sugars.

The area around the orifice in a curtain plate needs periodic cleaning (Figure 4.14). Electric fields created by charges on contaminating layers may disturb the electric field in front of the sampling tube or orifice and thus prevent ions from being entrained in the gas flow into the vacuum.

4.3.7 Ion Optics Between Sampling Orifice and Mass Analyzer

In the space between the sampling orifice (or outlet of sampling tube) and skimmer, ions are transported in the free-jet expansion of gas. Part of the beam of ions and gas passes through the skimmer into the next vacuum stage. Effective focusing of ions into a very narrow beam is not possible, but ions can be forced to remain closer to the axis of the beam if a tube or ring is positioned between nozzle orifice and skimmer.[41] A positive voltage on the tube lens or ring reduces the spreading of positive ions away from the axis. The gain in sensitivity is between twofold and fivefold. In a different arrangement, the loss of ions is prevented effectively by the positioning of an RF-only quadrupole right behind the sampling orifice. In this case the free-jet expansion takes place inside the quadrupole, and most of the ions are kept inside the quadrupole and are guided through a small orifice into the second vacuum stage.[45] Behind the skimmer the pressure is lower, but still not low enough for effective focusing with ion lenses. RF-only quadrupoles, hexapoles, and octopoles are preferred for the guiding of ions into the mass analyzer section. Douglas has observed that the efficiency of guiding ions is significantly increased if the pressure is raised to 10^{-2} torr.[43] Collisions between ions and background gas keep the ions on a path close to the quadrupole axis, in the same manner as ions are confined to the center of a quadrupole ion trap by collisions in 10^{-3} torr of helium. Ions are slowed down by collisions in the RF-only quadrupole and enter the next stage with a limited energy spread of a few electron volts. Ion transmission from the RF-only multipole into the mass analyzer becomes easier and focusing is much less dependent on mass. The occasional penetration of charged droplets into a quadrupole mass analyzer shows up as random spikes in mass spectra. Off-axis positioning of the sprayer can reduce this aerosol noise. In some mass spectrometers the skimmer is positioned slightly off-axis or at a 90° angle with respect to the flow from the ion sampling orifice. Since aerosol particles and heavy ion–solvent clusters follow a straight line on the axis of the ion transfer tube, they impinge on the outside wall of the off-axis skimmer. A sizable fraction of ions deviates from the center and is passed through the skimmer, which is located downstream from the Mach disk of the free-jet expansion. Ion counting detection is more immune to spikes of this nature when compared with analog current measurement, since a burst of *many* ions arriving together at the electron multiplier is recorded as *one* pulse by the ion counting detector, while the same burst is amplified to a large-voltage spike by the current to voltage converter of an analog detector.

4.4 UP-FRONT COLLISION-INDUCED DISSOCIATION

Electrospray mass spectra are usually devoid of fragment ions. Fragments can be formed in a triple quadrupole, in an ion trap, in an ICR cell, or in the field-free regions of magnetic sector mass spectrometers.

In atmospheric pressure ionization mass spectrometry, including electrospray mass spectrometry, fragmentation can easily be induced in one of the higher-pressure regions of the ion passageway from the source into the mass analyzer. Acceleration of ions between nozzle orifice (tube) and skimmer by $\Delta V(N - S)$ or between skimmer and RF-only multipole by $\Delta V(S - Q)$ results in collisions of ions with gas in which ions are entrained in their flow into the vacuum (Figure 4.15). Oftentimes this process is called "in source CID," which is clearly a misnomer. Further confusion is created by the use of different names for the same process by users and manufacturers: nozzle-skimmer fragmentation, cone voltage fragmentation, high orifice potential fragmentation, octopole fragmentation, to name a few.

Fragment ions produced by up-front collision induced dissociation are very efficiently transported into the mass analyzer. The major advantage of this poor man's MS/MS method is its simplicity: Only one voltage has to be increased, and no switching and adjustment of

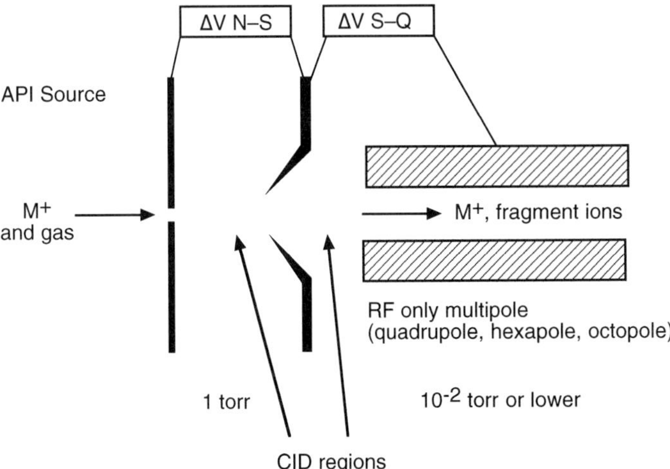

Figure 4.15. Collision-induced dissociation just outside the ion source, in the intermediate pressure regions between the atmospheric pressure ion source and high vacuum of the mass analyzer.

collision gas and retuning of ion optics is necessary. Of course, there is no parent ion selection.

In LC/MS, we can do one LC run without CID and do one with CID to obtain fragments of all components. An elegant application of up-front CID in LC/MS is stepping the CID (orifice) voltage during the course of each mass spectrum: A high voltage is used in the low mass region, to generate and detect low mass fragments, while the voltage is reduced again at, for example, *m/z* 200 and above, in order to collect unfragmented parent ions.[49] Using this approach, phosphorylated peptides were selectively detected in LC/MS of an enzymatic digest of a phosphoprotein.

Since up-front CID takes place just *outside* the ion source, it can be combined with any ionization technique that is applicable in atmospheric pressure ion sources.

In Figure 4.15 the accumulated effect of $\Delta V(N-S)$ and $\Delta V(S-Q)$ can lead to fragmentation.

CID takes place under the condition of sufficient translational energy and gas density. In a magnetic sector instrument the residual gas density in the accelerating region may be high enough for CID. An additional pumping stage is used by manufacturers to eliminate unwanted CID in the accelerating region.

Mild up-front CID reduces the abundance of background ions, a distinct advantage for LC/MS or CE/MS where the contribution of background ions to the baseline noise in the total ion current trace is a serious problem.[50]

In multiple-stage pumping instruments, the transmission of ions is improved if the voltage gradient in the first and second vacuum stages, such as $\Delta V(N-S)$, is increased–in particular for big protein ions. In fact, it is not too difficult to achieve good transmission of protein ions without inducing fragmentation since collision energy is partitioned over a very large number of vibrational degrees of freedom. It is not as easy to obtain good transmission of small molecules that fragment much more easily. A good test compound for unwanted CID is the drug naproxen. At low collision energy, the negative ion electrospray spectrum shows the $[M-H]^-$ ion at *m/z* 229 as the only peak. It should be easy to adjust $\Delta V(N-S)$ and $\Delta V(S-Q)$ for a spectrum free from fragment ions. Already at a moderate collision energy where most sample ions would survive, the $[M-H]^-$ ion loses CO_2 so easily that *m/z* 185 becomes the base peak. In the positive ion mode, the $[M+H]^+$ ion of dibutyl phthalate at *m/z* 279 can be used as a test ion. Fragmentation to *m/z* 149 is easy, but does not take place as readily as the loss of CO_2 from *m/z* 229 of naproxen.

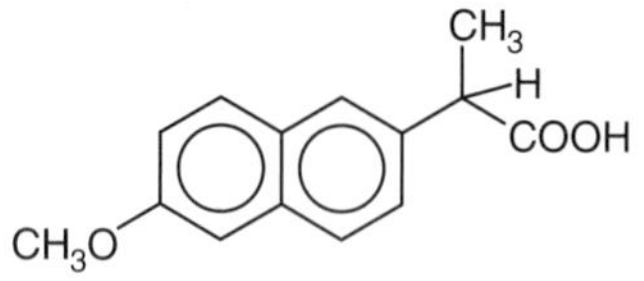

CH_3 H COOH CH_3O

naproxen

Removal of the last solvent molecule(s) from a sample ion by up-front CID may not be possible without unintended fragmentation of a labile sample ion.

4.5 MASS SCALE CALIBRATION

Calibration of the mass scale requires a series of ions evenly spaced throughout the mass range. Ideally, the calibrant should be easily removed from the source and leave no traces in the source or liquid handling system.

Mixtures of a protein (usually myoglobin) and some smaller peptides can be used; polyethylene glycols and polypropylene glycols are also widely used for calibration. Anacleto et al. have summarized different options and have proposed protonated water clusters and salt clusters[51] generated by pneumatically assisted electrospray. Water clusters provided a calibration range up to *m/z* 1000 in the SCIEX API III mass spectrometer. Alkali metal halides (sodium iodide) allow calibration on cluster ions $Na^+(NaI)_n$ or $I^-(NaI)_n$ up to at least *m/z* 2000, the full mass range of this instrument.

In principle, mass scale calibration is independent from the ionization technique. For many years we have routinely calibrated by the use of a combination of corona discharge and controlled cluster formation in the free-jet expansion. Corona discharge of the headspace of an aqueous ammonia solution creates $NH_4^+(H_2O)_n$ with $n = 0$–4 after passage through the gas curtain. If the curtain gas flow is reduced but not cut off completely, one can tune the gas flow to allow a controlled amount of moisture into the free-jet expansion so that a complete series of $NH_4^+(H_2O)_n$ ions is formed that extends all the way up to *m/z* 2000. The calibration mass series starts at *m/z* 54; and one out of every five clusters can be used, giving a calibration separation of 90 on the *m/z* scale. The relatively low abundance of deuterium and ^{17}O, along with the higher abundance of ^{18}O in the cluster ions, makes this calibrant suitable for resolution adjustment on isotope peaks.

REFERENCES

1. Bailey, A. G. *Electrostatic Spraying of Liquids*; Electrostatics and Electrostatic Applications Series, John Wiley & Sons, New York **1988**.
2. Special Issue: Electrosprays: Theory and Applications. *J. Aerosol Sci.* **1994**, *25*(6).
3. Special Issue: Electrohydrodynamic Atomization, *J. Aerosol Sci.* **1999**, *30*(7).
4. Parvin, L.; Galicia, M. C.; Gauntt, J. M.; Carney, L. M.; Nguyen, A. B., Park, E.; Heffernan, L.; Vertes, A. Electrospray diagnostics by Fourier analysis of current oscillations and fast imaging. *Anal. Chem.* **2005**, *77*, 3908–3915.
5. Smith, R. D.; Barinaga, C. J.; Udseth, H. R. Improved electrospray ionization interface for capillary zone electrophoresis-mass spectrometry. *Anal. Chem.* **1988**, *60*, 1948–1952.
6. Mylchreest, I. C.; Hail, M. E.United States Patent 5,122,670, June 16, **1992**.
7. Bruins, A. P.; Covey, T. R.; Henion, J. D. Ion spray interface for combined liquid chromatography/ atmospheric pressure ionization mass spectrometry. *Anal. Chem.* **1987**, *59*, 2642–2646.
8. Banks, J. F.; Shen, S.; Whitehouse, C. M.; Fenn, J. B. Ultrasonically assisted electrospray ionization for LC/MS determination of nucleosides from a transfer RNA digest. *Anal. Chem.* **1994**, *66*, 406–414.
9. Banks, J. F.; Quinn, J. P.; Whitehouse, C. M. LC/ESI-MS determination of proteins using conventional liquid chromatography and ultrasonically assisted electrospray. *Anal. Chem.* **1994**, *66*, 3688–3695.
10. Hail, M. E.; Mylchreest, I. C.United States Patent 5,170,053, December 8, **1992**.
11. Wahl, J. H.; Gale, D. C.; Smith, R. D. Sheathless capillary electrophoresis–electrospray ionization mass spectrometry using 10 μm I.D. capillaries: Analyses of tryptic digests of cytochrome c. *J. Chromatogr.* **1993**, *659*, 217–222.
12. Kriger, M. S.; Cook, K. D.; Ramsey, R. S. Durable gold-coated fused silica capillaries for use in electrospray mass spectrometry. *Anal. Chem.* **1995**, *67*, 385–389.
13. Gale, D. C.; Smith, R. D. Small volume and low flow-rate electrospray lonization mass spectrometry of aqueous samples. *Rapid Commun. Mass Spectrom.* **1993**, *7*, 1017–1021.
14. Emmet, M. R.; Caprioli, R. M. Micro-electrospray mass spectrometry: Ultra-high-sensitivity analysis of peptides and proteins. *J. Am. Soc. Mass Spectrom.* **1994**, *5*, 605–613.
15. Wilm, M. S.; Mann, M. Electrospray and Taylor–Cone theory, Dole's beam of macromolecules at last? *Int. J. Mass Spectrom. Ion Processes* **1994**, *136*, 167–180.
16. Yamashita, M.; Fenn, J. B. Electrospray ion source. Another variation on the free-jet theme. *J. Phys. Chem.* **1984**, *88*, 4451–4459.
17. Whitehouse, C. M.; Dreyer, R. N.; Yamashita, M.; Fenn, J. B. Electrospray interface for liquid chromatographs and mass spectrometers. *Anal. Chem.* **1985**, *57*, 675–679.
18. Larsen, B. S.; McEwen, C. N. An electrospray ion source for magnetic sector mass spectrometers. *J. Am. Soc. Mass Spectrom.* **1991**, *2*, 205–211.
19. Cody, R. B.; Tamura, J.; Musselman, B. D. Electrospray ionization/magnetic sector mass spectrometry:

Calibration, resolution, and accurate mass measurements. *Anal. Chem.* **1992**, *64*, 1561–1570.
20. Yamashita, M.; Fenn, J. B. Negative ion production with the electrospray ion source. *J. Phys. Chem.* **1984**, *88*, 4671–4675.
21. Cole, R. B.; Harrata, A. K. Charge-state distribution and electric-discharge suppression in negative-ion electrospray mass spectrometry using/chlorinated solvents. *Rapid Commun. Mass Spectrom.* **1992**, *6*, 536–539.
22. Hiraoka, K.; Kudaka, I. Negative-mode electrospray-mass spectrometry using nonaqueous solvents. *Rapid Commun. Mass Spectrom.* **1992**, *6*, 265–268.
23. Whitby, K. T.; Liu, B. Y. H. The electrical behaviour of aerosols. In Davies, C.N. (Ed.), *Aerosol Science*; Academic Press, London, **1966**, Chapter III.
24. Scalf, M.; Westphall, M. S.; Smith, L. M. Charge reduction electrospray mass spectrometry. *Anal. Chem.* **2000**, *72*, 52–60.
25. Ebeling, D. D.; Westphall, M. S.; Scalf, M.; Smith, L. M. Corona discharge in charge reduction electrospray mass spectrometry. *Anal. Chem.* **2000**, *72*, 5158–5161.
26. Schneider, B. B.; Hassan Javaheri, H.; Covey, T. R. Ion sampling effects under conditions of total solvent consumption. *Rapid Commun. Mass Spectrom.* **2006**, *20*, 1538–1544.
27. Covey, T. R.; Anacleto, J. F. United States Patent 5,412,208; May 2, **1995**.
28. Chowdhury, S. K.; Katta, V.; Chait, B. T. An electrospray–ionization mass spectrometer with new features. *Rapid Commun. Mass Spectrom.* **1990**, *4*, 81–87.
29. Anderson, J. B.; Andres, R. P.; Fenn, J. B. In Bates, D. R.; Estermann, I. (Eds.), *Advances in Atomic and Molecular Physics*, Academic Press, New York, **1965**, Chapter 8.
30. Campargue, R. Progress in overexpanded supersonic jets and skimmed molecular beams in free-jet zones of silence. *J. Phys. Chem.* **1984**, *88*, 4466–4474.
31. Jugroot, M., Groth, C. P. T., Thomson, B. A., Baranov, V., Collings, B. A., Numerical investigation of interface region flows in mass spectrometers: Ion transport. *J. Phys. D: Appl. Phys.* **2004**, *37*, 550–559.
32. Jugroot, M., Groth, C. P. T., Thomson, B. A., Baranov, V., Collings, B. A., Numerical investigation of interface region flows in mass spectrometers: Neutral gas transport. *J. Phys. D: Appl. Phys.* **2004**, *37*, 1289–1300.
33. Shahin, M. M. Mass-spectrometric studies of corona discharges in air at atmospheric pressures. *J. Chem. Phys.* **1966**, *45*, 2600–2605.
34. Kambara, H.; Kanomata, I. Determination of impurities in gases by atmospheric pressure ionization mass spectrometry. *Anal. Chem.* **1977**, *49*, 270–275.
35. Eisele, F. L. Direct tropospheric ion sampling and mass identification. *Int. J. Mass Spectrom. Ion Processes* **1983**, *54*, 119–126.
36. Busman, M. Sunner, Simulation method for potential and charge distributions in space charge dominated ion sources. *J. Int. J. Mass Spectrom. Ion Processes* **1991**, *108*, 165–178.
37. Busman, M.; Sunner, J.; Vogel, C. R. Space-charge-dominated mass spectrometry ion sources: Modeling and sensitivity. *J. Am. Soc. Mass Spectrom.* **1991**, *2*, 1–10.
38. Schneider, B. B.; Douglas, D. J.; Chen, D. D. Y. An atmospheric pressure ion lens to improve electrospray ionization at low solution flow-rates. *Rapid Commun. Mass Spectrom.* **2001**, *15*, 2168–2175.
39. Schneider, B. B.; Douglas, D. J.; Chen, David, D. Y. An atmospheric pressure ion lens that improves nebulizer assisted electrospray ion sources, *J. Am. Soc. Mass Spectrom.* **2002**, *13*, 906–913.
40. Bruins, A. P. Mass spectrometry with ion sources operating at atmospheric pressure. *Mass Spectrom. Rev.* **1991**, *10*, 53–77.
41. Mylchreest, I. C.; Hail, M. E.; Herron, J. R.United States Patent 5,157,260; October 20, **1992**.
42. Giles, K., Pringle, S. D., Worthington, K. R., Little, D., Wildgoose, J. L., Bateman, R. H. Applications of a traveling wave-based radio-frequency-only stacked ring ion guide, *Rapid Commun. Mass Spectrom.*, **2004**, *18*, 2401–2414.
43. Douglas, D. J.; French, J. B. Collisional focusing effects in radio frequency quadrupoles. *J. Am. Soc. Mass Spectrom.* **1992**, *3*, 398–408.
44. Ibrahim, Y., Tang, K., Tolmachev, A. V., Shvartsburg, A. A.; Smith, R. D. Improving mass spectrometer sensitivity using a high-pressure electrodynamic ion funnel interface. *J. Am. Soc. Mass Spectrom.* **2006**, *17*, 1299–1305.
45. Javaheri, H.; Thomson, B. A.; Groth, C. P. T.; Jugroot, M. In *Proceedings of the 53rd ASMS Conference on Mass Spectrometry*, San Antonio, TX, June 5–9, 2005.
46. Murphy, H. R., Miller, D. R. Effects of nozzle geometry on kinetics in free-jet expansions. *J. Phys. Chem.* **1984**, *88*, 4474–4478.
47. Lin, B.; Sunner, J. Ion transport by viscous gas flow through capillaries. *J. Am. Soc. Mass Spectrom.* **1994**, *5*, 873–885.
48. Guevremont, R.; Siu, K. W. M.; Wang, J.; LeBlanc, J. C. Y., Transport of electrospray ions through capillary tubes. In *Proceedings of the 42nd ASMS Conference on Mass Spectrometry and Allied Topics, May 29–June 3, 1994, Chicago, IL*, p. 999.
49. Carr, S. A.; Huddleston, M. J.; Bean, M. F., Selective identification and differentiation of N- and O-linked oligosaccharides in glycoproteins by liquid chromatography–mass spectrometry. *Protein Sci.* **1993**, *2*, 183–196.
50. Guo, X., Bruins, A. P., Covey, T. R. Characterization of typical chemical background interferences in atmospheric pressure ionization liquid chromatography–mass spectrometry. *Rapid Commun. Mass Spectrom.* **2006**, *20*, 3145–3150.
51. Anacleto, J. F.; Pleasance, S.; Boyd, R. K. Calibration of ion spray mass spectra using cluster ions. *Org. Mass Spectrom.* **1992**, *27*, 660–666.

Part II

Fundamentals of MALDI

Chapter 5

MALDI Ionization Mechanisms: an Overview

Richard Knochenmuss

Novartis Institute of Biomedical Research, Basel, Switzerland

Electrospray and MALDI Mass Spectrometry: Fundamentals, Instrumentation, Practicalities, and Biological Applications, Second Edition, Edited by Richard B. Cole

5.1 INTRODUCTION

It is interesting to compare the development of the two major modern ionization techniques: matrix-assisted laser desorption/ionization (MALDI)[1] and electrospray.[2,3] ESI emerged from the molecular beam community and therefore profited from an understanding of relatively well-defined predecessors such as condensation droplets in gas expansions.[4] While viewpoints continue to evolve regarding the final stages of ESI ion formation, there is a degree of consensus on much of the process.[5–7] In contrast, MALDI arose as a "soft" variant of laser ablation, a term encompassing widely varying fundamental phenomena, depending on the material and laser.[8] It is therefore perhaps not surprising that MALDI development initially lacked strong guiding principles, leading to much empirical work. This is illustrated by early surveys of hundreds of compounds as potential matrices.[9,10] Few have become widely used, and these were mostly identified very early.[10–15]

Clearly, it would be more efficient to have a conceptual framework to guide development, or at least show where fundamental limits might lie. A complete theory of MALDI should quantitatively predict the observed mass spectrum as a function of variables such as matrix choice, analyte physical and chemical properties, concentrations, preparation method, laser characteristics (wavelength, spatial and temporal properties), local environment (such as ambient pressure or substrate temperature), and ion extraction method. Here we focus only on ionization mechanisms and do not address all factors affecting a MALDI experiment. Some of these are discussed in other contributions to this volume. Only mechanisms involving molecular matrices and laser excitation are included, methods that depend primarily on properties of the substrate, such as nanoparticles[16] or structured surfaces (such as DIOS)[17] are not Hybrid methods, such as laser ablation into electrosprays, are also out of our scope, but vacuum and higher-pressure (e.g., atmospheric pressure) MALDI are both considered to have the same underlying mechanisms discussed here.

Various reviews and summary articles have addressed MALDI ionization. Some relevant concepts pre-date the advent of modern MALDI[1]; see, for example, Hillenkamp's discussion of LDI in 1983.[18] Significant early MALDI reviews[19,20] addressed a variety of possible mechanisms. Ehring, Karas, and Hillenkamp[21] elaborated three excitation schemes leading to seven ion types. The key intermediate step for all involved highly excited matrix. Liao and Allison[22] contributed an extensive discussion of protonated and sodiated adduct ions. This work emphasized the central role of ion–molecule reactions in the MALDI desorption plume. The consequent relevance of gas-phase thermodynamics to observed mass spectra has remained a central theme ever since.

Karas, Bahr, and Stahl-Zeng emphasized the multiple functions of the matrix in 1996,[23] especially proton transfer reactions of the matrix with analyte. A focused review of MALDI ionization mechanisms appeared in 1998,[24] which also explicitly introduced the concept of distinct primary and secondary ion formation processes. The former are rapid, early events creating the first ions. Initial ions may later undergo secondary reactions in the expanding plume, in a manner similar to the SIMS "selvedge" concept.[25]

An issue of *Chemical Reviews* (volume 103, number 2, 2003) covered "laser ablation of molecular substrates" and included several articles on MALDI. A microscopic view of desorption/ablation was provided by the molecular dynamics work of the Zhigilei and Garrison groups,[26] while Dreisewerd presented an overview of MALDI desorption,[27] and one recently developed cluster model was summarized and further developed by Karas and Krüger.[28] Secondary mechanisms were examined in detail by Knochenmuss and Zenobi.[29]

Several MALDI ionization-relevant papers have appeared in a special issue of the European *Journal of Mass Spectrometry* (volume 12, 2006): Proton transfer reactions in the plume were the object of experimental work by the Kinsel group[30] and calculations by Beran and co-workers.[31] Hoteling et al. considered electron transfer secondary reactions,[32] which are also of relevance for the fullerenes studied with solvent-free methods by Drewello's group.[33] Finally, cluster ionization and desolavtion processes were investigated by Tabet and co-workers[34] Also appearing in 2006 was a mechanisms review by Knochenmuss in *The Analyst*.[35]

5.2 THE TWO-STEP FRAMEWORK AND PRIMARY VERSUS SECONDARY MECHANISMS

As described below, there is ongoing discussion about initial (primary) ion formation, which may vary with laser wavelength or other variables. Secondary reactions of these ions with neutrals in the desorption/ablation plume are generally either invoked or not ruled out, making this aspect of MALDI uncontroversial since the early work of Liao and Allison.[22] In addition, it is now rarely challenged that local thermal equilibrium is often approached in the plume. This allows secondary reactions to be treated with conventional thermodynamics and has motivated extensive efforts to measure or calculate the corresponding thermodynamic quantities for MALDI-relevant molecules. Since the two-step model is fundamentally a consequence of the characteristics of the MALDI ablation/desorption event, this will be briefly discussed next.

5.2.1 Physics of Desorption/Ablation

The desorption/ablation aspect of MALDI has been the object of an excellent review by Dreisewerd.[27] Among the most basic functions of the matrix is absorption of the laser energy and conversion of most of it to heat. Subsequent matrix vaporization is sufficiently forceful that it entrains and ejects analyte that has been cocrystallized in or on the matrix. The time scale is not as short as in fast atom bombardment (FAB) or secondary ionization mass spectrometry (SIMS) because the typical MALDI lasers used emit pulses of a few to hundreds of nanoseconds duration (e.g., N_2 337 nm, 3 ns or tripled Nd:YAG, 355 nm, 4–7 ns, Er:YAG, 2.98 μm, 200 ns, although a range of pulse lengths has been studied[36–40]). This is slow compared to intramolecular motions. In UV MALDI, the energy conversion step is also temporally limited by storage of laser energy in electronic excited states of the matrix. Excited-state lifetimes of free, gas-phase, matrix molecules are on the order of tens of nanoseconds, but they drop to 1 ns or a few hundred picoseconds in the solid state.[41–43] This places a limit on the heating rate, independent of laser pulse length.

This period when the matrix is highly energized, but has not yet significantly expanded, is when most ionization can occur—the first step of the two-step framework. In the case of UV-MALDI, electronically excited molecules are abundant and can interact. In IR MALDI, where electronic states are probably rarely excited, physical forces are highest, including high-energy collisions and/or large-scale material disintegration. The initial pressure is very high, so the environment of the sample (vacuum or atmospheric pressure) is largely irrelevant during primary ionization. The expansion of the vaporizing matrix is comparatively slow, up to several microseconds. This is the timescale of secondary reactions, the second step. Primary processes that create separated ion pairs slow down, because energy is being diluted by conversion to physical expansion. The material begins to relax, physically and chemically, but full relaxation is never attained because of the expansion and because ions may be extracted from the plume by external fields. A schematic illustration of the timescales in MALDI is shown in Figure 5.1.

The phase change may be characterized as "desorption" or "ablation," mostly depending on the laser-deposited energy density.[26,44–52] Desorption is characterized

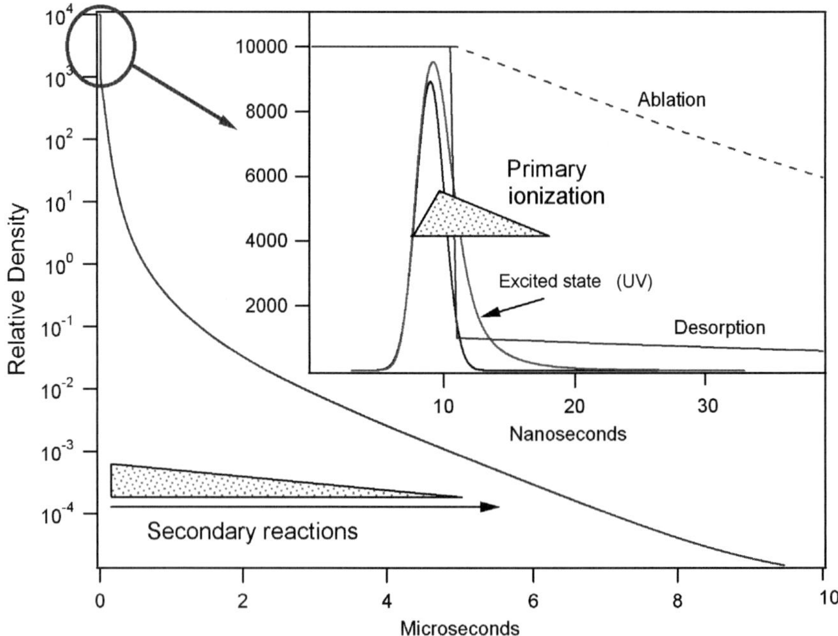

Figure 5.1. The origin of the two-step model: Expansion of the MALDI plume compared to the laser pulse and excited state decay. Density is plotted versus time for a plume expanding as an adiabatic free jet. A density of 1 is approximately that of a gas at 1 atm. The expansion stops when the gas reaches the environmental background pressure, which may vary from 1 atm. to high vacuum. The inset shows the early behavior, along with a 3-ns N_2 laser pulse (blue) and the typical lifetime of matrix excited states (green). Only during the time when the energy and material densities are high can significant ionization occur. The solid line represents the density if the sample vaporizes smoothly. The dashed line represents an explosive phase change for which a well-defined solid–gas boundary may not exist at short times (See color insert).

by a smooth transition from solid to gas, at the surface of the sample. In ablation, the sample is overheated and subsurface nucleation occurs, causing a "phase explosion." This ejects condensed material that can be captured on a cold plate and directly imaged,[50] or measured by particle-sizing methods.[51,52] Recent time-resolved imaging of IR-generated plumes show abundant micrometer-sized particles, which continue to be emitted long after ion generation is finished.[53,54] This is consistent with post-ionization of the plume by a delayed second laser pulse.[55] These clusters and particles are an important part of some models, and MALDI is often performed in the ablation regime. However, the collectable particle material and any ions within are lost for analytical purposes. Examples of these two regimes are shown in snapshots of molecular dynamics simulations in Figure 5.2. The amount of ablated material has been found to be well-described by a quasi-thermal vaporization model, having an exponential dependence on the energy deposited by the laser.[55]

Short laser pulses, faster than the thermal, mechanical relaxation of the excited matrix volume, can induce "stress confinement," along with spallation of large chunks or layers of material.[44] Typical UV MALDI experiments are not in this regime, since it requires either fast excitation (tens of picoseconds) or very intense laser pulses (to excite larger volumes). In IR MALDI the laser penetration depth is orders of magnitude larger, so stress confinement is much more common, especially if pulse lengths of a few nanoseconds or less are used. Recent detailed photoacoustic studies have examined this in detail for both liquid and solid IR matrices.[56] Ions were detected starting at the laser fluence where cluster ablation begins.

A range of peak temperatures in MALDI have been reported, probably because the effect of laser pulse energy has not always been accounted for. Early measurements gave values around 500 K.[55,57] More recently, unimolecular thermal fragmentation of "thermometer" molecules with known kinetics[58,59] suggests higher peak temperatures of around 1000 K. Time-resolved infrared emission measurements appear consistent with

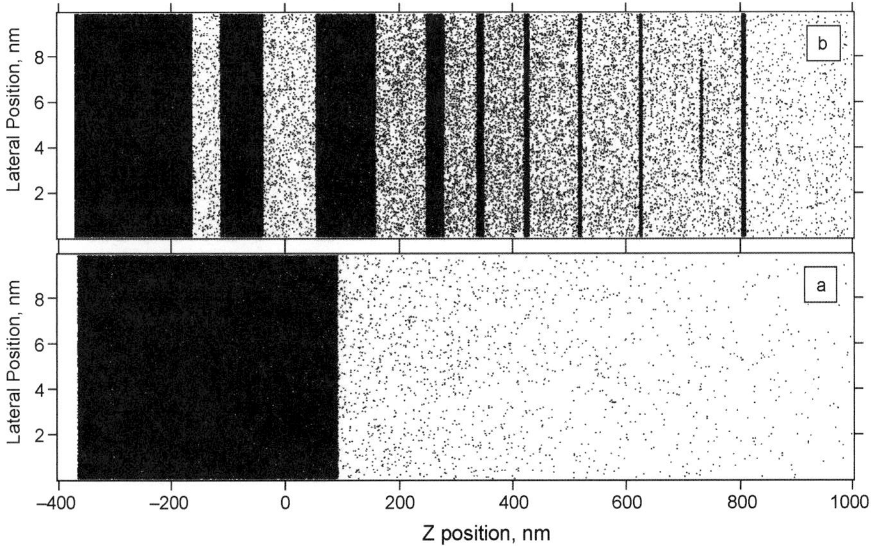

Figure 5.2. Snapshots of simulated MALDI events in the desorption (**a**) and ablation (**b**) regimes. Both generate ions, but the latter also generates condensed fragments of the original solid (clusters). These cool in the entraining gas and do not evaporate in the simulation. (Adapted from Ref. 62.)

this,[60] but suffer from lack of emissivity data for matrix materials, especially at elevated temperatures. An extensive study of leucine enkephalin fragmentation versus laser fluence[61] inferred values in the range of 500–750 K, but multiple processes were observed and in some cases imply up to 960 K. Models predict ranges between 600 and 1200 K, for typical MALDI fluences.[62,63] The material cools at a decreasing rate as the gas expands, and has a strong temperature gradient, due to laser attenuation by absorption in the upper layers of the sample. The predicted values also depend on the type of disintegration occuring, desorption versus ablation, and depth within the sample.

Most matrices rapidly vaporize at moderate temperatures, a few hundred degrees Celsius is typical.[64] The energy required from the laser varies with the initial sample temperature,[65–67] consistent with a requirement for reaching a characteristic minimal temperature to induce sufficient material ablation. The necessary laser-deposited energy density must be more associated with phase explosion than simple vaporization, since the heat of matrix sublimation is not correlated with MALDI efficacy.[68] Smaller or defect-rich crystals are more easily vaporized than larger, more ideal crystals, leading to qualitative changes in the sample as a function of number of laser shots.[69,70]

The plume begins at solid density and expands to high vacuum (or 1 atm in some instruments), a very wide density range. Because the starting density is so high, the plume is essentially unaffected by the local environment almost to the end, and it can be described as a free adiabatic expansion.[71] Methods developed for investigation of isolated molecules and clusters can be applied.[63] One important result is that cooling in this expansion is not dramatic, contrary to what is often assumed. Adiabatic expansion cooling depends critically on the heat capacity ratio, C_p/C_v, which is largest for monatomic gases, and approaches 1 for poylatomics such as MALDI matrix molecules.[4] In some cases, thermal decomposition to low-molecular-weight neutral fragments (such as CO_2) may aid cooling, as discussed for the pneumatic assistance model below.

Macroscopic models have been applied to the MALDI plume, including a hydrodynamic approach[72] and that of a pre-accelerated adiabatic expansion.[63] These cannot directly account for mixed gas/clusters in the ablation regime, but are still useful as a first approximation. They also cannot account for changes in plume development as the desorption/ablation crater shape changes.[73,74] In contrast, the mixed-phase aspect of ablation is a natural part of molecular dynamics, even though it is not computationally possible to include the full temporal and spatial extent of a real experiment. Nevertheless, molecular dynamics has illuminated numerous aspects of the phase transition aspect of MALDI.[26,47,75–78]

5.2.2 Ion Energetics, Matrix Collisions, and the Plume

It is important to examine the energetics of MALDI ion generation. Breaking a covalent C–H or O–H bond to yield R^- and H^+ requires about 14 eV or 1350 kJ/mol. An example is the O–H bond energy in phenol: 14.65 eV.[79] At 355 nm (tripled Nd:YAG laser), this would be a 4.2-photon process and hence is not very probable. However, MALDI ionization occurs in the condensed phase or dense plume. Neutral matrix is present in abundance, allowing protonated matrix to be readily formed in the above phenol example (m = neutral matrix). The total ionization process then requires only around 5 eV or 480 kJ/mol:

$$\mathrm{ROH + m \rightarrow RO^- + mH^+}$$

Such reactions will be discussed further below, but this illustrates how neutral matrix is involved in initial charge separation. It is also responsible for final, full isolation of the charges so they can be observed. After the first step, an ion pair may still be close enough to have a few electron volts of Coulomb attraction, and will eventually reneutralize. The matrix prevents this, at least sometimes, by driving them apart with collisions. Plume temperatures are moderately high, and therefore so are collision energies. Consider, for example, axial plume velocities: Mean values are a few hundred meters per second, but the distribution extends past 1000 m/s,[71,80–85] (although analytes may be somewhat slower than matrix ions[83,85,86] and the distributions can be nonthermal). Collision energies can then exceed 1 eV, readily achieving the final separation step.

5.2.3 Matrix Ions

Since matrix is nearly always a key reactant in the plume, it is particularly important to understand matrix ions and their reactivity. Matrix can be involved in proton, electron, and cation transfer secondary reactions. As a result, matrix ions associated with all these reactions can often be observed in the mass spectra. Thermodynamic data for several matrices and their various clusters, fragments, and ionic forms have been accumulating.[30–32,87–103] Experiment and theory are today generally in quite good agreement, see Table 5.1.

Matrix ionization potentials are mostly over 775 kJ/mol. Electron affinities are much lower, 100 kJ/mol or less.[95] Proton affinities range from 850 to more than 900 kJ/mol ($mH^+ \rightarrow m + H^+$). Electrostatic interactions make the acidities much higher, around 1300 kJ/mol ($m \rightarrow (m-H)^- + H^+$). Radical cation acidities ($m^{+\bullet} \rightarrow (m-H)^\bullet + H^+$) are, like the proton affinities, about 850 kJ/mol.

Alkali cation affinities of neutral matrices are in the range of 95–105 kJ/mol for K^+[104] and 140–170 kJ/mol for Na^+.[105,106] This trend of higher affinities for smaller, "harder" cations is as expected.

The energetics of matrix ion autoionization reactions are important reference values in MALDI. Some calculated examples are[95] (DHB = 2,5 dihydroxybenzoic acid, 3HPA = 3 hydroxypicolinic acid, NA = nicotinic acid, G = glycerol):

$2m \rightarrow m^{+\bullet} + m^{-\bullet}$	
DHB	703 kJ/mol
3HPA	766
NA	829
G	961

$2m \rightarrow mH^+ + (m-H)^-$	
DHB	510 kJ/mol
3HPA	499
NA	492
G	624

Table 5.1. Selected Thermodynamic Quantites Relevant to MALDI Primary and Secondary Ionization

Substance	Number of Prior Protons	PA (kJ/mol)	GB (kJ/mol)	Reference
Gramicidin S	$n=1$		$916.7 \pm 11.7 > 1018.0$	232
	$n=0$			232
Bradykinin	$n=1$		$968.2 > 1025.1$	233
	$n=0$			233
Leucine-enkephalin	$n=0$		967.8 ± 2.1	233
Cytochrome C	$n=5$		734.7	234
	$n=4$		722.2	234
	$n=3$		673.2	234
Gly		885 ± 13		235
		859.9	863.3	235
His		955 ± 9		235
		969.2	936.6	171
Arg		$>1016.$		235
		1025	992.3	171
		1037 (H-bonded)	171	
Gly-Gly-Pro			907.9	171
Gly-Pro-Gly			907.9	171
Pro-Gly-Gly			916.3	171
Gly-Gly-His			947.7/959.8	171
Gly-His-Gly			942.7/946.4	171
His-Gly-Gly			947.7/953.1	171
Gly-Gly-Arg			1015	171
Gly-Lys-Lys-Gly-Gly			995	171
Gly-Lys-Gly-Lys-Gly			997.9	171
Lys-Gly-Gly-Gly-Lys			1015	171

Matrix	PA (kJ/mol)	GB (kJ/mol)	GB((M-H)-) (kJ/mol)	Reference
α-Cyano-4-hydroxycinnamic acid (4HCCA)	841			88
	765.7 ± 8.4			90
	933 ± 9	900.5 ± 8.5		91
	841.5			99
4HCCA-H_2O	854 ± 14	822.5 ± 15.5		91
4HCCA-CO_2	894.5 ± 13.5	860.5 ± 11.5		91
Dihydroxybenzoic acid isomers:				
2,5-DHB	855 ± 8	822 ± 8		87
	853.5 ± 16.7			90
	854 ± 14	822.5 ± 15.5		91
	850.4	819.8		98
	855.8			99
			1329	92
				98
2,6-DHB	864 ± 6	830 ± 6		87
	855.2	823.6		98
			1284	96
Ferulic acid (FA)	879			88
	765.7 ± 8.4			90
	896.0	862.4		103

Table 5.1. (*Continued*)

Substance	Number of Prior Protons	PA (kJ/mol)	GB (kJ/mol)	Reference
Sinapinic acid (SA)	887			88
	894.5 ± 13.5	860.5 ± 11.5		91
	875.9			99
	903.4	869.0		103
SA-H_2O	933 ± 9	900.5 ± 8.5		91
SA-H_2O-$HOCH_3$	854 ± 14	822.5 ± 15.5		91
Nicotinic acid	907			88
	899.6 ± 16.7			90
3-Hydroxypicolinic acid (3HPA)	896			88
	898.5			99
			1365	92
2-(4-Hydroxyphenylazo) benzoic acid (HABA)	943			88
	765.7 ± 8.4			90
	950			99
trans-3-Indoleacrylic acid (IAA)	899.6 ± 16.7			90
	893.9			99
Dithranol (1,8-dihydroxyanthrone)	874.5 ± 8.4			90
	885.5			99
3,5-Dimethoxyhydroxy-cinnamic acid	853.5 ± 16.7			90
2,4,6-Trihydroxyaceto-phenone (THAP)	882			92
	893			99
			1324	92
Glycerol		874.8	820	236

Matrix	IP(eV)	Reference
2,5-DHB	8.054	100
	8.14	98
	8.19	32
	7.86	95
2,6-DHB	8.3	98
HABA	8.32	32
IAA	7.75	32
Dithranol (1,8-dihydroxyanthrone)	6.94	32
4HCCA	8.50	32
SA	7.72	32
	7.54	103
FA	7.82	103
THAP	8.44	32
Nicotinic acid	9.38	32
	9.21	95
3HPA	8.95	95

The first three are UV matrices, and the last one is an IR matrix. Two or three photons of typical MALDI UV lasers (2×337 nm = 710 kJ/mol, 2×355 nm = 673 kJ/mol) are sufficient for autoionization of the UV matrices. The IR matrix requires many IR photons for ionization (15 photons at $3000\ cm^{-1} = 598$ kJ/mol). This is not meant to imply a direct photoinduced process in either IR or UV, but rather to point out the energetic relationships between laser excitation and ionization.

Most matrices can simultaneously yield strong $m^{+\bullet}$, mH^+, mNa^+, mK^+, $m^{-\bullet}$, and $(m-H)^-$ signals (in the appropriate polarity). Sometimes decay products of the matrix are found, such as m-H_2O, or m-CO_2. Small amounts of more surprising matrix products may appear, such as $(m-nH)^{+\bullet}$, or $(m+2H)^{+\bullet}$. Radicals like $(m-H)^{\bullet}$ are likely to exist, although these are probably also generated by dissociative electron capture.[107,108]

The various matrix ions, which seem at first glance to be independent classes, may interconvert by ion–molecule reactions. In fact, this seems to be necessary for the widely observed matrix suppression effect discussed below. While not extensively investigated, for DHB it has been shown that interconversion of $m^{+\bullet}$, mH^+, and mNa^+ is thermally possible under typical plume conditions.[94,109] Protonated DHB is the most stable, but the other ion species are within 30 kJ/mol. Facile interconversion reactions are consistent with the observation of different ions in the mass spectrum for nearly all matrices.

5.3 PRIMARY IONIZATION MECHANISMS

How the first ions arise remains one of the more most controversial aspects of MALDI. Many different proposals have been made over the years, spanning a wide variety of physical effects. While cluster and photoionization models have received the most attention recently, it is highly informative to examine several of the others first.

a. Polar Fluid Model. The polar fluid model[67,110] is based on an attractive idea. It suggests that the vaporizing matrix behaves like a bulk polar solvent, in which ionic materials can dissociate. Crudely speaking, maybe the early MALDI plume is somewhat like a beaker of water. Many matrices are carboxylic acids, and so they might liberate solvated protons, thereby protonating analytes. Simple salts could analogously donate cations for adduct formation. A rich world of familiar solution-like chemistry could be active.

Early motivation for this model was the qualitative similarity of MALDI spectra regardless of whether UV or IR lasers were used. Another was observation of mass spectra dominated by analyte ions, and with no or very few matrix ions. As discussed below, these phenomena can be readily explained as a consequence of extensive secondary ion–molecule reactions in the plume; a common UV and IR mechanism is not, in fact, required.

Even if these early observations are now explained differently, a high transient plume density is expected, as described above. The model runs into difficulty mainly in that plume properties are not sufficiently "polar" for extensive ion separation. Liquid, substituted aromatics somewhat similar to MALDI matrices typically have dielectric constants around or below 10 at room temperature (phenol 9.8, methyl salicylate 9.4, acetic acid 6.1).[111] In addition, at least a factor-of-two decrease could be expected at MALDI plume temperatures. For example, 1-butanol has a dielectric constant of 15 at room temperature, but 7 at 400°C.[111] Assuming the MALDI plume is a similarly solvating fluid, separation of ionic substances (including matrix itself) will be extremely limited.

Another way to evaluate the polar fluid model is to consider its consequences for IR versus UV MALDI ion yields. Because the matrix is generally in large excess, its autoionization should dominate the total ion yield (ions/neutrals), in both MALDI variants. Instead, IR yields are orders of magnitude lower than in UV.[110,112] In addition, this mechanism cannot apply to those MALDI matrices that are nonpolar or have no ionizing functional groups.

While not a major mechanism of primary ionization, plume solvation should not be regarded as irrelevant in MALDI. Even moderate dielectric screening reduces the range at which electrostatic forces dominate over collisions, slowing ion recombination. Some indirect "polar fluid" effect may therefore be active for many matrices.

b. Excited-State Proton Transfer (ESPT). If simple ionic dissociation is not significant for typical matrices in the MALDI plume, one strategy might be to increase the driving force. Certain molecules show a large p*K* difference between ground and low electronic excited states, up to 9 units.[113]Excited-state proton transfer (ESPT) to suitable proton acceptors can therefore be transiently induced by laser excitation. Some MALDI matrices are related to (putative) ESPT molecules like salicyclic acid, leading many investigators to examine the contribution of ESPT to MALDI.[114–120] Because the relevant electronic excited states can be reached with one UV photon, this is an attractive mechanism. Obviously it is much less likely with IR excitation, although these states might be weakly accessible by multiphoton ladder climbing.

Perhaps unfortunately, ESPT does not seem to be active with the common matrices. This is because ESPT is highly dependent on efficient charge stabilization by the local environment, even if less so than ground-state ionization. Well-investigated ESPT systems tend to be active only in water or amine environments. Again the matrix plume is not likely to be sufficiently favorable. ESPT activity has also not been independently demonstrated for any MALDI matrix in any environment. For example, indicators of ESPT activity such as characteristic fluorescence shifts were not found for DHB in either solution or clusters.[100] But it need not be the matrix that is ESPT active in MALDI. Other, more strongly active substances can be added to the sample, but this approach has so far been unsuccessful.[21]

Experience to date and the above considerations indicate that ESPT in MALDI is probably very rare. If it exists, it may be limited to particular matrix–analyte complexes that are predisposed to proton transfer via strongly asymmetric hydrogen bonds and with stabilizing neighbor substituents.

5.3.1 Preformed Ions and Mechanical Separation Models

5.3.1.1 Preformed Ions

If models involving in-plume charge separation by solution-like mechanisms are inadequate, a next logical step might be to ask what happens to the ions that do exist in a typical sample preparation solution. MALDI samples are often prepared from polar or aqueous solution, using acidic matrices. Formic or trifluoroacetic acids may be added. In such solutions many analytes, such as typical proteins or peptides, will be protonated. If these survive the drying process intact as "preformed" ions, perhaps they merely need to be liberated in the desorption/ablation event.

Preformed ions might be an important ion source for many MALDI applications. Very interesting experimental evidence for preformed ions was found in MALDI samples of pH indicators, prepared at varying pH.[121] The dried samples retained the approximate color of the solution, indicating the same ionic state as in solution. While these ions are indeed preformed, this doesn't mean that they are free and available for release and observation in the mass spectrometer. As for other ionic substances prepared from solution, the protonated pH indicators are very probably crystallized with counterions. The resulting ion pairs are bound by considerable electrostatic forces. The same is true for any analytes that are charged in solution, either naturally or "tagged" by derivatization.[122–131] The problem of charge separation therefore remains also for the preformed ion models, even if ionization has, in a limited sense, already been initiated.

In addition, the color of a dopant in a host crystal may not correlate as expected with solution behavior. A particularly MALDI-relevant example is methyl red doped in DHB,[132] which exhibited different colors in different crystals or different regions of single crystals. This casts doubt on the interpretation of pH indicator colors in MALDI samples.

In the preformed model, MALDI results should depend strongly on the sample solution, and there is some evidence for this.[133] On the other hand, preformed ions might not be expected at all for nonpolar analytes and/or matrices or for samples prepared in low-polarity solutions. Since MALDI is clearly not limited to preparation methods and matrices where preformed ions are likely, this model would seem to have limited applicability.

5.3.1.2 The Cluster Model

A widely discussed cluster model was proposed and largely developed by the Karas group. Originally it was called the "lucky survivors" model.[134] Ions are taken to be largely preformed and are released by desorption/ablation. Neutralization in the plume was proposed to be very extensive, so those ions detected are "lucky survivors." Much of the neutralization of positive ions was initially suggested to take place with photoelectrons, but a more general cluster model was later developed, with more weight on other anions (secondary reactions).[28] In this form, it has also been taken up by others, and the key processes of most cluster models are sketched in Figure 5.3.

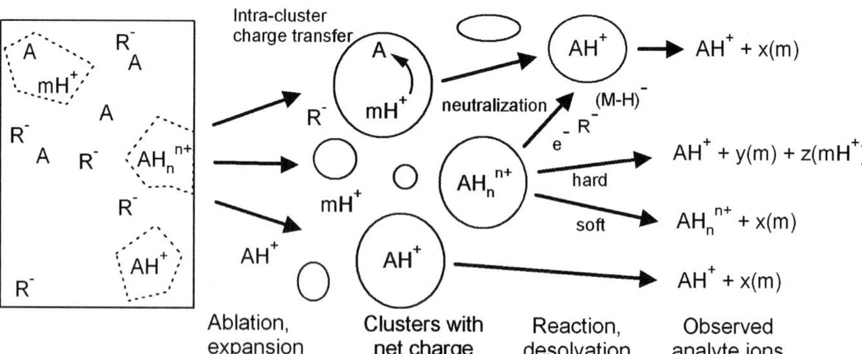

Figure 5.3. Sketch of the major processes proposed in cluster models of MALDI ionization. A, analyte; m, matrix; R^-, generic counterion. Preformed ions, separated in the preparation solution, are contained in clusters ablated from the initial solid material. Some clusters contain a net excess of positive charge, others net negative (not shown). If analyte is already charged, here by protonation, cluster evaporation may free the ion. In other clusters, charge may need to migrate from its initial location (e.g., on matrix) to the more favorable location on analyte (secondary reaction). For multiply charged analytes, hard and soft desolvation processes may lead to different free ions. Neutralization by electrons or counterions takes place to some degree but is not complete.

Charge separation in the cluster model is due to mechanical forces during ablation. Rapid appearance of gas–liquid boundaries leads to random excess of positive or negative charge on some chunks of emitted material. Either these clusters must evaporate to free analyte ions, or the ions must be ejected from the clusters (reminiscent of some electrospray models). This kind of charge separation requires a strongly ablative event. Under smooth photothermal vaporization conditions (desorption), the number of released free ions would be negligible. Since MALDI ions can be observed at fluences where clusters are not found (as shown by collecting plate experiments), the full range of MALDI conditions is not covered by the model. In ablated clusters, charge may migrate via proton, cation, or electron transfer. Depending on the mobilities of these charge carriers, secondary reactions may equilibrate, yielding the most thermodynamically favorable ions (again secondary reactions). Charge mobility means that intracluster neutralization will also be facile.

A less important charge separation process in the cluster model is photoionization and escape of electrons, leading to a net positive charge on the sample. This can only take place in a thin layer at the top of the sample. From the electron capture cross sections,[107,135] it can be shown that only a thin (few nanometers) layer emits electrons. Below this they are captured to form matrix anions (leading to further secondary products).[136] A large net positive charge excess cannot build up in the bulk of the sample.

Although this cluster model is relatively recent, clusters have been a MALDI topic for some time. Kinsel et. al.[137,138] were among the earliest to investigate this, for the case of late-appearing insulin ions. These were assigned to either (a) release of ions entrained in the plume or (b) ions generated later by downstream charge transfer reactions. Entrainment means either that the ions were carried on clusters or that the mean force on an ion due to collisions was at least as large as the force due to external extraction fields. Under such conditions, ion motion is dominated by the plume dynamics. Since collisions decrease as the plume expands, ions are progressively exposed to the field and accelerated to the detector. Note that this effect is larger for initially deeper ions (which must pass through more of the plume) than for surface ions.

Quite direct evidence for cluster ions in MALDI was found by Krutchinsky and Chait[139] in the "chemical noise" of MALDI spectra. By performing MS/MS on low-intensity peaks away from known matrix and analyte peaks, they showed that the precursor ions were clusters containing matrix. The fragment ions appeared at intervals corresponding to loss of neutral matrix molecules.

Late ions that may originate in clusters have been revisited by the Tabet group. With appropriate pulse sequences, they tried to select ions originating in heavy clusters.[140,141] Light ions were forced back into the sample, but the potential was switched before heavy ions were lost. The flight time distributions suggested that these ions were initially carried in massive clusters of up to 50,000 Da. To release the ions, evaporation of the hot clusters was invoked, possibly assisted by acceleration due to the extraction voltage.

The same group considered evaporation of clusters in some detail.[34] "Hard" and "soft" desolvation pathways were differentiated. Soft refers to sequential evaporation of neutral matrix, so no change of charge on the cluster occurs. A high yield of multiply charged analytes results, because all charge originally on the cluster gets concentrated on the analyte that remains. In hard desolvation, charged matrix is ejected, so low analyte charge states result. Charge migrates to analyte in the soft pathway, and it migrates to surface matrix in the hard pathway. Different kinetics are therefore expected, depending on the characteristics of the associated transition states. Since this depends on matrix, analyte charge states can be modulated somewhat by matrix choice.

The entrainment and massive cluster models are not necessarily mutually exclusive. Possibly they occur together in some cases, Since delayed ions are not limited to conditions of very large cluster formation, entrainment may be more likely or common. Also, while MALDI simulations show ion entrainment, the clusters usually are not hot enough to fully evaporate.[62]

5.3.1.3 Pneumatic Assistance

The pneumatic assistance or expanding bubble model[142–147] is essentially a variant of the cluster model. Its main feature is also mechanical separation of preformed ions during ablation, by gas bubbles forming below the surface. When these burst, small droplets or fragments are ejected, which then evolve to give individual ions as in other cluster models. Subsurface nucleation driving ablation of overlying melt has also been noted in molecular dynamics simulations, and the "bubble chamber" FAB/SIMS model is very similar.[148,149]

Perhaps the most insightful aspect of the pneumatic assistance model is the recognition that some matrices can thermally decompose to low-molecular-weight species that create significant overpressures at MALDI melt temperatures. In particular, many UV MALDI matrices are carboxylic acids that are known to decompose at moderate temperatures. Consistent with this, laser-induced CO_2 emission from such matrices *in vacuo* has been found to rise with increasing laser pulse energy.[144]

Many UV matrices are not thermally labile, so the model is limited in potential application. In IR MALDI, however, it could be more generally relevant. Entrapped residual low-molecular-weight solvent might be directly vaporized by the laser.[150] However, this would mean that IR MALDI is highly dependent on sample preparation in order to obtain the optimum amount and distribution of trapped solvent. It seems unlikely that this is consistent with experiment, but needs to be further explored.

Very interesting results from a rate equation implementation of the model have been reported,[144,146,147] but the equations have not appeared. Bubble pressures of 10–100 atm, depending on estimated decomposition rate parameters, were calculated. Such pressures could change an otherwise desorptive event into an ablation, increasing ion yield. On the other hand, if the pulse energy is high enough that ablation would happen already without decomposition, this mechanism increases the gas pressure by only about a factor of two. Larger pressure ranges can be sampled by varying the laser pulse length, but MALDI yields are insensitive to pulse length.[36–40] This suggest that low-molecular-weight gas bubbles are not decisive for MALDI ion formation.

5.3.2 Photoexcitation and Pooling Models

For IR MALDI, models that do not require direct electronic excitation by the laser, such as those above involving preformed ions, are the most attractive. Upconversion of many IR photons into molecular electronic excitations is not efficient, even with relatively intense lasers. In contrast, gas-phase photoionization by two-photon UV excitation is routinely used as a sensitive and selective detection method, so it is natural to explore this and related processes in the MALDI context.

5.3.2.1 Direct Multiphoton Ionization of Matrix or Matrix–Analyte Complexes

The most intuitive ionization pathway in UV MALDI might be direct matrix photoionization. The ionization potentials (IPs) of free, isolated matrix molecules have been found to lie in the three-photon region for typical UV-MALDI lasers. Compare the IP of DHB, at 8.054 eV,[100] with the tripled Nd:YAG laser at 3.5 eV per photon (355 nm) and the nitrogen laser at 3.7 eV per photon (337 nm). Relatively few matrix IPs have been measured, but modern *ab initio* methods are accurate to about 0.2 eV.[32,94–96,98]

Single-molecule three-photon processes are normally inefficient at MALDI-like laser intensities, but matrix and analyte are in the condensed phase, not free. Collective effects reduce IPs compared to isolated molecules, suggesting that more efficient two-photon ionization might still be possible in MALDI. IPs of matrix clusters are not well-studied, but only small reductions (a few tenths of an electron volt) were found for DHB clusters.[101] Clusters of a few molecules and bulk matrices are not the same, but the clusters show that even large IP reductions would not be important in practice, because ionization efficiency drops dramatically at low two-photon energies.[101] So even if a bulk MALDI matrix has a two-photon IP threshold for typical lasers, this might be a very inefficient method of creating matrix ions. A path around this limitation appears to exist in pooling, which will be discussed below.

More important for UV-MALDI might be IP reductions due to matrix–analyte interactions. Matrix–analyte interactions can be stronger than those between matrix molecules, and there is no *a priori* reason to expect low two-photon efficiencies for such complexes. Kinsel and colleagues have reported several experimental and theoretical studies of this effect in clusters.[97,102,151–153] In one example, strongly reduced IPs for DHB–proline complexes were found, down to 7 eV.[102] In addition, fragmentation of some complexes after ionization produced protonated analytes.[151,152] This is an efficient two-photon process that is probably active for certain matrix–analyte combinations. How often it contributes to MALDI is not yet clear, and this cannot be easily predicted in advance.

Even when facile cluster photoionization is possible in a particular system, the effect is currently not often exploited. The matrix–analyte mole ratio is generally 1000 or higher, so the laser energy is absorbed almost entirely by noncomplexed matrix. Even if relatively inefficient, matrix-only mechanisms will then generate the large majority of ions. Most analyte ions will be generated via secondary reactions with matrix primary ions.[136] Better results can be obtained in this case if the matrix–analyte ratio is reduced and the laser intensity is increased.

5.3.2.2 Pooling and Indirect Photoionization

Even if direct two-photon ionization is only an occasional or weak contributor to MALDI ionization, there are indirect processes in the condensed phase which make photoexcitation a strong mechanistic candidate. The most important of these is pooling, which is the basis for the second of the two most widely discussed modern MALDI ionization models.

This model invokes both mobile electronic excitations in the bulk matrix, as well as concentration of this energy in pooling events.[43,63,100] Pooling is a term for redistribution of electronic excitation energy initially distributed over more than one molecule. For example, two neighboring matrix molecules may be each raised to the first excited singlet state (S_1) by the laser. If these molecules are in contact in the solid sample, they may distribute their net two-photon energy in various ways. In addition to the original one quantum per molecule (which may be represented as S_1:S_1), the same energy can be concentrated on one molecule, which is raised to a higher excited state. The other molecule is deactivated, so the state description is S_n:S_0. Pooling makes high-energy processes such as ionization possible, even when the laser photon energy is well below the ionization energy.

Pooling was one of the many possibilities imagined in early work,[21] but its importance for MALDI requires more than excited neighbor pairs randomly created by the laser. These are not sufficiently numerous except at very high intensities.[43] However, excitations can be mobile in the solid state, greatly increasing their interaction probability.[154] Mobile excitations can be treated as pseudo-particles and are denoted "excitons."

Exciton motion and pooling are related phenomena, since both depend on the dynamics of electronic states as induced by intermolecular interactions.[155,156] Exciton pooling can be investigated by fluorescence quenching as a function of laser intensity, since it prevents S_1 excitations from radiatively decaying. This indicator has been examined for a few MALDI matrix materials,[42,43,157,158] but only 2,5-dihydroxybenzoic acid (DHB) has been studied in detail.[42,43] For DHB the time per exciton hop between molecules is approximately 50 ps, so the 0.5- to 1.5-ns excited-state lifetime allows for significant diffusion. (The lifetime of free DHB is near 30 ns.) Another indicator of exciton motion is trapping by dopants in the crystal with excited states energetically below those of the matrix.[43] This effect might be of some practical importance for certain analyte–matrix combinations, since it can reduce ionization efficiency.

As discussed above, two 337-nm or 355-nm photons probably can not efficiently ionize DHB or many other matrices in the solid state. If not, a $S_1 + S_1$ pooling event is also not sufficient. In that case, an additional $S_1 + S_n$ pooling event is required for ionization. This stepwise pooling model has been expressed as a set of differential equations and has been numerically integrated to provide quantitative MALDI predictions.[63] Figure 5.4 shows the intra- and intermolecular processes involved. The proposed ionizing pooling step was investigated with a time-delayed two-pulse experiment, and the results were consistent with the model.[159] Two weak laser pulses can be combined to give a strong MALDI signal, although neither alone does so. Changing the time delay between them provides information

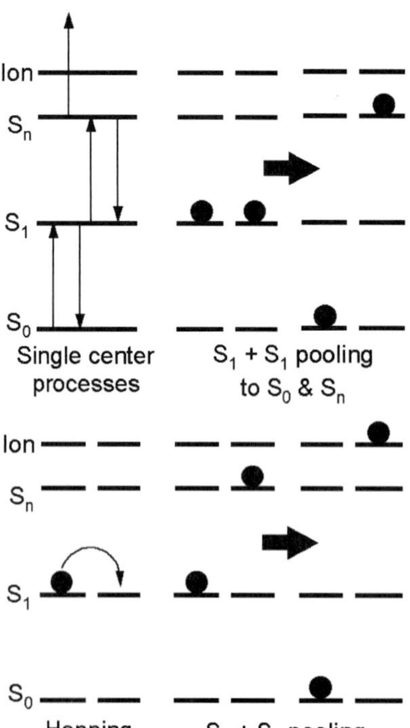

Figure 5.4. Unimolecular and biomolecular matrix processes included in the continuum MALDI ionization model of Refs. 63 and 225. Pooling reactions of matrix excited states are key steps in energy concentration and ionization.

about the dynamics of ionization. A delay of 2–3 ns gave the maximum DHB ion yield, followed by a gradual decrease over several nanoseconds.

Because exciton hopping, pooling, quenching, and recombination are second-order processes, they are strongly dependent on the intermolecular collision rates, and the local temperatures and pressures in the plume. These in turn depend on energy conversion from excited electronic states (including ions) to heat. This feedback makes inclusion of the plume expansion necessary for any model. Taking the plume to be a free jet seems to be a good approximation, but may not be correct at the earliest times, if nonequilibrium conditions temporarily exist. Molecular dynamics does not require specific models for the expansion, so the pooling model of ionization was added to the breathing-sphere molecular dynamics method of Zhigilei and Garrison, giving a molecular-level UV-MALDI model.[62] The results so far confirm the earlier pooling model assumptions, but provide many more details of the coupling of ionization to the expansion in desorption, ablation, and spallation regimes. All of these can even coexist in different layers of the same sample.

An example UV-MALDI model calculation is shown in Figure 5.5. Apparent are the buildup of excited states, followed by ion generation via pooling. After the phase change, recombination is initially rapid, but slows as the expansion dilutes the gas. These results assume that all ejected material is fully vaporized. Clusters create no free ions, so any cluster mass fraction proportionally reduces the free ion production.

The photoexcitation/pooling rate equation model allows quantitative comparison with a variety of experimental data. While the only sufficiently characterized matrix at the moment is DHB, many observed MALDI phenomena are very similar for others, suggesting a degree of generality for the model. The phenomena reproduced for DHB span a considerable range as summarized next[63]:

The ion yield is a fundamental observable and is in the range of 10^{-4} to 10^{-3}, depending on fluence. This matches well with experiment.[49,55,160–162]

Laser characteristics are among the most easily varied parameters. The model correctly predicts that UV MALDI ionization is fluence (J/cm^2)-dependent rather than irradiance (W/cm^2)-dependent[36,39,163–165] This stems from the energy storage effect of excited states and is a strong argument for the importance of those states.

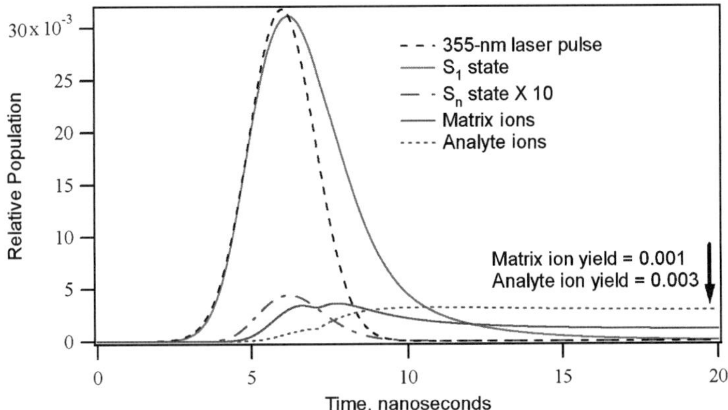

Figure 5.5. Example of the time-dependent evolution of a MALDI sample in the photoionization/pooling model. Pooling of abundant S_1 excitations leads to higher excited S_n matrix molecules. Pooling of S_1 and S_n leads to matrix ions (primary ionization). These react with analyte neutrals to yield analyte ions (secondary ionization), depleting the matrix ion signal.

MALDI ionization has a high-order (6 or higher has been reported) empirical dependence on laser fluence.[55,162] This leads to a perceived fluence threshold, above which ions are easily observed in a given instrument. The model predicts a rapid increase in signal, an apparent threshold, at a fluence where the vaporization of the sample becomes rapid. The threshold phenomenon is therefore not associated with ionization itself. Ions are formed at low fluence, but are not released if vaporization is poor. Because the of this, positive and negative ions should have the same threshold, as found experimentally.[166]

The apparent fluence threshold is wavelength-dependent and is inversely correlated with the absorption coefficient of the matrix.[1,167] This is reproduced by the model and is again a vaporization effect. A higher absorption coefficient heats the upper sample layers more efficiently, so they vaporize at lower fluence.

The degree of laser focusing affects MALDI yield in a nonlinear manner.[55,163,168] This initially seems surprising, if the fluence in the spot is held constant. However, it is readily interpreted in terms of the expanding jet plume model. The rate of approach to the collision-free regime is determined by the "orifice" from which a jet expands—in this case, the irradiated spot. A smaller spot means faster expansion, so bimolecular processes stop earlier in a plume emitted from a smaller spot. The model shows that this affects recombination more than ion formation, so yield per ejected volume is higher for a smaller spot at the same fluence. This would seem to explain the recently reported yield improvement for a structured rather than uniform laser spot.[169] Other associated physical properties of the plume are correctly reproduced. Temperatures are 600–1000 K, while velocities range from 500 to over 1000 m/s.

The MALDI yield at a given fluence depends on the initial sample temperature.[65,66] Again the coupling of vaporization and ionization clarifies this. The difference between the initial temperature and the matrix sublimation temperature determines how much laser heating is needed. More heating requires decay of more excited matrix, in addition to a longer time in the condensed phase before ion release. If ion formation is faster than recombination, this reduces yield. Higher temperatures therefore improve yield, up to the point where the sample is so rapidly vaporized that pooling does not have time to generate many ions. The yield therefore has a maximum versus initial temperature. Reproduction of these effects suggests that the jet model of plume–ionization interaction is basically correct.

5.4 SECONDARY IONIZATION PROCESSES

Depending on the mechanism, primary ions may be matrix or analyte, or both. Primary ions are created in a dense environment, perhaps even clusters and particles, but must reach a low-density regime to become available for mass analysis. During the expansion, primary ions react with neutrals to make the secondary ions observed in the mass spectrum. These secondary reactions are the key to understanding many MALDI phenomena and are the final theme to be discussed here. There are three types of charge transfer secondary reactions known in MALDI: proton, electron, and cation transfer. The characteristics of each will be discussed below, but all lead to qualitatively similar effects in MALDI mass spectra.

These similarities are once again a consequence of the plume expansion. During the expansion, collisions become less frequent and the rates of ion–molecule reactions decrease. Because matrix is normally in considerable excess, the last reactions of any ion will nearly always be with matrix neutrals. This is fortunate for understanding MALDI spectra, because simple bimolecular matrix–analyte reactions are the limiting reactions. Quantitative understanding of complex processes in dynamic cluster or dense plume environments is not necessary, at least to a good first approximation.

5.4.1 Kinetic Versus Thermodynamic Control of Secondary Reactions

Combined with the concept of limiting bimolecular reactions, approach to local thermal equilibrium in the plume allows straightforward prediction or interpretation of the mass spectrum. The observed ions should be determined more by the thermodynamics than the kinetics of reaction with matrix. One example might be

$$\mathrm{m}^{\bullet +} + \mathrm{A} \leftrightarrow \mathrm{m} + \mathrm{A}^{\bullet +}$$

for which the usual relationship between free energy and equilibrium constant applies:

$$\Delta G = -RT\ln(K)$$

Assuming uniform detection efficiency, the equilibrium constant, K, is given by the relative abundances of matrix and analyte ions in the spectrum. Since free energies are known or can be readily calculated for some matrices and analytes, this relationship can be tested. Kinsel et al.[30] have recently done so for proton transfer reactions of matrix to amino acids. Their Boltzmann plots were nicely linear, verifying the local equilibrium model in these cases.

Several other phenomena in MALDI can best be interpreted as more thermodynamically than kinetically limited, including the suppression effects discussed in detail below. It is interesting, however, to ask if conditions can exist in which MALDI spectra become kinetically dominated. This is possible at very low laser intensities, where so little material is emitted that it expands before secondary reactions can run to completion. As a result, a low-fluence spectrum can differ from a high-fluence one. Varying the fluence allows observation of and transition between kinetic and thermodynamic regimes, as seen in Figure 5.6.[170] This is a simple example, involving only two different ions. More complex cases with multiple matrix ions have also been documented.[170]

To obtain strong signal with relatively few laser pulses, MALDI fluences are nearly always high enough to reach the high-plume-density, thermodynamically limited regime for typical secondary reactions. Other situations where kinetic limitations may appear are at high initial sample temperature,[66] or for certain highly exothermic electron transfer reactions.[32]

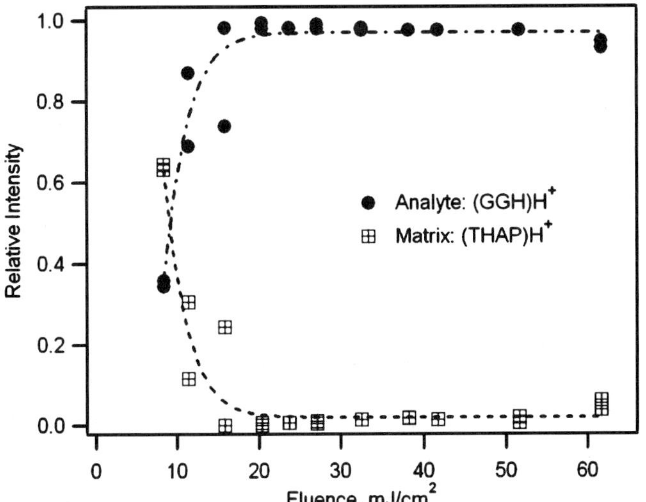

Figure 5.6. Normalized intensities of positive MALDI ions obtained from a sample of glycyl–glycyl–histidine in 2,4,6-trihydroxyacetophenone matrix (1 : 1 molar ratio) versus laser fluence. Matrix suppression is essentially complete at most fluences, but not near threshold. This is a consequence of incomplete secondary reactions of primary matrix ions with neutral analytes. (Adapted from Ref. 170.)

5.4.2 Proton Transfer Secondary Reaction Energetics

Analytes are very frequently observed as protonated molecules in MALDI. This is particularly so in biological applications such as protein and peptide analysis. If approach to local thermal equilibrium is extensive, we expect the spectrum to be defined by the gas-phase proton affinities (PAs) of matrix and analyte. Consider simple biomolecular analytes such as amino acids, which have PAs from 885 kJ/mol for glycine to 1025 kJ/mol for arginine.[171] Comparing with matrix PAs, proton transfer from typical matrices to the least basic amino acid, glycine, is found to be weakly exothermic to endothermic, while more basic arginine can easily abstract a proton from the common matrices.

The proton transfer reactivity of individual amino acids with matrix is the largest factor contributing to the rule of thumb that peptides containing basic residues are preferentially observed. A more refined analysis takes into account the higher basicity of polypeptide chains that better stabilize protons by multiple coordination.[171–174] While PAs may not be the sole determining factor in peptide analysis,[175–178] they are clearly the largest in typical practice. In contrast, oligonucleotides are weakly basic and are more amenable to MALDI analysis in negative polarity.[179] An appropriate matrix choice for these analytes has a high proton affinity in deprotonated form.

Final ion abundances may be largely predictable from the gas-phase thermodynamics of charge transfer if the molecules are stable, but the heat of reaction is localized in the participants for a certain time, which can lead to fragmentation. The extent of analyte decay or internal activation is therefore correlated with the exothermicity of reaction with protonated matrix.[64,180,181]

5.4.3 Cationization Transfer Secondary Reaction Energetics

The proton is a small, hard ion, so matrix and analyte PAs are high. Protonated species are therefore the most favorable secondary MALDI ions for many analyte classes. Nevertheless, there are major applications, such as analysis of synthetic polymers, where this is not the case. These are better analyzed as cation adducts because proton affinities or acidities are below those of most matrices, but affinities for some cations are comparatively high. While transition metal ions such as Cu^+ and Ag^+ have important applications, more ubiquitous are the alkali cations Na^+ and K^+.

Sodium affinities of common matrices are in the range of 140–170 kJ/mol, compared to >150 kJ/mol for amino acids and >160 for dipeptides.[182–184] Nucleobases and carbohydrates have even higher affinities, 164–190 and >160 kJ/mol, respectively.[182] Differences in matrix and analyte Na^+ affinities are evidently smaller than for proton transfer or even unfavorable, so adduct formation often needs to be optimized by choosing a matrix with a particularly low cation affinity. Dithranol (1,8-dihydroxyanthrone) is one such matrix and has become preferred over common matrices like sinapinic acid, DHB, or THAP for this application. It was discovered empirically, but its performance is now understood to be consistent with its cation transfer thermodynamics. The same is true for other matrices, as shown in various cationization studies.[22,185]

Because cations must exist in the sample prior to laser irradiation, preformed cations may be thought to be predominant. Interestingly, several studies instead support a predominantly gas-phase process.[186–190] Attempts to separate preparations of analyte and cationizing reagent are particularly convincing[189–191]; and solvent-free methods producing finely mixed, but not cocrystallized, sample are similar.[70,192]

Commercial matrix is often contaminated with salts,[193] as may be the analyte. Unfortunately, this is uncontrolled, leading to inconsistent cationization results. Controlled addition of cations to MALDI samples may be desirable both to increase signal and to simplify the spectrum by dominant ionization with the selected cation. Excess salt is deleterious, by adversely affecting cocrystallization and by matrix dilution.[193,194] A desirable salt quantity would be one analyte equivalent, but because affinities of matrix and analyte may be rather close, competition for cations occurs. Above-stoichiometric salt quantities often therefore give best results. Matrix cation adducts may be prominent in the spectrum, graphically demonstrating competition.[193] Clearly any anion, notably the often abundant deprotonated matrix primary ion, will have a much higher cation affinity than neutral analytes, reducing the analyte adduct signal.[195]

5.4.4 Electron Transfer Secondary Reaction Energetics

Third in the list of importance for MALDI applications are radical cations or anions.[196–201] These tend to be observed only for low polarity molecules with few polar functional groups, but make MALDI an effective analytical method for these substances.

Since electron transfer (ET) reactions between neutral analytes and primary matrix ions are involved,[197,198] the thermodynamic quantity of interest in positive polarity is the ionization potential (IP) of matrix and analyte.[198] In negative polarity it is the difference in electron affinities, as has been demonstrated for radical anions of fullerene derivatives in LDI and MALDI.[199,202] Once again the evidence points to thermodynamic rather than kinetic limitations in most cases. The ET excess energy has been strongly correlated with abundance of fluorofullerene fragments, for example.[199] Thermodynamic values needed for evaluation of electron transfer reactions are not yet widely available, although Lippa et al.[203] have recently measured electron affinities of matrix molecules. These lie roughly between 0.4 and 1.2 eV for the free molecules, rising significantly to over 1.5 eV for the dimers.

Multivalent cations are an important special case of ET in MALDI. Divalent salts of Ca^{2+}, Cu^{2+} or other cationizing agent are sometimes used, but only singly charged adducts are observed. The metal ions are reduced to $+1$,[123,204–207] or protons are ejected to reduce the net charge.[208,209] This can be understood from the thermodynamics of multiply charged gas-phase ions. The energy required for $Cu \rightarrow Cu^{+}$ is 7.7 eV, and $Cu^{+} \rightarrow Cu^{2+}$ requires 20.3 eV.[210] Because matrix IPs are near 8 eV, they can reduce divalent metal to $+1$ state, but not neutralize it. Proton ejection also effectively transfers an electron to the M^{2+} site and is strongly exothermic because of the high second ionization potential of the metal center.

5.4.5 Matrix–Analyte Reactions, Analyte–Analyte Reactions and Suppression Effects

5.4.5.1 Matrix Suppression Effect

The most important secondary reaction is that of matrix with analyte. In the photoionization model, analyte ions are only formed by reaction of analyte neutrals with primary matrix ions. Regardless of how they are formed, analyte ions in the plume will continue to undergo collisions, and hence possibly react with matrix neutrals until extracted to the detector. To understand the role and extent of matrix–analyte reactions, it is useful to consider one of the more dramatic MALDI phenomena: the matrix suppression effect (MSE). Although sometimes noted in the literature,[211–213] it has only been systematically studied after its importance for the two-step model was recognized.[214,215]

Empirically, the MSE is the loss of matrix ion signal as more analyte is added to the sample. At sufficient analyte concentration, matrix ions can be completely suppressed, leading to a very clean spectrum containing only analyte ions. An example is shown in Figure 5.7. It is a rather general effect, demonstrated for many analytes and matrices,[216] and has begun to be applied analytically.[217] If MSE does not appear in one polarity, it generally does in the other. Suppression includes all matrix ion types, not only a single type

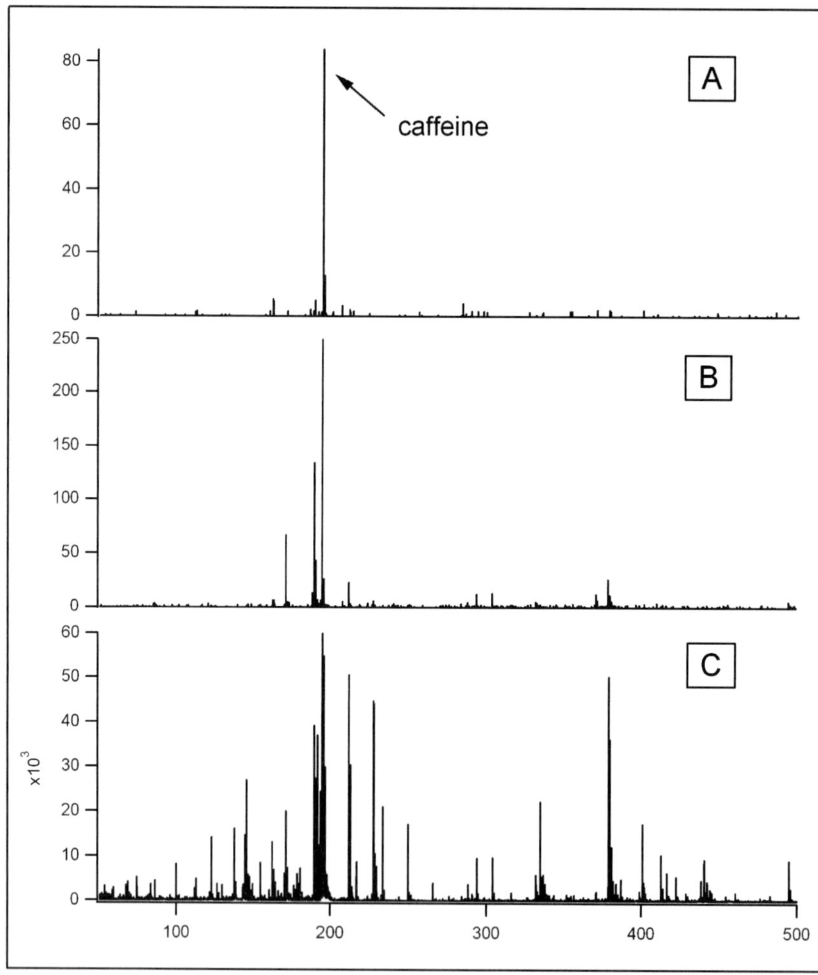

Figure 5.7. MALDI-TOF mass spectra of caffeine, in CHCA matrix. Matrix suppression is nearly complete in spectrum **A**, where the matrix–analyte ratio is low (3). At higher ratio, 27, more matrix signals appear, as seen in spectrum **B**. In panel **C**, many more matrix signals appear although the matrix:analyte mole ratio was again 3, as a result of much higher laser pulse energy. When more analyte than primary ions is present, suppression occurs as a consequence of secondary reactions. This ratio is affected by both the analyte concentration and the laser intensity. (Adapted from Ref. 216.)

such as protonated, cationized, or radical cations. In addition, all ion types are suppressed by analytes of any type.[109,214,215] This requires the interconversion reactions noted above, which are obviously also a kind of secondary plume reaction.

The MSE appears to point toward photoionization/pooling primary ionization models, because of inconsistency with preformed/cluster models. Although secondary matrix–analyte reactions occur in clusters,[28] observation of complete matrix suppression would require that there are *never* more preformed ions than analyte molecules in *any* randomly formed cluster. Statistics would seem to exclude this, especially since the analyte is often inhomogeneously distributed.[218–221] Diffusion of ions in the plume is better able to explain the MSE, since ion mobility and mixing is much better than in clusters.

The polarity of MSE is a natural consequence of secondary reaction thermodynamics. Creation of analyte ions from the respective matrix ion will typically be favorable only in one direction, positive or negative. Reaction in the opposite polarity is generally unfavorable, giving MSE a preferred polarity. Chemical intuition can generally be used to predict this. For example, basic analytes deplete protonated matrix, and suppression occurs in positive mode. Because matrix suppression is depletion of all primary matrix ions, it is a function of analyte concentration and laser fluence (concentration of primary ions). If matrix suppression is desired, an excess of analyte over matrix ions should be sought, by using a concentrated sample and low laser power. If it is considered undesirable, lower concentrations and higher laser intensity will be preferred.

5.4.6 Analyte Suppression Effect

Most MALDI samples contain two or more analytes. Often it is the relative intensities of these signals which are of most interest to the analyst, especially if one or more is an internal standard. The objective may be qualitative, such as in protein identification where maximal coverage of proteolytic peptides is desired, or quantitative, where sample concentrations are to be determined. In either case, inter-analyte effects must be understood to design or evaluate a MALDI experiment.

Analytes may influence each other indirectly via competitive reaction with matrix ions, or directly by reaction with each other:

$$\mathrm{m^+ + A \rightarrow m + A^+}$$
$$\mathrm{m^+ + B \rightarrow m + B^+}$$
$$\mathrm{A^+ + B \rightarrow A + B^+}$$

Whether due to matrix–analyte or analyte–analyte reactions, the analyte with the most favorable charge transfer thermodynamics will appear with greatest intensity in the mass spectrum. Since the thermodynamic picture of secondary plume reactions is not specific to any charge transfer reaction type, the same kind of suppression phenomena are expected for protonation, cationization, and electron transfer. As for the MSE, an increase in primary ions due to higher laser intensity reduces inter-analyte suppression since the matrix–analyte reactions are forced to the right. Also parallel to the MSE, ASE is favored by high analyte concentration. It is therefore normally preceded and accompanied by MSE. An example of both effects is shown in Figure 5.8. Because MSE and ASE are connected, suppression of one type of analyte ion by another can occur, and has been observed.[109]

The parameters that lead to and control the ASE are obviously also those that control relative sensitivity factors for any mixture. The relevant secondary plume reactions can be modulated by matrix choice, analyte concentration, and laser intensity. If analyte reactivity

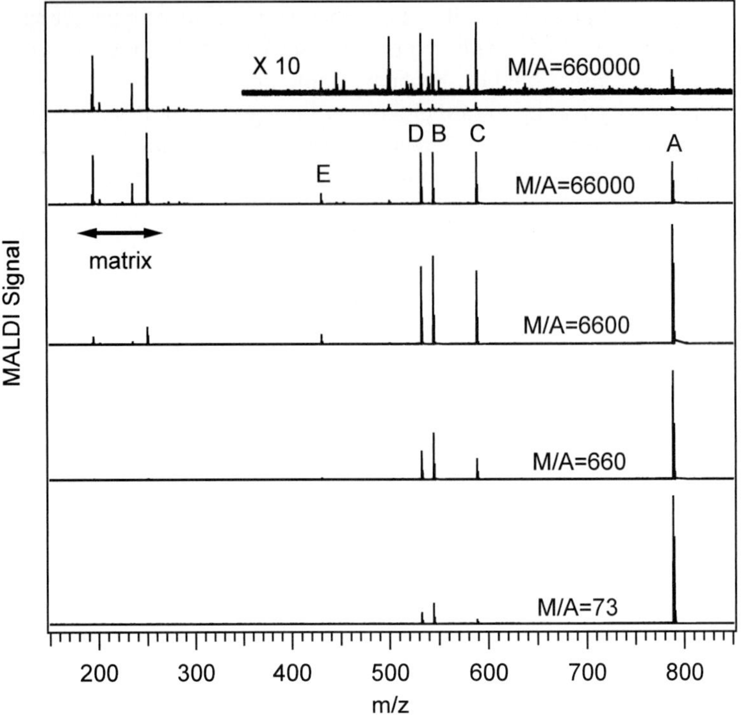

Figure 5.8. Positive-mode MALDI spectra versus matrix–analyte mole ratio (DCTB matrix) for an equimolar five-component mixture. A, M-T data; ionization potential (IP), 6.04 eV (CAS number: 124729-98-2); B, TTB; IP, 6.28 eV (76185-65-4); C, NPB; IP, 6.45 eV (123847-85-8); D, rubrene; IP, 6.50 eV (104751-29-9); E, D2NA; IP, 7.06 eV (122648-99-1). The molar mixing ratios of matrix to analyte are indicated for each spectrum. These analytes are observed exclusively as radical cations, and they exhibit matrix and analyte suppression effects analogous to those known from proton or cation transfer secondary reactions. Low ionization potential (IP) analytes suppress high IP analytes and matrix. (Adapted from Ref. 32.)

is expected to be highly variable, the secondary reaction picture suggests that good sensitivity for all analytes cannot be obtained with one set of conditions. A separation step may therefore be necessary to achieve full analyte coverage.

5.4.6.1 Multiply Charged Analytes

One important type of secondary reaction concerns the relative contribution of multiply charged ions to the mass spectrum. This is qualitatively predictable from potential plume reactions. Ions of higher charge state have higher internal electrostatic potential energy. This apparent from the basicities of biomolecules, which drop steadily versus prior protonation.[222] As a result, the removal of charge in a collision with neutral matrix is more favorable for higher charge states. The maximum charge state of a given species in MALDI is that at which matrix can exothermically abstract the next added charge, but not the present one. Since the internal electrostatic energy of an ion is a function of charge separation, the MALDI tendency to observe higher charge states increases with molecular size. This concept is consistent with the available data for protonation.[109,173,223,224] Multiple charging may also be kinetically limited by cluster desolvation,[34] even if thermodynamically favorable, as noted above.

5.4.7 The Quantitative Two-Step Model Including Analyte

The matrix-only pooling model described above has been extended to include analytes. It is currently the only model that is capable of quantitative predictions of matrix and analyte ion yields and that includes experimentally relevant factors such as laser characteristics, matrix/analyte ratio, and physical and chemical properties of matrix and analyte.[225] Primary ions are derived from matrix, so analyte signal is solely a consequence of the reactions that also lead to the MSE and ASE:

$$\mathrm{mH^+ + A \leftrightarrow m^+ AH^+}$$
$$\mathrm{AH^+ + B \leftrightarrow A^+ BH^+}$$

An Arrhenius form was assumed for the rates, with activation energies given by a nonlinear function of the reaction free energy, the only aspect of the model specific for ion type. Activation energies are substantial for free energies below 60–90 kJ/mol. The collisional prefactor introduces a dependence on the analyte size and, hence, molecular weight. Effective molecular radii can be estimated from ion mobility studies. In this form the secondary reaction model has no adjustable parameters beyond those necessary for primary ionization.

The correlation with experimental phenomena is good. The matrix and analyte suppression effects are well reproduced, including all the characteristics noted above such as dependence on secondary reaction exothermicities, concentration, and laser fluence.[109,214,215] The model shows that analyte intensity ratios can approach the original concentration ratios if more primary ions are created at higher laser fluence, but this ratio does not reach the correct values at any fluence.[32] Even analytes of similar reactivity but different molecular weight may exhibit different intensities, because of the size-dependent collision probabilities.[178]

Favorable plume reactions yielding the desired analyte ions depend on matrix choice. It should be emphasized that there is now considerable thermodynamic information available for many matrices, so it should seldom be necessary to try them at random. This includes experiments where cationization is the desired ionization pathway.

5.5 SURFACE EFFECTS

MALDI mechanisms have been discussed above with a focus on bulk samples. Recently, it has become apparent that the sample substrate can have an effect for thin samples in UV MALDI, via the ionization of the matrix. Some reports suggested that photoelectrons from the metal beneath thin samples reduce positive ion yield,[226] and they also suggested that yield is better from a gold than stainless-steel substrate, due to differences in the photoelectron energies compared to the matrix electron capture cross section.[227,228] From the cross sections,[108] the mean free path of low-energy electrons is 10–50 nm, defining the range of such a substrate effect. Since the samples in these studies were made by the dried droplet method, it is not clear what fraction of the samples was sufficiently thin, making comparisons difficult.

Better control of sample thickness is achieved by electrospray deposition. Recent studies with three matrices have shown dramatic enhancements (not reductions) of matrix and analyte signals for the bottommost layers on a stainless steel substrate surface.[229,230] Enhancement was greatest in the thinnest parts of the samples. Example images are shown in Figure 5.9.

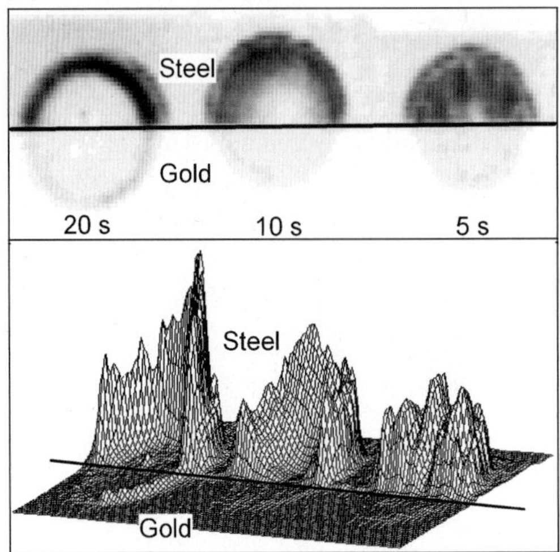

Figure 5.9. Electrosprayed MALDI spots of sinapinic acid matrix on stainless steel that is half-coated with 40 nm of gold. The thinnest spot is on the right, with the thickest spot is on the left, and the spray times are indicated in seconds. The images represent the sum of all SA matrix positive ions. Darker pixels indicate stronger signal. The substantial signal enhancement of thin sample regions on steel is apparent, as is the much weaker effect on gold.The thickness was estimated to be 100 nm per second of spray time, and the distance between laser craters is 0.15 mm. (See also Ref. 230.)

For two matrices, DHB and sinapinic acid, the enhancement was smaller on a gold substrate. HCCA matrix gave a similar enhancement on both metals. This pattern is consistent with a two-photon ionization model based on interaction of matrix orbitals with the conduction band of the metal.

The metal–matrix interface can also be at the top rather than the bottom of the sample. Thin metal layers can be sputter-coated with little difficulty.[231] The MALDI benefits of a top layer may extend beyond reducing surface charging of insulating samples to a moderate enhancement effect.[229]

5.6 CONCLUSIONS

MALDI ionization is a complex phenomenon because of the strong interaction between the physical and chemical components of the process. Desorption and ablation by phase explosion and spallation may all be active in a single MALDI event. Primary ionization models are now focused on two main approaches: (a) the thermal, mainly mechanical, picture involving preformed ions and (b) the indirect pooling/photoionization model (for UV-MALDI). The photochemical picture has made good quantitative predictions for many observables. Only one matrix is currently modeled in detail, but other matrices show many similarities, suggesting that the mechanisms could be fairly general. Acceptance and understanding of secondary reactions has advanced considerably. Recognition of the approach to local equilibrium has motivated measurement and application of gas-phase thermodynamic properties of matrices and analytes.

Mechanistic theory and models now can be applied to everyday analysis. MALDI applications can often be usefully optimized by consideration of the gas-phase thermodynamics of the possible reactions in the plume, and how much of the reactants are available. Experimental choices that favor formation of the desired analyte ions over other potential competitors can be made.

At the same time. MALDI mechanisms are not completely understood. Only a few matrix molecules have been well-characterized, and it remains unclear why one matrix is good and another not. Unusual species probably exist in the plume, and the chemical physics of the expanding, reacting, hot plume are still under active investigation. The very complex systems encountered in polymer analysis (matrix, polydisperse analyte, and cationizing reagent) are also a continuing challenge. There is still much to be learned about MALDI ionization.

REFERENCES

1. Karas, M.; Bachmann, D.; Hillenkamp, F. Influence of the wavelength in high-irradiance ultraviolet laser desorption mass spectrometry of organic molecules. *Anal. Chem.* **1985**, *57*, 2935–2939.
2. Yamashita, M.; Fenn, J. B. Negative ion production with the electrospray ion Source. *J. Phys. Chem.* **1984**, *88*, 4671–4675.
3. Yamashita, M.; Fenn, J. B. Electrospray ion source. Another variation on the free-jet theme. *J. Phys. Chem.* **1984**, *88*, 4451–4459.
4. Miller, D. R. Free jet sources. *Atom. and Mol. Beam Met.* **1988**, *1*, 14–43.
5. Cole, R. B. Some tenets pertaining to electrospray ionization mass spectrometry *J. Mass Spectrom.* **2000**, *35*, 763–772.
6. Kebarle, P. A brief overview of the present status of the mechanisms involved in electrospray mass spectrometry. *J. Mass Spectrom.* **2000**, *35*, 804–817.
7. Sunner, J.; Beech, I. B.; Hiraoka, K. On the distributions of ion/neutral molecule clusters in electrospray and laser spray—A cluster division model for the electrospray process. *J. Am. Soc. Mass Spectrom.* **2006**, *17*, 151–162.
8. Georgiou, S.; Hillenkamp, F. Introduction: Laser ablation of molecular substrates. *Chem. Rev.* **2003**, *103*, 317–319.
9. Fitzgerald, M. C.; Parr, G. R.; Smith, L. M. Basic matrices for the matrix-assisted laser-desorption/ionization mass-spectrometry of proteins and oligonucleotides. *Anal. Chem.* **1993**, *65*, 3204–3211.
10. Wu, K. J.; Steding, A.; Becker, C. H. Matrix-assisted laser desorption time-of-flight mass spectrometry of oligunucleotides using 3-hydroxypicolinic acid as an ultraviolet-sensitive matrix. *Rapid Commun. Mass Spectrom.* **1993**, *7*, 142–146.
11. Beavis, R. C.; Chait, B. T. Factors affecting the ultraviolet laser desorption of proteins. *Rapid Commun. Mass Spectrom.* **1989**, *3*, 233–237.
12. Juhasz, P.; Costello, C. E.; Biemann, K. Matrix-assisted laser desorption/ionization mass-spectrometry with 2-(4-hydroxyphenylazo)benzoic acid matrix. *J. Am. Soc. Mass Spectrom.* **1993**, *4*, 399–409.
13. Pieles, U.; Zürcher, W.; Schär, M.; Moser, H. Matrix-assisted laser desorption/ionization time-of-flight mass spectrometry: A powerful tool for the mass and sequence analysis of natural and modified oligunucleotides. *Nucl. Acid Res.* **1993**, *21*, 3191–3196.
14. Strupat, K.; Karas, M.; Hillenkamp, F. 2,5-Dihydroxybenzoic acid: A new matrix for MALDI-MS. *Int. J. Mass Spectrom. Ion Proc.* **1991**, *111*, 89–102.
15. Beavis, R. C.; Chaudary, T.; Chait, B. T. Alpha-cyano-4-hydroxycinnamic acid as a matrix for MALDI. *Org. Mass Spectrom.* **1992**, *27*, 156–158.
16. Tanaka, K.; Waki, H.; Ido, Y.; Akita, S.; Yoshida, Y.; Yoshida, T. Protein and polymer analyses of up to *m/z* 100,000 by laser ionization time-of-flight mass spectrometry. *Rapid Commun. Mass Spectrom.* **1988**, *2*, 151–153.
17. Wei, J.; Burlak, J. M.; Siuzdak, G. Desorption-ionization mass spectrometry on porous silicon. *Nature* **1999**, *399*, 243–246.
18. Hillenkamp, F. Laser induced ion formation from organic solids. *Sprin. Ser. in Chem. Phys.* **1983**, *25*, 190–205.
19. Hillenkamp, F.; Karas, M.; Beavis, R. C.; Chait, B. T. Matrix-assisted laser desorption/ionization mass spectrometry of bioploymers. *Anal. Chem.* **1991**, *63*, 1193A–1203A.
20. Karas, M.; Bahr, U.; Giessmann, U. Matrix-assisted laser desorption ionization mass spectrometry. *Mass Spectrom. Rev.* **1991**, *10*, 335–357.
21. Ehring, H.; Karas, M.; Hillenkamp, F. Role of photoionization and photochemistry in ionization processes of organic molecules and relevance for matrix-assisted laser desorption/ionization mass spectrometry. *Org. Mass Spectrom.* **1992**, *27*, 427–480.
22. Liao, P.-C.; Allison, J. Ionization processes in matrix-assisted laser desorption/ionization mass spectrometry: Matrix-dependent formation of $[M + H]^+$ vs $[M + Na]^+$ Ions of small peptides and some mechanistic comments. *J. Mass Spectrom.* **1995**, *30*, 408–423.
23. Karas, M.; Bahr, U.; Stahl-Zeng, J.-R. Steps toward a more refined picture of the matrix function in UV MALDI. In Baer, T.; Ng, C.-Y.; Powis, I. (Eds.), *Large Ions: Their Vaporization, Detection and Structural Analysis*, John Wiley & Sons, West Sussex, England, **1996**, pp. 27–48.
24. Zenobi, R.; Knochenmuss, R. Ion formation in MALDI mass spectrometry. *Mass Spectrom. Rev.* **1998**, *17*, 337.
25. Cooks, R. G.; Busch, K. L. Matrix effects, internal energies and MS/MS spectra of molecular ions sputtered from surfaces. *Int. J. Mass Spectrom. Ion Phys.* **1983**, *53*, 111–124.
26. Zhigilei, L. V.; Leveugle, E.; Garrison, B. J.; Yingling, Y. G.; Zeifman, M. I. Computer simulations of laser ablation of molecular substrates. *Chem. Rev.* **2003**, *103*, 321–348.

27. Dreisewerd, K. The desorption process in MALDI *Chem. Rev.* **2003**, *103*, 395–426.
28. Karas, M.; Krüger, R. Ion formation in MALDI: The cluster ionization mechanism. *Chem. Rev.* **2003**, *103*, 427–439.
29. Knochenmuss, R.; Zenobi, R. MALDI ionization: The role of in-plume processes. *Chem. Rev.* **2003**, *103*, 441–452.
30. Kinsel, G. R.; Yao, D.; Yassin, F. H.; Marynick, D. S. Equilibrium conditions in laser desorbed plumes: Thermodynamic propertues of alpha-cyano-4-hydroxycinnamic acid and protonation of amino acids. *Eur. J. Mass Spectrom.* **2006**, *12*, 359–367.
31. Rebber, B. L.; Halfacre, J. A.; Beran, K. A.; Gomez, M.; Beller, N. R.; Bashir, S.; Gianakopulow, A. E.; Derrick, P. J. Theoretical investigations of the proton affinity and gas-phase basicity of neutral *x,y*-dihydroxybenzoic acid (DHB) and its derivatives. *Eur. J. Mass Spectrom.* **2006**, *12*, 385–396.
32. Hoteling, A. J.; Nichols, W. F.; Giesen, D. J.; Lenhard, J. R.; Knochenmuss, R. Electron transfer reactions in LDI and MALDI: Factors influencing matrix and analyte ion intensities. *Eur. J. Mass Spectrom.* **2006**, *12*, 345–358.
33. Kotsiris, S. G.; Vasil'ev, Y. V.; Streletskii, A. V.; Han, M.; Mark, L. P.; Boltalina, O. V.; Chronakis, N.; Orfanopoulos, M.; Hungerbühler, H.; Drewello, T. Application and evaluation of solvent-free MALDI/MS for the analyis of derivatized fullerenes. *Eur. J. Mass Spectrom.* **2006**, *12*, 397–408.
34. Alves, S.; Fournier, I.; Afonso, C.; Wind, F.; Tabet, J.-C. Gas-phase ionization/desolvation processes and their effect on protein charge state distributions under MALDI conditions. *Eur. J. Mass Spectrom.* **2006**, *12*, 369–383.
35. Knochenmuss, R. Ion formation mechanisms in UV-MALDI. *The Analyst.* **2006**, *131*, 966–986.
36. Demirev, P.; Westman, A.; Reimann, C. T.; Håkansson, P.; Barofsky, D.; Sundqvist, B. U. R.; Cheng, Y. D.; Seibt, W.; Siegbahn, K. Matrix-assisted laser desorption with ultra-short laser pulses. *Rapid Commun. Mass Spectrom.* **1992**, *6*, 187–191.
37. Dreisewerd, K.; Schürenberg, M.; Karas, M.; Hillenkamp, F. Matrix-assisted laser desorption/ionization with nitrogen lasers of different pulse widths. *Int. J. Mass Spectrom. Ion Proc.* **1996**, *154*, 171–178.
38. Chevrier, M. R.; Cotter, R. J. A matrix-assisted laser desorption time-of-flight mass spectrometer based on a 600 ps, 1.2 mJ nitrogen laser. *Rapid Commun. Mass Spectrom.* **1991**, *5*, 611–617.
39. Papantonakis, M. R.; Kim, J.; Hess, W. P.; Haglund, R. F. What Do MALDI Mass Spectra Reveal About Ionization Mechanisms? *J. Mass Spectrom.* **2002**, *37*, 639–647.
40. Meffert, A.; Grotemeyer, J. Dissociative proton transfer in cluster ions: Clusters of aromatic carboxylic acids with amino acids. *Int. J. Mass Spectrom.* **2001**, *210/211*, 521–530.
41. Zechmann, C.; Muskat, T.; Grotemeyer, J. Wavelength- and time-resolved luminescence spectroscopy for investigation of the MALDI process. *Eur. J. Mass Spectrom.* **2002**, *8*, 287–293.
42. Lüdemann, H.-C.; Redmond, R. W.; Hillenkamp, F. Singlet–singlet annihilation in ultraviolet MALDI studied by fluorescence spectroscopy. *Rapid Comm. Mass Spectrom.* **2002**, *16*, 1287–1294.
43. Setz, P.; Knochenmuss, R. Exciton mobility and trapping in a UV-MALDI matrix. *J. Phys. Chem. A* **2005**, *109*, 4030–4037.
44. Zhigilei, L. V.; Kodali, P. B. S.; Garrison, B. J. Molecular dynamics model for laser ablation and desorption of organic solids. *J. Phys. Chem. B.* **1997**, *101*, 2028–2037.
45. Zhigilei, L. V.; Kodali, P. B. S.; Garrison, B. J. A microscopic view of laser ablation. *J. Phys. Chem. B* **1998**, *102*, 2845–2853.
46. Zhigilei, L. V.; Garrison, B. J. Microscopic mechanism of laser ablation of organic solids in the thermal stress confinement irradiation regime. *J. Appl. Phys.* **2000**, *88*, 1–18.
47. Zhigilei, L. V. Dynamics of the plume formation and parameters of the ejected clusters in short-pulse ablation. *Appl. Phys. A.* **2003**, *76*, 339–350.
48. Kristyan, S.; Bencsura, A.; Vertes, A. Modeling the cluster formation during infrared and ultraviolet MALDI of oligonucleotides in succinic acid matrix with molecular mechanics. *Theor. Chem. Acc.* **2002**, *107*, 319–325.
49. Puretzky, A. A.; Geohegan, D. B. Gas-phase diagnostics and LIF imaging of 3-hydroxypicolinic acid MALDI matrix plumes. *Chem. Phys. Lett.* **1997**, *286*, 425–432.
50. Handschuh, M.; Nettesheim, S.; Zenobi, R. Laser-induced molecular desorption and particle ejection from organic films. *Appl. Surf. Sci.* **1998**, *137*, 125–135.
51. Jackson, S. N.; Mishra, S.; Murray, K. K. Characterization of coarse particles formed by laser ablation of MALDI matrixes. *J. Phys. Chem. B.* **2003**, *107*, 13106–13110.
52. Alves, S.; Kalberer, M.; Zenobi, R. Direct detection of particles formed by laser ablation of matrices during MALDI. *Rapid Commun. Mass Spectrom.* **2003**, *17*, 2034–2038.
53. Leisner, A.; Rohlfing, A.; Röhling, U.; Dreisewerd, K.; Hillenkamp, F. Time-resolved imaging if the plume dynamics in infrared MALDI with a glycerol matrix. *J. Phys. Chem. B.* **2005**, *109*, 11661–11666.
54. Menzel, C.; Dreisewerd, K.; Berkenkamp, S.; Hillenkamp, F. The role of the laser pulse duration in infrared MALDI-MS. *J. Am. Soc. Mass Spectrom.* **2002**, *13*, 975–984.
55. Dreisewerd, K.; Schürenberg, M.; Karas, M.; Hillenkamp, F. Influence of the laser intensity and spot size on the desorption of molecules and ions in matrix-assisted laser-desorption/ionization with a uniform

beam profile. *Int. J. Mass Spectom. Ion Proc.* **1995**, *141*, 127–148.
56. Rohlfing, A.; Menzel, C.; Kukreja, L. M.; Hillenkamp, F.; Dreisewerd, K. Photoacoustic analysis of MALDI processes with pulsed infrared lasers. *J. Phys. Chem. B.* **2003**, *107*, 12275–12286.
57. Mowry, C. D.; Johnston, M. V. Internal energy of neutral molecules ejected by matrix-assisted laser-desorption. *J. Phys. Chem.* **1994**, *98*, 1904–1909.
58. Luo, G.; Marginean, I.; Vertes, A. Internal energy of ions generated by MALDI. *Anal. Chem.* **2002**, *74*, 6185–6190.
59. Gabelica, V.; Schulz, E; Karas, M. Internal energy build-up in MALDI. *J. Mass Spectrom.* **2004**, *39*, 579–593.
60. Koubenakis, A.; Frankevich, V.; Zhang, J.; Zenobi, R. Time-resolved surface temperature measurements of MALDI matrices under pulsed UV laser irradiation. *J. Phys. Chem. A.* **2004**, *108*, 2405–2410.
61. Campbell, J. M.; Vestal, M. L.; Blank, P. S.; Stein, S. E.; Epstein, J. A. Fragmentation of leucine enkephalin as a function of laser fluence in MALDI TOF-TOF. *J. Am. Soc. Mass Spectrom.* **2007**, *18*, 607–616.
62. Knochenmuss, R.; Zhigilei, L. V. A molecular dynamics model of UV-MALDI including ionization processes. *J. Phys. Chem. B.* **2005**, *109*, 22947–229957.
63. Knochenmuss, R. A quantitative model of ultraviolet matrix-assisted laser desorption and ionization. *J. Mass Spectrom.* **2002**, *37*, 867–877.
64. Stevenson, E.; Breuker, K.; Zenobi, R. Internal energies of analyte ions generated from different matrix-assisted desorption/ionization matrices. *J. Mass Spectrom.* **2000**, *35*, 1035–1041.
65. Schürenberg, M.; Dreisewerd, K.; Kamanabrou, S.; Hillenkamp, F. Influence of the sample temperature on the desorption of matrix molecules and ions in MALDI. *Int. J. Mass Spectrom.* **1998**, *172*, 89.
66. Wallace, W. E.; Arnould, M. A.; Knochenmuss, R. 2,5 dihydroxybenzoic acid: Laser desorption/ionization as a function of elevated temperature. *Int. J. Mass Spectrom.* **2005**, *242*, 13–22.
67. Chen, X.; Carroll, J. A.; Beavis, R. C. Near-ultraviolet-induced matrix-assisted laser desorption/ionization as a function of wavelength. *J. Am. Soc. Mass Spectrom.* **1998**, *9*, 885–891.
68. Price, D. M.; Bashir, S.; Derrick, P. R. Sublimation properties of *X,Y*-dihydroxybenzoic acid isomers as model MALDI matrices. *Thermochim. Acta* **1999**, *327*, 167–171.
69. Sadeghi, M.; Vertes, A. Crystallite size dependence of volatilization in matrix-assisted laser desorption ionization. *Appl. Surf. Sci.* **1998**, *127/129*, 226–234.
70. Trimpin, S.; Räder, H. J.; Müllen, K. Investigations of theoretical principles for MALDI-MS derived from solvent-free sample preparation: Part I. Preorganization. *Int. J. Mass Spectrom.* **2006**, *253*, 13–21.
71. Beavis, R. C.; Chait, B. T. Velocity distibutions of intact high mass polypeptide molecule ions produced by matrix-assisted laser desorpion. *Chem. Phys. Lett.* **1991**, *181*, 479–484.
72. Vertes, A.; Irinyi, G.; Gijbels, R. Hydrodynamic model of matrix-assisted laser Desorption Mass Spectrometry. *Anal. Chem.* **1993**, *65*, 2389–2393.
73. Fournier, I.; Marinach, C.; Tabet, J.-C.; Bolbach, G. Irradiation effects in MALDI, ablation, ion production and surface modifications. Part II: 2,5-Dihydroxybenzoic acid monocrystals. *J. Am. Soc. Mass Spectrom.* **2003**, *14*, 893–899.
74. Fournier, I.; Tabet, J.-C.; Bolbach, G. Irradiation effects in MALDI and surface modifications. Part I: Sinapinic acid monocrystals. *Int. J. Mass Spectrom.* **2002**, *219*, 515–523.
75. Leveugle, E.; Ivanov, D. S.; Zhigilei, L. V. Photomechanical spallation of molecular and metal targets: Molecular dynamics study. *Appl. Phys. A.* **2004**, *79*, 1643–1655.
76. Yingling, Y. G.; Zhigilei, L. V.; Garrison, B. J. The role of photochemical fragmentation in laser ablation: A molecular dynamics study. *J. Photochem. Photobiol. A* **2001**, *145*, 173–181.
77. Zhigilei, L. V.; Yingling, Y. G.; Itina, T. E.; Schoolcraft, T. A.; Garrison, B. J. Molecular dynamics simulation soft matrix-assisted laser desorption—connections to experiment. *Int. J. Mass Spectrom.* **2003**, *226*, 85–106.
78. Yingling, Y. G.; Garrison, B. J. Coarse-grained chemical reaction model. *J. Phys. Chem. B.* **2004**, *108*, 1815–1821.
79. Jouvet, C.; Lardeux-Dedonder, C.; Richard-Viard, M.; Solgadi, D.; Tramer, A. Reactivity of molecular clusters in the gas phase. Proton-transfer reactions in Neutral phenol-$(NH_3)_n$ and phenol-$(C_2H_5NH_2)_n$. *J. Phys. Chem.* **1990**, *94*, 5041–5048.
80. Huth-Fehre, T.; Becker, C. H. Energetics of gramicidin S after UV laser desorption from a ferrulic acid matrix. *Rapid Commun. Mass Spectrom.* **1991**, *5*, 378–382.
81. Pan, Y.; Cotter, R. J. Measurement of the initial translational energies of peptide ions in laser desorption/ionization mass spectrometry. *Org. Mass Spectrom.* **1992**, *27*, 3–8.
82. Spengler, B.; Bökelmann, V., Angular and time-resolved intensity distributions of laser-desorbed matrix ions *Nucl. Instr. Meth. Phys. Res. B* **1993**, *82*, 379–385.
83. Juhasz, P.; Vestal, M. L.; Martin, S. A. On the initial velocity of ions generated by matrix-assisted laser-desorption/ionization and its effect on the calibration of delayed extraction time-of-flight mass-spectra. *J. Am. Soc. Mass Spectrom.* **1997**, *8*, 209–217.
84. Glückmann, M.; Karas, M. The initial ion velocity as a marker for different desorption characteristics in MALDI. In *47th ASMS Conference on Mass Spectrometry and Allied Topics*, **1999**, p. 2232.

85. Glückmann, M.; Karas, M. The initial ion velocity and its dependence on matrix, analyte and preparation method in ultraviolet matrix-assisted laser desorption/ionization. *J. Mass Spectrom.* **1999**, *34*, 467–477.
86. Zhou, J.; Ens, W.; Standing, K. G.; Verentchikov, A. Kinetic energy measurement of molecular ions ejected into an electric field by matrix-assisted laser desorption. *Rapid Commun. Mass Spectrom.* **1992**, *6*, 671–678.
87. Mormann, M.; Bashir, S. M.; Derrick, P. J.; Kuck, D. Gas-phase basicities of the isomeric dihydroxybenzoic acids and gas-phase acidities of their radical cations. *J. Am. Soc. Mass Spectrom.* **2000**, *11*, 544–552.
88. Jørgensen, T. J. D.; Bojesen, G.; Rahbek-Nielsen, H. The proton affinities of seven matrix-assisted laser desorption/ionization matrices correlated with the formation of multiply charged ions. *Eur. Mass Spectrom.* **1998**, *4*, 39–45.
89. Jørgensen, T. J. D.; Vulpius, T.; Bojesen, G. The order of proton affinities of five common MALDI matrices. In *13th International Mass Spectrometry Conference*, **1994**, p. 175.
90. Burton, R. D.; Watson, C. H.; Eyler, J. R.; Lang, G. L.; Powell, D. H.; Avery, M. Y. Proton affinities of 8 matrices used for matrix-assisted laser desorption/ionization. *Rapid Commun. Mass Spectrom.* **1997**, *11*, 443–446.
91. Steenvoorden, R. J. J. M.; Breuker, K.; Zenobi, R. The gas-phase basicity of MALDI matrices. *Eur. Mass Spectrom.* **1997**, *3*, 339–346.
92. Breuker, K.; Knochenmuss, R.; Zenobi, R. The gas-phase basicities of deprotonated MALDI matrix molecules. *Int. J. Mass Spectrom.* **1999**, *184*, 25.
93. Breuker, K.; Knochenmuss, R.; Zenobi, R. Proton transfer reactions of matrix-assisted laser desorption/ionization matrix monomers and dimers. *J. Am. Soc. Mass Spectrom.* **1999**, *10*, 1111–1123.
94. Bourcier, S.; Bouchonnet, S.; Hoppilliard, Y. Ionization of 2,5-dihydroxybenzoic acid (DHB) MALDI experiment and theoretical study. *Int. J. Mass Spectrom.* **2001**, *210/211*, 59–69.
95. Bourcier, S.; Hoppilliard, Y. B3LYP DFT molecular orbital approach, an efficient method to evaluate the thermochemical properties of MALDI Matrices. *Int. J. Mass Spectrom.* **2002**, *217*, 231–244.
96. Yassin, F.; Marynick, D. S. Computational estimates of the gas-phase acidities of dihydroxybenzoic acid radical cations and their corresponding neutral species. *J. Mol. Struct.* **2003**, *629*, 223–235.
97. Kinsel, G. R.; Zhao, Q.; Narayanasamy, J.; Yassin, F.; Rasika Dias, H. V.; Niesner, B.; Prater, K.; St. Marie, C.; Ly, L.; Marynik, D. Arginine/2,5-dihydroxybenzoic acid clusters: An experimental and theoretical study of the gas-phase and solid-state systems. *J. Phys. Chem. A* **2004**, *108*, 3153–3161.
98. Yassin, F. H.; Marynick, D. S. Computational estimates of the gas-phase basicities, proton affinities and ionization potentials of the six isomers of dihydroxybenzoic. *Acid. Mol. Phys.* **2005**, *103*, 183–189.
99. Mizra, S. P.; Raju, N. P.; Vairamani, M. Estimation of the proton affinity values of fifteen MALDI matrices under electrospray ionization conditions using the kinetic method. *J. Am. Soc. Mass Spectrom.* **2004**, *15*, 431–435.
100. Karbach, V.; Knochenmuss, R. Do single matrix molecules generate primary ions in ultraviolet matrix-assisted laser desorption/ionization? *Rapid Commun. Mass Spectrom.* **1998**, *12*, 968–974.
101. Lin, Q.; Knochenmuss, R. Two-photon ionization thresholds of matrix-assisted laser desorption/ionization matrix clusters. *Rapid Comm. Mass Spectrom.* **2001**, *15*, 1422–1426.
102. Kinsel, G.; Knochenmuss, R.; Setz, P.; Land, C. M.; Goh, S.-K.; Archibong, E. F.; Hardesty, J. H.; Marynick, D. Ionization energy reductions in small 2,5-dihydroxybenzoic acid–proline clusters. *J. Mass Spectrom.* **2002**, *37*, 1131–1140.
103. Yassin, F. H.; Marynick, D. S. A computational study of the thermodynamic properties of sinapic and ferulic acids and their corresponding radical cations. *J. Mol. Struct.* **2006**, *766*, 137–141.
104. Zhang, J.; Dyachova, E.; Ha, T.-K.; Knochenmuss, R.; Zenobi, R. Gas-phase potassium binding energies of MALDI matrices, an experimental and theoretical study. *J. Phys. Chem. A* **2003**, *107*, 6891.
105. Zhang, J.; Ha, T.-K.; Knochenmuss, R.; Zenobi, R. Theoretical calculation of gas-phase sodium binding energies of common MALDI matrices. *J. Phys. Chem. A* **2002**, *106*, 6610–6617.
106. Zhang, J.; Knochenmuss, R.; Stevenson, E.; Zenobi, R. The gas phase sodium basicities of common MALDI matrices. *J. Mass Spectrom.* **2002**, *213*, 237–250.
107. Pshenichnyuk, S. A.; Asfandiarov, N. L.; Fal'ko, V. S.; Lukin, V. G. Temperature dependence of dissociative electron attachment to molecules of gentisic acid, hydroquinone and *p*-benzoquinone. *Int. J. Mass Spectrom.* **2003**, *227*, 281–288.
108. Asfandiarov, N. L.; Pshenichnyuk, S. A.; Forkin, A. I.; Lukin, V. G.; Fal'ko, V. S. Electron capture negative ion mass spectra of some typical MALDI matrices. *Rapid Commun. Mass Spectrom.* **2002**, *16*, 1760–1765.
109. Knochenmuss, R.; Stortelder, A.; Breuker, K.; Zenobi, R. Secondary ion–molecule reactions in MALDI. *J. Mass Spectrom.* **2000**, *35*, 1237–1245.
110. Niu, S.; Zhang, W.; Chait, B. T. Direct comparison of infrared and ultraviolet wavelength matrix-assisted laser desorption/ionization mass spectrometry of proteins. *J. Am. Soc. Mass Spectrom.* **1998**, *9*, 1–7.
111. Barthel; J.; Neuder; R.; Schröder; P. Chemistry Data Series, DCHEMA, Frankfurt, **1994**.
112. Cramer, R.; Haglund, R. F. Jr.; Hillenkamp, F. Matrix-assisted laser desorption and ionization in the O–H and C=O absorption bands of aliphatic and aromatic matrices: Dependence on laser wavelength and

temporal beam profile. *Int. J. Mass Spectrom. Ion Proc.* **1997**, *169/170*, 51–67.

113. Ireland, J. F.; Wyatt, P. A. H. Acid–base properties of electronically excited states of organic molecules. *Adv. Phys. Org. Chem.* **1976**, *12*, 131.
114. Karas, M.; Bachmann, D.; Bahr, U.; Hillenkamp, F. Matrix-assisted ultraviolet laser desorption of non-volatile compounds. *Int. J. Mass Spectrom. Ion Proc.* **1987**, *78*, 53–68.
115. Gimon, M. E.; Preston, L. M.; Solouki, T.; White, M. A.; Russell, D. H. Are proton transfer reactions of excited states involved in UV laser desorption/ionization? *Org. Mass Spectrom.* **1992**, *27*, 827–830.
116. Spengler, B.; Kaufmann, R. Gentle probe for tough molecules: Matrix-assisted laser desorption mass spectrometry. *Analusis* **1992**, *20*, 91–101.
117. Chiarelli, M. P.; Sharkey, A. G.; Hercules, D. M. Excited-state proton transfer in laser mass spectrometry. *Anal. Chem.* **1993**, *65*, 307–311.
118. Krause, J.; Stoeckli, M.; Schlunegger, U. P. Studies on the selection of new matrices for ultraviolet matrix-assisted laser desorption/ionization time-of-flight mass-spectrometry. *Rapid Commun. Mass Spectrom.* **1996**, *10*, 1927–1933.
119. Yong, H. Photochemistry and proton transfer reaction chemistry of selected cinnamic acid derivatives in hydrogen bonded environments. *Int. J. Mass Spectrom. Ion Proc.* **1998**, *175*, 187–204.
120. Lüdemann, H.-C.; Hillenkamp, F.; Redmond, R. Photoinduced hydrogen atom transfer in salicylate acid derivatives used as matrix-assisted laser desorption/ionization matrices. *J. Phys. Chem. A* **2000**, *104*, 3884–3893.
121. Krüger, R.; Pfenninger, A.; Fournier, I.; Glückmann, M.; Karas, M. Analyte incorporation and ionization in matrix-assisted laser desorption/ionization visualized by pH indicator molecular probes. *Anal. Chem.* **2001**, *73*, 5812–5821.
122. Liao, P.-C.; Allison, J. Enhanced detection of peptides in matrix-assisted laser desorption/ionization mass spectrometry through use of charge-localized derivatives. *J. Mass Spectrom.* **1995**, *30*, 511–512.
123. Lehmann, E.; Knochenmuss, R.; Zenobi, R. Ionization mechanisms in matrix-assisted laser-desorption/ionization mass-spectrometry—Contribution of preformed ions. *Rapid Commun. Mass Spectrom.* **1997**, *11*, 1483–1492.
124. Claereboudt, J.; Claeys, M.; Geise, H.; Gijbels, R.; Vertes, A. Laser microprobe mass spectrometry of quarternary phosphonium salts: direct versus matrix-assisted laser desorption. *J. Am. Soc. Mass Spectrom.* **1993**, *4*, 798–812.
125. Gut, I. G.; Jeffery, W. A.; Pappin, D. J. C.; Beck, S. Analysis of DNA by 'charge tagging' and matrix-assisted laser desorption/ionization mass spectrometry. *Rapid Commun. Mass Spectrom.* **1997**, *11*, 43–50.
126. Naven, T. J. P.; Harvey, D. J. Cationic derivatization of oligosaccharides with Girard's T reagent for improved performance in matrix-assisted laser desorption/ionization and electrospray mass spectrometry. *Rapid Commun. Mass Spectrom.* **1996**, *10*, 829–834.
127. Spengler, B.; Lützenkirchen, F.; Metzger, S.; Chaurand, P.; Kaufmann, R.; Jeffery, W.; Bartlet-Jones, M.; Pappin, D. J. C. Peptide sequencing of charged derivatives by postsource decay MALDI mass spectrometry. *Int. J. Mass Spectrom. Ion Proc.* **1997**, *169/170*, 127–140.
128. Asara, J. M.; Allison, J. Enhanced detection of oligonucleotides in UV MALDI MS using the tetraamine spermine as a matrix additive. *Anal. Chem.* **1999**, *71*, 2866–2870.
129. Strahler, J. R.; Smelyanskiy, Y.; Lavine, G.; Allison, J. Development of methods for the charge derivatization of peptides in polyacrylamide gels and membranes for their direct analysis using matrix-assisted laser desorption/ionization mass spectrometry. *Int. J. Mass Spectrom. Ion Proc.* **1997**, *169/170*, 111–126.
130. Berlin, K.; Gut, I. G. Analysis of negatively "charge tagged" DNA by MALDI TOF MS. *Rapid Comm. Mass Spectrom.* **1999**, *13*, 1739–1743.
131. Brancia, F. L.; Oliver, S. G.; Gaskell, S. J. Improved MALDI MS analysis of tryptic hydrolysates of proteins following guanidation of lysine-containing peptides. *Rapid Commun. Mass Spectrom.* **2000**, *14*, 2070–2073.
132. Cohen, D. E.; Benedict, J. B.; Morlan, B.; Chiu, D. T.; Kahr, B. Dyeing polymorphs: The MALDI host 2,5-dihydroxybenzoic acid. *Cryst. Growth Des.* **2007**, *7*, 492–495.
133. Krüger, R.; Karas, M. Formation and fate of ion pairs during MALDI analysis: Anion adduct generation as an indicative tool to determine ionization processes. *J. Am. Soc. Mass Spectrom.* **2002**, *13*, 1218–1226.
134. Karas, M.; Glückmann, M.; Schäfer, J. Ionization in MALDI: Singly charged molecular ions are the lucky survivors. *J. Mass Spectrom.* **2000**, *35*, 1–12.
135. Pshenichnyuk, S. A.; Asfandiarov, N. L.; Fal'ko, V. S.; Lukin, V. G. Temperature dependencies of negative ion formation by capture of low-energy electrons for some typical MALDI matrices. *Int. J. Mass Spectrom.* **2003**, *227*, 259–272.
136. Knochenmuss, R. Photoionization pathways and free electrons in UV-MALDI. *Anal. Chem.* **2004**, *76*, 3179–3184.
137. Kinsel, G. R.; Edmondson, R. D.; Russell, D. H. Profile and flight time analysis of bovine insulin clusters as a probe of matrix-assisted laser desorption/ionization ion formation dynamics. *J. Mass Spectrom.* **1997**, *32*, 714–722.
138. Kinsel, G. R.; Gimon-Kinsel, M. E.; Gillig, K. J.; Russell, D. H. Investigation of the dynamics of matrix-assisted laser desorption/ionization ion formation using an electrostatic analyzer/time-of-flight mass spectrometer. *J. Mass Spectrom.* **1999**, *34*, 684–690.
139. Krutchinsky, A. N.; Chait, B. T. On the nature of chemical noise in MALDI mass spectra. *J. Am. Soc. Mass Spectrom.* **2002**, *13*, 129–134.

140. Fournier, I.; Brunot, A.; Tabet, J.-C.; Bolbach, G. Delayed extraction experiments using a repulsive potential before ion extraction: Evidence of clusters as ion precursors in UV-MALDI. Part I: Dynamical effects with the matrix 2,5-dihydroxybenzoic acid. *Int. J. Mass Spectrom.* **2002**, *213*, 203–215.
141. Fournier, I.; Brunot, A.; Tabet, J.-C.; Bolbach, G. Delayed extraction experiments using a repulsive potential before ion extraction: Evidence of non-covalent clusters as ion precursors in UV MALDI. Part II—Dynamic effects with alpha-cyano-4-hydroxycinnamic acid matrix. *J. Mass Spectrom.* **2005**, *40*, 50–59.
142. Baldwin, M. A.; Talroze, V. L.; Burlingame, A. L.; Leipunsky, I. O. AP-MALDI and gas release in the MALDI process. In *50th ASMS Conference on Mass Spectrometry and Allied Topics,* American Society of Mass Spectrometry, Santa Fe, NM, **2002**.
143. Talroze, V. L.; Jacob, R. L.; Burlingame, A. L.; Baldwin, M. A. Experimental examination of the thermodynamic basis of MALDI. In *48th ASMS Conference on Mass Spectrometry and Allied Topics,* American Society of Mass Spectrometry, Santa Fe, NM, **2000**.
144. Talroze, V. L.; Jacob, R. L.; Burlingame, A. L.; Leipunsky, I. O.; Baldwin, M. A. A New theory of MALDI: Laser induced soft thermal dissociation of the matrix plays a decisive role in plume formation. In *49th ASMS Conference on Mass Spectrometry and Allied Topics,* American Society of Mass Spectrometry, Santa Fe, NM, **2001**.
145. Talroze, V. L.; Jacob, R. L.; Burlingame, A. L.; Baldwin, M. A., Insight into the MALDI Mechanism: Matrix decomposition and pneumatic assistance in plume formation. *Adv. Mass Sectrom.* **2001**, *15*, 481–482.
146. Talroze, V. L.; Leipunsky, I. O.; Burlingame, A. L.; Baldwin, M. A. Modeling nanosecond non-activated thermal decay of MALDI matrices containing preformed analyte ions and partially ionized matrix molecules as the origin of pneumatic assistance. In *51st ASMS Conference on Mass Spectrometry and Allied Topics,* American Society of Mass Spectrometry, Santa Fe, NM, **2003**.
147. Talroze, V. L.; Leipunsky, I. O.; Burlingame, A. L.; Baldwin, M. A. The thermal mechanism of MALDI—A refined quantitative model with experimental validation. In *52nd ASMS Conference on Mass Spectrometry and Allied Topics,* American Society of Mass Spectrometry, Santa Fe, New Mexico, **2004**.
148. Kosevich, M. V.; Shelkovsky, V. S.; Boryak, O. A.; Orlov, V. V. 'Bubble chamber mode' of fast atom bombardment induced processes. *Rapid Comm. Mass Spectrom.* **2003**, *17*, 1781–1792.
149. Vestal, M. L.; Juhasz, P.; Martin, S. A. Delayed extraction matrix-assisted laser-desorption time-of-flight mass-spectrometry. *Rapid Commun. Mass Spectrom.* **1995**, *9*, 1044–1050.
150. Talroze, V. L.; Person, M. L.; Whittal, R. M.; Walls, F. C.; Burlingame, A. L.; Baldwin, M. A. Insight into absorption of radiation/energy transfer in infrared matrix-assisted laser desorption/ionization: The roles of matrices, water and metal substrates. *Rapid Commun. Mass Spectrom.* **1999**, *13*, 21914–22198.
151. Land, C. M.; Kinsel, G. R. Investigation of the mechanism of intracluster proton transfer from sinapinic acid to biomolecular analytes. *J. Am. Soc. Mass Spectrom.* **1998**, *9*, 1060–1067.
152. Land, C. M.; Kinsel, G. R. The mechanism of matrix to analyte proton transfer in clusters of 2,5 dihydroxybenzoic acid and the tripeptide VPL. *J. Am. Soc. Mass Spectrom.* **2001**, *12*, 726–731.
153. Yassin, F. H.; Marynick, D. S. Computational study of matrix–peptide interactions in MALDI mass spectrometry: Interactions of 2,5- and 3,5-dihydroxybenzoic acid with the tripeptide valine–proline–leucine. *J. Phys. Chem. A* **2006**, *110*, 3820–3825.
154. Dexter, D. L.; Knox, R. S. *Excitons*, John Wiley & Sons, New York, 1965.
155. Birks; J. B. *Photophysics of Aromatic Compounds*; Wiley Interscience, London, 1970.
156. Birks; J. B. *Organic Molecular Photophysics*, John Wiley & Sons, New York, 1973.
157. Ehring, H.; Sundqvist, B. U. R. Studies of the MALDI process by luminescence spectroscopy. *J. Mass Spectrom.* **1995**, *30*, 1303–1310.
158. Ehring, H.; Sundqvist, B. U. R. Excited-state relaxation processes of MALDI-matrices studied by luminescence spectroscopy. *Appl. Surf. Sci.* **1996**, *96-8*, 577–580.
159. Knochenmuss, R.; Vertes, A. Time-delayed 2-pulse studies of MALDI matrix ionization mechanisms. *J. Phys. Chem. B.* **2000**, *104*, 5406–5410.
160. Mowry, C.; Johnston, M. Simultaneous detection of ions and neutrals produced by matrix-assisted laser desorption. *Rapid Commun. Mass Spectrom.* **1993**, *7*, 569–575.
161. Quist, A. P.; Huth-Fehre, T.; Sunqvist, B. U. R. Total yield measurements in matrix-assisted laser desorption using a quartz crystal microbalance. *Rapid Commun. Mass Spectrom.* **1994**, *8*, 149–154.
162. Ens, W.; Mao, Y.; Mayer, F.; Standing, K. G. Properties of matrix-assisted laser desorption. Measurements with a time-to-digital converter. *Rapid Commun. Mass Spectrom.* **1991**, *5*, 117–123.
163. Riahi, K.; Bolbach, G.; Brunot, A.; Breton, F.; Spiro, M.; Blais, J.-C. Influence of laser focusing in matrix-assisted laser desorption/ionization. *Rapid Commun. Mass Spectrom.* **1994**, *8*, 242–247.
164. Ens, W.; Schürenberg, M.; Hillenkamp, F. Dependence of ion yields on laser fluence in matrix-assisted laser desorption/ionization. In *45th ASMS Conference on Mass Spectrometry and Allied Topics,* American Society of Mass Spectrometry, Palm Springs, CA, **1997**, p. 1099.

165. Beavis, R. C., Phenomenological models for matrix-assisted laser desorption ion yields near the threshold fluence. *Org. Mass Spectrom.* **1992**, *27*, 864–868.
166. Yau, P. Y.; Chan, T. W. D.; Cullis, P. G.; Colburn, A. W.; Derrick, P. J. Threshold fluences for production of positive and negative ions in matrix-assisted laser desorption/ionization using liquid and solid matrices. *Chem. Phys. Lett.* **1993**, *202*, 93–100.
167. Heise, T. W.; Yeung, E. S. Dynamics of matrix-assisted laser desorption as revealed by the associated acoustic signal. *Anal. Chim. Acta.* **1995**, *199*, 377–385.
168. Feldhaus, D.; Menzel, C.; Berkenkamp, S.; Hillenkamp, F. Influence of the laser fluence in infrared MALDI with a 2.94 mm Er:YAG laser and a flat-top beam profile. *J. Mass Spectrom.* **2000**, *35*, 1320–1328.
169. Holle, A.; Haase, A.; Kayser, M.; Höhndorf, J. Optimizing UV laser focus profiles for improved MALDI performance. *J. Mass Spectrom.* **2006**, *41*, 705–716.
170. Breuker, K.; Knochenmuss, R.; Zhang, J.; Stortelder, A.; Zenobi, R. Thermodynamic control of final ion distributions in MALDI: In-plume proton transfer reactions. *Int. J. Mass Spectrom.* **2003**, *226*, 211–222.
171. Harrison, A. G. The gas-phase basicities and proton affinities of amino acids and peptides. *Mass Spectrom. Rev.* **1997**, *16*, 201–217.
172. Carr, S. R.; Cassidy, C. J. Gas-phase basicities of histidine and lysine and their selected di- and tripeptides. *J. Am. Soc. Mass Spectrom.* **1996**, *7*, 1203–1210.
173. Carr, S. R.; Cassady, C. J. Reactivity and gas-phase acidity determinations of small peptide ions consisting of 11 to 14 amino acid residues. *J. Mass Spectrom.* **1997**, *32*, 959–967.
174. Olumee, Z.; Vertes, A. Protonation of Glyn homologues in matrix-assisted laser desorption ionization. *J. Phys. Chem. B* **1998**, *102*, 6118–6122.
175. Baumgart, S.; Lindner, Y.; Kuehne, R.; Oberemn, A.; Wenschuh, H.; Krause, E. Peptide sensitivities. *Rapid Comm. Mass Spectrom.* **2004**, *18*, 863.
176. Wenschuh, H.; Halada, P.; Lamer, S.; Jungblut, P.; Krause, E. The ease of peptide detection by matrix-assisted laser desorption/ionization mass spectrometry: The Effect of secondary structure on signal intensity. *Rapid Commun. Mass Spectrom.* **1998**, *12*, 115–119.
177. Nishikaze, T.; Takayama, M. Cooperative effect of factors governing molecular ion yields in desorption/ionization mass spectrometry. *Rapid Commun. Mass Spectrom.* **2006**, *20*, 376–382.
178. Burkitt, W. I.; Giannakopulos, A. E.; Sideridou, F.; Bashir, S.; Derrick, P. J. Discrimination effects in MALDI-MS of mixtures of peptides—Analysis of the proteome. *Aust. J. Chem.* **2003**, *56*, 369–377.
179. Chou, C.-W.; Williams, P.; Limbach, P. A. Matrix influence on the formation of positively charged oligonucleotides in MALDI MS. *Int. J. Mass Spectrom.* **1999**, *193*, 15–27.
180. Gross, J.; Leisner, A.; Hillenkamp, F.; Hahner, S.; Karas, M.; Schäfer, J.; Lützenkirchen, F.; Nordhoff, E. Investigations of the metastable decay of DNA under ultraviolet matrix-assisted laser desorption/ionization conditions with post-source-decay analysis and hydrogen/deuterium exchange. *J. Am. Soc. Mass Spectrom.* **1998**, *9*, 866–878.
181. Schulz, E.; Karas, M.; Rosu, F.; Gabelica, V. Influence of the matrix on analyte fragmentation in atmospheric pressure MALDI. *J. Am. Soc. Mass Spectrom.* **2006**, *17*, 1005–1013.
182. Klassen, J. S.; Anderson, S. G.; Blades, A. T.; Kebarle, P. Reaction enthalpies for $M^+L = M^+ + L$, where $M^+ = Na^+$ and K^+ and $L =$ acetamide, *N*-methylacetamide, *N,N*-dimethylacetamide, glycine, and glycylglycine, from determinations of the collision-induced dissociation thresholds. *J. Phys. Chem.* **1996**, *100*, 14218–14227.
183. Hoyau, S.; Norrman, K.; McMahon, T. B.; Ohanessian, G. A quantitative basis for a scale of Na^+ affinities of organic and small biological molecule in the gas phase. *J. Am. Chem. Soc.* **1999**, *121*, 8864.
184. Hoyau, S.; Ohanessian, G. Interaction of alkali metal cations (Li–Cs) with glycine in the gas phase: A theoretical study. *Chem. Eur. J.* **1998**, *8*, 1561–1569.
185. Zhang, J.; Zenobi, R., Matrix-dependent cationization in MALDI mass spectrometry. *J. Mass Spectrom.* **2004**, *39*, 808–816.
186. Hoberg, A.-M.; Haddleton, D. M.; Derrick, P. M. Evidence for cationization of polymers in the gas phase during matrix-assisted laser desorption/ionization. *Eur. Mass Spectrom.* **1997**, *3*, 471–473.
187. Burgers, P. C.; Terlouw, J. K. Monoisotopic $^{65}Cu^+$ attachment to polystyrene. *Rapid Commun. Mass Spectrom.* **1998**, *12*, 801–804.
188. Rashidezadeh, H.; Guo, B. Investigation of metal attachment to polystyrenes in matrix-assisted laser desorption/ionization. *J. Am. Soc. Mass Spectrom.* **1998**, *9*, 724–730.
189. Wang, B. H.; Dreisewerd, K.; Bahr, U.; Karas, M.; Hillenkamp, F. Gas-phase cationization and protonation of neutrals generated by matrix-assisted laser desorption. *J. Am. Soc. Mass Spectrom.* **1993**, *4*, 393–398.
190. Erb, W. J.; Owens, K. G.; Hanton, S. D. A study of gas phase cationization in MALDI TOF MS. In *52nd ASMS Conference on Mass Spectrometry and Allied Topics,* American Society of Mass Spectrometry, Santa Fe, NM, **2004**.
191. Erb, W. J.; Hanton, S. D.; Owens, K. G. A study of gas-phase cationization in MALDI MS. *Rapid Commun. Mass Spectrom.* **2006**, *20*, 2165–2219.
192. Trimpin, S.; Rouhanipour, A.; Az, R.; Rader, H. J.; Mullen, K. New aspects in MALDI-TOF MS: A universal solvent-free sample preparation. *Rapid Commun. Mass Spectrom.* **2001**, *15*, 1364–1373.
193. Hoteling, A. J.; Owens, K. G. Cationization of polymers in MALDI/TOFMF-investigations of salt-to-

analyte ratio. In *53rd ASMS Conference on Mass Spectrometry and Allied Topics,* American Society of Mass Spectrometry, Santa Fe, NM, **2005**.

194. Hanton, S. D.; Owens, K. G.; Blair, W.; Hyder, I. Z.; Stets, J. R.; Guttman, C. M.; Giuseppi, A. Investigations of electrospray sample deposition for polymer MALDI. *J. Am. Soc. Mass Spectrom.* **2004**, *15*, 168–179.
195. Lecchi, P.; Olson, M.; Brancia, F. L. The role of esterification on detection of protonated and deprotonated peptide ions in MALDI MS. *J. Am. Soc. Mass Spectrom.* **2005**, *16*, 1269–1274.
196. Juhasz, P.; Costello, C. E. Generation of large radical ions from oligometallocenes by matrix-assisted laser desorption/ionization. *Rapid Commun. Mass Spectrom.* **1993**, *7*, 343–351.
197. McCarley, T. D.; McCarley, R. L.; Limbach, P. A. Electron-transfer ionization in matrix-assisted laser desorption/ionization mass spectrometry. *Anal. Chem.* **1998**, *70*, 4376–4379.
198. Macha, S. F.; McCarley, T. D.; Limbach, P. A. Influence of ionization energy on charge-transfer ionization in matrix-assisted laser desorption/ionization mass spectrometry. *Anal. Chim. Acta* **1999**, *397*, 235–245.
199. Streletskii, A.; Ioffe, I. N.; Kotsiris, S. G.; Barrow, M. P.; Drewello, T.; Strauss, S. H.; Boltalina, O. G. In-plume thermodynamics of the MALDI generation of fullerene anions. *J. Phys. Chem. A* **2005**, *109*, 714–719.
200. Lidgard, R.; Duncan, M. W. Utility of matrix-assisted laser-desorption/ionization time-of-flight mass-spectrometry for the analysis of low-molecular-weight compounds. *Rapid Commun. Mass Spectrom.* **1995**, *9*, 128–132.
201. Ulmer, L.; Mattay, J.; Torres-Garcia, H. G.; Luftmann, H. The use of 2-[(2e)-3-(4-*tert*-butylphenyl)-2-methylprop-2,2-enylidene] malononitrile as a matrix for matrix-assisted laser desorption/ionization mass spectrometry. *Eur. J. Mass Spectrom.* **2000**, *6*, 49–52.
202. Brown, T.; Clipston, N. L.; Simjee, N.; Luftmann, H.; Hungerbuhler, H.; Drewello, T. Matrix-assisted laser desorption/ionization of amphiphilic fullerene derivatives. *Int. J. Mass Spectrom.* **2001**, *210/211*, 249–263.
203. Lippa, T. P.; Eustis, S. N.; Wang, D.; Bowen, K. H. Electrophilic properties of common MALDI matrix molecules. *Int. J. Mass Spectrom.* **2007**, *268*, 1–7.
204. Deery, M. J.; K., J.; Jasieczek, C. B.; Haddleton, D. M.; Jackson, A. T.; Yates, H. T.; Scrivens, J. H. A study of cation attachment to polystyrene by means of matrix-assisted laser desorption/ionization and electrospray ionization mass spectrometry. *Rapid Commun. Mass Spectrom.* **1997**, *11*, 57–62.
205. Knochenmuss, R.; Lehmann, E.; Zenobi, R. Polymer cationization in matrix-assisted laser desorption/ionization. *Eur. Mass Spectrom.* **1998**, *4*, 421–427.
206. Yalcin, T.; Schriemer, D. C.; Li, L. Matrix-assisted laser-desorption/ionization time-of-flight mass-spectrometry for the analysis of polydienes. *J. Am. Soc. Mass Spectrom.* **1997**, *8*, 1220–1229.
207. Zhang, J.; Frankevich, V.; Knochenmuss, R.; Friess, S. D.; Zenobi, R. Reduction of Cu(II) in MALDI MS. *J. Am. Soc. Mass Spectrom.* **2002**, *14*, 42–50.
208. Nelson, R. W.; Hutchens, T. W. Mass Spectrometric Analysis of a Transition-Metal-Binding Peptide Using Matrix-Assisted Laser-Desorption Time-of-Flight Mass Spectrometry. A Demonstration of Probe Tip Chemistry. *Rapid Commun. Mass Spectrom.* **1992**, *6*, 4–8.
209. Woods, A. S.; Buchsbaum, J. C.; Worall, T. A.; Berg, J. M.; Cotter, R. J. Matrix-assisted laser desorption/ionization of noncovalently bound compounds. *Anal. Chem.* **1995**, *67*, 4462–4465.
210. Lide, D. R.(Ed.). *CRC Handbook of Chemistry and Physics*, CRC Press, Boca Raton, FL, **1992**.
211. Chan, T.-W. D.; Colburn, A. W.; Derrick, P. J. Matrix-assisted laser desorption/ionization using a liquid matrix: Formation of high-mass cluster ions from proteins. *Org. Mass Spectrom.* **1992**, *27*, 53–56.
212. Chan, T.-W. D.; Colburn, A. W.; Derrick, P. J., Matrix assisted UV laser desorption. Suppresion of the matrix peaks. *Org. Mass Spectrom.* **1991**, *26*, 342.
213. Bökelmann, V.; Spengler, B.; Kaufmann, R. Dynamical parameters of ion ejection and ion formation in matrix-assisted laser desorption/ionization. *Eur. Mass Spectrom.* **1995**, *1*, 81–93.
214. Knochenmuss, R.; Dubois, F.; Dale, M. J.; Zenobi, R. The Matrix Suppression Effect and Ionization Mechanisms in Matrix-Assisted Laser Desorption/Ionization. *Rapid Commun. Mass Spectrom.* **1996**, *10*, 871–877.
215. Knochenmuss, R.; Karbach, V.; Wiesli, U.; Breuker, K.; Zenobi, R. The matrix suppression effect in matrix-assisted laser desorption/ionization: Application to negative ions and further characteristics. *Rapid Commun. Mass Spectrom.* **1998**, *12*, 529–534.
216. McCombie, G.; Knochenmuss, R. Small molecule MALDI using the matrix suppression effect to reduce or eliminate background interferences. *Anal. Chem.* **2005**, *76*, 4990.
217. Vaidyanathan, S.; Gaskell, S.; Goodacre, R. Matrix-suppressed laser desorption/ionization mass spectrometry and its suitability for metabolome analyses. *Rapid Commun. Mass Spectrom.* **2006**, *20*, 1192–1198.
218. Dai, Y.; Whittal, R. M.; Li, L. Confocal fluorescence microscopic imaging for investigating the analyte distribution in MALDI matrices. *Anal. Chem.* **1996**, *68*, 2494.
219. Horneffer, V.; Forsmann, A.; Strupat, K.; Hillenkamp, F.; Kubitscheck, U. Localization of analyte molecules in MALDI preparations by confocal laser scanning microscopy. *Anal. Chem.* **2001**, *73*, 1016–1022.

220. Garden, R. W.; Sweedler, J. V. Heterogeneity within MALDI samples as revealed by mass spectrometric imaging. *Anal. Chem.* **2000**, *72*, 30–36.
221. Luxembourg, S. L.; McDonnell, L. A.; Duursma, M. C.; Guo, X.; Heeren, R. M. A. Effect of local matrix crystal variations in matrix-assisted ionization techniques for mass spectrometry. *Anal. Chem.* **2003**, *75*, 2333–2341.
222. Williams, E. R. Proton transfer reactivity of large multiply charged ions. *J. Mass Spectrom.* **1996**, *31*, 831–842.
223. Zhang, X.; Cassady, C. J. Apparent gas-phase acidities of multiply charged protonated peptide ions: Ubiquitin, insulin B, and renin substrate. *J. Am. Soc. Mass Spectrom.* **1996**, *7*, 1211–1218.
224. Pallante, G. A.; Cassady, C. J. Effects of peptide chain length on the gas-phase proton transfer properties of doubly protonated ions from bradykinin and its N-terminal peptides. *Int. J. Mass Spectrom.* **2002**, *12075*, 1–17.
225. Knochenmuss, R. A quantitative model of UV-MALDI including analyte ion generation. *Anal. Chem.* **2003**, *75*, 2199.
226. Frankevich, V.; Zhang, J.; Dashtiev, M.; Zenobi, R. Production and fragmentation of multiply charged ions in 'electron free' MALDI. *Rapid Commun. Mass Spectrom.* **2003**, *17*, 2343–2348.
227. Frankevich, V.; Zhang, J.; Friess, S. D.; Dashtiev, M.; Zenobi, R. Role of electrons in laser desorption/ionization mass spectrometry. *Anal. Chem.* **2003**, *75*, 6063–6067.
228. Dashtiev, M.; Frankevich, V.; Zenobi, R. Kinetic energy of free electrons affects MALDI positive ion yield via capture cross-section. *J. Phys. Chem. A.* **2006**, *110*, 926–930.
229. McCombie, G.; Knochenmuss, R. Enhanced MALDI ionization efficiency at the metal–matrix interface: Practical and mechanistic consequences of sample thickness and preparation method. *J. Am. Soc. Mass Spectrom.* **2006**, *17*, 737–745.
230. Knochenmuss, R.; McCombie, G.; Faderl, M. The Dependence of MALDI ion yield on metal substrates: Photoelectrons from the metal vs. surface-enhanced matrix photoionization. *J. Phys. Chem. A.* **2006**, *110*, 12728–12733.
231. Altelaar, A. F.; Klinkert, I.; Jalink, K.; deLange, R. P. J.; Adan, R. A. H.; Heeren, R. M. A.; Piersma, S. R. Gold-enhanced biomolecule surface imaging of cells and tissue by SIMS and MALDI mass spectrometry. *Anal. Chem.* **2006**, *78*, 734–742.
232. Gross, D. S.; Williams, E. R. Structure of gramicidin S $(M + H + X)^{2+}$ Ions (X = Li, Na, K) probed by proton transfer reactions. *J. Am. Chem. Soc.* **1996**, *118*, 202–204.
233. Kaltashov, I. A.; Fabris, D.; Fenselau, C. Assessment of gas phase basicities of protonated peptides by the kinetic method. *J. Phys. Chem.* **1995**, *99*, 10046–10051.
234. Schnier, P. D.; Gross, D. S.; Williams, E. R. Electrostatic forces and dielectric polarizability of multiply protonated gas-phase cytochrome-C Ions probed by ion/molecule Chemistry. *J. Am. Chem. Soc.* **1995**, *117*, 6747–6757.
235. Gorman, G. S.; Speir, J. P.; Turner, C. A.; Amster, I. J. Proton affinities of the 20 common alpha amino acids. *J. Am. Chem. Soc.* **1992**, *114*, 3986.
236. Hunter, E. P.; Lias, S. G. Evaluated gas phase basicities and proton affinities of molecules: An update. *J. Phys. Chem. Ref. Data.* **1998**, *27*, 413–656.
237. Lias, S. G.; Rosenstock, H. M.; Draxl, K.; Steiner, B. W.; Herron, J. T.; Holmes, J. L.; Levin, R. D.; Liebman, J. F.; Kafafi, S. A.Ionization energetics data. *NIST Standard Reference Database No. 69*, 1998.

Chapter 6

The Development of Matrix-Assisted Laser Desorption/Ionization Sources

Peter B. O'Connor

Department of Chemistry, University of Warwick, Coventry, United Kingdom

6.1 INTRODUCTION

Mass spectrometers do not measure mass. They measure mass–charge ratio. Therefore, to determine the mass of a molecular species, one must (1) ionize that species and (2) know the charge state to which it is ionized. Knowing the charge state is usually a straightforward process because charge is quantized in integral units of the charge on a proton or electron,

Electrospray and MALDI Mass Spectrometry: Fundamentals, Instrumentation, Practicalities, and Biological Applications, Second Edition, Edited by Richard B. Cole

$e \approx 1.6 \times 10^{-19}$ coulombs (the coulomb is the SI unit, so that 1 amp = 1 coulomb/second). In general, most ionization methods produce ionized molecular species with only a few low charge states that are easy to differentiate. The electrospray method discussed elsewhere in this book usually generates higher charge states than do the laser-based ionization methods; and thus, for electrospray, determining the charge state is frequently more difficult. The laser desorption/ionization (LDI) and matrix-assisted laser desorption/ionization (MALDI) methods that will be discussed in this chapter usually generate very low charge states with the 1+ or 1− charge state being dominant. Thus, determining the charge state of an ionized species is usually trivial for such methods. However, many alternative designs have been explored for ionization sources that efficiently generate ions with the appropriate velocities and internal vibrational energies for convenient analysis.

Laser desorption is the process of using a pulse of high-fluence laser light to lift (or desorb) a molecular species from a surface. The specific mechanisms are discussed in Chapter 5, but the basic idea is that the laser light must be absorbed into the electronic and rotational/vibrational states of some of the molecules on that surface. This adds energy to these molecules and detaches them from the surface. Usually, sufficient energy is added that a large number (10^6–10^8) of molecules are desorbed with a single pulse, leaving behind a miniature crater where the laser hit.

Laser desorption does not guarantee that a molecule will be ionized. The ionization process is usually considered to be separated from the desorption process, but this depends on the wavelength of the laser. Typical organic molecules have ionization potentials >10 eV (1 eV $\approx 1.6 \times 10^{-19}$ Joules; the joule is the SI unit, but the electron volt is a more convenient unit when working at the molecular scale), but most of the lasers used have single photon energies of <6 eV (see Table 6.1). Therefore, for direct laser ionization to be possible, the molecule must absorb several (or many) photons, which requires a high laser fluence. Furthermore, for a photon to be absorbed into a molecule, it must have an energy that is equal to the difference between the molecule's ground (or current) state and some higher excited electronic or vibrational state. For ultraviolet (UV) lasers, this generally refers to absorption in the electronically excited, but ground vibrational, states; but for infrared (IR) lasers, the vibrational states are usually the ones that are accessible.

Because the energy distance between states changes as the molecule's energy (either electronic or vibrational) increases, it's only in rare cases that the same photon energy can be

Table 6.1. Typical Lasers Used in MALDI

Laser	Wavelengths	Single-Photon Energy
Excimer		
ArF	193 nm	6.3 eV
KrF	248 nm	4.9 eV
XeCI	308 nm	3.9 eV
Nd:YAG		
	1064 nm	1.1 eV
3ω	355 nm	3.3 eV
4ω	266 nm	4.4 eV
N_2	337 nm	3.6 eV
Er:YAG	2.94 μm	0.4 eV
OPO lasers	2–6 μm	0.2–0.6 eV
CO_2	10.6 μm	0.1 eV

absorbed multiple times. Thus, direct laser ionization either uses very high energy photons via one or two photon processes that liberate an electron by exciting it to an unbound energy state, or by using two or more tuned lasers that specifically excite the molecule through a series of known energy transitions before liberating an electron.

This is true for free molecules in the gas phase, but in solid phase, most of the energy states of a molecule are perturbed by neighboring molecules, and the effective energy absorption window for particular states broadens. For any particular molecule, the energy absorption window is still as sharp in the solid phase as in the gas phase, but the net absorption that is observed experimentally is the average over the millions of individual molecules in the laser beam path, each of which has a slightly different local environment. Thus, it was long ago observed that, along with laser desorption, a small fraction of the desorbed molecules (perhaps 0.01%) were also ionized, with the fraction dependent on the molecular states available, contaminants, laser wavelength and fluence, and so on. This is the process known as direct laser desorption/ionization (Figure 6.1).[1–3]

This small fraction that was ionized could be easily separated from the neutrals using electric and magnetic fields and then sent into a mass spectrometer for measurement of its mass–charge ratio. Direct laser desorption/ionization is usually incredibly inefficient at converting neutral molecules to ions. The 0.01% estimate given above is a very optimistic number. However, in some cases, where the molecule has a strong fundamental absorption at the laser wavelength used, efficiencies can run up into the 1% range. Good examples include desorption of buckminsterfullerene (C_{60})[4,5] or chlorophyll[6] with a 337-nm nitrogen laser.

However, this high variability in ionization efficiencies implies that LDI is a very selective ionization method. In some cases, this selectivity is advantageous, for example, when one wishes to observe the presence or concentration of one, known, select molecule of interest.[7–12] However, in mass spectrometry, one usually is interested in detecting all molecules that are present, including unknown species. Thus, many methods were explored to ionize the molecules after they were desorbed by the laser. These methods are collectively known as laser desorption post-ionization methods, and the post-ionization techniques include electron impact (EI, diagrammed in Figure 6.2), chemical ionization (CI), photoionization (PI), resonant-enhanced multiphoton ionization (REMPI), and many others.

One of the more interesting laser desorption ion sources involved laser desorption of aerosol dust particles directly sampled from the air.[13–17] This instrumental design used an

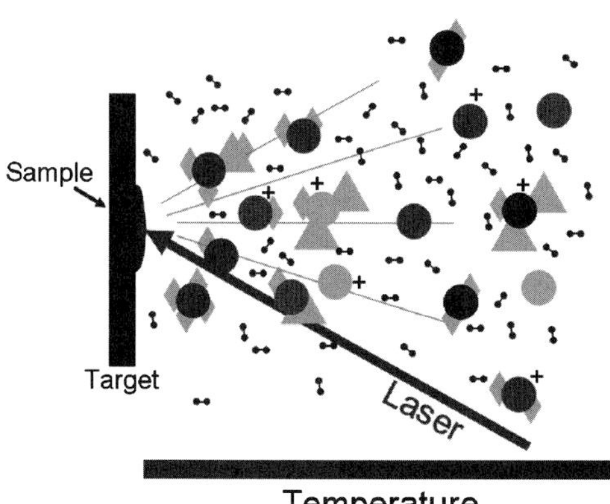

Figure 6.1. Laser desorption/ionization (See color insert).

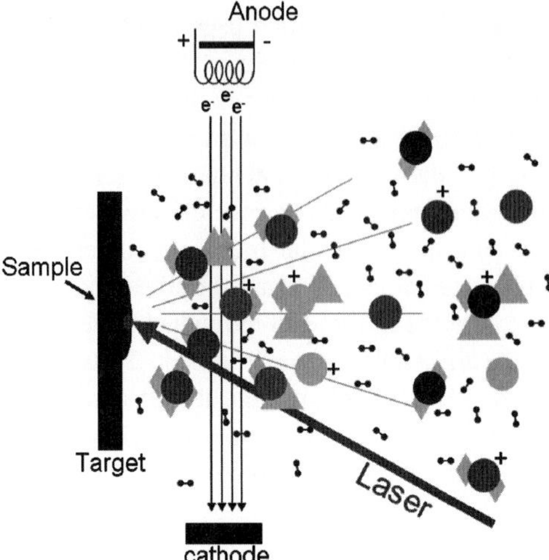

Figure 6.2. Laser desorption plus electron impact ionization (See color insert).

atmospheric sampling interface (several different interfaces have been explored, including use of an aerodynamic lens) that would sample aerosol particles continuously from the air. Light scattering from small continuous diode lasers was used to detect the presence and velocity of the aerosol particles, and then a high-power Nd:YAG laser was used to shatter and ionize these particles, with the ions then extracted and analyzed using time-of-flight mass spectrometry. A particularly interesting example of the utility of this approach involved sampling and detection of metal ions formed from the atmospheric explosion of fireworks during a July 4th celebration.[18]

The typical wavelengths of lasers used in LDI experiments are shown in Table 6.1. Generally speaking, unless unusual resonances are available in the molecule of interest, the UV lasers are more amenable to LDI, while the IR lasers are amenable to desorption, but not ionization. By far, the most common lasers are the nitrogen laser at 337 nm and the frequency-tripled Nd:YAG laser at 355 nm, because these lasers efficiently ionize molecules using the matrix-assisted laser desorption/ionization method discussed in Section 6.4.

The typical laser desorption ion source uses one laser that is focused through a window into a vacuum system onto a metal target that contains the sample spot of interest. Clearly, the focusing optics and window must be chosen for the laser of interest and the optics system, and the position of focusing lenses determines the laser spot size and shape on the target. Lenses and focusing elements are always wavelength-dependent, so, for optics that can pass a wide range of wavelengths, the optimal focusing positions will change with wavelength.

Finally, the mass analyzers used for laser desorption mass spectrometry experiments usually need to be synchronized to the laser's emission. The synchronization accuracy required varies from instrument to instrument. With time-of-flight instruments, the laser emission triggers the oscilloscope trace and must be known to within a few picoseconds. With ion trap, orthogonal time-of-flight, and Fourier transform mass spectrometers, usually the laser emission must be known to within a few microseconds at best. Thus, the electronics that are used to control the instruments can vary considerably.

For time-of-flight instruments, the laser emission is used to trigger the mass analyzer. In this arrangement, a small fraction of the laser light is diverted to a fast photodiode, and the output of the photodiode is used to trigger the oscilloscope that records the ion arrival times at the detector. Because the photodiode's output is noisy, the output signal from the photodiode can sometimes transition past the threshold triggering voltage several times over the course of a spectrum acquisition. In order to prevent re-triggering the oscilloscope, a simple and fast electronic device, known as a Schmitt trigger,[19] is inserted between the photodiode and the oscilloscope, so that only one TTL trigger is sent to the oscilloscope, on the rising edge of the laser output pulse. Thus, while the laser pulse may be nanoseconds long, the accuracy of the trigger is often in the tens of picoseconds on the rising edge of the laser light pulse. The actual time-position of this trigger is a parameter that can be addressed in the calibration of the spectrum, but the "jitter" of this trigger (its variance in time) determines the reproducibility of the mass spectrum and thus the ability to signal average.

For other instruments, the laser timing accuracy is not nearly so critical. For quadrupole and sector instruments, where the detection is essentially continuous, the laser is usually set to fire as fast as possible. For ion traps and Fourier transform mass spectrometers the accuracy needs to be controlled to within a few microseconds or milliseconds, depending on the arrangement. This can usually be achieved by letting the mass spectrometer send a trigger to the laser, which will then fire and emit its laser pulse a few hundred nanoseconds later. The jitter in this timing can be tens of nanoseconds, but it is sufficiently low for the instruments involved.

In addition to triggering the laser, the electronics of the mass spectrometer need to control the voltages on the sample plate and ion optics inside the source. Again, time-of-flight instruments have special requirements here. Other instruments generally set the voltages using computer controlled digital-to-analog converters (DACs) that are followed by amplifiers that bump those voltages up to whatever voltages are necessary. With time-of-flight instruments, the laser trigger is also often used to trigger the switching of a high-voltage pulse on the sample plate or extraction ion optics, in a method known as "delayed-extraction." These switches must trigger pulses of several kilovolts at a specific time (usually tens of nanoseconds after the rising edge of the laser pulse), controlled to within a few tens of picoseconds. Most of the major improvements in performance of these instruments have come about because of improvements in the power, timing, and accuracy of these switches.

6.2 LASER DESORPTION ANALYZERS

Once ions are produced, they are typically directed into a mass analyzer, where the ions are separated by their mass–charge ratios and detected. The types of mass analyzers vary widely.[20] By far, the most common is the time-of-flight mass analyzer due to its simplicity and the pulsed nature of ion extraction from the ion source, which makes it especially compatible with pulsed laser excitation, but laser desorption/ionization sources have been coupled to quadrupoles, electromagnetic sector instruments, orthogonal time-of-flight mass spectrometers, quadrupole ion traps, Fourier transform mass spectrometers, and even the recently introduced Orbitrap.[21] Detailed discussion of these mass analyzers is beyond the scope of this chapter but is not out of the scope of other chapters in this volume and elsewhere.[20] However, laser desorption ion sources produce ions with a particular momentum and temperature, and great care must be taken in coupling these ion sources to the mass spectrometers so that the ions' momentum and temperature parameters are compatible with the instrument at hand.

Again, the ubiquitous time-of-flight (TOF) mass analyzer is the special case. Since time-of-flight mass analyzers measure the velocity of ions, the initial velocity of desorption, and the initial position of the ions when the acceleration pulse is applied are critical parameters. In a typical axial MALDI-TOF ion source, the sample plate (at $>$ 10 kV) is positioned ~5 mm away from extraction ion optics (at ground potential), resulting in an electric field of 2 V/μm. Thus a 10-μm difference in desorption position (the height of a typical sample) results in a 0.2% difference in kinetic energy and a ~0.1% difference in arrival time at the detector. A ~0.1% difference in arrival time means that the maximum possible resolving power possible under such conditions is $\sim 10^3$. Thus, researchers developed reflectron TOF instruments[22–26] to refocus the kinetic energy distribution somewhat at the detector and the delayed extraction method[27,28] to squeeze the kinetic energy distribution. These two methods allowed TOF mass analyzers to exceed 10^4 resolving power for flat samples on flat surfaces, but they are insufficient to correct for more irregular surfaces such as thin-layer chromatography plates[29,30] or electrophoresis gels.[31] Again, for TOF mass analyzers, the kinetic energy of the ions is a matter that can be addressed by calibration, but the kinetic energy distribution and the initial spatial distribution (deviation from flatness) are critical because they affect the possible resolving power of the instrument.

For other types of mass spectrometers, it is necessary to couple the kinetic energy distribution to the instrument at hand. For example, with quadrupoles, quadrupole ion traps, and Fourier transform mass spectrometers, excessive kinetic energy of the ions can prevent them from being stably trapped. Thus, these instruments either tune their trapping and analyzing conditions to the kinetic energy distribution of the ions or desorb the ions into a region filled with a gas which cools and narrows the ions' kinetic energy distributions (see Section 6.7).

Furthermore, laser desorption/ionization usually increases the internal (ro)vibrational energy of the desorbed molecules, essentially heating the molecules up to internal temperatures of hundreds of degrees (°C) above room temperature. Depending on the molecule, this temperature is often sufficient to induce unimolecular dissociation. This is particularly true for molecules with highly labile moieties such as sulfate/phosphate or sialic acid groups. In order to measure the intact mass of such molecules, either (a) the internal vibrational energy must be cooled via collisions with cold neutral background gas or (b) the mass–charge ratios must be separated before the molecule dissociates (again, see Section 6.7).

In TOF mass analyzers, if the molecule dissociates after the grounded source extraction grid, all fragment ions will have the same velocity as the precursor molecule and will arrive at the detector at the same time. If the molecule dissociates in the source during the acceleration period, the fragment ion masses will be spread out over the whole range of acceleration potentials in the source depending on where, exactly, they were formed. Finally, if the molecule dissociates in the source before acceleration, the fragments will have a flight time consistent with their (lower) fragmentation mass. The first situation is called "post-source decay,"[32–34] the second is rather generically called "metastable decay," and the third is called "in-source decay." For linear, axial TOF mass analyzers, post-source decay does not distort the mass spectrum, but metastable decay during acceleration does distort it and can result in anything from a clean spectrum of the decay fragments to broad baseline humps where the fragment ion velocities are spread out, thereby generating a wide range of arrival times. For reflectron TOF instruments, post-source decay fragments usually result in these broad baseline humps as both the kinetic energies and arrival times for a fragment ion of a given *m/z* value are spread out, but they can be refocused onto the detector by lowering the reflectron voltage or by using a curved-field reflectron. However, usually the mass accuracy of the post-source decay fragments in reflectron mode is substantially lower than the mass accuracy of the precursor, and the sensitivity of the instrument is always lowered because some ions are lost and the signal of the ions that are observed is distributed over more detector channels.

Because the other types of mass analyzers are slower than TOF mass spectrometers, it is usually necessary to either (a) accept the fragments that are observed or (b) cool the ions before they dissociate. This necessity led to the development of laser desorption ion sources that operate at background gas pressures that are "elevated" from the usual high vacuum of $<10^{-6}$ mbar. These ion source designs are discussed in Section 6.7.

6.3 LASER MICROPROBE FOR MASS ANALYSIS

The laser microprobe for mass analysis (LaMMA) was simply a standard LDI source with a high resolution spatial translation stage for the sample target.[35–41] This allowed for a pixel-by-pixel generation of an image of the sample, where each pixel was the LDI mass spectrum of an individual position in the spectrum, generating a four-dimensional data set: two spatial dimensions, one mass dimension, and the peak intensity dimension. Such data allow for the generation of images that show the spatial distribution of any particular *m/z* value or, by summing the intensities of a series of masses, the spatial distribution of the total-mass intensity for a given ion and its fragment masses (for example).

The typical LaMMA source design is shown in Figure 6.3, and it combines the laser with an optical viewing system and camera that assisted the user in aligning the optical image with the mass spectrometry images. The main limitations of the LaMMA systems were due to (a) the selectivity of the ionization and (b) the propensity for decomposition of large, thermally labile species. Since direct laser desorption/ionization was used as the ionization source, clearly those molecules with strong chromophores (high extinction coefficient) at the desorption laser's wavelength would ionize much better than others. Furthermore, most imaging samples are extremely complex—for example, tissue samples or minerals. The spectra can be extremely complex, essentially impossible to analyze, and biased toward the low mass range. However, the spatial information provided by some small molecules of known mass, such as the distribution of a drug molecule in a tissue sample, could be very useful. Also, the LaMMA instrument's spatial distribution was not usually limited by the laser focus, but rather by the "interaction volume" of the laser with the sample. Due to energy coupling, the desorption plume's diameter is always larger than the laser spot

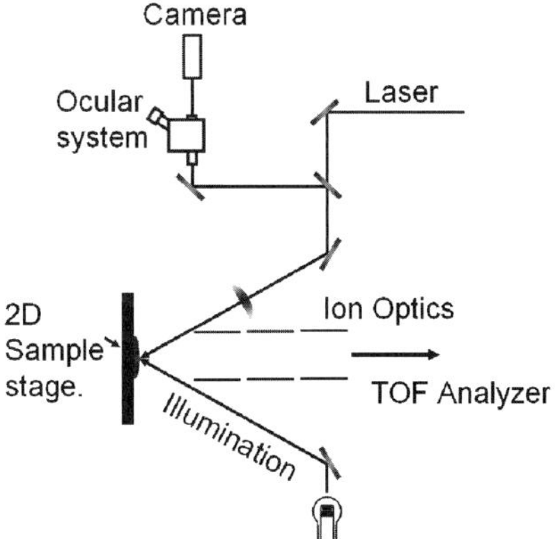

Figure 6.3. Laser microprobe for mass analysis (LaMMA) configuration.

size by several microns, depending on the laser fluence. The LaMMA instrument is the precursor for current imaging LDI mass spectrometers discussed in Chapter 8.

6.4 MATRIX-ASSISTED LASER DESORPTION

Ionization of large biomolecules, such as proteins and DNA, with LDI or any of the modified post-ionization methods was limited to very small fragments, typically under 1000 Da due to decomposition of the relatively fragile larger molecules upon ionization. Fragmentation during ionization isn't always a problem, but it's extremely useful to know the mass of the precursor species first, and these ions weren't usually observed. In 1985, two significant developments were made which led directly to the development of electrospray ionization and matrix-assisted laser desorption/ionization (MALDI), for which the Nobel prize was (jointly) awarded in 2002. Electrospray ionization is covered extensively elsewhere in this book, but the development of MALDI is to be addressed here.

The "matrix-assisted" effect refers to enhanced ionization efficiency of molecules that is observed when they are desorbed with another molecule in the mixture that acts as a primary chromophore as compared to their direct laser desorption ionization efficiency as a pure compound. Thus, in a matrix-assisted laser desorption/ionization (MALDI) experiment, one must always add the matrix to enhance the ionization efficiency of the analytes. Good matrices generally have high absorption coefficients for the laser wavelength of interest and are usually acidic to be able to donate a proton (positive ion mode) in the plume. It is also often desirable for a matrix to form large, flat crystals upon evaporation of solvent. This latter requirement is particularly true for time-of-flight mass analyzers because the resolving power is dependent on the flatness of the sample surface.

One of the first observations of this effect was reported in 1985 by Karas et al.,[42] and the relevant data from this manuscript is reproduced as Figure 6.4A, accompanied by the following description:

> *"Matrix-Assisted" Laser Desorption.* The mass spectrum of a mixture of alanine (Ala) and [tryptophan] Trp taken at Trp threshold irradiance is shown in Figure [6.4A]. A strong signal of the Ala quasi-molecular ion was observed in addition to that of Trp. It is important to note that its desorption took place at an irradiance of about a tenth of that necessary for obtaining spectra of alanine alone.
>
> Tryptophan thus must be regarded as an absorbing matrix resulting in molecular ion formation of the nonabsorbing alanine. This kind of "matrix-assisted LD" has also successfully been applied to reproducibly desorb other nonvolatiles, e.g., stachyose [Bahr, U., et al., unpublished work].

Thus, alanine laser desorption ionization efficiency increases when mixed with tryptophan, which was the first matrix. Acting on this clue, Karas et al.[43] continued this work testing various matrices using a q-switched frequency quadrupled Nd:YAG laser (266 nm, 10 ns) and testing the desorption of larger and larger, nonvolatile compounds ending with the spectrum shown in Figure 6.4B of a small protein, mellitin using nicotinic acid as the matrix. The mellitin spectrum showed no fragmentation and has essentially no absorption at 266 nm. This work was later extended to larger proteins up to bovine serum albumin at 67 kDa and higher.[44–46]

Following this work, Tanaka et al.[47] developed a similar method that used ultrafine (~300 Å) cobalt particles in glycerol as a matrix and showed the formation of lysozyme clusters up to 100 kDa. Tanaka et al. used a less expensive nitrogen laser (337 nm) for this

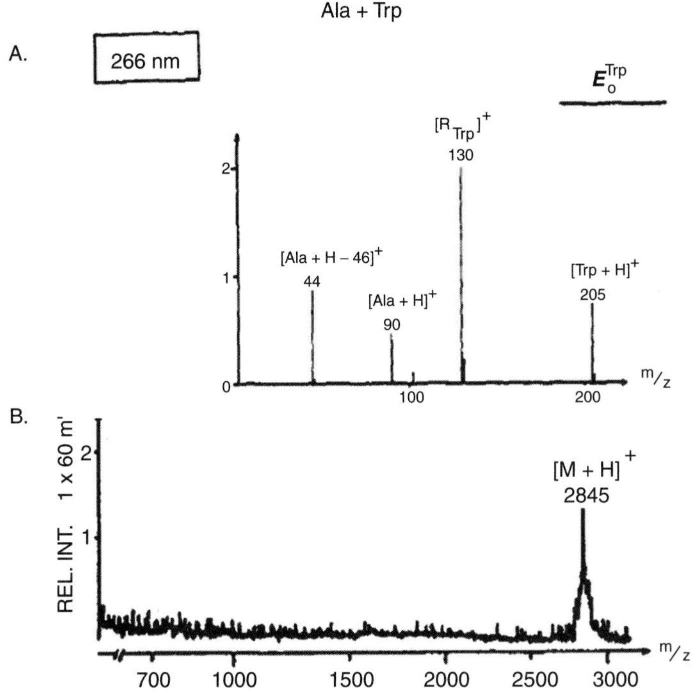

Figure 6.4. Two of the first MALDI spectra. (**A**) The matrix effect, desorbing alanine mixed with tryptophan, allowed observation of the alanine peaks at 10 × lower laser fluence than possible with pure alanine. (Reproduced with permission from reference 42). (**B**) The first MALDI of a protein, showing that desorption/ionization could occur with no observable fragmentation. (Reproduced with permission from reference 43).

work, but other than that convenience, the method was less reproducible and more difficult to use than the method above because glycerol is a liquid, constantly evaporating even under vacuum at room temperature, and because the ultrafine cobalt particles were not readily available.

Regardless of the differences, the two bodies of work clearly showed that, using a matrix and threshold laser irradiance, ions could be generated from peptides and proteins without their dissociation. Since the time that mass spectrometry was invented in the first half of the twentieth century, this capability had been sought after because the fragmentation of molecules upon ionization was what limited mass spectrometry to the analysis of small, stable, volatile organic molecules. The ability to ionize peptides and proteins without dissociation is the key technology that led to proteomics and allowed mass spectrometry to finally tackle the more difficult and intriguing questions in human health by providing new ways to sequence proteins, RNA, DNA, glycolipids, glycans, and oligosaccharides; identify proteins from their mass spectra; and track the post-translational modifications that are thought to be responsible for many (if not most) of the signal transduction cascades that form the backbone of the modern biotechnological industry and our current understanding of biology and medicine. The development of electrospray and MALDI was truly revolutionary.

The whole trick to success with MALDI is finding the right matrix. Some matrices are extremely good at ionizing peptides, others are preferred for proteins, others for DNA, still others for oligosaccharides. In addition to Chapter 7 in this volume, there's a large but diffuse body of literature published in diverse journal articles and book chapters where the matrices are explored; a few papers are referenced here.[48–63] Primarily, when starting a new MALDI project, it is recommended to try the sample preparation methods and matrices that have been published, but it is not uncommon to find that the published methods and matrices are not the best for any particular analyte. As mentioned above, MALDI matrices usually form large, flat crystals, have strong absorption at the laser wavelength of interest, and are acidic. The first criterion is important when MALDI is coupled to time-of-flight analyzers because a molecule's initial kinetic energy is determined by its position in the extraction field. If the surface is not flat, then ions start at different "heights" in the field and thus have slightly different kinetic energies, which broadens the peak and lowers the mass resolution. The second criterion is obvious since the primary job of the matrix is to lift the analyte into the gas phase; this is done by the matrix absorbing the laser light and exploding off the surface into a "plume," incidentally carrying along the embedded analyte. The third criterion provides for the hot, gaseous matrix molecules in the plume to protonate the desorbed analytes. Clearly, this also implies that the excited-state matrix molecules must have lower proton affinities than the analyte (this is essentially the definition of acidity in the solid state or gas phase) for positive-ion-mode acquisitions. Ionization in the negative mode is, of course, the reverse. The matrix must be able to accept a proton from or donate an electron to the analyte; additionally, cation and anion attachment strategies have also been reported. Accepting a proton from the analyte generates an even-electron species, donating an electron to the analyte generates an odd-electron species. Since even-electron species are almost always more stable than odd-electron species, most matrices that are currently used produce even-electron species by donation/extraction of a proton rather than donation/extraction of an electron. However, there are reports of odd-electron chemistry occurring in MALDI plumes which are probably related to the latter.[64]

The most commonly used matrices are α-cyano-hydroxycinnamic acid (CHCA, for peptides on MALDI-TOF instruments), sinapinic acid (also called sinapic acid, for peptides on MALDI-TOF instruments), 2,5-dihydroxy-benzoic acid (DHB, for proteins on MALDI-TOF and for proteins/peptides/labile molecules on decoupled instruments), hydroxypicolinic acid (for DNA), and succinic acid (for IR-MALDI; see below). The primary criteria for evaluating matrices is signal intensity, peak shape, extent of fragmentation upon ionization, and the amount of background matrix clusters that are co-ionized with the sample contributing to the abundance of contamination ions in the spectrum, which is often called "chemical noise."[65] For MALDI-TOF instruments, because the initial height of the crystals is a factor in determining the initial kinetic energy of the resulting ions, the best matrices generate crystals that are flat. DHB, for example, generates more ions than CHCA or sinapinic acid, but DHB, when dried down, forms sharp, vertical crystals that are not good for MALDI-TOF experiments and result in low-resolution spectra. DHB, however, is one of the preferred matrices for MALDI-ion trap, MALDI-qTOF, and MALDI-FTMS instruments for which the sample morphology is less important.

Most MALDI experiments operate with <10-ns UV laser pulses (at 337, 355, or sometimes 266 nm). However, MALDI can also be performed using infrared lasers at 1.064, 2.79, 2.93, and 10.6 μm, with typically 50- to 100-ns pulse durations.[66–75] The advantage of operating MALDI using infrared lasers is usually touted to be twofold: lower fragmentation and the ability to use aqueous buffers (even pure water ice) as the matrix. There has been some investigation of these methods, but the preferred lasers have much lower reliability and stability than do the UV lasers used, so the progress of IR MALDI has been somewhat slower than UV MALDI. The most dramatic IR-MALDI data to date is the mass spectra of

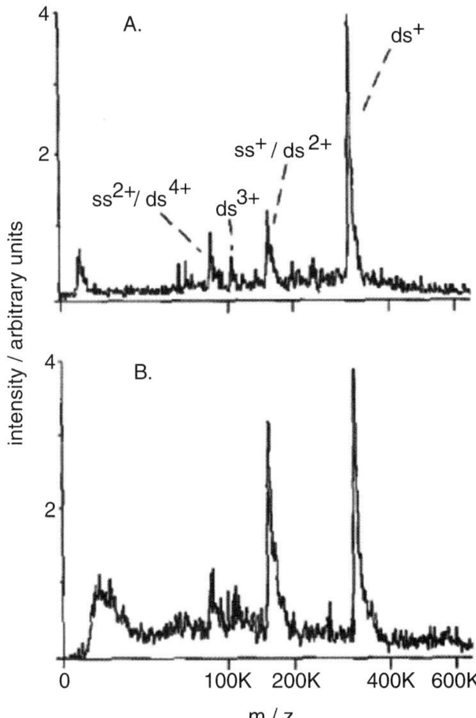

Figure 6.5. Infrared MALDI of double-stranded DNA. (Reproduced with permission from reference 76).

>400-kDa fragments of double-stranded DNA (Figure 6.5),[76] but it is also being explored for ionization of labile molecules such as gangliosides and sulfated oligosaccharides.

6.5 COUPLING CHROMATOGRAPHIC SEPARATION TO LDI

Combining chromatographic separation with mass spectrometry makes a lot of sense. Mass spectrometry runs into a number of problems when trying to ionize mixtures of molecules with different chemical and physical properties. The most important such problem is called the "signal suppression effect" or "matrix effect" in which analyte molecules, in the presence of other sample constituents, ionize with vastly different efficiencies so that MS peak intensity has little correlation with a molecule's concentration in the sample. This effect is most clearly observed in the mass spectra of enzymatic digests of proteins. In this case, all digest peptides have the exact same concentration (assuming the digestion reaction went to completion), but you commonly have >100-fold difference in peak intensities and some peptides aren't observed at all. This is true no matter what ionization source you use, and in spite of similarities in the structure of the peptides (particularly when trypsin is used which leaves a charged arginine or lysine on the C-terminus). To get around this problem, one approach is to separate the molecules prior to MS so that only one or a few molecules are analyzed at the same time. This substantially increases the effective dynamic range and reliability of the experiment.

However, LDI methods use solid-phase, dried-down samples, whereas chromatographic separations use solution phase samples. Thus, the liquid stream from the chromatograph must be deposited onto a sample target and dried prior to transfer to the LDI ion source. There are many methods to do this.

Since the LDI ion source is under vacuum (usually $\sim 10^{-6}$ mbar), the eluting end of the chromatography column cannot usually be inside the source, otherwise the pull of the vacuum will distort the chromatographic separation and possibly destroy the column packing material. Thus, various methods were explored to deposit the effluent on a target and transfer it into the source. Two such methods are diagrammed in Figure 6.6.[77,78] The left figure involves a direct transfer of the sample onto a wheel that rotates at a constant speed to transfer the sample into the desorption region. The obvious problem with such a source is contamination at the point where the wheel touches the repeller plate. This causes memory effects and sample crossover which, for chromatography, is mostly observed as peak-spreading. The right figure, the moving belt interface, was more effective because it allowed the deposition to proceed under vacuum via an atmospheric pressure liquid junction coupling. However, it has also not really panned out due to the need for high pumping speed and poor chromatographic performance.

Direct coupling of separations to LDI sources was therefore replaced by off-line coupling of various forms. One such form used a continuous liquid junction between the chromatographic effluent and a MALDI target to generate a continuous thin streak, or "snail-trail," of separated components.[79] This method works well, particularly for capillary electrophoretic separations since the elution buffer is constant. It works less well for traditional reversed-phase HPLC due to the gradient in hydrophobic/hydrophilic buffers which changes the evaporation rate and thus changes the width/shape of the eluent track. Additionally, high concentrations of organic molecules, such as MALDI matrices (see Section 6.4), if added to the eluent, tend to crystallize on the tip, distorting the eluent streak.

The most successful form of coupling of a chromatographic separation to LDI sources involves spotting the eluent into a series of sequential, separated droplets that evaporate prior to insertion of the target plate into the source. The droplet blotting technique takes many forms including direct touching of the droplet to the target (which requires z-axis motion), electroblotting (which uses a quick voltage pulse to make the droplet jump from the eluent tip to the target), and pneumatic blotting (which uses a small puff of air to do the same thing). The electroblotting method is shown schematically in Figure 6.7, using a system

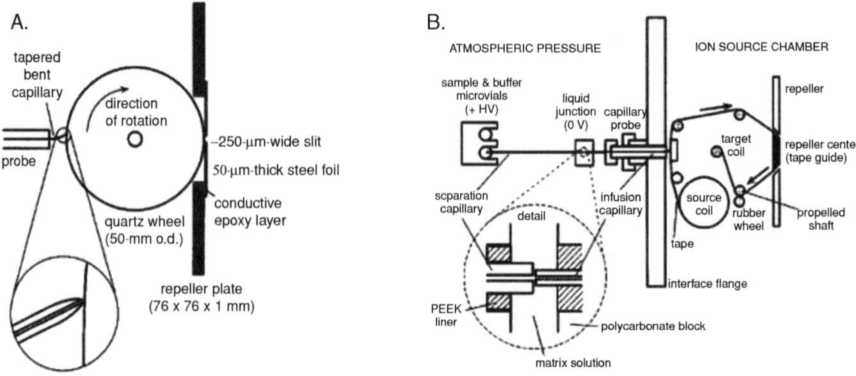

Figure 6.6. Two methods of direct coupling of chromatography with laser desorption/ionization. (**A**) direct coupling HPLC effluent at atmospheric using a wheel. (**B**) *In vacuo* coupling of capillary electrophoresis using a tape. (Figures reproduced from Refs. 77 and 78, with permission.)

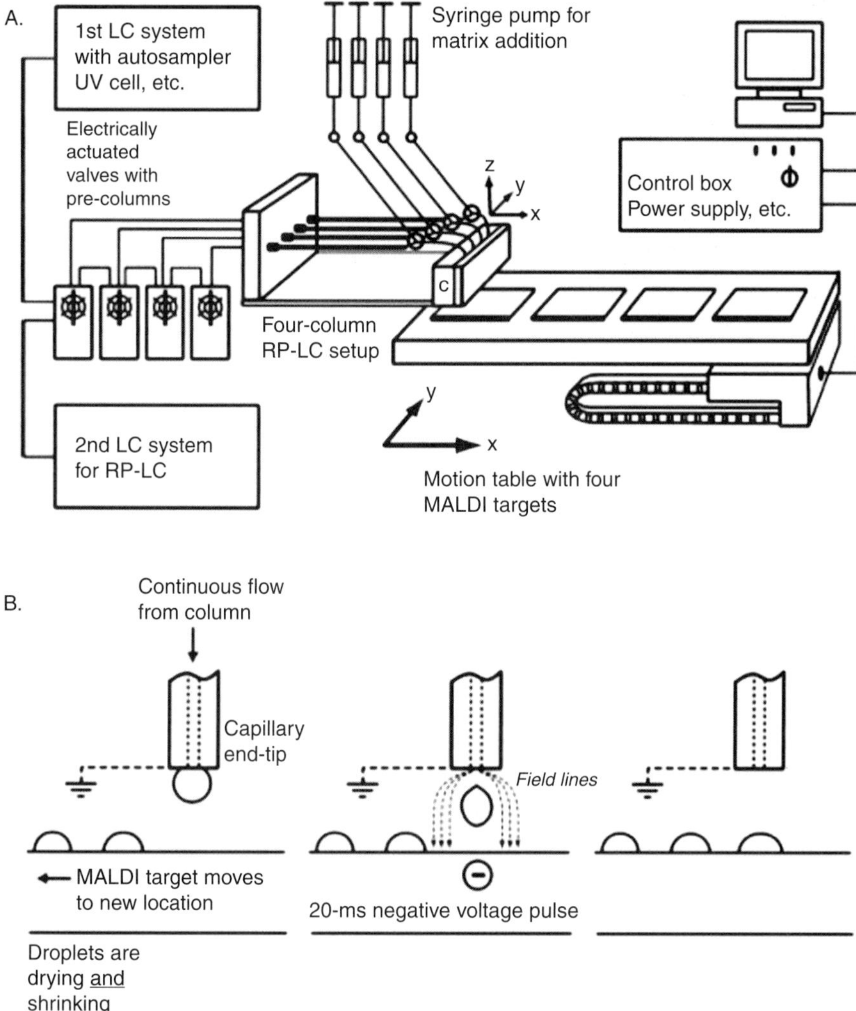

Figure 6.7. Coupling LC effluent to MALDI using an electroblotting drop interface (Reproduced from Ref. 80, with permission.)

where the matrix is mixed with the eluent prior to blotting, although a system where the matrix is spotted from a different eluent tip is clearly possible.[80] For HPLC, a blotting system still has the downside that the changes in evaporation rate along the gradient cause a change in spot size along the target traces, but many of the prior instabilities are avoided and blotting methods are generally preferred at this point.

6.6 MALDI FROM VARIOUS SURFACES

Matrix-assisted laser desorption/ionization is a surface technique involving use of laser light to desorb and ionize molecules from surfaces (Figure 6.8). For the typical proteomics experiments, the samples are proteins and peptide mixtures in a high concentration of matrix

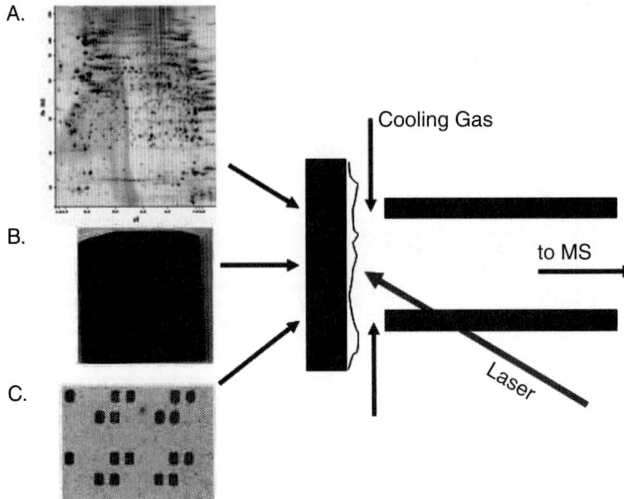

Figure 6.8. Coupling surface techniques to MALDI mass spectrometry. (**A**) 2D-page gel, (**B**) DIOS, (**C**) SPR array.

molecules dried onto a flat stainless steel target plate. However, many other surfaces and sample preparation methods have been explored. The chemistry occurring at a surface can be studied using auxiliary techniques that are combined with MALDI analysis. For example, surface plasmon resonance experiments can be performed to determine binding kinetics of molecules followed by MALDI mass spectrometry to characterize the binding partners (Figure 6.8C). When considering surfaces for use in MALDI, surface roughness is important, as is the chemistry of the surface.

One of the clearest cases of the importance of surface roughness is in the coupling of thin-layer chromatography (TLC) with MALDI. TLC-MALDI-TOF was explored in the mid-1990s by Hercules' group.[29,30,81,82] While the ability to separate molecules by TLC, followed by MALDI-TOF analysis was clearly demonstrated, the resolution apparent in the mass spectra was very low. This was expected because MALDI-TOF mass spectrometers are dependent on having a narrow kinetic energy distribution of the desorbed molecules, and the kinetic energy distribution is a function of the initial desorption "height" distribution of the molecules on the surface. Since TLC plates are porous, the desorbed molecules have a wide initial kinetic energy distribution. For example: consider a MALDI-TOFMS source with 24-kV extraction potential over 4 mm (a typical example from the literature[83]). In this case, with this electric field of 6 V/μm, ions desorbed from spots differing in height on the surface by 10 μm will experience a 60-V difference in acceleration potential (0.25%), and ions separated by 100 μm will experience a 600-V difference (2.5%). Achieving better than 0.1% mass accuracy or 1000 resolving power with such initial kinetic energy differences is extremely difficult. Delayed extraction can correct for modest potential differences but is limited by the field-free flight distance and the initial kinetic energy of the desorbed ions. In addition to use of instrumental improvements such as delayed extraction, the Hercules group used a sophisticated extraction scheme (Figure 6.9) to transfer the molecules in the TLC plate to a flat, stainless-steel MALDI target.

However, when TLC is coupled to a MALDI-qTOF[84] or MALDI-FTMS mass spectrometer,[85] the height variance does not greatly influence the resulting spectra because (1) the extraction potentials are a few volts, rather than 10+ kilovolts, and (2) the ions are trapped or focused before analysis, which cools and sharpens the kinetic energy distribution.

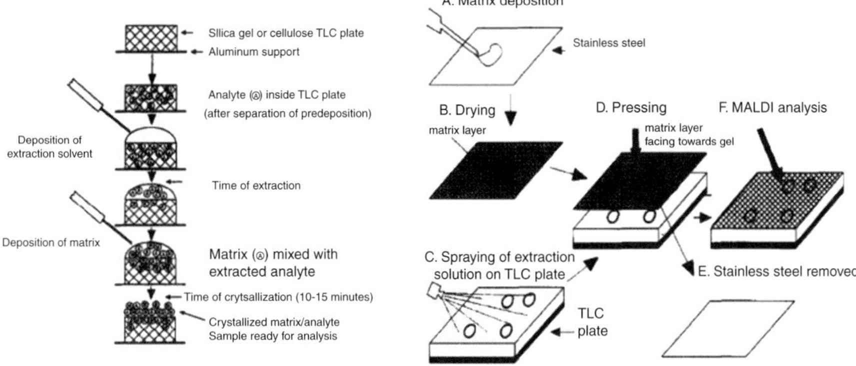

Figure 6.9. Extraction procedure for transferring TLC-separated molecules to a stainless-steel target while preserving their spatial separation. (Reproduced from Refs. 29 and 30, with permission.)

The result is the ability to couple TLC separations with the high resolution, high mass accuracy, and tandem mass spectrometry capabilities of modern mass spectrometers. It is important to note that the sensitivity of these TLC MALDI methods varies with the laser used. Since the IR lasers have a larger interaction volume with the sample, they showed improved sensitivity compared to UV-MALDI.[66]

Another high visibility surface for laser desorption ionization experiments is the use of porous silicon, a technique that has been called desorption/ionization on silicon (DIOS, Figure 6.10).[86,87] Again, the surface has a highly irregular topology, so that the use of uncoupled mass analyzers is advisable, and the method appears to yield ions that are less vibrationally activated than the same ions desorbed from stainless steel, although they are also more vibrationally activated than the same ions desorbed using a MALDI matrix. Currently, DIOS appears to be most useful for low-molecular-weight species because in

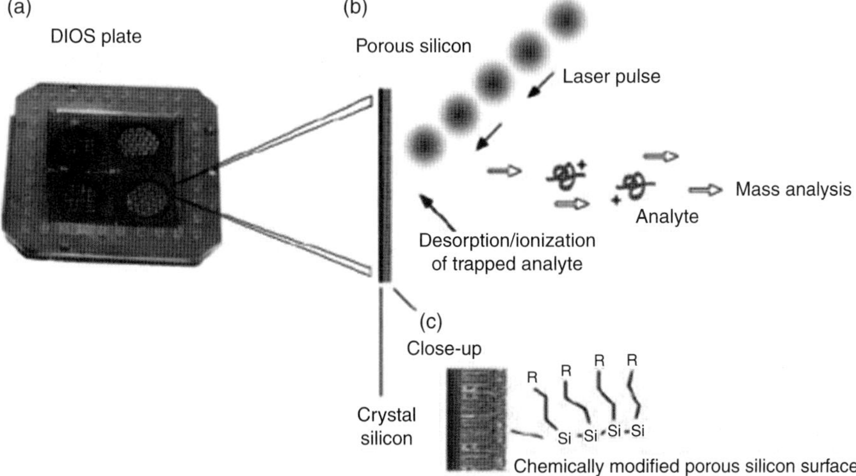

Figure 6.10. The DIOS experiment. (Reproduced with permission from reference 86).

MALDI, these low-mass ions often appear in the same mass range as the matrix clusters, confusing their analysis.[88]

A wide variety of other surfaces have been explored for MALDI analysis from polyacrylamide electrophoresis gels[31,89] to a range of polymeric materials.[90–99] The use of hydrophobic surfaces can greatly enhance sensitivity by forcing droplets to evaporate into a pinpoint sample spot, increasing the analyte concentration. When this trick is also combined with prestructured sample targets (Figure 6.11)[100] where tiny hydrophilic spots are created on a generally hydrophobic surface, one not only can increase analyte concentration and the sensitivity of a MALDI experiment, but also can know exactly where the spot will be, eliminating the need to search the plate optically prior to MALDI analysis.

6.7 MALDI AT ELEVATED PRESSURES

Electron impact ionization largely created the field of mass spectrometry by demonstrating that, with appropriate instrumentation, a chemist could both determine the molecular weight of a molecule and also obtain some very useful structural clues from its fragment

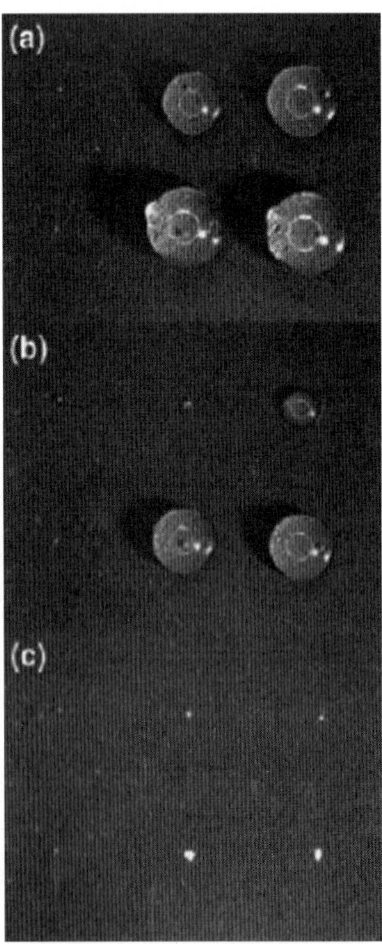

Figure 6.11. Anchor chips, showing the concentration of large droplets (**a**, **b**) into small spots (**c**) of known position. (Reproduced with permission from reference 100).

ions.[101] However, it quickly became clear that the requirement that the samples be vaporized prior to ionization and the extensive fragmentation resulting from electron impact hindered the analysis of larger, more labile biomolecules such as proteins. Therefore, mass spectrometrists embarked on a quest to develop ionization methods that could launch a molecule into the gas phase and ionize it without causing fragmentation. This search for "cooler" ionization techniques is directly responsible for the development of, among others, chemical ionization,[102] field desorption,[103] plasma desorption,[104] fast atom bombardment,[105] matrix-assisted laser desorption/ionization (MALDI),[43] and electrospray ionization (ESI).[106] The last two, MALDI and ESI, were revolutionary in scope because they allowed the ionization of large proteins and DNA molecules, as well as other biomolecules, while keeping the molecules intact and substantially improving sensitivity. It is not an exaggeration to say that MALDI and ESI are transforming biochemistry.

MALDI has had particularly widespread impact because of the simplicity of the time-of-flight (TOF) mass analyzer with which it is typically coupled. Biologists are cheerfully replacing gel electrophoresis results with the much more accurate MALDI-MS data, and the in-gel digest and protein sequence tag methods of searching protein databases using MALDI-MS data have made assignment of protein identities much more reliable and rapid than previously possible.[107] The term "proteomics" was coined to describe these exciting new methods. Although "proteomics" is now broadly applied to many types of MS-based protein analyses, MALDI remains one of the simplest and best methods available.

However, compared to ESI, MALDI has always been limited by the fact that desorbed ions become vibrationally excited on desorption (i.e., they are "hot"). This metastable activation is often sufficient to cause them to fragment, often on a >1-μs timescale. In 1994, Spengler et al.,[32] first took advantage of this metastable activation to sequence peptides and coined the two terms "post-source decay" and "in-source decay". The differentiation between fragmentation "post-source" and "in-source" is instrument-dependent because the two terms simply refers to metastable fragmentation that occurs after the ion leaves the MALDI source and enters the mass analyzer or before the ion extraction voltage pulse is applied.

6.7.1 Metastable Decay

Post-source decay occurs on a 0.1- to 10-μs timescale and is detectable in the MALDI-TOF mass spectrometer by comparing the linear mode spectrum with the reflectron mode spectrum. On such comparison, the ions that appear in the reflectron spectra (usually as broad humps in the baseline, but sometimes as fairly high intensity, broad peaks) are caused by metastable fragmentation after the ion leaves the source, but before it enters the electrostatic reflection mirror. In-source decay operates on a shorter timescale, typically <100 ns, and if the structure of the ion is not previously known, there is no simple way to determine whether a particular ion is an in-source fragment or an independent molecular species in the sample. Thus, while the ability to sequence peptides in this way is advantageous, mass spectrometric analysis of more fragile species is limited by fragmentation prior to detection of a molecular ion species. In a mixture of closely related species fragments on ionization, it is frequently impossible to determine whether a given ion represents a molecular species that is initially present or whether it merely originated by fragmentation of a higher-molecular-weight component, and quantitation under such conditions is not feasible.

Metastable ions become a particular problem in instruments that can only detect the ions on a timescale that is long compared to their metastable lifetime. Depending on the molecular species involved, all ion traps, ion cyclotron resonance, and quadrupole TOF mass spectrometers frequently run into this problem. For most molecules, metastables lifetimes are $<100\ \mu s$ whereas detection timescales of milliseconds to seconds are common for these instruments. Therefore, the ions detected are either (a) stable molecular ions or (b) fragments. High-pressure MALDI solves this problem by cooling the metastable ions, removing their excess vibrational energy, and thereby preventing fragmentation.

MALDI generates ions by evaporating their parent molecules, usually from a solid crystalline surface, using a fast laser pulse. The molecules of interest are mixed with a matrix, a small organic molecule that strongly absorbs photons of the laser's wavelength, and the mixture is deposited on the MALDI target surface and allowed to dry into small crystals. (Although a few liquids have been used, most MALDI matrices are solids.) Figure 6.1 is a sketch of what happens in a high-pressure MALDI source after the laser shot. When the laser pulse hits the surface, the matrix absorbs the laser light and is desorbed from the irradiated region with a velocity of several hundred meters per second propelling the analyte along with it. The laser shot generally heats up the ions and matrix, but the N_2 gas molecules (pulsed into the source during desorption) colliding with these clusters absorb both their kinetic and vibrational energy. Generally, the cooling of the ions' initial kinetic energy is observed first, at lower pressures, and the observation of vibrational energy cooling is observed later, at higher pressures. The pressure regimes involved are discussed below.

It is, in retrospect, obvious that increasing the pressure in the MALDI plume region during desorption would cool the desorbed ions. At 10 mbar, a slow-moving peptide molecule will experience ~100 gas collisions per microsecond; and since the metastable ions typically fragment on the ~1- to 100-μs timescale, they will have many collisions with background gas in this time. Additionally, since the background gas is usually adiabatically cooled nitrogen introduced via a leak valve or a pulse valve, the temperature of the background gas can be substantially lower than room temperature. The cool nitrogen absorbs the kinetic and vibrational energy of the ions.

As obvious as this seems now, this cooling effect was first reported by Loboda et al.[108] in 2000 with their development of a new MALDI source for the orthogonal injection quadrupole time-of-flight (QoTOF) mass spectrometer. This instrument increased the pressure in the MALDI source region up to 10 torr (13 mbar), collected the desorbed ions in a quadrupole, and focused them down into the mass analyzer. The results by Loboda et al. (Figure 6.12) demonstrated that by cooling the ions on desorption at high pressure, the myoglobin ions desorbed from the "hot" matrix, α-cyano-4-hydroxy cinnamic acid, were actually cooler (as evidenced by reduced "small molecule" losses on the low mass side of the main peak) than these same ions generated from the "cool" matrix, 2,5-dihydroxy benzoic acid at 10 mtorr (13×10^{-3} mbar). Thus, high-pressure MALDI with cooling gas pressures in the 10-mbar range had the potential for generating stable molecular ions from a wide variety of labile molecules.

"High-pressure MALDI" is unfortunately a rather vague term and usually requires definition of the pressure regime being used. Figure 6.13 shows a sketch of the energy regimes involved in the various types of MALDI sources in use today. On the left of the figure, at high vacuum, desorbed ions are not cooled, resulting in ions with relatively high initial velocities (~600 m/s)[109,110] and high vibrational activation.[111] On the right are the newly developed atmospheric pressure[112] and pneumatically assisted[113] MALDI sources in which the ions are formed under a high-pressure environment (typically ~1 bar), cooled to

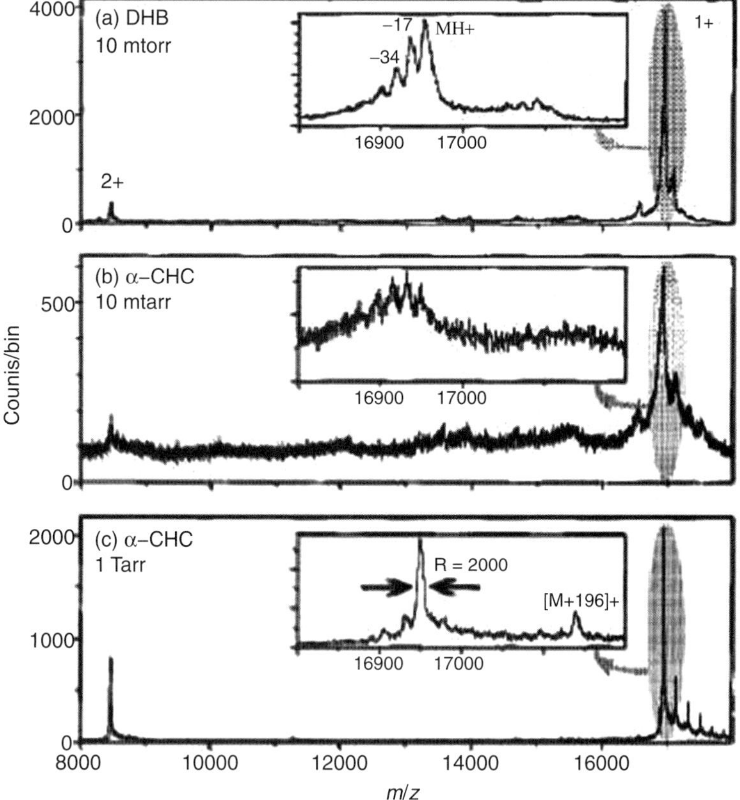

Figure 6.12. The first exploration of MALDI at elevated pressure showed reduced fragmentation of myoglobin and increased matrix adduction. (Reproduced from Ref. 121, with permission.)

room temperature in nanoseconds, and pulled down into the high-vacuum region of the mass spectrometer with a differentially pumped interface similar to that used in an electrospray ion source. In the middle of the graph are the two distinct regions in which several newly developed MALDI sources operate. For convenience sake, these are denoted the "intermediate-pressure"[114,115] and "high-pressure"[108,113,116–118] regions (the latter has been renamed "vibrational cooling" MALDI to remove some of the ambiguity). The intermediate-pressure region is used by commercially available MALDI-QTOF and

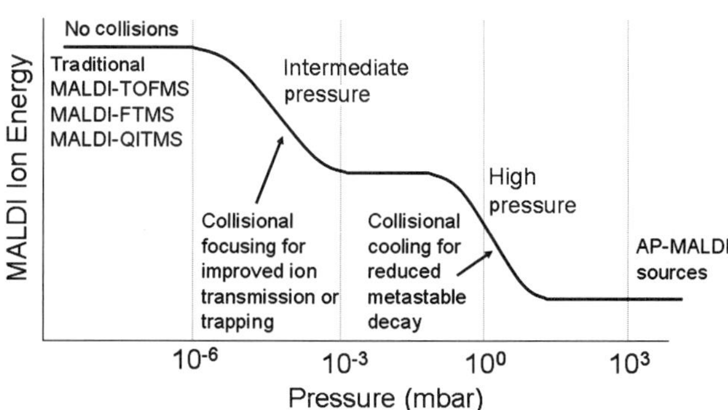

Figure 6.13. A sketch of the relevant pressure regimes in MALDI mass spectrometry.

MALDI-FT mass spectrometers, and it is convenient for removing the initial kinetic energy of the ions as well as collisionally focusing the ions in a multipole ion guide for improved ion transmission and sensitivity. Ion sources operating in the high-pressure region are now becoming commercially available.

6.7.2 Minimizing Metastable Decay

The advantage of operation in the high-pressure region is well illustrated by the ability to generate intact MALDI molecular ions from polysialylated gangliosides,[84,85,117] a task that had only previously been accomplished using MALDI on a linear TOF instrument. Gangliosides are important in cell surface interactions, and their structural variations play important roles in nervous system development and carcinogenesis, for example. In natural samples, they are usually composed of heterogeneous mixtures of very similar glycolipids typically varying only in the number and position of sialic acid residues and the compositions of their lipid-rich ceramide moieties. Figure 6.14 shows two gangliosides, G_{D1a} and

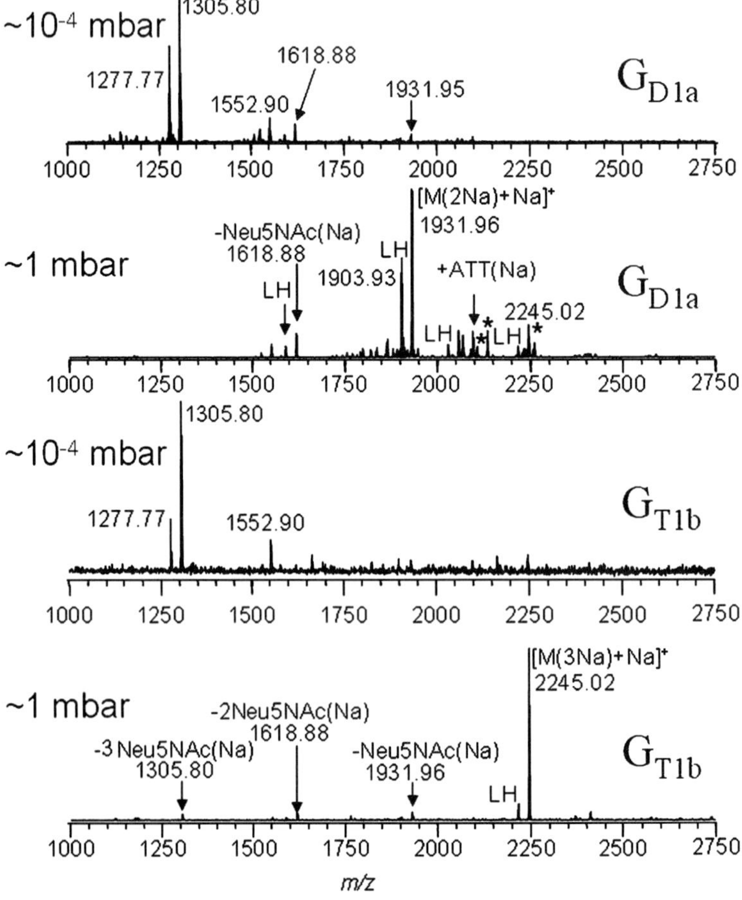

Figure 6.14. The vibrational cooling effect of MALDI at elevated pressures. (Reproduced from Ref. 117, with permission.)

G_{T1b}, having two and three sialic acid units, respectively. The ion pattern generated from desorbing the gangliosides shows very different ions depending on the background pressure during desorption. If MALDI occurs at $\sim 10^{-4}$ mbar, the most abundant peaks in the spectra are the two asialo peaks at *m/z* 1278 and 1306 (differing by C_2H_4 in the lipid) reflecting the loss of all sialic acid units, but when desorption occurs at $\sim$1 mbar, the ions' vibrational motion is stabilized and the loss of the sialic acid units is prevented. Even more interesting is the appearance of a previously undetected G_{T1} contaminant (*m/z* 2245.02) in the G_{D1a} sample; while its presence is not surprising biochemically, it was not observed when the sample was ionized at 10^{-4} mbar.

Cooling the vibrational energy of the desorbed ions also aids in generation of intact high-molecular-weight ions. Figure 6.15 shows a series of proteins from 8.5 kDa to 39 kDa that are ionized using a high-pressure MALDI source on a Fourier transform mass spectrometer (FTMS).[116,117] Although to date, isotopic resolving power has been achieved only up to myoglobin ($\sim$17.5 kDa), clearly the ion source can generate and stabilize these high-molecular-weight species. An additional advantage is that the increased multiple charging observed in the high-pressure MALDI-FTMS instrument[116] shifts the MW-related peaks into a spectral region where FTICR performance is improved. Furthermore, the Manitoba group recently reported singly charged ion signals generated from proteins exceeding 150 kDa on their high-pressure MALDI-QoTOFMS.[119]

6.7.3 The Matrix Adduction Problem

Figures 6.14 and 6.15, however, clearly demonstrate one of the limiting factors of high-pressure MALDI as well. As the desorbed ions are stabilized, the number of noncovalent matrix adducts increases. The prevailing picture of MALDI is that ions are blasted from the (usually) crystalline matrix in chunks that shake themselves apart because of residual

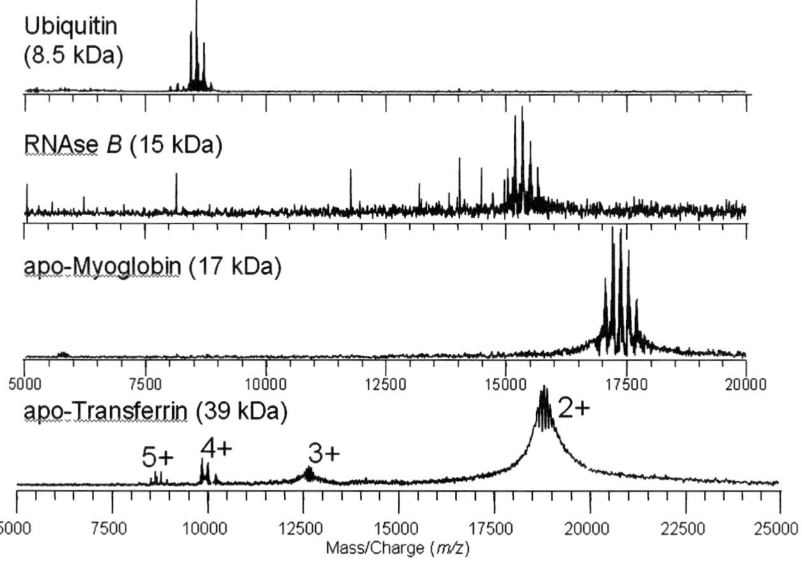

Figure 6.15. Vibrationally cooled MALDI-FTMS of proteins. (Reproduced with permission from reference 43).

vibrational energy in the clusters, according to the so-called "spallation" model.[120] Therefore, cooling the clusters before they can dissociate is a double-edged sword with which one can prevent labile covalent bonds from dissociating, but simultaneously one also prevents strong noncovalent bonds from dissociating. Cooling the desorbed matrix clusters also leads to the stabilization of matrix clusters which can increase the "chemical noise" of the high-pressure MALDI instruments.[65] The next stages of development in this technology clearly need to address this limitation.

The methods for solving the matrix adduction problem require either preventing the formation of the adducts or developing techniques to remove them without dissociating the covalent bonds. The Manitoba group has had some success with this by heating the front ion optics of the system, warming the cooled clusters up with black-body radiation to cause the matrix adducts to drop off.[121] They have also shown recently that the extent of matrix adduction is a strong function of laser fluence and sample preparation technique. In addition, the implementation of high-pressure MALDI on an FTMS also allows use of low-energy collisional activation of the matrix adducted ions in order to detach the matrix molecules.[116] Other possible methods for minimizing the presence of matrix adducts in the spectra include various forms of collisional activation or infrared multiphoton dissociation from a CO_2 laser. These techniques are currently being investigated.

The ability to desorb and cool ions even from such a hot matrix as α-cyano-4-hydroxycinnamic acid suggests that another approach may also work. While only a half dozen matrix preparation methods are routinely used in MALDI, all of these techniques have been developed initially for MALDI-TOF instruments. Since MALDI-TOFMS requires very flat desorption surfaces for the best resolving power, many potential matrices have not been explored. However, systems where the external ion source is decoupled from the mass analyzer, as is the case with MALDI-QTOF, MALDI-QIT, and MALDI-FTMS instruments, do not have this limitation. Indeed, some of the best signals from MALDI-FTMS systems are generated from large, irregular DHB crystals. Thus, there is a high probability that exploration of matrices with the decoupled MALDI sources will identify new matrices with favorable properties, such as improved ion abundance or reduced matrix adduction.

Additionally, one promising technology with regard to minimization of matrix clustering is the use of porous silicon wafers as a laser desorption surface.[122] The etched silicon oxide surface is reported to be able to generate ions that are considerably cooler than the same ions desorbed from a steel surface, particularly if an infrared Er:YAG (2.94 μm) laser is used. Because this technique uses no matrix, there is a strong chance that the results equivalent to those shown in Figures 6.14 and 6.15 can be generated on a high-pressure MALDI source, but without the matrix adducts. This concept has not yet been tested, however.

Although electrospray ionization already produces extremely cool ions, developing methods for cooling MALDI ions has certain potential advantages. While electrospray can, indeed, generate vibrationally cool ions, even to the point that noncovalently adducted species may be detected, ESI produces the highest ion signals from "clean" solutions, due to significant suppression effects from other species when they are present. Conversely, MALDI analysis is less affected when samples are relatively highly contaminated with salts and other small molecules. Even more importantly, MALDI ionizes samples from surfaces. Thus, two-dimensional surface analysis is feasible; and, provided that the surface is chemically stable, the surfaces can be stored and re-analyzed minutes (or months!) later. Furthermore, MALDI mass spectrometry is relatively straightforward to scale up to high-throughput operation.

6.7.4 Ion–Molecule Reactions

Witt et al.[115] reported H/D exchange reactions at $\sim 8.3 \times 10^{-6}$ mbar in their intermediate pressure MALDI source. These ion–molecule reactions on MALDI ions were used to determine surface accessibility and kinetics of exchangeable hydrogen atoms on peptide ions in the gas phase without disturbing the high vacuum of the FTICR cell. The possibilities demonstrated by this experiment are exciting. One could consider a large number of gas-phase chemical reactions that could be undertaken while still maintaining high performance in the mass analyzer. At the very least, the matrix clustering problem could be investigated by H/D exchange with the hope of determining structure of the matrix clusters.

6.7.5 High-Pressure MALDI and Ion Mobility/oTOF Spectrometry

Gillig et al.[123] have incorporated a MALDI ion source into a high pressure ($\sim$10 mbar) ion mobility spectrometer that is also coupled to an orthogonal TOF instrument. Although this instrument has not yet been applied to the analysis of fragile molecules, the source appears to meet the pressure requirements for collisional cooling. This system has the potential advantage of using ion mobility to separate conformational isomers while using the TOF to orthogonally generate *m/z* values for the desorbed ions. Since the ion mobility spectrometer can be used as a low-resolution mass spectrometer, this intriguing instrument can potentially also generate tandem mass spectrometry data.

6.7.6 AP-MALDI Versus HP-MALDI

The atmospheric pressure MALDI sources (Figure 6.16)[112,113,124] have many of the same advantages as the high pressure MALDI sources.[108,116,117] The expected trade-off between the two source geometries is likely to be one of sensitivity versus extent of collisional cooling; however, neither source geometry has yet been extensively tested with respect to either parameter. Early indications, however, seem to point in the direction of better collisional cooling but lower sensitivity for the atmospheric pressure sources. A prototype intermediate-pressure MALDI source demonstrated 10-attomole sensitivity,[114] compared

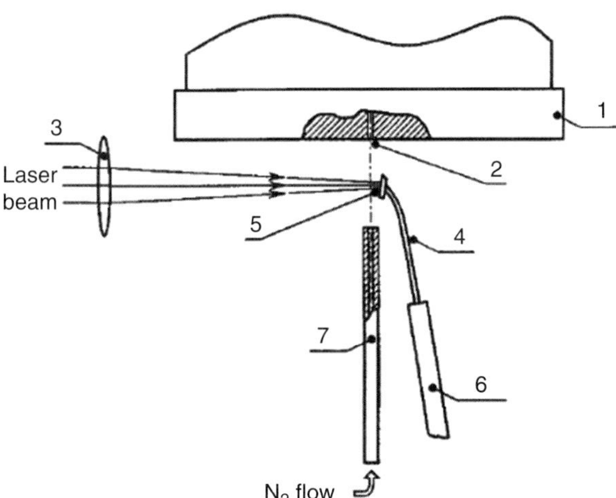

Figure 6.16. First pneumatically assisted atmospheric pressure MALDI source design. (Reproduced from Ref. 113, with permission.)

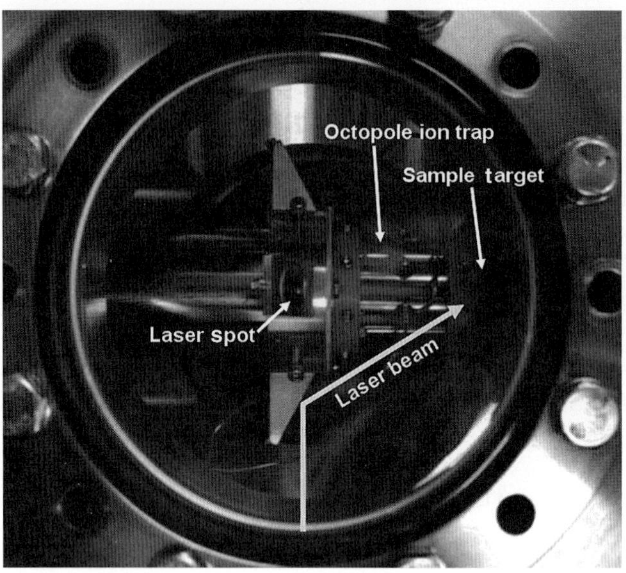

Figure 6.17. The plasma that develops if pressure goes too high in vibrationally cooled MALDI sources (See color insert).

to the initially reported picomole sensitivities of the first atmospheric-pressure MALDI source.[113] However, the atmospheric-pressure MALDI sources have already demonstrated intact desorption of noncovalent complexes and the ability to evaporate matrix adducts from the parent ion in the differential pumping stages,[113] both of which are still tasks to be accomplished in "high-pressure" MALDI sources, although early indications suggest methods for solving this problem. The AP MALDI sources appear to be much better at removing matrix adducts than their HP MALDI counterparts, presumably because they use existing electrospray desolvation inlets.

While the HP MALDI sources currently appear to have sensitivity advantages over the AP MALDI sources, they are not without their problems. With the current ion optics design in the high-pressure MALDI sources, strict control of the pressure applied is necessary to prevent arcing and formation of a plasma in the octopole. At the very least, such a plasma eliminates ion signal (presumably by a space-charge-related mechanism), and in some cases it can even destroy the electronics used to provide the trapping potentials. Thus, new methods that will focus a broad range of molecular ions at subatmospheric pressures could greatly improve the performance of these instruments. Interesting technologies that could potentially be used include the aerodynamic lens[125] and the ion funnel.[126]

While the real trade-off between HP and AP MALDI modes is not yet clear, the future significant role of MALDI at elevated pressures is unambiguous. The ability to desorb labile ions such as sialylated glycolipids, as well as ions with labile side groups such as glycopeptides and multiply phosphorylated peptides, greatly simplifies the analysis of these fragile molecules. Although the comparatively simple proteomics problem of identifying and quantifying proteins by MALDI mass spectrometry is not substantially affected, high-pressure MALDI is clearly an important tool for determination of the structures of post-translational modifications. Additionally, the ability to cool MALDI ions provides the possibility of studying noncovalent interactions, particularly with respect to relatively strong drug–protein interactions. Furthermore, the burgeoning field of glycomics crucially needs the ability to ionize labile species intact. Thus, the new technology of high-pressure MALDI is, indeed, very promising.

REFERENCES

1. Conzemius, R. J.; Capellen, J. M. A review of the applications to solids of the laser ion source in mass spectrometry. *Int. J. Mass Spectrom. Ion Processes* **1980**, *34*, 197–271.
2. Darke, S. A.; Tyson, J. F. Interaction of laser-radiation with solid materials and its significance to analytical spectrometry—A review. *J. Anal. At. Spectrom.* **1993**, *8*, 145–209.
3. Radziemski, L. J. Review of selected analytical applications of laser plasmas and laser-ablation, 1987–1994. *Microchem. J.* **1994**, *50*, 218–234.
4. Campbell, E. E. B.; Ulmer, G.; Hasselberger, B.; Busmann, H. -G.; Hertel, I. V. TOF laser ablation/fullerenes. *J. Chem. Phys.* **1990**, *93*, 6300–6307.
5. So, H.; Wilkins, C. L. First observation of carbon aggregate ions > C600 + by laser desorption FTMS. *J. Phys. Chem.* **1989**, *93*, 1187–1189.
6. O'Connor, P. B.; Heeren, R. M. A.Unpublished data, 1997.
7. Fountain, S. T.; Lubman, D. M. Wavelength-specific resonance-enhanced multiphoton ionization for isomer discrimination via fragmentation and metastable analysis. *Anal. Chem.* **1993**, *65*, 1257–1266.
8. Lustig, D. A.; Lubman, D. M. Selective resonance enhanced multiphoton ionization of aromatic polymers in supersonic beam mass-spectrometry. *Int. J. Mass Spectrom. Ion Processes* **1991**, *107*, 265–280.
9. Li, L.; Lubman, D. M. Resonance-enhanced multiphoton ionization jet spectroscopy and mass-spectrometry of tyrosine-containing dipeptides using a pulsed laser desorption-volatilization method. *Applied Spectroscopy* **1989**, *43*, 543–549.
10. Li, L.; Lubman, D. M. Resonance enhanced multiphoton ionization of nucleosides by using pulsed-laser desorption in supersonic beam mass-spectrometry. *Int. J. Mass Spectrom. Ion Processes* **1989**, *88*, 197–210.
11. Tembreull, R.; Sin, C. H.; Pang, H. M.; Lubman, D. M. Resonance enhanced multiphoton ionization spectroscopy for detection of azabenzenes in supersonic beam mass-spectrometry. *Anal. Chem.* **1985**, *57*, 2911–2917.
12. Lubman, D. M.; Kronick, M. N. Resonance-enhanced 2-photon ionization spectroscopy in plasma chromatography. *Anal. Chem.* **1983**, *55*, 1486–1492.
13. Morrical, B. D.; Fergenson, D. P.; Prather, K. A. Coupling two-step laser desorption/ionization with aerosol time-of-flight mass spectrometry for the analysis of individual organic particles. *J. Am. Soc. Mass Spectrom.* **1998**, *9*, 1068–1073.
14. Wood, S. H.; Prather, K. A. Time-of-flight mass spectrometry methods for real time analysis of individual aerosol particles. *Trends Anal. Chem.* **1998**, *17*, 346–356.
15. Gard, E.; Mayer, J. E.; Morrical, B. D.; Dienes, T.; Fergenson, D. P.; Prather, K. A. Real-time analysis of individual atmospheric aerosol particles: Design and performance of a portable ATOFMS. *Anal. Chem.* **1997**, *69*, 4083–4091.
16. Nordmeyer, T.; Prather, K. A. Real-time measurement capabilities using aerosol time-of-flight mass-spectrometry. *Anal. Chem.* **1994**, *66*, 3540–3542.
17. Prather, K. A.; Nordmeyer, T.; Salt, K. Real-time characterization of individual aerosol-particles using time-of-flight mass-spectrometry. *Anal. Chem.* **1994**, *66*, 1403–1407.
18. Liu, D. Y.; Rutherford, D.; Kinsey, M.; Prather, K. A. Real-time monitoring of pyrotechnically derived aerosol particles in the troposphere. *Anal. Chem.* **1997**, *69*, 1808–1814.
19. Horowitz, P.; Hill, W. *The Art of Electronics*, Cambridge University Press, Cambridge, 1989.
20. O'Connor, P. B.; Hillenkamp, F.In: Hillenkamp, F., Peter-Katalinic, J. (Eds.), *MALDI-MS*, Wiley-VCH Verlag GmbH & Co. KGaA, Weinheim, Germany, 2007, pp. 29–82.
21. Makarov, A. Electrostatic axially harmonic orbital trapping: A high-performance technique of mass analysis. *Anal. Chem.* **2000**, *72*, 1156–1162.
22. Mamyrin, B. A.; Karataev, V. I.; Schmikk, D. B.; Zagulin, B. A. The mass-reflectron, a new nonmagnetic time-of-flight mass spectrometer with high resolution. *Sov. Phys. JETP* **1973**, *37*, 45–48.
23. Ivanov, M. A.; Kozlov, B. N.; Mamyrin, B. A.; Shmikk, D. V.; Shchebelin, V. G. Mass-reflectron for the study of laser irradiation interaction processes with molecules in an ultrasonic gas-jet. *Zh. Tekh. Fiz.* **1983**, *53*, 2039–2044.
24. Niehuis, E. In Benninghoven, A. (Ed.), *Secondary Ion Mass Spectrometry (SIMS VII)*, Wiley-Interscience, Chichester, UK, 1989, pp. 299–304.
25. Bergmann, T.; Martin, T. P.; Schaber, H. High-resolution TOF mass spectrometer. *Rev. Sci. Instrum.* **1989**, *60*, 792–793.
26. Cornish, T. J.; Cotter, R. J. A curved-field reflectron for improved energy focusing of product ions in time-of-flight mass-spectrometry. *Rapid Commun. Mass Spectrom.* **1993**, *7*, 1037–1040.
27. Wiley, W. C.; McLaren, I. H. Time-of-flight mass spectrometer with improved resolution. *Rev. Sci. Inst.* **1955**, *29*, 1150–1157.
28. Takach, E. J.; Hines, W. M.; Patterson, D. H.; Juhasz, P.; Falick, A. M.; Vestal, M. L.; Martin, S. A. Accurate mass measurements using MALDI-TOF with delayed extraction. *J. Protein Chem.* **1997**, *16*, 363–369.
29. Gusev, A. I.; Proctor, A.; Rabinovich, Y. I.; Hercules, D. M. Thin-layer chromatography combined with matrix-assisted laser desorption ionization mass spectrometry. *Anal. Chem.* **1995**, *67*, 1805–1814.

30. Gusev, A. I.; Vasseur, O. J.; Proctor, A.; Sharkey, A. G.; Hercules, D. M. Imaging of thin-layer chromatograms using matrix/assisted laser desorption/ionization mass spectrometry. *Anal. Chem.* **1995**, *67*, 4565–4570.
31. Loo, R. R. O.; Stevenson, T. I.; Mitchell, C.; Loo, J. A.; Andrews, P. C. Mass spectrometry of proteins directly from polyacrylamide gels. *Anal. Chem.* **1996**, *68*, 1910–1917.
32. Spengler, B.; Kirsch, D.; Kaufmann, R.; Lemoine, J. Structure analysis of branched oligosaccharides using post-source decay in matrix-assisted laser desorption ionization mass spectrometry. *Org. Mass Spectrom.* **1994**, *29*, 782–787.
33. Cordero, M. M.; Cornish, T. J.; Cotter, R. J.; Lys, I. A. Sequencing peptides without scanning the reflectron—post-source decay with a curved-field reflectron time-of-flight mass-spectrometer. *Rapid Commun. Mass Spectrom.* **1995**, *9*, 1356–1361.
34. Kaufmann, R.; Chaurand, P.; Kirsch, D.; Spengler, B. Post-source decay and delayed extraction in matrix-assisted laser desorption/ionization/reflectron time-of-flight mass spectrometry—Are there trade-offs? *Rapid Commun. Mass Spectrom.* **1996**, *10*, 1199–1208.
35. Hercules, D. M.; Day, R. J.; Balasanmugam, K.; Dang, T. A.; Li, L. P. Laser microprobe mass analysis. *Anal. Chem.* **1982**, *54*, 280A–305A.
36. Van Vaeck, L.; Van Roy, W.; Struyf, H.; Adams, F.; Caravatti, P. Development of a laser microprobe Fourier transform mass spectrometer with external ion source. *Rapid Commun. Mass Spectrom.* **1993**, *7*, 323–331.
37. Hillenkamp, F.; Unsoeld, E.; Kaufmann, R.; Nitsche, R. Laser microprobe mass analysis of organic materials. *Nature (London, United Kingdom)* **1975**, *256*, 119–120.
38. Feigl, P.; Schueler, B.; Hillenkamp, F. Lamma-1000, a new instrument for bulk microprobe mass analysis by pulsed laser irradiation. *Int. J. Mass Spectrom. Ion Processes* **1983**, *47*, 15–18.
39. Guest, W. H. Recent developments of laser microprobe mass analyzers, lamma-500 and lamma-1000. *Int. J. Mass Spectrom. Ion Processes* **1984**, *60*, 189–199.
40. Verbueken, A. H.; Bruynseels, F. J.; Vangrieken, R. E. Laser microprobe mass analysis—A review of applications in the life sciences. *Biomed. Mass Spectrom.* **1985**, *12*, 438–463.
41. Hercules, D. M. Laser microprobe mass-spectrometry—The past, present, and future. *Microchem. J.* **1988**, *38*, 3–23.
42. Karas, M.; Bachmann, D.; Hillenkamp, F. Influence of the wavelength in high-irradiance ultraviolet laser desorption mass spectrometry of organic molecules. *Anal. Chem.* **1985**, *57*, 2935–2939.
43. Karas, M.; Bachmann, D.; Bahr, U.; Hillenkamp, F. Matrix-assisted ultraviolet laser desorption of non-volatile compounds. *Int. J. Mass Spectrom. Ion Processes* **1987**, *78*, 53–68.
44. Karas, M.; Hillenkamp, F. Laser desorption ionization of proteins with molecular masses exceeding 10,000 daltons. *Anal. Chem.* **1988**, *60*, 2299–2301.
45. Karas, M.; Bahr, U.; Hillenkamp, F. UV laser matrix desorption/ionization mass spectrometry of proteins in the 100000 dalton range. *Int. J. Mass Spectrom. Ion Processes* **1989**, *92*, 231–242.
46. Karas, M.; Bahr, U.; Ingendoh, A.; Hillenkamp, F. Laser desorption ionization mass-spectrometry of proteins of mass 100 000 to 250 000 dalton. *Angew. Chem. (International Edition in English)* **1989**, *28*, 760–761.
47. Tanaka, K.; Waki, H.; Ido, Y.; Akita, S.; Yoshida, Y.; Yoshida, T. Protein and polymer analyses up to *m/z* 100000 by laser ionization time-of-flight mass spectrometry. *Rapid Commun. Mass Spectrom.* **1988**, *2*, 151–153.
48. Strupat, K.; Karas, M.; Hillenkamp, F. 2,5-Dihydroxybenzoic acid: A new matrix for laser desorption-ionization mass spectrometry. *Int. J. Mass Spectrom. Ion Processes* **1991**, *111*, 89–102.
49. Karas, M.; Bahr, U.; Ingendoh, A.; Nordhoff, E.; Stahl, B.; Strupat, K.; Hillenkamp, F. Principles and applications of matrix-assisted UV laser desorption ionization mass-spectrometry. *Anal. Chim. Acta* **1990**, *241*, 175–185.
50. Karas, M.; Bahr, U.; Strupat, K.; Hillenkamp, F.; Tsarbopoulos, A.; Pramanik, B. N. Matrix dependence of metastable fragmentation of glycoproteins in MALDI TOF mass spectrometry. *Anal. Chem.* **1995**, *67*, 675–679.
51. Hillenkamp, F.; Karas, M.; Beavis, R. C.; Chait, B. T. Matrix-assisted laser desorption/ionization mass spectrometry of biopolymers. *Anal. Chem.* **1991**, *63*, 1193A–1203A.
52. Hillenkamp, F.; Karas, M. Mass spectrometry of peptides and proteins by matrix-assisted ultraviolet laser desorption/ionization. *Methods Enzymol.* **1990**, *193*, 280–295.
53. Tholey, A.; Heinzle, E. Ionic (liquid) matrices for matrix-assisted laser desorption/ionization mass spectrometry—Applications and perspectives. *Anal. Bioanal. Chem.* **2006**, *386*, 24–37.
54. Xu, S. Y.; Ye, M. L.; Xu, D. K.; Li, X.; Pan, C. S.; Zou, H. F. Matrix with high salt tolerance for the analysis of peptide and protein samples by desorption/ionization time-of-flight mass spectrometry. *Anal. Chem.* **2006**, *78*, 2593–2599.
55. Erra-Balsells, R.; Nonami, H. UV-matrix-assisted laser desorption/ionization time-of-flight mass spectrometry analysis of synthetic polymers by using nor-harmane as matrix. *Arkivoc* **2003**, 517–537.
56. Mirza, S. P.; Raju, N. P.; Vairamani, M. Estimation of the proton affinity values of fifteen matrix-assisted laser desorption/ionization matrices under electrospray ionization conditions using the kinetic method. *J. Am. Soc. Mass Spectrom.* **2004**, *15*, 431–435.

57. Eskinja, M.; Zollner, P.; Schmid, E. R. Determination of mercapturic acids using 1,4-dihydroxynaphthalene, a new matrix for matrix-assisted UV laser desorption/ionization mass spectrometry. *Eur. Mass Spectrom.* **1998**, *4*, 157–162.
58. Xu, N. X.; Huang, Z. H.; Watson, J. T.; Gage, D. A. Mercaptobenzothiazoles: A new class of matrices for laser desorption ionization mass spectrometry. *J. Am. Soc. Mass Spectrom.* **1997**, *8*, 116–124.
59. Krause, J.; Stoeckli, M.; Schlunegger, U. P. Studies on the selection of new matrices for ultraviolet matrix-assisted laser desorption/ionization time-of-flight mass spectrometry. *Rapid Commun. Mass Spectrom.* **1996**, *10*, 1927–1933.
60. Zollner, P.; Schmid, E. R.; Allmaier, G. K-4[Fe(CN)-(6)]/glycerol—A new liquid matrix system for matrix-assisted laser desorption/ionization mass spectrometry of hydrophobic compounds. *Rapid Commun. Mass Spectrom.* **1996**, *10*, 1278–1282.
61. Bai, J.; Liang, X.; Liu, Y. H.; Zhu, Y.; Lubman, D. M. Characterization of two new matrices for matrix-assisted laser desorption/ionization mass spectrometry. *Rapid Commun. Mass Spectrom.* **1996**, *10*, 839–844.
62. Tang, X. D.; Dreifuss, P. A.; Vertes, A. New matrices and accelerating voltage effects in matrix-assisted laser-desorption ionization of synthetic-polymers. *Rapid Commun. Mass Spectrom.* **1995**, *9*, 1141–1147.
63. Metzger, J. O.; Woisch, R.; Tuszynski, W.; Angermann, R. New-type of matrix for matrix-assisted laser-desorption mass-spectrometry of polysaccharides and proteins. *Fresenius J. Anal. Chem.* **1994**, *349*, 473–474.
64. Kocher, T.; Engstrom, A.; Zubarev, R. A. Fragmentation of peptides in MALDI in-source decay mediated by hydrogen radicals. *Anal. Chem.* **2005**, *77*, 172–177.
65. Krutchinsky, A. N.; Chait, B. T. On the nature of the chemical noise in MALDI mass spectra. *J. Am. Soc. Mass Spectrom.* **2002**, *13*, 129–134.
66. Dreisewerd, K.; Muthing, J.; Rohlfing, A.; Meisen, I.; Vukelic, Z.; Peter-Katalinic, J.; Hillenkamp, F.; Berkenkamp, S. Analysis of gangliosides directly from thin-layer chromatography plates by infrared matrix-assisted laser desorption/ionization orthogonal time-of-flight mass spectrometry with a glycerol matrix. *Anal. Chem.* **2005**, *77*, 4098–4107.
67. Dreisewerd, K. The desorption process in MALDI. *Chem. Rev.* **2003**, *103*, 395–425.
68. Dreisewerd, K.; Berkenkamp, S.; Leisner, A.; Rohlfing, A.; Menzel, C. Fundamentals of matrix-assisted laser desorption/ionization mass spectrometry with pulsed infrared lasers. *Int. J. Mass Spectrom. Ion Processes* **2003**, *226*, 189–209.
69. Berkenkamp, S.; Menzel, C.; Hillenkamp, F.; Dreisewerd, K. Measurements of mean initial velocities of analyte and matrix ions in infrared matrix-assisted laser desorption ionization mass spectrometry. *J. Am. Soc. Mass Spectrom.* **2002**, *13*, 209–220.
70. Menzel, C.; Dreisewerd, K.; Berkenkamp, S.; Hillenkamp, F. Mechanisms of energy deposition in infrared matrix-assisted laser desorption/ionization mass spectrometry. *Int. J. Mass Spectrom. Ion Processes* **2001**, *207*, 73–96.
71. Budnik, B. A.; Jensen, K. B.; Jorgensen, T. J. D.; Haase, A.; Zubarev, R. A. Benefits of 2.94 mu m infrared matrix-assisted laser desorption/ionization for analysis of labile molecules by Fourier transform mass spectrometry. *Rapid Commun. Mass Spectrom.* **2000**, *14*, 578–584.
72. Talrose, V. L.; Person, M. D.; Whittal, R. M.; Walls, F. C.; Burlingame, A. L.; Baldwin, M. A. Insight into absorption of radiation/energy transfer in infrared matrix-assisted laser desorption/ionization: The roles of matrices, water and metal substrates. *Rapid Commun. Mass Spectrom.* **1999**, *13*, 2191–2198.
73. Berkenkamp, S.; Menzel, C.; Karas, M.; Hillenkamp, F. Performance of infrared matrix-assisted laser desorption/ionization mass spectrometry with lasers emitting in the 3-μm wavelength range. *Rapid Commun. Mass Spectrom.* **1997**, *11*, 1399–1406.
74. Cramer, R.; Hillenkamp, F.; Haglund, R. F. Infrared matrix-assisted laser desorption and ionization by using a tunable mid-infrared free-electron laser. *J. Am. Soc. Mass Spectrom.* **1996**, *7*, 1187–1193.
75. Nordhoff, E.; Ingendoh, A.; Cramer, R.; Overberg, A.; Stahl, B.; Karas, M.; Hillenkamp, F.; Crain, P. F. Matrix-assisted laser desorption/ionization mass spectrometry of nucleic acids with wavelengths in the ultraviolet and infrared. *Rapid Commun. Mass Spectrom.* **1992**, *6*, 771–776.
76. Berkenkamp, S.; Kirpekar, F.; Hillenkamp, F. Infrared MALDI mass spectrometry of large nucleic acids. *Science* **1998**, *281*, 260–262.
77. Preisler, J.; Foret, F.; Karger, B. L. Online MALDI-TOF MS using a continuous vacuum deposition interface. *Anal. Chem.* **1998**, *70*, 5278–5287.
78. Preisler, J.; Hu, P.; Rejtar, T.; Karger, B. L. Capillary electrophoresis-matrix-assisted laser desorption/ionization time-of-flight mass spectrometry using a vacuum deposition interface. *Anal. Chem.* **2000**, *72*, 4785–4795.
79. Rejtar, T.; Hu, P.; Juhasz, P.; Campbell, J. M.; Vestal, M. L.; Preisler, J.; Karger, B. L. Off-line coupling of high-resolution capillary electrophoresis to MALDI-TOF and TOF/TOF MS. *J. Proteome Res.* **2002**, *1*, 171–179.
80. Ericson, C.; Phung, Q. T.; Horn, D. M.; Peters, E. C.; Fitchett, J. R.; Ficarro, S. B.; Saloma, A. R.; Brill, L. M.; Brock, A. An automated noncontact deposition interface for liquid chromatography matrix-assisted laser desorption/ionization mass spectrometry. *Anal. Chem.* **2003**, *75*, 2309–2315.
81. Nicola, A. J.; Gusev, A. I.; Hercules, D. M. Direct quantitative analysis from thin-layer chromatography plates using matrix-assisted laser desorption/ionization

mass spectrometry. *Appl. Spectrosc.* **1996**, *50*, 1479–1482.

82. Mehl, J. T.; Gusev, A. I.; Hercules, D. M. Coupling protocol for thin layer chromatography matrix-assisted laser desorption ionization. *Chromatographia* **1997**, *46*, 358–364.

83. Brown, R. S.; Lennon, J. J. Mass resolution improvement by incorporation of pulsed ion extraction in a matrix-assisted laser desorption/ionization linear time-of-flight mass spectrometer. *Anal. Chem.* **1995**, *67*, 1998–2003.

84. Ivleva, V. B.; Sapp, L. M.; O'Connor, P. B.; Costello, C. E. Ganglioside analysis by thin-layer chromatography matrix-assisted laser desorption/ionization orthogonal time-of-flight mass spectrometry. *J. Am. Soc. Mass Spectrom.* **2005**, *16*, 1552–1560.

85. Ivleva, V. B.; Elkin, Y. N.; Budnik, B. A.; Moyer, S. C.; O'Connor, P. B.; Costello, C. E. Coupling thin layer chromatography with vibrational cooling matrix assisted laser desorption/ionization Fourier transform mass spectrometry for the analysis of ganglioside mixtures. *Anal. Chem.* **2004**, *76*, 6484–6491.

86. Wei, J.; Burlak, J. M.; Siuzdak, G. Desorption–ionization mass spectrometry on porous silicon. *Nature* **1999**, *399*, 243–246.

87. Thomas, J. J.; Shen, Z. X.; Crowell, J. E.; Finn, M. G.; Siuzdak, G. Desorption/ionization on silicon (DIOS): A diverse mass spectrometry platform for protein characterization. *Proc. Nat. Acad. Sci. USA* **2001**, *98*, 4932–4937.

88. Lewis, W. G.; Shen, Z. X.; Finn, M. G.; Siuzdak, G. Desorption/ionization on silicon (DIOS) mass spectrometry: Background and applications. *Int. J. Mass Spectrom. Ion Processes* **2003**, *226*, 107–116.

89. Loo, R. R. O.; Mitchell, C.; Stevenson, T. I.; Martin, S. A.; Hines, W. M.; Juhasz, P.; Patterson, D. H.; Peltier, J. M.; Loo, J. A.; Andrews, P. C. Sensitivity and mass accuracy for proteins analyzed directly from polyacrylamide gels—Implications for proteome mapping. *Electrophoresis* **1997**, *18*, 382–390.

90. McComb, M. E.; Oleschuk, R. D.; Manley, D. M.; Donald, L.; Chow, A.; Oneil, J. D. J.; Ens, W.; Standing, K. G.; Perreault, H. Use of a non-porous polyurethane membrane as a sample support for matrix-assisted laser desorption ionization time-of-eight mass spectrometry of peptides and proteins. *Rapid Commun. Mass Spectrom.* **1997**, *11*, 1716–1722.

91. McComb, M. E.; Oleschuk, R. D.; Chow, A.; Ens, W.; Standing, K. G.; Perreault, H.; Smith, M. Characterization of hemoglobin variants by MALDI-TOF MS using a polyurethane membrane as the sample support. *Anal. Chem.* **1998**, *70*, 5142–5149.

92. McComb, M. E.; Oleschuk, R. D.; Chow, A.; Perreault, H.; Dworschak, R. G.; Znamirowski, M.; Ens, W.; Standing, K. G.; Preston, K. R. Application of nonporous polyurethane (PU) membranes and porous PU thin films as sample supports for MALDI-MS of wheat proteins. *Can. J. Chem.* **2001**, *79*, 437–447.

93. Bai, J.; Liu, Y. H.; Cain, T. C.; Lubman, D. M. Matrix-assisted laser desorption/ionization using an active perfluorosulfonated ionomer film substrate. *Anal. Chem.* **1994**, *66*, 3423–3430.

94. Hung, K. C.; Rashidzadeh, H.; Wang, Y.; Guo, B. Use of paraffin wax film in MALDI-TOF analysis of DNA. *Anal. Chem.* **1998**, *70*, 3088–3093.

95. Hung, K. C.; Ding, H.; Guo, B. Use of poly-(tetrafluoroethylene)s as a sample support for the MALDI-TOF analysis of DNA and proteins. *Anal. Chem.* **1999**, *71*, 518–521.

96. Liu, Y. H.; Bai, J.; Liang, X.; Lubman, D. M.; Venta, P. J. Use of a nitrocellulose film substrate in matrix-assisted laser desorption/ionization mass spectrometry for DNA mapping and screening. *Anal. Chem.* **1995**, *67*, 3482–3490.

97. Mock, K. K.; Sutton, C. W.; Cotrell, J. S. Sample immobilization protocols for matrix assisted laser desorption mass spectrometry. *Rapid Commun. Mass Spectrom.* **1992**, *6*, 233–238.

98. Walker, A. K.; Qiu, H.; Wu, Y.; Timmons, R. B.; Kinsel, G. R. Studies of peptide binding to allyl amine and vinyl acetic acid- modified polymers using matrix-assisted laser desorption/ionization mass spectrometry. *Anal. Biochem.* **1999**, *271*, 123–130.

99. Worrall, T. A.; Lin, S. H.; Cotter, R. J.; Woods, A. S. On-probe sample purification of lipids for matrix-assisted laser desorption/ionization time-of-flight mass spectrometry. *J. Mass Spectrom.* **2000**, *35*, 647–650.

100. Nordhoff, E.; Schurenberg, M.; Thiele, G.; Lubbert, C.; Kloeppel, K. D.; Theiss, D.; Lehrach, H.; Gobom, J. Sample preparation protocols for MALDI-MS of peptides and oligonucleotides using prestructured sample supports. *Int. J. Mass Spectrom. Ion Processes* **2003**, *226*, 163–180.

101. McLafferty, F. W.; Turecek, F. *Interpretation of Mass Spectra*, 4th ed., University Science Books, Mill Valley, CA, 1993.

102. Munson, M. S. B.; Field, F. H. Chemical ionization mass spectrometry. I. General introduction. *J. Am. Chem. Soc.* **1966**, *88*, 2621–2630.

103. Beckey, H. D. Field desorption mass spectrometry. *Int. J. Mass Spectrom. Ion Phys.* **1969**, *2*, 500.

104. Macfarlane, R. D.; Torgerson, D. F. 252Cf-plasma desorption time-of-flight mass spectrometry. *Int. J. Mass Spectrom. Ion Phys.* **1976**, *21*, 81–92.

105. Lattimer, R. P. Fast atom bombardment mass spectrometry of polyglycols. *Int. J. Mass Spectrom. Ion Processes* **1984**, *55*, 221–232.

106. Fenn, J. B.; Mann, M.; Meng, C. K.; Wong, S. F.; Whitehouse, C. M. Electrospray ionization for mass spectrometry of large biomolecules. *Science* **1989**, *246*, 64–71.

107. Jensen, O. N.; Podtelejnikov, A. V.; Mann, M. Identification of the components of simple protein

mixtures by high accuracy peptide mass mapping and database searching. *Anal. Chem.* **1997**, *69*, 4741–4750.

108. Loboda, A. V.; Ackloo, S.; Chernushevich, I. V. A high-performance matrix-assisted laser desorption/ionization orthogonal time-of-flight mass spectrometer with collisional cooling. *Rapid Commun. Mass Spectrom.* **2003**, *17*, 2508–2516.
109. Beavis, R. C.; Chait, B. T. Velocity distribution of intact high mass polypeptide molecule ions produced by matrix-assisted laser desorption. *Chem. Phys. Lett.* **1991**, *181*, 479–484.
110. Pan, Y.; Cotter, R. J. Measurement of initial translational energies of peptide ions in laser desorption/ionization mass spectrometry. *Org. Mass Spectrom.* **1992**, *27*, 3–8.
111. Spengler, B. Post-source decay analysis in matrix-assisted laser desorption/ionization mass spectrometry of biomolecules. *J. Mass Spectrom.* **1997**, *32*, 1019–1036.
112. Laiko, V. V.; Moyer, S. C.; Cotter, R. J. Atmospheric pressure MALDI/ion trap mass spectrometry. *Anal. Chem.* **2000**, *72*, 5239–5243.
113. Laiko, V. V.; Baldwin, M. A.; Burlingame, A. L. Atmospheric pressure matrix-assisted laser desorption/ionization mass spectrometry. *Anal. Chem.* **2000**, *72*, 652–657.
114. Baykut, G.; Jertz, R.; Witt, M. Matrix-assisted laser desorption/ionization Fourier transform ion cyclotron resonance mass spectrometry with pulsed in-source collision gas and in-source ion accumulation. *Rapid Commun. Mass Spectrom.* **2000**, *14*, 1238–1247.
115. Witt, M.; Fuchser, J.; Baykut, G. In-source H/D exchange and ion-molecule reactions using matrix assisted laser desorption/ionization Fourier transform ion cyclotron resonance mass spectrometry with pulsed collision and reaction gases. *J. Am. Soc. Mass Spectrom.* **2002**, *13*, 308–317.
116. O'Connor, P. B.; Costello, C. E. A high pressure matrix-assisted laser desorption/ionization Fourier transform mass spectrometry ion source for thermal stabilization of labile biomolecules. *Rapid Commun. Mass Spectrom.* **2001**, *15*, 1862–1868.
117. O'Connor, P. B.; Mirgorodskaya, E.; Costello, C. E. High pressure matrix-assisted laser desorption/ionization Fourier transform mass spectrometry for minimization of ganglioside fragmentation. *J. Am. Soc. Mass Spectrom.* **2002**, *13*, 402–407.
118. Brock, A.; Horn, D. M.; Peters, E. C.; Shaw, C. M.; Ericson, C.; Phung, Q. T.; Saloma, A. R. An automated matrix-assisted laser desorption/ionization quadrupole Fourier transform ion cyclotron resonance mass spectrometer for "bottom-up" proteomics. *Anal. Chem.* **2003**, *75*, 3419–3428.
119. Donald, L. J.; Stokell, D. J.; Holliday, N. J.; Ens, W.; Standing, K. G.; Duckworth, H. W. Multiple equilibria of the *Escherichia coli* chaperonin GroES revealed by mass spectrometry. *Protein Sci.* **2005**, *14*, 1375–1379.
120. Cramer, R.; Haglund, R. F. Jr.; Hillenkamp, F. Matrix-assisted laser desorption and ionization in the O–H and C=O absorption bands of aliphatic and aromatic matrixes: Dependence on laser wavelength and temporal beam profile. *Int. J. Mass Spectrom. Ion Processes* **1997**, *169/170*, 51–67.
121. Loboda, A. V.; Krutchinsky, A. N.; Bromirski, M.; Ens, W.; Standing, K. G. A tandem quadrupole/time-of-flight mass spectrometer with a matrix-assisted laser desorption/ionization source: Design and performance. *Rapid Commun. Mass Spectrom.* **2000**, *14*, 1047–1057.
122. Shen, Z. X.; Thomas, J. J.; Averbuj, C.; Broo, K. M.; Engelhard, M.; Crowell, J. E.; Finn, M. G.; Siuzdak, G. Porous silicon as a versatile platform for laser desorption/ionization mass spectrometry. *Anal. Chem.* **2001**, *73*, 612–619.
123. Gillig, K. J.; Ruotolo, B.; Stone, E. G.; Russell, D. H.; Fuhrer, K.; Gonin, M.; Schultz, A. J. Coupling high-pressure MALDI with ion mobility/orthogonal time-of-flight mass spectrometry. *Anal. Chem.* **2000**, *72*, 3965–3971.
124. Laiko, V. V.; Taranenko, N. I.; Berkout, V. D.; Musselman, B. D.; Doroshenko, V. M. Atmospheric pressure laser desorption/ionization on porous silicon. *Rapid Commun. Mass Spectrom.* **2002**, *16*, 1737–1742.
125. Schreiner, J.; Voigt, C.; Mauersberger, K.; McMurry, P.; Ziemann, P. Aerodynamic lens system for producing particle beams at stratospheric pressures. *Aerosol Sci. Technol.* **1998**, *29*, 50–56.
126. Shaffer, S. A.; Prior, D. C.; Anderson, G. A.; Udseth, H. R.; Smith, R. D. An ion funnel interface for improved ion focusing and sensitivity using electrospray ionization mass spectrometry. *Anal. Chem.* **1998**, *70*, 4111–4119.

Chapter 7

A Comparison of MALDI Matrices

Mahmud Hossain and Patrick A. Limbach

Rieveschl Laboratories for Mass Spectrometry, Department of Chemistry, University of Cincinnati, Cincinnati, Ohio

Electrospray and MALDI Mass Spectrometry: Fundamentals, Instrumentation, Practicalities, and Biological Applications, Second Edition, Edited by Richard B. Cole

7.1 INTRODUCTION

Since the development of the laser as a high-intensity source of light, there have been several investigations of the evaporation and ionization of solid materials by focused laser light.[1] Laser desorption (LD) became an effective method for producing gas-phase ions from nonvolatile low-mass thermally labile organic salts during the 1970s[2–4] and from biomolecules during the 1980s.[5,6] In LD, the probability of obtaining a useful mass spectrum depends critically on the various physical properties of the analyte (e.g., photo-absorption, volatility, etc.). Furthermore, molecules with masses over ~1000 Da almost always produce only fragment ions.

LD-based approaches were not regarded by the analytical community as a viable technique for examining high-molecular-mass compounds until 1987.[7] The introduction of an excess of a light-absorbing compound, the matrix, combined with the analyte of interest, facilitated this change in opinion. Specifically, the addition of ultrafine metal powder to analyte solutions in glycerol[8] and the cocrystallization of the analyte with an organic small molecule solution[9] revolutionized the area of mass spectrometry by producing mass spectra of proteins of ~100,000 Da.

7.1.1 Suspension Matrices

Tanaka et al.[8] changed the perception that LD cannot be used for high-mass biomolecules without fragmentation.[8] A sample was made into a solution having a concentration of about 10 μg/μL using distilled water as the solvent. About 10 μL of this sample solution was dripped onto the sample holder. An ultrafine metal powder (UFMP, cobalt powder of about 300-Å diameter) and glycerol were dissolved with organic solvents (e.g., ethanol and acetone), with 10 μL of this solution also being dripped onto the sample holder. This mixture was then vacuum-dried for a short time to remove volatile compounds of the solution. The sample holder, with sample, was then introduced into the mass spectrometer for analysis.[8]

UFMP, also known as "Japanese powder," is an extremely fine metallic powder with particle diameters measuring a few tens of nanometers. The particle diameter of UFMP is about the same as one wavelength in the UV–visible range, thus when compared to bulk metal, there is an increased possibility that scattered light could be taken up internally and light energy could be absorbed more efficiently.[10] A nitrogen laser (wavelength 337 nm, pulse width about 15 ns, pulse energy 4 mJ max.) was used for sample ionization. By this method, Tanaka et al.[8,11] were able to detect several high-mass molecular ions (Figure 7.1)—for example, lysozyme (mol. wt. 14.3 kDa), chymotrypsinogen (mol. wt. 25.7 kDa), poly(propylene glycol) (PPG) (avg. mol. wt. 4 kDa), carboxypeptidase A (mol. wt. 34.4 kDa), bovine insulin (mol. wt 5.7 kDa), cytochrome C (mol. wt 12.3 kDa), and PEG20K (avg. mol. wt. ~20 kDa). This particle suspension matrix approach was further pursued by only a few groups[12–15] and was shown to be useful for the analysis of some compounds that are less compatible with the normal MALDI chemical matrices,[16] because of limitations in the mass range, analytical sensitivity, and/or spectra quality.[15]

7.1.2 Chemical Matrices

The method of sample preparation chosen by Karas and Hillenkamp was fundamentally different. Their long experience with laser desorption led them to believe that often it was the substrate under a layer of organic material on a surface that absorbed the laser light, not the organic molecules themselves. In 1985, Hillenkamp and co-workers[17,18] began to publish papers containing the hypothesis that large molecule ions could be produced by mixing the analyte with a matrix material that was chosen for its ability to absorb light. This matrix material would absorb light, resulting in its ablation and, hopefully, a coupled desorption of analyte molecule ions. The profound role of the matrix in driving the desorption process led to the designation of this approach as matrix-assisted laser desorption/ionization (MALDI). A variety of matrices were investigated, including tryptophan,

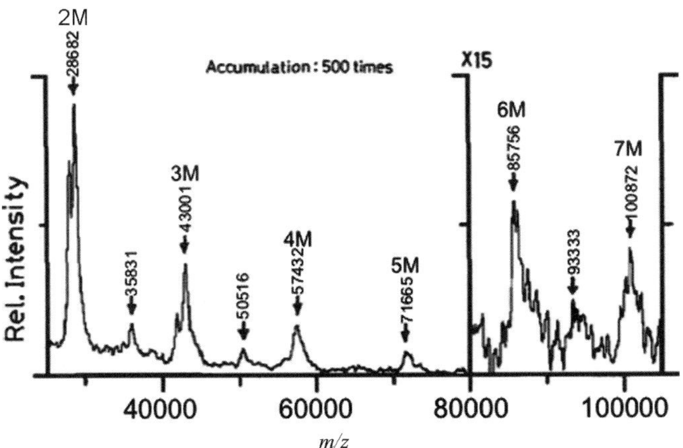

Figure 7.1. Laser ionization mass spectrum of lysozyme from chicken egg white, mol. wt. 14,306 Da. (Reproduced with permission from Tanaka, K.; Waki, H.; Ido, Y.; Akita, S.; Yoshida, Y.; Yoshida, T. Protein and polymer analyses up to *m/z* 100,000 by laser ionization time-of-flight mass spectrometry. *Rapid Commun. Mass Spectrom.* **1988**, 2, 151–153.)

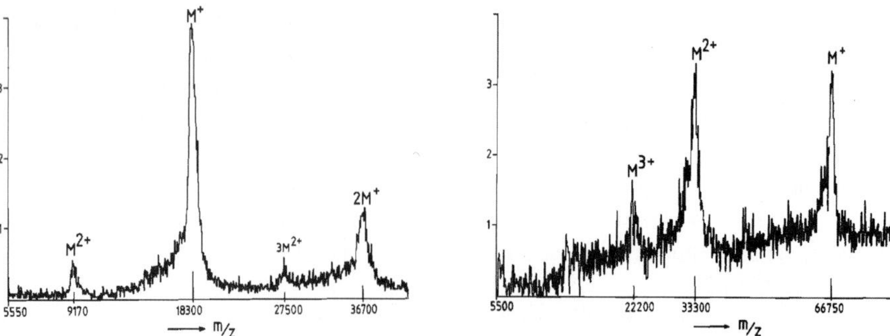

Figure 7.2. (**a**) Matrix-UVLD spectrum of β-lactoglubulin A, mass range of molecular ion region, 50 accumulated individual spectra. (**b**) Matrix-UVLD spectrum of bovine albumin, mass range of molecular ion region, 100 accumulated individual spectra. (Reproduced with permission from Karas, M.; Hillenkamp, F., Laser desorption ionization of proteins with molecular masses exceeding 10,000 Da. *Anal. Chem.* **1988**, *60*, 2299–2301.)

nicotinic acid, 3-nitrobenzyl alcohol, and 2-nitrophenyloctylether, and the results published were promising but not spectacular.[7]

The breakthrough for MALDI as applied to polypeptides came with the discovery that nicotinic acid had special properties for large polypeptides.[9,18,19] An Nd:YAG laser at 266 nm was used. Nicotinic acid is a solid at room temperature, with a low vapor pressure (but a tendency to sublime) and a high absorption coefficient at 266 nm. When aqueous solutions of a protein and nicotinic acid were mixed and a droplet of this mixture was dried, the deposit formed had the properties they had sought. They reported the ability of MALDI-MS to generate a large number of intact molecular ions as well as dimers and doubly charged molecular ions of proteins in the mass range above 10,000 Da, stable at least up to time lengths of about a millisecond. In all cases, singly charged molecular ions were the base peaks and no fragmentation was observed above 1000 Da (Figure 7.2).[9]

The Hillenkamp–Karas approach rapidly underwent essential developments over the subsequent years, resulting mainly in the change of the preferred laser wavelength from the originally used 266 nm to 337 nm, as well as the discovery of at least a dozen new good "chemical matrices." Hillenkamp defined the "chemical matrix" as a preferentially organic compound that is a liquid or that dissolves homogeneously in a suitable solvent and that *simultaneously* acts as the light absorber and interacts with the analyte in the ion-formation process.[15] A large variety of compounds has been empirically tested in the past for their suitability as a matrix. Today, analysts commonly make their choice from a relatively small number of established "chemical matrices"—for example, from benzoic or cinnamic acid derivatives. The actual choice depends on the type of the sample being analyzed, with only a few matrix compounds [e.g., 2,5-dihydroxbenzoic acid (2,5-DHB)] proving useful for a number of different compound classes.[20]

7.1.3 Technique Comparison

The major difference between these two approaches (suspension matrices versus chemical matrices) is sensitivity, with chemical matrices exhibiting greater sensitivity than suspension matrices. The signals obtained from the Hillenkamp–Karas method were more intense and had a higher signal-to-noise ratio. Laser desorption from suspension matrices was accompanied by a higher degree of analyte fragmentation compared to chemical matrices. These factors indicated that, for an analytical method, the chemical matrices were perceived to be more promising. The consequence of this perception is that the Hillenkamp–Karas method has become the focus of international research, while the Tanaka method has received less attention,[7] though two-phase matrix,[13] graphite-assisted,[16] and surface-assisted laser desorption ionization (SALDI)[14] have been introduced as alternative notations for Tanaka's approach.

7.2 MALDI MATRIX PROPERTIES AND PREPARATIONS

7.2.1 Matrix Properties

The MALDI technique involves the process of desorption, dissociation, and ionization of the analyte and the matrix under high laser energy. The essence of MALDI-MS is the matrix. Generally, analyte compounds are embedded in a surplus of matrix (~1000-fold molar excess) and are co-desorbed upon laser excitation. Despite the importance of the matrix, very little is known about what makes for a good matrix. Reviewing the matrices that have been discovered, we can identify some common qualities.[7,21–23] These features are:

1. ***Solubility.*** This condition can be expanded to include any solvent system in which the analyte of interest will co-dissolve with the matrix.
2. ***Absorptivity.*** This allows the energy to be deposited in the matrix, not in the analyte. In UV-MALDI, the matrix is typically an organic compound that has a relatively high molar absorptivity (ε_λ) at the wavelength of the laser and whose structure is based on an aromatic core (a chromophore) suitably functionalized to achieve the desired properties. In the case of IR-MALDI, fewer restrictions apply because wavelengths around 3 μm are effectively absorbed by O–H and N–H stretch vibrations, while wavelengths around 10 μm cause excitation of C–O stretch and O–H bending vibrations. Therefore, nonaromatic compounds can perform well as matrices in IR-MALDI.
3. ***Reactivity.*** The matrix itself is chemically inert toward the analyte. Matrices that covalently modify analytes cannot be used. A matrix can serve as protonating or deprotonating agent or as electron-donating or -accepting agent.
4. ***Volatility.*** The matrix should have a sufficiently low vapor pressure to minimize volatilization in the ion source vacuum, yet should sublime relatively easily.
5. ***Desorption.*** The matrix should facilitate co-desorption of the analyte upon laser irradiation.

7.2.2 Matrix Preparations

Sample preparation/handling is often considered the most important step in MALDI-MS. The quality of the preparation has a direct impact on the quality of the analytical results. Different types of sample preparations are available for various types of analytes and purposes.[24,25] Sample preparation methods can be divided into three major categories: solid matrix, liquid matrix, and special preparations.

7.2.2.1 Solid Matrices

a. Dried Droplet. Introduced in 1988 by Karas, and Hillenkamp,[9] this is the oldest and original sample preparation method that has remained, with minor modifications, in place for over two decades. A freshly prepared saturated solution of matrix compound is mixed with analyte solution and dried at ambient temperature. This method tolerates the presence of certain salts and buffers. However, aggregation of higher amounts of analyte–matrix crystals in a ring around the edge of the drop yield inhomogeneous and irregularly distributed crystals on the MALDI target; this requires the search for "hot" spots to generate good-quality spectra.[26–28] Vacuum drying was introduced as a modification to the dried-droplet method. Instead of using ambient temperature, the analyte-–matrix mixture is rapidly dried in a vacuum chamber that increased crystal homogeneity by reducing the size of analyte–matrix crystals.[29]

b. Crushed Crystal. This is an extension of the dried-droplet method developed to allow for the growth of analyte-doped matrix crystals in the presence of high concentrations of nonvolatile solvents: glycerol, 6 M urea, DMSO, and so on, without any purification steps.[30] The films produced are more uniform, with respect to ion production and spot-to-spot reproducibility, than dried-droplet deposits. The disadvantage of the crushed-crystal method is the increase in sample preparation time.

c. Fast Evaporation. Mann and co-workers[31] introduced fast evaporation with the main goal to improve the resolution and mass accuracy of MALDI measurements. Matrix and sample handling are completely decoupled in this simple procedure. The procedure allows matrix surfaces to be prepared in advance and, most importantly, results in a homogeneous distribution of matrix and analyte that enables easier and faster data acquisition (Figure 7.3).

d. Overlayer and Sandwich Methods. The overlayer (or two-layer) method was developed on the basis of two existing methods: crushed crystals and fast evaporation.[27] This method provides high detection sensitivity and excellent spot-to-spot reproducibility, which is mainly due to the increase of the matrix–analyte ratio and improved isolation between analyte molecules as a result of analyte–matrix deposition on top of matrix microcrystals.[28,32] The sandwich method was introduced as a blend of the overlayer and the fast-evaporation methods.[24,33]

e. Spin-Coating. Kantartzoglou and co-workers[34] first reported the preparation of near-homogeneous samples of large biomolecules (>150 kDa) by spin-coating sample substrates. Samples were prepared by spin coating the 2-mm-diameter metal sample probes (1-in.-diameter stainless steel and quartz substrate for larger volume) using premixed matrix and analyte solutions. This method is highly reproducible and generates enhanced molecular ions in the mass range of 5–150 kDa.

f. Electrospray. In the electrospray technique, the sample–analyte mixture is electrosprayed from a high-voltage (3–5 kV) stainless-steel capillary onto a grounded metal sample plate, mounted 0.5–3 cm away from the tip of the capillary (Figure 7.4).[35,36] It creates homogeneous layers of equally sized microcrystals wherein the analytes are evenly distributed. Later, Caprioli et al.[37] used this sample preparation technique for MALDI imaging of tissue samples, and Clench and co-workers[38] used it to spray matrix-only solutions on thin-layer chromatography plates.

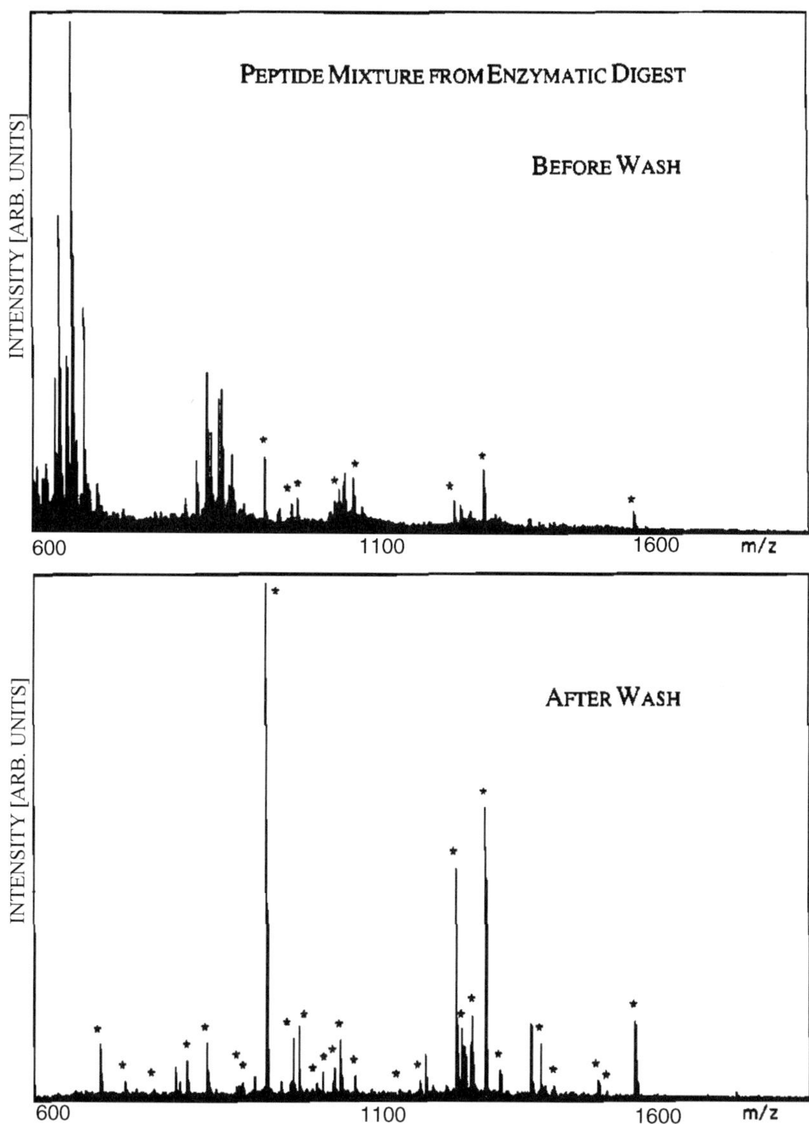

Figure 7.3. Reflector mass spectra of a peptide mixture before (**upper panel**) and after washing (**lower panel**). The matrix surface was made of α-cyano-4-hydroxycinnamic acid (06. μL 20 μg/μL in acetone with 1% water). Subsequently, 0.3 μL of peptide solution in 20% acetonitrile was applied. The wash was carried out by depositing 10 μL of 0.1% TFA on the sample spot, and the liquid was shaken off after ~5 s. The washing procedure was repeated once. Peaks from tryptic peptides are marked by asterisks. All peptide peaks are isotopically resolved. (Reproduced with permission from Vorm, O.; Roepstorff, P.; Mann, M. Improved resolution and very high sensitivity in MALDI TOF of matrix surfaces made by fast evaporation. *Anal. Chem.* **1994**, *66*, 3281–3287.)

g. Matrix-Precoated Layers. A drop of undiluted peptide and protein sample is added to a precoated target spot, which makes this method faster and more sensitive than many of the other methods. It also provides an opportunity to interface the MALDI sample preparation to capillary electrophoresis and liquid chromatography outputs. Many researchers have focused on the development of thin-layer matrix-precoated membranes—for example, nylon, PVDF, nitrocellulose, anion- and cation-modified cellulose, regenerated cellulose, or regenerated cellulose dialysis membranes.[39,40]

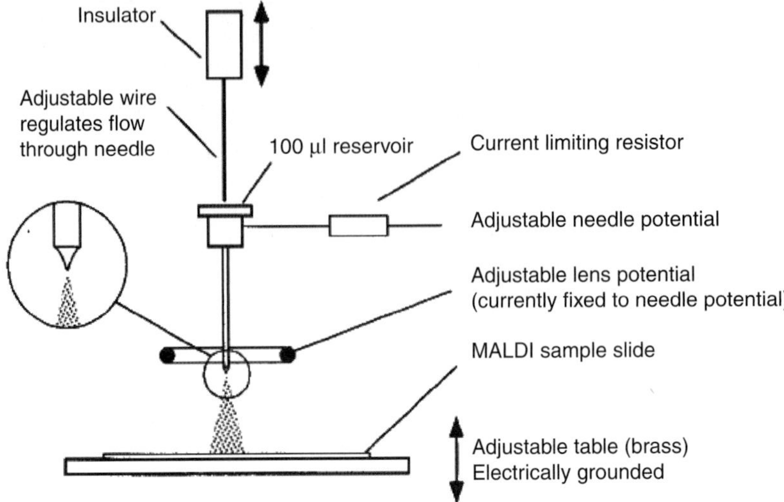

Figure 7.4. Electrospray sample deposition setup employed for preparation of MALDI sample slides. (Reproduced with permission from Axelsson, J.; Hoberg, A.-M.; Waterson, C.; Myatt, P.; Chield, G.; Varney, J.; Haddleton, D.; Derrick, J. Improved reproducibility and increased signal intensity in MALDI as a result of electrospray sample preparation. *Rapid Commun. Mass Spectrom.* **1997**, *11*, 209–213.)

7.2.2.2 Liquid Matrices

a. Chemical Liquid. In this straightforward method, samples are prepared by first dripping 0.5 μL of liquid matrix (a molar ratio of chemical liquids) onto a 1-mm^2 area of fibrous paper (Kimwipe™) followed by 0.5 μL of the sample solution. The paper is attached to the sample holder with double-sided sticky tape.[41] In the case of two-phase matrices, fine metal powder or graphite particles are suspended on a low-volatility solvent where the fine particles absorb most of the energy from the laser beam and the liquid molecules provide the charge for ionization. The analyte is mixed with the matrix suspension. After solvent evaporation, the remaining paste on the sample holder is introduced into the ion source.[12,15,42]

b. Chemical-Doped Liquid. This method is based on the dissolution of the solid matrix in a liquid support of low volatility such as glycerol. An appropriate solubilizing reagent is added to promote the dissolution of the matrix materials into the liquid support. Selection of the solubilizing reagent is empirically related to an acid–base relationship. This technique has wider scope and can convert most of the solid matrices to the corresponding solution forms to suit different classes of analytes.[43]

7.2.2.3 Special Preparations

a. Solid Supports. This method includes various solid supports that function as a matrix. Best known is the use of porous Si as a sample support, without the need for any other matrix (desorption/ionization on silicon or DIOS). The concept of an "active" MALDI support is very promising for practical applications but not without limitations including a limited mass range, poor reproducibility of the fabrication process, and effect of surface contamination from earlier experiments.[44]

b. Insoluble Samples. For insoluble samples, it is almost impossible to embed them in a matrix environment by any of the traditional methods. It has been found, however, that by pressing a mixture of finely ground matrix and analyte, it is possible to obtain MALDI data from insoluble compounds—for example, insoluble or high-molecular-weight synthetic polymers. A thick sample is produced by this method, resulting in long-lived samples that yield signals for thousands of laser shots impinging on one location.[45]

7.3 MATRIX CLASSES

7.3.1 Proteins/Peptides

MALDI-MS owes much of its popularity to its usefulness as an analytical tool for the characterization of peptides and proteins. While an extremely large number of matrix candidates have been investigated for their applicability in peptide and protein analysis, a relatively small group has proven reliable over the past 15 + years. The members of this small group include 2,5-dihydroxybenzoic acid (2,5-DHB), α-cyanohydroxycinnamic acid (CHCA or HCCA), and sinapinic acid (SA).

The performance of 2,5-DHB as a matrix for MALDI of proteins (chicken egg albumin, horse heart cytochrome c) at 337-nm-wavelength laser radiation was described by Strupat et al.[46] Samples from protein–matrix mixtures with mass ratios between 1.5×10^{-2} and 1×10^{-4} were prepared by drying solutions onto metal substrates as well as by growing single crystals. The detection limit was found to 1 fmol, and good shot-to-shot reproducibility was obtained. 2,5-DHB was found to be insensitive to contaminations by inorganic salts, buffers, and detergents, even up to 10% sodium dodecylsulfate (SDS).

Selected benzoic acid derivatives and related substances were used as additives to 2,5-DHB, and the MALDI performance of these mixtures was investigated.[47] Using 14 different benzoic acid derivatives substituted at position 2 and/or 5, or using related substances as a co-matrix in the 1–10% range with 2,5-DHB, improved ion yields and signal-to-noise ratios of analyte molecules, especially for the high-mass range. The enhanced performance was prominent with 2-hydroxy-5-methoxybenzoic acid. The authors suggested that the improved performance resulted from disorder in the 2,5-DHB crystal lattice allowing "softer" desorption.

Another popular class of matrices useful in peptide and protein analysis are the derivatives of cinnamic acid.[48] The most effective of these derivatives are 3-methoxy-4-hydroxycinnamic acid (ferulic acid), 3,4-dihydroxycinnamic acid (caffeic acid), and 3,5-dimethoxy-4-hydroxycinnamic acid (sinapinic acid, SA). The initial report of SA noted improved resolution, mass accuracy, and reproducibility.[49] SA can selectively ionize proteins in the presence of high concentrations of contaminating materials, such as lipids, carbohydrates, and salts. It is relatively nonselective in its behavior toward proteins of significantly different primary structures and modifications (such as phosphorylation and glycosylation). The general affinity of this matrix for proteins, coupled with its ability to tolerate high concentrations of various contaminants (buffering agents, acidification agents, and denaturing agents), allows the application of mass spectrometry to mixtures of crude biological extracts, including egg white, human blood cell lysates, *E. coli* cell lysates, crude seed extracts, enzyme digest mixtures, and commercial preparations of proteins containing many impurity proteins and peptides.

The matrix–analyte morphology and substrate interactions of insulin with vanillic, nicotinic, and sinapinic acid have been systematically investigated by optical microscopy.[50]

That study revealed that vanillic acid and, to a lesser extent, nicotinic acid matrices formed prominent crystalline rings around the dried sample spot whereas sinapinic acid forms a more uniform dispersion of the crystallized matrix. Optical microscopy of sample preparations revealed that sample morphology can differ dramatically, depending on choice of matrix. Samples were found to be heterogeneous containing two separate phases: a crystalline, birefringent matrix phase and an isotropic phase that contains analyte. Sinapinic acid, while constituting an improvement over nicotinic acid as a MALDI matrix, has two undesirable properties: (a) the formation of an adduct of a photodehydration product of sinapinic acid with proteins, having a mass 206 Da higher than the protonated protein, and (b) the production of an intense background of ions below m/z 1000, limiting the utility of the technique for the measurement of low–mass peptides.

The other successful protein and peptide matrix is CHCA, which was reported by Beavis et al.[51] CHCA was found to be a highly effective and efficient matrix for peptides and glycopeptides in the molecular mass range 500–5000. The MALDI mass spectra of proteins obtained from CHCA showed an increased tendency for multiple protonation compared with other widely used matrices. Several peptides were shown to undergo fragmentation during MALDI to a degree that complicated their analysis using CHCA as a matrix, even at threshold laser irradiance.[52] These peptides included synthetic peptides, peptides isolated from viral proteins, and a phosphopeptide from β-casein (residue 33–48). The excessive fragmentation occurred usually as a post-source phenomenon; however in–source fragmentation was also observed. The phosphopeptide studied exhibited a high degree of β-elimination of phosphates. It was demonstrated that the fragility exhibited by these peptides in CHCA, including β-elimination of phosphate from serine, was not evident with a matrix comprised of dihydroxyacetophenone (DHAP) and diammonium hydrogen citrate (DAHC). Table 7.1 summarizes information regarding matrix suitability for peptides and proteins.

Sample-matrix preparation procedures have been shown to greatly influence the quality of MALDI data from peptides and proteins. In particular, dramatic mass discrimination effects are observed when the matrix CHCA is used for analyzing complex mixtures of peptides and proteins.[29] The discrimination effects are found to be strongly dependent on the sample-matrix solution composition, pH, and the rates at which the sample-matrix co-crystals are grown. These findings demonstrated the need to employ great care in performing and interpreting the MALDI analysis of biological samples. The choice of the matrix solvent system and the rate of matrix crystal growth add a certain chromatographic-like feature in these sample-matrix preparation procedures that can be exploited to optimize analysis.

It has been shown that carbohydrate-containing co-matrices gave improved results over single-component matrices.[53] Of those studied, fucose plus 2,5-DHB produced a signal for 96% of the tryptic peptides from 280 fmol of unfractionated recombinant human growth hormone (rhGH) in a single mass spectrum. The incorporation of 5-methoxysalicylic acid (5-MSA) as a co-matrix with fucose and 2,5-DHB was shown to improve the results obtained from a recombinant human tissue plasminogen activator (rt-PA) sample. The fucose/2,5-DHB and 5-MSA/2,5-DHB co-matrices produced significant enhancements in spectral quality over 2,5-DHB alone, including suppression of matrix peaks, increased ion signal, improved resolution, an increased number of useful laser shots per crystal, and minimization of baseline slope.

9-Aminoacridine (9-AA) has been introduced as a matrix for negative-mode MALDI.[54] While most traditional MALDI matrices have been acidic in character, 9-AA is a moderately strong base. The mechanism by which 9-AA brings about ionization in the negative mode appears to involve abstraction of a labile proton in an acid–base reaction.

Table 7.1. Positive (Quasi) Molecular Ion Formation of Selected Organic Compounds and Their Usability as Matrix[a]

$M^{+\bullet}$		$[M-H]^+/[M+2H]^{+\bullet}$	
Indole	–	**1-Aminoindane**	0
Indole-3-carboxylic acid	–	**3-Aminopyrazine-2-carboxylic acid**	+
Indole-3-acetic acid	–	4,6-Dihydroxy-2-mercaptopyrimidine	
Indole-3-propionic acid			
Indole-3-butylic acid	–		
Indole-2-carboxylic acid (266 nm)		**Indole-3-aldehyde**	0
	–	**Dimethylaminomethyl indole (gramine)**	–
4-Aminobenzene sulfonic acid			
(Sulfanilic acid)	–	**Pyrazine-2-carboxylic acid**	+
1,4-Dihydroxybenzene (hydroquinone)		**Tetrahydroxy-*p*-benzoquinone**	+
4-Hydroxy-3-methoxyphenylacetic acid (homovanillic acid)			
3-(*p*-Hydroxyphenyl)propionic acid	–		
1-Naphthol	–		
3-Nitrobenzyl alcohol[b]	–		
3-Phenylpropionic acid	+		
(Hydrocinnamic acid)	–		
Pyrrole-2-carboxylic acid			
Tryptophol	–		
$[M+H]^+$		$[M+H]^+/M^{+\bullet}$	
Adenine	–	**2-Aminobenzoic acid**	+
2-Aminobenzimiazole	–	**2-Amino-3-hydroxybenzoic acid**	–
Aminobenzothiazole		**2-Amino-5-hydroxybenzoic acid**	+
Aminonicotinic acid[b]	0	**3-Amino-4-hydroxybenzoic acid**	0
***o*-Arsanilic acid**	–	**2,5-Dihydroxybenzoic acid**	+ +
***p*-arsanilic acid**	–	**2-Hydroxy-5-methoxybenzoic acid**	0
Chelidonic acid	–	**4-Hydroxy-3-methoxybenzoic acid (vanillic acid)**	+
Coumalic acid[b]	+		
Cytosine	–	**3,4,5-Trihydroxybenzoic acid**	0
2,3-Dihydroxybenzoic acid	+		
2,4-Dihydroxybenzoic acid	+	**3,4-Dihydroxycinnamic acid (caffeic acid)**	+
Guanine			
Hypoxanthine		**2,4-Dimethoxycinnamic acid**	0
Nicotinic acid[b]	+ +	**3,5-Dimethoxy-4-hydroxycinnamic acid (Sinapinic acid)**	+
Nicotinic acid methyl ester	–		
2-Nitrophenol[b]	0	**3-Hydroxy-4-methoxycinnamic acid**	0

(*continued*)

Table 7.1. (*Continued*)

$[M + H]^+$		$[M + H]^+/M^{+\bullet}$	
Phenylalanine	–	**4-Hydroxy-3-methoxycinnamic acid (ferulic acid)**	+
Quinoline-4-carboxylic acid	–		
Thymine[b]	0		
Tryptophan	–	**6-Aminoindazole**	–
Tyrosine	–	7-Azaindole	
Uracil	0	**3-Hydroxy-2-naphthoic acid**	+
Xanthine	–	Indazole	
		Indoleacetic acid hydrazide	
		3-Indoleacetone	
		Indole-2-carboxylic acid (337 nm)	+
		7,7,8,8-Tetracyanoquinodimethane	
		1,3,5-Trihydroxybenzene	0
		Tryptamine	+
		Tryptophanamide	–
		Tryptophan methyl ester	+

Source: Adapted with permission from Ehring, H.; Karas, M.; Hillen Kamp, F. Rote of photoionization and photochemistry in ionization process of organic molecules and relevance for matrix-assisted laser desorption and ionization mass spectrometry. *Org. Mass-Spectrom.* **1992**, *27*, 472–480. Copyright John Wiley & Sons.

[a]Matrix quality: all tested molecules are printed in bold type. *Key*: + +, very strong protein signals; +, strong protein signals; 0, weak protein signals;–, no protein signals observed.

[b]Radical fragments/adduct ions.

Most MALDI matrices readily donate protons; 9-AA readily accepts protons, leading to the formation of $[M-H]^-$ species. 9-AA was first examined as a matrix for low-molecular-weight compounds having acidic protons, such as phenols, carboxylic acids, sulfonates, amines, and alcohols, and has been extended to larger compounds including oligonucleotides, oligoamides, and proteins.

7.3.2 Carbohydrates

MALDI-MS is a particularly valuable analysis tool for carbohydrates because it enables underivatized, as well as derivatized, compounds to be examined. MALDI has been applied to the analysis of carbohydrates since the earliest reports of this technique. Hillenkamp et al.,[18] analyzed stachyose, a 666-Da tetrasaccharide, with and without tryptophan and nicotinic acid as matrices. Though sodium-adducted quasi-molecular ions could be obtained without the matrix, addition of matrix increased molecular ion abundances threefold and improved reproducibility.

These matrices were supplanted by 2,5-DHB within a short period of time.[26] 2,5-DHB facilitated the determination of molecular masses and size distribution of oligosaccharides without any noticeable fragmentation. A linear and a cyclic α-glucan, maltoheptaose, and cycloheptaose were analyzed separately and as a mixture. Glucan species were generally detected as monosodium and monopotassium carbohydrate adduct ions without any noticeable fragmentation, and protonated molecules were totally absent.

2,5-DHB remains one of the most successful matrices for carbohydrate analysis. Additives may affect crystallization, embedding the analyte more homogeneously.[55] Using 2-hydroxy-5-methoxybenzoic acid as a co-matrix in the 10% range with 2,5-DHB, producing a mixture known as "super-DHB," resulted in improved ion yields and a 2 to 3-fold increase of signal-to-noise ratio for the analysis of a dextran standard (Dextran 1000), presumably through a disordering in the 2,5-DHB crystal lattice allowing for "softer" desorption.[47] A mixture of 2,5-DHB and 1-hydroxy isoquinoline (HIQ) in a weight ratio of 3:1 was best suited for the analysis of β-cyclodextrin, maltoheptaose, sucrose octaacetate, chitotetraose, and raffinose-5-hydrate, providing enhanced molecular ion abundances, mass resolution, and suppression of unwanted matrix peaks.[55]

Acidic oligosaccharides are more difficult to analyze using conventional matrices. 6-Aza-2-thiothymine (ATT) and 2,4,6-trihydroxyacetophenone (THAP) were found to be particularly useful for acidic oligosaccharides, with THAP preferred because it gave less fragmentation.[56] In addition, THAP offered improved sensitivity for detection of acidic glycopeptides over CHCA. Among different substituted acetophenones, 2,5-dihydroxyacetophenone (2,5-DHAP) produced relatively strong signals from neutral carbohydrates in positive ion mode which was comparable to 2,5-DHB.[57] Esculetin (6,7-dihydroxycoumarin) efficiently ionized neutral glycans with better resolution than 2,5-DHB in both MALDI and AP-MALDI.[57] 5-Chloro-2-mercaptobenzothiazole (CMBT), an analog of MBT (2-mercapto-benzothiazole), has been found to be not only effective for the analyses of glycolipids, but also superior to conventional matrices for the analysis of some oligosaccharides.[58]

Commercially available β-carbolines (harmane, nor-harmane, harmine, harmol, harmaline, and harmalol) with their acidic NH-indolic group and basic *N*-pyridinic group have been shown to be useful matrices at 337 nm, for cyclic and acyclic oligosaccharides, in both positive and negative mode. β-Carbolines provide the same level of sensitivity and resolution in positive mode, but yield higher sensitivity and resolution in negative mode compared to conventional matrices (DHB or DHB/HIC). β-Carbolines have been used to generate deprotonated molecular ions of sulfated saccharides in negative ion mode and protonated and deprotonated molecular ions from neutral saccharides.[59–63]

a. Glycopeptides and Glycoproteins. Glycopeptides and glycoproteins contain free amino groups that can be protonated efficiently, and they exhibit frequently more abundant mass spectral ions than saccharides that are ionized by sodium addition. The ratio of the protonated to sodium adduct for glycopeptides depends on the percent of peptides in the molecule.[57] 2,5-DHB was found to be a preferred matrix for glycopeptides and glycoproteins compared to CHCA and SA by the dried-droplet sample preparation technique, although CHCA with the sandwich sample preparation technique can yield better results for the analysis of glycopeptides.[24] 2,6-DHAP containing diammonium hydrogen citrate (DAHC) exhibited superior spectra of sialylated glycopeptides with better resolution and less fragmentation compared to conventional matrices.[64] Protonated molecular ions were the prominent products and minimal cationic adducts were present during analysis.

7.3.3 Nucleic Acids

As with other biomolecule compound classes, a large number of candidate matrices have been investigated for the MALDI-MS analysis of nucleic acids. Over the years, only a handful have become standards within the field: THAP, 3-hydroxypicolinic acid (3-HPA), picolinic acid (PA), ATT, and 2,5-DHB. A variety of synthetic oligodeoxyribonucleotides ranging from 6 to 30 nucleotides were analyzed in MALDI with 2,5-DHB.[65] Molecular ions were observed almost exclusively, with little or no fragmentation and generally minor contributions from multiply charged species. Mixtures of poly $d(T)_{12\text{-}30}$ with low picomole amounts of each component gave well-resolved peaks. It was found that poly-G and poly-A were much more difficult to detect by MALDI. The failure to detect larger DNA segments may be due to the weak glycosidic bond of A and G.[66]

MALDI-TOF MS was used to produce quasi-molecular ion signals from underivatized mixed-base single-stranded DNA oligomers ranging from 10 to 67 nucleotides in length.[67] These results were obtained with 3-HPA, which showed significant improvement over many previously reported matrices studied in terms of mass range available, signal-to-noise ratio, and the ability to analyze mixed-base oligomers. The desorption/ionization was studied at 266, 308, and 355 nm. Spectra taken at 266 nm provided the smallest amounts of doubly charged and dimer ions—characteristics desirable for DNA sequencing by this technology. Negative-ion spectra were uniformly superior to positive-ion spectra.

PA was found to be suitable for oligonucleotides and tRNA.[68] The efficiency of MALDI of oligonucleotides using PA was superior to that found using 3-HPA. MALDI-MS of $tRNA^{Phe}$, a 76-base ribonucleic acid, was detected with a signal-to-noise ratio of >10. The more significant development was using a combination of 3-HPA and PA to analyze DNA fragments as large as 500 nucleotides.[69] ATT, cocrystallized with ammonium citrate, has been shown to be a suitable matrix for the analysis of oligonucleotides and short DNA fragments.[70] The major advantages of ATT over other conventional matrices (e.g., THAP and 3-HPA), were improved resolution and mass accuracy, easy sample preparation, and applicability to crude or partially purified samples.

The de facto standard for small oligonucleotides is THAP together with di- and triammonium salts of organic or inorganic acids.[71] A mixture of THAP in acetonitrile and aqueous triammonium citrate in a 1:1 molar proportion was found to be a good matrix for the detection of synthetic oligodeoxynucleotide samples.[72] A high proportion of volatile solvent as well as the high salt content ensured fast cocrystallization of the matrix, co-matrix, and analyte molecules. MALDI spectra obtained in negative ion reflectron mode from samples prepared with this protocol showed deprotonated molecules even in the presence of excess sodium salt. The matrix was found effective for low-mass modified single nucleotides as well as for longer oligodeoxynucleotides (up to 18-mers).

A MALDI-MS study of DNA analysis using 2,4,6-THAP and 2,3,4-trihydroxyacetophenone (2,3,4-THAP), separately and in combination, was performed.[73] The results showed that a mixture of 2,3,4-THAP, 2,4,6-THAP, and ammonium citrate with molar ratios of 2:1:1 serves as a good matrix for the detection of DNA, especially for samples containing a small quantity of DNA such as PCR products. The resolution and shot-to-shot reproducibility using this matrix combination were better than, and the MALDI sensitivity comparable to, that obtained when using 3-HPA, PA, and ammonium citrate matrix (9:1:1).

A unique aspect of MALDI-MS analysis of nucleic acid components is the necessity of a co-matrix to minimize cation adduction and improve oligonucleotide signals. The effect of ammonium salt in the detection of oligonucleotides was systematically investigated using several matrices with ammonium salt additives.[74] The results showed that the presence of ammonium salt in the matrix had a beneficial effect on protonation and deprotonation of oligonucleotides in addition to suppressing alkali-ion adducts. These observations indicated that a good matrix for DNA detection could be made from two components: one component for desorption, which has a relatively large absorption coefficient at the incident laser wavelength, and another component for ionization, which has good protonation and deprotonation ability. Experimental results showed that ammonium citrate and ammonium tartrate were good ionization components.

The addition of a co-matrix of sufficiently high proton affinity can serve as a proton sink to reduce oligonucleotide fragmentation in MALDI-MS.[75,76] A direct correlation between the co-matrix proton affinity and the oligonucleotide molecular ion stability was found. It was proposed that upon laser desorption/ionization, the co-matrix competed with the nucleobases for additional protons from the matrix. Titration of standard oligonucleotide matrices with several co-matrices of differing proton affinities demonstrated that the co-matrix mole fraction was an important factor in oligonucleotide molecular ion stability. The polyamine co-matrices—spermine tetrahydrochloride, spermine, spermidine trihydrochloride, and spermidine—were evaluated for their effectiveness at enhancing the mass spectral quality of oligonucleotides.[77,78] In general, polyamines co-matrices were found to be more effective than monofunctional amines for improved mass spectral data.

Binary or ternary matrices provide the capability to independently optimize the different properties required for the matrix by varying the identity and/or concentration of the mixture component. A mixed matrix of 3-HPA and pyrazinecarboxylic acid (PCA, a structural analogue of PA) was used for analysis of a variety of synthetic oligodeoxynucleotides.[79] The experimental results showed that DNA segments with masses in the range of 5000–10,500 Da can be analyzed with high resolution when 3-HPA/PCA was prepared by mixing saturated 3-HPA solution and saturated PCA solution (in 50% of a 0.5 mol/L solution of diammonium hydrogen citrate plus 50% acetonitrile) at the volume ratio of 4 : 1. Moreover, when this mixed matrix was used to analyze two mixtures containing two 23-mer DNA segments with a 9-Da difference (A and T) and the other with a 7-Da difference (AA and TG), the two 23-mer ion peaks were well-separated and an isotopically resolved spectrum of each component was found. When the 3-HPA/PCA mixture was compared to 3-HPA alone for the analysis of DNA segments, improved sample-to-sample reproducibility, resolution, and signal-to-noise ratio were observed.

Although 3,4-diaminobenzophenone (DABP) was originally introduced as a novel matrix for the analysis of peptides and proteins by MALDI-MS, it was also demonstrated to be useful in the analysis of oligonucleotides.[80] Intact oligonucleotide ions with this matrix can be readily produced with lower laser powers, resulting in better detection limits, less fragmentation, and fewer alkali metal ion adducts compared to conventional matrices (e.g., 3-HPA and THAP). This matrix alone was particularly useful for the analysis of small oligonucleotides (<30-mers) without any co-matrices. Besides, minimal fragmentation and fewer alkali metal ion adducts were seen even at low concentrations of oligonucleotides. It was also found that samples prepared with DABP are highly homogeneous, which resulted in excellent shot-to-shot reproducibility, better resolution, and a higher signal-to-noise ratio.

7.3.4 Lipids

Because lipids are low-molecular-weight compounds, MALDI analysis can be limited due to matrix background products. Thus, although a significant number of studies have been conducted, by far 2,5-DHB has been found to be the most effective matrix for various types of lipids and phospholipids (Table 7.2).[81–90]

Table 7.2. Matrix Compounds Used to Date in MALDI-MS of Lipids and Phospholipids

Chemical Structure	Common Name	Remarks
	2,5-Dihydroxybenzoic acid	So far, the most widely used matrix for lipids and PL. Gives only very little matrix peaks. May be used in positive and negative ion mode.
	α-Cyano-hydroxycinnamic acid	Strongly interfering with peaks of interest (oligo-merization). Scarcely soluble in methanol.
	Sinapinic acid	Strongly interfering with peaks of interest (oligo merization). Scarcely soluble in methanol.
	6,7-Dihydroxycoumarine	Scarcely used
	5-Ethyl-2-mercaptothiazole	Scarcely used
	para-Nitro-aniline (PNA)	Useful in complex mixture analysis. Provides intense negative ions signals of PE and PC. Only recently introduced.
C_x	Graphite	Seems only suitable for very apolar compounds. So far scarcely used.

Source: Reproduced with permission from Schiller, J.; SÜb, R.; Arnhold, J.; Fuchs, B.; Leßig, J.; MÜller, M.; PetkoviĆ, M.; Spalteholz; ZschÖrnig, O.; Arnold, K. Matrix-assisted laser desorption and ionization time-of-flight (MALDI-TOF) mass spectrometry in lipid and phospholipid research. *Prog. Lipid Res.* **2004**, *43*, 449–488.

b. *Glycolipids.* Underivatized and permethylated gangliosides from commercial sources and highly purified gangliosides from the human brain containing up to five sialic acid residues have been analyzed by MALDI.[91] A variety of matrix and wavelength combinations were tested in both the positive- and negative-ion modes. The best results were obtained with the matrices 2,5-DHB, 4-hydrazinobenzoic acid, 1,5-diaminonaphthalene, and ATT. Negative-ion mass spectra of the underivatized gangliosides exhibited spectra of better quality than did the positive-ion mass spectra with higher signal-to-noise ratios, improved resolution, reduced fragmentation, and, as expected, fewer sodium or potassium adducts.

Sphingolipids and both neutral and acidic glycosphingolipids exhibited good signals in the positive-ion mode but with the nature and abundance of the molecular ion and extent of fragmentation strongly dependent on the matrix.[90] 2,5-DHB, CHCA, and esculetin were found to provide the best signals. All of the compounds predominantly gave sodium adduct ions, whereas only sphingosine gave abundant protonated molecules. Fragmentation of the oligosaccharide chain of glycolipids containing HexNAc tended to be more random than that seen when HexNAc was absent, and it provided considerable sequence information.

7.3.5 Synthetic Polymers

MALDI MS has been demonstrated to be a powerful technique for the analysis of polymeric materials. The advantages of this technology include low sample consumption, ease of sample preparation, short analysis time, and soft ionization, which leads to negligible or no fragmentation of analytes. It provides absolute, fast, and accurate molecular masses for polymers characterized by narrow polydispersity. It gives masses for the entire polymer distribution, thereby providing molecular mass information that can be used to obtain (a) the mass of the end-groups (b) the mass of the repeat unit chemical modifications on the polymer if oligomer resolution is attained.[92]

One of the most useful studies was the division of synthetic polymers into four categories based upon solubility: water-soluble polymers, polar organic soluble polymers, nonpolar organic soluble polymers, and polymers soluble only in "difficult" solvents such as dimethylsulfoxide, hot 1,2-dichlorobenzene, or sulfuric acid.[93] This subdivision facilitates identification of appropriate MALDI matrices, depending upon the characteristics of the polymer. For water-soluble polymers, matrix conditions are similar to peptides or proteins. 2,5-DHB was used to analyze PEG, PPG, and PMMA at a concentration of 10 g/L in water/ethanol (9 : 1 ratio).[94] Various metal salts were added to increase cation adduction.

For polar organic-soluble polymers, such as acrylics, poly(hydroxybutanoate), and poly(vinyl acetate), use of common matrices with a solvent, such as acetone, tetrahydrofuran, or methanol, yields better spectra. Poly(methylmethacrylate) (PMMA) mixtures of molecular weight 28 kDa, 59 kDa, 127 kDa, and 260 kDa were analyzed by using t-3-indoleacrylic acid (IAA) as a matrix.[95] Both the samples and matrix were dissolved in acetone. In positive-ion mode, the singly charged molecular ions as well as multiply charged states were observed whereas in negative-ion mode the multiple charging was somewhat reduced.

Nonpolar organic-soluble polymers, such as hydrocarbons, have been addressed in a variety of ways. Bahr et al.[94] analyzed polystyrene (PS) with nitrophenyloctylether doped with silver trifluoroacetate to produce silver-cationized molecular ions. Danis et al.[93] used silver acetoacetonate with the matrices IAA or 1,4-di-(2-(5-phenyloxazolyl))benzene (POPOP) for the analysis of various hydrocarbons, PS, polybutadiene (PBD), and *cis*-1,4-polyisoprene (PIP). Silver-adduct molecular ions were produced for polymers up to 125 kDa. Silver salt clusters formed in MALDI analyses of nonpolar polymers is a problem. Silver cluster ions were observed from *m/z* 1500–7000 when acidic, polar matrices, such as 2,5-DHB, *all-trans*-retinoic acid (RTA), or 2-(4-hydroxyphenylazo)benzoic acid (HABA), were used to analyze nonpolar polymers. These background signals could be greatly reduced or eliminated by the use of nonpolar matrices, such as anthracene or pyrene.[92] Nonpolar polymers analyzed with acidic, polar matrices (e.g., RTA) and silver cationization reagents could yield lower-quality mass spectral results when interferences due to silver clusters were present. Replacing the polar matrices with nonpolar matrices, or the silver salts with copper salts, substantially improved the quality of the analytical results.[96]

Analysis of polymers from the fourth group is the most challenging. A solvent-less preparation technique was developed by Skelton et al.[45] to analyze polymers (such as polyamides) that are insoluble or poorly soluble in common organic solvents. This method consisted of pressing a pellet from a solid mixture of finely grounded powder of both polymer and matrix and was compared to the common dried droplet sample preparation method. This solid–solid sample preparation was found to give spectra that are reproducible, show good signal-to-noise ratios, and are not sensitive to changes in sample composition. It is applicable to completely insoluble materials and fragmentation was not observed for polyamides.

Solvent-free sample preparation was later extended to PS and PMMA in a mass range from 2 to 100 kDa with similar or better results compared to solvent-based methods, especially in terms of reproducibility and mass accuracy. Several problems during sample preparation such as immiscibility, segregation and suppression effects, solubility and incompatibility restrictions, and so on, can also be minimized with dithranol, 7,7,8,8-tetracyanoquinodimethane (TCNQ), and CHCA as matrices.[97]

7.4 NONTRADITIONAL MATRICES

7.4.1 DIOS (Desorption/Ionization on Silicon)

Just as the matrix serves to trap analyte molecules and absorb UV radiation, nanoporous materials such as porous silicon were found by Siuzdak and co-workers to be an effective medium for desorbing compounds as well as generating intact ions in the gas phase.[44,93–95] This is due to (a) the ability of its nanometer-sized pores to trap analyte molecules and (b) the silicon surface, with its photoactive and photoluminescent characteristics, which effectively absorbs UV light. This approach, desorption/ionization on porous silicon (DIOS), has demonstrated characteristics similar to MALDI in that intact molecules are observed at the femtomole to attomole levels with little or no fragmentation (Figure 7.5).[44,98] The structure and physiochemical properties of the porous silicon surfaces are vital for DIOS MS performance and are controlled by the selection of silicon and the electrochemical etching conditions.[99] DIOS surfaces are produced through a simple galvanostatic etching procedure that yields a UV-absorbing porous silicon exterior with high surface area and thermal insulating properties.

The observation of a variety of intact ions directly desorbed from these surfaces suggests that porous silicon has properties (such as high surface area and strong ultraviolet absorption) that may enhance pulsed laser desorption and ionization. The energy for analyte release from the surface may be transferred from silicon to the trapped analyte through vibrational pathways, or as a result of the rapid heating of porous silicon, producing hydrogen that releases the analyte which was defended by a study of Ogata et al.[100] on possible changes in the environment of hydrogen in porous silicon with thermal annealing. It has also been noted that porous silicon absorbs hydrocarbons from outside air, or while under vacuum, thus rapid heating/vaporization of trapped hydrocarbon contaminants or solvent molecules may increase the vaporization and ionization of the analyte embedded in the porous silicon. DIOS-MS offers good sensitivity, high tolerance to contaminants, no matrix interference, and more control over biological mass spectrometry as the surface properties of porous silicon can easily be customized.[44]

Different applications have been carried out by DIOS-MS including the characterization of polymers in the forensic field,[101] as well as studies of biological samples such as

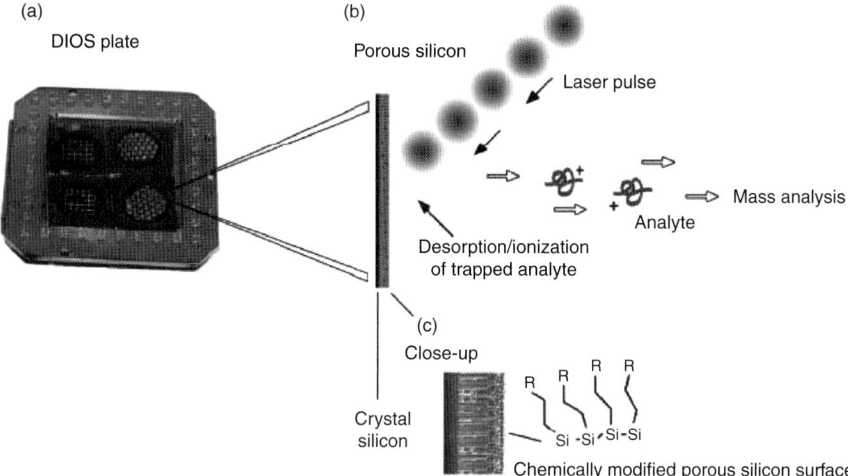

Figure 7.5. Experimental configuration for the DIOS-MS experiments. (**a**) Four porous silicon plates are placed on a MALDI plate. Each of the porous silicon plates contains photopatterned spots or grids prepared through illumination of *n*-type silicon with a 300-S tungsten filament through a mask and an *f*/50 reducing lens. (**b**) The silicon-based laser desorption/ionization process, in which the sample is placed on the porous silicon plate and allowed to dry, followed by laser-induced desorption/ionization mass spectrometry. (**c**) Cross section of porous silicon, and the surface functionalities after hydrosilylation; R represents phenyl or alkyl chains. (Reproduced with permission from Wei, J.; Buriak, J.; Siuzdak, G. Desorption/ionization mass spectrometry on porous silicon. *Nature* **1999**, *399*, 243–246.)

tissues (e.g., *Aplysia californica*) or cultured invertebrate neurons.[102] Recent efforts have investigated silicon modifications[103] and different applications such as metabolomics,[104] fatty acid analysis,[105] and surfactant-aided desorption/ionization[106] and its combination with an atmospheric-pressure MALDI source.[107]

7.4.2 Room-Temperature Ionic Liquid Matrices

Room-temperature ionic liquids (RTILs) are salts with melting points close to or below room temperature. They look like a classical liquid, but they do not contain any molecules: They are made of ions. Typically, an RTIL consists of nitrogen- or phosphorus-containing organic cations and large organic or inorganic anions.[108] They have good thermal stabilities, remain liquid over a range of 200°C to 300°C, and have practically no vapor pressure.[109] The general features of effective MALDI matrices are already known: They must dissolve (liquid matrix) or cocrystallize (solid matrix) with the sample, strongly absorb the laser light, remain in the condensed phase under high-vacuum conditions, stifle both chemical and thermal degradation of the sample, and promote the ionization of the sample via any of a number of mechanisms.[110–113] These requirements make ionic liquids possible candidates for MALDI matrices. They produce homogeneous solutions with great vacuum stabilities and are good solvents for a variety of organic, inorganic, and polymeric substances.[114]

It has been demonstrated that ionic liquids and solids make useful MALDI matrices.[115–121] With ionic matrices, it is possible to combine the beneficial qualities of liquid and solid matrices. Ionic liquids produce a much more homogeneous sample solution (as do all liquid matrices), yet they have greater vacuum stability than most solid matrices. In most cases, an ionic matrix can be found that produces greater spectral peak intensities and lower limits of detection than comparable solid matrices. Most ionic liquids readily dissolve biological oligomers, proteins, and polymers. However, ionic liquids can vary tremendously in their ability to promote analyte ionization. Both the cationic and anionic portion of the ionic matrix must be chosen with a consideration for the special requirements of UV MALDI detection. The ionic matrix must have significant absorbance at the desired wavelength, but also available protons. Most conventional ionic liquids that lack these properties are ineffective as MALDI matrices (Figure 7.6).

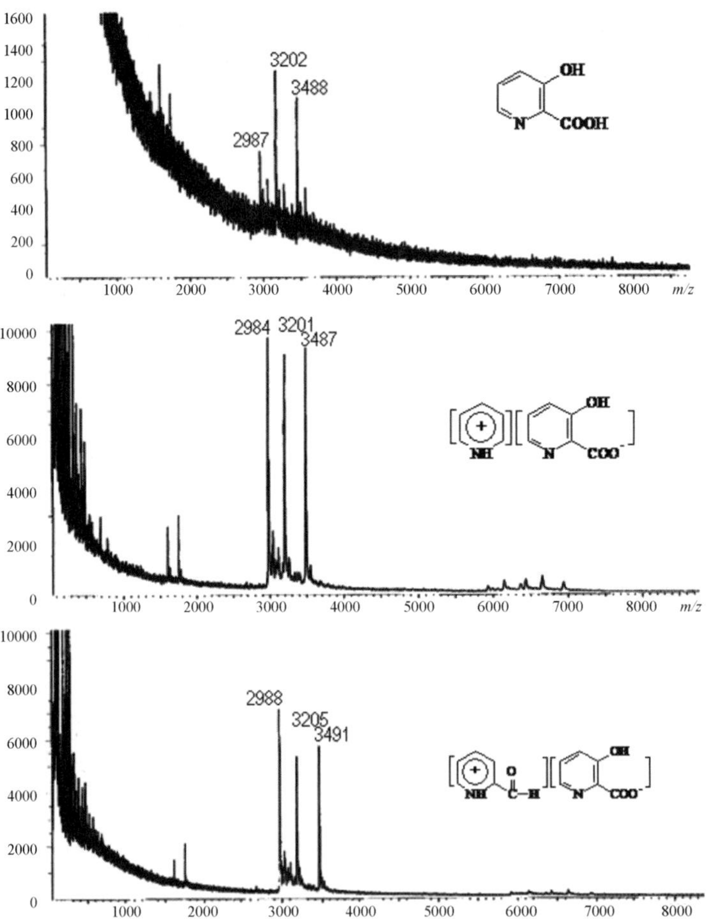

Figure 7.6. MALDI mass spectra of the three oligonucleotides [d(pT)10, d(pC),11 and d(pC)12] in different matrices: (**a**) 3-HPA + 10% ammonium citrate, (**b**) Ionic solid 21, and (**c**) ionic solid 26. Spectra obtained by accumulating 100 UV 237-nm laser shots. For the three experiments, the oligonucleotide–matrix molar ratio was 1:500,000 and the laser fluence was the same (attenuation 10). The signal strength is expressed in arbitrary units corresponding to the accumulation of 100 shots on a good spot. The 3-HPA scale (**top spectrum**) differs eightfold from that for the two salts (**middle and bottom spectra**). (Reproduced with permission from Carda-Broch, S.; Berthod, A.; Armstrong, D. Ionic matrices for matrix-assisted laser desorption/ionization time-of-flight detection of DNA oligomers. *Rapid Commun. Mass Spectrom.* **2003**, *17*, 553–560.)

Thirty-eight different ionic matrices having excellent solubilizing properties and vacuum stability were synthesized and tested with peptides, proteins, and poly(ethylene glycol) (PEG-2000) analytes.[115] Out of 38 ionic matrices tested, 20 ionic matrices produced homogeneous solutions of greater vacuum stability, higher ion peak intensity, and equivalent or lower detection limits than conventional solid matrices.

Zabet-Moghaddam et al.[120] characterized five different ionic liquids by LDI and by MALDI-MS. Signals of both anions and cations of the ionic liquids could be observed both in LDI- and in MALDI-MS without any metal adducts. Low-molecular-mass compounds and peptides could be analyzed best in the presence of water-immiscible ionic liquids, whereas proteins gave the best results in water-miscible ionic liquids. Optimal analytical conditions depend on the molar ratio of matrix-to-analyte and ionic liquid-to-matrix, although the homogeneity of samples in the presence of ionic liquids was reduced compared with classical MALDI preparations.

Recently, direct tissue profiling of peptides with ionic liquids made of matrix mixtures [CHCA/2-amino-4-methyl-5-nitropyridine (CHCA/2A4M5NP) and CHCA/*N*,*N*-dimethylaniline (CHCA/DANI)] was reported.[122] The ionic matrices were found to have several advantageous features for direct tissue analysis, especially CHCA/aniline as compared to CHCA, 2,5-DHB, and SA. These advantages included improved signal intensity and sensitivity, homogeneous crystallization, the possibility of *in situ* partial fragmentations, a high resistance to laser ablation especially using a high repetition rate laser, and the possibility to analyze compounds in positive and negative modes using the same slice for MALDI-MS high-quality imaging.

7.4.3 Nonpolar Matrices

Several nonpolar matrices are suitable for MALDI-MS analysis of low-molecular-weight nonpolar analytes. Terthiophene and anthracene, both of which in positive mode form only molecular radical cations ($M^{+\bullet}$) upon laser irradiation ($\lambda = 337$ nm) at near-threshold laser powers, have been used as MALDI matrices to promote the electron-transfer ionization of metallocenes (1,2-diferrocenylethane, ferrocene, and decamethylferrocene).[123] Analysis of several nonpolar analytes, including ferrocene and ferrocene derivatives, *trans*-stilbene, triphenylphosphine, 2,2′-methylenebis(6-*tert*-butyl-4-methylphenol), biphenyl and 1,4-bis (methylthio)benzene with different nonpolar matrices (acenaphthene, anthracene, perylene, pyrene, terthiophene, and triphenylene), demonstrated that the formation of the radical molecular cation depended on the difference in ionization energies between the matrix and the analyte.[124] Nonpolar matrices have also been shown to be effective as matrices for the analysis of low-molecular-weight hydrocarbon synthetic polymers. Anthracene, pyrene, and acenaphthene were utilized for the analysis of polybutadiene, polyisoprene, and polystyrene analytes of various average molecular weights ranging from ~700 to 5000 Da. The nonpolar matrices used in the study were shown to be equally effective with increased reproducibility compared to conventional acidic matrices. Silver salts were found to be the best cationization reagents for all of the cases studied, whereas copper salts worked well for polystyrene, poorly for polyisoprene, and not at all for polybutadiene analytes.[96,125]

A series of nonpolar matrices were evaluated for the analysis of synthetic polymers and compared to that of conventional ones. Among substituted polycyclic aromatic hydrocarbons 1,4-dihydroxy-2-naphthoic acid, 9-anthracenecarboxylic acid (9-ACA), and its mixtures with 5-methoxysalicylic acid ("super" 9-ACA) provided better signal for analysis of synthetic polymers – poly(ethylene glycol), poly(propylene glycol), Jeffamine, polybutadiene, poly(methyl methacrylate), poly(dimethyl siloxane), and some of their mixtures. 9-ACA provided better performance and higher effective mass resolution with the formation of $[M + Na]^+$ ions while the "super" 9-ACA composite matrix was found to be more suitable for larger polymer molecules (>8000 Da).[126] Two nonpolar matrices, anthracene and 9-cyanoanthracene, were found suitable for the analysis of bottom total liquid products (BTLP) resulting from crude oil distribution with either linear or reflectron mode MALDI-TOF-MS, compare to Field Ionization-MS (FI-MS) or laser-desorption ionization-MS (LDI-MS). These matrices were relatively insensitive to sample preparation conditions and yielded molecular mass distributions ranging from 200 to 4000 Da (linear mode) and from 200 to 600 Da (reflectron mode). They do not interfere with the analysis of lower-molecular components of the BTLP analytes because the matrices do not fragment or cluster.[127]

7.4.4 Carbon Nanotubes

Carbon nanotubes, prepared from coal by an arc discharge method, were investigated as the matrix for analyses of small molecules by MALDI-MS.[128] It was observed that the carbon nanotube layer on the probe before deposition of sample solution appears to have web morphology, which was masked by formation of analyte microcrystals after deposition of the sample solution. However, there is almost no change in configuration and shape for a single piece of carbon nanotube before and after the deposition of sample solution on the carbon nanotube layer (Figure 7.7). Nanotubes provide a lower laser power threshold for

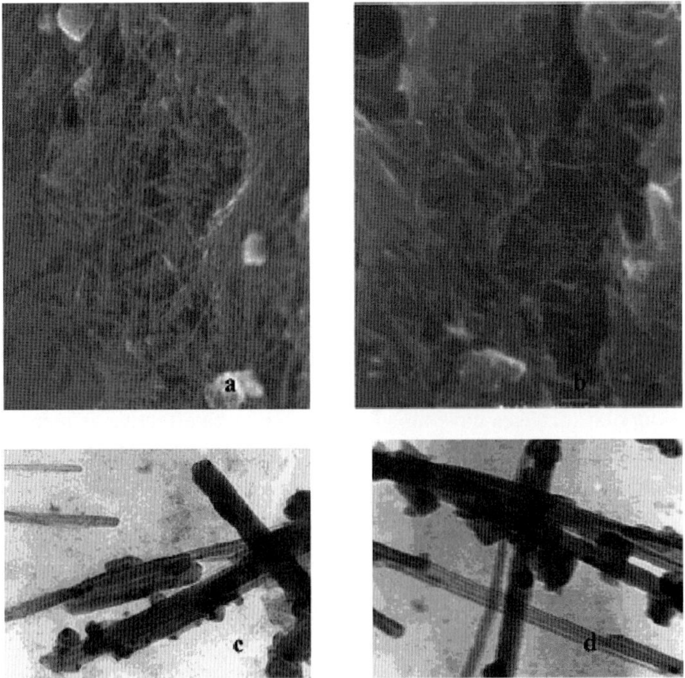

Figure 7.7. Morphology of carbon nanotube layer on probe well before and after dropping sample solution. (**a,b**) SEM image (×5000) of carbon nanotube layer before and after dropping sample solution. (**c,d**) TEM image (×100,000) of carbon nanotube layer before and after deposition of sample solution. (Reproduced with permission from Xu, S.; Li, Y.; Zou, H.; Qiu, J.; Gau, Z.; Guo, B. Carbon nanotubes as assisted matrix for laser desorption/ionization time-of-flight mass spectrometry. *Anal. Chem.* **2003**, *75*, 6191–6195.)

desorption/ionization and higher detection sensitivity and mass resolution of small molecules than organic matrices. Carbon nanotubes may greatly simplify sample preparation, eliminate interference of matrix background ions, and improve shot-to-shot reproducibility. A type of polyurethane adhesive, NIPPOLAN-DC 205, was employed to improve the applicability of carbon nanotubes as matrices for analyses of low-mass-weight compounds.[129]

A functionalized carbon nanotube (CNT), CNT 2,5-dihydroxybenzyl hydrazine derivative, was synthesized and used both as a pH-adjustable enriching reagent and as a matrix in MALDI-MS analysis of trace peptides.[130] Functionalized CNT provided twofold advantages over pristine CNT. The efficiency of enrichment of trace peptides was greatly enhanced as a result of increased surface area to volume ratio for adsorption in basic conditions. It also provided a convenient method for quick isolation. Compared with CHCA without enrichment, the detection limit for analytes can be enhanced about 10- to 100-fold after enrichment by the functionalized CNT (Figure 7.8).

7.4.5 Combined Detergent/MALDI Matrix

A functional cleavable detergent designed specifically for use in MALDI mass spectrometry was first reported by Norris et al.[131] (Figure 7.9). These combined detergents/MALDI matrices have two distinct functions: solubilize hydrophobic biomolecules and enhance MALDI mass analysis. The matrix can be generated *in situ* by adjusting the pH of the solution. Analysis of the membrane protein diacylglycerol kinase was accomplished using

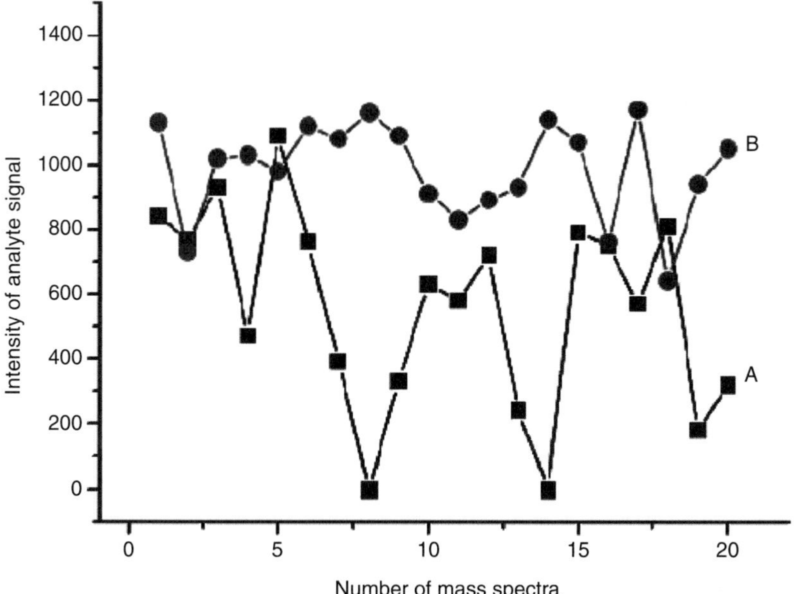

Figure 7.8. Comparison of intensity of analyte ion signal in the discrete location of probe well with the matrixes of (**A**) CCA and (**B**) carbon nanotubes. Twenty pulsed laser shots were applied under laser power set at 205 μJ. Equal concentration of analyte BAEE was used for MS experiments in both cases. (Reprinted with permission from Xu, S.; Li, Y.; Zou, H.; Qiu, J.; Gau, Z.; Guo, B. Carbon nanotubes as assisted matrix for laser desorption/ionization time-of-flight mass spectrometry. *Anal. Chem.* **2003**, *75*, 6191–6195. Copyright 2003 American Chemical Society.)

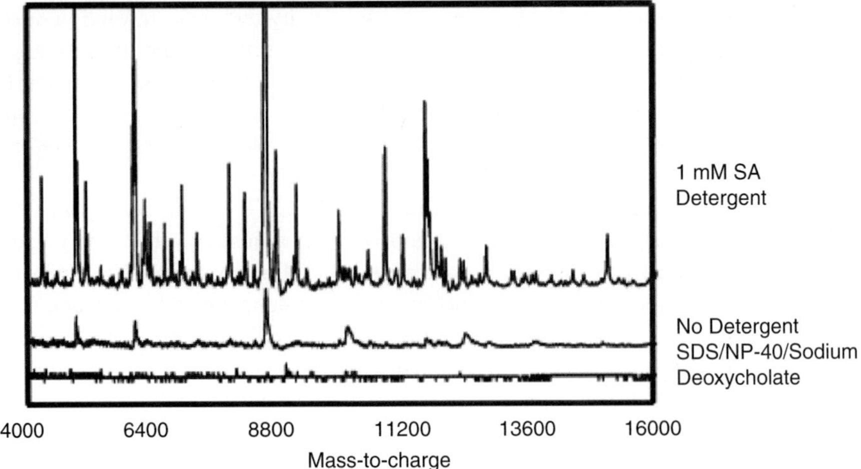

Figure 7.9. MALDI-MS analysis of RKO cell lysates using a combination detergent/MALDI matrix. RKO cell extracts prepared using **2** (1 mM) (**top**), without detergent (**middle**), and using noncleavalble detergents (0.1% SDS, 1% Nonidet P-40, 0.5% sodium deoxycholate) (**bottom**) were analyzed using MALDI-MS. The samples were mixed on target with an acidic solution of sinnapinic acid (20 mg/mL, pH < 1). Compound **2** facilitates detection of greater number of ions and increases signal-to-noise peaks common to controls. (Reprinted with permission from Norris, J.; Porter, N.; Caprioli, R. Combination detergent/MALDI matrix: Functional cleavage detergents for mass spectrometry. *Anal. Chem.* **2005**, *77*, 5036–5040. Copyright 2005 American Chemical Society.)

the combination detergent/MALDI matrix. When it is used to prepare membrane proteins and whole-cell lysates, these samples can be profiled using MALDI-MS. Applications of the functional cleavable detergents to the profiling of whole-cell lysates results in increased signal-to-noise ratios of many ions and the detection of additional proteins previously not observed.

7.5 HOT VERSUS COLD MATRICES

A better understanding of the influence of the matrix on ion fragmentation could lead to "tailor-made" sample preparations corresponding to a particular application that would produce higher-quality MALDI-MS data. MALDI-MS relies on the formation of intact molecular ions to achieve high-accuracy molecular weight determination. Problems arising from post-source decay (PSD)—ion dissociation due to unimolecular and bimolecular collision-induced reactions—strongly affect MALDI results in reflectron TOF instruments. It was found that the efficiency of collision-induced fragmentation is dependent not only on the collision energy and particle size, but also, to a considerable extent, on the internal energy of the molecules prior to a collision.[132]

It can be assumed that the matrix will transfer energy, taken up from the laser, into the analyte molecules to a varying extent, depending on the physiochemical properties of the matrix–analyte system. Sinapinic acid has been found to transfer more energy to analyte molecules compared to 2,5-DHB.[132] The strongly different unimolecular decay rates and observed different relative stabilities of parent molecular ions of cytochome c against a wide range of residual gas pressure [10^{-7}–10^{-4} mbar] obtained for these two matrices also support the notion that different energy is transferred from these matrices to the analyte. Matrix-induced internal cooling or heating of the analyte molecules might additionally take place in the dense plume just above the surface due to collisions between matrix and analyte. The kinetics of the jet of matrix molecules as well as the acceleration field strength will then affect this cooling or heating process.

MALDI matrices were initially characterized as "hot" or "cold," depending on their propensity to induce fragmentation of glycoproteins.[133] PSD efficiency was found to decrease in the order CHCA > SA > 2,5-DHB > DHBs ~ 3-HPA. The reported data showed that the matrix plays a key role in the stability and thus the fate of the ions produced by the MALDI process. The simplest explanation would be that different matrices yield a variably dense ablation plume due to their different ablation characteristics. It thus appears that even though the excitation energy stems from the same source, it would be higher in a denser plume due to a larger number of collisions. According to the photochemical MALDI ionization model, ionization occurs in the dense ablation plume via a bimolecular reaction between the matrix and analyte molecules, and the excess energy transferred to the ionized analyte molecule is determined by the gas-phase proton affinity difference between matrix and analyte. This result was also supported by other studies.[134] Zhu et al.[135] found that fragmentation pattern of oligodeoxynucleotides in UV-MALDI at 355 nm is different for 3-HPA compared to 2,5-DHB as a matrix. Ivleva et al.[136] have tested a set of 20 matrices for gangliosides and for phosphopeptide samples desorbed from TLC plate as well as from a standard stainless steel target. The mass spectral fragmentation patterns of the analyte were recorded as a function of a laser power that enabled marking each matrix in hot or cold terms.

Vertes reported the first quantification of ion internal energies produced in MALDI by using a set of substituted benzylpyridinium salts as thermometer molecules to probe the energy transferred from the matrix to the analyte.[137] The experiments were performed in the continuous extraction linear mode, and consequently the detected fragments are formed very shortly after laser excitation. Contrary to prior results, they consider CHCA, SA, and 2,5-DHB as cool, intermediate, and hot matrices, respectively, at the laser pulse energies that induce fragmentation in thermometer molecules. Gabelica et al.[138] reported studies on the internal energy of ions formed in vacuum MALDI-MS and AP MALDI-MS using benzylpyridinium (BzPy) cations as internal energy probes. Their study reveals three distinct contributions to internal energy, each having different effects on ion fragmentation. The relative "hotness" of the matrix (CHCA > 2,5-DHB > SA) with benzylpyridinium ions correlates well with the matrix sublimation temperature (455 > 443 > 426).

REFERENCES

1. Fenner, N.; Daly, N. Laser used for mass analysis. *Rev. Sci. Instrum.* **1966**, *37*, 1068–1070.
2. Vastola, F.; Pirone, A. Ionization of organic solids by laser irradiation. *Adv. Mass. Spectrom.* **1968**, *4*, 107–111.
3. Vastola, F.; Mumma, R.; Pirone, A. Analysis of organic salts by laser ionization. *Org. Mass. Spectrom.* **1970**, *3*, 101–104.
4. Mumma, R.; Vastola, F. Analysis of organic salts by laser ionization mass spectrometry. Sulfonates, sulfates and thiosulfates. *Org. Mass Spectrom.* **1972**, *6*, 1373–1376.
5. Posthumus, M.; Kistemaker, P.; Meuzelaar, H.; Ten Noever de Brauw, M. Laser desorption–mass spectrometry of polar nonvolatile bioorganic molecules. *Anal. Chem.* **1978**, *50*, 985–991.
6. Coates, M.; Wilkins, C. Laser desorption fourier transform mass spectra of malto-oligosaccharides. *Biomed. Mass. Spectrom.* **1985**, *12*, 424–428.
7. Beavis, R. Matrix-assisted ultraviolet laser desorption: evolution and principles. *Org. Mass Spectrom.* **1992**, *27*, 653–659.

8. Tanaka, K.; Waki, H.; Ido, Y.; Akita, S.; Yoshida, Y.; Yoshida, T. Protein and polymer analyses up to *m/z* 100, 000 by laser ionization time-of-flight mass spectrometry. *Rapid. Commun. Mass Spectrom.* **1988**, *2*, 151–153.
9. Karas, M.; Hillenkamp, F. Laser desorption ionization of proteins with molecular masses exceeding 10000 Da. *Anal. Chem.* **1988**, *60*, 2299–2301.
10. Tanaka, K. The origin of macromolecule ionization by laser irradiation (nobel lecture). *Angew. Chem. Int. Ed.* **2003**, *42*, 3861–3870.
11. Tanaka, K.; Ido, Y.; Akita, S.; Yoshida, Y.; Yoshida, T. Detection of high mass molecules by laser desorption time of flight mass spectrometry. In *Proceedings of the Second Japan–China Joint Symposium on Mass Spectrometry*, **1987**, pp. 185–188.
12. Dale, M.; Knochenmus, R.; Zenobi, R. Graphite/liquid mixed matrices for laser desorption/ionization mass spectrometry. *Anal Chem* **1996**, *68*, 3321–3329.
13. Dale, M.; Knochenmuss, R.; Zenobi, R. Two-phase matrix-assisted laser desorption/ionization: matrix selection and sample pretreatment for complex anionic analytes. *Rapid Commun. Mass Spectrom.* **1997**, *11*, 136–142.
14. Kraft, P.; Alimpiev, S.; Dratz, E.; Sunner, J. Infrared, surface-assisted laser desorption ionization mass spectrometry on frozen aqueous solutions of proteins and peptides using suspensions of organic solids. *J. Am. Soc. Mass Spectrom.* **1998**, *9*, 912–924.
15. Schurenberg, M.; Dreisewerd, K.; Hillenkamp, F. Laser desorption/ionization mass spectrometry of peptides and proteins with particle suspension matrices. *Anal. Chem.* **1999**, *71*, 221–229.
16. Dietemann, P.; Edelmann, M.; Meisterhans, C.; Pfeiffer, C.; Zumbuhl, S.; Knochenmuss, R.; Zenobi, R. Artificial photoaging of triterpenes studied by graphite-assisted laser desorption/ionization mass spectrometry. *Helv. Chim. Acta* **2000**, *83*, 1766–1776.
17. Karas, M.; Bachmann, D.; Hillenkamp, F. Influence of the wavelength in high-irradiance ultraviolet laser desorption mass spectrometry of organic molecules. *Anal. Chem.* **1985**, *57*, 2935–2939.
18. Karas, M.; Bachmann, D.; Bahr, U.; Hillenkamp, F. Matrix-assisted ultraviolet laser desorption of non-volatile compounds. *Int. J. Mass Spectrom. Ion Processes* **1987**, *78*, 53–68.
19. Hillenkamp, F.; Karas, M. Matrix-assisted laser desorption/ionization, an experience. *Int. J. Mass Spectrom.* **2000**, *200*, 71–77.
20. Dreisewerd, K. The desorption process in MALDI. *Chem. Rev.* **2003**, *103*, 395–425.
21. Jagtap, R.; Ambre, A. Overview literature on matrix assisted laser desorption ionization mass spectrometry (MALDI MS): Basics and its applications in characterizing polymeric materials. *Bull. Mater. Sci.* **2005**, *28*, 515–528.
22. Karas, M.; Bahr, U.; Stahl-Zeng, J. Steps towards a more refined picture of the matrix function in UV MALDI. In Baer, T.; Ng, C.; Powis, I. (Eds.), *Large Ions: Their vaporization, detection and Structural Analysis*, John Wiley & Sons, New York, **1996**, pp. 27–48.
23. Limbach, P. Matrix-assisted laser desorption-ionization mass spectrometry: An overview. *Spectroscopy* **1998**, *13*, 16–27.
24. Kussmann, M.; Nordhoff, E.; Rahbek-Nielsen, H.; Haebel, S.; Rossel-Larsen, M.; Jakobsen, L.; Gobom, J.; Mirgorodskaya, E.; Kroll-Kristensen, A.; Palm, L.; Roepstorff, P. Matrix-assisted laser desorption/ionization mass spectrometry sample preparation techniques designed for various peptide and protein analytes. *J. Mass Spectrom.* **1997**, *32*, 593–601.
25. Nordhoff, E.; Schürenberg, M.; Thiele, G.; Lübbert, C.; Kloeppel, K.-D.; Theiss, D.; Lehrach, H.; Gobom, J. *Int. J. Mass Spectrom.* **2003**, *226*, 163–180.
26. Stahl, B.; Steup, M.; Karas, M.; Hillenkamp, F. Analysis of neutral oligosaccharides by matrix-assisted laser desorption/ionization mass spectrometry. *Anal. Chem.* **1991**, *63*, 1463–1466.
27. Dai, Y.; Whittal, R.; Li, L. Confocal fluorescence microscopic imaging for investigating the analyte distribution in MALDI matrices. *Anal. Chem.* **1996**, *68*, 2494–2500.
28. Zheng, J.; Li, N.; Ridyard, M.; Dai, H.; Robbins, S.; Li, L. Simple and robust two-layer matrix/sample preparation method for MALDI MS/MS analysis of peptides. *J. Proteome Res.* **2005**, *4*, 1709–1716.
29. Cohen, S.; Chait, B. Influence of matrix solution conditions on the MALDI-MS analysis of peptides and proteins. *Anal. Chem.* **1996**, *68*, 31–37.
30. Xiang, F.; Beavis, R. A method to increase contaminant tolerance in protein matrix-assisted laser desorption/ionization by the fabrication of thin protein-doped polycrystalline films. *Rapid Commun. Mass Spectrom.* **1994**, *8*, 199–204.
31. Vorm, O.; Roepstorff, P.; Mann, M. Improved resolution and very high sensitivity in MALDI TOF of matrix surfaces made by fast evaporation. *Anal. Chem.* **1994**, *66*, 3281–3287.
32. Dai, Y.; Whittal, R.; Li, L. Two-layer sample preparation: A method for MALDI-MS analysis of complex peptide and protein mixtures. *Anal. Chem.* **1999**, *71*, 1087–1091.
33. Li, L.; Golding, R.; Whittal, R. Analysis of single mammalian cell lysates by mass spectrometry. *J. Am. Chem. Soc.* **1996**, *118*, 11662–11663.
34. Perera, I.; Perkins, J.; Kantartzoglou, S. Spin-coated samples for high resolution matrix-assisted laser desorption/ionization time-of-flight mass spectrometry of large proteins. *Rapid Commun. Mass Spectrom.* **1995**, *9*, 180–187.
35. Hensel, R.; King, R.; Owens, K. Electrospray sample preparation for improved quantitation in MALDI TOF MS. *Rapid Commun. Mass Spectrom.* **1997**, *11*, 1785–1793.
36. Axelsson, J.; Hoberg, A. -M.; Waterson, C.; Myatt, P.; Chield, G.; Varney, J.; Haddleton, D.; Derrick, J.

Improved reproducibility and increased signal intensity in MALDI as a result of electrospray sample preparation. *Rapid Commun. Mass Spectrom.* **1997**, *11*, 209–213.

37. Caprioli, R.; Farmer, T.; Gile, J. Molecular imaging of biological samples: Localization of peptides and proteins using MALDI-TOF MS. *Anal Chem* **1997**, *69*, 4751–4760.
38. Mowthorpe, S.; Clench, M.; Cricelius, A.; Richards, D.; Parr, V.; Tetler, L. MALDI TOF/Thin layer chromatography/MS—a rapid method for impurity testing. *Rapid Commun. Mass Spectrom.* **1999**, *13*, 264–270.
39. Zhang, H.; Caprioli, R. Direct analysis of aqueous samples by matrix-assisted laser desorption ionization mass spectrometry using membrane targets precoated with matrix. *J. Mass Spectrom.* **1996**, *31*, 690–692.
40. Preston, L.; Murray, K.; Russell, D. Reproducibility and quantitation of MALDI-MS: Effects of nitrocellulose on peptide ion yields. *Biol. Mass Spectrom.* **1993**, *22*, 544–550.
41. Zhao, S.; Somayajula, K.; Sharkey, A.; Hercules, D.; Hillenkamp, F.; Karas, M.; Ingendoh, A. Novel methods for matrix-assisted mass spectrometry of proteins. *Anal. Chem.* **1991**, *63*, 450–453.
42. Sunner, J.; Dratz, E.; Chen, Y. Graphite surface-assisted laser desorption/ionization time-of-flight mass spectrometry of peptides and proteins from liquid solutions. *Anal. Chem.* **1995**, *67*, 4335–4342.
43. Sze, E.; Chan, T.-W.; Wang, G. Formulation of matrix solutions for use in matrix-assisted laser desorption/ ionization of biomolucules. *J. Am. Soc. Mass Spectrom.* **1998**, *9*, 166–174.
44. Wei, J.; Buriak, J.; Siuzdak, G. Desorption/ionization mass spectrometry on porous silicon. *Nature* **1999**, *399*, 243–246.
45. Skelton, R.; Dubois, F.; Zenobi, R. MALDI sample preparation method suitable for insoluble polymers. *Anal. Chem.* **2000**, *72*, 1707–1710.
46. Strupat, K.; Karas, M.; Hillenkamp, F. 2,5-Dihydroxybenzoic acid: A new matrix for laser desorption/ionization mass spectrometry. *Int. J. Mass Spectrom. Ion Processes* **1991**, *11*, 89–102.
47. Karas, M.; Ehring, H.; Nordhoff, E.; Stahl, B.; Strupat, K.; Hillenkamp, F.; Grehl, M.; Krebs, B. Matrix-assisted laser desorption/ionization mass spectrometry with additives to 2,5-dihydroxybenzoic acid. *Org. Mass Spectrom.* **1993**, *28*, 1476–1481.
48. Beavis, R.; Chait, B. Cinnamic acid derivatives as matrices for ultraviolet laser desorption mass spectrometry of proteins. *Rapid Commun. Mass Spectrom.* **1989**, *3*, 432–435.
49. Beavis, R.; Chait, B. Rapid, sensitive analysis of protein mixtures by mass spectrometry. *Proc. Natl. Acad. Sci. USA* **1990**, *87*, 6873–6877.
50. Doktycz, S.; Savickas, P.; Frueger, D. Matrix/sample interactions in ultraviolet laser-desorption of proteins. *Rapid Commun. Mass Spectrom.* **1991**, *5*, 145–148.
51. Beavis, R.; Chaudhary, T.; Chait, B. α-Cyano-4-hydroxycinnamic acid as a matrix-assisted desorption mass spectrometry. *Org. Mass Spectrom.* **1992**, *27*, 156–158.
52. Gorman, J.; Ferguson, B.; Nguyen, T. Use of 2,6-dihydroxyacetophenone for analysis of fragile peptides, disulphide bonding and small proteins by matrix-assisted laser desorption/ionization. *Rapid Commun. Mass Spectrom.* **1996**, *10*, 529–536.
53. Billeci, T.; Stults, J. Tryptic mapping of recombinant proteins by matrix-assisted laser desorption/ionization mass spectrometry. *Anal. Chem.* **1993**, *65*, 1709–1718.
54. Vermillion-Salsbury, R.; Hercules, D. 9-Aminoacridine as a matrix for negative mode matrix-assisted laser desorption/ionization. *Rapid Commun. Mass Spectrom.* **2002**, *16*, 1575–1581.
55. Mohr, M.; Bornsen, K.; Widmer, H. Matrix-assisted laser desorption/ionization mass spectrometry: Improved matrix for oligosaccharides. *Rapid Commun. Mass Spectrom.* **1995**, *9*, 809–814.
56. Papac, D.; Wong, A.; Jones, A. Analysis of acidic oligosaccharides and glycopeptides by matrix-assisted laser desorption/ionization time-of-flight mass spectrometry. *Anal. Chem.* **1996**, *68*, 3215–3223.
57. Harvey, D. Matrix-assisted laser desorption/ionization mass spectrometry of carbohydrates and glycoconjugates. *Int. J. Mass Spectrom.* **2003**, *226*, 1–35.
58. Xu, N.; Huang, Z.-H.; Watson, J.; Gage, D. Mercaptobenzothiazoles: A new class of matrices for laser desorption/ionization mass spectrometry. *J. Am. Soc. Mass Spectrom.* **1997**, *8*, 116–124.
59. Nonami, H.; Fukui, S.; Erra-Balsells, R. β-Carboline alkaloids as matrices for matrix-assisted ultraviolet laser desorption time-of-flight mass spectrometry of proteins and sulfated oligosaccharides: a comparative study using phenylcarbonyl compounds, carbazoles and classical matrices. *J. Mass Spectrom.* **1997**, *32*, 287–296.
60. Nonami, H.; Tanaka, K.; Fukuyama, Y.; Erra-Balsells, R. β-Carboline alkaloids as matrices for UV-matrix-assisted laser desorption/ionization time-of-flight mass spectrometry in positive and negative ion modes. Analysis of proteins of high molecular mass, and of cyclic and acyclic oligosaccharides. *Rapid Commun. Mass Spectrum.* **1998**, *12*, 285–296.
61. Nonami, H.; Wu, F.; Thummel, R.; Fukuyama, Y.; Yamaoka, H.; Erra-Balsells, R. Evaluation of pyridoindoles, pyridylindoles and pyridylpyridoindoles as matrices for ultraviolet matrix-assisted laser desorption/ionization time-of-flight mass spectrometry. *Rapid Commun. Mass Spectrom.* **2001**, *15*, 2354–2373.
62. Wong, A.; Cancilla, M.; Voss, L.; Lebrilla, C. Anion dopant for oligosaccharides in matrix-assisted laser desorption/ionization mass spectrometry. *Anal. Chem.* **1999**, *71*, 205–211.
63. Cai, Y.; Jiang, Y.; Cole, R. Anionic adducts of oligosaccharides by matrix-assisted laser desorption/

ionization time-of-flight (MALDI-TOF) mass spectrometry. *Anal. Chem.* **2003**, *75*, 1638–1644.

64. Pitt, J.; Gorman, J. Matrix-assisted laser desorption/ionization time-of-flight mass spectrometry of sialylated glycopeptides and proteins using 2,6-dihydroxyacetophenone as a matrix. *Rapid Commun. Mass Spectrom.* **1996**, *10*, 1786–1788.
65. Parr, G.; Fitzgerald, M.; Smith, L. Matrix-assisted laser desorption/ionization mass spectrometry of synthetic oligodeoxyribonucleotides. *Rapid Commun. Mass Spectrom.* **1992**, *6*, 369–372.
66. Tang, K.; Allman, S.; Chen, C. Matrix-assisted laser desorption ionization of oligonucleotides with various matrices. *Rapid Commun. Mass Spectrom.* **1993**, *7*, 943–948.
67. Wu, K.; Steding, A.; Becker, C. Matrix-assisted laser desorption time-of-flight mass spectrometry of oligonucleotides using 3-hydroxypicolinic acid as an ultraviolet-sensitive matrix. *Rapid Commun Mass Spectrom* **1993**, *7*, 142–146.
68. Tang, K.; Taranenko, N.; Allman, S.; Chen, C.; Chang, L.; Jacobson, K. Picolinic acid as a matrix for laser mass spectrometry of nucleic acids and proteins. *Rapid Commun. Mass Spectrom.* **1994**, *8*, 673–677.
69. Tang, K.; Taranenko, N.; Allman, S.; Chang, L.; Chen, C. Detection of 500-nucleotide DNA by laser desorption mass spectrometry. *Rapid Commun. Mass Spectrom.* **1994**, *8*, 727–730.
70. Lecchi, P.; Le, H.; Pannell, L. 6-Aza-2-thiothymine: A matrix for MALDI spectra of oligonucleotides. *Nucleic Acids Res.* **1995**, *23*, 1276–1277.
71. Pieles, U.; Zürcher, W.; Schär, M.; Moser, H. Matrix-assisted laser desorption ionization time-of-flight mass spectrometry: A powerful tool for the mass and sequence analysis of natural and modified oligonucleotides. *Nucleic Acids Res.* **1993**, *21*, 3191–3196.
72. Lavanant, H.; Lange, C. Sodium-tolerant matrix for matrix-assisted laser desorption/ionization mass spectrometry and post-source decay of oligonucleotides. *Rapid Commun. Mass Spectrom.* **2002**, *16*, 1928–1933.
73. Zhu, Y.; Chung, C.; Taranenko, N.; Allman, S.; Martin, S.; Haff, L.; Chen, C. The study of 2,3,4-trihydroxyacetophenone and 2,4,6-trihydroxyacetophenone as matrices for DNA detection in matrix-assisted laser desorption/ionization time-of-flight mass spectrometry. *Rapid Commun. Mass Spectrom.* **1996**, *10*, 383–388.
74. Zhu, Y.; Tarnenko, N.; Allman, S.; Martin, S.; Haff, L.; Chen, C. The effect of ammonium salt and matrix in the detection of DNA by matrix-assisted laser desorption/ionization time-of-flight mass spectrometry. *Rapid Commun. Mass Spectrom.* **1996**, *10*, 1591–1596.
75. Simmons, T.; Limbach, P. Influence of co-matrix proton affinity on oligonucleotide ion stability in matrix-assisted laser desorption/ionization time-of-flight mass spectrometry. *J Am. Soc. Mass Spectrom.* **1998**, *9*, 668–675.
76. Simmons, T.; Limbach, P. The use of a co-matrix for improved analysis of oligonucleotides by matrix-assisted laser desorption/ionization mtime-of-flight mass spectrometry. *Rapid Commun. Mass Spectrom.* **1997**, *11*, 567–572.
77. Vandell, V.; Limbach, P. Polyamine co-matrices for matrix-assisted laser desorption/ionization mass spectrometry of oligonucleotides. *Rapid Commun. Mass Spectrom.* **1999**, *13*, 2014–2021.
78. Asara, J.; Allison, J. Enhanced detection of oligonucleotides in UV MALDI MS using the tetraamine spermine as a matrix additive. *Anal. Chem.* **1999**, *71*, 2866–2870.
79. Zhou, L.; Deng, H.; Deng, Q.; Zhao, S. A mixed matrix of 3-hydroxypicilinic acid and pyrazinecarboxylic acid for matrix-assisted laser desorption/ionization time-of-flight mass spectrometry of oligodeoxynucleotides. *Rapid Commun. Mass Spectrom.* **2004**, *18*, 787–794.
80. Fu, Y.; Xu, S.; Pan, C.; Ye, M.; Zou, H.; Guo, B. A matrix of 3,4-diaminobenzophenone for the analysis of oligonucleotides by matrix-assisted laser desorption/ionization time-of-flight mass spectrometry. *Nucleic Acids Res* **2006**, *34*, e94.
81. Schiller, J.; Süb, R.; Arnhold, J.; Fuchs, B.; Leßig, J.; Müller, M.; Petković, M.; Spalteholz; Zschörnig O.; Arnold, K. Matrix-assisted laser desorption and ionization time-of-flight (MALDI-TOF) mass spectrometry in lipid and phospholipid research. *Prog. Lipid Res.* **2004**, *43*, 449–488.
82. Benard, A.; Arnhold, J.; Lehnert, M.; Schiller, J.; Arnold, K. Experiments towards quantification of saturated and polyunsaturated diacylglycerols by matrix-assisted laser desorption and ionization time-of-flight mass spectrometry. *Chem. Phys. Lipids* **1999**, *100*, 115–125.
83. Petkovic, M.; Schiller, J.; Muller, M.; Benard, S.; Reichl, S.; Arnold, K.; Arnhold, J. Detection of individual phospholipids in lipid mixtures by matrix-assisted laser desorption/ionization time-of-flight mass spectrometry: Phosphotidylcholine prevents the detection of further species. *Anal. Biochem.* **2001**, *289*, 202–216.
84. Schiller, J.; Süb, R.; Petković, M.; Hilbert, N.; Arnhold, J.; Müller, M.; Zschörnig, O.; Arnhold, J.; Arnold, K. CsCl as an auxiliary reagent for the analysis of phosphatidylcholine mixtures by matrix-assisted laser desorption and ionization time-of-flight mass spectrometry (MALDI-TOF MS). *Chem. Phys. Lipids* **2001**, *113*, 123–131.
85. Zabrouskov, V.; Al-Saad, K.; Siems, W.; Hill, H.; Knowles, N. Analysis of plant phosphatidylcholines by matrix-assisted laser desorption/ionization time-of-flight mass spectrometry. *Rapid Commun. Mass Spectrom.* **2001**, *15*, 935–940.

86. Müller, M.; Schiller, J.; Petkovi, M.; Oehrl, W.; Heinze, R.; Wetzker, R.; Arnold, K.; J, A. Limits for the detection of (poly-)phosphoinositides by matrix-assisted laser desorption and ionization time-of-flight mass spectrometry (MALDI-TOF MS). *Chem. Phys. Lipids* **2001**, *110*, 151–164.
87. Al-Saad, K.; Siems, W.; Hill, H.; Zabrouskov, V.; Knowles, N. Structural analysis of phosphatidylcholines by post-source decay matrix-assisted laser desorption/ionization time-of-flight mass spectrometry. *J. Am. Soc. Mass Spectrom.* **2003**, *14*, 373–382.
88. Petkovi, M.; Schiller, J.; Müller, J.; Müller, M.; Arnold, K.; J, A. The signal-to-noise ratio as the measure for the quantification of lysophospholipids by matrix-assisted laser desorption/ionization time-of-flight mass spectrometry. *Analyst* **2001**, *126*, 1042–1050.
89. Schiller, J.; Arnhold, J.; Glander, H.; Arnold, K. Lipid analysis of human spermatozoa and seminal plasma by MALDI-TOF mass spectrometry and NMR spectroscopy—Effects of freezing and thawing. *Chem. Phys. Lipids* **2000**, *106*, 1450156.
90. Harvey, D. Matrix-assisted laser desorption/ionization mass spectrometry of phospholipids. *J. Mass Spectrom.* **1995**, *30*, 1333–1346.
91. Juhasz, P.; Costello, C. Matrix-assisted laser desorption ionization time-of-flight mass spectrometry of underivatized and permethylated gangliosides. *J. Am. Soc. Mass Spectrom.* **1992**, *3*, 785–796.
92. Macha, S.; Limbach, P. Matrix-assisted desorption/ionization (MALDI) mass spectrometry of polymers. *Curr. Opin. Solid State Mater. Sci.* **2002**, *6*, 213–220.
93. Danis, P.; Karr, D.; Xiong, Y.; Owens, K. Methods for the analysis of hydrocarbon polymers by matrix-assisted laser desorption/ionization time-of-flight mass spectrometry. *Rapid Commun. Mass Spectrom.* **1996**, *10*, 862–868.
94. Bahr, U.; Depee, A.; Karas, M.; Hillenkamp, F.; Giessman, U. Mass spectrometry of synthetic polymers by UV-matrix-assisted laser desorption/ionization. *Anal. Chem.* **1992**, *64*, 2866–2869.
95. Danis, P.; Karr, D. A facile sample preparation for the analysis of synthetic organic polymers by matrix-assisted laser desorption/ionization. *Org. Mass Spectrom.* **1993**, *28*, 923–925.
96. Macha, S.; Hanton, S.; Owens, K.; Limbach, P. Silver cluster interferences in matrix-assisted laser desorption/ionization (MALDI) mass spectrometry of non-polar polymers. *J. Am. Soc. Mass Spectrom.* **2001**, *12*, 732–743.
97. Trimpin, S.; Rouhanjpour, A.; Az, R.; Rader, H.; Mullen, K. New aspects in matrix-assisted laser desorption/ionization time-of-flight mass spectrometry: A universal solvent-free sample preparation. *Rapid Commun. Mass Spectrom.* **2001**, *15*, 1364–1373.
98. Lewis, W.; Shen, Z.; Finn, M.; Siuzdak, G. Desorption/ionization on silicon (DIOS) mass spectrometry: background and applications. *Int. J. Mass Spectrom.* **2003**, *226*, 107–116.
99. Thomas, J.; Shen, Z.; Crowell, J.; Siuzdak, G. Desorption/ionization on silicon (DIOS): A diverse mass spectrometry platform for protein characterization. *Proc. Natl. Acad. Sci. USA* **2001**, *98*, 4932–4937.
100. Ogata, Y.; Kato, F.; Tsuboi, T.; Sakka, T. Changes in the environment of hydrogen in porous silicon with thermal annealing. *Electrochem. Soc.* **1998**, *145*, 2439–2444.
101. Thomas, J.; Shen, Z.; Blackledge, R.; Siuzdak, G. Desorption/ionization on silicon of mass spectrometry: an application in forensics. *Anal. Chim. Acta* **2001**, *442*, 183–190.
102. Kruse, R.; Rubakhin, S.; Romanova, E.; Hohn, P.; Sweedler, J. Direct assay of *Aplysia* tissues and cells with laser desorption/ionization mass spectrometry on porous silicon. *J Mass Spectrom.* **2001**, *36*, 1317–1322.
103. Trauger, S.; Go, E.; Shen, Z.; Apon, J.; Compton, B.; Bouvier, E.; Finn, M.; Siuzdak, G. High sensitivity and analyte capture with desorption/ionization mass spectrometry on silylated porous silicon. *Anal. Chem.* **2004**, 4484–4489.
104. Kraj, A.; Dylag, T.; Gorecka-Drzazga, A.; Bargiel, S.; Dziuban, J.; Silberring, J. Desorption/ionization on silicon for small molecules: a promising alternative to MALDI TOF. *Acta Biochim. Pol.* **2003**, *50*, 783–787.
105. Budimir, N.; Blais, J.; Fournier, F.; Tabet, J. The use of desorption/ionization on porous silicon mass spectrometry for the detection of negative ions for fatty acids. *Rapid Commun. Mass Spectrom.* **2006**, *20*, 680–684.
106. Nordstrom, A.; Apon, J.; Uritboonthai, W.; Go, E.; Siuzdak, G. Surfactant-enhanced desorption/ionization on silicon mass spectrometry. *Anal. Chem.* **2006**, *78*, 272–278.
107. Laiko, V.; Taranenko, N.; Berkout, V.; Musselman, B.; Doroshenko, V. Atmospheric pressure laser desorption/ionization on pororous silicon. *Rapid Commun. Mass Spectrom.* **2002**, *16*, 1737–1742.
108. Wasserscheid, P.; Keim, W.; *Angew. Chem. Int. Ed.* **2000**, *39*, 3772–3789.
109. Welton, T.; *Chem. Rev.* **1999**, *99*, 2071–2083.
110. Cornett, D.; Duncan, M.; Amster, I.; *Anal. Chem.* **1993**, *65*, 2608–2613.
111. Williams, J.; Gusev, A.; Hercules, D. Use of liquid matrices for matrix-assisted laser desorption ionization of polyglycols and poly(dimethylsiloxanes). *Macromolecules* **1996**, *29*, 8144–8150.
112. Caldwell, K.; Murray, K.; *Appl. Surf. Sci.* **1998**, *127–129*, 242–47.
113. Zenobi, R.; Knockenmuss, R.; *Mass Spectrom. Rev.* **1998**, *17*, 337–366.
114. Armstrong, D.; Zhang, L.; He, L.; Gross, M. Ionic liquids as matrixes for matrix-assisted laser desorption/

ionization mass spectrometry. *Anal. Chem.* **2001**, *73*, 3679–3686.

115. Carda-Broch, S.; Berthod, A.; Armstrong, D. Ionic matrices for matrix-assisted laser sedorption/ionization time-of-flight detection of DNA oligomers. *Rapid Commun. Mass Spectrom.* **2003**, *17*, 553–560.
116. Zabet-Moghaddam, M.; Heinzle, E.; Lasaosa, M.; Tholey, A. Pyridinimum-based ionic liquid matrices can improve the identification of proteins by peptide mass-fingerprint analysis with with matrix-assisted laser desorption/ionization mass spectrometry. *Anal. Bioanal. Chem.* **2006**, *384*, 215–224.
117. Jones, J.; Batoy, S.; Wilkins, C.; Liyanage, R.; Lay, J., Jr. Ionic liquid matrix-induced metastable decay of peptides and oligonucleotides ans stabilization of phospholipids in MALDI FTMS analyses. *J. Am. Soc. Mass Spectrom.* **2005**, *16*, 2000–2008.
118. Mank, M.; Stahl, B.; Boehm, G. 2,5-Dihydroxybenzoic acid butylamine and other ionic liquid matrixes for enhanced MALDI-MS analysis of biomolecules. *Anal. Chem.* **2004**, *76*, 2938–2950.
119. Bungert, D.; Bastian, S.; Heckmann-Pohl, D.; Gifthorn, F.; Heinzle, E.; Tholey, A. Screening of sugar converting enzymes using quantitative MALDI-ToF mass spectrometry. *Biotechnol. Lett.* **2004**, *26*, 1025–1030.
120. Zabet-Moghaddam, M.; Heinzle, E.; Tholey, A. Qualitative and quantitative analysis of low molecular weight compounds by ultraviolet matrix-assisted laser desorption/ionization mass spectrometry using ionic liquid matrices. *Rapid Commun. Mass Spectrom.* **2004**, *18*, 141–148.
121. Ham, B.; Jacob, J.; Cole, R. MALDI-TOF MS of phosphorylated lipids in biological fluids using immobilized metal affinity chromatography and a solid ionic crystal matrix. *Anal. Chem.* **2005**, *77*, 4439–4447.
122. Lemaire, R.; Tabet, J.; Ducoroy, P.; Hendra, J.; Salzet, M.; Fournier, I. Solid ionic matrixes for direct tissue analysis and MALDI imaging. *Anal. Chem.* **2006**, *78*, 809–819.
123. McCarley, T.; McCarley, R.; Limbach, P. Electron-transfer ionization in matrix-assisted laser desorption/ionization mass spectrometry. *Anal. Chem.* **1998**, *70*, 4376–4379.
124. Macha, S.; McCarley, T.; Limbach, P. Influence of ionization energy on charge-tranfer ionization in matrix-assisted laser desorption/ionization mass spectrometry. *Anal. Chim. Acta* **1999**, *397*, 235–245.
125. Macha, S.; Limbach, P.; Savickas, P. Application of nonpolar matrices for the analysis of low molecular weight nonpolar synthetic polymers by matrix-assisted laser desorption/ionization time-of-flight mass spectrometry. *J. Am. Soc. Mass Spectrom.* **2000**, *11*, 731–737.
126. Tang, X.; Dreifuss, P.; Vertes, A. New matrices and accelerating voltage effects in matrix-assisted laser desorption/ionization of synthetic polymers. *Rapid Commun. Mass Spectrom.* **1995**, *9*, 1141–1147.
127. Robins, C.; Limbach, P. The use of nonpolar matrices for matrix-assisted laser desorption/ionization mass spectrometric analysis of high boiling crude oil fractions. *Rapid Commun. Mass Spectrom.* **2003**, *17*, 2839–2845.
128. Xu, S.; Li, Y.; Zou, H.; Qiu, J.; Gau, Z.; Guo, B. Carbon nanotubes as assisted matrix for laser desorption/ionization time-of-flight mass spectrometry. *Anal. Chem.* **2003**, *75*, 6191–6195.
129. Ren, S.; Zhang, L.; Cheng, Z.; Guo, Y. Immobilized carbon nanotubes as matrix for MALDI-TOF-MS analysis: applications to neutral small carbohydrates. *J. Am. Soc. Mass Spectrom.* **2005**, *16*, 333–339.
130. Ren, S.; Guo, Y. Carbon nanotubes (2,5-dihydroxybenzoyl hydrazine) derivatives as pH adjustables enriching reagent and matrix for MALDI analysis of trace peptides. *J. Am. Soc. Mass Spectrom.* **2006**, *17*, 1023–1027.
131. Norris, J.; Porter, N.; Caprioli, R. Combination detergent/MALDI matrix: Functional cleavage detergents for mass spectrometry. *Anal. Chem.* **2005**, *77*, 5036–5040.
132. Spengler, B.; Kirsch, D.; Kaufmann, R. Fundamental aspects of postsource decay in matrix-assisted laser desorption mass spectrometry. 1. Residual gas effects. *J. Phys. Chem.* **1992**, *96*, 9678–9684.
133. Karas, M.; Bahr, U.; Strupat, K.; Hillenkamp, F.; Tsarbopoulos, A.; Pramanik, B. Matrix dependence of metastable fragmentation of glycoproteins in MALDI TOF mass spectrometry. *Anal. Chem.* **1995**, *67*, 675–679.
134. Kinsel, G.; Preston, L.; Russell, D. *Biol. Mass Spectrom.* **1994**, *23*, 205.
135. Zhu, L.; Parr, G.; Fitzgerald, M.; Nelson, C.; Smith, L. Oligodeoxynucleotide fragmentation in MALDI/TOF mass spectrometry using 355-nm radiation. *J. Am. Soc. Mass Spectrom.* **1995**, *7*, 6048.
136. Ivleva, V.; Elkin, Y.; Budnik, B.; Moyer, S.; O'Connor, P.; Costello, C. Coupling thin-layer chromatography with vibrational cooling matrix-assisted laser desorption/ionization fourier transform mass spectrometry for the analysis of ganglioside mixtures. *Anal. Chem.* **2004**, *76*, 6484–6491.
137. Luo, G.; Marginean, I.; Vertes, A. Internal energy of ions generated by matrix-assisted laser desorption/ionization. *Anal. Chem.* **2002**, *74*, 6185–6190.
138. Gabelica, V.; Schulz, E.; Karas, M. Internal energy build-up in matrix-assisted desorption/ionization. *J. Mass Spectrom.* **2004**, *39*, 579–593.

Chapter 8

MALDI Imaging Mass Spectrometry

Alain Brunelle and Olivier Laprévote

Institut de Chimie des Substances Naturelles, CNRS, Gif-sur-Yvette cedex, France

8.1 INTRODUCTION

One of the major interests in science, in particular in the field of human health, consists of developing powerful tools to make it possible to visualize the precise location of a given molecule of interest within biological tissues and biopsies. Various techniques of spectroscopy and spectrometry were adapted during the past century in order to improve the tools of biological imaging.

In parallel, many efforts in the proteomic and metabolomic areas were devoted to the development of analytical protocols making it possible to individualize, within biological samples, differentiated compartments. For instance, the histological analysis of a biological tissue section followed by the laser microdissection of relevant regions can be proposed as a preliminary stage to an "omic" analysis intended to distinguish the chemical content of close tissue zones. In addition, approaches developed for the study of well-defined cellular compartments (such as membranes or mitochondria) enable one to differentiate subcellular

Electrospray and MALDI Mass Spectrometry: Fundamentals, Instrumentation, Practicalities, and Biological Applications, Second Edition, Edited by Richard B. Cole

proteomes starting from whole cells. However, all these techniques require different stages of extraction, purification, and separation of the concerned molecular families (proteins, metabolites) starting from a crude and complex biological material. These processes are often tainted with artifacts or sample loss. A technique that would render it possible to avoid these disadvantages by visualizing a great number of compounds directly on a biological tissue, with limited preliminary treatments, would thus be extremely welcome. Most of the biological imaging techniques used currently were developed with this end in mind. Their spatial resolution can vary from the nanometer to the millimeter scale, and they apply to both *in vivo* and *ex vivo* analyses. However, their common characteristic is the low number of molecules that can be detected simultaneously, owing as much to their weak resolving power as to their lack of sensitivity. In addition, the majority of these techniques are based on an *a priori* knowledge of the molecular target. Of course, this is not appropriate in a search for unknown molecules which could be characteristic of a given cellular state (i.e., biomarkers).

Compared to these constraints, and in view of these limitations, MALDI mass spectrometry can bring some answers. Indeed, it has been known for quite some time that the analysis of biological substrates by MALDI makes it possible to obtain ions characteristic of compounds present in the samples. The efficiency of MALDI for this purpose has been demonstrated in the cases of single cells,[1] bacteria,[2] and tissues.[3] Thus, without any preliminary knowledge of the detailed chemical composition of the biological target, it is possible to obtain characteristic "profiles" for each cellular type. This can be exploited, for example, for the identification of bacterial strains. By applying such an approach to a biological tissue section, it is possible to (a) distinguish various zones on the basis of a histological examination and (b) correlate them with their specific spectral patterns. This approach is at the origin of MALDI "profiling" studies. By extending this type of analysis to the whole surface of a tissue slice, which can be analyzed systematically at regularly spaced intervals, it becomes possible to correlate the intensity of a given ion peak with its location of origin. With all spectra being positioned on x,y coordinates, the intensity of each ion related to a z axis can be translated into a color code, thereby generating visual contrast. This is the guiding principle of the MALDI imaging mass spectrometry (IMS) technique (Figure 8.1).

This method has to be compared with the other preexisting mass spectrometric imaging methods. Time-of-flight-secondary ion mass spectrometry (TOF-SIMS) imaging has already existed for more than 40 years,[4] and recent improvements in polyatomic ion

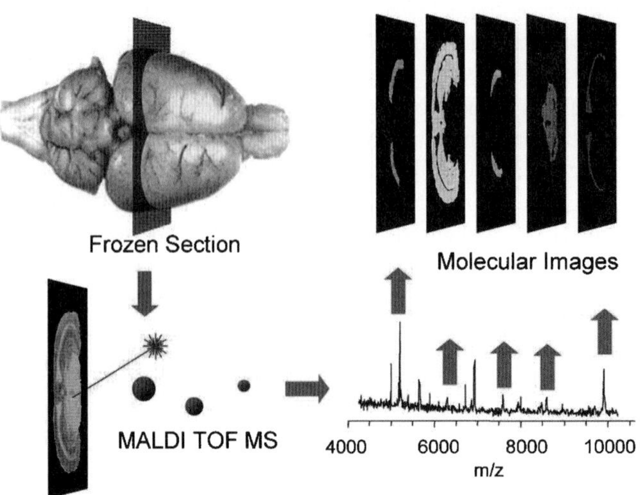

Figure 8.1. MALDI imaging mass spectrometry process. Reprinted from reference 42 with permission from Elsevier (See color insert).

sources are pushing SIMS imaging toward biological applications.[5] The nano-SIMS technique, which provides extreme spatial resolution of a few tens of nanometers, is also an exciting IMS method.[6] Instead of being competitive, these three techniques are highly complementary. Indeed, nano-SIMS presents the best resolution but accounts only for elemental analysis. The TOF-SIMS imaging instruments are able to map, at a micrometer scale, biomolecular compounds in a mass range limited to approximately 1000 Da whereas MALDI-IMS, with a 50-μm lateral resolution, brings the possibility to obtain images of peptides and proteins up to 30 kDa or more (Figure 8.2).

8.2 MALDI PROFILING VERSUS MALDI IMAGING

It is important to rigorously distinguish and define the profiling and imaging approaches that are often merged, unfortunately, in published papers and hence in the minds of potential users. These methods each have their own legitimacy in view of their biological applications.

In profiling studies, the MALDI matrix solution is deposited as a limited number of droplets on the tissue slice, thus delineating a well-defined analysis area. These particular zones are often chosen as the result of a preliminary examination of the tissue histology by optical microscopy. After crystallization of the matrix, MALDI mass spectra are recorded in each particular zone. It is expected that each spectrum would reflect the local chemical composition and be representative of a particular sample area. Such an approach presents several advantages over a scanning of the entire surface as is performed in the imaging mode. At first, the study is limited to only the regions of interest (often called ROI), with the mass spectra being correlated to other characteristics of the tissue (histological features, targeted zones by different types of chemical labeling, etc.).

Another interesting aspect of the profiling method is related to the problem of molecular delocalization on the biological target surface when the matrix solution is deposited. An inhomogeneous crystallization, or concentration variations of the matrix on the tissue section, can be responsible for a loss of reproducibility of the spectra and a lower lateral resolution. This is typically the case when 2,5-dihydroxybenzoic acid (DHB), which is known to crystallize as long and fine needles, is utilized as a matrix. In the profiling approach, such events occur only in the area on which the droplet was deposited. This allows for a much larger choice of matrix than in the imaging mode, which requires complex and problematic methods of matrix deposition. Furthermore, a large number of spectra can be acquired from a single matrix deposition area, thus increasing the sensitivity and reproducibility of the measurements. For the same reasons, the time required for a profiling

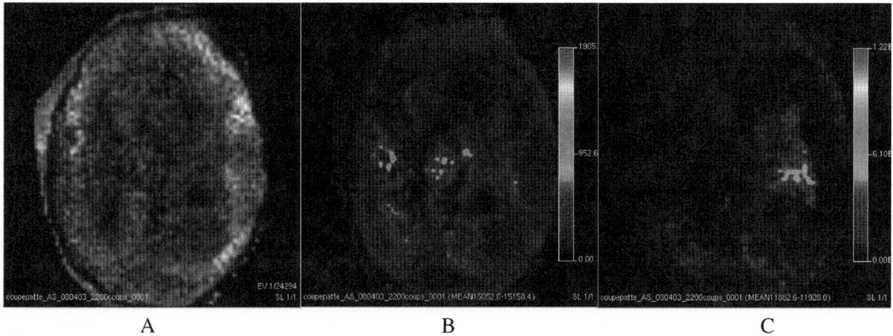

Figure 8.2. Differential localization of proteins on a mouse muscle section by MALDI imaging mass spectrometry. A: total ion image; B: MS image of the ion at *m/z* 15100; C: MS image of the ion at *m/z* 11950. (See color insert).

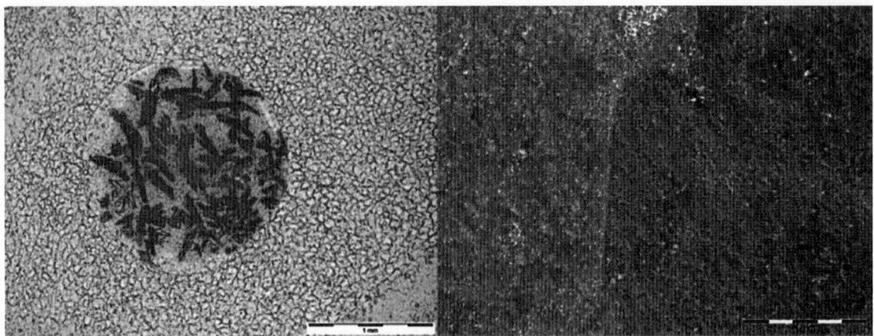

Figure 8.3. Left: the dried droplet deposited for a profiling study on a mouse brain section displays crystals of sinapinic acid. Right: the mouse brain section is homogeneously coated by sinapinic acid for an IMS experiment (air-spray deposition procedure) (See color insert).

analysis on a few points is much shorter than that necessary for the acquisition of spectra on thousands of pixels, as in the imaging mode. Consequently, a profiling study can precede, or even replace, a full IMS analysis. This has been successfully accomplished by different groups on single neuronal cells[7] and biological tissue sections analyzed directly[8] or using a blotting procedure.[9,10]

MALDI imaging mass spectrometry was described, in parallel to the MALDI profiling procedure, by the laboratory of R. M. Caprioli in a pioneering paper published in 1997.[11] This work was followed by a number of application studies, and a comprehensive review was published by the same group in 2004.[12] Contrary to the profiling approach, imaging mass spectrometry needs a complete and homogeneous coating of the biological target by the MALDI matrix while restricting, as much as possible, delocalization of the biomolecules on the analysed surface. This coating procedure is accompanied by some technical difficulties that will be discussed later in this chapter. Furthermore, the point-by-point analysis of the whole surface of a tissue section generates a huge amount of data. Considering, for example, a sample surface covering an area of 1 square centimeter at a lateral resolution limited to 100 micrometers, 10,000 points (pixels) have to be analysed, each of them corresponding to tens to hundreds of laser shots. The analysis time is thus particularly long even with high repetition rate lasers and rapid mechanical displacement of the target. Moreover, the ion density maps obtained by this method must be visualized in form of color-encoded images, necessitating appropriate software that is not commercially available from all mass spectrometry manufacturers. In some cases, the software tools must be written in-house and designed for a specific instrument. Consequently, the MALDI imaging approach must be used with discernment. MALDI-IMS is particularly justified in two different main cases: (i) when no clear histological patterns are observed on the biological tissues and (ii) when the localization of a pre-targeted diagnostic molecule or metabolite cannot be known *a priori*. In most of the other cases, the MALDI profiling approach would be preferred, particularly when the problem is to differentiate two (or more) well-defined areas on the tissue section (Figure 8.3).

8.2.1 Microprobe Imaging

MALDI-IMS was developed initially with the commercial instruments available at the end of the 1990s. At that time, the MALDI ionization sources were almost always coupled to time-of-flight (TOF) mass analyzers. The first experiments were thus carried out with commercial MALDI-TOF instruments. These mass spectrometers, which were not initially designed for IMS applications, suffered from a number of failures: The displacement of sample plates was slow, the signal acquisition and data treatment necessitated long downtimes between acquisitions, and they were equipped with nitrogen lasers irradiating at 337 nm and presenting a repetition rate of only 20 Hz. The focusing efficiency of the laser optics was also limited to 30- to 60- μm spot sizes. In order to illustrate the capabilities of this type of instrumentation for providing an IMS experiment, some constraints must be taken into account. Considering a surface covering 100×100 pixels and the use of a nitrogen laser at 20 Hz employing 150 laser shots on each pixel, the time required only to acquire the mass spectra is 20 h and 50 min. The time for moving the sample plate from one point to the next, as well as the time for data treatment, is approximately 2.4 s/point (pixel). For the 10,000 pixels, this supplementary time is about 6 h and 40 min. Consequently, such an IMS data acquisition would require 27.5 h!

Therefore, each of the parameters must be optimized and refined. First, the number of pixels, which depends upon the tissue surface and the desired resolution, must be carefully chosen. For a given surface, an increase of the lateral resolution by a certain factor (e.g., two) increases the pixel number and the acquisition time by its square (i.e., four). The number of shots in each pixel must be adequate for the expected sensitivity. It depends on the nature of the tissue, on the matrix, and on the type of ions to be analyzed. This parameter needs be optimized prior to any IMS experiment. Increasing the laser repetition rate is of great interest. For this purpose, tripled-frequency Nd:YAG lasers (355 nm) able to run at 200-Hz frequency are very desirable as compared to the slower nitrogen lasers. The use of such lasers and of fast digitizers leads to a significant reduction of the IMS acquisition time. Some improvements in the sample holder mechanics which accelerate the sample plate displacement will decrease again the total time of the experiment. With the present know-how and depending on the employed combination of all of these parameters, the total acquisition time required for an IMS experiment of 10,000 pixels can be reduced to 6–8 h.

Another analytical challenge is to increase the lateral resolution in order to achieve subcellular ion images. The Spengler group built an experimental setup designed to achieve a laser focus on the target of about 1 μm.[13] The scan rate could reach 50 pixels per second, and small areas ($100 \times 100\,\mu m^2$) were effectively analyzed in less than 50 min. This experiment was shown to be efficient for elemental analysis and promising for biomolecule imaging. However, the reduction of the laser spot size is clearly accompanied by a loss of sensitivity and a reduction of the useful mass range. It has been shown previously that for spot sizes lower than 10 μm, the ion signals generated from the matrix DHB were considerably weaker, involving laser ablation in competition with MALDI processes (Figure 8.4).[14]

For the last several years, the availability of commercial tandem time-of-flight (TOF-TOF) instruments opened the way for profiling and imaging mass spectrometry in the MS/MS analysis mode. Different manufacturers present adapted machines fitted with 200-Hz pulsed lasers and possessing fast sample moving capabilities which seem particularly well suited to the MALDI imaging constraints. It is interesting to note that before the advent of such TOF-TOF tandem analyzers, post-source decay (PSD) of metastable ions was studied with a conventional MALDI-TOF mass spectrometer in order to gain sequence information from peptide digests of imaged proteins.[9] Other types of mass analyzers that can be combined with MALDI ion sources, such as quadrupole/time-of-flight tandem mass spectrometers, can be used for imaging purposes in the MS/MS mode. These instruments showed interesting capabilities for detecting and imaging low-mass pharmaceutical drugs in mouse tumor tissue[15] or in rat brain.[16] With these mass spectrometers, the laser beam is

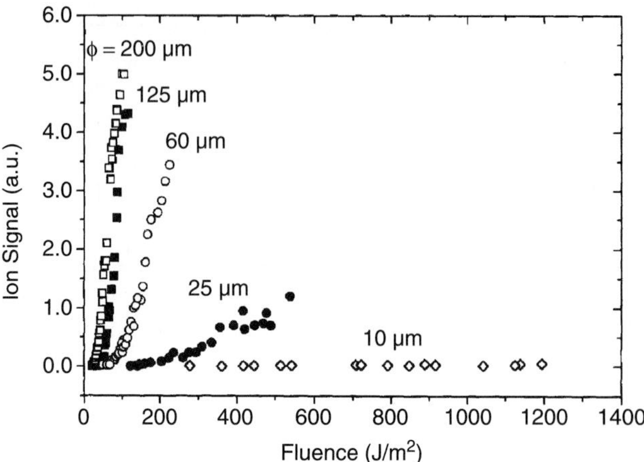

Figure 8.4. Ion signals of photoionized 2,5-dihydroxybenzoic acid molecules $M^{+\bullet}$ as a function of laser fluence for different laser spot diameters (10, 25, 60, 125 and 200 μm). Reprinted from reference 14 with permission from Elsevier.

transmitted to the sample surface by a fiber optic, thereby leading to spot diameters of hundreds of microns and, hence, to low lateral resolution. More recently, an intermediate-pressure MALDI source coupled to a linear traping quadrupole was used for investigating lipids from tissue sections.[17] Other instrumental improvements have been implemented, such as the high mass detection capability offered by the cryogenic detectors whose sensitivity is independent of mass-to-charge ratio.[18] Although shown to be efficient for improving the sensitivity in the high mass range, the usefulness of this detection mode has yet to be proven by real applications of imaging mass spectrometry.

8.2.2 Microscope Imaging

All the instrumental methods discussed up to this point are based on a microprobe imaging mode in which the mass spectra are collected from an array of predefined positions and are used to build molecular images using appropriate data treatment. A completely different approach is based on the concept of ion microscopy (Figure 8.5). In this imaging mode, the ions are produced by a single laser shot irradiating 200 μm on the target. The entire desorbed ion beam is passed through an ion optics system while maintaining the spatial integrity of the desorbed species such that an ion distribution representative of the initial spatial distribution on the surface is recorded by a two-dimensional detector. Data manipulation is then used to generate images of specific targeted ions. The time-of-flight separation of ion images by a triple electrostatic sector analyzer is achieved shot-by-shot at a rate of 12 Hz corresponding to the laser repetition rate. In a preliminary study, Luxembourg et al.[19] demonstrated that the addition of 20–80 images of a given ion could lead to lateral resolutions of approximatively 4 μm. This method appears to be very promising for the high-resolution mass spectrometry imaging of small tissue areas (e.g., 500 × 500 μm or less), but it remains to be validated with applications to biological samples.

8.3 SAMPLE PREPARATION

Sample preparation methods are a critical aspect of both profiling and imaging MALDI-MS studies.[20] They remain an open field in which each laboratory develops its own procedure.

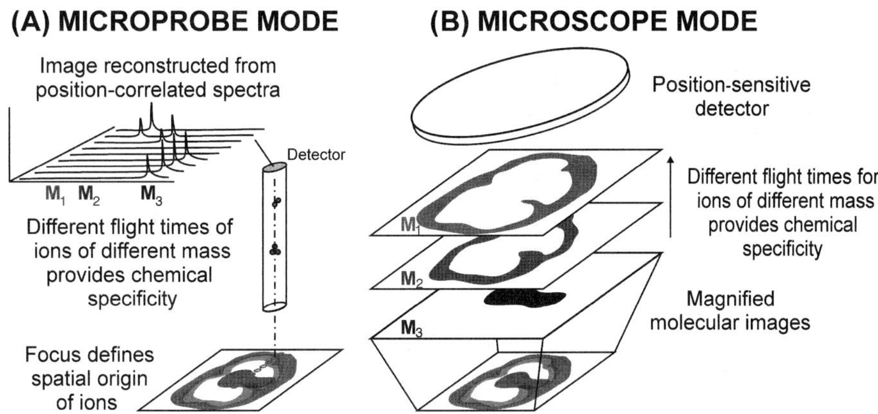

Figure 8.5. Schematics of the two different modes of molecular imaging mass spectrometry. Microprobe mode imaging (A) collects mass spectra from an array of designated positions to reconstruct a molecular image after completion of the experiment. In microscope imaging (B), magnified images of the ion distributions are directly acquired using a two-dimensional detector. Reprinted with permission from reference 19, Copyright 2004 American Chemical Society (See color insert).

However, a careful examination of the literature helps to derive some guidelines for avoiding some of the numerous problems encountered by the novice. Here, the matrix deposition procedures will be described in a practical manner.

Two main different protocols can be proposed for tissue analyses. The first one corresponds to the direct analysis of a tissue slice on a MALDI target plate, whereas the second one proceeds from the transfer of the tissue content onto a capture membrane (blotting). The latter method is less often used than the direct imaging approach. In an early experiment, blotting of the tissue was carried out on a target covered with C_{18}-coated resin beads.[11] More recently, an alternative procedure was suggested by Chaurand et al.[9] in a comparative study of different parts of the mouse intestine. The capture membrane was made of conductive polyethylene imbedded with carbon and fixed on the sample plate by double-sided conductive tape. After tissue blotting onto the membrane, water rinsing, and drying, the matrix is deposited on the membrane prior to mass analysis. If protein or peptide ions are desired, it is possible to perform a trypsin digestion of proteins on the target membrane.

For the direct analysis of tissues, the first practical aspect concerns the choice of the sample plate on which the tissue slice has to be deposited. In most of the published experiments, a conventional metallic MALDI plate is used. However, if correlations between mass spectrometric images and histological features of the tissue sections are needed, the use of optically transparent and noninsulating sample plates is necessary. Conductive quartz surfaces were used some years ago,[21] and they were shown to be compatible with MALDI analysis. More recently, glass slides coated by a thin layer of indium-tin oxide (ITO) have been proposed as convenient supports for MALDI. These conductive sample plates are also compatible with optical microscopy. Interestingly, various tissue staining methods that are commonly used for the histopathological examination of biological tissues have been adapted to MALDI mass spectrometry, and they have shown relatively good spectral quality.[22]

The preparation of biological tissue sections must be optimized in order to limit as far as possible any mechanical or chemical alteration. The tissues are generally frozen in liquid nitrogen just before dissection and conserved at −80°C. The samples are cut into 10- to 20-μm sections at −20°C using a microtome and are mounted directly onto the target plate; they are then stored at −80°C. Before matrix deposition, different treatment steps can be used. The sample can be fixed in a manner similar to that used in histology. The target plates bearing the tissue slices are immersed in 70% ethanol (aqueous) in order to precipitate the protein content of the tissue and remove some salts and impurities.[20] This preliminary washing step seems to improve the crystallization process when α-cyano-4-hydroxycinnamic acid (CHCA) is used as the MALDI matrix, but loss of alcohol-soluble compounds can be expected from this procedure. Other groups suggest the removal of lipids on the tissue slices in order to increase the mass spectrometric signal of peptides and proteins.

Lipid removal can be accomplished by successive step-wise rinsing of the tissue surface by organic solvents such as chloroform, hexane, acetone, toluene, or xylene.[23] Significant sensitivity improvements have been demonstrated in mass spectra leading to better contrasts of peptide and protein ion images. However, there is no rigorous evaluation of the chemical modifications, or damage, or of molecule losses due to these treatments. Moreover, there are no validated procedures for evaluating the level of molecule delocalization on a biological tissue submitted to a particular treatment. The use of immunohistochemistry could achieve this goal by comparing the localization of a targeted biomolecule of interest in treated and untreated vicinal tissue sections.[23] This approach cannot be applied, however, in a general and systematic manner because it requires antibodies specific to the proteins detected by MALDI.

The question of molecule delocalization on the biological surface is particularly relevant when considering matrix deposition procedures. It is widely admitted that the deposition of a matrix solution can induce a delocalization of the compounds in the area covered by the droplet. In MALDI profiling methodology, all common matrices can be used, with the choice being governed by the molecule type under investigation. In the imaging mode, for which a homogeneous matrix deposition and crystallization must be achieved, the matrix nature and the deposition process play a preeminent role in the success level of the experiment. The most generally used matrices are CHCA (for the analysis of peptides and small proteins in the reflectron mode) and sinapinic acid for higher-molecular-weight ions in the linear TOF analysis mode. They are diluted at 20–30 mg mL^{-1} in a mixed solvent water/acetonitrile/trifluoroacetic acid (50 : 50 : 0.1, V/V/V). Other matrices that are used for other MALDI applications, such as 2,5-dihydroxybenzoic acid (DHB) and 3,4,5-trihydroxyacetophenone (THAP), must be avoided for IMS experiments because they crystallize in the form of long and fine needles, thereby inducing a very significant delocalization of the molecules. The discovery of more efficient matrices for MALDI imaging mass spectrometry, in particular for increasing the sensitivity in the high mass range, is the object of continuing research. The so-called ionic matrices seem to provide an interesting solution.[24] These compounds are salts combining deprotonated conventional acidic matrices and substituted ammonium cations. One of them, composed of CHCA and aniline has proved to increase significantly the sensitivity especially for peptides and small proteins amenable to analysis in the reflectron mode.[25] This matrix is dissolved in water/acetonitrile/trifluroacetic acid (35/70/0.1, V:V:V) and the solution is directly applied on the tissue section by using a micropipette. The crystallization appears to be uniform; and, in addition to the advantageous increase of the image contrast, this matrix seems to be very stable under vacuum and more resistant to the laser irradiation. No craters are visible on the sample surface after the laser shots, and several acquisitions can be performed on the same sample.

The simplest means to cover the tissue section by the matrix is to deposit a single drop of matrix solution on the surface. However, because of the risk of molecule delocalization and inhomogeneity, other techniques have been proposed. One approach makes use of a deactivated glass spray nebulizer. The matrix solution is placed in a reservoir and sprayed by nitrogen stream nebulization onto the sample surface. The matrix coating is carried out by deposition in successive layers. In our experience, a volume of about 5 mL of matrix solution is necessary to adequately cover a MALDI sample plate of a few square centimeters. The most critical point of this method is the drying time between two successive spraying steps. Short drying times lead to the formation of small droplets that increase in size at each spraying stage, causing extensive molecular delocalization at the sample surface and the loss of spatial resolution. The matrix solution can also be distributed by using an electrospray device (Figure 8.6).[26] This technique was considered as the method of choice by Kruse and Sweedler[27] for the mass spectrometric profiling of *Aplysia californica* peptides from neuronal and exocrine gland tissues.

Very recently, a great deal of interest has been paid to novel matrix application techniques that could share the advantages of the profiling method (good sensitivity and limitation of the delocalization effects of the droplet scale) and the homogeneous tissue-coating approach (necessary for the IMS studies). The best solution was perhaps found by Aerni et al.[28] An automatic spotter delivering microdroplets of matrix solution on the tissue surface ensures both whole surface coverage and reproducibility of spot-to-spot crystallization. This leads to reliable ion images obtained with a 10-fold increase of sensitivity and an enrichment of diagnostic ions in the mass spectra. Due to the droplet volume, the diameter of the covered zone is about 200 μm, a value that corresponds to the ultimate lateral resolution. However, this idea has significantly evolved with the development of many new spotting devices including an inkjet printing system[29] that is dedicated to this particular application and delivers nanodroplets compatible with high lateral resolutions (Figure 8.7).

Even when using conductive sample holders, the tissue section itself remains insulating. Charge can accumulate on the sample surface, thereby disturbing the local extraction field with deleterious consequences on mass resolution. Recently, Altelaar et al.[30] circumvented this drawback by sputter coating the sample surface, after matrix deposition, with a few angtroms of gold. A previously published similar approach, which made use of heavy gold cluster implantation on the tissue, suggested an improved signal detection although no ion images were shown.[31]

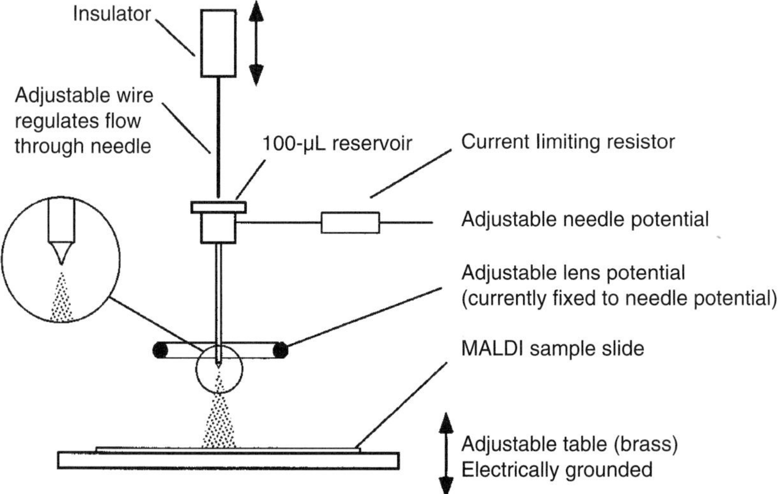

Figure 8.6. Electrospray deposition of a MALDI matrix solution on a sample plate. Reproduced with permission from reference 26, © 1997, John Wiley & Sons Limited.

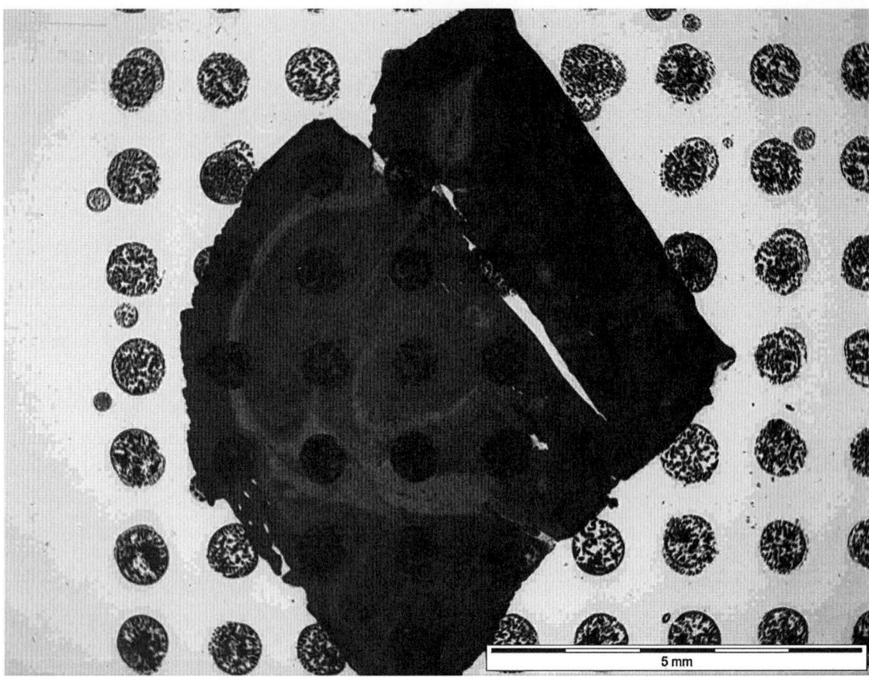

Figure 8.7. Automatic spotting of matrix onto a tissue section (See color insert).

8.4 DATA PROCESSING

Data treatment is a key issue in MALDI-MS imaging. The question of target displacement is not as trivial as it may seem, and it must be addressed in view of an instrument's performance. The huge amount of data acquired during one tissue investigation requires appropriate data treatment and storage procedures. Furthermore, the goal of any imaging experiment is to (a) create ion density maps from a bidimensional matrix of spectra and (b) translate them into readable images via the use of appropriate software.

Although acquisition and data treatment software are plentiful, the real availability of imaging tools remains limited. The acquisition software MMSIT and the imaging software BioMap developed by Stoeckli[32] are used by Applied Biosystems for their TOF and TOF-TOF mass spectrometers. Thermo Electron also offers a complete software package for the acquisition and visualization of mass spectrometry images acquired directly from tissue samples with its *v*MALDI ion source/LTQ linear trap LTQ instrument. Another software, FlexImaging, is patented by Bruker Daltonics and can be used with the MALDI-TOF and TOF-TOF spectrometers of this manufacturer. The data generated by a single MALDI imaging mass spectrometry experiment can attain hundreds of megabytes. However, since the beginning of IMS, faster processors have steadily appeared and the cost of hard disks and RAM have decreased in parallel. This evolution allows the efficient treatment of a full MALDI-IMS data set with a simple desktop computer.

Other than these general considerations, the MALDI imaging approach presents two very important characteristics with regard to the other biological imaging techniques. First, tens or even hundreds of ionic species can be detected simultaneously, and the same number of ion images can be theoretically obtained from a single IMS experiment. Second, most of the tissue investigations by IMS are carried out without any preliminary knowledge of the nature and quantity of the biomolecules that will be detected on the biological surface. Up to this point, it is in an empirical manner and only by careful inspection of the significant mass spectra that the ions and their corresponding images are identified. Consequently, the correlations of ion abundances of different biomolecules (positive in the case of co-localization, negative if the compounds are spatially separated) are deduced only in a second step from a subjective peak selection. It is thus of primary importance to have at one's disposal statistical data treatment tools capable of objectively deducing positive and negative correlations between different ion signals. A combination of clustering and multivariate methods was recently suggested to extract and efficiently unmask useful information from MALDI imaging experiments.[33] In a first step, cluster analysis is used to identify the spatial correlations of the mass spectra and to define the regions of interest within the tissue section. In a second step, principal component and discriminant analyses (PCA/DA) are performed in order to characterize the detailed changes of spectra between the different regions. In this way, the knowledge of the sample under investigation can be significantly increased, and an enhancement of the image contrast is obtained. Such tools, which will certainly undergo further improvement and be made widely available, are of very high importance for the future relevance and efficiency of MALDI-MS imaging as a biological investigation method.

8.5 MOLECULAR STRUCTURE IDENTIFICATION

A successful MALDI imaging experiment will offer a two-dimensional map of the distribution of a selected *m/z* ion on the surface of sample. But before giving any conclusion based on either mass spectra or ion images, the reproducibility of the experiment should be demonstrated. The question of reproducibility is often neglected, mostly because of the destructive nature of the analysis. Furthermore, the structure of the ions considered to be reliably detected by MALDI imaging MS should be unambiguously characterized independently by a robust method. Considering the case of the low-molecular-weight organic compounds, MS/MS can be the method of choice. Some tandem mass spectrometers have been used for that purpose (see above). Higher-molecular-weight protein ions are, by contrast, much more difficult to unambiguously identify. In many cases, conventional proteomic methods must be used after protein detection by MALDI imaging. The most often used procedures involve the extraction of the compounds of interest (peptides or proteins) and purification by liquid chromatography in order to yield enriched fractions of the analyte molecules. The fractions are then digested and the resulting peptides are analyzed by MS/MS on-line or off-line, after LC separation.[8,34] The major compounds can thus be identified unambiguously. By contrast, the minor components or the protein fragments detected in a preliminary IMS study are not always readily characterized by this technique. It is then useful to correlate the MALDI imaging results with histology, laser microdissection of the targeted area, and proteomics.[35] In that respect, MALDI-IMS can be considered as a "pre-proteomics" method, similar to 2-D gel electrophoresis or other separation methods, with the particular advantage that it gives directly the first structural information—that is, the molecular weight measurement. At first glance, even though hundreds of ion peaks can be observed in the mass spectra over a mass range extending to 30 kDa, they are often dominated by a few very abundant species.

In particular, histones, ubiquitin, myosin, actin, α- and β-globins, and so on, are systematically recovered on most biological tissue sections. Some of these proteins (human) are listed in the paper by Sköld et al.[36]

Because of the high chemical noise and the low signal-to-noise ratios of analyte proteins, it is relatively difficult to reliably observe subtle differences in protein abundances on biological tissue sections, especially at high mass. Owing to these limitations, the MALDI IMS method is currently the most relevant for large peptides or small proteins—that is, in the mass range 1000–10,000 Da. As an example, one of the most convincing MALDI imaging mass spectrometry analyses published so far in proteomics involves the small protein PEP-19 (6.7 kDa) whose expression is reduced in a mouse model of Parkinson's disease.[36] It is noteworthy that, in this case, confirmation of the protein identity required its extraction from the brain followed by purification, exact mass measurement using Fourier transform–ion cyclotron resonance (FT-ICR) mass spectrometry, and the analysis of its digestion peptides by MS/MS. Even so, the efficiency with which a small protein can be detected directly on tissue sections remains impressive when comparing the IMS approach with other separation techniques used in proteomics.

A novel alternative strategy to the conventional protocol of extraction and identification by digestion and MS/MS, has been proposed recently by Rohner et al.[37] In the "molecular scanner approach" the idea is to carry out the MALDI-IMS analysis of the digestion peptides rather than the proteins from which they arise (Figure 8.8). In this experiment, the tissue is blotted onto a first membrane containing immobilized trypsin. The digestion occurs *in situ* while preserving the initial location of the precursor proteins and the resulting peptides are captured by a second membrane. This capture membrane is subsequently coated with α-CHCA matrix and directly analyzed by MALDI-IMS. This method looks promising for localized peptide mapping even when the number of digestion peptides from a biological tissue is expected to be high. However, the acquisition of reliable tandem mass spectra requires sufficiently intense peptide ions or the accumulation of numerous spectra. Because of the destructive nature of MALDI, it seems difficult to acquire more than a few peptide sequences on a single target which, of course, will limit the number of proteins that can be reliably identified.

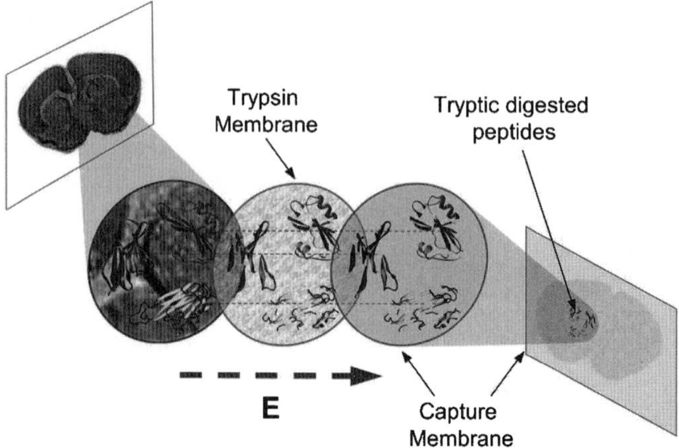

Figure 8.8. Descriptive scheme of the molecular scanner approach in imaging mass spectrometry. Reprinted from reference 35 with permission from Elsevier (See color insert).

8.6 BIOLOGICAL APPLICATIONS

It is important to recognize that up to the present, only a few biological applications have been published in which MALDI-MS imaging proved to be essential for obtaining molecular identifications. Because MALDI is able to produce ions from proteins at high mass, MALDI imaging mass spectrometry was first developed to locate specific proteins on the surfaces of biological tissue. The highly active Caprioli group performed a comparative study of the protein content at various locations along the mouse epididymis organ.[8] Some proteins were shown to be regionalized in specific areas, whereas others showed a concentration gradient from the *caput* to the *cauda* section. Identification of the main components was accomplished owing to laser capture microdissection of relevant tissue parts followed by HPLC and MS/MS.[8] Similar profiling studies of various cancer tissues showed protein content specific to each tumor type such as those of the brain[38] or lung.[39] MALDI profiling proved useful for the detection of proteomic changes in mouse mammary tumors due to the treatment by various active drugs.[40] Monitoring the variation of the protein profiles in relation to the disease progression is important for both prognosis and evaluation of treatment efficiency. As mentioned above, the assessment of neurodegenerative diseases can also be valuable target studies for MALDI imaging mass spectrometry. The discovery of a biomarker of Parkinson's disease in a mouse model was achieved mainly by MALDI IMS whereby a characteristic distribution of this protein was shown in the brain.[36] More recently, MALDI imaging was used to localize the eye lens alpha-crystallins and their modified forms in calf and bovine eye lenses. Distinct distribution patterns were observed for alphaA-crystallin and alphaB-crystallin, indicative of various degrees of modification including protein truncation and phosphorylation. In particular, a relationship between the phosphorylation level of alpha-crystallins and their location in the tissue was clearly established, giving new insight into the physiological significance of the modified forms of these proteins.[41] Because peptide ion peaks are generally more intense and more resolved than those of high-mass proteins, peptides may constitute targets of choice for MALDI-IMS investigations. An excellent example was provided by Stoeckli et al.,[42] who showed the specific distribution of amyloid peptides on brain slices obtained from a mouse model of Alzheimer's disease. Neuropeptides also drew attention of I. Fournier et al.[43] in a MALDI profiling study of the rat brain that confirmed the specific location of the small peptide vasopressin in the supraoptical nucleus. A solution of lipopolysaccharides injected into the rats 9 h before mass spectrometric analysis showed a strong effect on vasopressin release, thus demonstrating the usefulness of MALDI profiling for studying local peptidome modifications due to drug administration. The neuropeptides of *Aplysia californica*, a neuronal model, were also extensively studied by the group of J. Sweedler using MALDI profiling on single isolated neurons[44] and on the exocrine gland.[27] Provided that appropriate experimental protocols are used, it is possible to distinguish different peptide patterns according to the irradiated area of the cell.[7] In another peptide study, MALDI-IMS was performed to detect and locate novel bradykinin-related peptides in the skin of the amphibian *Phyllomedusa hypochondrialis*.[45]

It is noteworthy that most of these studies have been performed using MALDI profiling for which the spectral quality in terms of signal-to-noise ratio and mass resolution is much better than in the imaging mode. In that respect, it is important to mention the recent paper of G. Maddalo et al.[46] in which the question of mass spectral reproducibility of MALDI profiling studies was addressed. Based on a comparison of three different cancer models (salivary gland tumor, prostate tumor, and mammary gland carcinoma), the spectral variability for a given animal was estimated to be 30% (peak relative intensity), whereas the interindividual variability was shown to be 60%.[46] Such statistical evaluations of the reproducibility of data obtained by profiling or imaging mass spectrometry are necessary for validating the use of these methods in clinical molecular diagnosis.

In parallel to peptides and proteins, small molecules were also subjected to imaging mass spectrometric investigations. Attention has been paid to the lipid composition of mouse muscles in a model of Duchenne muscular dystrophy.[47] It was shown that depending on the degeneration state of the tissue, the degree of unsaturation of fatty acids in phospholipids was significantly modified and could be used as a marker of tissue regeneration within necrotic areas. Further studies gave confirmation of the reliability of MALDI mass spectrometry for localized lipidomics studies carried out directly on tissue sections.[17,48] In the field of pharmaceutical drugs, the distribution of certain therapeutic agents in skin was investigated by a MALDI imaging method.[49] An antitumor compound was mapped directly on mouse tumor tissue sections by using MALDI imaging in the MS/MS mode.[15] The specificity of detection was assessed by monitoring a characteristic fragmentation pathway of this compound. A similar approach showed a good correlation of MALDI IMS and autoradiographic methods for locating the antipsychotic drug clozapine in rat brain.[16] This type of analysis was shown to be sensitive to the experimental conditions, in particular the matrix solvent, and also to the type of tissue under investigation.[50] The applicability of MALDI imaging mass spectrometry to the detection and imaging of small organic compounds in tissues was also demonstrated by the study of the distribution of herbicide molecules in soya leaves and stems.[51] All these results are very encouraging and should prompt new applications of MALDI IMS in an increasing number of scientific fields including medical, pharmaceutical, and environmental analyses.

8.7 CONCLUSION AND PERSPECTIVES

MALDI imaging mass spectrometry is now a mature method that has been the subject of several reviews.[20,34,52–54] Despite this abundant and exciting literature, it is surprising to observe the relatively low number of laboratories that are significantly involved in the field. One explanation for this may be found in the inherent difficulties of sample treatment which necessitate a strong investment in methodological development and optimization. It is also necessary to recognize that the imaging approach was promoted more than the profiling method. The latter, however, is much easier to use, and up to this point in time, it has given the most significant results in various biological applications. The choice between profiling and imaging mass spectrometry must therefore be made cautiously, depending on the goal of the study under consideration. However, thanks to the pioneering work of a few research groups, many of the experimental drawbacks revealed in the early days of MALDI IMS have been attenuated or circumvented. Several validated methods are now described in detail[53,54] and are available to new users, thus facilitating initial efforts. New efficient matrix deposition protocols, continuing improvement of imaging software, and the decision of several manufacturers to include MALDI IMS in their instrumental specifications should lead, in the near future, to efficient and reliable tools for routine use of this technique. The specificity in detection of imaging mass spectrometry with regard to the other biological imaging techniques, along with the convincing results published to date, should ensure its future dissemination. The capability of IMS to simultaneously detect hundreds of molecules in a single acquisition without the need for any chemical or biological probe remains a unique feature within the arsenal of the imaging methods. It justifies a continuing effort for improving the methodologies and enlarging the fields of application. MALDI imaging mass spectrometry is now reaching a new era in its young existence.

REFERENCES

1. Li, L.; Garden, R. W.; Sweedler, J. V. Single cell MALDI: A new tool for direct peptide profiling. *Trends Biotechnol.* **2000**, *18*, 151–160.
2. Krishnamurthy, T.; Ross, P. L. Rapid identification of bacteria by direct matrix-assisted laser desorption/ionization mass spectrometric analysis of whole cells. *Rapid Commun. Mass Spectrom.* **1996**, *10*, 1992–1996.
3. Redeker, V.; Toullec, J.-Y.; Vinh, J.; Rossier, J.; Soyez, D. Combination of peptide profiling by matrix-assisted desorption/ionization Time-of-flight mass spectrometry and immuno detection on single glands or cells. *Anal. Chem.* **1998**, *70*, 1805–1811.
4. Castaing, R.; Slodzian, G. Microanalyse par emission ionique secondaire. *J. Microscopie* **1962**, *1*, 395–410.
5. Brunelle, A.; Touboul, D.; Laprévote, O. Biological tissue imaging with time-of-flight secondary ion mass spectrometry and cluster ion sources. *J. Mass Spectrom.* **2005**, *40*, 985–999.
6. Guerquin-Kern, J.-L.; Wu, T.-D.; Quintana1, C.; Croisy, A. Progress in analytical imaging of the cell by dynamic secondary ion mass spectrometry (SIMS spectroscopy). *Biochim. Biophys. Acta* **2005**, *1724*, 228–238.
7. Rubakhin, S. S.; Greenough, W. T.; Sweedler, J. V. Spatial profiling with MALDI MS: distribution of neuropeptides within single neurons. *Anal. Chem.* **2003**, *75*, 5374–5380.
8. Chaurand, P.; Fouchécourt, S.; Dague, B. B.; Xu, B. J.; Reyzer, M.; Orgebin-Crist, M.-C.; Caprioli, R. M. Profiling and imaging proteins in the mouse epididymis by imaging mass spectrometry. *Proteomics* **2003**, *3*, 2221–2239.
9. Chaurand, P.; Stoeckli, M.; Caprioli, R. M. Direct profiling of proteins in biological tissue sections by MALDI mass spectrometry. *Anal. Chem.* **1999**, *71*, 5263–5270.
10. Masumori, M.; Thomas, T. Z.; Chaurand, P.; Case, T.; Paul, M.; Kasper, S.; Caprioli, R. M.; Tsukamoto, T.; Shapell, S. B.; Matusik, R. J. A probasin-large T antigen transgenic mouse line develops prostate adenocarcinoma and neuroendocrine carcinoma with metastatic potential. *Cancer Res.* **2001**, *61*, 2239–2249.
11. Caprioli, R. M.; Farmer, T. B.; Gile, J. Molecular imaging of biological samples: Localization of peptides and proteins using MALDI-TOF MS. *Anal. Chem.* **1997**, *69*, 4751–4760.
12. Chaurand, P.; Schwartz, S. A.; Caprioli, R. M. Profiling and imaging proteins. *Anal. Chem.* **2004**, *76*, 86A–93A.
13. Spengler, B.; Hubert, M. Scanning microprobe matrix-assisted laser desorption ionization (SMALDI) mass spectrometry: Instrumentation for sub-micrometer resolved LDI and MALDI surface analysis. *J. Am. Soc. Mass Spectrom.* **2002**, *13*, 735–748.
14. Dreisewerd, K.; Schürenberg, M.; Karas, M.; Hillenkamp, F. Influence of the laser intensity and spot size on the desorption of molecules and ions in matrix-assisted laser desorption/ionization with a uniform beam profile. *Int. J. Mass Spectrom. Ion Proc.* **1995**, *141*, 127–148.
15. Reyzer, M. L.; Hsieh, Y.; Ng, K.; Korfmacher, W. A.; Caprioli, R. M. Direct analysis of drug candidates in tissue by matrix-assisted laser desorption/ionization mass spectrometry. *J. Mass Spectrom.* **2003**, *38*, 1081–1092.
16. Hsieh, Y.; Casale, R.; Fukuda, E.; Chen, J.; Knemeyer, I.; Wingate, J.; Morrison, R.; Korfmacher, W. Matrix-assisted laser desorption/ionization imaging mass spectrometry for direct measurement of clozapine in rat brain tissue. *Rapid Commun. Mass Spectrom.* **2006**, *20*, 965–972.
17. Garett, T. J.; Yost, R. A. Analysis of intact tissue by intermediate-pressure MALDI on a linear ion trap mass spectrometer. *Anal. Chem.* **2006**, *78*, 2465–2469.
18. Frank, M.; Labov, S. E.; Westmacott, G.; Benner, W. H. Energy-sensitive cryogenic detectors for high-mass biomolecule mass spectrometry. *Mass Spectrom. Rev.* **1999**, *18*, 155–186.
19. Luxembourg, S. L.; Mize, T. H.; McDonnell, L. A.; Heeren, R. M. A. High spatial resolution mass spectrometric imaging of peptide and protein distributions on a surface. *Anal. Chem.* **2004**, *76*, 5339–5344.
20. Schwartz, S. A.; Reyzer, M. L.; Caprioli, R. M. Direct tissue analysis using matrix-assisted laser desorption/ionization mass spectrometry: Practical aspects of sample preparation. *J. Mass Spectrom.* **2003**, *38*, 699–708.
21. Galicia, M. C.; Vertes, A.; Callahan, J. H. Atmospheric pressure matrix-assisted laser desorption/ionization in transmission geometry. *Anal. Chem.* **2002**, *74*, 1891–1895.
22. Chaurand, P.; Schwartz, S. A.; Billheimer, D.; Xu, B. J.; Crecelius, A.; Caprioli, R. M. Integrating histology and imaging mass spectrometry. *Anal. Chem.* **2004**, *76*, 86A–93A.
23. Lemaire, R.; Wisztorski, M.; Desmons, A.; Tabet, J.-C.; Day, R.; Salzet, M.; Fournier, I. MALDI-MS direct tissue analysis of proteins: Improving signal sensitivity using organic treatments. *Anal. Chem.* **2006**, *78*, 7145–7153.
24. Armstrong, D. W.; Zhang, L.-K.; He, L.; Gross, M. L. Ionic liquids as matrixes for matrix-assisted laser desorption/ionization mass spectrometry. *Anal. Chem.* **2001**, *73*, 3679–3686.
25. Lemaire, R.; Tabet, J.-C.; Hendra, J. B.; Salzet, M.; Fournier, I. Solid ionic matrixes for direct tissue analysis and MALDI imaging. *Anal. Chem.* **2006**, *78*, 809–819.
26. Axelsson, J.; Hoberg, A.-M.; Waterson, C.; Myatt, P.; Shield, G. L.; Varney, J.; Haddleton, D. M.; Derrick,

P. J. Improved reproducibility and increased signal intensity in matrix-assisted laser desorption/ionization as a result of electrospray sample preparation. *Rapid Commun. Mass Spectrom.* **1997**, *11*, 209–213.

27. Kruse, R.; Sweedler, J. V. Spatial profiling invertebrate ganglia using MALDI-MS. *J. Am. Soc. Mass Spectrom.* **2003**, *14*, 752–759.
28. Aerni, H.-R.; Cornett, D. S.; Caprioli, R. M. Automated acoustic matrix deposition for MALDI sample preparation. *Anal. Chem.* **2006**, *78*, 827–834.
29. Patel, S.; Goodacre, R.; Sloan, P.; Thakker, N.; Barnes, A.; Loftus, N. Characterising oral cancer tissue using MALDI MS with chemical printing. In *54th ASMS Conference on Mass Spectrometry*, Seattle, WA, ThP18, poster 327, **2006**.
30. Altelaar, F. M.; Klinkert, I.; Jalink, K.; de Lange, R. P. J.; Adan, R. A. H.; Heeren, R. M. A.; Pirsma, S. R. Gold-enhanced biomolecular surface imaging of cells and tissue by SIMS and MALDI mass spectrometry. *Anal. Chem.* **2006**, *78*, 734–742.
31. Novikov, A.; Caroff, M.; Della-Negra, S.; Lebeyec, Y.; Pautrat, M.; Schultz, J. A.; Tempez, A.; Wang, H.-Y. J.; Jackson, S. N.; Woods, A. Matrix-implanted laser desorption/ionization mass spectrometry. *Anal. Chem.* **2004**, *76*, 7288–7293.
32. Stoeckli, M.; Farmer, T. B.; Caprioli, R. M. Automated mass spectrometry imaging with a matrix-assisted laser desorption ionization time-of-flight instrument. *J. Am. Soc. Mass Spectrom.* **1999**, *10*, 67–71.
33. Mac Combie, G.; Staab, D.; Stoeckli, M.; Knochenmuss, R. Spatial and spectral correlations in MALDI mass spectrometry images by clustering and multivariate analysis. *Anal. Chem.* **2005**, *77*, 6118–6124.
34. Stoeckli, M.; Chaurand, P.; Hallahan, D. E.; Caprioli, R. M. Imaging mass spectrometry: A new technology for the analysis of protein expression in mammalian tissues. *Nature Med.* **2001**, *7*, 493–496.
35. Xu, B. J.; Caprioli, R. M.; Sanders, M. E.; Jensen, R. A. Direct analysis of laser capture microdissected cells by MALDI mass spectrometry. *J. Am. Soc. Mass Spectrom.* **2002**, *13*, 1292–1297.
36. Sköld, K.; Svensson, M.; Nilsson, A.; Zhang, X.; Nydahl, K.; Caprioli, R. M.; Svenningsson, P.; Andrén, P. E. Decreased striatal levels of PEP-19 following MPTP lesion in the mouse. *J. Proteome Res.* **2006**, *5*, 262–269.
37. Rohner, T. C.; Staab, D.; Stoeckli, M. MALDI mass spectrometric imaging of biological tissue sections. *Mechanisms of ageing and development*. **2005**, *126*, 177–185.
38. Schwartz, S. A.; Weil, R. J.; Johnson, M. D.; Toms, S. A.; Caprioli, R. M. Protein profiling in brain tumors using mass spectrometry: Feasibility of a new technique for the analysis of protein expression. *Clin. Cancer Res.* **2004**, *10*, 981–987.
39. Chaurand, P.; Schwartz, S. A.; Caprioli, R. M. Assessing protein patterns in disease using imaging mass spectrometry. *J. Proteome Res.* **2004**, *3*, 245–252.
40. Reyzer, M. L.; Caldwell, R. L.; Dugger, T. C.; Forbes, J. T.; Ritter, C. A.; Guix, M.; Arteaga, C. L.; Caprioli, R. M. Early changes in protein expression detected by mass spectrometry predict tumor response to molecular therapeutics. *Cancer Res.* **2004**, *64*, 9093–9100.
41. Han, J.; Schey, K. L. MALDI tissue imaging of ocular lens {alpha}-crystallin. *Invest. Ophtbalmol. Vis. Sci.* **2006**, *47*, 2990–2996.
42. Stoeckli, M.; Staab, D.; Staufenbiel, M.; Wiederhold, K.-H.; Signor, L. Molecular imaging of amyloid β peptides in mouse brain sections using mass spectrometry. *Anal. Biochem.* **2002**, *311*, 33–39.
43. Fournier, I.; Day, R.; Salzet, M. Direct analysis of neuropeptides by *in situ* MALDI-TOF mass spectrometry in the rat brain. *Neuroendocrinol. Lett.* **2003**, *24*, 9–14.
44. Garden, R. W.; Sweedler, J. V. Heterogeneity within MALDI samples as revealed by mass spectrometric imaging. *Anal. Chem.* **2000**, *72*, 30–36.
45. Brand, G. D.; Krause, F. C.; Silva, L. P.; Leite, J. R. S. A.; Melo, J. A. T.; Prates, M. V.; Pesquero, J. B.; Santos, E. L.; Nakaie, C. R.; Costa-Neto, C. M.; Bloch, C., Jr. Bradykinin-related peptides from *Phyllomedusa hypochondrialis*. *Peptides*. **2006**, *27*, 2137–2146.
46. Maddalo, G.; Petrucci, F.; Iezzi, M.; Pannellini, T.; Del Boccio, P.; Ciavardelli, D.; Biroccio, A.; Forli, F.; Di Ilio, C.; Ballone, E.; Urbani, A.; Federici, G. Analytical assessment of MALDI-TOF imaging mass spectrometry on thin histological samples. An insight in proteome investigation. *Clin. Chim Acta.* **2005**, *357*, 201–218.
47. Touboul, D.; Piednoël, H.; Voisin, V.; De La Porte, S.; Brunelle, A.; Halgand, F.; Laprévote, O. Changes in phospholipid composition within the dystrophic muscle by matrix-assisted laser desorption/ionization mass spectrometry and mass spectrometry imaging. *Eur. J. Mass Spectrom.* **2004**, *10*, 657–664.
48. Jackson, S. N.; Wang, H. Y.; Woods, A. S. Direct profiling of lipid distribution in brain and tissue using MALDI-TOFMS. *Anal. Chem.* **2005**, *77*, 4523–4527.
49. Bunch, J.; Clench, M. R.; Richards, D. S. Determination of pharmaceutical compounds in skin by imaging matrix-assisted laser desorption/ionisation mass spectrometry. *Rapid Commun. Mass Spectrom.* **2004**, *18*, 3051–3060.
50. Crossman, L.; McHugh, N. A.; Hsieh, Y.; Korfmacher, W.; Chen, J. Investigation of the profiling depth in matrix-assisted laser desorption/ionization imaging mass spectrometry. *Rapid Commun. Mass Spectrom.* **2006**, *20*, 284–290.
51. Mullen, A. K.; Clench, M. R.; Crosland, S.; Sharples, K. R. Determination of agrochemical compounds in

soya plants by imaging matrix-assisted laser desorption/ ionization mass spectrometry. *Rapid Commun. Mass Spectrom.* **2005**, *19*, 2507–2516.

52. Todd, P. J.; Schaaff, T. G.; Chaurand, P.; Caprioli, R. M. Organic ion imaging of biological tissue with secondary ion mass spectrometry and matrix-assisted laser desorption/ionization. *J. Mass Spectrom.* **2001**, *36*, 355–369.
53. Chaurand, P.; Schwartz, S. A.; Reyzer, M.; Caprioli, R. M. Imaging Mass Spectrometry: Principles and potentials. *Toxicologic Pathology.* **2005**, *33*, 92–101.
54. Rubakhin, S. S.; Jurchen, J. C.; Monroe, E. B.; Sweedler, J. V. Imaging mass spectrometry: Fundamentals and applications to drug discovery. *Drug Discovery Today* **2005**, *10*, 823–837.

Part III

ES and MALDI Coupling to Mass Spectrometry Instrumentation

Chapter 9

Coupling ESI and MALDI Sources to the Quadrupole Mass Filter, Quadrupole Ion Trap, Linear Quadrupole Ion Trap, and Orbitrap Mass Analyzers

Mark E. Bier

Center for Molecular Analysis, Department of Chemistry, Mellon College of Science, Carnegie Mellon University, Pittsburgh, Pennsylvania

Electrospray and MALDI Mass Spectrometry: Fundamentals, Instrumentation, Practicalities, and Biological Applications, Second Edition, Edited by Richard B. Cole

9.1 INTRODUCTION

The power of electrospray and matrix-assisted laser desorption techniques to ionize nonvolatile molecules, including macromolecules, and allow the flight of these ions in the gas phase has thrust mass spectrometry (MS) into the limelight of many new research laboratories over the last two decades. In concert with the development of these two ion sources has been the improvement of existing mass analyzers and significant inventions of new ones. The combination of these two essential components of a mass spectrometer, the ion source and the mass analyzer, is at the heart of this chapter. This chapter discusses the instrumental aspects of coupling electrospray ionization (ESI) and matrix-assisted laser desorption ionization (MALDI) sources to the four types of mass analyzers shown in Figure 9.1: (a) the well-known quadrupole mass filter (QMF), (b) the 3D-quadrupole field ion trap or "Paul" trap (QIT), (c) the 2D-quadrupole field ion trap, the popular geometry of which is a linear quadrupole ion trap (LQIT),[1,2] and, finally, the latest newcomer to the field of mass analyzer types, (d) the orbitrap.[3,4] Electrospray ionization (ESI) of "macroions"

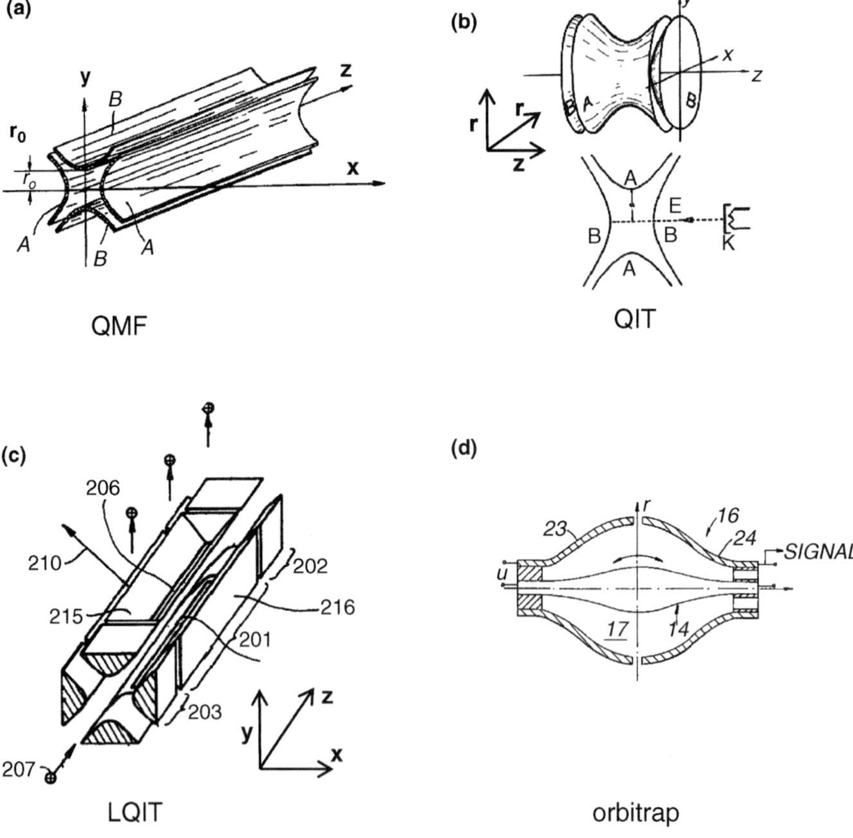

Figure 9.1. The four mass analyzers discussed in this chapter: (**a**) A 2D-quadrupole field mass filter (QMF). (**b**) A 3D-quadruople field ion trap (QIT). Parts a and b are from Paul, W.; Steinwedel, H. German Patent 944,900, 1956. (**c**) Linear 2D-quadrupole field ion trap (LQIT). (Bier, M. E., and Syka, J. E. P. U.S. Patent 5,420,425, 1995). (**d**) Orbitrap (Makarov, A. A. U.S. Patent 5,886,346, 1996). All figures were obtained from the original patents and partially modified for clarity.

was pioneered 30 years ago by the efforts of Malcolm Dole and colleagues.[5] Dole's group electrosprayed polystyrene molecules with molecular weights of 51 kilodaltons (kDa) and 411 kDa, and later they even electrosprayed the proteins lysozyme (MW 14.4 kDa) and zein (MW 38 kDa).[6–8] A simple energy analyzer was used to detect dimers and trimers of the 51 kDa polystyrene and lysozyme by measuring stopping potentials for these ions. Remarkably, they found that the larger polystyrene ion (411 kDa) carried up to five charges. In 1973, Dole et al.[7] discussed the use of a quadrupole mass filter (QMF) and a time-of-flight (TOF) mass spectrometer to measure the mass-to-charge ratios of these ions, but it was not until Fenn and co-workers at Yale University[9,10] and Aleksandrov and co-workers in Russia[11] coupled an ESI source to a mass spectrometer in 1984 that the field of ESI mass spectrometry began. MS took a leap forward when Fenn's group discovered that electrosprayed proteins may indeed carry multiple charges,[12] from 2 to greater than 100, and they further showed that the distribution of these multiply charged ions could be used to calculate accurate molecular weights. Fenn shared the 2002 Nobel Prize in Chemistry for making these remarkable electrospray ionization MS discoveries and he is still active in research today.

Koichi Tanaka and co-workers of Shimadzu Corporation in Japan initiated the first MALDI studies in 1987, and for this work Tanaka shared the 2002 Nobel Prize with Fenn. In Tanaka's research, metal powder dispersed in glycerol was used as the matrix to lift and mass analyze proteins up to 34 kDa by TOF MS.[13] MALDI was improved by Karas and Hillenkamp[14] who used UV absorbing nicotinic acid as the matrix to analyze proteins including serum albumin at 66.4 kDa.

Wolfgang Paul and H. Steinwedel at the University of Bonn publicly described quadrupole field mass spectrometers in 1953,[15] although Richard Post of the University of California at Berkeley independently carried out similar research at about the same time.[16,17] Three years later, Paul and Steinwedel were issued a German patent[18] that included the four electrode quadrupole mass filter (QMF) and the three-electrode quadrupole ion trap (QIT). In 1989, Paul shared the Nobel Prize in Physics[19] with H. G. Dehmelt at the University of Washington and with N. F. Ramsey at Harvard University, for their work with ion traps. The award was given to Paul for his work on radio-frequency (RF) ion traps and to Dehmelt and Ramsey for their work on atomic precision spectroscopy. Since conception to the present day, over 50 years of research activity has involved these interesting quadrupole field devices. The QMF popularity came first during the 1960s with the introduction of commercial instruments by Wade Fite's group at Extranuclear Corporation in Pittsburgh, PA and Robert Finnigan's team, first at Electronics Associates Inc. in Palo Alto, CA, and later at Finnigan Corporation in Sunnyvale, CA.

The quadrupole mass filter has seen continued performance improvements, but it is still operated fundamentally as it was when it was first introduced. The QMF is in demand today in the form of single-quadrupole, triple-quadrupole, or hybrid mass spectrometers. The birth of the QIT as a scanning mass spectrometer in 1983 by Finnigan Corporation occurred later because of the need for the often overlooked computer control which was essential for implementation of the scan function and the mass selective instability scan.[20,21]

Major ion trap developments have occurred over the last 15 years. In 1995, Bier and Syka patented the use of the mass selective instability scan from high-charge-capacity ion trap geometries such as the linear quadrupole ion trap (LQIT), toroidal trap (TQIT), curved or "banana" traps (CQIT), and elliptical traps (EQIT).[1] The LQIT with radial ejection was eventually commercialized in 2002, and it is used as a stand-alone mass analyzer and has been combined with QMFs, Fourier transform ion cyclotron resonance (FT ICR), and the orbitrap to form hybrid instruments. The LQIT analyzer has eclipsed the "conventional"

QIT in performance; as a result, the commercial price tag almost doubled! An almost simultaneous development by Hager and co-workers[2,108] at MDS Sciex Corporation in Toronto, Canada, was a high-charge-capacity LQIT that is operated using axial ejection. The Sciex LQIT was commercialized as a triple quadrupole hybrid. High-charge-capacity ion traps with unique scanning methods and field modifications were applied to the QIT and used by Varian,[22] Agilent Technologies, and Bruker Daltonics (see Appendix, Table 9.A). In 2001, a team at Shimadzu Corp. revisited and implemented the rectangular wave driven "digital" QIT.[23,24] The most recent mass analyzer innovation was the development of the high-resolution orbital trap, pioneered by Makarov[25] and discussed in 2000. Finally, the application of ion–ion reactions, electron capture dissociation (ECD), and electron transfer dissociation (ETD) to ion traps has added additional tools for top-down proteomics.

Our discussion starts with a brief section about some fundamentals and capabilities of the QMF, QIT, LQIT, and orbital trap or orbitrap (OT), followed by the instrumentation requirements necessary to couple MALDI and ESI to these mass analyzers, and ends with proteomic and genomic applications. For more fundamental knowledge of the QMF, LQIT, QIT, and orbital trap, several review papers are highly recommended. Detailed fundamentals about QMFs, most of which can be applied to the LQIT and the QIT, can be found in the invaluable resource by Dawson.[26] General QIT reviews,[27–31] including high-pressure ion sources,[32] and two focused on ESI[33,34] help to understand the fundamental ion trap developments. Douglas et al.[35] have written a general review about LQIT, and Hu et al.[36] have written an informative article about the orbitrap. An extensive general mass spectrometry instrumentation review is also available by McLuckey and Wells.[37]

Although this chapter covers certain aspects of the ESI and MALDI sources, earlier chapters of this book are recommended for details about these ion sources and especially for coverage of the new development of desorption ESI (DESI) by Cooks and co-workers[38] in 2004.

9.2 QUADRUPOLE FIELD MASS FILTER

9.2.1 Geometry of the Quadrupole Mass Filter

The quadrupole mass filter (QMF) is a four-electrode device. The four rod-like electrodes are positioned symmetrically around the ion axis and parallel to one another as shown in Figure 9.1a. The inner surfaces of the four electrodes are ideally hyperbolic in shape so as to create a pure quadrupole field inside the device; but many manufacturers of the quadrupole mass filter use round rod electrodes to create a high-quality quadrupole field near the ion axis. The rods are often held in place with ceramic rings or spacers and are typically about 12–20 cm long, although lengths have varied from 12.7 mm[39] to several meters.[40,41] The machining of the rods and the relative position of these rods to one another is extremely important in determining the overall quality of the mass filter. For this reason, the errors in the manufacturing of these devices are typically measured in microns. The distance from the center of the ion axis to the apex of a rod is defined as r_0. A QMF with a large r_0 will actually have a small mass range due to the inverse relationship of r_0 with m/z.

9.2.2 Operation of the Quadrupole Mass Filter

A two-dimensional (2D) quadrupole field is typically formed by applying a suitable RF and DC waveform to the four electrodes shown in Figure 9.1a. The same RF and DC waveform is applied to opposite pairs of (A or +) rods, but the RF is 180° out of phase and the DC is opposite in potential on the adjacent (B or −) rods. Ions enter the QMF through a small aperture in an entrance lens element at one end of the quadrupole and transmit through the device near the z axis. No quadrupole field exists in the z dimension and thus the field is two-dimensional (2D) rather than three-dimensional (3D) as seen within the QIT. The quadrupole field continuously forces ions toward the center of the QMF. For a stable ion, the force is linearly proportional to the distance the ion is from the center of the device. Ions follow trajectories described by the second-order Mathieu differential equation. However, there is some error in the predicted ion trajectory, because ion motion is affected by space-charge repulsion, gaseous collisions, and the imperfect quadrupole field, which are unaccounted for in the differential equation. Solutions to the differential equation are in terms of the Mathieu parameters a and q as shown in Eqs. (9.1) and (9.2) which are proportional to the DC and RF amplitudes, respectively. U is the DC amplitude, V is the RF amplitude, e is the charge on an ion, m is the mass of an ion, and $\Omega = 2\pi f_{RF}$, where f_{RF} is the frequency of the main RF voltage.

$$a_u = a_x = -a_y = \frac{8eU}{mr_0^2\Omega^2} \tag{9.1}$$

$$q_u = q_x = -q_y = \frac{4\,eV}{mr_0^2\Omega^2} \tag{9.2}$$

The entire QMF is electrically isolated and offset with a few negative volts respect to an ion source near ground potential so that a positive ion has a few electron volts of kinetic energy. If an ion of the proper m/z is not filtered, it will exit out of the QMF and go to a detector, a collision cell, or a second mass analyzer as shown in Figure 9.2a. Ions with the incorrect m/z undergo trajectories that eventually cause them to presumably strike the rods where they are annihilated, never making it to a detector or a second mass analyzer. The ion beam is thus filtered, typically allowing only a 1-Th-wide mass range through at any one time.

9.2.3 The 2D-Quadrupole Mass Filter Stability Diagram

The QMF stability diagram is shown in Figure 9.2d. Plotting and overlapping the solutions to the Mathieu equation in (a, q) space for the x and y dimensions forms the diagram. The Mathieu parameters, a and q, are indicative of the stability and motion of an ion in the 2D quadrupole field. The operating or scan line defined by the RF/DC ratio represents the unstable (a_u, q_u) points where ions are filtered and the stable (a_u, q_u) points in the stability diagram where ions can pass through the QMF. The operating line starts at $(a_u, q_u) = (0,0)$ and intersects the plot near the upper stability apex of the diagram. Theoretically, as the scan line (RF/DC slope) approaches the apex at $(a,q) = (0.23699, 0.70600)$ the resolution (R) would increase to infinity,[26] but in reality experimental values are typically in the thousands ($R = m/\Delta m$ at FWHM) since higher resolution results in a loss in sensitivity. Since q is inversely proportional to mass, high m/z ions have a lower q_x value than low m/z ions as represented by the different-sized circles on the diagram. In this case, m_4^+ has a larger m/z than m_3^+. By increasing U and V, the DC and RF amplitude applied to the rods, respectively, ions positioned along the QMF operating line shown in Figure 9.2d move to higher (a_u, q_u) values. Near, but below, the apex of the stability diagram, only ions over a narrow m/z can pass through the device (e.g., m_2^+) while unstable ions collide with the electrode rods. When scanning the RF and DC signals from

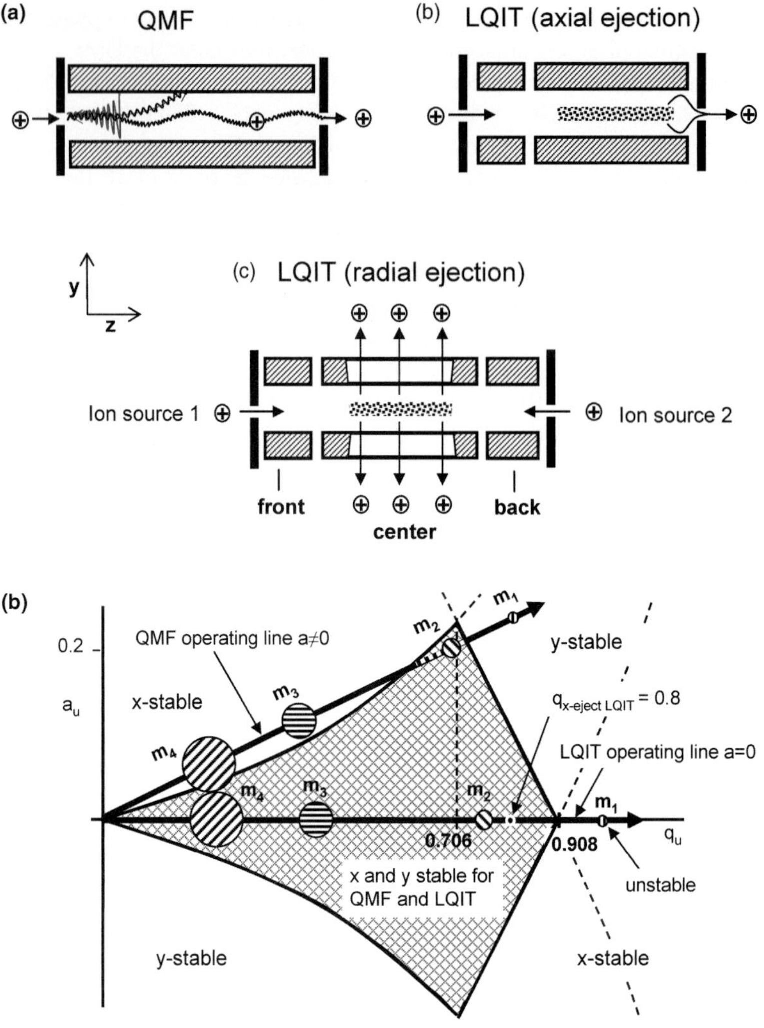

Figure 9.2. (**a**) Cross-sectional view of the QMF in the y–z plane with simplified ion trajectories for ion injection and mass filtering. (**b**) Cross section of the LQIT in the y–z plane with simplified ion trajectories for ion injection and mass analysis by axial resonance ejection. (**c**) Cross section of the LQIT in the y–z plane with simplified ion trajectories for ion injection from two ion sources and mass analysis by radial resonance ejection. Note the large ion trapping volumes in parts b and c. (**d**) The stability diagram for the 2D-quadrupole field QMF and LQIT with the operating line for the QMF filtering mode (RF/DC) and the trapping mode supplemented with resonance ejection (RF-only). The crossed circles represent ions of different m/z values (e.g., where $m_4 > m_3 > m_2 > m_1$, for $z = 1$).

low to high amplitudes, the lowest m/z ions are transmitted first, followed sequentially by successively higher m/z ions.

By setting the DC amplitude, U, to zero ($a_u = 0$), the QMF no longer acts as a mass filter, but instead as a RF-only quadrupole (q). These devices, termed "RF ion guides" or collision cells (see below), are very important in mass spectrometry because they are used to bunch and focus ions at high pressure over a broad mass range and guide them to the next device. Ion guides can also be constructed as higher multipole devices such as a hexapole, octopole, or decapole, RF ion guides are used in most of the instrumentation discussed in this chapter.

9.2.4 Scan Modes of the Triple Quadrupole Mass Spectrometer

A significant advantage of the QMF is that it can be coupled with many of the same or different mass analyzers. In one powerful configuration, a QMF is followed by a RF collision chamber or cell (RF ion guide inside a conductance limited cylinder) and then by a second mass analyzer to form a triple-quadrupole mass spectrometer (QqQ) (note that Q = QMF) for "tandem in space" mass spectrometry as shown in the figure included in Table 9.1. This combination for mass spectrometry/mass spectrometry (MS/MS) further improves the overall signal-to-noise (S/N) ratio by reducing the chemical noise. Several scan modes exist for this setup, including *selected or single-ion monitoring (SIM), product scan, precursor scan, neutral loss scan, and selected reaction monitoring* (SRM) as listed in Table 9.1 and described below:

- **Full -scan MS.** Either Q1 or Q3 RF/DC amplitude is scanned over a *m/z* range of 5–4000.
- **Single (or selected) ion monitoring (SIM):** Park Q1 RF/DC amplitudes at one $(m/z)_1$ with an ion isolation window of 1–10 Da. Q3 is set to RF-only mode to transmit all ions. Q1 and Q3 functions are interchangeable in SIM while q2 is RF-only, always. Multiple ions may be analyzed by jumping the parked RF/DC to various mass windows, but each jump will decrease the duty cycle for each ion accordingly.

Table 9.1. Scan Modes of the Triple-Quadrupole Mass Spectrometer

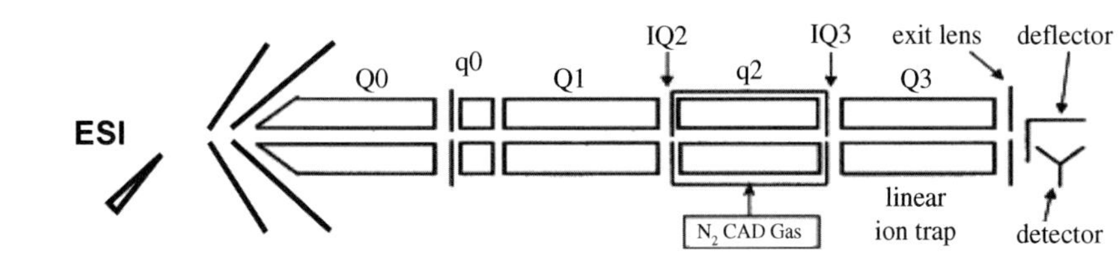

Scan Mode	Q0	Q1 RF/DC	q2	Q3 RF/DC	Duty Cycle[a]	Comments
Full scan	RF only	Scanned	RF only	RF only	Low	Q1 and Q3 functions are interchangeable.
SIM	RF only	Parked	RF only	RF only	High	Multiple ions are possible.
MS^n						
ion activation						PID, SID, (q2 gas required for CAD.)
Product scan	RF only	Parked	RF only	Scanned	Low	Q1 low res. for isotopes
Precursor scan	RF only	Scanned	RF only	Parked	Low	Look for homologs
Neutral loss scan	RF only	Scanned	RF only	Scanned	Low	Q1 & Q3 are synchronized
SRM	RF only	Parked	RF only	Parked	High	Multiple transitions are possible.

[a]Duty cycle is calculated using the scan time for ion one (in a QMF, but the total 'ion time' for ion traps.). SIM, selected ion monitoring; SRM, selected reaction monitoring; PID, photo induced dissociation; SID, surface induced dissociation.

Note: The triple-quadrupole picture is a hybrid mass spectrometer featuring an LQIT. Both q2 and Q3 have been used as LQIT (MDS Sciex Corp.) The LQIT scan modes are not covered in this table. Figure reprinted with permission from *J. Am. Soc. Mass Spectrom.* **2003**, *14*, 1130–1147.

- **Product scan** (daughter scan): Pass only the precursor ion $(m/z)_1$ in Q1 (RF/DC parked), collisionally activate and fragment $(m/z)_1$ inside the collision cell (q2, RF only), and scan Q3 (RF/DC amplitude scanned) to produce a mass spectrum of the product ions and unfragmented $(m/z)_1$.
- **Precursor scan** (parent scan): Scan all precursor ions $(m/z)_1, (m/z)_2, (m/z)_3, \ldots$, etc. over a large mass range in Q1 (RF/DC scanned), fragment precursors $(m/z)_1, (m/z)_2, (m/z)_3, \ldots$, sequentially inside the collision cell (q2, RF-only), and pass only a specific product ion $(m/z)_{\text{fragment}}$ through Q3 (RF/DC parked) to produce a mass spectrum of all the precursors ions that produce $(m/z)_{\text{fragment}}$.
- **Neutral loss scan:** Scan Q1 and Q3 together (RF/DC scanned), but with a set mass difference between them (e.g., *m/z* 98) so that all precursor ions that fragment in the collision cell (q2, RF-only) to produce the same $\Delta m/z$ loss will determine the mass spectrum. For example, show all ions that fragment to produce H_2O (18 Da), or H_3PO_4 (98 Da). The neutral loss scan can give information about functional groups attached to the precursor ions.
- **Selected reaction monitoring (SRM):** Pass only the precursor ion $(m/z)_1$ in Q1 (RF/DC parked), fragment $(m/z)_1$ inside the collision cell (q2, RF only) and pass only the fragment $(m/z)_{\text{fragment}}$ in Q3 (RF/DC amplitude parked). For multiple reaction monitoring (*MRM*), several transitions are chosen by jumping the RF/DC amplitudes of Q1 and Q3 simultaneously to the various ion transitions of interest. Each jump reduces the duty cycle and overall sensitivity accordingly. The SRM scan requires that both Q1 and Q3, which are parked at unique (*m/z*)s, pass the precursor and fragment from a selected transition for a signal to be registered at the detector. The highly sensitive *SRM or MRM* scan modes allow the operator to have a greater confidence in the identity of the target molecule over a SIM scan.

9.2.5 Performance of the Quadrupole Mass Filter

The quadrupole mass filter is a versatile and sensitive mass analyzer for the mass range *m/z* 10–4000. Typical QMFs operate at a RF frequency of 1–2 MHz and allow for isotope resolution throughout this mass range. The mass range is limited by the maximum RF and DC amplitudes that are electrically practical given the frequency that is required for adequate resolution at the high end of the mass range. The detection limit of a sample while operating the QMF at low resolution in a SIM or SRM mode ("parked" RF/DC amplitudes) is unrivaled. However, when operated in the scanning mode with unit resolution (all peaks separated by 1 amu) over a large mass range, the duty cycle increases by an order of magnitude and the detection efficiency is accordingly reduced by about the same order of magnitude. The MS/MS scan modes are powerful. For example, a neutral loss scan can be used to determine with high confidence what peptides are phosphorylated in a mixture in one experiment by looking for the neutral loss of 98 (H_3PO_4). Mass accuracies are at ~0.01%.

9.3 QUADRUPOLE FIELD ION TRAPS

Limited quadrupole ion trap theory is presented here, but a thorough coverage of the QIT can be found in March and Hughes,[42] in a three-volume book edited by March and Todd,[43] and in a book by March and Todd on QIT which includes a brief chapter on the LQIT.[44]

Quadrupole fundamentals that can be applied to 3D- and 2D-quadrupole field ion traps can be found in the book by Dawson.[26] We will use the acronym XQIT in this text, where X is a variable that refers to all quadrupole field ion traps whether they have 2D- or 3D-quadrupole fields. For the 2D traps, X will be replaced by L for "linear," RL for "rectilinear," T for "toroidal," and C for "curved," allowing for a nomenclature system for future geometries.

9.3.1 The Three-Dimensional Quadrupole Field Ion Trap

9.3.1.1 The History of Performance Enhancements for the 3D-Quadrupole Field Ion Trap

Over the past 20-plus years, a large number of performance enhancements have been developed for the 3D-quadrupole field QIT, which have allowed the device to compete with the QMF. These advances have been both fortuitous and by design. The analyzer performance enhancements include the use of helium dampening gas,[12] variable ionization times[45], resonance ejection[46], RF/DC isolation,[47] MS^n-scans,[15,48,49] ion injection,[50] precursor and neutral loss scanning,[51,52] mass range extension,[53–55] high-resolution scanning,[56–59] rapid high-sensitivity scanning[60] and tailored waveforms (TWF).[61–65] In addition to these improvements, the QIT was used as an electrodynamic vessel for carrying out gas-phase ion/molecule reactions.[66,67] Many types of ion sources and inlets have been coupled to the QIT. They include chemical ionization,[68,69] liquid secondary ion mass spectrometry (LSIMS),[19] thermospray,[70,71] glow discharge,[72] membrane,[73] particle beam,[39,74] laser desorption,[75–77]matrix-assisted laser desorption ionization (MALDI),[78–81]atmospheric pressure chemical ionization (APCI),[82] ESI and DESI.[38] While the union of ESI and MALDI with the QIT in the early 1990s formed extremely powerful analytical instruments,[78–81,83,84] only the ESI QIT was commercialized and made popular at that time. Now that the 2D-quadrupole field LQIT has been implemented and commercialized, many of the performance enhancements mentioned above have been adapted directly to this high-capacity ion trap as well as ESI and MALDI. The 3D Paul trap, however, is still alive and well with the development of the high-charge-capacity methodologies discussed later in this section.[22]

9.3.1.2 The Geometry of the 3D-Quadrupole Field Ion Trap

The QIT or "Paul trap" is a three-electrode device as shown in Figures 9.1b and 9.3a. The basic shape of the device can be imagined by rotating the hyperbolic electrode cross section shown in Figure 9.1b in space around the *z* axis (ion axis). Two of the electrodes formed are identical and are called *end caps*, while the third electrode is called the *ring electrode*. The ring electrode is often sandwiched between the two end caps at a precise distance typically maintained by ceramic or quartz spacers. In Figure 9.3a the QIT is shown configured to store ions from an ESI or MALDI source. A hole is machined in the entrance end cap for *ion injection* into the device from an ion source and, a second hole is located in the exit end cap for *ion ejection* out of the device to a detector. To allow for the injection and trapping of positive ions formed near ground potential from an ESI source or MALDI ionization source, ions "must go downhill." In one variation of this scheme, the DC potential on the electrodes can be set to a voltage of negative 3–15 V for positive ions and helium fills the trap at ~1 mtorr.

A three-dimensional quadrupole field is formed by the three electrodes when a suitable RF voltage ($V_{RF}\cos(2\pi f_{RF}t)$) is applied to the ring electrode as shown in Figure 9.3a. The

(a)

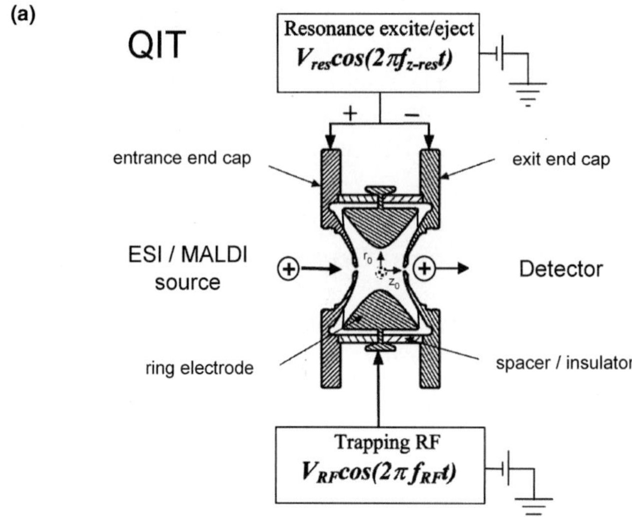

(b)

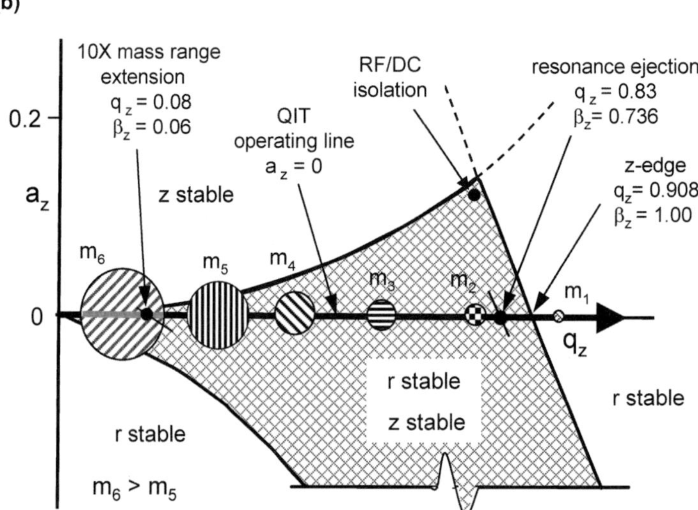

Figure 9.3. (**a**) A three-dimensional (3D) quadrupole field ion trap (QIT) consisting of two hyperbolic end caps and a central hyperbolic ring electrode positioned between an ESI or MALDI source and a detector. In this example the trapping RF voltage is applied to the ring electrode and the supplementary resonance excitation/ejection RF voltage is applied in dipolar fashion to the end cap electrodes. Ions are ejected toward the end caps during mass analysis. (**b**) The top portion of the QIT stability diagram (cross hatched region) with the $a_z = 0$ mass-selective instability operating line. The crossed circles represent ions of different m/z values (e.g., where $m_6 > m_5$) on the operating line with different q_z values. The mass selective instability scan is supplemented with resonance ejection at a $q_z = 0.8$ for a normal scan or at $q_z = 0.08$ to extend the mass range by a factor of 10×.

quadrupole field traps ions by continuously forcing them toward the center of the device. The force on an ion is linearly proportional to the distance of the ion from the center of the QIT. Ions stored inside the trap follow trajectories described by the second-order Mathieu differential equation. However, there is some error in the predicted ion trajectory, because ion motion is again affected by space-charge repulsion, the imperfect quadrupole trapping field, and collisions with gases (e.g., helium) which are unaccounted for in the differential equation. Solutions to the differential equation are in terms of the Mathieu parameters a_z and q_z as shown in Eqs. (9.3) and (9.4) for the three-dimensional ion trap case[42]:

$$a_z = -2a_r = \frac{-16eU_{DC}}{m(r_0^2 + 2z_0^2)\Omega^2} \tag{9.3}$$

$$q_z = -2q_r = \frac{-8eV_{RF}}{m(r_0^2 + 2z_0^2)\Omega^2} \tag{9.4}$$

In Eqs. (9.3) and (9.4), r represents the radial direction, z represents the axial direction, U_{DC} is the DC amplitude, V_{RF} is the RF amplitude, e is the charge on an ion, m is the mass of an ion, r_0 is the inner radius of the ring electrode, z_0 is the axial distance from the center of the device to the nearest point on one of the end-cap electrodes, and $\Omega = 2\pi f_{RF}$, where f_{RF} is the frequency of the main RF voltage.

One specific, but often cited geometry of the ion trap is defined by Eq. (9.5). Solving for z_0^2 in Eq. (9.5) and substituting the result into Eqs. (9.3) and (9.4) above, simplifies these Mathieu parameter equations.

$$r_0^2 = 2z_0^2 \tag{9.5}$$

During the 1980s, researchers at Finnigan Corporation observed that ion traps built with the specific geometry of Eq. (9.5) demonstrated poor mass accuracy. Mass assignments were shifted and the errors were compound-dependent. Syka[43] suggested that these mass shifts were due to a difference in ion radial distributions that occur because of the different ion collision cross sections, the high pressure of helium in the trap, and the effects of higher-order fields[85] that are caused by field imperfections resulting from the entrance and exit holes. Mass shifts were reduced to negligible deviations by spacing the end caps outward by a factor of $0.11z_0$.[86] This alteration of the geometry improves the homogeneity of the quadrupole field at the center on the trap where the ions are stored. To account for the "stretched" ion trap geometry, the z_0 characteristic dimension must be used in the complete form of Eqs. (9.3) and (9.4).

9.3.1.3 The Quadrupole Ion Trap Stability Diagram and the Resonant Frequencies of Ions

Plotting and overlapping the solutions to the Mathieu equation in (a, q) space for the r and z dimensions forms the QIT stability diagram. A portion of the stability diagram including the solutions where $a_z = 0$ is shown in Figure 9.3b. The Mathieu parameters, a and q, are indicative of the stability and motion of an ion in the 3D-quadrupole field.

The operating or scan line represents the (a_z, q_z) points in a stability diagram that ions pass through until they are scanned out of the trap. For the basic mass selective instability scan, the operating line starts at $(a_z, q_z) = (0,0)$ and passes through $(a_z, q_z) = (0,0.908)$ as seen in the diagram. Since q_z is inversely proportional to mass (m), high m/z ions have a lower q_z value than low m/z ions as shown by the different size representative circles on the diagram. In this case, m_5^+ has a larger m/z than m_4^+. By increasing V, the RF amplitude applied to the ring electrode, ions positioned along the operating line shown in Figure 9.3b move to higher q_z values. At the edge of the stability diagram at $q_{z\text{-edge}} = 0.908$, instability ensues in the z dimension and the ions leave the QIT through the holes in the end caps. The lowest m/z *ion* becomes unstable first, followed sequentially by the next higher m/z ion, and so on.

Ions of a specific *m/z* have a fundamental frequency of motion unique to their q_z value and this unique frequency is often used to resonate these ions. The frequencies of ion motion in the z dimension are defined by Eq. (9.6), where f_h is a frequency component of the ion motion, f_{RF} is the frequency of the RF applied to the ring electrode, and β_z is a parameter used to relate the frequency of an ion to its position on the stability diagram.[26]

$$f_h = (hf_{RF} \pm \beta_z f_{RF}/2), \text{ where } h = \{0, \pm1, \pm2, \pm3, \ldots\} \tag{9.6}$$

By knowing the q_z of an ion, one can calculate the β_z value[26,43] and thus the resonant frequencies of an ion following Eq. (9.6). At low β_z, the frequency components ($|h| > 0$) have a negligible effect on ion motion; and unless large voltages are applied to the end caps, these higher-order frequencies are usually insignificant and, thus, not utilized. In general, the fundamental resonant frequency of an ion is defined as the frequency where $h = 0$ (i.e., $f_{z\text{-res}} = f_0 = \beta_z f_{RF}/2$).

9.3.1.4 Resonance Ejection in the QIT

A dipole signal ($V_{res} \cos(2\pi f_{z\text{-res}} t)$) is applied to the end caps as shown in Figure 9.3a, for *resonance excitation* and/or *resonance ejection.* For either of these processes to occur efficiently, however, the frequency used must equal the fundamental resonant frequency of the ion, $f_{z\text{-res}}$. For example, during the mass selective instability scan supplemented with resonance ejection (see Figure 9.3b), ions are ejected from the trap axially as they are scanned into resonance at a frequency that corresponds to a $q_{z\text{-eject}} = 0.83$, $\beta_{z\text{-eject}} = 0.736$. Resolution, peak height, and sensitivity are greatly improved when using the mass selective instability scan mode supplemented with resonance ejection.

9.3.1.5 Effects of Space Charge

Space-charge effects are caused by the distortion of an electrostatic field, namely the applied trapping quadrupole field, due to the presence of the electrostatic fields from one or more ions. The greater the number of ions stored inside a XQIT, the greater the effects of space charge. This perturbation of the applied field forces ions to follow nonideal trajectories and is a fundamental limitation for all types of ion traps. Most notably, space charge causes degradation in resolution, a reduction in peak height, and a shift in mass assignments.[87,88] At severe space-charge conditions, the mass peaks are further broadened and reduced in peak height to the point where they flatten into the baseline. Beyond the extreme limit of space charge, the ion density becomes so large that additional ions injected into the quadrupole field may not be trapped at all or previously trapped ions may be displaced. Ion displacement should occur in descending order of *m/z* based on storage forces at various q_z values.[26] In a simplified analogy, an ion trap acts like a bucket. At "10% full," a bucket works well at carrying water and water may be poured out of the bucket with little or no spillage. However, filling a bucket full to the brim makes it difficult to carry or pour water. And, of course, any attempt to fill a bucket above the brim immediately results in spillage of water out of the bucket. To avoid the effects of space charge, the number of ions in an ion trap chamber is regulated at the "10% level," below the spectral ion charge limit, to achieve an optimum *m/z* analysis. This control is achieved by the prescan and a variable ionization time for the analytical scan.

9.3.1.6 High Ion Charge Capacity QIT

Several manufacturers (Varian, Bruker Daltonics, and Agilent Technologies) have increased the spectral ion charge capacity of the conventional 3D-QIT without making major physical changes to the electrode geometry. Varian modified the trapping field of their 500-MS Ion Trap mechanically to add octopole resonances, and electrically to add hexapole resonances to the trapping field.[22] These modifications result in a significant increase in the number of ions that can be efficiently scanned out without the deleterious effects of space charge. The method involves applying an AC voltage out of phase to each end-cap electrode at the same frequency as the main RF frequency. This causes the center of the trapping field to shift from the center to the end cap with the AC component in phase with the main RF voltage while adding a hexapole field. The resulting trapping field has a nonlinear resonance at the operating point of $\beta_z = 2/3$. Since the ions are displaced from the center of the trap by the asymmetric trapping field, the hexapole resonance and a parametric resonance resulting from a supplementary quadrupole field will also have a nonzero value. Finally, the addition of a supplementary dipole field at $\beta_z = 2/3$ causes dipolar resonant excitation. All three fields have nonzero values at the operating point and cause a triple resonance condition. An ion moved to $\beta_z = 2/3$ will be in resonance with, and absorb power from, all three fields simultaneously. The power absorption and axial ion motion by the ions will be nonlinear, resulting in rapid ejection from the QIT and improved resolution. In addition to the improved ion capacity, displacement of ions further from the trapping center toward the exit end cap causes the ions to be ejected through this electrode only, thus doubling the number of ions detected. Varian refers to this resonance method of ion ejection as enhanced charge capacity (ECC). A similar charge storage enhancement is found in the Bruker Daltonics (HCT™ Ion Trap) and the Agilent Technologies (1100 series Ion Trap), which also use nonlinear scanning techniques. These changes to the QIT geometry and scanning methods have remarkably resulted in an approximately 10-fold increase in spectral charge storage capacity over the more traditional QITs (Finnigan ITS40).[89,90] As shown in Appendix, Table 9.A, both the Bruker Daltonics and Agilent Technologies QITs also have an improved scan rate of 26,000 Th/s, which is about five times faster than the older Finnigan ITS40 or LCQ "classic." Care should be taken in comparing the improvement in spectral ion storage capacity between XQITs, because it is the *absolute number* of ions that can be scanned out and detected with good spectral quality that counts!

9.3.1.7 Effects of Helium Damping Gas

In the early 1980s, the first gas chromatograph (GC) was interfaced to a QIT. Because this QIT utilized internal ionization, the GC column was inserted directly into the QIT chamber. Serendipitously, they found that the resolution and sensitivity of the instrument were greatly improved when approximately 1 mtorr of helium was present in the QIT chamber.[20,91] This was especially evident for ions with higher m/z values. The improvement resulted from a collisional dampening effect of the ions by the low-molecular-weight helium atoms. The ions are kinetically cooled or damped to the center of the trap and occupy an area less than 2 mm in diameter as measured by tomography experiments.[92] Ions are damped kinetically over a period of a few milliseconds, which is pressure-dependent, thereby allowing ions to be ejected from the ion trap in dense ion packets during the mass-analysis step.

Successful trapping of injected ions into a QIT from an external source such as an ESI ion source is extremely dependent on the helium pressure. Without helium, or another suitable damping gas, only a few ions are trapped. At a helium pressure of 1 mtorr, ions are readily trapped. At helium pressures greater than 1 mtorr trapping efficiency is improved; however, the mass resolution is reduced at too high of a pressure. Typically, the helium pressure in the ion trap is between 1 and 4 mtorr[93] for mass analysis.

9.3.1.8 The Rectangular Wave-Driven Quadrupole Ion Trap

A QMF or QIT can be driven with an adjustable rectangular waveform rather than a sinusoidal waveform as used on the more conventional quadrupole mass analyzers.[26] Sheretov and co-workers[94,95] and Richards et al.[96] conducted experiments applying a RF rectangular waveform to a QMF and a QIT. In 1973, Richards and co-workers pointed out that creation of a rectangularly time-varying function is technologically easier than keeping the ratio of two voltages constant as was done for a typical QMF. The "digital" quadrupole ion trap (dQIT) using an adjustable rectangular wave[23] is being pursued by the Shimadzu Corporation in Manchester, UK. The name "digital" is used to describe the on/off nature of the waveform switching between two discrete voltage levels at precise times and is not due to the fact that the waveform is digitally produced. In both the dQIT and the dQMF cases, a quadrupole field is still formed inside the ion trap. The dQIT can have the same basic electrode geometry as the QIT (a ring and two end-cap electrodes), but the stability diagram is modified when driving the trap with a rectangular waveform, allowing for stability only out to a q_z of 0.712 rather than the more familiar q_z of 0.908. The ion motion in the dQIT was simulated by Ding and Kumashiro[97] to understand how this waveform might impact the ion motion. It is still early to determine all the advantages and disadvantage of driving a trap with a rectangular waveform, but some advantages noted[98] include: (i) the ability to easily generate the waveform, (ii) the ability to start and stop the waveform instantly without a complicated RF generator, (iii) the ease of tailoring the waveform, (iv) the ability to use rectangular waveforms at low amplitudes, and (v) the ability to keep precise timing parameters. Other XQITs should also be able to be driven by a rectangular wave.

9.3.1.9 Scan Modes of the QIT

See Section **9.3.3**, "**Ion Trap Scan Modes**." which covers all quadrupole ion traps.

9.3.1.10 QIT Performance

The QIT has a commercial mass range typically from ~50–2000 Th, although some manufacturers provide two or three mass ranges up to ~4000 Th. Scan rates vary from 500 Th/s to 27,000 Th/s and allow for resolution specifications of up to 5000. The typical mass accuracy is 50 ppm to 100 ppm at *m/z* 1000. Detection limits can be at the attomole level. Scan modes include the full scan, SIM, SRM, rapid high sensitivity and high resolution, and MS^n. Some QITs have a high charge capacity, which has increased the sensitivity by 10-fold. See Table 9.A in the Appendix.

9.3.2 The Linear Two-Dimensional Quadrupole Field Ion Trap

As with the QMF, the linear quadrupole ion trap (LQIT) provides a two-dimensional quadrupole field, so much of the theory of ion motion in the QMF can be applied directly to LQIT. In both cases there is no axial micromotion of the ions.

9.3.2.1 Geometry of the Linear Quadrupole Ion Trap Mass Analyzer

Linear quadrupole ion traps (LQIT) are shown in Figures 9.1c, 9.2b and 9.2c with geometries that are very similar to the QMF, but the LQIT described herein both traps ions and mass analyzes them. The LQIT is a versatile device like the QIT. The trap can be operated to trap, cool, isolate, react, dissociate, and, if so desired, mass analyze ions. Not all LQITs, however, are operated as mass analyzers and instead may be used as ion accumulators, as ion reaction vessels, or as pulsed ion sources. To mass analyze ions, several design considerations are made to account for this mode of operation. First, a trapping field must be applied along the z axis of the LQIT, which is absent in the QMF; second, a damping gas must be used to reduce the initial kinetic energy of the injected ions; and, finally, the LQIT must incorporate a method and/or geometry modification necessary to eject ions from the device for mass analysis. To trap ions, lens elements or quadrupole segments are used to generate a trapping field along the z axis. A segmented LQIT is shown in Figure 9.1c and in a cross-sectional view in Figure 9.2c. Both figures show the front-, center-, and back-quadrupole sections that can form a potential well along the z axis for trapping when the proper DC potentials are applied. The "front" and "back" denotations for the quadrupole sections is arbitrary since the unique QLIT can be readily coupled to two or more ion sources. In Figure 9.2c, the location for two ion sources are shown on either side of the trap. A less ideal QLIT design (not shown) can use a single lens element at the front and back sections of the device, but the central quadrupole field would be less ideal near the ends of the trapping section. Helium, not used in a QMF or orbitrap, is added into the device directly as done with the QIT to damp the ions by removing unwanted kinetic energy from the injection or a reaction process, and it can be used as an ion collision partner for collision-activated dissociation (CAD). Finally, to eject ions radially, exit slots are machined into two or more opposite rod pairs. MDS Sciex Corporation has taken a different approach to ion ejection by using the fringing field at the end of the LQIT assembly, and thus exit slots are not required. In the ThermoFisher LQIT (LTQ^{TM}), the two ejection electrodes (x dimension) are spaced apart an additional 10% to reduce the effect of the multipole resonances due to the 30-mm-long, 0.25-mm-wide slots. As with the "stretched" geometry QIT, this correction factor must also be incorporated into Eqs. (9.1) and (9.2).

In 2004, the Cooks' group at Purdue built a rectilinear-LQIT (RLQIT) with rectangular electrodes,[99] and they previously built cylindrical ion traps (CIT).[100–102] Both of these traps are attractive alternatives to XQITs that use hyperbolic electrodes because they are easier to build.

9.3.2.2 The Linear Ion Trap Stability Diagram and the Resonant Frequencies of Ions

The stability diagram for the LQIT is identical to the diagram for the QMF shown in Figure 9.2a, but in this case, the portion of the *operating line* that intersects the crossed-hatched region of the diagram represents ions with Mathieu parameters (a,q) that will be stably trapped rather than transmitted through the device. For the LQIT mass-selective instability scan, the operating line starts at $(a_u,q_u) = (0,0)$ and passes through $(a_u,q_u) =$ (0,0.908) as shown in the stability diagram of Figure 9.2d. Ions are located in the stable region with $q_u > 0$, but < 0.908. The ions that are outside of the stable region (i.e., where $q_u > 0.908$) are not stable and as a result are annihilated at the electrode rods. Again, since q_u is inversely proportional to m/z, high-mass ions of the same charge, z, have a lower q value than low m/z ions as shown by the different-sized circles on the diagram. In this case,

$(m/z)_4^+$ is greater than $(m/z)_3^+$. In an *RF amplitude scan*, or *full scan mode*, the ions positioned along the operating line shown in Figure 9.2d move to higher q_u and thus are at a higher frequency as the amplitude is increased. Eventually, the lowest $(m/z)_1$ ion becomes unstable first at the $\beta_u = 1$ edge of the stability diagram, followed sequentially by the next higher $(m/z)_2$ ion, and so on. In this case, without the supplementary resonance ejection waveform applied to the x rods, ions would be ejected in four directions, radially, out the xz and yz planes.

Ions of a specific m/z in the 2D quadrupole field have fundamental frequencies of motion unique to their q_u values, and these secular frequencies can be used to resonate these ions as in the QIT [see Eq. (9.6)]. However, in the LQIT case, the frequencies of ion motion will be defined in the x and y dimension (see Figure 9.1c). Again, by knowing the q_x of an ion, one can calculate the β_x value and thus the resonant frequencies of an ion ($f_{x\text{-res}} = f_0 = \beta_x f_{RF}/2$ for the x-dimension and likewise for the y-dimension).

9.3.2.3 Resonance Ejection from an LQIT

A dipolar excitation signal ($V_{res} \cos(2\pi f_{x\text{-res}} t)$) can be applied to the x rods (and/or y rods), for *resonance excitation* and/or *resonance ejection*. For either of these processes to occur efficiently, however, the frequency used must equal the fundamental resonant frequency of the ion, $f_{x\text{-res}}$. For example, during the mass selective instability scan supplemented with resonance ejection, ions can be ejected from the LQIT trap radially in the x dimension as they are scanned into resonance at a frequency that corresponds to a $q_{x\text{-eject}} = 0.8$. As with the QIT, resolution, peak height, and sensitivity are greatly improved when using the mass selective instability scan mode supplemented with resonance ejection. Ions can also be resonantly ejected during an amplitude scan by applying a quadrupolar excitation to the x (and/or y) rods of the QLIT at $2f_x$.[103] In all cases, as ions are excited, they undergo increasing excursions into the radial direction until they are ejected out of the slots in the x and/or y slotted electrodes. Note that ions could be ejected out all four rods if they were all slotted.

9.3.2.4 The Development of the LQIT with Radial Ejection

The invention of trapping ions in an ion trap with an increased or extended 2D-quadrupole field (whether linear or not) and scanning these ions out of the device radially using mass-selective excitation, with or without resonance ejection, was patented by Bier and Syka[1] in 1995. One of the patent figures used to describe the LQIT is shown in Figure 9.1c. Shown are three isolated segments totaling 12 electrode rods; of these, two of the center opposing y rods show slots for radial ejection. Welling, Thompson, and Walther[104] were the first to implement the LQIT as a mass spectrometer using radial ejection in 1996. They used three adjacent quadrupole segments to form the linear ion trap, and each segment was electrically isolated to allow for the application of three independent z-axis trapping potentials (front, center, and back). Ion motion in the axial direction could be controlled and thus the ions were trapped. They generated MgC_{60}^+ complexes inside an LQIT via ion–molecule reactions from Mg^+ formed from electron ionization and oven generated C_{60}. Photodissociation was used to break apart the complex. For mass analysis, an additional oscillating quadrupole field excitation frequency was scanned to radially eject ions, and an electron multiplier positioned perpendicular to the trap axis was used for detection from 20 Th to more than 1000 Th. In 1998, Walther and co-workers[105] further developed their earlier design by implementing RF/DC isolation, a low-resolution "q-scanning mode" in which ions are electrostatically pulled out of the trap radially at low well depths and they again used a "secular scanning mode" or parametric excitation allowing a mass range of 20 Th to over 4000 Th. As described in the patent,[1] ions could also be resonantly ejected radially through slots cut into the rod electrodes. This experiment was implemented in 2002 on an LQIT using RF amplitude scanning supplemented with a resonance ejection.[106]

9.3.2.5 The Development of the LQIT with Axial Ejection

During the same time period as the development of the radial ejection method, Hager at MDS Sciex Corporation developed an "axial" method of ion ejection as shown in Figure 9.2b. It is interesting to note that the radial ejection method was patented in the "nick of time" because researchers at MDS Sciex Corporation in Toronto were diligently working on this method until the patent was issued in 1995, at which time the Sciex group shelved the idea.[107] In the axial method of ejection, ions are scanned out of the LQIT in an "axial" direction using resonance ejection in combination with the usually detrimental, quadrupole fringing fields.[2,108,109] For the mass-selective axial method of ion ejection, the fringing fields at the ends of the quadrupole convert radial excursions into axial ones.[2,108,109] Axial ejection is accomplished by resonantly exciting the ions radially near the quadrupole exit where the ions can overcome the exit DC barrier for mass analysis. Londry and Hager[109] use the term, *cone of reflection,* to describe and determine what ions can be drawn past this barrier for mass analysis.[109] The MDS Sciex team built a powerful hybrid triple QMF/q2/LQIT mass spectrometer, which is shown above Table 9.1, using the axial mode of ion ejection while still taking advantage of the mass filtering ability of Q1.

9.3.2.6 High Ion Charge Capacity of the LQIT

The 2D-quadrupole field ion traps such as the LQIT achieve a higher charge capacity with no loss in spectral quality by trapping more ions in a larger ion-occupied trapping volume. That is, the ion trapping volume is increased by physically increasing the geometry of the trap (a bigger bucket!). Prestage et al.[110], Bier and Syka[1] and Campbell *et al.*[111] have all calculated the amount the charge capacity would increase in a LQIT versus a QIT. A simple model[1] is presented again here and the results are supported by current experimental findings. For example, let the number of ions in a QIT be defined by the equation $n = \rho v$, where ρ is the average ion charge density that can be scanned out of an ion trap with good spectral quality (i.e., no mass shift or loss in resolution) and v is the ion occupied volume (not the physical trapping chamber) under gas damped conditions. Based on tomography studies in a QIT by Hemberger and co-workers,[112,113] an assumption was made that when 95% of the ions are stored within a sphere (ion cloud) with a radius of 0.70 mm, the ion occupied volume would be 1.4 mm^3. If ρ is limited by space charge to 1000 ions/mm^3 (*Note:* Fischer trapped krypton ions at densities of 2000–4000 ions/mm^3 in non-helium-damped conditions.[114]), a QIT with this volume would store 1400 ions. As shown in "speckled" ion clouds of Figure 9.2b, 9.2c and 9.3a, the total volume of the LQIT would be considerably larger if one assumes a simplified geometry of the total ion occupied volume (ion cloud) to be $v = \pi r^2 L$ (i.e. a cylinder). With the same quadrupole field strength and assuming a similar ion density in the axial direction by the trapping DC field, we would expect to have the same ion density in an ion volume of 154 mm^3, for $L = 100$ mm and $r = 0.70$ mm. Thus, the larger ion volume in the LQIT would store 2.2×10^5 ions at the same charge density without the corresponding increase in space charge! Trapping more ions improves sensitivity, dynamic range, mass accuracy, and signal-to-noise (S/N) ratio, and the LQIT has other inherent advantages as discussed herein.

9.3.2.7 Scan Modes of the LQIT

The basic scan modes that can be used in a LQIT are essentially the same as those developed over the years for the QIT, and they are discussed in more detail in Section **9.3.3, "Ion Trap Scan Modes.**"

9.3.2.8 Performance of the LQIT

The MDS Sciex 4000 Q Trap, which is a triple-quadrupole instrument, has a mass range of *m/z* 5–2800 in Q1/Q3 RF/DC mode and has *m/z* 50–2800 in the LQIT mode. The scan speed is user-settable from 250 Th/s, 1000 Th/s, and 4000 Th/s in LQIT mode with a maximum of 2400 Th/s in RF/DC mode.[115] The mass resolution is >3000 at *m/z* 609 at 250Th/s. The LQIT from ThermoFisher (LTQ XL) has three mass ranges from 15–200, 50–2000, and 200–4000 Th. The resolution is unit (~2000 at *m/z* 1000) at the normal scan rate to 20,000 in the high-resolution scan mode (mass range not defined). See Tables 9.2 and 9.3 and Appendix Table A for mass accuracy and detection limits.

9.3.2.9 The Advantages of the LQIT over the QIT

Although the LQIT assembly is more complex to build due to the additional electrodes, insulators, and the precise tolerances that must be maintained along the *z* axis, there are a number of performance improvements gained by using a LQIT rather than a QIT. The LQIT advantages include:

i. **Higher ion charge capacity.** The ion trapping volume of a LQIT has been increased by more than an order of magnitude over the low charge capacity QIT.

ii. **Higher ion injection and trapping efficiency.** The ability to trap ions and fill the LQIT is improved by 10-fold or more. Trapping ions is more efficient because there is a longer axial stopping distance in which collisions with helium can damp the ion kinetic energy. The fill time is improved because there is no quadrupole field axially and the entrance aperture can be made larger with no deleterious effects. As a result, ion gate times are reduced.

iii. **Multidetector capability (radial ejection).** Those familiar with the classical QIT operation have realized that half of the ions are lost during ejection because only one end cap is followed by a detector. (Note that some manufacturers have solved this problem.) The LQIT, however, ejects ions radially and thus allows for the use of one or more radial ion detectors.

iv. **Dual ion source capability (radial ejection).** The LQIT can be operated so that ions may be trapped efficiently from two ion sources by transferring ions into both ends of the device. This allows for convenient ion/ion (−/+) reaction experiments.

v. **Excellent ion/molecule and ion/ion reactions capability.** The LQIT readily allows for ion/molecule reactions and electron transfer dissociation (ETD) experiments because of the advantages discussed in i, ii, and iv above, and because of the unique ability of XQITs to trap both positive and negative ions simultaneously.

Table 9.2. Mass Accuracy for the QMF, QIT, LQIT, Orbitrap

Analyte	Mass Analyzer	Theoretical MW (Da)	Experimental MW (Da)	Mass Accuracy (ppm) (Ref.)
		Proteins		
Unknown, from *E. coli*	LQIT	116718	116703[e]	128[a]
Lysozyme (chicken egg)	QIT	14306.2	14305.0[e]	60[a]
Carbonic anhydrase (bovine)	QIT	29024.6	29025.2[e]	20[a]
Cytochrome c (bovine)	QIT	12230.9	12231.5[e]	50[a]
B-lactoglobulin A (bovine milk)	QIT	18363.3	18364.5[e]	60[a]
Apomyoglobin (equine)	QIT	16951.5	16951.9[e]	20[a]
Hemoglobin-alpha chain (bovine)	QIT	15053.2	15053.7[e]	40[a]
Hemoglobin-beta chain (bovine)	QIT	15953.3	15053.5[e]	10[a]
		Peptides		
Ubiquitin (bovine)	QIT	8564.9	8565.0[e]	20[a]
Interleukin-8 (rat)	QIT	7840.09[d]	7840.38[d]	40[b]
Angiotensin I	QIT	1295.68[d]	1295.63[d]	80[c]
Angiotensin I	QIT	1295.68[d]	1295.63[d]	40[b]
Insulin	Orbitrap	1149 (+5)[d]		2.9 (158)
Insulin	Orbitrap	1436 (+4)[d]		1.5
Insulin	Orbitrap	1915 (+3)[d]		4.9
Calciseptine	QIT	7031.2[d]	7030.8[d]	60[b]
Bradykinin	QIT	1059.56[d]	1059.62[d]	60[b]
		Cytochrome c Peptides		
TGQAPGFTYTDANK	QIT	1469.68[d]	1469.66[d]	10[b]
KTEREDLIAYLK	QIT	1477.82[d]	1477.86[d]	30[b]
GITWKEETLMEYLENPKK	QIT	2208.11[d]	2208.19[d]	40[b]
		Oligonucleotides		
10 mer	QMF	2993.0	2993.1	33.4[a] (225)
20 mer	QMF	6168.1	6167.6	81.1[a] (225)
40 mer	QMF	12300.0	12299.7	24.4[a] (225)
60 mer	QMF	18410.0	18410.1	5.3[a] (225)
80 mer	QMF	24661.1	24661. 0	4.1[a] (225)
100 mer	QMF	30951.1	30950.6	16.2[a] (225)
120 mer	QMF	37031.0	37030.0	27.0[a] (225)
120 mer	LQIT	37031	37031	0[a] (328)

[a]Deconvolution method.

[b]Isotope resolution method (high resolution).

[c]Normal scan.

[d]Monoisotopic mass.

[e]Average mass.

Source: Data acquired by Gale, D.; Vasconcellos, T.; Sanders, M; Sweeney, M.; Chaudhary, T; Schwartz, J.; Land, A., Finnigan Corporation; Hail, M., Novatia LLC; Bier, M., Carnegie Mellon University.

vi. **Ion transportation.** Ions can be shuffled easily between trapping segments or other ion devices. This has led to a series of LQITs and the hybridization of the LQIT technology with other trapping technologies such as FTMS and the orbitrap (LQIT/QIT, Q/q/LQIT, LQIT/FTMS, LQIT/TOF, and LQIT/orbitrap).

Table 9.3. Detection Limits of Various Analytes Using QMF, QIT, LQIT and Orbitrap Mass Analyzers

Analyte	Mass Analyzer	Intro. Technique	Scan Type (*m*/*z*)	Ion(s) (*m*/*z*)	Detection Limit	S/N (Ref.)
Octaethylporphyrin	QIT	FIA/ESI	300–600	535^{+}	18 fmol	4 (329)
RVYVHPI	QIT	μLC/ESI	MS^2	442^{+2}	10 fmol	20 (74)
RVYVHPI	QIT	μLC/ESI	Rapid-MS^2	442^{+2}	550 amol	20 (67)
Bradykinin	QIT	μCE	100–1300	TIE	180 fmol	20 (84)
Leu-enkephalin	QIT	μCE	100–1200	556^{+}	650 amol	25 (175)
Leu-enkaphalin	QIT	μCE	100–1200	TIE	650 amol	6 (175)
Renin substrate	QIT	μCE	100–2000	587.3^{+3}	190 fmol	15 (169)
Substance P	QIT	μCE	100–2000	450.7^{+3} 456.0^{+3}	190 fmol	8 (169)
T-T (sodiated)	QIT	FIA	100–350	275^{+}	730 fmol	4 (330)
Berberine	QIT	LC	150–400	336^{+}	19 fmol	3 (331)
Berberine	QIT	μCE	140–400	TIE	51 fmol	20 (173)
Berberine	QIT	μCE	140–400	336^{+}	90–130 amol	3 (173)
Magnoflorine	QIT	μCE	140–400	TIE	37 fmol	15 (173)
Magnoflorine	QIT	μCE	140–400	343^{+}	90 amol	3 (173)
Reserpine	QIT	FIA	SIM	609^{+}	8 fmol	7
Reserpine	QIT	FIA	SRM^2	397^{+}, 436^{+}, 448^{+}	8 fmol	20
Reserpine	QIT	FIA	200–700	609^{+}	82 fmol	10
Apotransferrin	QIT	AMALDI	FS	$[M + H]^{+}$	50 fmol	(332)
Catalase	QIT	AMALDI	FS	$[M + H]^{+}$	50 fmol	(332)
Conalbumin	QIT	AMALDI	FS	$[M + H]^{+}$	100 fmol	(332)
Neurotensin	QIT	capLC-ESI	FS	559^{+3}	1 amol	8 (333)
Neurotensin	QIT	capLC-ESI	MS^2	M^{+3}	<1 amol	3 (333)
Angiotensin	QIT	capLC-ESI	FS	1297^{+}	1 amol	20 (333)
Aniline	QIT	MI-ESI	SIM	$[M + H]^{+}$	75 pptr	8 (334)
Pyridine	QIT	MI-ESI	SIM	$[M + H]^{+}$	1.0 ppb	14 (334)
Atrazine	QIT	MI-ESI	SIM	$[M + H]^{+}$	250 pptr	22 (334)
Simazine	QIT	MI-ESI	SIM	$[M + H]^{+}$	1.0 ppb	24 (334)
Pentachlorophenol	QIT	MI-ESI	SIM	$[M - H]^{-}$	100 pptr	10 (334)
Reserpine	LQIT	ESI-FIA 2 μL inj at 400 μL/min	SRM^3 (165–615 Da)	$609^{+} \rightarrow 397^{+} \rightarrow 365^{+}$	250 fmol	25 (150)

TIE, total ion electropherogram; μLC, capillary LC (100-μm i.d.); flow rate, ~1 μL/min, FIA, flow injection analysis (200 μL/min); LC, 2-mm-i.d. column zorbax; μCE, 20- to 25-μm-i.d. column., 50-μm-i.d. column (175); μCE, amino-propyl-silated column (169); T–T, thymine–thymine; MI, membrane inlet MS

9.3.3 Ion Trap Scan Modes

This section discusses the scan capabilities of the XQITs. The discussion starts with an example of a normal full scan function and evolves this analytical scan to cover the high-sensitivity (low-resolution), high-resolution (low-sensitivity), SIM, SRM^n, and MS^n scan modes.

9.3.3.1 The 2D- and 3D-Quadrupole Field Ion Trap Scan Functions and Timing Diagrams

Unlike triple quadrupole field mass spectrometers, where each operation on the ion beam is separated in space (i.e., in either Q1, q2, or Q3), ion traps operate on the ions over a period of time ("tandem in time"), but within the same analyzer. For example, an ion trap MS/MS scan consists of time periods for ion injection/cool, isolation, excitation/cool, and mass analysis. These steps are diagrammed in Figure 9.4, which shows a plot of the RF amplitude on the ring electrode (in the QIT case) or the quadrupole rods (in the LQIT case) versus time. This plot is referred to as an *ion trap scan function*. In order to avoid the effects of space charge resulting from trapping too many ions, a *prescan* can precede the *analytical scan*. A prescan rapidly determines the proper analytical *ion injection* time. Based on the prescan ion count measurement, the data system calculates a suitable ion injection time that does not inject an overabundance of ions that would cause the deleterious effects of space charge. For example, if the ion source signal level doubles, as measured by a prescan, the ion injection period during the analytical scan is shortened to one-half of the time used in the previous analytical scan and the data system scales up the recorded counts by a factor of two. The prescan may include many of the same steps found in the analytical scan, but the prescan ion injection time is often much shorter in duration (~10 ms) and the prescan mass analysis step is rapid since only an ion current measurement is required. The use of variable ion gate times has extended the linear dynamic range of the ion trap to five orders of magnitude in GC/MS applications.[116] A major advantage of the LQIT is that it can have an ion trapping volume at least ~40 times larger than a QIT and so the prescan calculation can scale up the total number of ions to trap, accordingly, with no deleterious effects. Another advantage of significantly shorter ion gating times is also realized because of the high efficiency of ion injection and trapping in the LQIT.

During the *ion injection* step (step 1), ions are gated using a pulsed electrode from an external ESI source or a pulsed laser which is part of the MALDI source. The ions are trapped in the quadrupole field by maintaining a suitable RF voltage on the ring or rod electrodes for the QIT or LQIT, respectively, and by using a partial pressure of helium in the trapping chamber of about 1–3 mtorr to cool ions. For an MS/MS scan, an *isolation* step (step 2) must be performed to select a precursor ion using one of a variety of isolation techniques that will be explained in more detail in a section that follows. After isolation, energy is deposited into the ion during an *excitation* step (step 3). Excitation of the ion is accomplished by several means such as through collisions with a gas, photon absorption, proton transfer, electron capture, and electron transfer. If the process deposits enough internal energy, dissociation will ensue. Finally, the last step (step 4) of a typical scan function involves *mass analysis*. During the mass analysis step, ions are ejected sequentially out of the trapping chamber through the holes in the QIT end caps, radially through the slots in the LQIT quadrupole rods or axially along the LQIT z axis. These ions can next collide with a post-acceleration dynode/electron multiplier detector. These four steps—ion injection/cool, isolation, excitation, and mass analysis—typically require 0.001–1000 ms, 5–30 ms, 5–30 ms, and 10–500 ms, respectively. The results of the scan are then sent to the data system and the scan function can then be repeated. One complete scan function ending in mass analysis has been termed a *microscan*. Note that after steps 1, 2, or 3, trapped ions could be transmitted to an orbitrap for high resolution mass analysis.

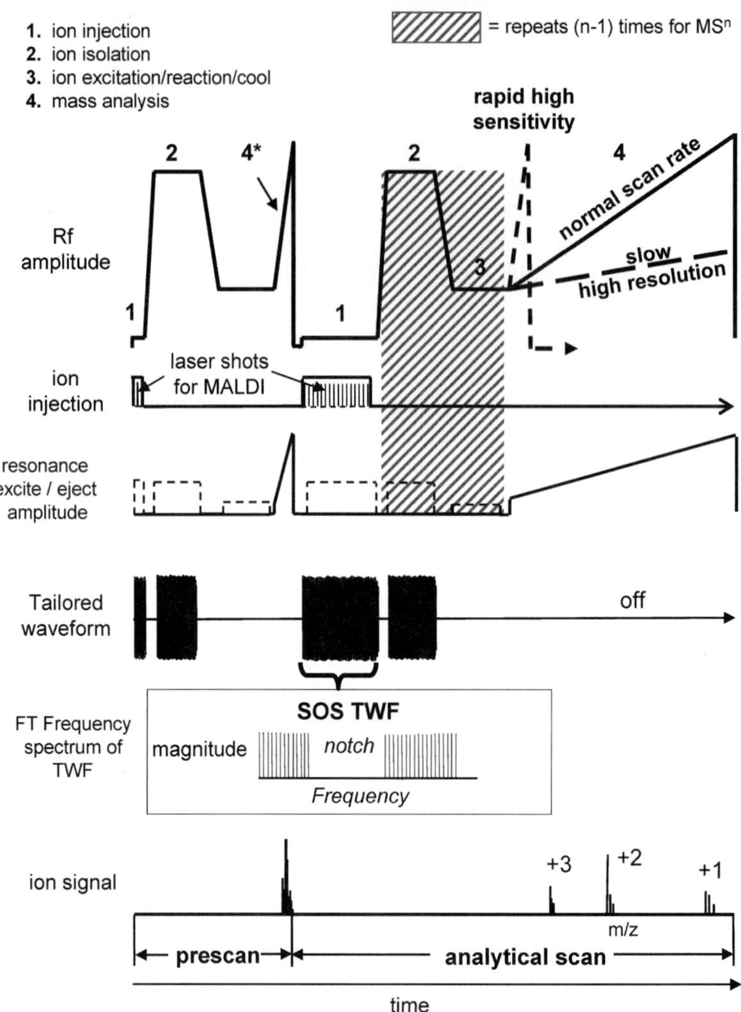

Figure 9.4. Scan functions for the XQIT showing the rapid prescan and the analytical scan which makes up one microscan. The four steps of the QIT operation are (1) ion injection, (2) ion isolation, (3) ion excitation, and (4) mass analysis. During step 1, ions are injected into the trap for storage. An optional isolation waveform can be applied to eject unwanted ions. Optional steps 2 and 3 require an additional tailored waveform (TWF) for ion isolation and a dipolar resonance signal for excitation, respectively (e.g., for MS^2). In one example, a frequency-domain spectrum from a Fourier transform (FFT) of a sum-of-sines (SOS) TWF applied during injection reveals the notch as shown in the boxed inset. The ion signal is acquired during the analytical mass analysis segment of the scan function (step 4), and the resonance ejection amplitude is increased for optimum resolution and linearity during mass analysis. The multiplier, not shown, can be turned on during mass analysis, step 4, to detect the ion signal.

9.3.3.2 Operation in the Full-Scan Mode

As discussed above, the operation of ion traps is frequently described using scan functions. Figure 9.4 shows the RF scan function used to generate a full-scan MS and includes example waveforms from other dynamic devices (see Figure 9.11a). Several additional features of the scan function are noteworthy. For the QIT, trapping of injected ions is highly dependent on the level of the RF amplitude applied to the ring electrode and this segment (1) can be divided up into different RF levels to allow for a more uniform capture of ions across the mass range. Ion injection efficiency is not as sensitive to the RF level in the LQIT[106] as expected, because there is no axial-RF along the ion injection axis. The potential of various lens elements can also be adjusted during each successive ion injection period to improve the transmission of higher *m*/*z* ions. Figure 9.4 also shows the use of an optional tailored-waveform (TWF) that can be applied to the end caps or to two opposing rods during ion injection in the QIT or LQIT, respectively. The TWFs can consist of a "sum-of-sines" (SOS),[63,64,117] filtered noise field (FNF) or stored waveform inverse Fourier transform (SWIFT) waveform to isolate a mass range of interest during the ion injection periods. The inset shows the frequency-domain spectrum from a Fourier transform (FT) analysis of the SOS TWF revealing the discrete frequencies used to eject all ions except those from within the notch. For full-scan MS, the notch can be quite wide. Tailored waveforms are discussed in more detail in the SIM section that follows. The resonance ejection voltage is increased with *m*/*z* for optimum resolution throughout the mass range and for a linear mass calibration.[26,118] A lens element such as an interoctopole lens (see Figure 9.11a) that permits ion transmission during ion injection periods can be used to stop the transmission of the supposed large charged particles during mass analysis. The RF amplitude applied to the octopole ion guides can be turned off after ion injection to ensure that a higher-order frequency (1–3 MHz) is not induced onto the entrance end cap during ion storage and to reduce charged particle noise.[119] Later QIT designs have added additional optics in front of the entrance end cap. The electron multiplier can be turned off during the ion injection period to protect it from the ion current due to untrapped ions. The ion signal plot in Figure 9.4 shows an example of a hypothetical ion signal from this scan.

An example of a full-scan ESI mass spectrum was collected using a QIT (Finnigan LCQ "classic"). The sample contained a mixture of Ultramark 1621, a mixture of perfluorinated alkyl phosphazenes (PCR, Gainesville, FL), the tetrapeptide MRFA (MW 523.3), and caffeine (MW 194.1). The sample mixture was dissolved in a 1% acetic acid solution of 50 : 25 : 25 acetonitrile : methanol : H_2O, and infused at 3 μL/min into the ESI ion source. The positive-ion spectrum showed the $(M + H)^+$ ion from caffeine at *m*/*z* 195.2, the $(M + H)^+$ from MRFA at *m*/*z* 524.3, the doubly charged ion from MRFA $(M + 2H)^{+2}$ at *m*/*z* 262.6, and the singly protonated fluorinated phosphazenes $(M_n + H)^+$ observed between *m*/*z* 800 and *m*/*z* 2000. Moini reported Ultramark 1621 to be an exceptional mixture to calibrate an ESI or APCI mass spectrometers because the molecular ions are readily formed in both positive- and negative-ion modes, singly charged, and evenly spaced every 100 Da over a large mass range.[120] A full-scan mass spectrum can also be used to analyze the product ions formed from precursor ions that are activated by collisions in the ESI interface. In this mode of operation, the precursor ion presumably collides vigorously with solvent molecules and dissociates, provided that there is a large potential gradient in the tube lens/skimmer region (~1 torr). A tube-lens collision activated dissociation (CAD) spectrum obtained from angiotensin I ($MW_{avg} = 1296.5$) showed intense b, a, y, sodiated, and doubly charged ions. Tube lens CAD MS can be complimentary to MS/MS because a shorter scan time is used and the $q_{z\text{-inject}}$ value may be set lower to trap the low mass fragments. In contrast, a relatively high $q_{z\text{-excite}}$ value of 0.25 is often required in MS/MS scans, and some low-mass fragments are lost if they have a $q_z > 0.908$ (see MS^n section to follow). A new pulse dissociation method at high *q* (PDQ) scan shows promising results in eliminating the low *m*/*z* cutoff.[121] The major disadvantage of the tube-lens CAD experiment is that it has no selectivity except by the separation technique and therefore may result in a product spectrum consisting of a mixture of various background analyte fragments compared to a selective MS^n scan.

9.3.3.3 The Effects of Scan Rate: The Normal, Rapid, and Slow Scan Rates

Scan rate (Th/s) refers to the rate of ejection of ions out of an ion trap during the mass analysis step. Scan rates can be decreased to improve resolution or increased to improve detection limits.

a. Rapid Scan Rates for High Sensitivity. Scan rates were first increased during the development of the prescan for variable ionization times.[45] Later, at Purdue University, the scan rate was increased when the mass range was extended.[53] Rapid scan rates can greatly improve detection limits because the ion current increases, that is, the signal height increases by 5.5-fold and the resulting shorter scan functions allow for improved signal averaging.[60,122] A rapid rate of ion ejection, however, causes mass peaks to broaden on the mass scale even though the peaks are narrower in time. The decreased resolution occurs when using these high scan rates because the ion undergoes a smaller number of resonance cycles before ejection. Rapid scan rates are less sensitive to the effects of space charge; hence, more than an order of magnitude of ions can be trapped without causing mass shifts or further degradation in mass resolution.[60] An example of a *normal scan* and a rapid *high sensitivity scan* (12 × normal rate) for an ESI mass spectrum of 40 fmol/µL apomyoglobin in 50:50 acetonitrile/water and 0.1% acetic acid is shown in Figure 9.5 where panel 9.5a is a normal scan (3 µscans over 5.6 s) and panel 9.5b is a rapid scan (3 µscans over 3.2 s). The Finnigan LCQ was set to a full-scan MS mode with a mass range from *m/z* 150 to 2000. The insets show the deconvoluted mass which only revealed noise in the normal scan. However, with the rapid scan mode it was possible to determine the molecular weight in 3.2 s. The mass error of 0.08% is due to several reasons: the LCQ Xcalibur software

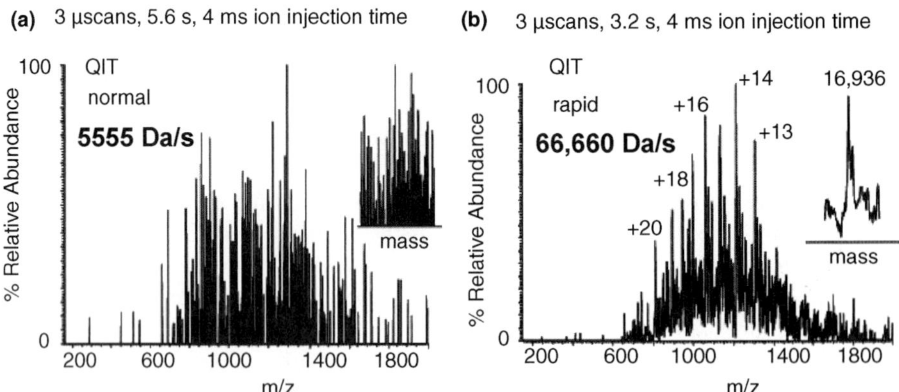

Figure 9.5. Electrospray full-scan mass spectra of 40 fmol/µL (10 µL/min) apomyoglobin in 50 : 50 acetonitrile/water and 0.1% acetic acid, with 3 µscans averaged each, where (**a**) is at the normal scan rate and (**b**) is at a rapid scan rate. The insets show the deconvoluted mass where at the normal scan rate, collected over 5.6 s, the molecular weight (MW) could not be determined, but at a rapid scan rate a MW was determined in 3.2 s with an improved S/N. (Reprinted with permission from Yang, C. G.; Bier, M. E. *Anal. Chem.* **2005**, *77*, 1663–1671.)

(Ver. 1.2) incorrectly uses monoisotopic masses rather than the average masses for the calibration procedure, the digitizer uses too few points to define the mass peaks accurately and, finally, the number of ions used for centroiding the peaks was low. The S/N improvement is clear in the comparison and suggests that this scan could be used as a survey scan in the full-scan mode for low-level samples. In my lab, we have found that rapid scanning is an excellent method of surveying the full-scan spectrum for low abundant ions where obtaining unit resolution is not critical. We have missed ions that were not observed above the noise before switching to the rapid scanning "turbo" mode. Varian (500-MS Ion Trap), Agilent (6340 Ion Trap), and Bruker Daltonics (HCT Ultra PTM Discovery) have all increased the scan rate on their QIT instruments to 27,000, 27,000, and 15,000 Th/s, respectively, and they have maintained a peak width at FWHM of 0.5–0.7. This improved resolution at higher rates is due to the improved method of resonance ejection which also allows for increased charge capacity as listed in the Appendix Table A (commercial XQITs specifications).

b. Slow Scan Rates for High Resolution. When the scan rate is decreased,[55,56] the resolution improves because ions can undergo an increased number of resonant cycles before ejection occurs. The scan rate can be reduced by 10-fold or more in a high-resolution scan mode, but the mass range used is typically reduced to a width of 10 Th wide to maintain a reasonable data acquisition rate. The high-resolution scan is, however, about 10 times more susceptible to the effects of space charge than a normal scan; as a result, a reduced number of ions must be trapped. The lower sensitivity and the necessity of a small mass range typically results in limited utility of this scan on the classic QIT. In any case, each scan rate, whether fast or slow, requires a proper calibration of the resonance ejection amplitude versus m/z.

9.3.3.4 Ion Isolation Techniques

Over the years, several methods of isolating ions in a QIT have been implemented. Methods include RF/DC isolation,[49,123] forward and reverse RF resonance ejection isolation,[124–126] and various forms of TWF isolation.[52,63,118,127–132] The RF/DC isolation methods position the ion of interest near the boundaries of the stability diagram for isolation. In this case, the parameter a_u is set to a nonzero value (see RF/DC isolation point in Figure 9.3b). The RF resonance ejection method sweeps the main RF amplitude and/or the resonance ejection frequency in both the forward and/or reverse directions to eject all but the ion of interest. This technique has been shown to yield high-resolution isolations and can be used to analyze multiply charged ions.[125]

A sum-of-sines (SOS) TWF consisting of many discrete frequencies can be applied to the end caps for a QIT or opposite rods of a LQIT in dipolar fashion to isolate a narrow m/z window (Figures 9.3a and 9.6). Ideally, one would like to apply a continuous band of frequencies at all β values except at β_{isolate} [refer to Eq. (9.6)] for the ion of interest, but, in practice, an SOS approach[130] is simpler to implement. Sine waves are added together so that each discrete frequency is applied at the desired power for efficient resonant ejection across the entire mass range. The waveform has a notch in the frequency domain at f_{res}, as shown in the boxed inset of Figure 9.4 that allows these ions to remain stable. For systems with a main RF near 1 MHz, the discrete frequencies can be calculated at a spacing of 250, 500, or 1000 Hz[117,133] or at every integer mass.[128] A high q_{isolate} (e.g., 0.83) for the isolation step provides the optimum resolution because of the high degree of frequency dispersion and the strong trapping forces.

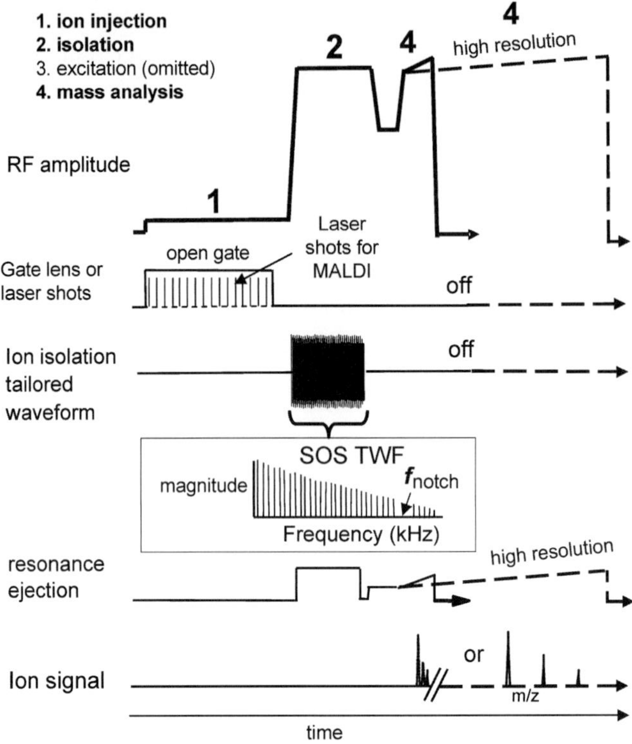

Figure 9.6. Scan functions for the SIM and high-resolution SIM modes of operation for the XQIT. Only the (1) ion injection, (2) ion isolation, and (3) mass analysis steps are required. In this example a SOS-TWF is used during high q isolation step to ejected unwanted ions from the trap. The optional *ion injection* TWF has been turned off in this example. A frequency-domain spectrum (FFT) of the SOS-TWF reveals the notch in frequencies applied as shown in the inset. The resonance ejection amplitude is increased during mass analysis for optimum resolution. The multiplier is turned on during the mass analysis segment, step 4, to collect the ion signal.

The SOS-TWF approach is somewhat similar to the SWIFT-TWF, introduced by Marshall.[134,135] Both create a defined time-domain waveform consisting of many frequencies. The SWIFT-TWF, however, is calculated by taking the inverse Fourier transform of a specified excitation or ejection frequency-domain spectrum. This synthesized time-domain spectrum can be applied to the end caps of the QIT for dipolar excitation similar to the SOS TWF discussed above. Guan and Marshall[136] as well as Julian et al.[137] have discussed the use of SWIFT-TWF in the QIT in detail.

Isolation of electrosprayed ions may also occur outside of the QIT. Jonscher and Yates built a hybrid linear QMF/QIT mass spectrometer. The QMF was used to filter out all but a 10-Th mass window of ions which contains the analyte m/z of interest. The 10-Th mass window of ions was then injected into the QIT. Preliminary results showed a reduction in space-charge effects for the analysis of peptides by using this prefiltering technique.[138] MDS Sciex has taken advantage of this coupling scheme and built a 4000 QTRAP with a QMF for filtering (Q1) and a q2 for collisional activation and trapping, followed by a LQIT.

9.3.3.5 Operation in the SIM and the High-Resolution Scan Modes

As with ion beam QMFs, quadrupole ion traps have improved ion signal-to-noise ratios in the SIM scan mode versus the full-scan mode.[139] The longer a mass spectrometer spends time detecting a selected ion, the lower the detection limit will be for that ion. QMF mass spectrometers have an important advantage over the QIT in the SIM scan mode. In QMF instruments, the duty cycle can increase from less than 0.1% for one ion in a scan from m/z 100–2000, to nearly 100% in a SIM scan [*duty cycle* = (ion collection or detection time/total scan time) × 100%]. However, the XQIT mass spectrometers have a duty cycle advantage in the full-scan mode of operation over QMF instruments (see Table 9.1). Since in a XQIT the duty cycle is dependent on the sample amount, unlike in the QMF case, low-level ion signals can reach duty cycles levels > 50% in the normal full-scan mode and can increase to levels >90% for the SIM scan mode.

A SIM scan function is shown in Figure 9.6. Although using an isolation TWF during injection is optional, and not shown in this figure, a TWF or some type of isolation step after the ion injection period is required for the SIM scan mode. This high q_z isolation step provides a significant advantage when operating in the SIM mode. In this case, the trap is filled beyond the "quality spectrum level" by 10 × or more with a variety of ions including the ion of interest (refer to Section 9.3.1, "Effects of Space Charge"). After a TWF has been applied, unwanted ions are ejected and leave behind a greater abundance of the isolated ion of interest at the desirable "quality spectrum level." In effect, the selected ion is purified and concentrated.

Molecular weight calculations of electrosprayed ions often require the determination of the ion charge state. High-resolution SIM mass analysis ($m/\Delta m \sim$ 20,000 at m/z 1000) achieved by reducing the scan rate (Th s^{-1}) an ion trap and supplementing this scan with an appropriate resonance ejection voltage[56,58] can be used to ascertain the charge state of ions. A high-resolution scan (over a 10-Th mass range) is shown in the mass analysis part of the SIM scan function in Figure 9.6. The mass resolution is improved because the ion spends more time in resonance during the ejection process.[26] High resolution in a QIT has been reported[140,141] at levels greater than 1×10^6; however, this has not become routine and the mass accuracy has not been shown to be commensurable.[142] At a resolution of 8000, an XQIT can be used to determine a charge state of up to ±4 at m/z 2000 by measuring the difference in the m/z of two adjacent ^{13}C isotope peaks ($\Delta m/z$). For example, if $\Delta m/z = 1$, the charge state is 1; if $\Delta m/z = 0.5$, the charge state is 2; if $\Delta m/z = 0.33$ the charge state is 3; and so on. Charge states much greater than 4 require even higher mass resolution. A mass resolution of approximately 25,000 would be required to resolve isotopes of charge state 25 at m/z 1000 for a 25-kDa protein. This resolving power has been demonstrated in FTMS[143,144] and orbitraps. High-resolution scans of multiply charged ions in a QIT have been limited to <7 charges because these slow scans (e.g., ~200 Th/s) are more susceptible to space-charge and fragmentation effects. An example of SIM scans was acquired at 66,600 Th/s, 5555 Th/s, and 200 Th/s for the +2 charge state of angiotensin I. The peak at the rapid scan rate was 5.5-fold more intense, but at a resolution of 300 no isotope peak separation was observed. The isotope peaks at the normal scan rate (5555 Th/s) were resolved with a 50% valley and the +2 charge state can be determined; and, finally, at a scan rate of 200 Th/s the isotope separation was easily determined to be ($\Delta m/z = 0.5$) at a resolution of 10,000 and the charge state can be determined as +2.

In a flow injection analysis (FIA) SIM scan using a TWF for isolation step 2, 1-μL injections of reserpine, m/z 609, were made into a 200-μL/min flow rate consisting of a solvent system of 1% acetic acid in a 50 : 50 methanol:water. A detection limit of 8 fmol (5 pg) with a signal-to-noise ratio of 7 was measured.

9.3.3.6 Operation with Multiple Stages of MS: The MS^n and the SRM^n Scan Modes

In MS/MS ($MS^2 = MS^n$, where $n = 2$), there are two stages of mass spectrometry. The first stage of ion trap mass spectrometry occurs during the isolation step of an ion (or ions), while the second stage of mass spectrometry occurs during the mass analysis step. The scan function for MS^2 (Figure 9.7) requires one more step than the SIM scan function,

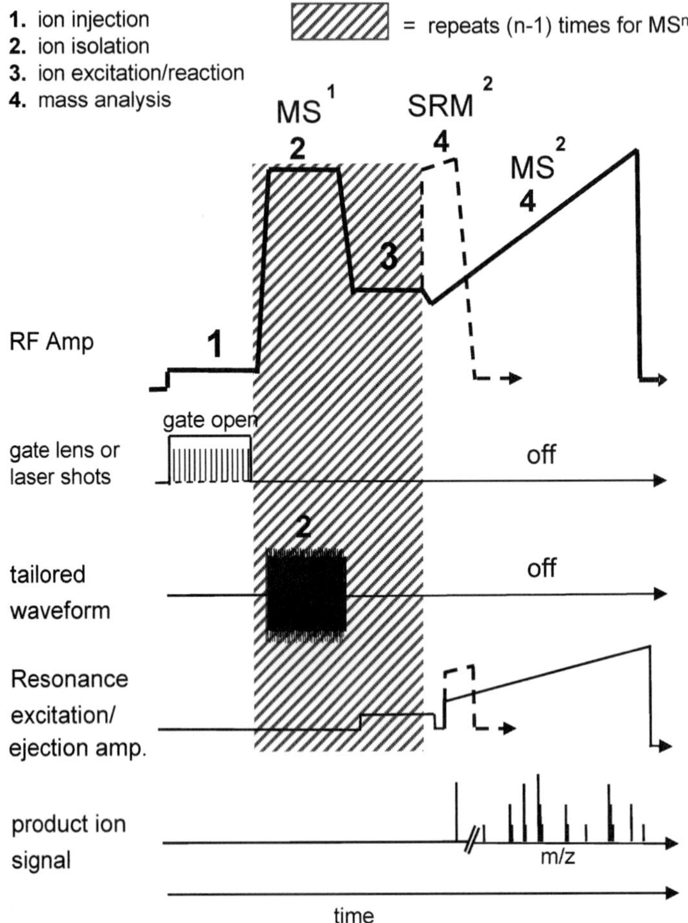

Figure 9.7. Scan functions for the SRM^n and MS^n modes of operation for the ion trap mass spectrometer. All four operations are executed: (1) ion injection, (2) ion isolation, (3) excitation, and (4) mass analysis. In this example an SOS TWF is used to isolate only the precursor ion(s) of interest. Resonance excitation, step 3, dissociates the precursor ion by CAD. The multiplier is turned on and the resonance ejection amplitude is increased during the mass analysis segment (4), allowing detection of the product ions.

namely, the excitation or activation step. This step could alternatively be an ion/molecule or ion/ion reaction step. A resonance excitation period is typically 5–30 ms in duration and requires an excitation voltage of less than 5 V_{pp} measured differentially, end cap to end cap. For the QITs, the resonance excitation frequency is calculated using Eq. (9.6) for a specified $q_{z\text{-excite}}$ value. During the resonance excitation step, ions undergo CAD in the ion trapping chamber where the resonant precursor ion repeatedly collides with helium buffer gas. A heavier collision gas partner can be mixed with the helium for improved internal energy deposition.[145–147] Given that enough internal energy is deposited by this multiple collision process, the precursor ion undergoes fragmentation. In 1987, Louris et al.[148] studied energy deposition and CAD efficiency in a QIT for the resonance excitation process. For the systems studied—tetraethylsilane, *n*-butylbenzene, and nitrobenzene—the CAD process was nearly 100% efficient when the internal energy deposited was less than 2 eV. In the CAD process described above, once the product ions are formed, they fall out of resonance unless a broadband excitation method is used over many frequencies.[149]

Typically, the q_{excite} value is chosen between 0.2 and 0.3 for resonance excitation. This allows the precursor ion to gain enough internal energy to fragment yet remain trapped in a strong RF field during resonance excitation. It also allows for a majority of the product ions to fall within the stability boundary at $q_u = 0.908$. For example, in a XQIT, if a doubly charged precursor ion [$(M + 2H)^{+2}$, having *m/z* 500 (MW 998 Da)] is resonantly activated at a $q_{\text{excite}} = 0.2$, product ions with a $q > 0.908$ are not trapped. In this case, no product ion of $m/z < (q_{\text{excite}}/q_{\text{edge}}) \times (\text{precursor } m/z) = (0.2/0.908) \times 500\text{ Th} = 110\text{ Th}$ will be stored. This "low *m/z* cutoff" (LMCO) is a fundamental limitation of this type of CAD experiment in a XQIT. The loss of product ion information below the LMCO can be revealed by using additional isolation and excitation steps on lower-mass fragments (MS^n-scans), but this requires additional scan time. A pulse dissociation method at high *q* (PDQ) has been introduced that involves a precursor ion resonance excitation pulse at a higher q_{excite} for 100 μs, followed by a 100 μs delay and then a 100-μsec return of the RF to a low *q* to trap the otherwise lost low-*m/z* fragment ions. The PDQ method allows trapping of ions with *m/z* below the LMCO and thus improves fragmentation coverage of, for example, the essential immonium ions in peptide sequencing.[150] In any regard, additional MS^n scans can provide a wealth of information regarding ion genealogy and structure not obtainable from MS^2 scans.[151]

The power of the MS^n scan is demonstrated in Figure 9.8a–9.8f. The results of six stages of MS (five stages of resonance CAD) are shown for an oleanolic acid glycoconjugate (MW = 1250). Each stage of MS shows a loss of one or two fragments allowing for a straightforward genealogical interpretation. Loss of the trifluoroacetic acid adduct (TFAH) was followed by losses of either a glucose (Glc) or rhamnose (Rha) group.[152] MS^n scans, however, do have limitations in that the ion current available for the mass analysis step is diminished as *n* increases and scan functions can become long in duration because SRM^n or MS^n scans require $(n - 1)$ additional isolation and excitation steps.

The mass analysis step after isolation and excitation may cover one ion of interest or a narrow range of ions for a SRM^n scan with no significant loss in duty cycle. This was demonstrated in a flow injection analysis (FIA) experiment where the detection limit of reserpine using an SRM scan (50-Da product mass range) was 8 fmol (5 pg) with a signal-to-noise ratio of 20. This detection limit was obtained by injecting 1 μL of 5 pg/μL reserpine into a 200 μL/min flow of 1% acetic acid in 50:50 methanol:water and summing three product ions in the positive mode at *m/z* 397, 436, and 448.

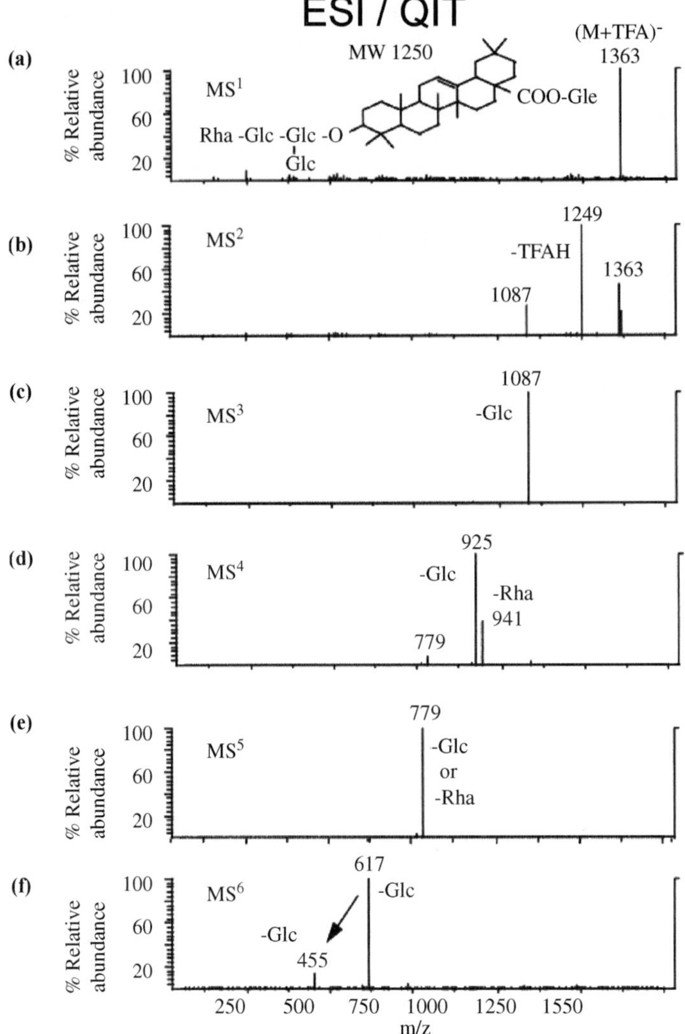

Figure 9.8. MS^6 of oleanolic acid glycoconjugate (MW 1250 Da). Figures (**a**) through (**f**) show six stages of MS resulting in the loss of the trifluoroacetic acid adduct and either a glucose or rhamnose group as noted. (Land, A. Finnigan Corporation 1995. Sample provided by the Université de Lausaune, Lausaune, Switzerland.)

9.4 THE ORBITRAP

9.4.1 A Brief History

In the early 1920s, at the suggestion of Irving Langmuir, K. H. Kingdon[153] at the General Electric Company studied the behavior of "imprisoned positive ions" orbiting in devices that consisted of a 2- or 5-cm-diameter cylinder (anode), a hot axial 0.1-mm-diameter wire (cathode), and two "guard" end caps that consisted of two sets of six radial wire spokes

(2-cm trap) or two metal disks (5-cm trap). A voltage was applied between the wire and cylinder (35 V) to allow positive ions formed by the filament to be trapped. The trapping occurs because positive ions from He, H, Ne, or Hg formed near the anode have an initial velocity transverse to the radius which prevents them from being annihilated at the cathode. The trapping times were increased by applying potentials to the guard end caps which caused trapped ions to, in Kingdon's words, "execute orbits to and fro across the tube." Remarkably, at a pressure near 10^{-5} torr, a helium ion was measured to orbit inside this simple device 175 times before losing enough energy to discharge at the cathode. Vane et al.[154] maintained the basic kingdon trap geometry with a cylindrical electrode to study electron capture of Ne^{+10}, and they added a quadrupole to mass analyze the ions inside the trap. Prior and Wang[155] also used an electrostatic ion trap with cylindrical geometry to study the hyperfine structure of the 2*s* state of $^{3}He^{+}$.

In 1981, Knight[156] described an improved Kingdon-type ion trap based on suggestions by M. H. Prior. The outer electrode was designed to give a harmonic axial potential or axial quadrupole term, and this non-cylindrically shaped electrode was split at the midplane to allow an RF voltage to be applied across it. The Knight trap was of limited utility primarily because a thin wire was still used as the axial electrode, and this geometry distorts the harmonic nature of the potential. In any case, ions were stored for up to 1 s at 10^{-8} torr. Mass analysis was performed by applying an RF resonance frequency to match the frequency of the ion, and detection was accomplished using an axial electron multiplier or radial Faraday plate. The resonances measured were weaker than those found in a Dehmelt's QIT,[157] and numerical solution of the Laplace equation for the Knight trap showed that indeed, the inner wire electrode distorted the field.

9.4.2 Geometry of the Orbitrap

Makarov at HD Technologies (Manchester, U.K.) improved on the Kingdon/Knight trap by developing an electrostatic (no RF) mass analyzer named the orbitrap.[25] A major advantage of this device is that no RF field or magnetic field is used to trap the ions. The orbitrap is a combination of quadrupolar and logarithmic potentials just like the Knight trap, but both the inner and outer electrodes have been modified. The modified axial symmetric electrodes as shown in Figure 9.1d create a more ideal "quadrologarithmic" electrostatic potential than what would be found in the Knight trap. A modern orbitrap consists of a precisely machined central spindle electrode, with a typical largest diameter of 8 mm inside an outer barrel-like electrode, with a typical largest inside diameter of 20 mm. The barrel electrode is split at midplane just as with the Knight trap, but instead of applying an AC signal, the barrel halves are used to measure the induced signal due to the ion axial motion. It is interesting to note that Knight stated that electrostatic ion traps "avoid the need for accurately machined" electrodes, but this is false. High performance from a modern orbitrap, QMF, LQIT, and the QIT (to a lesser extent) requires precision machining.

9.4.3 Ion Injection, Ion Motion, and m/z Analysis in the Orbitrap

A critical step in the operation of an orbitrap is the ability to inject and trap ions efficiently. Ions are injected at one end of the device away from the midplane, but perpendicular and off-axis of the device, between the spindle and the barrel electrode. As shown with the original Kingdon trap, ions with the proper *m/z* and kinetic energy will be trapped in orbits around the spindle. The ThermoFisher LTQ™ orbitrap uses a curved quadrupole ion trap or C-trap (CQIT); but rather than radially ejecting the ions with a mass selective instability scan, the cooled ions are instead pulsed out in $<0.3\ \mu s$ with DC potentials. The ions ejected from the CQIT are focused at the entrance to the orbitrap. Once trapped, the ions oscillate "to and fro" from one end of the barrel assembly to the other, while continuing to orbit the spindle as shown in Figure 9.9a. The frequency of the back-and-forth harmonic ion oscillations is measured by the acquisition of time-domain image current transients as graphed in Figure 9.9b, followed by a fast Fourier transform (FFT) to determine the *m/z* of the ions as shown in the mass spectrum of Figure 9.9c. The specially shaped electrodes allow the potential in the axial direction to be exclusively quadratic as in a harmonic oscillator. The *m/z* is related to the frequency of ion oscillation as shown in Eq. (9.7):

$$m/z = k/\omega^2 \tag{9.7}$$

The orbital motion of the ion is important because it is this motion that traps the ions radially. Ions of various *m/z* can be trapped as long as they enter the trap with the proper range of kinetic energies. The ion motion is directed along a path that is a balance of the centrifugal and centripetal forces just as found in an electrostatic analyzer.[158] A range of ion kinetic energies is an advantage in this trap because it radially increases the trapping volume. It is the frequency of the axial motion, not the orbital frequency, that is used to determine *m/z*. Ions of the same *m/z* form thin rings that oscillate back and forth axially. The net signal is detected using a differential amplifier so that the signals between the two halves of the outer barrel electrode do not cancel each other. With time, field imperfections of the orbitrap and collisions with residual gas cause the thin rings to broaden and the signal to diminish.

9.4.4 Orbitrap Scan Modes

Although the orbitrap is a trapping mass spectrometer, currently it has only been used in the full-scan mode or as the final stage of mass analysis for the other scan modes. For example, ion activation is first done in an LQIT (ThermoFisher LTQ Orbitrap), and then the products are injected into the orbitrap for the final MS^n scan of the product at high resolution. The orbitrap mass analysis scan thus replaces step 4 in the scan functions discussed earlier. Higher-resolution scans require longer scan times. Figure 9.9c shows a spectrum of bovine insulin collected at a 5-MHz sampling rate to produce 8 million data points and the resulting high-resolution spectrum ($R = 100{,}000$)[158]

9.4.5 Orbitrap Performance

Remarkably, orbitraps are capable of high mass resolution of $R = 150{,}000$, high mass accuracy (2–5 ppm), a mass range of at least *m/z* 6000, and a dynamic range greater than 10^3.[158] Makarov has reported that the resolution of the current orbitrap is 150,000 @ *m/z* 100, 50,000 @ *m/z* 1000, and 15,000 @ *m/z* 10,000 compared to the 7-tesla FT ICR which is 200,000 @ *m/z* 200, 40,000 @ *m/z* 1000, and 5000 @ *m/z* 10,000 for a 1-s acquisition; and mass accuracies can be better than 1 ppm at *m/z* 1522 when using an internal calibration.[159]

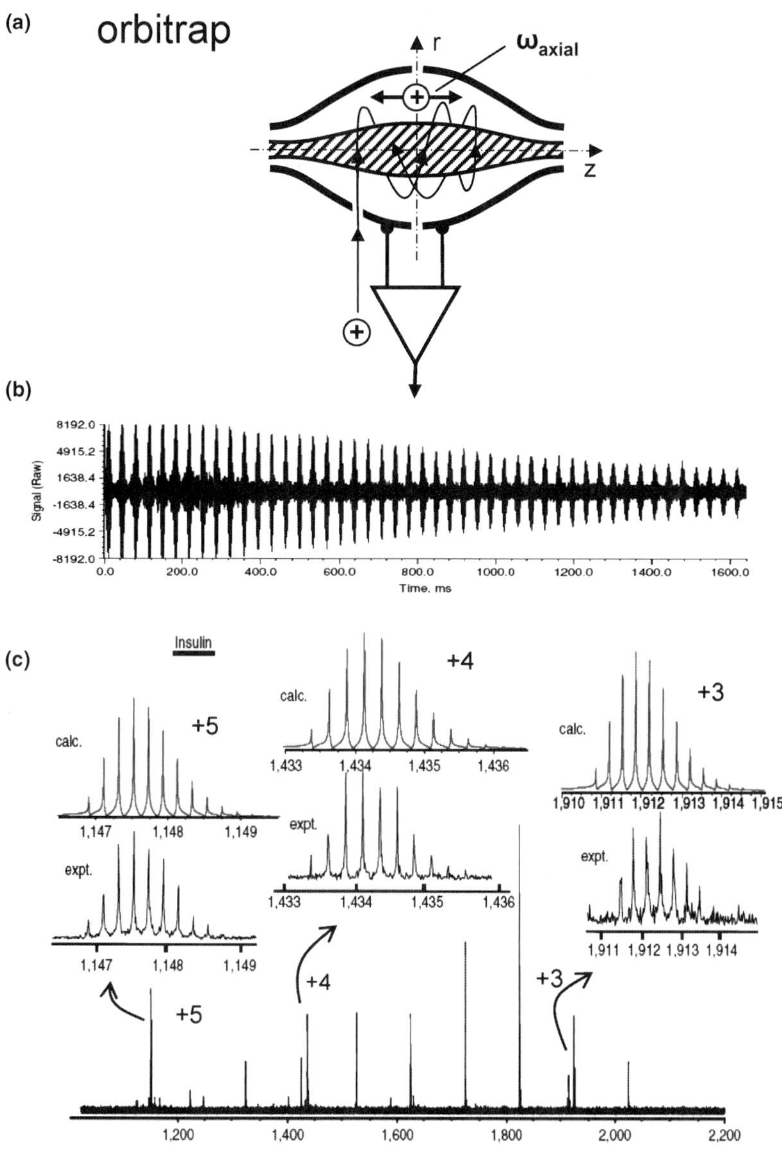

Figure 9.9. (**a**) An orbitrap (OT) mass analyzer showing the path of an ion during ion injection and while orbiting the spindle. The frequency of the axial ion motion is measured for mass analysis. (**b**) Typical transient acquired to record the mass spectrum of bovine insulin. (**c**) ESI mass spectrum of bovine insulin. Data acquisition parameters include a data sampling rate of 5 MHz, record length was 8 million data points, and the Fourier transform was performed with no apodization function or zero-filling. The lower spectrum shows a wide-range mass spectrum including the internal mass calibrant Ultramark 1621 whose oligomers are spaced by 100 mass/charge unit intervals. Lower traces in the close-ups show experimentally obtained isotopic distributions for each charge state. Upper traces in the close-ups show the theoretically expected isotopic distributions. The calculated isotope distributions were obtained from IsoPro 3.0 using Gaussian peak shapes with resolution of 100,000. (Reprinted with permission from Hu *et al. J. Mass Spectrom.* **2005**, *40*, 430.)

9.5 COUPLING ESI AND MALDI TO THE QMF, QIT, LQIT, AND ORBITRAP

A critical problem encountered when coupling an atmospheric ionization source, like ESI, to a mass spectrometer is that the analyte is introduced at atmospheric pressure, but the mass analyzer functions at low pressures. A difference of more than eight orders of magnitude can lie between the atmospheric pressure of the ESI source and the ambient pressure of a mass analyzer! The coupling problem is solved by using multipole RF ion guides[160] and several stages of differential pumping.

To couple a low-pressure MALDI source ($<1 \times 10^{-6}$ torr) to these analyzers is relatively straightforward because fewer stages of differential pumping are required and RF ion guides are not essential. However, for excellent MALDI performance, the photon and ion optics must be well-designed and aligned. Since a pulsed MALDI source is best coupled to mass analyzers that operate in a pulsing mode, a QMF is not a good candidate whereas the LQIT, QIT, and orbitrap are all well-suited.

9.5.1 Coupling the ESI to QMF, QIT, LQIT, and Orbitrap

Despite the significant pressure differences, Fenn's group successfully coupled an ESI source to a QMF in 1984.[9,10] The Fenn group at Yale along with Whitehouse at Analytica of Branford eventually designed a source that separated the atmospheric-pressure ion generation chamber from the QMF by using a glass capillary that was metal-coated on both ends to allow for the isolation of the applied potentials. Their ESI source coupled to the QMF is shown in Figure 9.10a. After the capillary, the expanding gas and ions were skimmed in the first pumping stage and introduced into the QMF in the main vacuum chamber. Fenn et al.[10] collected mass spectra of polyethylene glycol $M_n = 1000$ and 1450, carbonic anhydrase II, and alcohol dehydrogenase from this early ESI system.

Several laboratories successfully coupled the ESI source to a QIT mass spectrometer in the 1990s.[51,83,84,161] Van Berkel, Glish, and McLuckey at Oak Ridge National Laboratory were the first to couple an ESI ion source to a QIT in 1990.[83] In 1992, Mordehai and co-workers at Cornell University coupled their own ESI source to the first bench-top QIT (Saturn II™, Varian Instruments, Palo Alto, CA, USA) for use in LC and CE studies.[84] Bruker–Franzen of Germany also introduced an ESI/QIT instrument (ESQUIRE™) at that time. ThermoFisher Corporation introduced an ESI/QIT mass spectrometer (LCQ™) in 1995.[162,163] In 2002, Hager at MDS Sciex described the first ESI LQIT system.[108]

The design developed by the Oak Ridge group to couple an ESI source to a QIT is shown in Figure 9.10b. They modified an atmospheric glow discharge ionization source[72,83] for use as an ESI source by adding two lens elements. Sample solutions were sprayed from a 120-μm-i.d. "electrospray" needle positioned at the entrance orifice. The orifice (100 μm) in lens element *A1* samples the electrosprayed liquid from the needle and introduces the analyte into the first of two differentially pumped chambers. Chamber one was maintained at a pressure of 0.3 torr while chamber two, the QIT chamber, was maintained at a pressure of 0.01 mtorr (uncorrected). The ions were electrostaticly focused by a three-element lens stack (*L1*, *L2*, *L3*) onto a hole in the entrance end cap. Ion packets were injected into the QIT at the appropriate time by pulsing one of two semicircular plates (*L2*). A potential difference of 300 eV between the two half-plates deflected the ion beam away from the QIT entrance orifice when not injecting ions. Ions ejected from the QIT were detected with a dynode/electron multiplier detector.

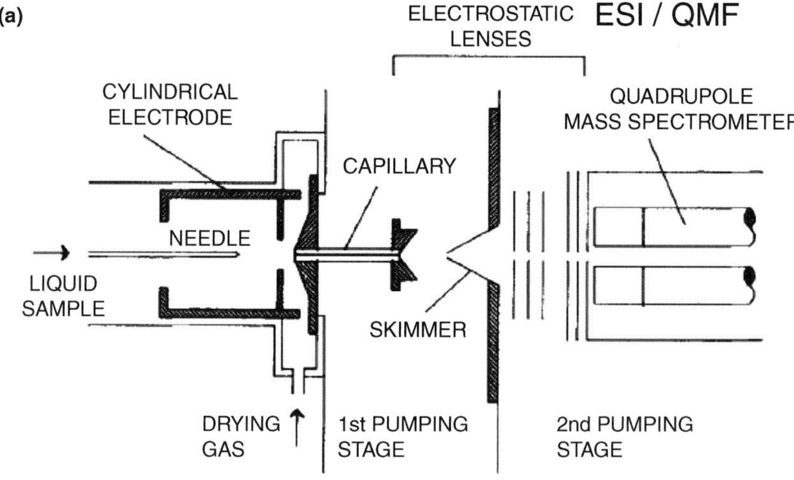

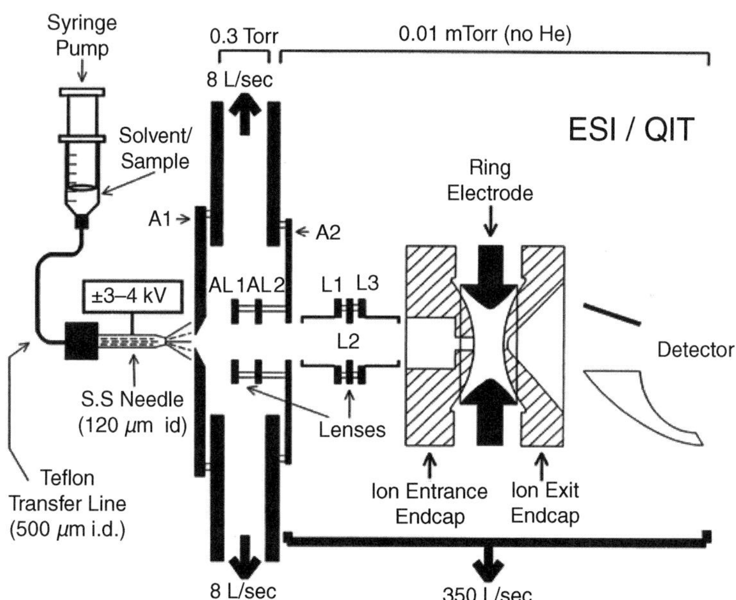

Figure 9.10. (**a**) Cross-sectional view of one of the first ESI QMF mass spectrometers. (Reprinted with permission from Fenn, J. B.; Mann, M.; Meng, C. K.; Wong, S. F. Whitehouse, C. M. *Mass Spectrom. Rev.* **1990**, *9*, 37.) (**b**) The first ESI QIT. (Reprinted with permission from Van Berkel, G. J.; Glish, G. L.; McLuckey, S. A. *Anal. Chem.* **1990**, *62*, 1284–1295.)

Most ESI interfaces desolvate solvent ion clusters by gaseous collisions and/or by heat. The ESI source developed by MDS Sciex Corporation (Toronto, Canada) coupled to a QMF used a nitrogen "curtain" near the orifice to break up ESI solvent clusters[164] (see the figure in Table 9.1). Chait's group at Rockefeller University developed the use of a heated metal capillary.[165] The Fenn–Whitehouse ESI source in Figure 9.10a (Analytica of Branford, Branford, CT) used a counter–current flow of hot nitrogen as a drying gas,[166] while Henion's group at Cornell used a heated lens.[84] The ESI source constructed by Oak Ridge was not heated and was prone to produce highly solvated biomolecular ions that were then trapped. To assist in desolvating these ions, they extended the trapping time and collisionally activated the solvated ions by resonance excitation prior to mass analysis.[83] The early Oak Ridge ESI source led to an interesting publication studying protonated water and methanol clusters.[167] They found that highly clustered solvent $(H_2O)_nH^{+n}$ and $(CH_3OH)_nH^{+n}$ ions underwent rapid desolvation when $n > 6$. The Oak Ridge group has also pointed out that a softer ESI source and ion transport process is desirable when studying weakly bound complexes by ESI/MS/MS in a QIT.[32]

A diagram of the first commercial ESI/QIT mass spectrometer system (LCQ™) is shown in Figure 9.11a and discussed herein. The vacuum system consisted of a rough pump and a dual-port, turbomolecular pump to maintain operating pressures P_2 through P_4 at 1 torr, 1×10^{-3} torr, and 2.5×10^{-5} torr (uncorrected), respectively. The solution is electrosprayed from a stainless steel needle held at ~4.5 kV, in the positive ion mode, and is sampled by the capillary which is typically heated to a temperature of 200°C. The capillary (400-μm i.d., 11.5 cm long) helps complete the desolvation process and serves as the nozzle for the supersonic expansion of the gas into the next chamber. The supersonic free jet of solvent molecules exiting the heated capillary forms a barrel shock that includes a Mach disk downstream.[168] A metal skimmer is positioned inside the barrel shock and is "attached" to the Mach disk for optimum ion transmission. Ions are transmitted through the skimmer using the tube lens as a gating element. Positive ions are pulsed through the skimmer by applying 0 to +200 V to the tube lens, while – 150 V stops transmission of the ions through the skimmer. The sampled ions are collected by the first RF octopole and are transmitted to the second RF octopole through an interoctopole lens. Each octopole is 5 cm long ($r_0 = 3.3$ mm) and both are operated at 2.5 MHz and 400 V_{pp}. Octopole RF ion guides were used to efficiently transmit ions through a region of relatively high pressure where ion scattering would otherwise occur.[169–171] It has been shown that low-mass ions are scattered easily by gases in static lens systems, whereas the RF octopole reduces these effects and improves transmission as was found with RF-collision cells in triple-stage quadrupole instruments.[172] The octopoles also allow for efficient differential pumping; and while each can have a separate DC offset potential for optimum ion transmission, only one RF supply is required. ESI/QMF, XQIT and orbitrap instruments have also used square quadrupoles and hexapoles (Waters QTOF2) to do the ion bunching and ion transmission.

The LCQ interoctopole lens has a 2.5-mm-i.d. aperture and it serves three functions: (i) to transmit ions during the ion injection period, (ii) to act as the conductance limit for differential pumping, and (iii) to act as a potential barrier against the transmission of large charged particles formed by the ESI source. Many users of ESI sources have noted the appearance of large and random noise spikes in the mass spectrum. These spikes are attributed to large multiply charged particles that apparently strike the detector and result in noise spikes of high intensity in the mass spectrum. To stop these particles, several hundred positive volts can be applied to an interoctopole lens or to deflection plates during the mass analysis period. This applied voltage dramatically reduces the particle noise during the subsequent mass analysis part of the scan function.[162] The reduction of background ESI noise has also been studied by Ramsey et al.[173] by the use of resonance excitation techniques.

Unique to the LCQ design was the use of an octopole to transmit ions directly to the ion trap chamber by placement of the second octopole inside the entrance end cap of the QIT

(a)

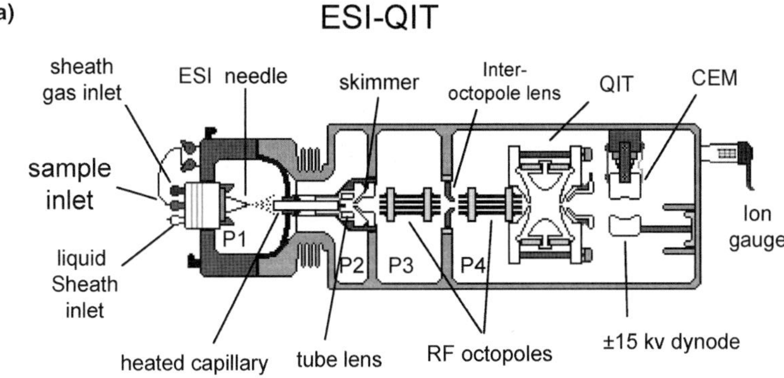

(b)

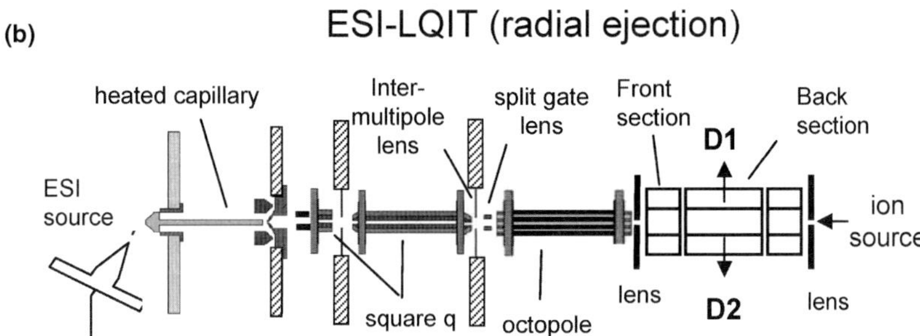

(c)

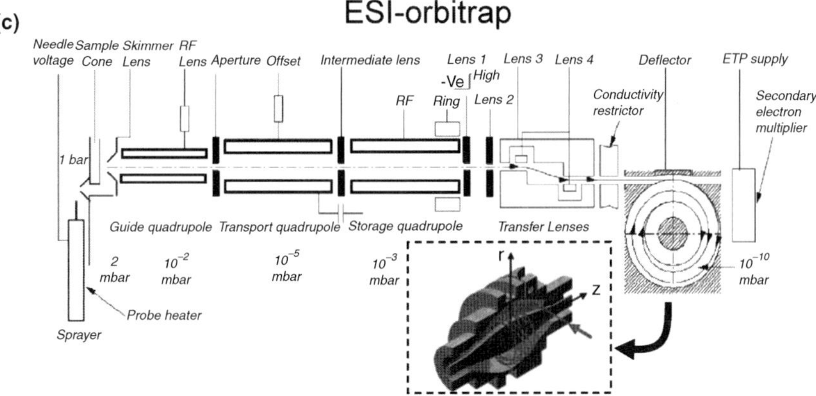

Figure 9.11. (**a**) Electrospray quadrupole ion trap mass spectrometer (LCQ™) featuring a heated capillary and dual octopoles RF ion guides to inject ions directly into the trapping chamber. P1 = 760 torr, P2 = 1 torr, P3 = 1×10^{-3} torr, P4 = 2.5×10^{-5} torr, and $P_{trap} = 1\text{–}2 \times 10^{-3}$ torr. (**b**) ThermoFisher LTQ™ with an additional pumping stage and the LQIT mass analyzer. (**c**) An experimental orbitrap mass spectrometer with multiple stages of differential pumping. Ions from the ESI source proceed through the collision quadrupole and selection quadrupole and then pass into the storage quadrupole. The storage quadrupole serves as an ion accumulator and buncher. After accumulation and bunching in the storage quadrupole, the exit lens ('Lens 1') is pulsed low, the ion bunches traverse the ion transfer lens system and are injected into the orbitrap mass analyzer. Inset shows a cut-away of the orbitrap. (part c is a reprint with permission from Hu, Q.; Noll, R. J.; LI, H.; Makarov, A.; Hardmanc, M.; Cooks, R. G. *J. Mass Spectrom.* **2005**, *40*, 430–443.)

(see Figure 9.11a). An octopole/3D-trap coupling scheme was used to reduce parts and power supplies wherein the entrance end cap was used both as a trapping electrode and a shield of the octopole RF once ions arrive inside the trap. Since then, others have added additional lenses in this region to offset the effects of nonlinear resonances created by the entrance aperture and to offer additional ion focusing.

Coupling an ESI source to a LQIT as shown in Figure 9.11b is straightforward given the previous design worked out for the first-generation ESI QITs. The LQIT offers the additional advantage of a larger acceptance aperture (a commercial LQIT uses a 2-mm aperture for the front lens), a non-RF electrostatic barrier along the z axis from a rear quadrupole segment or rear lens element, and a longer "runway" in which to trap the heavy ions. MDX Sciex simply replaced q2 and/or Q3 in their ESI/triple quadrupole system with a LQIT to build their 4000 QTRAP (see Table 9.1).

9.5.2 Coupling the ESI Source to the Orbitrap

Coupling an electrospray ionization source (ESI) to the orbitrap requires considerable differential pumping to reduce the solvent and atmospheric gas pressures to the low levels required by the mass analyzer. The orbitrap requires a low pressure ($<2 \times 10^{-7}$ torr) so that the ion orbits and axial oscillations can be maintained for an extended period. During this time the image current measurement is acquired, followed by Fourier transform analysis. In addition, ions must be injected into the orbitrap precisely for efficient trapping. One experimental design coupled ESI to the orbitrap by using several stages of differential pumping with an ion guide quadrupole, transport quadrupole, and storage quadrupole used to bunch and cool ions, followed by transfer lenses and a conductivity restrictor to focus ions into the orbitrap as shown in Figure 9.11c.[158] The storage quadrupole had a unique ring electrode used to create a small axial potential well. Ions pool in the well before being pulsed to the orbitrap. Using this storage quadrupole, an "energy lift" is created in the LQIT where the potential of the ion is raised while maintaining complete trapping of all ions. Compact packets of ions are formed from the additional cooling that occurs in the LQIT. When pulsed out of the LQIT, ions "roll down" the potential surface and enter the near ground potential orbitrap for ion injection and trapping. Commercial orbitrap instruments now follow a LQIT and/or a CQIT to similarly cool, elevate, and pulse out ions into the orbitrap. The ion injection process involves electrodynamic squeezing of the ions once they have entered the orbitrap.

9.5.3 Coupling MALDI to the QIT, LQIT, and Orbitrap

To couple a low-pressure MALDI source to a QIT, LQIT, or orbitrap mass analyzer, the primary concern other than having an efficient MALDI source is to be able to trap the gas-phase ions efficiently inside the analyzer. In the low-pressure MALDI experiment, ions can be formed inside the ion trap[80,174] or they can be generated outside the device and injected into the analyzer.[78,79,175–178] Ions are lifted by the matrix plume at a velocity of ~500 m/s with several electron volts of energy and may undergo additional reactions in the plume. Once in the gas phase, the ions are accelerated as a result of either (a) an RF field in the case of internal ionization or (b) an external lens system for the external ion formation/injection scheme. Ions can be trapped efficiently in a QIT or LQIT if they have low kinetic energies (typically <20 eV), but the trap must be set to the appropriate RF trapping potentials and filled with helium buffer gas. In two similar configurations from 1993,[78,79] the QIT was electrically isolated at −15 Volts while the MALDI probe tip was at ground potential as shown in Figure 9.12a. A three-element lens stack provided the extraction potentials to inject ions into a QIT. A nitrogen laser at 337 nm was triggered by the normal ionization gate trigger pulse and the packets of ions were injected into the QIT.[79] The QIT could be filled with ions using multiple laser shots if needed. A 20-kV dynode was used for detection of the higher-mass ions. For the internal or external ionization schemes, the use of helium pulsed to pressures greater than 1 mtorr helium improves the trapping ability and phase-locking of the laser with the main RF is important for the QIT case.[81,179,180]

Although most MALDI has been done at low pressures inside a vacuum chamber of the mass spectrometer, others have tried atmospheric pressure MALDI (APMALDI). For APMALDI,[181,182] Laiko et al.[183] removed the ESI source from a QIT and added a target holder external to the heated capillary. They were able to observe 20–50 fmol of peptide deposited. Keough et al.[184] used an AMALDI source (Mass Tech) and coupled it directly to a QIT to look at sulfonic derivatized peptides. Laiko et al.[185] used an IR laser for APMALDI of aqueous samples of peptides into a QIT and have also done APMALDI from porous silicon[186] into a QIT. A report by Coon and Harrison[187] has shown data where the IR APMALDI signal from a QIT can be increased 10-fold by using an additional corona discharge.

Miller, Yi, and Perkins of Agilent Technologies Corp. group have shown impressive results with an APMALDI source which uses a countercurrent flow of gas at the entrance tip. A tryptic digest of 125 amol of bovine serum albumin was identified by this technique combined with an Agilent Technologies 1100 series MSD trap.[188] Tan et al.[189] used pulsed dynamic focusing in the ion source to improve the efficiency of transmitting APMALDI ions into the vacuum chamber. In this method, a delay of the pusher voltage is generated to allow the ions to enter the heated capillary by the pressure gradient.

Several other MALDI/ESI trap schemes have been tried. A unique coupling scheme of a MALDI source to a QIT[190] was built by Harris, Reilly, and Whitten for the direct analysis of biomolecules from individual particles as shown in Figure 9.12b. The group used an aerodynamic lens system to introduce the preformed airborne particles into a QIT. Detection lasers signaled when to fire the ablation/ionization laser for the analysis. MALDI has also been coupled to a LQIT for the direct analysis of intact tissue.[191] In this case, the MALDI source was operated at 0.17 torr. The advantage of the dual source capability of the LQIT was demonstrated by Smith et al.,[192] who took advantage of the efficiency of coupling an ESI ion source to the other end of their LQIT system which was already coupled to MALDI as shown in Figure 9.12c.

Since a QMF is not operated to trap ions, but rather to transmit them while scanning over a large mass range (e.g., 2000 Th in 1 s), it is not ideal as a stand-alone mass analyzer for the pulsed MALDI technique. QMFs are, however, used as front-end precursor ion selectors for orthogonal TOF MS. In this chapter, the MALDI ions are all first trapped before mass analysis. Cotter's group and others have coupled a MALDI source to a QIT for time-of-flight mass analysis.[193–198] The QIT was used as a storage/cooling cell, not as the mass analyzer in these experiments.

9.5.4 Trapping Ions in the QIT or LQIT

Ions are typically injected axially through the entrance end-cap aperture of a QIT or axially through a lens element of a LQIT. It has been shown that ions can be injected and efficiently trapped directly from an octopole ion guide, despite having a significant

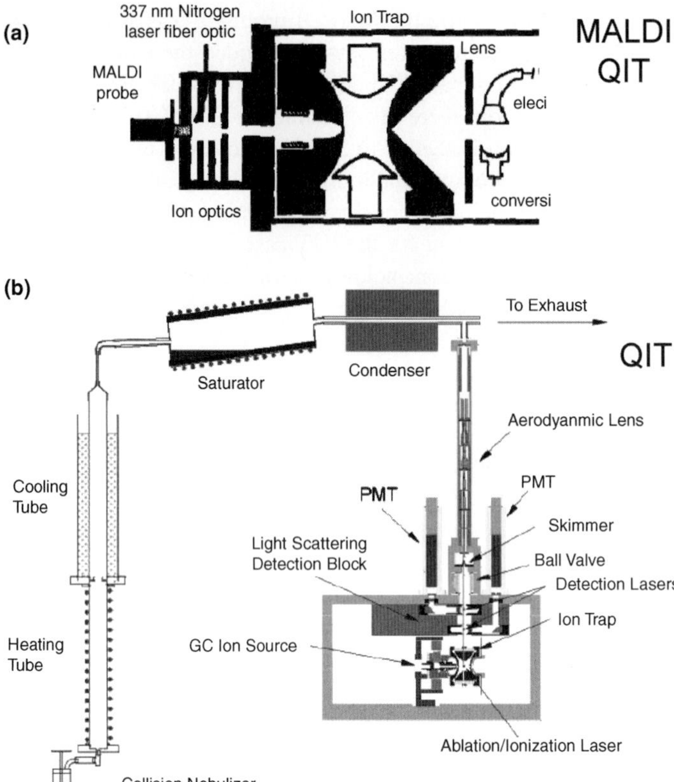

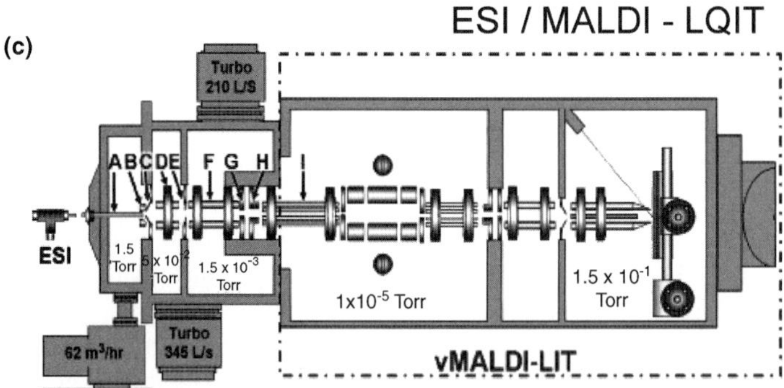

Figure 9.12. (**a**) External MALDI/QIT mass spectrometer with fiber-optic delivery of the laser light. (Reprinted with permission from Jonscher, K.; Currie, G.; McCormack, A. L.; Yates, J.R. III. *Rapid Commun. Mass Spectrom.* **1993**, *7*, 20–26.) (**b**) Schematic of aerosol generator, matrix applicator, and aerosol ion trap mass spectrometer. (Reprinted with permission from Harris, W. A.; Reilly, P. T. A. and Whitten, W. B. *Anal. Chem.*, **2005**, *77*, 4042.) (**c**) A dual ESI and MALDI LQIT mass spectrometer. (Reprinted with permission from Smith, F. A.; Blake, T. A.; Ifa, D. R.; Cooks, R. G.; Ouyang, Z. *J. Proteome Res.* **2007**, *6*, 837–845.)

transverse ion kinetic energy.[162] Improvements can be made with additional optics. In order to efficiently trap the axially injected ions, it is crucial to have an appropriate RF amplitude applied to the ring electrode of the QIT or the rods of a LQIT and to have the partial pressure of helium in the trap at a minimum of 1 mtorr. The injected ions collide with helium, and the ion radial trajectories are reduced while the trapping field continuously forces the ions toward the center (QIT) or *z* axis (LQIT) of the device. Louris et al.[50] originally plotted the intensity of injected ions of different *m/z*'s from perfluorotributylamine versus the RF amplitude on the ring electrode of a QIT. They observed that different *m/z* ions had different optimum RF amplitudes. Since then, it has been determined that, to a first approximation, a linear relationship exists between the optimum trapping RF level and *m/z*[199] and the slopes of these lines increase with ion injection energy.[162] The higher RF amplitudes are believed to reflect the need for stronger RF trapping fields at the higher ion kinetic energies. As expected, the LQIT has a much higher acceptance of ions over a much broader range of RF amplitudes.[106] Ion injection energies in most of the work described here are typically between 4 and 20 eV for both MALDI- and ESI-generated ions.

9.5.5 Unwanted Ion–Molecule Reactions

One of the limitations of ion traps that first used electron ionization (EI) is that given the long storage times (e.g., 10 ms and greater) some ions react with neutral molecules.[200] These ion/molecule reactions are dependent on the analyte ion reactivity, the partial pressure of the neutral reactant (e.g., water, methanol, nitrogen and the analyte neutral), and the duration of the reaction time. Little reactivity of the protonated ions is observed in ESI/QIT because they are even-electron ions. In general, even-electron ions are far less reactive than odd-electron radical ions formed by EI sources. In addition, ions formed by electrospray have already been exposed to the high partial pressures of reactants from solvents such as water, methanol, and acetonitrile, or acids like acetic or trifluoroacetic in the capillary/skimmer region. In the low-pressure MALDI case ($<1 \times 10^{-6}$ torr), the ions are also primarily even-electron ions and undergo collisions with many matrix neutrals in the plume above the surface before acceleration. In AMALDI the ions would additionally undergo high-pressure gas-phase collisions above the MALDI plate, in the introduction optics and/or at the expansion region in front of the skimmer. As a result, an ion/molecule reaction should have already taken place in the ESI or MALDI interface prior to trapping. Product ions formed from resonance excitation in the trap chamber, however, are not exposed to the high pressure of gases found in these sources. Usually, these product ions are also even-electron ions, but the relatively long ion trap storage times increase the probability that an ion/molecule reaction may occur. Routine ESI analysis has shown that unwanted ion/molecule reactions are indeed uncommon and also that when they do occur, they often are structurally informative.[82]

In the case of the ESI QMF interface, the same arguments above apply with the exception that the ions are not stored in the device, so no additional reaction along the flight path is expected unless generated by the operator in the collision cell (q2). Since the orbitrap is operated at low pressures, ion–molecule reactions are unlikely and if ion/neutral collisions do occur during mass analysis, a reduction in resolution would occur.

9.5.6 Mass Range Extension

9.5.6.1 Mass Range Extension in Quadrupole Ion Traps

The mass range of the first commercial QIT was 650 Th. Since ESI and MALDI sources allow the formation of ions greater than m/z 650, a method was needed to extend the mass range of the XQIT. This is even more of an issue for the analysis of low-charge-state macromolecular ions formed in the MALDI source. As indicated by Eq. (9.4), if one uses *RF amplitude scanning*, one could increase the mass range by lowering the fundamental RF frequency, reducing r_0, and/or increasing V_{MAX}. The RF amplitude (V_{MAX}), has a practical working limit of approximately 8.5 kV$_{0\text{-p}}$, above which arcing may occur. In the *RF frequency scanning* case, one can scan the frequency of the RF to sequentially eject ions at higher mass while maintaining the RF amplitude constant.

In 1990, the RF amplitude scan was used to extend the mass range of a QIT to over 70,000 Th by decreasing the $q_{z\text{-eject}}$ point at which ions are resonantly ejected from the trapping chamber.[53] For example, if all other parameters remain constant, by decreasing $q_{z\text{-eject}}$ by a factor of 10, from $q_{z\text{-edge}} = 0.800$ to $q_{z\text{-eject}} = 0.080$, the maximum mass range increases from 650 to 6500 Th as defined by Eq. (9.8) and indicated in Figure 9.3b. For example, now $m_6^+ = 5000$ can be resonantly ejected as shown in this illustration.

$$(m/z)_{MAX} = \frac{8V_{MAX}}{(r_0^2 + 2z_0{}^2)\Omega^2 q_{z-\text{eject}}} \tag{9.8}$$

In Eq. (9.8), V_{MAX} is the maximum amplitude of the RF and $q_{z\text{-eject}}$ is the point on the stability diagram where the ions are resonantly ejected. By using a lower $q_{z\text{-eject}}$, however, the scan rate (Th s^{-1}) increases, but resolution is reduced. In the example above, the scan rate would be increased to 55,550 Th s^{-1}. This rapid scan rate will also result in fewer points acquired across a mass peak unless a higher data acquisition rate is used (see Section **9.3.3.3, "*The Effects of* Scan Rate: The Normal, Rapid, and Slow Scan Rates"**).[60] A solution to this problem would be to reduce the scan rate back to 5555 Th s^{-1} and to limit the mass range to avoid long scan times. Mass range extension in the LQIT can be accomplished by applying the same methods as described above.

9.5.6.2 Ghost Peaks in Ion Traps

When the q_{eject} point is reduced for mass range extension, artifact peaks or "ghost peaks" can be observed. Ghost peaks are the result of ions that have fallen between the q_{eject} point and $q_{\text{edge}} = 0.908$ of the stability diagram at some time before or during the mass analysis scan. During the mass-selective instability scan, these ions are ejected at the q_{edge} rather than at the resonance point and appear as broad underresolved peaks. To avoid ghost peaks, rather than working at a low q_{eject} value, the mass range can be extended to 2500 Th by using a $V_{0\text{-}p}$ of 8.5 kV, an r_0 of 6.5 mm, an f_{RF} of 700 kHz, and a slightly reduced q_{eject} (to less than $q_{\text{edge}} = 0.908$). For the XQITs, one can also apply the main RF in bipolar fashion so that the mass range will be extended by a factor of two.

9.5.6.3 Mass Range Extension in the QMF

For the QMF, Eqs. (9.1) and (9.2) can be rearranged to create Eq. (9.9), which is similar to Eq. (9.8), but since the QMF operates at a defined $q_{\text{transmit}} = 0.706$ and the V_{RF} is already maximized, that leaves only the reduction of the main RF frequency and the reduction of r_0 as possible parameters to modify.[26]

$$(m/z)_{\mathrm{MAX}} = \frac{7 \times 10^6 V_{\mathrm{MAX}}}{r_0^2 \Omega^2} \tag{9.9}$$

Since the use of a QMF is not practical for a pulsed MALDI technique and because ESI often produces multiply charged ions, thus decreasing the ion's m/z, there has been little attempt commercially to increase the mass range of the QMF beyond m/z 4000. However, several researchers have increased the QMF to m/z 9000[201] and 45,000.[202] Sobott and co-workers[203] showed that it was possible to analyze m/z 22,000 within an isolation window of 22 Th on a quadrupole time-of-flight (QTOF) tandem mass spectrometer by lowering the QMF to 300 kHz.

9.5.6.4 Mass Range Extension in the Orbitrap

The current upper mass range limitation of one orbitrap was found to be m/z 7000 (Ref. [158]) for a multiply charged alcohol dehydrogenase, but this was believed to be a limitation of the poor transmission at high m/z in the double orthogonal geometry ion source. To trap ions over a wide mass range, electrodynamic squeezing[25,204] is used by reducing the field on the spindle during injection.

9.5.7 Detection of Ions

In the QIT and XQITs, ions are damped to the center of the trap by helium prior to mass analysis so that they may be ejected in a dense stream. Ion ejection at a high q_{eject} value will cause the kinetic energies of high mass ions to reach several kilo-electron volts.[162,205] The kinetic energy is imparted to the ions by the high RF amplitude used during mass analysis. Despite these high ion axial ejection energies, off-axis $\pm$15-kV dynode(s) can deflect and focus these high-mass ions onto its surface for detection. When higher helium pressures are used to trap injected ions more efficiently, the detector may be differentially pumped to avoid the detrimental effects of ion scattering or arcing.[206] For some of the early QITs, one-half of the ions were lost back toward the ion source, whereas in the LQIT with radial ejection, ions may be scanned out towards multiple detectors. In the QMF and the LQIT with axial ejection, one detector is placed off-axis to capture axially ejected ions. Ion detection is accomplished in the orbitrap by measuring the image current from the outer electrodes and applying Fourier transform analysis.

9.6 APPLICATIONS OF MALDI AND ESI COUPLED TO QMF, QIT, LQIT, AND HYBRID MASS SPECTROMETERS

The focus in this section is on proteomics and genomics and the use of ion traps as chemical reactors just like a traditional test tube. Ion activation is at the heart of these reactions and these activation methods will be covered later in this section.

9.6.1 Analysis of Peptides and Proteins

Biochemists are interested in determining the molecular weights, amino acid sequences, and/or post-translational modifications (PTM) of peptides and proteins. Due to the high sensitivity of ESI and MALDI mass spectrometry, they have become the techniques of choice. The information gained can be used to identify proteins,[207] identify PTMs sites, sequence epitopes, study a biological process and/or diagnose disease. A molecular weight can be determined by resolving the isotopes and calculating the mass based on charge state or by calculating the molecular weight based on several charge states. In MALDI, the matrix is typically chosen to produce a charge state of +1 or +2, so a deconvolution algorithm is not typically used.

An example of an ESI QIT high-resolution molecular weight determination of a quadruply charged peptide, interleukin-8 (rat), is shown in Figure 9.13. The isotope peaks in Figure 9.13b are separated by 0.25 Da, which indicates a +4 charge state. The exact molecular weight of the ^{12}C monoisotopic peak of interleukin-8 was measured to be

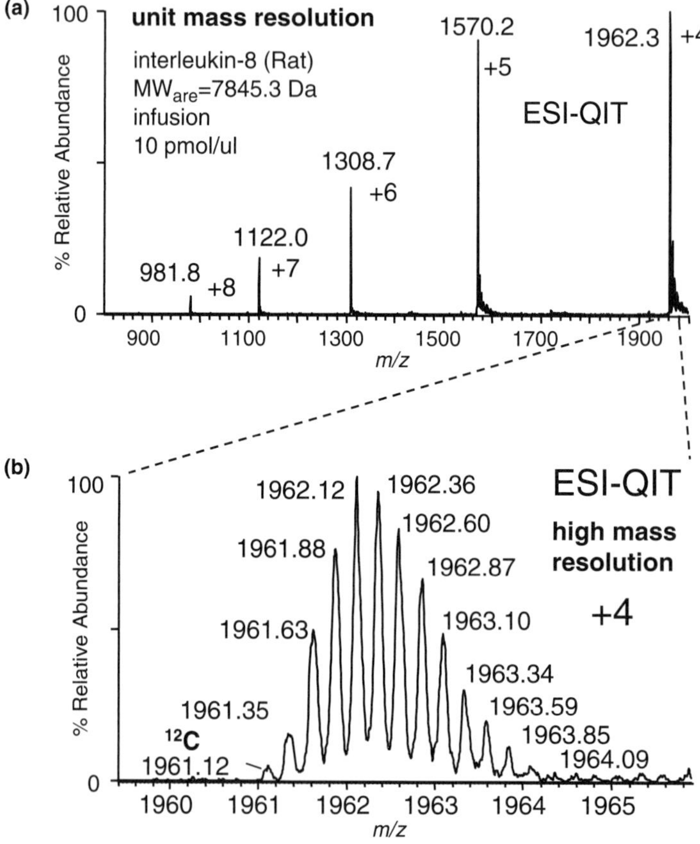

Figure 9.13. Molecular weight determination of interleukin-8 (rat) +4 ion using a QIT. The spectrum in (**a**) shows several of the multiply charged interleukin-8 ions while (**b**) shows the high-resolution spectrum of the +4 charge state used to determine the ^{12}C monoisotopic MW 7840.38 (theoretical MW 7840.09 Da). (M. Sanders, Finnigan Corporation 1995.)

7840.38 ± 0.14 Da using the *m/z*'s of all the +4 isotopes that each carry four protons. The theoretical exact mass of the ^{12}C monoisotopic peak of interleukin-8 is 7840.09 Da, indicating that four of the -SH groups were oxidized to two disulfide bonds. This was a high-resolution SIM method over a 10-Da window that determined the monoisotopic molecular weight of interleukin-8 to within 0.3 Da (0.004%). For an example of a multiply charged molecular weight determination of a peptide less than 10,000 Da, 1-μL injection (20 pmol/μL) of bovine ubiquitin (MW_{avg} 8564.9 Da) was collected on a ESI QIT and gave a deconvoluted average molecular weight of 8565.0 Da.[208] The deconvolution method determined the average molecular weight of bovine ubiquitin to within 0.1 Da (0.002%).

One of the powerful applications of mass spectrometry is the analysis of larger proteins (>10 kDa) by ESI MS.[209] For speed and accuracy, deconvolution algorithms are used to calculate molecular weights since there may be tens to hundreds of multiply charged mass peaks. Because this is a multiplicative method that uses all of the charge states for the molecular weight determination, the result can be highly accurate. Figure 9.14 shows the capillary LC ESI mass spectra of four proteins, 12-kDa cytochrome c (equine), 17-kDa apomyoglobin (equine) and 66-kDa serum albumin (bovine) (BSA) that were acquired using a QIT (Finnigan LCQ™), and an unknown 117-kDa protein that was collected on a LQIT (ThermoFisher LTQ™). Deconvolution algorithms (BIOMASS for Figures 9.14a–c. and ProMass™ [210] for Figure 9.14d) were applied to the multiply charged spectra and the deconvoluted plots are shown in the insets. The mass accuracy for these proteins was found to be 0.01% or better. Note that the bovine serum albumin spectrum shows significant adduct tailing, which suggests that the BSA, a major carrier protein in blood, is not pure. As a result of collecting this cap-LC ESI QIT spectrum, the author stopped using this "98% pure" BSA as a calibrant.

The ESI LQIT spectrum of a protein with an expected MW of 116,718 Da (over-expressed in *E. coli*) was eluted from a polymer column (2- × 50-mm PLRP-s 4000A) using a gradient from 10–65% ACN in 20 min (solvents A & B contained 0.01% TFA) and is shown in Figure 9.14d. It is particularly interesting to note the 101 different positive-charge states observed from +59 to +159 and the bimodal distribution of charge states. The bimodal distribution is an indication of two conformations, the more unfolded conformation being more highly charged. These molecular weight results are tabulated in Table 9.2 with other peptide and protein examples.

Examples of protein molecular weight determinations by MALDI QIT are seen in Figure 9.15. In 1993, two groups coupled MALDI externally to a QIT.[78,79] In both cases, fiber optics were used to bring 337-nm nitrogen laser light onto the MALDI plate, but without focusing. The ions formed were extracted from the plate and injected through the entrance end cap of the two experimental QITs. As displayed in Figure 9.15a, Jonscher et al.[78] showed a signal-to-noise ratio of ~10 for 250 fmol of renin substrate in a 1 : 2000 ratio (analyte:matrix) with 2,5-dihydroxybenzoic acid and with resonance ejection at 89 kHz (9 V_{pp}). Figure 9.15b shows an intense $[M + H]^+$ peak and the doubly charged peak at $[M + 2H]^{2+}$ from 3 pmol of alpha-lactalbumin (MW 14,175 Da) collected with 15 laser shots on a MALDI QIT.[79] To obtain this spectrum, the resonance ejection frequency was set below 10 kHz for mass range extension. Schlunegger and Caprioli[179] have further extended the mass range of the QIT using an internal MALDI ionization method with pulsed helium and a frequency scan. Figure 9.15c shows a MALDI mass spectrum of the singly charged molecular ion of 10 pmol IgG using sinapinic acid as the matrix from an average of 100 scans.[179] The frequency was scanned from 20 to 10 kHz in 500 ms.

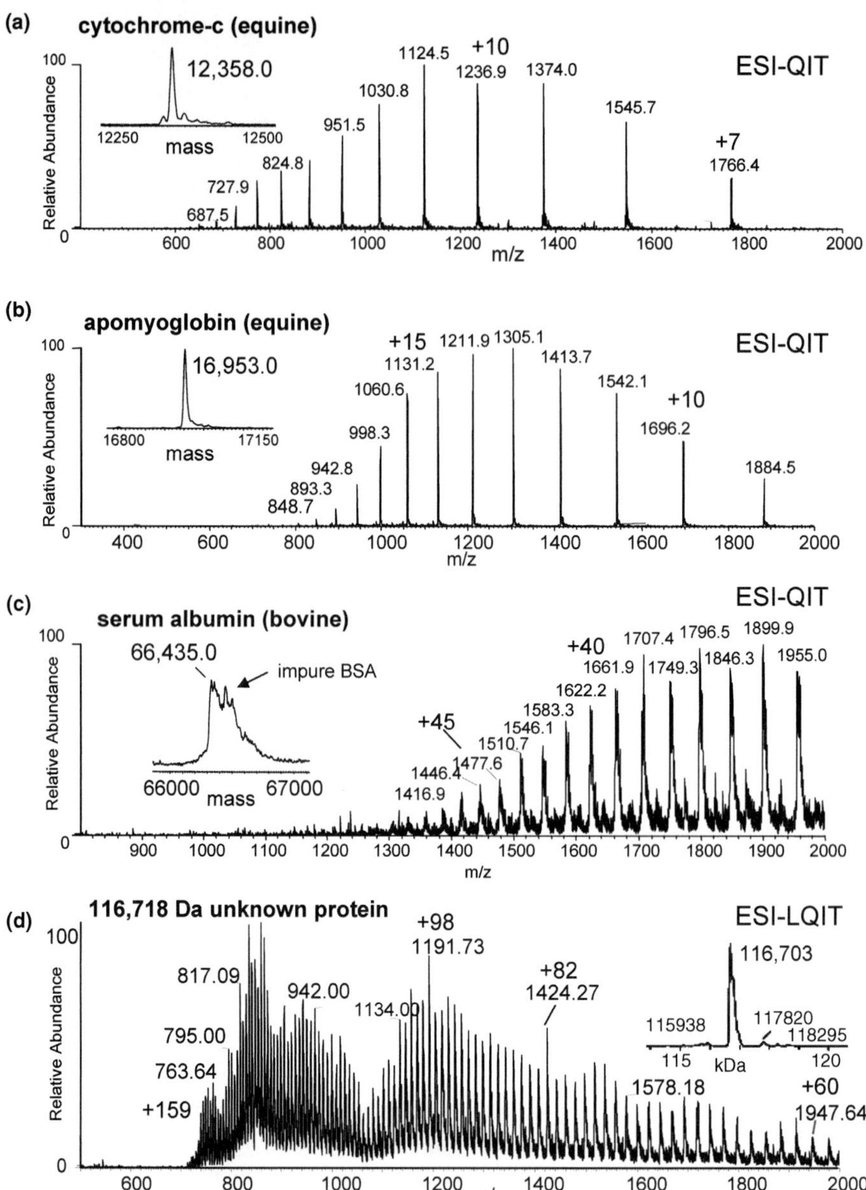

Figure 9.14. Protein ESI ion trap mass spectra of (**a**) cytochrome-C (equine), MW_{ave} 12,360.1 (expected); (**b**) apomyoglobin (equine), MW_{ave} 16, 951.6 (expected); and (**c**) 8 pmol serum albumin (bovine), MW_{ave} 66,430.0 (expected) using a QIT. Spectra **b** and **c** are from capillary-LC-MS with the final gradient concentration of 98% acetonitrile, 1.9% water, and 0.1% acetic acid. (**d**) ESI LQIT (ThermoFisher LTQ™) spectrum of a protein with an expected MW of 116,718 Da (overexpressed in *E. coli*) run off of a polymer column shows 101 different charge states. Deconvolution algorithms were applied to the multiply charged spectra, and the deconvoluted plots are shown in the insets. Note that the serum albumin (bovine) sample in spectrum **c** is modified (as expected) as shown by the significant adduct tailing seen in the deconvoluted mass inset. (**a–c:** M. E. Bier, Carnegie Mellon University, 2000. **d:** With permission from Mark Hail, Novatia, LLC.)

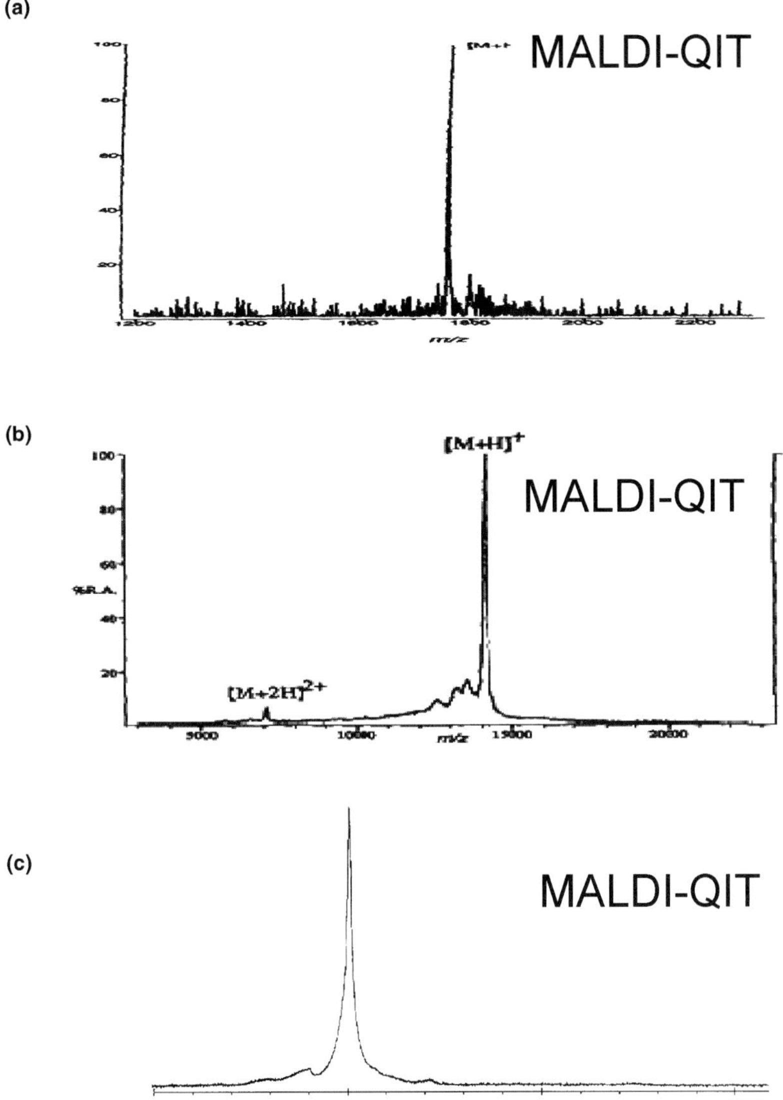

Figure 9.15. MALDI QIT MS from three different QITs. (**a**) Mass spectrum of 250 fmol of tetradecapeptide renin substrate in a 1:2000 ratio with 2,5-dihydroxybenzoic acid $[M + H]^+ = 1758.9$ (average) using external injection. The exclusion limit was 95 Da and resonant ejection occurred at a frequency of 89 kHz. (With permission from Jonscher, K.; Yates, J. *Rapid Commun. Mass Spectrom.* **1993**, *7*, 20–26.) (**b**) External-MALDI QIT mass spectrum of alpha-lactalbumin with a MW 14175, 3 pmol, 15 laser shots. (Reprinted with permission from Schwartz, J.; Bier, M. E. *Rapid Comm. in Mass Spectrom.* **1993**, 7, 27–32) (**c**) Internal–MALDI QIT mass spectrum of IgG (MW ~148,500Da) with 10 pmol on target. (Reprinted with permission from Schlunegger, U.; Caprioli, R. *Rapid Commun. Mass Spectrom.* **1999**, *13*, 1792–1796.)

9.6.2 Protein Folding

Many researchers have used ESI to examine protein folding with various analyzer types. Konermann and co-workers have published a nice series of papers in which they studied the denaturation of proteins in real-time. In one such work, Sogbein et al.[211] denatured holomyoglobin at high pH and examined this reaction over time. The holomyoglobin was in a water–methanol mixture at pH 11.2 and was monitored in negative-ion mode by ESI QMF MS. As shown in Figure 9.16, the heme group is separated from the apo group over time and the two forms apo- and holo- are represented in the mass spectra with the apo-form being more highly charged than the folded holo- form. Multiply charged holomyoglobin, which can retain the prosthetic heme group, has also been analyzed by CAD in a QIT.[212] Isolation of the +8 charge state between *m/z* 2190–2310 followed by resonance excitation yielded the heme group at *m/z* 616 and the +7 charge state of apomyoglobin. In the absence of resonance excitation, myoglobin was stored for 1 s at a pressure of 1 mtorr helium with no dissociation to heme and apomyoglobin species. This study suggests that other biologically important noncovalently bound complexes may be studied with an ion trap using the soft ionization process of ESI.

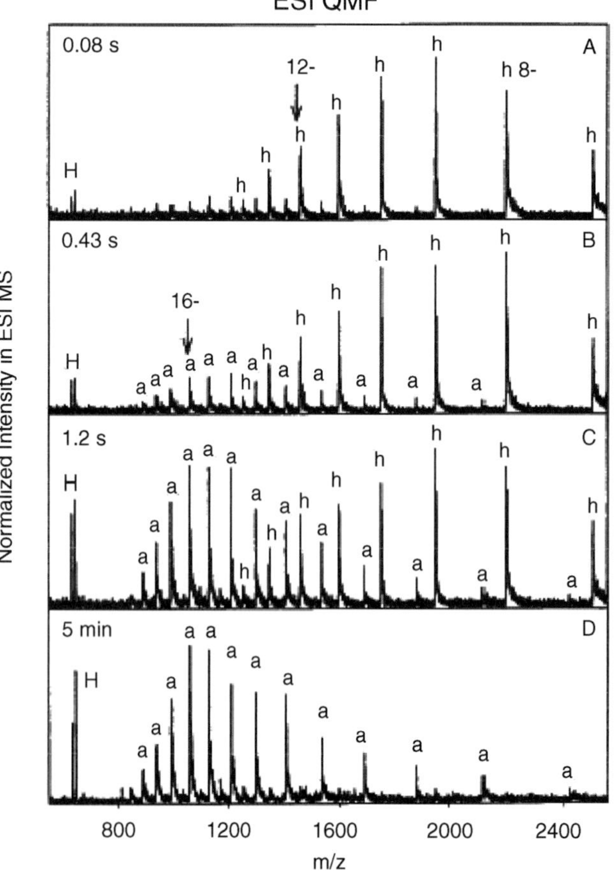

Figure 9.16. Denaturation of holomyoglobin (hMb) in water/methanol (75 : 25 v/v) at pH 11.2 monitored by time-resolved ESI MS in the negative-ion mode. These spectra were recorded 0.08 s (**A**), 0.43 s (**B**), 1.2 s (**C**), and 5 min (**D**) after initiation of denaturation. Notation: h is hMb; a is apomyoglobin (aMb), H is heme. Also indicated are the charge states of some protein ions. (Reprinted with permission from Sogbein, O.; Simmons, D. A.; Konermann, L. J. *Am. Soc. Mass Spectrom.* **2000**, *11*, 312–319.)

9.6.3 Analyses of Oligonucleotides

Prior to 1995, few QIT instruments were configured with negative ion formation and detection capabilities.[213–218] Now with MALDI and ESI coupled to many QITs, LQITs, and hybrid traps, there has been a renewed interest in biomolecular ions that are negatively charged in solution, but could not be ionized intact by most other methods. Oligonucleotides are one class of biomolecules that are readily ionized by ESI in the negative-ion mode because of the ability of the phosphodiester linkages and phosphate groups to stabilize a negative charge.[219] Sodium and other metal counterions are commonly found bound to these sites forming salts with the negative-ion formula $(M - nNa^+)^{-n}$ and $(M - nNa^+ + mH^+)^{-(n\text{-}m)}$.[220] Figure 9.17a shows an ESI mass spectrum of 5′-dTAGTCTAG-3′ with little evidence of bound alkali counterions because the oligonucleotide was desalted prior to analysis by an ammonium acetate/ethanol precipitation. The desalted oligonucleotide was infused in a 50 : 50, water:isopropanol solution at 3 μL/min. The charge state, and therefore the molecular weight, may be determined by the ($\Delta m/z$) spacing of the −2 anion isotopes as shown in the higher-resolution inset. Datta et al.[221] formed interesting four stranded quadruplex complexes using a guanine-rich PNA dodecamer having the sequence $H\text{-}G_4T_4G_4\text{-}Lys\text{-}NH_2$ (G_4-PNA) and examined these forms to be both dimeric and tetrameric four-stranded quadruplexes by ESI QIT.

ESI work on large duplex DNA has an exciting potential application for diagnosing genetic diseases. In 1995, Doktycz et al.[222] showed ESI/QIT mass spectra from 72- and 75-base single-stranded oligonucleotides and a duplex DNA consisting of complementary strands with 72 base pairs. Their ESI mass spectrum was acquired from a 20 μM duplex DNA solution containing 10 mM ammonium acetate. The average charge on this duplex DNA was determined to be −31. Although analytes electrosprayed with high salt concentrations in aqueous solutions show reduced sensitivity, their data still demonstrated the utility of an ESI/QIT instrument for the analysis of large double-stranded DNA. A 76-mer oligonucleotide was electrosprayed and analyzed by Smith et al.[223] using a QMF. Figure 9.17b shows the negative-ion ESI mass spectrum from a 120-mer oligonucleotide of MW 37,031-Da run on a LQIT by Hail and co-workers at Novatia LLC[224]. This is an excellent example of how sample clean-up can produce impressive results. The sample was first desalted at 1 mL/min on 1- × 10-mm C18 trap, washed for a short period at a high flow rate, and eluted off with 60/40 A/B at 0.2 mL/min, where A contained water and B contained 90% methanol. Both A and B solvents contained 0.75% hexafluoroisopropanol, 0.0375% triethylamine, and 10 μM ethylenediaminetetraacetic acid (EDTA) (pH 7.3 for solvent A), thereby forming an ion-pairing mobile phase, adapted from the work of Apffel et al.[225] A deconvoluted spectrum[210] gave a MW of 37,031 Da as predicted. In general, they found that an ESI QMF or an ESI LQIT allowed for superior mass accuracy over MALDI TOF MS at high mass.[224]

9.6.4 Ion Activation

9.6.4.1 Variations of the Collisional Activated Dissociation Theme

In sections covering the MS/MS scan, we discussed collisional-activated dissociation (CAD) in a triple quadrupole and in a quadrupole ion traps instruments. Other ion trap CAD schemes have been developed to increase energy deposition and/or the degree of fragmentation. For example, a characteristic of using single-frequency resonance excitation is that when the product ions are formed, they are not activated because they have a

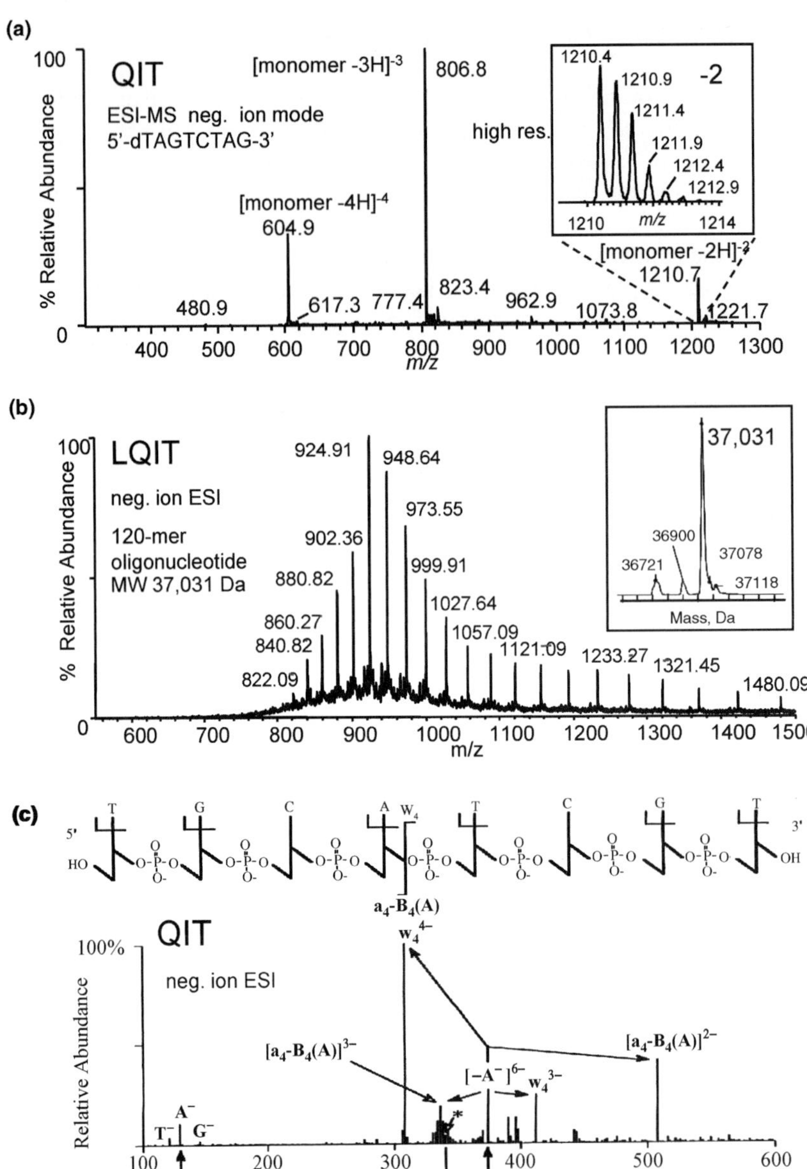

Figure 9.17. (**a**) ESI/QIT MS of 5′-TAGTCTAG-3′ under normal resolution conditions showing only minor alkali ion addition due to the use of a desalting procedure prior to analysis. The inset shows the high-resolution mass spectrum of the −2 anion. (D. C. Gale, Finnigan Corporation 1996. Sample provided by R. Griffey, ISIS Pharmaceuticals, Inc.) (**b**) A 120-mer oligonucleotide with a MW 37,031 Da acquired on a LQIT. The sample was first desalted on a 1 × 10-mm C_{18} trap. A deconvoluted spectrum gave a MW of 37,031 Da. (Reprinted with permission from M. Hail, Novatia, LLC.) (**c**) ESI QIT MS/MS spectrum of $(M - 7H^+)^{-7}$ ion, where M = 5′-d(TGCATCGT)-3′. The asterisk (*) indicates the *m/z* of the precursor ion. Closed arrowheads indicate the genealogy of some complementary pairs of fragments. (Reprinted with permission from McLuckey, S. A.; Habibi-Goudarzi, S. *J. Am. Chem. Soc.* **1993**, *115*, 12085. Copyright 1993 American Chemical Society.)

resonance frequency that is different from the applied frequency (i.e., they are stored at a different β). Broadband excitation is a variation of the resonance excitation method that can be used to activate these product ions. In this case, a broadband of frequencies is applied to the end caps to excite the precursor ion as well as any product ions formed from the first CAD step.[149] A low-frequency activation technique has also been studied in ion traps by applying a square wave to the end caps.[226] This activation process is somewhat similar to the nonresonance collision process in the multipole collision cell used in a triple quadrupole mass spectrometer, and the process has shown the ability to produce high-energy deposition. Qin and Chait[227] used a red-shifted, off-resonance excitation technique[227] where they applied a large excitation amplitude (21 V_{pp}) and observed efficient fragmentation of the singly charged protonated peptide, substance P, into structurally significant b_3 through b_{11} product ions. The high-q excitation method (PDQ) was also described previously in the MS/MS scan mode section.

9.6.4.2 MS/MS of Oligonucleotides

Tandem mass spectrometry of multiply charged oligonucleotide anions by ESI was demonstrated in a QIT by McLuckey et al.[220,228] These researchers proposed a nomenclature for different oligonucleotide fragment types analogous to the one developed for peptides.[229] Sequencing of small oligonucleotides was possible using MS/MS, but MS^n was recommended for larger oligomers. An example of the product ions observed from collisional activation of an oligonucleotide is shown in Figure 9.17c.[230] The oligomer (M), where M = 5′-d(TGCATCGT)-3′, was electrosprayed to form $(M-7H^+)^{-7}$, isolated, and then resonantly excited for 20–40 ms to form product ions. The major route of decomposition involved the loss of an adenine anion (A^-) as was seen with other adenine containing oligonucleotides in their work. Complementary fragment ions can be located in the MS/MS spectrum to aid in fragment identification. For example, $A^-/(M-7H^+-A^-)^{-6}$ are complementary ions produced from the activated precursor, while $w_4^{-4}/(a_4-B_4(A))^{-2}$ and $w_4^{-3}/(a_4-B_4(A))^{-3}$ are complementary ions of activated $(M-7H^+-A^-)^{-6}$.

Duplex DNA has also been collisionally activated in a QIT. The precursor ion, a duplex decamer consisting of 5′-d(ACATTCTGGC)-3′ and 5′-d(GCCAGAATGT)-3′, was isolated at low resolution and resonantly excited.[231] Single-strand fragments from the excitation of the −5 duplex were formed. The data demonstrated that a peak resonantly excited in a large mass window could indeed be identified as a duplex ion by noting the complementary product ions formed from the dissociation of the −5 precursor duplex DNA. These data suggest that an ion trap with high ion capacity and improved resolution may be an ideal instrument for elucidating structural changes on DNA, such as sites of alkylation, by using MS^n.

9.6.4.3 Photoinduced Dissociation

Most ion activation analyses with the QIT and LQIT have used CAD; however, photoinduced dissociation (PID) has shown promise as a means for high internal energy deposition. In addition, it is also a means to avoid the low-mass cutoff (LMCO) limitation since the precursor ion does not have to reside at a relatively high q for excitation as in the CAD process. PID has been demonstrated in ion cyclotron resonance (ICR)[232–234] as well as in the QIT.[235–238] Louris et al.[239] used a fiber optic to introduce light from an Nd:YAG laser into the QIT chamber to photodissociate protonated benzaldehyde, butylbenzene, and perfluoropropylene.[239] The spectra collected in this study suggested that PID deposits more energy than CAD. Stephenson et al.[240–242] built an ESI/QIT which incorporated PID. This instrument included mirrored light optics on the ring electrode. Their design focuses the laser light into a 0.3-cm^2 aperture of the ring electrode and allows for eight passes of the laser beam. The longer path length improves the PID efficiencies. A comparison was made between CAD and PID product spectra acquired in a QIT from the trisaccharide raffinose using a pulsed CO_2 laser with a maximum energy of 1.1 J. Both the CAD and the PID spectra showed the loss of a single monosaccharide at m/z 343 from the protonated molecular ion at m/z 505, and the PID spectrum showed more extensive fragmentation leading to the formation of the protonated monosaccharide ion at m/z 164. The CAD and PID data were collected with the precursor ion at a $q_{z\text{-excite}} = 0.1$,[243] but dissociation was more extensive in the PID experiment, apparently because of the higher internal energy deposited. Energy deposition at low $q_{z\text{-excite}}$ resonance CAD is limited because ion ejection of the precursor ion will result before dissociation if the resonance excitation amplitude is excessive. PID, however, appears to offer higher activation at lower $q_{z\text{-excite}}$ values, which allows lower m/z fragments to be observed. Extensive fragmentation from photodissociation has also been demonstrated in the activation of angiotensin I. In addition to the amount of energy deposited, PID offers a narrow energy distribution of activation that should allow for more selective fragmentation.

A drawback to the PID technique is that it requires the added expense and maintenance of the laser and light optics. In addition, in the QIT, the ring electrode may require modifications. A limitation of the PID process is that the energy deposition is dependent on the molecular structure. The latter problem is becoming less restrictive, however, because more powerful lasers with wider wavelengths of excitation are now available. Tae-Young Kim et al.[244] have explored using atmospheric MALDI PID for peptides at 157 nm in an LQIT and compared the fragmentation efficiency and low-mass cutoff (LMCO) results to data acquired with CAD by resonance excitation. The laser beam was focused axially through the rear of the LQIT. Figure 9.18 shows the LQIT PID of the peptide FSWGAEFQR by PID in panal 9.18a and by CAD in 9.18b. Although the spectrum was blown up 5-fold more for the PID spectrum, it is fragment-rich and allows for sequencing of the peptide. The PID technique does not have the same LMCO disadvantage as CAD on an XQIT, which is particularly critical for MALDI since this ionization technique produces primarily the $+1$ charge state for peptides and proteins.

9.6.4.4 IRMPD of Oligonucleotides

Several researchers have investigated infrared multiphoton photodissociation (IRMPD) inside a QIT.[241,245–254] Recently, Wilson and Brodbelt[255] showed an infrared multiphoton photodissociation (IRMPD) spectrum of Duplex DNA by ESI QIT MS/MS. They excited the precursor, which was a complex of the duplex d(GCGGGGATGGGGCG)/(CGCCCCATCCCCGC) $[M - 5H^+]^{-5}$ and the drug actinomycin. Figure 9.19 shows their results in a comparison between CAD and IRMPD where part a shows CAD at 0.86 V for 30 ms and part b shows IRMPD at 50 W for 1.5 ms with 73% and 6% of the product ions coming from uninformative base losses, respectively. Fragment labels were given a subscript G or C if they pertained specifically to the G-rich or C-rich strand of the duplex. Note that the IRMPD product spectrum appears far more fragment-rich than the CAD spectrum and includes ions below the CAD LMCO.

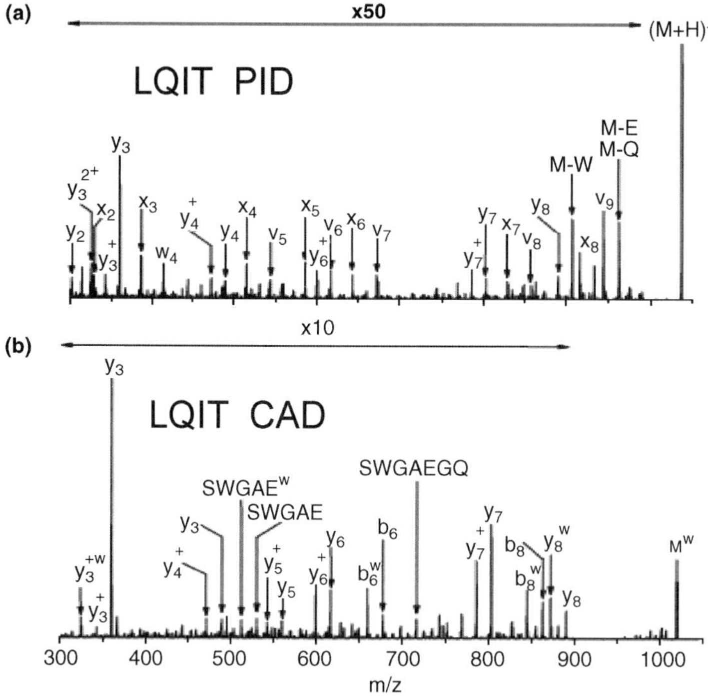

Figure 9.18. MALDI LQIT MS^2 spectra of singly charged FSWGAEGQR obtained (**a**) by 157-nm photodissociation and (**b**) by CID. Ions arising from loss of NH_3, H_2O are labeled with superscripts of *w, respectively, and loss of amino acid side chains from $[M + H]^+$ is denoted as the single-letter code of the amino acid involved with a minus sign. Internal fragment ions are labeled with their sequences. (Reprinted with permission Kim, T.-Y.; Thompson, M. S.; Reilly, J. P. *Rapid Commun. Mass Spectrom.* **2005**, *19*, 1657–1665.)

9.6.4.5 Ion Reactions in XQITs

A plethora of experiments have been done using ion/ion and ion–molecule reactions in ion traps since 1995.[256–265] McLuckey and Stephenson[266] have written a review of ion/ion chemistry that covers high-mass and highly charged ions in QMFs and QITs. Many of these reactions can be used to reduce ion charge, study adduction, determine charge state, and produce extensive fragmentation similar to ECD. In addition to the high-resolution method demonstrated in the peptide section above, the next sections give interesting examples of charge-state determinations by the use of ion/molecule reactions and other ion/electron and ion/ion reactions at higher charge states.

a. Ion–Molecule Reactions and Charge-State Determinations. Macromolecules have broad isotopic distributions which increase with mass. McLuckey and Goeringer[267] showed that ion traps could be used to determine the charge states of highly charged positive or negative macromolecules by using ion–molecule reactions. Real samples are also never pure, both in the mixture of counterions that can be present and in the macromolecular heterogeneity. This can result in an overlap of adjacent *m/z* distributions. By isolating a narrow window of the broad overlapping distributions and then allowing a gas-phase reagent to react with these ions, the charge state can readily be determined. This was demonstrated using *E. coli* t-RNA which was electrosprayed into a

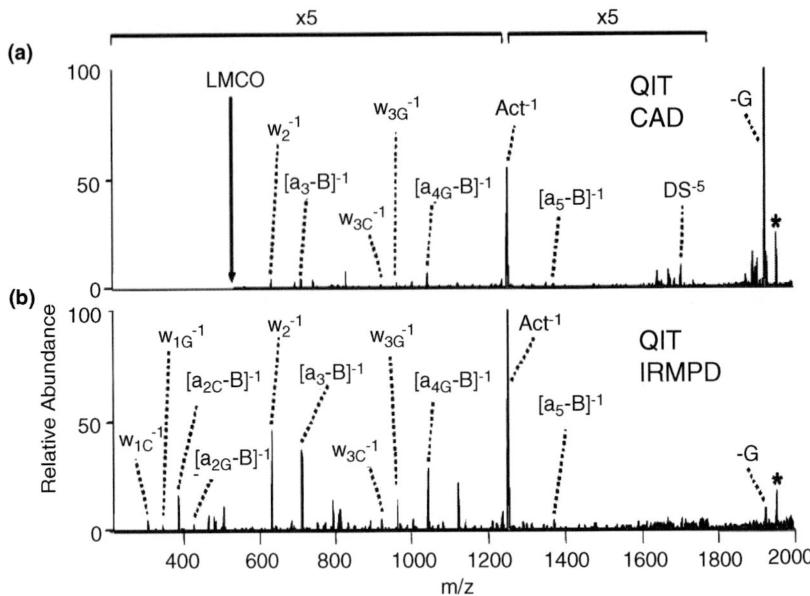

Figure 9.19. QIT ESI-MS/MS spectra of the $[M - 5H]^{5-}$ complex containing duplex d(GCGGGGATGGGGCG)/(CGCCCCATCCCCGC) and actinomycin by (**a**) CAD (0.86 V, 30 ms) (73%) and (**b**) IRMPD (50 W, 1.5 ms) (6%) with the portion of uninformative base loss ions given in parentheses. Fragment labels include a subscript G or C if they pertain specifically to the G-rich or C-rich strand of the duplex, respectively. The magnification scale applies to both spectra. Precursor ion is denoted with an asterisk (*). (Reprinted with permission from Wilson, J. J.; Brodbelt, J. S. *Anal. Chem.* **2007**, *79*, 2067.)

QIT, and a narrow mass range was selected from the anions. The narrow mass range was next reacted with trifluoroacetic acid for 200 ms. The trifluoroacetic acid transfers protons to the isolated distribution and two higher *m/z* distributions were observed. Assuming that the newly formed product distributions at higher *m/z* differs by one negative charge and in mass by 1 Da due to the addition of a proton [i.e., $m/z = (m + 1)/|z + 1|$], the charge of the isolated precursor can be determined.

Similarly, an ion/molecule reaction procedure can be used to determine the charge state of CAD products.[268] The $(M + 4H)^{+4}$ cation of melittin was mass-selected and resonantly excited to produce the $y_{13}{}^{+3}$ at *m/z* 542. Next, this ion was isolated for a MS^3 reaction step with the strong base 1,6-diaminohexane. The charge state of this ion was determined by the newly observed product ion formed from the single-proton transfer reaction producing $y_{13}{}^{+2}$ at *m/z* 812.

A known clustering reaction can also be used to determine charge states.[268] By reacting 1,6-diaminohexane (MW 116) with the $(M + 4H)^{+4}$ ion of bovine insulin, several cluster species incorporating one, two, and three bases are formed. When the precursor ion $(M + 4H)^{+4}$ was isolated and fragmented, a product ion at *m/z* 1430 could be isolated and reacted with 1,6-diaminohexane and observed in a MS^3 spectrum. Again one, two, and three bases are attached to this product ion. Given the assumption that two adjacent peaks differ by one base molecule, the charge state (*n*) for both species is determined to be + 4. To make this determination, the amount of increase in *m/z* of the adjacent peak was divided into m_B, the molecular weight of the base molecule as shown by equation $n = m_B/|(m/z)_1 - (m/z)_2|$.

b. Ion/Ion Reactions. Although there has been considerable effort directed at studying ion/molecule reactions in a QIT, only in the last decade has there been a recent surge in studies involving reactions between positive and negative ions and in the use of the LQIT. Quadrupole ion traps and Fourier transform ion cyclotron resonance mass spectrometers are unique in that both positive and negative ions can be stored simultaneously[269–272] prior to reactive annihilation or charge reduction. The polarity of the ions may also be selected in a quadrupole ion trap by adjusting the a_u value to an appropriate point in the stability diagram. The parameter a_u can be set to a calculated nonzero value by applying a DC potential to the ring electrode at some time prior to the mass-selective instability scan. Additionally, a second ion source can be used to make the oppositely charge ions.

Herron, Goeringer, and McLuckey[273] have electrosprayed and isolated triply and doubly charged anions from the single-stranded deoxynucleotide, 5′-d(AAAA)-3, and allowed the anions to undergo ion–ion proton transfer reactions with protonated pyridine cations. Pyridine vapor was introduced into the vacuum chamber at $1–3 \times 10^{-7}$ torr (uncorrected), and the protonated cations were produced by internally ionizing pyridine by electron ionization using a radially injected electron beam. This pressure of pyridine was sufficient to cause complete protonation of the nucleotide molecular ion through self-chemical ionization (self-CI)[274–276] in 10 ms. Next, the single-stranded DNA anions were injected into the QIT and simultaneously stored with the pyridine cations for up to one second during which time the ion–ion proton transfer reactions occurred. Figure 9.20a shows the product ions formed from the reaction of isolated $d(A_4)^{-3}$ with protonated pyridine. The $d(A_4)^{-3}$ ion undergoes ion–ion proton transfer with the protonated pyridine to form the doubly charged species $d(A_4 + H)^{-2}$. Figure 9.20b shows the result from further reacting the doubly charged product anion $d(A_4 + H)^{-2}$ from the first reaction with protonated pyridine to form the singly charged anion $d(A_4 + 2H)^{-1}$. Very little fragmentation was reported for this exothermic process. The Oak Ridge group has also shown ion-charge-state determinations in MS^3 experiments where pyridine reacts with product anions of 5′-d(AAAA)-3′ and the A-chain of bovine insulin.[277] In these examples, a reaction

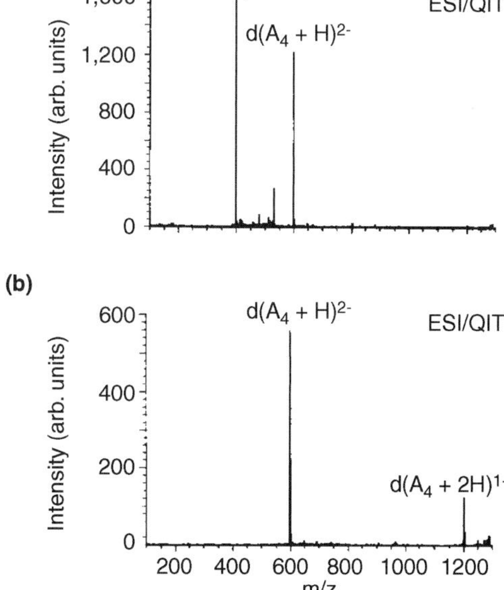

Figure 9.20. (**a**) MS/MS product ions formed from the reaction of isolated $d(A_4)^{-3}$ with protonated pyridine. The $d(A_4)^{-3}$ ion undergoes ion–ion proton transfer with the protonated pyridine to form the doubly charged species $d(A_4 + H)^{-2}$. (**b**) The result from reacting the doubly charged product anion $d(A_4 + H)^{-2}$ from the first reaction with the protonated pyridine to form the singly charged anion $d(A_4 + 2H)^{-1}$. (Reprinted with permission from Herron, W. J.; Goeringer, D. E.; McLuckey, S. A. *J. Am. Soc. Mass Spectrom.* **1995,** *6*, 529–532.)

time of 200 ms was used for the proton transfer reaction to proceed. The isolated product ions produced by CAD were further reacted with protonated pyridine to produce protonated MS^3 product ions. Again, by assuming a change in mass of 1 Da, the charge state was readily obtained.

One of the limitations of all three of the charge-state determination techniques discussed above is that the long ion trap reaction time necessary (200 ms) lengthens the overall scan time. As a result, fewer scans would be acquired across an eluting peak from an LC or CE column. However, reducing the mass range to two or three isotope distributions, can help shorten the scan time. An additional limitation of the ion/molecule charge state determination method is that it relies on the reactivity of the ion with a given neutral and all ions do not react to the same degree, if at all. Also, the neutral reactant is always present in the QIT chamber in the ion/molecule experiment unless a pulsed valve is used to introduce the gas only during the reaction period. The ion/ion reaction method is far more universal in reactivity and is more easily controlled,[277] but it does require the ionization and capturing of both positive and negative ions.

c. Electron Capture Dissociation (ECD). In 1998, Zubarev, Kelleher, and McLafferty introduced electron capture dissociation (ECD) as a new form of ion activation used in FT ICR ion trap mass analyzers.[278,279] ECD involves capturing an electron preferably by a multiply charge positive ion. ECD can be used for top-down protein analysis because of the high-energy deposition (~6 eV) that can allow for the extensive fragmentation of large peptides (>3 kDa) along the backbone in a nonergodic process.[279] Because ECD is highly exothermic and nonergodic, fragmentation is sequence-independent compared to CAD or IRMPD, and this makes it a useful tool for the structural elucidation of post-translation modifications (PTM).[278]

Since the reaction cross sections of electrons are small (10^{-15} m^2) and maximum at kinetic energies <1 eV, ECD experiments have primarily been carried out in FT ICR MS (FTMS) instruments because the magnetic field can trap low-energy electrons over long periods of time.[281–283] Because ECD causes fragmentation at c and z sites to allow for improved fragmentation of large peptides/proteins and easier interpretation of the MS/MS spectrum, it is also desirable to implement the process in XQITs. Unfortunately, heating of electrons in the RF field makes ECD in XQITs difficult.[284] Baba et al.[285] were initially not successful in observing ECD using a 3D Penning trap, but they have now succeeded in demonstrating ECD in a LQIT for the first time using an electron confining axial 50-mT magnetic field.[285] The electrons are maintained at <1eV by injecting them along the z axis of the LQIT and are confined by the magnetic field. They fragmented substance P over a 10-min period with a dissociation efficiency of 4% for the $c_4{}^+$, $c_5{}^+$, $c_6{}^+$, and $c_7{}^+$ fragments, which is comparable to the FT ICR case.

d. Electron Transfer Dissociation (ETD). Because it is difficult to provide thermal electrons for ECD in RF traps, an alternative ion/ion electron transfer dissociation (ETD) process was pursued to deliver an electron to a multiply charged cation by Hunt and co-workers.[286] The ETD process is highly exothermic by ~5 eV and nonergodic, so a similar fragmentation pattern to ECD is expected. They used an LQIT with an ESI source to make the positive protein cations, and a negative chemical ionization (CI) source with methane was coupled to the opposite side (rear) of the QLIT to inject anthracene anions, the electron donors. Their steps in the ETD process on the LQIT were as follows: (1) Trap injected precursor ions in the center section of the trap and then store them in the front section; (2) inject and trap anthracene anions in the center section; (3) isolate the precursor cations with charge ≥2 and anthracene anions with a 3-Da window; (4) allow the reaction to occur in the center section; and (5) mass analyze the products. The resulting spectrum contains c- and z-type fragments and b- and y-type fragments from electron transfer and proton abstraction reactions that can be followed by CAD if desired. Many of the unique advantages of the LQIT were used in this experiment (see Section 9.3.2.9, "Advantages of the LQIT over the QIT"). An example of an ETD/PTR/MS/MS of a 70S ribosomal subunit protein L32 is shown in Figure 9.21a for the intact protein form $[M + 12H]^{+12}$ at *m/z* 527.3 (MW 6316). In this experiment, ETD is followed by a proton transfer/charge reduction (PTR) reaction with benzoate anions to reduce the charge state to $+1$ and $+2$ for sequence reading. Almost complete fragment coverage from c_3 to c_{17} and complete coverage from z_3 to z_{17} is shown for the normal form. Coon et al.[287] have also examined the ion/ion reactions of multiply deprotonated peptides with a xenon cation.

ETD has been implemented on a hybrid QMF/LQIT/LQIT mass spectrometer (modified Applied BioSystems/MDS Sciex 4000 QTRAP) by two related methods.[288] In method one, the cation analyte is stored in LQIT1 while reagent anions are continuously transmitted into the trapped cloud of cations. The newly formed cation products are then trapped. In method two, the reagent anions are trapped in LQIT1 and analyte cations are transmitted into the trap and the products are either collected or transmitted into the second LQIT2 for mass analysis. Both methods use a pulsed dual ion source: (a) APCI for the production of azobenzene radical anions and (b) ESI for the production of multiply charged peptides. This "transmission mode" ETD does not require the superposition of RF to the containment lenses of the LQIT. The scan function for method two was as follows: (1) ESI positive-ion injection and trapping in LQIT1 (15 ms), (2) anion mass filtering in QMF1 and then transmission of a specific reactant through LQIT1 (80 ms), and (3) transfer of ion/ion reaction products to LQIT2 for mass analysis (50 ms). Figure 9.21b shows the resulting ETD MS/MS spectrum of $[M + H]^{+3}$ of the phosphopeptide TRDIpYETDYYRK and product ions trapped in LQIT2 after reacting with azobenzene radical anions for 100 ms. The resulting c- and z-type product ions from every inter-residue bond, except the bond between two tyrosines, were observed. The site of phosphorylation was determined by c-type fragment ions (c_4–c_5) and z-type fragment ions (z_7–z_8) that are 80 mass units higher than the corresponding unmodified peptide (not shown).

Finally, ETD has also been implemented on a modified hybrid LQIT/OT where the orbitrap serves as the high-resolution ($R \sim 60{,}000$) mass analyzer,[289] while the LQIT serves as a high-charge-capacity reactor. An RF trapping voltage was applied to the end lenses of the LQIT to trap both polarities of ions, simultaneously. A dual ESI source was used for cation production of a multiply charged peptide and 9-anthracenecarboxylic acid which undergoes CAD in the LQIT to produce the ETD reactive decarboxylated anion. Figure 9.21c shows an ETD MS/MS spectrum from 10 averaged scans of ATCH peptide (SYSMEHFRWGKPVGKK-RRPVKVYP) reaction product which was mass-analyzed on the orbitrap. Again a rich fragmentation series of c- and z-type product ions are observed. Two drawbacks to their method were the longer cycle times and the lower ETD efficiency found compared to "conventional" ETD instrumentation.

9.6.5 Separation Techniques Coupled to Mass Analyzers

9.6.5.1 Capillary LC for Peptide Sequencing

A mixture of peptides can be sequenced by first separating them using LC or by direct mixture analysis and then analyzing the individual peptides by ESI/MS^n. The first on-line

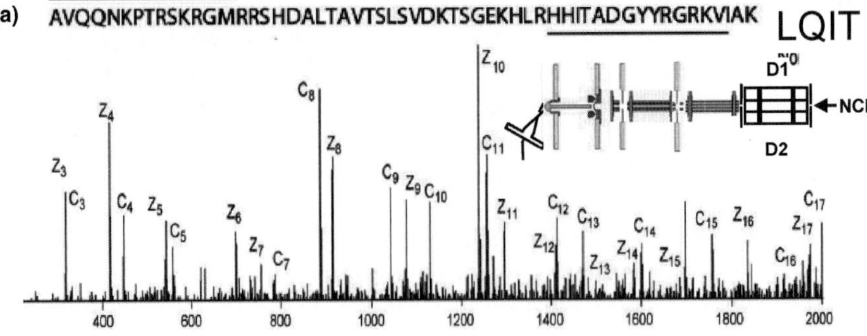

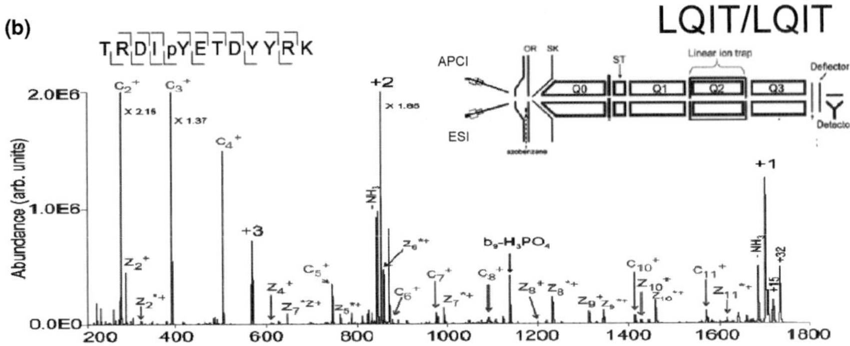

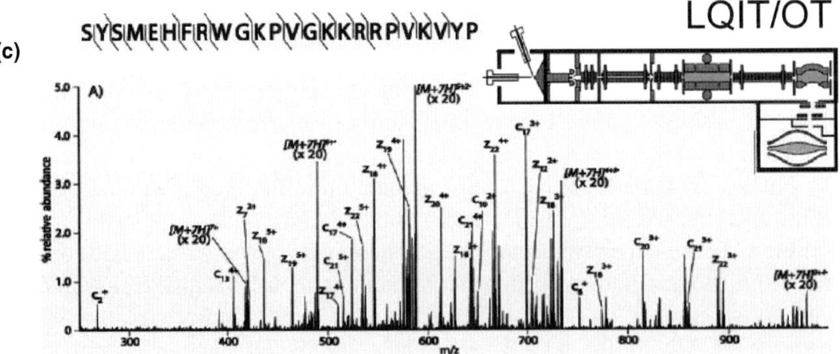

Figure 9.21. (**a**) ETD/PTR-MS/MS spectra recorded on the *E. coli* 70S ribosomal protein, L32. Lines above and below the sequence represent the coverage defined by ions of types c and z. (Reprinted with permission from Chi, A.; Bai, D. L.; Geer, L. Y.; Shabanowitz, J.; Hunt. D. F. *Int. J. Mass Spectrom.* **2007**, *259*, 197–203.) (**b**) Mass spectrum derived from transmission mode ion/ion electron-transfer reaction of triply protonated phosphopeptide TRDIpYETDYYRK trapped in Q2 LQIT while passing azobenzene radical anions through it for 100 ms. (Reprinted with permission from Liang, X.; Hager, J. W.; McLuckey, S. A. *Anal. Chem.* **2007**, *79*, 3363–3370.) (**c**) ETD spectra of the ATCH peptide performed using orbitrap mass analysis. (Reprinted with permission from McAlister, G. C.; Phanstiel, D.; Good, D. M.; Berggren, W. T.; Coon J. J. *Anal. Chem.* **2007**, *79*, 3525–3534.)

coupling of a 1-mm-i.d. LC column to an ESI/QIT showed useful ion currents at the 2.5 pmol level of injected tryptic peptides and at the 300 fmol level for human serum albumin (MW_{avg} 66 kDa).[290] Since then, countless labs have coupled capillary LC to a ESI/ion traps for peptide sequencing[161,163] using the short column capillary LC technique adapted from Kennedy and Jorgenson.[291] In using this method, a capillary column (100-μm i.d. and 15 cm long) was packed with 10-μm beads of POROS R2™ (Applied Biosystems,[122] Framingham, MA) and coupled to ESI in a way that is similar to the diagram in Figure 9.22a. This configuration reduces dead volume and thus avoids peak broadening. In one experiment, a standard mixture of five angiotensins (Michrom BioResources, Inc., Auburn, CA) was hydrostatically loaded onto the head of the column and separated within a 15-min gradient elution. The flow rate (200 μL/min) from a Michrom micro-HPLC pump (Auburn, CA) was split 200 : 1 and delivered to the capillary column. Figure 9.22b shows the MS/MS product ion spectrum of the isolated doubly charged ion (442.3^{+2}) and its isotopes, from 10 fmol of the peptide RVYVHPI ($MW_{monoisotopic} = 882.6$ Da). At the 10-fmol level, ions can readily be assigned as the y''^{+1}_{2}, b_5^{+2},a_7^{+2}, b_4^{+1}, a_5^{+1}, b_5^{+1}, and b_5^{+1} product ions. Similar data have been collected using triple-quadrupole mass spectrometers (QqQ)[292]; however, a decrease in the resolution of the quadrupoles to give 2- to 3-Da wide peaks was required to achieve the necessary 10-fold improvement in signal-to-noise level. The ESI/QIT MS/MS data in figure 9.22b were collected at unit mass resolution. The detection limit of peptides can be improved by operating the QIT using rapid scan rates.[60,122] Figure 9.22c shows a MS/MS spectrum of 550 amol of RVYVHPI analyzed in a QIT at 16× the normal scan rate of 5555 Th s^{-1}. The faster scan rate of 88,880 Th s^{-1} improved the signal height by ~6×, and additional scans were collected for improved signal averaging. Haskin et al.[293] also used an improved capillary LC setup on a QIT to achieve a 4-amol detection limit, and they discovered endogenous peptides from rats. They used a fused-silica capillary LC column (25-μm i.d.) with 3-μm particles and formed a polymer frit cured by UV light. Flow rates during the mass analysis were 20 nL/min, and the sample loop volume was 1.8 μL. Figure 9.22d shows their MS^2 spectrum with excellent S/N of 600 pM (1 femtomole loaded) of the synthetic peptide SPQLEDEAKE which was used to confirm the sequence of an *in vivo* analysis.

9.6.5.2 Capillary Electrophoresis

Capillary electrophoresis (CE) is an ion mobility separation technique that is rapid and efficient at resolving biomolecules such as peptides and proteins [294–298] as well as pharmaceutical analytes. Coupling CE to a mass spectrometer is desirable due to the ability of CE to separate nonvolatile analytes rapidly and the ability of an ESI source to ionize them; however, an incompatibility exists. Often CE separations are done with buffer systems that can cause poor ESI performance, for example, phosphate or borate buffer systems should be avoided and the atmospheric-pressure photoionization technique should be considered.[299] CE has also been coupled to MALDI in an off-line approach.[300] A review and special issue of CE/MS is available for further reading.[301–303]

CE/ESI was first done in 1987 by Olivares and co-workers in Smith's group at Pacific Northwest Laboratory.[304] In their work they used an ESI source coupled to a QMF which was manufactured by Extrel Corp. (Pittsburgh, PA). They reported excellent detection capabilities for quaternary ammonium salts at 14–17 fmol injected and even observed these analytes at the level of 1 fmol injected. Since then, several other laboratories have coupled CE to a ESI/QIT[305–307] and attomole detection limits have been achieved, although often

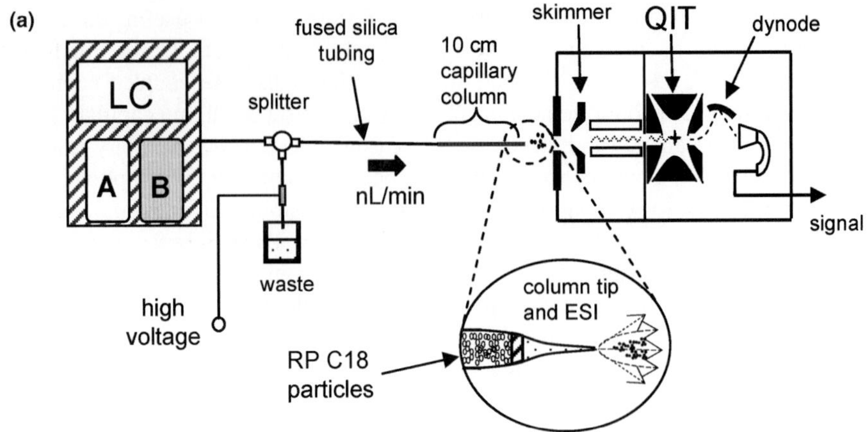
(a)
LC
A
B
splitter
fused silica tubing
nL/min
10 cm capillary column
skimmer
QIT
dynode
signal
waste
high voltage
column tip and ESI
RP C18 particles

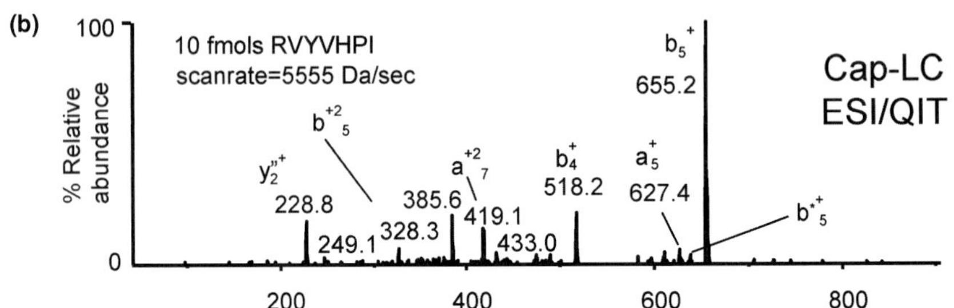
(b)
10 fmols RVYVHPI
scanrate=5555 Da/sec
Cap-LC
ESI/QIT
% Relative abundance
100
0
228.8
249.1
328.3
385.6
419.1
433.0
518.2
627.4
655.2
200
400
600
800

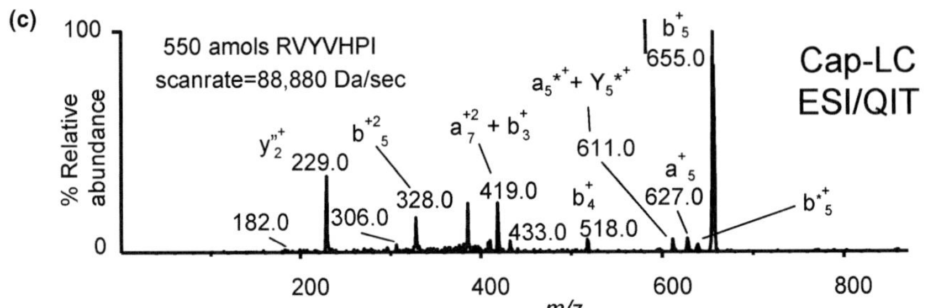
(c)
550 amols RVYVHPI
scanrate=88,880 Da/sec
Cap-LC
ESI/QIT
% Relative abundance
100
0
182.0
229.0
306.0
328.0
419.0
433.0
518.0
611.0
627.0
655.0
200
400
600
800
m/z

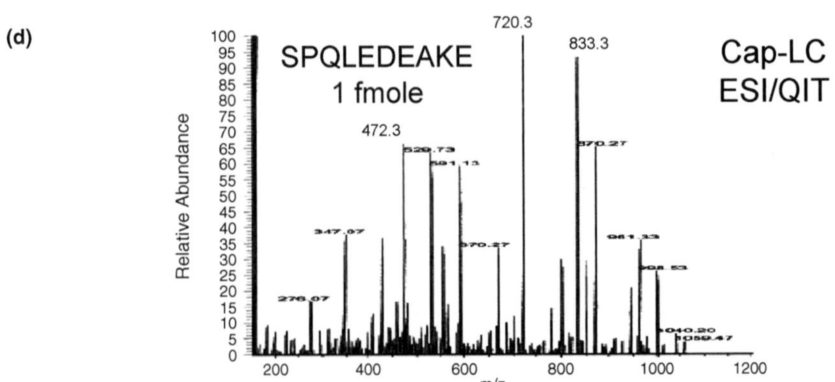
(d)
SPQLEDEAKE
1 fmole
Cap-LC
ESI/QIT
Relative Abundance
720.3
833.3
472.3
200
400
600
800
1000
1200
m/z

with concentrated samples. Moseley et al.[307] has recommended a special amino-propyl-silated fused silica column so that the separations can take place at an acidic pH for optimum ESI.

Henion et al.[306] reported the first use of CE/ESI/QIT to quantitatively determine the amounts of the isoquinoline alkaloids berberine and palmatine in the bark of *Phellodendron wilsonii*. Dried *P. wilsonii* tree bark is used in China and other countries to treat dysentery and jaundice.[309] Relative migration times and ESI source CAD was used to identify these compounds. Signal-to-noise ratios greater than 10 were demonstrated for mass chromatograms of a synthetic mixture of these compounds at the 400- to 500-atmol level. At the picogram level, the calibration curves for berberine and palmatine were linear. Berberine and palmatine were determined in the bark at 13.95 μg/mg and 0.234 μg/mg, respectively.

The attainment of mid-attomole detection limits for the peptide leucine-enkephalin in total ion electropherograms using 20-μm-i.d. capillaries without using a liquid sheath has been achieved by Ramsey et al.[310] The 20-μm-i.d. capillaries were tapered down to 3-μm i.d. to form the ESI tip and coated with gold to make the electrical connection. A 650-atmol injection resulted in a signal-to-noise ratio of 6:1 in the total ion electropherogram (TIE) and 25 : 1 in the reconstructed mass electopherogram of $(M + H)^+$ at m/z 556. Overall, the non-sheath capillary ESI interface showed a gain in signal of 3–4 over the sheath flow ESI source.

9.6.6 Limits of Detection

Detection limits of analytes measured in ESI and MALDI using the mass analyzers covered in this chapter are quite noteworthy. Table 9.3 lists detection limits for various molecules analyzed by LC, CE, and FIA coupled to ESI QIT, LQIT, and the orbitrap or with the use of MALDI. The lowest limits of detection are measured when an isolation step is used to accumulate and/or isolate only the ion(s) of interest. For example, in Table 9.3, the detection limits for reserpine in SIM and SRM scan modes are better than in the full-scan mode. Although ion trap full MS scans have a higher duty cycle over QMF ion beam instruments, detection limits can be reduced in the full-scan MS mode due to matrix effects. In this latter case, at low analyte amounts, the QIT is filled up primarily with unwanted matrix ions, leaving little space to store the ions of interest. The advantage of the high-charge-capacity

Figure 9.22. (**a**) Capillary liquid chromatography electrospray ionization mass spectrometry (cap-LC-ESI-MS) used to concentrate, separate and analyze femtomole to attomole levels of peptides and proteins. The packed capillary column length is typically 2–10 cm long and the column flow rate is maintained at 20–500 nL/min. The inset shows the electrospray tip formed at the end of a 100-μm-i.d. fused silica capillary column. (**b**) Capillary-LC ESI/MS/MS spectra of the doubly charged peptide RVYVHPI ($MW_{monoiso}$ 882.6 Da) at m/z 442.3 loaded on a 100-μm-i.d. capillary column. Spectrum **b** shows the results at 10 fmol using a scan rate of 5555 Da s^{-1}, while **c** shows approximately the same S/N ratio at 550 amol using a scan rate of 88,880 Da s^{-1}, but with reduced mass resolution. (From Ref. 122) (**d**) MS^2 spectrum using an ESI/QIT of 600 pM (1 fmol) of the synthetic peptide SPQLEDEAKE which was used to confirm the sequence of an *in vivo* analysis. The LC consisted of a fused-silica capillary columns (25-μm i.d.) with 3-μm particles and homemade polymer frit cured by UV light. Flow rates were 20 nL/min and the sample loop volume was 1.8 μL. (Reprinted with permission from Haskins, W. E.; Wang, Z.; Watson, C. J.; Rostand, R. R.; Witowski, S. R.; Powell, D. H.; Kennedy, R. T. *Anal. Chem.* **2001**, *73*, 5005–5014.)

ion traps, whether of the 2D or 3D type, is clear in this example, that is, more of the analyte ions can be trapped. It should be noted, however, that just because the current LQIT has a high charge capacity, it does not necessarily mean that it will perform better than a high-charge-capacity QIT if the samples are not identical. A 10× purer sample run on the QIT would most likely have a lower detection limit than the impure sample run on a commercial LQIT. In any case, detection limits are outstanding in the full-scan MS/MS mode for the ion traps because the duty cycle is high and the effects of the matrix are reduced during isolation.

In review of the data in Table 9.3, it is clear that MS is a sensitive technique; and with a reasonable introduction technique, whether at atmospheric pressure (cap-LC/ESI, AMALDI) or not (vacuum-MALDI), the ion source mass analyzer combinations mentioned in this chapter offer single-digit femtomole detection in all but the most difficult samples; with special separation techniques, low-attomole detection levels are achievable.

9.6.7 Automated Data-Dependent Scanning

A data-dependent scan utilizes an algorithm to change the acquisition mode or system parameters in real-time, based on the current set of MS data. Essentially, the data system executes the next scan type based on previously collected data.[311,312] For example, a MS system can be programmed to implement the following experiment: (i) Acquire a full-scan mass spectrum; (ii) select a peak from the mass spectrum that meets a preset signal threshold, if no peak meets this criteria go back to (i); (iii) acquire a MS/MS scan with the appropriate isolation width and product ion mass range for that precursor ion; and (iv) acquire an MS^3 scan with the appropriate product mass range for that precursor and a selected product ion. Alternatively, with a hybrid mass analyzer instrument, step iii could read, isolate, and collisionally activate the precursor ion and collect the fragments in an LQIT and send the fragments to an orbitrap or FT ICR for high-resolution analysis.[313] Data-dependent scans provide system automation to rapidly make changes that a human could not do in the same time frame. It can greatly increase the amount of information gained from a single LC analysis. Quadrupole ion traps have the advantage over ion-beam mass spectrometers because they can immediately switch between a full-scan MS and MS/MS in the presence of helium collision gas with no loss in signal, CAD efficiency, or time to switch gases on or off. Triple-stage QMFs would have to reduce the gas pressure in the collision cell to achieve a similar full-scan MS result. Data-dependent scanning on a QIT has been used to indentify proteins associated with hypertrophy of cardiomyocytes,[314] analyze digests from proteins separated by two-dimensional polyacrylamide gel electrophoresis,[315] identify proteins,[207] and identified phosphorylation sites of peptides by an MS/high-resolution/MS/MS sequence of scans.[316] In a real example of a data-dependent scan discussed above, a myoglobin tryptic digest was acquired by LC QIT MS. In Figure 9.23a, an MS scan of 10 pmol of an unknown digest constituent is shown. The charge state and thus fragment mass was determined for *m/z* 368.5 as shown in the inset by a high-resolution scan. The charge state was determined to be +2 and the mass range was set for the MS/MS scan shown in Figure 9.23b. This peptide was searched for and identified in a protein database by a MS/MS cross-correlation algorithm (SEQUEST)[317,318] and found to be the sub-sequence HKIPIK (MW 734) from horse heart myoglobin. Protein identification using MS^3 and MS^4 scans have also been used with successful results.[319]

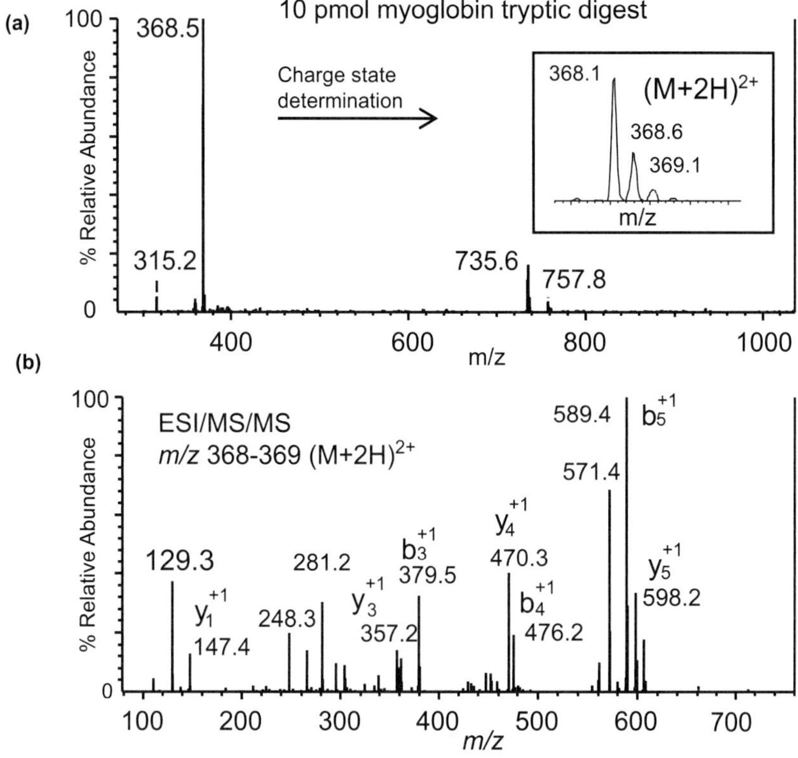

Figure 9.23. An example of a data-dependent scanning during an LC separation of a myoglobin tryptic digest where panel **a** shows a full scan MS of a component of the digest eluting from the column. This scan is followed by a charge state determination of the largest peak in the MS spectrum by a higher-resolution scan shown in the inset. Finally, an MS/MS scan of that ion is shown in panel **b** providing product ion information for automated protein identification. (A. Land, Finnigan Corporation.)

9.6.8 Diagnosis of Disease

There are several areas not covered in this chapter due to limited space that already have tremendous utility or are likely to become a major analytical method in the future. Twenty years ago the idea of using mass spectra from proteins or oligonucleotides to diagnose disease was not a reality. With the development of these higher-performance mass analyzers and separation techniques, we are now at the brink of making a major impact in medicine for early stage disease diagnosis. New biomarkers have been discovered and new MS methodologies have been invented in the last 10 years like isotope-coded affinity tag (ICAT),[320]isobaric tag for relative and absolute quantitation (iTRAQ),[321] and stable isotope labeling by amino acids in cell culture (SILAC)[322] to determine the relative abundances of different proteins from diseased versus normal cells that are helping to pave the way down this path. In addition, experiments that can lay down pure biological samples on surfaces[323] or lift biomarkers from surfaces using the techniques of desorption electrospray ionization (DESI),[38,324–326] direct real-time analysis (DART),[327] laser ablation electrospray ionization (LAESI),[328] and other desorption methodologies have the potential to add thrust to these efforts.

9.7 SUMMARY

Over 50 years have past since Wolfgang Paul's invention of the QMF and the QIT, and 30 years have past since Malcolm Dole's ESI experiments. Both mass analyzers and ESI and MALDI have improved over the years and we now add the LQIT and orbitrap to our mass analyzer family. Together these components can be assembled into remarkably high performance single analyzers or hybrid mass spectrometers. The marriage of ESI to a mass spectrometer has made the coupling of LC and CE separation techniques routine. The analyst can readily employ an assortment of scan types (e.g., MS^n, SIM, SRM^n) for real-time detailed structural and/or quantitative analyses. A major advancement has been the increase in ion charge capacity of $\sim 10\times$ by electrode modifications/resonance techniques in the QIT and of $\sim 40\times$ by larger ion trapping volumes such as in the linear quadrupole ion trap. Resolution and scan rates have improved and the orbitrap can achieve a resolution of over 100,000 for lower-*m*/*z* ions. Detection limits determined with these instruments are exemplary, especially when coupled to nano-flow capillary separation techniques. We have now shifted from femtomole detection levels of a decade ago into the low-attomole levels. The QMF has been a workhorse since it was introduced commercially in the early 1960s, and it is widely used today in single-quadrupole or triple-quadrupole mass spectrometers. The QMF has also found a new vocation as the front-end analyzer for several hybrid instruments such as the QTOF. Quadrupole ion traps can be used as extremely useful electrostatic reaction vessels for electrosprayed ion/molecule and ion/ion reactions, and this has facilitated the important ETD and proton transfer (PT) reactions especially on the LQIT. This area of study will no doubt see future growth and may lead to some enlightening biochemical discoveries. As the capabilities continue to expand in the hands of many more scientists, new niches of the modern mass spectrometer featuring these components will be discovered and the traditional ones should flourish. I can assure you that more inventions of both new instruments and methodologies will be revealed in the future to keep the innovations continuing well into the next decade. I hope that you get to share in them!

ACKNOWLEDGMENTS

I thank Drs. David Sipe and FMD for review of the manuscript. I thank Julia Sheehy for help with the references and permissions. I thank Dr. Mark Hail of Novatia Corp. for the unpublished protein and DNA data. I thank Dr. Alex Mordhai at Agilent Technologies Tech. for his helpful comments about their high-capacity QIT. I especially thank my Ph.D. advisor, Dr. Robert Graham Cooks and Dr. Jonathan Amy for their mentorship and friendship.

APPENDIX

Parameters for various commercial quadrupole field ion-trap mass spectrometers are listed in Table 9.A.

Table 9.A. Select Commercial Quadrupole Field Ion-Trap Mass Spectrometers

Product (Company)	QF	Ion Source	Ion Activity	Mass Range (m/z)	Scan Speed (m/z)	Resolution ($m/\Delta m$) at FWHM at m/z 1000)	Detection Limit(s)	Ion Charge Capacity	Mass Accuracy (ppm)
4000 QTrap (Applied Biosystems/MDS Sciex)	2D	ESI APCI APPI	CID	Model 4000: 50–2800 Model 3200: 50–1700	250 1000 4000	5000 2500 1600	~1-fmol peptide MS2 protein id.; 100-amol peptide quantitation	High[a]	~50 ppm 1000 (m/z)/sec
LTQ XL (Thermo-Fisher)	2D	ESI	CID	50–2000; 100–4000	1100.	20000	250 fg of reserpine at S/N 100:1 (2-μL injection) in MS/MS	High[a]	~50 ppm
		APPI APCI MALDI	ETD		16,700 125,000	5000 1100			
340 Ion Trap (Agilent Tech.)	3D	ESI APCI MALDI APPI	ETD CID	2200 4000	Up to 27,000	2000 at 27,000	250 fg reserpine at S/N 50:1 in MS/MS	High[b]	~ 50 ppm externally calibrated
HCT Ultra PTM Discovery (Bruker Daltonics)	3D	ESI APCI MALDI APPI	ETD CID	3000 6000	27,000	2000 at 27,000	250 fg reserpine at S/N 50:1 in MS/MS	High[b]	~ 50 ppm externally calibrated

(*continued*)

Table 9.A. (*Continued*)

Product (Company)	QF	Ion Source	Ion Activity	Mass Range (m/z)	Scan Speed (m/z)	Resolution ($m/\Delta m$) at FWHM at m/z 1000)	Detection Limit(s)	Ion Charge Capacity	Mass Accuracy (ppm)
DECA XP MAX (Thermo-Fisher)	3D	ESI APPI APCI	CID	15–200 50–2000 100–4000	500 5500	5000 1400	2 pg reserpine at S/N 50 : 1 in MS/MS	Standard	~ 50 ppm normal resolution, full-scan mode
500-MS LC (Varian, Inc.)	3D	ESI APCI	CID	50–2000	5000 15,000	2000	250 fg reserpine MS/MS	High[b]	Not specified

Acronyms:
2D, two-dimensional quadrupole field; 3D, three-dimensional quadrupole field; APCI, atmospheric pressure chemical ionization; APPI, atmospheric-pressure photoionization; ESI, electrospray ionization; MALDI, matrix-assisted laser desorption ionization; CID, collision-induced dissociation: ETD, electron transfer dissociation.

[a]Due primarily to a larger ion trapping volume.

[b]Due to unique mode of resonance ejection.

Source: Parts of this table came from Ref. [336]: Jonchers, K. Not your run-of-the-mill test tube. Product review. *Anal. Chem.* **2006**, Oct. 1, 6713.

REFERENCES

1. Bier, M. E.; Syka, J. E. P. Ion trap mass spectrometer system and method. U.S. Patent 5,420,425, 1995.
2. Hager, J. W. Axial ejection in a multipole mass spectrometer. U.S. Patent 6,177,668, 2001.
3. Makarov, A. A. Mass spectrometer. U.S. Patent 5,886,346, 1996.
4. Hardman, M.; Makarov, A. A. Interfacing the orbitrap mass analyzer to an electrospray ion source. *Anal. Chem.* **2003**, *75*(7), 1699–1705.
5. Dole, M.; Mack, L. L.; Hines, R. L.; Mobley, R. C.; Ferguson, L. D.; Alice, M. B. Molecular beams of macroions. *J. Chem. Phys.* **1968**, *49*, 2240.
6. Clegg, G. A.; Dole, M. Molecular beams of macroions. *Biopolymers* **1971**, *10*, 821–826.
7. Dole, M.; Cox, H. L.; Gieniec, J. Electrospray mass spectroscopy. M. Erzin (Ed.), *Advances in Chemistry Series*, No. 125, Polymer Molecular Weight Methods, Chapter 7, American Chemical Society, Washington D.C., **1973**, pp. 73–84.
8. Gieniec, J.; Mack, L. L.; Nakamae, K.; Gupta, C.; Kumar, V.; Dole, M. Electrospray mass spectroscopy of macromolecules: Application of an ion-drift spectrometer. *Biomed. Mass Spectrom.* **1984**, *11*, 259–268.
9. Yamashita, M.; Fenn, J. B. Application of electrospray mass spectrometry in medicine and biochemistry. *Iyo Masu Kenkyukai Koenshu* **1984**, *9*, 203.
10. Fenn, J. B.; Mann, M.; Meng, C. K.; Wong, S. F.; Whitehouse, C. M. Electrospray ionization—Principles and practice. *Mass Spectrom. Rev.* **1990**, *9*, 37.
11. Aleksandrov, M. L.; Gall, L. N.; Krasnov, N. V.; Nikolaev, V. I.; Pavlenko, V. A.; Shkurov, V. A. Ion extraction from solutions at atmospheric pressure - a method for mass-spectrometric analysis of bioorganic substances. *Doklady Akademii Nauk SSSR* **1984**, *277* (2), 379–83.
12. Meng, C.K.; Mann, M..; Fenn, J.B. Of protons or proteins. *Z. Phys. D.* **1988**, *10*(2–30), 361–368.
13. Tanaka, K.; Ido, Y.; Akita, S.; Yoshida, T. Detection of high mass molecules by laser desorption time-of-flight mass spectrometry. In *Second Japan-China Joint Symposium on Mass Spectrometry*, Osaka, Japan; September **1987**, pp. 185–188.
14. Karas, M.; Hillenkamp, F. Laser desorption ionization of proteins with molecular masses exceeding 10,000 daltons. *Anal. Chem.* **1988** *60*(20), 2299–2301.
15. Paul, W.; Steinwedel, H. A new mass spectrometer without magnetic field. *Z. Naturforsch.* **1953**, *8a*, 448–450.
16. Post, R. F. Report. No. UCRL-2209, University of California Radiation Laboratory, Berkeley, CA, 1953.
17. Finnigan, R.E. Quadrupole mass spectrometers. *Anal. Chem.* **1994**, *66*(19), 969A–975A.
18. Paul, W.; Steinwedel, H. Apparatus for separating charged particles of different specific charges, German Patent 944,900, 1956. U.S. Patent 2,939,952, 1960.
19. Paul, W. Electromagnetic cages for charged and neutral particles (Nobel lecture). *Angew. Chem. Int.* **1990**, *29*, 739–748.
20. Stafford, G. C.; Kelley, P.; Stephens, D. R. Method of mass analyzing a sample by use of a quadrupole ion trap U.S. Patent 4,540,884, 1985.
21. Stafford, G. C.; Kelley, P. E.; Syka, J. E. P.; Reynolds, W. E.; Todd, J. F. J. Recent improvements in and analytical applications of advanced ion trap technology. *Int. J. Mass Spectrom. Ion Proc.* **1984**, *60*, 85–98.
22. Splendore, M.; Marquette, E.; Oppenheimer, J.; Huston, C.; Wells, G. A new ion ejection method employing an asymmetric trapping field to improve the mass scanning performance of an electrodynamic ion trap. *Int. J. Mass Spectrom.* **1999**, *190/191*, 129–143.
23. Ding, L.; Nuttall, J. E. Methods and apparatus for driving a quadrupole ion trap device. U.S. Patent 7,193,207, 2007.
24. Ding, L.; Sudakov, M.; Brancia, F. L.; Giles, R.; Kumashiro, S. A digital ion trap mass spectrometer coupled with atmospheric pressure ion sources. *J. Mass Spectrom.* **2004**, *39*(5), 471–484.
25. Makarov, A. Electrostatic axially harmonic orbital trapping: A high-performance technique of mass analysis. *Anal. Chem.* **2000**, *72*(6), 1156–1162.
26. Dawson, P. H., (Ed.). *Quadrupole Mass Spectrometry and Its Applications*, Elsevier, Amsterdam, **1976**. (republished by the American Institute of Physics, Woodbury, NY, 1995).
27. March, R. E. Ion trap mass spectrometry. *Int. J. Mass Spectrom. Ion Proc.* **1992**, *118/119*, 71–135.
28. McLuckey, S. A.; Van Berkel, G. J.; Goeringer, D. E.; Glish, G. L. Ion trap mass spectrometry using high-pressure ionization. *Anal. Chem.* **1994**, *66*, 689A–696A.
29. Todd, J. F. J. Ion trap mass spectrometer—Past, present, and future (?) *Mass Spectrom. Rev.*, **1991**, *10*, 3.
30. Todd, J. F. J.; Penman, A. D. The recent evolution of the quadrupole ion trap mass spectrometer—An overview. *Int. J. Mass Spectrom. Ion Process.* **1991**, *106*, 1.
31. Cooks, R. G.; Glish, G. L.; McLuckey, S. A.; Kaiser, R. E. Ion trap mass spectrometry. *Chem. Eng. News* **1991**, *69*, 26.
32. McLuckey, S. A.; Van Berkel, G. J.; Goeringer, D. E.; Glish, G. L. Ion trap mass spectrometry. Using high-pressure ionization. *Anal. Chem.* **1994**, *66*, 737A–743A.
33. McLuckey, S. A.; Van Berkel, G. J.; Glish, G. L.; Schwartz, J. C. In March, R. E.; Todd, J. F. J., (Eds.), *Practical Aspects of Ion Trap Mass Spectrometry*, Vol. II, CRC Series Modern Mass Spectrometry, New York, **1995**, Chapter 3, pp. 89–141.
34. Schwartz, J. C.; Jardine, I. In Karger, B.; Hancock, W., (Eds.), *High Resolution Separation and Analysis of*

Biological Macromolecules, Part A: Fundamentals, Vol. 270, Methods in Enzymology, Spectrum Publication Services, York, PA, 1996, 552–586.
35. Douglas, D. J.; Frank, A. J.; Mao, D. Linear ion traps in mass spectrometry. *Mass Spectrom. Rev.* **2005**, *24* (1), 1–29.
36. Hu, Q.; Noll, R. J.; Li, H.; Makarov, A.; Hardman, M.; Cooks, R. G.; The orbitrap: A new mass spectrometer. *J. Mass Spectrom.* **2005**, *40*(4), 430–443.
37. McLuckey, S. A.; Wells, J. M. Mass analysis at the advent of the 21st century. *Chem. Rev.* **2001**, *101*(2), 571–606.
38. Takats, Z.; Wiseman, J. M.; Gologan, B.; Cooks, R. G. Mass spectrometry sampling under ambient conditions with desorption electrospray ionization. *Science* **2004**, *306*(5695), 471–473.
39. Holkeboer, D. H.; Karandy, T. L.; Currier, F. C.; Frees, L. C.; Ellefson, R. E. Miniature quadrupole residual gas analyzer for process monitoring at milliTorr pressures. *J. Vac. Sci. Technol. A.* **1998** *16* (3, Pt. 1), 1157–1162.
40. Zahn, U. von. Precision mass determinations with the electric mass filter. *Z. Phys.* **1962**, *168*, 129–142.
41. Zahn, U. von.; Gebauer, S.; Paul, W. *10th Annual Conference on Mass Spectrometry*, New Orleans, LA, June 3, 1962.
42. March, R. E.; Hughes, R. J. In Winefordner, J. D. (Ed.), *Quadrupole Storage Mass Spectrometry*, Vol. 102, Analytical Chemistry and Its Applications, John Wiley & Sons, New York, 1989.
43. March, R. E.; Todd, J. F. J. (Eds.), *Practical Aspects of Ion Trap Mass Spectrometry*, Vols. 1–3, CRC Press, Boca Raton, FL, **1995** vol. 1, 2, 3.
44. March, R. E.; Todd, J. F. J. In Winefordner, J. D, (Ed.), *Quadrupole Ion Trap Mass Spectrometry 2nd ed.*, John Wiley & Sons, New York, **2005**.
45. Stafford, G. C.; Taylor, D. M.; Bradshaw, S. C. Method of increasing the dynamic range and sensitivity of a quadrupole ion trap mass spectrometer. U.S. Patent 5,107,109, 1992.
46. Syka, J. E. P.; Louris, J. N.; Kelley, P. E.; Stafford, G. C.; Reynolds, W. E. Method of operating ion trap detector in MS/MS mode. U.S. Patent 4,736,101, 1988.
47. Weber-Grabau, M. Method of operating a three-dimensional ion trap with enhanced sensitivity. U.S. Patent 4,818,869, 1989.
48. Louris, J. N.; Cooks, R. G.; Syka, J. E. P.; Kelley, P. E.; Stafford, G. C.; Todd, J. F. Instrumentation, applications, and energy deposition in quadrupole ion-trap tandem mass spectrometry. *Anal. Chem.* **1987**, *59*, 1677–1685.
49. Louris, J. N.; Brodbelt-Lustig, J. S.; Cooks, R. G.; Glish, G. L.; Van Berkel, G. J.; McLuckey, S. A. Ion isolation and sequential stages of mass spectrometry in a quadrupole ion trap mass spectrometer. *Int. J. Mass Spectrom. Ion Proc.* **1990**, *96*, 117–137.
50. Louris, J. N.; Amy, J. W.; Ridley, T. Y.; Cooks, R. G. Injection of ions into a quadrupole ion trap mass spectrometer. *Int. J. Mass Spectrom. Ion Proc.* **1989**, *88*, 97–111.
51. Johnson, J. V.; Pedder, R. E.; Yost, R. A. MS–MS parent scans on a quadrupole ion trap mass spectrometer by simultaneous resonant excitation of multiple ions. *Int. J. Mass Spectrom. Ion Proc.* **1991**, *106*, 197–212.
52. Johnson, J. V.; Pedder, R. E.; Yost, R. A.; Story, M. S. Quadrupole ion trap mass spectrometer having two pulsed axial excitation. U.S. Patent 5,075,547, **1991**.
53. Kaiser, R. E.; Louris, J. N.; Amy, J. W.; Cooks, R. G. Extending the mass range of the quadrupole ion trap using axial modulation. *Rapid Commun. Mass Spec.* **1989**, *3*, 225–229.
54. Kaiser, R. E.; Cooks, R. G.; Moss, J.; Hemberger, P. H. Mass range extension in a quadrupole ion-trap mass spectrometer. *Rapid Commun. Mass Spec.* **1989**, *3*, 50–53.
55. Kaiser, R. E.; Cooks, R. G.; Stafford, G. C.; Syka, J. E. P.; Hemberger, P. H. Operation of a quadrupole ion trap mass spectrometer to achieve high mass/charge ratios. *Int. J. Mass Spectrom. Ion Proc.* **1991**, *106*, 79–115.
56. Schwartz, J. C.; Syka, J. E. P.; Jardine, I. High resolution on a quadrupole ion trap mass spectrometer. *J. Am. Soc. Mass Spectrom.* **1991**, *2*, 198–204.
57. Goeringer, D. E.; McLuckey, S. A.; Glish, G. L. Enhancement of mass resolution in the quadrupole ion trap via resonance ejection, *Proceedings of the 39th Annual conference of Mass Spectrometry and Allied Topics*, Nashville, TN, May 19–24, **1991**, pp. 532–533.
58. Kaiser, R. E. Ph.D. thesis, Purdue University, West Lafayette, IN, 1990, pp. 120–123.
59. Londry, F. A.; Wells, G. J.; March, R. E. Enhanced mass resolution in a quadrupole ion trap. *Rapid Commun. Mass Spec.* **1993**, *7*, 43–45.
60. Yang, G.; Bier, M. E. Investigation of a rapid scan on an electrospray ion trap mass spectrometer. *Anal. Chem.* **2005**, *77*, 1663–1671.
61. Kelley, P. E. Mass spectrometry method using notch filter. U.S. Patent 5,134,286, 1992.
62. Kelley, P. E. Mass spectrometry method using filtered noise signal. U.S. Patent 5,206,507, 1993.
63. Hoekman, D. J.; Kelley, P. E. Method for generating filtered noise signal and broadband signal having reduced dynamic range for use in mass spectrometry. U.S. Patent 5,256,875, 1993.
64. Goeringer, D. E.; Asano, K. G.; McLuckey, S. A.; Hoekman, D.; Stiller, S. W. Filtered noise field signals for mass-selective accumulation of externally formed ions in a quadrupole ion trap. *Anal. Chem.* **1994**, *66*, 313–318.
65. Guan, S.; Marshall, A. G. Stored waveform inverse Fourier transform axial excitation/ejection for quadrupole ion trap mass spectrometry. *Anal. Chem.* **1993**, *65*, 1288–1294.
66. Brodbelt, J. S.; Cooks, R. G. Ion/molecule reactions investigated by using tandem mass spectrometry in an

ion trap: Halomethylation. *Anal. Chim. Acta*, **1988**, *206*, 239–251.

67. McLuckey, S. A.; Van Berkel, G. J.; Glish, G. L. Reactions of dimethylamine with multiply charged ions of cytochrome c. *J. Am. Chem. Soc.* **1990**, *112*, 5668.
68. Louris, J. N.; Syka, J. E. P.; Kelley, P. E. Method of operating quadrupole ion trap chemical ionization mass spectrometry. U.S. Patent 4,686,367, 1987.
69. Weber-Grabau, M.; Bradshaw, S. C.; Syka, J. E. P. Method of increasing the dynamic range and sensitivity of a quadrupole ion trap mass spectrometer operating in the chemical ionization mode. U.S. Patent 4,771,172, 1988.
70. Kaiser, R. E.; Williams, J. D.; Schwartz, J. C.; Lammert, S. A.; Cooks, R. G. Thermospray LC/MS with a quadrupole ion trap and recent application of the ion injection. *Proceedings of the 37th Annual conference of Mass Spectrometry and Allied Topics*, Miami, FL, May 21–26, **1989**, 369–370.
71. Bier, M. E.; Hartford, R. E.; Herron, J. R.; Stafford, G. C. Coupling a particle beam and a thermospray interface to a quadrupole ion trap mass spectrometer. In *Proceedings of the 39th ASMS Conference on Mass Spectrometry and Allied Topics*, Nashville, TN, May 21–24, **1991**, 538–539.
72. McLuckey, S. A.; Glish, G. L.; Asano, K. G. Coupling of an atmospheric-sampling ion source with an ion-trap mass spectrometer. *Anal. Chim. Acta.* **1989**, *225*, 25–35.
73. Lister, A. K.; Wood, K. V.; Cooks, R. G.; Noon, K. R. Direct detection of organic compounds in water at parts-per-billion levels using a simple membrane probe and a quadrupole ion trap. *Biomed. Environ. Mass Spectrom.* **1989**, *18*, 1063.
74. Bier, M. E.; Winkler, P. C.; Herron, J. R. Coupling a particle beam interface directly to a quadrupole ion trap mass spectrometer. *J. Am. Soc. for Mass Spectrom.* **1993**, *4*, 38–46.
75. Heller, D. N.; Lys, I.; Cotter, R.; Uy, O. M. Laser desorption from a probe in the cavity of a quadrupole ion storage mass spectrometer. *Anal. Chem.* **1989**, *61*, 1083–1086.
76. Louris, J. N.; Amy, J. W.; Ridley, T. Y.; Cooks, R. G. Injection of ions into a quadrupole ion trap mass spectrometer. *Int. J. Mass Spectrom. Ion Proc.* **1989**, *88*, 97–111.
77. Cox, K. A.; Williams, J. D.; Cooks, R. G.; Kaiser, R. E. Quadrupole ion trap mass spectrometry: Current applications and future directions for peptide analysis. *Biol. Mass Spectrom.* **1992**, *21*, 226.
78. Jonscher, K.; Currie, G.; McCormack, A. L.; Yates, J. R. III. Matrix-assisted laser desorption of peptides and proteins on a quadrupole ion trap mass spectrometer. *Rapid Commun. Mass Spectrom.* **1993**, *7*, 20–26.
79. Schwartz, J. C.; Bier, M. E. Matrix-assisted laser desorption of peptides and proteins using a quadrupole ion trap mass spectrometer. *Rapid Commun. Mass Spectrom.* **1993**, *7*, 27–32.
80. Doroshenko, V. M.; Cotter, R. J. High-performance collision-induced dissociation of peptide ions formed by matrix-assisted laser desorption/ionization in a quadrupole ion trap mass spectrometer. *Anal. Chem.* **1995**, *67*(13), 2180–2187.
81. Doroshenko, V. M.; Cotter, R. J. Pulsed gas introduction for increasing peptide CID efficiency in a MALDI/quadrupole ion trap mass spectrometer. *Anal. Chem.* **1996**, *68*(3), 463–472.
82. Taylor, L. C. E.; Singh, R.; Cahng, S. Y.; Johnson, R. L.; Schwartz, J. The identification of *in vitro* metabolites of bupropion by ion-trap mass spectrometry. *Rapid Commun. Mass Spectrom.* **1995**, *9*, 902–910.
83. Van Berkel, G. J.; Glish, G. L.; McLuckey, S. A. Electrospray ionization combined with ion trap mass spectrometry. *Anal. Chem.* **1990**, *62*, 1284–1295.
84. Mordehai, A. V.; Hopfgartner, G.; Huggins, T. G.; Henion, J. D. Atmospheric-pressure ionization interface for a bench-top quadrupole ion trap. *Rapid Commun. Mass Spectrom.* **1992**, *6*, 508–516.
85. Wang, Y; Franzen, J. J. The non-linear resonance QUISTOR Part 1. Potential distribution in hyperboloidal QUISTORs. *Int. J. Mass Spectrom Ion Proc.* **1992**, *112*, 167–178.
86. Louris, J.; Schwartz, J.; Stafford, G.; Syka, J.; Taylor, D. The Paul ion trap mass selective instability scan: Trap geometry and resolution. In *Proceedings of the 40th ASMS Conference on Mass Spectrometry and Allied Topics*, Washington, D.C., May 31–June 5, 1992, pp. 1003–1004.
87. Cleven, C. D.; Cox, K. A.; Cooks, R. G.; Bier, M. E. Mass shifts due to ion/ion interactions in a quadrupole ion-trap mass spectrometer. *Rapid Commun. Mass Spectrom.* **1994**, *8*, 451–454.
88. Cox, K. A., Cleven, C. D.; Cooks, R. G. Mass shifts and local space charge effects observed in the quadrupole ion trap at higher resolution. *Int. J. Mass Spectrom. Ion Proc.* **1995**, *144*, 47–65.
89. Mordehai, A.; Miller, B.; Bai, J.; Brekenfeld, A.; Baessmann, C. Schubert, M.; Hosea, K. In *Improved 3D Ion Trap–Ion Detector Coupling & Techniques for Evaluating Ion Trap Capacity, Proceedings of the 43rd ASMS Conference on Mass Spectrometry and Allied Topics*, Nashville, TN, May 23–27, 2004.
90. Personal communications with Dr. Alex Mordehai, Agilent Technologies, Inc., 2007.
91. Stafford, G. C.; Kelley, P. E.; Syka, J. E. P.; Reynolds, W. E.; Todd, J. F. J. Recent improvements in and analytical applications of advanced ion trap technology. *Int. J. Mass Spectrom. Ion Phys.* **1984**, *60*, 85–98.
92. Williams, J. D.; Cooks, R. G.; Syka, J. E. P.; Hemberger, P. H.; Nogar, N. S. Determination of positions, velocities, and kinetic energies of resonantly excited ions in the quadrupole ion trap mass

spectrometer by laser photodissociation. *J. Am. Soc. Mass Spectrom.* **1993**, *4*, 792–797.
93. Cleven, Curtis D.; Cox, Kathleen A.; Cooks, R. Graham; Bier, M. E. Mass shifts due to ion/ion interactions in a quadrupole ion-trap mass spectrometer. *Rapid Commun. Mass Spectrom.* **1994**, *8* (6), 451–454.
94. Sheretov, E. P.; Terent'ev, V. I. Bases of the theory of quadrupole mass spectrometers during pulsed feeding. *Zh.Tech. Fiz.* **1972**, *42*(5), 953.
95. Sheretov, E. P.; Gurov, V.; Safonov, M.; Philippov, I. W. Hyperboloid mass spectrometers for space exploration. *Int. J. Mass Spectrom.* **1999**, *189*, 9.
96. Richards, A.; Huey, M.; Hiller, J. New operating mode for the quadrupole mass filter. *Int. J. Mass Spectrom. Ion Phys.* **1973**, *12*, 317.
97. Ding, L.; Kumashiro, S. Ion motion in the rectangular wave quadrupole field and digital operation mode of a quadrupole ion trap mass spectrometer. *Rapid Commun. Mass Spectrom.* **2006**, *20*(1), 3–8.
98. Ding, L.; Sudakov, M.; Kumashiro, S. A simulation study of the digital ion trap mass spectrometer. *Int. J. Mass Spectrom.* **2002**, *221*(2), 117–138.
99. Ouyang, Z.; Wu, G.; Song, Y.; Li, H.; Plass, W. R.; Cooks, R. G. Rectilinear ion trap: Concepts, calculations, and analytical performance of a new mass analyzer. *Anal. Chem.* **2004**, *76*(16), 4595–4605.
100. Wells, J. M.; Badman, E. R.; Cooks, R. G. A quadrupole ion trap with cylindrical geometry operated in the mass-selective instability mode. *Anal. Chem.* **1998**, *70*(3), 438–444.
101. Badman, E. R.; Wells, J. M.; Bui, H. A.; Cooks, R. G. Fourier transform detection in a cylindrical quadrupole ion trap. *Anal. Chem.* **1998**, *70*(17), 3545–3547.
102. Badman, E. R.; Johnson, R. C.; Plass, W. R.; Cooks, R. G. A miniature cylindrical quadrupole ion trap: simulation and experiment. *Anal. Chem.* **1998**, *70*(23), 4896–4901.
103. Langmuir, D. B.; Langmuir, R. V.; Sheldton, H.; Weurker, R. F. Containment device. U.S. Patent 3,065,640, 1962.
104. Welling, M.; Thompson, R. I.; Walther, H. Photodissociation of $MgC_{60}{}^{+}$ complexes generated and stored in a linear ion trap. *Chem. Phys. Lett.* **1996**, *253*(1,2), 37–42.
105. Welling, M.; Schuessler, H. A.; Thompson, R. I.; Walther, H. Ion/molecule reactions, mass spectrometry and optical spectroscopy in a linear ion trap. *Int. J. Mass Spectrom. Ion Proc.* **1998**, *172*, 95–114.
106. Schwartz, J.; Senko, M.; Syka, John, E. A two-dimensional quadrupole ion trap mass spectrometer. *J. Am. Soc. Mass Spectrom.* **2002**, *13*(6), 659–669.
107. Personal communication by researchers at MDS Sciex, ca. 1998.
108. Hager, J. W. A new linear ion trap mass spectrometer. *Rapid Commun. Mass Spectrom.* **2002**, *16*(6), 512–526.
109. Londry, F. A.; Hager, J. W. Mass selective axial ion ejection from a linear quadrupole ion trap. *J. Am. Soc. Mass Spectrom.* **2003**, *14*(10), 1130–1147.
110. Prestage, J. D.; Dick, G. J.; Maleki, L. New ion trap for frequency standard applications. *J. Appl. Phys.* **1989**, *66*, 1013–1017.
111. Campbell, J. M.; Collings, B. A.,et al. A new linear ion trap time-of-flight system with tandem mass spectrometry capabilities. *Rapid Commun. Mass Spec.* **1998**, *12*, 1463–1474.
112. Hemberger, P. H.; Nogar, N. S.; Williams, J. D.; Cooks, R. G.; Syka, J. E. P. Laser photodissociation probe for ion tomography studies in a quadrupole ion-trap mass spectrometer. *Chem. Phys. Lett.* **1992**, *191*(5), 405–410.
113. Williams, J. D.; Cooks, R. G.; Syka, J. E. P.; Hemberger, P. H.; Nogar, N. S. Determination of positions, velocities, and kinetic energies of resonantly excited ions in the quadrupole ion trap mass spectrometer by laser photodissociation. *J. Am. Soc. Mass Spectrom.* **1993**, *4*, 792–797.
114. Fischer, E. Thermodynamic explanation of large periods in crystalline high polymers. *Z. Naturforsch., A: Phys. Sci.* **1959**, *156*, 26.
115. http://www.mdssciex.com, Technical Specification of 4000 Q TRAP. Searched July 15, **2007**.
116. Yost, R. A.; McClennen, W.; Snyder, A. P. Picogram to microgram analysis gas chromatography/ion trap mass spectrometry. In *Proceedings of the 35th Annual Conference of Mass Spectrometry and Allied Topics*, Denver, CO, May 24–29, **1987**, pp. 789–790.
117. Louris, J. N.; Taylor, D. M. Method and apparatus for ejecting unwanted ions in an ion trap mass spectrometer. U. S. Patent 5,324,939, 1994.
118. Doroshenko, V. M.; Cotter, R. J. Linear mass calibration in the quadrupole ion-trap mass spectrometer. *Rapid Commun. Mass Spectrom.* **1994**, *8*, 766–776.
119. Bier, M. E. Method of reducing noise in an ion trap mass spectrometer coupled to an atmospheric pressure ionization source. U.S. Patent 5,750,993, 1998.
120. Moini, M. Ultramark 1621 as a calibration/reference compound for mass spectrometry. *Rapid Commun. Mass Spectrom.* **1994**, *8*, 711–714.
121. ThermoFisher LTQ Brochure, 2007.
122. Bier, M. E.; Schwartz, J. C.; Zhou, J.; Syka, J. E. P.; Taylor, D.; Land, A.; James, M.; Fies, B.; Peptide sequencing using fast MS and MS/MS scan rates on a μLC/quadrupole ion trap. In *Proceedings of the 43rd ASMS Conference on Mass Spectrometry and Allied Topics*, Atlanta, GA, May 21–26, **1995**, p. 988.
123. Yates, N. A.; Yost, R. A.; Bradshaw, S. C.; Tucker, D. B. A comparative study of Selective Mass Storage and two-step Isolation in the quadrupole In trap. *Proceedings of the 39th Annual conference of Mass Spectrometry and Allied Topics*, Nashville, TN, May 19–24, **1991**, pp. 1489–1490.

124. KAISER, R. E.; COOKS, R. G.; SYKA, J. E. P.; STAFFORD, G. C. Collisionally activated dissociation of peptides using a quadrupole ion-trap mass spectrometer. *Rapid Commun. Mass Spectrom.* **1990**, *4*, 30–33.
125. SCHWARTZ, J. C.; JARDINE, I. High-resolution parent-ion selection/isolation using a quadrupole ion-trap mass spectrometer. *Rapid Commun. Mass Spectrom.* **1992**, *6*, 313–317.
126. MCLUCKEY, S. A.; GOERINGER, D. E.; GLISH, G. L. Selective ion isolation/rejection over a broad mass range in the quadrupole ion trap. *J. Am. Soc. Mass Spectrom.* **1991**, *2*, 11–21.
127. DOROSHENKO, V. M.; COTTER, R. J. Advance swift techniques for MALDI/quadrupole ion trap mass spectrometry. In *Proceedings of the 43rd Conference on Mass Spectrometry and Allied Topics*, Atlanta, GA, May 21–26, **1995**, 1102.
128. SHAFFER, B. A.; KARNICKY, J.; BUTTRILL, S. E. Development of automated selected ion storage for quadrupole ion trap mass spectrometry. In *Proceedings of the 41st Conference on Mass Spectrometry and Allied Topics*, San Francisco, CA, May 30–June 4, 1993, p. 802.
129. GOERINGER, D. E.; ASANO, K. G.; MCLUCKEY, S. A. Filtered noise field signals for mass-selective accumulation of externally formed ions in a quadrupole ion trap. *Anal. Chem.* **1994**, *66*, 313–318.
130. TAYLOR, D.; SCHWARTZ, J.; ZHOU, J.; JAMES, M.; BIER, M.; KORSAK, A.; STAFFORD, G. Application of tailored waveform generation to the quadrupole ion trap. In *Proceedings of the 43rd Conference on Mass Spectrometry and Allied Topics*, Atlanta, GA, May 21–26, **1995**, p. 1103.
131. SCHUBERT, M.; NAGEL, M.; WANG, Y.; FRANZEN, J. Exciting waveform generation for ion traps. In *Proceedings of the 43rd Conference on Mass Spectrometry and Allied Topics*, Atlanta, GA, May 21–26, 1995, p. 1107.
132. WELLS, G.; HUSTON, C. Field-modulated selective ion storage in a quadrupole ion trap. *J. Am. Soc. Mass Spectrom.* **1995**, *6*, 928–935.
133. GOERINGER, D. E.; ASANO, K. G.; MCLUCKEY, S. A.; HOEKMAN, D; STILLER, S. W. Filtered noise field signals for mass-selective accumulation of externally formed ions in a quadrupole ion trap. *Anal. Chem.* **1994**, *66*, 313–318.
134. MARSHALL, A. G.; WANG, T.-C. L.; RICCA, T. L. Tailored excitation for Fourier transform ion cyclotron mass spectrometry. *J. Am. Chem. Soc.* **1985**, *107*, 7893–7897.
135. CHEN, L.; WANG, T. C. L.; RICCA, T.; MARSHALL, A. G. Phase-modulated stored waveform inverse Fourier transform excitation for trapped ion mass spectrometry. *Anal. Chem.* **1987**, *59*, 449–454.
136. GUAN, S.; MARSHALL, A. G. Stored waveform inverse Fourier transform axial excitation/ejection for quadrupole ion trap mass spectrometry. *Anal. Chem.* **1993**, *65*, 1288–1294.
137. JULIAN, R. K.; COX, K.; COOKS, R. G. Broadband excitation in the quadrupole ion trap mass spectrometer using shaped pulses (inverse Fourier transform). In *Proceedings of the 40th Conference on Mass Spectrometry and Allied Topics*, Washington, D.C., May 31–June 5, **1992**, pp. 943–944.
138. JONSCHER, K. R.; YATES, J. R. Analyzing mixtures with a hybrid quadrupole/ion trap mass spectrometer. In *Proceedings of the 43rd Conference on Mass Spectrometry and Allied Topics*, Atlanta, GA, May 21–26, **1995**, p. 990.
139. WELLS, G.; HUSTON, C. High-resolution selected ion monitoring in a quadrupole ion trap mass spectrometer. *Anal. Chem.* **1995**, *67*, 3650–3655.
140. WILLIAMS, J. D.; COX, K.; MORAND, K. L.; JULIAN, R. K.; JULIAN, R. K.; KAISER, R. E. High mass-resolution using a quadrupole ion trap mass spectrometer. In *Proceedings of the 39th ASMS Conference on Mass Spectrometry and Allied Topics*, Nashville, TN, **1991**, pp. 1481–1482.
141. LONDRY, F. A.; WELLS, G. J.; MARCH, R. E. Enhanced mass resolution in a quadrupole ion trap. *Rapid Commun. Mass Spectrom.* **1993**, *7*, 43–45.
142. WILLIAMS, J. D.; COOKS, R. G. Improved accuracy of mass measurement with a quadrupole ion-trap mass spectrometer. *Rapid Commun. Mass Spectrom.* **1992**, *6*, 524–527.
143. BEU, S. C.; SENKO, M. W.; QUINN, J. P.; WAMPLER III, F. M.; MCLAFFERTY, F. W. Fourier-transform electrospray instrumentation for tandem high-resolution mass spectrometry of large molecules. *J. Am. Soc. Mass Spectrom.* **1993**, *4*, 557–565.
144. WINGER, B. E.; HOFSTADLER, S. A.; BRUCE, J. E.; UDSETH, H. R.; SMITH, R. D. High-resolution accurate mass measurements of biomolecules using a new electrospray ionization ion cyclotron resonance mass spectrometer. *J. Am. Soc. Mass Spectrom.* **1993**, *4*, 566–577.
145. MORAND, K. L.; COX, K. A.; COOKS, R. G. Efficient trapping and collision-induced dissociation of high-mass cluster ions using mixed target gases in the quadrupole ion trap. *Rapid Commun. Mass Spectrom.* **1992**, *6*, 520–523.
146. MCLUCKEY, S. A.; GLISH, G. L.; ASANO, K. G. Coupling of an atmospheric-sampling ion source with an ion-trap mass spectrometer. *Anal. Chim. Acta* **1989**, *22*, 25–35.
147. DOROSHENKO, V. M.; COTTER, R. Pulsed gas introduction for increasing peptide CID efficiency n a MALDI/quadrupole ion trap mass spectrometer. *Anal. Chem.* **1996**, *68*, 463–472.
148. LOURIS, J. N.; COOKS, R. G.; SYKA, J. E. P.; KELLEY, P. E.; STAFFORD, G. C.; TODD, J. F. Instrumentation, applications, and energy deposition in quadrupole ion-trap tandem mass spectrometry. *Anal. Chem.* **1987**, *59*, 1677–1685.
149. MCLUCKEY, S. A.; GOERINGER, D. E.; GLISH, G. L. Collisional activation with random noise in ion trap mass spectrometry. *Anal. Chem.* **1992**, *64*, 1455–1460.

150. LTQ XL Product Specification, ThermoScientific. San Jose, CA, **2007**.
151. Strife, R. J.; Schwartz, J.; Bier, M. E.; Zhou, J. Electrospray/ion trap MS^n for structure elucidation of surfactants. In *Proceedings of the 43rd Conference on Mass Spectrometry and Allied Topics*, Atlanta, GA, May 21–26, **1995**, p. 160.
152. Wolfender, J. L.; Rodriguez, S.; Hostettmann, K.; Winfried, W. R. Comparison of liquid chromatography/electrospray, atmospheric pressure chemical ionization, thermospray and continuous-flow fast atom bombardment mass spectrometry for the determination of secondary metabolites in crude plant extracts. *J. Mass Spectrom.* **1995**, S35–S46.
153. Kingdon, K. H. A method for the Neutralization of electron space charge by positive ionization at very low gas pressures, *Phys. Rev.* **1923**, *21*, 408.
154. Vane, C. R.; Prior, M. H.; Marrus, Richard. Electron capture by neon(10 +) trapped at very low energies. *Phys. Rev. Lett.* **1981**, *46*(2), 107–110.
155. Prior, M. H.; Wang, E. C. Hyperfine structure of 2*s* helium-3(+) ion by an ion-storage technique. *Phys. Rev. Lett.* **1975**, *35*(1), 29–32.
156. Knight, R. D. Storage of ions from laser-produced plasmas. *Appl. Phys. Lett.* **1981**, *38*(4), 221–223.
157. Dehmelt, H. G. Radiofrequency spectroscopy of stored ions. II. Spectroscopy. *Adv. At. Mol. Phys.* **1969**, *5*, 109.
158. Hu, Q.; Noll, R. J.; Li, H.; Makarov, A.; Hardman, M.; Cooks, R. G. The orbitrap: A new mass spectrometer. *J. Mass Spectrom.* **2005**, *40*(4), 430–443.
159. Makarov, A. Theory and practice of the orbitrap mass analyzer. In *Proceedings of the 54th ASMS Conference on Mass Spectrometry and Allied Topics*, Seattle, WA, May 28–June 1, 2006.
160. Lunney, M. D.; Moore, R. B. Cooling of mass-separated beams using a radiofrequency quadrupole ion guide. *Int. J. Mass Spectrom.* **1999**, *190/191*, 153–160.
161. Yates, N.; Kottmeier, D.; Shabanowitz, J.; Hunt, D. Multi-stage collision activated dissociation of protonated peptides by electrospray/quadrupole ion trap mass spectrometry. *Proceedings of the 42nd ASMS Conference on Mass Spectrometry and Allied Topics*, Chicago, IL, May 29–June 3, 1994, p. 212.
162. Bier, M. E.; Schwartz, J. C.; Zhou, J.; Taylor, D. M.; Syka, J. E. P.; James, M. S.; Fies, B.; Stafford, G. C. Ion optics and design considerations for a new LC/quadrupole ion trap. In *Proceedings of the 43rd Conference on Mass Spectrometry and Allied Topics*, Atlanta, GA, May 21–26, 1995, p. 1117.
163. Schwartz, J. C.; Bier, M. E.; Taylor, D. M.; Zhou, J.; Syka, J. E. P.; James, M. S.; Stafford, G. C. Technological aspects of a new LC/MS^n quadrupole ion trap. In *Proceedings of the 43rd Conference on Mass Spectrometry and Allied Topics*, Atlanta, GA, May 21–26, 1995, p. 1114.
164. Bruins, A. P.; Covey, T. R.; Henion, J. D. Ion spray interface for combined liquid chromatography/atmospheric pressure ionization mass spectrometry. *Anal. Chem.* **1987**, *59*, 2642–2646.
165. Chowdhury, S. K.; Katta, V.; Beavis, R. C.; Chait, B. T. Peptide adducts in electrospray ionization. *J. Am. Soc. Mass Spectrom.* **1990**, *1*, 382–388.
166. Whitehouse, C. M.; Dreyer, R. N.; Yamashita, M.; Fenn, J. B. Electrospray interface for liquid chromatographs and mass spectrometers. *Anal. Chem.* **1985**, *57*, 675–679.
167. McLuckey, S. A.; Glish, G. L.; Asano, K. G.; Bartmess, J. E. Protonated water and protonated methanol cluster decompositions in a quadrupole ion trap. *Int. J. Mass Spectrom. Ion Proc.* **1991**, *109*, 171–186.
168. Ashkenas, H.; Sherman, F. S. In De Leeuw, J. (Ed.), *Rarefied Gas Dynamics*, Vol. 2, Academic Press, New York, 1966, p. 84.
169. Teloy, E.; Gerlich, D. Integral cross sections for ion–molecule reactions. I. Guide beam technique. *Chem. Phys.* **1974**, *4*, 417.
170. Tosi, P.; Fontana, G.; Longano, S.; Bassi, D. Transport of an ion beam through an octopole guide operating in the rf-only mode. *Int. J. Mass Spectrom. Ion Processes* **1989**, *93*, 95–105.
171. Szabo, I. New ion-optical devices utilizing oscillatory electric fields. I. Principle of operation and analytical theory of multipole devices with two-dimensional electric fields. *Int. J. Mass Spectrom. and Ion Processes* **1986**, *73*(3), 197–235.
172. Yost, R. A.; Enke, C. G. Triple quadrupole mass spectrometry for direct mixture analysis and structure elucidation. *Anal. Chem.* **1979**, *51*(12), 1251A–1264A.
173. Ramsey, R. S.; Goeringer, D. E.; McLuckey, S. A. Active chemical background and noise reduction in capillary electrophoresis/ion trap mass spectrometry. *Anal. Chem.* **1993**, *65*, 3521–3524.
174. Doroshenko, V. M.; Cotter, R. J. A new method of trapping ions produced by matrix-assisted laser desorption ionization in a quadrupole ion trap. *Rapid Commun. Mass Spectrom.* **1993**, *7*(9), 822–827.
175. Qin, J.; Chait, B. T. Matrix-assisted laser desorption ion trap mass spectrometry: Efficient trapping and ejection of ions. *Anal. Chem.* **1996**, *68*(13), 2102–2107.
176. Qin, J.; Chait, B. T. Matrix-assisted laser desorption ion trap mass spectrometry: Efficient isolation and effective fragmentation of peptide ions. *Anal. Chem.* **1996**, *68*(13), 2108–2112.
177. Doroshenko, V. M.; Cotter, R. J. Injection of externally generated ions into an increasing trapping field of a quadrupole ion trap mass spectrometer. *J. Mass Spectrom.* **1997**, *32*(6), 602–615.
178. Krutchinsky, A. N.; Kalkum, M.; Chait, B. T. Automatic identification of proteins with a MALDI-quadrupole ion trap mass spectrometer. *Anal. Chem.* **2001**, *73*(21), 5066–5077.
179. Schlunegger, U.; Caprioli, R. Frequency scan for the analysis of high mass ions generated by matrix-assisted

laser desorption/ionization in a Paul trap. *Rapid Commun. Mass Spectrom.* **1999**, *13*, 1792–1796.
180. Robb, D. B.; Blades, M. W. Optimum phase angle for laser desorption ion trap mass spectrometry is dependent on the number of ions produced. *Int. J. Mass Spectrom.* **1999**, *190/191*, 69–80.
181. Laiko, V. V.; Burlingame, A. L. Atmospheric pressure matrix assisted laser desorption. U.S. Patent 5,965,884, **1999**.
182. Laiko, V. V.; Baldwin, M. A.; Burlingame, A. L. Atmospheric pressure matrix-assisted laser desorption/ionization mass spectrometry. *Anal. Chem.* **2000**, *72* (4), 652–657.
183. Laiko, V. V.; Moyer, S. C.; Cotter, R. J. Atmospheric pressure MALDI/ion trap mass spectrometry. *Anal. Chem.* **2000**, *72*(21), 5239–5243.
184. Keough, T.; Lacey, M. P.; Strife, R. J. Atmospheric pressure matrix-assisted laser desorption/ionization Ion trap mass spectrometry of sulfonic acid derivatized tryptic peptides. *Rapid Commun. Mass Spectrom.* **2001**, *15*(23), 2227–2239.
185. Laiko, V. V.; Taranenko, N. I.; Berkout, V. D.; Yakshin, M. A.; Prasad, C. R.; Lee, H. S.; Doroshenko, V. M. Desorption/ionization of biomolecules from aqueous solutions at atmospheric pressure using an infrared laser at 3 microns. *J. Am. Soc. Mass Spectrom.* **2002**, *13*(4), 354–361.
186. Laiko, V. V; Taranenko, N. I.; Berkout, V. D.; Musselman, B. D.; Doroshenko, V. M. Atmospheric pressure laser desorption/ionization on porous silicon. *Rapid Commun. Mass Spectrom.* **2002**, *16*(18), 1737–1742.
187. Coon, J. J; Harrison, W. W. Laser desorption-atmospheric pressure chemical ionization mass spectrometry for the analysis of peptides from aqueous solutions. *Anal. Chem.* **2002**, *74*(21), 5600–5605.
188. Miller, C. A.; Yi, D.; Perkins, P. D. An atmospheric pressure matrix-assisted laser desorption/ionization ion trap with enhanced sensitivity. *Rapid Commun. Mass Spectrom.* **2003**, *17*(8), 860–868.
189. Tan, P. V.; Laiko, V. V.; Doroshenko, V. M. Atmospheric pressure MALDI with pulsed dynamic focusing for high-efficiency transmission of ions into a mass spectrometer. *Anal. Chem.* **2004**, *76*(9), 2462–2469.
190. Harris, W. A.; Reilly, P. T. A.; Whitten, W. B. MALDI of individual biomolecule-containing airborne particles in an ion trap mass spectrometer. *Anal. Chem.* **2005**, *77*(13), 4042–4050.
191. Garrett, T. J.; Yost, R. A. Analysis of intact tissue by intermediate-pressure MALDI on a linear ion trap mass spectrometer. *Anal. Chem.* **2006**, *78*(7), 2465–2469.
192. Smith, S. A.; Blake, T. A.; Ifa, D. R.; Cooks, R. G.; Ouyang, Z. Dual-source mass spectrometer with MALDI -LQIT-ESI configuration. *J. Proteome Res.* **2007**, *6*(2), 837–845.
193. Fountain, S. T.; Lee, H.; Lubman, D. M. Ion fragmentation activated by matrix-assisted laser desorption/ionization in an ion-trap/reflectron time-of-flight device. *Rapid Commun. Mass Spectrom.* **1994**, *8*(5), 407–416.
194. Doroshenko, V. M.; Cotter, R. J. A quadrupole ion trap/time-of-flight mass spectrometer with a parabolic reflectron. *J. Mass Spectrom.* **1998**, *33*(4), 305–318.
195. Lee, H.; Lubman, D. M. Sequence-specific fragmentation generated by matrix-assisted laser desorption/ionization in a quadrupole ion trap/reflectron time-of-flight device. *Anal. Chem.* **1995**, *67* (8), 1400–1408.
196. Chen, Y.; Jin, X.; Misek, D.; Hinderer, R.; Hanash, S. M.; Lubman, D. M. Identification of proteins from two-dimensional gel electrophoresis of human erythroleukemia cells using capillary high performance liquid chromatography/electrospray–ion trap–reflectron time-of-flight mass spectrometry with two-dimensional topographic map analysis of in-gel tryptic digest products. *Rapid Commun. Mass Spectrom.* **1999**, *13*(19), 1907–1916.
197. Martin, R. L.; Brancia, F. L. Analysis of high mass peptides using a novel matrix-assisted laser desorption/ionisation quadrupole ion trap time-of-flight mass spectrometer. *Rapid Commun. Mass Spectrom.* **2003**, *17*(12), 1358–1365.
198. Warscheid, B.; Jackson, K.; Sutton, C.; Fenselau, C. MALDI analysis of Bacilli in spore mixtures by applying a quadrupole ion trap time-of-flight tandem mass spectrometer. *Anal. Chem.* **2003**, *75*(20), 5608–5617.
199. Kaiser, R. Ph.D. Thesis, Purdue University, West Lafayette, IN, 1990, pp. 120–23.
200. Eichelberger, J. W.; Budde, W. L.; Slivon, L. E. Existence of self chemical ionization in the ion trap detector. *Anal. Chem.* **1987**, *59*, 2730–2732.
201. Labastie, P.; Doy, M. A high mass range quadrupole spectrometer for cluster studies. *Int. J. Mass Spectrom. Ion Proc.* **1989**, *91*, 105–112.
202. Winger, B. E.; Light-Wahl, K. J.; Ogorzalek-Loo, R. R.; Udseth, H. R.; Smith, R. D. Observation and implications of high mass-to-charge ratio ions from electrospray ionization mass spectrometry. *J. Am. Soc. Mass Spectrom.* **1993**, *4*, 536–545.
203. Sobott, F.; Hernandez, H.; McCammon, M. G.; Tito, M. A.; Robinson, C. V. A tandem mass spectrometer for improved transmission and analysis of large macromolecular assemblies. *Anal. Chem.* **2002**, *74*, 1402–1407.
204. Makarov, A. A. Mass spectrometer. U.S. Patent 5,886,346, 1999.
205. Reiser, H. P.; Kaiser, R. E.; Savickas, P. J.; Cooks, R. G. Measurement of kinetic energies of ions ejected from a quadrupole ion trap. *Int. J. Mass Spectrom. Ion Proc.* **1991**, *106*, 237–247.
206. Mordehai, A. V.; Henion, J. A novel differentially pumped design for atmospheric pressure ionization–ion trap mass spectrometry. *Rapid Commun. Mass Spectrom.* **1993**, *7*, 205–209.

207. Washburn, M. P.; Wolters, D.; Yates, J. R. Large-scale analysis of the yeast proteome by multidimensional protein identification technology. *Nat. Biotechnol.* **2001**, *19*(3), 242–247.
208. BIOMASS™ deconvolution program, Finnigan Corp.
209. Bier, M. E. In Brown, W. E.; Howard, G. C. (Eds.), *Analysis of Proteins by Mass Spectrometry. Modern Protein Chemistry: Practical Aspects*, Catalog Number 9453; CRC Press, Boca Raton, FL; 2002, pp. 71–101.
210. ProMass, Novatia, LLC, Monmouth Junction, NJ, 2007.
211. Sogbein, O. O.; Simmons, D. A. Konermann, L. Effects of pH on the kinetic reaction mechanism of myoglobin unfolding studied by time-resolved electrospray ionization mass spectrometry, *J. Am. Soc. Mass Spectrom.* **2000**, *11*, 312–319.
212. McLuckey, S. A.; Ramsey, R. S. Gaseous myoglobin ions stored at greater than 300 K. *J. Am. Soc. Mass Spectrom.* **1994**, *5*, 324–327.
213. McLuckey, S. A.; Glish, G. L.; Kelley, P. E. Collision-activated dissociation of negative ions in an ion trap mass spectrometer. *Anal. Chem.* **1987**, *59*, 1670–1674.
214. Berberich, D. W.; Yost, R. A. Negative chemical ionization in quadrupole ion trap mass spectrometry: effects of applied voltages and reaction times. *J. Am. Soc. Mass Spectrom.* **1994**, *5*, 757–764.
215. March, R. E.; Hughes, R. *Practical Organic Mass Spectrometry*, John Wiley & Sons, New York, 1989.
216. Harrison, A. G. *Chemical Ionization Mass Spectrometry*, 2nd ed., CRC Press, Boca Raton, FL, 1992, p. 1.
217. Mather, R. E.; Todd, J. F. J. The quadrupole ion store (QUISTOR). Part VII. Simultaneous positive/negative ion mass spectrometry. *Int. J. Mass Spectrom. Ion Phys.* **1980**, *33*, 159.
218. Schermann, J. P.; Major, F. G. Characteristics of electron-free plasma confinement in an rf quadrupole field. *Appl. Phys.* **1978**, *16*, 225.
219. Covey, T. R.; Bonner, R. F.; Shushan, B. I.; Henion, J. The determination of protein, oligonucleotide and peptide molecular weights by ion-spray mass spectrometry. *Rapid Commun. Mass Spectrom.* **1988**, *2*, 249–256.
220. McLuckey, S. A.; Van Berkel, G. J. V.; Glish, G. L. Tandem mass spectrometry of small, multiply charged oligonucleotides. *J. Am. Soc. Mass Spectrom.* **1992**, *3*, 60–70.
221. Datta, B.; Bier, M. E.; Roy, S.; Armitage, B. A. Quadruplex formation by a guanine-rich PNA oligomer. *J. Am. Chem. Soc.* **2005**, *127*(12), 4199–4207.
222. Doktycz, M. J.; Hurst, G. B.; Habibi-Goudarzi, S.; McLuckey, S. A.; Tang, K.; Chen, C. H.; Uziel, M.; Jacobson, K. B.; Woychik, R. P.; Buchanan, M. V. *Anal. Biochem.* **1995**, *230*, 205–214.
223. Smith, R. D.; Loo, J. A.; Edmonds, C. G.; Barinaga, C. J.; Udseth, H. R. New developments in biochemical mass spectrometry: electrospray ionization. *Anal. Chem.*, **1990**, *62*, 882–899.
224. Hail, M. E.; Elliott, B.; Anderson, K. High-throughput analysis of oligonucleotides using automated electrospray ionization mass spectrometry. *Am. Biotech. Lab.* **2004**, *22*, 12–14.
225. Apffel, A.; Chakel, J. A.; Fischer, S.; Lichtenwalter, K.; Hancock, W. S. Analysis of oligonucleotides by HPLC-electrospray ionization mass spectrometry. *Anal. Chem.* **1997**, *69*, 1320–1325.
226. Wang, M.; Wells, G. Non-resonance excitation and ejection in ion trap. In *Proceedings of the 41st Annual Conference of Mass Spectrometry and Allied Topics*, San Francisco, CA, **1993**, p. 463.
227. Qin, J.; Chait, B. T. Matrix-assisted laser desorption ionization ion trap tandem mass spectrometry of peptides: A new excitation scheme for CID. In *Proceedings of the 43rd Conference on Mass Spectrometry and Allied Topics*, Atlanta, GA, May 21–26, **1995**, p. 1100.
228. McLuckey, S. A.; Habibi-Goudarzi, S. Ion trap tandem mass spectrometry applied to small multiply charged oligodeoxyribonucleotides with a modified base. *J. Am. Soc. Mass Spectrom.* **1994**, *5*, 740.
229. Roepstorff, P. Proposal for a common nomenclature for sequence ions in mass spectra of peptides. *Biomed. Mass Spectrom.* **1984**, *11*(11), 601.
230. McLuckey, S. A.; Habibi-Goudarzi, S. Decompositions of multiply charged oligonucleotide anions. *J. Am. Chem. Soc.* **1993**, *115*, 12085.
231. Doktycz, M. J.; Habibi-Goudarzi, S.; McLuckey, S. A. Accumulation and storage of ionized duplex DNA molecules in a quadrupole ion trap. *Anal. Chem.* **1994**, *66*, 3416–3422.
232. Williams, E. R.; Furlong, J. J. P.; McLafferty, F. W. Efficiency of collisionally-activated dissociation and 193-nm photodissociation of peptide ions in Fourier-transform mass spectrometry. *J. Am. Soc. Mass Spectrom.* **1990**, *1*, 288–294.
233. Little, D. P.; Speir, P. J.; Senko, M. W.; O'Connor, P. B. McLafferty, F. W. Infrared multiphoton dissociation of large multiply charged ions for biomolecule sequencing. *Anal. Chem.* **1994**, *66*, 2809–2815.
234. Castro, J. A.; Nuwaysir, L. M.; Ijames, C. F.; Wilkins, C. L. Comparative study of photodissociation and surface-induced dissociation by laser desorption Fourier transform mass spectrometry. *Anal. Chem.* **1992**, *64*, 2238–2243.
235. Hughes, R. J.; March, R. E.; Young, A. B. Multiphoton dissociation of ions derived from 2-propanol in a QUISTOR with low-power CW infrared laser radiation. *Int. J. Mass Spectrom. Ion Phys.* **1982**, *42*, 255–263.
236. Hughes, R. J.; March, R. E.; Young, A. B. Multiphoton dissociation of gaseous ions in a quadrupole ion store (QUISTOR). *Int. J. Mass Spectrom. Phys.* **1983**, *47*, 85–88.

237. Hughes, R. J.; March, R. E.; Young, A. B. Optimization of ion trapping characteristics for studies of ion photodissociation. *Can. J. Chem.* **1983**, *61*, 824.
238. Ensberg, E. S.; Jefferts, K. B. Visible photodissociation spectrum of ionized methane. *Astrophys. J.* **1975**, *195*, L89.
239. Louris, J. N.; Brodbelt, J. S.; Cooks, R. G. Photodissociation in a quadrupole ion trap mass spectrometer using a fiber optic interface. *Int. J. Mass Spectrom. Ion Proc.* **1987**, *75*, 345–352.
240. Stephenson, J. L.; Booth, M. M.; Boue, S. M.; Eyler, J. R.; Yost, R. A. Analysis of biomolecules using electrospray ionization-ion-trap mass spectrometry and laser photodissociation. ACS Symposium Series 1996, 619 (Biochemical and Biotechnological Applications of Electrospray Ionization Mass Spectrometry), 512–564.
241. Stephenson, J. L.; Booth, M. M.; Shalosky, J. A.; Eyler, J. R.; Yost, R. A. Infrared multiple photon dissociation in the quadrupole ion trap via a multipass optical arrangement. *J. Am. Soc. Mass Spectrom.* **1994**, *5*, 886–893.
242. Boue, S. M., Stephenson, J. L.; Yost, R. A. Photodissociation of electrosprayed in the quadrupole ion trap. In *Proceedings of the 43rd ASMS Conference on Mass Spectrometry and Allied Topics*, Atlanta, GA, May 21–26, **1995**, p. 408.
243. Stephenson, J. L. Personal communication.
244. Kim, T.- Y.; Thompson, M. S.; Reilly, J. P. Peptide photodissociation at 157 nm in a linear ion trap mass spectrometer. *Rapid Commun. Mass Spectrom.* **2005**, *19*, 1657–1665.
245. Payne, A. H.; Glish, G. L. Thermally assisted infrared multiphoton photodissociation in a quadrupole ion trap. *Anal. Chem.* **2001**, *73*, 3542–3548.
246. Crowe, M. C.; Brodbelt, J. S. Infrared multiphoton dissociation (IRMPD) and collisionally activated dissociation of peptides in a quadrupole ion trap with selective IRMPD of phosphopeptides. *J. Am. Soc. Mass Spectrom.* **2004**, *15*, 1581–1592.
247. Shen, J.; Brodbelt, J. S. Characterization of ionophore–metal complexes by infrared multiphoton photodissociation and collision activated dissociation in a quadrupole ion trap mass spectrometer. *Analyst* **2000**, *125*, 641–650.
248. Keller, K. M.; Brodbelt, J. S. Collisionally activated dissociation and infrared multiphoton dissociation of oligonucleotides in a quadrupole ion trap. *Anal. Biochem.* **2004**, *326*, 200–210.
249. Goolsby, B. J.; Brodbelt, J. S. Characterization of β-lactams by photodissociation and collision-activated dissociation in a quadrupole ion trap. *J. Mass Spectrom.* **1998**, *33*, 705–712.
250. Colorado, A.; Shen, J. X. X.; Vartanian, V. H.; Brodbelt, J. Use of infrared multiphoton photodissociation with SWIFT for electrospray ionization and laser desorption applications in a quadrupole ion trap mass spectrometer. *Anal. Chem.*, **1996**, *68*, 4033–4043.
251. Hashimoto, Y.; Hasegawa, H.; Yoshinari, K.; Waki, I. Collision-activated infrared multiphoton dissociation in a quadrupole ion trap mass spectrometer. *Anal. Chem.*, **2003**, *75*, 420–425.
252. Crowe, M. C.; Brodbelt, J. S.; Goolsby, B. J.; Hergenrother, P. *J. Am. Soc. Mass Spectrom.* **2002**, *13*, 630–649.
253. Crowe, M. C.; Brodbelt, J. S. Differentiation of phosphorylated and unphosphorylated peptides by high-performance liquid chromatography–electrospray ionization–infrared multiphoton dissociation in a quadrupole ion trap. *Anal. Chem.* **2005**, *77*, 5726–5734.
254. Murrell, J.; Despeyroux, D.; Lammert, S. A.; Stephenson, J. L.; Goeringer, D. E. "Fast excitation" CID in a quadrupole ion trap mass spectrometer. *J. Am. Soc. Mass Spectrom.* **2003**, *14*, 785–789.
255. Wilson, Jeffrey, J., Brodbelt, Jennifer, S. Infrared Multiphoton Dissociation of Dulex DNA/Drug Complexes in a Quadrupole Ion Trap. *Analytical Chemistry* **2007**, *79*(5), 2067–2077.
256. Herron, W. J.; Goeringer, D. E.; McLuckey, S. A. Product ion charge state determination via ion/ion proton transfer reactions. *Anal. Chem.* **1996**, *68*(2), 257–262.
257. McLuckey, S. A.; Herron, W. J.; Stephenson, J. L., Jr.; Goeringer, D. E. Cation attachment to multiply charged anions of oxidized bovine insulin A-chain. *J. Mass Spectrom.* **1996**, *31*(10), 1093–1100.
258. Stephenson, J. L., Jr.; McLuckey, S. A. Ion/ion proton transfer reactions for protein mixture analysis. *Anal. Chem.* **1996**, *68*(22), 4026–4032.
259. Stephenson, J. L., Jr.; McLuckey, S. A. Counting basic sites in oligopeptides via gas-phase ion chemistry. *Anal. Chem.* **1997**, *69*(3), 281–285.
260. Stephenson, J. L., Jr.; McLuckey, S. A. Charge reduction of oligonucleotide anions via gas-phase electron transfer to xenon cations. *Rapid Commun. Mass Spectrom.* **1997**, *11*(8), 875–880.
261. McLuckey, S. A.; Stephenson, J. L., Jr.; Asano, K. G. Ion/ion proton-transfer kinetics: implications for analysis of ions derived from electrospray of protein mixtures. *Anal. Chem.* **1998**, *70*(6), 1198–1202.
262. Stephenson, J. L., Jr.; McLuckey, S. A. Charge manipulation for improved mass determination of high-mass species and mixture components by electrospray mass spectrometry. *J. Mass Spectrom.* **1998**, *33*(7), 664–672.
263. Stephenson, J. L., Jr.; McLuckey, S. A. Ion/ion reactions for oligopeptide mixture analysis: Application to mixtures comprised of 0.5–100 kDa components. *J. Am. Soc. Mass Spectrom.* **1998**, *9*(6), 585–596.
264. Stephenson, J. L., Jr.; Schaaff, T. G.; McLuckey, S. A. Hydroiodic acid attachment kinetics as a chemical probe of gaseous protein ion structure: bovine

pancreatic trypsin inhibitor. *J. Am. Soc. Mass Spectrom.* **1999**, *10*(6), 552–556.

265. Stephenson, J. L., Jr; McLuckey, S. A. Simplification of product ion spectra derived from multiply charged parent ions via ion/ion chemistry. *Anal. Chem.* **1998**, *70* (17), 3533–3544.
266. McLuckey, S. A.; Stephenson, J. L., Jr. Ion/ion chemistry of high-mass multiply charged ions. *Mass spectrometry reviews* **1998**, *17*(6), 369–407.
267. McLuckey, S. A.; Goeringer, D. E. Ion/molecule reactions for improved effective mass resolution in electrospray mass spectrometry. *Anal. Chem.* **1995**, *67*, 2493–2497.
268. McLuckey, S. A.; Glish, G. L.; Van Berkel, G. J. Charge determination of product ions formed from collision-induced dissociation of multiply protonated molecules via ion/molecule reactions. *Anal Chem.* **1991**, *63*, 1971–1978.
269. Schermann, J. P.; Major, F. G. Characteristics of electron-free plasma confinement in an rf quadrupole field. *Appl. Phys.* **1978**, *16*, 225.
270. Mather, R. E.; Todd, J. F. J. The quadrupole ion store (QUISTOR). Part VII. Simultaneous positive/negative ion mass spectrometry. *Int. J. Mass Spectrom. Ion Phys.* **1980**, *33*, 159–165.
271. Williams, J. D.; Cooks, R. G. Reduction of space-charging in the quadrupole ion trap by sequential injection and simultaneous storage of positively and negatively charged ions. *Rapid Commun. Mass Spectrom.* **1993**, *7*, 380–382.
272. Gorshkov, M. V.; Guan, S.; Marshall, A. G. Dynamic ion trapping for Fourier-transform ion cyclotron resonance mass spectrometry: Simultaneous positive- and negative-ion detection. *Rapid Commun. Mass Spectrom.* **1992**, *6*, 166.
273. Herron, W. J.; Goeringer, D. E.; McLuckey, S. A. Ion–ion reactions in the gas phase: Proton transfer reactions of protonated pyridine with multiply charged oligodeoxyribonucleotide anions. *J. Am. Soc. Mass Spectrom.* **1995**, *6*, 529–532.
274. McLuckey, S. A.; Glish, G. L.; Asano, K. G.; Van Berkel, G. J. Self chemical ionization in an ion trap mass spectrometer. *Anal. Chem.* **1988**, *60*, 2312–2314.
275. Pannell, L. K.; Pu, Q. L.; Fales, H. M.; Mason, R. T.; Stephenson, J. L. Intermolecular processes in the ion trap mass spectrometer. *Anal. Chem.* **1989**, *61*, 2500–2503.
276. Ratnayake, W. M. N.; Timmins, A.; Ohshima, T.; Ackman, R. G. Mass spectra of fatty acid derivatives, of isopropylidenes of novel glyceryl ethers of cod muscle and of phenolic acetates obtained with the Finnigan MAT ion trap detector. *Lipids.* **1986**, *21*, 518–524.
277. Herron, W. J.; Goeringer, D. E.; McLuckey, S. A. Product ion charge state determination via ion/ion proton transfer reactions. *Anal. Chem.* **1996**, *68*(2), 257–262.
278. Zubarev, R. A.; Kelleher, N. L.; McLafferty, F. W. Electron capture dissociation of multiply charged protein cations. A nonergodic process. *J. Am. Chem. Soc.* **1998**, *120*, 3265–3266.
279. Zubarev, R. A.; Kruger, N. A.; Fridriksson, E. K.; Lewis, M. A.; Horn, D. M.; Carpenter, B. K.; McLafferty, F. W. Electron capture dissociation of gaseous multiply-charged proteins is favored at disulfide bonds and other sites of high hydrogen atom Affinity. *J. Am. Chem. Soc.* **1999**, *121*, 2857–2862.
280. Breuker, K.; Oh, H. B.; Cheng, L.; Barry, K. C.; and McLafferty, F. W. Nonergodic and conformational control of the electron capture dissociation of protein cations. *Proceedings of the National Academy of Sciences*, **2004**, *101*, 14011.
281. Taybin, Y. O.; Hakanson, P.; Budnik, B. A.; Haselmann, K. F.; Kjeldsen, F.; Gorshkov, M.; Zubarev, R. A. Improved low-energy electron injection systems for high rate electron capture dissociation in Fourier transform ion cyclotron resonance mass spectrometry. *Rapid Commun. Mass Spectrom.* **2001**, *15*, 1849–854.
282. Zubarev, R. A.; Horn, D. M.; Fridriksson, E. K.; Kelleher, N. L.; Kruger, N. A.; Lewis, M. A.; Carpenter, B. K.; McLafferty, F. W. Electron capture dissociation for structural characterization of multiply charged protein cations. *Anal. Chem.* **2000**, *72*, 563.
283. Zubarev, R. A. Electron-capture dissociation tandem mass spectrometry. *Curr. Opin. Biotechnol.* **2004**, *15*, 12–16.
284. Vachet, R. W.; Clark, S. D.; Glish, G. L. Electron capture by ions in a quadrupole ion trap. In *Proceedings of the 43rd ASMS Conference on Mass Spectrometry and Allied Topics*, **1995**, p. 1111.
285. Baba, T.; Hashimoto, Y.; Hasegawa, H.; Hirabayashi, A.; Waki, I. Electron capture dissociation in a radio frequency ion trap. *Anal. Chem.* **2004**, *76*(15), 4263–4266.
286. Syka, J. E. P.; Coon, J. J.; Schroeder, M. J.; Shabanowitz, J.; Hunt, D. F. Peptide and protein sequence analysis by electron transfer dissociation mass spectrometry. *Proc. Nat. Acad. Sci. USA* **2004**, *101*(26), 9529.
287. Coon, J.; Schabanowitz, J.; Syka, J. E. P.; Hunt, D. F. Electron transfer dissociation of peptide anions. *J. Am. Soc. Mass Spectrom.* **2005**, *16*, 880–882.
288. Liang, X.; Hager, J. W.; McLuckey, S. A. Transmission mode ion/ion electron-transfer dissociation in a linear ion trap. *Anal. Chem.* **2007**, *79*, 3363–3370.
289. McAlister, G. C.; Phanstiel, D.; Good, D. M.; Berggren, W. T.; Coon, J. J. Implementation of electron-transfer dissociation on a hybrid linear ion trap-orbitrap mass spectrometer. *Anal. Chem.* **2007**, *79*, 3525–3534.
290. McLuckey, S. A.; Van Berkel, G. J.; Glish, G. L.; Huang, E. C.; Henion, J. D. Ion spray liquid chromatography/ion trap mass spectrometry

determination of biomolecules. *Anal. Chem.* **1991**, *63*, 375–383.

291. Kennedy, R. T.; Jorgenson, J. W. Preparation and evaluation of packed capillary liquid chromatography columns with inner diameters from 20 to 50 micrometers. *Anal. Chem.* **1989**, *61*, 1128–1135.

292. Cox, A. L.; Skipper, J.; Chen, Y.; Henderson, R. A.; Darrow, T. L.; Shabanowitz, J.; Engelhard, V. H.; Hunt, D. F.; Slingluff, C. L., Jr. Identification of a peptide recognized by five melanoma-specific human cytotoxic T cell lines. *Science (New York, N.Y.)* **1994**, *264*(5159), 716–719.

293. Haskins, W. E.; Wang, Z.; Watson, C. J.; Rostand, R. R.; Witowski, S. R.; Powell, D. H.; Kennedy, R. T. Capillary LC-MS2 at the attomole level for monitoring and discovering endogenous peptides in microdialysis samples collected *in vivo*. *Anal. Chem.* **2001**, *73*(21), 5005–5014.

294. Stutz, H. Advances in the analysis of proteins and peptides by capillary electrophoresis with matrix-assisted laser desorption/ionization and electrospray–mass spectrometry detection. *Electrophoresis* **2005**, *26*, 1254–1290.

295. Simpson, D. C.; Smith, R. D. Combining capillary electrophoresis with mass spectrometry for applications in proteomics. *Electrophoresis* **2005**, *26*, 1291–1305.

296. Schmitt-Kopplin, P., Englmann, M., Capillary electrophoresis–mass spectrometry: Survey on developments and applications 2003–2004. *Electrophoresis* **2005**, *26*, 1209–1220.

297. Hu, S., Zhang, L., Newitt, R., Aebersold, R., Kraly, J. R., Jones, M., Dovichi, N. J. Identification of proteins in single-cell capillary electrophoresis fingerprints based on comigration with standard proteins. *Anal. Chem.* **2003**, *75*, 3502–3505.

298. Hernandez-Borges, J., Neusüss, C., Cifuentes, A., Pelzing, M. On-line capillary electrophoresis-mass spectrometry for the analysis of biomolecules. *Electrophoresis* **2004**, *25*, 2257–2281.

299. Hommerson, P.; Khan, A. M.; de Jong, G. J.; Somsen, G. W. Comparison of atmospheric pressure photoionization and ESI for CZE-MS of drugs. *Electrophoresis* **2007**, *28*, 1444–1453.

300. Monton, M. M. N.; Terabe, S. Recent developments in capillary electrophoresis–mass spectrometry of proteins and peptides. *Anal. Sci.* **2005**, *21*, 5–13.

301. Schmitt-Kopplin, P.; (Ed.). Special Issue: Capillary Electrophoresis–Mass spectrometry 2007. *Electrophoresis* **2007**, *28*(9), 1303.

302. Hofstadler, S. A.; Wahl, J. H.; Bruce, J. E.; Smith, R. D. On-line capillary electrophoresis with Fourier transform ion cyclotron resonance mass spectrometry. *J. Am Chem. Soc.* **1993**, *115*, 6983.

303. Vladislav Dolník Capillary electrophoresis of proteins 2003–2005. *Electrophoresis* **2006**, *27*, 126–141.

304. Olivares, J. A., Nguyen, N. T., Yonker, C. R., Smith, R. D. Measurement of isotope ratios by Doppler-free laser spectroscopy applying semiconductor diode lasers and thermionic diode detection. *Anal. Chem.* **1987**, *59*, 1230–1232.

305. Ramsey, R. S.; Goeringer, D. E.; McLuckey, S. A. Active chemical background and noise reduction in capillary electrophoresis/ion-trap mass spectrometry. *Anal. Chem.* **1993**, *65*, 3521–3524.

306. Henion, J. D.; Mordehai, A. V.; Cai, J. Quantitative capillary electrophoresis–ion spray mass spectrometry on a benchtop ion trap for the determination of isoquinoline alkaloids. *Anal. Chem.* **1994**, *66*, 2103–2109.

307. Ramsey, R. S.; McLuckey, S. A. Gaseous myoglobin ions stored at greater than 300 K. *J. Am. Soc. Mass Spectrom.* **1994**, *5*, 324.

308. Moseley, M. A.; Jorgenson, J. W.; Shabanowitz, J.; Hunt, D. F.; Komer, K. B. Optimization of capillary zone electrophoresis/electrospray ionization parameters for the mass spectrometry and tandem mass spectrometry analysis of peptides. *J. Am. Soc. Mass Spectrom.*, **1992**, *3*, 289–300.

309. Li, C. P. Chinese Herbal Medicine; U.S. Department of Health, Education, and Welfare, Public Health Service, National Institute of Health, **1974**.

310. Ramsey, R. S.; McLuckey, S. A. Capillary electrophoresis/electrospray ionization ion trap mass spectrometry using a sheathless interface. *J. Microcolumn Separations* **1995**, *7*, 461–469.

311. Stahl, D. C.; Martino, P. A.; Swiderek, K. M.; Davis, M. T.; Lee, T. D. In *Proceedings of the 40th ASMS Conference on Mass Spectrometry and Allied Topics*, Washington DC, May 31–June 5, **1992**, pp. 1801–1802.

312. Mylchreest, I.; Campbell, C.; Wheeler, K.; Wakefield, M. Data dependent automated MS/MS switching Methods for routine LC/MS/MS. *Proceedings of the 43rd Conference on Mass Spectrometry and Allied Topics*, Atlanta, GA, May 21–26, **1995**, p. 436.

313. Syka, J. E. P.; Marto, J. A; Bai, D. L; Horning, S.; Senko, M. W; Schwartz, J. C.; Ueberheide, B.; Garcia, B.; Busby, S.; Muratore, T.; Shabanowitz, J.; Hunt, D. F. Novel linear quadrupole ion trap/FT mass spectrometer: performance characterization and use in the comparative analysis of histone H3 post-translational modifications. *J. Proteome Res.* **2004**, *3* (3), 621–626.

314. Arnott, D.; King, K.; Bier, M.; Land, A.; Stults, J. Identification of proteins associated with hypertrophy of cardiomycytes by 2D gel electrophoresis combined with Microcapillary LCMS and tandem MS. In *Proceedings of the 43rd ASMS Conference on Mass Spectrometry and Allied Topics*, Atlanta, GA, May 21–26, **1995**, p. 31.

315. Kanai, M.; Seta, K.; Nakayama, H.; Isobe, T.; Land, A. P.; Bier, M. E. HPLC-ESI ion trap MS/MS analysis

of proteins separated by 2D-PAGE. In *Proceedings of the 43rd Conference on Mass Spectrometry and Allied Topics*, Atlanta, GA, May 21–26, **1995**, 616.

316. GILLECE-CASTRO, B. L.; ARNOTT, D. P.; BIER, M. E.; LAND, A. P.; STULTS, J. T. Peptide phosphorylation site identification: A new method based on ion trap mass spectrometry. In *Proceedings of the 43rd Conference on Mass Spectrometry and Allied Topics*, Atlanta, GA, May 21–26, **1995**, p. 302.
317. ENG, J. K.; MCCORMACK, A. L.; YATES, J. R. An approach to correlate tandem mass spectral data of peptides with amino acid sequences in a protein database. *J. Am. Soc. Mass Spectrom.* **1994**, *5*, 976–989.
318. YATES, J. R.; ENG, J. K.; MCCORMACK, A. L.; SCHIELTZ, D. Method to correlate tandem mass spectra of modified peptides to amino acid sequences in the protein database. *Anal. Chem.* **1995**, *67*, 1426–1436.
319. ENG, J. K.; YATES, J. R.; SCHIELTZ, D. M. Protein database searching with ion trap MS^n data. *Proceedings of the 43rd Conference on Mass Spectrometry and Allied Topics*, Atlanta, GA, May 21–26, **1995**, p. 641.
320. GYGI, S. P.; RIST, B.; GERBER, S. A.; TURECEK, F.; GELB, M. H.; AEBERSOLD, R. Quantitative analysis of complex protein mixtures using isotope-coded affinity tags. *Nat. Biotechnol.* **1999**, *17*(10), 994–999.
321. DESOUZA, L.; DIEHL, G.; RODRIGUES, M. J.; GUO, J.; ROMASCHIN, A. D.; COLGAN, T.J; SIU, K. W. M. Search for cancer markers from endometrial tissues using differentially labeled tags iTRAQ and cICAT with multidimensional liquid chromatography and tandem mass spectrometry. *J. Proteome Res.* **2005**, *4*(2), 377–386.
322. ONG, S.-E.; BLAGOEV, B.; KRATCHMAROVA, I.; KRISTENSEN, D. B.; STEEN, H.; PANDEY, A.; MANN, M. Stable isotope labeling by amino acids in cell culture, SILAC, as a simple and accurate approach to expression proteomics. *Mol. Cell. Proteomics* **2002**, *1* (5), 376–386.
323. GOLOGAN, B.; TAKATS, Z.; ALVAREZ, J.; WISEMAN, J. M.; TALATY, N.; OUYANG, Z.; COOKS, R. G. Ion soft-landing into liquids: Protein identification, separation, and purification with retention of biological activity. *J. Am. Soc. Mass Spectrom.* **2004**, *15*(12), 1874–1884.
324. TAKATS, Z.; WISEMAN, J. M.; COOKS, R. G. Ambient mass spectrometry using desorption electrospray ionization (DESI): Instrumentation, mechanisms and applications in forensics, chemistry, and biology. *J. Mass Spectrom.* **2005**, *40*(10), 1261–1275.
325. COOKS, R. G.; OUYANG, Z.; TAKATS, Z.; WISEMAN, J. M. Ambient mass spectrometry. *Science*, **2006**, *311* (5767), 1566–1570.
326. PAN, Z.; GU, H.; TALATY, N.; CHEN, H.; SHANAIAH, N.; HAINLINE, B. E.; COOKS, R. G.; RAFTERY, D. Principal component analysis of urine metabolites detected by NMR and DESI - MS in patients with inborn errors of metabolism. *Anal. Bioanal. Chem.* **2007**, *387*(2), 539–549.
327. CODY, R. B.; LARAMEE, J. A.; DURST, H. D. Versatile new ion source for the analysis of materials in open air under ambient conditions. *Anal. Chem.* **2005**, *77*(8), 2297–2302.
328. NEMES, P.; VERTES, A. Laser ablation electrospray ionization for atmospheric pressure, *in vivo*, and imaging mass spectrometry. *Anal. Chem.* **2007**, *79* (21), 8098–8106.
329. Personal communication with Mark Hail, Novatia LLC, **2007**.
330. Van BERKEL, G. J.; MCLUCKEY, S. A.; GLISH, G. L. Electrospray ionization of porphyrins using a quadrupole ion trap for mass analysis. *Anal. Chem.* **1991**, *63*, 1098–1109.
331. RAMSEY, R. S.; Van BERKEL, G. J.; MCLUCKEY, S. A.; GLISH, G. L. Determination of pyrimidine cyclobutane dimers by electrospray ionization/ion trap mass spectrometry. *Biol. Mass Spectrom.* **1992**, *21*, 347–352.
332. LIM, H. K.; WU, W-N.; MORDEHAI, A.; HENION, J. D. Determination of quaternary isoquinoline alkaloids by ion spray ion trap mass spectrometer (IS/ITMS). In *Proceedings of the 41st Conference on Mass Spectrometry and Allied Topics*, San Francisco, CA, May 31–June 4, **1993**, p. 51.
333. MEHL, J. T.; CUMMINGS, J. J.; ROHDE, E.; YATES, N. Automated protein identification using atmospheric-pressure matrix-assisted laser desorption/ionization. *Rapid Commun. Mass Spectrom.* **2003**, *17*(14), 1600–1610.
334. YI, E. C.; LEE, H.; AEBERSOLD, R.; GOODLETT, D. R. A microcapillary trap cartridge-microcapillary high-performance liquid chromatography electrospray ionization emitter device capable of peptide tandem mass spectrometry at the attomole level on an ion trap mass spectrometer with automated routine operation. *Rapid Commun. Mass Spectrom.* **2003**, *17*(18), 2093–2098.
335. VANHASSEL, E.; BIER, M. E. An electrospray membrane probe for the analysis of volatile and semivolatile organic compounds in water. *Rapid Commun. Mass Spectrom.* **2007**, *21*, 413–420.
336. JONCHERS, K. Not your run-of-the-mill test tube. Product review. *Anal. Chem.* **2006**, Oct. 1, 6713.

Chapter 10

Time-of-Flight Mass Spectrometry

Robert J. Cotter

Middle Atlantic Mass Spectrometry Laboratory, Department of Pharmacology and Molecular Sciences, Johns Hopkins School of Medicine, Baltimore, Maryland

Electrospray and MALDI Mass Spectrometry: Fundamentals, Instrumentation, Practicalities, and Biological Applications, Second Edition, Edited by Richard B. Cole

10.1 THEN AND NOW

What began in the 1950s as an instrument with a mass range of 400 Da now routinely records ions with masses of 40 kDa, or even 400 kDa! MALDI[1,2] is, of course, well-recognized as the reason for the dramatic increase in mass range. But, both MALDI and electrospray[3] changed our perceptions about the size of molecules that could be lofted intact and in ionized form into the gas phase, and precipitated other changes to these instruments. For example, in the 1970s the Model 2000 (CVC Products, Rochester, NY), the LAMMA 1000 (Leybold-Hereaus, Cologne, Germany) and the LIMA (Cambridge Instruments, Cambridge, UK) used accelerating voltages in the 3- to 5-kV range, limiting the detectable mass range using conventional particle multipliers. Both continuous dynode and multichannel plate (MCP) detectors rely upon secondary electron emission for detection, which decreases with increasing mass and decreasing ion velocity.[4] One approach is to increase the ion kinetic energy; thus, in the 1980s the BIO ION Nordic (Uppsala, Sweden) plasma desorption mass spectrometer used an accelerating voltage of 10 kV to record ions as high as 35 kDa.[5] Today, voltages of 20–30 kV are typical for matrix-assisted laser desorption/ionization (MALDI) instruments. There have been other significant improvements in detector design as well: faster channel plates, conversion dynodes, and inductive detectors.[6] Cryocooled detectors,[7,8] such as the superconducting tunnel junction (STJ) detector (Figure 10.1) available on the Comet (Flammatt, Switzerland) MACROMIZER, can easily detect ions in the 100-kDa to several-megadalton range. Knowing that we can make such large ions has been a powerful driving force for development.

MALDI has always been well-matched to the time-of-flight (TOF) mass spectrometer, an instrument that can record all the ions made by each short (subnanosecond) laser pulse. Though we generally take the *multichannel recording advantage* of the TOF for granted, it was not always the case. Instruments commercialized in the 1950s by the Bendix Corporation (Rochester, NY) used an oscilloscope to produce repetitive displays from each ionizing cycle, but permanent recording required a *boxcar* approach that Holland et al.[9] have described as *time-slice recording*, the recording of 10-ns time segments from each 100-μs spectral transient. In addition to the obviously paltry duty cycle, there were other consequences. The recording cycle was determined by the high repetition rate (10 kHz) required to produce a "scanned" mass spectrum, which, in turn, limited the mass range to around 400 Da.

Thus, a key element of today's TOF mass spectrometers is the availability of high-speed, high-bandwidth digitizers: *time-to-digital converters* (TDCs), *waveform* or *transient recorders*, and *digital oscilloscopes*. In the 1970s, waveform recorders manufactured by Biomation (Ontario, Canada) were used on the laser microprobe instruments, but these 6- to 8-bit digitizers had sampling rates of 20 Msamples/s and effective dynamic range at full recording speed of around 3.5 bits. Today, 8-Gsample/s 8-bit, high-bandwidth digital scopes have a far better dynamic range and longer record lengths for recording the wide mass ranges of protein/peptide spectra.

The early inventors of TOF mass spectrometers provided remarkable insights into the mass resolution problem, with approaches to addressing this aspect of TOF instruments that are pertinent to this day. Developing a focusing method compatible with an electron impact ionization source, they devised an approach using pulsed ion extraction which, when applied after a suitable time delay, would improve mass resolution by compensating for differences in initial kinetic energy. Today, the subnanosecond pulsed lasers used in laser desorption and MALDI provide a more compatible ionization source that has seemed for many years to not require this complex approach. *Delayed extraction* has of course now been revisited[10–13] to

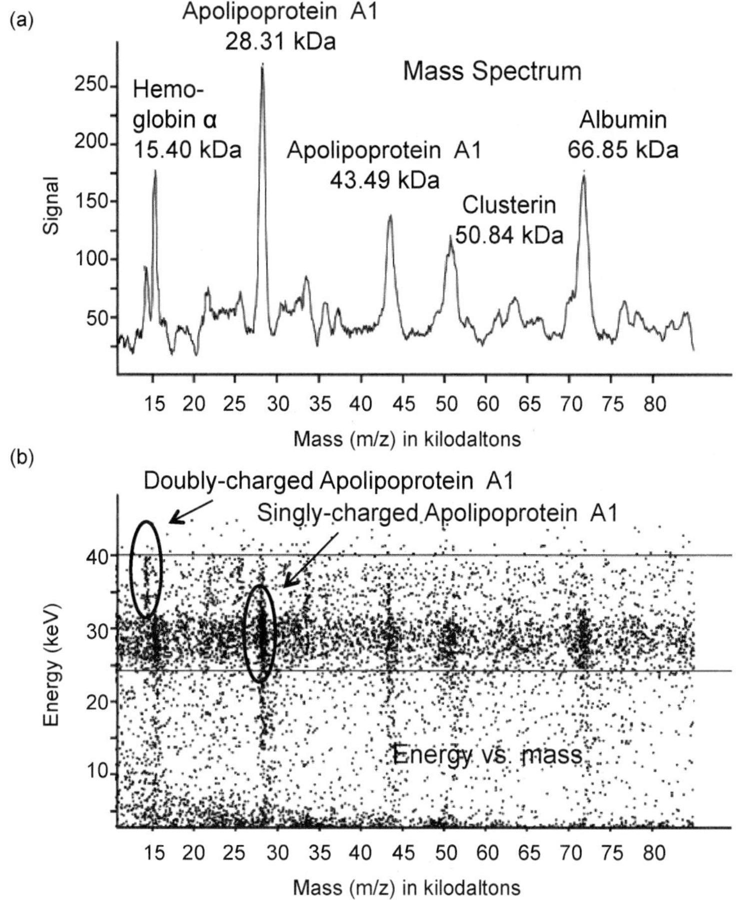

Figure 10.1. High-mass MALDI spectrum recorded using a superconducting tunnel junction detector. In addition to the mass spectrum (**a**), the detector records ion kinetic energies (**b**), enabling one to distinguish multiply charged ions and metastables.

provide extraordinarily high mass resolution with high-speed electronics providing high-voltage extraction pulses with very short rise times. Compatibility with continuous ionization sources has also been revisited, primarily with electrospray. Perhaps one of the most significant recent innovations in TOF technology has been the development of the *orthogonal acceleration* TOF (or *oaTOF*), which enables use of virtually any ionization technique, including those carried out at atmospheric pressure.[14,15] Though very different from classical TOF geometry, oaTOF design addresses the same spatial and kinetic energy questions (described below) to enable good focusing and high mass resolution.

10.2 THE QUESTION OF MASS RESOLUTION

In 1946 W. E. Stephens suggested that a mass spectrometer could be constructed without a magnetic field, measuring velocities from ions formed in a pulsed source.[16] Two years later,

Cameron and Eggers[17] introduced their *velocitron*, an instrument with a Nier electron impact source, a 317-cm drift tube, and oscilloscope recording. Ions were accelerated to a constant energy of 500 eV, so that their velocities would be inversely proportional to the square root of their masses. The 1955 instrument by Katzenstein and Friedland[18] pulsed both the ionizing electron beam and the ion extraction field, using a *drawout* (or *pushout*) *pulse*. A pulsed extraction approach was also developed by Wiley and McLaren[19] as a means to improve mass resolution. Their instrument was commercialized, and their approach is the forerunner of the current *delayed extraction* methods used in MALDI.

10.2.1 Time-of-Flight Basics

The basic TOF mass spectrometer consists of (a) a short source region s with a high electric field E and (b) a longer drift region D which is field-free (Figure 10.2). Typically, the source length may be around 1 cm, while the drift length is 1 m or more. An accelerating voltage V defines the field $E = V/s$, and on many different instruments it has ranged from 3 to 20 kV. If a positive ion is formed on the source *backing plate* (Figure 10.2a), then it is accelerated toward the drift region to a kinetic energy $\frac{1}{2}mv^2 = eV$, where m is the ion mass, v is its velocity, and e is the charge. The time t for the ion to travel across the drift region, $t = D/v$, is then

$$t = \left(\frac{m}{2eV}\right)^{1/2} D \tag{10.1}$$

While this is the general form of the TOF equation used to calibrate TOF instruments, it is an approximation. First, the ions do spend a finite time in the source region and, if the ions are formed in the space above the backing plate (Figure 10.2b), then the flight time is more accurately expressed as

$$t = \left(\frac{m}{2eEs}\right)^{1/2} (2s + D) \tag{10.2}$$

where s may be a distribution of positions: $s = s_0 + \Delta s$ around a mean distance s_0 from the source exit. This so-called *initial spatial distribution* Δs results in a distribution of flight

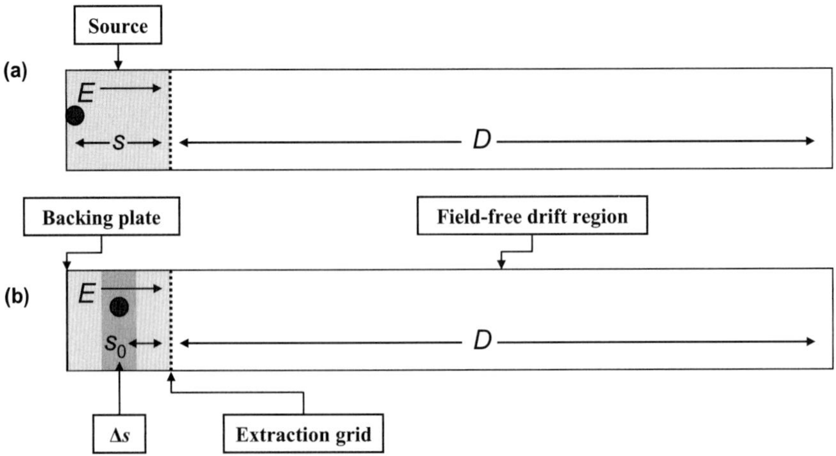

Figure 10.2. Basic linear time-of-flight mass spectrometer. (**a**) Formation of ions on a surface; (**b**) formation of ions in the gas phase.

times Δt and reduces mass resolution: $m/\Delta m = t/2\Delta t$. Generally, ions will also have an *initial kinetic energy distribution*. In that case, a more general TOF expression would be[20]

$$t = \frac{(2m)^{1/2}}{eE}\left[(U_0 + eEs)^{1/2} \mp U_0^{1/2}\right] + \frac{(2m)^{1/2}D}{2(U_0 + eEs)^{1/2}} + t_0 \tag{10.3}$$

where U_0 is the initial kinetic energy of an ion, the expression $\mp U_0$ is the so-called *turn-around time* and depends upon the initial direction of velocity (along the time-of-flight axis), and t_0 is the uncertainty in the time of ion formation. While the latter is not a particular problem for the short pulsed lasers used in MALDI instruments, it was significant in early pulsed electron impact (EI) sources. The effects of initial spatial and kinetic energy distributions on mass resolution are difficult to compensate simultaneously. Increasing the accelerating voltage V will minimize the effects of the initial kinetic energy as $eEs \gg U_0$, but this in turn increases the contribution from the spatial distribution.

It should also be noted that in addition to the axial velocities, the initial kinetic energy spread also includes a velocity distribution in the radial direction. While this does not affect ion flight times, a spread in radial velocities does affect ion transmission, so that some form of ion optics is generally used to focus the ion beam through longer flight lengths. Radial velocities can have a pronounced effect on ion transmission through reflectrons (see below), in some cases resulting in a significant loss in the observed ion signal (compared with linear analyzers), particularly in the higher mass ranges.

10.2.2 Spatial Distribution

Approaches to addressing the spatial distribution in the source are shown in Figure 10.3. Ions formed toward the back of the source will be in the field longer and leave the source with higher velocities than those formed closer to the source exit. At some point in the drift tube,

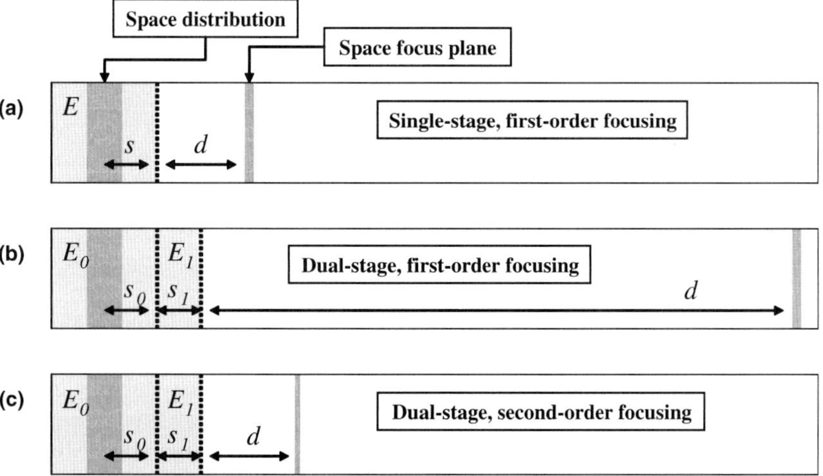

Figure 10.3. Methods for focusing the initial spatial distribution. (**a**) Single-stage, first-order focusing. (**b**) Dual-stage, first-order focusing. (**c**) Dual-stage, second-order focusing.

known as the *space-focus plane*, these faster ions will catch up with the slower ions. In the simplest case, the *single-stage* source shown in Figure 10.3a, the difference in time spent in the source between an ion formed at s_0 and one at $s_0 + \Delta s$ is

$$\Delta t_s = \left(\frac{m}{2eE}\right)^{1/2} 2\left[(s_0 + \Delta s)^{1/2} - s_0^{1/2}\right] \tag{10.4}$$

while the difference in time to reach the *space focus plane d* in the drift region would be

$$\Delta t_d = \left(\frac{m}{2eE}\right)^{1/2} \left[\frac{1}{(s_0 + \Delta s)^{1/2}} - \frac{1}{s_0^{1/2}}\right] d \tag{10.5}$$

Space focusing will occur at the point d in the flight tube where the difference in flight time $\Delta t = \Delta t_s + \Delta t_d$ is zero. Solution of these equations requires the expansion of the term

$$(s_0 + \Delta s)^{1/2} = s_0^{1/2} + \frac{\Delta s}{2s_0^{1/2}} - \frac{\Delta s^2}{8s_0^{3/2}} + \frac{3\Delta s^3}{48s_0^{5/2}} + \text{etc.} \tag{10.6}$$

where the terms shown represent the zero-, first-, second-, and third-order terms of the expansion. In the *single-stage* extraction source, only a first-order solution is possible, and space-focusing occurs at the point where $d = 2s_0$.

The problem with single-stage extraction is that ions are not focused to the end of the flight tube. Generally, longer flight tubes are advantageous because they increase the flight time t, thereby increasing the mass resolution $t/2\Delta t$. Thus, the 1955 Wiley and McLaren instrument[19,21] used a *dual-stage* source to push the *space-focus plane* to the detector end of the flight tube (Figure 10.3b). In this case, the two source/extraction stages s_0 and s_1 have different field strengths E_0 and E_1, respectively, where generally $E_1 \gg E_0$. Similarly, the space focus plane (and the best location for the detector) can be found when $\Delta t = \Delta t_0 + \Delta t_1 + \Delta t_d = 0$ for ions formed at s_0 and $s_0 + \Delta s$. Expressions for the time differences in the first (source) and drift regions (Δt_0 and Δt_d) are similar to those in Eqs. (10.6) and (10.7), respectively, while the time difference in the second region is

$$\Delta t = \left(\frac{m}{2eE_0}\right)^{1/2} \left[\frac{1}{(\sigma + \Delta s)^{1/2} + (s_0 + \Delta s)^{1/2}} - \frac{1}{\sigma^{1/2} + s_0^{1/2}}\right] 2s_1 \tag{10.7}$$

where $\sigma = s_0 + (E_1/E_0)s_1$. Two terms in this equation can be expanded, giving rise to first- or second-order solutions. A first-order solution locates the focal plane at

$$d = 2\sigma^{3/2}\left[\frac{1}{s_0^{1/2}} - \frac{2s_1}{s_0^{1/2}(\sigma^{1/2} + s_0^{1/2})^2}\right] \tag{10.8}$$

a more complex expression that depends upon the values of s_0, s_1, E_0 and E_1. In the original Wiley and McLaren instrument,[19] these values were 0.2 cm, 1.2 cm, 320 V/cm, and 1280 V/cm, respectively, resulting in a drift length of around 40 cm. Most instruments today use much higher voltages, from 5 to 20 kV, though the field strength in the second extraction region is similarly much larger than the first to give a drift length of around 1 m.

Second-order solutions[22] are also possible for the dual-stage source (Figure 10.3c), and they result in a much shorter optimal drift length. This configuration is rarely used in a simple linear spectrometer, but can be used to form an intermediate virtual point source for a reflectron (see below). A shorter, higher-order focal plane has also been used in the development of instruments of compact size.

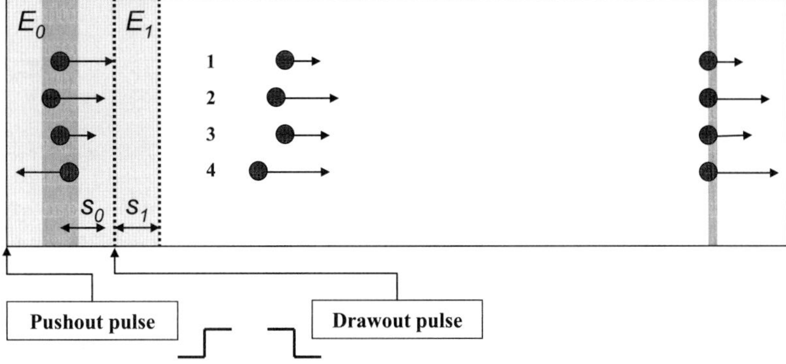

Figure 10.4. Time-lag focusing scheme. Ion 2 is formed behind ion 1 and is accelerated to higher energy, catching up at the space focus point. Ion 3 has less initial energy than ion 1, drifts more slowly during the delay period, and is accelerated to higher energy. Ion 4 has the same initial energy as ion 1, but its velocity is reversed. During the drift period it travels toward the back of the source. It leaves the source late, but with much higher energy.

10.2.3 Kinetic Energy Distribution and Time-Lag Focusing

An ion formed by any ionization method will have an initial kinetic energy U_0 so that (after acceleration) it will enter the drift region with an energy $eV + U_0$. This additional energy will reduce its flight time, and a distribution of such energies will reduce the resolution as well. The distribution of initial kinetic energies is also a distribution in initial velocities. While the initial direction of motion does not change the time in the drift region, it does change the time in the source, giving rise to the term for the *turnaround time* in Eq. (10.3).

The 1955 Wiley and McLaren instrument[19] addressed the initial kinetic energy problem with a technique they called *time-lag focusing*. In the simplest terms, ions with initial velocities are allowed to drift in a field-free source for a period, before being extracted by a *drawout pulse* applied to the source exit grid and/or a *pushout pulse* applied to the backing plate. During this time lag (see Figure 10.4), ions with a velocity component in the direction of the backing plate move toward the back of the source. When the extraction field is applied, they will travel a longer distance in the field than ions that have moved toward the front of the source, will have higher velocities, and will catch up at some point in the flight tube (generally at the detector) in much the same way that ions are space focused. In addition to direction, the correction is of course proportional to the magnitude of the velocity, giving a greater boost from the field to ions with initially low velocities.

The Wiley and McLaren instrument combined *time-lag focusing* with a two-stage source to provide both space and energy focusing, though this arrangement cannot provide optimal focusing of these two distributions simultaneously. Moreover, the spatial and kinetic energy distributions are uncorrelated and the correction for initial kinetic energy is mass-dependent. This latter point means that the optimal length of the time delay between ionization and ion extraction increases with mass. Thus, modern MALDI instruments generally require some tuning of the *delayed extraction* to match the mass range of interest.

10.3 REFLECTRON TIME-OF-FLIGHT (RTOF) INSTRUMENTS

In 1975 Boris Mamyrin developed an instrument for mass-independent kinetic energy focusing.[23] In the so-called *reflectron,* a retarding electric field added at the end of the drift

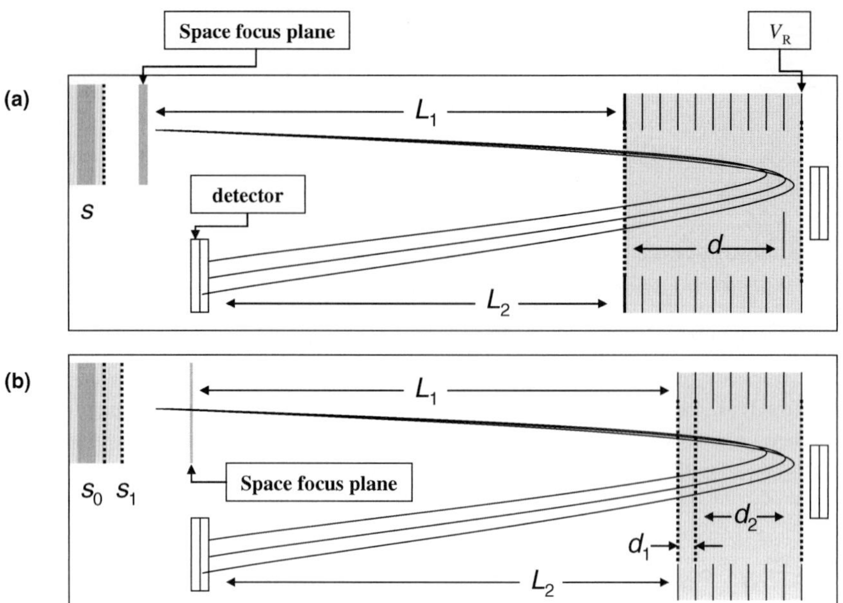

Figure 10.5. The *reflectron* mass analyzer: (**a**) single-stage and (**b**) dual-stage.

region slows down and then reverses the direction of the ions to return them back through the drift region. Ions of a particular m/z value but with higher energies $eV + U_0$ will penetrate this field more deeply and will thereby have a longer flight path than less energetic ions of that mass. The idea is again that they arrive at the detector simultaneously.

The simplest embodiment of the reflectron is shown in Figure 10.5a along with a single-stage ion source. The source space-focus plane $d = 2s$ is used here as the focal point for the reflectron as well. The ions from this point travel through the drift region L_1, turn around in the reflectron at an average depth of d, and travel back through the drift region L_2 to the detector. The reflecting field is defined by an entrance grid, a series of rings or lenses, and a backing grid that permits detection of ions in the "linear" mode when the reflectron is turned off. The field in this *single-stage reflectron* is generally constant, determined by a linear voltage gradient $V_x = ax$, where x is the depth from the front of the reflectron. In most cases this is accomplished by a resistive divider connected to the lens elements. The maximum voltage V_x at the back of the reflectron V_R is generally 5–10% higher than the source accelerating voltage to ensure that all ions will be reflected before hitting the rear grid. The flight time for an ion in a reflectron is

$$t = \left(\frac{m}{2eV}\right)^{1/2} [L_1 + L_2 + 4d] \tag{10.9}$$

For an ion with higher kinetic energy $eV + U_0$, the time that the ion spends in the linear regions (where the ions are traveling much faster) is decreased. However, the time spent in the reflecting field is longer as a result of the increase in the effective field penetration depth: $d(eV + U_0)/eV$. The focusing expression for the reflectron is obtained by expanding the term $(eV + U_0)$, where the first-order solution $L_1 + L_2 = 4d$ is similar to that of the space-focusing from the source, $d = 2s$, but covers the forward and reverse passes through the reflectron.

10.3.1 Higher-Order Reflectrons

It is of course possible to devise a *dual-stage reflectron*, and that is shown in Figure 10.5b along with its analog two-stage source. Similar to the space focus plane expression given in Eq. (10.8), the focal length for the dual-stage reflectron is given by

$$L = 2\delta^{3/2}\left[\frac{1}{d_2^{1/2}} - \frac{2d_1}{d_2^{1/2}(\delta^{1/2}+d_2^{1/2})^2}\right] \tag{10.10}$$

where $L=(L_1+L_2)/2$, $\delta = d_2 + (\Delta V_1 d_2/\Delta V_2 d_1)d_1$, ΔV_1 is the voltage drop across the first stage d_1, and ΔV_2 is the voltage drop across the second stage d_2. While this is a first-order solution, the original Mamyrin reflectron[23] was in fact a second-order reflectron. In general, the dual-stage reflectron provides second-order focusing when the second region is about 10 times the depth of the first ($d_2 = 10d_1$), and the voltage drop in the first stage is about two to three times that of the second stage.[24]

Infinite-order focusing is theoretically possible with a *quadratic reflectron*, in which the voltage is given by $V = ax^2$. In that case the flight time

$$t = 2\int_0^{x_{\max}} \frac{dx}{\left[\frac{2e}{m}(V_x - ax^2)\right]^{1/2}} = \pi\left(\frac{m}{2ea}\right)^{1/2} \tag{10.11}$$

is independent of ion kinetic energy. Quadratic reflectrons have not been widely used because it is difficult to provide a true quadratic potential along the axis using a series of lenses and because the off-axis field defocuses the ion beam and reduces transmission. Both of these problems have generally dictated the use of long, narrow-bore reflectrons with very closely spaced lens elements. A quadratic reflectron designed in 1984 by Yoshida[25] was a coaxial geometry constructed with 100 closely spaced 40-mm-i.d. lenses, and used by Tanaka et al.[2] to obtain the first mass spectra above 100 kDa. A tandem TOF mass spectrometer developed by Raptakis and co-workers[26] uses a quadratic reflectron in the second stage (see below). Cornish and Cotter[27] have developed an *endcap reflectron* geometry for miniaturized instruments that approximates a quadratic field as well as a *curved-field reflectron*[28] for focusing product ions in a tandem TOF instrument (see below). Even high-transmission grids, when used as field boundaries in a reflectron, will significantly reduce ion transmission due to the localized "lensing" effects at the grid holes. Thus, many reflectrons are now designed to be *gridless*.

10.4 IONIZING BIOMOLECULES DIRECTLY FROM SURFACES: LD AND MALDI

Beginning in the 1970s with the laser microprobes,[29] a succession of new ionization techniques were developed that desorbed ions directly from surfaces. These included plasma desorption,[30] laser desorption,[31,32] and secondary ion mass spectrometry, or SIMS, using keV primary ion beams.[33,34] Formation of ions directly from a conductive surface effectively eliminated an initial spatial distribution, so that time-lag focusing was generally abandoned in favor of *prompt extraction* and mass-independent reflectron focusing of the kinetic energy spread. However, during that period our laboratory developed a *time-resolved laser desorption* technique[35,36] to improve mass resolution for ions desorbed by a pulsed infrared CO_2 laser—in this case, thermally induced desorption extending beyond the duration of the laser pulse.

10.4.1 Space–Velocity-Correlated Focusing

By the 1990s the widespread use of MALDI for protein structure analysis and proteomics continued to motivate the development of higher-performance instruments. At that time, several groups reported dramatic increases in mass resolution using MALDI in conjunction with time-delayed pulsed ion extraction.[10–12] This is not entirely surprising since (in fact) desorption from a surface does not produce a pure energy distribution, but one that is accompanied by a time distribution resulting from the different times ions of differing energy spend in the source. The source is not in actuality a point source, and reflectron focusing of the energy cannot produce a peak width at the detector narrower than the peak emerging from the source. Thus, a pulsed extraction scheme can indeed improve the focus at the detector, or more often a focal point that acts as a virtual point source for the reflectron. While a number of terms were used to describe the use of pulsed extraction with MALDI, the term *delayed extraction* coined by PerSeptive BioSystems (now Applied BioSystems, Framingham, MA) appears to be the term generally used. Using delayed extraction with MALDI has produced mass resolutions of 15,000 or more in TOF instruments.

The rationale for the extraordinary improvement in mass resolution in MALDI has been offered by Colby and Reilly.[13] While in one sense this is simply *time-lag focusing* in which Δs is zero, a spatial distribution is in fact generated after the delay time, where the location of each ion depends upon its initial velocity. Focusing ions when the spatial and velocity distributions are correlated is very different from the case in which the two distributions are uncorrelated, and their method is called *space–velocity correlation focusing*.[13] It should be noted, however, that *delayed extraction* or *space–velocity correlation focusing* is still mass-dependent. On many commercial MALDI TOF instruments the delay time is tuned to optimize mass resolution for a particular mass range. In some cases, it is simply set around *m*/*z* 5000 (for example) to focus a range covering peptides and small proteins. Figure 10.6 shows a MALDI mass spectrum of a tryptic digest of histone H3 obtained using both delayed extraction and a reflectron. Expansion of the molecular ion region of one of the peptides provides sufficient mass resolution to distinguish the isotopic distributions of labeled and unlabeled peptide in a quantitation experiment.[37]

10.4.2 Mass-Correlated Acceleration

One of the ways to understand this mass-dependence is shown in Figure 10.7a, which depicts the positions *after the delay time* of three sets of ions with different masses. The scheme uses a dual-stage extraction source, so that *space–velocity correlation focusing* at the detector depends upon E_0, E_1, s_0, s_1, and of course the delay time. It also depends upon mass. Because their velocities are greater, lighter ions will distribute differently than heavier ions during the delay time. In particular, they will be distributed about a different mean position s_0 than the heavier ions and will focus further down the flight tube. One obvious solution is to increase the delay time so that the heavier ions have drifted to the optimal position prior to extraction. This is, of course, what is often done to focus higher mass ranges. However, if one wishes to focus the entire mass range in each TOF cycle, then the delay time will be fixed. Because distances are also fixed, then the only opportunity is to change the extraction fields in a time-dependent manor. One such scheme was developed by Kovtoun et al.[38] and is known as *mass-correlated acceleration* and is shown in Figure 10.7b. Because lighter ions reach the second extraction region sooner than heavier ions, it is

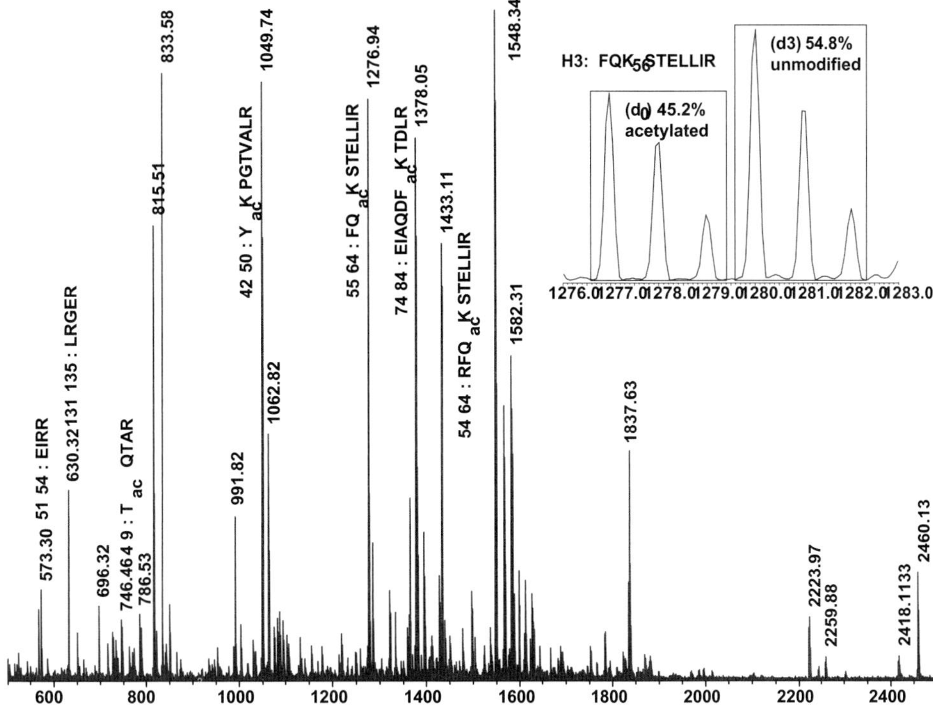

Figure 10.6. MALDI mass spectrum of the tryptic digest of histone H3. The protein was deutero-acetylated prior to digestion to enable quantitation of acetylation at each lysine site.[37] Inset shows the mass resolution of the isotopic cluster for one of the peptides.

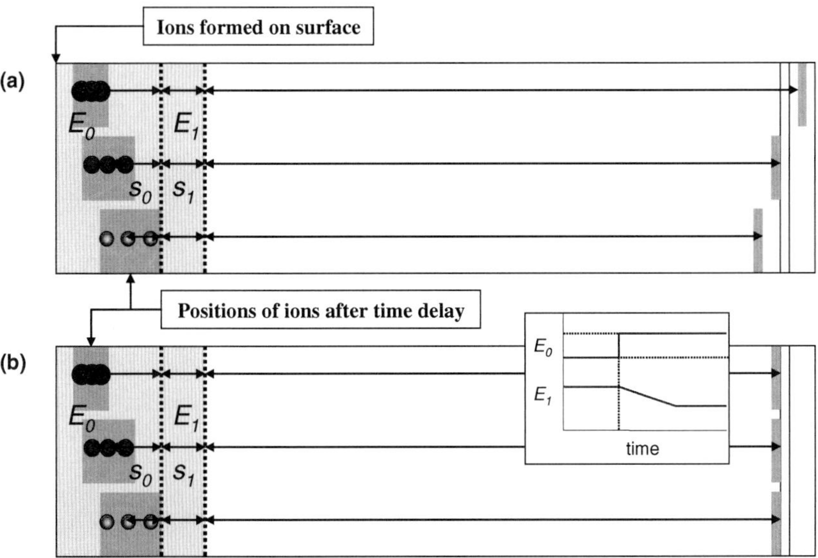

Figure 10.7. (**a**) Positions of ions following *delayed extraction* showing normal mass dependence of *space–velocity-correlated focusing*. (**b**) Positions of ions after correction using *mass-correlated acceleration*.

possible to change the field in the second region, so that the ratio E_1/E_0 focusing the ions compensates for the effective difference in s_1/s_0 for each mass.

10.5 FRAGMENTATION AND TANDEM TOF MASS SPECTROMETRY

Much of proteomics is concerned with the identification of proteins and/or changes in their levels of expression in response to disease, condition, or genetics. When purified using high-performance liquid chromatography (HPLC) or when obtained from a spot on 2D polyacrylimide gels, proteins can be identified by mapping the molecular masses of their tryptic peptides and comparing these to a database, a method known as *protein mass fingerprinting* (PMF). When not completely fractionated, however, proteins are generally identified from the amino acid sequences of their tryptic peptides, which in turn requires some form of tandem mass spectrometry.

10.5.1 The Important Role of the Reflectron

Before the recent commercial offerings of TOF/TOF instruments there had been a number of reports of mass spectrometers composed of two TOF mass analyzers (Table 10.1). Perhaps the simplest configuration was reported by Jardine et al.[39] and consisted of two drift lengths separated by a voltage step. Because product ions formed in a drift region have the same velocities as their parents, the additional electric field is able to distinguish their flight times. However, it is more common to use a reflectron as the second mass analyzer (MS2), the case for the remaining instruments in Table 10.1.[26,40–45] Using this notation, the commercial MALDI tandem TOF mass spectrometers are TOF/RTOF instruments, in which the first mass analyzer is a linear drift region with resolution provided by delayed extraction focusing.

Table 10.1. History of Tandem Time-of-Flight Mass Spectrometers

Year	Type	Description	Reference
1989	TOF/RTOF	Linear mass analyzer with an orthogonal reflectron mass analyzer	Schey et al.[40]
1992	TOF/TOF	Two in-line linear mass analyzers with a voltage step	Jardine et al.[39]
1992	TOF/RTOF	MS1 is second-order space focus. Ions are reaccelerated into a second-order reflectron	Boesl et al.[41]
1993	RTOF/RTOF	ArF excimer photodissociation in floated chamber between two reflectron analyzers	Seeterlin et al.[42]
1993	RTOF/RTOF	Pulsed gas, floated collision chamber between two dual-stage reflectrons	Cotter and Cornish[43]
1993	RTOF/RTOF	CID in floated collision region with two single-stage reflectrons	Cornish and Cotter[44]
1994	RTOF/RTOF	Single stage as MS1; curved-field reflectron (CFR) as MS2	Cornish and Cotter[45]
2002	TOF/RTOF	MS2 is a quadratic reflectron	Giannakopulos et al.[26]

In 1993 two laboratories produced instruments with two reflectron mass analyzers—that is, RTOF/RTOF configurations, both using two dual-stage reflectrons. The instrument developed by Enke et al.[42] used photodissociation, while that by Cotter and Cornish[43,44] used a pulsed gas valve to provide collision-induced dissociation. Both of these instruments were able to retard the precursor ions before dissociation. This reduces the kinetic energy spread of the product ions, which are then reaccelerated before entering the reflectron. A later version of the Cornish and Cotter instrument[45] used a *curved-field reflectron* to accommodate the wide energy spread, while Derrick and co-workers[26] used a *quadratic reflectron.*

Figure 10.8 is a somewhat generalized schematic of a TOF/RTOF configuration which, for simplicity, uses a single-stage reflectron as MS2. The total flight time of an ion is given by the time spent in the first and second mass analyzers: $t = t_1 + t_2$. For all ions m_a formed in the ion source, this is

$$t = \left(\frac{m_a}{2eV}\right)^{1/2} [D + L_1 + L_2 + 4d] \tag{10.12}$$

essentially the same result as Eq. (10.9). Figure 10.8a shows that these ions are focused (by delayed extraction) as they cross from MS1 to MS2, and then they are refocused at the detector. This scheme results in very high mass resolution in the MS mode. The first focus point is also used as the location of a mass selection gate. In the MS/MS mode (Figure 10.8b), ions of only one m/z value are selected. After dissociation the resultant product ions m_b will have the same velocity as their precursors, but will have lower kinetic energies: $(m_b/m_a)eV$ and will penetrate less deeply into the reflectron. Their flight times will be

$$t = \left(\frac{m_a}{2eV}\right)^{1/2} \left[D + L_1 + L_2 + 4\frac{m_b}{m_a}d\right] \tag{10.13}$$

and will define a *linear* product ion mass scale. However, the ions will not all be in focus at the detector, because this range of energies exceeds the reflectron bandwidth. One approach is to

(a)

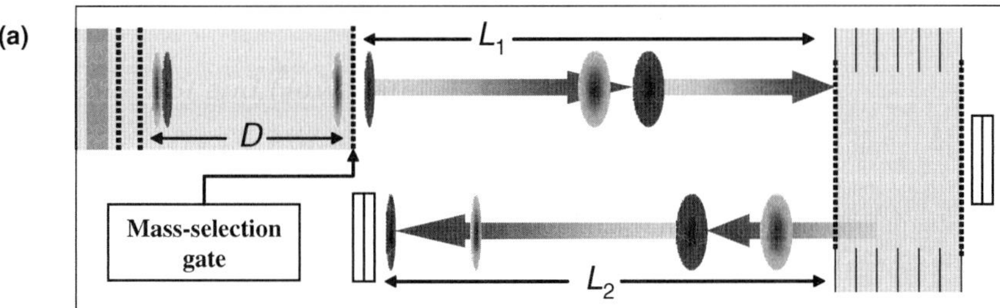

(b)

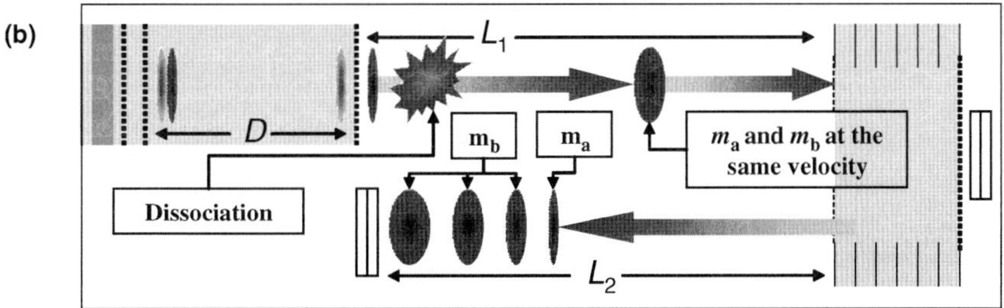

Figure 10.8. TOF/RTOF tandem time-of-flight mass spectrometer: (**a**) MS mode with mass selection gate OFF. (**b**) MS/MS mode (See color insert).

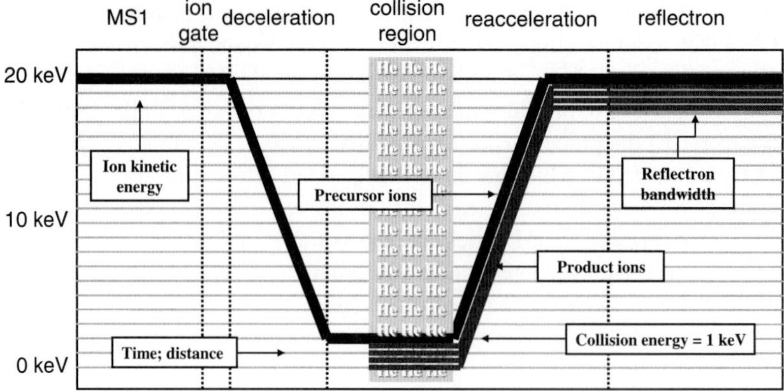

Figure 10.9. Scheme for the deceleration and reacceleration of ions to match the reflectron focusing range.

decelerate the precursor ions and reaccelerate the product ions (Figure 10.9). For example, a precursor ion beam whose ions have energies of 20 keV can be decelerated to 1 keV, so that their product ions will cover the energy range from 0 to 1 keV. The product ions are then reaccelerated to energies ranging from 19 to 20 keV. Thus, this approach compresses the product energy range to accommodate the reflectron.

10.5.2 Commercial TOF/TOF Mass Spectrometers

Variations of this approach are used on some commercial instruments. The model 4700 and 4800 instruments offered by Applied Biosystems (Framingham, MA) decelerate the ions to 1–2 keV using a series of retarding lenses (Figure 10.10a). The ions are mass-selected and

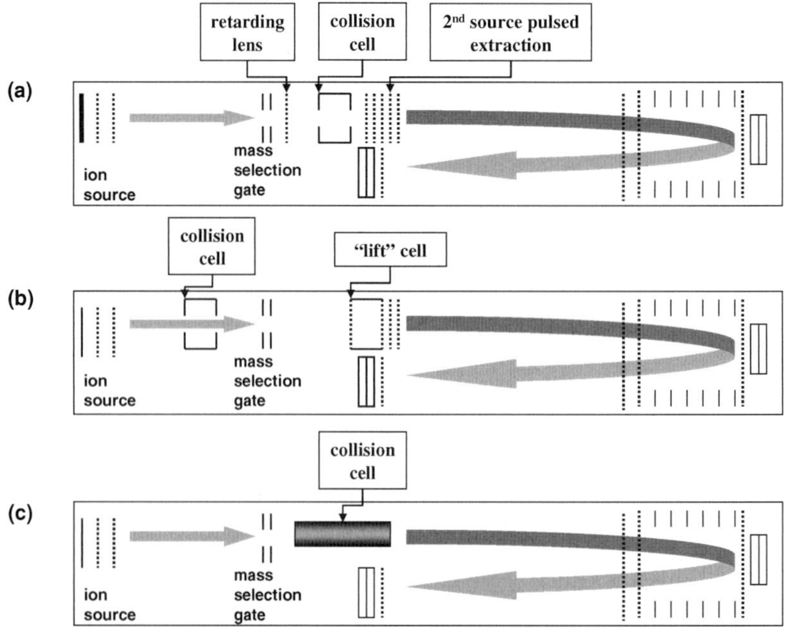

Figure 10.10. Schematic comparison of current commercial tandem TOF instruments.

then dissociated in a collision chamber. They enter a *second source* from which they are reaccelerated and refocused by pulsed extraction.[46–48] In the Bruker Daltonics (Billerica, MA) *Ultraflex* instrument, 8-keV ions are not decelerated, so that collision energies are much higher (Figure 10.10b). The product ions enter a *lift cell* that sends them into the reflectron with an energy range of 17–25 keV.[49]

Alternatively, the Shimadzu Corporation (Manchester, UK) *AXIMA TOF*^{2} tandem uses a curved field reflectron (Figure 10.10c). This reflectron uses a nonlinear field defined by potentials on the lens elements that approximates the arc of a circle: $R^2 = V^2 + x^2$, where V is the voltage, x is the distance from the reflectron entrance, and R is a constant. Because the reflectron can accommodate a wide range of product ion energies, the ions are not decelerated or reaccelerated.[50,51] Collision energies can then be as high as 20 keV.

10.6 ORTHOGONAL ACCELERATION AND HYBRID INSTRUMENTS

One of the challenges has been to develop TOF mass spectrometers, which are basically pulsed instruments that can be used with continuous ionization sources. Since the major interest here is electrospray, then the requirement is also compatibility with ionization at atmospheric or intermediate pressure.

In 1989 Dawson and Guilhaus[14] introduced their *orthogonal acceleration* time-of-flight (oaTOF) mass spectrometer. In their instrument, ions formed by a continuous EI source were focused into a thin beam, injected (along the x-direction) into a storage volume (Figure 10.11a), and extracted by a *pushout* pulse applied orthogonal to the ion beam direction. Focusing is possible because the thin beam provides minimal spatial distribution in the extraction direction, and the initial velocity distribution is almost entirely in the x-direction. At the *12th International Mass Spectrometry Conference* in 1991 in Amsterdam, Dodenov and co-workers reported, and later published,[15] an *orthogonal extraction* instrument with electrospray ionization. This instrument also used a dual-stage reflectron, while later instruments by Verentchikov et al.[52] used a single-stage reflectron, or *ion mirror*. In 1998, RF-only quadrupole ion guides were introduced as interfaces between the electrospray source and the high-vacuum region of the TOF.[53,54] Quadrupole or other multiple guides are now used in nearly all oaTOF instruments and hybrids and are generally located in an intermediate-pressure region of a differentially pumped system (Figure 10.11b). Collisions of ions with neutrals not only cool the ions, but tightly collimate the ion beam for injection through a small, low-conductance orifice into the orthogonal extraction region. This scheme has also been used in a miniaturized oaTOF with a vacuum EI source,[55] where the addition of 4.0 mtorr of helium into the RF-only ion guide improves the mass resolution. An interesting property of the oaTOF/ion guide configuration is that TOF resolution no longer depends upon the initial conditions of energy and space. Indeed, ions entering the oaTOF mass analyzer have essentially no “memory” of the type of ionization source used. Thus, such instruments can be used with interchangeable sources operating in vacuum, at atmospheric pressure, or at an intermediate pressure.

10.6.1 Hybrid Quadrupole/TOF Mass Spectrometers

A major benefit of the oaTOF technology has been the development of hybrid quadrupole/TOF mass spectrometers. In these instruments a number of RF quadrupole (or other multipole) devices are used to provide the transition from atmospheric pressure sources,

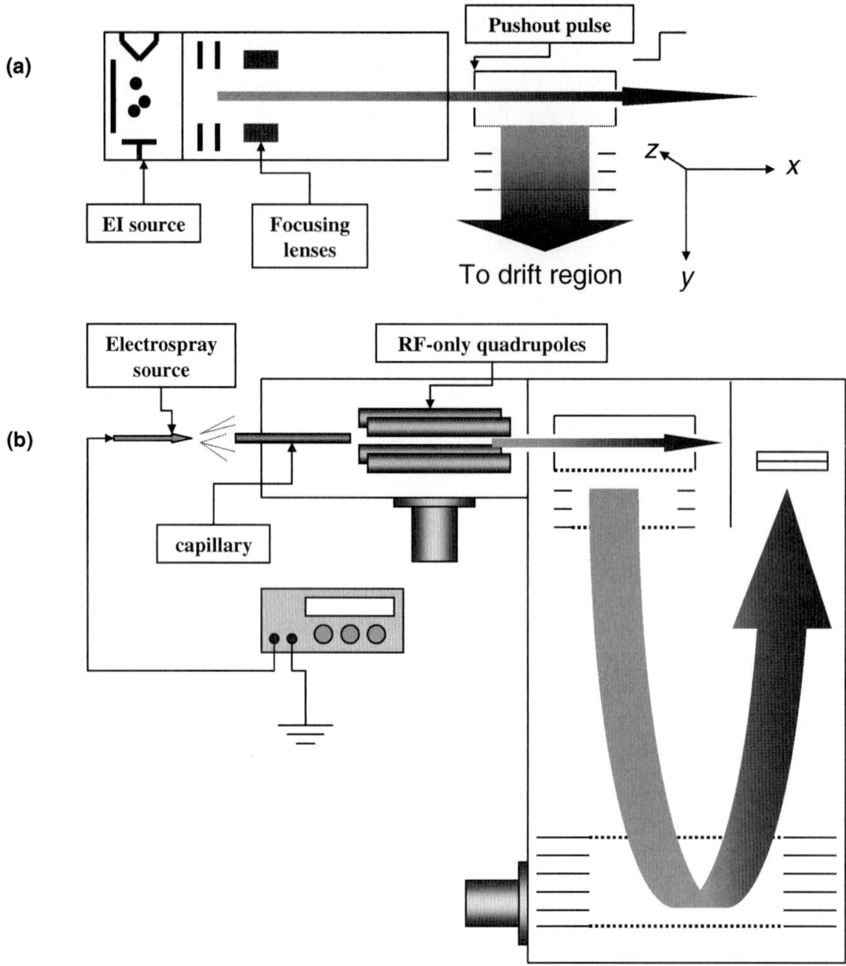

Figure 10.11. Orthogonal acceleration TOF mass spectrometers. (**a**) Original design of Dawson and Guilhaus[14] using lens focusing. (**b**) An oaTOF shown with an electrospray source and RF ion guide.

collimation of the ion beam, mass selection (MS1), and collisional activation. Shown in Figure 10.12 is an RF-only quadrupole (Q_1) interfaced to the capillary inlet of an electrospray source, a quadrupole mass filter (Q_2) that is the first mass analyzer in this tandem instrument, and a second RF-only quadrupole (Q_3) used as collision chamber and for final collimation of the beam entering the ion storage area. Commercial hybrid quadrupole/TOF mass spectrometers include the QTOF from Waters-Micromass (Milford, MA) and the QSTAR from Applied BioSystems.

10.6.2 Quadrupole Ion Trap/TOF Mass Spectrometers

Time-of-flight mass spectrometers have also been used with quadrupole ion traps, where they record the final mass spectrum in any MS^n experiment. One of the first such instruments by Chien and Lubman[56] used an electrospray source. A MALDI trapTOF was reported by Doroshenko and Cotter[57] using a quadratic reflectron as the TOF mass analyzer. In this

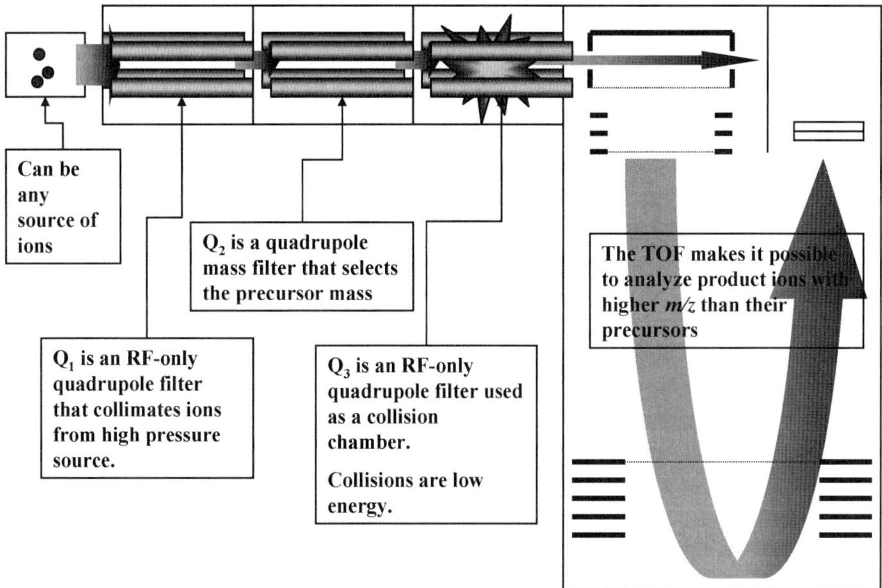

Figure 10.12. Schematic of a quadrupole/time-of-flight hybrid mass spectrometer.

instrument, ions produced by MALDI outside the trap were captured by ramping the trapping field.[58] A MALDI QIT/TOF mass spectrometer was designed by Ding et al.[59] Available commercially from Kratos Analytical, this instrument captures external MALDI ions by rapidly gating the trap's fundamental RF amplitude. Because mass spectra are acquired by extracting ions from the trap in a single pulse and recording mass spectra in the TOF mass analyzer (rather than scanning the fundamental frequency), ions with m/z up to 60 kDa can be recorded.

10.7 SUMMARY AND FUTURE PROSPECTS

A wide array of MALDI-TOF, hybrid quadrupole/TOF, trapTOF, and TOF/TOF instruments are now available commercially and found in most proteomics laboratories. But, the development of TOF technology goes much further. Miniaturization is one important direction, which includes instruments for homeland security, environmental monitoring, and point-of-care diagnostics. An advantage of the TOF is that relatively compact instruments can be assembled that retain high mass ranges. One approach (Figure 10.13) is to utilize ion source dimensions that can sustain high accelerating voltages while making use of high-order space-velocity focusing in a very short drift length.[60] English et al.[61] have described the use of a 3-in. TOF mass analyzer for the identification of *Bacillus* spores from their small acid-soluble proteins (SASPs). TOF mass spectrometers *on a chip* have also been constructed, but with a considerably smaller mass range.[62]

Caprioli et al[63] have pioneered an approach for molecular imaging of biological tissue using MALDI time-of-flight mass spectrometry. This is a MALDI *microprobe* approach that generally uses the XY sample stage to scan the tissue. Alternatively, Heeren and co-workers[64] have developed a SIMS *microscope* TOF that irradiates an area of the tissue and images the ions on an array detector. This instrument, known as the TRIFT, utilizes three isochronous electrostatic energy analyzers to preserve both TOF and spatial information.[65]

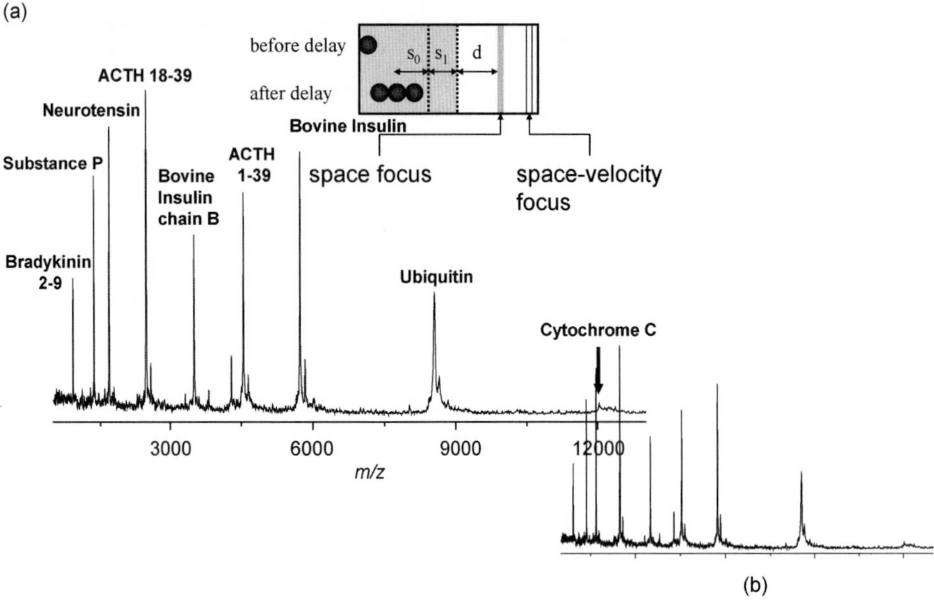

Figure 10.13. (**a**) Compact mass spectrometer using high-order space–velocity focusing. (**b**) Mass spectrum of a peptide mixture on a 3-in. TOF mass analyzer.

A number of laboratories have developed TOF mass spectrometers for the analysis of aerosol particles[66,67] and aerosolized bacterial spores.[68] In these instruments, aerosol particles are brought from the atmosphere directly into the vacuum region by means of an aerodynamic lens, skimmers, and differentially pumped regions. They are then ionized directly in flight by a pulsed laser beam. While most of these studies have involved laser desorption, the addition of matrix coatings to aerosol particles in flight has been reported[69] in a MALDI bioaerosol mass spectrometer.

No doubt, the prospects for continued development of the TOF mass spectrometer are very good indeed!

REFERENCES

1. Karas, M.; Hillenkamp, F. Laser desorption ionization of proteins with molecular masses exceeding 10,000 daltons. *Anal. Chem.* **1988**, *60*, 2299.
2. Tanaka, K.; Waki, H.; Ido, Y.; Akita, S.; Yoshida, Y.; Yoshida, T. Protein and polymer analyses up to m/z 100,000 by laser ionization time-of-flight mass spectrometry. *Rapid Commun. Mass Spectrom.* **1988**, *2*, 151.
3. Yamashita, M.; Fenn, J. B. Electrospray ion source. Another variation on the free-jet theme. *J. Phys. Chem.* **1984**, *88*, 4451.
4. Frank, M.; Labov, S. E.; Westmacott, G.; Benner, W. H. Energy-sensitive cryogenic detectors for high-mass biomolecule mass spectrometry. *Mass Spectrom. Rev.* **1999**, *18*, 155–186 (and references therein).
5. Cotter, R. J. Plasma desorption mass spectrometry: Coming of age. *Anal. Chem.* **1988**, *60*, 781A–793A.
6. Park, M. A.; Callahan, J. H.; Vertes, A. An inductive detector for time-of-flight mass spectrometry. *Rapid Commun. Mass Spectrom.* **1994**, *8*, 317–322.
7. Frank, M.; Labov, S. E.; Westmacott, G.; Benner, W. H. Energy-sensitive cryogenic detectors for high-mass biomolecule mass spectrometry. *Mass Spectrom. Rev.* **1999**, *18*, 155–186.
8. Hilton, G. C.; Martinis, J. M.; Wollman, D. A.; Irwin, K. D.; Dulcie, L. L.; Gerber, D.; Gillevet, P. M.; Twerenbold, D. Impact energy measurement in time-of-flight mass spectrometry with cryogenic microcalorimeters. *Nature* **1998**, *391*, 672–675.
9. Holland, J. F.; Enke, C. G.; Allison, J.; Stults, J. T.; Pinkston, J. D.; Newcombe, B. N.; Watson, J. T. Mass

spectrometry in the chromatographic time frame. *Anal. Chem.* **1983**, *65*, 997A–1112A.
10. WHITTAL, R. M.; LI, L. High-resolution matrix-assisted laser desorption/ionization in a linear time-of-flight mass spectrometer. *Anal. Chem.* **1995**, *67*, 1950–1954.
11. BROWN, R. S.; LENNON, J. J. Mass resolution improvement by incorporation of pulsed ion extraction in a matrix-assisted laser desorption/ionization linear time-of-flight mass spectrometer. *Anal. Chem.* **1995**, *67*, 1998–2003.
12. VESTAL, M. L.; JUHASZ, P.; MARTIN, S. A. Delayed extraction matrix-assisted laser desorption time-of-flight mass spectrometry. *Rapid Commun. Mass Spectrom.* **1995**, *9*, 1044–1050.
13. COLBY, S. M.; REILLY, J. P. Space–velocity correlation focusing. *Anal. Chem.* **1996**, *68*, 1419–1428.
14. DAWSON, J. H. J.; GUILHAUS, M. Orthogonal-acceleration time-of-flight mass spectrometer. *Rapid Commun. Mass Spectrom.* **1989**, *3*, 155.
15. MIRGORODSKAYA, O. A.; SHEVCHENKO, A. A.; CHERNUSHEVICH, I. V.; DODENOV, A. F.; MIROSHNIKOV, A. I. Electrospray-ionization time-of-flight mass spectrometry in protein chemistry. *Anal. Chem.* **1994**, *66*, 99–107.
16. STEPHENS, W. E. *Phys. Rev.* **1946**, *69*, 691.
17. CAMERON, A. E.; EGGERS, D. F. An ion "Velocitron." *Rev. Sci. Instr.* **1948**, *19*, 605.
18. KATZENSTEIN, H. S.; FRIEDLAND, S. S. New time-of-flight mass spectrometer. *Rev. Sci. Instr.* **1955**, *26*, 324.
19. WILEY, W. C.; MCLAREN, I. H. Time-of-flight mass spectrometer with improved resolution. *Rev. Sci. Instrum.* **1955**, *26*, 1150–1157.
20. COTTER, R. J. *Time-of-Flight Mass Spectrometry: Instrumentation and Applications in Biological Research*, American Chemical Society, Washington DC, 1997.
21. WILEY, W. C.; MCLAREN, I. H. Bendix time-of-flight mass spectrometer. *Science* **1956**, *124*, 817–820.
22. BOESL, U.; WEINKAUF, R.; SCHLAG, E. W. Reflectron time-of-flight mass spectrometry and laser excitation for the analysis of neutrals, ionized molecules and secondary fragments. *Int. J. Mass Spectrom. Ion Processes* **1992**, *112*, 121–166.
23. MAMYRIN, B. A.; KARATAEV, V. I.; SHMIKK, D. V.; ZAGULIN, V. A. *Sov. Phys. JETP* **1973**, *37*, 45.
24. BRUNELLE, A.; DELLA-NEGRA, S.; DEPAUW, J.; JORET, H.; LEBEYEC, Y. Time-of-flight mass spectrometry with a compact two-stage electrostatic mirror: Metastable-ion studies with high mass resolution and ion emission from thick insulators. *Rapid Commun. Mass Spectrom.* **1991**, *5*, 40–43.
25. YOSHIDA, Y. US Patent 4,625,112, 1984.
26. GIANNAKOPOULOS, A. E.; THOMAS, B.; COLBURN, A. W.; REYNOLDS, D. J.; RAPTAKIS, E. N.; MAKAROV, A. A.; DERRICK, P. J. Tandem time-of-flight mass spectrometer (TOF-TOF) with a quadratic-field ion mirror. *Rev. Sci. Instrumen.* **2002**, *73*, 2115–2123.
27. CORNISH, T. J.; COTTER, R. J. High order kinetic energy focusing in endcap reflectron time-of-flight mass spectrometer. *Anal. Chem.* **1997**, *69*, 4615–4618.
28. CORNISH, T. J.; COTTER, R. J. A curved field reflectron for improved energy focusing of product ions in time-of-flight mass spectrometry. *Rapid Commun. Mass Spectrom.* **1993**, *7*, 1037–1040.
29. HILLENKAMP, F.; UNSOLD, E.; KAUFMANN, R.; NITSCHE, R. Laser microprobe mass analysis of organic materials. *Nature* **1975**, *256*, 119–120.
30. MACFARLANE, R. D.; TORGERSON, D. F. Californium-252 plasma desorption mass spectroscopy. *Science* **1976**, *191*, 920–925.
31. POSTHUMUS, M. A.; KISTEMAKER, P. G.; MEUZELAAR, H. L. C.; M. C.; Ten Noever DEBRAUW,. Laser desorption–mass spectrometry of polar nonvolatile bio-organic molecules. *Anal. Chem.* **1978**, *50*, 985.
32. COTTER, R. J. Lasers and mass spectrometers (Review). *Anal. Chem.* **1984**, *56*, 485A.
33. BENNINGHOVEN, A.; JASPERS, D.; SICHTERMANN, W. *Appl. Phys.* **1976**, *11*, 35.
34. TANG, X.; BEAVIS, R.; ENS, W.; LAFORTUNE, F.; SCHUELER, B.; STANDING, K. G. A secondary ion time-of-flight mass spectrometer with an ion mirror. *Int. J. Mass Spectrom. Ion Processes* **1988**, *85*, 43–67.
35. TABET, J.-C.; JABLONSKI, M.; COTTER, R. J.; HUNT, J. E. Time resolved laser Desorption: III. The metastable decomposition of chlorophyll a and some derivatives. *Int. J. Mass Spectrom. Ion Processes* **1985**, *65*, 105–117.
36. TABET, J.-C.; COTTER R. J. Laser desorption time-of-flight mass spectrometry of high mass molecules. *Anal. Chem.* **1984**, *56*, 1662.
37. CELIC, I.; MASUMOTO, H.; GRIFFITH, W. P.; MELUH, P.; COTTER, R. J.; BOEKE, J. D.; VERREAULT, A. The sirtuins Hst3 and Hst4p preserve genome integrity by controlling histone h3 lysine 56 deacetylation. *Curr. Biol.* **2006**, *16*, 1280–1289.
38. KOVTOUN, S. V.; ENGLISH, R. D.; COTTER, R. J. Mass correlated acceleration in a reflectron MALDI TOF mass spectrometer: An approach for enhanced resolution over a broad mass range. *J. Am. Soc. Mass Spectrom.* **2002**, *13*, 135–143.
39. JARDINE, D. T.; MORGAN, J.; ALDERDICE, D. S.; DERRICK, P. J. A tandem time-of-flight mass spectrometer. *Org. Mass Spectrom.* **1992**, *27*, 1077–1083.
40. SCHEY, K. L.; COOKS, R. G.; KRAFT, A.; GRIX, R.; WOLLNIK, H. Ion/surface collision phenomena in an improved tandem time-of-flight instrument. *Int. J. Mass Spectrom. Ion Processes* **1989**, *94*, 1–14.
41. BOESL, U.; WEINKAUF, R.; SCHLAG, E. W. *Int. J. Mass Spectrom. Ion Processes* **1992**, *112*, 121–166.
42. SEETERLIN, M. A.; VLASAK, P. R.; BEUSSMAN, D. J.; MCLANE, R. D.; ENKE, C. G. High efficiency photo-induced dissociation of precursor ions in a tandem time-of-flight mass spectrometer. *J. Am. Soc. Mass Spectrom.* **1993**, *4*, 751–754.

43. Cotter, R. J.; Cornish, T. J. A tandem time-of-flight (TOF/TOF) mass spectrometer. *Anal. Chem.* **1993**, *65*, 1043–1047.
44. Cornish, T. J.; Cotter, R. J. Collision-induced dissociation in a tandem time-of-flight mass spectrometer with two single-stage reflectrons. *Org. Mass Spectrom.* **1993**, *28*, 1129–1134.
45. Cornish, T. J.; Cotter, R. J. A curved field reflectron time-of-flight mass spectrometer for the simultaneous focusing of metastable product ions. *Rapid Commun. Mass Spectrom.* **1994**, *8*, 781–785.
46. Medzihradszky, K. F.; Campbell, J. M.; Baldwin, M. A.; Falik, A. M.; Juhasz, P.; Vestal, M. L.; Burlingame, A. L. The characteristics of peptide collision-induced dissociation using a high performance Maldi-TOF/TOF tandem mass spectrometer. *Anal. Chem.* **2000**, *72*, 552–558.
47. Yergey, A. L.; Coorssen, J. R.; Backlund, P. S., Jr.; Blank, P. S.; Humphrey, G. A.; Zimmerberg, J.; Campbell, J. M.; Vestal, M. L. De novo sequencing of peptides using MALDI/TOF-TOF. *J. Am. Soc. Mass Spectrom.* **2002**, *13*, 784–791.
48. Barofsky, D. F.; Håkansson, P.; Katz, D. L.; Piyadasa, C. K. G. Tandem time-of-flight mass spectrometer. US Patent No 6,489,610, (2002).
49. Schnaible, V.; Wefing, S.; Resemann, A.; Suckau, D.; Bücker, A.; Wolf-Kümmeth, S.; Hoffman, D. Screening for disulfide bonds in proteins by MALDI in-source decay and LIFT-TOF/TOF-MS. *Anal. Chem.* **2002**, *74*, 4980–4988.
50. Cotter, R. J.; Gardner, B.; Iltchenko, S.; English, R. D. Tandem time-of-flight mass spectrometry with a curved field reflectron. *Anal. Chem.* **2004**, *76*, 1976–1981.
51. Pittenauer, E.; Zehl, M.; Belgacem, O.; Raptakis, E.; Mistrik, R.; Allmaier, G. Comparison of CID spectra of singly charged polypeptide antibiotic precursor ions obtained by positive-ion vacuum MALDI IT/RTOF and TOF/RTOF, AP-MALDI-IT and ESI-IT mass spectrometry. *J. Mass Spectrom.* **2006**, *41*, 421–447.
52. Verentchikov, A. N.; Ens, W.; Standing, K. G. Reflecting time-of-flight mass spectrometer with an electrospray ion source and orthogonal extraction. *Anal. Chem.* **1994**, *66*, 126–133.
53. Koslovsky, V.; Fuhrer, K.; Tolmachev, A.; Dodonov, A.; Raznikov, V.; Wollnik, H. Cooling of direct current beams of low mass ions. *Int. J. Mass Spectrom.* **1998**, *181*, 27–30.
54. Krutchinsky, A. N.; Chernushevich, I. V.; Spicer, V. L.; Ens, W.; Standing, K. G. Collisional damping interface for an electrospray ionization time-of-flight mass spectrometer. *J. Am. Soc. Mass Spectrom.* **1998**, *9*, 569–579.
55. Berkout, V. D.; Segers, D. P.; Cotter, R. J. Miniaturized EI/Q/oa TOF mass spectrometer. *J. Am. Soc. Mass Spectrom.* **2001**, *12*, 641–647.
56. Chien, B. M.; Lubman, D. M. Analysis of the fragments from collision-induced dissociation of electrospray-produced peptide ions using a quadrupole ion trap storage/reflectron time-of-flight mass spectrometer. *Anal. Chem.* **1994**, *66*, 1630–1636.
57. Doroshenko, V. M.; Cotter, R. J. A quadrupole ion trap/time-of-flight mass spectrometer with a parabolic reflectron. *J. Mass Spectrom.* **1998**, *33*, 305–318.
58. Doroshenko, V. M.; Cotter, R. J. A new method of trapping ions produced by matrix-assisted laser desorption ionization in a quadrupole ion trap. *Rapid Commun. Mass Spectrom.* **1993**, *7*, 822–827.
59. Ding, L.; Kawatoh, E.; Tanaka, K.; Smith, A. J.; Kumashiro, S. High efficiency MALDI-QIT-ToF mass spectrometer. *Proc. SPIE* **1999**, *3777*, 144.
60. Prieto, M. C.; Kovtoun, V. V.; Cotter, R. J. Miniaturized linear time-of-flight mass spectrometer with pulsed extraction. *J. Mass Spectrom.* **2002**, *37*, 1158–1162.
61. English, R. D.; Warscheid, B.; Fenselau, C.; Cotter, R. J. Bacillus spore identification using proteolytic mapping and a miniaturized MALDI TOF mass spectrometer. *Anal. Chem.* **2003**, *75*, 6886–6893.
62. Yoon, H. Y.; Kim, J. H.; Choi, E. S.; Yang, S. S.; Jung, K. W. Fabrication of a novel micro time-of-flight mass spectrometer. *Sensors and Actuators A* **2002**, *97–98*, 441–447.
63. Caprioli, R. M.; Farmer, T. B.; Gile, J. Molecular imaging of biological samples: Localization of peptides and proteins using MALDI-TOF MS. *Anal. Chem.* **1997**, *69*, 4751–4760.
64. McDonnell, L. A.; Heeren, R. M. A.; de Lange, R. P. J.; Fletcher, I. W. Higher sensitivity secondary ion mass spectrometry of biological molecules for high resolution, chemically specific imaging. *J. Am. Soc. Mass Spectrom.* **2006**, *17*, 1195–1202.
65. Schueler, B. *Microsc. Microanal. Microstruct.* **1992**, *3*, 1–21.
66. Salt, K.; Noble, C. A.; Prather, K. A. Aerodynamic particle sizing versus light scattering intensity measurement as methods for real-time particle sizing coupled with time-of-flight mass spectrometry. *Anal. Chem.* **1996**, *68*, 230–234.
67. Wang, S.; Zordan, C. A.; Johnston, M. V. Chemical characterization of individual, airborne sub-10-nm particles and molecules. *Anal. Chem.* **2006**, *78*, 1750–1754.
68. Czerwieniec, G. A.; Russell, S. C.; Tobias, H. J.; Pitesky, M. E.; Fergenson, D. P.; Steele, P.; Srivastava, A.; Horn, J. M.; Frank, M.; Gard, E. E.; Lebrilla, C. B. Stable isotope labeling of Entire Bacillus atrophaeus Spores and Vegetative Cells Using Bioaerosol Mass spectrometry. *Anal. Chem.* **2005**, *77*, 1081–1087.
69. Stowers, M. A.; van Wuijckhuijse, A. L.; Marijnissen, J. C. M.; Scarlett, B.; Van Baar, B. L. M.; Kientz, C. E. Application of matrix-assisted laser desorption to on-line aerosol time-of-flight mass spectrometry. *Rapid Commun. Mass Spectrom.* **2000**, *14*, 829–833.

Chapter 11

Fourier Transform Ion Cyclotron Resonance and Magnetic Sector Analyzers for ESI and MALDI

Jeremy J. Wolff and I. Jonathan Amster

Department of Chemistry, University of Georgia, Athens, Georgia

Electrospray and MALDI Mass Spectrometry: Fundamentals, Instrumentation, Practicalities, and Biological Applications, Second Edition, Edited by Richard B. Cole

11.1 FOURIER TRANSFORM MASS SPECTROMETRY

11.1.1 FTMS Overview

Fourier transform mass spectrometry (FTMS) offers the highest mass resolution and mass measurement accuracy of all mass analyzers. FTMS, also referred to as Fourier transform ion cyclotron resonance mass spectrometry (FTICR MS), can be adapted to a wide variety of ion sources[1–8] and ion dissociation methods.[9–17] The fundamental behavior of ions in FTMS instruments is based on the principle of ion cyclotron resonance (ICR), conceived and developed by E. O. Lawrence in the 1930s to build ion accelerators for nuclear physics experiments.[18] ICR was first implemented in mass spectrometry in an instrument called the omegatron, developed by scientists at the National Bureau of Standards in the 1950s.[19] Advances such as the application of Fourier transform methods to ICR spectrometry[20,21] and the trapped analyzer cell[22] resulted in the development of a powerful analytical instrument.

At the present time, there are a wide variety of both custom and commercial FTMS instruments throughout the world. Regardless of the type of FTMS instrument, the FTMS instrument consists of five basic components. These components are the magnet, analyzer cell, vacuum system, data system, and an ion source. The FTMS instrument utilizes either a permanent magnet, an electromagnet, or a superconducting magnet. Permanent magnets are rarely found in FTMS instruments due to their low field strength. However, because permanent magnets do not require any maintenance, there has been recent interest in developing a commercial FTMS instrument with such magnets. With magnetic field strengths ranging up to 1 T, FTMS instruments using permanent magnets are ideally suited for analysis of low-molecular-weight species, with applications to gas analysis, for example. Several early FTMS instruments used electromagnets, with field strengths of 1–2T. Because the performance of the FTMS instrument increases with field strength, electromagnets were replaced by superconducting magnets to achieve higher magnetic fields. Many properties of FTMS scale linearly or quadratically with magnetic field strength, including mass resolving power, mass accuracy, dynamic range, and the upper *m/z* limit.[23] The first commercial FTMS instruments utilized superconducting magnets, and these are the favored choice of FTMS instruments today. Superconducting FTMS magnets are similar to NMR magnets but have a much wider bore (100–160 mm) necessary to accommodate the analyzer cell and the vacuum system. Because of the large bore and homogeneous magnetic field necessary for FTMS, magnetic field strengths for commercial FTMS instruments are typically lower than those found on commercial NMR instruments. Superconducting magnetic field strengths of 4.7 T, 7 T, 9.4 T, and 12 T can currently be found in commercial FTMS instruments, with some research FTMS instruments using 14- to 15-T magnets. The technology exists to build a 21-T magnet, equivalent to the field strength of a 900-MHz NMR, and construction of such a large magnet is under consideration by some of the large national laboratories.

A second component of the FTMS instrument is the analyzer cell. It is the heart of the FTMS instrument where ions are stored, translationally excited, mass selectively isolated or ejected, and detected. The analyzer cell may also be used to dissociate ions for tandem mass spectrometry. The original ICR analyzer cell was cubic in design, consisting of six plates arranged as shown in Figure 11.1A. This cell design efficiently utilized the space between the flat pole caps of an electromagnet. Other types of analyzer cells are better adapted to the cylindrical bore of a solenoidal superconducting magnet. Commercial FTMS instruments utilize cylindrical cells, as shown in Figures 11.1B, 11.1C, and 11.1D. Figure 11.1B shows the open-ended cylindrical cell. The middle cylinder is divided into four plates, two for

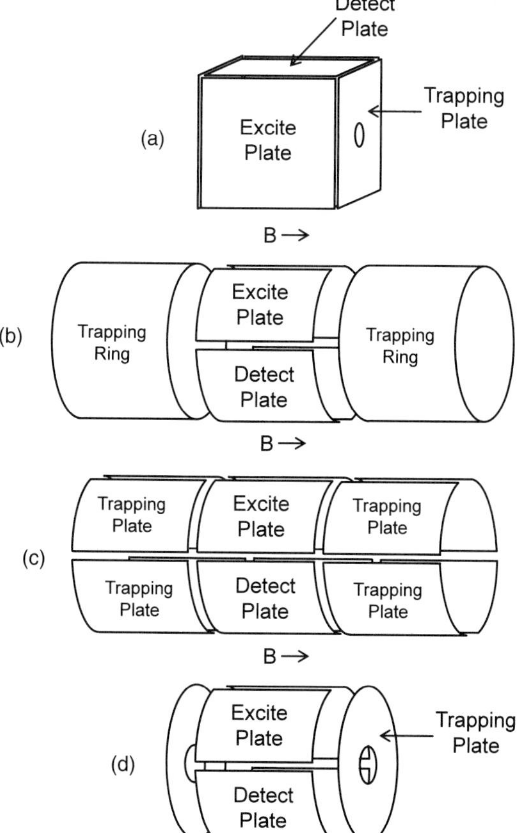

Figure 11.1. Four different types of analyzer cells commonly found in FTMS instruments. (**A**) Cubic cell, (**B**, **C**) open-cylindrical cell, and (**D**) closed-cylindrical cell. The cell diagrams are not drawn to scale. The cells are oriented so that their principal axis is aligned with the magnetic field, as indicated by the vector, **B**.

excitation of ion motion and two for detection of the trapped ions, and the outer rings are trapping electrodes used to contain the ions within the analyzer cell. The outer rings may also be divided into four segments similar to the central section as shown in Figure 11.1C, so that the excitation signal can be capacitively coupled to the electrodes adjacent to the excitation plates. The trapping electrodes of the open-ended cell can be replaced with flat plates that are oriented perpendicular to the magnetic field axis, giving rise to the closed-cylindrical cell, shown in Figure 11.1D.

A third component of the FTMS instrument is the vacuum system. During detection, ions travel several kilometers as they orbit the analyzer cell, and it is important that they do not undergo collisions with background neutral gases. FTMS instruments require ultrahigh vacuum (UHV)—that is, 10^{-9}–10^{-10} torr—in the analyzer region to provide a very long mean free path between collisions. In order to achieve ultrahigh vacuum (UHV), cryopumps or turbopumps are commonly used. It is worth noting that UHV is required only in the analyzer region and only during the excitation and detection of ions. Prior to excitation and detection, the pressure can be elevated in the analyzer region, followed by a delay in the experimental sequence to pump away the gas.

A fourth component of the FTMS is the data system. While the mechanical assembly of a FTMS system is relatively simple compared to other mass spectrometers, the data system is quite sophisticated. The FTMS data system consists of a pulse sequence generator to control the timing of the various events in a measurement cycle, a frequency synthesizer, a wideband excitation amplifier for ion isolation and ion detection, and a transient digitizer. A computer controls all of these components and is used to acquire and process the data.

The fifth and final component of the FTMS system is the ion source. All mass spectrometers have an ion source. With early FTMS instruments, the ion source was an integral part of the analyzer cell, or located adjacent to the analyzer cell, within the homogeneous magnetic field region of the instrument. In the 1980s, instruments were developed with ion sources located outside the magnetic field. This configuration is called an external ion source. By placing the source outside the magnetic field where it can be differentially pumped, almost any type of ion source can be interfaced to a FTMS instrument. When the source is located outside the magnet, ion optics such as a quadrupole or a series of electrostatic lenses are used to transfer the ions from the source to the analyzer cell, as will be discussed in more detail below.

11.1.1.1 Ion Motion in FTMS

Two fundamental forces act on ions in the analyzer cell, from interaction of their charge with the magnetic and electric fields. The magnetic field is homogeneous and unidirectional in the analyzer cell. The electric fields arise from the DC voltage applied to the trapping plates, as well as from the RF voltages applied to the excitation plates. The magnetic field causes an ion to move in a circular orbit, perpendicular to the direction of the magnetic field. The forces acting on an ion in a purely magnetic field are illustrated in Figure 11.2A. The Lorentz force, qvB (q is charge, v is velocity perpendicular to the magnetic field, B is magnetic field strength), is opposed by centrifugal force, mv^2/r (r is radius of the ions orbit). Equating these two forces and solving for v/r (angular frequency) leads to the fundamental equation:

$$\frac{v}{r} = \omega_c = \frac{qB}{m} \tag{11.1}$$

$$f_c = \frac{\omega_c}{2\pi} = \frac{qB}{2\pi m} \tag{11.2}$$

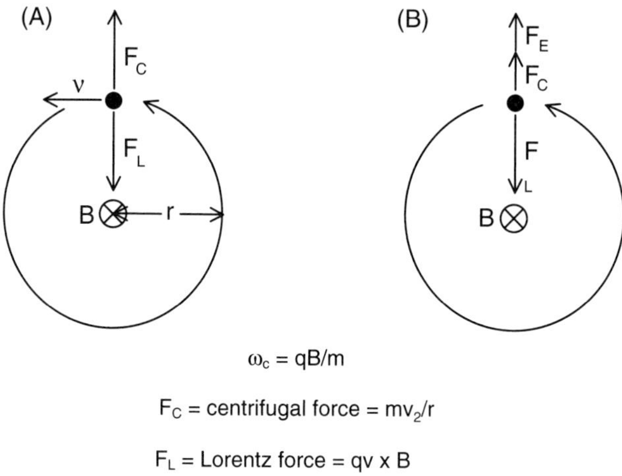

Figure 11.2. (**A**) The Lorentz force, F_L, causes the ion to undergo a circular orbit in the presence of a magnetic field, B. The Lorentz force is opposed by the centrifugal force of the orbiting ion, F_C. This ion orbits with a cyclotron frequency, ω_c, and with a radius, r. (**B**) The trapping potential produces a radially repulsive force, F_E, which, along with the centrifugal force, F_C, opposes the Lorentz force, F_L. (Adapted from Ref. 161.)

f_c is called the cyclotron frequency and has values in the range of 10 kHz to 1 MHz. Note that the cyclotron frequency is dependent only on the m/z of the ion and the strength of the magnetic field. The kinetic energy, and therefore the velocity of the ion, has no effect on the cyclotron frequency. The velocity of the ion does influence the radius of the cyclotron orbit, as can be seen in Eq. (11.1). Increasing the kinetic energy of an ion will increase the radius of the constant frequency cyclotron orbit. This is used to detect the ions and will be described in further detail below.

As a result of the Lorentz force, ion motion is constrained to a circular orbit perpendicular to the principal axis of the magnetic field, denoted here as the Z axis. However, the Lorentz force does not influence ion motion parallel to the magnetic field, so that ions can move freely along the Z axis. In order to contain the ions within the analyzer cell, a DC voltage is applied to the trapping plates or rings (Figure 11.1). The trapping potential is a small, symmetric voltage, approximately 1 V in magnitude. A schematic of the potential well created by the trapping plates of a cubic analyzer cell for storing positive ions is shown in Figure 11.3. A positive voltage is applied to trap positive ions, and a negative voltage is applied in order to trap negative ions. Ions trapped along the Z axis within the potential well oscillate freely between the two trapping plates, allowing ions to be stored from seconds to hours.

In addition to cyclotron and trapping motions, a third fundamental mode of behavior for ions in a FTMS analyzer cell is magnetron motion. This mode is a result of the combined electric and magnetic forces acting on an ion. In addition to trapping ions along the magnetic field axis, the trapping voltage creates a radially repulsive electric field perpendicular to the magnetic field. This results from the fact that the electric potential in the center of the analyzer cell is not zero, but is one-third of the trapping voltage for a cubic cell and is approximately this voltage for cylindrical cells. One can envision an extra force vector F_E (as shown in Figure 11.2B) that adds to the outward directed centrifugal force. The

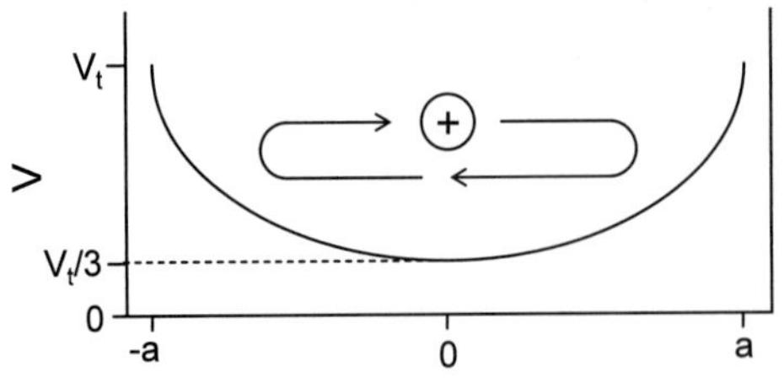

Figure 11.3. Plot of the electric potential along the central axis of the analyzer cell used to trap ions. The distance between the center of the cell and each trapping plates is a, and the distance between each trapping plate is $2a$. A trapping voltage (V_t) of $+1$ V is applied to each trapping plate. The trapped ion freely travels between the trapping plates ($-a$ to a), parallel to the magnetic field.

magnitude of this force is the product of charge and the radial electric field, qE_r. This additional force produces magnetron motion, which is an orbiting motion of the center of the cyclotron orbit around the central axis of the analyzer cell. The frequency of this motion is mass-independent, but is proportional to the ratio of the trapping potential to the magnetic field strength, V_T/B. Magnetron frequencies are on the order of 1–100 Hz, much lower than cyclotron frequencies. The calculated ion trajectory in Figure 11.4 illustrates the combination of cyclotron and magnetron motion on an ion within the analyzer cell. An ion undergoing cyclotron motion (the smaller radius) will also precess along an isopotential field line (the larger radius) around the cell. For Figure 11.4 the magnitude of the magnetron orbit relative to the cyclotron motion has been greatly exaggerated. Under normal conditions, the cyclotron orbit radius is much greater than the magnetron orbit radius. The greatest influence on the magnitude of magnetron orbit is the initial displacement of the ion when entering the cell. For this reason, ions should ideally enter the cell along the central axis of the analyzer cell. There is no analytical utility to magnetron motion. In addition, it affects mass measurement, reducing the observed cyclotron frequency by an amount equal to the magnetron frequency [Eq. (11.3)]. Under normal conditions, the observed frequency (ω_o) is reduced by 10–100 ppm from the unperturbed cyclotron frequency (ω_c) of Eq. (11.1).

$$\omega_o = \omega_c - \omega_m \tag{11.3}$$

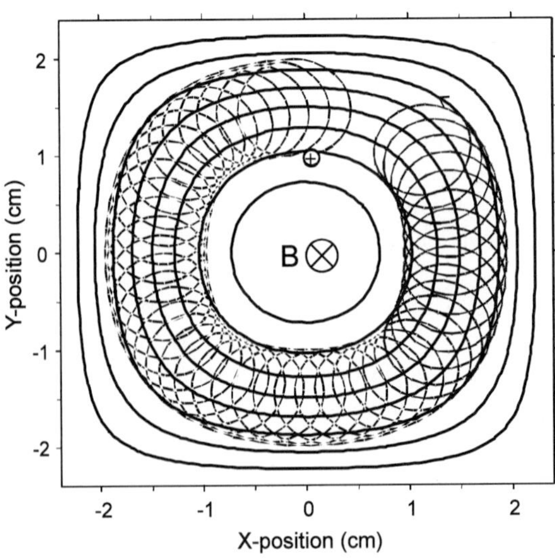

Figure 11.4. Plot of an ion undergoing both magnetron and cyclotron motion in the XY plane perpendicular to the magnetic field. For this plot, the slow precession of the magnetron motion (large orbit) is greatly exaggerated over the much higher frequency cyclotron motion (small orbit). (Reprinted from Ref. 161 with permission.)

11.1.1.2 Event Sequences in the FTMS

The operation of the FTMS instrument is different compared to beam-type mass spectrometers such as magnetic sector, time-of-flight, or quadrupole instruments. With beam instruments, ion generation, mass analysis, and ion detection are separated from each other in *space*. In contrast to this arrangement, the operations of an FTMS instrument are separated in *time*. The events, such as ionization, mass analysis, and detection, can all occur within the analyzer cell, but at different times. In more modern FTMS instruments, ionization occurs in a separate region, but the time-dependent nature of the experiment is still significant. This series of events is referred to as an experimental sequence. In FTMS, the data system controls the execution of events within the experimental sequence. Many early ICR instruments used the analyzer cell for ionization and detection. In comparison to these early instruments, modern FTMS instruments also operate in *space*. Modern instruments can ionize, isolate, and dissociate outside the analyzer cell, but the latter is still required for detection. Because the FTMS instrument separates its operations in time, it is classified as a *pulsed* mass spectrometer. In a pulsed mass spectrometer, the ions are formed in small packets that are, in turn, analyzed as a group. This is in contrast to a *scanning* instrument (e.g., magnetic sector and quadrupole) in which ions are made continuously, and the mass analyzer is scanned, detecting ions one mass-to-charge value at a time. A typical FTMS experimental sequence is shown in Figure 11.5. Because detection in the FTMS analyzer is nondestructive (see below), ions remain in the analyzer cell after they are detected. Before the experimental sequence can begin, ions are generally removed from the cell by a process referred to as a cell quench. After all ions have been removed, an ionization pulse places a new group of ions into the analyzer cell. Ions are then excited and, shortly after, detected. The timing of events in the FTMS experiment is important; Figure 11.5 also shows a delay between ionization and detection, where the length of the delay is controlled by the instrument operator. Such delays can be used for a variety of purposes—for example, to either (a) allow excess gas to be removed from the instrument or (b) control the number of ions that enter the analyzer cell.

11.1.1.3 Ion Excitation and Detection

As previously discussed, the principal motion of ions in a magnetic field is the cyclotron orbit. Cyclotron frequency is inversely proportional to mass-to-charge ratio, as shown in Eq. (11.1). Therefore, the mass-to-charge ratio can be determined from the strength of the magnetic field and the cyclotron frequency of the ion. To measure cyclotron frequency, ions are excited into coherent motion within the analyzer cell. Ions are introduced into the analyzer cell with low kinetic energies, typically below 1 eV. A notable exception

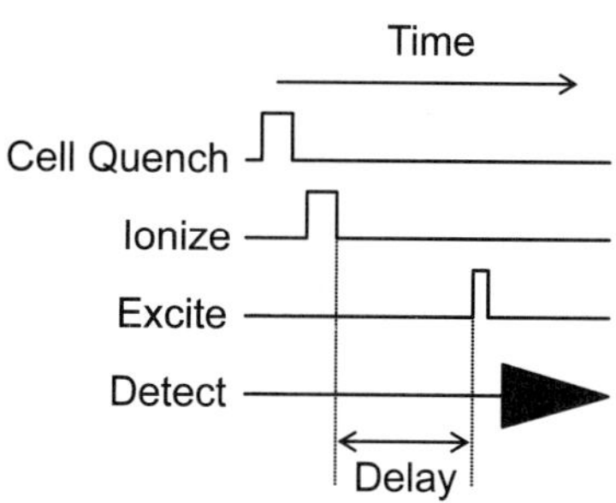

Figure 11.5. Example of an experimental sequence used in FTMS. Events such as the cell quench, ionization, and detection occur at different times during the experimental sequence. Events can be spaced by user-defined delays, allowing the timing of the FTMS experiment to be precisely controlled.

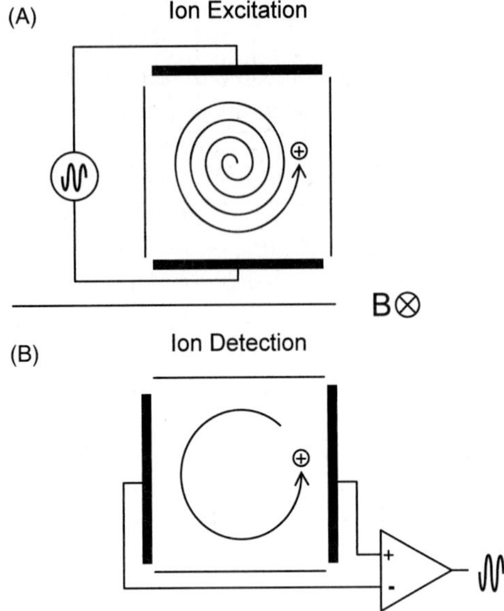

Figure 11.6. Sequence of events for ion excitation and detection in the analyzer cell. (**A**) Ions trapped within the analyzer cell are excited using dipolar excitation, resulting in expansion of the cyclotron radius while maintaining the same cyclotron frequency. The excitation is terminated before the ions collide with the analyzer cell, when the cyclotron radius is approximately three-quarters of the radius of the cell. (**B**) An image current of the orbiting ions is produced on the detection plates, and this signal is converted into a voltage, amplified, and acquired by the data system. (Adapted from Ref. 161.)

is for ions generated by internal matrix-assisted laser desorption/ionization (MALDI), which will be discussed below. Because of the low kinetic energy of ions upon entering the analyzer cell, their initial cyclotron radius is usually less than 1 mm, much smaller than the dimensions of the cell. To detect the ions, the radius of their cyclotron orbit is increased before detection by the application of a dipolar electric field, as indicated by the ion trajectory shown in Figure 11.6A. For dipolar excitation, a sinusoidal waveform is applied to an opposing pair of electrodes, called the excitation plates, with opposite polarity applied to the two opposing plates. If the frequency of the excitation signal equals the cyclotron frequency of ions in the cell, a resonance condition exists that causes acceleration of the ion and expansion of the radius of its cyclotron orbit. The cyclotron frequency remains constant as ions accelerate, and ions that are not in resonance with the excitation signal remain at the center of the analyzer cell. By controlling the excitation amplitude and duration, ions can be accelerated sufficiently so that the cyclotron radius will expand until the ion collides with the cell plates and is neutralized. This is referred to as radial ejection and is a method that is used to mass-selectively remove unwanted ions from the analyzer cell—for example, to isolate ions of a single mass-to-charge value as the first step in an MS/MS experiment. For detection, however, ions are accelerated to a radius that is smaller than that of the cell, generally around 75% of the analyzer cell radius. Orbiting ions produce an image current in an external circuit that is connected to a second opposing pair of electrodes, called the detection plates, as shown in Figure 11.6B. All ions of the same mass-to-charge travel in a spatially coherent packet. As the packet of ions passes one of the detection plates, electrons are drawn toward this detection plate and away from the opposite detection plate. The attraction is reversed when the orbiting ion packet reaches the other detection plate. This back-and-forth motion of the electrons in the external circuit that joins the two detection plates is referred to as the image current. It has a frequency that is identical to the cyclotron frequency of the orbiting ions in the cell. The image current is converted into a voltage and then amplified, digitized, and stored by the FTMS data system. Note that detection of the ions via image current is nondestructive, and ions remain in the trap after excitation and detection. This feature is unique to FTMS instruments, including ICR and the orbitrap, because all other mass spectrometers rely on destructive collisions with an electron multiplier for ion detection.

FTMS instruments detect ions of all mass-to-charge values within the analyzer cell at the same time. In order to accomplish this, ions in the analyzer cell of varying mass-to-charge must be excited near simultaneously. This is achieved by exciting a range of frequencies over a very short period of time. For example, the excitation signal can sweep all frequencies from 10 kHz to 10 MHz over a period of 1 ms. This is known as a broadband excitation and is generally accomplished with an RF "chirp," which is a waveform in which one steps through a sequence of frequencies, spending an equal amount of time at each frequency. When the frequency of the RF chirp is resonant with an ion's cyclotron frequency, it is excited and its cyclotron radius expands. As the excitation signal steps through the frequencies specified for the RF chirp, all ions are excited to a similar cyclotron orbit regardless of mass-to-charge. After the excitation waveform is executed, a collective image current is detected for all the ions (simultaneous detection) within the analyzer cell, producing a signal referred to as a transient, also known as the time-domain signal. An example of a typical transient from FTMS is shown in Figure 11.7. The summation of the sinusoidal frequencies of each ion in the cell produces the transient, where the amplitude of each sinusoidal component is proportional to the number of ions with that cyclotron frequency. A Fourier transform is applied to the time-domain signal, thus extracting individual frequency components and their amplitudes. A mass calibration equation is then applied to convert the observed frequencies to mass-to-charge values as shown in Figure 11.7. A more in-depth discussion of signal acquisition and Fourier analysis can be found elsewhere.[24–26]

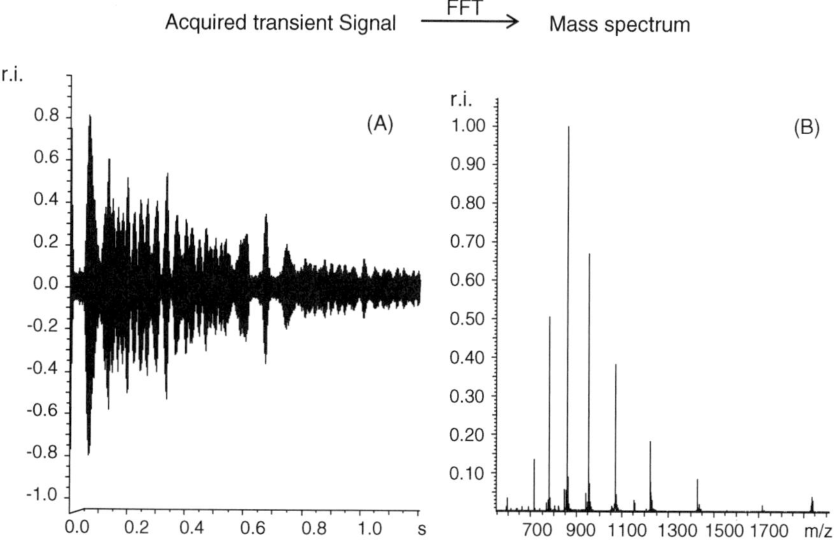

Figure 11.7. (**A**) After excitation, the summation of all ions trapped within the analyzer cell produces a transient signal that is digitized and stored by the data system. A fast Fourier transform (FFT) is applied to the transient (**B**), producing a frequency spectrum that is then converted into a mass spectrum via a calibration equation. (Reprinted from Ref. 161 with permission.)

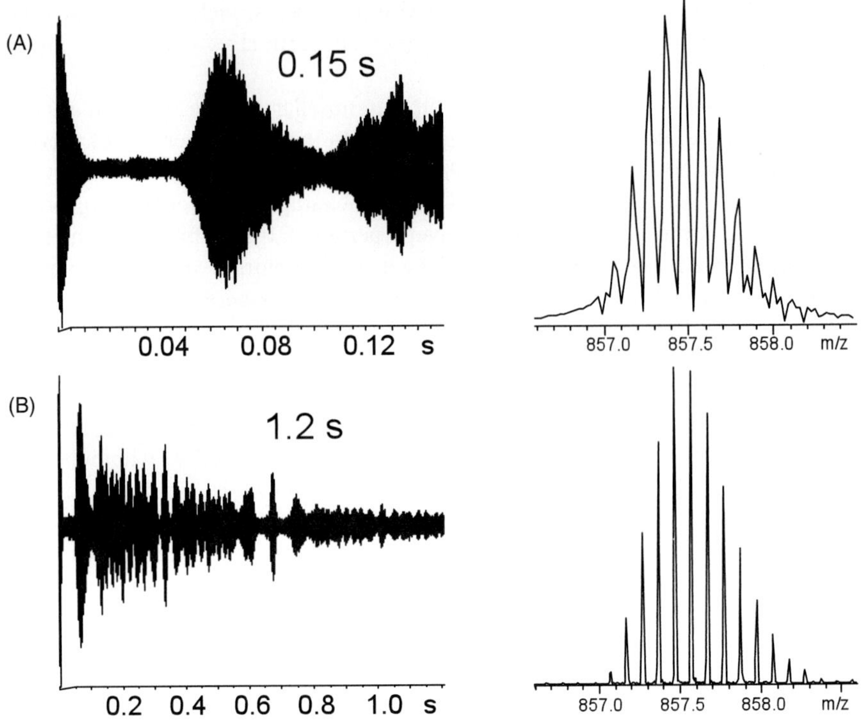

Figure 11.8. Peak resolution increases in FTMS as the transient length increases. (**A**) A 0.15-s transient yields a resolving power of 9000 at *m/z* 857 in a 7-T instrument. (**B**) A resolving power of 75,000 is achieved for the same ion, but with a 1.2-s transient. (Adapted from Ref. 161.)

Assuming coherent ion motions continue for the duration of the acquired signal, the maximum obtainable resolution is Fourier-limited and is defined by the equation

$$\mathrm{RP} = \frac{f_c T}{2} = \frac{2\pi BT}{m/q} \tag{11.4}$$

where RP is resolving power, f_c is cyclotron frequency in hertz, and T is the acquisition time in seconds. For example, an ion of *m/z* 1400 has a cyclotron frequency of 100 kHz at 9.4 T. The expected mass resolving power for a 1-s transient is 50,000. Increasing the data acquisition time improves mass resolution, as shown in Figure 11.8. The resolving power is clearly much lower for data acquired with the shorter transient (Figure 11.8A) when compared to the same data acquired with a longer transient (Figure 11.8B). FTMS is a powerful technique because resolution depends solely on the length of time for which a coherent signal is acquired. Since the cyclotron frequency of an ion decreases as its mass-to-charge increases, resolving power decreases at higher mass-to-charge. Conversely, increasing the strength of the magnetic field will increase mass resolving power. While increasing charge on an ion will increase the resolving power for a given mass-to-charge, it has no practical benefit for mass resolution, because isotope peaks are more closely spaced by the same factor for which resolving power increases.

Factors that improve mass resolution generally improve mass accuracy as well. Narrower peak width implies a smaller standard deviation in determining the peak's centroid. However, mass accuracy depends on a number of other factors, as well. Proper mass calibration is essential to accurate mass measurement. Internal calibration provides the best mass accuracy; better than 1 ppm is easily achieved using internal calibration in modern instruments with magnetic field strengths of 7 T or higher. However, internal calibration is not always practical to use, particularly for LC/MS and tandem mass spectrometry experiments. With external calibration, it is important to ensure that all experimental parameters are kept identical between the acquisition of the mass spectra of the calibrant and the analyte, particularly the trapping potential and the excitation waveform. Space–charge limits mass accuracy in most trapping mass analyzers, as the electric field of the trapped ions scales with the number of ions in the analyzer, and influences ion motion and mass measurement. For FTMS analyzers, space-charge effects on mass accuracy are understood fairly well and can be controlled so that accurate mass measurement can be obtained with external calibration.[27–31] For example, with one commercial instrument, the number of ions that are injected into the analyzer cell can be controlled dynamically, and is referred to as automatic gain control (AGC).[32] This can be used to compensate for the large variations in ion production that occur during an LC/MS run, for example.

11.1.1.4 Limits of Ion Detection at Higher Mass-to-Charge

There are a number of physical parameters that contribute to the upper mass limit and the upper mass-to-charge limit of FTMS instruments. The upper mass limit of an FTMS is limited by the magnetic field strength, trapping potential, and cell geometry. One can consider both axial and radial limits for stable trapping of high-mass ions in the analyzer cell. The axial limit refers to the ability of the trapping plates to confine the ions within the analyzer cell parallel to the magnetic field. Axial motion affects the upper mass limit in MALDI, where ions are singly charged for the most part, resulting in ion *m/z* scaling directly with analyte mass. As discussed below (see Section 11.1.2.2, "Internal MALDI"), MALDI ions are generated with similar velocities regardless of mass, and as a result, the kinetic energy of ions formed under vacuum MALDI conditions scales with analyte mass. It is important to note that the kinetic energy of the analyte is dependent on velocity and mass, but is independent of B_0. Therefore, increasing the trapping voltage increases the axial limit by allowing ions with higher KE, and thus higher masses, to be trapped. The maximum axial limit is theoretically unlimited, but depends on the maximum voltage than can be applied to the analyzer cell trapping plates. The kinetic energy along the Z axis (parallel to the magnetic field) of a trapped ion can be reduced prior to ion detection so that trapping voltage can be lowered to a value that provides good resolution (e.g., ≤ 1 V).

The radial limit refers to the stable cyclotron orbit of the ion. Ions stored in the cell must have a cyclotron radius that is less than the radius of the analyzer cell, otherwise they will strike the cell plates and be neutralized. Recall that the cyclotron radius is proportional to the velocity of the ion as defined by the equation

$$r = \frac{v}{\omega_c} \tag{11.5}$$

Substituting the equation for kinetic energy ($E = \frac{1}{2}mv^2$) as solved for the ion velocity, along with the cyclotron equation as solved for ω_c, results in the equation

$$r = \frac{\sqrt{2Em}}{qB_0} \tag{11.6}$$

The cyclotron orbit of an ion increases as the square root of the analyte mass. Therefore, as the mass of the analyte increases, the cyclotron radius will eventually exceed the radius of the trap, causing loss of the ion. In contrast to the axial limit, the radial limit is inversely proportional to B_0.

Another limiting factor is that for stable ion motion, the cyclotron frequency must exceed the magnetron frequency of an ion. Since magnetron frequency is independent of mass to a first approximation, one can see that as mass-to-charge increases, cyclotron and magnetron frequencies will converge.[33] Since magnetron frequency is directly proportional to trapping potential, the convergence point decreases in mass-to-charge as the trapping potential is raised. Taking both the axial and radial limits into account, the upper mass limit must be balanced between a high or low trapping voltage, where a lower voltage results in axial ion loss and a higher voltage results in radial ion loss. For a cylindrical trap and $B_0 = 7.0$ T, the calculated optimal trapping voltage for high-mass operation is 17.13 V, resulting in an upper limit of m/z 12,400.[34] However, such a high trapping potential has deleterious effects on signal shape and is not practical. While these calculations were demonstrated with $B_0 = 7.0$ T, current FTMS instrument configurations use higher magnetic fields that will decrease the cyclotron radius, allowing for an increase in the upper mass limit. The highest mass detected by MALDI-FTMS is the transferrin protein dimer at ~157 kDa, albeit at low resolution.[35] The highest mass detected by ESI-FTMS was an individual T4 DNA ion, with a nominal molecular mass of 1.1 MDa.[36]

The influence of isotope peaks subtly influences the practical upper mass limit, independent of charge. This limit has greatest application to ions produced by ESI, which brings high-mass ions into the usable m/z range of an FTMS by multiply charging the ion, but it also influences both internal and external MALDI. Because of multiple charging, the isotopes of these high-mass ions will be closely spaced in mass-to-charge and will therefore also have closely spaced cyclotron frequencies. During detection, these closely spaced cyclotron frequencies exhibit destructive interference, leading to "beating" effects as illustrated in Figure 11.9. The destructive interference of two cyclotron frequencies spaced by Δf is shown in Figure 11.9A. Increasing the number of cyclotron frequencies, each spaced by Δf, from 2 to 16 (Figures 11.9A to 11.9D) leads to a reduction in the measurable signal with "beats" occurring with a frequency equal to Δf. As shown in Figure 11.9, the duration of these beats becomes narrower as the number of added cyclotron frequencies increases, decreasing the amount of measurable signal. The duration between the signal beats increases with increasing analyte mass, as the frequency spacing between isotope peaks decreases with increasing mass. The challenge of analyzing high-mass molecules by ESI is that as the mass of the analyte increases, the occurrence of the second signal "beat" may exceed the transient acquisition time. When only the first beat at $t = 0$ will be observed, a low-resolution signal is produced. In order to obtain isotopic resolution, the second beat must be acquired,[37] and this becomes more challenging as the beat spacing increases. Moreover, the signal-to-noise ratio decreases, because one is averaging a transient that is mostly noise (between beats) with very little usable signal (short beats). The challenge of high-mass analysis by ESI-FTMS can be seen in Figure 11.10. ESI of a non-denatured 44-kDa protein dimer in a 7-T FTMS[38] produces three charge states, 10^+–12^+, as shown in Figure 11.10A. The transient (Figure 11.10B) shows that measurable signal appears in short (tens of milliseconds) bursts that are separated by almost 2 s. For a 44-kDa molecule at 7 T, the frequency difference between isotopes for the 11^+ ion is 0.61 Hz, and the beat period is predicted to be 1/0.61 Hz = 1.6s, as observed.

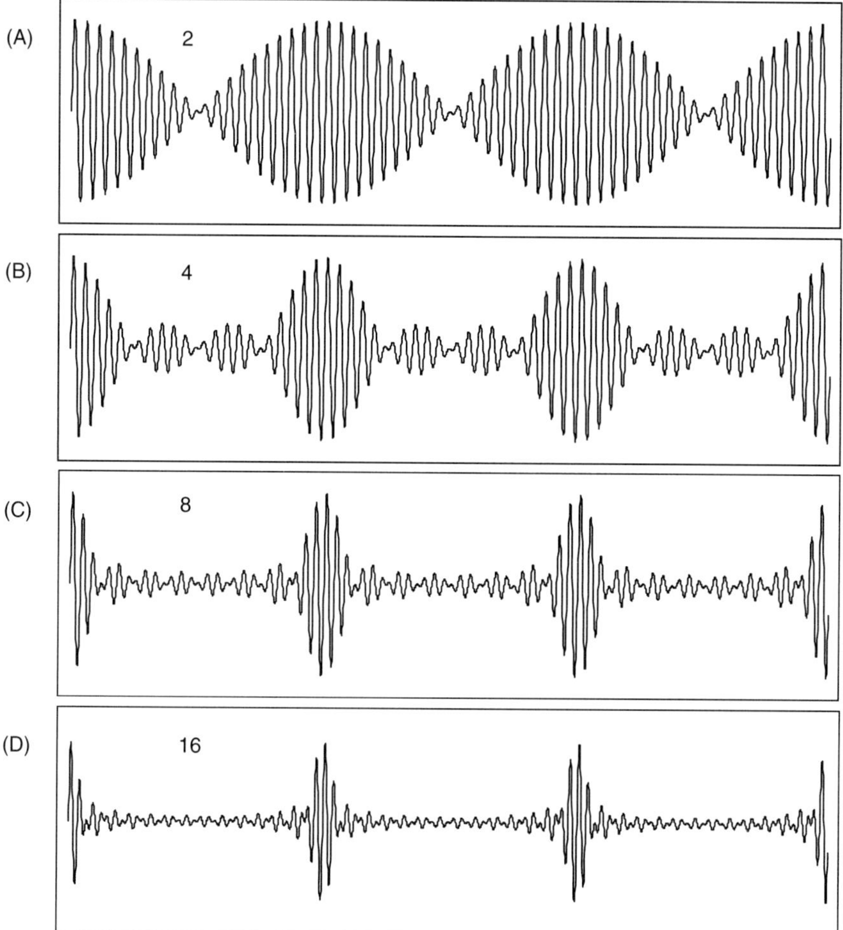

Figure 11.9. Illustration of the destructive interference that results from closely spaced frequencies, and the resulting "beats" that appear in the summed signal. The summation of (**A**) 2, (**B**) 4, (**C**) 8, and (**D**) 16 sine waves, equally spaced in frequency, produces the signals shown in **A–D**, respectively. Destructive interference reduces the measurable signal as the number of co-added sine waves increases. The spacing between beats is the inverse of the difference in frequency between the co-added components.

11.1.2 MALDI Combined with FTMS

11.1.2.1 Chronology of Developments in MALDI-FTMS

Shortly after MALDI was first reported in 1988,[39–41] researchers became interested in interfacing MALDI to FTMS. Because MALDI can be performed in vacuum, it was easy to conceive of an internal MALDI experiment, in which the sample target was placed adjacent to the analyzer cell. Moreover, the pulsed nature of MALDI ion production lends itself well to the pulsed nature of FTMS analysis. Because MALDI produces predominately singly charged ions, analyte ion *m/z* increases linearly with analyte mass. Early MALDI developments focused on increasing the upper mass limit and improving resolution and

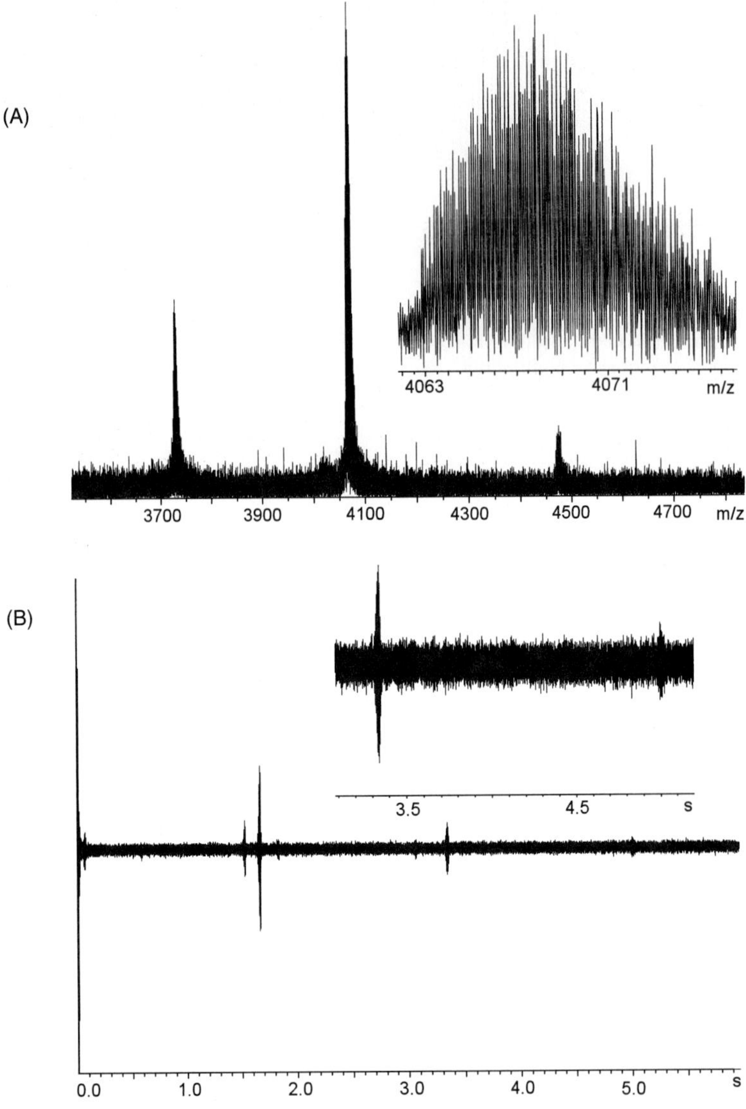

Figure 11.10. (**A**) Nondenaturing ESI-FTMS of nigerythrin. A 44-kDa protein dimer produces the 10^+–12^+ charge states. **Inset**: Zoom showing the resolution of the 11^+ charge state. (**B**) The signal in the transient from the data shown in **A** appears in short bursts separated by nearly 2 s. (Reproduced from Ref. 38 with permission.)

mass accuracy. Also, techniques to decrease metastable decompositions resulting from MALDI ionization were investigated. Increasing the upper mass limit was achieved by altering the MALDI matrix and modification of how ions are trapped within the analyzer cell. Increasing the pressure of the MALDI source during ionization through the development of intermediate-pressure and atmospheric-pressure MALDI has also greatly decreased the amount of metastable decomposition.

11.1.2.2 Internal MALDI

Internal MALDI is performed by placing the MALDI target in close proximity (1 mm to 1 cm) to the analyzer cell. As a result, the MALDI target is within the UHV region of the FTMS as well as inside the magnetic field. The earliest experiments used a direct probe and sliding seal to bring a sample from a pumpdown chamber to an opening in a trapping plate, adjacent to the analyzer cell. An illustration of this arrangement is shown in Figure 11.11. The focused laser beam enters the vacuum system via a window and passes through the cell before striking the MALDI target. The ions that are generated desorb normal to the MALDI target (constrained by the magnetic field) and pass into the analyzer cell. While this design was mechanically simple, the one-stage seal between atmosphere and the analyzer region limited the vacuum that could be achieved to 10^{-8} torr. In later designs, the direct probe was replaced by a sample transfer mechanism that could allow MALDI targets to be evacuated in a pumpdown chamber and then transferred via a transport mechanism that was entirely within the analyzer vacuum system.[42] This allowed UHV (10^{-8}–10^{-10} torr) pressures to be achieved.

With internal MALDI, ions desorbed from the target pass into the analyzer cell where they are trapped for subsequent excitation and detection. Upon ionization, a plume of ions and neutral molecules rapidly expands along the magnetic field axis. Entrapment of the analyte within the rapidly expanding MALDI plume results in analyte ions having similar velocities as the matrix regardless of mass. Reported initial velocities of matrix ions vary from a few hundred meters to second to greater than 1000 m/s,[43–48] depending upon the matrix used, laser fluence, and measurement method. Because analyte ions have similar initial velocity distributions independent of their masses, the kinetic energy of an analyte ion scales directly with its molecular mass. The dependence of analyte kinetic energy on analyte velocity and mass is illustrated in Figure 11.12.[42] Kinetic energies start as low as 1 eV for molecules below *m/z* 100, reaching $\geq$100 eV for high-mass molecules, with the average kinetic energy increasing in direct proportion to analyte mass. The large distribution in analyte kinetic energy influences the ability to capture ions, because the trapping potential

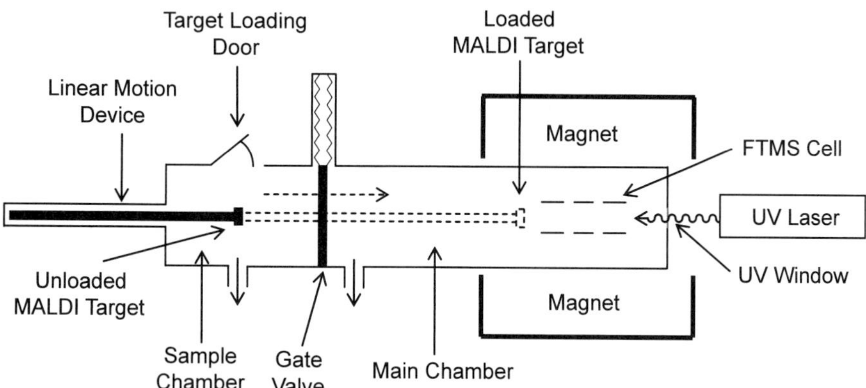

Figure 11.11. Schematic of an internal-MALDI FTMS. The sample target is first loaded into an antechamber that is then evacuated to a pressure of approximately 10^{-8} torr. The sample target then passes into the main chamber, where it transferred to a holder adjacent to the analyzer cell. The antechamber and main chamber are isolated from each other with a gate valve after the sample is transferred, to achieve ultrahigh vacuum in the analyzer region.

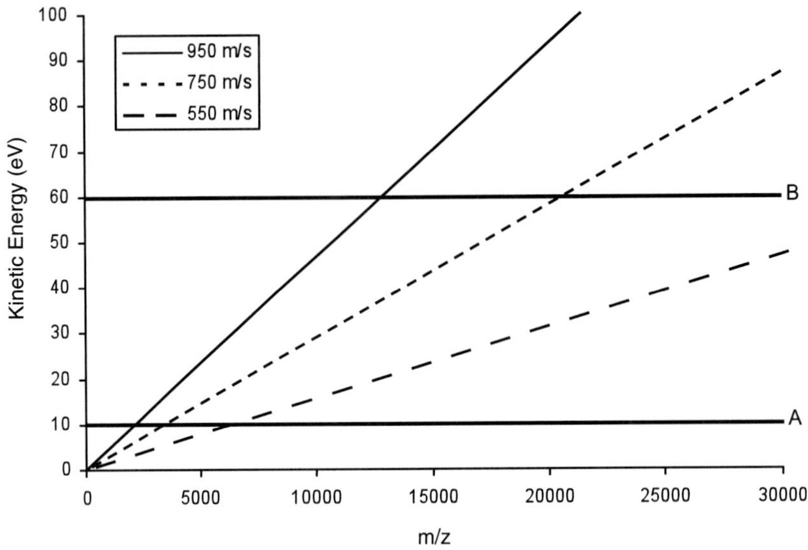

Figure 11.12. Relationship of analyte ion kinetic energy to mass analyte mass for ions produced by vacuum MALDI. Ions below line A may be captured by a 10-V trapping potential, while ions below line B require a 60-V trapping potential to be captured. (Adapted from Ref. 42.)

must be set greater than the kinetic energy of the ions to be analyzed. The trapping plates act like a kinetic energy filter, where only ions with kinetic energies below the trapping plate potential are stopped, and those of higher energies pass. Ions must pass the first trapping plate and must be stopped by the second trapping plate in order to be captured. For example, if the trapping plates are set to 1 V, a favorable voltage for detection, Figure 11.12 indicates that only analytes with masses below 1000 Da have a low enough kinetic energy to be efficiently captured. If the trapping voltage increases to 10 V (Figure 11.12, line A), the maximum analyte mass that can be trapped increases to approximately 4000 Da. Similarly, a trapping voltage of 60 V (Figure 11.12, line B) allows analytes up to 20 kDa to be trapped. Due to the wide range in analyte kinetic energies produced from vacuum MALDI, the trapping voltage must be adjusted according to the mass of the molecules of interest.

In order to trap ions with the wide range of kinetic energies generated in MALDI experiments and to provide optimal trapping voltages for excitation and detection, a number of methods have been developed. The trapping methods are described here with application to internal MALDI, but some methods are also applicable to external MALDI and ESI. Depending on the design of the analyzer cell, ions may be trapped by rings (an open cylindrical cell), plates (a closed cell), or a combination of the two. To simplify the description of trapping techniques, the analyzer cell will be described with trapping plates. A diagram of the analyzer cell used to describe these trapping methods, with the trapping and excite/detect plates annotated, is shown in Figure 11.13A. The trapping plate closest to the source (internal MALDI target, or external ESI or MALDI source) will be defined as the front trapping plate; the trapping plate farthest from the source will be referred to as the rear trapping plate. In order to capture and store ions, the trapping voltage of the analyzer cell must be equal to or higher than the product of an ions kinetic energy and its charge. The earliest MALDI experiments were conducted with static trapping voltages. In order to be captured in the analyzer cell, ions had enough energy to pass over the potential barrier of the

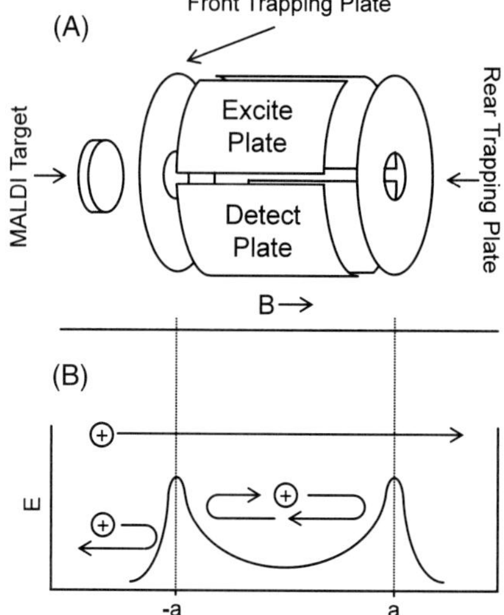

Figure 11.13. Illustration of the challenge of capturing externally formed ions using a fixed trapping voltage. (**A**) Diagram of the analyzer cell showing position of MALDI target. (**B**) Ions with kinetic energy less than trapping potential are reflected by the front trapping plate and do not enter the cell, while ions with kinetic energy greater than the trapping potential pass through the cell. Only ions with enough kinetic energy to pass the front trapping plate, but with less energy than the rear trapping plate, are captured. This requires ions to lose kinetic energy while in the analyzer region (e.g., by a collision with a neutral) or to convert some of the axial energy into radial energy as they pass through the cell.

front trapping plate, yet not collide with the rear trapping plate. As shown in Figure 11.13B, ions with kinetic energy below the trapping potential are reflected and do not enter the analyzer cell, while ions with kinetic energy greater than the trapping potential pass through the analyzer cell. Only ions that have adequate kinetic energy to pass the first trapping potential but not the second trapping potential are captured. The ions that were captured lost translational energy while passing through the cell via a collision (aided by the pulsed addition of a collision gas during the MALDI event) or by conversion of the axial energy into cyclotron or magnetron motion. This type of trapping is very inefficient.

Improvements in FTMS instrumentation allowed the potentials applied to the trapping plates to be altered during the course of the experiment. Altering the potential applied to the trapping plates during an experimental sequence is generally referred to as dynamic trapping. Depending on how the trapping potentials are altered, dynamic trapping can be more specifically identified as gated trapping or as adiabatic cooling. Gated trapping is a commonly used trapping method for FTMS and involves altering the voltage of the front trapping plate during the FTMS experiment. The experimental sequence of gated trapping experiment is shown in Figures 11.14B–11.14E. The front trapping plate is held at a voltage at or near ground during MALDI ionization, while the rear trapping plate is held at a potential greater than the kinetic energy of the analyte ions. Ions generated by MALDI pass through the cell until they encounter the higher potential of the rear trapping plate (Figure 11.14B) and are reflected in the opposite direction (Figure 11.14C). Before the ions can exit the cell, the front trapping plate is increased to the same voltage as the rear trapping plate (Figure 11.14D). The voltages applied to the trapping plates are typically greater than the trapping voltages used for optimal excitation/detection conditions. Therefore, the trapping voltages are symmetrically lowered to approximately 1 V prior to excitation/detection as shown in Figure 11.14E. The symmetric lowering of the trapping potential is referred to as adiabatic cooling.[49] The kinetic energy of the trapped ions is reduced by this method. Due to the short distance the ions travel from the MALDI target to the entrance of the analyzer cell, *m/z* discrimination is minimized compared to trapping of ions generated from external sources, thus providing the means to trap wide mass distributions—for example, for internal MALDI of polymer samples.[50,51]

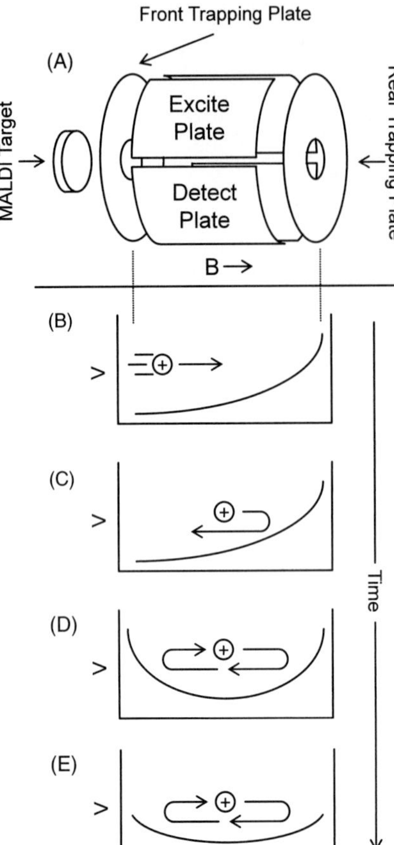

Figure 11.14. Voltages changes for a gated trapping event. (**A**) Schematic of the analyzer cell showing position of MALDI target and trapping plates (see text for description). Ions enter the analyzer cell (**B**) where they are reflected by the rear trapping plate and reverse direction (**C**). The front trapping plate is brought up to a higher voltage (**D**) to prevent ions from exiting the cell. (**E**) The trapping voltage is symmetrically lowered before excite/detect. The gated trapping events occur on a timescale of tens of microseconds. The lowering of the trapping potential after gated trapping occurs on the tens-of-milliseconds timescale.

Another form of dynamic trapping that is found on some commercial FTMS instruments uses deflection electrodes, and is referred to as Sidekick™.[52] The cell is modified with a pair of cylindrical electrodes centered around the axis (*Z* axis) of the cell, adjacent to the front trapping plate, as shown in Figure 11.15A. Using this cell configuration, both the front and rear trapping plates are held at a potential higher than the kinetic energy of the high mass ions. The potential on the front trapping plate must be overcome for the ions to enter the analyzer cell. To allow ions to pass into the analyzer cell, the Sidekick electrodes are set to a voltage of opposite polarity to that of the trapping plates. This creates a "hole" in the electric field of the front trapping plate, allowing ions to pass into the analyzer cell. In order to prevent ions that have reflected from the rear trapping plate from exiting the analyzer cell, a small differential voltage is applied to the Sidekick electrodes, pushing the ions off-axis as they enter the cell, as shown in Figure 11.15B. Prior to ion excitation and detection, the Sidekick electrodes are brought to the same voltage as the front trap plate to ensure a uniform electric trapping field during the detection of the image current.

While these gated trapping methods are effective for lower-mass analytes, trapping high-mass molecules is still difficult due to the large kinetic energies of these molecules. The addition of a sugar co-matrix (D-fructose) was used in early internal MALDI-FTMS experiments to improve the trapping efficiency of high-mass ions.[4] This improvement

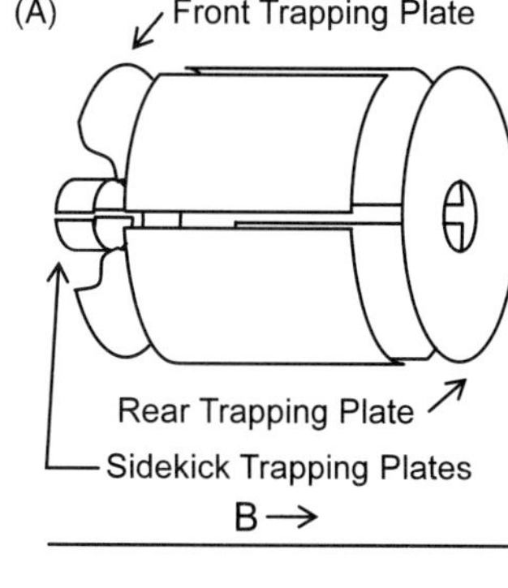

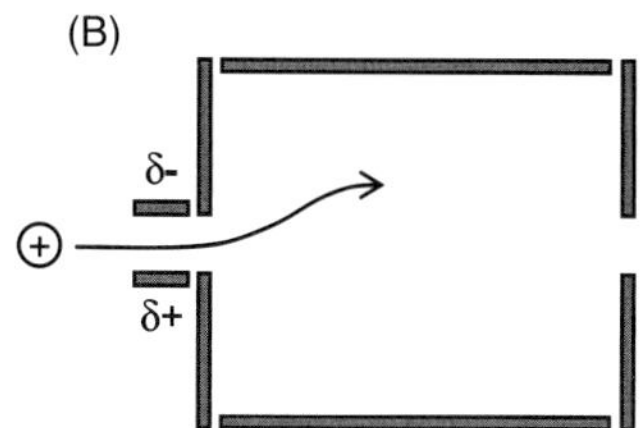

Figure 11.15. Configuration and operation of an analyzer cell equipped with electrodes for Sidekick™ trapping. (**A**) For clarity, a small section of the front trapping plate has been removed so that the Sidekick electrodes can be seen. (**B**) Ions entering the analyzer cell are shifted off-axis due to a small differential voltage applied to the sidekick electrodes.

was attributed to collisional cooling of the analyte ions by CO_2 and H_2O gases formed from the decomposition of the sugar co-matrix. The trapping of high-mass analytes (bovine serum albumin, MW ~66,000 Da) was achieved using a modified gated trapping method. A large (50 V) voltage was applied to the rear trapping plate, and a lower voltage (35 V) was applied to both the excite/detect plates, while the front trapping plate was held at ground potential.[53] Simulations revealed that the five-plate trapping method formed a "cuplike" electrostatic field that is more conducive to trapping ions by reducing the radially repulsive electric field and reducing magnetron expansion. As with more conventional gated trapping experiments, the front trapping plate is raised from ground to a higher voltage to trap the ions within the analyzer cell. Prior to excitation/detection, the front and rear trapping plates are lowered to a more favorable voltage and the excite/detect plates are brought to 0 V.

11.1.2.3 Quadrupolar Excitation and Axialization

Capturing of ions in the analyzer cell may result in a spatial distribution for ions which is nonoptimal for tandem mass spectrometry and excitation/detection. Ions can be manipulated by the application of the appropriate electric fields to improve their spatial distribution. Quadrupolar excitation (QE) is an ion excitation technique that causes the ions to interconvert between magnetron and cyclotron motion.[54–56] If a collision gas is present, the cyclotron motion dampens much more rapidly than the magnetron motion, causing the ions to move to the center of the cell, a process known as axialization (QEA). The difference between the polarity of the excitation for dipolar excitation and QE is illustrated in Figure 11.16. During dipolar excitation, the excitation waveform is applied to the excite plates, with equal amplitude but opposite polarity applied to opposing plates as shown in Figure 11.16A. As originally described, the resonance QE waveform is applied to all four plates as shown in Figure 11.16B. Opposing pairs of plates receive the same polarity of the

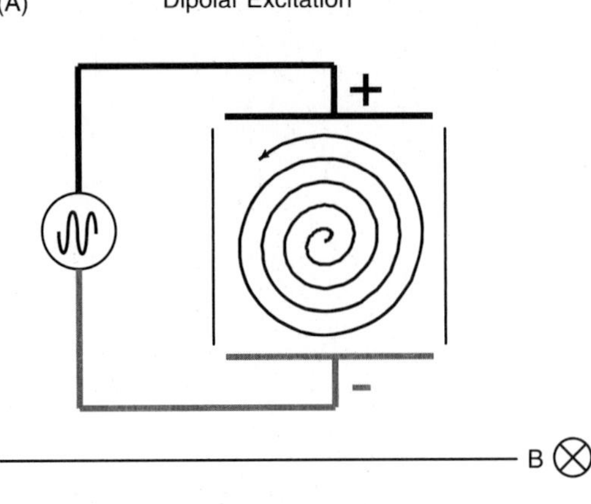

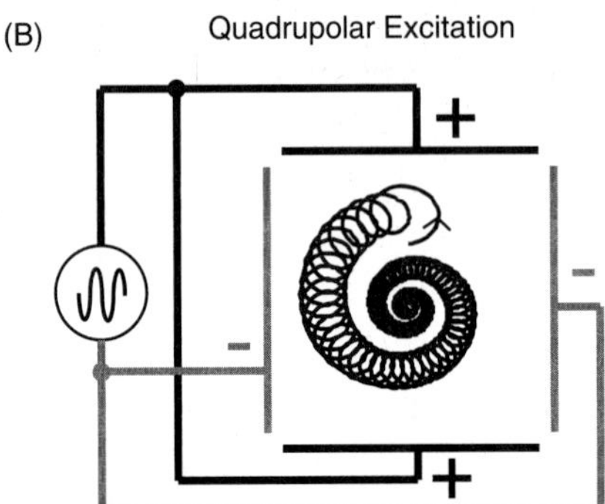

Figure 11.16. Illustration of the applied waveforms for dipolar versus quadrupolar excitation. (**A**) For dipolar excitation, opposite phases of the excitation waveform is applied to opposing excite plates. (**B**) Four excite plates are used for quadrupolar excitation, with opposing pairs of plates applying the same polarity but with adjacent plates having opposite polarity. The symmetry of the resulting electric field is dipolar in **A** and quadrupolar in **B**. (Adapted from Ref. 161.)

applied waveform, and adjacent plates receive the opposite polarity. Later implementations of QE used a single phase of the excitation waveform applied to both excite plates while leaving the detect plates grounded.[57]

QEA provides a large improvement in transient duration and mass resolution. Because ion detection in FTMS is nondestructive, QE can be used to refocus ions after excitation and detection while allowing the same ions to be excited and detected over and over again, a process known as remeasurement. Remeasurement of the same population of ions up to 500 times (i.e., one ionization event, but 500 measurement events) has been demonstrated at 1 T,[56] and use of remeasurement and signal averaging have been used to achieve attomole sensitivity of biomolecules from a single laser shot at 3 T.[58] The efficiency of axialization by QEA decreases with increasing magnetic field strength; and although it has been used at 7 T,[29] it is more difficult to perform this experiment with higher field magnets. The reason for this adverse dependence on magnetic field strength is that the kinetic energy of ions of a given orbital radius increases as the square of the magnetic field; at higher magnetic field strengths, ions that are excited to a radius suitable for detection can dissociate from the collisions that are used QE axialization. Even if dissociation were not an issue, the number of collisions necessary to relax ions to the center of the cell scales in direct proportion to the kinetic energy of the ions, and thus to the square of the magnetic field strength. Thus, each doubling in magnetic field strength requires ions to be exposed to a four times higher pressure of collision gas, or a four times longer period of collisional relaxation.

11.1.2.4 Metastable Decomposition in MALDI-FTMS

Ions that are produced in the MALDI ionization process can accumulate internal energy that is sufficient to break bonds. This fragmentation is a slow process, and in time-of-flight instruments, dissociation can occur after the ions have exited the ion source, referred to as metastable fragmentation. The much longer timescale of an FTMS measurement (seconds versus hundreds of microseconds for time-of-flight) allows such fragmentation to be a more abundant process. Typical products observed from metastable decomposition include loss of small molecules such as H_2O, NH_3, or small side chains[59,60] and decomposition of labile post-translation modifications of amino acids in peptides, such as phosphorylation[61] or sulfation[62] sites. If enough energy is imparted to the analyte, metastable decomposition can result in fragmentation of the peptide backbone. This is particularly problematic with internal MALDI-FTMS, because the ions are desorbed into high vacuum and do not have the opportunity to undergo collisional relaxation of the excess internal energy on the timescale of metastable decompositions. This reduces the applicability of internal MALDI-FTMS for the analysis of peptides. On the other hand, internal MALDI-FTMS has been demonstrated to be useful for analyzing synthetic polymers, because these are more rugged molecules.[50] The development of higher pressure MALDI and external ion sources has led to FTMS instruments that are better suited for peptide analysis.

11.1.2.5 External MALDI

External MALDI-FTMS places the sample target outside the magnetic field, away from the analyzer cell. A diagram of a simple external MALDI- FTMS instrument is shown in Figure 11.17A. The MALDI source is placed external to the magnetic field and is held at a higher pressure than the analyzer cell (typically 10^{-6} torr versus 10^{-10} torr). Because the source operates at a higher pressure than for internal MALDI, transfer of the MALDI target from the laboratory to the vacuum system is more rapid as less pump down time is necessary. Placement of the MALDI source external to the magnetic field results in the MALDI target and the analyzer cell being separated by distances of up of 1 m or greater. Ion guides are used to transfer ions from the source to the analyzer cell, with differential pumping used to incrementally lower the pressure between the source and the analyzer cell.[63–66]

11.1.2.6 Ion Guides in FTMS Instruments

When the ion source is placed outside the magnetic field, ions must be transferred through the steep magnetic field gradient in order to be delivered to the analyzer cell. Ideally, ions produced in the external source should be transferred without ion loss, mass discrimination, or fragmentation. Transmitting the ions through the magnetic field gradient is difficult due to the magnetic mirror effect. As ions pass from low to high magnetic field, their cyclotron frequencies increase, causing ions to decelerate. The increase in kinetic energy in the plane perpendicular to the magnetic field draws energy from the translational mode parallel to the magnetic field, slowing the velocity of the ion. Without any focusing elements, a majority of the ions would be slowed to a standstill, and then they reverse direction and exit the magnetic field. In order to overcome this problem, the ion beam must be well-focused parallel to the magnetic field axis. Two solutions have been devised to transmit ions through the magnetic field gradient. An RF-only multipole can be used to focus and move the ions through the magnetic field and into the analyzer cell.[66] An external MALDI-FTMS arrangement using a RF-only quadrupole is shown in Figure 11.17A. Because the multipole is operated in RF-only mode, it acts as an ion guide and not a mass filter. Alternatively, a series of

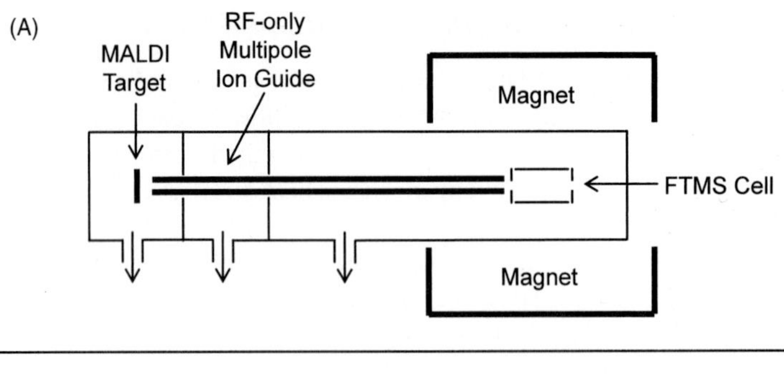

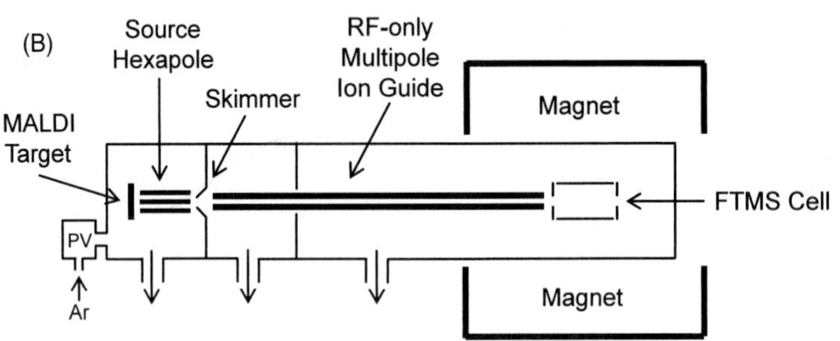

Figure 11.17. Schematic representations of external MALDI-FTMS systems. (**A**) vacuum MALDI-FTMS system where ions are generated in the external source and transferred to the analyzer cell via an RF-only multipole ion guide. (**B**) IP-MALDI-FTMS system. Argon is introduced in a short burst via a pulsed valve to briefly elevate the pressure within the source region. Ions are first stored in the source hexapole and undergo collisional cooling before transfer to the analyzer cell via an RF-only ion guide.

high-voltage electrostatic ion optics can be used to accelerate ions to high kinetic energy, from 1 to 4 keV. Electrostatic ion optics then focus the ions into a parallel beam that is transmitted through the fringe magnetic field without reflection.[67] Once the ions enter the region of homogeneous magnetic field, they are decelerated by an electric field to permit trapping in the analyzer cell. While both methods effectively transfer ions from the source into the analyzer cell, both are susceptible to a mass discrimination problem. When ions of varying *m/z* are accelerated to similar kinetic energies, lower *m/z* ions will have higher velocities than high *m/z* ions and therefore will arrive at the analyzer cell first. The timing of the dynamic trapping event relative to the time at which ions leave the source can cause mass discrimination in the ions that are trapped by the analyzer cell. For practical FTMS analysis, careful consideration of the *m/z* range of the analyte ions must be taken into account when setting the timing sequence for dynamic trapping.

11.1.2.7 Intermediate/Atmospheric-Pressure MALDI

Placement of the MALDI target outside the magnetic field allows desorption to occur at higher pressures, with an observed decrease in metastable decomposition.[5] With the external MALDI source held at an elevated pressure, collisions with neutral gas molecules occur during ionization which decreases the internal energy of the ions. This collisional relaxation leads to a decrease in products resulting from metastable decomposition. This stands in sharp contrast to internal MALDI, in which very few collisions with neutral gas molecules occur during ionization in the UHV region of the analyzer. Instruments have been

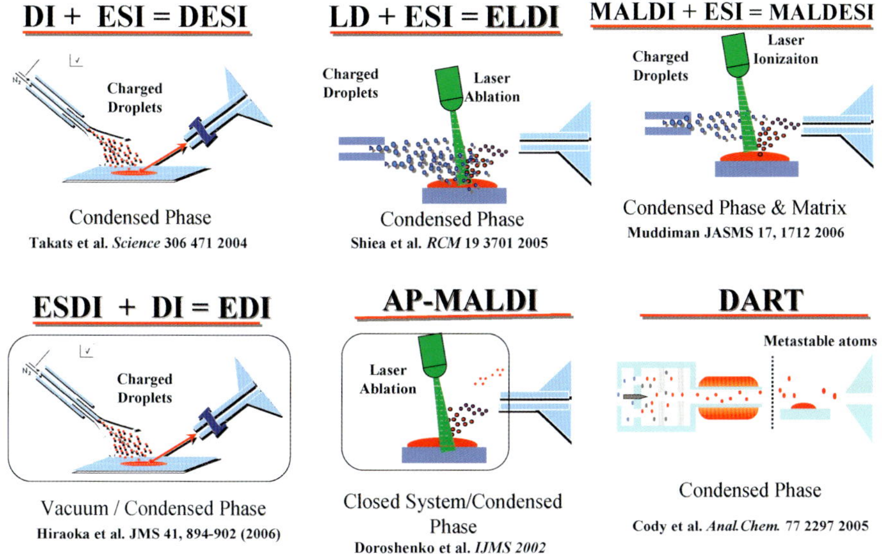

Figure F.1. Ambient ionization methods for ionizing condensed phase samples on surfaces using droplets, atom beams, electrical discharges and laser beams. For comparison, some vacuum methods are also shown.

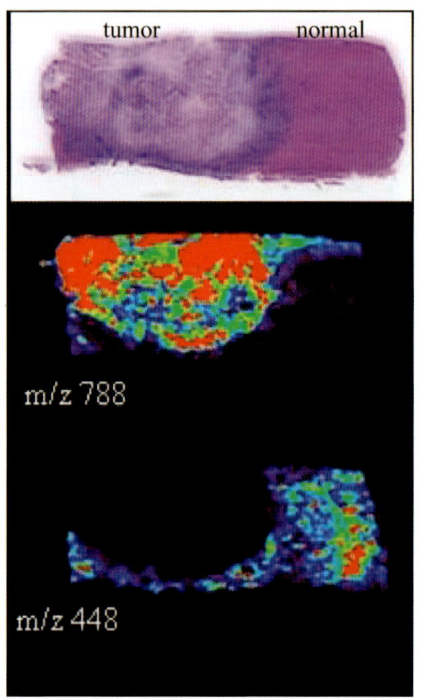

Figure F.2. Human liver sample. (**a**) H&E (hematoxylin and eosin) stain. (**b**) Negative ion images of phosphatidylserine (ion of m/z 788). (**c**) Chenodeoxycholic acid (ion of m/z 448). (Unpublished DESI imaging data of Wiseman, Caprioli, Puolitaival, and Cooks; false color intensity; tissue shown is 17 mm by 8 mm by 12 μm.)

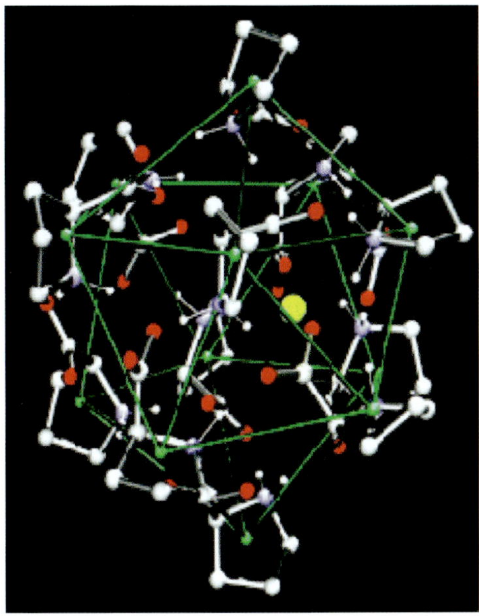

Figure F.3. Calculated structure of the protonated form of the homochiral proline-12 icosahedral cluster. (From Clemmer, D. E., et al. *J. Am. Chem. Soc.* **2006**, *128*, 15988–15989.)

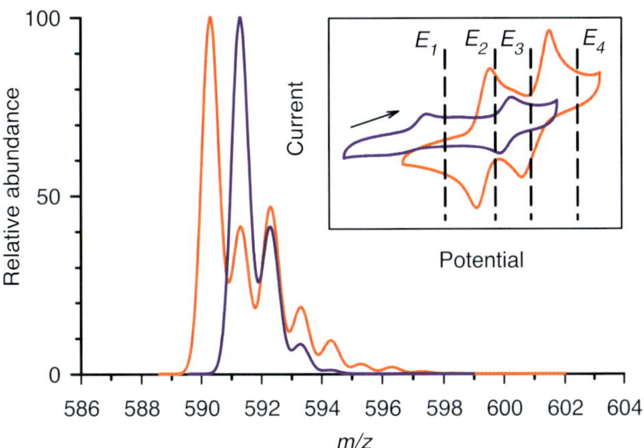

Figure 3.15. Theoretical isotope patterns for the single-charge cations of NiOEP (6, red line) and CoOEP (7, blue line). **Inset**: Off-line cyclic voltammograms of 0.5 mM NiOEP (6, red curve) and 0.5 mM CoOEP (7, blue curve) (50/50 (v/v) acetonitrile/methylene chloride containing 50 mM lithium triflate) at a scan rate of 50 mV/s. E_1–E_4 represent the potential conditions under which the data shown in Figure 3.16 were collected. (Adapted with permission from Ref. 89. Copyright 2005, American Chemical Society.)

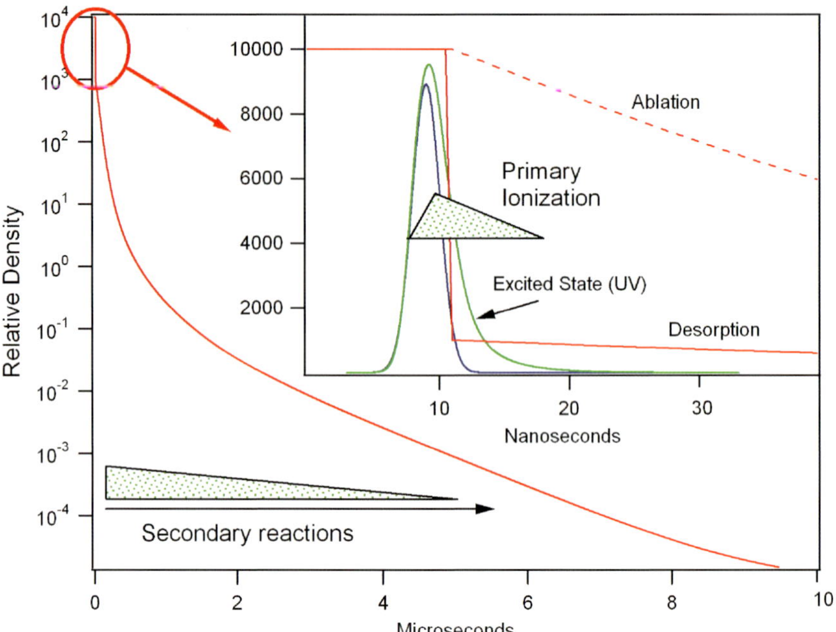

Figure 5.1. The origin of the two-step model: Expansion of the MALDI plume compared to the laser pulse and excited state decay. Density is plotted versus time for a plume expanding as an adiabatic free jet. A density of 1 is approximately that of a gas at 1 atm. The expansion stops when the gas reaches the environmental background pressure, which may vary from 1 atm. to high vacuum. The inset shows the early behavior, along with a 3-ns N_2 laser pulse (blue) and the typical lifetime of matrix excited states (green). Only during the time when the energy and material densities are high can significant ionization occur. The solid line represents the density if the sample vaporizes smoothly. The dashed line represents an explosive phase change for which a well-defined solid–gas boundary may not exist at short times.

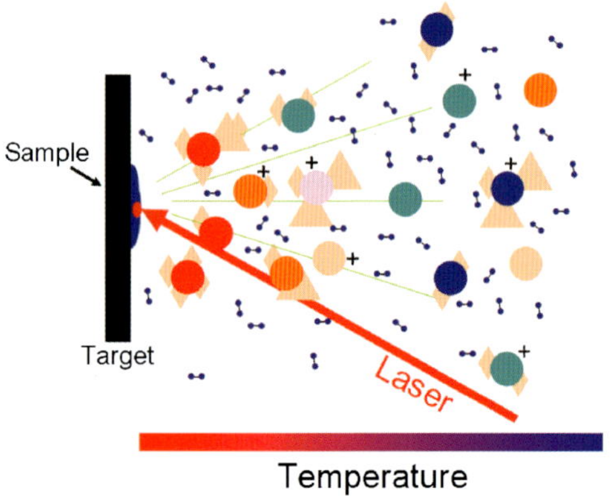

Figure 6.1. Laser desorption/ionization.

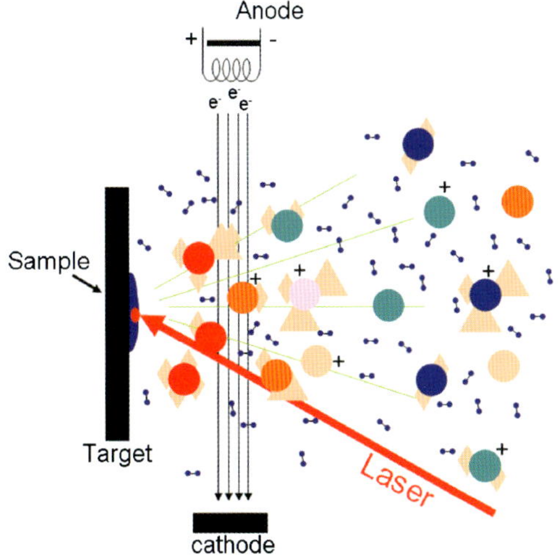

Figure 6.2. Laser desorption plus electron impact ionization.

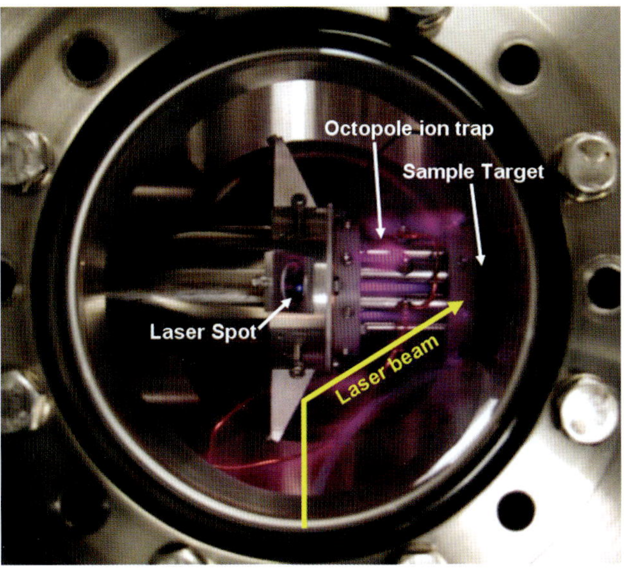

Figure 6.17. The plasma that develops if pressure goes too high in vibrationally cooled MALDI sources.

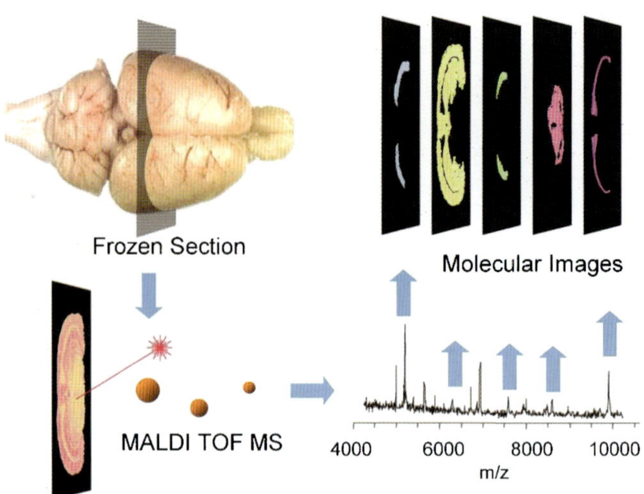

Figure 8.1. MALDI imaging mass spectrometry process. Reprinted from reference 42 with permission from Elsevier.

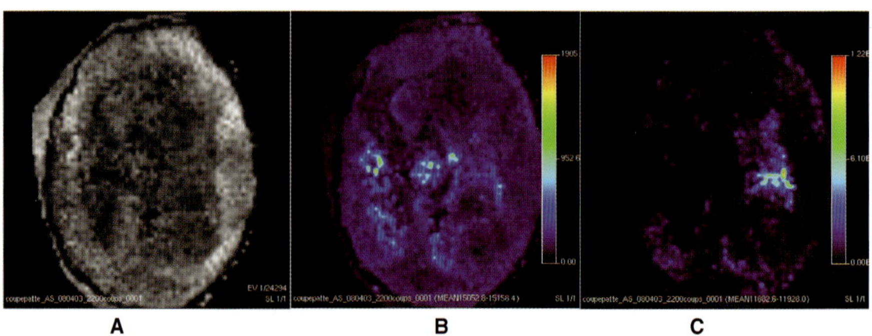

Figure 8.2. Differential localization of proteins on a mouse muscle section by MALDI imaging mass spectrometry. A: total ion image; B: MS image of the ion at *m/z* 15100; C: MS image of the ion at *m/z* 11950

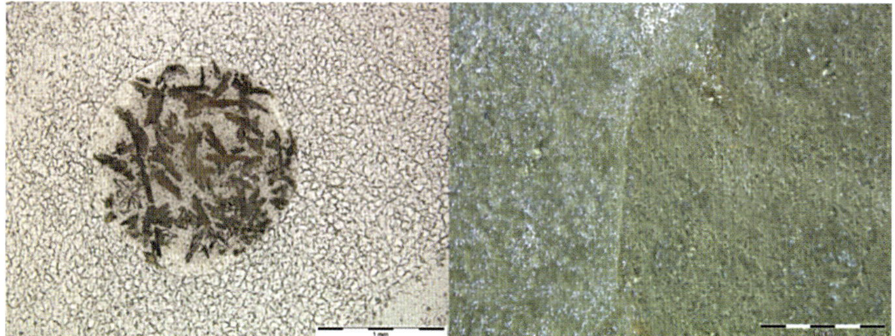

Figure 8.3. Left: the dried droplet deposited for a profiling study on a mouse brain section displays crystals of sinapinic acid. Right: the mouse brain section is homogeneously coated by sinapinic acid for an IMS experiment (air-spray deposition procedure).

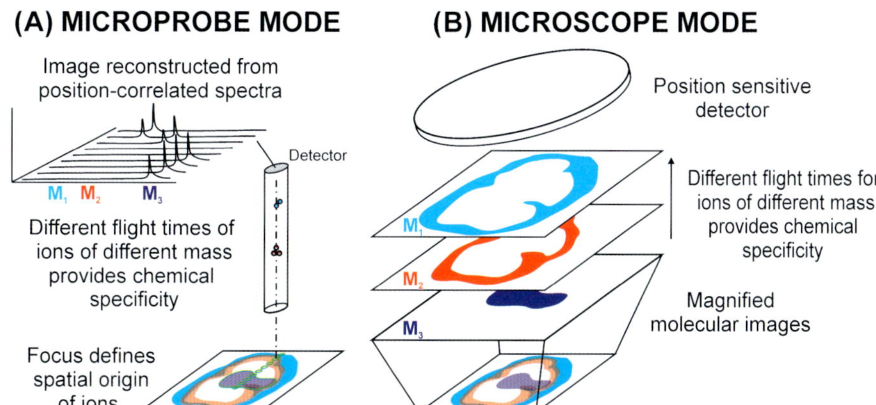

Figure 8.5. Schematics of the two different modes of molecular imaging mass spectrometry. Microprobe mode imaging (A) collects mass spectra from an array of designated positions to reconstruct a molecular image after completion of the experiment. In microscope imaging (B), magnified images of the ion distributions are directly acquired using a two-dimensional detector. Reprinted with permission from reference 19, Copyright 2004 American Chemical Society.

Figure 8.7. Automatic spotting of matrix onto a tissue section.

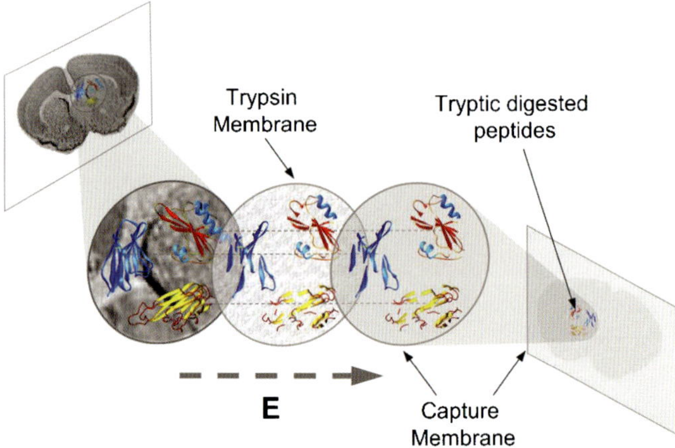

Figure 8.8. Descriptive scheme of the molecular scanner approach in imaging mass spectrometry. Reprinted from reference 35 with permission from Elsevier.

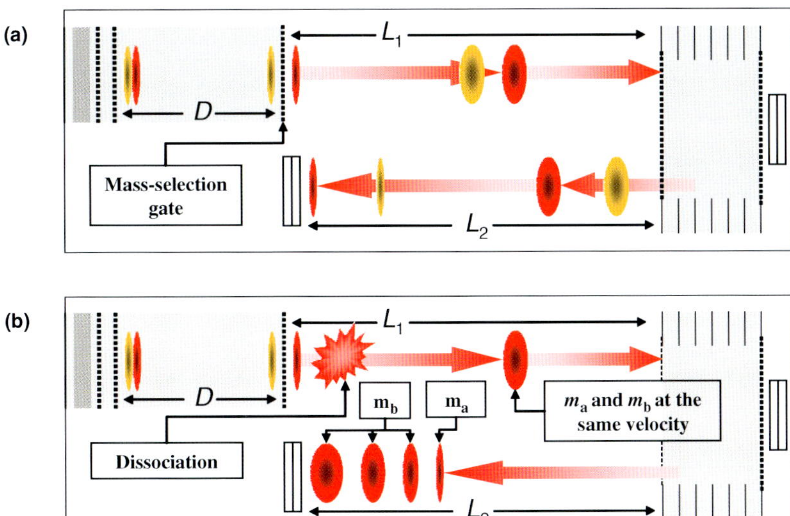

Figure 10.8. TOF/RTOF tandem time-of-flight mass spectrometer: (**a**) MS mode with mass selection gate OFF. (**b**) MS/MS mode.

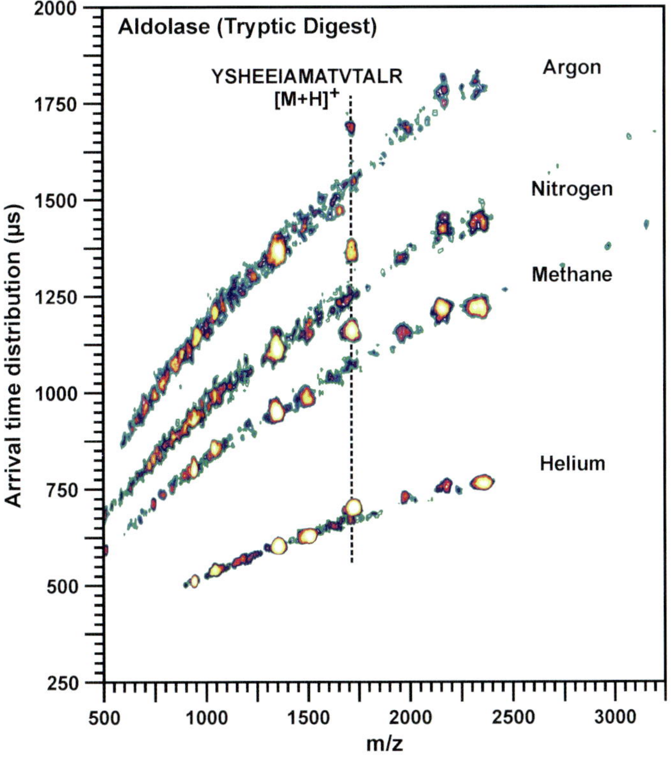

Figure 12.6. Four overlaid IM-MS conformation space plots for a tryptic digest of aldolase (rabbit muscle), for separations performed in each of four drift gases (Ar, N_2, CH_4, and He). The vertical dashed line is provided to assist in visualizing the relative shift of the peptide YSHEEIAMATVTALR from the other tryptic peptides as the mass of the drift gas is increased from He to Ar. (Adapted with permission from *J. Mass Spectrom.* **2004**, *39*, 361–367.)

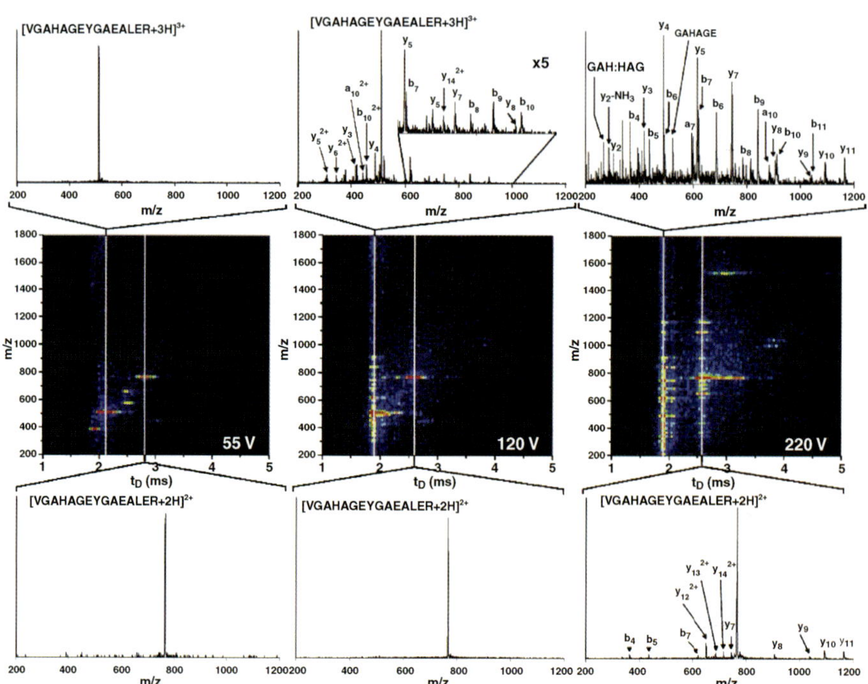

Figure 12.10. A comparison of correlated peptide sequencing data obtained by LC-IM-MS and LC-IM-CID-MS obtained sequentially using a field-modulated ESI-IM-TOFMS as depicted in Figure 12.8b. Mass spectra correspond to integrating across the IM arrival times as indicated by white lines. (Adapted with permission from *J. Proteome Res.* **2005**, *4*, 25–35. Copyright 2005 American Chemical Society.)

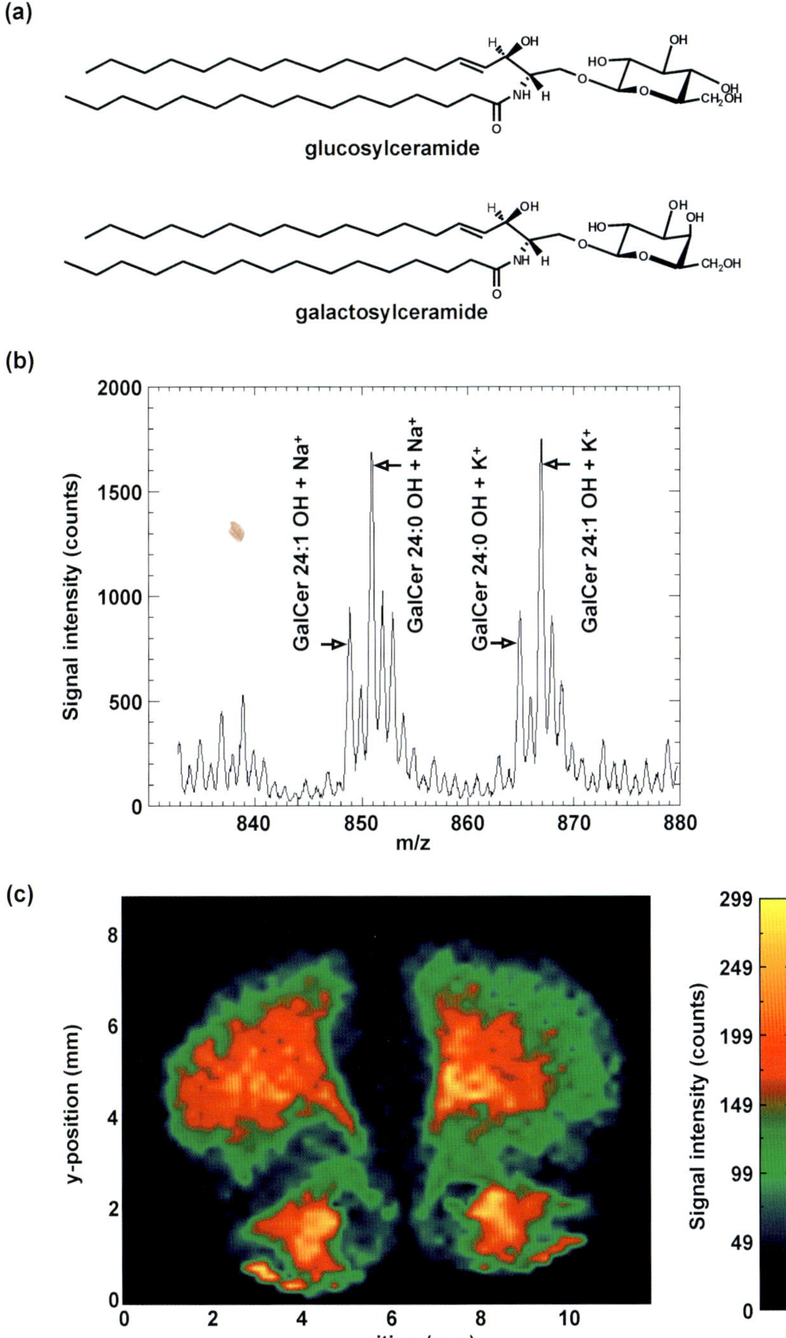

Figure 12.11. Imaging-mode MALDI-IM-MS of a thin tissue section (16 μm) of a rat brain. (**a**) Structures of glucosylceramide and galactosylceramide, the two major classes of cerebroside molecules. Both cerebrosides are glycosphingolipids, which are important components in nerve cell membranes. (**b**) Integrated mass spectrum obtained at the specific IM-ATD for these cerebrosides directly from rat brain tissue. (**c**) False color image of signal intensity for where these cerebrosides occur spatially in the context of the thin tissue section. Imaging MALDI-IM-MS was performed at 349 nm using a 2-nm Au nanoparticle matrix, with an imaging spatial resolution of 200 μm.

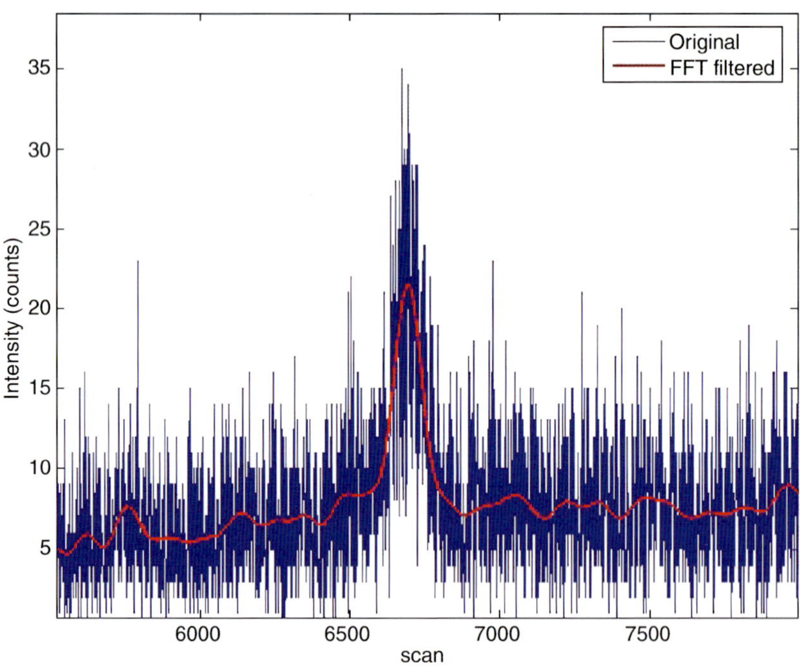

Figure 13.10. Removal of high-frequency noise with FFT. Low-frequency noise is revealed in the baseline of the FFT-processed chromatogram.

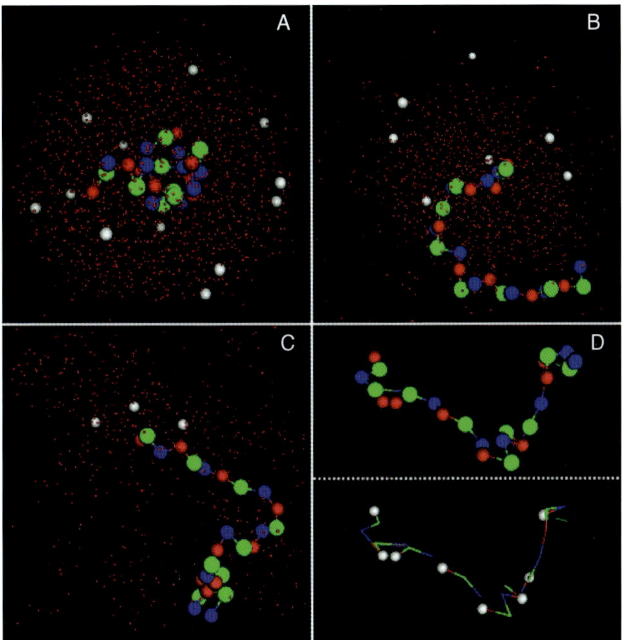

Figure 14.4. Progressive disintegration of a charged nanodroplet containing an unfolded protein. The time points and charge states are (**A**) 3 ps, 2+; (**B**) 87 ps, 7+; (**C**) 97 ps, 7+; (**D**) 126 ps, 8+. (Reprinted from Ref. 68 with permission.)

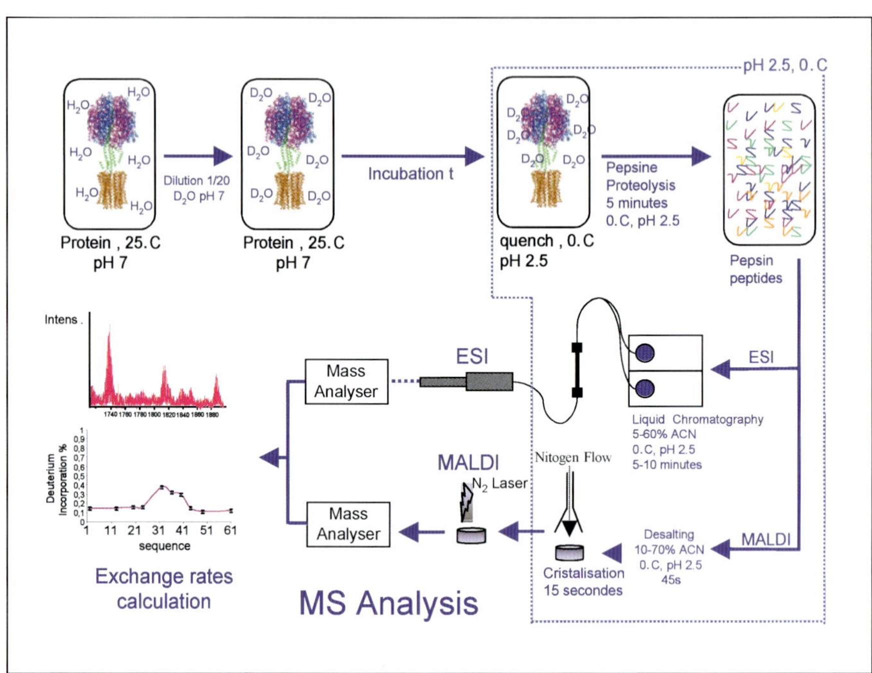

Figure 15.4. Principle of HDX-MS analysis: The protein complex is diluted in deuterium oxide to start the exchange reaction. After the incubation time at room temperature, the exchange reaction is quenched by lowering both pH and temperature. To study the topology of deuterium incorporation, the protein complex is digested with pepsin under quenched conditions. The peptides generated can be analyzed using ESI or MALDI ionization. During the mass spectrometric analysis the number of deuterium atoms incorporated is determined to calculate the exchange rate along the sequence of the protein.

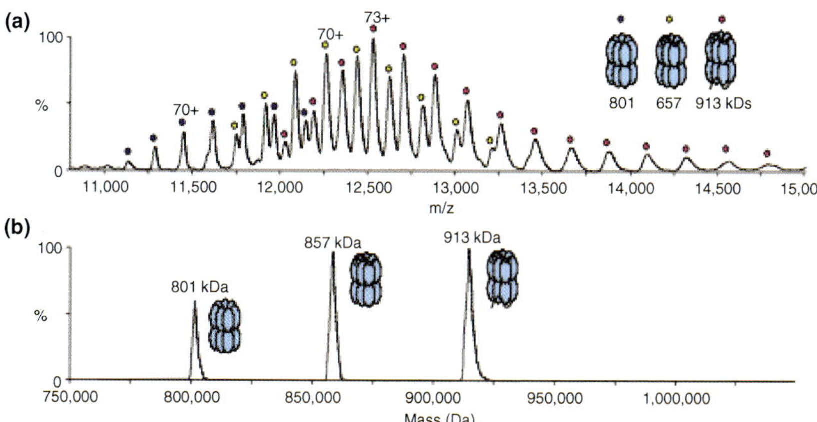

Figure 15.8. Stoichiometry of GroEL:gp23 binding and characterization of the complexes by native nano-ESI MS. (**a**) Mass spectrum of a 1 : 4 mixture of GroEl and unfolded polypeptide gp23. The charge state distribution of free GroEl (800 kDa, blue circles) GroEL with one (856 kDa, yellow circles), and two (912 kDa, red circles) gp23 substrate molecules bound can be discerned. (**b**) Deconvoluted spectrum of panel **a**, revealing the presence of the three different complex stoichiometries. (Reproduced from Ref. 90 with permission. Copyright © 2005 Nature Publishing group.)

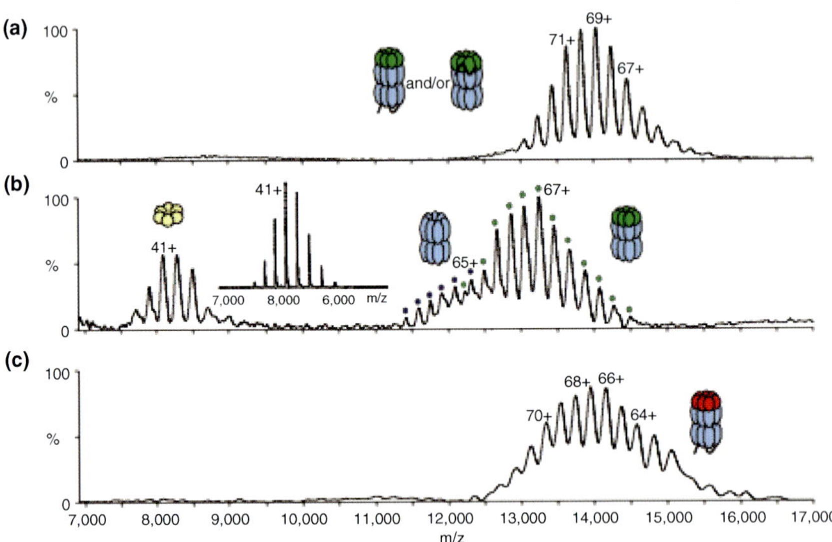

Figure 15.9. *In vivo* refolding of gp23 monitored by mass spectrometry. (**a**) Nano-ESI mass spectrum of a solution containing GroEl, gp31, gp23, ADP, and Mg^{2+}. Only one charge-state distribution is observed, corresponding to a complex with a mass of 944 kDa, which closely matches the expected mass of a ternary GroEl–gp31-ADP-Mg^{2+} complex with one gp23 substrate molecule bound. (**b**) Same as panel **a**, except that ATP was used instead of ADP. This resulted in formation of gp23 hexamer assemblies (336 kDa, yellow charge-state distribution centered around *m/z* values of 8000), revealing *in vivo* refolding of gp23. The GroEL–gp31 complex gives rise to a charge-state distribution centered around *m/z* values of 13,000 (green circles). A small amount of tetradecameric GroEL is also present (blue circles). The inset shows the mass spectrum obtained for hexameric gp23 as purified from *E. coli*. (**c**) Same as panel **b**, except that GroES was used instead of gp31. No gp23 refolding was observed. (Reproduced from Ref. 90. with permission. Copyright © 2005 Nature Publishing group.)

designed to utilize MALDI at pressures ranging from 10^{-6} torr to atmospheric pressure. Different names are used in order to distinguish the pressure regime under which MALDI ionization occurs. Pressures ranging from 10^{-6} to $\leq 10^{-10}$ torr are referred to as vacuum MALDI. MALDI occurring at pressures around 10^{-4}–10^{-1} torr is known as intermediate-pressure MALDI (IP-MALDI). Atmospheric-pressure MALDI (AP-MALDI) is used to describe MALDI ionization occurring outside the vacuum system.

IP-MALDI was developed before AP-MALDI and was first reported using an o-TOF mass spectrometer.[68] An elevated pressure in the MALDI source region was maintained by admitting the collisional cooling gas into the source region with a leak valve. IP-MALDI FTMS was first reported in 2000 using a modified external vacuum MALDI source.[69] The arrangement of the IP-MALDI FTMS instrument is illustrated in Figure 11.17B. This MALDI source contained an RF-only hexapole positioned directly in front of the MALDI target. Ions desorbed from the MALDI target can be stored in the source hexapole before transfer to the analyzer cell by using the MALDI target and a skimmer on the opposite side of the hexapole as trapping electrodes. For IP-MALDI FTMS, the source pressure is briefly brought from vacuum MALDI pressures ($\leq 10^{-6}$ torr) to IP-MALDI (10^{-1} torr) pressures using a pulsed valve located in the source region. The pulsed valve introduces gas into the source region just prior to firing the laser, and ions from one or more laser shots can be accumulated in the source hexapole. The accumulation of ions from several laser shots allows one to deliver a more uniform number of ions to the analyzer cell than one gets from a single laser shot. Collisional cooling of the ions produced by MALDI occurs because of the elevated source pressure. Adequate time is allowed for the removal of excess gas by the vacuum pumps before ions are released from the source hexapole and transferred to the analyzer cell.

AP-MALDI was first developed in 2000 by focusing a pulsed laser beam onto a target placed directly in front of the capillary at the entrance of an electrospray ionization (ESI) orthogonal quadrupole time of flight (QoTOF) instrument.[70] No instrument or sample modifications were required other than (a) the placement of the probe in front of the ESI capillary and (b) the use of a buffer gas to direct the MALDI generated ions toward the ESI capillary inlet. Application of AP-MALDI to FTMS was first reported in 2004.[71] Similar to AP-MALDI on the QoTOF, AP-MALDI FTMS was performed by placing the MALDI target directly in front of the ESI capillary. For AP-MALDI, cooling of the ions occurs due to the enormous number of thermalizing collisions that occur at atmospheric pressure before the ions enter the instrument via the ESI capillary.

Increasing the pressure during MALDI ionization increases the number of collisions between the ions and neutral gas molecules within the MALDI plume. These collisions are responsible for reducing the internal energy of the ions via collisional thermalization. This results in a substantial decrease in the number of observed products of metastable decomposition.[72,73] Typically, "cool" matrices such as DHB are used in vacuum MALDI because they produce optimal signal-to-noise ratios and reduce (but do not eliminate) metastable losses of neutrals such as H_2O or NH_3. The use of "hot" matrices such as α-cyano-4-hydroxy cinnamic acid (α-cyano) in vacuum MALDI is not common.[73] Interestingly, IP- and AP-MALDI allow the use of "hot" matrices (e.g., α-cyano) that typically do not work well for vacuum MALDI-FTMS. These "hot" matrices can be used in IP- and AP-MALDI with minimal metastable decomposition of the analyte ions. IP- and AP-MALDI may result in the formation of analyte–matrix adducts or matrix clusters. An example of matrix clusters and analyte–matrix adducts is shown in Figure 11.18A, where MALDI of the peptide angiotensin II produces abundant DHB clusters as well as peptide–DHB adducts. The presence of these labile species is viewed as evidence for the decrease in internal energy

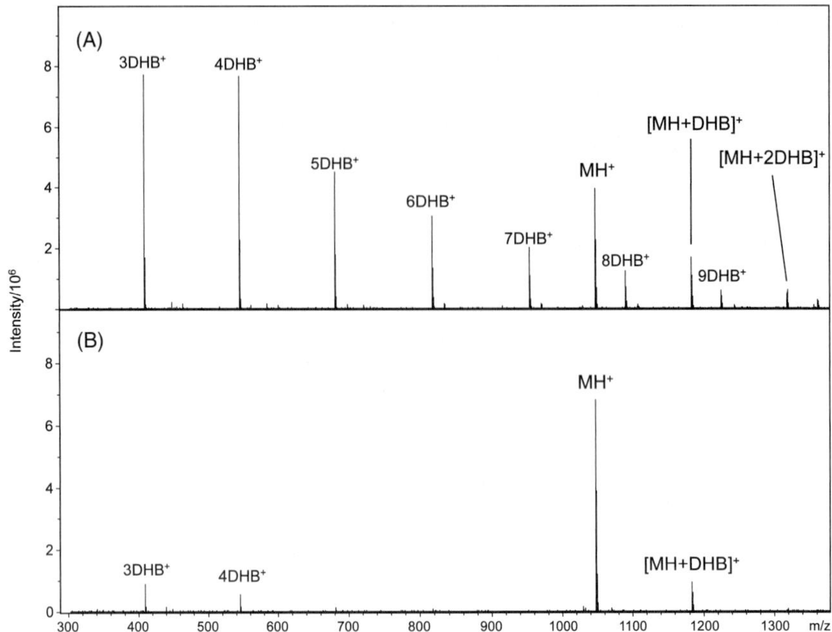

Figure 11.18. Influence of low-energy collisions on elimination of analyte–matrix clusters. (**A**) IP-MALDI FTMS of the peptide angiotensin II produces the molecular ion, the molecular ion plus DHB, and DHB clusters. (**B**) Many of these DHB clusters can be eliminated by application of low-energy collisions. Low-energy collisions were produced in the hexapole collision cell of a QhFTMS instrument.

during ionization by collisional cooling. These noncovalent complexes can easily be removed by gently exciting the ions within the analyzer cell.[73] Alternatively, low-energy collisional activation prior to transmission of the ions (e.g., CAD in a hybrid FTMS instrument; see below) into the magnetic field can be used to remove many of these clusters as shown in Figure 11.18B.

11.1.2.8 MALDI-FTMS Applications

In the past 20 years, MALDI-FTMS has been widely applied to the analysis of a variety of compounds, and a substantial number of papers can be found in the literature. A few of these applications will be mentioned briefly. The characterization of complex polymer samples has been performed on an internal MALDI-FTMS instrument.[74] Because the internal MALDI arrangement minimizes TOF discrimination, it is well-suited for the analysis of samples with large mass distributions. Polymer ions produced from internal MALDI-FTMS were observed from *m/z* 500–3500 over a wide dynamic range. Internal MALDI-FTMS analysis of copolymers produces many isobaric ions and is further complicated by the presence of both sodium and potassium cationized species. The high mass accuracy allows assignment of polymer repeat units, and the high resolving power is used to characterize polymer end groups as well as distinguish linear from cyclic polymer cations.

The analysis of oligosaccharides is particularly challenging due to the lability and complexity of these molecules. While vacuum MALDI-FTMS of oligosaccharides may produce some metastable decomposition of the labile oligosaccharides, it occurs with far less abundance and without the chemical noise typically observed for FAB ionization. Metastable decay of oligosaccharides can be further reduced when IP-MALDI is used.[73] Similarly, the addition of alkali metals to the matrix has been shown to decrease metastable decomposition.[75] MALDI-FTMS has been widely applied to the analysis of peptides and proteins. For example, identification of proteins by peptide mass fingerprinting (PMF) using MALDI-FTMS improves protein identification compared to MALDI-TOF because of the high-mass accuracy obtainable for peptide analysis.[76] The extremely high resolving power of MALDI-FTMS has been used to distinguish deamidated from nondeamidated peptides. With resolutions from 350,000–400,000, the A + 1 isotope peak of a non-deamidated peptide was distinguished from the deamidated peptide; moreover, the ^{13}C isotope peak was distinguished from the ^{34}S isotope peak.[77] Modern MALDI-TOF instruments can operate with mass accuracies slightly higher than FTMS instruments, typically ≤10 ppm. However, high mass accuracy must often be accompanied by high resolution to ensure proper identifications. Thus, isobaric peptides cannot always be distinguished by MALDI-TOF due to the relatively low resolving power of these instruments, possibly resulting in incorrect peptide assignment. To resolve the isobaric peptides shown in Figure 11.19, the high resolving power of FTMS is necessary, or the monoisotopic peak may be misidentified. MALDI-FTMS can be also be used for the analysis of complex peptide mixtures (e.g., digestion of a solution of proteins) by coupling MALDI to off-line HPLC with the use of a robotic sample spotter.[78] The combination of MALDI-FTMS and tandem mass spectrometry will be discussed later.

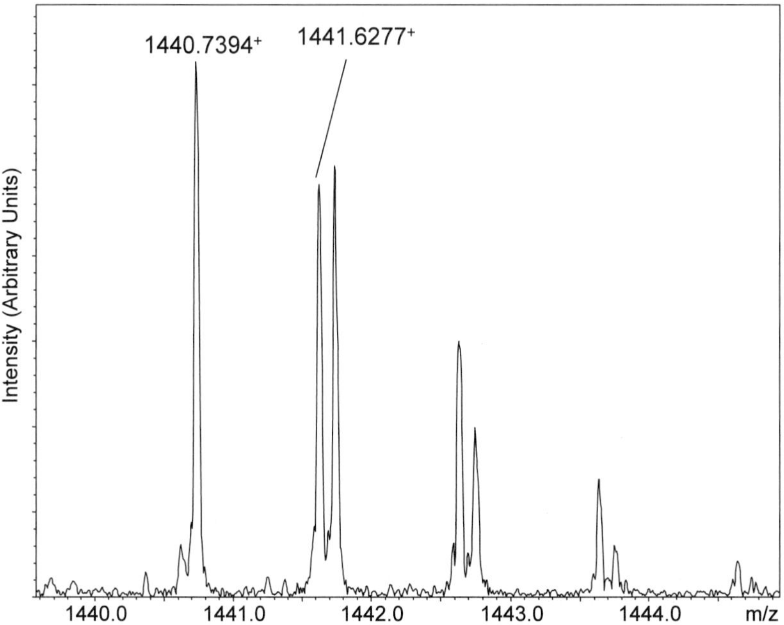

Figure 11.19. Tryptic peptides from a protein digestion analyzed by MALDI-FTMS produce isobaric peptide ions. The monoisotopic peaks for these overlapping ions are resolved by using the high resolving power ($m/\Delta m = 70{,}000$) of FTMS.

11.1.3 ESI Combined with FTMS

11.1.3.1 Chronology of Developments in ESI-FTMS

Electrospray ionization (ESI) was first combined with mass spectrometry with applications to biomolecules by John Fenn in 1988.[79,80] For this contribution to the field of mass spectrometry, John Fenn was awarded the Nobel Prize in Chemistry in 2002. Fenn's original ESI source was interfaced to a quadrupole mass spectrometer.[81] Its combination with FTMS offers many mutual advantages. The multiple-charging that is characteristic of ESI compresses peaks into a narrower range of m/z values, and the high resolution of FTMS allows simple identification of ion charge state. Interfacing an ESI source to FTMS was first reported in 1989 by McLafferty and co-workers.[2] The combination of ESI and FTMS produces a powerful analytical tool, allowing the ionization and rapid analysis of many labile molecules. The combination of ESI with FTMS has led to the widespread deployment of FTMS, particularly for protein analysis and proteomics.

11.1.3.2 ESI-FTMS Instrumentation

There are a number of challenges that must be overcome in order to interface ESI with FTMS. ESI is performed at atmospheric pressure, and considerable differential pumping is required to lower the pressure in the analyzer cell to $\leq 10^{-9}$ torr. The ESI source is placed outside the magnetic field and requires ion guides to transfer the ions from the source to the analyzer cell. The first ESI-FTMS instrument was configured in this manner and utilized a series of RF-only quadrupole ion guides to focus ions from the source into the homogeneous region of the magnetic field.[2] Another interfacing challenge is that ESI produces a continuous beam of ions, while FTMS is a pulsed detection technique. In the first coupling of ESI to FTMS, a continuous beam of ions passed from the source to the analyzer cell and the ions were captured by gated trapping. Inefficient transfer of the ions from the source to the analyzer cell by the quadrupole ion guides resulted in poor sensitivity, which was attributed to the broad distribution of ion kinetic energies generated in the ESI source. To overcome the problems associated with ion transfer, an internal ESI-FTMS system was investigated.[3,82] This arrangement is similar to an internal MALDI instrument, where the source is located within the bore of the magnet and in close proximity to the analyzer cell. Concentric vacuum chambers were used to lower the pressure from atmosphere to 10^{-7} torr. The internal ESI-FTMS setup allowed for abundant ion current compared to external ESI-FTMS instruments at that time. However, the difficulty in achieving low vacuum pressures, along with improvements in ion transmission using multipoles or electrostatic ion optics, has rendered the internal ESI-FTMS impractical.

Improvements in both signal-to-noise (S/N) ratio and duty cycle were necessary to make ESI-FTMS a viable analytical technique. Modern ESI-FTMS instruments have a storage multipole (typically an RF-only hexapole) located between the ESI source and the ion transfer optics, outside the magnetic field. In this location, ions are stored in a higher-pressure region of the instrument, typically 10^{-3}–10^{-6} torr. Ions produced continuously by the ESI source are stored in the source multipole, creating an ion "packet." The packet of ions is then moved from the multipole, down the ion guide, and into the analyzer cell. There are a number of advantages in storing ions in a multipole external to the analyzer cell. The amount of time that ions are stored in the source multipole is variable and user-controlled, and increasing the accumulation time increases overall S/N.[83] However, the multipole will eventually reach its maximum charge capacity and not store additional ions. The utilization of the continuous beam of ions from ESI is more efficient as ions may be accumulated externally while the analyzer detects ions. The interweaving of accumulation and ion detection provides almost 100% duty cycle for FTMS analysis of ions formed by ESI. Finally, if selective ion accumulation is used, the dynamic range of the instrument will be increased for the mass-to-charge of interest.[84]

11.1.3.3 Hybrid ESI-FTMS Instruments

FTMS instruments have much longer acquisition times when compared to quadrupole, ion trap, or TOF mass spectrometers. For LC-MS experiments, the experimental sequence can be modified to acquire data every 0.5–1 s, sacrificing mass resolution in order to increase chromatographic resolution. Experimental cycle time further increases when performing collisional activated dissociation (CAD) in the analyzer cell, because a small amount of gas is admitted into the analyzer region in order for dissociative collisions to occur (for a more detailed description of SORI-CAD see Section 11.4.1). The amount of time required to perform an in-cell CAD experiment also must include a significant delay to remove excess gas from the analyzer region before detection. Due to the long cycle time, performing multiple in-cell MS/MS experiments is difficult over the typical length of a chromatographic peak in a LC-MS/MS experiment. In order to increase the duty cycle of FTMS, hybrid FTMS instruments have been developed. The term hybrid FTMS instrument refers to the combination of an additional mass analyzer and ion dissociation region between the ion source and the analyzer cell. Typically, this mass analyzer is a quadrupole mass filter[85] or linear ion trap (LIT).[32] Hybrid FTMS instruments containing a quadrupole are often referred to as "Q-FTMS" instruments. The Q-FTMS is designed with the quadrupole mass spectrometer placed between the ESI source and the ion transfer optics. The Q-FTMS instrument shown in Figure 11.20A may more specifically be described as a QhFTMS, with the "h" referring to the RF-only hexapole collision cell located after "Q," the mass filtering quadrupole. Different acronyms are used to refine the description of Q-FTMS instruments. For example, a QqFTMS contains an RF-only quadrupole collision cell after the mass filtering quadrupole, while a QoFTMS contains an RF-only octopole collision cell in place of the hexapole.[86] Regardless of the configuration, the collision multipole is operated in the RF-only mode and is used to accumulate, fragment, or transfer ions in a non-mass-selective fashion. The quadrupole mass filter that precedes the collision cell can be set to pass all ions, thus operating as an ion guide. Alternatively, the mass-selective quadrupole can mass-filter and pass only a user-defined *m/z* range of ions. Typically, the S/N of mass-selected ions will increase with mass-selection because the space-charge limited storage capacity of the multipole collision cell is devoted to a single species of ion. The multipole collision cell is kept at an elevated pressure relative to the surrounding vacuum system and can also be used to dissociate ions simply by accelerating the ions as they enter the collision region. Performing CAD in the hexapole instead of in the analyzer cell eliminates analyzer pumpdown time (see below) and decreases the instrument experimental cycle time. As a result, the Q-FTMS is well-suited for performing LC-MS/MS experiments.

Following the development of the hybrid Q-FTMS instruments, a linear ion trap (LIT) was used in place of the mass-filtering quadrupole and storage multipole.[32] This hybrid-FTMS configuration is found in some commercial FTMS instruments, and an example of this instrument configuration is shown in Figure 11.20C. Ions pass from the ESI source into the LIT where they are stored, and then they are passed down to the analyzer cell. With the LIT, ion accumulation, mass selection, and dissociation (including MS^n) can be performed within the LIT instead of the analyzer cell. CAD is typically performed within the LIT, while other methods of ion activation, infrared multiphoton dissociation (IRMPD), and electron capture dissociation (ECD) (vide infra) are performed inside the analyzer cell. One unique

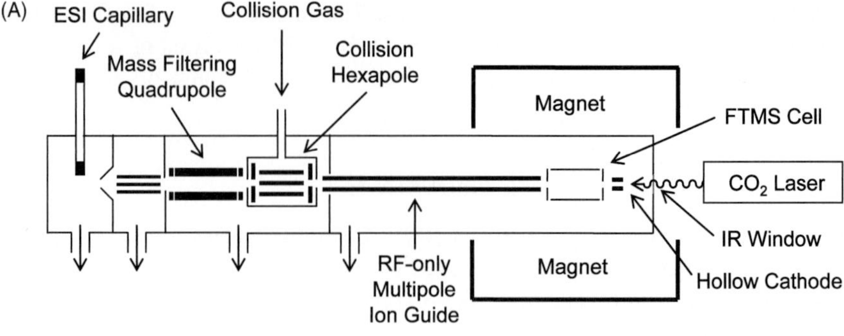

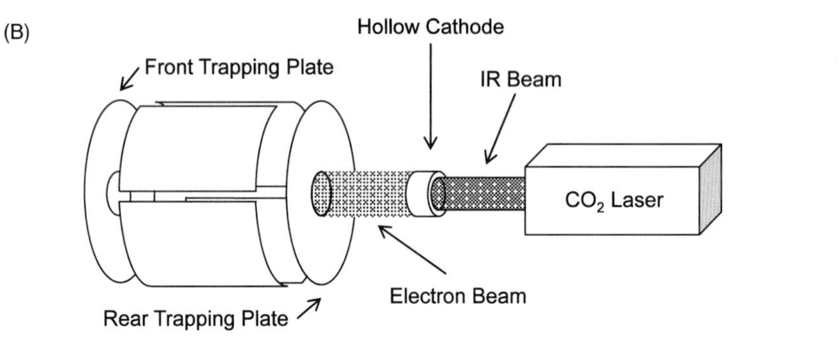

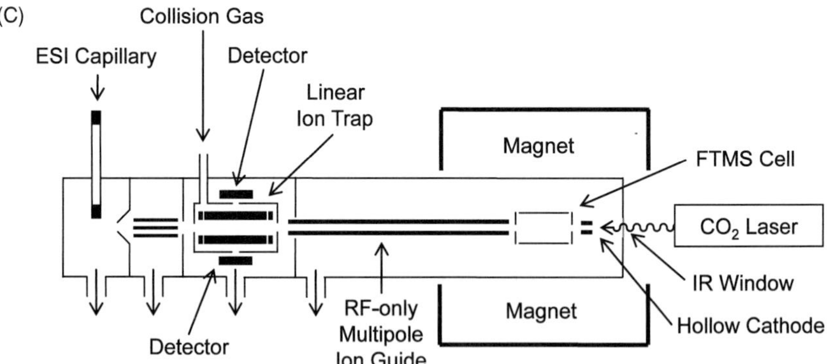

Figure 11.20. (**A**) Schematic of a hybrid ESI-FTMS, a Q-FTMS, with a mass-selective quadrupole and a hexapole collision cell. The position of the hollow cathode and CO_2 laser relative to the vacuum system is also shown. (**B**) Schematic of analyzer cell showing the implementation of the hollow cathode and CO_2 laser that allows both electron activation and photodissociation experiments to be performed. (**C**) Schematic of a hybrid ESI-FTMS using a linear ion trap. In this configuration, low-resolution detection can occur by ejecting the ions between the LIT rods, striking the adjacent detector. For FTMS detection, ions exit the end of the LIT and are passed to the analyzer cell via an ion guide.

feature of this hybrid instrument is that the LIT can also be used for ion detection and be used for rapid, low-resolution analysis. For non-FTMS detection using the LIT, ions are ejected between the rods of the LIT and strike a pair of electron multipliers for detection. MS^n can be performed outside the analyzer cell, with a very short experimental cycle time compared to FTMS. In a typical mode of operation, the FTMS analyzer is used to make highly accurate mass measurements of peptide precursor ions in an on-line LC-MS experiment, and the subsequent MS/MS analysis of interesting peptides is made by the LIT between survey scans. However, the LIT has a low *m/z* cutoff that is approximately one-third of the *m/z* of the precursor ion, making it possible for low *m/z* product ions of interest to be lost. A commercial LIT-FTMS instrument also features automatic gain control (AGC).[32,84] In order to minimize space charge effects on mass accuracy, the instrument is programmed to admit a constant number of ions into the FTMS analyzer cell for each scan by controlling the duration of ion accumulation time in the LIT. A short prescan using the LIT allows an instantaneous measurement of the ion current from the ESI source, and this data is then used to determine the accumulation time needed to capture a predetermined number of ions in the LIT. Thus a relatively constant population of ions can be passed from the LIT to the FTMS analyzer for each mass scan. With a similar number of ions in the analyzer cell for each analysis, space charge is constant and mass accuracy is improved. AGC is particularly useful for maintaining low-ppm mass accuracy during an LC-MS run, where the number of ions produced by ESI can fluctuate dramatically during the course of chromatographic elution.

11.1.3.4 ESI-FTMS Applications

ESI-FTMS is well-suited for accurate mass measurement of large molecules. The superb mass accuracy and resolving power allow the determination of the molecular weights of large proteins. Unit mass resolution has been demonstrated for the protein chondroitinase I and II, assigning the molecular weights as 112,509 Da and 111,714 Da, respectively.[87] These measurements are within 1 Da of the calculated mass. The high resolving power of FTMS has been used to analyze complex mixtures such as heavy crude oil and humic substances. The high resolving power was necessary because approximately 11,000 unique species were present in a single mass spectrum of crude oil extract, >75% of which could be identified based on accurate mass measurement.[88] ESI-FTMS has also been used to analyze humic and fulvic acid samples.[89–91] Similar to ESI-FTMS of oil extracts, humic and fulvic acid samples produce thousands of unique peaks from *m/z* 200–3000 that could only be separated and identified with the high resolving power of FTMS.

Because ESI is a soft-ionization technique, it can be used to analyze labile molecules that dissociate through other ionization methods. Non-denaturing ESI-FTMS has been used to determine the molecular weights and oxidation states of the metal centers in metalloproteins.[38,92,93] For example, the oxidation state of the metal center in the copper-containing metalloprotein azurin was altered using chemical or electrochemical methods.[93] Non-denaturing ESI-FTMS of the protein, shown in Figure 11.21, was used to distinguish between the Cu^{1+} and Cu^{2+} oxidation states. Applications from the combination of ESI-FTMS and tandem mass spectrometry will be discussed below.

11.1.4 Tandem Mass Spectrometry in FTMS

The ability to store ions for long periods of time in the low-pressure analyzer cell allows a wide variety of in-cell tandem mass spectrometry (MS/MS) techniques to be applied to

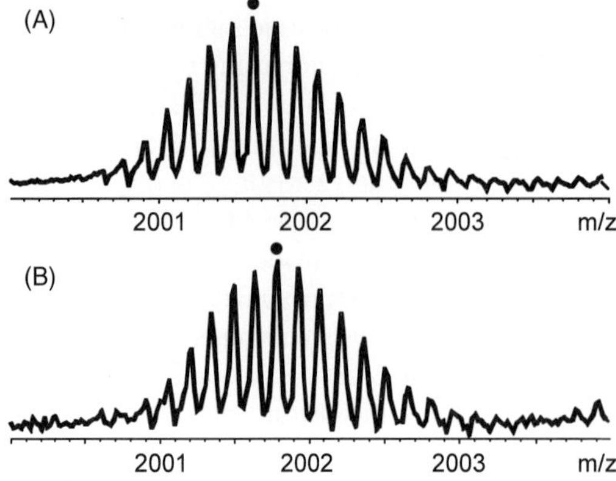

Figure 11.21. Non-denaturing ESI-FTMS used to determine the oxidation state of the metalloprotein azurin. (**A**) Oxidized azurin with Cu^{2+}. (**B**) Reduced form of azurin with Cu^{1+}. The shift of the most intense peak in the mass spectrum, labeled with a •, corresponds to the change from Cu^{2+} to Cu^{1+}. An extra proton is attached to the reduced form of the protein to yield an ion of the same charge state as the oxidized form, shifting the observed mass of the ion by one mass unit. (Adapted from Ref. 93.)

FTMS. Ion activation by collisions with neutral molecules, photoexcitation, and electron interactions have all been applied to FTMS. The recent development of hybrid FTMS instrumentation allows MS/MS to be performed outside of the cell, thus improving the duty cycle. Discussion of MS/MS capabilities of FTMS will be divided into two sections, in-cell (or internal) MS/MS and external MS/MS.

11.1.4.1 Internal MS/MS

Prior to the development of hybrid FTMS instruments, MS/MS was performed in the analyzer cell. Because FTMS performs MS/MS experiments "in time," precursor ion selection, ion activation, and product ion detection can all occur within the analyzer cell, in the proper sequence, as shown in Figure 11.22. Multiple ion isolation and activation events can be put together to achieve MS^n, requiring only modification of the event sequence via software control. This is shown by the bracketing of the ion isolation and precursor ion excitation events in Figure 11.22. For in-cell MS/MS, mass-selected ions are activated by one of a number of methods causing dissociation. The products and any remaining precursor ions are then detected using the standard method for broadband excitation and detection. In order to perform MS^n, additional ion isolation and activation events are placed after the first ion activation and before excitation and detection in the event sequence.

Mass selection of ions can be achieved within the analyzer cell by application of an excitation waveform that causes radial ejection of all unwanted ions, except the ions of

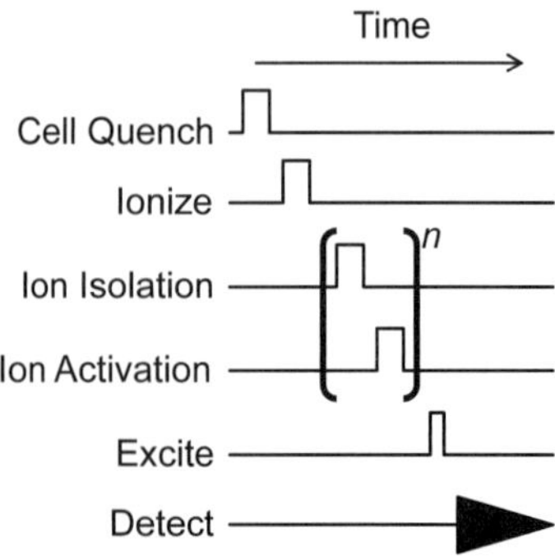

Figure 11.22. An experimental sequence demonstrating MS/MS for FTMS. Precursor ion selection and dissociation (bracketed section) occurs prior to the final excitation and detection. To perform MS^n, the bracketed section is repeated as needed.

interest. In its simplest form, the excitation waveform is an RF chirp that covers all frequencies except the frequency of the precursor ion. The precursor ion is not excited and remains at the center of the analyzer cell. For high-resolution ion isolation, other excitation waveforms are used to isolate ions, such as a coherent harmonic excitation frequency (CHEF)[94] or a stored waveform inverse Fourier transform (SWIFT) excitation.[95] CHEF requires no additional hardware to execute. It is a type of RF chirp in which the duration of each frequency step is equal to the inverse of the frequency step size. For example, an RF chirp that excites over a range of 10 kHz to 1 MHz, using 100 frequency steps, with a frequency increment of 10 kHz per step would satisfy the CHEF criteria if each step lasted 0.1 kHz, that is, 100 μs. The CHEF excitation skips the frequency of the ion to be retained in the analyzer cell. In this fashion, all ions can be ejected from the cell, leaving only the mass-selected precursor ions. The smaller the frequency steps (and the longer the CHEF waveform duration), the higher the resolving power of the isolation step. Isobaric species have been separated by this method. SWIFT excitation can also be used to achieve high-resolution ion isolation, but this method requires additional hardware (an arbitrary waveform generator) and software.

Following ion isolation, the mass-selected precursors are then activated and undergo dissociation. A wide variety of ion activation methods have been applied for in-cell MS/MS in FTMS. Collisionally activated decomposition (CAD) imparts energy into the molecule via inelastic collisions of precursor ions with neutral gas molecules. In-cell CAD is achieved by applying an excitation waveform that increases the kinetic energy of the precursor ion, thus allowing the mass-selected ion to undergo energetic collisions. Because of the low pressure required for the FTMS analysis step, it necessary to raise the pressure in the analyzer cell transiently by addition of the collision gas via a pulsed valve. After the short ($\sim$1 ms) introduction of gas into the analyzer, the pressure rises to $\sim 10^{-4}$ torr and is then pumped back to $\sim 10^{-9}$ torr. Adequate pumpdown time (1–10 s) is required to remove excess neutral gas from the analyzer cell before excitation and detection. Until the early 1990s, an on-resonance excitation pulse was used to increase the ions' translational energy prior to the addition of the collision gas. The ions would be excited to a radius of $\sim$1/3 the cell dimensions, giving them hundreds of electron volts of kinetic energy. The on-resonance ion excitation technique is no longer used because of the challenge of detecting product ions that are formed in different regions of the analyzer cell, well away from the center axis.[96,97]

In the early 1990s, Jacobson and coworkers developed the method of sustained off-resonance excitation (SORI) for CAD.[13] With SORI, precursor ions are excited with a waveform that is 1 kHz off resonance. This causes the alternate expansion and contraction of the cyclotron orbit of the precursor ion, with a cycle time of 1 ms, as shown in Figure 11.23A. The excitation is applied for 1 s or more, allowing thousands of cycles of expansion/contraction of the cyclotron orbit, shown in Figure 11.23B. The low amplitude of the excitation signal ensures that the expansion of the cyclotron orbit is small, approximately 10% of the radius of the cell, and that the kinetic energy gained by the precursor ions is small. However, many hundreds of collisions occur during the SORI excitation, and the cumulative effect is that the ion is activated sufficiently to undergo dissociation. With SORI, product ions are produced near the central axis of the cell, facilitating their detection by standard means. As the excitation is off-resonance, precursor ions are not ejected from the analyzer cell. However, any product ions formed during the SORI pulse that have the same frequency as the SORI pulse may be ejected from cell, creating a "blind spot" in the mass spectrum where no product ions will be present. The width of these "blind spots" depends on the intensity and duration of the SORI pulse and decreases at lower *m/z*. In order to ensure that

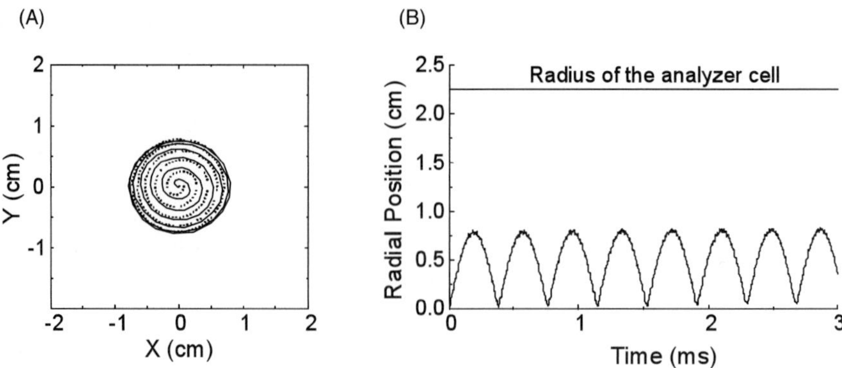

Figure 11.23. Illustration of ion cyclotron orbit during SORI excitation. (**A**) For one cycle, the ion cyclotron radius will expand (solid line) and contract (dashed line). (**B**) For the duration of the SORI excitation, the ion cyclotron radius will expand and contract many times.

all product ions are detected, two separate SORI-CAD experiments can be performed: In one experiment the SORI excitation occurs above the frequency of the precursor ion, and in the second experiment the SORI excitation occurs below the frequency of the precursor ion.[98] Because only the precursor is undergoing SORI excitation, this excitation event can be sustained long enough to drive all precursors into products without any concern about activating product ions and creating secondary product ions.

There are a number of different methods that are used to activate ions by irradiation with light, commonly referred to as photodissociation. Of these techniques, infrared multiphoton dissociation (IRMPD) is the most common for FTMS.[16,99] Irradiation is typically achieved by using the infrared light from a CO_2 laser, although simple heating of the cell can lead to the infrared heating of the ions by blackbody infrared irradiation (BIRD).[100] The light can be introduced along the central axis of the magnet, to irradiate ions located at the center of the analyzer cell, as shown in Figure 11.20A. The rear trapping plate of the analyzer cell must contain a hole large enough to permit the light to enter the cell and to interact with the stored ions (Figure 11.20B). For an IRMPD experiment, precursor ions are first mass-selected, and then these ions are irradiated by the IR (10.6 μm) output of a CO_2 laser. The time of irradiation depends on the photon flux provided by the laser. Under typical conditions, 10–15 W of power are applied for 100 ms to dissociate peptide ions. IRMPD performs best when the IR beam is directed along the central axis of the analyzer cell, thus maximizing overlap of the IR beam with the stored ions, as illustrated in Figure 11.24A. However, injection/isolation of ions within the analyzer cell may shift the precursor ions off-axis, reducing their overlap with the IR beam, illustrated by the solid cyclotron orbit in Figure 11.24B. In this case, an excitation pulse can be combined with the IRMPD experiment to increase overlap of the precursor ion with the IR beam.[101] The excitation pulse causes the ions' cyclotron orbit to expand, increasing the ions' overlap with the IR beam, as shown by the dashed cyclotron orbit in Figure 11.24B. While it has not been demonstrated, axialization by QE could be used to move the ions to the center of the cell to maximize beam overlap. Because IRMPD and SORI-CAD are both threshold dissociation methods, similar products are observed for the two ion activation methods.

IRMPD has a few advantages over SORI-CAD for tandem mass spectrometry in FTMS. First, the use of a collision gas is not necessary in IRMPD MS/MS; this maintains optimal pressure in the analyzer cell and eliminates pumpdown time, facilitating its use for on-line

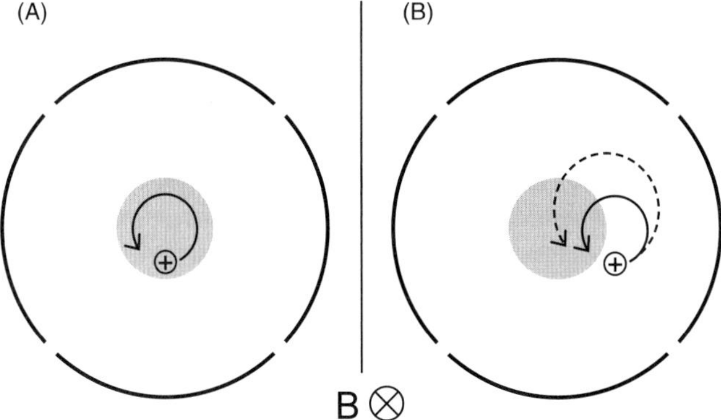

Figure 11.24. (**A**) IRMPD performs best when the IR beam (grey circle) overlaps the ion cyclotron orbit (solid arrow). (**B**) Ions with a large magnetron radius (solid arrow) will not favorably overlap the IR beam. The ion can be excited to a larger cyclotron radius to increase overlap with the IR beam (dashed arrow), improving IRMPD performance. These ion trajectories are not drawn to scale. (Adapted from Ref. 101.)

HPLC MS/MS. Unlike SORI-CAD, there is no issue with "blind spots" in the IRMPD product ion spectrum. However, product ions produced on-axis may also absorb the IR radiation to produce secondary fragments. Care must be taken when performing IRMPD to eliminate or minimize the formation of secondary product ions, which are of lower analytical utility than the primary products. This requires that the duration of the event be selected so that no more than half of the precursor ions undergo fragmentation. In contrast with SORI-CAD, 100% of the precursor ions can be converted into products.

Photodissociation using ultraviolet (UV) light is also used in FTMS, but it is not as common as IRMPD. UV photodissociation (UVPD) has been performed with an excimer laser, operated at either 193 nm[9,102,103] or 157 nm.[104] The arrangement of the excimer laser and the FTMS instrument is similar to the arrangement shown in Figure 11.20A, but with a UV transmitting window in place of the IR window. UVPD of peptides at 193 nm produces some backbone cleavage, whereas UVPD of proteins produces low abundance and uninformative backbone cleavages.[9,102,103] UVPD with 157-nm light results in backbone as well as disulfide bond cleavage, without charge reduction of the precursor ion.[104] However, because IRMPD is more reliable and reproducible compared to UVPD, IRMPD is more widely used.

Both SORI-CAD and IRMPD have been used to provide sequence information for proteins in ESI-FTMS.[105,106] Generally, similar products were observed for both SORI-CAD and IRMPD activation of ubiquitin ions, but more uninformative H_2O and NH_3 loss products occurred in SORI-CAD. Secondary fragmentation, which gives undesirable internal fragmentations, is observed for longer irradiation times only in IRMPD due to activation of product ions in the IR beam, but no secondary product ions are observed in SORI-CAD. Photodissociation of a peptide by IRMPD and 193-nm UVPD have been compared.[105] Dissociation of the peptide by UVPD produced more unassignable product ions, suggesting that IRMPD is more useful, in general, for peptide analysis. Overall, IRMPD is favored over other dissociation methods due to its ease of implementation, high efficiency, faster activation (100 ms), lack of products accompanied by H_2O and NH_3 loss, and the avoidance of introducing gas into the analyzer with its requisite long pumpdown time.

11.1.4.2 ECD, EDD, and EID

A number of different ion activation methods can be performed using electrons. While the previously mentioned dissociation methods can be accomplished with other types of mass spectrometers, electron-ion dissociation methods are best performed by FTMS. Generally, these methods reduce the charge of an ion and thus require a multiply charged precursor. The most widely applied of these methods is electron capture dissociation (ECD). ECD was discovered serendipitously by Zubarev et al.[17] when a misaligned UV laser irradiated a cell plate, releasing electrons that reduced and fragmented a multiply charged positive ion. Zubarev was able to show that similar fragmentation resulted when electrons were produced by a directly heated filament located within the vacuum chamber just outside the analyzer cell, adjacent to the rear trapping plate. This arrangement produces a low flux of electrons, necessitating an irradiation time of the order of tens of seconds to yield an adequate number of product ions. The use of an indirectly heated cathode (the cathode is electrically isolated from a resistive heater element, but is in thermal contact with the heater) has been shown to provide a much higher flux of electrons, enabling electron-ion experiments to be performed on much shorter timescales, typically 1 s or less.[107] The configuration of FTMS instruments with an indirectly heated cathode is shown in Figures 11.20A–11.20C. One advantage to performing electron-ion experiments is that the energy of the electrons can be precisely controlled, permitting a wide range of dissociation energies to be explored. The experimental sequence of events for performing electron-ion experiments in FTMS is similar to those for an IRMPD experiment. Mass-selected ions are stored in the analyzer cell where they are subsequently irradiated by electrons. The product ions and remaining precursor ions are then excited and detected. No SORI excitation or gas pulse is required to perform these experiments, but SORI excitation, on-resonance excitation, or QE can be used to ensure overlap of the electron beam with the ion cloud.[108–110] Similar to IRMPD, electrons pass from the filament or cathode, through a hole in the rear trapping plate, where they interact with the mass-selected ions (Figure 11.20B).

ECD involves the recombination of low-energy ($\leq$1.5 eV) electrons with a multiply charged positive precursor ions.[17] This recombination creates a radical site that initiates decomposition, producing a different set of products than are obtained by CAD or IRMPD. ECD dissociation is believed to be nonergodic; that is, fragmentation occurs before internal energy distributes itself throughout a molecule.[111,112] For peptide and protein analysis, the result is that the C_α–N bond is broken along the backbone of the peptide chain, yet labile modifications such as glycosylation, phosphorylation, and sulfation are retained.[113–118] ECD can also be performed with higher electron energies varying from 3 eV to 13 eV and is referred to as "hot" ECD (HECD).[119] Using a heated filament, electron irradiation times ranging from 1 s to >10 s were necessary to fragment the molecule. Another option is to combine other means of activating ions with ECD. For example, protein ions can be "warmed" by low-energy CAD while they are undergoing ECD to facilitate their dissociation,[120] a process known as activated ion ECD (aiECD). This is found to give far more sequence coverage for proteins above 10 kDa in molecular weight.

Electron detachment dissociation (EDD)[11] is the negative-ion complement of ECD.[121] For EDD, multiply charged negative ions are irradiated with moderate energy (19 eV) electrons. An electron is detached from the ion, creating a radical site that can undergo rearrangement and fragmentation. EDD was originally performed using a heated filament and required irradiation times of 10–100 s. For EDD using an indirectly heated cathode, electron irradiation times are typically $\leq$1 s. EDD has shown great potential for fragmenting acidic molecules that readily form negative ions but not positive ions, such as oligonucleotides[122] and sulfated carbohydrates.[123–125] Activation of ions by electrons without electron capture was originally referred to as electron impact exciation of ions from organics (EIEIO)[12,126] but is now referred to as electron-induced dissociation (EID).[10] Generally, product ions resulting from EID are similar to those observed by collisional or photoactivation (CAD and IRMPD)[127] and thus was not widely used. However, recent work on carbohydrate cations,[10] phosphorylated metabolites,[128] and amino acids[129] has shown promise for the characterization of these small molecules by EID. EID can be performed on either positive or negative ions that are either singly or multiply charged, and therefore it can be used both with MALDI or ESI. Many FTMS instruments are equipped with the capability to perform both ECD/EDD and IRMPD experiments. These dissociation methods perform best when the light or electrons are irradiated on-axis (in the center) of the analyzer cell. Because a solid indirectly heated cathode would block incoming IR light, a hollow cathode has been developed, as shown in Figure 11.20B. The indirectly heated hollow cathode is annular, producing an on-axis, hollow cylindrical electron beam (the lighter trace). For dissociation experiments using infrared light, light passes through the center of the hollow cathode, entering the FTMS analyzer on-axis (the darker trace).

11.1.4.3 External MS/MS

The development of hybrid-FTMS instruments allows mass selection and/or CAD to be performed outside the analyzer cell. There are a number of advantages to performing tandem mass spectrometry outside the analyzer cell. Placing the collision region away from the analyzer eliminates the need for a pumpdown cycle as with in-cell CAD. Thus MS/MS using CAD outside the analyzer cell occurs much faster, decreasing the duration of the experiment. Compared to performing MS/MS in the analyzer cell, precursor and product ions are all kept on-axis, maintaining optimum conditions for excitation and detection. Also, because isolation and mass-selected ion accumulation occur before dissociation, S/N is generally improved because the storage multipole, whose capacity is limited by space charge, contains only the precursor ion. Hybrid instruments equipped with either a quadrupole or LIT have the ability to perform CAD outside the analyzer cell. Because the quadrupole is a scanning MS, and not a trapping MS, only one stage of MS/MS can be performed. Additional stages of MS/MS can be performed using in-cell MS/MS. In contrast, the hybrid-FTMS instrument equipped with a LIT can perform MS^n outside the analyzer cell, using the FTMS for detection of the final products. Ion isolation and fragmentation can be performed many times (e.g., MS^n) within the LIT before products are transferred to the FTMS for analysis. Electron transfer dissociation (ETD) has recently been applied to FTMS.[130] ETD produces product ions similar to ECD, and it can be performed much faster than ECD.

11.1.4.4 Tandem Mass Spectrometry Applications in FTMS

The ability to perform multiple stages of MS/MS makes FTMS a powerful tool for the characterization of complex molecules. Typically, tandem mass spectrometry is performed using ESI-FTMS and MALDI-FTMS, but dissociation techniques that reduce precursor ion charge (e.g., ECD, ETD, and EDD) must be performed on multiply charged ions and are therefore used with ESI.

The application of tandem mass spectrometry to the characterization of proteins and protein mixtures has greatly improved the analysis of these complex molecules. Tandem mass spectrometry analysis of proteins or protein mixtures are distinguished by the method used to characterize them and are referred to as either top-down or bottom-up protein analysis.[131] Top-down protein analysis is ideally suited for characterization of unknown proteins as well as location of the sites of post-translational modification. For top-down protein analyses, proteins are dissociated (via MS^n) to determine the primary sequence of the protein. Threshold dissociation methods such as CAD or IRMPD can be used to determine the primary sequence, but ECD has been shown to provide more sequence coverage than IRMPD or CAD.[132] Also, ECD can be used for top-down sequencing of the protein and localization of labile post-translational modifications. This has been demonstrated for histone acetylation[133] and phosphorylation sites on carbonic anhydrase.[134] The use of activated ion ECD (AI-ECD) has been shown to increase the number of cleavages observed in ECD, allowing more complete coverage from a single tandem mass spectrometry experiment.[120] ESI-FTMS coupled with HPLC allows for rapid, bottom-up analysis of proteomes. Bottom-up protein analysis involves the digestion of a complex mixture of proteins, followed by ESI-FTMS/MS of the peptide mixture. Many proteins can be identified by ESI-FTMS using only accurate mass measurement of the peptide mixture. Tandem mass spectrometry of individual peptides provides additional sequence information that aids in peptide identification.

Many different tandem mass spectrometry methods have been applied to the analysis of oligosaccharides and peptides. FTMS is highly suited for the analysis of these complex biomolecules because high resolution can be used to distinguish isobaric products while multiple stages of MS^n can be used to characterize the protein, peptides, and oligosaccharides. The combination of protein digestion, HPLC, and ESI-FTMS was used to determine the location and type of post-translational modification on a bovine milk protein.[135] In this analysis, the protein was digested and the resulting mixture was separated and analyzed by HPLC ESI-FTMS. ECD of the peptides was used to determine the site of modification, while IRMPD was used to characterize the modification. This analysis confirmed eight known protein modifications and one unknown *O*-glycosylation.[135] ECD,[10] EDD,[123,136] and IRMPD[75,137] have also been used to characterize oligosaccharides using a single tandem mass spectrometry experiment.

11.2 MAGNETIC SECTOR MASS SPECTROMETRY

11.2.1 Magnetic Sector Overview

Prior to the advent of Fourier transform mass spectrometers, double focusing magnetic sector mass spectrometers provided the highest mass resolution and the highest mass accuracy of any type of mass spectrometer. High-end double-focusing instruments can provide resolving powers of 50,000 and mass accuracy of 5 ppm or better for ions up to *m/z* 10,000. Moreover, such instruments can be used for tandem mass spectrometry measurements by collisionally activated decomposition (CAD) at high energy ($>$1 keV). In the mid-1980s, four sector magnetic sector instruments combined with fast atom bombardment sources were considered to be the premier high-performance mass spectrometers for biomolecule analysis. Shortly thereafter, as electrospray ionization and MALDI became the ionization methods of choice for analyzing biomolecules, magnetic sector technology began to be challenged by other types of mass spectrometers. Ultimately, technological developments in instrumentation led to the ascendancy of other mass analyzers for biological analysis, and the use of magnetic sector mass spectrometers for biological analysis has all but disappeared. Prior to these developments, there was a short period of time in which there was interest in combining the high-performance features of magnetic sector mass spectrometers with MALDI and electrospray ionization for biomolecule analysis. For completeness, these developments are presented below.

An important consideration in combining a sector mass spectrometer with an ion source for biomolecules is the mass-to-charge range of the mass spectrometer, which is determined by the magnetic field strength (B), the radius of curvature of the magnetic sector (r), unit charge (e), and the acceleration voltage of the source (V) as shown in Eq. (11.7). To extend the upper mass-to-charge limit, the radius of curvature of the magnetic sector can be increased, adding considerably to the size, weight, and expense of the magnet. Increasing the field strength of the magnet also increases the upper mass-to-charge limit; however, the ferrous core of the electromagnets used for these instruments saturate around 1 T, providing an upper limit to the field strength that can be achieved. The upper mass-to-charge limit can also be extended by operating the instrument at reduced acceleration energy; however, this has deleterious effects on the mass resolution and sensitivity. A typical upper mass-to-charge limit for a high-end double-focusing mass spectrometer operated at full acceleration energy is m/z 10,000. MALDI applications will thus be limited to biomolecules of relatively moderate size, given the propensity of this ionization method to form singly charged species. On the other hand, electrospray ionization is well-suited for this analyzer, because the multiple-charging that is typical of electrospray produces ions of low mass-to-charge even for very high-molecular-weight samples.

$$\frac{m}{z} = \frac{r^2 B^2 e}{2V} \tag{11.7}$$

Two other features of magnetic sector mass spectrometers bear on the interfacing of MALDI or electrospray ionization sources. One is that ions must be accelerated to high kinetic energy in order to achieve high mass resolution, typically 1–10 keV. Generally, this is achieved by operating the ion source at high electric potential (1–10 kV) and converting the ions' potential energy to kinetic energy as they move toward the mass analyzer that is held at ground potential. A second feature is that for high-resolution performance, magnetic sectors instruments are usually operated as a scanning analyzer with a point detector. Ions of a narrow range of mass-to-charge values are focused onto the detector, and a mass spectrum is recorded by ramping the magnetic field to scan over the range of mass values of interest.

11.2.2 MALDI Combined with Magnetic Sector Mass Spectrometry

The high acceleration voltage utilized by sector mass spectrometers provides no deterrent for MALDI applications. MALDI sources that operate at 10–30 kV are commonly used in time-of-flight mass spectrometers, and similar designs can be implemented in a sector instrument. The MALDI sample target is held at the acceleration potential of the mass spectrometer, and the desorbed ions are accelerated and focused using the standard ion optics for an ion source of a magnetic sector mass spectrometer. The implementation of MALDI with a sector mass spectrometer is particularly challenged by the second consideration mentioned above, namely the scanning mass analyzer. With MALDI, ions are produced in short pulses of nanosecond duration, with a typical repetition rate of 10 Hz. This yields a very low duty cycle for ion production that renders it impossible to utilize a slow scanning mass analyzer. In order to enable the implementation of MALDI, focal plane detection is used to provide a means for simultaneous detection over a range of masses. Focal plane detection was used in some of the very first mass spectrometers, for example, in the instrument designed by Dempster circa 1918. In the Dempster instrument, photographic film was used to detect a wide mass range of ions simultaneously. For MALDI applications, an integrating focal plane array detector has been used to electronically record a range of masses simultaneously.[138] This provided subpicomole detection limits. However, the mass resolving power that could be achieved with this detector is only 5000. When this instrument was first described in 1994, this mass resolution surpassed that of time-of-flight (TOF) mass spectrometers. Such instruments were used to examine oligosaccharides,[139] sphingolipids,[140] and phospholipids[141] with higher resolution than TOF could provide at that time. However, innovations with delayed extraction ion sources led shortly thereafter to the development of TOF instruments with comparable mass resolution. The new TOF mass spectrometers were smaller in scale, less expensive to produce, and less complex to operate, and they eventually eliminated the rationale for a magnetic-sector-based MALDI instrument for accurate mass measurement.

MALDI and magnetic sector mass spectrometers have also been combined in a novel instrument for tandem mass spectrometry measurements. With this instrument, the magnetic sector mass spectrometer was used to provide high-resolution precursor ion selection, along with orthogonal acceleration into a TOF mass spectrometer that was used to mass analyze the product ions formed by collisionally activated decomposition in a gas collision cell at the interface of the two mass spectrometers.[142] The TOF mass spectrometer is well-suited to mass analyze the product ions that are produced in brief pulses. This arrangement also uses an optional array detector to use the magnetic sector mass spectrometer to mass analyze the ions produced by the MALDI event. This instrument provided the means to perform CAD on ions formed by MALDI, at a collision energy of 800 eV. A similar instrument with a reflectron TOF mass spectrometer has been demonstrated to provide subfemtomole detection limits for peptide analysis.[143,144] The capabilities of these instruments were later matched or surpassed by the development of tandem time-of-flight (TOF–TOF) mass spectrometers.

11.2.3 ESI Combined with Magnetic Sector Mass Spectrometry

In the early 1990s, several researchers worked to combine electrospray ionization with the high-resolution capabilities of a double-focusing magnetic sector mass spectrometer. The multiple charging that is characteristic of electrospray allows magnetic sector instruments to be applied to high-molecular-weight samples, extending the molecular weight range of analysis. The high resolution that can be obtained with double-focusing mass spectrometers extends the upper range of sample molecular weight that can be analyzed compared to a quadrupole mass spectrometer, by providing the means to resolve the more closely spaced charge states that occur for high-molecular-weight compounds.

The combination of electrospray ionization with magnetic sector mass spectrometers proved to be less technically challenging than the interfacing of MALDI. Electrospray ionization produces ions continuously, and thus it combines easily to scanning mass analyzers such as the magnetic sector. Thus, no modifications of the mass analyzer or detector are required for electrospray ionization. However, the ions produced by electrospray must be accelerated to kilo-electron-volt energies, and this requires some innovation with the design of the ion source. One solution is to float the entire electrospray source and its power supply at the acceleration potential of the sector mass spectrometer.[145,146] However, this complicates the introduction of the sample—for example, by on-line HPLC—because the sample must also float at the acceleration potential. The use of a glass capillary for the air–vacuum interface of the ESI source allows independent control of the potentials applied at the atmosphere and vacuum ends of the capillary.[147–150] Thus, one can float the interior components of the electrospray source (e.g., the skimmer and ion optics) at the source potential while operating the electrospray needle and inlet end of the capillary at typical voltages for ESI (e.g., needle at 0 V and electrospray inlet at −3.5 kV for positive ions). In the elevated pressure that exists in the first region of differential pumping within the ion source, the high acceleration voltage can form a discharge that is electrically conducting and can charge the mechanical pump that evacuates this region, so that its housing reaches the acceleration potential of the ion source. This pump must be electrically isolated from its surroundings in order to operate under these conditions.

ESI combined with double-focusing mass spectrometers has been used to examine high-molecular-weight proteins,[151–154] noncovalent peptide complexes,[155] and protein–protein complexes.[156,157] Negative-ion mode has been used to examine anionic compounds.[158] Capillary electrophoresis has been interfaced to an ESI-magnetic sector instrument.[159] Although a point detector has been used for most applications, focal plane detection with array detectors can provide an enhancement in sensitivity.[160] Despite the high-resolution capabilities offered by magnetic sector instruments, the combination of electrospray with FTMS and with TOF mass spectrometry produced instruments with comparable or superior mass resolution and mass accuracy, and thus these have now replaced magnetic sector instruments for ESI analysis.

REFERENCES

1. Amster, I. J.; Loo, J. A.; Furlong, J. J. P.; McLafferty, F. W. Cesium ion desorption ionization with Fourier-transform mass-spectrometry. *Anal. Chem.* **1987**, *59*, 313–317.
2. Henry, K. D.; Williams, E. R.; Wang, B. H.; McLafferty, F. W.; Shabanowitz, J.; Hunt, D. F. Fourier-transform mass-spectrometry of large molecules by electrospray ionization. *Proc. Natl. Acad. Sci. USA* **1989**, *86*, 9075–9078.
3. Hofstadler, S. A.; Laude, D. A. Electrospray ionization in the strong magnetic field of a Fourier transform ion cyclotron resonance mass spectrometer. *Anal. Chem.* **1992**, *64*, 569–572.
4. Koster, C.; Castoro, J. A.; Wilkins, C. L. High-resolution matrix-assisted laser desorption ionization of biomolecules by Fourier-transform mass-spectrometry. *J. Am. Chem. Soc.* **1992**, *114*, 7572–7574.
5. Li, Y.; Tang, K.; Little, D. P.; Koster, H.; Hunter, R. L.; McIver, R. T. High-resolution MALDI Fourier transform mass spectrometry of oligonucleotides. *Anal. Chem.* **1996**, *68*, 2090–2096.
6. Loo, J. A.; Williams, E. R.; Amster, I. J.; Furlong, J. J. P.; Wang, B. H.; McLafferty, F. W.; Chait, B. T.; Field, F. H. Cf-252 plasma desorption with Fourier-transform mass-spectrometry. *Anal. Chem.* **1987**, *59*, 1880–1882.
7. Watson, C. H.; Barshick, C. M.; Wronka, J.; Laukien, F. H.; Eyler, J. R. Pulsed gas glow discharge for ultrahigh mass resolution measurements with Fourier transform ion cyclotron resonance mass spectrometry. *Anal. Chem.* **1996**, *68*, 573–575.
8. Winger, B. E.; Hofstadler, S. A.; Bruce, J. E.; Udseth, H. R.; Smith, R. D. High-resolution accurate mass measurements of biomolecules using a new electrospray-ionization ion-cyclotron resonance mass-spectrometer. *J. Am. Soc. Mass. Spectrom.* **1993**, *4*, 566–577.
9. Beu, S. C.; Senko, M. W.; Quinn, J. P.; Wampler, F. M.; McLafferty, F. W. Fourier-transform electrospray instrumentation for tandem high-resolution mass spectrometry of large molecules. *J. Am. Soc. Mass. Spectrom.* **1993**, *4*, 557–565.
10. Budnik, B. A.; Haselmann, K. F.; Elkin, Y. N.; Gorbach, V. I.; Zubarev, R. A. Applications of electron-ion dissociation reactions for analysis of polycationic chitooligosaccharides in Fourier transform mass spectrometry. *Anal. Chem.* **2003**, *75*, 5994–6001.
11. Budnik, B. A.; Haselmann, K. F.; Zubarev, R. A. Electron detachment dissociation of peptide di-anions: An electron-hole recombination phenomenon. *Chem. Phys. Lett.* **2001**, *342*, 299–302.

12. Cody, R. B.; Freiser, B. S. Electron impact excitation of ions from organics: An alternative to collision induced dissociation. *Anal. Chem.* **1979**, *51*, 547–551.
13. Gauthier, J. W.; Trautman, T. R.; Jacobson, D. B. Sustained off-resonance irradiation for collision-activated dissociation involving Fourier-transform mass-spectrometry–collision-activated dissociation technique that emulates infrared multiphoton dissociation. *Anal. Chim. Acta* **1991**, *246*, 211–225.
14. Ijames, C. F.; Wilkins, C. L. Surface-induced dissociation by Fourier-transform mass-spectrometry. *Anal. Chem.* **1990**, *62*, 1295–1299.
15. Williams, E. R.; Henry, K. D.; McLafferty, F. W.; Shabanowitz, J.; Hunt, D. F. Surface-induced dissociation of peptide ions in Fourier-transform mass-spectrometry. *J. Am. Soc. Mass. Spectrom.* **1990**, *1*, 413–416.
16. Woodin, R. L.; Bomse, D. S.; Beauchamp, J. L. Multiphoton dissociation of molecules with low-power continuous wave infrared-laser radiation. *J. Am. Chem. Soc.* **1978**, *100*, 3248–3250.
17. Zubarev, R. A.; Kelleher, N. L.; McLafferty, F. W. Electron capture dissociation of multiply charged protein cations. A nonergodic process. *J. Am. Chem. Soc.* **1998**, *120*, 3265–3266.
18. Lawrence, E. O.; Livingston, M. S. The production of high speed light ions without the use of high voltages. *Phy. Rev.* **1932**, *40*, 19.
19. Hipple, J. A.; Sommer, H.; Thomas, H. A. A precise method of determining the faraday by magnetic resonance. *Phys. Rev.* **1949**, *76*, 1877–1878.
20. Marshall, A. G.; Comisarow, M. B.; Parisod, G. Theory of Fourier-transform ion-cyclotron resonance mass spectroscopy.1. Relaxation and spectral-line shape in Fourier-transform ion resonance spectroscopy. *J. Chem. Phys.* **1979**, *71*, 4434–4444.
21. Comisaro, M. B.; Marshall, A. G. Fourier-transform ion-cyclotron resonance spectroscopy. *Chem. Phys. Lett.* **1974**, *25*, 282–283.
22. McIver, R. T. A trapped ion analyzer cell for ion cyclotron resonance spectroscopy. *Rev. Sci. Instrum.* **1970**, *41*, 555.
23. Marshall, A. G.; Guan, S. H. Advantages of high magnetic field for Fourier transform ion cyclotron resonance mass spectrometry. *Rapid Commun. Mass Spectrom.* **1996**, *10*, 1819–1823.
24. Marshall, A. G.; Schweikhard, L. Fourier-transform ion-cyclotron resonance mass-spectrometry—Technique developments. *Int. J. Mass Spectrom. Ion Processes* **1992**, *118*, 37–70.
25. Marshall, A. G.; Verdun, F. R. *Fourier Transforms in NMR, Optical, and Mass Spectrometry: A User's Handbook*, Elsevier, New York, **1990**.
26. Vartanian, V. H.; Anderson, J. S.; Laude, D. A. Advances in trapped ion cells for Fourier-transform ion-cyclotron resonance mass-spectrometry. *Mass Spectrom. Rev.* **1995**, *14*, 1–19.
27. Easterling, M. L.; Mize, T. H.; Amster, I. J. Routine part-per-million mass accuracy for high-mass ions: Space-charge effects in MALDI FT-ICR. *Anal. Chem.* **1999**, *71*, 624–632.
28. Jing, L.; Li, C. Y.; Wong, R. L.; Kaplan, D. A.; Amster, I. J. Improved mass accuracy for higher mass peptides by using SWIFT excitation for MALDI-FTICR mass spectrometry. *J. Am. Soc. Mass. Spectrom.* **2008**, *19*, 76–81.
29. Taylor, P. K.; Amster, I. J. Space charge effects on mass accuracy for multiply charged ions in ESI-FTICR. *Int. J. Mass spectrom.* **2003**, *222*, 351–361.
30. Wong, R. L.; Amster, I. J. Sub part-per-million mass accuracy by using stepwise-external calibration in Fourier transform ion cyclotron resonance mass spectrometry. *J. Am. Soc. Mass. Spectrom.* **2006**, *17*, 1681–1691.
31. Wong, R. L.; Amster, I. J. Experimental evidence for space-charge effects between ions of the same mass-to-charge in Fourier-transform ion cyclotron resonance mass spectrometry. *Int. J. Mass spectrom.* **2007**, *265*, 99–105.
32. Syka, J. E. P.; Marto, J. A.; Bai, D. L.; Horning, S.; Senko, M. W.; Schwartz, J. C.; Ueberheide, B.; Garcia, B.; Busby, S.; Muratore, T.; Shabanowitz, J.; Hunt, D. F. Novel linear quadrupole ion trap/FT mass spectrometer: Performance characterization and use in the comparative analysis of histone H3 post-translational modifications. *J. Proteome Res.* **2004**, *3*, 621–626.
33. Ledford, E. B.; Rempel, D. L.; Gross, M. L. Space charge effects in Fourier transform mass spectrometry. II. Mass calibration. *Anal. Chem.* **1984**, *56*, 2744–2748.
34. Wood, T. D.; Schweikhard, L.; Marshall, A. G. Mass-to-charge ratio upper limits for matrix-assisted laser desorption Fourier transform ion cyclotron resonance mass spectrometry. *Anal. Chem.* **1992**, *64*, 1461–1469.
35. Solouki, T.; Gillig, K. J.; Russell, D. H. Detection of high-mass biomolecules in Fourier transform ion cyclotron resonance mass spectrometry: Theoretical and experimental investigations. *Anal. Chem.* **1994**, *66*, 1583–1587.
36. Chen, R.; Cheng, X.; Mitchell, D. W.; Hofstadler, S. A.; Wu, Q.; Rockwood, A. L.; Sherman, M. G.; Smith, R. D. Trapping, detection, and mass determination of coliphage T4 DNA ions by electrospray ionization Fourier transform ion cyclotron resonance mass spectrometry. *Anal. Chem.* **1995**, *67*, 1159–1163.
37. Hofstadler, S. A.; Bruce, J. E.; Rockwood, A. L.; Anderson, G. A.; Winger, B. E.; Smith, R. D. Isotopic beat patterns in Fourier-transform ion-cyclotron resonance mass-spectrometry—Implications for high-resolution mass measurements of large biopolymers. *Int. J. Mass Spectrom.* **1994**, *132*, 109–127.

38. Taylor, P. K.; Kurtz, D. M.; Amster, I. J. Electrospray ionization Fourier transform ion cyclotron resonance mass spectrometry of multimeric metalloproteins. *Int. J. Mass spectrom.* **2001**, *210*, 651–663.
39. Karas, M.; Bachmann, D.; Bahr, U.; Hillenkamp, F. Matrix-assisted ultraviolet laser desorption of non-volatile compounds. *Int. J. Mass Spectrom. Ion Processes* **1987**, *78*, 53–68.
40. Karas, M.; Hillenkamp, F. Laser desorption ionization of proteins with molecular masses exceeding 10,000 daltons. *Anal. Chem.* **1988**, *60*, 2299–2301.
41. Tanaka, K.; Waki, H.; Ido, Y.; Akita, S.; Yoshida, Y.; Yoshida, T.; Matsuo, T. Protein and polymer analyses up to *m/z* 100,000 by laser ionization time-of-flight mass spectrometry. *Rapid Commun. Mass Spectrom.* **1988**, *2*, 151–153.
42. Easterling, M. L.; Pitsenberger, C. C.; Kulkarni, S. S.; Taylor, P. K.; Amster, I. J. A 4.7 Tesla internal MALDI-FTICR instrument for high mass studies: Performance and methods. *Int. J. Mass Spectrom. Ion Processes* **1996**, *158*, 97–113.
43. Beavis, R. C.; Chait, B. T. Velocity distributions of intact high mass polypeptide molecule ions produced by matrix assisted laser desorption. *Chem. Phys. Lett.* **1991**, *181*, 479–484.
44. Chan, T. -W. D.; Thomas, I.; Colburn, A. W.; Derrick, P. J. Initial velocities of positive and negative protein molecule-ions produced in matrix-assisted ultraviolet laser desorption using a liquid matrix. *Chem. Phys. Lett.* **1994**, *222*, 579–585.
45. Huth-Fehre, T.; Becker, C. H.; Beavis, R. C. Energetics of gramicidin S after UV laser desorption from a ferulic acid matrix. *Rapid Commun. Mass Spectrom.* **1991**, *5*, 378–382.
46. Juhasz, P.; Vestal, M. L.; Martin, S. A. On the initial velocity of ions generated by matrix-assisted laser desorption ionization and its effect on the calibration of delayed extraction time-of-flight mass spectra. *J. Am. Soc. Mass. Spectrom.* **1997**, *8*, 209–217.
47. Pan, Y.; Cotter, R. J. Measurement of initial translational energies of peptide ions in laser desorption/ionization mass spectrometry. *Org. Mass Spectrom.* **1992**, *27*, 3–8.
48. Glückmann, M.; Karas, M. The initial ion velocity and its dependence on matrix, analyte and preparation method in ultraviolet matrix-assisted laser desorption/ionization. *J. Mass Spectrom.* **1999**, *34*, 467–477.
49. Gorshkov, M. V.; Masselon, C. D.; Anderson, G. A.; Udseth, H. R.; Harkewicz, R.; Smith, R. D. A dynamic ion cooling technique for FTICR mass spectrometry. *J. Am. Soc. Mass. Spectrom.* **2001**, *12*, 1169–1173.
50. Easterling, M. L.; Mize, T. H.; Amster, I. J. MALDI FTMS analysis of polymers: Improved performance using an open ended cylindrical analyzer cell. *Int. J. Mass Spectrom.* **1997**, *169*, 387–400.
51. Mize, T. H.; Amster, I. J. Broad-band ion accumulation with an internal source MALDI-FTICR-MS. *Anal. Chem.* **2000**, *72*, 5886–5891.
52. Caravatti, P.Spectrospin AG (CH) Method and apparatus for the accumulation of ions in a trap of an ion cyclotron resonance spectrometer, by transferring the kinetic energy of the motion parallel to the magnetic field into directions perpendicular to the magnetic field. Patent number 4,924,089, May 8 1990.
53. Yao, J.; Dey, M.; Pastor, S. J.; Wilkins, C. L. Analysis of high-mass biomolecules using electrostatic fields and matrix-assisted laser desorption/ionization in a Fourier transform mass spectrometer. *Anal. Chem.* **1995**, *67*, 3638–3642.
54. Savard, G.; Becker, S.; Bollen, G.; Kluge, H. J.; Moore, R. B.; Otto, T.; Schweikhard, L.; Stolzenberg, H.; Wiess, U. A new cooling technique for heavy-ions in a Penning trap. *Phys. Lett. A* **1991**, *158*, 247–252.
55. Schweikhard, L.; Guan, S. H.; Marshall, A. G. Quadrupolar excitation and collisional cooling for axialization and high-pressure trapping of ions in Fourier-transform ion-cyclotron resonance mass-spectrometry. *Int. J. Mass Spectrom. Ion Processes* **1992**, *120*, 71–83.
56. Speir, J. P.; Gorman, G. S.; Pitsenberger, C. C.; Turner, C. A.; Wang, P. P.; Amster, I. J. Remeasurement of ions using quadrupolar excitation Fourier-transform ion-cyclotron resonance spectrometry. *Anal. Chem.* **1993**, *65*, 1746–1752.
57. Hendrickson, C. L.; Drader, J. J.; Laude, D. A. Simplified application of quadrupolar excitation in Fourier transform ion cyclotron resonance mass spectrometry. *J. Am. Soc. Mass. Spectrom.* **1995**, *6*, 448–452.
58. Solouki, T.; Marto, J. A.; White, F. M.; Guan, S.; Marshall, A. G. Attomole biomolecule mass analysis by matrix-assisted laser desorption/ionization Fourier transform ion cyclotron resonance. *Anal. Chem.* **1995**, *67*, 4139–4144.
59. Cotter, R. J. Time-of-flight mass spectrometry for the structural analysis of biological molecules. *Anal. Chem.* **1992**, *64*, 1027A–1039A.
60. Spengler, B.; Kirsch, D.; Kaufmann, R.; Cotter, R. J. Metastable decay of peptides and proteins in matrix-assisted laser-desorption mass spectrometry. *Rapid Commun. Mass Spectrom.* **1991**, *5*, 198–202.
61. Annan, R. S.; Carr, S. A. Phosphopeptide analysis by matrix-assisted laser desorption time-of-flight mass spectrometry. *Anal. Chem.* **1996**, *68*, 3413–3421.
62. Laremore, T. N.; Murugesan, S.; Park, T. J.; Avci, F. Y.; Zagorevski, D. V.; Linhardt, R. J. Matrix-assisted laser desorption/ionization mass spectrometric analysis of uncomplexed highly sulfated oligosaccharides using ionic liquid matrices. *Anal. Chem.* **2006**, *78*, 1774–1779.
63. Heeren, R. M. A.; Boon, J. J. Rapid microscale analyses with an external ion source Fourier transform ion cyclotron resonance mass spectrometer. *Int. J. Mass Spectrom. Ion Processes* **1996**, *158*, 391–403.

64. LEBRILLA, C. B.; AMSTER, I. J.; McIVER, R. T. External ion-source FTMS instrument for analysis of high mass ions. *Int. J. Mass Spectrom. Ion Processes* **1989**, *87*, R7–R13.
65. LI, Y. Z.; McIVER, R. T.; HUNTER, R. L. High-accuracy molecular-mass determination for peptides and proteins by Fourier-transform mass-spectrometry. *Anal. Chem.* **1994**, *66*, 2077–2083.
66. McIVER, R. T.; LI, Y. Z.; HUNTER, R. L. High-resolution laser-desorption mass-spectrometry of peptides and small proteins. *Proc. National Acad. Sci. USA* **1994**, *91*, 4801–4805.
67. KOFEL, P.; ALLEMANN, M.; KELLERHALS, H.; WANCZEK, K. P. External generation of ions in ICR spectrometry. *Int. J. Mass Spectrom. Ion Processes* **1985**, *65*, 97–103.
68. KRUTCHINSKY, A. N.; LOBODA, A. V.; SPICER, V. L.; DWORSCHAK, R.; ENS, W.; STANDING, K. G. Orthogonal injection of matrix-assisted laser desorption/ionization ions into a time-of-flight spectrometer through a collisional damping interface. *Rapid Commun. Mass Spectrom.* **1998**, *12*, 508–518.
69. BAYKUT, G.; JERTZ, R.; WITT, M. Matrix-assisted laser desorption/ionization Fourier transform ion cyclotron resonance mass spectrometry with pulsed in-source collision gas and in-source ion accumulation. *Rapid Commun. Mass Spectrom.* **2000**, *14*, 1238–1247.
70. LAIKO, V. V.; BALDWIN, M. A.; BURLINGAME, A. L. Atmospheric pressure matrix-assisted laser desorption/ionization mass spectrometry. *Anal. Chem.* **2000**, *72*, 652–657.
71. KELLERSBERGER, K. A.; TAN, P. V.; LAIKO, V. V.; DOROSHENKO, V. M.; FABRIS, D. Atmospheric pressure MALDI-Fourier transform mass spectrometry. *Anal. Chem.* **2004**, *76*, 3930–3934.
72. HARVEY, D. J.; BATEMAN, R. H.; BORDOLI, R. S.; TYLDESLEY, R. Ionisation and fragmentation of complex glycans with a quadrupole time-of-flight mass spectrometer fitted with a matrix-assisted laser desorption/ionisation ion source. *Rapid Commun. Mass Spectrom.* **2000**, *14*, 2135–2142.
73. O'CONNOR, P. B.; COSTELLO, C. E. A high pressure matrix-assisted laser desorption/ionization Fourier transform mass spectrometry ion source for thermal stabilization of labile biomolecules. *Rapid Commun. Mass Spectrom.* **2001**, *15*, 1862–1868.
74. MIZE, T. H.; SIMONSICK, W. J.; AMSTER, I. J. Characterization of polyesters by matrix-assisted laser desorption/ionization and Fourier transform mass spectrometry. *Eur. J. Mass Spectrom.* **2003**, *9*, 473–486.
75. CANCILLA, M. T.; PENN, S. G.; CARROLL, J. A.; LEBRILLA, C. B. Coordination of alkali metals to oligosaccharides dictates fragmentation behavior in matrix assisted laser desorption ionization Fourier transform mass spectrometry. *J. Am. Chem. Soc.* **1996**, *118*, 6736–6745.
76. HORN, D. M.; PETERS, E. C.; KLOCK, H.; MEYERS, A.; BROCK, A. Improved protein identification using automated high mass measurement accuracy MALDI FT-ICR MS peptide mass fingerprinting. *Int. J. Mass Spectrom.* **2004**, *238*, 189–196.
77. STULTS, J. T. Minimizing peak coalescence: High resolution separation of isotope peaks in partially deamidated peptides by matrix assisted laser desorption/ionisation Fourier transform ion cyclotron resonance mass spectrometry. *Anal. Chem.* **1997**, *69*, 1815–1819.
78. PETERS, E. C.; BROCK, A.; HORN, D. M.; PHUNG, Q. T.; ERICSON, C.; SALOMON, A. R.; FICARRO, S. B.; BRILL, L. M. An automated LC-MALDI FT-ICR MS platform for high-throughput proteomics. *LC-GC Europe* **2002**, *15*, 423–428.
79. FENN, J. B.; MANN, M.; MENG, C. K.; WONG, S. F.; WHITEHOUSE, C. M. Electrospray ionization for mass-spectrometry of large biomolecules. *Science* **1989**, *246*, 64–71.
80. MENG, C. K.; MANN, M.; FENN, J. B. Of protons or proteins—A Beam's a Beam for a That. *Z. Phys. D—Atoms Molecules and Clusters* **1988**, *10*, 361–368.
81. YAMASHITA, M.; FENN, J. B. Electrospray ion-source—Another variation on the free-jet theme. *J. Phys. Chem.* **1984**, *88*, 4451–4459.
82. HOFSTADLER, S. A.; LAUDE, D. A. Trapping and detection of ions generated in a high magnetic field electrospray ionization Fourier transform ion cyclotron resonance mass spectrometer. *J. Am. Soc. Mass. Spectrom.* **1992**, *3*, 615–623.
83. SENKO, M. W.; HENDRICKSON, C. L.; EMMETT, M. R.; SHI, S. D. -H.; MARSHALL, A. G. External accumulation of ions for enhanced electrospray ionization Fourier transform ion cyclotron resonance mass spectrometry. *J. Am. Soc. Mass. Spectrom.* **1997**, *8*, 970–976.
84. BELOV, M. E.; RAKOV, V. S.; NIKOLAEV, E. N.; GOSHE, M. B.; ANDERSON, G. A.; SMITH, R. D. Initial implementation of external accumulation liquid chromatography/electrospray ionization Fourier transform ion cyclotron resonance with automated gain control. *Rapid Commun. Mass Spectrom.* **2003**, *17*, 627–636.
85. McIVER, R. T.; HUNTER, R. L.; BOWERS, W. D. Coupling a quadrupole mass-spectrometer and a Fourier-transform mass-spectrometer. *Int. J. Mass Spectrom. Ion Processes* **1985**, *64*, 67–77.
86. HAKANSSON, K.; CHALMERS, M. J.; QUINN, J. P.; McFARLAND, M. A.; HENDRICKSON, C. L.; MARSHALL, A. G. Combined electron capture and infrared multiphoton dissociation for multistage MS/MS in a Fourier transform ion cyclotron resonance mass spectrometer. *Anal. Chem.* **2003**, *75*, 3256–3262.
87. KELLEHER, N. L.; SENKO, M. W.; SIEGEL, M. M.; McLAFFERTY, F. W. Unit resolution mass spectra of 112 kDa molecules with 3 Da accuracy. *J. Am. Soc. Mass. Spectrom.* **1997**, *8*, 380–383.

88. Hughey, C. A.; Rodgers, R. P.; Marshall, A. G. Resolution of 11,000 compositionally distinct components in a single electrospray ionization Fourier transform ion cyclotron resonance mass spectrum of crude oil. *Anal. Chem.* **2002**, *74*, 4145–4149.
89. Brown, T. L.; Rice, J. A. Effect of experimental parameters on the ESI FT-ICR mass spectrum of fulvic acid. *Anal. Chem.* **2000**, *72*, 384–390.
90. Kujawinski, E. B.; Hatcher, P. G.; Freitas, M. A. High-resolution Fourier transform ion cyclotron resonance mass spectrometry of humic and fulvic acids: Improvements and Comparisons. *Anal. Chem.* **2002**, *74*, 413–419.
91. Stenson, A. C.; Landing, W. M.; Marshall, A. G.; Cooper, W. T. Ionization and fragmentation of humic substances in electrospray ionization Fourier transform-ion cyclotron resonance mass spectrometry. *Anal. Chem.* **2002**, *74*, 4397–4409.
92. Gao, H.; Leary, J.; Carroll, K. S.; Bertozzi, C. R.; Chen, H. Y. Noncovalent complexes of APS reductase from M-tuberculosis: Delineating a mechanistic model using ESI-FTICR MS. *J. Am. Soc. Mass. Spectrom.* **2007**, *18*, 167–178.
93. Johnson, K. A.; Shira, B. A.; Anderson, J. L.; Amster, I. J. Chemical and on-line electrochemical reduction of metalloproteins with high-resolution electrospray ionization mass spectrometry detection. *Anal. Chem.* **2001**, *73*, 803–808.
94. Heck, A. J. R.; de Koning, L. J.; Pinkse, F. A.; Nibbering, N. M. M. Mass-specific selection of ions in Fourier-transform ion cyclotron resonance mass spectrometry. Unintentional off-resonance cyclotron excitation of selected ions. *Rapid Commun. Mass Spectrom.* **1991**, *5*, 406–414.
95. Marshall, A. G.; Wang, T. C. L.; Ricca, T. L. Tailored excitation for Fourier transform ion cyclotron mass spectrometry. *J. Am. Chem. Soc.* **1985**, *107*, 7893–7897.
96. Haebel, S.; Gaumann, T. Difference measurements and shaped wave-forms applied to Hadamard-transform Ft/Icr. *Int. J. Mass Spectrom. Ion Processes* **1995**, *144*, 139–166.
97. Haebel, S.; Walser, M. E.; Gaumann, T. High front-end resolution collision-induced dissociation in Fourier transform ion cyclotron resonance mass spectrometry. *Int. J. Mass Spectrom. Ion Processes* **1995**, *151*, 97–115.
98. Hofstadler, S. A.; Wahl, J. H.; Bakhtiar, R.; Anderson, G. A.; Bruce, J. E.; Smith, R. D. Capillary electrophoresis Fourier transform ion cyclotron resonance mass spectrometry with sustained off-resonance irradiation for the characterization of protein and peptide mixtures. *J. Am. Soc. Mass. Spectrom.* **1994**, *5*, 894–899.
99. Bomse, D. S.; Woodin, R. L.; Beauchamp, J. L. Molecular activation with low-intensity Cw infrared-laser radiation—Multi-photon dissociation of ions derived from diethyl-ether. *J. Am. Chem. Soc.* **1979**, *101*, 5503–5512.
100. Schnier, P. D.; Price, W. D.; Jockusch, R. A.; Williams, E. R. Blackbody infrared radiative dissociation of bradykinin and its analogues: Energetics, dynamics, and evidence for salt-bridge structures in the gas phase. *J. Am. Chem. Soc.* **1996**, *118*, 7178–7189.
101. Hofstadler, S. A.; Sannes-Lowery, K. A.; Griffey, R. H. *m*/*z*-selective infrared multiphoton dissociation in a Penning trap using sidekick trapping and an rf-tickle pulse. *Rapid Commun. Mass Spectrom.* **2001**, *15*, 945–951.
102. Bowers, W. D.; Delbert, S. S.; Hunter, R. L.; McIver, R. T. Fragmentation of oligopeptide ions using ultraviolet-laser radiation and Fourier-transform mass-spectrometry. *J. Am. Chem. Soc.* **1984**, *106*, 7288–7289.
103. Bowers, W. D.; Delbert, S. S.; McIver, R. T. Consecutive laser-induced photodissociation as a probe of ion structure. *Anal. Chem.* **1986**, *58*, 969–972.
104. Fung, Y. M. E.; Kjeldsen, F.; Silivra, O. A.; Chan, T. W. D.; Zubarev, R. A. Facile disulfide bond cleavage in gaseous peptide and protein cations by ultraviolet photodissociation at 157 nm. *Angew. Chem. Int. Ed.* **2005**, *44*, 6399–6403.
105. Little, D. P.; Speir, J. P.; Senko, M. W.; Oconnor, P. B.; McLafferty, F. W. Infrared multiphoton dissociation of large multiply-charged ions for biomolecule sequencing. *Anal. Chem.* **1994**, *66*, 2809–2815.
106. Senko, M. W.; Speir, J. P.; McLafferty, F. W. Collisional activation of large multiply-charged ions using Fourier-transform mass-spectrometry. *Anal. Chem.* **1994**, *66*, 2801–2808.
107. Tsybin, Y. O.; Håkansson, P.; Budnik, B. A.; Haselmann, K. F.; Kjeldsen, F.; Gorshkov, M.; Zubarev, R. A. Improved low-energy electron injection systems for high rate electron capture dissociation in Fourier transform ion cyclotron resonance mass spectrometry. *Rapid Commun. Mass Spectrom.* **2001**, *15*, 1849–1854.
108. Gorshkov, M. V.; Masselon, C. D.; Nikolaev, E. N.; Udseth, H. R.; Pasa-Tolic, L.; Smith, R. D. Considerations for electron capture dissociation efficiency in FTICR mass spectrometry. *Int. J. Mass Spectrom.* **2004**, *234*, 131.
109. Mormann, M.; Peter-Katalinic, J. Improvement of electron capture efficiency by resonant excitation. *Rapid Commun. Mass Spectrom.* **2003**, *17*, 2208–2214.
110. McFarland, M. A.; Chalmers, M. J.; Quinn, J. P.; Hendrickson, C. L.; Marshall, A. G. Evaluation and optimization of electron capture dissociation efficiency in Fourier transform ion cyclotron resonance mass spectrometry. *J. Am. Soc. Mass. Spectrom.* **2005**, *16*, 1060–1066.
111. Haselmann, K. F.; Jorgensen, T. J. D.; Budnik, B. A.; Jensen, F.; Zubarev, R. A. Electron capture dissociation

of weakly bound polypeptide polycationic complexes. *Rapid Commun. Mass Spectrom.* **2002**, *16*, 2260–2265.

112. Turecek, F. N-C-alpha bond dissociation energies and kinetics in amide and peptide radicals. Is the dissociation a non-ergodic process? *J. Am. Chem. Soc.* **2003**, *125*, 5954–5963.
113. Hakansson, K.; Cooper, H. J.; Emmett, M. R.; Costello, C. E.; Marshall, A. G.; Nilsson, C. L. Electron capture dissociation and infrared multiphoton dissociation MS/MS of an *N*-glycosylated tryptic peptide to yield complementary sequence information. *Anal. Chem.* **2001**, *73*, 4530–4536.
114. Kelleher, N. L.; Zubarev, R. A.; Bush, K.; Furie, B.; Furie, B. C.; McLafferty, F. W.; Walsh, C. T. Localization of labile posttranslational modifications by electron capture dissociation: The case of gamma—Carboxyglutamic acid. *Anal. Chem.* **1999**, *71*, 4250–4253.
115. Mirgorodskaya, E.; Roepstorff, P.; Zubarev, R. A. Localization of *O*-glycosylation sites in peptides by electron capture dissociation in a Fourier transform mass spectrometer. *Anal. Chem.* **1999**, *71*, 4431–4436.
116. Niiranen, H.; Budnik, B. A.; Zubarev, R. A.; Auriola, S.; Lapinjoki, S. High-performance liquid chromatography–mass spectrometry and electron-capture dissociation tandem mass spectrometry of osteocalcin—Determination of gamma-carboxyglutamic acid residues. *J. Chromatogr. A* **2002**, *962*, 95–103.
117. Shi, S. D. H.; Hemling, M. E.; Carr, S. A.; Horn, D. M.; Lindh, I.; McLafferty, F. W. Phosphopeptide/phosphoprotein mapping by electron capture dissociation mass spectrometry. *Anal. Chem.* **2001**, *73*, 19–22.
118. Stensballe, A.; Jensen, O. N.; Olsen, J. V.; Haselmann, K. F.; Zubarev, R. A. Electron capture dissociation of singly and multiply phosphorylated peptides. *Rapid Commun. Mass Spectrom.* **2000**, *14*, 1793–1800.
119. Kjeldsen, F.; Haselmann, K. F.; Budnik, B. A.; Jensen, F.; Zubarev, R. A. Dissociative capture of hot (3-13 eV) electrons by polypeptide polycations: An efficient process accompanied by secondary fragmentation. *Chem. Phys. Lett.* **2002**, *356*, 201–206.
120. Horn, D. M.; Ge, Y.; McLafferty, F. W. Activated ion electron capture dissociation for mass spectral sequencing of larger (42 kDa) proteins. *Anal. Chem.* **2000**, *72*, 4778–4784.
121. Zubarev, R. A. Reactions of polypeptide ions with electrons in the gas phase. *Mass Spectrom. Rev.* **2003**, *22*, 57.
122. Yang, J.; Mo, J.; Adamson, J. T.; Hakansson, K. Characterization of oligodeoxynucleotides by electron detachment dissociation Fourier transform ion cyclotron resonance mass spectrometry. *Anal. Chem.* **2005**, *77*, 1876–1882.
123. Wolff, J. J.; Amster, I. J.; Chi, L.; Linhardt, R. J. Electron detachment dissociation of glycosaminoglycan tetrasaccharides. *J. Am. Soc. Mass. Spectrom.* **2007**, *18*, 234–244.
124. Wolff, J. J.; Chi, L. L.; Linhardt, R. J.; Amster, I. J. Distinguishing glucuronic from iduronic acid in glycosaminoglycan tetrasaccharides by using electron detachment dissociation. *Anal. Chem.* **2007**, *79*, 2015–2022.
125. Wolff, J. J.; Laremore, T. N.; Busch, A. M.; Linhardt, R. J.; Amster, I. J. Electron detachment dissociation of dermatan sulfate oligosaccharides. *J. Am. Soc. Mass. Spectrom.* **2008**, *19*, 294–304.
126. Cody, R. B.; Freiser, B. S. Electron impact excitation of ions in Fourier transform mass spectrometry. *Anal. Chem.* **1987**, *59*, 1054–1056.
127. Wang, B. -H.; McLafferty, F. W. Electron impact excitation of ions from larger organic molecules. *Org. Mass Spectrom.* **1990**, *25*, 554–556.
128. Yoo, H. J.; Liu, H.; Hakansson, K. Infrared multiphoton dissociation and electron-induced dissociation as alternative MS/MS strategies for metabolite identification. *Anal. Chem.* **2007**, *79*, 7858–7866.
129. Lioe, H.; O'Hair, R. Comparison of collision-induced dissociation and electron-induced dissociation of singly protonated aromatic amino acids, cystine and related simple peptides using a hybrid linear ion trap–FT-ICR mass spectrometer. *Analytical and Bioanalytical Chemistry* **2007**, *389*, 1429–1437.
130. Kaplan, D. A.; Hartmer, R.; Speir, J. P.; Stoermer, C.; Gumerov, D.; Easterling, M. L.; Brekenfeld, A.; Kim, T.; Laukien, F.; Park, M. A. Electron transfer dissociation in the hexapole collision cell of a hybrid quadrupole–hexapole Fourier transform ion cyclotron resonance mass spectrometer. *Rapid Commun. Mass Spectrom.* **2008**, *22*, 271–278.
131. Bogdanov, B.; Smith, R. D. Proteomics by FTICR mass spectrometry: Top down and bottom up. *Mass Spectrom. Rev.* **2005**, *24*, 168–200.
132. Ge, Y.; Lawhorn, B. G.; ElNaggar, M.; Strauss, E.; Park, J. H.; Begley, T. P.; McLafferty, F. W. Top down characterization of larger proteins (45 kDa) by electron capture dissociation mass spectrometry. *J. Am. Chem. Soc.* **2002**, *124*, 672–678.
133. Pesavento, J. J.; Kim, Y. B.; Taylor, G. K.; Kelleher, N. L. Shotgun annotation of histone modifications: A new approach for streamlined characterization of proteins by top down mass spectrometry. *J. Am. Chem. Soc.* **2004**, *126*, 3386–3387.
134. Sze, S. K.; Ge, Y.; Oh, H.; McLafferty, F. W. Top-down mass spectrometry of a 29-kDa protein for characterization of any posttranslational modification to within one residue. *Proc. Natl. Acad. Sci.* **2002**, 1774–1779.
135. Kjeldsen, F.; Haselmann, K. F.; Budnik, B. A.; Sorensen, E. S.; Zubarev, R. A. Complete characterization of posttranslational modification sites in the bovine milk protein PP3 by tandem mass

spectrometry with electron capture dissociation as the last stage. *Anal. Chem.* **2003**, *75*, 2355–2361.

136. Adamson, J. T.; Hakansson, K. Electron detachment dissociation of neutral and sialylated oligosaccharides. *J. Am. Soc. Mass. Spectrom.* **2007**, *18*, 2162–2172.
137. Penn, S. G.; Cancilla, M. T.; Lebrilla, C. B. Collision-induced dissociation of branched oligosaccharide ions with analysis and calculation of relative dissociation thresholds. *Anal. Chem.* **1996**, *68*, 2331–2339.
138. Bordoli, R. S.; Howes, K.; Vickers, R. G.; Bateman, R. H.; Harvey, D. J. Matrix-assisted laser-desorption mass-spectrometry on a magnetic-sector instrument fitted with an array detector. *Rapid Commun. Mass Spectrom.* **1994**, *8*, 585–589.
139. Harvey, D. J.; Rudd, P. M.; Bateman, R. H.; Bordoli, R. S.; Howes, K.; Hoyes, J. B.; Vickers, R. G. Examination of complex oligosaccharides by matrix-assisted laser-desorption ionization mass-spectrometry on time-of-flight and magnetic-sector instruments. *Org. Mass Spectrom.* **1994**, *29*, 753–766.
140. Harvey, D. J. Matrix-assisted laser-desorption ionization mass-spectrometry of sphingo-lipids and glycosphingo-lipids. *J. Mass Spectrom.* **1995**, *30*, 1311–1324.
141. Harvey, D. J. Matrix-assisted laser-desorption ionization mass-spectrometry of phospholipids. *J. Mass Spectrom.* **1995**, *30*, 1333–1346.
142. Bateman, R. H.; Green, M. R.; Scott, G.; Clayton, E. A combined magnetic sector–time-of-flight mass-spectrometer for structural determination studies by tandem mass-spectrometry. *Rapid Commun. Mass Spectrom.* **1995**, *9*, 1227–1233.
143. Strobel, F. H.; Russell, D. H. In *Time-of-Flight Mass Spectrometry*, ACS Symposium Series 549, 1993, pp. 73–94.
144. Strobel, F. H.; Solouki, T.; White, M. A.; Russell, D. H. Detection of femtomole and sub-femtomole levels of peptides by tandem magnetic-sector reflectron time-of-flight mass-spectrometry and matrix-assisted laser desorption ionization. *J. Am. Soc. Mass. Spectrom.* **1991**, *2*, 91–94.
145. Gallagher, R. T.; Chapman, J. R.; Mann, M. Design and performance of an electrospray ionization source for a doubly-focusing magnetic-sector mass-spectrometer. *Rapid Commun. Mass Spectrom.* **1990**, *4*, 369–372.
146. Dobberstein, P.; Schroeder, E. Accurate mass determination of a high-molecular-weight protein using electrospray-ionization with a magnetic-sector instrument. *Rapid Commun. Mass Spectrom.* **1993**, *7*, 861–864.
147. Meng, C. K.; McEwen, C. N.; Larsen, B. S. Electrospray ionization on a high-performance magnetic-sector mass-spectrometer. *Rapid Commun. Mass Spectrom.* **1990**, *4*, 147–150.
148. Meng, C. K.; McEwen, C. N.; Larsen, B. S. Peptide sequencing with electrospray ionization on a magnetic-sector mass-spectrometer. *Rapid Commun. Mass Spectrom.* **1990**, *4*, 151–155.
149. Larsen, B. S.; McEwen, C. N. An Electrospray ion-source for magnetic-sector mass spectrometers. *J. Am. Soc. Mass. Spectrom.* **1991**, *2*, 205–211.
150. Cody, R. B.; Tamura, J.; Musselman, B. D. Electrospray ionization magnetic-sector mass-spectrometry—calibration, resolution, and accurate mass measurements. *Anal. Chem.* **1992**, *64*, 1561–1570.
151. Chapman, J. R.; Gallagher, R. T.; Mann, M. Some biochemical applications of electrospray magnetic-sector mass-spectrometry. *Biochem. Soc. Trans.* **1991**, *19*, 940–943.
152. Cody, R. B.; Tamura, J.; Finch, J. W.; Musselman, B. D. Improved detection limits for electrospray-ionization on a magnetic-sector mass-spectrometer by using an array detector. *J. Am. Soc. Mass. Spectrom.* **1994**, *5*, 194–200.
153. McEwen, C. N.; Larsen, B. S. Accurate mass measurement of proteins using electrospray ionization on a magnetic-sector instrument. *Rapid Commun. Mass Spectrom.* **1992**, *6*, 173–178.
154. Wada, Y.; Tamura, J.; Musselman, B. D.; Kassel, D. B.; Sakurai, T.; Matsuo, T. Electrospray ionization mass-spectra of hemoglobin and transferrin by a magnetic-sector mass-spectrometer—Comparison with theoretical isotopic distributions. *Rapid Commun. Mass Spectrom.* **1992**, *6*, 9–13.
155. Loo, J. A.; Holsworth, D. D.; Rootbernstein, R. S. Use of electrospray-ionization mass-spectrometry to probe antisense peptide interactions. *Biol. Mass Spectrom.* **1994**, *23*, 6–12.
156. Loo, J. A. Observation of large subunit protein complexes by electrospray-ionization mass-spectrometry. *J. Mass Spectrom.* **1995**, *30*, 180–183.
157. Loo, J. A.; Loo, R. R. O.; Andrews, P. C. Primary to quaternary protein-structure determination with electrospray-ionization and magnetic-sector mass-spectrometry. *Org. Mass Spectrom.* **1993**, *28*, 1640–1649.
158. Ghardashkhani, S.; Gustavsson, M. L.; Breimer, M. E.; Larson, G.; Samuelsson, B. E. Negative electrospray-ionization mass-spectrometry analysis of gangliosides, sulfatides and cholesterol 3-sulfate. *Rapid Commun. Mass Spectrom.* **1995**, *9*, 491–494.
159. Perkins, J. R.; Tomer, K. B. Capillary electrophoresis electrospray mass-spectrometry using a high-performance magnetic-sector mass-spectrometer. *Anal. Chem.* **1994**, *66*, 2835–2840.
160. Loo, J. A.; Pesch, R. Sensitive and selective determination of proteins with electrospray-ionization magnetic-sector mass-spectrometry and array detection. *Anal. Chem.* **1994**, *66*, 3659–3663.
161. Amster, I. J. Fourier transform mass spectrometry. *J. Mass Spectrom.* **1996**, *31*, 1325–1337.

Chapter 12

Ion Mobility–Mass Spectrometry

John A. McLean, [*] ***J. Albert Schultz,*** [†] ***and Amina S. Woods*** [‡]

[*] *Department of Chemistry, Vanderbilt Institute of Chemical Biology and Vanderbilt Institute of Integrative Biosystems Research and Education, Vanderbilt University, Nashville, Tennessee*

[†] *Ionwerks Inc., Houston, Texas*

[‡] *Cellular Neurobiology, Intramural Research Program, National Institute on Drug Abuse, National Institutes of Health, Baltimore, Maryland*

Electrospray and MALDI Mass Spectrometry: Fundamentals, Instrumentation, Practicalities, and Biological Applications, Second Edition, Edited by Richard B. Cole

12.1 MULTIDIMENSIONAL SEPARATIONS BY ION MOBILITY–MASS SPECTROMETRY

Recent advances in mass spectrometry (MS) instrumentation, techniques, and data interrogation strategies have revolutionized the practice of protein separation, identification, and characterization. The primary means for increasing the information derived from contemporary proteomics experiments is through increasing the dimensionality of the data that are obtained. It is now routine to interface various liquid chromatography separations prior to MS, or MS/MS, detection for improved protein identification from complex samples (e.g., whole-cell lysates). In parallel with developments in LC-MS, recent and considerable progress has been made in the development of post-ionization gas-phase separations on the basis of ion mobility–MS (IM-MS). In contrast with LC-MS separations, separations by IM-MS are extraordinarily rapid (μs-ms versus min-h) and can be further combined with additional pre-ionization separation techniques. Moreover, IM separations provide enhanced data dimensionality in the form of direct analyte structural information, which is largely insensitive to the complexity of the sample from which the analyte is derived.

Ion mobility provides separations on the basis of the ion–neutral collision cross section, or the integral of analyte ion collisions ($\sim 10^5$ to 10^8) with neutral drift gas atoms or molecules. Briefly, ions are injected into the IM drift cell where they migrate under the influence of a weak electrostatic field and their progress is impeded by the neutral drift gas by an amount depending on the collision cross section, or apparent surface area, of the ion. Analyte structural information is inferred by comparing empirical IM-derived cross sections with complementary results from molecular modeling—that is, which ion conformation(s) correlate with the empirical results.

A primary aim of this chapter is to describe the motivation and information content that can be obtained by combining IM with MS. In this context, the IM drift cell provides great experimental flexibility by treating it as a compartmentalized tool that can be utilized in a variety of instrumental arrangements depending on the type of data desired. The instrumental challenges associated with combining IM with MS are discussed, because it is a marriage of apparently paradoxical traits—that is, ion–neutral collisions are promoted in IM separations, while they are undesirable in the MS dimension (pressure ranges of approximately 1–760 torr and approximately $\ll 10^{-7}$ torr in IM and MS, respectively). This is potentially one reason why there are relatively few groups involved in IM-MS development and few commercially available IM-MS instruments; however, this situation is quickly changing. Various forms of IM-MS instruments are now becoming commercially available; and due to the analytical advantages of combining IM with MS, they will likely become more commonplace tools in structural MS and biological laboratories.

To fully appreciate the capabilities of IM-MS, it is necessary to understand the principles of each dimension separately and the distinctive synergistic qualities resulting from their combination. To avoid duplication with other works in this edition, only the unique aspects of IM and the attributes of coupling IM-MS are described. Furthermore, the present chapter is not intended to be a comprehensive review of the field, but rather a survey of key points related to the practice of measuring ion structure, inherent advantages of the multidimensional nature of the data, and instrumental parameters and considerations in the construction of IM-MS devices. As such, selected examples of data and IM-MS arrangements have been chosen for illustrative purposes, rather than for a comprehensive description of the area. The target audience is for individuals interested in the present capabilities and limitations of IM-MS instrumentation, or to evaluate the potential of IM-MS in their research endeavors. Although IM has been and can be used for a wide variety of analytes, IM and IM-MS is discussed with a specific focus on biomolecular separations. A brief historical background, conceptual framework of the motivation to perform such separations, and practical aspects of determining structural analyte information from IM-MS measurements is presented in Section 12.1. Section 12.2 describes instrumental considerations and several instrumental arrangements that are used for biomolecular studies. Finally, an outlook to promising new IM-MS approaches on the horizon is the topic of Section 12.3. A number of outstanding texts are devoted to the detailed theoretical treatment of gas-phase ion–molecule collisions, different instrumental arrangements for performing IM, and applications of IM.[1–5] Furthermore, fundamental and applied aspects of IM and IM-MS are reviewed elsewhere.[6–15]

12.1.1 The Modern Renaissance of Ion Mobility

Gas-phase ion separations on the basis of migration and diffusion of ions through a neutral gas have existed for over a century, with quantitative studies of ionized gases being initiated by the discovery of X rays by Wilhelm Röntgen in 1895,[16] for which he was awarded the first Nobel Prize in Physics in 1901.[17] A detailed account on the historical development of ion mobility spectrometry and milestones in the progress of gas-phase IM techniques and small-molecule applications is presented elsewhere.[6,18] For many years, IMS separations were used for the fundamental study of atomic and small molecular ions in plasma physics. It is important to note that over the historical span of IMS development, the technique has been referred to by several names including plasma chromatography, ion chromatography, and the presently accepted (and more accurate) term ion mobility spectrometry (IMS). In the 1960s and 1970s, a transition occurred from using IMS primarily as a fundamental physics research tool, to using it as a separations device for analytical and physical chemistry applications. Notably, the early 1960s also marked the first reports of combining IMS with MS: first, with quadrupole mass spectrometry[19] and subsequently with time-of-flight mass spectrometry (TOFMS[20]).

Although the primary applications of stand alone IMS are for drug, explosives, and chemical warfare agent detection, the utility of IMS for the separation of multiply charged protein ions by electrospray ionization (ESI)-IMS was demonstrated in the mid-1980s to early 1990s.[21–23] A particularly important milestone in the development of IMS for biological applications occurred in the mid-1990s when Bowers and colleagues[24,25] first reported IM-MS structural studies for peptides, and studies on structural analysis of intact proteins were reported by Jarrold, Clemmer, and co-workers.[26,27] In large part, this work and that of others in the IMS and IM-MS fields denotes a period of rapid and expanding work over the past decade resulting in the development of biological IM-MS as a multidimensional separations tool for complex biological samples and for measuring fundamental biomolecular structural parameters in biophysics.

12.1.2 Matching Data Dimensionality with Desired Information

12.1.2.1 Information Derived by the Ordering of Separation Dimensions

The combination of IM-MS provides enhanced data dimensionality because it provides both ion structural information on the basis of ion–neutral collision cross section (or apparent surface area of the ion) and mass-to-charge (*m/z*) information determined by accurate mass measurement in MS. Note that in contrast with highly orthogonal multidimensional separations,[28] in IM-MS the two separation dimensions exhibit a high degree of correlation, which can be simultaneously a challenge and a significant advantage. The underlying reason for correlation can be rationalized by dimensional analysis. Given the relatively few types of atoms involved in the composition of most biomolecules (C, H, O, N, P, and S), the mass and volume of a molecule are largely related by a narrow range of density—that is, biomolecular mass scales as volume or length cubed. On the other hand, IM collision cross section, or surface area, scales as length squared. Nevertheless, the average density for different types of biomolecules (e.g., peptides versus lipids) can be quite different as discussed in Section 12.1.2.2, which provides great utility in the separation of complex samples. The apparent high correlation of IM-MS separation dimensions within a particular type of biomolecule results in the 2D region of separation space (referred to as "conformation" space) exhibiting relatively low peak capacity ($\sim 10^3$–10^4) when compared with more orthogonal multidimensional separation techniques.[7,29,30,31] However, this is offset by the fast rate of 2D IM-MS separations which yields an exceptional peak capacity production rate ($\sim 1 \times 10^6$ peaks s^{-1}).

An advantage of the apparent correlation between IM and MS is that it is relatively straightforward to utilize each dimension in different configurations, depending on the aims of a particular experiment. A conceptual diagram of how different separation dimensions are coupled in contemporary IM-MS instrumentation is illustrated in Figure 12.1. In the most common arrangement (Figure 12.1a), an ion source (typically ESI or MALDI) is used to inject analyte ions directly into the IM drift cell which is fluidly coupled to the mass analyzer for selection of a particular *m/z* (e.g., with a quadrupole MS) or for complete mass analysis (e.g., with an orthogonal TOFMS). For biomolecular IM-MS, the orthogonal TOFMS is typically the mass analyzer of choice, because the timescale of TOFMS separation (ca. tens of microseconds) is well-suited for sampling of the ion distribution eluting from the IM drift cell (ca. hundreds of microseconds to milliseconds). Note that for most typical multidimensional separation techniques, slower separation dimensions precede those of faster separation dimensions for reasons of instrumental and data acquisition practicability. Although, for convenience, IM separations are commonly performed at laboratory temperature, one significant advantage of varying the system temperature (e.g., 80–800 K, Figure 12.1b) is the ability to determine biophysical thermodynamic and kinetic data for the structural interconversion of analyte conformations.

A further advantage of the high correlation in IM and MS separations is the ability to obtain biomolecular structure information in the MS dimension by performing so-called

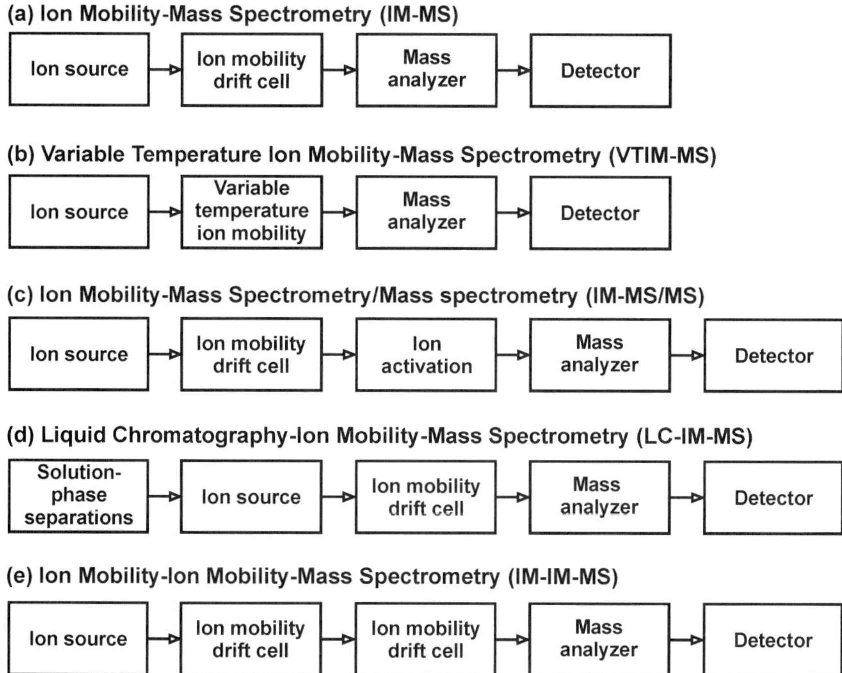

Figure 12.1. Diagram illustrating the versatility of IM-MS instrumentation. In the most common form of IM-MS, (**a**) the instrument consists of an ion source, a drift cell, a mass analyzer, and a detector. Additional information can be obtained by expanding on this framework, including: (**b**) VTIM-MS, for the determination of structurally resolved thermodynamic and kinetic information; (**c**) IM-MS/MS, for acquisition of nearly simultaneous precursor and fragment ion spectra; (**d**) LC-IM-MS, for three-dimensional separations of complex samples; and (**e**) IM-IM-MS, for further dimensionality in the structural information obtained.

IM-MS/MS (Figure 12.1c). In this analysis mode, the IM dimension is analogous to MS^1 in traditional tandem MS experiments; however, analyte ions are dispersed in time on the basis of their ion–neutral collision cross section rather than *m/z*. Ion activation following IM separation allows for both precursor and fragment ions to be determined in the MS dimension (i.e., MS^2). Note that in contrast with conventional scanning MS/MS approaches, where a particular *m/z* analyte is selected for fragmentation, IM-MS/MS provides a "multiplex advantage" in that nearly all ions can be selected for fragmentation, because the fragment ions are correlated to the precursor ion from which they are derived by the elution time from the IM drift cell. Furthermore, additional stages of pre-ionization separations can be coupled with IM-MS [e.g., high-performance liquid chromatography (HPLC), capillary electrophoresis (CE), etc. (Figure 12.1d)], or additional stages of IM, MS, MS/MS, or IM/IM can be performed, depending on the specific goals of the experiment (Figure 12.1e)]. Importantly, each of these experimental modes is not exclusive, and it is conceptually straightforward (although in some cases experimentally challenging) to perform combinations thereof.

12.1.2.2 Separation of Complex Biological Mixtures on the Basis of Molecular Class

Typical proteomic-scale experiments require one or more separation steps prior to MS analysis to remove concomitant species that result in ion suppression effects or chemical noise, to enrich the sample in the analytes of interest (i.e., peptides and proteins), and to simplify the mass spectra for higher confidence-level interpretation. Analytically, these aims correspond to improving the limits of detection, selectivity, and identification/quantitation, respectively. IM-MS also provides these attributes by providing pre-MS analyte structural separation, but at much faster separation rates in comparison to condensed-phase separations. Figure 12.2(i) illustrates a typical 2D plot of IM-MS conformation space for peptides determined from a proteolytic digestion of bovine hemoglobin.[32] Note that the 2D plot is a projection of 3D data, in that the relative abundance of specific signals is indicated in false coloring/shading of particular signals, similar to the representation of spectra obtained in 2D nuclear magnetic resonance techniques.

The IMS electropherogram in this example, or what would be observed by placing the detector at the end of the IM drift cell, is shown in panel (ii). Conversely, the mass spectrum that would be obtained in the absence of IM is illustrated in panel (iii). By integrating the mass spectra over small regions of IM-separation space (e.g., 1250–1350 μs), the integrated mass spectrum exhibits only those signals corresponding to species of a particular arrival time distribution (ATD) at the detector, or collision cross section, (Figure 12.2(v)]. By comparison of Figures 12.2(v) and 2(iii), the attenuation of chemical noise is demonstrated

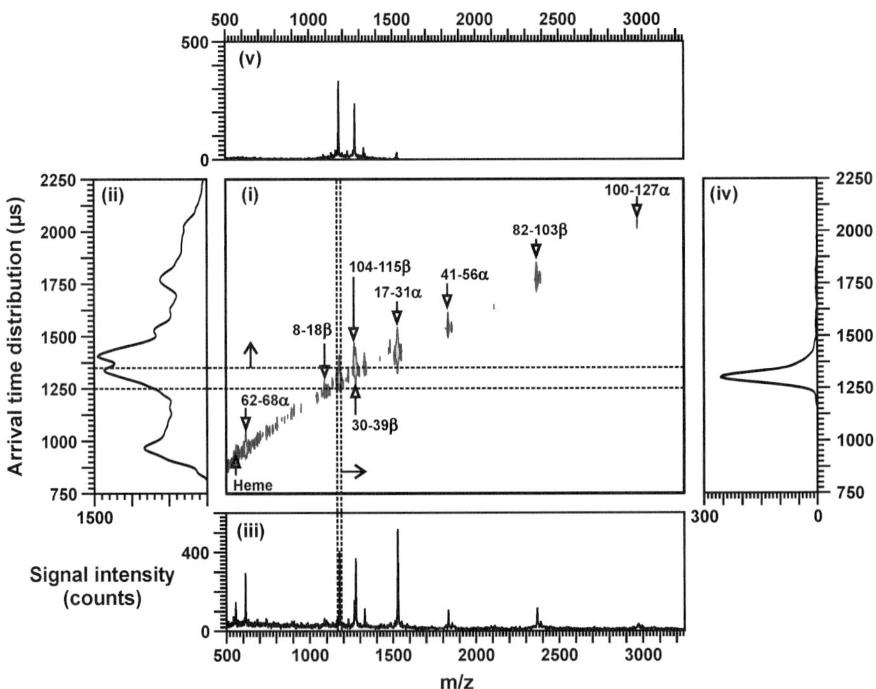

Figure 12.2. A 2D MALDI-IM-MS plot of conformation space illustrating the separation of peptides obtained from a tryptic digest of bovine hemoglobin (**i**). (**ii**) The ion mobility ATD integrated over all *m/z* space. (**iii**) The mass spectrum obtained by integrating over all ATD space. (**iv**) The ATD integrated over the *m/z* range of 1175–1185 (corresponding to the peptide VVAGVANALAHR, 132–143*β*), and (**v**) the mass spectrum obtained by integrating over the ATD of 1250–1350 μs (regions outlined by dashed-lines). MALDI was performed using *α*-cyano-4-hydroxycinnamic acid matrix with a frequency-tripled Nd:YAG laser (355 nm) operated at a repetition rate of 100 Hz. The IM drift cell was operated at 85 V cm^{-1} $torr^{-1}$ with argon gas. (Adapted from Ref. 32 with permission.)

by the reduced baseline and number of m/z signals. Complementary to this observation is the deconvolution of the IM profile by integrating over a narrow m/z range as indicated in Figure 12.2(iv), which yields a baseline-resolved IM profile for the singly protonated peptide.

In the analysis of complex samples—that is, those expected to contain analytes encompassing multiple molecular classes—it is empirically known that different molecular classes exhibit significant differences in their gas-phase packing efficiencies (i.e., oligonucleotides > carbohydrates > peptides > lipids; Figure 12.3).[33,34] Conceptually, this reflects the inherent degrees of freedom that the different biopolymers possess and is subsequently reflected in the overall preferred gas-phase conformation, or predicted collision cross section-m/z correlation that is generally observed. The separation of molecular classes in Figure 12.3 is illustrated by the observation of three distinct IM-MS correlations, for a lipid, peptide, oligonucleotide and their in-source decay fragments, respectively. The general trend in IM-MS correlation for molecular classes has been demonstrated for a wide variety of samples and complex mixtures, and it is noteworthy that a predicted IM-MS correlation is largely invariant for specific experimental parameters. Thus, the qualitative assignment of a particular signal to a molecular class is straightforward in the analysis of a complex biological sample. In proteomic analyses, integrating only the signals corresponding to peptides results in reduced chemical noise and significantly enhanced confidence levels in protein IDs obtained by database searching (i.e., elimination of concomitant nonpeptidic signals), relative to MS-only techniques.[35] Furthermore, the IM-MS separation of chemical noise results in an improved dynamic range for

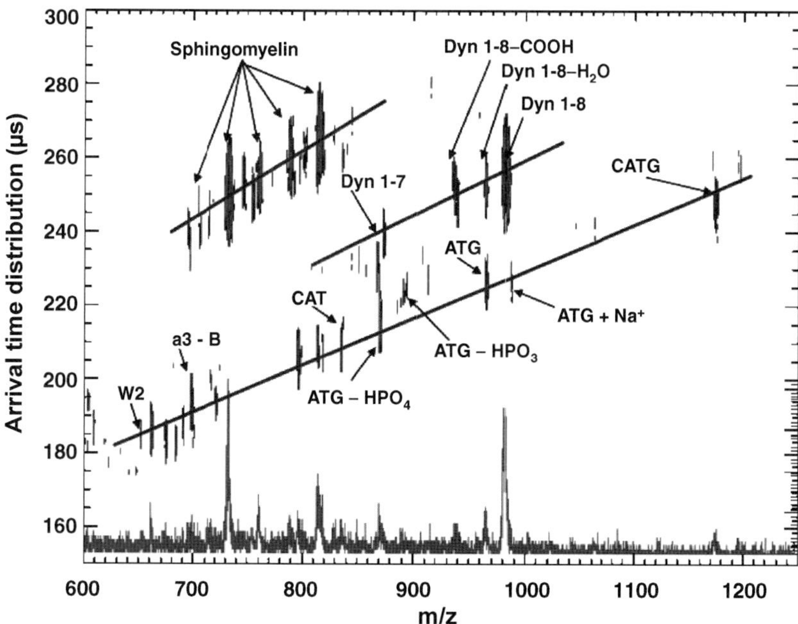

Figure 12.3. A 2D plot of MALDI-IM-MS conformation space for a short oligonucleotide (CATG), a peptide (dynorphin 1–8, YGGLFRRI), a lipid (sphingomyelin), and their respective in-source decay fragment ions. Lines are overlaid on signals arising for particular molecular classes for visualization purposes. The gray spectrum (**bottom**) illustrates the mass spectrum obtained by integrating over the entire IM dimension. (Adapted with permission from *Anal. Chem.* **2004**, *76*, 2187–2195. Copyright 2004 American Chemical Society.)

low-abundance proteins (including post-translationally modified ones), which are often the proteins of biological interest (e.g., signaling MAP kinases, ubiquinated proteins, sumoylated proteins, etc.).[35] Additional conformational differences within a particular molecular class (e.g., peptides/proteins), about the predicted correlation, are also observed owing to conformational isomers arising from peptide/protein secondary and tertiary structure,[36–40] intramolecular charge solvation,[25,41–43] small molecule–peptide or noncovalent complexes (i.e., quaternary structure),[39,44–48] and time-dependent structural reordering of the analyte ion.[49,50]

12.1.3 Two-Dimensional Measurements in Conformation Space

Several excellent texts and reviews describe the theory of IM and the derivation of ion–neutral collision cross-section measurements from IM profiles using the kinetic theory of gases.[4,51,52] This section summarizes several of the key equations and practical considerations for determining ion–neutral collision cross sections in uniform electrostatic-field IM instrumentation.

12.1.3.1 Transforming Drift Time to Collision Cross Section

The separation of ions in a weak electrostatic field (E) is measured as the ion drift velocity (v_d) and is related by the proportionality constant, K, which is the mobility of the ion in a particular neutral gas:

$$v_d = KE \tag{12.1}$$

Generally speaking, the drift cell is of a fixed length (L), and therefore the velocity of the ion packet is determined by measuring the drift time (t_d) of the packet across the drift cell. In practice, however, the parameter that is physically measured is the arrival time distribution (t_{ATD}) of the ion packet at the detector, which is the sum of both the drift time (t_d) of the ion packet through the IM cell and time the ion packet resides in other regions of the instrument (i.e., in the ion source, ion optics, and MS regions, etc.). The amount of time the ion packet spends outside of the drift cell can be empirically determined by performing the IM separation at several different electrostatic-field strengths [i.e., changing the potential (V) applied across the length of the drift cell]. As illustrated in Figure 12.4, the t_{ATD} for each of the separations is determined and plotted as a function of the inverse of the IM separation voltage. Provided that the separations are performed using sufficiently weak electrostatic fields (where K remains constant), the points can be fitted by a linear regression where the y intercept corresponds to the time the ions reside outside of the drift cell (i.e., at the limit of infinitely fast ion velocities across the IM cell or $t_d = 0$). By subtracting this time from t_{ATD}, the measured times are corrected and represent t_d across the IM cell. Note that for the most accurate results in IM-MS, the drift time correction (t_{corr}) should be evaluated for each component in the IM profile. The motivation for evaluating individual drift time corrections arises from additional ion–neutral collisions in the differential pumping regions at the entrance and/or exit of the IM drift cell. In these regions the gas dynamics typically transition from viscous to molecular flow—for example, at the exit aperture of the drift cell at 2–10 torr to the high vacuum ($\sim 10^{-8}$ torr) of the mass spectrometer, respectively.

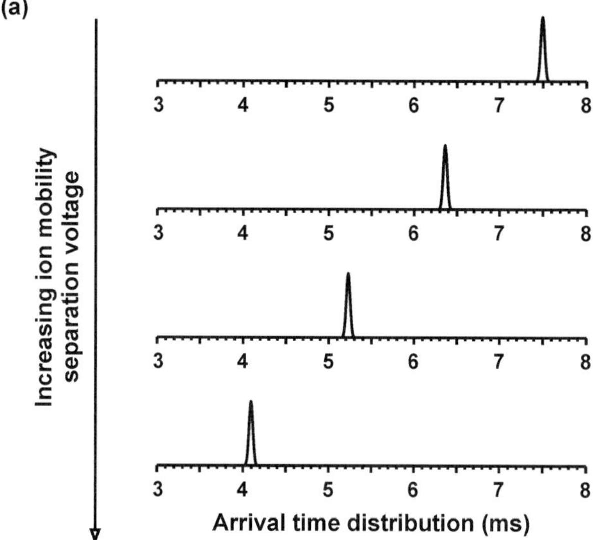

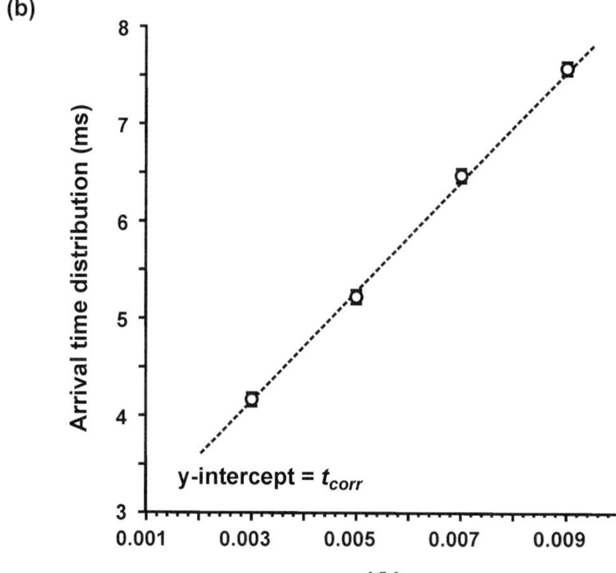

Figure 12.4. Procedure for transforming t_{ATD} for IM signals to t_d. (**a**) The arrival time distribution is measured sequentially at several increasing separation voltages. By selecting the time at the apex of the IM profile, the t_{ATD} values are plotted as a function of $1/V$, where V is the separation voltage applied across the drift cell (**b**). A linear best fit to these points indicates that the separation is performed under low-field conditions, where the y intercept represents the time the ion packet resides in regions of the instrument outside of the drift cell (t_{corr}).

In evaluating K, the drift velocity of the ion packet depends not only on the electrostatic field strength, but also on the pressure (p, torr) of the neutral drift gas and the temperature (T, kelvin) of separation. Therefore, it is conventional practice to report K as the standard or reduced mobility (K_0), which normalizes the results to standard temperature and pressure conditions (i.e., 0°C and 760 torr):

$$K_0 = K \frac{p}{760} \frac{273}{T} \tag{12.2}$$

For applications where IM is used to obtain structural information about the ion, such as those in structural proteomics and biophysics, the IM separations are performed using weak electrostatic fields. Provided that the field strength is sufficiently weak, a closed equation for the ion–neutral collision cross section can be expressed from the kinetic theory of gases. If a Maxwellian distribution function of molecular (or ion) velocities in thermodynamic equilibrium is assumed, then the mean thermal velocity of a gas is

$$v_{\text{mean}} = \left(\frac{8kT}{\pi M_r}\right)^{1/2} \tag{12.3}$$

When the electrostatic field is sufficiently weak, the ion velocity in the gas will be the random motion of ions at the temperature of the gas, on which a small velocity component in the direction of the electrostatic field is imposed. Provided that these separation parameters are met, the IM is performed under so-called "low-field" conditions. At higher electrostatic fields, the ion velocity distribution depends less strongly on the temperature of the separation and the mean ion energy increases as it traverses the drift region. Consequently, K is no longer constant and depends on the specific ratio of the electrostatic field to the gas number density (E/N). When the IM separations are performed in low-field conditions (i.e., constant K), the mobility is related to the collision cross section of the ion–neutral pair:

$$K_0 = \frac{(18\pi)^{1/2}}{16}\frac{ze}{(k_B T)^{1/2}}\left[\frac{1}{m_i}+\frac{1}{m_n}\right]^{1/2}\frac{760}{p}\frac{T}{273}\frac{1}{N_0}\frac{1}{\Omega} \tag{12.4}$$

where these parameters include the charge of the ion (ze), the number density of the drift gas at STP (N_0, $2.69 \times 10^{19}\ \text{cm}^{-3}$), the reduced mass of the ion–neutral collision pair (ion and neutral masses of m_i and m_n, respectively), Boltzmann's constant (k_B), and the ion–neutral collision cross section (Ω). Inspection of Eq. (12.4) shows that the mobility of an ion is inversely related to its collision cross section, or apparent surface area, which forms the basis for typical bioanalytical IM separations. Substituting for K_0 in Eq. (12.4) and rearranging to solve for the collision cross section yields

$$\Omega = \frac{(18\pi)^{1/2}}{16}\frac{ze}{(k_B T)^{1/2}}\left[\frac{1}{m_i}+\frac{1}{m_n}\right]^{1/2}\frac{t_d E}{L}\frac{760}{p}\frac{T}{273}\frac{1}{N_0} \tag{12.5}$$

which is the typical functional form of the equation used to solve for collision cross sections from IM data. Note that these equations are derived from classical electrodynamics, and as such, great care should be exercised in the dimensionality of the units used.*

In both Eqs. (12.4) and (12.5), the collisions of ions with neutrals are considered to be a completely elastic process. Thus, the collision cross section obtained is termed the "hard sphere" collision cross section because it emulates the scattering process of billiard balls; that is, only momentum is transferred between the two collision partners. By comparing empirically determined cross sections with theoretical results, it has been shown that the hard-sphere approximation is best suited for analytes larger than ~1000 Da, which is typically the size range in which many bioanalytical measurements are made.[53] However, as the size of the analyte approaches the size scale of the drift gases used for separation,

*Specifically, the units for E should be expressed in cgs Gaussian units—that is, statvolts cm^{-1}, where 1 statvolt equals 299.79 V. Note that statvolts cm^{-1} is equivalent to statcoulombs cm^{-2} and that elementary charge, e, is 4.80×10^{-10} statcoulombs.

long-range interaction potential between the ion and neutral must be considered for accurate results.[54–56]

12.1.4 Factors Affecting the Separation of Biomolecules in Ion Mobility

12.1.4.1 Influence of Gas Selection on Separations

From the above discussion, it should be noted that for structural studies using the hard-sphere approximation, the long-range interaction between the ion and neutral should be minimized. Accordingly, helium gas is typically used because (i) it has a low polarizability ($\sim 0.21 \times 10^{-24}\,\mathrm{cm}^3$) and (ii) ion transmission efficiency is enhanced, since it is relatively low mass ($\sim$4 Da); that is, scattering losses are minimized. Nevertheless, the selection of the neutral drift gas composition in IM can alter the ion separation selectivity and absolute drift times, which is conceptually similar to the selection and tuning of mobile-phase composition in HPLC separations. Alternatively, reactive gases can be used as a complementary probe of analyte ion structure, or as reagents for probing the structural consequences of ion–molecule reactions.

a. Controlling Separation Conditions Using Neutral Gases. As a separations tool, the drift gas composition used in an unreactive mode can be tuned to serve several purposes including, primarily: (i) to change the mobility of the analyte (i.e., for faster or slower drift times) and (ii) to alter selectivity for specific analytes on the basis of ion-induced dipole interactions. Inspection of Eq. (12.4) indicates that the mobility of an analyte decreases with increasing drift gas mass, which yields longer drift times for more massive neutrals (t_d, Ar > N_2 > He, etc.). In particular, this is the case in the limit of increasing analyte ion mass, because the reduced mass term approximates the mass of the neutral gas. For high-throughput separations, faster drift times are advantageous, but for accurate determination of collision cross sections, slower drift times are desirable (i.e., to minimize the relative magnitude of t_{corr} with respect to t_d). A further instrumental motivation for changing the rate at which analytes elute from the drift cell is to optimize the number of time points sampled across the IM peak for accuracy in the profiles obtained.

In the separation of small molecules, including small biologically relevant species, Hill and co-workers[57–60] have described the utility of tuning the selectivity of IM separations (elution order of analytes) on the basis of drift gas polarizability. In this case, the long-range potential between the analyte ion and drift gas is promoted in the form of ion-induced dipole interactions. The contribution of ion-induced dipole interaction to K_0 is defined as the polarization limit, or K_{pol}, which represents the mobility of an ion, in a gas of particular polarizability, in the limit of vanishing energy and temperature[51,61]:

$$K_{pol} \equiv K_0(E/N \to 0, T \to 0) = \frac{13.853}{(\alpha_d \mu)^{1/2}} \qquad (12.6)$$

where α_d is the dipole polarizability of the neutral and μ is the reduced mass of the ion–neutral collision pair. It is noteworthy that Eq. (12.6) implies a significant decrease in the influence of ion-induced dipole interactions on the mobility of analyte ions >~500 Da.[30,51,54,62] However, for smaller analyte ions, ion-induced dipole interactions can exhibit a pronounced effect on the elution order of analytes. For example, Figure 12.5 illustrates the IM profiles observed in the separation of chloroaniline ($M_r = 128$ Da) and

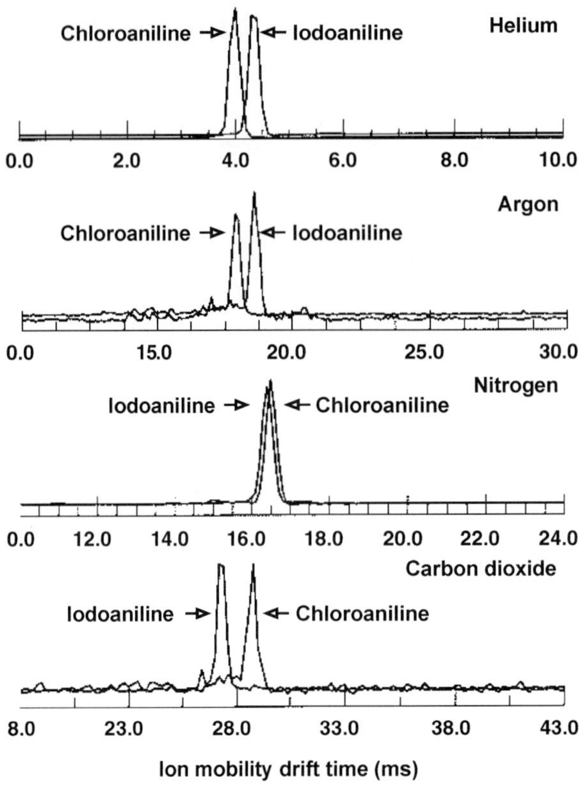

Figure 12.5. Ion mobility spectra of chloroaniline and iodoaniline in each of the four drift gases. As the polarizability increases (from 0.2 to $2.9 \times 10^{-24}\,cm^3$ for He and CO_2, respectively), the velocity of the chloroaniline ion decreases relative to the iodoaniline ion. This change allows one to use drift gas polarizability to change separation factors (α) in ion mobility spectrometry. (Adapted with permission from *Anal. Chem.* **2000**, *72*, 580–584. Copyright 2000 American Chemical Society.)

iodoaniline ($M_r = 220$) in increasing polarizable gases—that is, He, Ar, N_2, and CO_2, which have gas-phase polarizabilities of 0.21, 1.64, 1.74, and $2.91 \times 10^{-24}\,cm^3$, respectively [57]. Although iodoaniline is approximately twice the mass of chloroaniline, its mobility sequentially increases with the polarizability of the drift gas as indicated by the inversion of the IM profiles. Note that this is generally not observed for higher mass analytes as illustrated for tryptic peptide ions in Figure 12.6. Although the signal for $[M + H]^+$ of the peptide YSHEEIAMATVTALR appears to shift further from that predicted for tryptic peptides (i.e., the correlation of IM versus m/z), from 37 µs in He to 140 µs in Ar, the relative shift normalized for drift time remains constant, irrespective of drift gas polarizability.

In addition to neutral drift gas mass and polarizability, there is significant potential for utilizing other long-range ion–neutral interactions, such as ion–dipole, hydrogen bonding, and so on—for example, by doping the neutral drift gas with a partial pressure of a high gas-phase basicity reagent (e.g., acetonitrile, ammonia), because such gases have longer–lived ion–neutral collisions with peptides in gas-phase H/D exchange experiments.[63] A particularly interesting drift gas modification was recently described by Hill and co-workers[64] in which a chiral gas-phase modifier (*S*-(+)-2-butanol, or *R*-(−)-2-butanol) was added at ~10 ppm to the N_2 drift gas. This modification allowed for the gas-phase chiral separation of enantiomer species such as L- and D-enantiomers of various amino acids and small biomolecules (Figure 12.7). Although the general utility of this approach to large biomolecular analytes seems unlikely, it has significant potential for small chiral biomolecules often encountered in such areas as metabolomics or therapeutic natural product discovery.

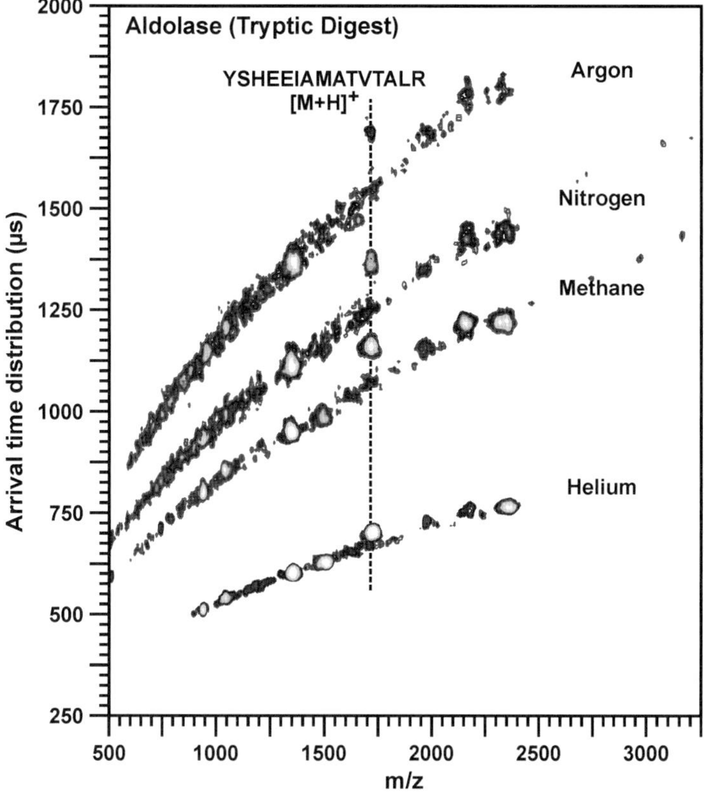

Figure 12.6. Four overlaid IM-MS conformation space plots for a tryptic digest of aldolase (rabbit muscle), for separations performed in each of four drift gases (Ar, N_2, CH_4, and He). The vertical dashed line is provided to assist in visualizing the relative shift of the peptide YSHEEIAMATVTALR from the other tryptic peptides as the mass of the drift gas is increased from He to Ar. (Adapted with permission from *J. Mass Spectrom.* **2004**, *39*, 361–367.)

b. Structural Separations Utilizing Reactive Collisions. The study of reactions between gas-phase ions and neutrals are important in many areas such as atmospheric, inorganic, and physical organic chemistry. Typically, such reactions are performed by adding a small amount of the reactive gas to an excess of inert drift gas. Assuming that the collision frequency with the reactive gas or conjugate product species is sufficiently low, which can be tailored depending on the partial pressure, then reverse reactions can be considered negligible. Although there are numerous studies on the reaction of atomic and small-molecule ions with reactive gases,[2,5] there are few studies utilizing reactive collisions for the study of biomolecules. Valentine and Clemmer[65,66] have demonstrated the utility of H/D exchange in the drift cell for interpreting the consequence of protein structure on the number of exchangeable hydrogens. Importantly, H/D exchange in the drift cell by adding a partial pressure of D_2O to the He drift gas provides complementary information regarding the analyte structure. Note that measurements of H/D exchange following drift cell elution (e.g., in an FT-ICR-MS[67]), or as a complementary MS measurement to collision cross-section profile determinations, have demonstrated great utility in structural interpretation (see, for example, Refs. [12] and [68–70]).

A second area of particular importance in biophysics is the determination of the structural consequences, or thermodynamics, of stepwise hydration of biomolecules. These

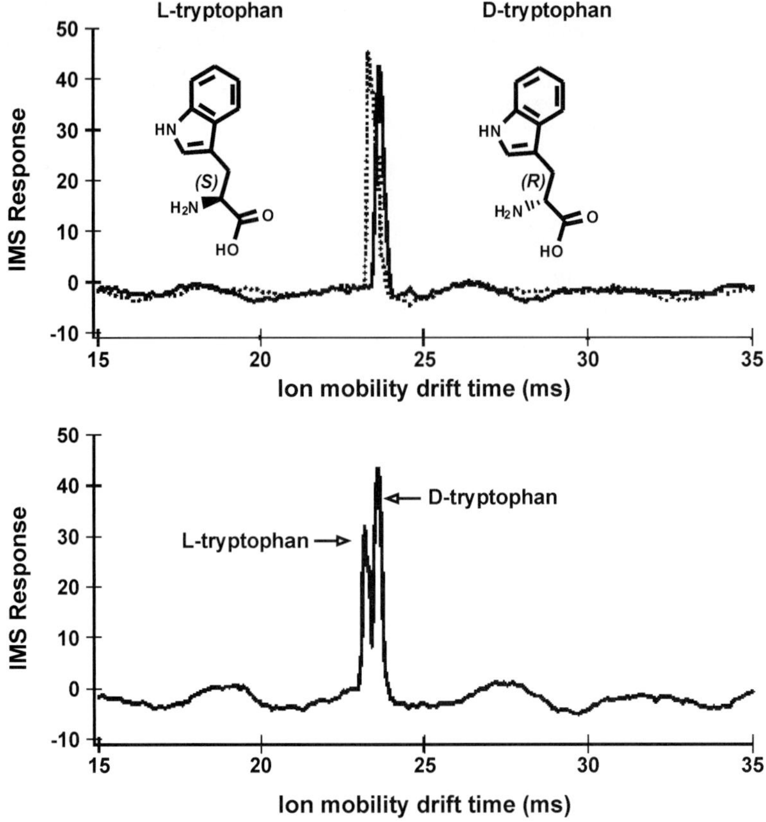

Figure 12.7. Ion mobility profiles demonstrating the chiral gas-phase separation of tryptophan enantiomers. (**Top**) Superimposed profiles of L- and D-tryptophan, dashed and solid lines, respectively, obtained by using 10 ppm of (S)-(+)-2-butanol as a chiral modifier in inert nitrogen drift gas. (**Bottom**) Separation of the enantiomeric tryptophan forms from a racemic mixture. (Adapted with permission from *Anal. Chem.* **2006**, *78*, 8200–8206. Copyright 2006 American Chemical Society.)

studies are typically accomplished by adding H_2O to a variable temperature drift cell, measuring an equilibrium constant for the analyte ion and analyte–$(H_2O)_n$ coordinated ion (by measuring relative intensities of the species and the partial pressure of H_2O) at different system temperatures, and constructing a van't Hoff plot to deduce the change in enthalpy and entropy upon water coordination. In a series of excellent studies, Jarrold and co-workers[71–77] have investigated the stepwise hydration of a number of small protein and peptide systems, and Bowers and colleagues have studied the effects of hydration on amino acids and peptides[78–80] and, more recently, on the hydration of mononucleotides to discern the competition between hydration and Watson–Crick base pairing in DNA.[81]

12.1.4.2 Influence of Electrostatic Field Strength on Separations

Independent of the gas type selected for IM separations, the electrostatic field strength applied across the drift cell can be used to tailor (i) ion drift velocity, (ii) IM resolution, and (iii) analyte separation selectivity. Complementary to the selection of drift gas for changing the mobility of an analyte ion, for a given K, the drift velocity is proportional to E (Eq. (12.1)).

a. Resolution Considerations. Typically the resolution in IM separations ($t_d/\Delta t_{d1/2}$) is limited by the ion longitudinal diffusion in the drift cell. For separations under low-field conditions, the diffusion-limited resolution for an initially narrow pulse of ions injected into the drift tube is described by[51]

$$\frac{t_d}{\Delta t_{d1/2}} = \frac{1}{4}\left(\frac{zeV}{k_B T \ln 2}\right)^{1/2} \tag{12.7}$$

where these parameters correspond to those in Section 12.1.3.1, with the exception of V, which is the voltage drop across the drift cell. Inspection of Eq. (12.7) indicates that IM resolution can be improved by increasing the voltage applied across the drift cell, which implies increasing the drift cell length or using higher drift cell pressures (i.e., to avoid gas breakdown at high voltages). High-resolution ($R \sim 100$–200) IM-MS instruments specifically designed for high-pressure and high-drift-voltage operation have been constructed.[82,83] Although pulsed ion sources such as MALDI inject a narrow pulse of ions (ca. <4 ns) into the drift cell, the modulation of continuous ion sources such as ESI should inject ions with pulse durations significantly shorter than the drift time across the cell. If the latter criterion is not met, the ion injection pulse width adds additional terms to the resolution equation presented above,[84,85] which ultimately result in impaired resolution relative to that predicted at the diffusion limit.

b. Low-Field Versus High-Field Mobility. In Section 12.1.3.1 it was emphasized that the closed equation for determining collision cross section in IM measurements [Eq. (12.5)] assumes that separations are made in the low-field limit—that is, that the energy imparted to the ion by virtue of the electrostatic-field is negligible in comparison with the thermal energy of the system. This is qualitatively described by reference 4

$$\left(\frac{m_n}{m_i} + \frac{m_i}{m_n}\right) zE\lambda \ll k_B T \tag{12.8}$$

where λ represents the mean free path of the ion in the direction of the electrostatic field. In turn, λ is inversely proportional to the product of collision cross section and pressure. Thus, specific E/p ratios that maintain the separation in the low-field regime increase with increasing Ω. In the separation of atomic ions (small Ω), the E/p ratio should be maintained at $<2\,\text{V cm}^{-1}\,\text{torr}^{-1}$ reference 4, whereas peptide ions are observed to retain low-field behavior even at E/p ratios up to $70\,\text{V cm}^{-1}\,\text{torr}^{-1}$ or greater.[86]

The concepts developed in the preceding sections are generally applicable for IM-MS separations in uniform electrostatic fields—that is, for separations performed at constant K and in low E-fields. A fundamentally different type of IMS separation, based on promoting differences in mobility by alternating between low and high E-field conditions, is termed high-field asymmetric waveform ion mobility spectrometry (FAIMS).[87–89] Atmospheric-pressure FAIMS devices have generated considerable interest in combination with a variety of MS instruments, because (i) they provide higher sensitivity by focusing ions at the MS interface and (ii) structural selectivity can be obtained. Guevremont[90] has published an excellent review of the historical development, fundamentals, and applications of FAIMS-MS. However, there is considerable ongoing research to understand the fundamental physical processes of ion separations in FAIMS,[91,92] and, as such, it is not presently possible to write a closed equation analogous to Eq. (12.4) for uniform-field IMS. Thus, collision cross-section measurements using FAIMS devices are presently derived by using literature values for Ω, obtained on uniform-field instruments, for calibration purposes.[93]

12.1.5 Structural Characterization by Correlation of Empirical and Computational Results

To infer analyte ion structural details, IM-MS-determined collision cross sections are typically supported by complementary computational studies. Although the specific procedural details for comparing empirical cross-section values with those obtained by computational techniques varies by laboratory, the general framework for comparisons consists of five steps: (i) generation of model *in silico* structures, (ii) exploration of the conformational landscape (e.g., by molecular mechanics/molecular dynamics (MM/MD) and simulated annealing protocols), (iii) determination of modeled structure collision cross sections (typically via MOBCAL developed by Jarrold and co-workers[53,54,94]), (iv) generation of scatter plots of relative energy versus collision cross section, and (v) inspection of the lowest-energy structures consistent with the IM-MS collision cross section(s). An outstanding tutorial of these procedures and discussion of fundamental considerations is provided by Bowers and colleagues.[95] For biomolecular structural studies, MM/MD approaches are typically used because of the large size of the molecules, but these techniques require that suitably parameterized force fields exist and these approaches generally have difficulty in sampling compact conformations.[96–98] It is important to recognize that comparing computationally derived structures consistent with the IM-MS experimental results does not afford high-resolution (e.g., atomic coordinate) structural information. Rather, it provides interpretive power for understanding the structural motifs that may prevail, or for eliminating structural conformations which are inconsistent with experimental results.

12.2 INSTRUMENTAL ARRANGEMENTS FOR ION MOBILITY–MS

Since the earliest reports of coupling IM with MS,[19,20] a wide variety of instrumental arrangements have been described utilizing a multitude of drift cell and mass analyzer arrangements. For example, IM drift cells have been interfaced with quadrupoles,[19,99–102] quadrupole ion traps,[103,104] double-focusing sector-fields,[105] Fourier transform ion cyclotron resonance,[106] and TOFMS,[20,107,108] among other arrangements. It is beyond the scope of this work to discuss the merits and pitfalls of each design; rather, several illustrative and contemporary examples are highlighted specifically for IM-MS applications in the analysis of complex biological samples. Nevertheless, it is important to underscore the use of IM drift cells as a modular component in the MS toolbox; that is, specific applications may favor a particular ordering of IM and MS as indicated in Figure 12.1. Although this discussion focuses on uniform electrostatic-field drift cells, several excellent recent works in segmented-quadrupole[109–111] and traveling-wave ion guide (TWIG[112,113]) IM-MS designs are presented elsewhere.

12.2.1 Fundamental Considerations for Ion Source Selection (MALDI Versus ESI)

For biological IM-MS instrumentation, both MALDI[114,115] and ESI[116] sources have been effectively coupled for gas-phase separations.[14,105–118] However, from an instrumental standpoint there are clearly fundamental differences that dictate the chosen source for particular experiments. To date, the majority of IM-MS designs have incorporated ESI,[119–125] which produces an envelope of different charge state ions depending on the chemical composition (e.g., number of basic sites) and size of the analyte. In contrast, MALDI typically produces singly charged ions. The charge state and number of charges are salient considerations for IM-MS, because (i) the mobility of the ion is directly proportional to the charge-state, (ii) in complex samples multiple charge states yield a multiplicity of signals which can result in congested spectra and reduced sensitivity by partitioning signals into multiple channels, and (iii) in the low dielectric of the drift cell, charge repulsion results in ion structures (Ωs) that are larger than those predicted from solution-phase studies. On the other hand, higher charge-state ions are advantageous when performing IM-MS/MS because they provide higher fragmentation efficiency and they also shift analyte ions to a lower *m*/*z* range that may be necessary for certain mass analyzers (e.g., quadrupoles) and for higher sensitivity, depending on the detection scheme that is used (e.g., image current or microchannel plates).

Note that it remains unclear whether the specific ion structures obtained by using MALDI or ESI are equivalent, or how these structures correspond to those obtained in solution. For model systems there are some indications based on hydration IM studies and by using complementary solution-phase structural probes, but these have largely been performed for small model systems. This remains an important area of ongoing research (see Sections 12.1.2.2 and 12.1.4.2.a above). It is likely an oversimplification to suggest that ions generated by ESI, directly from the solution-phase, retain solution-phase structural elements of the analyte more successfully than do those generated from the solid phase by MALDI. In both cases, at least mildly (to strongly) denaturing conditions are typically used in sample preparation, either (a) by the addition of organic modifiers or acids for ESI or (b) cocrystalization with small organic molecules in MALDI. Furthermore, in both cases the anhydrous ions that are probed should ultimately adopt structures corresponding to the energy minimum attained by intramolecular folding forces—that is, in the absence of intermolecular solvation. For both ionization techniques, it is also important to note that the internal energy imparted to the molecule from ionization can be mitigated by evaporative cooling, either from solution evaporation in ESI or from matrix cluster evaporation in MALDI. Although challenging, future studies aimed at delineating the differences in ESI- and MALDI-derived ion structures will be highly beneficial for biophysical studies utilizing MS-based approaches.[126,127]

12.2.2 Contemporary Instrumental Arrangements for Biomolecular IM-MS

Of the mass analyzers coupled with IM for complex biological samples, TOFMS and in particular orthogonal-TOFMS provide desirable attributes in comparison with scanning MS-instrumentation (e.g., quadrupoles). Following time-dispersive IM, it is relatively straightforward to perform a second time-dispersive measurement (e.g., TOFMS). Secondly, TOFMS provides nearly simultaneous measurement of all of the analytes eluting from the drift cell on a timescale (tens of microseconds) ideally suited to IM separations (hundreds of microseconds to milliseconds). Two types of IM-MS platforms, based on MALDI and ESI ion sources, are described.

12.2.2.1 MALDI-IM-Time-of-Flight-MS

Significant developmental effort has resulted in the evolution of a series of moderate pressure MALDI-IM-TOFMS instruments following early proof-of-concept experiments.[108] A schematic diagram of the most modern of these instruments is presented in Figure 12.8a. In this instrument, ions are produced by high repetition rate MALDI (up to 5000 Hz) at the pressure of the IM drift cell (3–10 torr). In contrast with earlier designs, high-repetition-rate MALDI provides significant advantages in terms of limits of detection (femtomole to attomole) and throughput,[128] particularly at moderate pressures where neutral plume stagnation occurs, resulting in higher ionization efficiencies from subsequent laser pulses. The ions are then directed into a dc-only ion guide drift cell,[129] which further improves sensitivity by refocusing ions to the center of the drift cell—that is, eliminating ion losses from radial diffusion as the ions traverse the drift cell. The ions are then focused through a differential pumping region terminated by a skimming cone and are directed into the source of a reflectron orthogonal TOFMS which is operated at pressures of $\sim 10^{-8}$ torr.

The arrival time distributions (ATDs) of ions in this IM-MS arrangement range from hundreds of microseconds (peptides) to ~1–2 ms (proteins). Such fast drift times are beneficial for high throughput IM-MS studies, such as those necessary for proteomics-scale research. The attributes of MALDI facilitate rapid and high sensitivity separations, because ions are produced in both a temporally (picosecond to nanosecond) and spatially focused region (micrometer squared). Under these conditions, resolution is not limited by temporally gating the ion injection,[57,130–132] and ion transmission efficiency is enhanced by better projection of the ion source onto the aperture plate and skimming cone at the terminus of the drift cell, which is analogous to ion projection in sector-field instrumentation.[133] By producing ions in a pulsed mode, the MALDI event itself provides a well-defined t_0 for the time-dispersive IM measurement.

In coupling two time-dispersive measurements, complications can arise in matching the time resolution of each separation dimension. For example, if the orthogonal TOFMS is operated at pulse-frequency of 10 kHz, sequential MS spectra would be collected every 100 μs, resulting in a 1-ms IM profile with 100-μs time resolution or composed of 10 points. Given a typical IM resolution of 30–50, this corresponds to peak profile widths (FWHM) of ~10–20 μs, and thus ion packets eluting in the IM dimension in the midst of an MS cycle may pass though the source of the TOFMS without detection. To overcome this challenge, a time interleaving scheme for data acquisition was developed which provides independent time resolution in both time-dispersive dimensions.[7,134] In the time interleaving mode, TOFMS spectra are acquired sequentially at the rate of the TOFMS, but are slightly offset by a desired amount for each subsequent MALDI event. Following a sufficient number of ion injections into the drift cell (t_0^{IM}), the offset TOFMS spectra are then stitched together (Figure 12.9), providing a 2D IM-MS plot of conformation space with independent control of the number of time points sampled in both the IM and MS dimensions, respectively.

12.2.2.2 ESI-IM-Time-of-Flight-MS

As indicated above, the majority of IM-MS instrument platforms have been constructed utilizing ESI ion sources. Initially, these were largely devoted to projects investigating

(a)

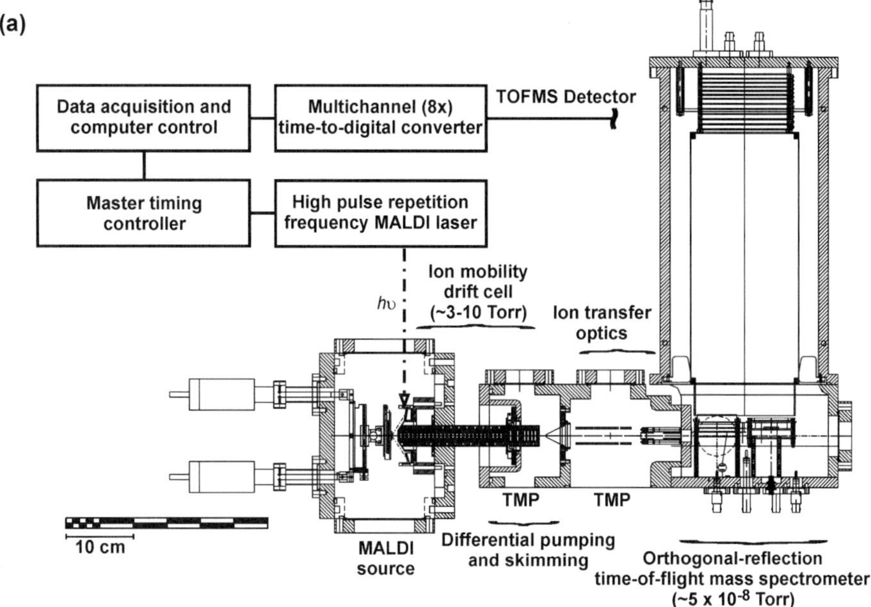

(b)

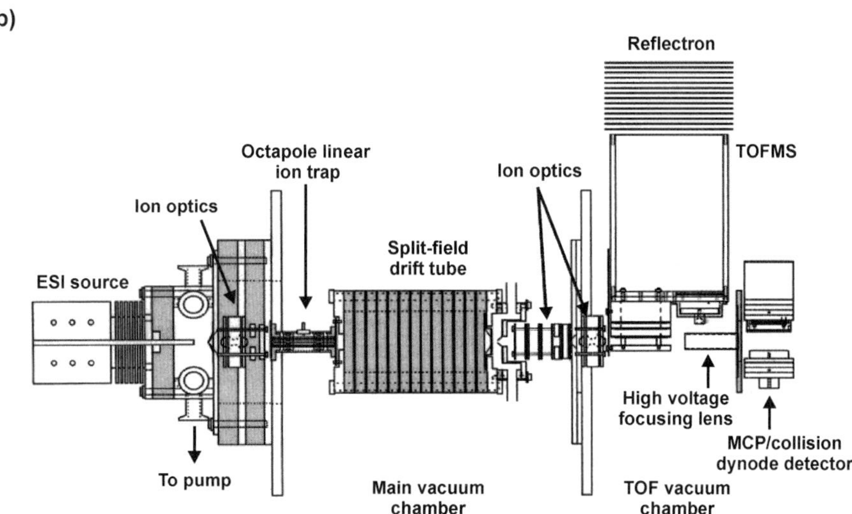

Figure 12.8. Instrumental arrangements for performing (**a**) MALDI-IM-TOFMS and (**b**) ESI-IM-TOFMS, respectively. (The instrument depiction in panel b is adapted with permission from *Anal. Chem.* **2003**, *75*, 6202–6208. Copyright 2003 American Chemical Society.)

gas-phase ion structures, but are increasingly being used for peptide characterization in proteomics. In contrast with MALDI, ESI ion sources are typically operated as continuous, rather than pulsed, sources of ions. Advances in utilizing pulsed ion traps[103,124] and ion funnels,[121,135,136] for ion storage and to gate the injection of ion packets into the drift cell, have resulted in devices which inject nearly 100% of the ions into the IM drift cell, despite the step of ion source modulation. Thus, sensitivity obtained on ESI-IM-MS instruments is commensurate with levels on commercial ESI-MS instruments.[135]

(a)

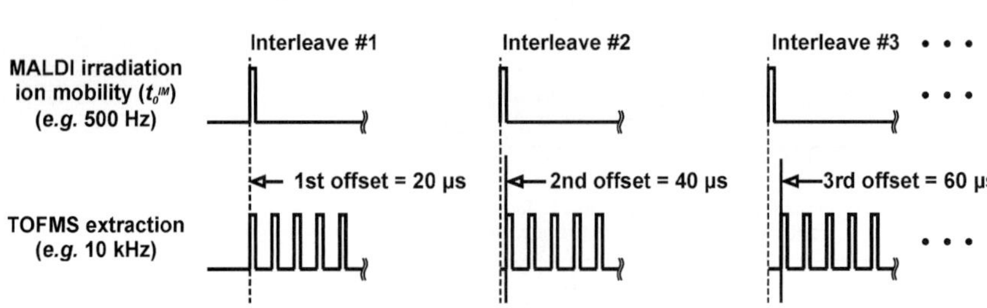

(b)

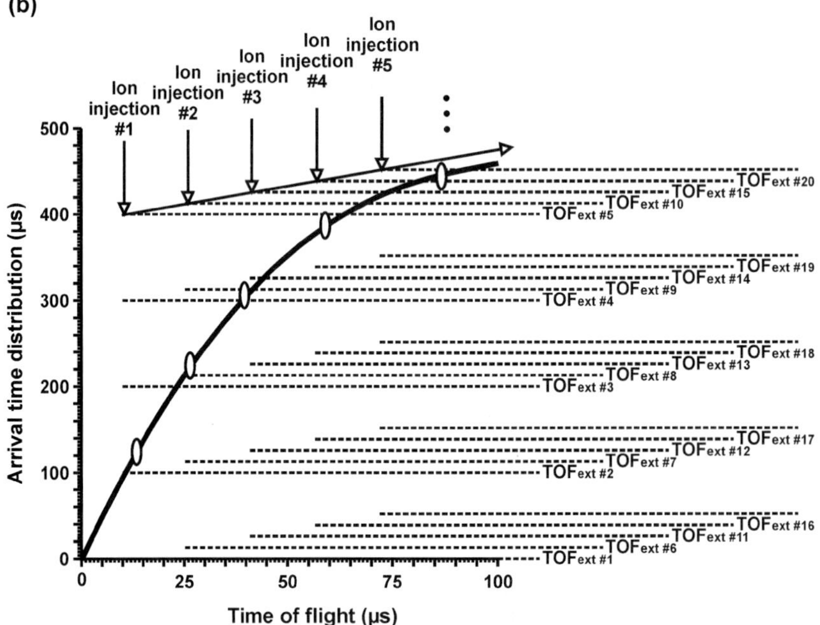

Figure 12.9. A hypothetical timing sequence diagram illustrating interleaved data acquisition methodology for independently defining the measurement time resolution in both the IM and MS dimensions. Following the initial ion injection event [i.e., t_0^{IM}, (**a**)] the number of TOFMS extractions that are performed define the total IM arrival time distribution space that is sampled. For subsequent ion injections, TOFMS extractions are offset relative to t_0^{IM} by a user definable extent (**b**). The interleaved TOFMS spectra are then stitched together, yielding a two-dimensional plot of ATD versus *m*/*z*.

A schematic diagram for an innovative ESI-IM-TOFMS is illustrated in Figure 12.8b. In this arrangement, ions generated by ESI are desolvated and focused into an octopole linear ion trap. Ions are accumulated in the ion trap for a desired duration (up to ~6 ms) prior to their injection into a moderate pressure (1.7 torr) IM drift cell. This particular design incorporates a unique region at the exit of the drift cell, whereby the final region in the drift cell can be raised to significantly higher field strengths to promote collision-induced dissociation (CID) fragmentation prior to mass analysis.[124] Termed a split-field drift tube, the utility of such a device is described in Section 12.2.3 below. An important advantage of ESI coupled with IM-MS is the relative ease with which additional liquid-phase separation techniques can be coupled to 2D post-ionization IM-MS separations in an on-line approach.

These techniques have largely been pioneered by Clemmer and colleagues for proteomic applications.[137–142] By increasing the data dimensionality of the IM-MS experiment with additional liquid-phase separations, exceptional information content can be obtained. For example, by combining LC-ESI-IM-CID-MS, Liu et al.[141] were able to identify $\sim 2 \times 10^4$ and 6×10^5 precursor and fragment ions, respectively, over the course of a 21-min analysis.[141]

12.2.3 Analyte Sequencing Using IM-MS/MS

In parallel with peptide mass fingerprinting techniques, tandem mass spectrometry (MS/MS) technology has become an indispensable tool for proteome analysis by providing additional peptide sequence information for higher confidence-level protein identification and validation of peptide mass fingerprinting strategies.[143] As briefly discussed above (Section 12.1.2.1), IM-MS/MS can be performed by activating the ions as they elute from the IM drift cell, where the fragment ions observed in the MS dimension are correlated to their precursor ion by the IM drift time. Thus, provided that there is a structural separation in the IM dimension, *all* parent ions can be activated in the same 2D separation. This is in contrast to typical MS/MS technology, which is operated in a scanning mode; that is, a mass filter is set to transmit a particular mass (or range of masses), and the selected ions are activated (e.g., by CID) and fragment to produce sequence-informative fragment ions that are subsequently analyzed by a second mass spectrometer. In this sense, IM-MS/MS approaches what would be desired in an ideal MS/MS experiment—that is, the *simultaneous* acquisition of parent ions and fragment ions with spatial or temporal correlation, to retain information content.[144]

Separations utilizing an ion activation region after the IM drift cell for IM-MS/MS data acquisition have been demonstrated using CID,[145]surface-induced dissociation (SID),[146] and UV photodissociation[147] ion activation strategies. An example illustrating ESI-IM-CID-TOFMS, utilizing the split-field instrument described in Section 12.2.2.2, is shown in Figure 12.10 for the $+2$ and $+3$ charge states of the peptide VGAHAGEYGAEALER.[148] This example illustrates a further advantage to utilizing precursor ion time dispersion in the IM dimension as MS^1, namely that the ion activation conditions can be modulated, within a single analysis, for purposes of optimizing the fragment ions that are obtained. In the present example, this is demonstrated by increasing the potentials applied to the final region of the drift cell in steps from 55 V (low-field) to 220 V (high-field), where more extensive fragmentation patterns are observed at increasing potentials, but they remain correlated to the precursor elution time from the IM drift cell. By providing comprehensive fragmentation data, IM-MS/MS methodologies thus afford advantages in terms of throughput, sample consumption, and instrumental duty cycle over scanning tandem mass spectrometry approaches.

12.3 EMERGING DIRECTIONS IN IM-MS TECHNOLOGY

Similar to the new directions taken in IM-MS for biotechnology in the mid-1990s to the present, IM-MS instrumentation is poised to become much more commonplace because it is now becoming commercially available in a variety of forms. This is significant, because technological developments now yield IM-MS figures of merit (in particular sensitivity) similar to commercially available high-performance MS-only instrumentation. Thus,

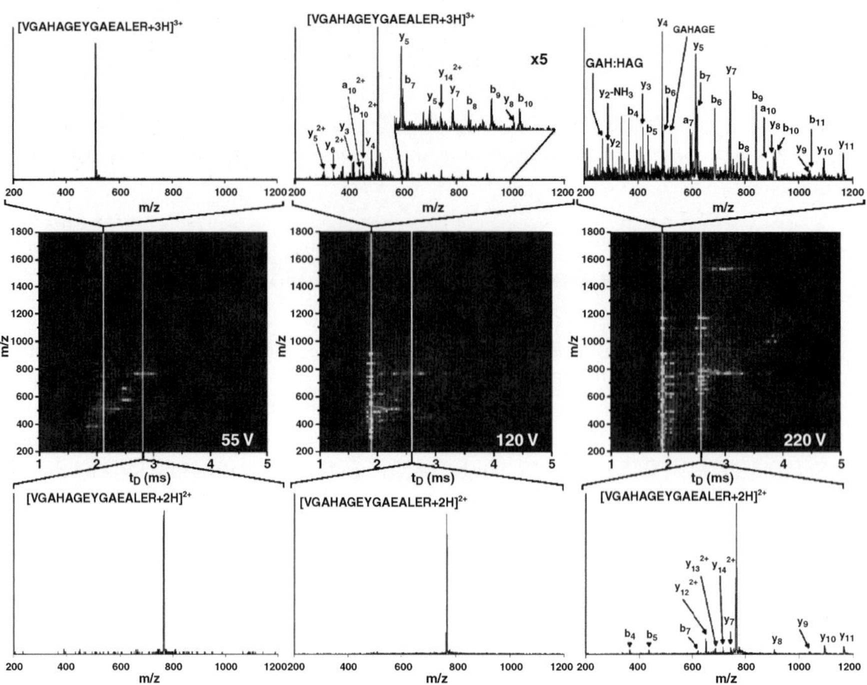

Figure 12.10. A comparison of correlated peptide sequencing data obtained by LC-IM-MS and LC-IM-CID-MS obtained sequentially using a field-modulated ESI-IM-TOFMS as depicted in Figure 12.8b. Mass spectra correspond to integrating across the IM arrival times as indicated by white lines. (Adapted with permission from *J. Proteome Res.* **2005**, *4*, 25–35. Copyright 2005 American Chemical Society.)

applications utilizing IM-MS approaches will inevitably increase. For example, imaging MS applications whereby MALDI-TOFMS spectra are collected in a spatially resolved manner have demonstrated enormous potential in life sciences and clinical research.[149,150] In imaging MS acquisitions, each image potentially contains thousands of molecular signals corresponding to potentially thousands of individual maps of spatially resolved biomolecular distribution—for example, in the context of a thin tissue section or biopsy. Imaging MS requires high sensitivity, because the total volume of sample is limited to the spatial coordinates from which the individual MS spectra are obtained. Nevertheless, the individual mass spectra can oftentimes be highly congested owing to the complexity of the samples, and it is challenging to image regions containing isobaric species (e.g., neuropeptides that are isobaric with lipids in brain tissues). Figure 12.11 illustrates the emerging technique of imaging MALDI-IM-MS,[151,152] which shows the spatial distribution of two cerebrosides as they occur in a transverse section (16 μm) of rat brain. Note that the spatial-resolved IM-MS image in Figure 12.11c corresponds to *m/z* values only from the IM-ATD region corresponding to these cerebrosides, which is analogous to integrating only those signals for sphingomyelin in Figure 12.3. Thus, the IM-MS image provides added data dimensionality in that it is selective to both analyte structure and *m/z*. Notably, in addition to biomolecular and biophysical measurements, it is likely to have significant impact in areas such as nanotechnology and bionanotechnology, where nano-architecture can be structurally probed in the IM dimension and characterized by MS in a second dimension.[32,153]

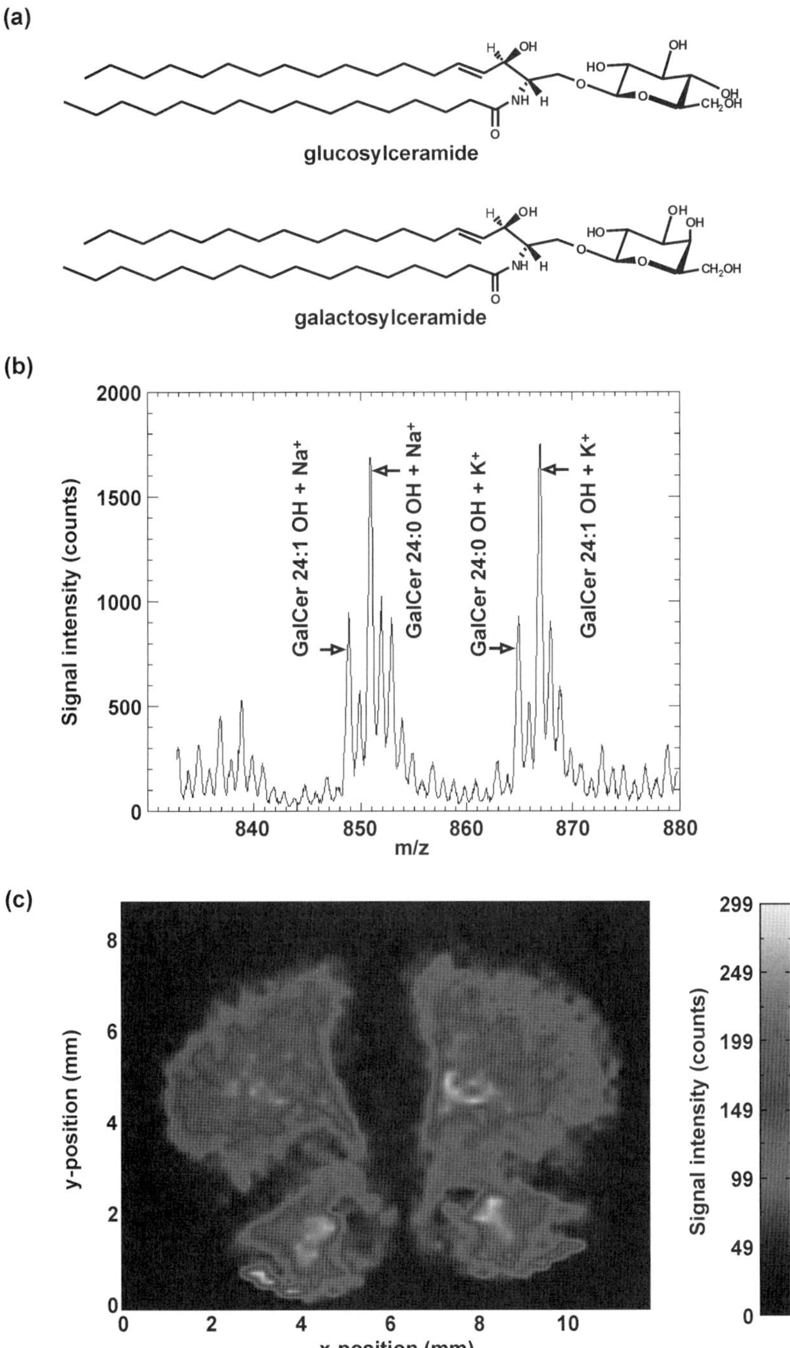

Figure 12.11. Imaging-mode MALDI-IM-MS of a thin tissue section (16 μm) of a rat brain. (**a**) Structures of glucosylceramide and galactosylceramide, the two major classes of cerebroside molecules. Both cerebrosides are glycosphingolipids, which are important components in nerve cell membranes. (**b**) Integrated mass spectrum obtained at the specific IM-ATD for these cerebrosides directly from rat brain tissue. (**c**) False color image of signal intensity for where these cerebrosides occur spatially in the context of the thin tissue section. Imaging MALDI-IM-MS was performed at 349 nm using a 2-nm Au nanoparticle matrix, with an imaging spatial resolution of 200 μm.

ACKNOWLEDGMENTS

We thank Larissa S. Fenn for assistance in the revision of this chapter. Financial support for this work was provided in part by the National Institutes of Health, the Vanderbilt University College of Arts & Sciences, the Vanderbilt Institute of Chemical Biology, the Vanderbilt Institute of Integrative Biosystems Research and Education, and Ionwerks Inc.

REFERENCES

1. McDaniel, E. W. *Collision Phenomena in Ionized Gases*, John Wiley & Sons, New York, **1964**.
2. McDaniel, E. W.; Mason, E. A. *The Mobility and Diffusion of Ions in Gases*, John Wiley & Sons, New York, **1973**.
3. Carr, T. W. (Ed.), *Plasma Chromatography*, Plenum Press, New York, **1984**.
4. Mason, E. A.; McDaniel, E. W. *Transport Properties of Ions in Gases*, John Wiley & Sons, New York, **1988**.
5. Eiceman, G. A.; Karpas, Z. *Ion Mobility Spectrometry*, 2nd ed., CRC Press, Boca Raton, FL, **2005**.
6. Borsdorf, H.; Eiceman, G. A. Ion mobility spectrometry: Principles and applications. *Appl. Spectrosc. Rev.* **2006**, *41*, 323–375.
7. McLean, J. A.; Ruotolo, B. T.; Gillig, K. J.; Russell, D. H. Ion mobility–mass spectrometry: A new paradigm for proteomics. *Int. J. Mass Spectrom.* **2005**, *240*, 301–315.
8. Valentine, S. J.; Liu, X.; Plasencia, M.; Hilderbrand, A. E.; Kurulugama, R.; Koeniger, S. L.; Clemmer, D. E. Developing liquid chromatography ion mobility mass spectrometry techniques. *Expert Rev. Proteomics* **2005**, *2*, 553–565.
9. Creaser, C. S.; Griffiths, J. R.; Bramwell, C. J.; Noreen, S.; Hill, C. A.; Thomas, C. L. P. Ion mobility spectrometry: A review. Part 1. Structural analysis by mobility measurement. *Analyst* **2004**, *129*, 984–994.
10. Wyttenbach, T.; Bowers, M. T. Gas-phase conformations: The ion mobility/ion chromatography method. *Top. Curr. Chem.* **2003**, *225*, 207–232.
11. Collins, D. C.; Lee, M. L. Developments in ion mobility spectrometry–mass spectrometry. *Anal. Bioanal. Chem.* **2002**, *372*, 66–73.
12. Jarrold, M. F. Peptides and proteins in the vapor phase. *Annu. Rev. Phys. Chem.* **2000**, *51*, 179–207.
13. Hoaglund-Hyzer, C. S.; Counterman, A. E.; Clemmer, D. E. Anhydrous protein ions. *Chem. Rev.* **1999**, *99*, 3037–3079.
14. Clemmer, D. E.; Jarrold, M. F. Ion mobility measurements and their applications to clusters and biomolecules. *J. Mass Spectrom.* **1997**, *32*, 577–592.
15. St. Louis, R. H.; Hill, H. H., Jr. Ion mobility spectrometry in analytical chemistry. *Crit. Rev. Anal. Chem.* **1990**, *21*, 321–355.
16. Röntgen, W. C. On a new kind of rays. *Nature* **1895**, *53*, 274–276.
17. *Nobel Lectures, Physics 1901–1921*, Elsevier Publishing Company, Amsterdam, **1967**.
18. Eiceman, G. A.; Karpas, Z. Introduction to ion mobility spectrometry. In *Ion Mobility Spectrometry*, 2nd ed., CRC Press, Boca Raton, **2005**, pp. 3–36.
19. Barnes, W. S.; Martin, D. W.; McDaniel, E. W. Mass spectrographic identification of the ion observed in hydrogen mobility experiments. *Phys. Rev. Lett.* **1961**, *6*, 110–111.
20. McAfee, K. B., Jr.; Edelson, D. Identification and mobility of ions in a Townsend discharge by time-resolved mass spectrometry. *Proc. Phys. Soc. London* **1963**, *81*, 382–384.
21. Gieniec, J.; Mack, L. L.; Nakamae, K.; Gupta, C.; Kumar, V.; Dole, M. Electrospray mass spectroscopy of macromoelcules: Application of an ion-drift spectrometer. *Biomed. Mass Spectrom.* **1984**, *11*, 259–268.
22. Wittmer, D.; Chen, Y. H.; Luckenbill, B. K.; Hill, H. H., Jr. Electrospray ionization ion mobility spectrometry. *Anal. Chem.* **1994**, *66*, 2348–2355.
23. Smith, R. D.; Loo, J. A.; Ogorzalek, R.; Busman, M.; Udesth, H. R. Principles and practice of electrospray ionization–mass spectrometry for large polypeptides and proteins. *Mass Spectrom. Rev.* **1991**, *10*, 359–452.
24. von Helden, G.; Wyttenbach, T.; Bowers, M. T. Conformation of macromolecules in the gas phase: Use of matrix-assisted laser desorption methods in ion chromatography. *Science* **1995**, *267*, 1483–1485.
25. Wyttenbach, T.; von Helden, G.; Bowers, M. T. Gas-phase conformation of biological molecules: Bradykinin. *J. Am. Chem. Soc.* **1996**, *118*, 8355–8364.
26. Clemmer, D. E.; Hudgins, R. R.; Jarrold, M. F. Naked protein conformations: Cytochrome c in the gas phase. *J. Am. Chem. Soc.* **1995**, *117*, 10141–10142.
27. Shelimov, K. B.; Clemmer, D. E.; Hudgins, R. R.; Jarrold, M. F. Protein structure *in vacuo*: The gas phase conformations of BPTI and cytochrome c. *J. Am. Chem. Soc.* **1997**, *119*, 2240–2248.
28. Giddings, J. C. Two-dimensional separations: Concept and promise. *Anal. Chem.* **1984**, *56*, 1258A–1270A.
29. Ruotolo, B. T.; Gillig, K. J.; Stone, E. G.; Russell, D. H. Peak capacity of ion mobility mass spectrometry: Separation of peptides in helium buffer gas. *J. Chrom. B* **2002**, *782*, 385–392.
30. Ruotolo, B. T.; McLean, J. A.; Gillig, K. J.; Russell, D. H. Peak capacity of ion mobility mass spectrometry: The utility of varying drift gas polarizability for the separation of tryptic peptides. *J. Mass Spectrom.* **2004**, *39*, 361–367.

31. McLean, J. A.; Russell, D. H. New vistas for mass spectrometry-based proteomics and biotechnology: Rapid two-dimensional separations using gas-phase electrophoresis/ion mobility-mass spectrometry. *Am. Biotech. Lab.* **2005**, *23*, 18–21.
32. McLean, J. A.; Schultz, J. A.; Ugarov, M. V.; Egan, T. F.; Russell, D. H. Ultrafast two-dimensional ion separations based on ion mobility-time-of-flight mass spectrometry: From biotechnology to Nanotechnology. *Braz. J. Vac. Appl.* **2005**, *24*, 3–9.
33. Koomen, J. M.; Ruotolo, B. T.; Gillig, K. J.; McLean, J. A.; Russell, D. H.; Kang, M.; Dunbar, K. R.; Fuhrer, K.; Gonin, M.; Schultz, J. A. Oligonucleotide analysis with MALDI–ion mobility–TOFMS. *Anal. Bioanal. Chem.* **2002**, *373*, 612–617.
34. Woods, A. S.; Ugarov, M.; Egan, T.; Koomen, J.; Gillig, K. J.; Fuhrer, K.; Gonin, M.; Schultz, J. A. lipid/peptide/nucleotide separation with MALDI–ion mobility–TOFMS. *Anal. Chem.* **2004**, *76*, 2187–2195.
35. McLean, J. A.; Tao, L.; Perkins, S. L.; Russell, D. H. Multidimensional proteomic analysis of *Escherichia Coli* whole cell lysates by HPLC–MALDI–ion mobility–MS: Extending dynamic range in protein analysis. *In the 53rd American Society for Mass Spectrometry Annual Meeting*, San Antonio, TX (June 2005).
36. Hudgins, R. R.; Ratner, M. A.; Jarrold, M. F. Design of helices that are stable *in vacuo*. *J. Am. Chem. Soc.* **1998**, *120*, 12974–12975.
37. Ruotolo, B. T.; Verbeck, G. F.; Thomson, L. M.; Gillig, K. J.; Russell, D. H. Observation of conserved solution-phase secondary structure in gas-phase tryptic peptides. *J. Am. Chem. Soc.* **2002**, *124*, 4214–4215.
38. Ruotolo, B. T.; Russell, D. H. Gas-phase conformations of proteolytically derived protein fragments: influence of solvent on peptide conformation. *J. Phys. Chem. B* **2004**, *108*, 15321–15331.
39. McLean, J. A.; McLean, J. R.; Russell, D. H. Investigation of anhydrous protein structures: The influence of post-translational modifications on tertiary and quaternary structure. In *19th Annual Symposium of the Protein Society*, Boston, MA (August 2005).
40. Counterman, A. E.; Clemmer, D. E. *Cis–trans* signatures of proline-containing tryptic peptides in the gas phase. *Anal. Chem.* **2002**, *74*, 1946–1951.
41. Ruotolo, B. T.; Verbeck, G. F.; Thomson, L. M.; Woods, A. S.; Gillig, K. J.; Russell, D. H. Distinguishing between phosphorylated and nonphosphorylated peptides with ion mobility–mass spectrometry. *J. Proteome Res.* **2002**, *1*, 303–306.
42. Ruotolo, B. T.; Gillig, K. J.; Woods, A. S.; Egan, T. F.; Ugarov, M. V.; Schultz, J. A.; Russell, D. H. Analysis of phosphorylated peptides by ion mobility–mass spectrometry. *Anal. Chem.* **2004**, *76*, 6727–6733.
43. Sawyer, H. A.; Marini, J. T.; Stone, E. G.; Ruotolo, B. T.; Gillig, K. J.; Russell, D. H. The structure of gas-phase bradykinin fragment 1-5 (RPPGF) ions: An ion mobility spectrometry and H/D exchange ion-molecule reaction chemistry study. *J. Am. Soc. Mass Spectrom.* **2005**, *16*, 893–905.
44. Woods, A. S.; Koomen, J. M.; Ruotolo, B. T.; Gillig, K. J.; Russell, D. H.; Fuhrer, K.; Gonin, M.; Egan, T. F.; Schultz, J. A. A study of Peptide–peptide interactions using MALDI ion mobility o-TOF and ESI mass spectrometry. *J. Am. Soc. Mass Spectrom.* **2002**, *13*, 166–169.
45. Kaleta, D. T.; Jarrold, M. F. Noncovalent interactions between unsolvated peptides. *J. Phys. Chem. A* **2002**, *106*, 9655–9664.
46. Bernstein, S. L.; Liu, D.; Wyttenbach, T.; Bowers, M. T.; Lee, J. C.; Gray, H. B.; Winkler, J. R. α-Synuclein: Stable compact and extended monomeric structures and pH dependence of dimer formation. *J. Am. Soc. Mass Spectrom.* **2004**, *15*, 1435–1443.
47. Bernstein, S. L.; Wyttenbach, T.; Baumketner, A.; Shea, J. -E.; Bitan, G.; Teplow, D. B.; Bowers, M. T. Amyloid β-protein: monomer structure and early aggregation states of Aβ42 and its pro19 alloform. *J. Am. Chem. Soc.* **2005**, *127*, 2075–2084.
48. Ruotolo, B. T.; Giles, K.; Campuzano, I.; Sandercock, A. M.; Bateman, R. H.; Robinson, C. V. Evidence for macromolecular protein rings in the absence of bulk water. *Science* **2005**, *310*, 1658–1661.
49. Badman, E. R.; Myung, S.; Clemmer, D. E. Evidence for unfolding and refolding of gas-phase cytochrome c ions in a Paul trap. *J. Am. Soc. Mass Spectrom.* **2005**, *16*, 1493–1497.
50. Myung, S.; Badman, E. R.; Lee, Y. J.; Clemmer, D. E. Structural transitions of electrosprayed ubiquitin ions stored in an ion trap over approximately 10 ms to 30 s. *J. Phys. Chem. A* **2002**, *106*, 9976–9982.
51. Revercomb, H. E.; Mason, E. A. Theory of plasma chromatography/gaseous electrophoresis—A review. *Anal. Chem.* **1975**, *47*, 970–983.
52. Mason, E. A. Ion mobilty: Its role in plasma chromatography. In: Carr, T. W. (Ed.), *Plasma Chromatography*, Plenum Press, New York, **1984**, pp. 43–93.
53. Shvartsburg, A. A.; Jarrold, M. F. An exact hard-spheres scattering model for the mobilities of polyatomic ions. *Chem. Phys. Lett.* **1996**, *261*, 86–91.
54. Mesleh, M. F.; Hunter, J. M.; Shvartsburg, A. A.; Schatz, G. C.; Jarrold, M. F. Structural information from ion mobility measurements: Effects of the long-range potential. *J. Phys. Chem.* **1996**, *100*, 16082–16086.
55. Wyttenbach, T.; von Helden, G.; Batka, J. J., Jr.; Carlat, D.; Bowers, M. T. Effect of the long-range interaction potential on ion mobility measurements. *J. Am. Soc. Mass Spectrom.* **1997**, *8*, 275–282.
56. Wyttenbach, T.; Witt, M.; Bowers, M. T. On the stability of amino acid zwitterions in the gas phase: The influence of derivatization, proton affinity, and alkali ion addition. *J. Am. Chem. Soc.* **2000**, *122*, 3458–3464.

57. Asbury, G. R.; Hill, H. H., Jr. Using different drift gases to change separation factors (α) in ion mobility spectrometry. *Anal. Chem.* **2000**, *72*, 580–584.
58. Asbury, G. R.; Hill, H. H., Jr. Evaluation of ultrahigh resolution ion mobility spectrometry as an analytical separation device in chromatographic terms. *J. Microcolumn Sep.* **2000**, *12*, 172–178.
59. Matz, L. M.; Hill, H. H.; Beegle, L. W.; Kanik, I. Investigation of drift gas selectivity in high resolution ion mobility spectrometry with mass spectrometry detection. *J. Am. Soc. Mass Spectrom.* **2002**, *13*, 300–307.
60. Beegle, L. W.; Kanik, I.; Matz, L.; Hill, H. H. Effects of drift-gas polarizability on glycine peptides in ion mobility spectrometry. *Int. J. Mass Spectrom.* **2002**, *216*, 257–268.
61. Mason, E. A.; McDaniel, E. W. Some accurate theoretical results. In *Transport Properties of Ions in Gases*, John Wiley & Sons, New York, **1988**, Chapter 6, pp. 225–381.
62. Tammet, H. Size and mobility of nanometer particles, clusters, and ions. *J. Aerosol Sci.* **1995**, *26*, 459–475.
63. Campbell, S.; Rodgers, M. T.; Marzluff, E. M.; Beauchamp, J. L. Structural and energetic constraints on gas phase hydrogen/deuterium exchange reactions of protonated peptides with D_2O, CD_3OD, CD_3CO_2D, and ND_3. *J. Am. Chem. Soc.* **1994**, *116*, 9765–9766.
64. Dwivedi, P.; Wu, C.; Matz, L. M.; Clowers, B. H.; Seims, W. F.; Hill, H. H., Jr. Gas-phase chiral separations by ion mobility spectrometry. *Anal. Chem.* **2006**, *78*, 8200–8206.
65. Valentine, S. J.; Clemmer, D. E. H/D exchange levels of shape-resolved cytochrome c conformers in the gas phase. *J. Am. Chem. Soc.* **1997**, *119*, 3558–3566.
66. Valentine, S. J.; Clemmer, D. E. Temperature-dependent H/D exchange of compact and elongated cytochrome c ions in the gas phase. *J. Am. Soc. Mass Spectrom.* **2002**, *13*, 506–517.
67. Robinson, E. W.; Williams, E. R. Multidimensional separations of ubiquitin conformers in the gas phase: Relating ion cross sections to H/D exchange measurements. *J. Am. Soc. Mass Spectrom.* **2005**, *16*, 1427–1437.
68. Wyttenbach, T.; Paizs, B.; Barran, P.; Breci, L.; Liu, D.; Suhai, S.; Wysocki, V. H.; Bowers, M. T. The effect of the initial water of hydration on the energetics, structures, and H/D exchange mechanism of a family of pentapeptides: An experimental and theoretical study. *J. Am. Chem. Soc.* **2003**, *125*, 13768–13775.
69. Ruotolo, B. T.; Russell, D. H. Gas-Phase Conformations of Proteolytically derived protein fragments: Influence of solvent on peptide conformation. *J. Phys. Chem. B* **2004**, *108*, 15321–15331.
70. Sawyer, H. A.; Marini, J. T.; Stone, E. G.; Ruotolo, B. T.; Gillig, K. J.; Russell, D. H. The Structure of gas-phase bradykinin fragment 1–5 (RPPGF) Ions: An ion mobility spectrometry and H/D exchange ion–molecule reaction chemistry study. *J. Am. Soc. Mass Spectrom.* **2005**, *16*, 893–905.
71. Woenckhaus, J.; Mao, Y.; Jarrold, M. F. Hydration of gas phase proteins: Folded +5 and unfolded +7 charge states of cytochrome c. *J. Phys. Chem. B* **1997**, *101*, 847–851.
72. Woenckhaus, J.; Hudgins, R. R.; Jarrold, M. F. Hydration of gas-phase proteins: A special hydration site on gas-phase BPTI (bovine pancreatic trypsin inhibitor). *J. Am. Chem. Soc.* **1997**, *119*, 9586–9587.
73. Fye, J. L.; Woenckhaus, J.; Jarrold, M. F. Hydration of folded and unfolded gas-phase proteins: saturation of cytochrome c and apomyoglobin. *J. Am. Chem. Soc.* **1998**, *120*, 1327–1328.
74. Jarrold, M. F. Unfolding, refolding, and hydration of proteins in the gas phase. *Acc. Chem. Res.* **1999**, *32*, 360–367.
75. Mao, Y.; Ratner, M. A.; Jarrold, M. F. One water molecule stiffens a protein. *J. Am. Chem. Soc.* **2000**, *122*, 2950–2951.
76. Kohtani, M.; Jarrold, M. F. The initial steps in the hydration of unsolvated peptides: Water molecule adsorption on alanine-based helices and globules. *J. Am. Chem. Soc.* **2002**, *124*, 11148–11158.
77. Kohtani, M.; Breaux, G. A.; Jarrold, M. F. Water molecule adsorption on protonated dipeptides. *J. Am. Chem. Soc.* **2004**, *126*, 1206–1213.
78. Liu, D.; Wyttenbach, T.; Barran, P. E.; Bowers, M. T. Sequential hydration of small protonated peptides. *J. Am. Chem. Soc.* **2003**, *125*, 8458–8464.
79. Liu, D.; Wyttenbach, T.; Carpenter, C. J.; Bowers, M. T. Investigation of noncovalent interactions in deprotonated peptides: Structural and energetic competition between aggregation and hydration. *J. Am. Chem. Soc.* **2004**, *126*, 3261–3270.
80. Wyttenbach, T.; Liu, D.; Bowers, M. T. Hydration of small peptides. *Int. J. Mass Spectrom.* **2005**, *240*, 221–232 and references therein.
81. Liu, D.; Wyttenbach, T.; Bowers, M. T. Hydration of mononucleotides. *J. Am. Chem. Soc.* **2006**, *128*, 15155–15163.
82. Dugourd, Ph.; Hudgins, R. R.; Clemmer, D. E.; Jarrold, M. F. High-resolution ion mobility measurements. *Rev. Sci. Instr.* **1997**, *68*, 1122–1129.
83. Wu, C.; Siems, W. F.; Asbury, G. R.; Hill, H. H., Jr. Electrospray ionization high-resolution ion mobility spectrometry–mass spectrometry. *Anal. Chem.* **1998**, *70*, 4929–4938.
84. Rokushika, S.; Hatano, H.; Baim, M. A.; Hill, H. H., Jr. Resolution measurement for ion mobility spectrometry. *Anal. Chem.* **1985**, *57*, 1902–1907.
85. Siems, W. F.; Wu, C.; Tarver, E. E.; Hill, H. H., Jr.; Larsen, P. R.; McMinn, D. G. Measuring the resolving power of ion mobility spectrometers. *Anal. Chem.* **1994**, *66*, 4195–4201.
86. Ruotolo, B. T.; McLean, J. A.; Gillig, K. J.; Russell, D. H. The influence and utility of varying field strength for the separation of tryptic peptides by ion mobility–

mass spectrometry. *J. Am. Soc. Mass Spectrom.* **2005**, *16*, 158–165.

87. Buryakov, I. A.; Krylov, E. V.; Nazarov, E. G.; Kh. Rasulev, U. A new method of separation of multi-atomic ions by mobility at atmospheric pressure using a high-frequency amplitude-asymmetric strong electric field. *Int. J. Mass Spectrom. Ion Proc.* **1993**, *128*, 143–148.
88. Purves, R. W.; Guevremont, R.; Day, S.; Pipich, C. W.; Matyjaszczyk, M. S. Mass spectrometric characterization of a high-field asymmetric waveform ion mobility spectrometer. *Rev. Sci. Inst.* **1998**, *69*, 4094–4105.
89. Guevremont, R.; Purves, R. W. Atmospheric pressure ion focusing in a high-field asymmetric waveform ion mobility spectrometer. *Rev. Sci. Inst.* **1999**, *70*, 1370–1383.
90. Guevremont, R. High-field asymmetric waveform ion mobility spectrometry: A new tool for mass spectrometry. *J. Chrom. A* **1058**, *2004*, 3–19.
91. Krylova, N.; Krylov, E.; Eiceman, G. A.; Stone, J. A. Effect of moisture on the field dependence of Mobility for gas-phase ions of organophosphorus compounds at atmospheric pressure with field asymmetric ion mobility spectrometry. *J. Phys. Chem. A* **2003**, *107*, 3648–3654.
92. Shvartsburg, A. A.; Tang, K.; Smith, R. D. Optimization of the design and operation of FAIMS analyzers. *J. Am. Soc. Mass Spectrom.* **2005**, *16*, 2–12.
93. Purves, R. W.; Barnett, D. A.; Ells, B.; Guevremont, R. Investigation of bovine ubiquitin conformers separated by high-field asymmetric waveform ion mobility spectrometry: Cross section measurements using energy-loss experiments with a triple quadrupole mass spectrometer. *J. Am. Soc. Mass Spectrom.* **2000**, *11*, 738–745.
94. MOBCAL is available from Prof. Martin F. Jarrold (Department of Chemistry, Indiana University) at http://nano.chem.indiana.edu/Software.html.
95. Prof. Michael T. Bowers (Department of Chemistry, University of California Santa Barbara), http://bowers.chem.ucsb.edu/theory_analysis/.
96. Shvartsburg, A. A.; Schatz, G. C.; Jarrold, M. F. Mobilities of carbon cluster ions: critical importance of the molecular attractive potential. *J. Chem. Phys.* **1998**, *108*, 2416–2423.
97. Hudgins, R. R.; Jarrold, M. F. Conformations of unsolvated glycine-based peptides. *J. Phys. Chem. B* **2000**, *104*, 2154–2158.
98. Bernstein, S. L.; Wyttenbach, T.; Baumketner, A.; Shea, J.-E.; Bitan, G.; Teplow, D. B.; Bowers, M. T. Amyloid β-protein: Monomer structure and early aggregation states of Aβ42 and its Pro19 alloform. *J. Am. Chem. Soc.* **2005**, *127*, 2075–2084.
99. Karasek, F. W.; Hill, H. H., Jr.; Kim, S. H. Plasma chromatography of heroin and cocaine with mass-identified mobility spectra. *J. Chromatog.* **1976**, *117*, 327–336.
100. Hudgins, R. R.; Woenckhaus, J.; Jarrold, M. F. High resolution ion mobility measurements for gas phase proteins: Correlation between solution phase and gas phase conformations. *Int. J. Mass Spectrom. Ion Proc.* **1997**, *165/166*, 497–507.
101. Liu, Y.; Valentine, S. J.; Counterman, A. E.; Hoaglund, C. S.; Clemmer, D. E. Injected-ion mobility analysis of biomolecules. *Anal. Chem.* **1997**, *69*, 728A–735A.
102. Wu, C.; Siems, W. F.; Asbury, G. R.; Hill, H. H., Jr. Electrospray ionization high-resolution ion mobility spectrometry–mass spectrometry. *Anal. Chem.* **1998**, *70*, 4929–4938.
103. Hoaglund, C. S.; Valentine, S. J.; Clemmer, D. E. An ion trap interface for esi-ion mobility experiments. *Anal. Chem.* **1997**, *69*, 4156–4161.
104. Creaser, C. S.; Benyezzar, M.; Griffiths, J. R.; Stygall, J. W. A tandem ion trap/ion mobility spectrometer. *Anal. Chem.* **2000**, *72*, 2724–2729.
105. von Helden, G.; Wyttenbach, T.; Bowers, M. T. Inclusion of a MALDI ion source in the ion chromatography technique: Conformational information on polymer and biomolecular ions. *Int. J. Mass Spectrom. Ion Proc.* **1995**, *146/147*, 349–364.
106. Bluhm, B. K.; Gillig, K. J.; Russell, D. H. Development of a Fourier-transform ion cyclotron resonance mass spectrometer-ion mobility spectrometer. *Rev. Sci. Instr.* **2000**, *71*, 4078–4086.
107. Hoaglund, C. S.; Valentine, S. J.; Sporleder, C. R.; Reilly, J. P.; Clemmer, D. E. Three-dimensional ion mobility/TOFMS analysis of electrosprayed biomolecules. *Anal. Chem.* **1998**, *70*, 2236–2242.
108. Gillig, K. J.; Ruotolo, B. T.; Stone, E. G.; Russell, D. H.; Fuhrer, K.; Gonin, M.; Schultz, J. A. Coupling high-pressure MALDI with ion mobility/orthogonal time-of-flight mass spectrometry. *Anal. Chem.* **2000**, *72*, 3965–3971.
109. Javahery, G.; Thomson, B. A Segmented radiofrequency-only quadrupole collision cell for measurements of ion collision cross section on a triple quadrupole mass spectrometer. *J. Am. Soc. Mass Spectrom.* **1997**, *8*, 697–702.
110. Guo, Y.; Wang, J.; Javahery, G.; Thomson, B. A.; Siu, K. W. M. Ion mobility spectrometer with radial collisional focusing. *Anal. Chem.* **2005**, *77*, 266–275.
111. Guo, Y.; Ling, Y.; Thomson, B. A.; Siu, K. W. M. Combined ion-mobility and mass-spectrometry investigations of metallothionein complexes using a tandem mass spectrometer with a segmented second quadrupole. *J. Am. Soc. Mass Spectrom.* **2005**, *16*, 1787–1794.
112. Giles, K.; Pringle, S. D.; Worthington, K. R.; Little, D.; Wildgoose, J. L.; Bateman, R. H. Applications of a traveling wave-based radiofrequency-only stacked ring ion guide. *Rapid Commun. Mass Spectrom.* **2004**, *18*, 2401–2414.
113. Ruotolo, B. T.; Giles, K.; Campuzano, I.; Sandercock, A. M.; Bateman, R. H.; Robinson, C. V. Evidence for

macromolecular protein rings in the absence of bulk water. *Science* **2005**, *310*, 1658–1661.
114. Tanaka, K.; Waki, H.; Ido, Y.; Akita, S.; Yoshida, Y.; Yohida, T. Protein and polymer analyses up to *m/z* 100,000 by laser ionization time-of-flight mass spectrometry. *Rapid Commun. Mass Spectrom.* **1988**, *2*, 151–153.
115. Karas, M.; Hillenkamp, F. Laser desorption ionization of proteins with molecular masses exceeding 10,000 daltons. *Anal. Chem.* **1988**, *60*, 2299–2301.
116. Fenn, J. B.; Mann, M.; Meng, C. K.; Wong, S. F.; Whitehouse, C. M. Electrospray ionization for mass spectrometry of large biomolecules. *Science* **1989**, *246*, 64–71.
117. Steiner, W. E.; Clowers, B. H.; English, W. A.; Hill, H. H., Jr. Atmospheric pressure matrix-assisted laser desorption ionization with analysis by ion mobility time-of-flight mass spectrometry. *Rap. Commun. Mass Spectrom.* **2004**, *18*, 882–888.
118. von Helden, G.; Hsu, M.-T.; Kemper, P. R.; Bowers, M. T. Structures of carbon cluster ions from 3 to 60 atoms: Linears to rings to fullerenes. *J. Chem. Phys.* **1991**, *95*, 3835–3837.
119. Mason, E. A.; McDaniel, E. W. Measurement of drift velocities and longitudinal diffusion coefficients. In *Transport Properties of Ions in Gases*, John Wiley & Sons, New York, **1988**, Chapter 2, pp. 31–102.
120. Shumate, C. B.; Hill, H. H., Jr. Coronaspray nebulization and ionization of liquid samples for ion mobility spectrometry. *Anal. Chem.* **1989**, *61*, 601–606.
121. Wyttenbach, T.; Kemper, P. R.; Bowers, M. T. Design of a new electrospray ion mobility mass spectrometer. *Int. J. Mass Spectrom.* **2001**, *212*, 13–23.
122. Kinnear, B. S.; Hartings, M. R.; Jarrold, M. F. Helix unfolding in unsolvated peptides. *J. Am. Chem. Soc.* **2001**, *123*, 5660–5667.
123. Hoaglund-Hyzer, C. S.; Lee, Y. J.; Counterman, A. E.; Clemmer, D. E. Coupling ion mobility separations, collisional activation techniques, and multiple stages of MS for analysis of complex peptide mixtures. *Anal. Chem.* **2002**, *74*, 992–1006.
124. Valentine, S. J.; Koeniger, S. L.; Clemmer, D. E. A split-field drift tube for separation and efficient fragmentation of biomolecular ions. *Anal. Chem.* **2003**, *75*, 6202–6208.
125. Clowers, B. H.; Hill, H. H., Jr. Mass analysis of mobility-selected ion populations using dual gate, ion mobility, quadrupole ion trap mass spectrometry. *Anal. Chem.* **2005**, *77*, 5877–5885.
126. Loo, J. A. Electrospray ionization mass spectrometry: A technology for studying noncovalent macromolecular complexes. *Int. J. Mass Spectrom.* **2000**, *200*, 175–186.
127. Winston, R. L.; Fitzgerald, M. C. Mass spectrometry as a readout of protein structure and function. *Mass Spectrom. Rev.* **1997**, *16*, 165–179.
128. McLean, J. A.; Russell, D. H. Sub-femtomole peptide detection in ion mobility–time-of-flight mass spectrometry measurements. *J. Proteome Res.* **2003**, *2*, 428–431.
129. Gillig, K. J.; Ruotolo, B. T.; Stone, E. G.; Russell, D. H. An electrostatic focusing ion guide for ion mobility-mass spectrometry. *Int. J. Mass Spectrom.* **2004**, *239*, 43–49.
130. Verbeck, G. F.; Ruotolo, B. T.; Gillig, K. J.; Russell, D. H. Resolution equations for high-field ion mobility. *J. Am. Soc. Mass Spectrom.* **2004**, *15*, 1320–1324.
131. Rokushika, S.; Hatano, H.; Baim, M. A.; Hill, H. H., Jr. Resolution measurement for ion mobility spectrometry. *Anal. Chem.* **1985**, *57*, 1902–1907.
132. Watts, P.; Wilder, A. On the resolution obtainable in practical ion mobility systems. *Int. J. Mass Spectrom. Ion Proc.* **1992**, *112*, 179–190.
133. Berry, C. E. Image curvature caused by fringing fields in magnetic sector mass spectrometers. *Rev. Sci. Instrum.* **1956**, *27*, 849–853.
134. Fuhrer, K.; Gonin, M.; McCully, M. I.; Egan, T.; Ulrich, S. R.; Vaughn, V. W.; Burton, W. D., Jr.; Schultz, J. A.; Gillig, K.; Russell, D. H. Monitoring of fast processes by TOFMS. In the 49th American Society for Mass Spectrometry Annual Meeting, Chicago, IL, May 2001.
135. Tang, K.; Shvartsburg, A. A.; Lee, H.-N.; Prior, D. C.; Buschbach, M. A.; Li, F.; Tolmachev, A. V.; Anderson, G. A.; Smith, R. D. High-sensitivity ion mobility spectrometry/mass spectrometry using electrodynamic ion funnel interfaces. *Anal. Chem.* **2005**, *77*, 3330–3339.
136. Koeniger, S. L.; Merenbloom, S. I.; Valentine, S. J.; Jarrold, M. F.; Udseth, H. R.; Smith, R. D.; Clemmer, D. E. An IMS-IMS Analogue of MS-MS. *Anal. Chem.* **2006**, *78*, 4161–4174.
137. Valentine, S. J.; Kulchania, M.; Srebalus Barnes, C. A.; Clemmer, D. E. Multidimensional separations of complex peptide mixtures: A combined high-performance liquid chromatography/ion mobility/time-of-flight mass spectrometry approach. *Int. J. Mass Spectrom.* **2001**, *212*, 97–109.
138. Lee, Y. J.; Hoaglund-Hyzer, C. S.; Srebalus Barnes, C. A.; Hilderbrand, A. E.; Valentine, S. J.; Clemmer, D. E. Development of high-throughput liquid chromatography injected ion mobility quadrupole time-of-flight techniques for analysis of complex peptide mixtures. *J. Chromatog.* **2002**, *782B*, 343–351.
139. Moon, M. H.; Myung, S.; Plasencia, M.; Hilderbrand, A. E.; Clemmer, D. E. Nanoflow LC/ion mobility/CID/TOF for proteomics: Analysis of a human urinary proteome. *J. Proteome Res.* **2003**, *2*, 589–597.
140. Myung, S.; Lee, Y. J.; Moon Myeong, M. H.; Taraszka, J.; Sowell, R.; Koeniger, S.; Hilderbrand, A. E.; Valentine, S. J.; Cherbas, L.; Cherbas, P.; Kaufmann, T. C.; Miller, D. F.; Mechref, Y.; Novotny, M. V.; Ewing, M. A.; Sporleder, C. R.; Clemmer, D. E. Development of high-sensitivity ion trap ion mobility

spectrometry time-of-flight techniques: A high-throughput nano-LC-IMS-TOF separation of peptides arising from a *Drosophila* protein extract. *Anal. Chem.* **2003**, *75*, 5137–5145.

141. Liu, X.; Plasencia, M.; Ragg, S.; Valentine, S. J.; Clemmer, D. E. Development of high throughput dispersive LC–ion mobility–TOFMS techniques for analysing the human plasma proteome. *Brief Funct. Genomes and Proteomes* **2004**, *3*, 177–186.
142. Taraszka, J. A.; Gao, X.; Valentine, S. J.; Sowell, R. A.; Koeniger, S. L.; Miller, D. F.; Kaufman, T. C.; Clemmer, D. E. Proteome profiling for assessing diversity: Analysis of individual heads of *Drosophila melanogaster* using LC–ion mobility–MS. *J. Proteome Res.* **2005**, *4*, 1238–1247.
143. Yates, J. R. III. Database searching using mass spectrometry data. *Electrophoresis* **1998**, *19*, 893–900.
144. McLafferty, F. W. (Ed.), *Tandem Mass Spectrometry*, Plenum Press, New York, **1983**.
145. Hoaglund-Hyzer, C. S.; Li, J.; Clemmer, D. E. Mobility labeling for parallel CID of ion mixtures. *Anal. Chem.* **2000**, *72*, 2737–2740.
146. Stone, E.; Gillig, K. J.; Ruotolo, B.; Fuhrer, K.; Gonin, M.; Schultz, A.; Russell, D. H. Surface-induced dissociation on a MALDI–ion mobility–orthogonal time-of-flight mass spectrometer: Sequencing peptides from an "in-solution" protein digest. *Anal. Chem.* **2001**, *73*, 2233–2238.
147. McLean, J. A.; Gillig, K. J.; Ruotolo, B. T.; Ugarov, M.; Bensaoula, H.; Egan, T.; Schultz, J. A.; Russell, D. H. Ion mobility-photodissociation (213 nm)-time-of-flight mass spectrometry for simultaneous peptide mass mapping and protein sequencing. In *Proceedings of the 51st Conference of the American Society for Mass Spectrometry, Montreal, Canada*, **2003**.
148. Koeniger, S. L.; Valentine, S. J.; Myung, S.; Plasencia, M.; Lee, Y. J.; Clemmer, D. E. Development of field modulation in a split-field drift tube for high-throughput multidimensional separations. *J. Proteome Res.* **2005**, *4*, 25–35.
149. Caprioli, R. M.; Farmer, T. B.; Gile, J. Molecular imaging of biological samples: Localization of peptides and proteins Using MALDI-TOFMS. *Anal. Chem.* **1997**, *69*, 4751–4760.
150. Chaurand, P.; Norris, J. L.; Cornett, S. D.; Mobley, J. A.; Caprioli, R. M. New developments in profiling and imaging of proteins from tissue sections by MALDI mass spectrometry. *J. Proteome Res.* **2006**, *5*, 2889–2900 and references therein.
151. Jackson, S. N.; Ugarov, M.; Egan, T.; Post, J. D.; Langlais, D.; Schultz, J. A.; Woods, A. S. MALDI–ion mobility–TOFMS imaging of lipids in rat brain tissue. *J. Mass Spectrom.* **2007**, *42*, 1093–1098.
152. McLean, J. A.; Ridenour, W. B.; Caprioli, R. M. Profiling and imaging of tissues by imaging ion mobility–mass spectrometry. *J. Mass Spectrom.* **2007**, *42*, 1099–1105.
153. Scott, C. D.; Ugarov, M. V.; Hauge, R. H.; Sosa, E. D.; Arepalli, S.; Schultz, J. A.; Yowell, L. Characterization of large fullerenes in single-wall carbon nanotube production by ion mobility mass spectrometry. *J. Phys. Chem.* **2007**, *111C*, 36–44.

Part IV

Practical Aspects of ES and MALDI

Chapter 13

ESI, APCI, and MALDI A Comparison of the Central Analytical Figures of Merit: Sensitivity, Reproducibility, and Speed

Thomas R. Covey, Bradley B. Schneider, Hassan Javaheri, J.C. Yves LeBlanc, Gordana Ivosev, Jay J. Corr, and Peter Kovarik

ABI/MDS-Sciex, Concord, Ontario, Canada

Electrospray and MALDI Mass Spectrometry: Fundamentals, Instrumentation, Practicalities, and Biological Applications, Second Edition, Edited by Richard B. Cole

13.1 INTRODUCTION

This chapter provides an analysis of three widely used ionization techniques examining, for each ion source, the three most important figures of merit for any analytical technique: sensitivity, reproducibility, and speed. Because this analysis was done primarily on a single type of mass analyzer common to all sources—namely, a triple quadrupole (QqQ)—the performance characteristics of the sources can be directly compared. To a large extent, the ion source behavior will be examined in the context of quantitative measurements where the QqQ, operating in multiple reaction monitoring (MRM) mode, is the analyzer and method of choice. In the field of quantitation, more attention is applied to statistically validating these figures of merit and maintaining them on a daily basis, than to qualitative methodologies such as the identification of unknowns. As a result, there is a large body of data upon which to draw accurate representations of past and current performance measurements.

In this chapter we will attempt to establish, within reasonable limits, the current performance levels of the three ion sources with regard to these three metrics. First to be considered for each metric will be the fundamental limits imposed by the instrumentation and ionization processes. The second consideration for each metric will be the single most important analytical parameter under control of the analyst that determines how close these fundamental limits can be approached. Third, after all aspects of an analytical methodology are considered, reference to practically achievable performance levels for these three metrics will be discussed. Hopefully, this discussion will provide some basis for predicting what improvements may or may not be possible in the future. Insight regarding which ionization technique would be most appropriate for a given application is also a consequence of such an analysis. Electrospray ionization (ESI),[1–3] particularly with its pneumatically assisted counterparts,[4,5] and atmospheric pressure chemical ionization[6–8] are generally chosen for quantitative bioanalytical measurements, depending on which demonstrates the lowest detection limits and best reproducibility. In most regards they are similar with respect to speed of analysis. Matrix-assisted laser desorption/ionization (MALDI)[9,10] is an emerging technique in the area of quantitation which has the potential to greatly exceed the other sources on the third figure of merit—that is, speed of analysis. Quantitative applications where speed is more important than sensitivity and reproducibility may find MALDI becoming an increasingly important tool.

13.2 SENSITIVITY

13.2.1 Sampling Efficiency

One means of quantifying the sensitivity of an ion source is to measure the sampling efficiency.[11] The sampling efficiency is the product of the ionization efficiency (efficiency of ion creation) and transfer efficiency (efficiency of transferring ions from atmosphere into vacuum). The sampling efficiency can be measured with reasonable accuracy, however, it is difficult to directly measure either the ionization efficiency or transfer efficiency. Ideal instrumental and sample inlet conditions can be established where the transfer efficiency is nearly 100% for ESI and MALDI, which allows the ionization efficiency to be inferred from the sampling efficiency measurement. Figure 13.1 diagrammatically represents these concepts for the three major ion sources. High-flow ESI and APCI are depicted together in Figure 13.1A because their characteristics are so similar with regard to sampling efficiencies. Nanoelectrospray (NanoESI) is treated as a special case in Figure 13.1B, and MALDI is reflected in Figure 13.1C.

The sampling efficiency is obtained by recording the number of ions that hit the detector from a known amount injected or deposited into the ion source. Experiments using an electrometer to measure ion current at various elements in the ion optics provide data on the losses that occur in the various segments of the path to the mass analyzer. This data can be used to calculate the number of ions entering the first stage of the ion optic path. When the measurements are done in the MRM mode, as discussed throughout this chapter, these losses include those at the mass filters, lenses, and detector, as well as those in the collision cell itself when the molecular ion is fragmented into several product ions and only one is monitored.[11] As an example, on an API 5000 under the conditions described here, 5% of the Reserpine precursor ion current that enters the first stage of the ion optics from the ion source arrives at the detector when monitoring the product ion *m/z* 195. In the work described throughout this chapter, the basic mass analyzer and vacuum system used for high-flow ESI and APCI was an API 4000™ or API 5000™ triple quadrupole with some modifications as noted for NanoESI[11,12] and MALDI.[10] The absolute numbers reported here for sensitivity, reproducibility, and speed will differ on different mass analyzers, instrument models, and ion sources of different designs, but the general trends and the relative numbers—that is, relative sensitivity of different compounds with a particular ionization technique and relative performance of the different sources compared to each other—should be similar regardless of the mass analyzer or instrument model.

Sections 13.2.1.1–13.2.1.3 describe sampling efficiency under conditions that provide the highest sensitivity currently achievable with this instrumentation or modifications to it. The construction of the sources and interfaces are such that the transfer efficiencies are as

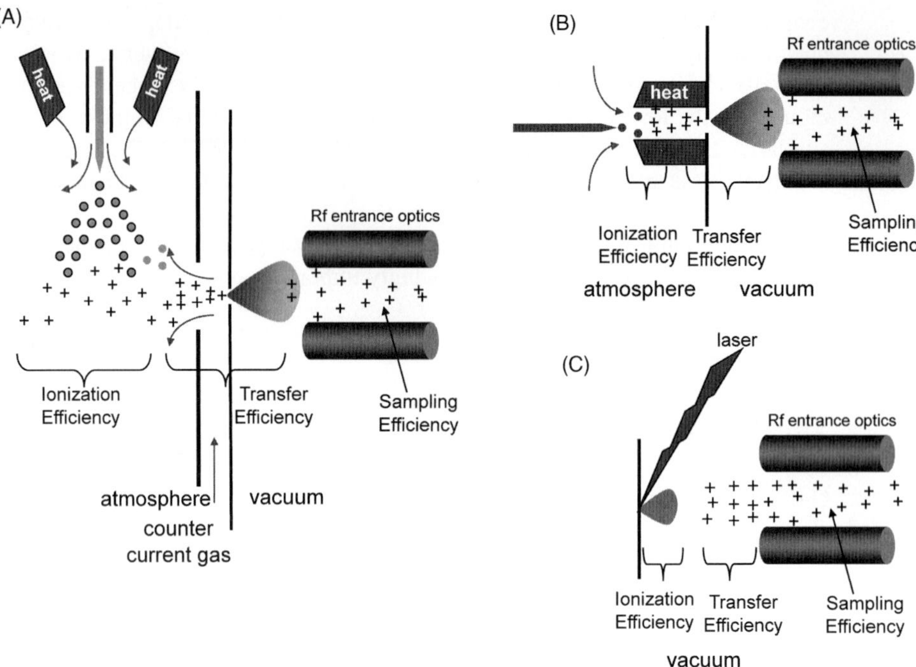

Figure 13.1. Diagram of the origin of ion losses from different ion sources expressed in terms of ionization and transfer efficiency. The product of these two is the sampling efficiency which represents the percentage of molecules that enter an ion source, either in the liquid phase with ESI and APCI or on a surface with MALDI, and are captured as ions in the Rf ion guide optics. The sampling efficiency can be calculated from the signal at the detector when the losses in the mass analyzer and collision cell are well-characterized. (**A**) Diagram of a high-flow-rate electrospray source and an APCI source where the transfer efficiency is low but the ionization efficiency is high. (**B**) Diagram of a NanoESI source where both the ionization and transfer efficiency are high. (**C**) Diagram of a vacuum MALDI source where the ionization efficiency is lower than the spray techniques but the transfer efficiency is high.

high as possible, approaching 100% in some of the cases described. Estimation of the ionization efficiency can then be made from the sampling efficiency measurement. For all three ionization techniques, the ionization efficiency varies over several orders of magnitude, depending on the chemical species being measured (see Section 13.2.1.4). It should be understood that sample introduction conditions providing close to 100% transfer efficiency are not always practical from an analytical methodology perspective, and variations in these conditions can have a profound effect on the transfer efficiencies. Section 13.2.2 will deal with variation in sampling efficiencies as a function of sample introduction conditions, primarily liquid flow rate. The final section of the sensitivity topic (Section 13.2.3) will deal with how all of the above translate into limits of detection in real-world situations and thus will consider signal-to-noise (S/N) as well as absolute sensitivity.

13.2.1.1 ESI

In the case of NanoESI (50- to 500-nL/min flow rates), very high sampling efficiencies have been observed for compounds that have high ionization efficiency. Under NanoESI conditions, the transfer efficiency is optimal. Sampling efficiencies greater than 85% have been measured at concentrations below source saturation; this indicates that, individually, both the ionization and transfer efficiencies must exceed this number.[13] It should be pointed out that achieving these very high sampling efficiencies requires very low flow rates and special experimental care, particularly with reference to sprayer geometry and positions, ensuring that all of the sprayed solvent is captured in the laminar flow of gas through the atmospheric aperture established by the vacuum drag of the analyzer pumps. Instrumentation with large apertures, capable of handling large gas loads, and simultaneously providing efficient desolvation conditions can achieve these peak efficiencies of 85% more easily, and with flows as high as 500 nL/min.[11] The instrument in reference 11 was designed to maximize the sampling efficiency in order to establish the theoretical sensitivity limits of ESI. This indicates that there is not much room for improvement; however such instrumentation is not commercially available today. More typical NanoESI sampling efficiencies with widely available instrumentation would be on the order of 5–20%,[10,13] with losses primarily occurring during the transfer of ions into the vacuum chamber optics; they are not due to reduced ionization efficiency in the atmospheric region. Even these efficiencies are quite high and can be achieved as long as the flow rates are less than approximately 500 nL/min, where the entire vaporized solvent stream can be "inhaled" by the vacuum system. The sensitivity ceiling for NanoESI is clearly in sight today. As will be described in Section 13.2.2.1 sampling efficiency is considerably different at higher flows where significant challenges and opportunities still lie.

13.2.1.2 APCI

APCI sampling efficiencies equivalent to NanoESI have not been observed, primarily because of the difficulty in creating conditions whereby the entire chemical ionization reaction region can be inhaled by the vacuum system. Discharge conditions create space charge and other effects that make total sample consumption difficult to achieve. APCI sampling efficiencies tend to be close to high-liquid-flow introduction ESI sources. The ionization efficiency of APCI can be close to that of ESI for compounds with high gas-phase proton affinities or basicities. The large number of collisions rapidly encountered between the vaporized analyte molecules and the reagent ions at atmospheric pressure ensures that reactions will proceed to a thermodynamic equilibrium and will not be kinetically limited. The ionization efficiencies for ESI and APCI are the highest observed for ion sources designed to create molecular ions from organic compounds. The majority of the losses occur during the transfer into the vacuum with APCI and high-flow ESI where ultimately sampling efficiencies of only a fraction of a percent are achieved. This is discussed further in Section 13.2.2.2, which deals with the effect of source conditions on sampling efficiencies; and, as mentioned earlier, further improvements remain to be harvested.

13.2.1.3 MALDI

In the case of MALDI the ion sampling efficiency is less than the maximum values observed for NanoESI but is not far off from the more typical values for the NanoESI operation cited above. MALDI sampling efficiency refers to the percentage of molecules deposited on the target that are converted to gas-phase ions and transferred into the initial focusing optics. Because MALDI is conducted inside the vacuum system, the transfer of ions created in the ablation plume into the optics can be nearly 100%; therefore, the sampling efficiency approaches the ionization efficiency. Measurements of this number for compounds with favorable ionization characteristics, and under conditions of nearly 100% transfer efficiency, have been reported to be on the order of 5%.[10] Since the ionization process is fundamental, within reasonable error this 5% value approximates the MALDI sensitivity limit today. Although lower than the upper theoretical limit demonstrated for ESI (>85%), it is still very high by organic mass spectrometry standards and, like the other techniques, will vary within a wide range below this number, depending on the chemical entity, matrix, sample deposition, and general source tuning conditions. To obtain the efficiency measurements cited here, a known amount of sample was deposited on the target and the entire spot was completely ablated by continuous rastering of the spot with the laser; the experimental approach is depicted in Figure 13.2. This means that to obtain these high efficiencies for MALDI, conditions must be such that the entire spot can be consumed in the time frame of the experiment. This aspect will be discussed further in the next section considering the effects of source conditions on sensitivity, but the implication is that for analytical scenarios requiring fast measurements, and where there is no time to completely raster an entire spot, MALDI sampling efficiencies will be considerably lower than the 5% cited above. In situations where time is not of the essence and continuous rastering of an entire spot is possible, efficiencies similar to typical nanoESI can be achieved. The signal averaging of product ion spectra of peptides for database identification is one scenario where this is commonly practiced.

13.2.1.4 Compound Dependency

Variations in sampling efficiency between diverse compounds are primarily due to differences in ionization efficiency and not transfer efficiency, the transfer efficiencies

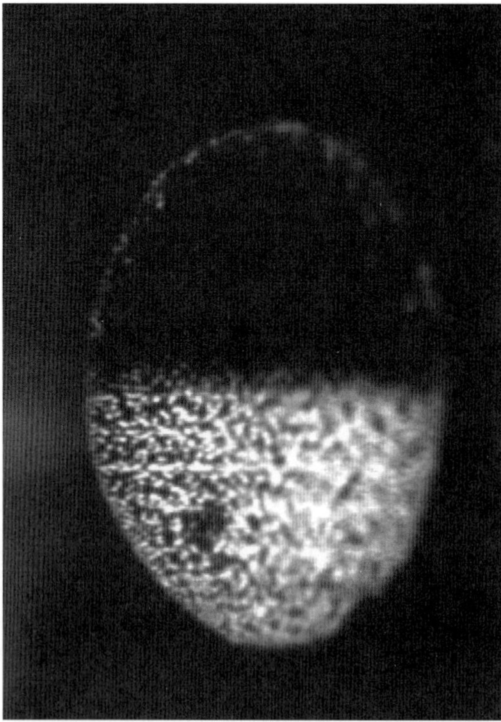

Figure 13.2. Image of a sample spot 50% ablated. The entire 0.5-μL spot was consumed in order to obtain a measure of the ion yield for a sample of Reserpine monitoring the 609 → 195 transition. (Reprinted from Ref. 10 with permission.)

being reasonably constant between different compounds with not widely disparate molecular weights. To establish some indication of the relative ionization efficiencies for different compounds with each source, a set of 31 molecules spanning chemical diversity relevant to the biomedical and pharmaceutical fields was chosen for study. In Tables 1–3 the data for each source is presented relative to the most efficiently ionized compound for that source. These data were obtained by comparing the sampling efficiencies of the different compounds, considering the total product ion current from the molecular ions reaching the detector, and accounting for all losses in the ion optics. The relative sampling and ionization efficiencies can be considered to be equivalent because the transfer efficiency does not vary

Table 13.1. Relative Sampling Efficiency and MRM Sensitivity with ESI for 31 Compounds [a]

Compound	Relative Sampling Efficiency (%)	Relative MRM Sensitivity (%)
Verapamil	100	100
Buscopan	65	40
Safranin Orange	58	24
Estradiol sulfate	49	16
Tamoxifen	45	53
Haloperidol	35	68
Morphine	16	4
Clenbuterol	10	24
Phenylbutazone	9.0	3
Bromocryptin	8.0	11
Morphine glucuronide	8	15
Succinyl choline	7	4
Aztreonam	5	2
Furosemide	4	5
Dianabol	3	3
Oxyphenylbutazone	3	2
Bentazon	2	2
Beclomethazone	2	2
Taurocholic acid	2	2
Naproxen	2	2
Fosinopril	2	3
Dexamethazone	2	1
Prednisolone	1	1
Testosterone glucuronide	1	1
5-Fluorouracil	1	1
Bromazepam	1	1
Taxol	1	0.3
Lovastatin	1	0.3
α-Cyclodextrin	0.3	0.2
Erythromycin	0.2	0.2
Cyclosporin	0.2	0.1

[a] The data are referenced to the most efficient molecule, in this case Verapamil. The MRM sensitivity is the signal obtained from one product ion which reflects typical assay conditions. The sampling efficiency refers to the number of ions entering the vacuum system optics. The relative sampling efficiency directly reflects the relative ionization efficiency because the transfer efficiency will be very similar for all compounds. The relative sampling and MRM efficiencies are similar but not identical because some compounds concentrate the majority of the ion current in a single product ion. Data acquired by flow injection analysis at 200 μL/min in 50/50 MeOH/H_2O, 0.1% formic acid. For the chemical structures see the Appendix or Ref. 57.

Table 13.2. Relative Sampling Efficiency and MRM Sensitivity with APCI for 31 Compounds [a]

Compound	Relative Sampling Efficiency (%)	Relative MRM Sensitivity (%)
Verapamil	100	100
Tamoxifen	27	31
Safranin Orange	16	6
Oxyphenylbutazone	15	9
Haloperidol	12	23
Morphine	7	2
Clenbuterol	6	14
Bromazepam	6	5
Dianabol	5	4
Beclomethazone	5	5
Phenylbutazone	5	2
Lovastatin	4	2
Dexamethazone	4	3
Prednisolone	2	1
5-Fluorouracil	1	1
Naproxen	0.3	0.5
Bentazon	0.3	0.3
Taxol	0.3	0.1
Estradiol sulfate	0.2	0.1
Buscopan	0.2	0.1
Bromocryptin	0.2	0.2
Cyclosporin	0.1	0.02
Furosemide	0.03	0.05
Fosinopril	0.02	0.04
Erythromycin	0.01	0.02
Taurocholic Acid	0.01	0.01
Testosterone glucuronide	0.004	0.003
Morphine glucuronide	0	0
Aztreonam	0	0
α-Cyclodextrin	0	0
Succinyl choline	0	0

[a] See caption for Table 13.1 for explanation of efficiencies. Data acquired by flow injection analysis at 1000 μL/min in 50/50 MeOH/H_2O. For the chemical structures see the Appendix or Ref. 57.

significantly within the narrow mass range studied here, and the losses in the optics are similar for all of the compounds. Large proteins were not considered. The relative MRM response is also shown in these tables because it is what is typically observed by the analyst. It considers only the signal obtained on the most abundant product ion and closely mirrors the sampling efficiency, but not exactly. This is because, among the different compounds, sensitivity is either boosted or diminished as a result of a large portion of the molecular ion current either being conserved in just one product ion or widely distributed to several product ions.

A few important generalities can be surmised from these data. The range of ESI is 500-fold and for the others much greater. The trend is similar for all sources: Highly sensitive or insensitive species for one source tend to behave similarly for the other sources, with some notable exceptions. These exceptions are primarily explained by the degree of thermal stability of a neutral species during the desorption and/or vaporization event, and the ability of a compound to withstand the energy transfer during the ionization event without fragmenting. ESI is by far the least energetic and is thus the most universal of the three

Table 13.3. Relative Sampling Efficiency and MRM Sensitivity with MALDI for 31 compounds [a]

Compound	Relative Sampling Efficiency (%)	Relative MRM Sensitivity (%)
Safranin Orange	100	100
Verapamil	33	80
Buscopan	27	39
Haloperidol	14	64
Tamoxifen	13	37
Bromazepam	5	9
Phenylbutazone	3	2
Morphine	2	1
Estradiol sulfate	2	1
Cyclosporine	2	0.5
Oxyphenylbutazone	0.3	0.4
Bromocryptin	0.2	0.6
Clenbuterol	0.2	1.0
Morphine glucuronide	0.2	0.7
Taurocholic acid	0.1	0.3
Testosterone glucuronide	0.1	0.2
Dianabol	0.1	0.2
Bentazon	0.1	0.2
Naproxen	0.04	0.1
Aztreonam	0.02	0.03
Dexamethazone	0.02	0.03
Furosemide	0.02	0.06
Lovastatin	0.013	0.02
Beclomethazone	0.01	0.02
Taxol	0.01	0.01
α-Cyclodextrin	0.02	0.01
5-Fluorouracil	0.004	0.01
Erythromycin	0.002	0.01
Fosinopril	0	0
Prednisolone	0	0
Succinyl choline	0	0

[a]See caption for Table 13.1 for explanation of efficiencies. Data acquired by ablating a portion of a 1-μL spot deposited on target. See text for further explanation of the method of data acquisition. For the chemical structures see the Appendix or Ref. 57.

ion sources. With ESI, the ionization event is mild, most often occurring in the final solution droplets, and the desorption event requires no sudden heating to launch the ion from the droplet. Even when heat is employed to aid desolvation, the droplet temperatures rise very little before the ion emission event. With regard to fragmentation, MALDI falls between ESI and APCI. Like APCI, MALDI is known to have a gas-phase chemical ionization component at play during the ionization step, which is more energetic than ESI. MALDI also depends to some degree on the thermal vaporization of neutral species, like APCI, but it is a more rapid heating process that is significantly dampened by the presence of excess MALDI matrix. Of the three ionization techniques, APCI shows these thermal fragmentation effects to the most severe degree followed by MALDI, both of which are relatively energetic processes compared to the ion evaporation process of ESI. MALDI and APCI are closer to each other than either is to ESI mechanistically and in the observed empirical results.

One good example of thermal effects is observed with Buscopan, which, based on its structure, would be expected to have high gas-phase proton affinity and ionization efficiency by APCI. Instead, its efficiency is very low relative to the other two sources. The energetics of the APCI process are too severe to maintain significant molecular ion current. In this case the culprit is more likely the thermal desorption than the ionization step. This observation derives by inference from the MALDI data. The energy transfer during the chemical ionization event in MALDI and APCI is likely similar with some consideration given to the proton affinity of the reagent ion population, both of which are different but difficult to characterize in MALDI. Yet with MALDI, this compound does not suffer the same amount of fragmentation as with APCI. The thermal vaporization step is more rapid and is also dampened with MALDI, leading to less fragmentation. A similar protective or thermal dampening effect is often observed with APCI when large quantities of materials exogenous to the analyte during the thermal desorption process are present. This will be shown in more detail in the discussion on reproducibility and the compound dependency observed with matrix effects (Section 13.3.2.3). Generally, drug conjugates such as glucuronides and sulfates also follow a behavior similar to that of Buscopan and exhibit thermal degradation with APCI and no degradation with ESI; their performance with MALDI falls somewhere between the two.

Azetreonam is another example of a very labile compound, but in this case, even the MALDI process is observed to be far too energetic, showing a 100-fold lower sampling efficiency relative to ESI. There are several cases of this scenario in the lower third of the list of compounds in Tables 13.1–13.3. In many cases, as can be seen toward the bottom of Tables 13.2 and 13.3, virtually no measurable signal can be obtained from these compounds with MALDI or APCI, whereas all compounds could be measured to some degree by ESI. Although there is roughly a 500-fold difference between the best and worst compounds in the ESI list, there remains sufficient sensitivity with those at the bottom of the list for many analytically useful applications, as will be shown in Section 13.2.3.1.

It should be pointed out that the methodology used to collect this relative ionization efficiency data was standardized to some degree. The expected molecular ions (protonated or deprotonated except in the case of quaternary amines) were used as the precursor ions, although adducts were considered but were not found to be the predominate ion in any of these cases. Fragment ions were not considered as precursor ions that may have been useful in some cases with APCI and MALDI; but in general when fragmentation occurred, it was widely distributed. Absolutely perfect conditions of solvent and matrix composition were not selected for each individual compound. Source inlet conditions were established to be relatively universal and remained fixed throughout. Temperatures, orifice potentials, and collision energies were the primary parameters tuned for each compound. For ESI, the flow was maintained at 200 μL/min and solvent composition at 1/1 MeOH/H_2O, 0.1% formic acid. Experiments were done to see if there were noticeable trends toward different relative ionization efficiencies under high-flow ESI conditions versus nanoESI conditions, and with careful source optimization, none were observed.[11] The ESI data of Table 13.1 should be reasonably representative over broad ranges of both flow and compound type. APCI was conducted at 1 mL/min with 1/1 MeOH/H_2O, and no attempt was made to adjust the proton affinity of the solvent relative to the analyte. All ESI and APCI data were generated by flow injection analysis where peak areas were used to obtain efficiency numbers. Extensive studies with a variety of matrices and matrix-free desorption surfaces concluded that α-cyano-4-hydroxycinnamic (CHCA) was the most universal (known to date) for small molecules.[14] Although CHCA is known to be a "hot" matrix, inducing fragmentation of glycopeptides in MALDI TOF experiments, under the relatively high pressure MALDI source conditions used here (1 torr) this effect was reduced.[10,15] MALDI laser power was also tuned for each compound. With MALDI, the general trend was increasing ionization efficiency with increasing power until a threshold was reached, typically a few microjoules per pulse on this system, above which fragmentation would predominate. There are also classes of compounds that give no response with ESI such as hydrocarbons which are highly nonpolar and have little opportunity to retain a charge using any of these techniques. These classes of compounds are not represented in this tabulation.

13.2.2 Sampling Efficiency Dependence on Source Conditions

When coupling liquid chromatography to mass spectrometry, depending on the application, there are good reasons to utilize separations that could span a flow range of four orders of magnitude, from hundreds of nanoliters per minute to milliliters per minute. Of the three ionization modes discussed here, the one whose sampling efficiency is most affected by flow rate is ESI. ESI will therefore be considered in most detail. When coupling off-line separations to MALDI via a liquid deposition device, small spot sizes with the concentration of the analyte as high as possible are desirable. For this reason, LC/MALDI is, and will likely remain, relegated to capillary LC separations. There is an optimum balance between sampling efficiency, laser illumination area, spot size, flow, column dimensions, and column capacity, which generally makes the most sense for MALDI; details of these will be addressed further below. APCI is the least affected by solvent flow, with no particular advantages at low flow rates. However, all of the ionization techniques are fundamentally mass-sensitive; that is, the number of ions created in the source is proportional to the number of analyte molecules entering the source and not the concentration of analyte in the liquid. APCI reflects this behavior with the fewest artifacts. The reason APCI has been implemented primarily as a high-flow-rate technique relates to these points, and this will also be elaborated on further below.

13.2.2.1 ESI and Flow Rate

Electrospray ion sources have evolved significantly since their first commercial inception in 1988 with the Sciex API 3 QqQ. Sampling efficiencies have improved across the flow range by greater than two orders of magnitude since that time. In addition, ion path transmission efficiencies have also improved.[16–18] To produce stable ion current across the entire flow range utilized by liquid chromatography, pneumatic nebulization[4] was implemented. At that time, sampling efficiency was observed to be reduced as one increased the liquid flow rate and column internal diameter. The decrease was roughly proportional to the difference in the cross-sectional area of the column (square of the ratio of the column diameters) used at its van Deempter optimal flow rate. For instance, the signal obtained from a given amount of substance injected on a 4.6-mm-i.d. column flowing at 1000 μL/min was 80–100× less than that observed on a 0.5-mm column at 10 μL/min. This behavior is similar to what may be observed with a detector responding to the concentration of the sample in the mobile phase and not the absolute mass. This, in turn, is similar to a UV detector that responds according to the Beer–Lambert Law and, as such, was the topic of considerable discussion,[5,19,20] although it has been generally understood that ESI was fundamentally a mass responsive phenomena demonstrated by some experiments conducted under more controlled conditions in the nanoflow regime.[3,21] The observation of a behavior similar to a concentration-sensitive detector was the result of a reduction in sampling efficiency as the flow rate increased and not due to some inherent property of the ionization process.

Improvements in sampling efficiencies over a broad flow range have addressed both ionization and transfer efficiency losses, to a significant degree, with the implementation of high-temperature gases and controlled gas dynamics.[5,12] Experiments to quantify these efficiencies are relatively simple to perform, and an example is summarized in the data in Figure 13.3, with calculations included. In this case, a sample of Reserpine was injected into a 2.1-mm column operating at 500 μL/min on an API 5000 QqQ and a Turbo IonSpray® source. A post-column split was inserted, reducing the mobile phase flow entering the source in a stepwise fashion by changing the split ratio. The amount of sample entering the source was also reduced in direct proportion to the split ratio. Split ratios were carefully measured to avoid large errors in the calculations, particularly in the nanoflow regime where the outflow from the sprayer was measured gravimetrically to ensure accuracy. The source

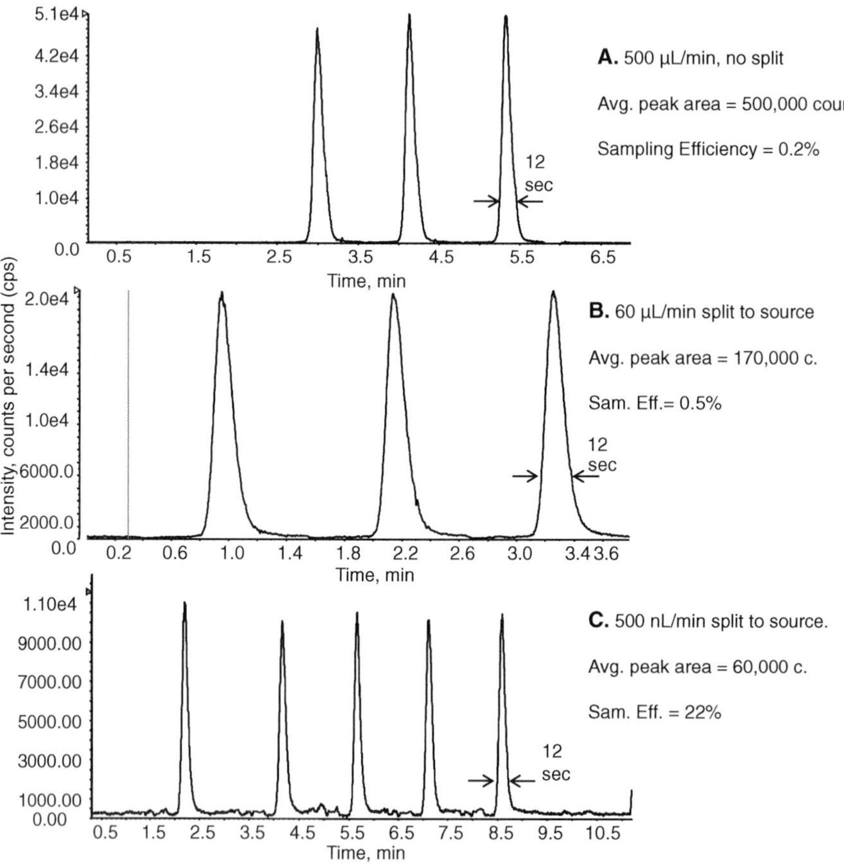

Figure 13.3. LC/ESI chromatograms and sampling efficiencies obtained at different split ratios from a 2.1-mm column operating at 500 μL/min and the resulting calculated sampling efficiency. The same amount of Reserpine was injected on column in each case. The amount entering the ion source was reduced proportionally as a result of the split. An example of the efficiency calculation for case C is as follows: (1) Ions hitting detector = 60,000. (2) Analyzer transmission efficiency on product ion m/z 195 = 5.12%. (3) Ions entering analyzer (Q0) = 60,000/.0512 = 1.1×10^6 ions. (4) 5 pg Reserpine = 4.9×10^9 molecules × 0.001 split ratio = 4.9×10^6 molecules. (5) Sampling efficiency = $1.1 \times 10^6 / 4.9 \times 10^6$ = 22%.

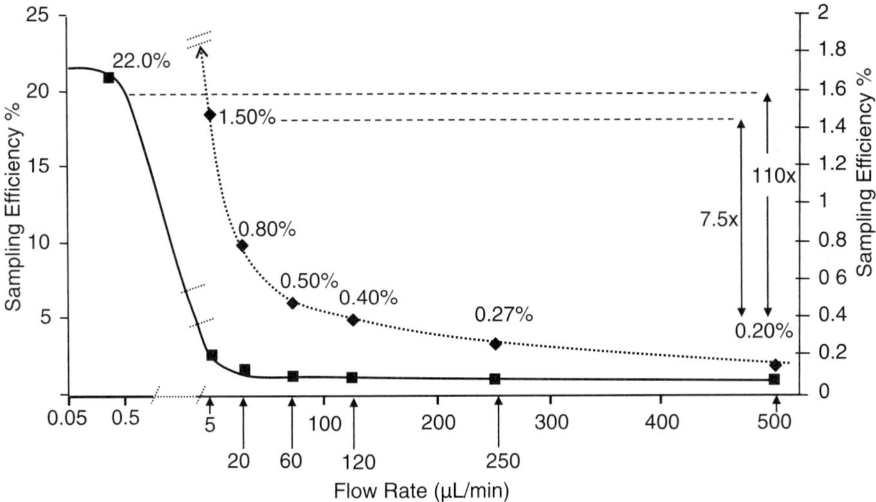

Figure 13.4. Sampling efficiency versus flow rate. The data in the dotted line are read from the right *Y* axis and are an expansion of the data in the solid line that are read from the left *Y* axis. For points above 1 μL/min, a TurboIonSpray™ source with standard interface was used. For points below 1 μL/min, a 15-μm aperture fused silica capillary was used to electrospray into a heated chamber interface.[12]

conditions were re-optimized at each flow rate to ensure optimal conditions. The data were acquired in the MRM mode, and the sampling efficiencies were calculated from the area counts as described previously[11,13] and plotted as a function of flow rate as shown in Figure 13.4. The difference in sampling efficiency in the flow regime between 500 and 5 μL/min was only 7.5× despite the fact that there was a 100-fold reduction in flow to the source. Within this flow range the ion sources require operation as shown in Figure 13.1A. The nebulized charged droplets were sprayed outside the immediate region of the orifice and vacuum draw of the instrument primarily because the gas load from the solvent was too great and the spray too divergent to be entirely consumed by the vacuum system. The majority of the factor of 7.5 loss was due to a lower transfer efficiency at high flows from an increasingly divergent spray rather than a loss in the ionization efficiency.

The points on the graph in Figure 13.4 that are below 1 μL/min represent the nanoflow regime or a mode of operation where most or all the sprayed solvent can be transferred into the vacuum system. As mentioned earlier, operation in this regime is optimally done by modifying the atmosphere-to-vacuum interface so proper desolvation will occur in the very short distance between the sprayer and vacuum system as shown in Figure 13.1B.[12] Operating in this mode, the transfer losses are largely eliminated and a significant jump in sampling efficiency is observed. Utilizing this interface offers no advantages in the higher flow regime above 5–10 μL/min because the entire solvent flow and spray cannot be inhaled. A 15× improvement in sampling efficiency over 5 μL/min, and a 100× improvement over 500 μL/min is seen. On this model of instrument, an API 5000, the maximum sampling efficiency into the Q0 region of the mass spectrometer is achieved at 22% at 500 nL/min. A reduction in flow rate below 500 nL/min provides no further gains. At 500 μL/min, where the sampling efficiency is 0.20% due to a reduced efficiency of transferring the ions generated in the source through the orifice, a significant gain is still achievable.

The analytical implication of using capillary columns that operate in the 100- to 500-nL/min flow regime (250-μm i.d.) is that the best possible conditions for absolute sensitivity can be achieved: on the order of 100-fold better than at 500 μL/min with a 2.1-mm-i.d. column. Capillary columns operating in the 5- to 30-μL/min flow range will exhibit absolute sensitivities on the order of 3–8 times better than at 500 μL/min for a given sample amount injected. However, this does not necessarily translate into lower detection limits, particularly if one is concerned about concentration detection limits which will be discussed in the detection limit section (Section 13.2.3.1.)

13.2.2.2 APCI and Flow Rate

The effect of flow rate on sampling efficiency with APCI is relatively small. Although the ionization efficiency can be as high as ESI, improvements in the transfer efficiency with flow reduction are minimal. With APCI, ionization occurs in a reaction zone located between a corona discharge needle and a counterelectrode located at a significant distance from the orifice outside the vacuum draw area. Space charge from the corona and other ionization effects prevent efficient operation within this critical area. The liquid is completely vaporized in a heated nebulizer device, and the volatilized neutral analyte molecules are injected into this region and then ionized by ion molecule reactions whose chemistry is dominated by the solvent. Drift voltages bring the ion population toward the vacuum aperture.[8] Because ionization occurs in the greater source region, APCI has transfer efficiencies similar to the high-flow ESI techniques, on the order of a few tenths of a percent, but are relatively constant across the flow range from tens of microliters to mL/min, provided that the heated nebulizer is tuned to the flow conditions. With APCI it is difficult to obtain the NanoESI advantage because of the requirements for a corona discharge, including (a) a relatively large reaction region outside the vacuum draw and (b) space-charge effects from the discharge.

13.2.2.3 MALDI Spot Size and Flow Rate

When measuring the sampling efficiency with LC/MALDI, as described earlier in this chapter, the efficiency measurements are entirely flow-independent. Whether a sample is concentrated in a small spot at low flow rates or spread out over the entire target with exceedingly high flows, the sampling efficiency numbers should be similar since the entire sample area is rastered regardless of size. The purpose of such measurements is to establish the sampling efficiency that sets the upper limits of MALDI sensitivity regardless of the sample introduction conditions. In practice, however, there are at least two good reasons to operate LC/MALDI at reduced flows and concentrate the sample to the highest degree possible. The first is the conservation of target real estate, particularly for LC/MALDI where the entire chromatogram is recorded on the target as a series of fractions collected at sufficient frequency to maintain chromatographic resolution. The second reason is to maintain, as much as possible, high sensitivity at high sampling rates. This is particularly important for high-speed quantitation. Both will be elaborated on further.

Capillary LC is naturally suited to LC/MALDI, particularly in the flow rate range of 0.5–10 μL/min using 200- to 360-μm-i.d. columns. A simple calculation illustrates target real estate conservation. At 10 μL/min with a 1-s/drop deposition frequency, sufficient to crudely profile chromatographic peaks, individual droplets on the order of 166 nL will be produced which can be spaced approximately 1 mm apart from center to center. At a deposition frequency of 1 Hz, roughly eight 20-min chromatograms could be recorded on a single microtiter-sized target. Some improvement of chromatographic density would be possible, depending on the chosen frequency and flow within these ranges. At higher flows and droplet volumes, surface tension forces confining droplets to small areas are quickly exceeded, requiring 5-mm to 1-cm spacing between fractions and a drastic reduction in the number of chromatograms per target. It can become difficult to fit a single chromatogram on one target.

For high-speed quantitation, maximizing target real estate is also a motivation for keeping the spot size and flow rates low. However, a more subtle reason has to do with the relationship of spot size to sampling efficiency when data are acquired under conditions where the minimum time per sample is spent. In this scenario, only a small fraction of a spot is ablated in order to obtain a quantifiable signal. Ideally, only an area equivalent to the illumination area of the laser will be ablated in approximately 100 microns. With spot sizes of 2 mm in diameter, only a fraction of a percent of the sample will be illuminated, which will result in a reduction in the sampling efficiency by this amount. Figure 13.5 is an example of data acquired in this fashion. To minimize sampling efficiency losses, in general, a smaller and more concentrated spot size is desirable, but there are practical limits to this as well. Spot sizes smaller in diameter than the laser illumination area provide no additional sensitivity advantage. For mass spectrometry, the practical limit to laser beam diameter is approximately 200 μm,[10] which corresponds to a spot volume of approximately 200 nL. Because MALDI is a surface ionization technique, layering samples on top of each other to increase the mass of material in a small-diameter spot offers diminishing sensitivity returns. For these reasons, from both a sensitivity and target real estate perspective, flows in the 0.5- to 10-μL/min range are optimal.

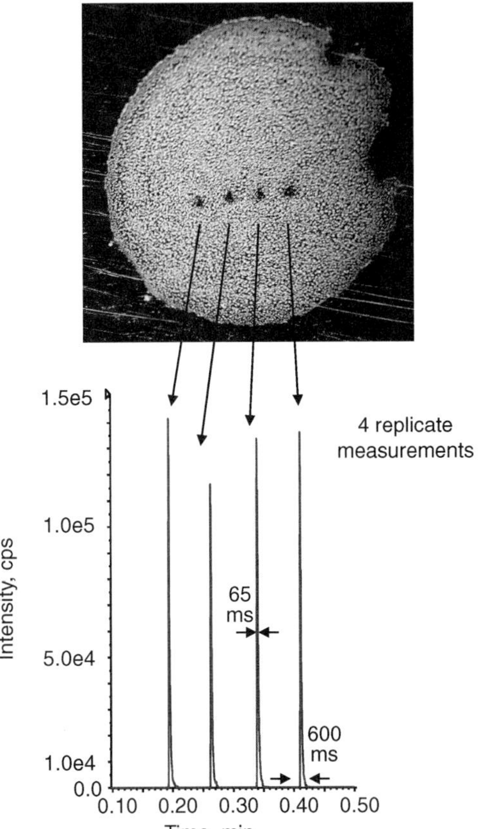

Figure 13.5. Signal transients from firing a laser at four different locations of a sample spot. The image is the spot showing the laser burn holes for the four separate replicate measurements. The laser remained firing until the sample was depleted in the illuminated area. Laser frequency was 1400 Hz; MRM dwell time was 10 ms. The analyte was Prazosin monitoring m/z 384 → 247 of a 10-pg/μL solution; 1-μL sample spots were used.

13.2.3 Detection Limits

The detection limit metric, as used here, is different from the sensitivity metric. Unlike sensitivity, which deals only with absolute signal, detection limits consider signal-to-noise ratio (S/N). Detection limits also consider the entire analytical method including sample extraction efficiencies and injection volumes in addition to the response characteristics and noise levels of the mass spectrometer signal. Detection limits are therefore analytically more relevant than sensitivity. Sensitivity measurements allow one to estimate the theoretical limits for the quantity of a compound that can be detected, whereas detection limits define what can actually be achieved in a practical setting to provide a meaningful result toward the solution of some application of mass spectrometry.

13.2.3.1 ESI

From the discussion on sampling efficiency and Figure 13.4, it is clear that as one drives the flow rate and column dimensions down, using instrumentation currently available, the sensitivity improves. An example of this is shown in Figure 13.6 comparing the signal obtained during an MRM analysis of Buspirone and metabolite at flow rates of 20 μL/min and 500 μL/min. In both instances, rates are set substantially above the van Deempter optimum flow for the column dimensions to increase analysis speed. The sensitivity and detection limit improvement is approximately four- to fivefold as predicted in Figure 13.4. However, the improvement in detection limits shown does not translate to improvements in concentration detection limits as described below.

Concentration detection limits refer to the measurement of the quantity of a target analyte in the original sample volume or mass. Some examples are the assay of drugs and metabolites in biological samples, typically measured as picograms of drug per milliliter of original blood plasma, or the measurement of environmental contaminants in water or soil samples in terms of parts per trillion. The vast majority of bioanalytical and environmental analyses are not severely sample limited, and methods generally have at least hundreds of microliters of extracts available. A typical strategy for achieving lower concentration detection limits is simply to inject more sample, because column capacities that scale with the cross-sectional area of the column must be considered. A 2.1-mm-i.d. column operating at 500 μL/min would have 100-fold greater injection volume and mass capacity than a 0.2-mm-i.d. column operating at 500 nL/min. The 100-fold reduction in sampling efficiency shown in Figure 13.4 is compensated for by column capacity resulting in equivalent concentration detection limits in this case.

For this reason, bioanalytical applications that are not sample-limited continue to prefer the higher flow rate LC/MS conditions, whereas those that are truly sample limited, such as peptide mapping of protein extracts from 2-D gels, benefit greatly from the capillary LC/MS

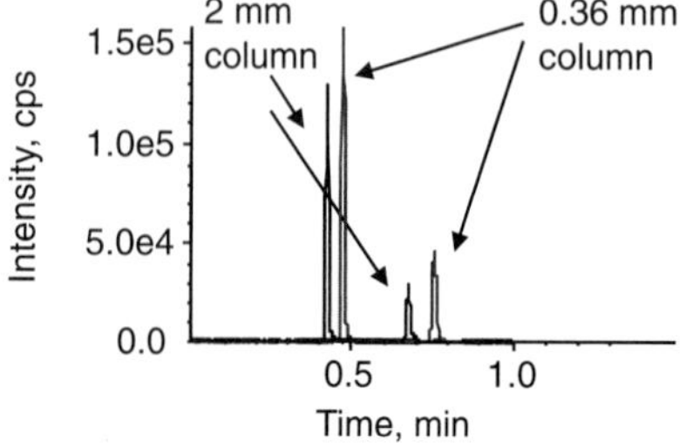

Figure 13.6. Comparison of sensitivity at 20 μL/min with 360-μm-i.d. column and 500 μL/min with a 2.1-mm-i.d. column. Two picograms of Buspirone and metabolite was injected. Fast gradient elution conditions were used in both cases.

approach.[22] For the latter type of analysis, mass detection limits are used as the primary metric and they are frequently measured as the absolute amount of material required to obtain sufficient sequence coverage of the original protein from a database search of the peptide product ion spectra. Detection limits in the low femtomoles range are relatively routine with trapping and TOF instrumentation and NanoLC/ESI. Analyses in the hundreds of attomoles are challenging, but achievable, when all aspects of the system are fully optimized.[23] One of the bottlenecks to performing routine NanoLC/ESI has been the availability of chromatography instrumentation designed for operation in the nanoflow regime. This situation has improved dramatically over the past few years, solving long-standing problems of chromatographic reproducibility, resolution, and general reliability. Pumping systems using pressurized gas fluid drives, different from conventional reciprocating pistons, have shown to have particular merit for delivering fluids in the nanoliter to tens of microliter/minute range.[24]

Because detection limits are dependent on every aspect of an analytical methodology, it is meaningful to provide examples of absolute numbers only if the conditions are well-defined. The case of Albuterol, a compound similar to Reserpine that has high ionization efficiency, MRM sensitivity, and extraction efficiency, is shown here as a point of reference representing close to the best values that can be achieved today. With a 2.1-mm column at 200 μL/min, a detection limit of 2.5 fg (11 attomoles) of Albuterol injected on column was achieved and is shown in Figure 13.7. With a liquid/liquid extraction efficiency of 50% from 1 mL of plasma and an injection of 20 μL from a 100-μL concentrated extract, a sample concentration detection limit of 25 fg/mL of plasma, or a limit of quantitation of 250 fg/mL, would be achievable. Because this assay was not validated, it is uncertain whether these low levels could be maintained over many samples and time, but it is not uncommon to see values for limits of quantitation for validated assays to be in the low picogram/milliliter plasma level.[25,26] It should be mentioned that although thousands of samples are run with such low limits of quantitation, all of the following are required: skilled method development, compounds with high ionization efficiency, high MRM efficiency, and instrumentation operating at peak performance.

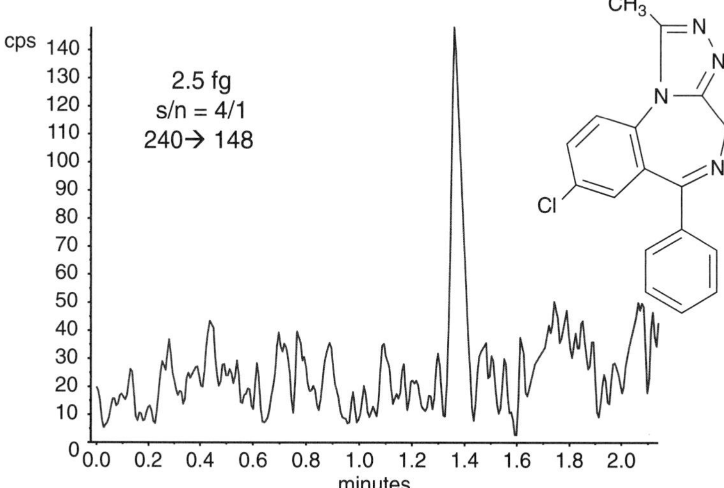

Figure 13.7. Limit of detection for Albuterol, 2.1-mm column at 200 μL/min.

The aforementioned data represents compounds that have high detection sensitivities with ESI. As shown in Table 13.1, there is a difference of more than 500× between the sampling efficiencies and MRM detection sensitivities across the chemical space studied using ESI. At first consideration, it would appear that those compounds at the bottom of the list have such low sensitivities that they cannot be assayed at biologically relevant levels, however, this is not the case. LC/MS/MS assays have been developed for the anti-tissue rejection drug Cyclosporin (bottom of Table 13.1) for therapeutic drug monitoring purposes with limits of quantitation of 1 ng/mL.[27] Indeed, many toxicology and *in vitro*-based assays, such as those observed in ADME property screening applications, operate in the 1- to 100-ng/mL (μM) range.[28,29] There are very few compounds of biological relevance that do not show at least this quantitation level with ESI. One report stated that greater than 95% of the compounds in a major pharmaceutical drug library could be assayed with ESI at these levels.[30]

13.2.3.2 APCI

Compounds with high gas-phase acidities or basicities and having thermal stability show similar detection limits to high-flow ESI. This is consistent with the understanding that the ionization efficiency is very high but the transfer efficiency is low because the ion creation zone is large and far removed from the inlet aperture of the mass spectrometer. This leads to sampling efficiencies on the order of a few tenths of a percent. Because the desolvation process is aided by impacting the nebulized droplets on very hot surfaces of a small-diameter tube, complete vaporization at very high flows and confinement of the spray is achieved.[8] APCI can therefore be used with 4.6-mm columns at 1–2 mL/min without major compromises, which is viewed as an important advantage over ESI. The sampling efficiency also remains relatively constant across an HPLC gradient offering a further advantage. Its behavior with regard to matrix effects is different from ESI, which is often viewed as an advantage as well and will be discussed further in Section 13.3.2. For these reasons, APCI is often considered to be the method of choice for many bioanalytical assays, and it is included in up to 30% of pharmaceutical methods employing mass spectrometry and has demonstrated the ability to achieve low picogram/milliliter concentration detection limits.[26] The first demonstrations of high-speed quantitative LC/MS/MS[31,32] and validated analytical methods[33,34] were done using the heated nebulizer APCI interface. This general approach has been widely used for 20 years and remains virtually unchanged.

The MRM data in Table 13.2 give some indication, however, that there is a wider swing in sensitivities across chemical space than with ESI. Unlike ESI, nearly 20% of the compounds tested provided no useable signal primarily due to thermal effects as highlighted in Section 13.2.1.4. For this reason, APCI is primarily relegated to targeted analysis and assay development where the opportunity exists to test a compound for its ionization behavior before analysis begins. It is rarely used for characterizing unknowns because of the uncertainty about whether these compounds can survive intact through the process of vaporization and chemical ionization.

13.2.3.3 MALDI

In order to meaningfully compare MALDI detection limits to the other techniques, specifically defined analytical conditions must be established. The case of peptide mapping is a good starting point because this class of chemical species ionizes well using both MALDI and ESI. Moreover, a relatively large body of work using both NanoLC/ESI and NanoLC/MALDI has been reported from which some initial conclusions may be drawn. As described in Section 13.2.3.1, NanoLC/ESI can produce detection limits, as measured by sequence coverage, on the order of low femtomoles. For LC/MALDI, under capillary LC conditions in the 0.5- to 10-μL/min range, spot volumes can be kept below 1 μL with corresponding spot diameters of approximately 1–2 mm. Spotting frequencies of less than 1 Hz, which strive to capture the majority of a peak in a single spot at the expense of chromatographic resolution, are frequently used for this application. With laser illumination dimensions of 0.1–0.2 mm, several passes of the chromatogram can be made. The first survey pass finds the molecular ions in the single MS mode. Because this mode is highly limited by chemical noise, sacrificing the sampling efficiency (by ablating only a small portion of the sample) has little effect on the S/N.

Each peptide-containing spot is then rastered and completely consumed while signal averaging the product spectra of the targeted peptides. In so doing, detection limits similar to those reported for NanoLC/ESI are being achieved (i.e., low femtomoles to hundreds of attomoles), which are required for good sequence coverage.[35] The results described in Sections 13.2.1.1 and 13.2.1.3 regarding NanoESI and MALDI sampling efficiencies are examples. In situations where the entire spot can be consumed in a MALDI experiment, sampling efficiencies of 5% were measured for favorable compounds, whereas they were in the range of 5–20% for NanoESI depending on the instrumentation.

The situation using LC/MALDI for quantitation of drugs is different from the one described above. This is because rastering an entire spot to achieve total consumption is not a practical alternative due to the amount of time that would be required to do this. Given conditions where the entire sample is concentrated in a single spot of reasonable volumes and dimensions (1–2 μL, 1- to 2-mm diameter) and currently employed MALDI laser beam dimensions of 0.1–0.2 mm, less than 1% of the sample can be ionized from a single illumination area. This method of acquisition is the fastest means of acquiring data for quantitation, and it produces S/N measurements equivalent to a slow rastering of the entire spot. Figure 13.5 is an example of this concept where it is clear that the sampling efficiency must be low—in this scenario approximately 0.025%. This is on the order of one-tenth of what would be observed from high flow rate ESI or APCI at several hundreds of microliter/min. High-speed quantitation by MALDI, as described here, will have detection limits that are considerably higher, on the order of 10×, than conventional ESI and APCI bioanalytical methods when comparing compounds with favorable ionization conditions in all cases.

13.2.3.4 Compound Dependency

Up to this point, the detection limit and sensitivity comparisons of the different sources have focused primarily on compounds that ionize efficiently with all the techniques. It is important to understand the coverage or scope of an ionization technique across the chemical space of general interest, particularly when confronted with unknown compounds, or compounds whose structures are known but whose ionization properties have not been tested, and there is no time to assess a variety of options. This situation occurs in many drug discovery laboratories measuring *in vitro* ADME properties (administration, distribution, metabolism, excretion) where many different chemical species need to be assayed quickly. The data used to generate the relative efficiency values within a source in Tables 1–3 were used to calculate the relative MRM efficiency between the three sources and are shown in Table 13.4. The MALDI data were acquired in the most practical fashion to obtain a quantitative measurement where only a small percentage of the sample spot was ablated with a single raster. The ESI and APCI data were obtained by flow injection analysis at 200 and 1000 μL/min, respectively. Electrospray is the most sensitive ion source in nearly all cases. The sensitivity of APCI is close in several cases, within 5× in 9 cases out of 31, which could lead to very similar detection limits depending on the chemical noise and other methodological parameters such as injection volumes which tend to be greater with APCI.

The top four compounds of the list in Table 13.4 would be considered to have favorable ionization characteristics by both MALDI and ESI. They exhibit MALDI sensitivities close to 10-fold lower compared to high flow ESI under these acquisition conditions. However, two compounds closely match or exceed ESI. For APCI, eight compounds were more than 900-fold less sensitive than ESI, implying that they produce almost no analytically useful signals, while three compounds fell into this category for MALDI. It is notable that Phase II drug conjugates like glucuronides and sulfates are particularly poor performers by APCI. This is primarily a thermal degradation issue where full-scan APCI spectra of such compounds show the aglycones of glucuronides as the base peaks.

Table 13.4. MRM Sensitivity of APCI and MALDI Relative to ESI for the 31 Compounds [a]

Compound	Highest Sensitivity Source	ESI/APCI	ESI/MALDI
Verapamil	ESI	4	8
Haloperidol	ESI	12	7
Tamoxifen	ESI	7	9
Buscopan	ESI	1342	6
Clenbuterol	ESI	7	156
Safranin Orange	ESI	15	1.5
Estradiol sulfate	ESI	978	81.0
Morphine Glucuronide	ESI	∞	123
Bromocryptin	ESI	174	108
Furosemide	ESI	439	520
Morphine	ESI	10	18
Succinyl choline	ESI	∞	∞
Phenylbutazone	ESI	7	9
Fosinopril	ESI	271	53,031
Dianabol	ESI	2	91
Aztreonam	ESI	∞	527
Bentazon	ESI	31	81
Taurocholic acid	ESI	907	49
Oxyphenylbutazone	APCI	0.8	32
Naproxen	ESI	20	108
Beclomethazone	ESI/APCI	1	479
Bromazepam	MALDI	0.5	0.4
Dexamethazone	ESI	1	205
5-Fluorouracil	ESI	4	657
Testosterone glucuronide	ESI	1165	32
Prednisolone	ESI	3	∞
Lovastatin	APCI	0.5	105
Erythromycin	ESI	71	356
Taxol	ESI	9	158
Alpha-cyclodextrin	ESI	∞	108
Cyclosporin	ESI	13	22

[a] The same data used to calculate Tables 13.1–13.3 were used to establish these numbers. A ratio of ∞ means ESI was infinitely more sensitive because no measurable MRM signal could be obtained from the other sources. See Tables 13.1–13.3 and text for method of data acquisition.

To conclude this section on sensitivity and limits of detection, it should be readily apparent that ESI is both the most universal and most sensitive of the ionization techniques discussed here and itself exhibits a broad range of response factors for different chemical species. Comparative studies of these three modes of ionization across chemical space are sparse in the literature. One important publication exists testing a representative subset of a drug library for the applicability of MALDI for the quantitation of *in vitro* ADME tests.[30] Approximately 85% of the compounds that worked by ESI provided sufficient results by MALDI to obtain reasonable quantitative numbers, at lower sensitivity but at much higher speed. This is the MALDI advantage for quantitation[36–38] to be discussed further in Section 13.4.2. NanoLC/MALDI can rival NanoLC/ESI for qualitative types of analyses where MS/MS signal averaging techniques can be utilized for the acquisition of full scan product ion spectra and where there is time to consume the entire sample spot by rastering the laser. The strength of APCI is its low concentration detection limits for a subset of chemical space, on first approximation 10–30% of a typical drug library,[26] primarily because of its ability to maintain reasonably high sensitivity at high flow rates and large column diameters. In general, APCI is a popular technique for steroid analysis due to the thermal stability and gas-phase chemistry of this class of compounds.

13.3 REPRODUCIBILITY

The discussion on reproducibility for the three sources will take a similar tack to the discussion on sensitivity starting with a discussion of the limits imposed by the instrumentation and the ionization process on ion current stability. Secondly, we will describe the most important methodological factor affecting reproducibility, which is under some control of the analyst through the use of sample preparation and chromatography. This is ion suppression and will be discussed in some detail. Thirdly, some real-world examples of assay precision and accuracy will be provided. In all of these categories there is less data available from MALDI than from the other sources because its implementation as a quantitative tool is relatively recent.

Reproducibility, as used here, refers to the ability to repeat measurements with a high degree of accuracy and precision. Invariably, quantitative measurements of this nature involve calibration of the system with the targeted analyte and referencing the signal to that of an internal standard. Of course, foregoing the use of an internal standard results in less precise and accurate measurements. Absolute quantitation—that is, the prediction of the molar amount of a substance from a signal without any calibration or predetermination of a response factor—has not been shown to be feasible to date with any of these three ionization techniques. That this is so should be apparent from the data in Tables 1–4 showing the broad range of ionization efficiencies across compound space that exists for all three sources.

13.3.1 Ion Current Stability

Reproducibility is fundamentally determined by the degree of ion current stability. We describe here three general categories of ion current stability or signal fluctuation that can affect reproducibility in different ways, each of which is controlled by different instrumental, ionization, and methodological conditions. All three will be considered here for the three sources; however, the characteristics of ESI and APCI are so similar that they will be discussed together. The three categories can be distinguished based upon the timescale that is being considered.

1. Long-term ion current fluctuations exhibited over periods of hours to days.
2. Short-term ion current fluctuations exhibited over periods of milliseconds.
3. Intermediate-term ion current fluctuations exhibited over periods of seconds.

The first category refers to the stability of ion current measurements over long periods of time (hours to days). Maintenance of instrument sensitivity on a daily basis is essential to sustain continued instrument use. As long as the drift is slow, however, and not greater than factors of 2–4× over periods of days, variations of this sort do not seriously affect accuracy, precision, or even limits of detection. A fourfold loss in sensitivity would result in a twofold detection limit increase if the measurement was limited by chemical noise, which is almost always the case. This type of long-term ion current instability is generally caused by slow contamination of the entrance optics of the mass spectrometer, and it can be addressed with periodic routine cleaning.

The second type of signal fluctuation is observed on a very short timescale—that is, fluctuations of ion current approximating the measurement time of the mass spectrometer. For MRM data, this measurement time, often referred to as dwell time, is typically on the order of tens to hundreds of milliseconds. The source of these fluctuations can be traced to a variety of aspects of instrument design including electronic, photon, and particle noise reaching the detector, stability of the signal measurement circuitry, the method of signal measurement (pulse counting versus analogue detection), and ion sampling and transport effects encountered with particular atmosphere to vacuum interface designs.[39] Fluctuations in the rate of ion production from the source play a very large role as well. There are limits to accuracy, precision, and detection imposed by these fluctuations, but they are not as severe as one might expect, primarily because the average frequency of this noise readily distinguishes it from a real analyte signal, which has a signal component equal to the width of the chromatographic peak. This will be discussed further in Section 13.3.1.

The third type of signal fluctuation to consider is the most severe with regard to its effect on reproducibility. Fluctuations of ion current of this type are observed on the timescale of a single sample analysis measured in seconds to minutes. Variations in sensitivity occurring during a chromatographic run can seriously affect measurements, particularly if analyte and internal standards do not co-elute. Shifts in the ion current baseline similar in width to a chromatographic peak also create a noise element that is difficult to distinguish from a real signal, particularly at the detection limit. This not only increases the limit of detection, but also can introduce precision errors during the process of peak integration when defining the start- and stop-points of a peak. The source of this type of noise can be traced to (a) instrument parameters in the ion source, such as sprayer plugging, and (b) sample introduction equipment, with stability of mobile phase flow being quite important. Methodological effects are also important, the most critical of which is often referred to as ionization suppression.

13.3.1.1 ESI and APCI

The long-term stability of an instrument is primarily related to the ability of a particular design to resist contamination. Manufacturers strive to minimize this effect by a variety of means such that instrument sensitivity will remain relatively constant over periods of days, weeks, or months under the most severe sample introduction loads. There are several means of achieving this, including the use of countercurrent flows of gas, heating of critical areas that become contaminated, off-axis sprayers, and the deliberate misalignment of inlet optics to block particle penetration to critical areas. Because the mean time required between cleanings is completely dependent on the type and number of samples introduced, it is difficult to quantify this condition in a meaningful way. The frequency of instrument cleaning will vary, depending on the nature of the instrument use. One useful strategy is to leave the ionizing voltage off, or, alternatively, divert the mobile phase flow, during the elution of the column void. Even with the most contaminating samples, such as protein-precipitated plasma or neat injection of *in vitro* buffers, taking this simple precaution will result in mean time between cleaning of many months. Suffice it to say that for both high-flow ESI and APCI, which behave similarly in this regard, with proper routine cleaning of the source, drifts of this sort will not seriously affect precision, accuracy, or detection limits.

NanoESI will contaminate an instrument more readily than the high-flow modes because vacuum drag is used to affect sample consumption with very little discrimination between neutrals and ions. Efforts to ameliorate this effect have to some degree improved this aspect of NanoESI,[12] taking advantage of the large momentum difference between gas-phase ions and neutral particles to sort them before entering the vacuum chamber and deposit them on atmospheric lenses to reduce their affect on instrument performance. Never the less, greater degrees of contamination continue to be the norm with nanoESI operation.

Short-term ion current instability is characterized by fluctuations that occur on the millisecond timescale and can be related to the specifics of a particular instrumental design. It is also related to acquisition speed and, with ESI, sample introduction flow rate. High-frequency noise of this type has some affect on precision, accuracy, and detection limits. Because it can be easily distinguished from the real signal, which is lower in frequency (i.e., chromatographic peaks are seconds, not milliseconds in duration), its effect is less important than sensitivity and baseline drift that occur over several seconds. Figure 13.8 is an example of data with high-frequency noise introduced by decreasing the dwell time from 200 to 5 ms. The high-frequency noise is revealed when the dwell period is reduced giving the data the appearance of low S/N. Simple smoothing algorithms, however, readily distinguish this noise from the real, low-frequency signal as shown in the inset in panel A of Figure 13.8. Application of smoothing algorithms or acquiring data at very long dwell times improves the data appearance but tends to increase the peak area variation measurements. Repetition of the analyses shown in Figure 13.8, conducted close to the limit of detection, generated peak area variations of 10–15% for the 5- to 20-ms dwell data and 15–20% for the 200-ms dwell data. This is primarily because the inflection points at the beginning and end of the peaks can be more accurately determined in the raw data acquired with short dwell times and/or data that are not smoothed. Smoothing or binning the data distorts the true position of this inflection point as illustrated in Figure 13.9. The use of frequency transformations of the data avoids this binning effect and reduces peak area integration errors. An example of a fast Fourier transformation (FFT) of high-flow-rate ESI data with substantial high-frequency noise is shown in Figure 13.10. What should be apparent from this data is that it is not the high-frequency noise that limits the peak detectability or area integration. As long as the fluctuations are uniform around a constant signal such that their effects average out over a few measurements, it is the low-frequency noise (very apparent in the baseline of the transformed data) which appears as signals indistinguishable from the analyte if they are close in retention time.

One source of this high-frequency noise is the massive charged particles known to originate in ESI sources which, when they impinge on the detector, cause large instantaneous ion current spikes. More of these particles and un-desolvated charged droplets are

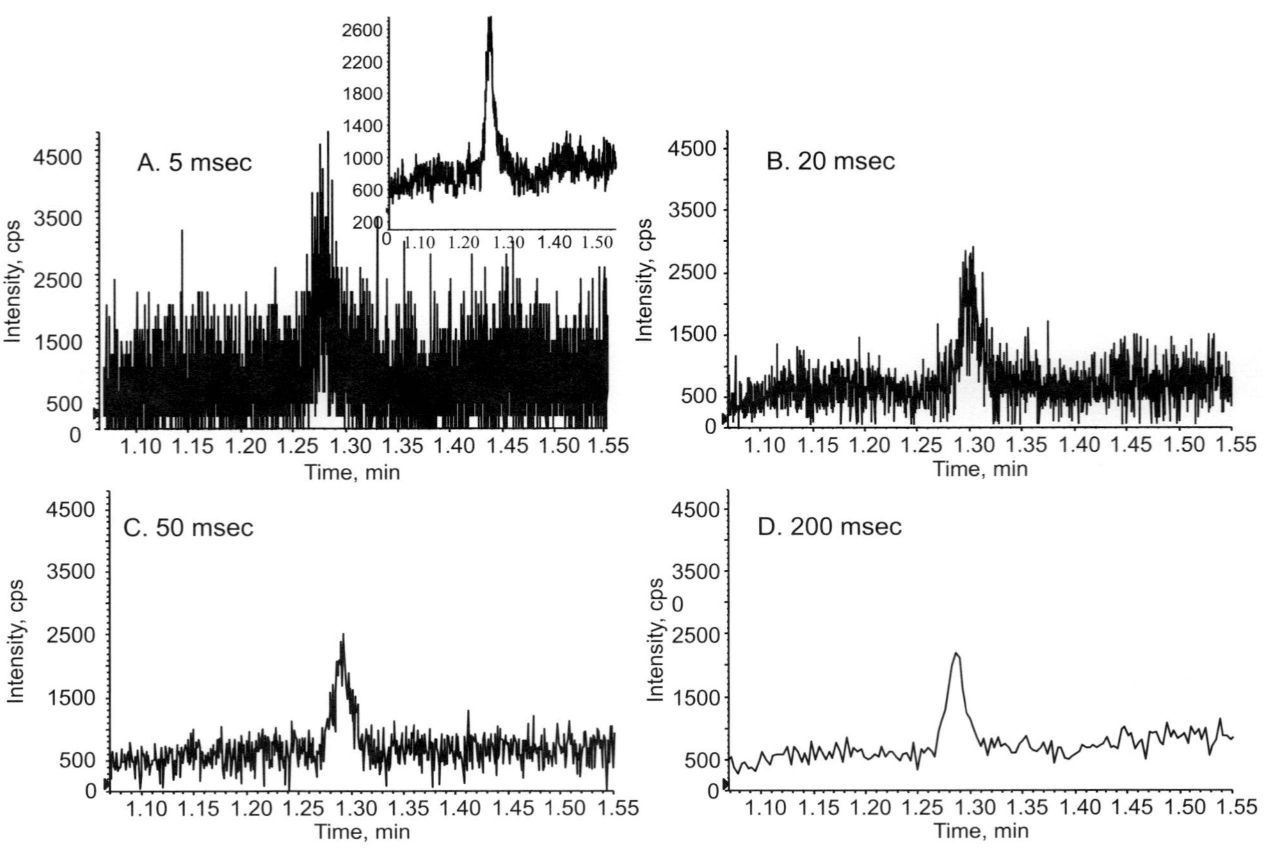

Figure 13.8. The effect of high-frequency noise introduced by increasing acquisition rates on limit of detection. Insert in panel A is a Savitsky-Golay 40-point smooth of the 5-ms dwell data. High-flow LC/ESI of Reserpine (30 fg on column) on a 2.1-mm column at 200 μL/min was acquired in MRM mode at different dwell times.

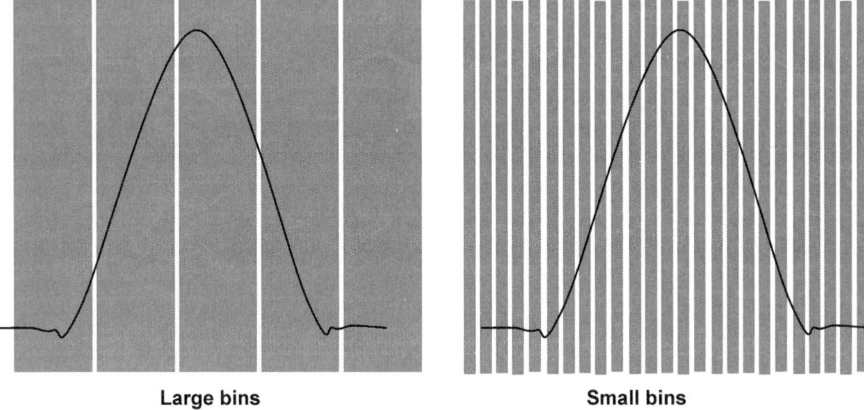

Figure 13.9. Illustration of the distortion of the peak start and end points that can occur when data are binned by either smoothing algorithms or long dwell times that can have a significant effect on peak area integration and reproducibility.

created at high flows than at low flows. These can be sorted from ions to some degree by off-axis ion sampling skimmers or entrance optics in the vacuum system,[40,41] or by using ion counting detectors that register these occurrences as imperceptible single ion events. The ion generation process is also more sporadic at high flows than at low flows, introducing additional high-frequency (signal carried, "shot") noise on the analyte signal.

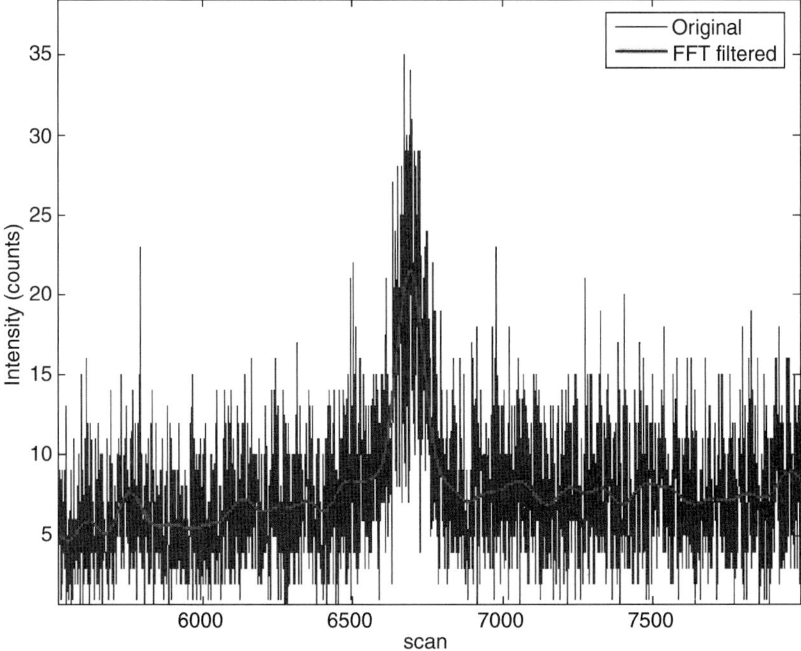

Figure 13.10. Removal of high-frequency noise with FFT. Low-frequency noise is revealed in the baseline of the FFT-processed chromatogram (See color insert).

A reduction of this high-frequency noise is observed when one operates at lower ESI flow rates.[42] This can be demonstrated, under idealized inlet conditions, by using a syringe pump to infuse an analyte at different flows to establish a baseline ion current where both the high-frequency noise and low-frequency baseline drifts can be observed. Figure 13.11 shows the results of such an experiment at the extremes of flows tested, although several intermediate flows were also tested to verify the trend. The nanoflow data shows a threefold improvement over the high-flow ESI data for both the high-frequency ion current fluctuations and the area count fluctuations from segments of the trace, indicative of the limits to peak area reproducibility at these signal levels. Although there is a significant difference between the two flows, under conditions that measure only variations generated by the ion source and mass analyzer, the area variations are small, <1%, for both. The high-frequency fluctuations average out over the time of a chromatographic peak. Typical analytical methods provide variations much greater than this, on the order of several percentage points, primarily due to varying conditions of liquid flow delivered to a sprayer (pumps), flow emitted from a sprayer (tip plugging), sample compositions, compounds adhering to surfaces, and injection reproducibility.

APCI provides similar stability results to the high-flow ESI data when either infusing a sample in a liquid to the nebulizer or when simply monitoring a background ion with the nebulizer turned off and without any liquid or substantial air flow into the source. This indicates that some of the stability difference between NanoESI and the high-flow techniques relates to the location of ion creation in the source, as well as the processes of droplet creation, evaporation, and ionization. With NanoESI the ions are created in the vacuum drag area where there is very little opportunity for sampling variability to occur as all substance is inhaled. With the other techniques the ions are created well outside this area, requiring several centimeters of drift at atmospheric pressure. Only a small, statistically varying percentage enter the orifice region. This source of the additional variability is fundamental and difficult to correct but, as described above, has a relatively small effect on peak detectability or integration errors compared to those of the overall method.

Most important to detection limit determination and peak area reproducibility are ion current variations of the third type involving shifts in the baseline that occur on a time scale

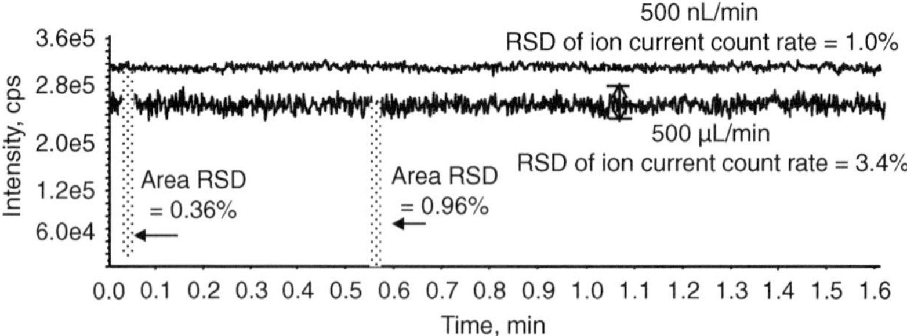

Figure 13.11. Comparison of ion current stability between NanoESI and high-flow ESI and APCI techniques. The RSD of the baseline count rate is a measure of the high-frequency ion current fluctuations. The area counts were calculated over 3-s windows, approximating chromatographic peak width, taken at 10 separate segments of the 1.6-min trace from beginning to end. Only one segment is shown for each trace to illustrate the measurement. This is a measure of the low-frequency noise reflected in baseline drift. Similar experiments at intermediate flows under the conditions of Figure 13.1A (not total consumption nanoESI conditions) verified the trend. 5 µL/min = 1.3% count rate RSD, 0.51% area RSD. Different sample concentrations were used to obtain similar ion count rates and counting statistics. MRM data from Reserpine acquired at 100-ms dwell times.

of seconds. These shifts resemble—and, as a result, interfere with—chromatographic peak detection and integration. This is illustrated in Figure 13.10 with some real data near the limit of detection. When the high-frequency noise is removed with a fast Fourier transform (FFT) filter, the noise that limits detectability is revealed as baseline deviations resembling chromatographic peaks. It is possible that some of these fluctuations are real chromatographic signals from unknown substances and others are not. Shifts in the baseline ion current in the middle of a peak will also alter peak area and therefore introduce significant precision measurement errors. Drift on this timescale is generally not related to the mass spectrometer itself. Drift of this nature, if not due to real substances eluting from a column or periodically bleeding from system components, is usually due to fluctuations in flow of the LC mobile phase coming from the pumps, flow variations from sprayer plugging, variations in column pressure altering flow, or variations in split ratios if splits are used. Under conditions where all of these variables are controlled, variations in baseline drift are maintained to $<1\%$ (Figure 13.11) for both NanoESI and high-flow ESI. This number should be a close approximation of the highest attainable reproducibility for peak areas, as limited by the mass spectrometer ion source and instrument. Realistically, variations in flow and sample components will make this difficult to attain. Precise control of flow conditions in the nanoflow regime is far more difficult to achieve than at high flows. For this reason, NanoLC/ESI has generally exhibited lower precision and accuracy than attainable at conventional flow rates. Improvements in pumping systems that eliminate many of these sources of error at low flow rates[23] should help in this regard.

The most common origin of this type of signal deviation is the sample itself, and it is due to variations in the composition of eluting peaks during the course of a chromatographic run which can affect reproducibility in the most dramatic fashion. This phenomenon is commonly referred to as ion suppression and, because of its importance, will be treated in more detail in Section 13.3.2.

13.3.1.2 MALDI

Since MALDI is a pulsed ionization technique whereas ESI is virtually continuous, it is difficult to compare, in a direct fashion, the ion current stability metrics discussed above. However, there are some meaningful analogies that can be drawn.

Long-term stability—that is, freedom from analyzer contamination under high sample load conditions—is a prerequisite for an instrument designed to perform high-speed quantitation. To date, most MALDI systems have found application in the field of protein and peptide analysis which, from an analyzer contamination perspective, is not indicative of small-molecule quantitation applications, which tend to be more demanding. There is not as large a body of historical field experience with MALDI under these conditions as there is with ESI or APCI. The MALDI process requires the ablation of large quantities of matrix material inside the vacuum system in the vicinity of critical lensing elements, which is reason for concern. However, efforts to enrich the ion-to-neutral particle ratio entering the lens elements of the mass spectrometer have demonstrated that more than 200,000 samples can be analyzed before a routine cleaning was required.[10]

Measurement of the high-frequency noise as described for ESI and APCI is more difficult with MALDI because the ionization event and the noise are on similar timescales. It is difficult to generate a continuous ion beam with a constant baseline from which to make comparable measurements. Since MALDI is a surface ionization technique, many parameters come into play, including depth of beam penetration, rate of raster, and crystal morphology. The most meaningful relationship that can be drawn between the ESI

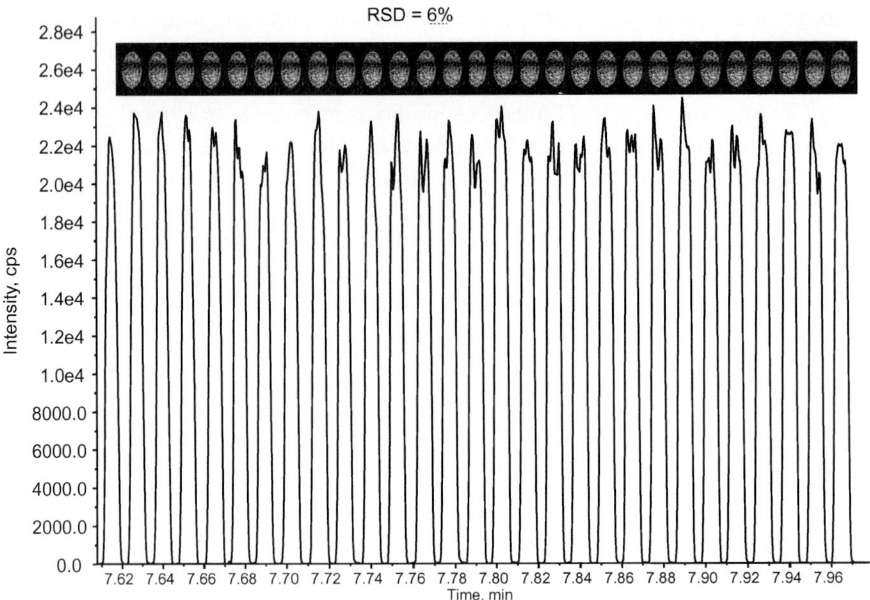

Figure 13.12. MALDI signal stability rastering over a series of 30 spots. Acquisition time approximately 30 s. The compound spotted was Propanolol.

discussion and MALDI regarding high-frequency noise is that the same principles apply. As long as the signal duration is longer than that of the noise, the two can be distinguished. Operation of MALDI at acquisition rates similar to the high-frequency noise, which is more of a possibility with MALDI than with ESI, will make detection increasingly difficult. Additional discussion of MALDI acquisition rates will be presented in Section 13.4.2, which deals with speed and throughput.

A comparison of the low-frequency sensitivity drift is also difficult to do directly, but a reasonable analogy can be drawn. A variation in sensitivity over a series of sample spots is a situation similar to infusion and observation of baseline drift with ESI. Using ideal conditions of sample deposition and crystallization with pure standards and solvents establishes some idea of reproducibility limits. An example is shown in Figure 13.12, where a series of spots from a sample of the drug Propanolol were created with an electrostatic deposition device.[43] The spots were rastered at a speed such that each spot was traversed in 600 ms. Absolute area reproducibility of approximately 6% was achieved—that is, worse than the infusion data of NanoESI and the high flow techniques, but still reasonable, particularly if referencing the signal to an internal standard (not done here) in the same spot that is ionized at the same time as the analyte. Suffice it to say that good quantitation is possible when an internal standard is cocrystallized with the analyte.[30] ESI or APCI do not require coelution of the internal standard and analyte.

13.3.2 Ionization Suppression

The most important methodological parameter effecting accuracy and precision is a sample matrix effect commonly referred to as ionization suppression.[44–47] Sample matrix effects are generally categorized as a change in the ionization efficiency of an analyte during the course of a chromatographic run as a result of the presence of other species at the time of ionization. Matrix effects can be manifested either as an increase in ionization efficiency or, more commonly, as a decrease in efficiency. Examples of both will be shown. All ionization techniques exhibit these effects to varying degrees, although the underlying mechanisms and causative agents are different. Some understanding of the mechanism provides a basis for predicting when these effects are most likely to occur during a chromatography run.

13.3.2.1 Mechanistic Aspects

ESI is a droplet surface ionization phenomenon; consequently, the mechanism of ion suppression can be understood from this perspective. Compounds that are surface-active and concentrate at the droplet surface are very effective at suppressing analyte signals. Surfactants such as SDS and lipids are good examples of this category of compounds. Materials of this sort tend to be highly retained on C18 columns. Analytes that have these same properties are least likely to be subject to suppression effects. Gas-phase acid–base characteristics also play a role, but are secondary to the surface effects.

APCI is dominated by the thermodynamics and kinetics of gas-phase acid–base ion chemistry. Ions with low proton affinities in the presence of compounds with high proton affinities will rapidly transfer their protons and return to the neutral state. Biological matrices have an abundance of compounds of widely varying gas-phase acidities and basicities; they tend to be polar and elute early from C18 columns. In general, APCI is considered to be less prone to matrix effects than ESI, but this is not always the case as will be seen in some of the following examples. This perception may have more to do with the predictability of when causative agents will elute from reversed-phase columns. It is also not uncommon to observe an increase in analyte signal with APCI when it coelutes with other sample constituents. This is not an ion chemistry effect; it typically occurs with thermally labile compounds and can be attributed to a decrease in thermal degradation resulting from a reduction in the extent of direct physical contact of the analyte with the hot vaporization surface when coeluting with an excess of sample materials.

MALDI ionization is the least understood of the three largely because the ionization mechanism is a combination of both a surface ionization process and gas-phase reactions in the ablation plume. Components of the sample that inhibit the formation of crystals decrease the energy transfer at the surface and reduce the ionization efficiency. Reactions of ions and neutrals in the gas phase will also determine the final ion population, similar to APCI. At this point, it is important to distinguish endogenous sample matrix from MALDI ionization matrix. Ionization matrix is the material used to form the crystals that absorb and transfer the photon energy. Endogenous sample matrix is the biological material in a sample.

13.3.2.2 Observation and Characterization of Matrix Effects

Suppression of analyte signal is a serious problem for any type of mass spectrometric analysis, but it is most serious and most easily observed in quantitative analysis as an unexpected deviation of standard curve points and quality control samples (QCs) from expected values.[44] Sample clean-up and chromatography are the most effective methods to control suppression. Fast chromatographic techniques are the most prone to exhibit these effects where compounds elute close to the column void in the vicinity of the majority of the sample matrix. Some components of biological matrices, particularly lipophilic compounds that exert the most serious effects in ESI, adhere strongly to C18 columns to such an extent that they can unpredictably elute from a column after several subsequent sample injections.

The best means to avoid this is gradient elution chromatography, but even this will not guarantee that a column will be completely cleared after every sample injection.

One effective experimental technique for observing this ion suppression effect involves a post-column infusion of analyte while chromatographing the sample matrix.[45,46,47] With this system, a "T" junction is installed between the column and the ion source, and a solution of the targeted analyte is infused at this point while monitoring the appropriate precursor/product transition. A blank matrix is injected into the column, and the chromatographic method chosen for the specified analyte is run. The result is a "matrix effect chromatogram" showing the retention times of the offending substances as a negative perturbation of the continuous ion current trace.

Figure 13.13 is an example of an electrospray matrix effect chromatogram of an injection of 10 μL of protein precipitated plasma (1/1 CH_3CN/plasma) with infusion of the corticosteroid beclomethazone post-column. The red trace is a superimposed chromatogram of standard beclomethazone and metabolite to indicate the point of elution under these conditions. The analyte is eluting at a position where suppression is occurring, which leads to highly variable peak area ratios, particularly if the background material changes, or shifts in retention time occur during the analysis. Also, the broad unresolved region between 1.5 and 3.5 min appears to contain late eluters—that is, compounds eluting from a previous injection showing broad unresolved peaks. These types of materials are particularly insidious because they do not have predictable retention times. A gradient with ample time spent at high organic content mobile phase flush would likely improve this situation by helping to clear the column with each injection.

Figure 13.14 is an example of two superimposed matrix effect chromatograms, one generated with ESI and the other with APCI. These data highlight a common observation that suppression effects are more serious and more commonly observed with ESI than APCI. The suppression effects are actually serious in both cases, but the retention time characteristics of the suppressing materials with APCI are more readily predicted:

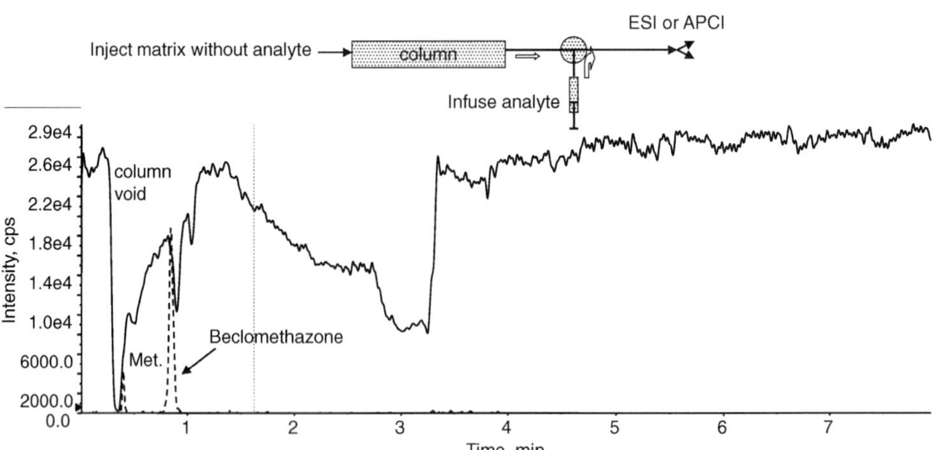

Figure 13.13. The continuous trace is an ESI matrix effect chromatogram of a 10-μL injection of protein-precipitated plasma while post-column infusing beclomethazone. The dashed chromatogram is an injection of standard beclomethazone and metabolite superimposed to indicate retention times of the analytes. MRM data 466 → 376, 2.1 mm × 5 cm C18 column, 200 μL/min, isocratic conditions 60/40 CH_3CN/H_2O 2 mM NH_4OAc.

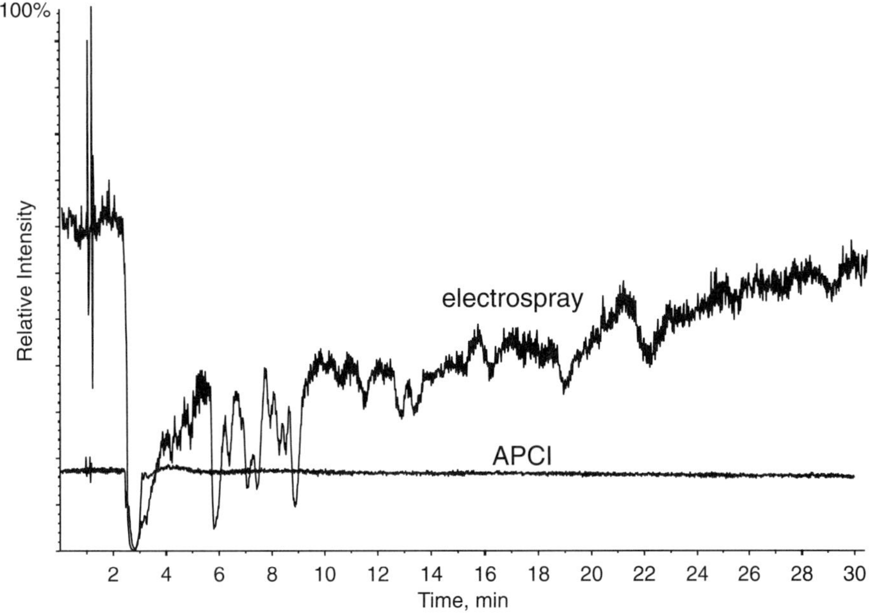

Figure 13.14. Electrospray and APCI matrix effect chromatograms of a 20-μL injection of rat urine while post-column infusing Minoxidil, MRM 210 → 195, 2.1-mm × 5-cm C18 column, 200 μL/min, isocratic conditions 70/30 CH_3CN/H_2O 2 mM NH_4OAc.

They elute in or near the column void. Highly polar nitrogen containing compounds with high gas-phase basicities populate this region of the chromatogram and exert suppression effects with APCI. Large quantities of proteins, lipids, and other surface-active compounds stick to columns and elute at later times and affect ESI more than APCI. In practice, it is clear why the problem manifests itself more in ESI than APCI. As will be seen in the next section, however, the degree to which ionization efficiency is decreased by various sample matrices is not drastically different between ESI, APCI, and MALDI. Different compounds are affected to different degrees in different sample matrices as one would expect.

13.3.2.3 Compound and Sample Matrix Dependency

It is difficult to develop any hard, fast rules to predict endogenous sample matrix effects and their relative importance to ESI, APCI, and MALDI, beyond the obvious. Using ESI, analytes permanently charged in solution with surfactant properties will likely experience minimal effects under even severe conditions. High-gas-phase proton affinity analytes will be similarly resistant with APCI, and MALDI because of the mechanistic similarities. The unpredictability comes from the unknown composition of the sample matrix. To get some empirical handle on the situation, we took the same set of 31 compounds used to measure sampling efficiencies and determined the suppression effects in two different common biological matrices for all three sources. A liquid/liquid extract of blood plasma was used to represent a relatively clean sample involving extraction with an equal volume of methyl *tert*-butyl ether, separation and evaporation of the organic phase, and reconstitution in an equal volume of mobile phase. Protein precipitated plasma, via the addition of an equal volume of acetonitrile and centrifugation, was used to represent a notoriously complex and dirty sample. This was done entirely without chromatographic separation, either using flow injection of analyte mixed with sample matrix with ESI and APCI or by direct spotting with MALDI. Without chromatography, this represents a worst-case scenario.

An example of the data collection process is shown in Figures 13.15–13.17 for the compound Clenbuterol. Blanks for all matrices were analyzed, and the neat compound was spiked into the extracts or pure solvent. Five microliters of sample were flow injected at 200 μL/min for ESI (Turbo Ion Spray®) and at 1 mL/min for APCI. For MALDI, an equivalent amount of sample was spotted on the target mixed 1/1 with CHCA matrix and a single lane was rastered across the spot. As described in the discussion on sampling efficiency, only a portion of the total spot was consumed, but in this case, this corresponded to more than a single laser illumination area.

In the case of Clenbuterol, the relative suppression effect is greatest with ESI and least with APCI, and MALDI lies in between. Although commonly observed, this trend was not entirely consistent across this chemical space. Figures 13.18–13.20 show the results with Erythromycin demonstrating the aforementioned signal enhancement effect observed with APCI and MALDI where thermal degradation effects play an important role in the sample vaporization step. Some amount of sample matrix can be beneficial; in this case it came from the liquid/liquid plasma extract. This effect has been observed since the earliest attempts at desorption chemical ionization in the early 1980s which involved drying a sample on a filament, rapidly heating to vaporize the sample, followed by chemical ionization with reagent ions from an ionized gas. For compounds that are thermally labile, the presence of material in addition to the analyte can serve to significantly minimize thermal degradation of

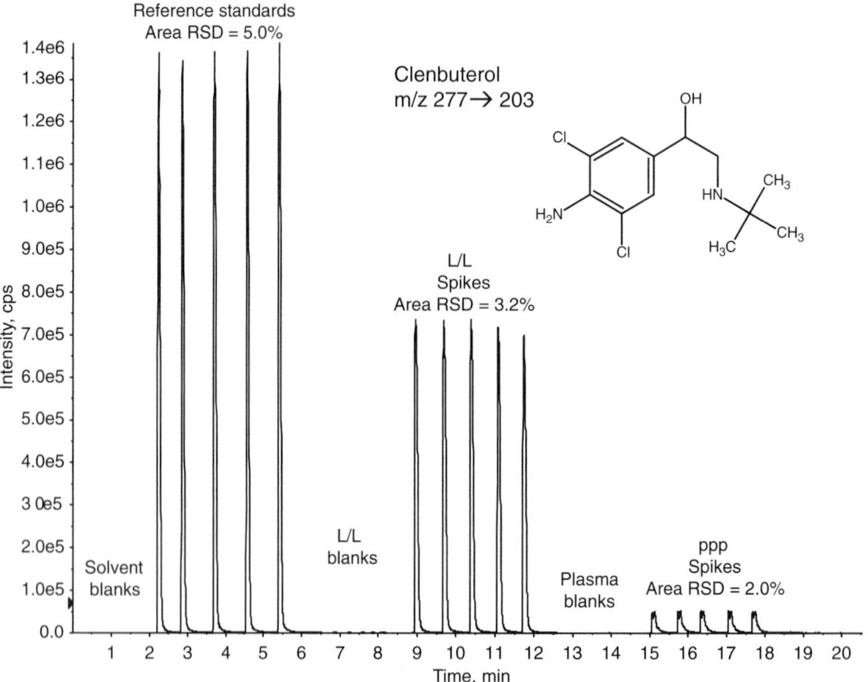

Figure 13.15. Matrix suppression experiment with ESI by flow injection analysis. Replicates of pure standard compounds are followed by spiked liquid/liquid extracts (L/L) and finally by spiked protein precipitated plasma (ppp) with unspiked blanks of each. Flow conditions were 200 μL/min, 50/50 MeOH/H_2O 1 mM NH_4OAc, 5 μL injected, 500 pg injected.

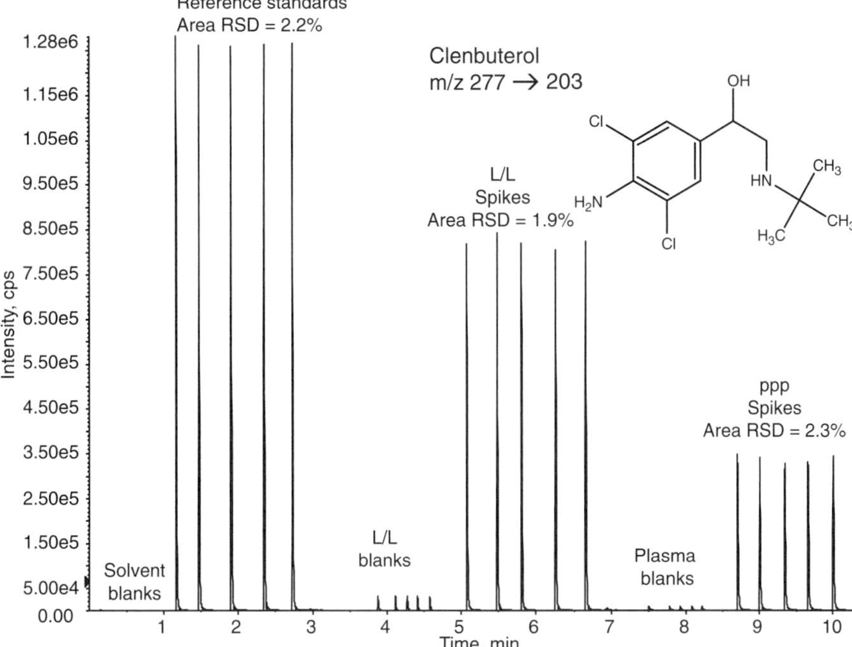

Figure 13.16. Matrix suppression experiment with APCI by flow injection analysis. Replicates of pure standard compounds are followed by spiked liquid/liquid extracts (L/L) and finally by spiked protein precipitated plasma (ppp) with unspiked blanks of each. Flow conditions were 1 mL/min, 50/50 MeOH/$H_2$0, 5 μL injected, 500 pg injected.

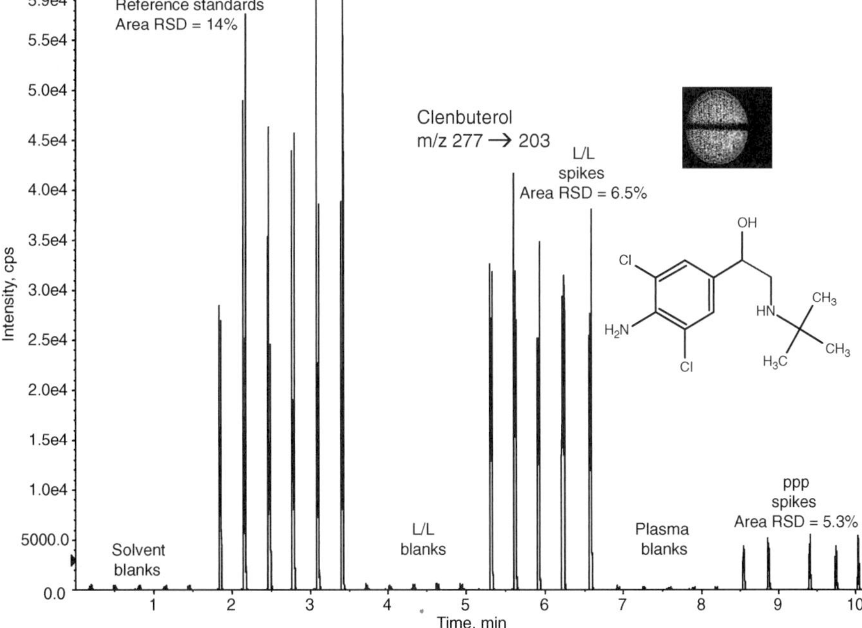

Figure 13.17. Matrix suppression experiment with MALDI with an across spot raster (inset image). Replicates of pure standard compounds are followed by spiked liquid/liquid extracts (L/L) and finally by spiked protein precipitated plasma (ppp) with unspiked blanks of each matrix. 500 pg on target (1-μL spot). Equal volume of CHCA matrix cocrystallized.

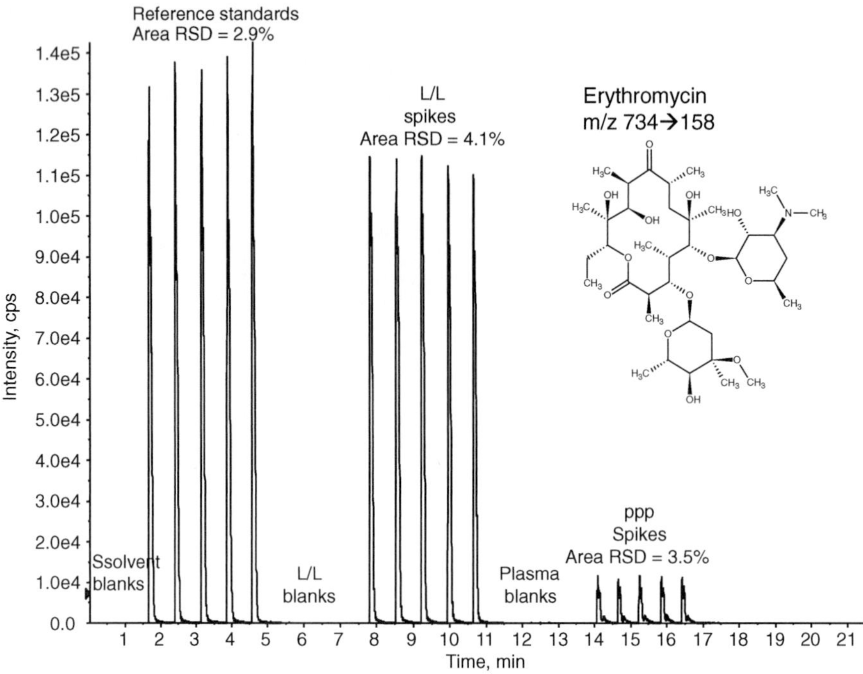

Figure 13.18. Matrix suppression experiment with ESI by flow injection analysis. Replicates of pure standard compounds are followed by spiked liquid/liquid extracts (L/L) and finally by spiked protein precipitated plasma (ppp) with unspiked blanks of each. Flow conditions were 200 μL/min, 50/50 MeOH/H_2O 1 mM NH_4OAc, 5 μL injected, 5 ng injected.

the compound on the hot surface. In this case, the matrix is preventing the compound from undergoing complete thermal fragmentation. The effect is also seen in the MALDI data for Erythromycin. The effect has not been observed to any significant degree with ESI, in any of the cases studied, and this is to be expected since there is no thermal desorption element inherent to the ESI ionization process. Even with the high gas temperatures used with the high-flow ESI sources, the temperatures of the liquid that the samples are exposed to do not exceed 45°C to 50°C and analytes never encounter a solid surface.[5] With MALDI, surface heating is required, and is much more difficult to measure or quantify than with APCI. But the ability of MALDI to produce results on thermally labile compounds like glucuronides would suggest that it is significantly less energetic than APCI.

The results for all compounds and sample matrices for each source are displayed in Table 13.5. The data for each sample matrix are presented relative to the signal obtained with a pure standard for that ion source. A value of 100% means that no suppression occurred, whereas a value less than 100% indicates suppression occurred to that degree; that is, a value of 20 means that the analyte produced 20% of the signal it did when present as a pure standard, and a number greater than 100% indicates that signal enhancement occurred owing to the sample matrix. A value of 800 means that the analyte produced eightfold more signal in the presence of matrix than it did as a pure standard. It is difficult to say conclusively that one ionization technique is generally more or less prone to suppression effects than another. It very much depends on the source, the compound being studied, and the chemical species involved in the suppression event for that source. Suffice it to say that no technique is immune from these phenomena and all sources behave differently,

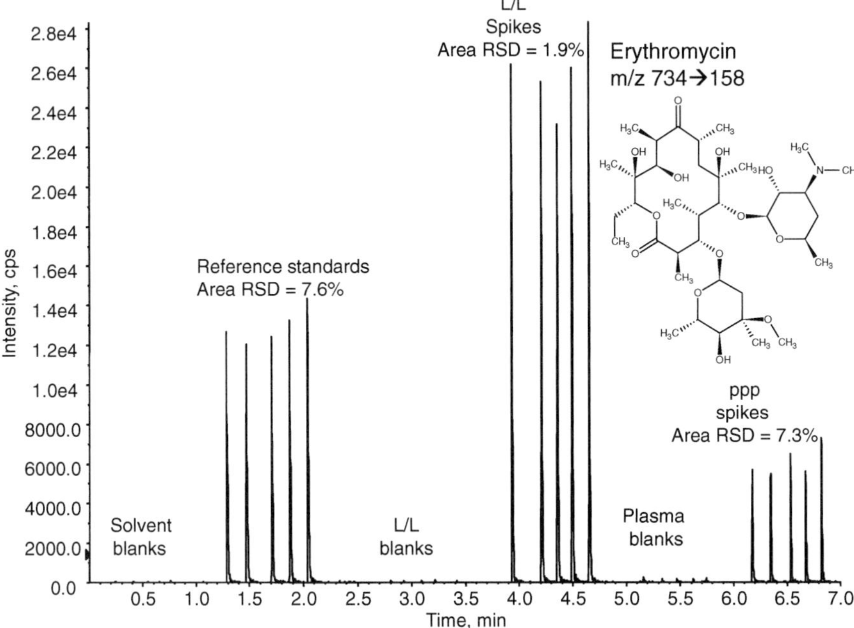

Figure 13.19. Matrix suppression experiment with APCI by flow injection analysis. Replicates of pure standard compounds are followed by spiked liquid/liquid extracts (L/L) and finally by spiked protein precipitated plasma (ppp) with unspiked blanks of each. Flow conditions were 1 mL/min, 50/50 $MeOH/H_2O$, 5 μL injected, 5 ng injected.

which is to be expected if the mechanism of ionization was different for the three ionization modes.

13.3.3 Assay Precision and Accuracy

It has been 15 years since publication of the first validated LC/MS/MS assays.[33,34] Since that time, LC/ESI and LC/APCI on triple quadrupole mass spectrometers have become the quantitative method of choice in both regulated and nonregulated laboratories.[25,28,29] Regulatory requirements place the most stringent specifications on reproducibility and accuracy; therefore these specifications are worth noting as a point of reference. These criteria can be summarized by the ±20/15 rule established by the U.S. Food and Drug Administration (FDA).[48] Precision (expressed as relative standard deviation or coefficient of variation CV) and accuracy (expressed as % deviation from the theoretical concentration) should be ≤15% for all calibrators except for the lower limit of quantification, which should be ≤20%. Nonregulated environments, such as drug discovery, can loosen these criteria as long as the results provide meaningful data. All measurements that we are describing here involve the use of an internal standard where the signal is expressed as a ratio of analyte to internal standard peak areas. It is possible to obtain quantitative measurements without an internal standard as long as the response factor of the targeted compound is known, but using this approach will make it unlikely that the ±20/15 criteria would be achieved, unless conditions are very carefully controlled.

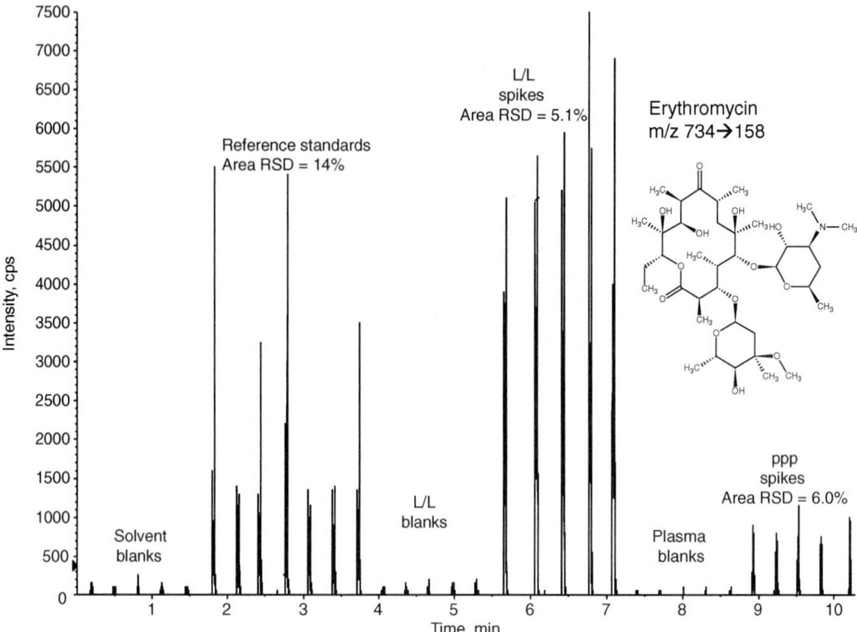

Figure 13.20. Matrix suppression experiment with MALDI with an across spot raster (inset image). Replicates of pure standard compounds are followed by spiked liquid/liquid extracts (L/L) and finally by spiked protein precipitated plasma (ppp) with unspiked blanks of each matrix. 5 ng on target (1-μL spot). Equal volume of CHCA matrix cocrystallized.

The literature abounds with assay statistics; but perhaps the most statistically interesting publication, based solely on the vast number of samples analyzed, is an epidemiological study of the effects of exposure to secondhand cigarette smoke conducted by the U.S. Center for Disease Control.[49] This assay employed APCI LC/MS/MS for the determination of cotinine, a nicotine metabolite, in serum using a stable isotope internal standard. At the time of the publication, data on >32,000 samples utilizing a single mass spectrometer had been amassed. Precision and accuracy of the method was approximately 6% except at the limit of quantitation (25 pg/mL), which had a CV of 12%. It is appropriate to point out here that these statistics provide the variation in the entire method. The previous discussion on reproducibility (see Section 13.3.1 and 13.3.2) dealt only with variability limits imposed by the mass spectrometer based on ion current stability measurements that establish the theoretical limits.

Electrospray has demonstrated results similar to those in the cotinine example and, as described in the sensitivity section, is more widely used and is frequently a more sensitive approach. But it is not always the method of choice when considering targeted specific assays. The ease of tuning APCI, its routine use with 4.6-mm columns, and more predictable susceptibility to suppression effects explain its continued acceptance. In some bioanalytical laboratories, APCI is utilized in up to 30% of the assays developed.[26]

A significant body of work describing MALDI as a quantitative tool exists in the literature.[50–52] In recent years, more attention has been paid to optimizing MALDI instrumentation[10] and matrices[14,37] in order to maximize its utility as a quantitative tool. The development of orthogonal MALDI, which decoupled the laser ionization event from the mass analysis measurement,[53] enabled the deployment of MALDI on mass analyzers other than TOF. Coupling MALDI to a triple quadrupole and the implementation of

Table 13.5. Degree of Signal Suppression for 31 Compounds in the Three Ion Sources in Two Different Sample Matrices [a]

Compound	ESI % Supp. in L/L	ESI % Supp. in ppp	APCI % Supp. in L/L	APCI % Supp. in ppp	MALDI % Supp. in L/L	MALDI % Supp. in ppp
5-Fluorouracil	105	20	147	14	112	36
Phenbutazone	103	8	22	1	48	1
Succinyl choline	101	3	NS	NS		NS
Furosemide	97	19	824	0	77	24
Fosinopril	92	7	609	484	62	90
Erythromycin	92	10	181	41	328	35
Buscopan	83	24	99	0	117	50
Safranin Orange	71	25	83	53	99	24
Clenbuterol	66	8	79	41	120	12
Haloperidol	64	7	90	58	137	18
Morphine	63	2	63	37	138	20
Tamoxifen	58	13	64	19	74	1
Bromocryptin	55	0.5	139	40	107	10
Morphine glucuronide	36	0.3	NS	NS	88	37
Aztreonam	28	0.3	NS	NS	110	62
Prednisolone	26	0.5	130	20		NS
Dianabol	25	3.7	66	12	101	10
Lovastatin	22	0.0	101	43	130	7
Beclomethazone	21	0.9	45	9	71	128
Oxyphenbutazone	17	0.8	66	2	32	2
Taxol	17	0.4	166	35	125	74
Naproxin	12	0.4	46	0.5	77	7
Verapamil	12	0.0	2	1	183	16
Testosterone glucuronide	7	0.0	517	43	90	14
Bentazon	7	0.0	47	1	74	4
Cyclosporin	1	1	116	89	90	13
Alpha-cyclodextrin	1	0.1	NS	NS	49	1

[a] Percent suppression (% Supp.) refers to the amount of MRM signal for the compound spiked in the sample matrix relative to the signal of a pure standard for that ion source. Numbers greater than 100% indicate a signal enhancement from the matrix. A value of 800% would indicate that the analyte generated 8× the signal in the matrix than it did as a pure standard. A value of 1% means that the analyte generated only 1% of the signal when in the matrix compared to a pure standard. L/L means liquid/liquid extract, ppp means protein precipitated plasma. NS means no measurable molecular ion signal with either pure compound or spiked extract.

kilohertz frequency solid-state lasers has the potential to bring quantitative capabilities to the technique that have not existed previously.[36,38]

Reports of accuracy and precision levels for MALDI that meet the FDA ± 20/15 criteria have appeared in the literature;[36,37] but as indicated in the previous sections, the parameters of accuracy, precision, and limit of quantitation generally will not be as good as with conventional ESI and APCI. There will be some instances where it could exceed the more traditional approaches, as mentioned in the sensitivity section, but in general this will not likely be the case—barring important improvements in sampling efficiencies which will allow, on a routine basis, a larger portion of the sample on the target to be ablated instantaneously. The important MALDI advantage is speed, which will be the next topic of discussion.

13.4 SPEED OF ANALYSIS

The difference in the potential for sample throughputs between MALDI and the liquid ionization techniques can be summarized as follows: ESI and APCI are limited by fluidic restrictions—that is, the velocity of fluids in tubes. MALDI is limited by the flight time of ions in a vacuum system. From first principle considerations the two differ by at least two orders of magnitude.

13.4.1 ESI and APCI

ESI and APCI techniques are limited by the rate of sample introduction into the ion source. Autosampler technology is the primary rate-determining factor; but in addition to this, fluid transfer rates and chromatographic separation can also be bottlenecks in the process. Two approaches, generally categorized as parallel sample introduction and fast serial sample introduction, have been taken to increase the throughput.

Parallel sample introduction is characterized by the concept of "indexing" whereby multiple parallel chromatographic systems are shunted into an ion source with multiple sprayers and each channel is toggled on and off rapidly in synchrony with the mass spectrometer acquisition. Samples are introduced into the system simultaneously but are analyzed by the mass spectrometer in a rapid serial fashion. Only a portion of a chromatographic peak can be analyzed with this approach, the fraction being determined by the number of parallel channels operating simultaneously. Because of the duty cycle loss, this approach has not proven particularly useful for high-speed quantitation, but is more applicable to improving throughput where the chromatography is relatively slow and peaks are broad, such as in purification applications.[54] A more in-depth description of the various approaches for construction of such ion sources can be found in Ref. 55.

Fast serial sample introduction, originally described as "stacked elution zone" chromatography, runs parallel chromatographic separations slightly offset in time and, through use of appropriately synchronized valves, toggles one channel into the mass spectrometer at a time. There is no duty cycle loss with the mass spectrometer measurement in this case, making it ideal for high-speed chromatography. This approach was pioneered at Pfizer in the late 1990s,[56,57] and since then the design and optimization of the hardware and software has been developed to achieve consistently high throughputs on a daily basis. For drug discovery ADME applications, this system has set the bar for the liquid introduction ion sources. Throughputs of 15 s per sample have been routinely achieved, and speeds as high as 8 s per sample have been demonstrated with dual spray sources. Tens of thousands of samples are analyzed on a weekly basis with this highly optimized system.

Various versions employing some of the components of this approach have appeared over the past several years. These have been primarily focused on fast chromatography and column switching,[58] as well as staggered parallel columns[59] and software to time and gate chromatographic runs into a mass spectrometer in a fast serial fashion.[60] Recently, a novel concept for a high-speed autosampler has been introduced that improves the duty cycle of the autosampler by the use of multiple independent injection modules.[61] Arrays of independent injection syringes, each with their own autonomous drive motors and RAM, are all engaged in different activities simultaneously—that is, sample aspiration, dispensation, injection, and washing. With this approach the duty cycle of the autosampler, which is traditionally very poor, can be driven to nearly 100% without compromising key performance parameters such as carryover or cross-contamination of analyte from sample to sample. This approach should be able to match the Pfizer injection system speed, but because system washing and sample injecting are occurring simultaneously rather than in parallel as with conventional auto samplers, lower carryover values will be achievable at these fast injection rates.

13.4.2 MALDI

The introductory statement that MALDI speed is limited only by the flight time of ions in the mass spectrometer is based on the premise that sample preparation will be conducted off-line in a highly parallel fashion and therefore will not "bottleneck" the process. Such sample preparation systems are not fully operational today, partially because there has been no driving force to justify their development. It is not difficult, however, to imagine that they could be constructed to keep up with such a high-speed analyzer, particularly if leveraged on the automation and fluid handling capabilities of the microtiter plate industry. Reasonable sample volumes, on the order of 0.2 μL, can be spotted on 1-mm centers resulting in sample densities of nearly 10,000 per microtiter-plate-sized target as shown in Figure 13.21. The speed of loading and unloading such targets in a vacuum system would not represent a throughput bottleneck with these sample densities. Pumpdown times of 10–20 s have been described.[10]

One alternative is conducting solid-phase extraction in parallel. Another would be high-speed capillary gradient chromatography with multiple separations occurring in parallel. The advantage to this approach is the generic nature of the clean-up step. A single-gradient

Figure 13.21. Photograph of a portion of a MALDI target spotted 1 mm on center, each spot 100 nL. Approximately 9600 spots per microtiter-dimensioned target.

method could be applied to nearly all compounds to provide high recoveries; high sample purity could be utilized to improve reproducibility, and highly concentrated sample spots could be employed to enhance sensitivity. This approach would also provide the lowest cost per sample analysis, since greater than 1000 injections per column could be done with ADME-type *in vitro* samples, as opposed to a single-use, solid-phase or membrane-type extraction approach. Multiple-use extraction cartridges would help in this regard. The disadvantage is that there would be a slower sample preparation step: More spots per sample would be required to profile the peaks, which would slow down the acquisition rate and require more target real estate. Ballistic or very rapid gradients with injection-to-injection times of 30 s have been demonstrated using 360 μm-i.d. columns and 5-μL injection volumes at flows of 15 μL/min, with use of pumping systems especially designed for this purpose. An example of an ADME-type analysis run under these conditions is shown in Figure 13.22. In this case, 5 spots per peak were deposited and analysis of the chromatographic region of interest was possible using a MALDI acquisition time of approximately 2 s for this 30-s chromatogram. Windowing the deposition to the chromatographic region of interest would require approximately 20 spots in this case, along with a raster time across each spot of 100 ms. This would allow for approximately 480 samples to be deposited on a microtiter-sized MALDI target using the spot density shown in Figure 13.22. Using a parallel eight-channel LC/MALDI deposition system, 23,040 samples could be prepared in 24 h on 48 reusable MALDI targets on a single instrument of this nature. Eight-channel capillary LCs with this capability exist today,[62] as do electrostatically driven deposition systems that readily handle multiple parallel channels;[41] therefore development of such equipment is not inconceivable.

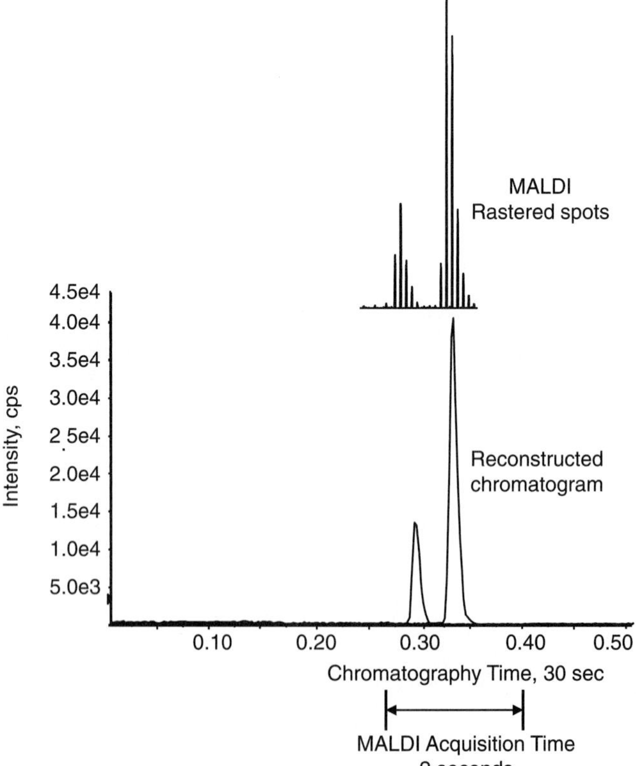

Figure 13.22. LC/MALDI data on Buspirone & metabolite (pyrimidinyl piperazine). On top is the raw data from the rastered spots. Below is the reconstructed chromatogram. 20 pg of Buspirone in Hank's Buffer (Caco cell matrix) injected onto a 360-mm-i.d. column, flow = 10 μL/min, 0.2 μL injected, 30-s gradient, 4-Hz spotting frequency. This quantity represents the approximate quantity predicted for Caco cell ADME tests where the initial concentration is 0.25 μM.

If the problems of sample preparation and data processing are solved, then the mass spectrometer itself becomes the speed limitation. More important than flight times of ions through the analyzer are the phenomena that cause ion beam spreading. There are two primary sources of this, the ablation event itself and the spreading of the beam in the analyzer optics resulting from collisions with background gas. There are two recently described developments that significantly reduce the beam spreading as a result of the ablation event.[10] The first is the use of 1- to 2-kHz high-repetition-rate solid-state lasers. Figure 13.5 demonstrates the speed of the signal transients using lasers with these specifications. Conventional nitrogen lasers used for MALDI typically operate at repetition rates of 1–30 Hz and would produce signal transients nearly two orders of magnitude longer.[10] The second consideration has been to keep the number of laser shots to the minimum required to obtain a quantitative measurement. To do this effectively, a high-repetition-rate laser needs to be gated. Ten milliseconds of illumination (approximately 10 laser shots) have been shown to produce good-quality, reproducible data.[10]

With the above-described source conditions, the limiting factor then becomes beam spreading in the optics of the analyzer. High-pressure regions of the analyzer have been implemented to maximize ion collisions with neutrals for the purpose of fragmentation (collision cell) or focusing (entrance Rf rods). These elements also introduce ion beam spreading because, statistically, some ions will encounter more collisions and lose more speed than others. On a triple-quadrupole-type system the net effect is the spreading of the ion beam, which can be on the order of 100 ms. This establishes the fundamental limit to sample throughput, at least on this instrument, to be 10 samples per second if each spot is a different sample and no carryover of signal between them can be tolerated. Fundamental limits are never reached in practice, as we have seen in the above discussions on sensitivity and reproducibility. However, it is not inconceivable, using conventional mechanical motion control systems, that speeds of one sample spot per 500 ms could be achieved in the future.

To gain perspective, this should be compared to current high-throughput screening systems used for drug activity assays. These systems are typically fluorescence-based and achieve high speed by parallel detection schemes that use light imaging techniques. Throughputs on the order of 100,000 samples per day are considered high-throughput technology with these benchmark systems. A single mass spectrometer capable of two samples per second (one sample per spot) in a serial fashion could achieve analyses in excess of 50,000 samples in 12 h. This might be excessive for today's needs but could potentially open up new drug discovery strategies not currently implemented, if throughputs near to this could be maintained on a daily basis.

13.5 CONCLUSIONS

This chapter is an attempt to provide a snapshot perspective of the three primary ionization sources used today in biological and bioanalytical mass spectrometry—ESI, APCI, and MALDI—and how they compare on three key analytical figures of merit: sensitivity, reproducibility, and speed. In addition to comparing the performance characteristics of the sources, some indication of the potential improvements for each of these three techniques is discussed. In the area of sensitivity, particularly the high-flow ESI techniques, there are significant improvements yet to be advanced, but they will be hard-gained. Reproducibility appears to be limited primarily by methodological factors and not instrumentation, so improvements in this area are difficult to predict and may be minimal. The limits to speed have been established for ESI and APCI and are currently being investigated with the development of new sample introduction equipment. A major advancement in analysis speed appears possible with MALDI, and this could open new doors of opportunity for mass spectrometry in the future.

REFERENCES

1. Whitehouse, C.; Dreyer, R.; Yamashita, M.; Fenn, J. Electrospray interface for liquid chromatography and mass spectrometers. *Anal. Chem.* **1985**, *57*(3), 675–679.
2. Wilm, M.; Mann, M. Analytical properties of the nanoelectrospray ion source. *Anal. Chem.* **1996**, *68*, 1–8.
3. Thomson, B. The heated nebulizer LC-MS interface. In Gross, M. L.; Caprioli, R. (Eds.), *The Encyclopedia of Mass Spectrometry*, Vol. 6, Elsevier, San Diego, CA, 2007, pp 366–370.
4. Bruins, A. P.; Covey, T. R.; Henion, J. D. Ion spray interface for combined liquid chromatography/atmospheric pressure ionization mass spectrometry. *Anal. Chem.* **1987**, *59*, 2642–2646.
5. Covey, T. Pneumatically assisted electrospray ionization. In Gross, M. L.; Caprioli, R. (Eds.), *The Encyclopedia of Mass Spectrometry*, Vol. 6, Elsevier, San Diego, CA, 2007, pp 426–435.
6. Horning, E.; Horning, M.; Carroll, D.; Dzidic, I.; Stillwell, R. New picogram detection system based on a mass spectrometer with an external ionization source at atmospheric pressure. *Anal. Chem.* **1973**, *45*, 936–943.
7. Thomson, B. A.; Danylewich-May, L. Design and performance of a new total liquid introduction LC/MS interface. In *Proceedings of the 31st Annual Conference on Mass Spectrometry and Allied Topics, Boston*, 1983.
8. Thomson, B. Micro and nano electrospray ionization techniques. In Gross, M. L.; Caprioli, R. (Eds.), *The Encyclopedia of Mass Spectrometry*, Vol. 6, Elsevier: San Diego, CA, 2007, 435–444.
9. Karas, M.; Hillencamp, M. Laser desorption ionization of proteins with molecular mass exceeding 10,000 daltons. *Anal. Chem.* **1988**, *60*, 2299.
10. Corr, J. J.; Kovarik, P.; Schneider, B. B.; Hendrikse, J.; Loboda, A.; Covey, T. R. Design considerations for high speed quantitative mass spectrometry with MALDI ionization. *J. Am. Soc. Mass Spectrom.* **2006**, *17*, 1129–1141.
11. Schneider, B. B.; Javaheri, H.; Covey, T. R. Ion sampling effects under conditions of total solvent consumption. *Rapid Commun. Mass Spectrom.* **2006**, *20*, 1538–1544.
12. Schneider, B.; Baranov, V.; Javaheri, H.; Covey, T. Particle discriminator interface for nanoflow ESI-MS. *J. Am. Soc. Mass Spectrom.* **2003**, *14*, 1236–1246.
13. El-Faramawy, A.; Michael, Siu, K. W.; Thomson, B. A. Efficiency of nano-electrospray ionization. *J. Am. Soc. Mass Spectrom.* **2005**, *16*, 1702–1707.
14. Corr, J.; Covey, T. R.; Chau, T.; Kovarik, P.; Fisher, W. MALDI MS/MS on a triple quadrupole mass spectrometer: A new technology for high throughput small molecule quantitation. In *Proceedings of the ASMS Conference on Mass Spectrom. and Allied Topics* 2003.
15. Verentchikov, A.; Smirnov, I.; Vestal, M. Collisional cooling and ion formation processes in orthogonal MALDI at intermediate gas pressures. In *Proceedings of the ASMS Conference on Mass Spectrometry and Allied Topics*, 1999.
16. Douglas, D.; French, J. Collisional focusing effects in radio frequency quadrupoles. *J. Amer. Soc. Mass Spectrom.* **1992**, *3*, 398–408.
17. Thomson, B. A.; Douglas, D. J.; Corr, J. J.; Hager, J. W.; Jolliffe, C. L. Improved collisionally activated dissociation efficiency and mass resolution on a triple quadrupole mass spectrometer system. *Anal.Chem.* **1995**, *67*(10), 1696–1704.
18. Lock, C.; Dyer, E. Characterization of high pressure quadrupole collision cells possessing direct axial fields. *Rapid Commun. Mass Spectrom.* **1999**, *13*, 432–448.
19. Hopfgartner, G.; Bean, K.; Henion, J. Ion spray mass spectrometric detection for liquid chromatrography: A concentration—or a mass-flow-sensitive device. *Journal of Chromatography*. **1993**, *647*, 51–61.
20. Covey, T. In Snyder, A. P. (Ed.), *Biochemical and Biotechnological Applications of Electrospray Ionization Mass Spectrometry*, American Chemical Society, Anaheim, CA, April 2–6, 1995, pp. 21–59.
21. Covey, T. R.; Pinto, D. In Pramanik, B.; Ganguly, A. K.; Gross, M. (Eds.), *Applied Electrospray Mass Spectrometry*, Marcel Dekker, New York, 2002, Chapter 2, pp. 105–148.
22. Valaskovic, G. A.; Kelleher, N. L. Miniaturized formats for efficient liquid chromatography mass spectrometry-based proteomics and therapeutic development. In Lee, M. S.; (Ed.), *Integrated Strategies for Drug Discovery Using Mass Spectrometry*, John Wiley & Sons, New York, 2005, pp. 1–26.
23. Schneider, B.; Guo, X.; Fell, L.; Covey, T. Stable gradient nanoflow LC-MS. *J. Am. Soc. Mass Spectrom.* **2005**, *16*, 1545–1551.
24. Harris, C. M. Shrinking the LC landscape. *Anal. Chem.* **2003**, *75*, 4A–69A.
25. Hopfgartner, G.; Bourgogne, E. Quantitative high-throughput analysis of drugs in biological matrices by mass spectrometry. *Mass Spectrom. Rev.* **2003**, *22*, 195–214.
26. Lagerwerf, Fija M.; van Dongen, William D.; Steenvoorden, Ruud J. J. M.; Honing, M.; Jonkman, J.

H. G. Exploring the boundaries of bioanalytical quantitative LC-MS-MS. *Trends Anal. Chem.* **2000**, *19*(7), 418–427.

27. Keevil, B. G.; Tierney, D. P.; Cooper, D. P.; Morris, M. R. Rapid liquid chromatography–tandem mass spectrometry method for routine analysis of cyclosporine A over an extended concentration range. *Clin. Chem.* **2002**, *48*(1), 69–76.
28. Lee, M. *LC/MS Applications in Drug Development.* John Wiley & Sons, New York, 2002, pp. 115–122.
29. Lee, M.; Kerns, E. H. LC/MS applications in drug development. *Mass Spectrom. Rev.* **1999**, *18*, 187–279.
30. Gobey, J.; Cole, M.; Janiszewski, J.; Covey, T.; Chau, T.; Kovarik, P.; Corr, J. Characterization and performance of MALDI on a triple quadropole mass spectrometer for analysis and quantification of small molecules. *Anal. Chem.* **2005**, *77*(17), 5643–5654.
31. Covey, T. R.; Lee, E. D.; Henion, J. D. High-speed liquid chromatography/tandem mass spectrometry for the determination of drugs in biological samples. *Anal. Chem.* **1986**, *58*, 2453–2460.
32. Covey, T. R.; Lee, E. D.; Bruins, A. P.; Henion, J. D. Liquid chromatography mass spectrometry, instrumentation review. *Anal. Chem.* **1986**, *58*(14), 1451A–1461A.
33. Fouda, H.; Nocerini, M.; Schneider, R.; Geduits, C. Quantitative analysis by high-performance liquid chromatography atmospheric pressure chemical ionization mass spectrometry: The determination of the renin inhibitor CP-80794 in human serum. *J. Am. Soc. Mass. Spectrom.* **1991**, *2*, 164.
34. Gilbert, J. D.; Hand, E. L.; Yuan, A. S.; Olah, T. V.; Covey, T. R. Determination of L-365,260, a new cholecystokinin receptor (CCK-B) antagonist, in plasma by liquid chromatography/atmospheric pressure chemical ionization mass spectrometry. *Biol. Mass Spectrom.* **1992**, *21*, 63–68.
35. Schneider, B.; Lock, C.; Covey, T. AP and vacuum MALDI on a QqLIT instrument. *J. Amer. Soc. Mass Spectrom.* **2005**, *16*, 176–182.
36. Sleno, Lekha; Volmer, Dietrich A. Some fundamental and technical aspects of the quantitative analysis of pharmaceutical drugs by matrix-assisted laser desorption/Ionization mass spectrometry. *Rapid Commun. Mass Spectrom.* **2005**, *19*, 1928–1936.
37. vanKampen, Jeroen J.A.; Burgers, Peter C.; deGroot, Ronald; Luider, Theo M. Qualitative and quantitative analysis of pharmaceutical compounds by MALDIi-TOF mass spectrometry. *Anal.Chem.* **2006**, *78*(15), 5403–5411.
38. Hopfgartner, G.; Varesio, E. New approaches for quantitative analysis in biological Fluids Using mass spectrometric detection. *Trends Anal. Chem.* **2005**, *24*, 583–589.
39. Bruins, A. P. Mass spectrometry with ion sources operating at atmospheric pressure. *Mass Spectrom. Rev.* **1991**, *10*, 53–77.
40. Mylchreest, I.; Hail, M.Electrospray ion source with reduced neutral noise. U.S. Patent 5,171,990, 1992.
41. Takada, Y.; Sakairi, M.; Hirabayashi, A.; Ose, Y.;Mass spectrometer with reduced charged droplet noise. U.S. Patent 5,481,107, 1996.
42. Schultz, G.; Pace, E.; Henion, J. Comparison of TurboIonSpray to Triversa with ESI chip for small molecule quantitation. In *Proceedings of the 54th American Society for Mass Spectrometry and Allied Topics*, 2006, A060082.
43. Ericson, C.; Phung, Q.; Horn, Peters, E.; Fitchett, J.; Ficarro, S.; Salomon, A.; Brill, L.; Brock, A. An automated noncontact deposition interface for liquid chromatography matrix-assisted laser desorption/ionization mass spectrometry. *Anal. Chem.* **2003**, *75*, 2309–2315.
44. Jessome, Lori Lee; Volmer, Dietrich A. Ion suppression: A major concern in mass spectrometry. *LCGC North Am.* **2006**, *24*(4), 2–7.
45. Bonfiglio, Ryan; King, Richard C.; Olah, Timothy V.; Merkle, Kara; The effects of sample preparation methods on the variability of the electrospray ionization response for model dug compounds. *Rapid Commun. Mass Spectrom.* **1999**, *13*, 1175–1185.
46. King, R.; Bonfiglio, R.; Fernandez-Metzler, C.; Miller-Stein, C.; Olah, T. Mechanistic investigation of ionization suppression in electrospray ionization. *J. Am. Soc. Mass. Spectrom.* **2000**, *11*, 942–950.
47. Polson, C.; Sarkar, P.; Incledon, B.; Raguvaran, V.; Grant, R. Optimization of protein precipitation based upon effectiveness of protein removal and ionization effect in liquid chromatography–tandem mass spectrometry. *J. Chromatogr. B* **2003**, *785*, 263–275.
48. FDA 2001 at http://www.fda.gov/cder/guidance/4252fnl.pdf.
49. Bernert, J. T., Jr.; Turner, W. E.; Pirkle, J. L.; Sosnoff, C. S.; Akins, J. R.; Waldrep, M. K.; Ann, Q.; Covey, T. R.; Whitfield, W. E.; Gunter, E. W.; Miller, B. B.; Patterson, D. G., Jr.; Needham, L. L.; Hannon, W. H.; Sampson, E. J. Development and validation of sensitive method for determination of serum cotinine in smokers and Nonsmokers by liquid chromatography/atmospheric pressure ionization tandem mass spectrometry. *Clin. Chem.* **1997**, *43*(12), 2281–2291.
50. Bucknall, M.; Fung, K. Y. C. Practical quantitative biomedical applications of MALDI-TOF mass spectrometry. *J. Am. Soc. Mass Spectrom.* **2002**, *13*, 1015–1027.
51. Cohen, L. H.; Gusev, A. I. Small molecule analysis by MALDI mass spectrometry. *Anal. Bioanal. Chem.* **2002**, *373*, 571–586.
52. Wu, J.; Chatman, K.; Harris, K.; Siuzdak, G. An automated MALDI mass spectrometry approach for optimizing cyclosporin extraction and quantitation. *Anal. Chem.* **1997**, *69*, 3767–3771.
53. Kruchinsky, A.; Loboda, A.; Spicer, V.; Dworschak, R.; Ens, W.; Standing, K. Orthogonal injection of matrix-assisted laser desorption/ionization ions into a

time-of-flight spectrometer through a collisional dampening interface. *Rap. Commun. Mass Spectrom.* **1998**, *12*, 508–518.

54. Zeng, L.; Kassel, D. B. Developments of a fully automated parallel HPLC/mass spectrometry system for the analytical characterization and preparative purification of combinatorial libraries. *Anal. Chem.* **1998**, *70*, 4380–4388.
55. Schneider, B.; Covey, T. In Gross, M. L.; Caprioli, R. (Eds.), *The Encyclopedia of Mass Spectrometry*, Vol. 6-7-9, Elsevier, San Diego, CA, 2006.
56. Cole, M. **1999**. High speed, high capacity analysis of metabolism screens. In *LC/MS and Related Techniques Workshop*, 47th ASMS Conference, June 1999, Dallas, TX.
57. Janiszewski, J. S.; Rogers, K. J.; Whalen, K. M.; Cole, M. J.; Liston, T. E.; Duchoslav, E.; Fouda, H. G. A high-capacity LC/MS system for the bioanalysis of samples generated from plate-based metabolic screening. *Anal. Chem.* **2001**, *73*, 1495–1501.
58. Ackermann, B.; Berna, M. J.; Murphy, A. T. Advances in high throughput quantitative drug discovery bioanalysis. In Lee, M. S. (Ed.), *Integrated Strategies for Drug Discovery Using Mass Spectrometry*, John Wiley & Sons, New York, pp. 315–358.
59. Van Pelt, C. K.; Corso, T. N.; Schultz, G. A.; Lowes, S.; Henion, J. A four-column parallel chromatography system for isocratic or gradient LC/MS analyses. *Anal. Chem.* **2001**, *73*, 582–588.
60. www.cohesivetech.com.
61. Kovarik, P.; LeBlanc, Y.; Aiello, M.; Covey, T.; Londo, T. Ultra high throughput autosampler technology using multiple parallel independent channels. *Proceedings of the 54th American Society Mass Spectrometry and Allied Topics*, 2006, A061331.
62. Welch, C.; Sajonz, P.; Biba, M.; Gouker, J.; Fairchild, J. Comparison of multifluidic HPLC instruments for high throughput analyses in support of pharmaceutical process research. *J. Liq. Chromatogr. Related Technol.* **2006**, *29*, 2185–2200.
63. NIST chemical structure data base, NIST Chemistry WebBook (http://webbook.nist.gov/chemistry/); Advanced Chemistry Development Inc. structure data base, ACDLabs/ChemSketch; e-Molecule data structure data base (http://www.emolecules.com/).

APPENDIX

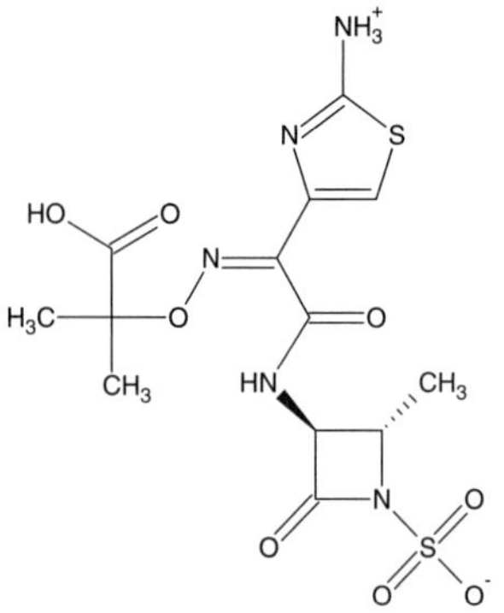

Azetreonam; $C_{13}H_{17}N_5O_8S_2$; MW = 435.1 $(M-H)^- =$ 434.1 → 96; CE = 32 eV; OR = 70 V

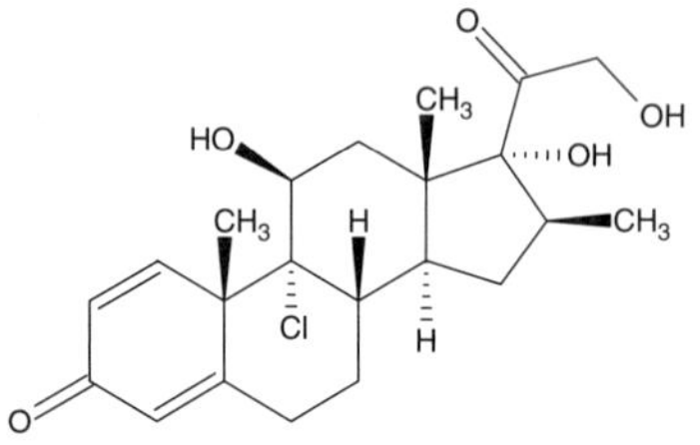

Beclomethazone; $C_{22}H_{29}ClO_5$; MW = 408.18 $(M+H)^+ = 409.2$ → 391.2; CE = 18 eV; OR = 50 V

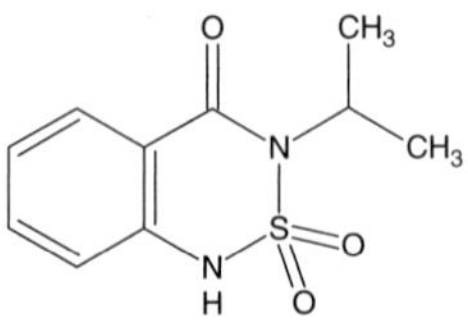

Bentazon; $C_{10}H_{12}N_2O_3S$; MW = 240.1 $(M+H)^+ =$ 241.1 → 199.1; CE = 17 eV; OR = 40 V

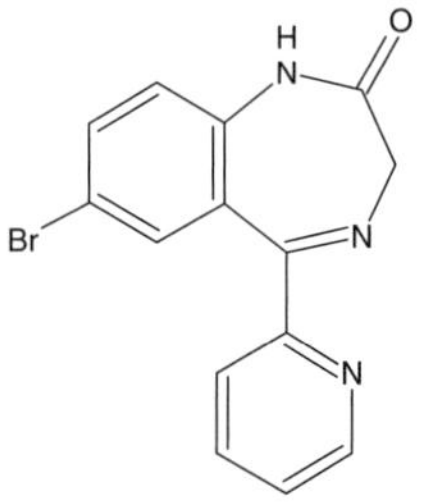

Bromazepam; $C_{14}H_{10}N_3OBr$; MW = 315.1 $(M+H)^+ = 316.1$ → 182.1; CE = 45 eV; OR = 80 V

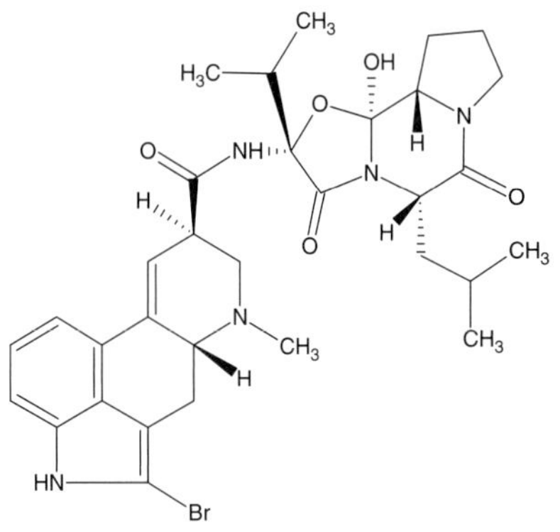

Bromocryptine; $C_{32}H_{40}BrN_5O_5$; MW = 653.3 $(M+H)^+$ = 654.3 → 346.2; CE = 40 eV; OR = 100 V

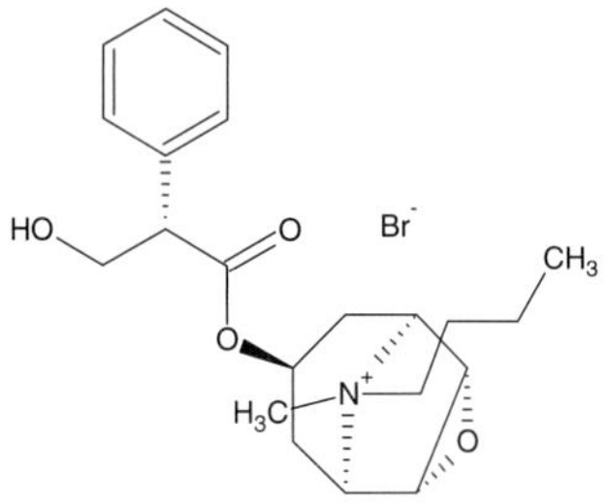

Buscopan; $C_{21}H_{30}NO_4$; MW = 360.2 $(M)^+$ = 360.2 → 194.1; CE = 35 eV; OR = 80 V

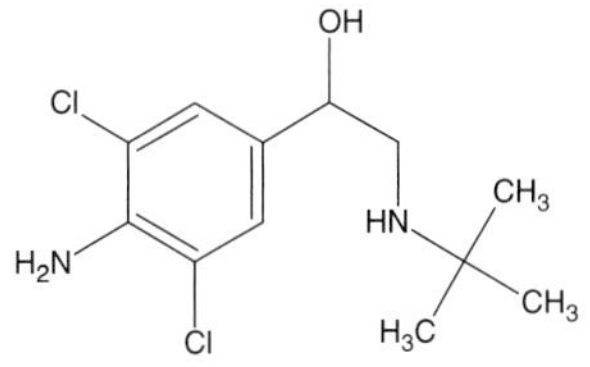

Clenbuterol; $C_{12}H_{18}Cl_2N_2O$; MW = 276.1 $(M+H)^+$ = 277.1 → 203.1; CE = 25 eV; OR = 50 V

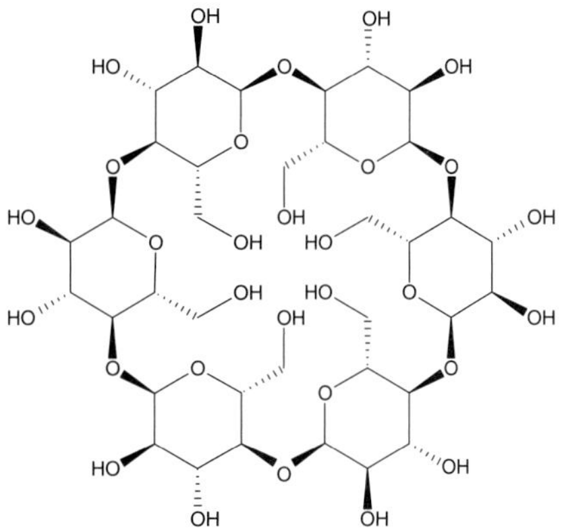

α-Cyclodextrin; $C_{36}H_{60}O_{30}$; MW = 972.4 $(M+H)^+$ = 973.4 → 325.1; CE = 40 eV; OR = 70 V

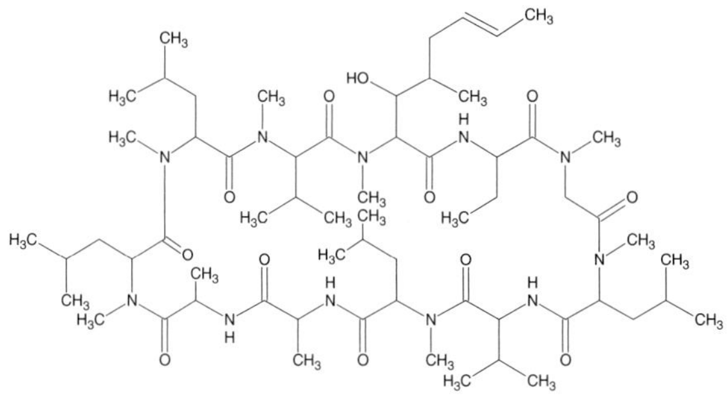

Cyclosporine; $C_{62}H_{111}N_{11}O_{12}$; MW = 1201.9 $(M+H)^+$ = 1202.9 → 286.2; CE = 50 eV; OR = 80 V

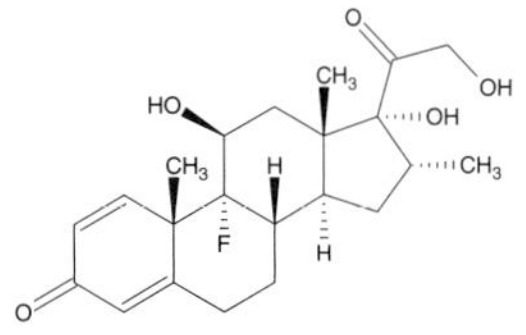

Dexamethazone; $C_{22}H_{29}FO_5$; MW = 392.2 $(M+H)^+$ = 393.2 → 373.2; CE = 13 eV; OR = 50 V

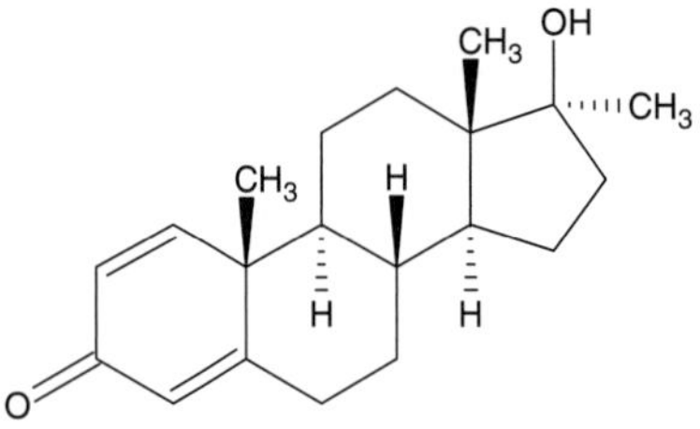

Dianabol; $C_{20}H_{28}O_2$; MW = 300.2 $(M+H)^+$ = 301.2 → 121.1; CE = 40 eV; OR = 60 V

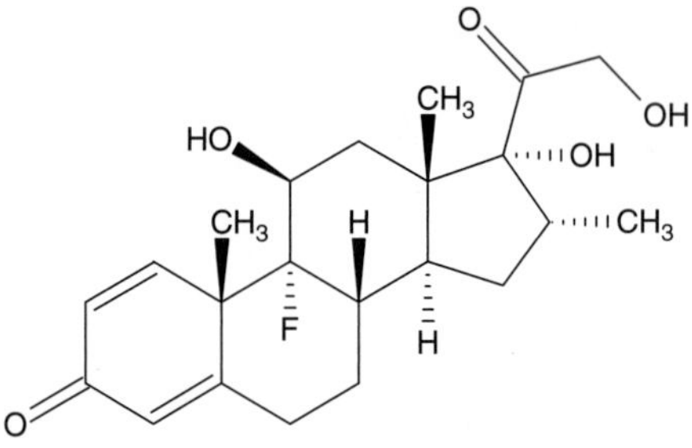

Erythromyacin; $C_{37}H_{67}NO_{13}$; MW = 732.4 $(M+H)^+$ = 733.4 → 158.1; CE = 43 eV; OR = 80 V

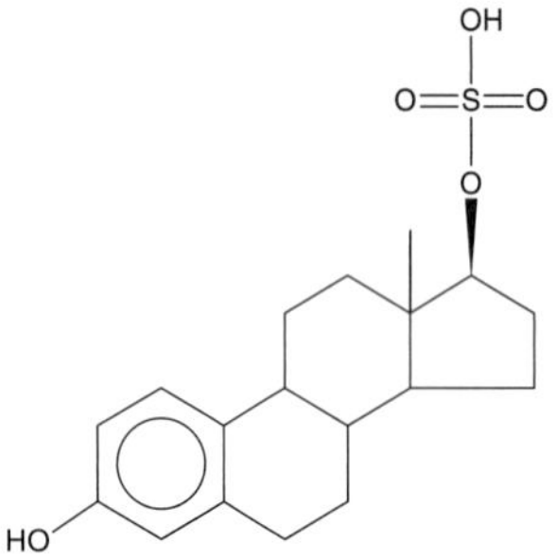

Estradiol-3-sulfate; $C_?H_?O_5S$; MW = 352.3 $(M-H)^-$ = 351.3 → 271.2; CE = 50 eV; OR = 100 V

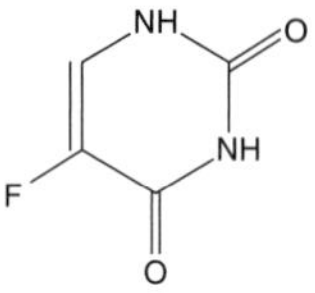

5 Fluorouracil; $C_4H_3FN_2O_2$; MW = 130.1 $(M-H)^-$ = 129.1 → 41; CE = 40 eV; OR = 60 V

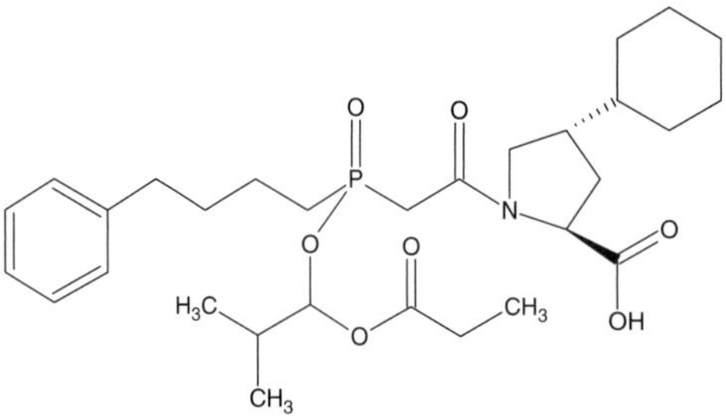

Fosinopril; $C_{30}H_{46}NO_7P$; MW = 563.3 $(M-H)^-$ = 562.3 → 416.2; CE = 18 eV; OR = 60 V

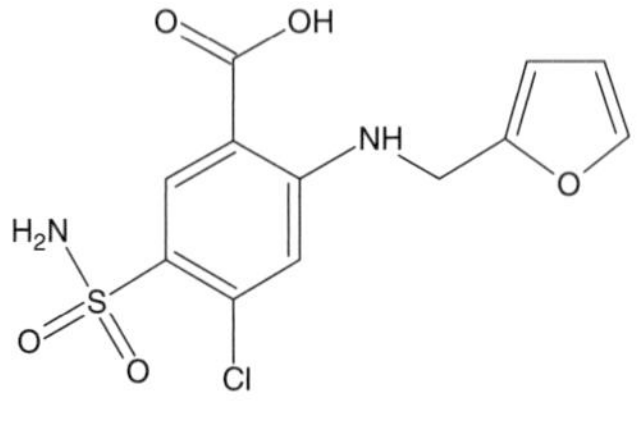

Furosemide; $C_{12}H_{11}ClN_2O_5S$; MW = 330.1 $(M-H)^-$ = 329.1 → 285; CE = 20 eV; OR = 70 V

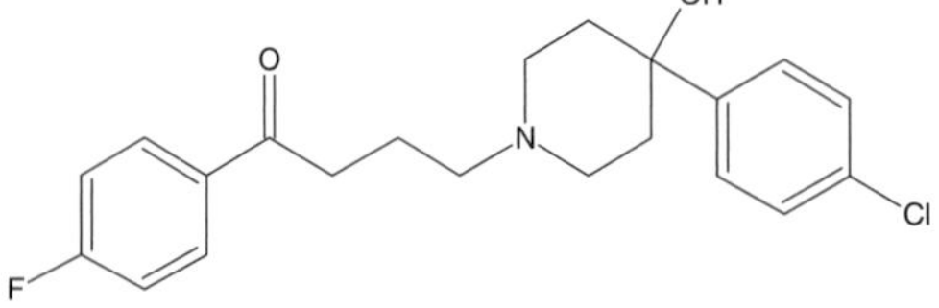

Haloperidol; $C_{21}H_{23}ClFNO_2$; MW = 375.2 $(M-H)^-$ = 376.2 → 165.1; CE = 35 eV; OR = 80 V

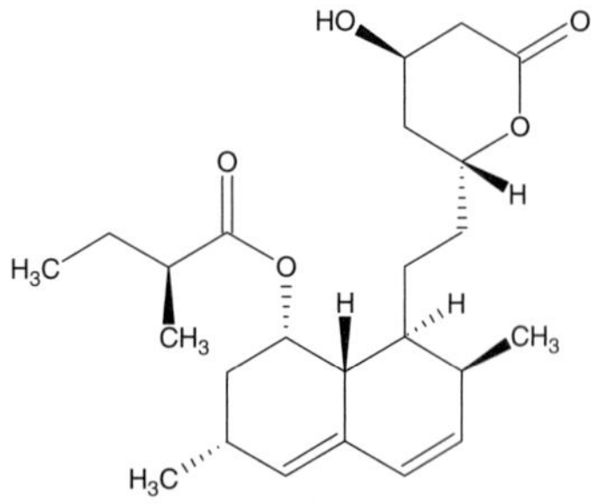

Lovastatin; $C_{24}H_{36}O_5$; MW = 404.3 $(M+H)^+$ = 405.2 → 373.2; CE = 13 eV; OR = 50 V

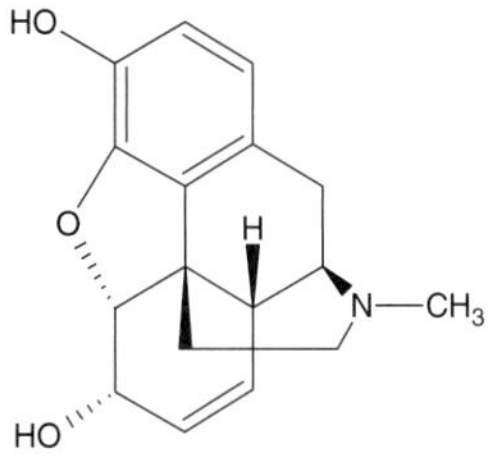

Morphine; $C_{17}H_{19}NO_3$; MW = 285.2 $(M+H)^+ = 286.2 \rightarrow 152.1$; CE = 85 eV; OR = 85 V

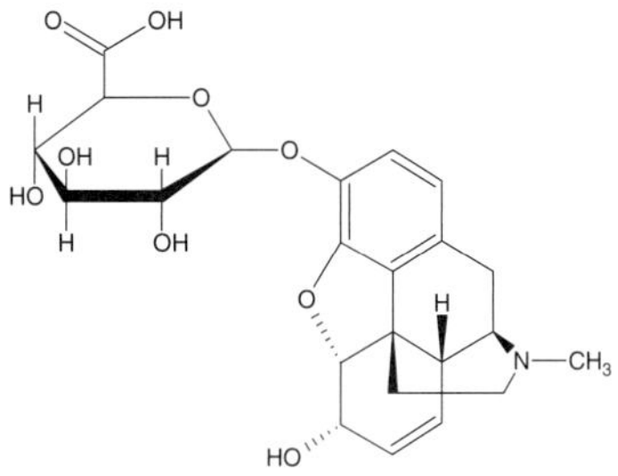

Morphine-3-Glucuronide; $C_{23}H_{27}NO_9$; MW = 461.2 $(M+H)^+ = 462.2 \rightarrow 286.2$; CE = 50 eV; OR = 80 V

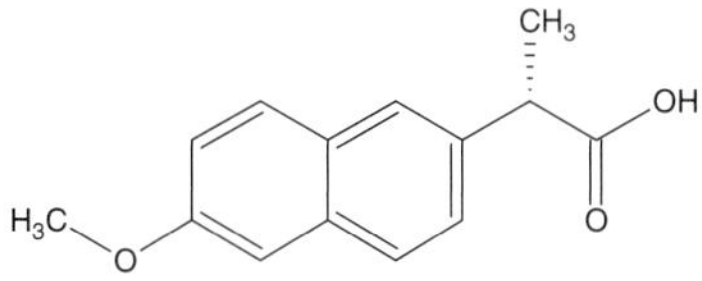

Naproxen; $C_{14}H_{14}O_3$; MW = 230.1 $(M+H)^+ = 231.1 \rightarrow 185.1$; CE = 18 eV; OR = 40 V

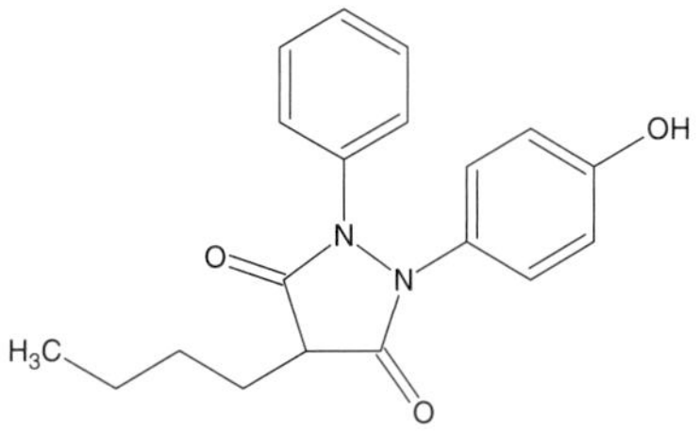

Oxyphenylbutazone; $C_{12}H_{18}Cl_2N_2O_3$; MW = 324.2 $(M+H)^+ = 325.2 \rightarrow 204.1$; CE = 20 eV; OR = 60 V

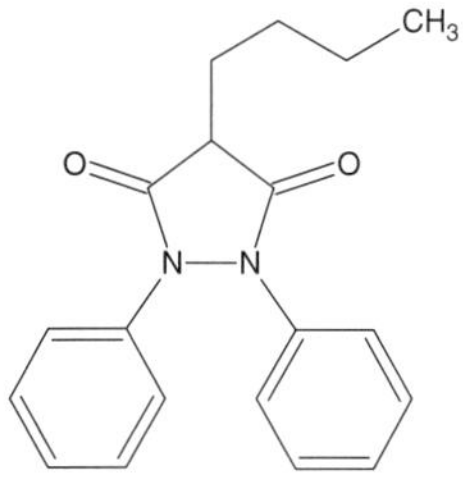

Phenylbutazone; $C_{19}H_{20}N_2O_2$; MW = 308.2 $(M+H)^+ = 309.2 \rightarrow 160.1$; CE = 30 eV; OR = 80 V

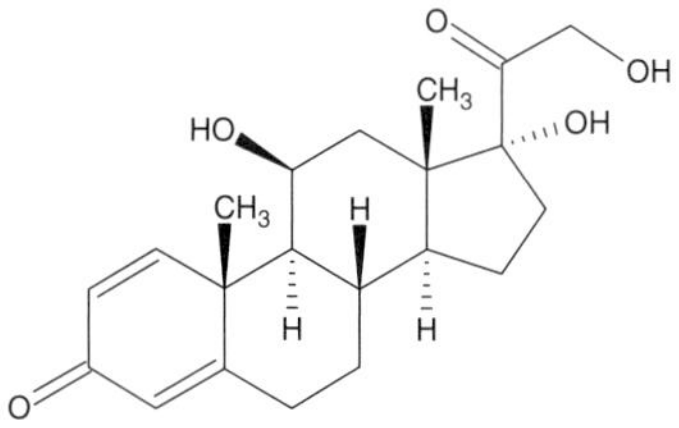

Prednisilone; $C_{21}H_{28}O_5$; MW = 360.2 $(M+H)^+ = 361.2 \rightarrow 147.1$; CE = 35 eV; OR = 50 V

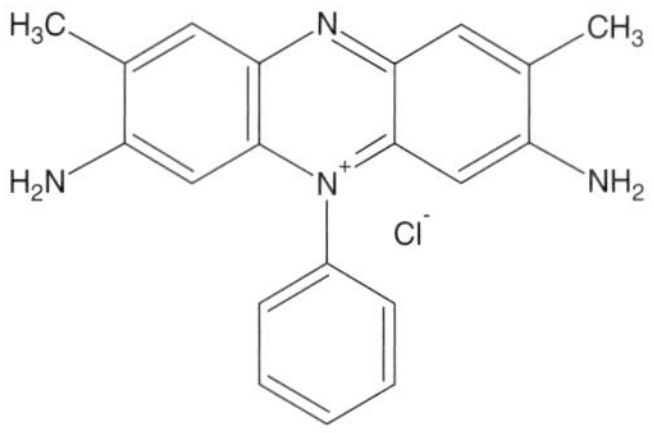

Safranin Orange; $C_{20}H_{19}N_4$; MW = 315.2$(M)^+ = 315.2$ 238.1; CE = 50 eV; OR100 V

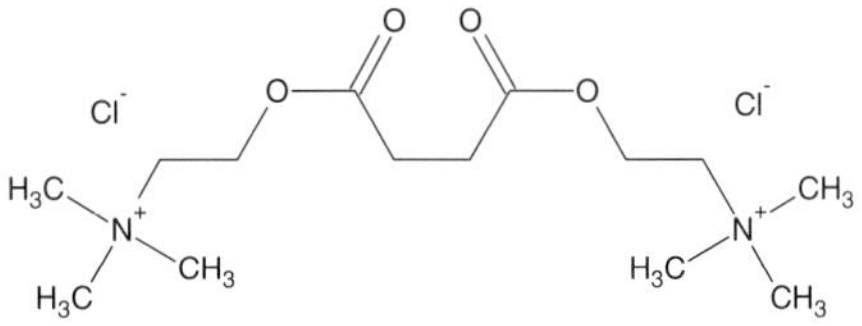

Succinyl Choline; $C_{14}H_{30}N_2O_4$; MW = 290.2 $(M)^{2+} = 145.1 \rightarrow 115.1$; CE = 18 eV; OR = 50 V

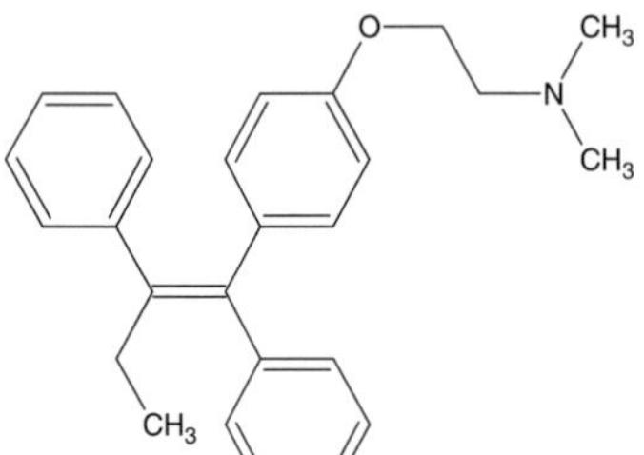

Tamoxifen; $C_{26}H_{29}NO$; MW = 371.2 $(M+H)^+$ = 372.3 → 72.1; CE = 50 eV; OR = 80 V

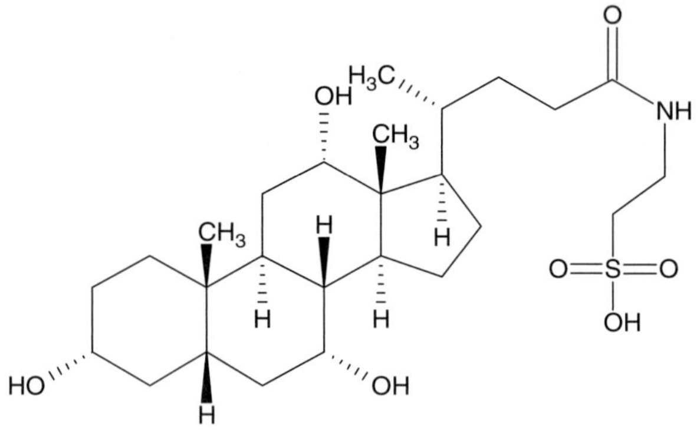

Taurocholic Acid; $C_{26}H_{45}NO_7S$; MW = 515.3 $(M-H)^-$ = 514.3 → 80; CE = 90 eV; OR = 200 V

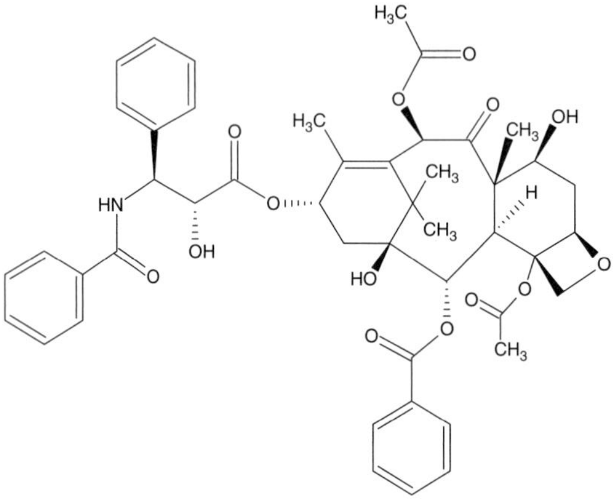

Taxol; $C_{47}H_{51}NO_{14}$; MW = 853.4 $(M+H)^+$ = 854.4 → 286.1; CE = 25 eV; OR = 55 V

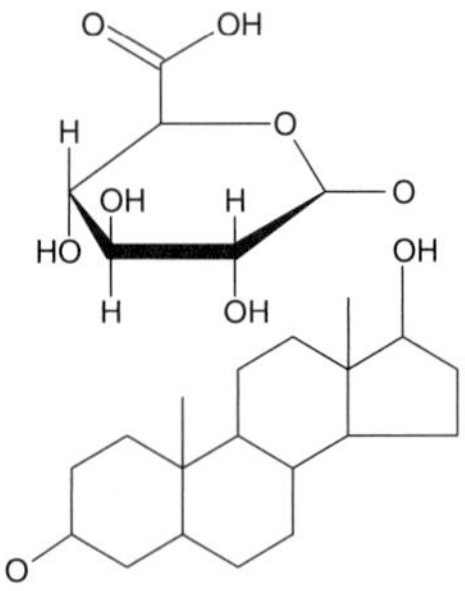

Testosterone Glucuronide; $C_{22}H_{29}ClO_5$; MW = 464.2 $(M+H)^+$ = 465.2 → 289.2; CE = 20 eV; OR = 50 V

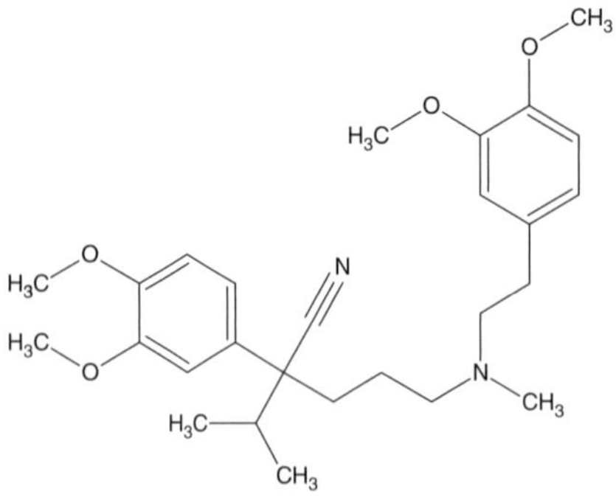

Verapimil; $C_{27}H_{38}N_2O_4$; MW = 454.3 $(M+H)^+$ = 455.3 → 165.1; CE = 40 eV; OR = 85 V

Chapter 14

Charge State Distributions in Electrospray and MALDI

Yan Li[†] *and Richard B. Cole*[*]

[†]*Protein Chemistry Technology Center, University of Texas Southwestern Medical Center, Dallas, Texas*
[*]*Department of Chemistry, University of New Orleans, New Orleans, Louisiana*

Electrospray and MALDI Mass Spectrometry: Fundamentals, Instrumentation, Practicalities, and Biological Applications, Second Edition, Edited by Richard B. Cole

Electrospray (ES) and matrix-assisted laser desorption/ionization (MALDI) may each yield ions corresponding to intact molecules of varying charge state in acquired mass spectra. The relative abundances of analyte ions of different charge state collectively constitute what is referred to as the charge state distribution. Knowledge concerning the factors that influence the charge state distribution is important for several reasons. Notably, if quantification relies upon monitoring the intensity of a certain peak corresponding to one selected charge state, then results will be skewed if other charge states become either favored or disfavored due to changing experimental conditions. Moreover, for experiments such as electron capture dissociation and electron transfer dissociation, higher charge states will offer improved results in terms of electron capture/transfer efficiency; thus, it is to the analyst's advantage to have some control over the charge states produced during ionization. Conversely, the interpretation of spectra is often simplified when only singly charged species are present. A last point is that variability in the charge state distribution may be exploited to obtain analytical information—for example, concerning the changing three-dimensional conformation of an analyte in response to certain imposed conditions. This chapter presents an overview of the factors that influence charge state distributions in ES and MALDI.

14.1 ELECTROSPRAY

A unique characteristic of electrospray (ES) ionization that distinguishes it from all other mass spectrometric ionization techniques is the tendency to form ions having extensive multiple charging. It was Malcolm Dole and co-workers[1,2] who first reported multiply charged macroions of polystyrene in the late 1960s. Dole envisioned that these multiply charged species arose when the excess charges of the initially formed droplets attached to polystyrene molecules as the last of the volatile solvent molecules departed. This portrayal of the birth of gas-phase ions from a residue of the charged droplet became known as the "charged-residue model" (CRM). In the next decade, an alternative ionization mechanism was put forth by Iribarne and Thomson,[3–6] whereby ion formation was proposed to occur via direct desorption of charged species from highly charged droplets. Conditions become favorable for this so-called "ion evaporation" as solvent vaporizes and the remaining charges on the shrinking droplet surface raise the local electric field to a high enough strength to lift an ion into the gas phase (possibly taking solvent molecules along with it). This early "ion evaporation model" (IEM), however, did not address the issue of multiple charging. An extension of the original Iribarne–Thomson IEM was later offered by Fenn et al.[7] to rationalize the phenomenon of multiple charging. They suggested that the charge-state distributions of polyethylene glycols and proteins reflect the rate of ion desorption, which depends upon droplet charge density. Ions of lower charge state start to desorb as the surface charge density of droplets increases with solvent evaporation. As evaporation continues, desorption of higher charge-state ions begins to occur at a faster rate. Eventually, however, a decrease in the flux of desorbed ions of even higher charge state occurs because the majority of ions have already left the droplet before collecting the additional charges characteristic of the highest observable charge states. This scenario was one of the earliest explanations given to account for the commonly observed "envelope" of charge state distributions arising from proteins and polynucleotides. However, despite being a proponent for the ion evaporation model, Fenn did concede that very large proteins, with linear dimensions larger than the charged droplets that hold them, are most likely to be ionized according to the charged residue model.[8]

Much of the electrospray literature has been dominated by protein and peptide studies, and it was recognized early on that basic sites (e.g., amine groups of arginine and of lysine amino acid residues) on these molecules would be the favored points of proton attachment in the positive-ion mode, while acidic sites (e.g., those on aspartic acid and glutamic acid amino acid residues) would logically be facile sites of proton removal. The number of basic amino acids in a protein was reported by Loo et al.[9,10] and Covey et al.[11] to be directly related to the upper limit of the charge state of the protein observed in ES-MS. For several disulfide-bond-containing proteins, the observed maximum charge states were considerably lower than those predicted by the number of basic residues. This disparity was rationalized as being attributable to the reduced accessibility of a number of basic sites owing to the tighter conformation imposed by the disulfide bonds.[10]

14.1.1 Solution Equilibrium—The Starting Place

Between the time that a solution containing macromolecules is introduced into the electrospray device and the time that the presence of gas-phase ions is detected, a complex series of events unfolds that ultimately results in a distribution of multiply charged species in the ES mass spectrum. Initially, in the original charge-balanced solution that is subjected to electrospray, one can rely on solution equilibrium theory[12,13] to describe the degree of salt dissociation and the average solution charge state of an analyte that has known acid–base properties. In the initial neutral solution, of course, an overall charge balance is maintained; that is, for every charge on a particular analyte (or other solution species), a counterion is present near that charge site.

14.1.2 Excess Charge—The Driving Force

The droplets that emanate from the filament or "jet" of the Taylor cone carry a charge imbalance—that is, an excess of charge that corresponds to the polarity of the imposed electric field. With oxidation reactions predominating at the ES emitter in the positive-ion mode, an excess of positive charges is present in formed droplets, while in the negative-ion mode, reduction reactions at the emitter result in an excess of negative charges present. It is these excess charges which directly or indirectly give rise to the gas-phase ions detected by the mass spectrometer.

The full implications of this assertion cannot be realized unless one considers the case of the so-called "wrong-way-round" ions. These are ions appearing in a positive or negative electrospray mass spectrum whose polarity is opposite to the analyte polarity (anionic or cationic, respectively) known to exist in a neutral solution at a specific pH. "Wrong-way-round" ions were first commented upon by Fenselau and coworkers,[14] who observed multiply protonated myoglobin (charge states from $+8$ to $+19$) originating from a pH 10 solution where myoglobin is known to exist almost entirely in anionic form. Conversely, at pH 3.5 where protonation of myoglobin is extensive, anionic (deprotonated) species were observed in the negative ion mode (-9 to -17 charge states). The same group reported similar results for lysozyme, and later Hiraoka et al.[15] reported on wrong-way-round ions arising from a series of amino acids.

In positive-mode experiments, Mirza and Chait[16] showed that the charge-state distribution was influenced by the type of anion (conjugate base) that was used to acidify the solution. These authors considered that the anion remained associated with the multiply protonated protein in solution and that charge removal could occur via dissociation of the neutral acid (anion plus proton in tow) in the later moments of the droplet lifetime. Shortly afterwards, LeBlanc et al.[17] proposed that neutral nitrogen bases that were noncovalently attached to gramicidin S peptide molecules also could serve to remove charge as the complex underwent collisions in the gas phase. These conclusions were reiterated by Hiraoka et al.[15] in their work with amino acids.

Several years later, clever experiments by Boyd and co-workers[18] offered an alternative route to the formation of wrong-way-round ions in the positive-ion mode. By using tetramethylammonium hydroxide (which is not a proton donor) to basicify solutions of histidine, they were able to rationalize the existence of protonated histidine at pH 10, which formed through the liberation of protons from water (requiring pairing of the complementary hydroxide ion with a tetramethylammonium cation). Analogous indirect pathways to liberate protons in basic solution, or liberate hydroxide ions (which could deprotonate acidic sites on analyte molecules) in acidic solution, can account for formation of the so-called wrong-way-round ions. To reiterate, it is the excess (unbalanced) charge in droplets which drives the production of gas-phase ions, be it directly or indirectly.

14.1.3 The Maximum Charge State

14.1.3.1 Intrinsic Basicity and Coulomb Repulsion

The pioneering experiments of Fenn and co-workers, investigating charge attachment to polyethylene glycols (PEGs),[19,20] led to an early model to define the maximum charge state obtainable in ES. They proposed that the capacity of a PEG molecule to retain charge reaches its upper limit when, because of Coulomb repulsion by other charges, the electrostatic potential energy of the centermost charge equals the energy that binds it to its site.[19] In other words, if the electrostatic repulsion surpasses the binding energy, this charge will not remain attached. In the case of polypeptides, the early observations of Loo et al.[9] and Covey et al.[11] led to the proposition of the existence of a direct relationship between the upper limit to the number of charges that a protein may carry in the gas phase and the number of basic residues contained in that protein. Building upon this rough approximation, a model for calculation of the maximum charge state of multiply protonated peptides and proteins was presented by Williams and co-workers.[21] "Apparent gas-phase basicities" of multiply charged ions from 13 proteins were calculated. Apparent gas-phase basicity values consisted of an estimate of intrinsic (Brønsted–Lowry) basicity of a nucleophilic site minus the accumulated electrostatic repulsion (Coulomb energy) of the positively charged protons already attached to the molecule. These values were then compared to the gas-phase basicity of the solvent employed, while reasoning that it is energetically and kinetically favorable for a multiply charged ion with an apparent gas-phase basicity inferior to that of the solvent to transfer an arriving proton to the solvent during the final moments of a droplet's lifetime. This procedure allowed calculation of a maximum charge state for a given protein. The maximum charge state corresponds to the charge state of the multiply protonated molecule whose calculated apparent gas-phase basicity is just below the gas-phase basicity of the solvent; all lower charge state ions will have higher apparent gas-phase basicities than the solvent. Calculated maximum charge state values compared favorably with experimental values reported in the literature, pointing to the conclusion that the maximum charge state for proteins is determined by the final competition for protons between the multiply charged analyte and the residual solvent molecules constituting the final charged cluster. However, for peptides comprised of few basic amino acids, the maximum charge state was found to correlate more closely with the solution charge state based upon a calculated number of amino acids that are protonated in solution.[21] A protein bearing the calculated maximum number of charges is postulated to be formed from a solvent–protein cluster which could conceivably carry more charge than that found on the protein in bulk solution, and also more than that of a "naked" gas-phase protein ion, due to charge stabilization afforded by the solvent dielectric. As the final solvent molecules depart, partitioning of the charge between analyte and solvent is proposed to be governed by the gas-phase chemistry of the species comprising the final cluster (Coulomb energy of the multiply charged analyte and intrinsic gas-phase basicities of analyte versus solvent).

14.1.3.2 Rayleigh Limit to Available Charge

For a large protein contained in a charged droplet about to undergo the final stages of solvent evaporation, Fernandez de la Mora[22] reasoned that the resultant gas-phase ion will carry as many charges as held by the droplet an instant earlier. That droplet has a charge determined by the Rayleigh limit, which in turn will be affected by the surface tension of the liquid subjected to electrospray. This report contends that, for studies of globular proteins and dendrimers having molecular weights of 6000 Da or larger, the observed charges are within 60–110% of the Rayleigh limit of charge (q_R)[22] given below in terms of surface tension (γ) and the droplet radius (R), with ε_0 being the permittivity of the vacuum.

$$q_R = 8\pi(\gamma\varepsilon_0 R^3)^{1/2} \qquad (14.1)$$

According to the above equation, the maximum charge available thus depends upon the solution parameters of surface tension and the final droplet radius.

14.1.4 Solution Effects

14.1.4.1 pH

Variation of solution pH has predictable effects on acid–base equilibria of analyte species and, most importantly in the context of this chapter, analyte charging in solution. It must not be forgotten, however, that in neutral solution, for every charge that a given analyte bears, a counterion will be present nearby. In a landmark experiment, the Chait group[23] first demonstrated that the lowering of initial solution pH beyond a critical value caused a dramatic, discontinuous shift toward higher values in observed ES charge states of proteins. This abrupt shift in the ES spectrum was attributed to an acid-induced change from the native conformation to a denatured state of the protein. As shown in Figure 14.1, bovine cytochrome c undergoes a dramatic change in charge-state distribution when the solution pH is lowered from 5.2 to 2.6. The most intense peak obtained at pH 5.2 corresponds to the native (folded) protein carrying 10 positive charges (Figure 14.1c). Lowering the pH to 3.0 causes the protein to partially unfold, and two discrete charge state distributions centered at 16+ and 8+ are observed (Figure 14.1b). Further lowering of the pH to 2.6 results in complete denaturation, and the obtained mass spectrum shows only one distribution of charge states centered at 16+ (Figure 14.1a). The bimodal nature of the charge-state distribution as a function of solution pH was attributed to two different degrees of solvent accessibility to ionizable side chains.[23] In the native state, some sites are sequestered in hydrophobic "pockets"; with unfolding, many more basic sites suddenly become available to accept protons.

Later, an attempt was made to decouple the above conformation effects from solution charging effects, in an examination of the effect of pH on analyte charging in ES-MS.[24] The small peptides bradykinin and gramicidin S were employed in order to avoid the complications caused by pH-induced conformational changes and thereby probe more directly the effect of pH alone on analyte charge state distributions. The range of hydronium ion concentration was varied from near 10^{-3} M to below 10^{-10} M. Over this seven-orders-of-magnitude variation in pH, only minor changes in the charge states of protonated peptides were observed in ES mass spectra, even though large shifts in *solution* protonation were expected. Thus, the level of peptide protonation in the gas phase as evidenced by the 2+/1+ charge state ratios in ES mass spectra was *not* commensurate with the degree of protonation in the initial solution. This study provided further evidence of changes to analyte protonation that can be incurred between the initial solution condition and the formation of gas-phase ions. In particular, when protons are virtually completely removed from acidic sites, multiply protonated gas-phase ions can still be observed.[14,24,25] Similarly, positively charged proteins that exist in solution can be converted to anionic gas-phase ions in the negative-ion mode of ES-MS operation. Still, the acid–base properties of a given potential charge site clearly affect the ability of this site to retain charge. This is well illustrated in an example comparing the ES mass spectra of two small molecules 4,4′-bipyridyl and 4,4′-bipiperidine,[26] whose size and conformation are nearly the same, hence, Coulomb repulsion in doubly protonated forms of the two molecules would be similar. While singly charged ions were observed for both compounds, only 4,4′-bipiperidine yielded a detectable doubly charged ion. The pK_{a2} for doubly protonated 4,4′-bipyridyl is extremely low in aqueous solution, making it virtually impossible to retain protons on both nitrogen atoms.[26]

During the course of the ES process, the actual concentration of hydronium ions may be changing substantially as first suggested by Gatlin and Turecek,[27] who estimated that acidity may be augmented by 1000- to 10,000-fold in the outermost surface layer of the charged

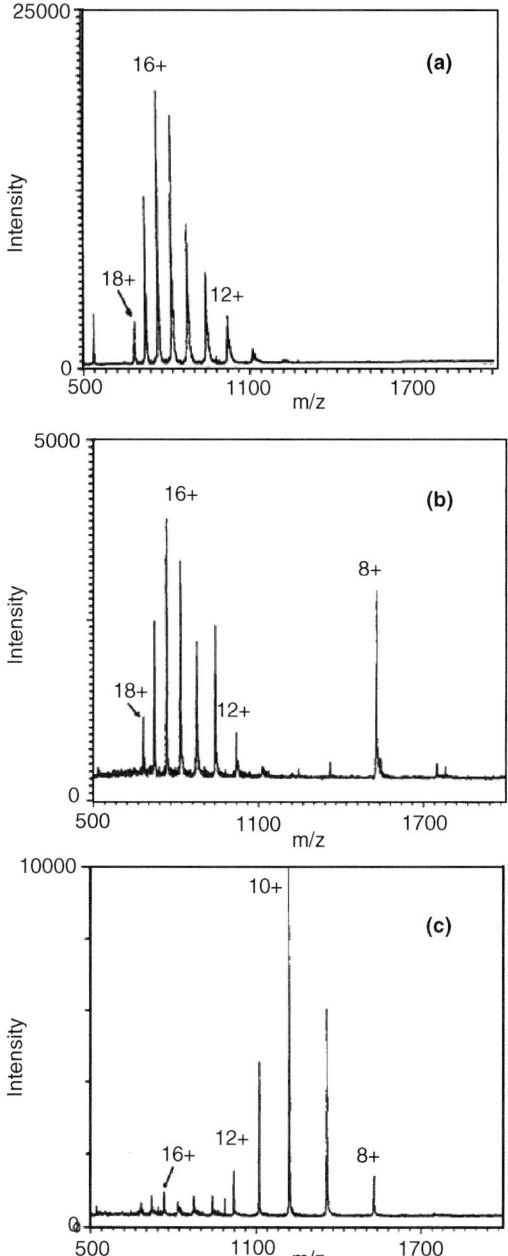

Figure 14.1. Positive-ion ES mass spectra of 10^{-5} M bovine cytochrome c obtained at different pH values (adjusted with acetic acid). (**a**) pH 2.6; (**b**) pH 3.0; (**c**) pH 5.2 (no acetic acid). The charge states on the peaks indicate the number of protons attached to the protein molecule. (Reprinted from Reference 23 with permission.)

droplets. Using an indicator compound that has different optical properties in basic and acidic form, Van Berkel et al.[28] showed that under particular ES conditions, solution pH may be reduced by as much as 4 pH units as a result of the electrolytic oxidation of water inherent to positive-mode ES of aqueous solutions. This production of hydronium ions will have the most impact on low-flow-rate systems, nonbuffered solutions, and systems employing electrical contacts made from difficult to oxidize materials. Conversely, ES droplets formed from aqueous solutions in the negative mode are likely to be enriched in hydroxide ion and thus may undergo considerable pH increase. In either case, the point is that the distribution of analyte charging as it nears the surface of the final droplet may be altered radically compared to what existed in the initial solution.

The discrepancy between solution-phase equilibria expectations based upon initial pH of the neutral solution and observed ES mass spectral charge state distributions was considered in terms of ion pairing and Coulomb repulsion phenomena.[24] The possibility was raised that attractive forces between multiply charged species and nearby counterions likely increase at later moments in the charged droplet's lifetime as fewer solvent molecules remain to provide dielectric shielding for charged species. The increased ion pairing can serve to remove charge to varying extents, depending upon the relative proton affinities of the analyte and the counterion. The simultaneously rising Coulomb repulsion between charge sites in the final moments of the charged droplet's lifetime will likewise tend to lower the number of protons ultimately retained on gas-phase analyte molecules relative to the initial starting solution.[24]

The ES-MS behavior of peptides and proteins in the presence of small, neutral nitrogen bases was also examined.[17] Beyond the subtle direct effect on initial solution pH, it was proposed that the desorbed peptides enter into the gas phase with the nitrogen bases still attached at specific protonated sites. These complexes could then dissociate in the gas phase with the fate of the central proton determined by the competition to capture it between the nitrogen base and the basic site on the molecule. Departure of the protonated nitrogen base, of course, will lower the peptide's charge state.

Other early ES-MS studies looked at the effect of hydronium ion concentration on the charge-state distributions of other types of biological polymers—that is, oligonucleotides[29] and oligodeoxynucleotides.[30] In these reports, the maximum charge states observed for the respective oligonucleotides or oligodeoxynucleotides in the negative-ion mode were reported to decrease with increasing hydronium ion concentration. Observed ES-MS charge states were correlated with the basicity of the nucleoside bases[30]; but unlike the case for proteins, pH variation induced a rather gradual and continual shift in charge states for the oligodeoxynucleotides. Addition of diamines was also effective at reducing the negative charge states of oligonucleotides.[29] Diammonium forms of the additive were proposed to initially interact with the phosphate sites forming complexes that subsequently dissociated yielding a net proton transfer to the oligonucleotide to neutralize the site(s).

14.1.4.2 Solvent Polarity

A pair of negative-ion studies[31,32] and a later positive-ion study[33] examined the dependence of charge-state distributions on the solvent employed in ES-MS. Each of the examined analytes bore two permanent charge sites that were either phosphate groups (negative mode) or quaternary alkylammonium groups (positive mode). Despite the difficulty in varying a single solvent characteristic without simultaneously changing others, experimental results clearly pointed to the conclusion that the polarity of the solvent is a decisive factor in determining charge state distributions of permanently charged analytes. Specifically, solvents of higher dielectric constant consistently shifted charge-state distributions toward higher values. This trend is illustrated in Figures 14.2a–e where the mass spectra of cardiolipin, a phospholipid with two readily ionizable phosphate groups that initially hold sodium conterions (Figure 14.2f), are shown in a series of solvents listed in descending order of polarity, as given by the dielectric constant. The lipid portion of cardiolipin consists of four hydrocarbon chains of varying chain length. This heterogeneity in the number of

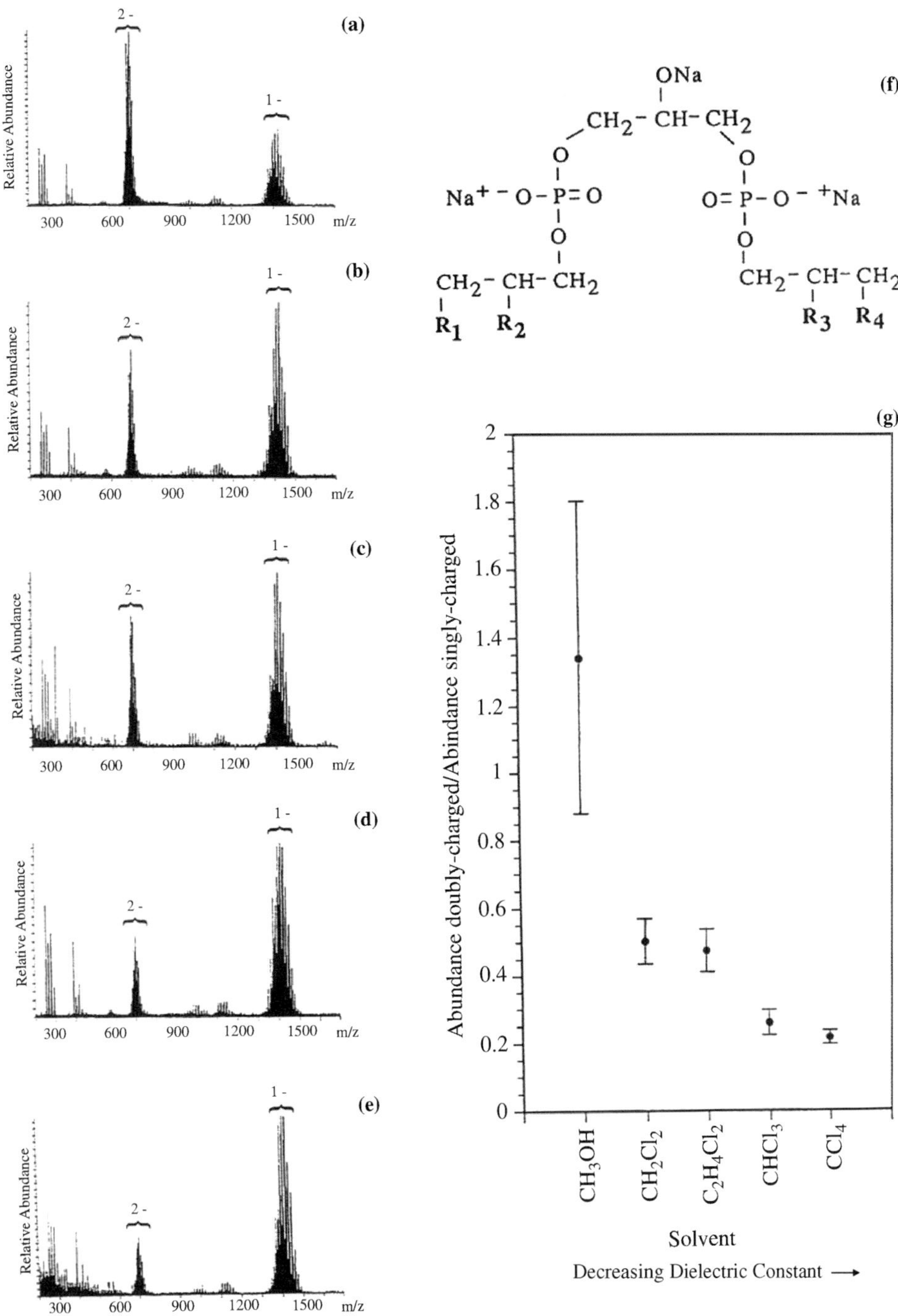

Figure 14.2. Negative-ion ES mass spectra of the disodium salt of cardiolipin dissolved in 10% chloroform and 90%: (**a**) Methanol, (**b**) dichloromethane, (**c**) 1,2-dichloroethane, (**d**) chloroform, and (**e**) carbon tetrachloride. Solvents are listed in order of decreasing polarity as determined by their dielectric constants. (**f**) Structure of cardiolipin; R_1–R_4 represent ester-linked fatty acyl chains of varying length (approximately 16 carbons each). (**g**) Ratios of the abundances of doubly charged/singly charged ions for the disodium salt of cardiolipin using data from **a–f**. The ratio diminishes as the dielectric constant of the solvent decreases. (Adapted from Ref. 32 with permission.)

$-CH_2-$ groups is responsible for the observed variety of cardiolipin species that appear at m/z values separated by 14 and 7 m/z units for singly and doubly charged species, respectively. Figure 14.2g shows the ratios of doubly charged to singly charged species and standard deviations based upon five measurements. A clear trend showing a decreased tendency to form ions having the higher (2−) charge state with decreasing solvent polarity is evident. Analogous negative-ion results were obtained from diphosphoryl lipid A from *E. coli* in series of chlorinated solvents, alcohols, and other solvents.

In the positive-ion mode, diquaternary salts exhibited a similar tendency to dissociate as a function of solvent polarity—that is, complete dissociation with formation of doubly charged cations was more efficient in solvents of higher polarity. This held true for ES mass spectra acquired using series of chlorinated solvents as well as homologous short-chain alcohols.[33] The increased tendency toward formation of higher-charged-state ions in ES mass spectra acquired from solvents of higher polarity in either the positive- or negative-ion modes was attributed to an increased ease of dissociation in solvents of higher dielectric constant. Of course, higher-polarity solvents are more effective at stabilizing multiply charged species, whereas lower-polarity solvents will disfavor dissociation and counterions will have an increased propensity to remain attached to at least one of the charge sites.

From a mechanistic standpoint, it seems reasonable to postulate that solvents with higher dielectric constants would be more effective at boosting the charge level in offspring droplets upon droplet fission, relative to their lower dielectric counterparts. It is known that more polar solutions can tolerate higher charge densities, which implies an increased conductivity due to augmented charge separation. In these high-polarity solutions, the excess charge will also be confined to a thinner layer at the surface, analogous to the effect of increased conductivity on the width of the double layer at a charged electrode's surface (see also Chapter 2 in this volume). In addition, owing to the fact that both cardiolipin and the employed diquaternary ammonium molecules have large nonpolar moieities, they are each expected to have substantially enhanced surface activity in the more polar solvents. The combination of increased surface activity along with increased charge separation (counter-ion departure) in higher dielectric solvents can account for an increased level of charging in offspring droplets upon droplet fission.

14.1.4.3 Electrolyte and Counterions

The nature of the particular counterion present in solution was first shown to exhibit an impact on the appearance of the ES mass spectrum by Mirza and Chait.[16] They found that the charge states of proteins and peptides were shifted to lower values in varying degrees by neutralization of positive charge sites by anions present in solution. The magnitude of the shift in the positive-mode charge-state distributions of tested peptides and proteins diminished in the order $CCl_3COO^- > CF_3COO^- > CH_3COO^- \approx Cl^-$, regardless of the salt form in which the ions were introduced. This ranking correlates with the electroselectivity of anions toward anion exchange resins.[34,35] In addition to the proteins and peptides, the trend also held for compounds that exist in salt form, such as the chloride salts of diquaternary ammonium cations.[33] When cesium trifluoroacetate was added to the initial solution, ES mass spectra revealed that trifluoroacetate ions readily displace chloride counterions on the diquaternary ammonium analytes and more effectively bind to the permanent charge sites as evidenced by the much higher abundance of $(M - 2Cl^- + CF_3COO^-)^+$ versus $(M - Cl^-)^+$ ions (here M denotes the initial diquaternary ammonium cation plus its two chloride counterions). These results imply that counterions with the greatest affinities for cationic species will be most effective at shifting charge state distributions toward lower values.

If salts consisting of anions of relatively low affinity for anion exchange resins (e.g., Cl^- from CsCl) are added to solutions containing proteins in preparation for ES-MS analysis, it was reported[36] that over a wide range of electrolyte concentration (i.e., 10^{-6}–10^{-2} M) the average charge state, z, of test proteins remained virtually unchanged. The average charge state is defined as

$$Z = \frac{I_{A+,ms} + 2(I_{A2+,ms}) + \cdots + i(I_{Ai+,ms})}{I_{A+,ms} + I_{A2+,ms} + \cdots + I_{Ai+,ms}} = \frac{\sum iI_{Ai+,ms}}{\sum I_{Ai+,ms}} \quad (14.2)$$

where $I_{Ai+,ms}$ is the signal of analyte A of charge state i+ detected by the mass spectrometer.[36] The point was illustrated for the proteins myoglobin and lysozyme, which exhibited nearly constant charge state distributions while at the same time being subjected to dramatic drops in overall signal intensities attributable to the signal suppression effect of the added electrolyte. Increasing concentrations of other varieties of electrolytes, such as ammonium acetate, caused a small shift in the ES-MS charge states of lysozyme toward lower values. The basicity of the acetate counterions likely favored removal of protons from the lowest basicity sites of lysozyme to increasing extents as the concentration of this electrolyte was raised.[36]

14.1.4.4 Surface Tension

In a series of papers devoted to investigating the effects of selected additives to solutions subjected to electrospray, Iavorone, Jurchen, and Williams[37–40] reported on a phenomenon that they termed "supercharging." Unlike the previous examples that employed analytes with two "fixed" charge sites,[31–33] these studies targeted effects on observed charge states of peptides and small proteins that were run in methanol/water mixtures containing 3% acetic acid. It was shown that the addition of *m*-nitrobenzyl alcohol or glycerol to these solutions resulted in the appearance of a higher maximum charge state and an overall shift in the charge state distribution toward higher values. While it was noted that these additives were of lower volatility than the solvent components and that under the employed conditions all of the protein analytes were highly denatured,[37] the "supercharging" phenomenon was ultimately attributed to the high surface tension of the additives. It was reasoned that the raised solution surface tension that resulted from introduction of the additive necessitated augmented droplet surface charging in order to achieve Rayleigh fission. The additional number of charges available in the presence of additives (e.g., *m*-nitrobenzyl alcohol) that raise the surface tension is what was deemed responsible for increasing the signals of higher charge states of gas-phase protein ions. Using polyethylene glycol (PEG) as the test analyte, this same *m*-nitrobenzyl alcohol additive was shown to increase the level of charging in both positive- and negative-ion modes of operation. Again, the raised surface tension was deemed responsible for the requirement of a higher degree of surface charging which led to a shift in charge states toward higher values in both positive and negative modes. Notably, the addition of glycerol to PEG solutions run in the positive mode actually lowered the average charge state. This was attributed to glycerol's high affinity for available sodium, thus allowing it to rob PEG of some of the available charge.[38]

In a later study employing dendrimers with multiple peripheral amino groups, Iavorone and Williams[39] found that addition of *m*-nitrobenzyl alcohol to methanol solutions increased the average charge state, and they attributed this shift to the resulting increase in surface tension that necessitated an increased level of droplet charging to obtain droplet fission. Conversely, however, they found that *m*-nitrobenzyl alcohol addition to pure aqueous solutions actually decreased the observed charge states of the tested dendrimers, an effect attributable to a lowering of the surface tension compared to that of pure water. In some cases, small peaks appeared at higher maximum charge states when *m*-nitrobenzyl alcohol was added to aqueous solutions, but the highest average charge states were actually achieved from solutions of pure water. Moreover, the most intense peaks in the spectrum (i.e., the base peaks) were also of higher charge for pure aqueous solutions as compared to their counterparts containing *m*-nitrobenzyl alcohol.[39]

The above results from Iavorone, Jurchen, and Williams linking higher charge states to higher surface tensions for solutions of peptides, small proteins, and functionalized dendrimers prompted us to go back to look at surface tension trends obtained in our earlier work. Unlike the mobile charges (protons) responsible for charging of the above samples, we had examined solvent-induced shifts in charge state distributions employing analytes with fixed phosphate groups (negative-ion mode[31,32]) or fixed quarternary ammonium groups (positive-ion mode[33]). Interestingly, in contrast to the results outlined above for compounds relying on protons for charging, for our analytes containing "permanent" charge sites, often there existed an *inverse* relationship between maximum charge state observed and the surface tension of the solvent. For example, in comparing methanol, 1,2-dichloroethane, dichloromethane, chloroform, and carbon tetrachloride [surface tensions (dynes/cm) of pure solvent at 20°C = 22.6, 23.4 (35°C), 26.0, 27.0, 27.0, respectively], when these pure solvents were mixed with 10% chloroform containing the analytes, the onset voltage for electrospray was observed to increase monotonically through the series as expected because of the increasing surface tensions.[32] But the 2− : 1− ratio (Figure 14.2) for the test analyte cardiolipin containing two fixed anionic sites rose in the opposite order: carbon tetrachloride < chloroform < 1,2-dichloroethane ≈ dichloromethane < methanol; that is, *higher surface tension led to less charging of cardiolipin*. Using the same solvent series minus 1,2-dichloroethane, the results were exactly mirrored in the positive-ion mode for two diquaternery ammonium salts; that is, 2+ : 1+ abundance decreased over the series methanol, dichloromethane, chloroform, and carbon tetrachloride, where surface tension is rising. Similarly, when comparing alcohols: methanol, ethanol, 1-propanol, 1-butanol [surface tensions (dynes/cm) of pure solvent at 20°C = 22.6, 22.8, 23.8, 24.6, respectively], in the negative mode, the 2− : 1− ratio of both cardiolipin and lipid A from *E. coli* decreased steadily as the surface tension rose,[32] and it did so again in the positive-ion mode for the diquaternary ammonium salts.[33]

To rationalize the contrasting results obtained for peptides, proteins, and amine-derivatized dendrimers versus those obtained for doubly phosphorylated and diquaternary molecules containing fixed charge sites, it is worthwhile to consider that the former acquire charge via the *acquisition of excess protons* while the latter acquire charge by *counterion removal*. In the case of the former, excess protons become available as a consequence of methanol/water oxidation[28] and possibly via removal/neutralization of acetate counterions originating from the added acetic acid. As the surface tension of the solvent increases, one can reason that the higher charge density required for droplet breakup will create a larger excess of available protons and ultimately lead to higher charge states on desorbed proteins, peptides, and other analytes bearing basic sites that accept protons. In the case of the diphosphorylated and diquaternary ammonium molecules bearing fixed charges, a higher solution surface tension also will produce a larger overall excess charge. However, an accompanying higher solvent dielectric will serve to simultaneously allow a higher level of dissociation of the counterions that accompany the fixed charge sites. While the precise fate of these counterions is not known, to account for the higher charge states observed, we propose that higher-polarity solvents can more efficiently boost the charge level in offspring droplets produced by droplet fission. Both charge density and conductivity can reach higher values in these more polar solutions, and the excess charge will tend to be present in a more compact layer at the surface. The charged species in this thin layer can more effectively slip over to the offspring droplet in a fission event, leading to increased charging of offspring droplets upon droplet fission relative to lower polarity solutions. The plausibility of this scenario is corroborated by the fact that the analytes (especially those examined in the negative-ion mode[31,32]) have substantial surfactant character.

Bearing in mind the relationship between the surface tension of the solution and the Rayleigh limit that charged droplets may carry as expressed above in Eq. (14.1) Samalikova and Grandori[41] set out to test whether protein charge state distributions responded to changes in surface tension caused by various additives. Addition of 1-propanol (e.g, 20%) did lower the surface tension of aqueous solutions, but the charge state distributions of four tested proteins did not vary as a result. In a later study by the same group,[42] 50% addition of 1-propanol or 2-propanol both lowered the surface tension of acidified aqueous protein solutions, but the vapor pressure of the latter is higher than that of water, while the vapor pressure of the former is below that of water. Again, charge-state distributions were not affected by the additives. Other additives such as DMSO (having a higher surface tension than aqueous protein solutions containing acetic acid) did cause shifts in charge-state distributions toward higher values in a manner consistent with the explanation of "supercharging" given by Williams and co-workers,[37–40] but Samalikova and Grandori[42] conclude that their results are not easily reconciled by considering only charge availability in droplets at the Rayleigh limit as determined by surface tension.

14.1.5 Analyte Effects

14.1.5.1 Analyte Concentration and Competition for Charge

The relationship between ES-MS signal intensity and analyte concentration in the initial sprayed solution has been studied extensively.[43–49] The linear dynamic range can extend up to about 10^{-5} M, beyond which the signal intensity levels off and may even decrease at extremely high concentrations. For multiply charged analytes, ions of a discrete charge state exhibit similar dependences on concentration; that is, ions of one particular m/z value have a linear dynamic range followed by a level-off region at higher concentrations. However, the slope of the linear range and the concentration where the transition into the flatter portion of the curve begins are variable for ions of different charge state. The consequence is that the resultant charge state distribution of the analyte is shifted toward progressively lower values as the concentration is raised.[48] The interpretation of the effect of analyte concentration on the charge-state distribution observed in ES-MS is intertwined with the perceived ionization mechanism. At elevated concentrations, the decrease in the relative abundances of analyte ions of higher charge states was hypothesized by Chowdhury et al.[50] to be a direct result of an increase in the competition for the limited number of available charges on the droplets. In a similar vein, Smith et al.[51] proposed that a different "saturation" limit exists for ions of various charge states, which could be responsible for the increase in the abundances of ions of lower charge state, whereas those of higher charge state level off, as concentration is raised.

The role of analyte concentration (and that of counterions present) in influencing ES-MS charge state distributions was modeled using analyte species originally present in salt form in both positive and negative modes.[48] ES mass spectra revealed that charge states were consistently shifted toward lower values as analyte concentrations were raised. In contemplating the underlying issues from a mechanistic standpoint, it was deemed useful to introduce the N/N_0 quotient—that is, the ratio of the total number of excess charges (N) to the total number of analyte molecules (N_0) in initially formed electrospray droplets. N is defined as the total number of elemental charges in the droplets that have no counterion. In the absence of corona discharge, the number of excess charges generated per unit time can be approximated as being equal to the ES current (I, amperes = C/s). Dividing this current (I) by the elemental charge (e, 1.602×10^{-19} C/charge) yields charges per second, that is, the rate at which positive charges leave the spray tip [Eq. (14.3)]. N_0 corresponds here to the totality of analyte molecules entering electrosprayed droplets; its value per unit time can deduced from the product of the analyte concentration (C, mol/L) and the solution flow rate (U, L/s) multiplied by Avogadro's number (A, 6.023×10^{23} molecules/mol):

$$\frac{N}{N_0} = \frac{I/e}{\text{ACU}} = \frac{(\text{charges/second})}{(\text{molecules/second})} \quad (14.3)$$

The significance of the N/N_0 quotient or ratio is that it indicates the number of excess charges available per analyte molecule present. The value of N/N_0 was compared with the value of the average analyte charge state as a function of rising analyte concentration. Because both N/N_0 and the average charge states decreased concomitantly as concentration rose, it was postulated that the diminishing N/N_0, indicative of an increasing amount of ion pairing for multiply charged analyte species with counterions in solution, was the underlying factor leading to the lowering of the average charge states [Eq. (14.2)] observed by ES-MS. Qualitatively speaking, as the number of analyte molecules, N_0, in the solution increases due to a concentration rise, N (the total number of excess charges) may also increase slightly, but not in proportion to the increase in N_0. This is attributable to the weak dependence of the electrospray current, I, on the solution conductivity, σ ($I = H\sigma^n$, $n \sim 0.2$–0.5,[44,45,52,53] H is a proportionality constant); thus, even with a slight increase in I, overall, N/N_0 will decrease. The detailed rationale provided for this model was tested using both diquaternary ammonium cations and doubly protonated diamines.[48]

14.1.5.2 Analyte Structure/Conformation

The three-dimensional structure of an analyte molcule in solution can have a profound effect on the charge state distribution of that compound observed in ES mass spectra. The abrupt shift in charge state toward higher values that occurs upon protein denaturation was first demonstrated in the Chait lab[23] (see Figure 14.1). The increased solvent accessibility to ionizable side chains that occurred upon unfolding of the protein was deemed to be responsible for allowing an increased level of protonation of previously 'buried' basic sites. Thus, a tightly folded conformation typically yields lower charge states in ES mass spectra. But if the proteins suddenly adopt a more extended form due to denaturation, basic sites now become exposed to the solvent and are thus susceptible to protonation, thereby shifting the charge state distribution to higher values (lower m/z values). The proposed limited access to ionizable sites was corroborated by the lower degree of hydrogen–deuterium exchange that occurred in solution for proteins in the folded conformation relative to the denatured conformation.[54–56]

Another means of denaturing proteins involves reduction of disulfide bridges to thiols in cysteine-containing proteins, thereby perturbing the tertiary structure. Using 1,4-dithiothreitol

(DTT) as the reducing agent, Loo et al.[10] compared ES mass spectra of several proteins that were either treated or untreated. All examined proteins showed a shift in the charge-state distribution toward higher values as a result of addition of DTT, which allowed the protein to unfold into a more extended conformation capable of accommodating more charges. Heat can also provoke denaturation of proteins as reported by Le Blanc et al.,[57] who found that globular proteins exhibited an increase in charge state as a result of thermal denaturation. Similar results were obtained by Mirza et al.[58] when their ES emitter was heated to 98°C. Other early experiments by Loo et al.[59] used organic solvents to induce conformation changes in proteins. The minimum amounts of solvent addition required to induce a noticeable shift in ES charge state distributions were largely consistent with previous studies of protein conformational changes resulting from addition of the same solvents as evaluated by circular dichroism and nuclear magnetic resonance.

Later studies by Konermann and Douglas[60] also combined the use of optical techniques (circular dichroism, absorption, and fluorescence) to examine the effects of denaturation of cytochrome c by addition of both organic solvent (methanol) and acid (HCl). Especially notable was their conclusion that the shift in charge state to higher values upon denaturation was attributable specifically to the breakdown in tertiary structure rather than secondary structure. This highly significant finding was complemented by a subsequent report from Konermann and Douglas[61] describing cases of cooperative (two-state) folding/unfolding behavior of proteins that could be distinguished from noncooperative unfolding involving multiple protein conformations. Conclusions were based upon observed pH-dependent shifts in charge state distributions in positive-ion ES mass spectra. Further evidence that tertiary structure is the predominant factor in determining the ES charge state distribution in the positive-ion mode was given in a third paper from this group.[62] Interestingly, negative-ion mode results from this paper demonstrated that only minor changes in the negative-ion charge state distribution occurred in response to denaturing conditions, leading the authors to preferentially endorse the use of positive-ion charge state distributions as a means of probing conformational changes.

The above studies ushered in the notion that charge-state distributions could be used as a tool to evaluate the conformational dynamics of proteins. A relatively recent report by the Kaltashov group[63] shows a nice example of an ES mass spectrum capturing the native state of the 58-kDa glycoprotein anti-thrombin III, plus at least three additional non-native states as shown by distinct charge state distributions centered around the $+20$, $+32$, and $+38$ charge states (Figure 14.3). The unfolding of individual domains of larger proteins does not necessarily occur simultaneously, as suggested by this example. The relative abundances of the ions associated with each charge state distribution change as the denaturing conditions pass from mild to strong, and higher charge states are successively favored in this progression.

Computer simulations have been used to account for, as well as exploit for analytical purposes, the variation in charge-state distributions with protein conformation. The Kaltashov group has used deconvolution strategies to correlate charge state with the conformer's overall shape and heterogeneity.[64–66] A central goal is to computationally derive, from multimodal ES mass spectral charge-state distributions, information concerning the relative quantities of the various conformational isomers of proteins that coexist in solution. The underlying assumptions are that (1) under a given set of experimental conditions, each conformer presents a typical distribution of charges that is characteristic of that conformer's overall shape and surface area and (2) the ES mass spectrum represents the ensemble (linear sum) of all such distributions. From experimental mass spectral data, the objective of the computations was to deconvolute the specific charge state distributions

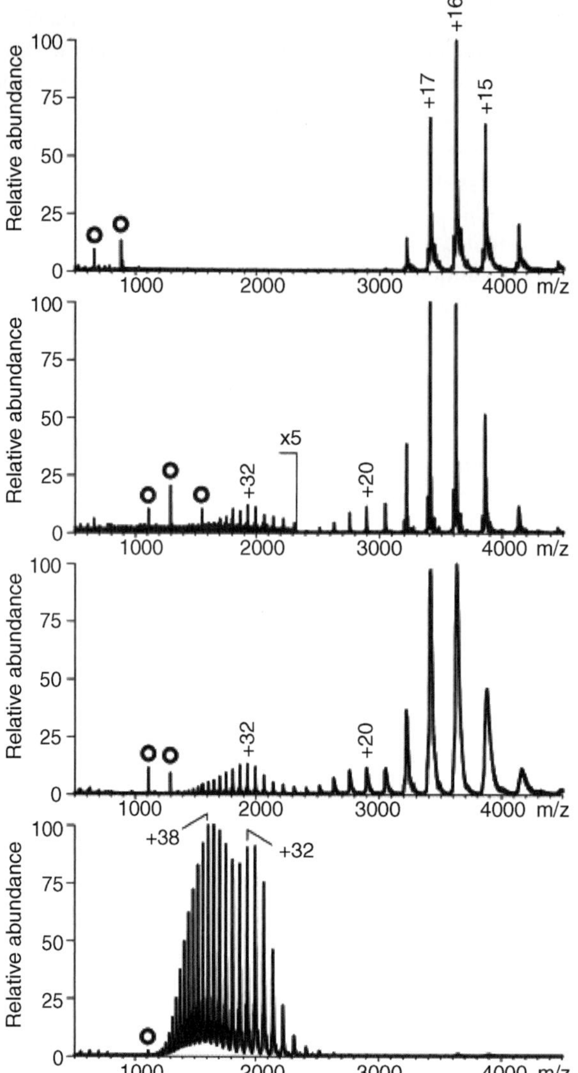

Figure 14.3. ES mass spectra of antithrombin III acquired under denaturing conditions of progressively increasing severity in moving from top to bottom. In addition to the native conformation, at least three distinct non-native states are revealed. Peaks labeled with circles indicate impurities. (Reprinted from Ref. 63 with permission.)

(referred to as "basis functions") corresponding to each conformer. The basis functions were first approximated to be Gaussian, and the ES mass spectrum was considered to exhibit the sum of the weighted basis functions. This approach necessitated the processing of each spectrum individually.[64] In later work, the data reduction process was expedited via a chemometric approach. Here, the multivariate technique of factor analysis was used to process an array of ES mass spectral data in view of determining the number of different protein conformations present, as well as describing how the distribution of conformers changed in response to altered experimental conditions (pH and alcohol content).[66] The authors point out that further refinements to the approach are certainly needed, such as the ability to better correlate the conformers' physical characteristics (e.g., size, shape, surface area and fractional concentration) with experimentally derived parameters. Nonetheless, the approach shows much promise for studies of protein conformational dynamics, especially in systems containing multiple proteins, mostly because the specific detection of ES mass spectrometry is superior to that of other techniques for such mixture analyses.[64,66]

Several years later, Kuprowski and Konermann[67] examined relative signal responses of various conformational states of lysozyme and myoglobin in an attempt to ascertain whether measured ion abundances accurately reflect solution-phase concentrations of the individual conformers. They found that non-native (unfolded) proteins give much stronger signals than the native (folded) conformation. The response of the unfolded conformation could be as much as two orders of magnitude higher than would be expected had the desorption/ionization efficiencies of all conformers been equal. This increased response for unfolded conformations is attributed to an increased hydrophobic character that augments surface activity and favors transfer of the unfolded protein into offspring droplets. The comparatively hydrophilic native conformation tends to remain more in the interior of the droplet; thus, folded conformers are less efficiently transferred to progeny droplets. The authors point out the consequences for quantitative comparisons of folded versus unfolded protein conformers, as well as for studies designed to determine ligand–protein binding affinities; that is, correction factors to compensate for different desorption/ionization efficiencies are likely warranted.[67]

At around the same time as his above study was performed, Konermann developed what he termed a "minimalist" molecular dynamics model for simulating protein charging during the final stages of the ES process.[68] Shown in Figure 14.4 is a temporal depiction of the final moments of the lifetime of a charged droplet containing a single unfolded protein as the last solvent molecules depart. Compared to the folded form that exhibited a charge-state distribution of 4+ to 7+, the unfolded conformers retained more charge—that is, a distribution of 6+ to 9+. While recognizing that there is room for future improvements in the model's descriptions of the protein, solvent, and proton dynamics, results from the simulation nonetheless lead to the conclusion that protein unfolding

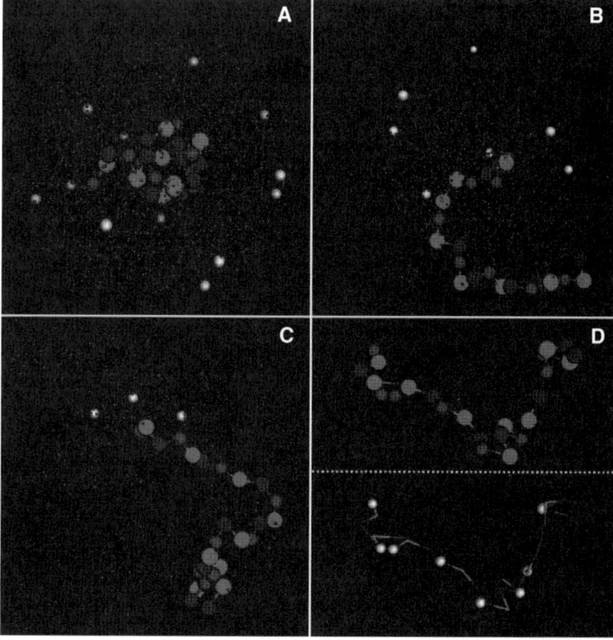

Figure 14.4. Progressive disintegration of a charged nanodroplet containing an unfolded protein. The time points and charge states are (**A**) 3 ps, 2+; (**B**) 87 ps, 7+; (**C**) 97 ps, 7+; (**D**) 126 ps, 8+. (Reprinted from Ref. 68 with permission.) (See color insert)

augments the steric accessibility and the electrostatic exposure of protonation sites, which allows an unfolded protein to capture a larger number of protons during droplet disintegration.[68]

While the vast majority of the work on conformational dynamics has been performed on proteins, an investigation into the relationship between charge-state distributions and secondary structure of oligonucleotide strands has also been reported.[69] In contrast to the behavior of proteins, these negative-ion mode studies revealed that for oligonucleotides of the same base composition but different sequence, the hairpin (folded) strands show somewhat higher charge states than the "linear" strands lacking the hairpin. The authors proposed that the added secondary structure of the hairpin strand serves to stabilize the more highly charged negative ions. It is also noted that the lower charge states likely arise from protonation of the bases, aided by the addition of ammonium acetate, rather than from neutralization of the phosphates constituting the backbone. This report[69] represents a foray toward the use of ES-MS charge state distributions as a probe of secondary structure within single oligonucleotide strands. It builds on previous evidence[70] that hydrogen bonding and base-stacking interactions are maintained in the gas-phase for ES-generated oligodoxy-nucleotide duplexes.

a. Reciprocal Stabilization of Opposite Charges and Conformation-Dependent Charge Neutralization. In an effort to elaborate on the underlying reasons that folded proteins carry less charges in electrospray mass spectra than those that have been denatured, Grandori and Samalikova[71,72] propose that one must consider the totality of charges (both positive and negative) borne by proteins. In folded proteins, charges opposite to that of the net (overall) charge are likely to be stabilized by nearby charge sites corresponding to the excess charges. Thus, in the positive-ion mode, negative charge sites (e.g., carboxylate groups) are prone to be situated near protonated amine sites and are thus less likely to add a proton to neutralize the carboxylate site relative to the same protein in an unfolded conformation. In an analogous way, in the negative-ion mode, positively charged basic sites are more likely to be stabilized by one of the more numerous negatively charged sites, thus impeding proton departure from that basic site. This so-called "reciprocal stabilization of opposite charges" is proposed to render sites of opposite polarity to the overall charge on folded proteins less susceptible to neutralization. This decreased tendency toward neutralization will thereby contribute to the lower charge states observed in ES mass spectra. In other words, a consequence of protein unfolding may well be that charge sites of opposite polarity to the overall charge (and ionization mode) are more susceptible to neutralization reactions. Accordingly, the difference in charge states observed in ES mass spectrometry would derive from a higher number of reciprocal stabilization sites in the folded conformation, rather than a higher degree of ionization in the unfolded form.[71] Thus, when considering folded proteins, in addition to the increased Coulomb repulsion due to the shorter distances between charges,[43] along with the reduced accessibility of polar solvent to hydrophobic sites,[23] intramolecular conformation-dependent charge neutralization can contribute to the lower charge states observed in ES-MS versus unfolded protein counter-parts.[72] Notably, in the limiting case (positive mode example) where every negative charge is protected against neutralization by reciprocal stabilization, the number of charge states appearing in the ES mass spectrum would directly correspond to the number of basic residues minus the acidic residues.[71]

A short time later, Nesatyy and Suter[73] commented directly on the conformation-dependent neutralization theory, claiming that while certain data do concur with the theory, many other examples do not. Their preference was for the Rayleigh limiting charge theory (see Section 14.1.3.2) which can be used to estimate the *maximum* charge state carried by an ES-generated protein (conformation-dependent neutralization theory predicts the *most abundant* charge state). Nesatyy and Suter's analysis of charge-state data from 79 individual proteins led them to conclude that observed ES maximum charge states correlate well with those predicted by Rayleigh charge limit theory and that simply multiplying this calculated maximum charge state by a coefficient (i.e., 0.85) was a good predictor of the most abundant charge state. They conceded that conformation-dependent neutralization theory does take into account chemical aspects of protein character; but in their view, the level of protein charging predicted by this theory will be correct only when it does not contradict physical constraints imposed by the Rayleigh limiting charge.[73]

b. Breadth in the Distribution of Charge States. In addition to the mean value of the charge state, the factors that influence the width of the Gaussian-like distribution of charge states is also of interest to comprehend. In fact, much less has been written about the breadth of observed charge states in the charge state distribution of a given protein. Halgand and Laprevote[74] noted that the distribution of charge states, as measured by the standard deviation about the mean charge state, was much narrower for folded proteins versus denatured varieties of the same proteins. Thus, the denaturation process not only shifted charge states toward higher values, but also led to a wider range of charge states. The broadening of the charge state distribution upon denaturing was particularly dramatic for RNase A and less so for lysozyme. The authors considered that perhaps additional stabilization may be provided by intramolecular ion pairing in the three-dimensional structures of the proteins; but upon examination, they did not find evidence for a higher gas-phase stability of lysozyme. On the point that folded proteins often give only one or two intense peaks whereas unfolded proteins typically exhibit a much broader charge state distribution, Grandori[71] accounts for this based upon the differing "history" of the ions. That is, the folded proteins exist as preformed ions in solution that are rather homogeneous in nature, and they are characterized by substantial intramolecular ion-pairing. The denatured proteins, on the other hand, have undergone varying degrees of neutralization during the ES process, with the exact level of neutralization depending upon a variety of factors including the conformational dynamics of the protein that has been denatured as well as the specific experimental conditions that ES spectra were acquired under.

14.1.6 Instrumental Effects

Even after closely following a specific protocol for sample solution preparation, ES mass spectra obtained under different circumstances can still have significantly varying appearances. This becomes obvious when comparing mass spectra acquired using different types of instrumentation. Differences can also be observed when comparing two sets of data acquired on the same type of instrument, perhaps of the same model constructed by the same manufacturer, when certain adjustable parameters are not duplicated. Moreover, ES mass spectra acquired on a single instrument at different points in time can exhibit some degree of variability, even when attempts are made to faithfully reproduce adjustable parameters, if other specific conditions (e.g., pressure or temperature in a given portion of the source region or instrument contamination) are not as well controlled.

14.1.6.1 Spray Tip Orifice Diameter

Over the course of studying charge state distributions of various compounds, we had noticed an inconsistency in obtained average charge states [Eq. (14.2)] for certain analytes when electrospray versus nanospray conditions were employed. This prompted us to launch a study[75] into the effect of ES capillary diameter on analyte charging. Using metal-coated silica electrospray tips and a constant electrospray current, the effect of ES spray tip orifice diameter on the charge-state distributions of peptides and a small protein were investigated. It was found that charge states consistently shifted toward higher values as the spray tip diameter became narrower. Shown in Figure 14.5 are three ES mass spectra obtained from a solution of 20 μM of angiotensin I using nanospray emitters of 1-, 2-, and 5-μm diameter. A fixed spray current of 50 nA was employed in acquiring all three mass spectra. This was achieved by adjusting the ES high voltage while monitoring the spray current; all other instrumental parameters were held constant. Analogous observations showing significant shifts in charge state toward higher values from narrower diameter spray tips were obtained from solutions of the 8560-Da protein ubiquitin.[75]

These results were rationalized by again considering the ratio of excess charges to analyte molecules as given by the N/N_0 quotient introduced above in Eq. (14.3).[48] As the tip orifice progressively increases from 1.0- to 2.2- to 5.1-μm diameter, the flow rate was also measured to increase from ~1.3 to ~2.0 to ~5.3 nL/s (electrostatic induced flow, i.e., no external solution pumping). In comparing the upper extreme in this series with the lower, N/N_0 has decreased by a factor of four. It was thereby deduced that the constant number of excess charges (arising from the fixed current) was being divided over an increasing number of analyte molecules as the spray tip diameter and flow rate became larger. With fewer charges available per molecule, the average charge state of the analytes decreases at the higher flows associated with wider-diameter capillaries.[75]

In situations where it is desirable to preferentially form analyte ions of higher charge state—for example, to obtain more reactive precursors for tandem mass spectrometry experiments or to improve the efficiency of electron capture in either electron capture dissociation (ECD) or electron transfer dissociation (ETD) experiments—then the use of narrow diameter capillaries is recommended. Conversely, if lower charge states are desirable, such as singly charged small molecule precursor ions that present more straightforward spectral interpretation, then wider diameter capillaries are preferable.[75]

14.1.6.2 Source Geometry and Gas Pressure

Early on it was recognized that ES source conditions can have a substantial bearing on the appearance of the overall mass spectrum including the charge state distribution. In work reported by Winger et al.,[76] two different ES source geometries were employed to gauge the effect of source geometry on the appearance of ES mass spectra. In an arrangement employing a countercurrent (nitrogen) gas, a higher level of protonation was observed as compared to a design using a heated capillary inlet without drying gas. The authors also remarked that with larger distances between the spray capillary and the sampling orifice, a discrimination against higher charge states resulted. These combined results imply that construction of an ES source that reduces the amount of solvent species entering the mass spectrometer by use of drying gas can result in the production of more highly charged ions.[76]

Further investigations of the influence of countercurrent gas (sometimes called bath or drying gas) as well as concurrent "cooling" gas were reported by Fenn.[43] In extending

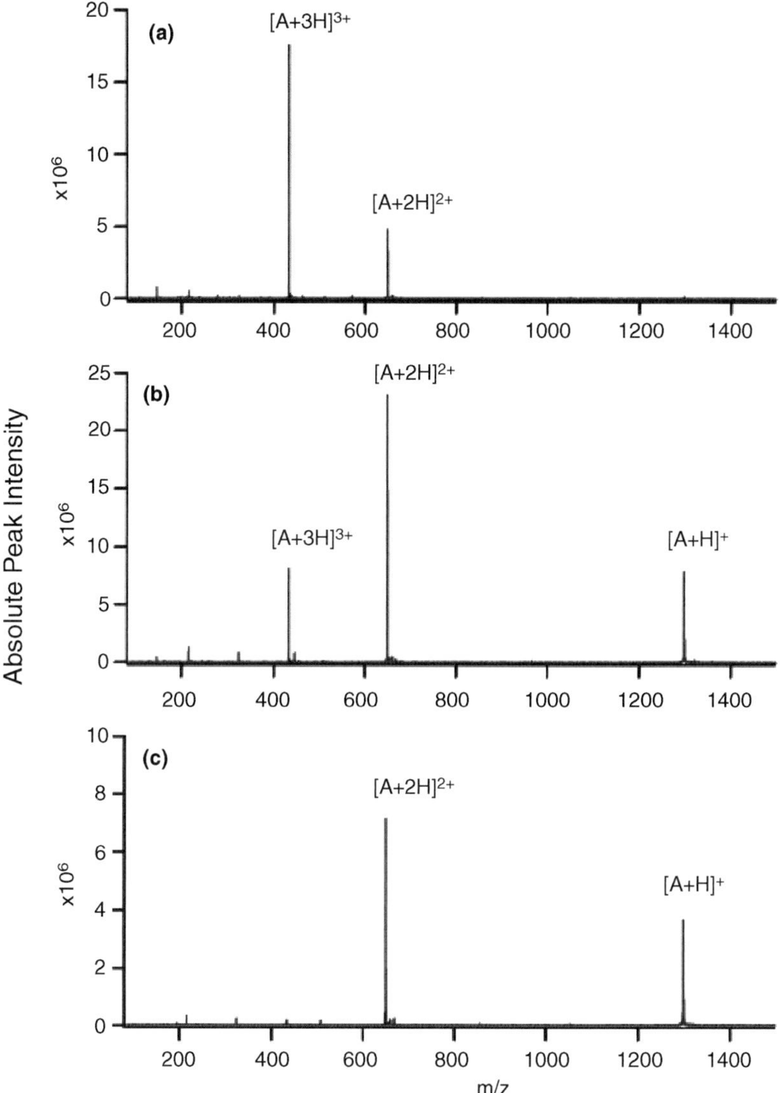

Figure 14.5. Nano-ES mass spectra of 20 μM angiontensin I loaded into (**a**) 1-, (**b**) 2-, and (**c**) 5-μm tips. The solution and employed experimental parameters (including distance between tip exit and counterelectrode) were held constant. Only the applied high voltage was adjusted to obtain a fixed current value of 50 nA at the emitter tip. (Reprinted from Ref. 75 with permission.)

concepts of ion evaporation theory, as charged droplets shrink, the surface charge density, and consequently, the magnitude of the local field on the surface of the droplet both increase. At a certain point in this process, the field acting on the surface charges becomes sufficiently high to allow an ion to overcome the activation energy barrier leading to escape from the droplet. Ions that are lifted into the gas phase deplete the surface of excess charges and Fenn[43] argues that the specific surface-charge density of the droplet at the moment of ion departure will exert an important influence upon the charge state of that ion as it is desorbed at a particular instant in time. More specifically, ions desorbed in the earlier stages of droplet evaporation (when the surface-charge density is relatively low) will carry fewer charges than ions desorbing at later stages, when the droplet is smaller and the surface charge density has increased.

It was further postulated[43] that a slowing of the evaporation rate, by a decreased drying gas flow, would prolong the time that the droplet spent in the low surface-charge density regime. This could lead to desorption of an increased proportion of lower charge state ions; that is, the charge state distribution would be shifted toward lower values relative to a higher drying gas flow. Experimental results did, in fact, show a higher distribution of charge states with higher drying gas flow. Raising the bath gas temperature to increase the level of drying had a more subtle, but corroborative, effect. Lastly, the flow of concurrent (same direction as solution flow) "cooling" gas was also varied while other parameters were held constant.[43] Increasing the flow of this cooling gas served to slow the solvent evaporation process, resulting in larger droplet radii and a shorter time period of desorption of ions into the mass spectrometer. Augmented cooling gas flow rates led to a steady shift in the charge state of the most abundant ion toward lower values. These results are consistent with a desorption scenario wherein ions desorbed during earlier stages of offspring droplet evaporation carry with them lower numbers of charges than those departing at later moments.

14.1.6.3 Nozzle-Skimmer Voltage, Adduct Formation, and Desolvation

The effect of varying the potential difference between the nozzle and skimmer, sometimes called the "cone" or "fragmentor" voltage and more recently the "declustering potential," to provoke fragmentations is very well known. The voltage applied to the skimmer may serve an electrostatic lens function as well as providing a means to provoke energetic collisions with residual gases. The latter phenomenon has been termed "in-source CID," "up front CID," or "nozzle-skimmer" decomposition, among other names. Collisions are most efficient—that is, they produce the highest yield of product ions—when the pressure is sufficiently low to allow a significant mean free path of movement before a collision occurs, but sufficiently high to result in multiple collision events. Increasing the potential difference between the nozzle and skimmer presumably augments the velocity of ions in the region in between these two elements. Higher translational energy in this zone, where a good balance between mean free path of movement and collisional cross section exists, results in increased fragmentations, although the maximum amount of energy uptake achievable for analyte ions appears to be significantly below that attainable in the collision cell of a triple quadrupole mass spectrometer.

Loo et al.[25] were the first to report that the charge state distribution of multiply charged proteins underwent a shift toward lower values as the nozzle–skimmer potential was raised. They rationalized this observation as being due to an improved efficiency of CID for higher charge state ions in the nozzle–skimmer region. They noted that higher-charge-state ions have greater translational energies and thus will be more susceptible to decompositions upon collision than ions of lower charge state.[25,77] Ashton et al.[78] also found a progressive shift in charge state distributions toward lower values with repeated increases in the nozzle–skimmer potential difference.

An investigation to probe the details underlying this shift in charge-state distributions toward lower values with increasing applied nozzle–skimmer potential was undertaken by Thomson.[79] He considered that there are at least four possible reasons to favor higher m/z ions as the potential is raised: (1) mass-dependent ion focusing; (2) collisional stripping of protons from "bare" proteins; (3) solvent molecules carrying protons are stripped from charged solvent–protein clusters; or (4) high-charge-state proteins are fragmented, while those of low charge state are more effectively declustered. A triple quadrupole was used to test these different possibilities; conditions were found which allowed the acquisition of qualitatively similar charge state distributions from a scan of the first quadrupole (conventional ES mass spectrum) versus that obtained by using Q1 as a nonresolving quadrupole and introducing low-pressure collision gas (7 mtorr) into the Q2 collision cell while scanning Q3. Because the charge state distributions in the two situations had similar initial appearances and the trends in the shift were the same when collision energies were augmented, the authors concluded that ion transmission is not an issue and that the weak signal observed at low nozzle–skimmer potentials is due to a low number of declustered ions rather than to poor ion focusing.

In addressing the possibility for CID loss of bare protons, Thomson[79] observed rather symmetric distributions of product ions having charge states both higher and lower than the selected multiply protonated precursor, but only when low nozzle–skimmer potentials were used. The presence of small quantities of multiply protonated *dimer* (and possibly higher-order multimer) precursors that have the same *m/z* value (proportionally higher mass and higher charge) as the monomeric precursors was proposed to explain the symmetrical distribution of product ion charge states. That is to say, multiply charged dimers carrying twice as many charges as the monomers of the same *m/z* are selected as precursors along with the monomers; the dimers dissociate into a pair of monomers, but the charges (protons) are distributed in a nonsymmetrical way, thus creating monomers of both higher and lower charge as compared to the monomeric precursors. This attribution of asymmetric charge distribution in product ion spectra was in agreement with results from Richard Smith's group, which had previously performed a detailed study of protein dimers[80]; this led Smith and coworkers to revise their original proposition that CID may provoke the dislodging of protons and/or electrons.[25,77,81] In the absence of evidence for direct loss of protons in CID of multiply protonated proteins, Thomson also concluded that proton "stripping" was not a significant process.

To test the third possibility above, product ion spectra were acquired while passing a wide *m/z* range of "undeclustered" apomyoglobin ions through Q1 and subjecting them to collisions in Q2 (Figure 14.6[79]). The absence of any ions corresponding to protonated solvent molecules or protonated solvent clusters in Q3 scans obtained under a variety of energies and gas pressures led to the conclusion that the shift in charge states to lower values at higher collision energies was not caused by departure of proton-bearing solvent molecules. Figure 14.6a shows the dense and noisy mass spectrum obtained with no

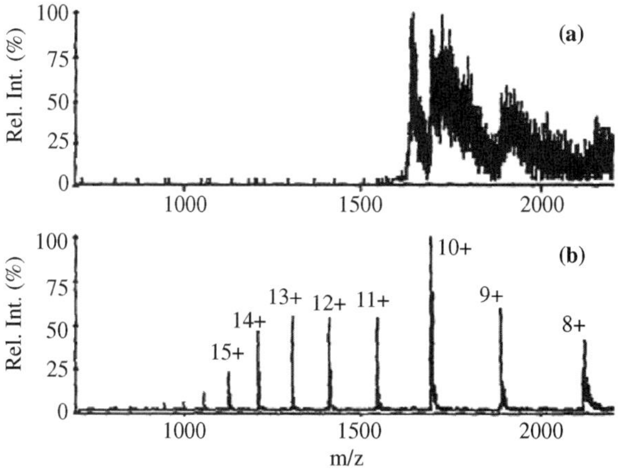

Figure 14.6. ES mass spectra of apomyoglobin obtained by Q3 scan on a triple quadrupole with Q1 set at *m/z* 2000 in rf-only mode (low mass cutoff ~1633); declustering in the ion source region was negligible. (**a**) No CID gas in Q2; (**b**) CID gas was added to Q2. (Reprinted from Ref. 79 with permission.)

nozzle–skimmer potential difference (i.e., Δnozzle–skimmer = 0 V) and Q1 set to transmit only peaks above *m/z* 1633. The displayed peaks correspond to multiply solvated apomyoglobin, mostly of charge states 10+ and 9+, although heavily solvated 11+ peaks and lightly solvated 8+ peaks probably are also present. When the CID collision gas is introduced into the Q2 collision cell (Figure 14.6b), ions corresponding to 10+, 9+, and 8+ charge states increase in intensity while ions of higher charge state appear at lower *m/z* values, leading to the inference that the highly charged ions appearing below *m/z* 1633 are derived from initially higher *m/z* species. This reasoning is consistent with the notion that among the ionic species are dimers (and possibly higher-order multimers), quite possibly solvated with varying numbers of solvent molecules, that fragment with asymmetric partitioning of protons to form the higher-charge-state species. In addition, highly clustered monomers undergo declustering (loss of neutral solvent molecules) upon collision that results in enhancement of Signals of higher *m/z* ions. While Figure 14.6 shows results obtained using CID in the Q2 collision cell, Thomson extrapolates findings back to the nozzle–skimmer region and concludes that the shift in charge state distribution toward lower values caused by elevated nozzle–skimmer voltage differences is due to increased fragmentation of higher charge state species plus improved declustering of lower-charge-state ions.[79]

14.1.7 Gas-Phase Modifications

14.1.7.1 Solution Additives and Proton Transfer

As the molecules composing a final droplet reach the last stages of evaporation, the mobile charges will experience increased electrostatic (ion–dipole) attraction to both analyte and solvent molecules. If additives are present, they too may offer stabilization to charge sites. The behavior of peptides and proteins in the presence of small nitrogen bases was investigated,[17] and it was proposed that protonated polypeptides may be desorbed into the gas phase with the small nitrogen bases attached to one or more protonated sites. These complexes can then dissociate in the gas phase with a partitioning of the adducted protons between the departing nitrogen base and the polypeptide, thus lowering the charge states of the polypeptide in ESI mass spectra.

The Kebarle group showed that if ammonium acetate or ammonium formate is present as an additive, then, upon desolvation of the final droplet, NH_4^+ will act largely as a protonating agent owing to the low basicity of ammonia in the gas phase relative to those of arginine, lysine, histidine, tryptophan, and proline amino acid residues.[82] Building on studies by the Williams group to quantify Coulomb energy in multiply protonated molecules including peptides,[83] proteins,[21,84] and diaminoalkanes,[85] as well as work by Gronert[86] who described a method for estimating internal electrostatic repulsion and potential energy surfaces for multiply charged ions, the Kebarle group refined their investigations of apparent gas-phase basicities in multiply charged molecules using doubly protonated alkylamines[87] and doubly deprotonated dicarboxylate anions[88] as simplified model compounds that could approximate proton transfer sites in polypeptides. They concluded that the decrease in positive-mode charge states observed with alkylammonium ions compared to ammonium ion is a result of the higher basicities of the conjugate alkylamines versus that of ammonia.

Ammonium actetate adjusted with acetic acid is a common "buffer" additive in ES experiments, and Kaltashov and co-workers[63,65] argue that at low pH (e.g., near the pK_a of acetic acid, ~4.8), complexation of acetate anions with multiply charged proteins via ion-pairing in solution is particularly efficient. Once formed, the complex, consisting of a multiply protonated protein with one or more acetate anions in tow, can be desorbed into the gas phase, and there it is proposed to dissociate via loss of (one or more) acetic acid neutral molecules. This process of attachment of acetate anion(s) followed by acetic acid loss(es) (and certainly *not* a tightening of protein structure) was proposed to account for the reduction in charge state observed to occur below a pH of about 4.5. In the negative mode, added acetic acid was cited by Blades et al.[88] as being responsible for protonating carboxylate sites on protein side chains, thereby lowering charge states.

Two other studies of solution effects on charge-state distributions of proteins are also worthy of mention here. In 2004, a report by Otto et al.[89] used partial least-squares modeling and multivariate analysis of variance to assess the significance of protein concentration, percent added butanol, flow rate, and buffer concentration on ES mass spectra. The first three factors were found to have a significant effect on charge-state distributions of apomyoglobin while the fourth did not. Additionally, Wood and co-workers[90] demonstrated that the presence of certain redox agents could serve to shift charge-state distributions of proteins to lower values. It was demonstrated that proton transfer, and not electron transfer, was responsible for the, at times, substantial charge-state reduction of proteins. These proton transfer reactions were presumably occurring in solution, as inferred from the relatively low gas-phase basicities of the additives.

In addition to proteins, the effects of added organic bases or acids on *oligonucleotides* have also been investigated. Greig and Griffey[91] reported that for negative-ion-mode ES-MS, addition of stronger bases, such as triethylamine or piperidine, was effective at diminishing the deleterious signals of sodium- or potassium-adducted oligonucleotides. Suppression of sodium and potassium adduct ions was accompanied by markedly improved sensitivities. Furthermore, a bimodal distribution of charge states was observed upon addition of high concentrations of imidazole; each distribution was hypothesized to result from a separate conformation of single-stranded oligonucleotide oligomers. In a later report from the same group,[92] ammonium acetate adjusted to pH 7 was used as an additive to solutions containing one of two tested deoxynucleotides. Negative-mode charge-state distributions were observed to shift in average charge state from -7.2 to -3.8, and then to -3.1 as the concentration of ammonium acetate was progressively raised. The shift in charge-state distributions was attributed to a *solution* equilibration of cations associated with the DNA strand, as opposed to *gas-phase* proton transfers. The authors found that the maximum percentage of neutralized phosphate groups on the investigated oligonucleotides was 79%, regardless of length and sequence. Muddiman et al.[93] confirmed the shift in charge state distributions toward lower charge states for negative-ion-mode ES-MS studies of oligonucleotides that employed various mixtures of imidazole and piperidine as additives. They also concurred that the use of imidazole and piperidine substantially suppressed sodium addition to oligonucleotides, and especially when combined with acetonitrile, resulted in cleaner, more intense signals.

14.1.7.2 Gas-Phase Reactivity of Multiply Charged Ions and Charge Reduction

McLuckey et al.[94] were the first to examine the effect of *gas-phase* additives on charge-state distributions. When dimethylamine was added at low pressure to a 3D quadrupole ion trap, it reacted with multiply protonated cytochrome c to diminish the degree of multiple charging by proton abstraction. The rate constants for deprotonation were shown to increase with increasing charge state of the protein. Subsequent MS/MS experiments by the same group[95] using 1,6-diaminohexane as the gaseous base demonstrated how the reduction in charge state resulting from proton abstraction could be used to identify the charge state (and thus the mass) of unknown peaks in low resolution mass spectra. In the course of this study, it was noted that the most reactive ions were those bearing the largest number of protons. Moreover, after a certain number of proton transfers to the gaseous 1,6-diaminohexane base, the remaining protein-bound protons became unreactive. This result was attributed to (a) a reduced Coulomb repulsion between neighboring charges and (b) an increased basicity of the remaining proton-bearing sites on the charged–reduced protein.

A short time afterwards, reports began to appear from Richard Smith's group that verified the notion that ions of higher charge states are indeed less stable than their less charged counterparts. It was demonstrated that activation energies for unimolecular dissociation reactions of multiply protonated melitten molecules were consistently lower for ions of higher charge state.[96] Using a capillary inlet/reactor coupled to a triple quadrupole, gaseous water (acting as a base) was shown to dramatically shift the charge-state distribution toward lower values.[76] The relative pressure of the reactive gas was found to be crucial in determining the extent of proton transfer; higher pressures led to lower analyte charge states, but also concomitantly reduced the signal intensity. In addition to Coulomb effects, it was suggested that entropic factors can also augment the proclivity for proton transfer under seemingly unfavorable conditions. If one considers that highly charged gas-phase ions are conformationally constrained owing to the high level of coulomb repulsion, then proton transfer can reduce this strain; the resulting gain in entropy can help to surmount a positive enthalpy of reaction. Related studies from the same group[97] reported the lowered reactivity of lower charge states of multiply protonated protein ions to be a general phenomenon for all proteins studied. When comparing relative reactivity (proton transfer to diethylamine) for ions of varying charge state, differences were more evident between species at the lower charge end, because ions of the highest charge states all reacted quite rapidly. Moreover, it was found that disulfide reduced forms of three examined proteins were not as reactive as their equally charged counterparts in native (nonreduced) form. Increased Coulomb repulsion present in the highly constrained native forms (containing several disulfide bridges) was proposed to contribute to the increased tendency for proton transfer. A follow-up study[98] showed that differences in reaction rates (proton transfer) of protein ions of lower charge state were more subtle than those of higher charge state in comparisons of disulfide intact (compact) versus disulfide reduced forms. The faster-than-expected (based upon relative proton affinities) reactions with trimethylamine were rationalized by Coulomb repulsion considerations. An ensuing study examined temperature dependences of gas-phase proton transfer reactions for bases reacting with multiply protonated proteins (positive ion mode), as well as acids reacting with multiply deprotonated proteins (negative ion mode).[99] Cassady et al.[100] reiterated findings of increased reactivity (i.e., higher rate constants for proton transfer) for more highly charged forms of ubiquitin. Moreover, the rate of proton transfer was demonstrated to increase steadily for gaseous amines of progressively higher gas-phase basicity. It was proposed that a multiply protonated protein in a given charge state may have different protonation sites with very similar basicities; the reaction rates for these various structures would be expected to be comparable.

An alternative approach to charge reduction in the gas phase was introduced by Lloyd Smith's group.[101–105] They used a radioactive ^{210}Po source to induce reactions between the ES-generated aerosol (on which the emitted alpha particles impinge) and the multiply charged analyte; the reactions result in charge neutralization as first demonstrated for proteins and oligonucleotides. The degree of charge reduction is heavily influenced by the particle flux from the radioactive source as well as the residence time of the aerosol in the neutralization chamber.[101] Charge reduction, potentially producing predominately the singly charged species, was attributed to proton transfer reactions in both positive- and negative-mode experiments on proteins and oligonucleotides.[102] Another variation of the concept employed a corona discharge electrode in place of the ^{210}Po emitter.[103,104]

14.1.7.3 Ion Parking

The McLuckey laboratory has been exploring ion–ion reactions in the gas phase, especially those of oppositely charged ions, using a three-dimensional quadrupole ion trap. Ions of charge opposite to the macroions of interest can be reacted with the latter to overcome the limitations experienced with neutral reagents to effect charge state reduction. Because all such ion–ion reactions are highly exothermic, charge states can be efficiently reduced to low values, including zero.[106] From these early experiments, the idea of converting an initially wide distribution of charge states to a narrow distribution, or even single charge state, via ion–ion reactions was born, and the concept was named "ion parking." It entails inhibiting ion–ion reactions of selected ion species such that these selected species preferentially survive in the trap for subsequent tandem mass spectrometry experiments. Initial ion parking studies[107] employed the application of single-frequency dipolar (AC) on-resonance excitation to the end caps of the 3D quadrupole ion trap. The frequency is tuned to increase the relative velocity of a specific ion–ion reaction pair which diminishes the cross section for ion–ion "capture" and thereby inhibits the reaction of this specific ion–ion pair. Reaction rates of higher and lower mass-to-charge ions are unaffected; hence, they will undergo transformations to other species, including conversion of higher-charge-state species to the "parked" ion. Thus, ions of a particular charge state can be preferentially retained in the ion trap owing to inhibition of the rate of ion–ion proton transfer reactions for that particular m/z ion. Higher-charge-state species will lose protons via transfer reactions to eventually transform into the charge state of the parked ion, while lower-charge-state species will eventually become neutralized. The higher charge state ions have greater probabilities of reacting in a given time period.[107]

A major repercussion of ion parking is that a single (lower)-charge-state ion may be accumulated at the expense of other (higher) charge states. Shown in Figure 14.7[107] is

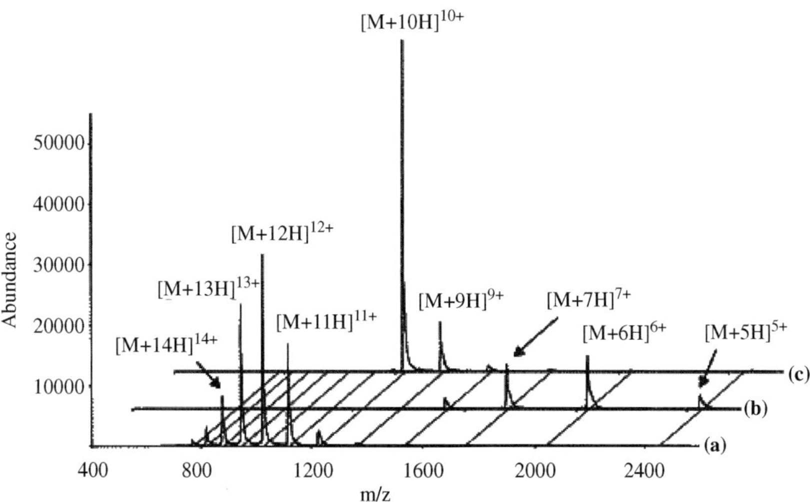

Figure 14.7. Mass spectra of bovine cytochrome c acquired in (**a**) pre-ion/ion, (**b**) post-ion/ion (no ion parking), and (**c**) ion parking modes. Approximately 90% of the ions in **a** are accounted for in **c**. Of the totality of ions in **c**, about 83% are in the 10+ charge state. (Reprinted from Ref. 107 with permission.)

the ion parking of the +10-charge-state ion of cytochrome c. Compared to the initial (pre-ion–ion reaction) distribution of charge states (Figure 14.7a), approximately 83% of the ions are accounted for in the signal of the +10 species after it was selected for ion parking (Figure 14.7c). Without ion parking, this +10 signal would have completely disappeared (Figure 14.7b). The other 17% of the initial signal is accounted for in the lower-charge-state species. Ion parking thus provides a means to "concentrate", into a single-charge-state, analyte ion signal that normally would be dispersed over a wide range of charge states.

An alternative approach to ion parking was introduced by Grosshans et al.[108,109] in a pair of patent filings wherein a dipolar DC voltage was applied across the end-cap electrodes of a quadrupole ion trap. The result is a *m/z*–dependent separation in space of cation and anion "clouds" that results in the parking of all ions above a certain *m/z* value.

In further developing their ion parking technique, McLuckey and co-workers[110] were able to expand from parking ions of a single *m/z* value to simultaneous multiple-ion parking, which they termed "parallel" ion parking. The approach of inhibiting ion–ion reaction rates by resonant excitation with the inherent increase in velocities of selected ions that was developed for single-ion parking was now extended to multiple ions of differing *m/z* value. In parallel-ion parking, a tailored waveform is applied that contains two or more frequencies that are distinct from those of the reactant ions. To enable this excitation, a filtered noise field (FNF)[111] may be used to resonantly excite ions over a broad *range* of *m/z* values. In this case, the ion–ion reaction rates over a wide swathe of *m/z* values were thus reduced, parking all ions in a certain "window." This method was used advantageously to boost the yield of "first-generation" products of electron transfer dissociation in a 3D-quadrupole ion trap.[110] In parallel-ion parking, a broad range of applied frequencies will result in a wide range of *m/z* values of parked ions. The broadest range of *m/z* values is parked when "notches" are included corresponding to the initial reactant ions, such that the latter are not accelerated, and hence they are not preserved.

A third means of ion parking based upon application of *high-amplitude*, low-frequency (HALF) auxiliary AC for parallel-ion parking was later developed by the McLuckey group.[112] This approach takes advantage of the fact that the width of *m/z* values affected in the the 3D quadrupole ion trap increases as the *amplitude* of a given AC signal imposed on the end-cap electrodes goes up. Thus, the broad "shoulders" of the AC signal are capable of exciting a wide range of *m/z* values. Similar to ion parking using the filtered noise field (but unlike the single-ion parking approach), precise frequency tuning is not required for HALF parallel ion parking because of the relatively high voltage employed and nonselective excitation involved.[112]

14.1.8 Electrospray Conclusion

Between the moment that an initially neutral sample solution enters the ES capillary and the time when the detected ions are displayed in the mass spectrum, a complex series of events has occurred during the ES process. The initial solvent conditions clearly define the starting condition and conformation for analyte molecules. As the solution responds to the imposed ES high voltage, redox products are formed at the metal–solution interface and they are likely to be present in the released liquid. Thereafter, the solution contains a charge imbalance. The emerging liquid is transformed into charged droplets wherein excess charges are located near the surface layer. Uneven fission of these initially formed droplets boosts the charge-to-mass ratio of produced progeny droplets from which detected ions are ultimately desorbed. Charged clusters containing analyte, solvent(s), counterions, additives, and possibly redox products are headed toward the vacuum region of the mass spectrometer. As the final solvent molecules evaporate, solvophobic effects disappear and electrostatic effects augment between charge sites and dipoles of the remaining surrounding molecules. These nearby molecules, especially if they are counterions, can have a profound influence in decreasing the charge state of analyte species in the gas phase as the last of the solvent molecules depart. The dissociation of the final cluster can lead to a partitioning of charge between the remaining gas-phase species. In the end, the charge borne by analyte species will reflect the gas-phase chemistry (relative gas-phase affinities) of the species in contact with the available charge (including other analyte molecules) and the size, shape, and surface area of the analyte. The precise description of events is complicated by thermochemical considerations such as the free-jet expansion of the sprayed solution and the kinetically-controlled collisional heating of the formed charged clusters that take place within poorly defined thermal and pressure gradients inside the mass spectrometer.

Despite these complications in the comprehension of the exact events that lead to a given charge state distribution in ES mass spectrometry, the study of charge states may provide some unique information that cannot be accessed by other analytical means. In the years since the previous edition of this volume,[26] an enormous amount of work has been dedicated to comprehending the relationship between protein conformational dynamics and ES charge-state distributions. In this regard, the efforts of Konermann and Douglas,[60–62] the Konermann group[67,68] and the Kaltashov group[63–66] will be highlighted here as indicative of the future direction of charge-state distribution research, while noting that the cited work of these groups hadn't yet emerged at the time of the previous volume. The finding that a breakdown in tertiary structure rather than secondary structure was responsible for the shift in charge state distributions of proteins to higher values[60] is key to the future use of ES-MS as an accepted method to evaluate changes in the conformation of proteins. The launching of ES-MS in the field of protein conformational dynamics hinges upon the notion that individual conformations yield unique charge-state distributions based upon the relationship between charge state and the physical dimensions of a protein molecule in a given conformation.[60,63] But advanced computational methods will be needed to "tease out" the contributions of each conformer's charge state distribution to the overall ES mass spectrum.[64–66] In addition, a more detailed comprehension of exactly how physical parameters (i.e., "compactness") of protein conformation translate into charge-state distributions will be needed.[64,68] Also to be factored in is the knowledge that solution parameters can alter charge-state distributions and that observed signal responses are not direct reflections of relative solution concentrations.[67] Nonetheless, the groundwork has been laid to analytically exploit charge-state distributions as a means to gauge protein conformational changes. Future prospects include possibilities for refined characterization of non-native protein states, improved characterization of conformational dynamics of larger protein assemblies of all types, including protein–DNA, protein–oligosaccharide interactions, and nonbiological systems such as protein–polymer and protein nanoparticle assemblies.[63] Studies into the utility of ES charge-state distributions to perform analogous conformational investigations of other biomolecules such as the polynucleotides[69,70,91,92,113] have begun, but they lag behind those of the proteins.

14.2 MALDI

Matrix-assisted laser desorption/ionization (MALDI) has been widely used in mass spectrometric analysis of macromolecules such as biochemical compounds since its invention in the late 1980s.[114–117] Ions formed in the MALDI process typically contain fewer charges compared to those generated by electrospray (ES). In this section, we shall focus on the factors that affect the charge-state distributions of MALDI-generated ions. The observed ion distribution in the MALDI spectrum is inherently related to how the ions are produced. It thus seems necessary to first discuss the mechanisms of ion formation in view of shedding light on the analyte ion charge state distribution.

Numerous reviews have been present in the mechanisms of ion formation and desorption in the MALDI process,[118–123] including Chapter 5 of this volume. Although the picture of MALDI ionization is still far from complete, a consensus does exist concerning several aspects of the process: A laser beam serves as the desorption and ionization source. It triggers the primary ionization and formation of a plume of desorbed matrix and analyte and other ionic species or clusters. The matrix absorbs most of the laser energy and is the major species that gets ionized during the primary ionization phase. The primary ionized molecules (mainly matrix ions) or clusters transfer their charges to other molecules or ions through secondary ion–molecule/ion–ion reactions in the expanding plume. Analyte ionization is thermodynamically determined; highly charged primary ions are more reactive, so their charge tends to drop.

During the MALDI process, analyte molecules become ionized while still being protected by the matrix from the intense energy of the laser, thus, they can be detected with little or no fragmentation. Thermodynamically favorable products have a better chance to reach the detector. Under typical MALDI conditions, singly charged ions are predominately produced. However, it is not rare to detect multiply charged ions. Here, we roughly summarize the factors that are considered to affect the charge-state distributions of MALDI-generated ions. Our examination looks into the following categories: matrix properties, laser fluence; sample preparation, analyte properties, and involvement of the electrospray process as an associated ionization tool.

14.2.1 Matrix Properties

The matrix plays a key role in the MALDI process. It is thought to serve several functions including absorption of photon energy from the laser light; isolation of analyte molecules from each other; assistance in the vaporization of analyte molecules; and facilitation of the ionization of the analyte.

With an appropriate choice of the matrix, various macromolecules (for example, peptides, proteins, synthetic polymers, oligonucleotides, and oligosaccharides) can be analyzed by the MALDI-MS technique. A large variety of matrices have been employed. The matrix originally used by Karas and Hillenkamp[114,116] is nicotinic acid, which exhibits strong absorption at 266 nm. Beavis and Chait[124–126] found that cinnamic acid derivatives, such as α-cyano-hydroxycinnamic acid (CHCA) and sinapinic acid (SA), produce mass spectra of proteins showing considerably reduced adduction compared with nicotinic acid. Moreover, cinnamic acid derivatives absorb strongly at longer wavelengths—for example, 355-nm radiation from ND(YAG) lasers and 337-nm radiation from nitrogen lasers. Beavis, Chaudhary, and Chait's study[126] revealed an enhancement of the degree of protonation when proteins were embedded in the CHCA matrix as compared to the SA matrix. Figure 14.8 shows the spectra they obtained for 1 pmol of protein bovine tranferrin (molecular weight 78 kDa). The highest charge state observed is $+9$ when using CHCA as the matrix (Figure 14.8, top), while it is $+4$ in the SA matrix (Figure 14.8, bottom). Another study[127] concurred that multiple charging is a more common occurrence with the CHCA matrix.

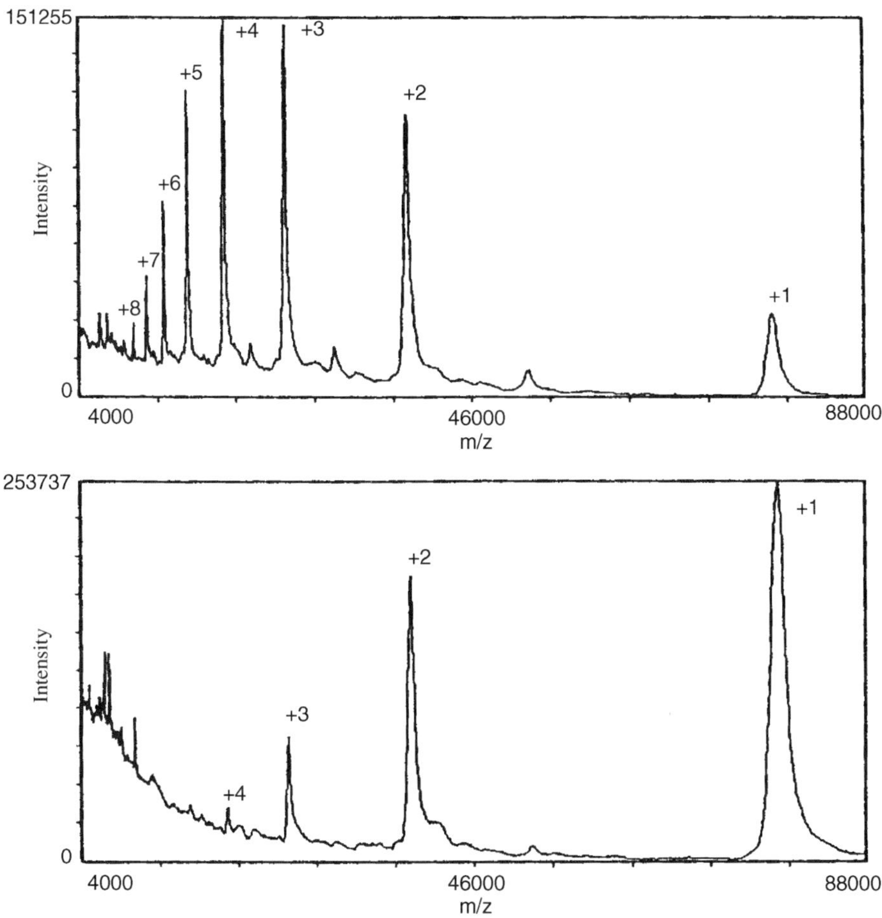

Figure 14.8. MALDI MS spectra of 1 pmol of protein bovine tranferrin obtained from the CHCA matrix (**top**) and the SA matrix (**bottom**). (Reproduced from Ref. 126 with permission.)

The tendency for a given matrix to favor or disfavor multiple charging of analytes has been considered in terms of gas-phase proton affinities (PAs) and ionization energies (IEs) of the matrix molecule. Jørgensen et al.[128] and Mirza et al.[129] used the kinetic method to measure the gas-phase proton affinities of several matrices commonly used for MALDI, including CHCA and SA. Table 14.1 shows the PA values obtained by three different groups. These data demonstrate that the PA of CHCA is lower than that of SA. A lower PA of the matrix will increase the tendency for available protons to be captured by the analyte, thus leading to higher charge states in the positive-ion mode. Jørgensen et al.[128] correlated the PAs of the seven matrices with the yield of multiply charged ions, using insulin, ribonuclease A, and myoglobin as analytes. They showed that for insulin and myoglobin, a low matrix PA was indeed associated with high yields of multiply charged analyte ions. Alves et al.[130] also showed that higher-charge-state ions were observed for apomyoglobin when CHCA was used as the matrix instead of SA in positive ion MALDI. However, the authors noted that HABA, a matrix with a relatively high PA, sometimes gave contradictory results, depending on the protein examined.

Table 14.1. Literature Values of Proton Affinity (PA) and Ionization Energy (IE) for Six Commonly Used MALDI Matrices

	PA Values			IE Values
Matrix	Jørgensen et al.[128] (kJ/mol)	Mirza et al.[129] (kcal/mol)	Burton et al.[135] (kcal/mol)	Frankevich et al.[131] (eV)
CHCA	841	201.0 ± 0.27	183 ± 2	8.3
DHB			204 ± 4	8.1
Caffeic acid	866			7.8
Ferulic acid	879		183 ± 2	7.6
SA	887	209.2 ± 0.38	204 ± 4	7.3
HABA	943	226.9 ± 0.26	183 ± 2	7.0

Instead of considering the PA value of the matrix, Frankevich et al.[131] reported a correlation between the theoretically calculated matrix ionization energy (IE) and the yield of the doubly charged human insulin ion (molecular weight 5807.6 Da) (Figure 14.9). They showed that the electron yield from dihydroxybenzoic acid (DHB) matrix increased as the laser pulse energy was raised while, concomitantly, the formation of multiply charged ions decreased. Thus, they proposed that at high laser energy, the number of photoelectrons in the plume increases; these photoelectrons can directly or indirectly neutralize multiply charged ions. Since a matrix with a higher IE has a lower tendency to produce free electrons, it is reasonable that matrices with high ionization energies, such as CHCA, show a higher yield of multiply charged ions. Furthermore, when a metal MALDI plate is employed, the yield of doubly charged ions decreases in favor of singly charged ions, due to a high yield of electrons from the metal–organic interface.

Based on data from Jørgensen et al.[128] and Frankevich et al.,[131] the relative ranking of proton affinities is CHCA < caffeic acid < ferulic acid < SA < HABA (2-(hydroxyphenylazo)benzoic acid) whereas the ranking of ionization energy thresholds is CHCA > DHB > caffeic acid > ferulic acid > SA > HABA; as shown in Table 14.1, both correlations are consistent with positive-ion studies that demonstrate that the observed charge state of analytes is higher in the CHCA matrix.[126,127,130,132,133] Reasonable secondary

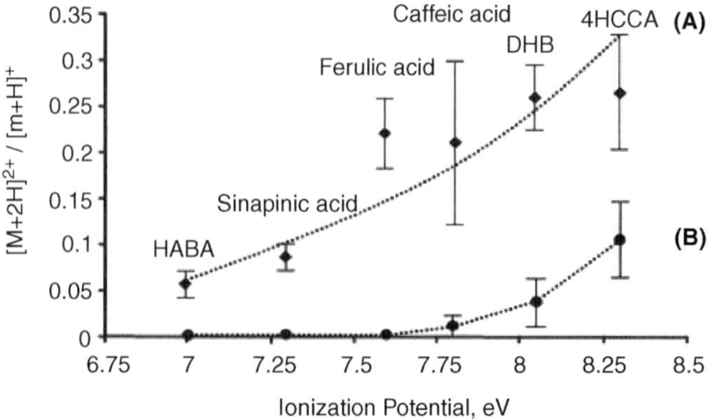

Figure 14.9. The peak intensity ratio of doubly to singly charged human insulin ions as a function of the calculated gas-phase ionization energy of the matrix. (**A**) A nonmetallic plate was used as the MALDI target. (**B**) A metal plate was employed. (Reproduced from Ref. 131 with permission.)

ion–molecule reactions can be proposed to support both correlations. However, the PA and IE models may be indicative of different ionization pathways, especially in the primary ionization process. The PA model is in accordance with the generally believed hypothesis that matrix ions are dominant during or immediately after the laser pulse and that gas-phase proton transfer reactions play a significant role in the formation of analyte ions under MALDI conditions. On the other hand, it was suggested that the analyte's charge state in the matrix is about the same as in solution[134]; so, it is possible that multiply charged ions are being generated during the primary ionization and are subsequently reduced by secondary ion–molecule reactions. The IE model reflects the ease of production of photoelectrons in the laser desorption/ionization process, which will be discussed further in the section examining the effect of laser fluence.

14.2.2 Laser Fluence

The tendency to form multiply charged ions in the MALDI process has been found to exhibit a dependence on the laser fluence. Although laser fluence (units of joules/cm^2) is used here, other dimensions such as laser power or laser energy (both in joules) or laser irradiance (in Watts/cm^2) are also seen in the literature. Using CHCA as the matrix, a 1995 study by Zhou and Lee[127] reported that the charge state of BSA shifted to lower values as the laser power was increased from 30% to 40% of full power. Later, Frankevich et al.[136] introduced SF_6 gas into their FTICR cell and assumed that the final amount of generated SF_6^- ions was proportional to the number of electrons present in the FTICR cell. They indicated that electrons are formed by the photoelectric effect on the metal–matrix interface. When a nonmetallic MALDI plate was used, or thick matrix layers were applied, the emission of photoelectrons from the support–matrix interface was inhibited, so far fewer electrons were in evidence.[136] In a subsequent study employing the FTICR instrument, after spotting a DHB–insulin mixture on a PEEK plate, Frankevich et al.[131] showed that although the electron yield was negligible at laser powers less than 0.15 mJ, substantial numbers of electrons were still emitted from the DHB matrix at laser powers larger than 0.2 mJ. Moreover, the abundance of doubly charged bovine insulin β-chain ions (molecular weight 3495.9 Da) was maximized at a laser power of 0.175 mJ, but the $2+/1+$ ratio rapidly decreased as the laser power was raised above 0.2 mJ (Figure 14.10).[131] Thus, a three-way correlation was established between higher laser fluence and increased electron yields and reduced charge states in positive-ion MALDI-MS. This general trend of a higher laser fluence disfavoring higher charge states was later confirmed by Alves et al.[130] for a wider variety of matrices including CHCA, SA, and DHB.

In a study of doubly charged porphyrin compounds, Schafer and Budzikiewicz[137] found that fast atom bombardment (FAB)-MS and MALDI-MS spectra of these inherently doubly charged compounds showed obvious analogies. They proposed that the formation of singly charged ions of the type $[A^{2+} + e^-]^{+\bullet}$ (A represents analyte) pointed toward an electron capture mechanism similar to that implicated in FAB-MS behavior. In the liquid secondary ion mass spectrometry (liquid SIMS)/FAB process, one-electron or two-electron reductions have been shown to contribute to the formation of singly charged ions.[138,139] Results mentioned above suggest that electrons generated from the MALDI process can directly or indirectly reduce or neutralize charges on multiply charged ions. The model proposed by Karas et al.[119] rationalized why singly charged molecular ions are dominant in MALDI. Among several proposed reactions leading to reduced charge states are: direct electron reduction of positively charged ions [Eq. (14.4)] and protons being abstracted from

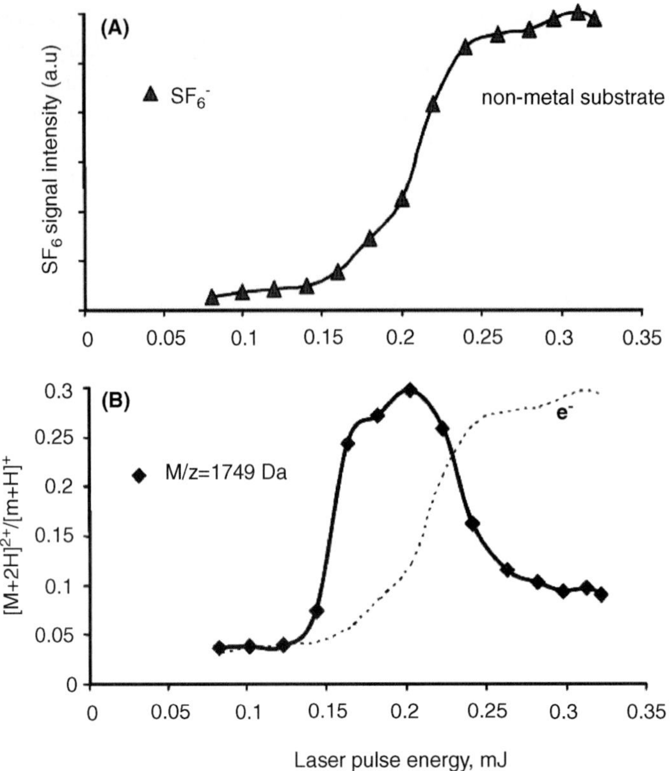

Figure 14.10. Influence of the laser pulse energy on analyte charge state distributions. (**A**) SF_6^- signal intensity as a function of laser pulse energy when insulin β chain in DHB matrix was spotted on a PEEK plate. (**B**) The peak intensity ratio of doubly to singly charged bovine insulin β chain as a function of laser pulse energy. (Reproduced from Ref. 136 with permission.)

multiply charged positive ions by anions [Eq. (14.5)].

$$(\mathrm{MH}_x^{x+}) + (x-1)e^- \leftrightarrows \mathrm{MH}_x^{+(x-1)\bullet} \rightarrow \text{fast fragmentation} \tag{14.4}$$

$$[n\mathrm{T} + \mathrm{MH}_x^{x+} + (x-1)\mathrm{B}^-]^+ \leftrightarrows \mathrm{MH}^+ + (x-1)\mathrm{HB} + n\mathrm{T} \tag{14.5}$$

where M represents an analyte molecule, H_X^{X+} represents addition of x protons, T represents a matrix molecule, and B^- represents an anion.[119]

A later model presented by Frankevich et al.[140] suggested an indirect reduction involving matrix anions such as

$$\mathrm{M} + e^- = [\mathrm{M-H}]^- + \mathrm{H}^\bullet \tag{14.6}$$

$$\mathrm{A}^{2+} + [\mathrm{M-H}]^- = [\mathrm{A-H}]^+ + \mathrm{M} \tag{14.7}$$

$$\mathrm{A}^{2+} + [\mathrm{M-H}]^- = \mathrm{A}^{+\bullet} + [\mathrm{M-H}]^\bullet \tag{14.8}$$

where M represents matrix and A represents analyte. Both models point to the same underlying phenomenon as being responsible for the reduced level of charging at higher laser fluences: Because more electrons are formed at higher laser fluences, their presence augments the degree of analyte reduction, thus higher analyte charge states are disfavored.

14.2.3 Sample Preparation

Two points relating directly to sample preparation have been mentioned in the laser energy section: (1) Electrons originating from the matrix that are generated in the MALDI process can reduce the number of charges carried by the analyte ions, and (2) fewer electrons were generated when using a nonmetallic plate or when applying thick matrix layers during sample preparation. Under these latter conditions, it makes good sense that the charge state distribution of the analyte shifts to higher values. Zhou and Lee[127] had noticed this effect when they spotted the same BSA/CHCA mixture using the same dried-droplet method onto two different MALDI targets. The first target consisted of a bare stainless steel plate, thereby yielding a thin matrix layer, whereas the second target held a matrix "pad" adhered to the stainless steel plate, resulting in a thick matrix layer; MALDI mass spectra were acquired with low laser irradiance. They reported that the BSA spectra obtained on the matrix pad exhibited a minor shift toward higher charge states; the dominant peak carried three charges compared to two charges when spotting directly on a stainless steel plate. Furthermore, as was shown by the work of Frankevich et al.[131] in Figure 14.9, when measured at the laser threshold energy for peak appearance, the peak intensity ratio of doubly to singly charged human insulin ions was much higher if a nonmetal plate was used instead of a metal plate. This result held true for six different tested matrices (i.e., CHCA, DHB, caffeic acid, ferullic acid, SA, and HABA).[131]

Returning now to the work of Zhou and Lee,[127] in addition to the above findings regarding target preparation, interestingly, they also found that when glycerol was added (20% by volume) to the BSA–CHCA mixture and spotted on the matrix pad, the charge state shifted to even higher values (the dominant peak carrying five charges) than those obtained on the matrix pad without glycerol (most abundant peak carried three charges). The authors noted that once glycerol was added to the sample-matrix solution, after a few minutes drying time, there was no obvious change in the spot volume; they also mentioned that any remaining liquid was removed from the matrix pad before starting the acquisition. The clear effect of the glycerol additive in shifting the charge state distribution of BSA toward higher values is shown in Figure 14.11.

More recently, König et al.[132] detected highly charged peptide/protein ions by AP-IR-MALDI-IT. Samples were prepared by adding 1 μL of glycerol and 0.5 μL of 0.1% TFA to 0.5 μL of aqueous analyte solution, and then removing the excess water. The matrix target, this time irradiated at atmospheric pressure, was characterized as a liquid. In the sample preparation protocol, König et al.[132] reported that the addition of acid was crucial for formation of multiply charged ions. Among more than 10 peptides/proteins studied, the smallest one, bradykinin (molecular weight = 1059.6 Da), was detected in the form of singly and doubly charged ions; whereas for the largest one, myoglobin (molecular weight 16,951 Da), species from $[M + 9H]^{9+}$ to $[M + 13H]^{13+}$ were detected, with $[M + 11H]^{11+}$ being the dominant peak. Even though the latter signal-to-noise ratios were rather weak, the numbers of charges on the peptides/proteins were almost at the same high level as those produced by ES-MS. Because their work used a liquid matrix, König et al.[132] speculated that plume expansion dynamics at atmospheric pressure, involving droplet breakup via collision, were contributing to the phenomenon of extensive multiple charging. In this regard, they expressed the view that the overall ionization mechanism appears to bridge the MALDI and ES processes.

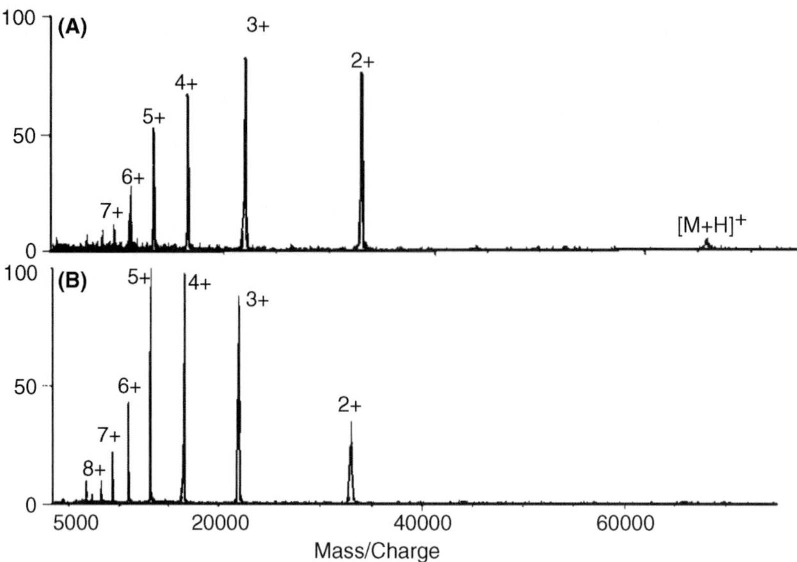

Figure 14.11. MALDI spectra of BSA acquired at 30% of full laser power acquired using the dried droplet method employing: (**A**) BSA/CHCA layers; (**B**) 20% glycerol (by volume) added to the BSA–CHCA mixture. (Reproduced from Ref. 127 with permission.)

It is worthy of note that in the above experiments, an Er:YAG IR laser with 2.94-μm wavelength was used.[132] Although most of the time the laser wavelengths used for MALDI-MS of peptides/proteins are in the UV range, it was actually shown early on that an IR laser is also capable of generating useful peptide/protein MALDI mass spectra.[141] Experimental data from two different labs indicated that the IR-MALDI and UV-MALDI spectra of a given protein are very similar, even though the corresponding interpretations of the ionization/desorption mechanism may be different.[142,143] However, compared to UV-MALDI, even less is known about the energy absorption and desorption/ionization mechanism of the IR-MALDI process, and the exact role of the matrix in IR-MALDI remains poorly understood; see Chapter 5–7 in this volume for additional information.

In the work of Kononikhin et al.,[144] bovine-insulin (molecular weight 5733 Da) was electrospray-deposited to form a thin layer on top of a previously prepared CHCA matrix layer. At 3-mW/cm^2 laser irradiance, peaks with charge states up to $+5$ were detected, which is comparable to those obtained by ES-MS. By contrast, using a conventional two-layer sample preparation approach (no ES deposition of analyte), the highest charge state detected was $+3$. In rationalizing the appearance of $+5$ charge state ions only when ES deposition was used, the authors suggest that desorption of multiply charged "preformed" ions may be occurring; in this case, preformed ions were generated by ES during sample preparation.

Yao et al.[145] investigated the effects of various buffers (six in total) on MALDI mass spectra of three selected proteins. Using DHB as the matrix and 26 μM BSA solution as the analyte, while keeping the DHB : BSA molar ratio at 40,000 : 1, BSA ions with charge states up to $+3$ were detected. However, for BSA in 200 mM BICINE (*N*,*N*-bis[2-hydroxyethyl] glycine) buffer solution, ions carrying five charges were observed, although the S/N quality of the spectrum was not exceptional. The authors conclude that a fine-tuning of the matrix–analyte–buffer ratio could be used to augment the degree of analyte charging in MALDI-MS experiments investigating analytes dissolved in buffer-containing solutions.

When testing the effect of the matrix–analyte ratio on charge-state distributions, contradictory results are noted in the literature. Zhong and Zhao[133] reported increased levels of doubly charged species relative to singly charged ions when a higher matrix–-analyte ratio (more dilute analyte on the target) was used. They demonstrate this effect showing MALDI mass spectra of the small protein cytochrome c (~12,000 Da) embedded in both SA and CHCA at various matrix–analyte ratios. By contrast, Alves et al.[130] report that an increase in matrix–analyte ratio (i.e., dilution of the analyte) results in a *decrease* in the desorption/ionization of higher charge state species. The latter investigation tested a medium-sized protein—that is, bovine serum albumin (~66,000 Da)—in SA.

14.2.4 Analyte Properties

The size of the analyte molecule has been shown to heavily influence the charge states of desorbed ions formed during the MALDI process. Using CHCA as the matrix, Beavis et al.[126] detected only singly charged ions for the small peptide "substance P" (molecular weight 1347.7 Da); whereas for a bovine transferrin (molecular weight 78 kDa), up to +9 charged ions were detected. It was shown that the number of charges carried on the analytes ions increases as the mass of the analyte molecules increases[126,146,147] when comparing masses within one class of analytes, that is, peptides/proteins. Employing the laser-induced fluorescence technique in combination with a quadrupole ion trap, Chang and co-workers[146] reported that the MALDI technique can ionize very high mass polystyrene particles (with use of 3-hydroxypicolinic acid as the matrix). They determined that the number of charges on particles of 27 nm and 1.1 μm in diameter (with masses in the 10^6-Da and 10^{12}-Da range, respectively) were 1–10 and 80–400, respectively. Zenobi and co-workers[147] utilized a multichannel array cryodetector to directly detect high-mass MALDI-generated ions. For a 150-kDa IgG sample, ions with up to three charges were detected; whereas for a 1 MDa IgM, ions with up to four charges were detected. Analogous to arguments used in the ES section of this chapter (i.e., within one class of analytes), the internal electrostatic repulsion between like charges will decrease as the size of the analyte increases; and if the surface exposed area of the larger molecule is indeed greater than that of the lower molecular weight counterpart, it then seems reasonable that the larger molecule would be able to accommodate a greater number of charges.

14.2.5 Involvement of the Electrospray Process as an Associated Ionization Tool for Generation of Highly Charged Ions

Shiea and co-workers[148] introduced an electrospray assisted laser desorption ionization source (ELDI) in 2005; their ELDI setup is shown schematically in Figure 14.12. They demonstrated that during the ELDI process, protein was desorbed from the stainless-steel sample support plate by laser irradiation and it became ionized by the electrospray droplets under ambient pressure. Using bovine cytochrome C as the analyte, they obtained mass spectra characteristic of electrospray. When only bovine cytochrome C, without matrix, was spotted on the plate, ions with up to 19 charges were detected. When mixing bovine cytochrome C with CHCA matrix, ions with up to 10 charges were detected. In 2006, Sampson, Hawkridge, and Muddiman[149] reported a matrix-assisted laser desorption electrospray ionization (MALDESI) source. The MALDESI source is, in principle, virtually identical to the ELDI source. A subtle difference lies in that MALDESI relies

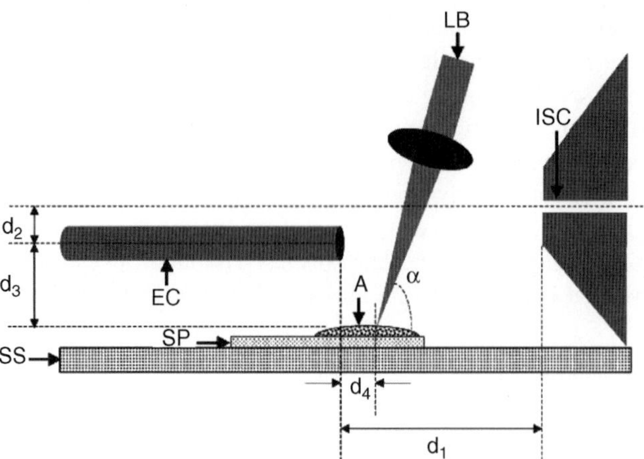

Figure 14.12. Schematic drawing of the ELDI setup. The analyte (A), deposited on the stainless steel sample support plate (SP), is positioned by the sliding sample stage (SS) and irradiated with a pulsed nitrogen laser (LB). The laser beam irradiates the at the incident angle $\alpha = 45°$. Laser-ablated material is ionized in the electrospray-generated solvent plume exiting the electrospray capillary (EC). Resulting ions enter into the mass spectrometer through the ion sampling capillary (ISC). (Reproduced from Ref. 148 with permission.)

upon use of a matrix, whereas for ELDI, the matrix seems optional. Upon coupling the MALDESI source to a FT-ICR MS,[149] it was found that the addition of matrix was necessary to obtain an appreciable signal. It was also reported that for bradykinin and ubiquitin (using SA as matrix), the average charge states observed by MALDESI-FTICR respectively (i.e., 1.97 and 9.58), are nearly the same as those observed by ESI-FTICR, 1.92 and 9.73, respectively. This result provided further evidence to corroborate prior conclusions made by Shiea et al.[148] that ionization is occurring via the ES event.

14.2.6 MALDI Conclusion

Investigations aimed at improving the understanding of the exact roles of the various components involved in the MALDI process that contribute to the final observed charge-state distribution in MALDI mass spectra have lagged behind parallel studies in ES ionization. This section has attempted to group the relevant MALDI studies investigating charge state distributions into the categories of matrix selection, laser fluence, sample preparation techniques, and analyte properties. At laser fluences well above the threshold for ion generation, the observed MALDI-generated ions are considered to largely reflect the thermodynamically favored products (those of highest stability); whereas just above the threshold, ion distributions appear to be kinetically limited.[118,150,151] It seems reasonable to assume that any aspect of the MALDI process that can affect the primary ionization and/or secondary ion–molecule/ion–ion reactions will play a role in determining the distribution of charge states arriving at the detector. In particular, in the positive-ion mode, high proton affinities of the matrix have been correlated with lower analyte charge states, which is clearly an issue of competition for available protons. Furthermore, a consensus has emerged that a high availability of electrons—occurring, for example, as a result of the use of a matrix with a relatively low ionization energy, or by use of higher laser powers—will disfavor the survival of high-charge-state ions. The advent of new combined techniques such as

ELDI and MALDESI promise to push the envelope of accessible charge states of laser desorbed species into much higher realms than were previously considered to be achievable.

REFERENCES

1. Dole, M.; Mack, L. L.; Hines, R. L.; Mobley, R. C.; Ferguson, L. D.; Alice, M. B. Molecular beams of macroions. *J. Chem. Phys.* **1968**, *49*, 2240–2249.
2. Dole, M.; Hines, R. L.; Mack, L. L.; Mobley, R. C.; Ferguson, L. D.; Alice, M. B. Gas phase macroions. *Macromolecules* **1968**, *1*, 96.
3. Iribarne, J. V.; Thomson, B. A. On the evaporation of small ions from charged droplets. *J. Chem. Phys.* **1976**, *64*, 2287–2294.
4. Thomson, B. A.; Iribarne, J. V. Field-induced ion evaporation from liquid surfaces at atmospheric pressure. *J. Chem. Phys.* **1979**, *71*, 4451–4463.
5. Thomson, B. A.; Iribarne, J. V.; Dziedzic, P. J. Liquid ion evaporation/mass spectrometry for the detection of polar and labile molecules. *Anal. Chem.* **1982**, *54*, 2219–2224.
6. Iribarne, J. V.; Dziedzic, P. J.; Thomson, B. A. Atmospheric pressure ion evaporation–mass spectrometry. *Int. J. Mass Spectrom. Ion Phys.* **1983**, *50*, 331–334.
7. Fenn, J. B.; Mann, M.; Meng, C. K.; Wong, S. K.; Whitehouse, C. Electrospray ionization—principles and practice. *Mass Spectrom. Rev.* **1990**, *9*, 37–70.
8. Fenn, J. B.; Rosell, J.; Nohmi, T.; Banks, F. J. In Snyder, A. P. (Ed.), *Biochemical and Biotechnological Applications of Electrospray Ionization Mass Spectrometry*, American Chemical Society, Washington, DC, 1995, pp. 60–80.
9. Loo, J. A.; Udseth, H. R.; Smith, R. D. Solvent effects on the charge distribution observed with electrospray ionization–mass spectrometry of large molecules. *Biomed. Environ. Mass Spectrom.* **1988**, *17*, 411–414.
10. Loo, J. A.; Edmonds, C. G.; Udseth, H. R.; Smith, R. D. Effect of reducing disulfide-containing proteins on electrospray ionization mass spectra. *Anal. Chem.* **1990**, *62*, 693–698.
11. Covey, T. R.; Bonner, R. F.; Shushan, B. I.; Henion, J. The determination of protein, oligonucleotide, and peptide molecular weights by ion-spray mass spectrometry. *Rapid Commun. Mass Spectrom.* **1988**, *2*, 249–256.
12. Creighton, T. E. *Proteins*, W. H. Freeman, New York, 1993.
13. Shaw, K. L.; Grimsley, G. R.; Yakovlev, G. I.; Makarov, A. A.; Pace, C. N. The effect of net charge on the solubility, activity, and stability of ribonuclease Sa. *Protein Sci.* **2001**, *10*, 1206–1215.
14. Kelly, M. A.; Vestling, M. M.; Fenselau, C. C.; Smith, P. B. Electrospray analysis of proteins: A comparison of positive-ion and negative-ion mass spectra at high and low pH. *Org. Mass Spectrom.* **1992**, *27*, 1143–1147.
15. Hiraoka, K.; Murata, K.; Kudaka, I. Do the electrospray mass spectra reflect the ion concentrations in sample solution? *J. Mass Spectrom. Soc. Japan* **1995**, *43*, 127–138.
16. Mirza, U. A.; Chait, B. T. Effects of anions on the positive ion electrospray ionization mass spectra of peptides and proteins. *Anal. Chem.* **1994**, *66*, 2898–2904.
17. Le Blanc, J. C. Y.; Wang, J.; Guevremont, R.; Siu, K. W. M. Electrospray mass spectra of protein cations formed in basic solutions. *Org. Mass Spectrom.* **1994**, *29*, 587–593.
18. Mansoori, B. A.; Volmer, D. A.; Boyd, R. K. Wrong-way-round electrospray ionization of amino acids. *Rapid Commun. Mass Spectrom.* **1997**, *11*, 1120–1130.
19. Wong, S. F.; Meng, C. K.; Fenn, J. B. Multiple charging in electrospray ionization of poly(ethylene glycols). *J. Phys. Chem.* **1988**, *92*, 546–550.
20. Nohmi, T.; Fenn, J. B. Electrospray Mass Spectrometry of poly(ethylene glycols) with molecular weights up to five million. *J. Am. Chem. Soc.* **1992**, *114*, 3241–3246.
21. Schnier, P. D.; Gross, D. S.; Williams, E. R. On the maximum charge state and proton transfer reactivity of peptide and protein ions formed by electrospray ionization. *J. Am. Soc. Mass Spectrom.* **1995**, *6*, 1086–1097.
22. Fernandez de la Mora, J. Electrospray ionization of large multiply charged species proceeds via Dole's charged residue mechanism. *Anal. Chim. Acta* **2000**, *406*, 93–104.
23. Chowdhury, S. K.; Katta, V.; Chait, B. T. Probing conformational changes in proteins by mass spectrometry. *J. Am. Chem. Soc.* **1990**, *112*, 9012–9013.
24. Wang, G.; Cole, R. B. Disparity between solution-phase equilibria and charge state distributions in positive-ion electrospray. mass spectrometry. *Org. Mass Spectrom.* **1994**, *29*, 419–427.
25. Loo, J. A.; Udseth, H. R.; Smith, R. D. Collisional Effects on the charge distribution of ions from large molecules, formed by electrospray-ionization mass spectrometry. *Rapid Commun. Mass Spectrom.* **1988**, *2*, 207–210.
26. Wang, G.; Cole, R. B. Solution, gas-phase, and instrumental parameter influences on charge state distributions in electrospray ionization mass spectrometry. In Cole, R. B. (Ed.), *Electrospray*

Ionization Mass Spectrometry: Fundamentals, Instrumentation, and Applications, Wiley-Interscience, New York, **1997**, pp. 137–174.

27. GATLIN, C. L.; TURECEK, F. Acidity determination in droplets formed by electrospraying methanol–water solutions. *Anal. Chem.* **1994**, *66*, 712–718.
28. Van BERKEL, G. J.; ZHOU, F.; ARONSON, J. T. Changes in bulk solution pH caused by the inherent controlled-current electrolytic process of an electrospray ion source. *Int. J. Mass Spectrom. Ion Proc.* **1997**, *162*, 55–67.
29. CHENG, X.; GALE, D. C.; UDSETH, H. R.; SMITH, R. D. Charge state reduction of oligonucleotide negative ions from electrospray ionization. *Anal. Chem.* **1995**, *67*, 586–593.
30. TONG, X.; HENION, J.; GANEM, B. Studies on the distribution of oligodeoxynucleotide charge states in the gas phase by ion spray mass spectrometry. *J. Mass Spectrom.* **1995**, *30*, 867–871.
31. COLE, R. B.; HARRATA, A. K. Charge-state distribution and electric-discharge suppression in negative-ion electrospray mass spectrometry using chlorinated solvents. *Rapid Commun. Mass Spectrom.* **1992**, *6*, 536–539.
32. COLE, R. B.; HARRATA, A. K. Solvent effect on analyte charge state, signal intensity, and stability in negative ion electrospray mass spectrometry; implications for the mechanism of negative ion formation. *J. Am. Soc. Mass Spectrom.* **1993**, *4*, 546–556.
33. WANG, G.; COLE, R. B. Effects of solvent and counterion on ion pairing and observed charge states of diquaternary ammonium salts in electrospray ionization mass spectrometry. *J. Am. Soc. Mass Spectrom.* **1996**, *10*, 1050–1058.
34. GREGOR, H. P.; BELLE, J.; MARCUS, R. A. Ion-exchange resins. XIII. Selectivity coefficients of quaternary base anion-exchange resins toward univalent anions. *J. Am. Chem. Soc.* **1955**, *77*, 2713–2719.
35. GJERDE, D. T.; SCHMUCKLER, G.; FRITZ, J. S. Anion chromatography with low-conductivity eluents. II. *J. Chrom.* **1980**, *187*, 35–45.
36. WANG, G.; COLE, R. B. Effect of solution ionic strength on analyte charge state distributions in positive and negative ion electrospray mass spectrometry. *Anal. Chem.* **1994**, *66*, 3702–3708.
37. IAVARONE, A. T.; JURCHEN, J. C.; WILLIAMS, E. R. Supercharged protein and peptide ions formed by electrospray ionization. *Anal. Chem.* **2001**, *73*, 1455–1460.
38. IAVARONE, A. T.; WILLIAMS, E. R. Supercharging in electrospray ionization: Effects on signal and charge. *Int. J. Mass Spectrom.* **2002**, *219*, 63–72.
39. IAVARONE, A. T.; WILLIAMS, E. R. Mechanism of charging and supercharging molecules in electrospray ionization. *J. Am. Chem. Soc.* **2003**, *125*, 2319–2327.
40. IAVARONE, A. T.; WILLIAMS, E. R. Collisionally activated dissociation of supercharged proteins formed by electrospray ionization. *Anal. Chem.* **2003**, *75*, 4525–4533.
41. SAMALIKOVA, M.; GRANDORI, R. Protein charge-state distributions in electrospray–ionization mass spectrometry do not appear to be limited by the surface tension of the solvent. *J. Am. Chem. Soc.* **2003**, *125*, 13352–13353.
42. SAMALIKOVA, M.; GRANDORI, R. Testing the role of solvent surface tension in protein ionization by electrospray. *J. Mass Spectrom.* **2005**, *40*, 503–510.
43. FENN, J. B. Ion Formation from charged droplets: Roles of geometry, energy, and time. *J. Am. Soc. Mass Specrom.* **1993**, *4*, 524–535.
44. KEBARLE, P.; TANG, L. From ions in solution to ions in the gas phase—The mechanism of electrospray mass spectrometry. *Anal. Chem.* **1993**, *65*, 972A–986A.
45. TANG, L.; KEBARLE, P. Dependence of ion intensity in electrospray mass spectrometry on the concentration of the analytes in the electrosprayed solution. *Anal. Chem.* **1993**, *65*, 3654–3668.
46. SUNNER, J.; NICOL, G.; KEBARLE, P. Factors determining relative sensitivity of analytes in positive mode atmospheric pressure ionization mass spectrometry. *Anal. Chem.* **1988**, *60*, 1300–1307.
47. KOSTIAINEN, R.; BRUINS, A. P. Effect of multiple sprayers on dynamic range and flow rate limitations in electrospray and ionspray mass spectrometry. *Rapid Commun. Mass Spectrom.* **1994**, *8*, 549–558.
48. WANG, G.; COLE, R. B. Mechanistic interpretation of the dependence of charge state distributions on analyte concentrations in electrospray ionization mass spectrometry. *Anal. Chem.* **1995**, *67*, 2892–2900.
49. CECH, N. B.; ENKE, C. G. Practical implications of some recent studies in electrospray ionization fundamentals. *Mass Spectrom. Rev.* **2001**, *20*, 362–387.
50. CHOWDHURY, S. K.; KATTA, V.; CHAIT, B. T. An electrospray–ionization mass spectrometer with new features. *Rapid Commun. Mass Spectrom.* **1990**, *4*, 81–87.
51. SMITH, R. D.; LOO, J. A.; EDMONDS, C. G.; BARINAGA, C. J.; UDSETH, H. R. New developments in biochemical mass spectrometry: electrospray ionization. *Anal. Chem.* **1990**, *62*, 882–899.
52. TANG, L.; KEBARLE, P. Effect of the conductivity of the electrosprayed solution on the electrospray current. *Anal. Chem.* **1991**, *63*, 2709–2715.
53. COLE, R. B. Special feature commentary: Some tenets pertaining to electrospray ionization mass spectrometry. *J. Mass Spectrom.* **2000**, *35*, 763–772.
54. KATTA, V.; CHAIT, B. T. Conformational changes in proteins probed by hydrogen-exchange electrospray–ionization mass spectrometry. *Rapid Commun. Mass Spectrom.* **1991**, *5*, 214–217.
55. KATTA, V.; CHAIT, B. T. Hydrogen/deuterium exchange electrospray ionization mass spectrometry: A method for probing protein conformational changes in solution. *J. Am. Chem. Soc.* **1993**, *115*, 6317–6321.

56. Suckau, D.; Shi, Y.; Beu, S. C.; Senko, M. W.; Quinn, J. P.; Wampler, F. W.; McLafferty, F. W. Coexisting stable conformations of gaseous protein ions. *Proc. Natl. Acad. Sci. USA* **1993**, *90*, 790–793.
57. Le Blanc, J. C. Y.; Beuchemin, D.; Siu, K. W. M.; Guevremont, R.; Berman, S. S. Thermal denaturation of some proteins and its effect on their electrospray mass spectra. *Org. Mass Spectrom.* **1991**, *26*, 831–839.
58. Mirza, U. A.; Cohen, S. L.; Chait, B. T. Heat-induced conformational changes in proteins studied by electrospray ionization mass spectrometry. *Anal. Chem.* **1993**, *65*, 1–6.
59. Loo, J. A.; Ogorzalek Loo, R. R.; Udseth, H. R.; Edmonds, C. G.; Smith, R. D. Solvent-induced conformational changes of polypeptides probed by elecrospray–ionization mass spectrometry. *Rapid Commun. Mass Spectrom.* **1991**, *5*, 101–105.
60. Konermann, L.; Douglas, D. J. Acid-induced unfolding of cytochrome c at different methanol concentrations: Electrospray ionization mass spectrometry specifically monitors changes in the tertiary structure. *Biochem.* **1997**, *36*, 12296–12302.
61. Konermann, L.; Douglas, D. J. Equilibrium unfolding of proteins monitored by electrospray ionization mass spectrometry: Distinguishing two-state from multi-state transitions. *Rapid Commun. Mass Spectrom.* **1998**, *12*, 435–442.
62. Konermann, L.; Douglas, D. J. Unfolding of proteins monitored by electrospray ionization mass spectrometry: A comparison of positive and negative ion modes. *J. Am. Soc. Mass Spectrom.* **1998**, *9*, 1248–1254.
63. Kaltashov, I. A.; Abzalimov, R. R. Do ionic charges in ESI MS provide useful information on macromolecular structure? *J. Am. Soc. Mass Spectrom.* **2008**, *19*, 1239–1246.
64. Dobo, A.; Kaltashov, I. A. Detection of multiple protein conformational ensembles in solution via deconvolution of charge-state distributions in ESI MS. *Anal. Chem.* **2001**, *73*, 4763–4773.
65. Gumerov, D. R.; Dobo, A.; Kaltashov, I. A. Protein-ion charge-state distributions in electrospray ionization mass spectrometry: Distinguishing conformational contributions from masking effects. *Eur. J. Mass Spectrom.* **2002**, *8*, 123–129.
66. Mohimen, A.; Dobo, A.; Hoerner, J. K.; Kaltashov, I. A. A chemometric approach to detection and characterization of multiple protein conformers in solution using electrospray ionization mass spectrometry. *Anal. Chem.* **2003**, *75*, 4139–4147.
67. Kuprowski, M. C.; Konermann, L. Signal response of coexisting protein conformers in electrospray mass spectrometry. *Anal. Chem.* **2007**, *79*, 2499–2506.
68. Konermann, L. A minimalist model for exploring conformational effects on the electrospray charge state distribution of proteins. *J. Phys. Chem. B* **2007**, *111*, 6534–6543.
69. Guo, X.; Bruist, M. F.; Davis, D. L.; Bentzley, C. M. Secondary structural characterization of oligonucleotide strands using electrospray ionization mass spectrometry. *Nucl. Acids Res.* **2005**, *33*, 3659–3666.
70. Gabelica, V.; De Pauw, E. Comparison between solution-phase stability and gas-phase kinetic stability of oligodeoxynucleotide duplexes. *J. Mass Spectrom.* **2001**, *36*, 397–402.
71. Grandori, R. Origin of the conformation dependence of protein charge-state distributions in electrospray ionization mass spectrometry. *J. Mass Spectrom.* **2003**, *38*, 11–15.
72. Samalikova, M.; Grandori, R. Role of opposite charges in protein electrospray ionization mass spectrometry. *J. Mass Spectrom.* **2003**, *38*, 941–947.
73. Nesatyy, V. J.; Suter, Marc J. -F. On the conformation-dependent neutralization theory and charging of individual proteins and their non-covalent complexes in the gas phase. *J. Mass Spectrom.* **2004**, *39*, 93–97.
74. Halgand, F.; Laprevote, O. Mean charge state and charge state distribution of proteins as structural probes. an electrospray ionisation mass spectrometry study of lysozyme and ribonuclease A. *Eur. J. Mass Spectrom.* **2001**, *7*, 433–439.
75. Li, Y.; Cole, R. B. Shifts in peptide and protein charge state distributions with varying spray tip orifice diameter in nano-electrospray FT-ICR mass spectrometry. *Anal. Chem.* **2003**, *75*, 5739–5746.
76. Winger, B. E.; Light-Wahl, K. J.; Smith, R. D. Gas-phase proton transfer reactions involving multiply charged cytochrome C ions and water under thermal conditions. *J. Am. Soc. Mass Spectrom.* **1992**, *3*, 624–630.
77. Smith, Richard D.; Loo, Joseph A.; Barinaga, Charles J.; Edmonds, Charles G.; Udseth, Harold R. Collisional activation and collision-activated dissociation of large multiply charged polypeptides and proteins produced by electrospray ionization. *J. Am. Soc. Mass Spectrom.* **1990**, *1*, 53–65.
78. Ashton, D. S.; Beddell, C. R.; Cooper, D. J.; Green, B. N.; Oliver, R. W. A. Mechanism of production of ions in electrospray mass spectrometry. *Org. Mass Spectrom.* **1993**, *28*, 721–728.
79. Thomson, B.A. Declustering and fragmentation of protein ions from an electrospray ion source. *J. Am. Soc. Mass Spectrom.* **1997**, *8*, 1053–1058.
80. Smith, R. D.; Light-Wahl, K. J.; Winger, B. E.; Loo, J. A. Preservation of non-covalent associations in electrospray ionization mass spectrometry: Multiply charged polypeptide and protein dimers. *Org. Mass Spectrom.* **1992**, *27*, 811–821.
81. Smith, R. D.; Barinaga, C. J.; Udseth, H. R. Tandem mass spectrometry of highly charged cytochrome C molecular ions produced by electrospray ionization. *J. Phys. Chem.* **1989**, *93*, 5019–5022.

82. Felitsyn, N.; Peschke, M.; Kebarle, P. Origin and Number of Charges Observed on Multiply-Protonated Native Proteins Produced by ESI. *Int. J. Mass Spectrom.* **2002**, *219*, 39–62.
83. Gross, D. S.; Williams, E. R. Experimental measurement of Coulomb energy and intrinsic dielectric polarizability of a multiply protonated peptide ion using electrospray ionization Fourier-transform mass spectrometry. *J. Am. Chem. Soc.* **1995**, *117*, 883–890.
84. Schnier, P. D.; Gross, D. S.; Williams, E. R. Electrostatic forces and dielectric polarizability of multiply protonated gas-phase cytochrome c ions probed by ion/molecule chemistry. *J. Am. Chem. Soc.* **1995**, *117*, 6747–6757.
85. Gross, D. S.; Rodriguez-Cruz, S. E.; Bock, S.; Williams, E. R. Measurement of Coulomb energy and dielectric polarizability of gas-phase diprotonated diaminoalkanes. *J. Phys. Chem.* **1995**, *99*, 4034–4038.
86. Gronert, S. Determining the gas-phase properties and reactivities of multiply charged ions. *J. Mass Spectrom.* **1999**, *34*, 787–796.
87. Peschke, M.; Blades, A.; Kebarle, P. Charged states of proteins. reactions of doubly protonated alkyldiamines with NH3: Solvation or deprotonation. extension of two proton cases to multiply protonated globular proteins observed in the gas phase. *J. Am. Chem. Soc.* **2002**, *124*, 11519–11530.
88. Blades, A.; Peschke, M.; Verkerk, U.; Kebarle, P. Rates of proton transfer from carboxylic acids to dianions, $CO_2(CH_2)pCO_2^{2-}$, and their significance to observed negative charge states of proteins in the gas phase. *J. Phys. Chem. A* **2002**, *106*, 10037–10042.
89. Otto, M.; Knabe, F.; Claussnitzer, U. Response surface study on protein charge-state distributions in electrospray ionization mass spectrometry related to multiple ion equilibria in solution. *Chemom. Intell. Lab. Syst.* **2004**, *72*, 245–252.
90. Zhao, C.; Johnson, R. W.; Bruckenstein, S.; Wood, T. D. FT-ICRMS distinguishes the mechanism of the charge state reduction for multiply charged protein cations admixed with redox reagents in ESI. *J. Mass Spectrom.* **2006**, *41*, 641–645.
91. Greig, M.; Griffey, R. H. Utility of organic bases for improved electrospray mass spectrometry of oligonucleotides. *Rapid Commun. Mass Spectrom.* **1995**, *9*, 97–102.
92. Griffey, R. H.; Sasmor, H.; Greig, M. J. Oligonucleotide charge states in negative ionization electrospray–mass spectrometry are a function of solution ammonium ion concentration. *J. Am. Soc. Mass Spectrom.* **1997**, *8*, 155–160.
93. Muddiman, D. C.; Cheng, X.; Udseth, H. R.; Smith, R. D. Charge-state reduction with improved signal intensity of oligonucleotides in electrospray ionization mass spectrometry. *J. Am. Soc. Mass Spectrom.* **1996**, *7*, 697–706.
94. McLuckey, S. A.; Van Berkel, G. J.; Glish, G. L. Reactions of dimethylamine with multiply charged ions of cytochrome C. *J. Am. Chem. Soc.* **1990**, *112*, 5668–5670.
95. McLuckey, S. A.; Glish, G. L.; Van Berkel, G. J. Charge determination of product ions formed from collision-induced dissociation of multiply protonated molecules via ion/molecule reactions. *Anal. Chem.* **1991**, *63*, 1971–1978.
96. Busman, M.; Rockwood, A. L.; Smith, R. D. Activation energies for gas-phase dissociations of multiply charged ions from electrospray ionization mass spectrometry. *J. Phys. Chem.* **1992**, *96*, 2397–2400.
97. Ogorzalek Loo, R. R.; Loo, J. A.; Udseth, H. R.; Fulton, J. L.; Smith, R. D. Protein structural effects in gas phase ion/molecule reactions with diethylamine. *Rapid Commun. Mass Spectrom.* **1992**, *6*, 159–165.
98. Ogorzalek Loo, Rachel R.; Winger, Brian E.; Smith, Richard D. Proton transfer reaction studies of multiply charged proteins in a high mass-to-charge ratio quadrupole mass spectrometer. *J. Am. Soc. Mass Spectrom.* **1994**, *5*, 1064–1071.
99. Ogorzalek Loo, R. R.; Smith, R. D. Proton transfer reactions of multiply charged peptide and protein cations and anions. *J. Mass Spectrom.* **1995**, *30*, 339–347.
100. Cassady, C. J.; Wronka, J.; Kruppa, G. H.; Laukien, F. H. Deprotonation reactions of multiply protonated ubiquitin ions. *Rapid Commun. Mass Spectrom.* **1994**, *8*, 394–400.
101. Scalf, M.; Westphall, M. S.; Krause, J.; Kaufman, S. L.; Smith, L. M. Controlling charge states of large ions. *Science* **1999**, *283*, 194–197.
102. Scalf, M.; Westphall, M. S.; Smith, L. M. Charge reduction electrospray mass spectrometry. *Anal. Chem.* **2000**, *72*, 52–60.
103. Ebeling, D. D.; Westphall, M. S.; Scalf, M.; Smith, L. M. Corona discharge in charge reduction electrospray mass spectrometry. *Anal. Chem.* **2000**, *72*, 5158–5161.
104. Frey, B. L.; Lin, Y.; Westphall, M. S.; Smith, L. M. Controlling gas-phase reactions for efficient charge reduction electrospray mass spectrometry of intact proteins. *J. Am. Soc. Mass Spectrom.* **2005**, *16*, 1876–1887.
105. Smith, L. M. Is charge reduction in ESI really necessary? *J. Am. Soc. Mass Spectrom.* **2008**, *19*, 629–631.
106. Stephenson, J. L., JJr.; McLuckey, S. A. Ion/ion reactions in the gas phase: Proton transfer reactions involving multiply-charged proteins. *J. Am. Chem. Soc.* **1996**, *118*, 7390–7397.
107. McLuckey, S. A.; Reid, G. E.; Wells, J. M. Ion parking during ion/ion reactions in electrodynamic ion traps. *Anal. Chem.* **2002**, *74*, 336–346.
108. Grosshans, P. B.; Ostrander, C. M.; Walla, C. A.; Methods and apparatus to control charge neutralization reactions in ion traps. U. S. Patent 6,570,151 B1, May 27, **2003**.

109. GROSSHANS, P. B.; OSTRANDER, C. M.; WALLA, C. A.; Methods and apparatus to control charge neutralization reactions in ion traps. U. S. Patent 6,674,067 B2, January 6, **2004**.
110. CHRISMAN, P. A.; PITTERI, S. J.; MCLUCKEY, S. A. Parallel ion parking: Improving conversion of parents to first-generation products in electron transfer dissociation. *Anal. Chem.* **2005**, *77*, 3411–3414.
111. GOERINGER, D. E.; ASANO, K. G.; MCLUCKEY, S. A.; HOEKMAN, D.; STILLER, S. W. Filtered noise field signals for mass-selective accumulation of externally formed ions in a quadrupole ion trap. *Anal. Chem.* **1994**, *66*, 313–18.
112. CHRISMAN, P. A.; PITTERI, S. J.; MCLUCKEY, S. A. Parallel ion parking of protein mixtures. *Anal. Chem.* **2006**, *78*, 310–316.
113. FAVRE, A.; GONNET, F.; TABET, J. -C. Location of the negative charge(s) on the backbone of single-stranded deoxyribonucleic acid in the gas phase. *Eur. J. Mass Spectrom.* **2000**, *6*, 389–396.
114. KARAS, M.; BACHMAN, D.; BAHR, U.; HILLENKAMP, F. Matrix-assisted ultraviolet laser desorption of non-volatile compounds. *Int. J. Mass Spectrom. Ion Proc.* **1987**, *78*, 53–68.
115. TANAKA, K.; IDO, Y.; AKITA, S.; YOSHIDA, Y.; YOSHIDA, T.In Matsuda, H.; Xiao-tian, L. (Eds.), *Proceedings, Second Japan–China Joint Symposium on Mass Spectrometry, Osaka, Japan, 15–18 September 1987*, pp. 185–188.
116. KARAS, M.; HILLENKAMP, F. Laser desorption ionization of proteins with molecular masses exceeding 10.000 daltons. *Anal. Chem.* **1988**, *60*, 2299–2301.
117. TANAKA, K.; WAKI, H.; IDO, Y.; AKITA, S.; YOSHIDA, Y.; YOSHIDA, T. Protein and polymer analysis up to *m/z* 100.000 by laser ionization time-of-flight mass spectrometry. *Rapid Commun. Mass Spectrom.* **1988**, *2*, 151–153.
118. ZENOBI, R.; KNOCHENMUSS, R. Ion formation in MALDI mass spectrometry. *Mass Spectrom. Rev.* **1998**, *17*, 337–366.
119. KARAS, M.; GLUCKMANN, M.; SCHAFER, J. Ionization in matrix-assisted laser desorption/ionization: Singly charged molecular ions are the lucky survivors. *J. Mass Spectrom.* **2000**, *35*, 1–12.
120. KNOCHENMUSS, R. A Quantitative model of ultraviolet matrix-assisted laser desorption/ionization. *J. Mass Spectrom.* **2002**, *37*, 867–877.
121. KNOCHENMUSS, R.; ZENOBI, R. MALDI ionization: The role of in-plume processes. *Chem. Rev.* **2003**, *103*, 441–452.
122. DREISEWERD, K. The desorption process in MALDI. *Chem. Rev.* **2003**, *103*, 395–425.
123. KARAS, M.; KRUGER, R. Ion formation in MALDI: The cluster ionization mechanism. *Chem. Rev.* **2003**, *103*, 427–439.
124. BEAVIS, R. C.; CHAIT, B. T.; FALES, H. M. Cinnamic acid derivatives as matrices for ultraviolet laser desorption mass spectrometry of proteins. *Rapid Commun. Mass Spectrom.* **1989**, *3*, 432–435.
125. BEAVIS, R. C.; CHAIT, B. T.; STANDING, K. G. Matrix-assisted laser- desorption mass spectrometry using 355 NM radiation. *Rapid Commun. Mass Spectrom.* **1989**, *3*, 436–439.
126. BEAVIS, R. C.; CHAUDHARY, T.; CHAIT, B. T. **α**-Cyano-4-hydroxycinnamic acid as a matrix for matrix-assisted laser desorption mass spectrometry. *Org. Mass Spectrom.* **1992**, *27*, 156–158.
127. ZHOU, J.; LEE, T. D. Charge state distribution shifting of protein ions observed in matrix-assisted laser desorption ionization mass spectrometry. *J. Am. Soc. Mass Spectrom.* **1995**, *6*, 1183–1189.
128. JØRGENSEN, T. J. D.; BOJESEN, G. The proton affinities of seven matrix-assisted laser desorption/ionization matrices correlated with the formation of multiply charged ions. *Eur. J. Mass Spectrom.* **1998**, *4*, 39–45.
129. MIRZA, S. P.; RAJU, N. P.; VAIRAMANI, M. Estimation of the proton affinity values of fifteen matrix-assisted laser desorption/ionization matrices under electrospray ionization conditions using the kinetic method. *J. Am. Soc. Mass Spectrom.* **2004**, *15*, 431–435.
130. ALVES, S.; FOURNIER, F.; AFONSO, C.; WIND, F.; TABET, J. C. Gas-phase ionization/desolvation processes and their effect on protein charge state distribution under matrix-assisted laser desorption/ionization conditions. *Eur. J. Mass Spectrom.* **2006**, *12*, 369–383.
131. FRANKEVICH, V.; ZHANG, J.; DASHTIEV, M.; ZENOBI, R. Production and fragmentation of multiply charged ions in "electron-free" matrix-assisted laser desorption/ionization. *Rapid Commun. Mass Spectrom.* **2003**, *17*, 2343–2348.
132. KÖNIG, S.; KOLLAS, O.; DREISEWERD, K. Generation of highly charged peptide and protein ions by atmospheric pressure matrix-assisted infrared laser desorption/ionization ion trap mass spectrometry. *Anal. Chem.* **2007**, *79*, 5484–5488.
133. ZHONG, F.; ZHAO, S. A study of factors influencing the formation of singly charged oligomers and multiply charged monomers by matrix-assisted laser desorption/ionization mass spectrometry. *Rapid Commun. Mass Spectrom.* **1995**, *9*, 570–572.
134. KRUGER, R.; PFENNINGER, A.; FOURNIER, I.; GLUCKMANN, M.; KARAS, M. Analyte incorporation and ionization in matrix-assisted laser desorption/ionization visualized by pH indicator molecular probes. *Anal. Chem.* **2001**, *73*, 5812–5821.
135. BURTON, R. D.; WATSON, C. H.; EYLER, J. R.; LANG, G. L.; POWELL, D. H.; AVERY, M. Y. Proton affinities of eight matrices used for matrix-assisted laser desorption/ionization. *Rapid Commun. Mass Spectrom.* **1997**, *11*, 443–436.
136. FRANKEVICH, V.; KNOCHENMUSS, R.; ZENOBI, R. The origin of electrons in MALDI and their use for

sympathetic cooling of negative ions in FTICR. *Int. J. Mass Spectrom.* **2002**, *202*, 11–19.

137. Schafer, M.; Budzikiewicz, H. A fast atom bombardment and matrix-assisted laser desorption/ionization mass spectrometry study of doubly charged porphyrins. *J. Mass Spectrom.* **2001**, *36*, 1062–1068.
138. Clayton, E.; Wakefield, A. J. C. Fast atom bombardment (FAB) mass spectrometry: mechanism of ionization. *J. Chem. Soc., Chem. Commun.* **1984**, *15*, 969–970.
139. Cerny, R. L.; Gross, M. L. Abundances of molecular ion species desorbed by fast atom bombardment: Observation of $(M + 2H)^{+\bullet}$ and $(M + 3H)^{+}$. *Anal. Chem.* **1985**, *57*, 1160–1163.
140. Frankevich, V.; Zhang, J.; Friess, S. D.; Dashtiev, M.; Zenobi, R. Role of electrons in laser desorption/ionization mass spectrometry. *Anal. Chem.* **2003**, *75*, 6063–6067.
141. Overberg, A.; Karas, M.; Bahr, U.; Kaufmann, R.; Hillenkamp, F. Matrix-assisted infrared-laser (2 μm) desorption/ionization mass spectrometry of large biomolecules. *Rapid Commun. Mass Spectrom.* **1990**, *4*, 293–296.
142. Overberg, A.; Karas, M.; Kaufmann, R.; Hillenkamp, F. Matrix-assisted laser desorption of large biomolecules with a TEA-carbon dioxide laser. *Rapid Commun. Mass Spectrom.* **1991**, *5*, 128–131.
143. Niu, S.; Zhang, W.; Chait, B. T. Direct Comparison of infrared and ultraviolet wavelength matrix-assisted laser desorption/ionization mass spectrometry of proteins. *J. Am. Soc. Mass Spectrom.* **1998**, *9*, 1–7.
144. Kononikhin, A. S.; Nikolaev, E. N.; Frankevich, V.; Zenobi, R. Multiply charged ions in matrix assisted laser desorption/ionization generated from electrosprayed sample layers. *Eur. J. Mass Spectrom.* **2005**, *11*, 257–259.
145. Yao, J.; Scott, J. R.; Young, M. K.; Wilkins, C. L. Importance of matrix:analyte ratio for buffer tolerance using 2,5-dihydroxybenzoic acid as a matrix in matrix-assisted laser desorption/ionization–Fourier transform mass spectrometry and matrix-assisted laser desorption/ionization–Time of flight. *J. Am. Soc. Mass Spectrom.* **1998**, *9*, 805–813.
146. Cai, Y.; Peng, W. P.; Kuo, S. J.; Sabu, S.; Han, C. C.; Chang, H. C. Optimal detection and charge-state analysis of MALDI-generated particles with molecular masses larger than 5MDa. *Anal. Chem.* **2002**, *74*, 4434–4440.
147. Wenzel, R. J.; Matter, U.; Schultheis, L.; Zenobi, R. Analysis of megadalton ions using cryodetection MALDI time-of-flight mass spectrometry. *Anal. Chem.* **2005**, *77*, 4329–4337.
148. Shiea, J.; Huang, M. Z.; Hsu, H. J.; Lee, C. Y.; Yuan, C. H.; Beech, I.; Sunner, J. Electrospray-assisted laser desorption/ionization mass spectrometry for direct ambient analysis of solids. *Rapid Commun. Mass Spectrom.* **2005**, *19*, 3701–3704.
149. Sampson, J. S.; Hawkridge, A. M.; Muddiman, D. C. Generation and detection of multiply-charged peptides and proteins by matrix-assisted laser desorption electrospray ionization (MALD-ESI) Fourier transform ion cyclotron resonance mass spectrometry. *J. Am. Soc. Mass Spectrom.* **2006**, *17*, 1712–1716.
150. Knochenmuss, R.; Stortelder, A.; Breuker, K.; Zenobi, R. Secondary Ion–molecule reactions in matrix-assisted laser desorption/ionization. *J. Mass Spectrom.* **2000**, *35*, 1237–1245.
151. Breuker, K.; Knochenmuss, R.; Zhang, J.; Stortelder, A.; Zenobi, R. Thermodynamic control of final ion distributions in MALDI: In-plume proton transfer reactions. *Int. J. Mass Spectrom.* **2003**, *226*, 211–222.

Chapter 15

Probing Noncovalent Interactions by Electrospray Ionization and Matrix-Assisted Laser Desorption/ Ionization

Sonal Mathur,[*] *Alexis Nazabal,*[†] *and Renato Zenobi*[‡]

[*]*Max Planck Institute for Chemical Ecology, Jena, Germany*
[†]*CovalX AG, Zurich, Switzerland*
[‡]*Department of Chemistry and Applied Biosciences, ETH Zurich, Zurich, Switzerland*

Electrospray and MALDI Mass Spectrometry: Fundamentals, Instrumentation, Practicalities, and Biological Applications, Second Edition, Edited by Richard B. Cole

15.1 INTRODUCTION

Soft ionization methods in mass spectrometry (MS)—in particular, electrospray ionization (ESI) and matrix-assisted laser desorption/ionization (MALDI)—have had an enormous impact in the field of proteomics[1] and simultaneously spurred the industrial development of excellent instrumentation that is robust and designed to be operated by nonspecialists. ESI and MALDI also allow, under appropriate conditions, the observation of noncovalently bound complexes in the gas phase.[2,3] Recent research suggests that bio-macromolecules taking part in such noncovalent interactions are still in their folded conformation in the gas phase, close to their native conformation in solution. This surprising hypothesis has not yet been rigorously proven, but—if it holds true—rationalizes many of the features of soft ionization mass spectra recorded from noncovalent complexes.

In the first part of this chapter we will briefly discuss the nature of noncovalent interactions, followed by an overview of standard methods for probing them (optical and fluorescence spectroscopy, isothermal titration calorimetry/microcalorimetry, differential scanning calorimetry, and surface plasmon resonance), with particular attention to their application to biomolecules. An overview of mass spectrometric methods suitable for detecting noncovalent complexes will then be presented. The advantages of mass spectrometry compared with conventional analytical methods are sensitivity, speed, and the ability to obtain stoichiometric information directly.

In the second part of the chapter, the application of MALDI mass spectrometry for probing noncovalent interactions will be discussed. While special precautions have to be taken to avoid dissociation of complexes during MALDI sample preparation and during the desorption/ionization step, there are some extremely useful features of MALDI that make it suitable for fast readout in solvent accessibility studies (via HDX) and even for the determination of noncovalent binding constants via the so-called SUPREX method.

In the third part of this chapter, the wealth of possibilities offered by electrospray ionization mass spectrometry for detecting noncovalent interactions and measuring their binding strengths is delineated. In the case of ESI, special conditions are required, too, for maintaining noncovalent complexes through the spray process—in particular, "native-like" aqueous buffers. The high surface tension of water renders the formation of a stable ESI spray more difficult; nevertheless, excellent progress has been made in recent years to overcome this difficulty.

The fourth part of this chapter will present a critical discussion of the use of mass spectrometry to determine noncovalent binding constants. The chapter is concluded with a summary and a list of references.

15.1.1 Nature of Noncovalent Interactions

Noncovalent interactions occurring between molecules that are charged, have a dipole moment, or are polarizable come in many different flavors. If both molecules carry an opposite charge, then the electrostatic interaction between them will be attractive and follow Coulomb's law. The Coulomb energy E_c depends on the charges (Q_1 and Q_2), the distance r between them, and the dielectric constant ε of the medium.

$$E_c = \frac{Q_1 Q_2}{4\pi\varepsilon\varepsilon_0 r} \tag{15.1}$$

A negative energy is equivalent to an attractive interaction. Coulombic interactions can of course be repulsive too, for the case of two molecular species carrying the charges of the same sign [Eq. (15.1)]. All other types of interactions are attractive because permanent or induced dipoles can orient themselves to accommodate the forces acting on them. As an example, let's consider the interaction energy between freely rotating dipoles with dipole moments u_1 and u_2, which is sometimes referred to as the Keesom energy E_K [Eq. (15.2)]:

$$E_K = -\frac{u_1^2 u_2^2}{3(4\pi\varepsilon\varepsilon_0)^2 kTr^6} \tag{15.2}$$

Interestingly, the dipole–nonpolar interaction energy (called Debye energy) for a freely rotating dipole takes almost exactly the same form, with the polarizability replacing one of the terms in the numerator. Both the Keesom energy and the Debye energy are important attractive forces for molecules in very close proximity, and they are called van der Waals interactions. Van der Waals energies exhibit the typical $1/r^6$ distance dependence. However, the dependence on the relative dielectric constant ε is of much greater importance in the realm of mass spectrometry. Consider, for example, going from an aqueous environment ($\varepsilon = 80$) to vacuum ($\varepsilon = 1$): Coulomb interaction energies will *increase* by a factor of 80, and the van der Waals interactions even by a factor of 6400! In other words, once a noncovalent complex has survived the critical stages of ion formation and desolvation, it is *less likely* to dissociate in the vacuum environment of the mass spectrometer compared to solution. At first glance, this prediction may be somewhat counterintuitive for novices in this field. Also, it does not hold for all types of noncovalent interactions. The notable exception is a hydrophobic interaction, which in solution is driven by the entropy gain when structurally organized solvent (usually water) molecules are liberated from a binding interface upon formation of a complex. In vacuum, the solvent is gone, and the only entropy gain is provided by the dissociation of the complex itself; that is, hydrophobic complexes are *more likely* to dissociate in the vacuum compared to the solution phase. These considerations also explain why MS experiments carried out in the vacuum to measure binding energies cannot give the same values as those found in solution.

Neither hydrophobic interactions nor hydrogen bonds can be expressed by a simple formula. Hydrophobic interactions depend critically on the size and the structure of the contact surfaces. Hydrogen bonding is a specific noncovalent interaction of primarily electrostatic nature and can vary in strength from very weak (1–2 kJ mol^{-1}) to strong (40 kJ mol^{-1}). Hydrogen bonds and hydrophobic interactions are prevalent in the folding of proteins as well as in the formation of noncovalent complexes.

15.1.2 Standard Methods for Studying Macromolecular Interactions

Before discussing MS-based methods for probing noncovalent interactions, we give an overview of conventional techniques and discuss their capabilities and limitations, in order to put the mass spectrometric methods into some perspective.

15.1.2.1 Optical Spectroscopy

Important contributions to the field of macromolecular interactions were made using spectroscopic approaches including scattering, fluorescence, and circular dichroism (CD).[4] Fluorescence is one of the most widespread optical methods for measuring protein–protein interactions because fluorophores are highly sensitive to their surroundings, and therefore the changes associated with binding can be effectively monitored.[5,6] New fluorescence-based techniques such as fluorescence resonance energy transfer (FRET)[7] and fluorescence correlation spectroscopy (FCS)[8] have been introduced for studying biomolecular interactions both *in vitro* and *in vivo*. An excellent recent study, aimed at the characterization of the FKBP (FK506 binding protein)–rapamycin–FRB (FKBP-rapamycin binding domain) ternary complex, was presented by Banaszynski et al.[9] using fluorescence polarization method. In this work, the authors report modest affinity ($K_d = 26 \pm 0.8\ \mu M$) for the rapamycin–FRB complex in the absence of FKBP, and a dramatic increase in the affinity (12 ± 0.8 nM) in the presence of FKBP demonstrating that the protein–protein interactions play a major role at the interface for the stabilization of the ternary complex. These methods are useful, but they are quite system-dependent and always require labeling with a fluorophore; controls are necessary to verify that this chemical modification does not affect the noncovalent binding interaction.

Optically active compounds show differential absorption of left- and right-circularly polarized light leading to a well-known phenomenon called circular dichroism (CD) that has also been used to study protein–ligand interactions. Dissociation constants can be determined in the cases where binding leads to some secondary structural changes, for example, ligand-induced changes in protein secondary (in far-UV region) or tertiary (in near-UV region) structures (for a review, see Ref. 10). A classical example is the determination of equilibrium dissociation constants of class I MHC–peptide complexes by measuring changes in melting temperature as recorded by a CD spectrometer.[11] The role of CD in studying noncovalent interactions is diminishing due to the following limitations: (a) Chirality or asymmetry in the molecule, or in its environment, is a prerequisite in order to get a CD signal; (b) commonly observed baseline artifacts are present in the CD spectrum; and (c) high sample concentrations are required. In contrast, mass spectrometric methods are applicable to all systems and provide reliable data at very low analyte concentrations. Affinities (K_ds) that can be studied using spectroscopic methods are in the range of 10^{-6}–10^{-11} M.[4]

15.1.2.2 Isothermal Titration Calorimetry

Isothermal titration calorimetry (ITC), a quantitative method for the determination of thermodynamic properties of biomolecular interactions, measures directly the heat evolved in a binding reaction (association of a ligand with its binding partner) at constant temperature. In a typical ITC protocol, one binding partner is titrated against its interaction partner in the sample cell of the calorimeter followed by the measurement of heat generated or absorbed during binding. The final outcome of these experiments yields information about association constant (K_a), stoichiometry (n), enthalpy of binding (ΔH_b), Gibbs's free energy (ΔG), entropy (ΔS), and the heat capacity (ΔC_p) if experiments are performed over a range of temperatures. The applications of ITC for studying biomolecular interactions have been the subject of many reviews in the recent past.[12,13] Velazquez-Campoy et al.[14] demonstrated the use of ITC for the determination of HIV-1 protease inhibitors. The authors have used a low-affinity inhibitor with an opposite sign of binding enthalpy in competition experiments with the high-affinity target compound that helped in amplifying the calorimetric signals in the displacement titration. Mathematical calculations involved in this study are described by Wang[15] and Sigurskjold.[16] In a more recent study, Keeble et al.[17] have identified enthalpically favorable lead compounds for HIV-1 protease using microcalorimetry. ITC is the only technique that can provide individual contributions of entropy and enthalpy to binding, which is essential for designing high-affinity ligands. ITC is a very simple method and provides abundant information on the thermodynamic parameters of the system in use; but because the observed heat change (ΔH_{app}) is a global property, it is often difficult to distinguish among the factors contributing to ΔH_{app}. Dissociation constants in the range of 10^{-6}–10^{-11} M can be measured using ITC.[4] The sensitivity of this method is frequently compromised due to problems such as baseline instability, and it requires large amounts of sample for analysis. MS methods are much more promising for dealing with issues like sensitivity and sample consumption.

15.1.2.3 Differential Scanning Calorimetry

Differential scanning calorimetry (DSC) is a technique in which the heat capacity change of a solution containing the macromolecule is monitored as a function of temperature, relative to a reference solution in order to study the conformational transitions of biomolecules—for example, folding and unfolding of proteins or single- or double-stranded DNA. Theoretically, the changes in transition entropy (ΔS) and free energy (ΔG) can also be observed from DSC measurements. The theory and the applications of DSC have been discussed extensively.[18–20] Recently, Michnik et al.[21] have presented an elegant piece of work to study kinetics of thermal denaturation of haemoglobin using DSC. A new variant of DSC, pressure perturbation calorimetry (PPC), which uses small pressure pulses in sample and reference cells to measure volumetric changes associated with protein unfolding as well as ligand binding, has recently been introduced by Cooper et al.[22] Interestingly, a combination of ITC and DSC can be used for characterizing the energy profile of the interacting systems to understand the energetics of structural changes in biomolecules and their specific complexes with binding partners.[23,24]

DSC delivers more precise information than ITC because it measures the partial heat capacity that gives information about the conformational state of a protein within the temperature range of the experiment. However, these advantages come at the expense of (a) limited throughput of conventional DSC instruments (5–10 measurements per day), (b) limited sensitivity (milligram amounts of purified sample required per experiment), and (c) nonspecific aggregation of macromolecules in the calorimeter cell during DSC measurements at higher concentrations. In short, utmost care is needed here while operating the instrument to get reliable data, which limits its application to some extent when compared to methods based on mass spectrometry. DSC is suitable only for tight binding complexes exhibiting affinities from 10^{-9} to 10^{-20} M.[4]

15.1.2.4 Surface Plasmon Resonance

Surface plasmon resonance (SPR) is a well-established technique for the determination as well as quantification of biomolecular interactions. A surface plasmon is a surface electromagnetic wave propagating along the interface of a metal and a dielectric. In SPR, a monochromatic *p*-polarized light source is employed and the interface between two optically dense media is coated with a thin metal film. Under conditions of total internal reflection, a component of incident light (the evanescent wave) still penetrates beyond the reflecting surface in the medium of lower refractive index and interacts with the free oscillating electrons (plasmons) in the metal film as shown in Figure 15.1A. At a specific angle of incidence, the momentum of incoming photons matches the momentum of surface plasmons, resulting in energy transfer and a reduction in the intensity of reflected light that can be detected by a photodiode or charge-coupled detectors (CCD). The resonance conditions are influenced by the chemical composition of the solution in contact with the thin metal film leading to changes in resonance angle. Changes in the resonance signal over time are displayed in a plot called a sensorgram (Figure 15.1B). The rate of change of the SPR signal is utilized to yield apparent rate constants for the association and dissociation phases of the reaction.

SPR measurements are usually performed on sensor chips, which are signal transducers providing sites for interaction and converting the mass change on the surface to an SPR signal. They are composed of a glass chip carrying a thin (50 nm) gold film on one side, and a dextran matrix (thickness 100 nm) is covalently attached to it through a linker layer on the

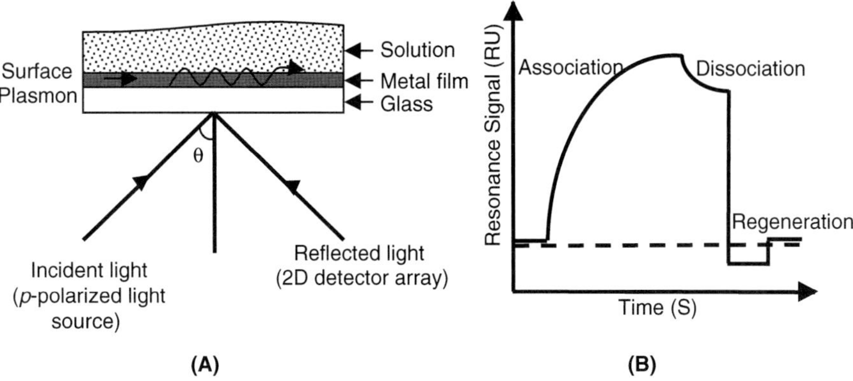

Figure 15.1. (**A**) Schematic of surface plasmon resonance based on Kretchmann configuration used in most of the commercial SPR instruments. (**B**) Typical sensorgram showing the binding of analyte with its immobilized ligand on the chip surface (association phase), separation of the two binding partners (dissociation phase), and, finally, regeneration of the surface.

gold film. The target molecule (called the ligand) is immobilized on the dextran layer, and its binding partner (analyte) is flown over the chip surface. The outstanding performance of SPR in the arena of biomolecular interactions has been the subject of a number of review articles.[25–27]

Biomolecular interaction analysis–mass spectrometry (BIA-MS), a combination of SPR with MS, was first explored by Krone et al. in 1996 where SPR was used to quantify interactions between proteins and MS was used to determine the structural features of the bound proteins. Review articles on SPR-MS describe recent progress in the field of SPR-MS, including future trends and possible applications.[26,28]

SPR provides real-time analysis of biomolecular interactions and can be efficiently applied to study bioaffinity interactions in a label-free manner even for complex protein mixtures. On the downside, it is relatively slow and in some cases immobilization processes interfere with the binding site of the biomolecule, thereby decreasing its activity. The affinity of systems that do not have a 1 : 1 stoichiometry is often difficult to obtain due to steric hindrance between partners during binding. MS methods are good vehicles for determining different stoichiometries, and they do not interfere with the binding sites of the interacting partners. A typical range of affinities determined by SPR is 10^{-5}–10^{-12} M.

15.1.3 MS Techniques for Detecting Noncovalent Complexes

We will next turn to the question how to detect noncovalent complexes by soft ionization mass spectrometry; the determination of the binding constant will be dealt with in later sections. The two most widely used soft ionization methods are certainly MALDI and ESI. Several comprehensive review articles, by our group,[29] Schalley,[30] Hofstadler and Griffey,[31] Hardouin and Lange,[32] Heck and van der Heuvel,[33] and Schermann et al.,[34] give good overviews of the application of these methods to the study of noncovalent interactions. MALDI is generally known to be somewhat less soft than ESI. It is in fact more difficult to preserve noncovalent complexes throughout the MALDI sample preparation and desorption/ionization steps. Sample preparation methods that avoid organic solvents and strongly acidic matrices are advantageous, as are moderate laser pulse energies. A compromise often has to be found, because these conditions are not optimal for detecting high-molecular-weight complexes. MALDI has thus not been employed very frequently to study noncovalent interactions. Zehl and Allmaier[35,36] have reviewed sample preparation methods and instrumental parameters for optimum detection of noncovalent peptide–peptide complexes by MALDI MS. One general problem is that MALDI is known to produce predominantly singly charged ions. High-mass complexes, (e.g., large protein assemblies) thus become very difficult to analyze and detect using MALDI and conventional mass spectrometers and detectors.

Older soft ionization methods that are related to MALDI—such as fast atom bombardment (FAB), liquid secondary ion mass spectrometry (LSIMS), or field desorption (FD)—in principle have potential for the generation of intact gas-phase ions from noncovalent complexes. In particular, the extensive clustering often found in FAB-MS, which is generally thought of as a nuisance, can be viewed as evidence for this. However, the limitation of these methods lies in their inability to ionize very large molecules. It is generally very difficult or impossible to obtain useful mass spectra from compounds with molecular weights above a few thousand daltons. For this reason, FAB-MS, LSIMS, and FD-MS only play a very minor role in this field.

The lion's share of MS-based research on noncovalent complexes uses electrospray ionization, due to a number of reasons. One is the limited mass range of many mass spectrometers. Since ESI produces multiply charged ions, it is often possible to use mass analyzers with limited m/z ranges such as quadrupole instruments, even for fairly high molecular weight complexes. However, as will be discussed below, it is found that for native-like conformations of proteins in the gas phase, the number of charges they carry is limited. In particular for very large multiprotein molecular assemblies, specialized equipment with an extended m/z range must be used even with ESI as an ionization source. Another reason for the predominance of ESI MS is that it is the softest of the modern "soft ionization" methods. However, in the case of noncovalent complexes, special precautions have to be taken even when using ESI: aqueous sample solutions buffered to pH values near the physiological range, as well as gentle desolvation and ion transfer conditions, are important for the successful detection of noncovalent complexes.

15.1.4 New Spray Methods on the Horizon

In recent years, several variants of ESI have appeared. Many of them are very attractive for probing noncovalent interactions. Yamaguchi and co-workers[37–40] have presented a new method, namely, cold-spray ionization mass spectrometry. Using a modified electrospray source operating at low temperature, the solution-phase structures of biomolecules, labile organic species, asymmetric catalysts, and supramolecular assemblies have been investigated. Among other impressive examples, the Japanese group presented mass spectra of intact cage-type platinum complexes, double-interlocking-type copper–palladium complexes, Grignard reagents, self-assembled porphyrin oligomers, and hyperstranded DNA architecture.

Nano-ESI is a variant of ESI that uses smaller spray capillaries (and hence smaller flows, often below 1 μL/min, thus the term "nano"), and it produces smaller initial droplets. This is beneficial for the investigation of noncovalent complexes. In particular for quantitative studies, the situation where less than one molecule is contained in a droplet, and thus the point where the spray and desolvation processes cannot distort an equilibrium any more, will be reached much faster for nano-ESI than for ESI. A variant of nano-ESI is a fully automated, chip-based nanoelectrospray system that has recently become available commercially. Only

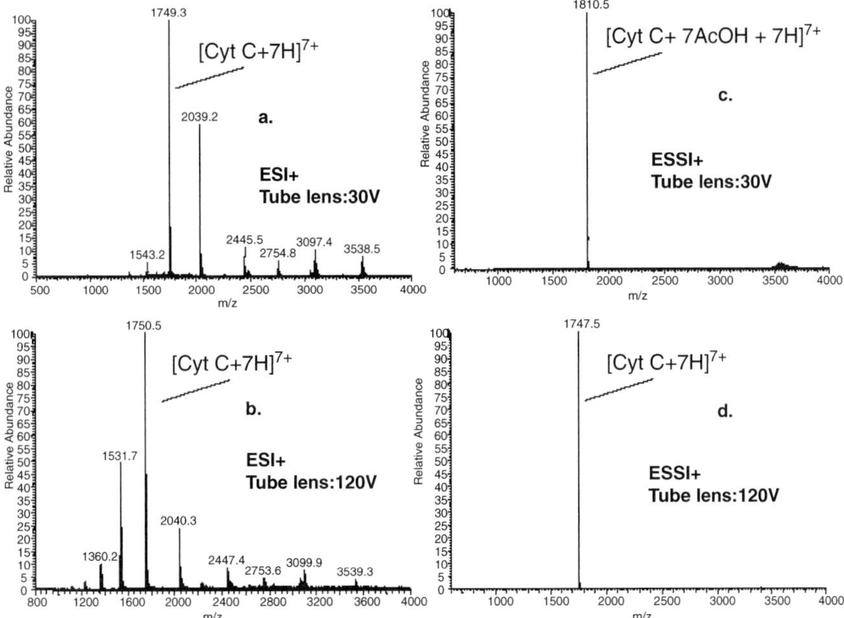

Figure 15.2. Spectra of bovine cytochrome c, 0.01 mg/mL in 10 mM aqueous ammonium acetate, taken under different conditions. Panels **a** and **b** show ESI spectra, while panels **c** and **d** show ESSI spectra of the same sample. The most striking difference is that using ESSI, the ion current is almost exclusively concentrated in a single charge state. (Reproduced from Ref. 43 with permission. Copyright © 2004 American Chemical Society.)

very few studies using this setup for the investigation of noncovalent interactions have appeared in the literature.[41,42] In a study published in 2003, Zhang et al.[42] presented measurements of the dissociation constants of complexes between proteins with oligosaccharide and cytidylic acid ligands. For one complex, K_d values determined with nano-ESI titration were compared with K_d determinations from the literature using conventional methods such as calorimetry and circular dichroism. K_d values in the low-micromolar range were obtained by all methods. The nano-ESI data was between 25% and a factor of four higher than the other values, which is, however, considered to be reasonable agreement.

A significant methodological contribution to this subject area was the introduction of a new variant on ESI, called electrosonic spray ionization (ESSI) by Cooks and co-workers (Figure 15.2). ESSI employs a traditional micro-ESI source with a supersonic nebulizing gas.[43–45] The use of nebulizing gas has been shown to provide efficient pneumatic spraying of the charged liquid and rapid breakup of larger into smaller droplets, while keeping noncovalent complexes intact. ESSI results in very narrow charge distributions, believed to represent near-native conformations of the biomolecules studied.

15.2 DETECTING NONCOVALENT COMPLEXES AND DETERMINING THEIR K_d BY MALDI MS

15.2.1 Introduction

The key issue when studying noncovalent complexes by MALDI ionization is to distinguish between specific complexes and (nonspecific) clusters formed during the ionization step. Nonspecific clusters of proteins can be generated in the plume, depending on the laser energy used or the concentration of the protein in the sample. In other words, what are the criteria that allow us to distinguish a nonspecific cluster from a specific complex? Several criteria have been proposed: (a) By far the most powerful criterion is a control experiment, in which one or several functionalities that are important for the formation of the complex are chemically modified. For example, the active site of an enzyme can be covalently blocked, thereby preventing binding of ligands or inhibitors.[46] If the putative complex signal disappears from the MALDI mass spectrum upon chemical modification, this is strong evidence that it was indeed due to a specific complex. (b) While nonspecific clusters exhibit an exponentially decaying intensity distribution, specific complexes are often found to "stick out" of this distribution, giving rise to much more intense signals. (c) Sometimes, the best strategy for observation of specific noncovalent complexes is to collect data from the first laser shot fired at a fresh sample spot only; this is called the "first shot phenomenon."[3,47] (d) The problem can be circumvented in the case of the analysis of protein complexes by combined cross-linking and high-mass MALDI by comparing the cross-linked spectra with a non-cross-linked control spectrum.[48]

Another challenge to study noncovalent complexes by MALDI is the dissociation that often occurs during sample preparation or during the desorption/ionization step. To circumvent this problem, special sample preparation methods have been developed using modified solvent or matrices,[36,49] combined with protocols such as first shot analysis.[50] Cross-linking chemistry is also used for the study of protein complexes to get rid of dissociation problems by linking covalently the intact protein complexes before ionization. Until now, most studies using MALDI ionization and cross-linking chemistry employ a proteolysis step before analysis, making use of the high accuracy of MALDI-MS to identify the covalently tagged peptides.

MALDI mass spectrometry has also been used for the study of protein complexes by hydrogen deuterium exchange (HDX). The main advantage of using MALDI is the high accuracy with which deuterium incorporation can be determined, along with the relative simplicity of the spectra obtained after protolysis of the protein complex analyzed. Because most of the species formed by MALDI are singly charged pseudomolecular ions, the spectra obtained are relatively simple, and a fractionation step is not required before analysis. Another advantage of using MALDI for hydrogen exchange experiments is the speed of the analysis, which is important to maintain deuterium within the protein during analysis.

15.2.2 MALDI Approaches for Detecting Noncovalent Complexes

15.2.2.1 Nonacidic Matrices for Maintaining Native Complexes

Initially, most or even all known MALDI matrices were acidic compounds, such as benzoic and cinnamic acid derivatives. However, the acidic medium is not well-suited to keeping noncovalently bound complexes intact. In the 1990s, a couple of groups therefore introduced a number of nonacidic MALDI matrices such as 2-amino-4-methyl-5-nitropyridine, *para*-nitroaniline, or 6-aza-2-thio-thymine.[51,52] These were shown to be beneficial for MALDI MS of noncovalent complexes—for example, to keep peptide–oligonucleotide complexes or DNA duplexes intact.[53,54] However, their performance at high molecular weight is not as good as those of other matrices—for example, sinapinic acid. For the detection of high-molecular-weight protein complexes by MALDI-MS, conventional matrices in aqueous solution are therefore often preferred, especially if very harsh conditions (organic solvents, acidification) during sample preparation can be avoided and if the sample crystallization time can be kept short enough that unfolding of a protein component of the complex is unlikely.

15.2.2.2 Stabilization by Cross-Linking

Cross-linkers are widely used in biochemistry to probe protein–protein interactions. When two or more proteins are involved in complexes or in transient interactions, cross-linking reagents can covalently bind together the complex formed by these proteins to allow further analysis. Chemical cross-linking can be used in conjunction with MALDI mass spectrometry to characterize protein interactions and complexes with high resolution and sensitivity. Compared with the analysis of protein complexes by electrospray ionization, the use of chemical cross-linking and MALDI mass spectrometry presents several advantages: (a) As chemical cross-linking reactions take place in solution, and cross-linking changes the mass, this approach avoids artifacts that can result from gas-phase processes (e.g., clustering). (b) There is tolerance for sample impurities and buffer conditions. (c) Most of the molecules are detected as singly charged ions. (d) As the cross-linking reactions generate covalent complexes, the sample can be analyzed without special instrumental settings.

Mass spectrometrists can use chemical cross-linking to characterize protein interactions and complexes at two different levels: (a) It can be used to determine complex stoichiometry by directly analyzing the intact cross-linked protein complex. (b) By analyzing protein complexes after proteolysis, structural information from the covalently tagged peptides can be obtained. The direct analysis of protein complexes using cross-linking chemistry and MALDI time-of-flight (TOF) mass spectrometry is a useful method to analyze stoichiometry and to characterize subunits belonging to a complex. This direct analysis, without proteolysis or fragmentation, was initiated in the early 1990s with the analysis of multimers in the 60- to 150-kDa mass range from purified samples.[48] In this study, the dimer and tetramer of avidin stabilized by glutaraldehyde were analyzed with a MALDI TOF mass spectrometer. The main hurdle to develop this method further was the limit in detection of high-mass ions by microchannel plate (MCP) detectors and the relatively poor specificity of the cross-linking protocol used.

Because most protein complexes are high-molecular-weight molecules, the analysis of intact protein complexes by MALDI MS requires detection in the high-mass range. Recently, this detection issue has been addressed by the development of new types of technology such as the superconducting tunnel junction (STJ), significantly improving detection sensitivity of high-molecular-weight ions in the MDa range.[55] Following this development, STJ detection and specific cross-linking protocols have been used to analyze intact immunocomplexes (Figure 15.3) and to develop immunochemistry applications by mass spectrometry.[56] In another study, an ion conversion dynode (ICD) detector, also allowing the detection of ions up to 1 MDa, has been used to analyze intact hetero-oligomeric protein complexes by MALDI mass spectrometry.[57] A great advantage of the ICD technology is that it is completely compatible with commercial TOF instrumentation, which in principle would allow retrofitting or upgrading of existing equipment in the field.

Chemical cross-linking can also be used to map the interactions between subunits within a protein complex. In this approach, a protein complex is subjected to

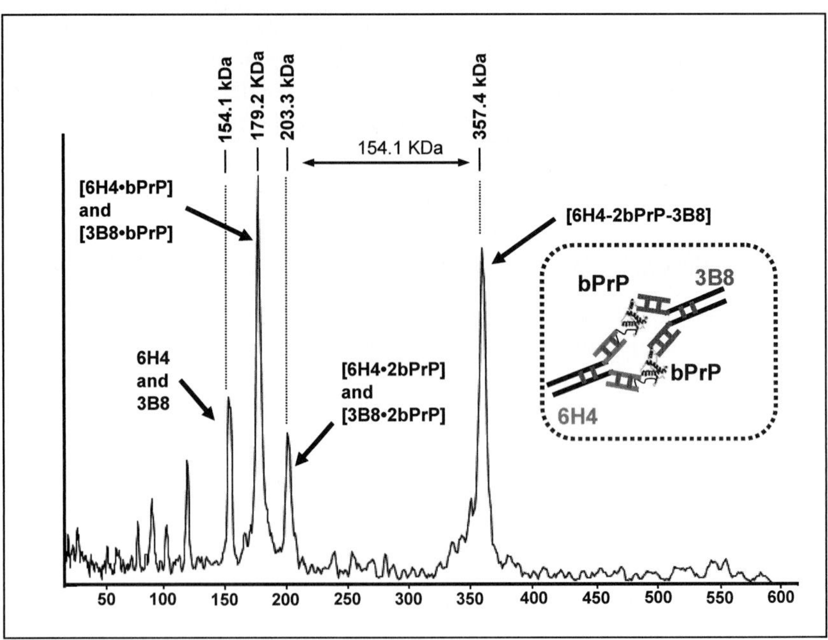

Figure 15.3. Sandwich immunocomplex analyzed by high-mass MALDI mass spectrometry: A complex formed by two monoclonal antibodies (6H4 and 3B8) against the bovine prion protein (bPrP) has been cross-linked and analyzed by high-mass MALDI TOF mass spectrometry. The specific protein complex [6H4·bPrP·bPrP·3B8] is detected at *m/z* 357.4 kDa.

cross-linking and enzymatic proteolysis prior to high-resolution mass spectrometric analysis.[58] The positions of the cross-links give information about distance constraints between the side chains of the amino acids involved in the interactions. These distance constraints allow one to deduce structural information concerning protein complex structures.[59] This method presents several advantages to map protein–protein interactions at the molecular level. Only a few picomoles of protein complexes are necessary to perform the analysis. The analysis is generally fast and, because of the proteolysis, the upper masses of protein complexes that can be analyzed are theoretically unlimited. A drawback of such analysis is the complexity of the peptide mixture generated by the proteolysis of the cross-linked complexes. Different strategies have been used to circumvent the problem of complexity. Affinity chromatography and tri-functional cross-linkers have been used to enrich cross-linked peptides. Isotopically labeled cross-linkers have also been used to facilitate the detection of cross-linked peptides by high-resolution mass spectrometry. The challenge of building up structural models for protein complexes from the mass spectrometric data generated has been address by improving the bioinformatics tools employed.[60]

The use of cross-linkers for the study of noncovalent interactions by mass spectrometry helps to circumvent one of the major issues of the analysis: dissociation of the interactions during analysis. Even if the use of chemical cross-linkers and mass spectrometry is a rapidly growing area, it is still not yet a generally applicable technique for characterizing protein complexes. The synthesis of new cross-linkers and the improvement of enrichment strategies for the cross-linked species are needed to further develop this promising technique.

15.2.2.3 Hydrogen/Deuterium Exchange

Hydrogen/deuterium exchange analyzed by mass spectrometry (HDX-MS) is a valuable tool for the investigation of protein conformation and dynamics. This technique consists of incubating a protein sample in a solvent containing deuterium oxide. During the incubation, the hydrogen atoms located on the backbone amides of the proteins exchange with the deuterium atoms present in the solvent.

The exchange of amide hydrogen atoms (protons) with deuterons in the backbone of a protein depends on solvent accessibility, hydrogen bonding, and the spatial arrangement of amino acids in structural conformations such as helices and ß-sheets. Thus, the solvent accessibility of a protein under different states, such as a soluble form and an aggregated form, may be monitored by HDX-MS (a typical protocol is schematically shown in Figure 15.4). For the study of noncovalent complexes, HDX is very useful because the binding interfaces of the interaction partners are usually much less solvent accessible and thus protected against HDX. An overall lower HDX rate will thus be observed for interaction partners involved in noncovalent complex formation compared to that of the isolated species. This information can be refined when an enzymatic cleavage is performed, in order to reveal the distribution of deuterons within the sequence of a protein having a known primary structure. However, reliable information can only be obtained when proteolysis occurs under conditions of quenched exchange—that is, at low temperature and pH. Indeed, at 0°C and pH 2.5, the HDX rate is slower by about five to six orders of magnitude than at

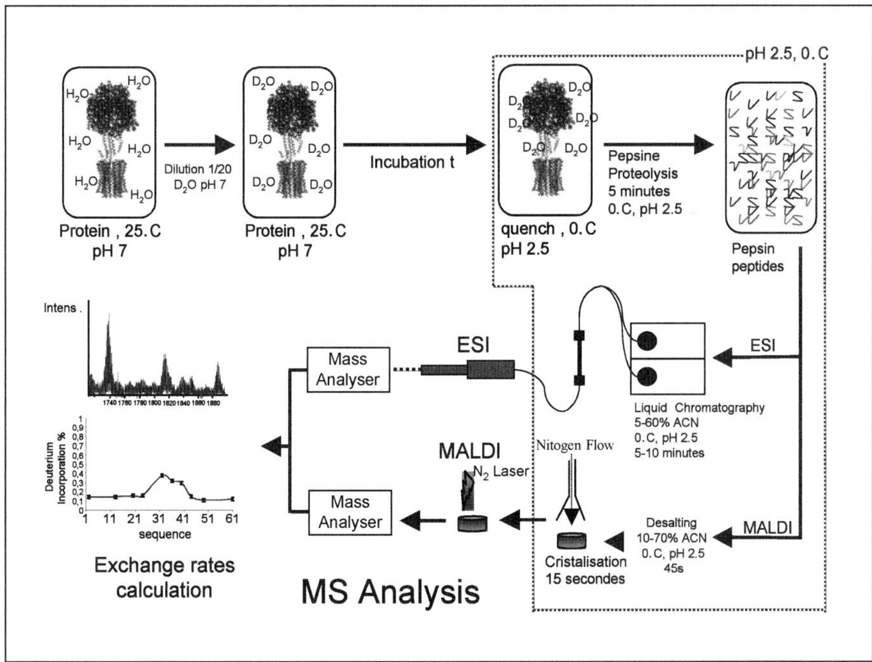

Figure 15.4. Principle of HDX-MS analysis: The protein complex is diluted in deuterium oxide to start the exchange reaction. After the incubation time at room temperature, the exchange reaction is quenched by lowering both pH and temperature. To study the topology of deuterium incorporation, the protein complex is digested with pepsin under quenched conditions. The peptides generated can be analyzed using ESI or MALDI ionization. During the mass spectrometric analysis the number of deuterium atoms incorporated is determined to calculate the exchange rate along the sequence of the protein (See color insert).

room temperature and pH 7.[61] In these conditions, the protease of choice is most often pepsin, in spite of its broad specificity.

HDX-MS studies are often conducted by ESI. In that case, peptide mapping is achieved by means of on-line coupling with a liquid chromatograph providing the required separation of peptides. Since both sample preparation and separation steps must be conducted under conditions of quenched exchange, experimental constraints on the chromatographic setup are rigorous. Furthermore, corrections for H/D back exchange must be carefully applied. Alternatively, MALDI can be used instead of ESI for HDX-MS analysis, as demonstrated for the first time by Mandell et al.[62] The matrix concentration (about 10 mM in aqueous acid solution) is usually 10,000 times higher than the sample concentration, and the sample-matrix mixing process leads to an efficient quenching of the HDX reaction, provided that the solvent evaporation is fast enough. Furthermore, specific characteristics of MALDI such as speed of spectral acquisition and sensitivity render this method well-suited for HDX-MS analysis. In the case of HDX analyzed by MALDI-MS, we observed back exchange levels in the 12–16% range. This is similar to the values that we and other authors have observed in the case of HDX analyzed by ESI-MS.[63]

One of the major differences between MALDI- and ESI-HDX-MS is the nature of the ions generated by these ionization methods. After MALDI ionization, most of the ions observed for peptides produced by proteolytic cleavage are singly charged, which strongly reduces the number of peaks observed in mass spectra. In ESI, however, multiply charged ions are generated, leading to a higher complexity of mass spectra observed for the same peptides. Most of the time, a chromatographic separation is required before the ionization step to reduce overlaps of peaks of multiply charged ions. This overlapping phenomenon of isotopic clusters increases after HDX and can lead to confusion in peak assignments. The use of MALDI ionization reduces the overlapping phenomenon and thus greatly improves the quality of information obtained by HDX-MS experiments.

Limitations of MALDI-HDX-MS are mostly related to selective desorption–ionization effects encountered for complex peptide mixtures. Although these selective desorption–ionization effects depend on several factors, a general trend is the dominance of arginine-containing peptides in MALDI mass spectra. As a result of this phenomenon, complete sequence coverage of a protein of interest might be difficult to obtain by this methodology alone. Microfractionation by step gradient elution of peptides from pipette tips packed with reversed-phase chromatographic support is an effective way to overcome this problem.[64]

15.2.2.4 Laser-Induced Liquid Beam Ionization/Desorption (LILBID)

A method related to MALDI, called laser-induced liquid beam ionization/desorption (LILBID),[65] has also shown promise for the detection of high-mass biomolecules, their clusters, and specific noncovalent complexes.[66–68] In this method, a laser beam desorbs species directly from a liquid jet of solution sprayed into a specially designed ion source. LILBID requires custom-designed instruments and is, until now, in use only in a small number of laboratories.[69,70]

15.2.3 Approaches for Determining Binding Affinities by MALDI-MS

Recently, a method called SUPREX (stability of unpurified proteins from rates of H/D exchange) was introduced by the Fitzgerald group[71] for the quantitative measurement of

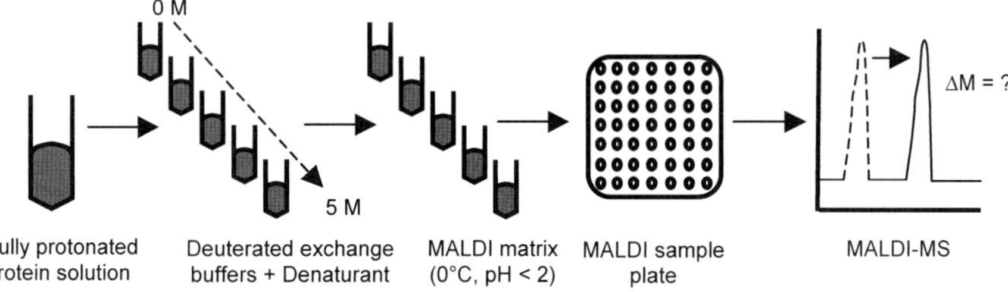

Figure 15.5. A typical SUPREX protocol. (Adapted from Ref. 78).

stability of unpurified proteins in complex biological matrices. The SUPREX technique uses HDX and relies on MALDI mass spectrometry, an approach previously described by Komives and co-workers.[62] In a polypeptide chain, some hydrogen atoms can exchange freely with the surrounding solvent; however, some of these hydrogen atoms are not solvent-accessible under native conditions.[72] Some of these protected protons can exchange only if the protein globally unfolds[73] and that is how the stability of a protein can be analyzed by monitoring the exchange rates of these "globally protected" hydrogens[74] using MS.[75] The pioneering experiments of the Fitzgerald group in this field led to some excellent articles describing the use of SUPREX to measure the thermodynamic stability of proteins both *in vitro* and *in vivo* with good accuracy and high precision.[76–78] More recently, SUPREX was employed to investigate the strength of protein–ligand binding interactions in solution.[79]

The SUPREX protocol (Figure 15.5) for the determination of K_d values includes initiation of HDX by adding 10-fold excess of deuterated exchange buffer to the sample containing the target protein usually at physiological pH (referred to as pD = 7.4). The exchange buffers contain varying concentrations of a chemical denaturant—for example, guanidinium chloride (GdmCl) or urea. The denaturant leads to unfolding of the protein and enhancement in HDX rate, thus increasing the rate of deuterium incorporation into the protein. At a given exchange time, a small aliquot of the exchange reaction is added to matrix (sinapinic acid) solution in a 10-fold excess and the sample is subjected to MALDI analysis. The change in mass relative to the fully protonated sample (Δmass) in the spectra is plotted as a function of denaturant concentration, and the data are fitted to a sigmoidal function to obtain the transition midpoint, the $C^{1/2}_{\text{SUPREX}}$. Solution-phase folding free energies (ΔG^0_f) in the presence and the absence of ligand can be calculated using Eq. (15.3):

$$-\Delta G^0_f = mC^{1/2}_{\text{SUPREX}} + RT\ln\frac{\left(\frac{\langle k_{\text{int}}\rangle t}{0.693} - 1\right)}{\left(\frac{n^n}{2^{n-1}}[P]^{n-1}\right)} \tag{15.3}$$

In Eq. (15.3), m is defined as $\delta\Delta G^0_f/\delta[\text{denaturant}]$, $C^{1/2}_{\text{SUPREX}}$ is the concentration of denaturant at the SUPREX transition midpoint, R is the gas constant, T is the temperature in kelvin, $\langle k_{\text{int}}\rangle$ is the average intrinsic exchange rate of an amide proton, t is the HDX time, n is the number of subunits in the protein, and $[P]$ is the protein concentration expressed in n-mer equivalents. The transition region in the SUPREX curve depicts the range of denaturant concentrations where both the native and the unfolded state of protein are significantly populated.

The folding free energies ΔG^0_f of a protein and a protein–ligand complex are different, and this difference, $\Delta\Delta G^0_f$, can be employed to obtain the protein–ligand binding constant.

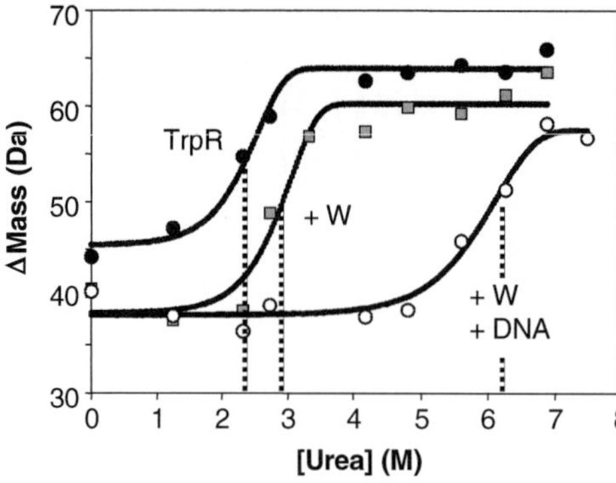

Figure 15.6. SUPREX data obtained on TrpR in the absence of ligand (●), in the presence of L-tryptophan (W) (■), and in the presence of W and cognate DNA (○). The dotted lines indicate the transition midpoints of each curve. (Adapted with permission from Ref. 80. Copyright © 2002 American Chemical Society.)

As an example of this approach, SUPREX curves generated for one of the four models studied by Powell et al.[79] are shown in Figure 15.6: the ternary protein–DNA complex composed of Trp repressor (TrpR), two molecules of L-tryptophan (W), and a 25-base-pair duplex of DNA-containing TrpR's cognate DNA sequence. The $C_{\mathrm{SUPREX}}^{1/2}$ values determined from the TrpR data as depicted in Figure 15.6 were found to be higher in the presence of W and DNA ligands, indicating binding-induced stabilization. The dissociation constant can be calculated using Eq. (15.4), where n is the number of independent binding sites and $[L]$ is the molar concentration of free ligand.

$$\Delta\Delta G_f^0 = -nRT \ln\left[1 + \frac{[L]}{[K_d]}\right] \tag{15.4}$$

The K_d values obtained in this study were in reasonable agreement (within a factor of 3) with the previously established K_d values for most of the complexes investigated. A prerequisite for SUPREX is that the protein must unfold in a two-state manner; that is, only the fully folded and the fully unfolded states of the protein should be populated at equilibrium, thus posing a limit on its application. SUPREX works in a high-throughput automated fashion, requires only minute amounts of sample, and performs analysis of purified as well as unpurified protein–ligand complexes.

15.3 DETECTING NONCOVALENT COMPLEXES AND DETERMINING THEIR K_d VALUES BY ESI MS

15.3.1 Introduction

The conformation of a protein in solution is highly dependent on the solvent system used. Biochemical sample preparation protocols usually require buffers that are not compatible with mass spectrometric analyses. The presence of nonvolatile salts, even in trace quantities, has a large impact on the quality of ESI mass spectra and can cause severe signal loss. On the other hand, organic modifiers such as methanol, acetonitrile, isopropanol, and organic acids (acetic acid and formic acid) used routinely in ESI-MS are known to destroy the noncovalent interactions. To avoid this, the sample has to be formulated in or dialyzed against a buffer that is compatible with ESI-MS and maintains the native conformation of the resulting complex. Ammonium salts offer fairly high volatility and are therefore suitable for the characterization of noncovalent complexes using ESI-MS as illustrated by Lemaire et al.[81]

The analysis of biomolecules under native conditions (aqueous medium, pH 6–9) leads to a decrease in the number of effective charges when compared to the detection under denaturing conditions because, under native conditions, biomolecules tend to assume a more compact conformation close to the folded solution conformation. Therefore, ions of higher *m/z* values are produced, which is a good indicator of proper operating conditions. Sometimes, bimodal charge distributions, representing folded as well as unfolded conformations of the same species, can be observed. Quadrupole mass analyzers have a limited *m/z* range (3000–4000), which allows the isolation of folded proteins or protein complexes only up to a molecular weight of 50–60 kDa. With the development of ESI instruments coupled to TOF mass analyzers this limitation has been overcome. Additional benefits include the virtually unlimited *m/z* range, the high sensitivity, and the speed of TOF mass analysis. Quadrupole TOF mass spectrometers are also appropriate for the study of large proteins and protein complexes, and they have the additional potential to perform tandem mass spectrometry experiments. Robinson and co-workers[82] have modified a quadrupole TOF hybrid instrument that has been designed to overcome mass range limitation with a custom-built quadrupole operating up to 22,000 *m/z*. Interface conditions in the source region play an important role in the detection of noncovalent complexes and need to be optimized to obtain high sensitivity, good spectrum quality, and prevention of dissociation of the complex. In order to preserve the specific noncovalent interactions, tuning of the cone voltage (V_c) and the interfacial pressure (P) is critical. The effect of different combinations of these parameters is explained in Figure 15.7. Additionally, the source temperature should also be optimized to achieve proper ion desolvation without thermal dissociation of the complex.[83]

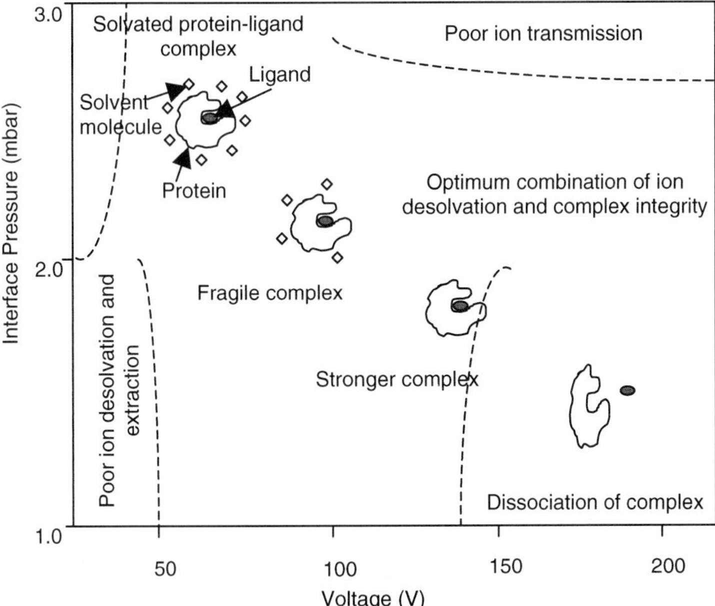

Figure 15.7. Effect of interface pressure (P) and the cone voltage (V) on the dissociation of the complex in the source region of mass spectrometer. The central region (dashed lines) in the diagram corresponds to the optimum conditions for the detection of intact complexes with appropriate desolvation of ions. (Adapted from Ref. 83.)

15.3.2 ESI Approaches for Detecting Noncovalent Complexes

15.3.2.1 Direct Monitoring of Protein–Ligand Interactions

The following illustrations of the capabilities of ESI for the detection of noncovalent complexes will focus mostly on studies carried out in recent years. These studies range from the detection of very high molecular weight multiprotein complexes to complexes between proteins and small molecules (drugs) and metal ions.

a. Protein–Protein Complexes. Robinson and co-workers[84–87] have published impressive work on the architecture of supra(bio)molecular complexes, in particular the ribosome and its protein complement. Loo and co-workers[88] have reported the analysis of the 20S proteasome complex using MS and gas-phase ion mobility. Heck's group continues to make excellent contributions in the area.[89–91]

One beautiful study involved mass spectrometric monitoring of the chaperonin-assisted protein folding cycle,[90] which shall be discussed in some more detail. The data are shown in Figures 15.8 and 15.9. Native ESI-MS was used to analyze the complexes involved in the GroEL–GroES chaperonine-assisted refolding of gp23, a major capsid protein of bacteriophage T4. This folding process is dependent on ATP, Mg^{2+}, and the presence of a co-chaperonin called gp31. GroEL was found to bind up to two unfolded gp23 molecules, but when GroEL has formed a complex with gp31, it binds exclusively one gp23. Figures 15.8 and 15.9 illustrate the nano-ESI characterization of the various complexes involved, as well as the *in vivo* refolding of gp23.

b. Protein–Peptide Complexes. For complexes involving peptides, the molecular weight is often fairly low, and MALDI as well as ESI have been used. Liu et al.[92] have examined noncovalent peptide–peptide and peptide–water interactions in small model systems using an ESI mass spectrometer equipped with a high-pressure drift cell. Woods et al.[93,94] published detailed studies of noncovalent peptide–peptide interactions. Using MALDI MS with

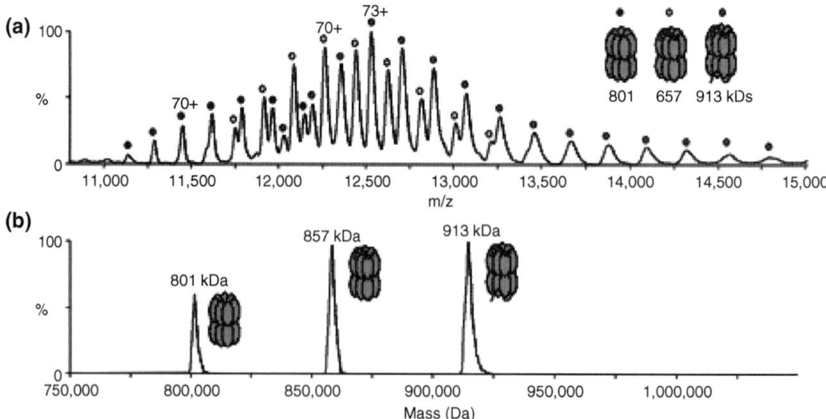

Figure 15.8. Stoichiometry of GroEL:gp23 binding and characterization of the complexes by native nano-ESI MS. (**a**) Mass spectrum of a 1 : 4 mixture of GroEl and unfolded polypeptide gp23. The charge state distribution of free GroEl (800 kDa, blue circles) GroEL with one (856 kDa, yellow circles), and two (912 kDa, red circles) gp23 substrate molecules bound can be discerned. (**b**) Deconvoluted spectrum of panel **a**, revealing the presence of the three different complex stoichiometries. (Reproduced from Ref. 90 with permission. Copyright © 2005 Nature Publishing group.) (See color insert).

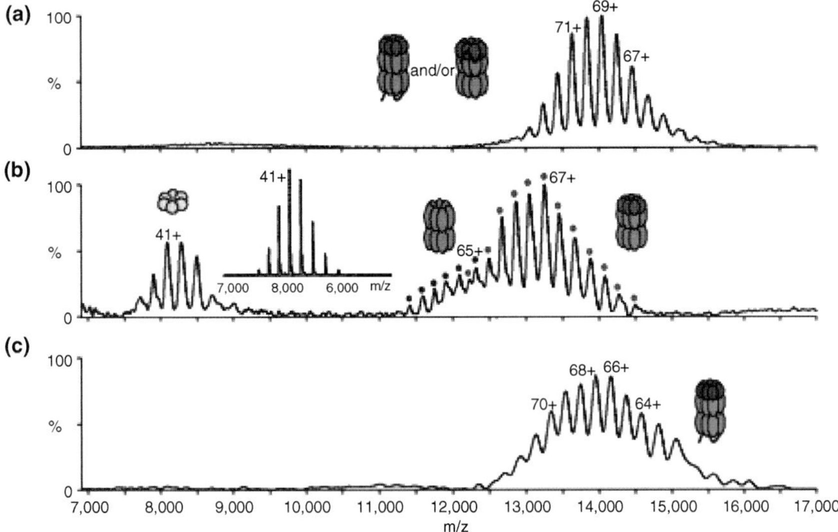

Figure 15.9. *In vivo* refolding of gp23 monitored by mass spectrometry. (**a**) Nano-ESI mass spectrum of a solution containing GroEl, gp31, gp23, ADP, and Mg^{2+}. Only one charge-state distribution is observed, corresponding to a complex with a mass of 944 kDa, which closely matches the expected mass of a ternary GroEl–gp31-ADP-Mg^{2+}complex with one gp23 substrate molecule bound. (**b**) Same as panel **a**, except that ATP was used instead of ADP. This resulted in formation of gp23 hexamer assemblies (336 kDa, yellow charge-state distribution centered around *m/z* values of 8000), revealing *in vivo* refolding of gp23. The GroEL–gp31 complex gives rise to a charge-state distribution centered around *m/z* values of 13,000 (green circles). A small amount of tetradecameric GroEL is also present (blue circles). The inset shows the mass spectrum obtained for hexameric gp23 as purified from *E. coli*. (**c**) Same as panel **b**, except that GroES was used instead of gp31. No gp23 refolding was observed. (Reproduced from Ref. 90. with permission. Copyright © 2005 Nature Publishing group.) (See color insert).

nonacidic matrices and time-of-flight as well as ion mobility mass analysis, the authors proposed a specific mode of interaction: Dynorphin, an opioid peptide, and five of its fragments that contain two adjacent Arg residues were all found to interact noncovalently with peptides that contain two to five adjacent acidic residues (Asp or Glu). However, peptides containing adjacent Lys or His did not form noncovalent complexes with acidic peptides.

c. Complexes Involving Carbohydrates Lipids, and Nucleic Acids. The group at Isis Pharmaceuticals is very active in the study of DNA/RNA–drug interactions.[31,95–98] Recently, this group has presented an automated high-throughput discovery platform to quantitatively investigate ligand binding to structured RNA drug targets.[98] Interestingly, K_ds similar to solution-phase values were obtained, despite the fact that no precautions were taken to account for varying ionization efficiencies.

Examples from other groups include gas-phase thermal dissociation experiments, implemented with blackbody infrared radiative dissociation (BIRD) and FT-ICR MS on a series of protein–carbohydrate complexes,[99,100] and the detection of fusion peptide–phospholipid noncovalent interactions using nano-ESI FTICR-MS.[101] An interesting example of protein–DNA interaction studied by ESI-MS is the *trp* repressor–DNA operator complex.[102] *Escherichia coli trp* apo-repressor (TrpR), a homodimer, is a DNA-binding protein that binds to two molecules of co-repressor L-tryptophan to form a holorepressor complex at higher salt concentrations. The mass spectrum of noncovalent

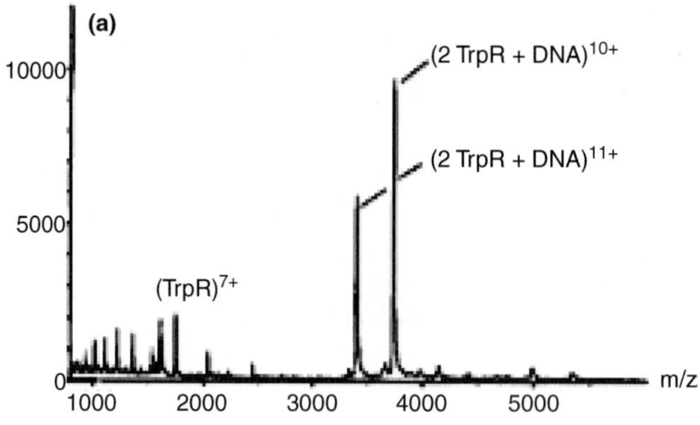

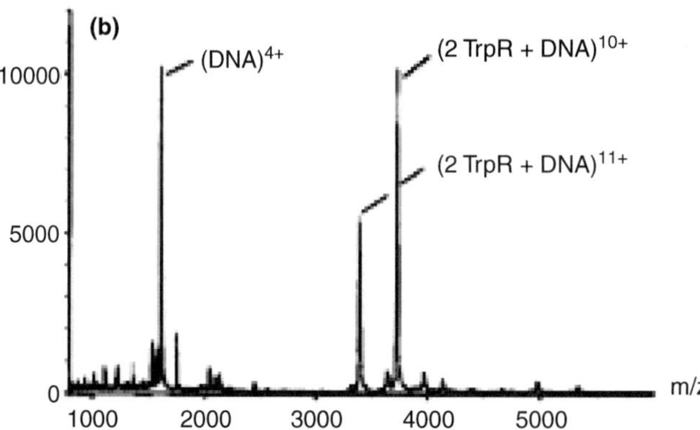

Figure 15.10. Characterization of the TrpR-DNA complex. ESI *m/z* spectra of the *trp* repressor (20 μM) in the presence of (**A**) 10 μM of its specific operator DNA and (**B**) 30 μM DNA in 5 mM NH_4OAc, pH 6.0. A molecular mass of 37,266 ± 8 Da was measured for the TrpR dimer–DNA complex, which has the expected stoichiometry of [protein dimer: dsDNA] = [1:1]. The declustering voltage was 100 V. (Reproduced with permission from Ref. 102. Copyright © 1998 The Protein Society.)

complex of TrpR dimer and double-stranded (ds) DNA with 1 : 1 stoichiometry in equimolar amounts is shown in Figure 15.10A. An incubation with threefold excess of dsDNA leads to the same charge envelope as in the previous case but with additional peaks of single-stranded (ss) DNA (Figure 15.10B). Notably, no complex of TrpR and ssDNA or complexes of other stoichiometries are detected, thus demonstrating the specificity of the complexes formed. Additional controls were used, such as incubating the protein with three different DNAs having slight modifications in their base-pair sequences.

d. Protein–Small Molecule Complexes. Van Dorsselaer's group is very active in studying noncovalent protein–ligand interactions by ESI-MS.[103–106] In one recent study, they obtained direct evidence of the formation of nuclear receptor complexes and monitored the effect of ligand binding on the stability of such complexes.[105] This example is displayed in Figure 15.11; details are given in the figure caption. A particular technical problem presented by this complex system was the delicate balance between keeping the complex intact and obtaining sufficient desolvation in order to distinguish different species in the mass spectra. This was accomplished by careful optimization of the acceleration voltage Vc between sampling cone and skimmer in the ESI source.

e. Protein–Metal Ion Complexes. The challenge posed by these systems is often that the ionization efficiency of free and metal-bound protein can be quite different. In one recent example, Gao et al.[107] have performed interesting experiments for the measurement

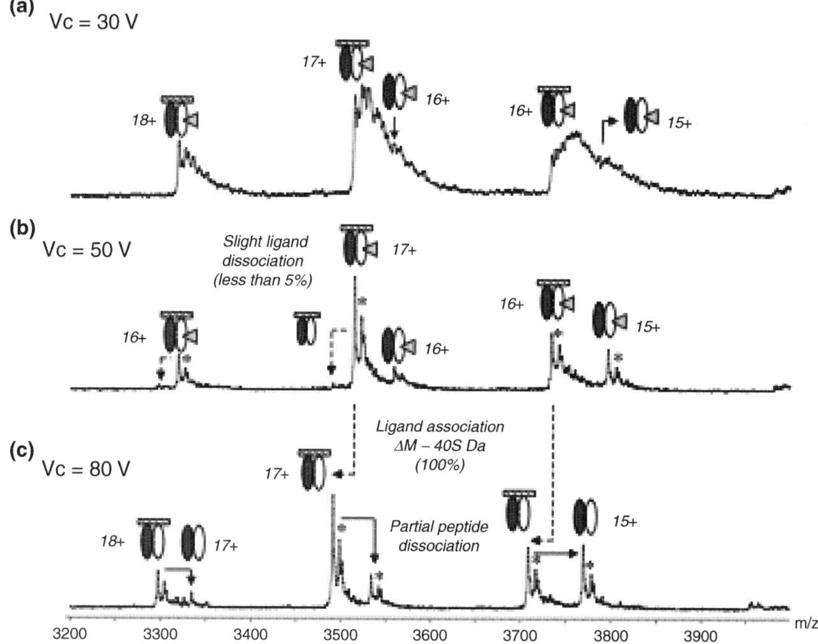

Figure 15.11. Optimization of the ESI operating conditions on detecting the retinoid acid receptor (RAR)–retinoid X receptor (RXR) heterodimer (HD) binding with a corepressor peptide (CoRNR1) and an inverse pan-RAR agonist called BMS493. (**A**) At low acceleration voltages, the HD-BMS493-CoRAR1 noncovalent comlex is detected as the major component, but the peaks are broad, due to insufficient desolvation. (**B**) Slight dissociation of the ligand (less than 5%). (**C**) One hundred percent dissociation of the ligand from the HD-CoRNR1 comples, as well as partial dissociation of the corepressor peptide. Peaks labeled with * correspond to a species with an additional N-terminal methionine. (Reproduced with permission from Ref. 105. Copyright 2004 © Blackwell Publishing.)

of dissociation constants of phosphomannose and a zinc complex including the effect of zinc binding on the metalloenzyme using ESI-FTICR MS. Our laboratory presented a study where ESI-MS was used to measure *de novo* the dissociation constant of minute amounts of synthetic b-peptides with Zn.[108] The b-peptides were designed to mimic Zn finger behavior; available quantities were far too small to allow K_d determination by any method other than mass spectrometry. The problem of the potentially different ionization efficiencies of complexed versus free b-peptide was solved by quantifying only the free b-peptide against a noninteracting peptide used as an internal standard.

15.3.2.2 Diffusion Measurements

A different approach for studying noncovalent interactions between potential ligands and biological macromolecules in solution was introduced by Konermann and co-workers in 2003.[109] It takes into account the translational diffusion of biomolecules in bulk solution. The diffusion coefficient D of a molecule is determined by the Stokes radius R_s as depicted in Eq. (15.5), where k is the Boltzmann constant, T is the temperature, and η is the viscosity of the solution.

$$D = \frac{kT}{6\pi\eta R_s} \tag{15.5}$$

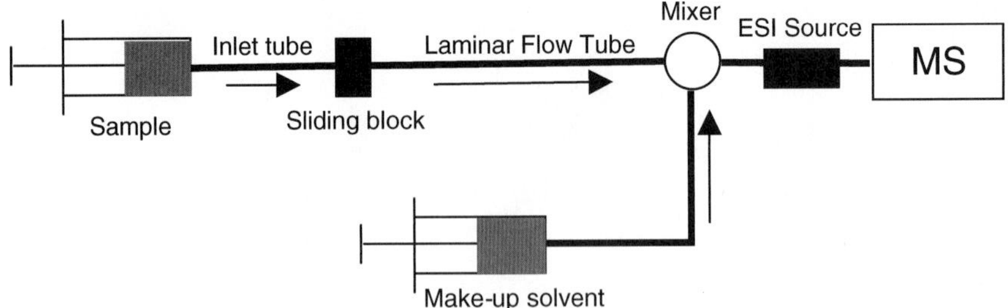

Figure 15.12. Schematic representation of the experimental setup used for dispersion experiments. (Adapted from Ref. 111.)

The molecules having smaller Stokes radii diffuse rapidly, whereas the macromolecules with larger Stokes radii diffuse at a slower rate. Therefore, the binding of a ligand to a macromolecule reduces the rate of diffusion of the ligand. Diffusion coefficients of a macromolecule and its low-molecular-weight ligand are determined by generating an initially sharp boundary between two solutions of different concentration in a laminar flow tube as described.[110] The boundary tends to be distorted in a parabolic form when these solutions are pumped towards the ESI mass spectrometer mounted at the end of the capillary. This is due to the laminar flow inside the capillary, which generates high flow velocity in the center and a zero velocity at the walls of the capillary. Typically, a similar laminar flow tube setup (Figure 15.12) has been employed to record dispersion profiles of the binding partners.[111]

ESI-MS offers an effective means of monitoring in parallel dispersion profiles of multiple analytes in parallel to predict binding of protein–ligand complexes on the basis of their diffusion coefficients. The effective diffusion coefficient for each analyte can be obtained by fitting to the individual profiles. An elegant study has been performed by the same group[112] using the binding of heme to myoglobin as a model system. Simulated ESI-MS-based dispersion profiles showing the signal intensity of selected analytes at the end of a laminar flow tube under different conditions are shown in Figure 15.13.

Some of the main advantages of this technique are as follows: (a) It allows the investigation of noncovalent ligand–macromolecule interactions directly in solution; (b) changes in the nature of protein–ligand interactions can be detected, which are undetectable by ESI-MS; and (c) it works well for systems that do not survive the ESI process. This method could be very useful for membrane receptors that are undetectable by direct ESI measurements. Interestingly, the dissociation constant of protein–ligand complexes with 1 : 1 binding can also be determined from the dispersion profiles obtained by this method.[113]

15.3.2.3 H/D Exchange

The first study using HDX-MS was performed using FAB ionization to study the microenvironment of amino acids of human interleukin-2.[114] After this pioneering work, the introduction of ESI in the early 1990s allowed significant improvements, making determination of deuterium incorporation more precise and accurate.[115] An important step in the development of the method was achieved by the use of enzymatic proteolysis allowing topological analysis of deuterium incorporation in proteins.[116] For this proteolysis, pepsin was preferred for its ability to generate peptides from proteins in exchange-quenched conditions (0°C, pH 2.5).

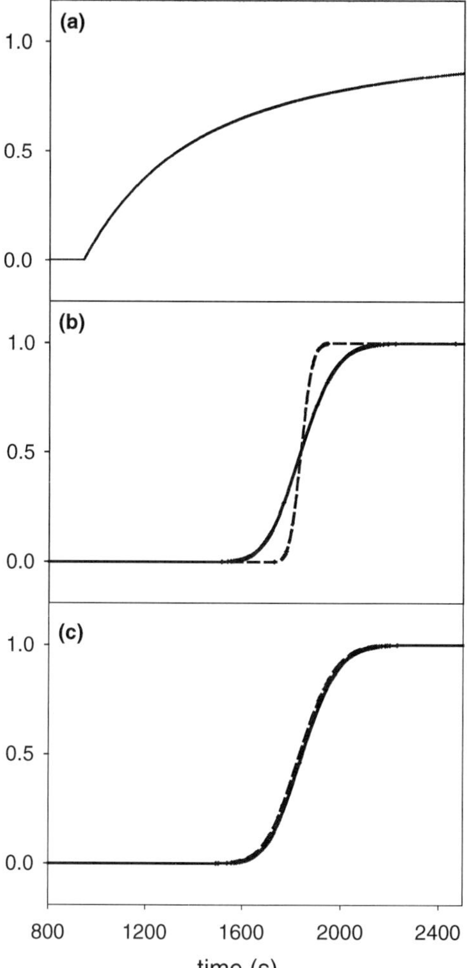

Figure 15.13. Simulated ESI-MS dispersion profiles, representing the signal intensity of selected analytes at the end of a laminar flow tube under different conditions. (**a**) Dispersion profile expected for a single analyte in the absence of diffusion, that is, $D = 0$. (**b**) Dispersion profiles expected for a macromolecule ($D = 1 \times 10^{-10}\,m^2/s$, solid curve), and a small molecule ($D = 10 \times 10^{-10}\,m^2/s$, dashed curve). It is assumed that the two analytes do *not* interact in solution. (**c**) Dispersion profiles as in panel **b**, but under the assumption of tight noncovalent binding between the two analytes. Under these conditions the profiles measured for the two species will be identical (for the purpose of presentation, one of the profiles has been slightly shifted). Parameters used: Tube length $l = 3$ m, tube radius $R = 129.1\,\mu m$, flow rate $= 5\,\mu L/min$ (corresponding to $v_{max} = 3.18 \times 10^{-3}\,m/s$). The dispersion profile in panel **a** has been calculated based on Eq. (19) in Ref. 110; all the other profiles have been calculated based on Eq. (17) from the same reference. For simplicity, all dispersion profiles have been normalized to unity. (Reproduced with permission from Ref. 112. Copyright © 2003 Elsevier.)

Following these methodological improvements, HDX and ESI-MS have been used in numerous studies for analysis of protein–protein,[117] protein–ligand,[118] and protein–-membrane interactions.[119] HDX and ESI-MS have also been used for the study of protein conformational dynamics. In a study performed in the Gross group, four different conformations of insulin were detected using HDX and ESI-MS.[120]

The main advantage of electrospray ionization for monitoring hydrogen/deuterium exchange is the possibility to precede the mass spectrometric analysis with liquid chromatographic separation. This allows a fast and accurate analysis of deuterium incorporation and minimizes the rate of back exchange resulting from sample preparation. The liquid chromatography separation facilitates the determination of deuterium incorporation in exchanged peptides by decreasing peak overlap and by removing deuterium atoms present on the side chains of proteins.

However, most of the experiments using ESI and hydrogen exchange are performed with liquid chromatography columns of 1- to 2-mm diameter, requiring 500 pmol of sample for each analysis. This relatively large amount of sample and the loss of deuterium due to the liquid separation make the use of liquid chromatography difficult and subjected to time-consuming optimizations. To reduce the problem of deuterium loss, columns coated with pepsin have been used.[121] The development of nano-electrospray has significantly improved the sensitivity of HDX analysis, thereby reducing the volume necessary for liquid chromatography and, consequently, the amount of sample needed. By reducing the internal diameter of the columns from 0.1 to 0.5 mm, the amount of sample needed has been reduced by a factor of 100. Wang and Smith[122] have optimized the analysis conditions by using a column containing immobilized pepsin and a reversed-phase capillary. Despite the significant improvements made, further methodological advances are necessary to permit routine use of HDX, in particular, because of the difficulties related to back exchange and sample preparation.

15.3.3 Approaches for Determining Binding Affinities by ESI-MS

15.3.3.1 Titration Method

Solution-phase titration methods are among the most common methods of determining binding constants using ESI-MS. Generally, the concentration of the target biomolecule is kept constant and the concentration of its ligand is varied over a range of about two orders of magnitude and the intensity of the corresponding complex is then compared to the intensity of the binding partner for every ligand concentration. It is assumed here that the ionization process does not affect the equilibrium existing in the solution phase. Binding constants are determined using graphic linearization such as Scatchard plots where the ratio of bound ligand (PL) over free ligand (L_0) is plotted against the concentration of bound ligand. It gives a linear relationship with slope $-1/K_d$ as shown in Eq. (15.6):

$$\frac{[PL]}{[L_0][P_0]} = \frac{-1}{K_d} \times \frac{[PL]}{[P_0]} + \frac{n}{K_d} \tag{15.6}$$

In the above expression, n gives the number of ligands bound to one protein molecule (stoichiometry). Scatchard plots consider multiple binding sites as noncooperative (equivalent and independent) for the calculation of K_d. However, this linearization is not really necessary in the age of computer-based curve fitting, and in fact it is somewhat problematic because the ordinate and abscissa are not independent. Figure 15.15 provides an example where a nonlinear least-squares fit was used instead of linearization.

In one of the early attempts, Lim et al.[123] presented a beautiful study of the noncovalent binding of glycopeptides (vancomycin and ristocetin) to a peptide ligand (AC_2KAA) in solution. Titration experiments were performed by varying the ligand concentration with the glycopeptide concentration held constant; the MS intensities were measured in each case. The data were analyzed using Scatchard plots to obtain the binding constants—for example, 6.25×10^5 for ristocetin–Ac_2KAA complex, which is in a good agreement with the reported literature value (5.9×10^5) obtained from other solution-phase methods.[124]

Greig et al.[124] have identified two independent binding sites for the binding of oligonucleotides to bovine serum albumin (BSA). They have determined the respective dissociation constants by fitting the data acquired from titration experiments to a second-order polynomial. K_d values obtained by this method match well with the solution K_d values obtained using capillary electrophoresis (CE). Eckart and Spiess[125] have monitored the concentration dependent binding of biotin to streptavidin using ESI-MS. Loo et al.[126] have presented a nice piece of work taking the model system of phosphopeptides binding to SH2 motifs where the binding involves both electrostatic and hydrophobic interactions. Both phosphorylated and nonphosphorylated peptides were studied, but only the phosphorylated peptides were found to bind to the SH2 domain and the mass spectrometric data correlated well to the data derived from solution-based methods.

Similar investigations have been published by Griffey et al.[95] and Carte et al.[127]. Furthermore, Sannes-Lowery et al.[96] have compared different methods to determine dissociation constants of RNA-aminoglycoside complexes and found that the best strategy is to hold the RNA concentration constant below the expected K_d, followed by titration with the aminoglycoside. This comparison, however, shows that the K_d values obtained using the titration method for an unknown system should be treated with caution. Recently, we presented a study[46] on the detection and quantification of noncovalent complexes of chicken muscle adenylate kinase and two inhibitors, P1,P4-di(adenosine-5′)tetraphosphate (Ap4A) and P1,P5-di(adenosine-5′)pentaphosphate (Figure 15.14). Association constants ($9.0 \times 10^4\,M^{-1}$ for Ap4A and $4.0 \times 10^7\,M^{-1}$ for Ap5A) were determined using the titration method (Figure 15.15), and the results were in good agreement with literature data. Additional control experiments were performed to confirm the noncovalent nature and the specificity of the complexes under study, and a data evaluation method was also developed that is independent of the overall ionization efficiency of the complex as well as individual binding partners.

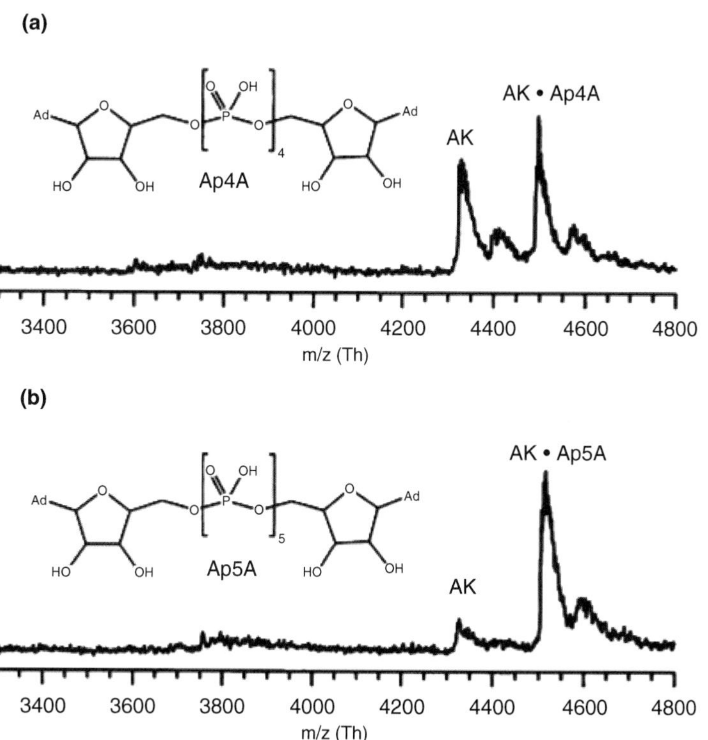

Figure 15.14. ESI mass spectra of 8.8 μM adenylate kinase (AK, +5 charge state) with two inhibitors: (**a**) 8 μM Ap4A and (**b**) 8 μM Ap5A in 50 mM $(HNEt_3)HCO_3$. The ionization and the desolvation are very soft, leading to broad peaks. Many salt and buffer adducts are observed. In the insets, the structures of Ap4A and Ap5A are shown where Ad denotes adenine. (Reproduced with permission from Ref. 46. Copyright © 2003 Elsevier.)

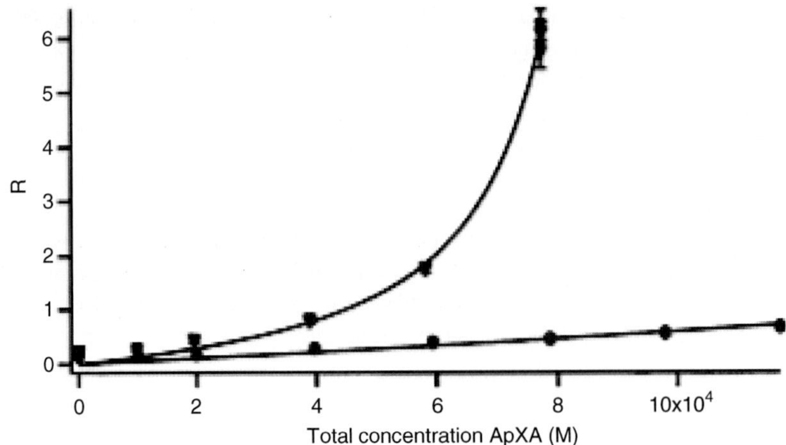

Figure 15.15. Result of the titration experiments of AK with Ap5A The lines are the fitted data with a nonlinear least-squares fit to the equation.

$$R = \frac{1}{2}\left(-1 - K_a \times [AK]_0 + K_a \times [ApXA]_0 + \sqrt{4 \times K_a \times [ApXA]_0 + (K_a \times [ApXA]_0 - K_a \times [AK]_0 - 1)^2} \right)$$

where R is the ratio of complex to free protein. The only adjustable parameter is K_a, the association constant for the formation of this 1 : 1 complex. Filled inverted triangles: AK with Ap5A, $K_a = 4.0 \times 10^7\ M^{-1}$. Filled circles: AK with Ap4A, $K_a = 9.0 \times 10^4\ M^{-1}$. (Reproduced with permission from Ref. 46. Copyright © 2003 Elsevier.)

15.3.3.2 Melting Curves

The strength of noncovalent interactions of nucleic acid complexes is routinely determined by their temperature-dependent "melting" behavior in solution where the fraction of intact complex ($\propto$) is determined as a function of the temperature of the solution. UV absorption, fluorescence, or circular dichroism is usually employed as the means of detection of biomolecular interactions in this type of experiments. A transition enthalpy can be obtained from the slope of the melting curve as described by Marky and Breslauer[128] and given in Eq. (15.7), where n is the molecularity of the association reaction (e.g., $n = 2$ for dimerization) and T_m is the melting temperature, normally defined for $\alpha = 50\%$.

$$\Delta H = (2 + 2n)RT_m^2 \left(\frac{\partial \alpha}{\partial T} \right)_{T=T_m} \tag{15.7}$$

ESI-MS has also been used for the detection of these complexes by controlling the temperature of the spray capillary in an ESI source. Typically, the temperature of the solution maintained at a constant temperature is raised gradually and α is determined using MS. It is essential to note that above-mentioned setup affects only the temperature of the spray solution, thereby giving a measure of solution-phase and not the gas-phase stability.

In one of the earlier attempts, Cheng et al.[129] showed an interesting comparison of the stabilities of natural DNA and 2′,5′-linked DNA duplexes using ESI-MS. The data, acquired using a capillary interface temperature of 90°C, clearly showed that the spectrum of natural DNA is mainly dominated by the peaks of duplex (heterodimer formed through Watson–-Crick base pairing), unlike 2′,5′-linked DNA. Moreover, a plot of mole fraction of single and double strands and the interface capillary temperature (melting curve) was also generated from the ESI data. A good agreement was shown between the relative stabilities of both of the DNA duplexes by ESI-MS and by solution-phase stability measurements. Fändrich et al.[130] have also shown the effects of higher temperatures on the structure of the chaperone GimC/prefolding homologue complex.

15.3.3.3 Competition Method

Competition among the different binding partners of a target molecule for its binding site can be a good source of information about the absolute and relative binding affinities of host–guest complexes. This was first demonstrated by Jorgensen et al.,[131] who considered the binding of vancomycin and risocetin with a number of guest molecules. The binding constants of vancomycin and three peptides were determined in a single ESI-MS experiment. This method relies on the measurement of relative peak intensities of free host and the complexes of the host with different guests. Based on this protocol, the equilibrium concentration of the host is given by

$$[H_a] = \frac{H_a[H_a]_0}{H + HG_x + HG_y + HG_z} \tag{15.8}$$

where H, HG_x, HG_y, and HG_z represent the peak intensities of the host and its complexes with three different guests. The initial concentration of host is denoted by $[H_a]_0$, and H_a applies to all forms of the host. Working with equimolar mixtures, affinities can be calculated by the following expression:

$$K_{HG_X} = \frac{[HG_x]}{[H][G_x]} = \frac{[HG_x]}{[H]([H] + [HG_y] + [HG_z])} \tag{15.9}$$

The fast speed of this method is a major advantage over other techniques. However, the ionization efficiencies of the host and the complexes are assumed to be identical, which is a limiting factor. The same group has also proposed a different approach for the determination of absolute binding constants.[132]

An indirect method to determine absolute binding constants using ESI-MS was described by Kempen and Brodbelt,[133] where only the signal intensity of a reference complex with a known binding constant is monitored. As a first step, a calibration curve for the intensity versus concentration of reference complex is constructed using ESI-MS. The concentration of the reference complex at each calibration point is given by.

$$K_R = \frac{[H_RM]^+}{[H_R]_F[M]_F^+} \tag{15.10}$$

$$[H_R]_F = [H_R]_T - [H_RM]^+ \tag{15.11}$$

$$[M]_F^+ = [R]_T - [H_RM]^+ \tag{15.12}$$

Here, subscript R denotes the reference complex, a subscript T stands for the total concentration of guest metal ion (M^+) or host (H) present in the solution. The determination of binding constants is based on the competition of two different hosts (reference *guest* and new *guest*) for one common guest or vice versa. A shift in the equilibrium of reference host–guest complex is observed as a result of competition for guest binding sites between the host molecules. The concentration of the reference complex decreases as a function of binding affinity of the second host/guest, thereby giving the binding constant of the new host–guest complex. Upon solving Eqs. (15.1) and (15.2) simultaneously as explained by Kempen et al.,[133] the binding constant of the complex of interest (K_N) can be simultaneously calculated:

$$K_N = \left[\frac{[H_RM]^+ + K_R([H_R]_T - [H_RM]^+)([H_RM]^+ - [M]_T^+)}{-[H_RM]^+ + K_R([H_N]_T + [H_RM]^+ - [M]_T^+)([H_RM]^+ - [H_R]_T)}\right]\left[\frac{K_R([H_R]_T - [H_RM]^+)}{[H_RM]^+}\right] \quad (15.13)$$

where G stands for guest, both for the known and unknown complexes. H_R and H_N denote reference and new hosts, respectively. A key assumption here is that the ionization efficiency of the reference complex is the same over the complete concentration range—that is, a negligible effect of the presence of guest or host on the ionization efficiency of the reference complex. A stability of the spray throughout the experiment is critical to obtaining high-quality data, but that is not always easy to obtain. This method is rapid, but it does not take into account the ionization efficiency of the unknown complex that can introduce some error, but it avoids the need for estimation of transfer coefficients.

Relative binding affinities can also be determined using the aformentioned method. Chen et al.[134] have implemented ESI-FTICR-MS for the determination of relative binding affinities of 16 inhibitors derived from *para*-substituted benzenesulfonamides towards bovine carbonic anhydrase II in a single experiment using the competition method.

15.3.3.4 PLIMSTEX

A method complementary to SUPREX to quantify *P*rotein-*LI*gand *I*nteractions in solution by *MS*, *T*itration and H/D *EX*change (PLIMSTEX) was described by Zhu et al. 2003.[135] Unlike SUPREX, this method does not require denaturants; instead it provides *in situ* desalting that is extensively needed for ESI-MS analyses. The main steps of the PLIMSTEX method are summarized in Figure 15.16. The protocol requires a high D/H ratio in the forward reaction and a high H/D ratio in the back-exchange. Back-exchange of the labile and non-amide protons of the protein immobilized on the column gives the sample an isotopic-exchange "signature" in its amide bonds, thus manifesting its state in the initial solution. Eventually, the sample is eluted either (a) by an isocratic run of mobile phase containing a high-percentage organic content or (b) with a fast pH 2.5 gradient in front of an ESI source for the determination of the molecular weight. The mass difference between the nondeuterated and deuterated sample (deuterium uptake) is plotted against the total ligand concentration to obtain the PLIMSTEX curve, which is

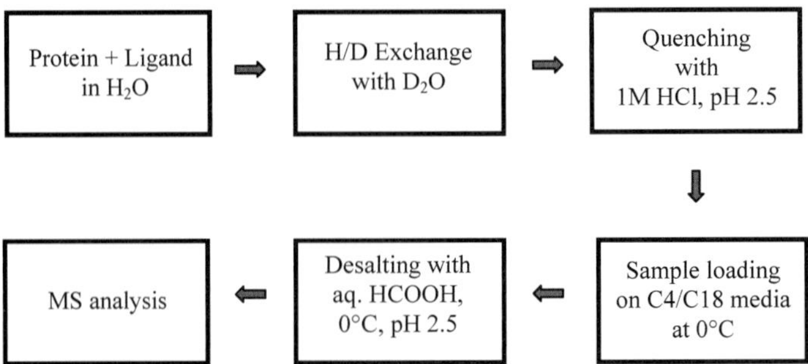

Figure 15.16. A general protocol used for sample preparation and analysis in PLIMSTEX.

then fitted to different binding models using Mathcad[136] to determine the binding affinity of the system under study.

Zhu and coworkers have applied PLIMSTEX to different test systems including the binding of small organic molecules, metal ions and peptides to proteins. The binding constants obtained by this method are within a factor of six of the literature values determined using conventional methods.[136] When the overall deuterium shift (ΔD) is positive, it gives a quantitative measure of increased protection from HDX thereby indicating some ligand-induced conformational change, which makes the protein less solvent accessible. On the other hand, negative ΔD values show a decreased protection and an opening of tertiary structure of protein upon ligand binding. PLIMSTEX curves are sensitive to the total protein concentration and are not reliable when performed at high concentrations ($\sim$100 times the K_d).

PLIMSTEX can be applied for the determination of conformational change, binding stoichiometry and binding affinities of protein-ligand complexes. It requires low quantities of sample and it works in biological media at high ionic strength with a high throughput. Contrarily, PLIMSTEX is limited in its use for proteins, which do not change their conformation after the binding of ligand. Complex systems involving different stoichiometries other than 1 : 1 binding are also not appropriate for this methodology.

15.4 CRITICAL DISCUSSION OF MS-BASED METHODS TO DETERMINE DISSOCIATION CONSTANTS

15.4.1 Do ESI-MS-Based K_d Measurements Accurately Represent the Solution Phase?

It is difficult to find comprehensive studies where comparisons were made between MS-based and conventional methods for K_d determination.[42,98,137,138] Generally, good to excellent agreement is found, in particular for cases where such comparisons were not made against older literature data, suggesting that ESI can indeed provide a "snapshot" of the solution phase composition.

In this context, a deeper understanding about processes within the electrospray plume is important. In particular, the implications of these processes for the ESI-MS quantification of species that are in chemical equilibrium with each other are subject to discussion. The key question is whether the ESI process has the capacity to grossly distort a solution phase equilibrium, or whether relative concentrations of interaction partners are left undisturbed.

Investigations of these phenomena directly in the electrospray plume have been published: Zhou et al.[139] used spatially resolved laser-induced fluorescence to observe pH changes within the electrospray plume by using fluorescent pH indicators. Although a small pH shift to more acidic values could be observed, the results remained unclear, due to the problem that no information about the decrease in droplet size was available. A similar study by Scott and co-workers[140] used the equilibrium between a fluorophore and a quencher to investigate equilibria within the spray. The authors could indeed monitor a higher quenching rate at later stages of the spray due to the concentration increase in the shrinking droplets. Another interesting strategy to investigate the shrinking droplets was presented Agnes and co-workers[141,142]: By using single droplets, levitated by a hydrodynamic balance, they were able to show crystallization of analytes within the evaporating droplet. Also, they landed droplets on a glass target and then measured the analyte distribution with confocal fluorescence microscopy. Konermann and co-workers[143] showed that HDX immediately prior to the spray process can be used to distinguish between (a) specific complexes that reflect the solution phase and (b) artificially created complexes as well as dissociation products created in the electrospray process.

Theoretical considerations have also added to our understanding of ESI processes. Experimental parameters that can have an influence on the investigation of a chemical equilibrium with ESI-MS were investigated by and co-workers[144] and Roeraade and co-workers[145]: it was shown that the molecular weight, the concentration of an analyte, and the binding mode involved can influence measurements of K_d if the titration method is used. Also, instrumental parameters such as the capillary-cone distance and hence the sampled droplet size can influence the outcome of the experiments. A mathematical approach to model the behavior of a host–guest system with two guests within electrospray droplets was presented by Brodbelt's group.[146] These authors find that the processes taken into account for their study have only little influence on the outcome of experiments.

A caveat is, of course, that it should not generally be assumed that relative peak intensities in an ESI mass spectrum can be directly related to relative concentrations in solution. The ionization efficiency and transfer coefficients for individual species have to be taken into account, which is often difficult.

15.4.2 Gas-Phase K_d Determinations and Their Relationship to Solution-Phase Measurements

MS methods that determine the dissociation energies of gas-phase complex ions—such as collision-induced dissociation, electron capture dissociation, or blackbody infrared radiative dissociation experiments—and MS-based methods used for monitoring solution equilibria must be clearly distinguished. For the latter, results generally agree well with known solution-phase values, although rigorous comparisons using many different kinds of interactions and experimental methods should be made in the future. On the other hand, gas-phase MS methods yield interaction energies that typically do not agree with the solution-phase value. It is not really surprising that a binding energy in solution has no direct relationship with the corresponding gas-phase value, because ionic, ion–dipole, and dipole–dipole interactions will become stronger in the absence of screening by the solvent, while some interactions such as hydrophobic interactions are destabilized in the absence of solvent. Theoretically, gas-phase and solution-phase binding energies can be related by subtracting energies of solvation; however, these are often unknown. This chapter focused mainly on the determination of binding interactions and dissociation constants in solution, using MS readout. These are certainly the relevant data in a biochemical context. However, it can also be very useful to determine values for gas-phase binding energies. These can be directly compared to quantum mechanical computations that are typically obtained without incorporating solvent. Again such comparisons have not been made in a systematic fashion.

15.5 SUMMARY

We predict that more and more noncovalent complexes will be analyzed by mass spectrometry in the future. Complexes with high mass and from complicated protein extracts will be characterized; the latter will require a high tolerance for impurities during the ionization step to be analyzed successfully. For this purpose, MALDI ionization with its simplicity, high speed, and high salt tolerance may be the ionization method of choice. ESI, as the softest of the modern soft ionization methods (and its variants), is well-suited to obtain mass spectra of intact complexes directly from native conditions, opening up many possibilities for detecting specific noncovalent complexes and for determining binding constants. We have highlighted (a) the criteria to identify specific complexes in ESI and MALDI mass spectra and (b) the limitations of the mass spectrometric approach to detect and quantify protein–protein and protein–ligand complexes. Overall, we believe that this field has a bright future, especially in the pharmaceutical industry. The promises of mass spectrometry are sensitivity, speed, and the possibility to obtain precise molecular weight and stoichiometric information.

ACKNOWLEDGMENTS

We would like to extend our sincere thanks to Prof. L. Konermann, Prof. R. G. Cooks, Dr. N. Potier, Prof. H. Duckworth, Prof. M.L. Fitzgerald, and Prof. A. Heck for providing figures for this chapter. Financial support from the Swiss National Science Foundation (grant no. 200020-111831/1) and the Commission for Technology and Innovation (grant no. CTI 8070.2LSPP) is gratefully acknowledged.

REFERENCES

1. Aebersold, R.; Mann, M. Mass spectrometry-based proteomics. *Nature* **2003**, *422*, 198–207.
2. Ganem, B.; Li, Y. T.; Henion, J. D. Observation of noncovalent enzyme–substrate and enzyme–product complexes by ion-spray mass spectrometry. *J. Am. Chem. Soc.* **1991**, *113*(20), 7818–7819.
3. Cohen, L. R. H.; Strupat, K.; Hillenkamp, F. Analysis of quaternary protein ensembles by matrix assisted laser desorption/ionization mass spectrometry. *J. Am. Soc. Mass Spectrom.* **1997**, *8*(10), 1046–1052.
4. Hensley, P. Defining the structure and stability of macromolecular assemblies in solution: The re-emergence of analytical ultracentrifugation as a practical tool. *Structure* **1996**, *4*, 367–373.
5. Ahmadian, M. R.; Hoffmann, U.; Goody, R. S.; Wittinghofer, A. Individual rate constants for the interaction of Ras proteins with Gtpase-activating proteins determined by fluorescence spectroscopy. *Biochemistry* **1997**, *36*(15), 4535–4535.
6. Li, W.; Dennis, C. A.; Moore, G. R.; James, R.; Kleanthous, C. Protein–protein interaction specificity of Im9 for the endonuclease toxin Colicin E9 defined by homologue-scanning mutagenesis. *J. Biol. Chem.* **1997**, *272*(35), 22253–22258.
7. Leifert, W.; Bailey, K.; Cooper, T.; Aloia, A.; Glatz, R.; McMurchie, E. measurement of heteromeric G-protein and regulators of G-protein signaling interactions by time-resolved fluorescence resonance energy transfer. *Anal. Biochem.* **2006**, *355*, 201–212.
8. Pramanik, A. Ligand–receptor interactions in live cells by fluorescence correlation spectroscopy. *Curr. Pharm. Biotechnol.* **2004**, *5*(2), 205–12.
9. Banaszynski, L. A.; Liu, C. W.; Wandless, T. J. Characterization of the Fkbp·Rapamycin·Frb ternary complex. *J. Am. Chem. Soc.* **2005**, *127*, 4715–4721.
10. Zsila, F.; Bikadi, Z.; Fitos, I.; Simonyi, M. Probing protein binding sites by circular dichroism spectroscopy. *Curr. Drug. Discovery Technol.* **2004**, *1*(2), 133–53.
11. Morgan, C. S.; Holton, J. M.; Olafson, B. D.; Bjorkman, P. J.; Mayo, S. L. Circular dichroism determination of Class I Mhc-peptide equilibrium dissociation constants. *Protein Sci.* **1997**, *6*(8), 1771–1773.
12. Weber, P. C.; Salemme, F. R. Applications of calorimetric methods to drug discovery and the study of protein interactions. *Curr. Opin. Struct. Biol.* **2003**, *13* (1), 115–121.
13. Perozzo, R.; Folkers, G.; Scapozza, L. Thermodynamics of protein–ligand interactions: History, presence, and future aspects. *J. Recept. Signal Transduct. Res.* **2004**, *24*(1–2), 1–52.
14. Velazquez-Campoy, A.; Kiso, Y.; Freire, E. The binding energetics of first- and second-generation Hiv-1 protease inhibitors: Implications for drug design. *Arch. Biochem. Biophys.* **2001**, *390*(2), 169–175.
15. Wang, Z. X. An exact mathematical expression for describing competitive binding of two different ligands to a protein molecule. *FEBS Lett.* **1995**, *360*(2), 111–114.
16. Sigurskjold, B. W. Exact analysis of competition ligand binding by displacement isothermal titration calorimetry. *Anal Biochem* **2000**, *277*(2), 260–206.
17. Keeble, A. H.; Kirkpatrick, N.; Shimizu, S.; Kleanthous, C. Calorimetric dissection of colicin Dnase—Immunity protein complex specificity. *Biochemistry* **2006**, *45*(10), 3243–3254.
18. Plum, G. E.; Breslauer, K. J. Calorimetry of proteins and nucleic acids. *Curr Opin. Struct. Biol.* **1995**, *5*(5), 682–690.

19. Freire, E. Differential scanning calorimetry. *Methods Mol. Biol.* **1995**, *40*, 191–218.
20. Bruylants, G.; Wouters, J.; Michaux, C. Differential scanning calorimetry in life science: Thermodynamics, stability, molecular recognition and application in drug design. *Curr. Med. Chem.* **2005**, *12* (17), 2011–2020.
21. Michnik, A.; Drzazga, Z.; Kluczewska, A.; Michalik, K. Differential scanning microcalorimetry study of the thermal denaturation of haemoglobin. *Biophys. Chem.* **2005**, *118*(2–3), 93–101.
22. Cooper, A.; Johnson, C. M.; Lakey, J. H.; Nollmann, M. Heat does not come in different colours: Entropy–enthalpy compensation, free energy windows, quantum confinement, pressure perturbation calorimetry, solvation and the multiple causes of heat capacity effects in biomolecular interactions. *Biophys. Chem.* **2001**, *93*(2–3), 215–230.
23. Waldron, T. T.; Schrift, G. L.; Murphy, K. P. The salt-dependence of a protein–ligand interaction: Ion–protein binding energetics. *J. Mol. Biol.* **2005**, *346* (3), 895–905.
24. Connelly, P. R. Acquisition and use of calorimetric data for prediction of the thermodynamics of ligand-binding and folding reactions of proteins. *Curr. Opin. Biotechnol.* **1994**, *5*(4), 381–388.
25. Jonsson, U.; Fagerstam, L.; Ivarsson, B.; Johnsson, B.; Karlsson, R.; Lundh, K.; Lofas, S.; Persson, B.; Roos, H.; Ronnberg, I.; Sjolander, S.; Stenberg, E.; Stahlberg, R.; Urbaniczky, C.; Ostlin, H.; Malmqvist, M. Real-time biospecific interaction analysis using surface-plasmon resonance and a sensor chip technology. *Biotechniques* **1991**, *11*(5), 620–627.
26. Nedelkov, D.; Nelson, R. W. Surface plasmon resonance mass spectrometry: Recent progress and outlooks. *Trends Biotechnol.* **2003**, *21*(7), 301–305.
27. Davies, J. Surface plasmon resonance—the technique and its applications to biomaterial processes. *Nanobiology* **1994**, *3*, 5–16.
28. Buijs, J.; Franklin, G. C. Spr-Ms in functional proteomics. *Brief Funct. Genomic Proteomic* **2005**, *4* (1), 39–47.
29. Daniel, J.; Friess, S.; Rajagopalan, S.; Wendt, S.; Zenobi, R. Quantitative determination of noncovalent binding interactions using soft ionization mass spectrometry. *Int. J. Mass Spectrom.* **2002**, *216*, 1–27.
30. Schalley, C. A. Molecular recognition and supromolecular chemistry in the gas phase. *Mass Spectrom. Rev.* **2002**, *20*(5), 253–309.
31. Hofstadler, S. A.; Griffey, R. H. Analysis of Noncovalent complexes of DNA and Rna by Mass Spectrometry. *Chem. Rev.* **2001**, *101*(2), 377–390.
32. Hardouin, J.; Lange, C. M. Biological noncovalent complexes by mass spectrometry. *Curr. Org. Chem.* **2005**, *9*(3), 317–324.
33. Heck, A. J. R.; van der Heuvel, R. H. Investigation of intact protein complexes by mass spectrometry. *Mass Spectrom. Rev.* **2004**, *23*(5), 368–389.
34. Schermann, S. M.; Simmons, D. A.; Konermann, L. Mass spectrometry-based approaches to protein–ligand interactions. *Expert Rev. Proteom.* **2005**, *2*(4), 475–485.
35. Zehl, M.; Allmaier, G. Investigation of sample preparation and instrumental parameters in the matrix-assisted laser desorption/ionization time-of-flight mass spectrometry of noncovalent peptide/peptide complexes. *Rapid Commun. Mass Spectrom.* **2003**, *17*, 1931–1940.
36. Zehl, M.; Allmaier, G. Instrumental parameters in the Maldi-Tof mass spectrometric analysis of quaternary protein structures. *Anal. Chem.* **2005**, *77*(1), 103–110.
37. Sakamoto, S.; Fujita, M.; Kim, K.; Yamaguchi, K. Characterization of self-assembling nano-sized structures by means of coldspray ionization mass spectrometry. *Tetrahedron* **2000**, *56*, 955.
38. Sakamoto, S.; Yoshizawa, M.; Kusukawa, T.; Fujita, M.; Yamaguchi, K. Characterization of encapsulated supramolecules by using Csi Ms with ionization-promoting reagents. *Org. Lett.* **2001**, *3*, 1601.
39. Sakamoto, S.; Yamaguchi, K.; Hyperstranded, DNA architectures observed by cold-spray ionization Ms. *Angew. Chem. Int. Ed.* **2003**, *42*, 905–908.
40. Yamaguchi, K. Cold-spray ionization mass spectrometry: Principle and applications. *J. Mass Spectrom.* **2003**, *38*, 473–490.
41. Zhang, S.; Van Pelt, C. K. Chip-based nanoelectrospray mass spectrometry for protein characterization. *Expert Rev. Proteomics* **2004**, *1*(4), 449–468.
42. Zhang, S.; Van Pelt, C. K.; Wilson, D. B. Quantitative determination of noncovalent binding interactions using automated Nanoesi Ms. *Anal. Chem.* **2003**, *75*, 3010–3018.
43. Takats, Z.; Wiseman, J. M.; Gologan, B.; Cooks, R. G. Electrosonic spray ionization. A gentle technique for generating folded proteins and protein complexes in the gas phase and for studying ion–molecule reactions at atmospheric pressure. *Anal. Chem.* **2004**, *76*, 4050–4058.
44. Wiseman, J. M.; Takats, Z.; Gologan, B.; Davisson, V. J.; Cooks, R. G. Direct characterization of enzyme–substrate complexes using electrosonic spray ionization mass spectrometry. *Angew. Chem.* **2005**, *44*, 913–916.
45. Yang, P.; Cooks, R. G.; Ouyang, Z.; Hawkridge, A. M.; Muddiman, D. C. Gentle protein ionization assisted by high velocity gas flow. *Anal. Chem.* **2005**, *77*, 6174–6183.
46. Daniel, J. M.; McCombie, G.; Wendt, S.; Zenobi, R. Mass spectrometric determination of association constants of adenylate kinase with two noncovalent inhibitors. *J. Am. Soc. Mass Spectrom.* **2003**, *14*, 442–448.
47. Rosinke, B.; Strupat, K.; Hillenkamp, F.; Rosenbusch, J.; Dencher, N.; Krüger, U.; Galla, H. J. Maldi Ms of

membrane proteins and noncovalent complexes. *J. Mass Spectrom.* **1995**, *30*, 1462–1468.

48. Farmer, T. B.; Caprioli, R. M. Assessing the multimeric states of proteins: studies using laser desorption Ms. *Biol. Mass Spectrom.* **1991**, *20*, 796–800.
49. Glocker, M. O.; Bauer, S. H. J.; Kast, J.; Volz, J.; Przybylski, M. Characterization of Specific Noncovalet Protein Complexes by Uv Maldi Mass Spectrometry. *J. Mass Spectrom.* **1996**, *31*, 1221–1227.
50. Rosinke, B.; Strupat, K.; Hillenkamp, F.; Rosenbusch, J.; Dencher, N.; Kruger, U.; Galla, H. J. Matrix-assisted laser desorption/ionization mass-spectrometry (Maldi-Ms) of membrane-Proteins and noncovalent complexes. *J. Mass Spectrom.* **1995**, *30*(10), 1462–1468.
51. Fitzgerald, M. C.; Parr, G. R.; Smith, L. M. Basic matrices for the matrix-assisted laser desorption/ionization mass spectrometry of proteins and oligonucleotides. *Anal. Chem.* **1993**, *65*, 3204–3211.
52. Jespersen, S.; Niessen, W. M. A.; Tjaden, U. R.; van der Greef, J. Basic matrices in the analysis of noncovalent complexes by Maldi Ms. *J. Mass Spectrom.* **1998**, *33*, 1088–1093.
53. Lehmann, E.; Zenobi, R. Detection of specific noncovalent zinc finger peptide–Oligodeoxynucleotide complexes by matrix-assisted laser desorption/ ionization mass spectrometry. *Angew. Chem. Int. Ed. Engl.* **1998**, *38*, 3430–3432.
54. Rajagopalan, S.; Zenobi, R. The detection and stability of DNA duplexes probed by Maldi mass spectrometry. *Helv. Chim. Acta* **2002**, *85*, 3136–3143.
55. Wenzel, R. J.; Matter, U.; Schultheis, L.; Zenobi, R. Analysis of megadalton ions using cryodetection Maldi time-of-flight mass spectrometry. *Anal. Chem.* **2005**, *77*, 4213–4219.
56. Nazabal, A.; Wenzel, R. J.; Zenobi, R. Immunoassays with direct mass spectrometric detection. *Anal. Chem.* **2006**, *78*(11), 3562–3570.
57. Nazabal, A.; Schlumberger, M.; Hardt, W.; Zenobi, R. Super-shifting: A new method for the analysis of protein complexes using high-mass Maldi Tof mass spectrometry. In *Proceedings of the 54th Conference of the American Society for Mass Spectrometry,* May 28–June, 1, 2006, Seattle, WA **2006**.
58. Young, M. M.; Tang, N.; Hempel, J. C.; Oshiro, C. M.; Taylor, E. W.; Kuntz, I. D.; Gibson, B. W.; Dollinger, C. High throughput protein fold identification by using experimental constraints derived from intramolecular cross-links and mass spectrometry. *Proc. Natl. Acad. Sci. USA* **2000**, *97*, 5802–5806.
59. Sinz, A. Chemical cross-linking and mass spectrometry for mapping three-dimensional structures of proteins and protein complexes. *J. Mass Spectrom.* **2003**, *38* (12), 1225–1237.
60. Van Dijk, A. D. J.; Boelens, R.; Bonvin, A. M. J. J. Data-driven docking for the study of biomolecular complexes. *FEBS J.* **2005**, *272*, 293–312.
61. Bai, Y.; Milne, J. S.; Mayne, L.; Englander, S. W. Primary structure effects on peptide group hydrogen exchange. *Proteins* **1993**, *17*(1), 75–86.
62. Mandell, J. G.; Falick, A. M.; Komives, E. A. Measurement of amide hydrogen exchange by Maldi-Tof mass spectrometry. *Anal. Chem.* **1998**, *70*(19), 3987–3995.
63. Resing, K. A.; Ahn, N. G. Deuterium exchange mass spectrometry as a probe of protein kinase activation. Analysis of wild-type and constitutively active mutants of map kinase kinase-1. *Biochemistry* **1998**, *37*(2), 463–475.
64. Belghazi, M.; Bathany, K.; Hountondji, C.; Grandier-Vazeille, X.; Manon, S.; Schmitter, J. M. Analysis of protein sequences and protein complexes by matrix-assisted laser desorption/ionization mass spectrometry. *Proteomics* **2001**, *1*(8), 946–954.
65. Kleinekofort, W.; Avdiev, J.; Brutschy, B. A new method of laser desorption mass spectrometry for the study of biological macromolecules. *Int. J. Mass Spectrom. Ion Processes* **1996**, *152*(2/3), 135–142.
66. Kleinekofort, W.; Pfenninger, A.; Plomer, T.; Griesinger, C.; Brutschy, B. Observation of noncovalent complexes using Lilbid. *Int. J. Mass Spectrom. Ion Processes* **1996**, *156*, 195–202.
67. Sobott, F.; Kleinekofort, W.; Brutschy, B. Cation selectivity of natural and synthetic ionophores probed with laser-induced liquid beam mass spectrometry. *Anal. Chem.* **1997**, *69*(17), 3587–3594.
68. Wattenberg, A.; Sobott, F.; Barth, H. -D.; Brutschy, B. Studying noncovalent protein complexes in aqueous solution with laser desorption Ms. *Int. J. Mass Spectrom.* **2000**, *203*, 49–57.
69. Charvat, A.; Lugovoj, E.; Faubel, M.; Abel, M. Analytical laser induced liquid beam desorption mass spectrometry of protonated amino acids and their non-covalently bound aggregates. *Eur. Phys. J.* **2002**, *7*(3), 573–582.
70. Kohno, J. T.; Toyama, N.; Buntine, M. A.; Mafune, F.; Kondow, T. Gas phase ion formation from a liquid beam of arginine in aqueous solution by Ir multiphoton excitation. *Chem. Phys. Lett.* **2006**, *420*(1–3), 18–23.
71. Ghaemmaghami, S.; Fitzgerald, M. C.; Oas, T. G. A quantitative, high-throughput screen for protein stability. *Proc. Natl. Acad. Sci. USA* **2000**, *87*(15), 8296–8301.
72. Hvidt, A.; Nielsen, S. O. Hydrogen exchange in proteins. *Adv. Protein Chem.* **1966**, *21*, 287–386.
73. Englander, S. W.; Sosnick, T. R.; Englander, J. J.; Mayne, L. Mechanisms and uses of hydrogen exchange. *Curr. Opin. Struct. Biol.* **1996**, *6*(1), 18–23.
74. Huyghues-Despointes, B. M.; Scholtz, J. M.; Pace, C. N. Protein conformational stabilities can be determined from hydrogen exchange rates. *Nat. Struct. Biol.* **1999**, *6*(10), 910–912.
75. Miranker, A.; Robinson, C. V.; Radford, S. E.; Dobson, C. M. Investigation of protein folding by mass spectrometry. *Faseb J.* **1996**, *10*(1), 93–101.

76. Powell, K. D.; Fitzgerald, M. C. Measurements of protein stability by H/D exchange and MALDI Ms using picomoles of material. *Anal. Chem.* **2001**, *73*, 3300–3304.
77. Ghaemmaghami, S.; Oas, T. G. Quantitative protein stability measurement *in vivo*. *Nat. Struct. Biol.* **2001**, *8* (10), 879–882.
78. Powell, K. D.; Fitzgerald, M. C. Accuracy and precision of a new H/D exchange- and mass spectrometry-based technique for measuring the thermodynamic properties of protein–peptide complexes. *Biochemistry* **2003**, *42*(17), 4962–4970.
79. Powell, K. D.; Ghaemmaghami, S.; Wang, M. Z.; Ma, L.; Oas, T. G.; Fitzgerald, M. C. A general mass spectrometry-based assay for the quantitation of protein–ligand binding interactions in solution. *J. Am. Chem. Soc.* **2002**, *124*, 10256–10257.
80. Powell, K. D.; Ghaemmaghami, S.; Wang, M. Z.; Ma, L. Y.; Oas, T. G.; Fitzgerald, M. C. A general mass spectrometry-based assay for the quantitation of protein–Ligand binding interactions in solution. *J. Am. Chem. Soc.* **2002**, *124*(35), 10256–10257.
81. Lemaire, D.; Marie, G.; Serani, L.; Laprevote, O. Stabilization of gas-phase noncovalent macromolecular complexes in electrospray mass spectrometry using aqueous triethylammonium bicarbonate buffer. *Anal. Chem.* **2001**, *73*(8), 1699–1706.
82. Sobott, F.; Hernandez, H.; McCammon, M. G.; Tito, M. A.; Robinson, C. V. A tandem mass spectrometer for improved transmission and analysis of large macromolecular assemblies. *Anal. Chem.* **2002**, *74*(6), 1402–1407.
83. Potier, N.; Rogniaux, H.; Chevreux, G.; Van Dorsselaer, A. Ligand–metal ion binding to proteins: Investigation by Esi mass spectrometry. *Methods Enzymol.* **2005**, *402*, 361–389.
84. Sobott, F.; Robinson, C. V. Characterising electrosprayed biomolecules using tandem-Ms—the noncovalent GroEL chaperonin assembly. *Int. J. Mass Spectrom.* **2004**, *236*(1–3), 25–32.
85. Videler, H.; Ilag, L. L.; McKay, A. R. C.; Hanson, C. L.; Robinson, C. V. Mass spectrometry of intact ribosomes. *FEBS Lett.* **2005**, *579*(4), 943–947.
86. Aquilina, J. A.; Benesch, J. L. P.; Ding, L. L.; Yaron, O.; Horowitz, J.; Robinson, C. V. Subunit exchange of polydisperse proteins—Mass spectrometry reveals consequences of alpha αA-crystallin truncation. *J. Biol. Chem.* **2005**, *280*(15), 14485–14491.
87. Callaghan, A. J.; Redko, Y.; Murphy, L. M.; Grossmann, J. G.; Yates, D.; Garman, E.; Ilag, L. L.; Robinson, C. V.; Symmons, M. F.; McDowall, K. J.; Luisi, B. F. "Zn-link": A metal-sharing interface that organizes the quaternary structure and catalytic site of the endoribonuclease, Rnase E. *Biochemistry* **2005**, *44* (12), 4667–4675.
88. Loo, J. A.; Berhane, B.; Kaddis, C. S.; Wooding, K. M.; Xie, Y.; Kaufman, S. L.; Chernushevich, I. V. Electrospray ionization mass spectrometry and ion mobility analysis of the 20s proteasome complex. *J. Am. Soc. Mass Spectrom.* **2005**, *16*(7), 998–1008.
89. Pinske, M. W. H.; Heck, A. J. R.; Rumpel, K.; Pullen, F. Probing noncovalent protein–ligand interactions of the Cgmp-dependent protein kinase using electrospray ionization time of flight mass spectrometry. *J. Am. Soc. Mass Spectrom.* **2004**, *15*(10), 1392–1399.
90. van Duijn, E.; Bakkes, P. J.; Heeren, R. M. A.; van den Heuvel, R. H. H.; van Heerikhuizen, H.; van der Vies, S. M.; Heck, A. J. R. Monitoring macromolecular complexes involved in the chaperonin-assisted protein folding cycle by mass spectrometry. *Nature Meth.* **2005**, *2*(5), 371–376.
91. van den Heuvel, R. H. H.; Heck, A. J. R. Native protein mass spectrometry: From intact oligomers to functional machineries. *Curr. Opin. Chem. Biol.* **2005**, *8*(5), 519–526.
92. Liu, D.; Wyttenbach, T.; Carpenter, C. J.; Bowers, M. T. Investigation of noncovalent interactions in deprotonated peptides: structural and energetic competition between aggregation and hydration. *J. Am. Chem. Soc.* **2004**, *126*, 3261–3270.
93. Woods, A. S.; Koomen, J. M.; Ruotolo, B. T.; Gillig, K. J.; Russel, D. H.; Fuhrer, K.; Gonin, M.; Egan, T. F.; Schultz, J. A. A study of peptide–peptide using Maldi ion mobility O-Tof and Esi mass spectrometry. *J. Am. Soc. Mass Spectrom.* **2002**, *13*(2), 166–169.
94. Woods, A. S.; Huestis, M. A. A study of peptide–peptide interaction by matrix-assisted laser desorption/ionization. *J. Am. Soc. Mass Spectrom.* **2001**, *12*(1), 88–96.
95. Griffey, R. H.; Hofstadler, S. A.; Scannes-Lowery, K. A.; Ecker, D. J.; Crooke, S. T. Determinants of aminoglycoside-binding specificity for Rrna by using mass spectrometry. *Proc. Natl. Acad. Sci. USA* **1999**, *96*, 10129–10133.
96. Sannes-Lowery, K. A.; Griffey, R. H.; Hofstadler, S. A. Measuring dissociation constants of Rna and aminoglycoside antibiotics by electrospray mass spectrometry. *Anal. Biochem.* **2000**, *280*, 264–271.
97. Ding, Y.; Hofstadler, S. A.; Swayze, E. E.; Risen, L.; Griffey, R. H. Design and synthesis of paromomycin-related heterocycle-substituted aminoglycoside mimetics based on a mass spectrometry Rna-binding assay. *Angew. Chem. Int. Ed.* **2003**, *42*, 3409–3412.
98. Sannes-Lowery, K. A.; Cummins, L. L.; Chen, S.; Drader, J. J.; Hofstadler, S. A. high throughput drug discovery with Esi-Fticr. *Int. J. Mass Spectrom.* **2004**, *238*(2), 197–206.
99. Wang, W.; Kitova, E. N.; Klassen, J. S. Nonspecific protein–carbohydrate complexes produced by nanoelectrospray ionization. Factors influencing their formation and stability. *Anal. Chem.* **2005**, *77*(10), 3060–3071.
100. Wang, W.; Kitova, E. N.; Sun, J.; Klassen, J. S. Blackbody infrared radiative dissociation of Nonspecific protein–carbohydrate complexes

produced by nanoelectrospray ionization: The nature of the noncovalent interactions. *J. Am. Soc. Mass Spectrom.* **2005**, *16* (10), 1583–1594.

101. Li, Y.; Heitz, F.; Le Grimellec, C.; Cole, R. B. Fusion peptide–phospholipid noncovalent interactions as observed by nanoelectrospray Fticr-Ms. *Anal. Chem.* **2005**, *77*(6), 1556–1565.
102. Potier, N.; Donald, L. J.; Chernushevich, I.; Ayed, A.; Ens, W.; Arrowsmith, C. H.; Standing, K. G.; Duckworth, H. W. Study of a noncovalent Trp repressor:DNA operator complex by electrospray ionization time-of-flight mass spectrometry. *Protein Sci.* **1998**, *7*(6), 1388–1395.
103. Liang, Y.; Du, F.; Sanglier, S.; Zhou, B. R.; Xia, Y.; Van Dorsselaer, A.; Maechling, C.; Kilhoffer, M. C.; Haiech, J. Unfolding of rabbit muscle creatine kinase induced by acid. A study using electrospray ionization mass spectrometry, isothermal titration calorimetry, and fluorescence spectroscopy. *J. Biol. Chem.* **2003**, *278*(32), 30098–30105.
104. Darmanin, C.; Chevreux, G.; Potier, N.; Van Dorsselaer, A.; Hazemann, I.; Podjarny, A.; El-Kabbani, O. Probing the ultra-high resolution structure of aldose reductase with molecular modelling and noncovalent mass spectrometry. *Bioorg. Med. Chem.* **2004**, *12*(14), 3787–3806.
105. Sanglier, S.; Bourguet, W.; Germain, P.; Chavant, V.; Moras, D.; Gronemeyer, H.; Potier, N.; Van Dorsselaer, A. Monitoring ligand-mediated nuclear receptor–coregulator interactions by noncovalent mass spectrometry. *Eur. J. Biochem.* **2004**, *271*(23–24), 4958–4967.
106. Chevreux, G.; Potier, N.; Van Dorsselaer, A.; Bahloul, A.; Houdusse, A.; Wells, A.; Sweeney, H. L. Electrospray ionization mass spectrometry studies of noncovalent myosin Vi complexes reveal a new specific calmodulin binding site. *J. Am. Soc. Mass Spectrom.* **2005**, *16*(8), 1367–1376.
107. Gao, H.; Yu, Y.; Leary, J. A. Mechanism and kinetics of metalloenzyme phosphomannose isomerase: Measurement of dissociation constants and effect of zinc binding using Esi-Fticr mass spectrometry. *Anal. Chem.* **2005**, *77*(17), 5596–5603.
108. Wortmann, A.; Rossi, F.; Lelais, G.; Zenobi, R. Determination of zinc to beta-peptide binding constants with electrospray ionization mass spectrometry. *J. Mass Spectrom.* **2005**, *40*, 777–784.
109. Clark, S. M.; Konermann, L. Diffusion measurements by electrospray mass spectrometry for studying solution-phase noncovalent interactions. *J. Am. Soc. Mass Spectrom.* **2003**, *14*, 430–441.
110. Clark, S. M.; Leaist, D. G.; Konermann, L. Taylor dispersion monitored by electrospray mass spectrometry: A novel approach for studying diffusion in solution. *Rapid Commun. Mass Spectrom* **2002**, *16* (15), 1454–1462.
111. Clark, S. M.; Konermann, L. Screening for noncovalent ligand–receptor interactions by electrospray ionization mass spectrometry-based diffusion measurements. *Anal. Chem.* **2004**, *76*, 1257–1263.
112. Clark, S. M.; Konermann, L. Diffusion measurements by electrospray mass spectrometry for studying solution-phase noncovalent interactions. *J. Am. Soc. Mass Spectrom.* **2003**, *14*(5), 430–431.
113. Clark, S. M.; Konermann, L. Determination of ligand–protein dissociation constants by electrospray mass spectrometry-based diffusion measurements. *Anal. Chem.* **2004**, *76*(23), 7077–7083.
114. Miyano, H.; Suzuki, E.; Akashi, S.; Furuya, M.; Tsuji, T.; Hirayama, K.; Nagashima, N.; *Anal. Sci.* **1989**, *5*, 759–761.
115. Thevenon-Emeric, G.; Kozlowski, J.; Zhang, Z.; Smith, D. L. Determination of amide hydrogen exchange rates in peptides and Ms. *Anal. Chem.* **1992**, *64*, 2456–2458.
116. Zhang, Z.; Smith, D. L. Determination of amide hydrogen exchange by mass spectrometry: A new tool for protein structure elucidation. *Protein Sci.* **1993**, *2* (4), 522–531.
117. Kraus, M.; Bienert, M.; Krause, E. Hydrogen exchange studies on Alzheimer's amyloid-beta peptides by mass spectrometry using Maldi and Esi. *Rapid Commun. Mass Spectrom.* **2003**, *17*, 222–228.
118. Gonzalez de Peredo, A.; Saint-Pierre, C.; Latour, J. M.; Michaud-Soret, I.; Forest, E. Conformational changes of the ferric uptake regulation protein upon metal activation and DNA binding; First evidence of structural homologies with the diphtheria toxin repressor. *J. Mol. Biol.* **2001**, *310*(1), 83–91.
119. Demmers, J. A.; van Duijn, E.; Haverkamp, J.; Greathouse, D. V.; Koeppe, R. E., 2nd; Heck A. J.; Killian, J. A. Interfacial positioning and stability of Transmembrane Peptides in Lipid bilayers studied by combining hydrogen/deuterium exchange and mass spectrometry. *J. Biol. Chem.* **2001**, *276*(37), 34501–34508.
120. Ramanathan, R.; Gross, M. L.; Zielinski, W. L.; Layloff, T. P. Monitoring recombinant protein drugs: A study of insulin by H/D exchange and electrospray ionization mass spectrometry. *Anal. Chem.* **1997**, *69* (24), 5142–5145.
121. Wang, L.; Pan, H.; Smith, D. L. Hydrogen exchange–mass spectrometry: Optimization of digestion conditions. *Mol. Cell Proteomics* **2002**, *1*(2), 132–138.
122. Wang, L.; Smith, D. L. Downsizing improves sensitivity 100-fold for hydrogen exchange-mass spectrometry. *Anal. Biochem.* **2003**, *314*(1), 46–53.
123. Lim, H. -K.; Hsieh, Y. L.; Ganem, B.; Henion, J. Recognition of cell-wall peptide ligands by vancomycin group antibiotics: Studies using ion spray mass spectrometry. *J. Mass Spectrom.* **1995**, *30*(5), 708–714.
124. Greig, M. J.; Gaus, H.; Cummins, L. L.; Sasmor, H.; Griffey, R. H. Measurement of macromolecular binding using electrospray mass spectrometry.

Determination of dissociation constants from oligonucleotide—Serum albumin complexes. *J. Am. Chem. Soc.* **1995**, *117*, 10765–10766.
125. Eckart, K.; Spiess, J. Electrospray ionization mass spectrometry of biotin binding to Streptavidin. *J. Am. Soc. Mass Spectrom.* **1995**, *6*(10), 912–919.
126. Loo, J. A.; Hu, P.; McConnell, P.; Mueller, W. T. A study of Src Sh2 domain protein–phosphopeptide binding interactions by electrospray ionization mass spectrometry. *J. Am. Soc. Mass Spectrom.* **1997**, *8*(3), 234–243.
127. Carte, N.; Legendre, F.; Leize, E.; Potier, N.; Reeder, F.; Chottard, J. -C.; Van Dorsselaer, A. Determination by electrospray mass spectrometry of the outersphere association constants of DNA/ platinum complexes using 20-Mer oligonucleotids and ($[Pt(Nh_3)_4]^{2+}$, $2cl^-$) or ($[Pt(Py)_4]^{2+}$, $2cl^-$). *Anal. Biochem.* **2000**, *284*, 77–86.
128. Marky, L. A.; Breslauer, K. J. Calculating thermodynamic data for transitions of any molecularity from equilibrium melting curves. *Biopolymers* **1987**, *26*, 1601–1620.
129. Cheng, X.; Gao, Q.; Smith, R. D.; Jung, K.-E.; Switzer, C. Comparison of 3′,5′- and 2′,5′-linked DNA duplex stabilities by electrospray ionization mass spectrometry. *Chem. Commun.* **1996**, 747–748.
130. Fändrich, M.; Tito, M. A.; Leroux, M. R.; Rostom, A. A.; Hartl, F. U.; Dobson, C. M.; Robinson, C. V. Observation of the noncovalent assembly and disassembly pathways of the chaperone complex Mtgimc by mass spectrometry. *Proc. Natl. Acad. Sci. USA* **2000**, *97*(26), 14151–14155.
131. Jorgensen, T. J. D.; Roepstorff, P.; Heck, A. J. R. Direct determination of solution binding constants for noncovalent complexes between bacterial cell wall peptide analogs and vancomycin group antibiotics by electrospray ionization mass spectrometry. *Anal. Chem.* **1998**, *70*(20), 4427–4432.
132. Jorgensen, T. J. D.; Staroske, T.; Roepstorff, P.; Williams, D. H.; Heck, A. J. R. Subtle differences in molecular recognition between modified glycopeptide antibiotics and bacterial receptor peptides identified by Esi Ms. *J. Chem. Soc. Perkin Trans.* **1999**, *2*, 1859–1863.
133. Kempen, E. C.; Brodbelt, J. S. A method for the determination of binding constants by electrospray ionization mass spectrometry. *Anal. Chem.* **2000**, *72*, 5411–5416.
134. Cheng, X.; Chen, R.; Bruce, J. E.; Schwartz, B. L.; Anderson, G. A.; Hofstadler, S. A.; Gale, D. C.; Smith, R. D. Using electrospray ionization Fticr mass spectrometry to study competitive binding of inhibitors to carbonic anhydrase. *J. Am. Chem. Soc.* **1995**, *117*, 8859–8860.
135. Zhu, M. M.; Rempel, D. L.; Du, Z.; Gross, M. L. Quantification of protein–ligand interactions by mass spectrometry, titration, and H/D exchange: Plimstex. *J. Am. Chem. Soc.* **2003**, *125*(18), 5252–5253.
136. Zhu, M. M.; Chitta, R.; Gross, M. L. Plimstex: A novel mass spectrometric method for the quantification of protein–ligand interactions in solution. *Int. J. Mass Spectrom.* **2005**, *240*, 213–240.
137. Wendt, S.; McCombie, G.; Daniel, J.; Kienhöfer, A.; Hilvert, D.; Zenobi, R. Quantitative evaluation of noncovalent chorismate mutase—inhibitor binding by Esi-Ms. *J. Am. Soc. Mass Spectrom.* **2003**, *14*, 1470–1476.
138. Tjernberg, A.; Carnö, S.; Oliv, F.; Benkestock, K.; Edlung, P. -O.; Griffiths, W. J.; Hallen, D. Determination of dissociation constants for protein–ligand complexes by electrospray ionization mass spectrometry. *Anal. Chem.* **2004**, *76*, 4325–4331.
139. Zhou, S.; Prebyl, B. S.; Cook, K. D. Profiling Ph changes in the electrospray plume. *Anal. Chem.* **2002**, *74*, 4885–4888.
140. Ham, J. E.; Durham, B.; Scott, J. R. Design of instrumentation for probing changes in electrospray droplets via the Stern–Volmer relationship. *Rev. Sci. Instrum.* **2005**, *76*, 014101-1–014101-6.
141. Bakhoum, S. F. W.; Agnes, G. R. Study of chemistry in droplets with net charge before and after Coulomb explosion: Ion-induced nucleation in solution and implications for ion production in an electrospray. *Anal. Chem.* **2005**, *77*(10), 3189–3197.
142. Haddrell, A. E.; Agnes, G. R. Organic cation distributions in the residues of levitated droplets with net charge: Validity of the partition theory for droplets produced by an electrospray. *Anal. Chem.* **2004**, *76*, 53–61.
143. Hossain, B. M.; Simmons, D. A.; Konermann, L. Do electrospray mass spectra reflect the ligand binding state of proteins in solution?. *Can. J. Chem.* **2005**, *83*, 1953–1960.
144. Peschke, M.; Verkerk, U. H.; Kebarle, P. Features of the Esi mechanism that affect the observation of multiply charged noncovalent protein complexes and the determination of the association constant by the titration method. *J. Am. Soc. Mass Spectrom.* **2004**, *15*, 1424–1443.
145. Benkestock, K.; Sundqvist, G.; Edlung, P. -O.; Roeraade, J. Influence of droplet size, capillary-cone distance and selected instrumental parameters for the analysis of noncovalent protein-ligand complexes by nano-electrospray ionization mass spectrometry. *J. Mass Spectrom.* **2004**, *39*, 1059–1067.
146. Sherman, C. L.; Brodbelt, J. S. Partitioning model for competitive host–guest complexation in Esi-Ms. *Anal. Chem.* **2005**, *77*, 2512–2523.

Chapter 16

Ion Activation Methods for Tandem Mass Spectrometry

Kristina Håkansson [*] ***and John S. Klassen*** [†]

[*] *Department of Chemistry, University of Michigan, Ann Arbor, Michigan*
[†] *Department of Chemistry, University of Alberta, Edmonton, Alberta, Canada*

Electrospray and MALDI Mass Spectrometry: Fundamentals, Instrumentation, Practicalities, and Biological Applications, Second Edition, Edited by Richard B. Cole

16.1 INTRODUCTION

Tandem mass spectrometry (MS/MS) refers to any mass spectrometric method that involves at least two stages of mass analysis.[1] The first-stage mass-selectively isolates precursor (parent) ions of interest. These mass-selected ions can then undergo, for example, a gas-phase chemical reaction that can either decrease or increase the ion mass. Finally, the second stage of mass analysis detects products of the reaction. In the most common MS/MS implementation, the step between the two stages of mass analysis involves activation of ions, either by increasing their internal energies, through radical-driven reactions, or a combination of both. Activation results in the rupture of chemical bonds, either homolytically (to create two radical products from an even-electron precursor ion) or heterolytically (to create two even-electron species from an even-electron precursor ion). The resulting product (fragment) ions are generally representative of the structure of precursor ions and are highly valuable for, for example, structural elucidation and biomolecule identification. Use of MS/MS rather than MS alone can also provide more accurate quantification of molecules of interest from complex matrices because precursor ions with similar masses often have different fragmentation patterns.

A plethora of different MS/MS activation methods are available, including low- and high-energy collision-induced dissociation (CID),[2] surface-induced dissociation (SID),[3–5] infrared multiphoton dissociation (IRMPD),[6,7] blackbody infrared radiative dissociation (BIRD),[8,9] ultraviolet photodissociation (UVPD),[10,11] and electron-based methods such as electron-induced dissociation [EID; also termed electron impact excitation of ions from organics (EIEIO)][12] and the more modern techniques of, for example, electron capture dissociation (ECD)[13] and electron transfer dissociation (ETD).[14] Each method involves a different physical principle for which the timescale of activation, the amount of energy imparted into precursor ions, and the mode of activation (e.g., vibrational versus electronic excitation) is different. An excellent summary of the timescale differences between activation methods, published by McLuckey and Goeringer,[15] is shown in Figure 16.1. Because this scheme dates to 1997, more recent techniques such as ECD

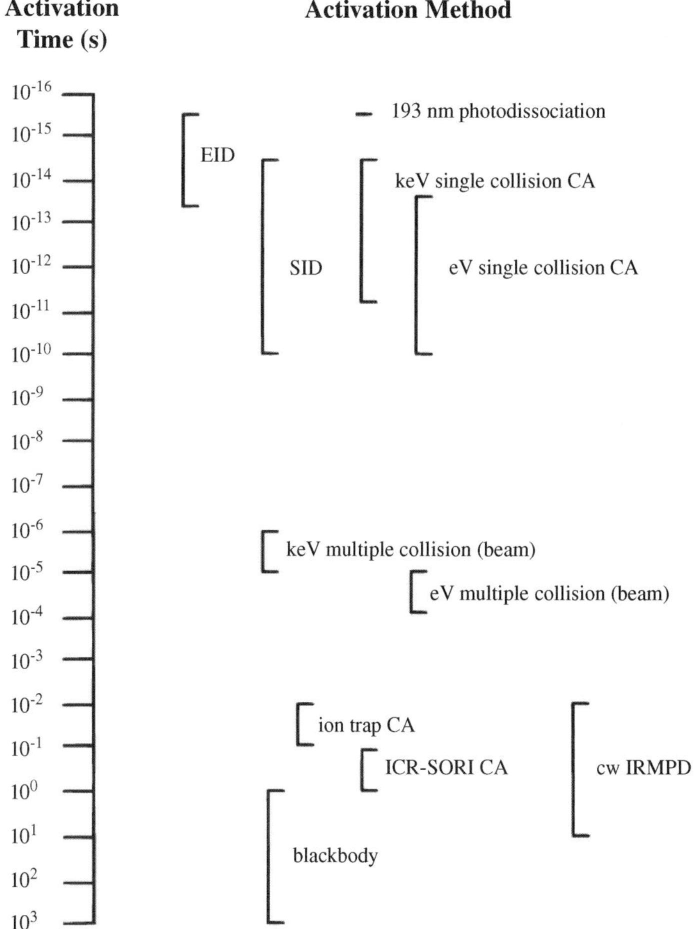

Figure 16.1. Typical ranges of activation times associated with various activation methods in tandem mass spectrometry. EID, electron-induced dissociation; SID, surface-induced dissociation; CA, collisional activation; (beam), typical conditions used in a beam-type tandem mass spectrometer; ICR-SORI, ion cyclotron resonance sustained off-resonance irradiation; cw IRMPD, continuous-wave infrared multiphoton dissociation. (Reproduced from Ref. 15 with permission from John Wiley & Sons.)

and ETD are not included. However, these techniques are difficult to directly compare to the techniques shown in Figure 16.1 because they involve radical-driven chemistry and therefore *routinely* produce radical products from even-electron precursor ions, a process that is not typical for other techniques. Furthermore, the timescale—that is, whether the imparted energy is distributed over all vibrational degrees of freedom prior to bond cleavage (ergodic behavior) or not (nonergodic behavior)—of ECD is still heavily debated in the literature,[16–19] as further discussed below. Due to the highly varying principles of the different MS/MS activation methods, the fragmentation outcome for a particular analyte ion can vary drastically between techniques, thereby resulting in complementary information. The technique chosen also depends on the type of mass spectrometer used—that is, triple quadrupole versus quadruple ion trap versus time-of-flight, versus Fourier transform ion cyclotron resonance (FT-ICR), versus hybrid instruments, and so on—because optimum implementation of the various activation methods may require different geometries. For example, photodissociation and electron-based methods are most straightforwardly implemented in trapping type mass analyzers. This chapter will discuss the physical principles of MS/MS activation methods and describe their utility (pros and cons) for specific applications in small and large molecule analysis. For a more comprehensive overview of the fundamentals of ion activation and dissociation, readers are referred to, for example, the Wiley book *Principles of Mass Spectrometry Applied to Biomolecules*.[20]

16.2 COLLISIONAL ACTIVATION METHODS

Energetic ion–neutral collisions (collision-induced dissociation) is the most common means of ion activation in tandem mass spectrometry. This activation method can be achieved in a variety of tandem mass spectrometers, including triple quadrupoles, hybrid quadrupole/time-of-flight (TOF) instruments, quadrupole ion traps, FT-ICR mass spectrometers, hybrid quadrupole ion trap/orbitrap, TOF/TOF instruments, and hybrid sector instruments. In CID, the neutrals typically correspond to inert gases such as helium, nitrogen, or argon, although heavier gases (e.g., xenon) may be used to transfer more energy into precursor ions[21] [see Eq. (16.1) below]. CID is traditionally divided into two energy regimes termed "low-energy CID" and "high-energy CID," where the divider between the two is around 100 eV. Each regime is described separately in the text below, followed by a section on ion–surface collisions.

16.2.1 Low-Energy Collision-Induced Dissociation (CID)

16.2.1.1 Mechanism

Low-energy CID is frequently further divided into "beam-type CID" and "ion trap-type CID" where the latter type involves lower collision energies than the former. This division is clarified in Figure 16.1 (reproduced from an excellent CID review by McLuckey and Goeringer[15]), which shows that ion trap collisional activation (CA) and ICR-sustained off-resonance irradiation (SORI)[22] CA (both of which involve multiple collisions) occur over much longer timescales than CA from <100-eV multiple collisions in a beam-type instrument (e.g., a triple quadrupole or hybrid quadrupole/TOF configuration). Thus, each collision imparts less energy in ion trap-type CID compared to beam-type CID. This behavior can be understood from the fact that *collisions alone* are not sufficient to raise the internal energy of precursor ions to a level that will cause bond rupture: In order to achieve energetic overall inelastic collisions, the *kinetic energy* of precursor ions also needs to be raised.

In beam-type instruments, precursor kinetic energy is determined by the potential difference between the ion source and the collision cell; the latter is located between the two mass analyzers responsible for the two stages of mass analysis. This potential difference can typically be varied in the 0- to 100-V range, depending on instrument design and vacuum conditions. Beam-type instruments are also referred to as tandem-in-space mass spectrometers because precursor ion selection and product ion detection occur at different spatial locations and are performed by two separate mass analyzers. In ion trapping instruments (also termed tandem-in-time mass spectrometers), precursor ion trajectories must be confined within the dimensions of the trap and, hence, precursor ion kinetic energy can only be raised to a level where stable ion motion still occurs. Consequently, the achievable precursor kinetic energy is lower than that in a beam-type instrument, typically $<15\,\mathrm{eV}$.[23] In quadrupole ion traps, kinetic energy is raised by introducing a radio-frequency (rf) waveform in resonance with the axial frequency of selected precursor ions. In FT-ICR mass analyzers, an rf waveform either on- or slightly ($\sim$1000 Hz) off-resonance with the ICR frequency of selected precursor ions is applied. For on-resonance excitation, the rf waveform has to have either low amplitude or short duration to avoid ejection of precursor ions. The advantage of off-resonance excitation is that precursor ions remain close to the center of the ICR cell due to periodic (period equal to the difference between the excitation frequency and the ICR frequency)[22] acceleration/deceleration of precursor ions. Longer activation times are therefore allowed and product ions are generated close to the center, which facilitates their detection.

Regardless of the precise low-energy CID implementation, the fraction of precursor kinetic energy that can be converted to internal energy in a collision is determined by the collision energy in the center-of-mass frame of reference[2]:

$$E = \frac{N}{m_p + N} E_k \tag{16.1}$$

where N is the atomic/molecular weight of target gas, m_p is the molecular weight of precursor ion, and E_k is the precursor ion kinetic energy. In the $<$100-eV regime (i.e., low-energy CID), the resulting internal energy increase is generally sufficient to vibrationally, but not electronically, excite precursor ions. The internal energy increase is distributed over all analyte vibrational degrees of freedom (ergodic process) and therefore results in the cleavage of the most labile chemical bonds. It is important to note, however, that the internal energy increase described by Eq. (16.1) is not a single value but an energy transfer distribution, which broadens with increasing precursor ion kinetic energy.[2] This characteristic of CID generally results in observation of product ions associated with various energy barriers with more complex spectra for higher kinetic energies—for example, in beam-type instruments versus ion traps. As an example, a direct comparison between beam-type and ion-trap-type CID of protonated ubiquitin ions in a linear quadrupole ion trap, performed by McLuckey and co-workers,[24] is shown in Figure 16.2. It is clear that much more selective fragmentation (likely due to discrimination against higher-energy channels[25]) occurs in ion-trap-type CID (left panels) with abundant products corresponding to backbone cleavages on the C-terminal side of aspartic acid and the N-terminal side

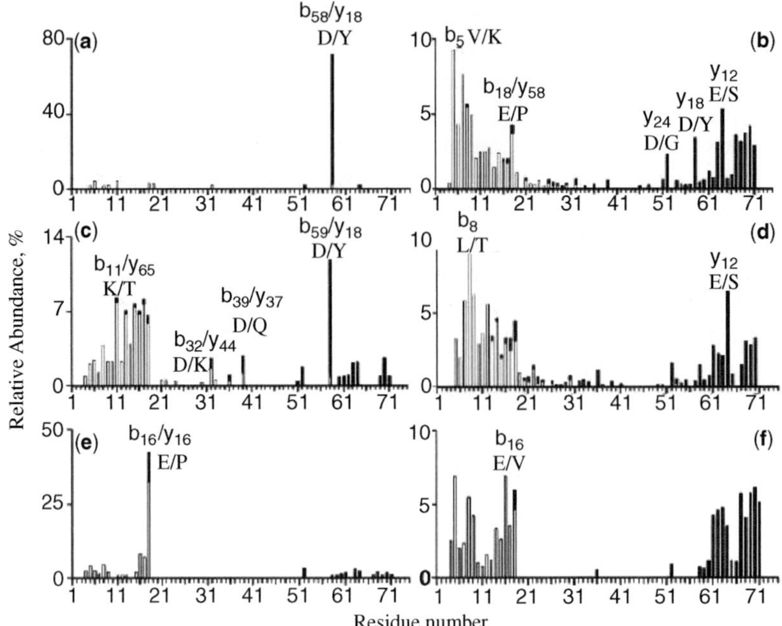

Figure 16.2. Relative abundances of *b*-type (white column) and *y*-type (black column) product ions resulting from cleavage of the ubiquitin backbone with ion trap CID for (**a**) $[M + 6H]^{6+}$, (**c**) $[M + 8H]^{8+}$, and (**e**) $[M + 11H]^{11+}$ ions; and with beam-type CID for (**b**) $[M + 6H]^{6+}$, (**d**) $[M + 8H]^{8+}$, and (**f**) $[M + 11H]^{11+}$ ions. (Reproduced from Ref. 24 with permission from the American Chemical Society.)

of proline. These peptide linkages are known to be particularly labile in peptide and protein CID.[26–28]

In addition to the above-mentioned adaptability to various types of mass spectrometers, another great advantage of low-energy CID is the degree to which it can be driven in terms of fragmentation efficiency, which translates into analytical sensitivity. In beam-type CID, an analogy to Beer's law can be made[2]:

$$I = I_0 e^{-\sigma n l} \tag{16.2}$$

where I is the precursor ion abundance following CID, I_0 is the initial precursor ion abundance, σ is the cross section of all loss processes for the precursor ion, including CID, scattering, and charge transfer to the target gas (the cross section for CID is dependent on the energy transfer distribution, the precursor ion internal energy prior to CID, and the rate constants for dissociation: in practice this value amounts to $\sim$10–100 Å^2),[2] n is the target gas number density, and l is the path length. From this analogy, it is clear that the fragmentation can be driven toward completeness by increasing the precursor ion kinetic energy (which affects the energy transfer distribution) or by changing the target gas pressure (changing the path length is generally not practical). Both these changes result in increased scattering, which reduces the quality of MS/MS spectra. Alternatively, ion internal energy can be raised to increase both fragmentation efficiency and the number of fragmentation pathways. For example, Glish and co-workers have demonstrated thermally assisted CID in a quadrupole ion trap.[29]

Further advantages of CID include its applicability to both cations and anions as well as to both singly and multiply charged ions. Activation of multiply charged ions is advantageous for larger molecules due to their significantly higher number of vibrational degrees of freedom compared to smaller molecules; therefore, the former need higher energy input to reach dissociation thresholds. Multiply charged ions are accelerated to higher kinetic energies through a fixed potential difference compared to singly charged ions and therefore can undergo more facile dissociation. Furthermore, the presence of more than one charge of equal polarity in a gas-phase analyte ion introduces Coulomb repulsion, which renders precursor ions less stable and therefore further contributes to more facile dissociation. Due to these advantages of multiple charging, electrospray ionization (ESI, which typically generates multiply charged ions for large molecules)[30] is more commonly used than matrix-assisted laser desorption/ionization (MALDI) (which typically generates singly charged ions)[31,32] to ionize large analytes prior to low-energy CID. Precursor ion polarity in ESI is typically chosen according to the analyte pK_a, although positive ion mode is more commonly employed than negative ion mode for CID MS/MS. One reason for this bias is that the fragmentation behaviors of protonated species is better understood than those of deprotonated species. Also, more analytically useful fragmentation pathways are observed in positive ion mode for common analyte types such as peptides.[33–35]

One disadvantage of low-energy CID is its limited reproducibility (in terms of both product ion types and relative abundances) on different types of mass spectrometers due to the sensitivity of energy transfer distributions to small changes in experimental conditions. This reproducibility issue has precluded generation of universal (instrument-independent) CID spectral libraries for identification of unknowns. However, significant efforts have been made, and are currently being made, toward this goal. For example, a recent report by Hopley et al. compared CID spectra of 48 compounds on 11 mass spectrometers (located in different laboratories): six ion traps, two triple quadrupoles, a hybrid triple quadrupole, and two quadrupole time-of-flight instruments.[36] Their results show that use of a spectral calibration point for each instrument allows a degree of reproducibility across all instrument types. They further state that reproducibility of CID product ion spectra is increased when comparing tandem-in-time-type instruments and tandem-in-space-type instruments as two separate groups. Another example of an extensive effort to catalogue CID MS/MS spectra is the web-based METLIN metabolite database.[37] A second potential disadvantage of low-energy CID is the preferential cleavage of the most labile chemical bonds. Such cleavages may not always provide the maximum amount of analytical information. For example, a key feature of phosphopeptide characterization is to identify the presence but also the precise location of a phosphate group. Because phosphate groups are labile, H_3PO_4 and/or HPO_3 loss generally dominates in low-energy CID of cationic phosphopeptides. This neutral loss is highly useful for phosphopeptide identification[38]; however, because the phosphate is lost, information on its location within the peptide can be difficult to ascertain with low-energy CID.

16.2.1.2 Low-Energy CID Application

Because of its highly versatile nature, low-energy CID has been applied to virtually every class of ionizable compound, including small molecules and biomacromolecules. A complete description of low-energy CID applications is beyond the scope of this chapter, and only a few examples will be given. First, low-energy CID, frequently in combination with on-line liquid chromatography (LC), plays a significant role in drug metabolism analysis, including pharmacokinetics.[39–42] Most analyses of this type involve quantification of a drug and its key metabolites in a biological fluid. Due to the complexity of the matrix involved, there is a high probability that compounds having similar retention times and molecular weights as the analytes of interest are present and therefore may result in false positives. However, use of LC/MS/MS (with low-energy CID as the activation method) rather than LC/MS can ensure more accurate quantitation because fragmentation pathways of background compounds may be different from those of analytes of interest. Many LC/MS/MS protocols involve so-called selected (single) ion reaction monitoring (SRM) or multiple ion reaction monitoring (MRM) with a triple quadrupole mass spectrometer. Here, quantification is based on selected CID product ions associated with a specific precursor ion mass. As an example, Figure 16.3 shows a comparison between LC/MS (on a TOF instrument) and SRM LC/MS/MS (on a triple quadrupole instrument detecting a CID product ion at 97.9 amu). This work, by Zhang and Henion,[43] seeks to quantify idoxifene in human plasma. As can be seen, there is background from endogenous compounds in LC/MS but not in LC/MS/MS. Furthermore, an improved limit of detection was obtained with LC/MS/MS. Similar LC/MS/MS protocols are frequently applied in environmental,[44–46] toxicological,[47,48] and forensics[49] applications.

In addition to the above-described small molecule analyses, which typically involve singly charged ions, low-energy CID is tremendously valuable for analysis of large multiply charged biomolecular ions. In particular, CID MS/MS in combination with two-dimensional chromatography of peptides is one of the preferred methods of protein identification in proteomics.[50,51] The key factor behind the success of low-energy CID for this purpose is the highly selective cleavage of backbone amide bonds of peptides

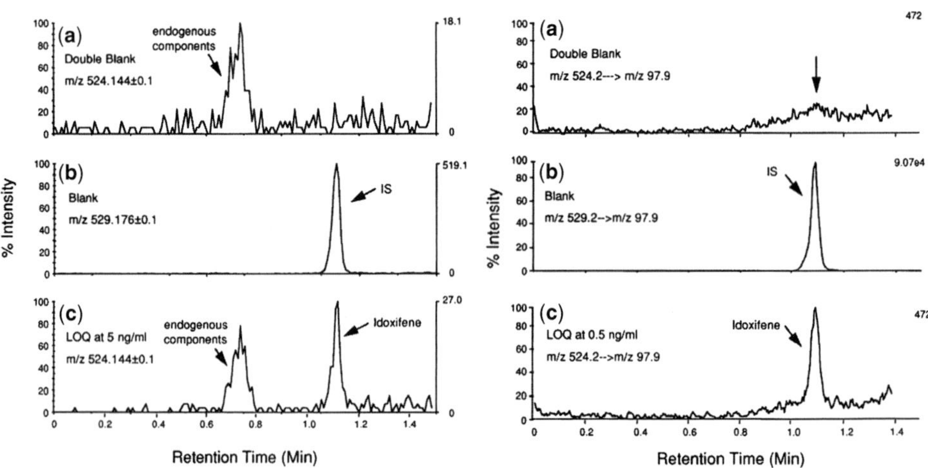

Figure 16.3. LC-TOF-MS (**left panels**) and SRM LC-triple quadrupole (TQ)-MS (**right panels**) extracted ion chromatograms of idoxifene and d5-idoxifene. (**A**) Double blank human plasma extracts for idoxifene. (**B**) Blank human plasma extracts for d5-idoxifene. (**C**) Control blank human plasma extract spiked at limit of quantification of the method (5 ng/mL for LC-TOF-MS and 0.5 ng/mL for SRM LC-TQ-MS) for idoxifene. (Adapted from Ref. 43 with permission from Elsevier Science.)

that provides ladder-like product ion spectra in which mass differences between peaks correspond to amino acid residue masses, thus providing peptide "fingerprints" that can be searched against genomic data for protein identification.[34,52] This method has some disadvantages though, including altered fragmentation pathways, mainly involving side chain losses, when labile post-translational or chemical modifications are present.[53] To partially circumvent this problem, low-energy CID of intact proteins (so-called top-down proteomics)[54] is also employed. However, limited fragmentation is typically observed due to the large number of vibrational degrees of freedom at higher molecular weights.

Although peptide and protein analyses dominate in biomacromolecular mass spectrometry, low-energy CID is also highly useful for characterizing other types of large molecules, including nucleic acids and carbohydrates. Oligonucleotides undergo selective fragmentation in CID: For DNA, neutral nucleobase loss is followed by cleavage of the backbone C–O bond 3′ to the nucleotide from which the nucleobase was lost[55] and, for RNA, direct backbone P–O bond cleavage on the 3′ side of phosphorous occurs.[56,57] In both cases, ladder-like product ion spectra are generated, similar to peptide low-energy CID, which can be interpreted to yield oligonucleotide sequence information.[58] Such sequence information can be valuable for the analysis of genetic markers, such as short tandem repeats and single-nucleotide polymorphisms.[59] Furthermore, low-energy CID can provide information on the presence and location of chemical modifications, including modified nucleobases and alterations of sugar rings and the phosphate backbone.[60–63] Chemically modified oligonucleotides, which play an important role in biomedical/pharmaceutical research, particularly for antisense applications,[64] often evade characterization by classical enzymatic techniques due to their altered chemical nature.

Carbohydrates are a much more challenging class of biomacromolecules to characterize than proteins and nucleic acids due to their frequently branched nature. This class of molecules fragments readily in low-energy CID, particularly at glycosidic linkages between monosaccharide residues. However, the corresponding product ion spectra cannot be interpreted back to a single structure because there are many possibilities as to how the fragments go back together. One solution to this problem is carbohydrate permethylation prior to CID. This chemical modification alters carbohydrate low-energy CID fragmentation pathways and enhances sugar cross-ring fragmentation; that is, breakage of two covalent bond across a sugar ring is typically observed.[65–67] Such product ions contain information about the precise connectivity between monosaccharide residues and are crucial to branched glycan structural elucidation. Low-energy CID of permethylated glycans is most commonly performed in quadrupole ion traps such that several stages of MS (MS^n) can be utilized to obtain maximum structural information.[65–67]

16.2.2 High-Energy CID

16.2.2.1 Mechanism

The kilo-electron-volt precursor ion kinetic energies required to reach the high-energy CID regime are obtainable in beam-type instruments void of quadrupoles (which require lower ion kinetic energies for optimal mass analysis), including hybrid-sector and TOF-type instruments. The main body of high-energy CID analyses has been carried out on sector-type instruments, which have been available much longer than TOF/TOF mass analyzers. The introduction of MS/MS as a structural tool is attributed to McLafferty[68] and to Jennings.[69] Both groups used sector-type instruments with high-energy CID as the ion activation method. In these types of instruments, precursor ions (selected by a first nonscanning sector) can straightforwardly be accelerated through several kilovolts under high vacuum conditions. Collisions take place in a collision cell located between sectors at a pressure of $\sim 10^{-6}$ torr, sufficient to result in 1–10 collisions. This collision number is in stark contrast to low-energy CID, which typically occurs at much higher pressure ($\sim 10^{-3}$ torr) and can involve as many as 10^6 collisions in a quadrupole ion trap.

At kilo-electron-volt collision energies, the energy transfer distribution extends into an energy range where electronic excitation of precursor ions becomes possible. Direct vertical Franck–Condon transitions may occur if ion–neutral interaction times are very short ($<10^{-14}$ s), which generally is not the case for medium- to large-size molecules undergoing high-energy CID.[2] Instead, a curve-crossing mechanism is more likely in which, for example, a crossover from the first electronically excited state to a higher-level vibrational state of the electronic ground state occurs. Both direct electronic excitation and curve-crossing are likely to result in different fragmentation patterns compared to the slow vibrational excitation occurring in low-energy CID because higher analyte energy levels are sampled. Consequently, high-energy CID spectra are typically more complex than beam-type low-energy CID spectra (which are more complex than ion-trap-type low-energy CID spectra). For example, backbone cleavages other than rupture of amide bonds and side-chain cleavages are typically observed for peptides in addition to the backbone amide bond cleavages that dominate in low-energy CID.[70] High-energy CID is known to promote so-called charge-remote fragmentation—that is, cleavage of chemical bonds located remotely from the charge.[71] This type of fragmentation requires a nonmobile charge and typically results in very different cleavage patterns than low-energy CID in which fragmentation is typically charge driven (e.g., a mobile proton can weaken a particular chemical bond to cause selective cleavage).[35]

A major advantage of high-energy CID is the diverse fragmentation pathways observed due to the broad energy transfer distribution. The resulting rich product ion spectra contain a maximum of structural information. On the other hand, such complex spectra may be difficult to interpret. A second drawback is the significant scattering of ions after multiple collisions that are necessary for sufficient excitation of large analytes. Examples of applications for which high-energy CID can provide unique structural information compared to low-energy CID are given in the following section.

16.2.2.2 High-Energy CID Application

As alluded to in Section 16.2.1.2, proteomics is currently a vast and still expanding field in which MS/MS plays a major role. In bottom-up proteomics, proteins are digested into peptides that can be introduced into a mass spectrometer via either ESI or MALDI. For MS/MS analysis, ESI is advantageous due to its straightforward coupling to liquid chromatography but also due to the typical generation of multiply charged ions, which undergo more facile dissociation (see Section 16.2.1.1). However, MALDI has the advantages of being more tolerable to sample contaminants such as salts and detergents, as well as providing complementary ionization due to its completely different ionization mechanism. For example, peptides that are not detected with ESI can be detected with MALDI and vice versa. Furthermore, MALDI may provide higher sensitivity because the total signal for a particular peptide is not spread out over multiple charges states such as in ESI. Due to these advantages, MALDI may be the preferred ionization method. However, because MALDI generally produces singly charged ions, low-energy CID often provides only limited peptide fragmentation. The driving forces behind the development of the TOF/TOF mass spectrometer are its straightforward coupling to MALDI (TOF analyzers have virtually unlimited mass-to-charge (*m/z*) range, which is compatible with large singly charged ions) and the possibility of performing high-energy CID to dissociate MALDI-generated peptide ions for improved protein identification.

Section 16.2.1.2 also discussed carbohydrate MS/MS and the importance of obtaining cross-ring fragmentation. Such fragmentation is greatly enhanced in high-energy CID compared to low-energy CID.[72,73] High-energy CID allows analysis of underivatized oligosaccharides (versus permethylated ones for low-energy CID), which avoids challenges associated with incomplete reactions, side reactions, and losses in derivatization and clean-up processes. Such challenges are of particular concern when sample amounts are scarce. Another compound class that can greatly benefit from high-energy CID is fatty acids. Such compounds can be ionized to contain a nonmobile charge—for example, metal ions bound to the carboxylate end. Fixed-charge-containing fatty acids dissociate via charge remote fragmentation pathways, and relative abundances of product ions from alkyl chains can provide information on chemical functionalities such as double bonds and oxo groups.[71] An example, by Cheng and Gross,[71] of isomeric lithium cationized oxyfatty acid, $[M - H + 2Li]^+$ (where M denotes the neutral-molecular-weight), charge-remote fragmentation by high-energy CID, is shown in Figure 16.4. Here, the respective locations of the oxy groups are revealed from the charge-remote fragmentation patterns.

16.2.3 Surface-Induced Dissociation (SID)

16.2.3.1 Mechanism

Equation (16.1) states that, in order to vary the energy transfer distribution in an ion-neutral collision, the precursor ion kinetic energy can be varied. The second parameter that can be varied is the mass of the target. In surface-induced dissociation,[5,74,75] a surface rather than a neutral gas molecule is used as collision target, which extends the target mass. Thus, very high precursor ion internal energies should be reachable, which may result in unique fragmentation patterns. Another advantage of SID is that no collision gas is needed. Thus, this activation method may be more compatible than CID with high-vacuum mass analyzers such as FT-ICR. Furthermore, it may provide more reproducible MS/MS spectra because a specific target pressure does not need to be maintained.

The use of a surface as collision target was first demonstrated by Cooks et al.[76] in a sector-type mass spectrometer by dissociation of CO_2^{2+}. These authors concluded that SID resulted in a fragmentation spectrum very similar to that of high-energy CID of the same species, and they highlighted experimental convenience as the major advantage of SID. Since this first implementation, SID has been demonstrated in a variety of mass spectrometers, including quadrupole,[77] TOF,[78] hybrid Q-TOF,[79] and FT-ICR.[74] A crucial aspect of

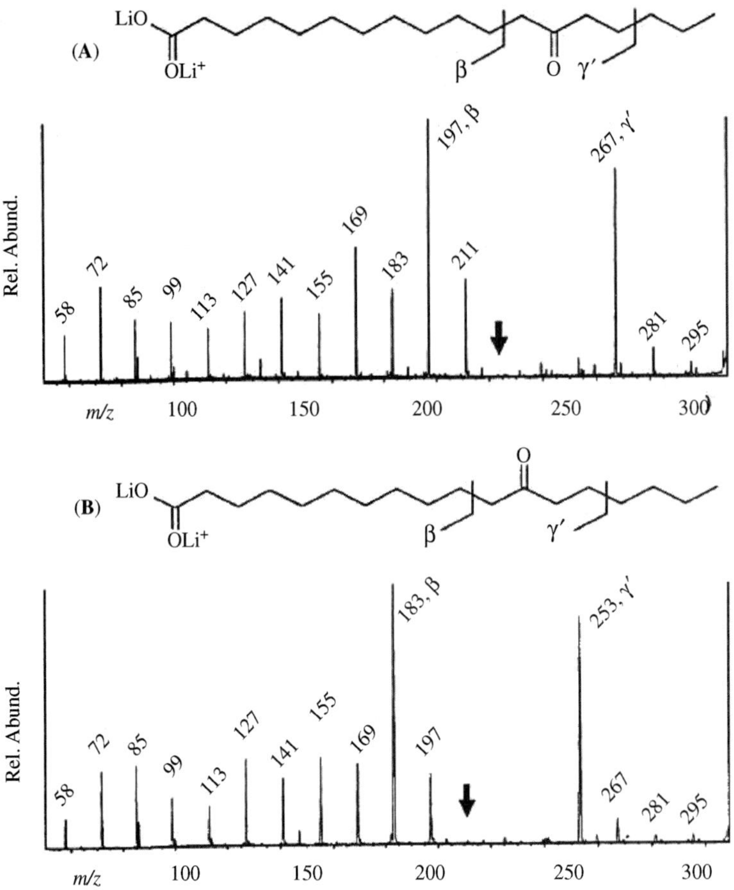

Figure 16.4. High-energy CAD spectra of isomeric oxofatty acids charged with Li^+ as [M − H + 2Li]. The arrows indicate a point in the spectrum that corresponds to the location of the oxo group. (**A**) 13-oxo- and (**B**) 12-oxooctadecanoic acid (*m/z* 311). (Adapted from Ref. 71 with permission from John Wiley & Sons.)

SID is the choice of target surface. Different surfaces provide differences in parameters such as neutralization probability (highly important because mass spectrometers cannot detect neutrals), sticking probability, internal energy uptake, and the nature of any reactive encounters.[5] Metal surfaces result in a high (>99%) degree of projectile ion neutralization[74] and have to be thoroughly cleaned prior to use.[5] Thus, other surfaces with high ionization energies are preferred, such as organic thin films on metal substrates. Self-assembled monolayers (SAMs) are popular choices,[80–82] and use of Langmuir-Blodgett films has also been demonstrated.[83] Other surfaces used for SID include, for example, graphite[84] and metal halides or diamond films.[85]

Another choice in implementing SID is the geometry of the ion–surface collision with respect to the mass analyzer—that is, whether the surface should be mounted at a 90° or smaller angle with respect to the incoming ion beam, or whether grazing incidence should be used. Molecular dynamics simulations[86,87] and experimental results[88] have shown that precursor ion internal energy increase is almost independent of the incidence angle with a slight maximum at 45°. However, Jungclas and co-workers[89] have shown that grazing incidence ion–surface interactions can yield particularly high (40–70%) kinetic to internal energy conversion efficiencies, proposed to be due to a different excitation mechanism than for nongrazing SID. Following ion–surface interaction, product ion yields and types may also show a scattering angular dependence such that the detection angle needs to be chosen carefully. Most instruments with a 45° incidence angle also use 45° for detection, although some systems have been built that allow selection of both incidence and scattering angles.[90–92] In all types of SID instruments, the target surface is usually mounted on a translator/rotator such that product ion abundance can be optimized at the detector.

As stated above, major advantages of SID are the high level of control over the collision energy, the high-energy fragmentation pathways that can be accessed (some not accessible with high-energy CID), and the absence of a need for a collision gas. The main drawback is signal loss due to partial neutralization.

16.2.3.2 SID Application

Ion–surface collisions have been vastly employed to study fundamentals of this process (e.g., chemical sputtering, elastic and reactive ion scattering, and soft landing)[5,93] and to analyze surface composition. Here, only examples of SID as an activation method to yield structural information about the precursor ions themselves will be provided. One advantage of SID over CID is the absence of a pressure dependence, which allows for a high level of control over the collision energy. This level of control is important for gaining detailed information on the energetics of precursor ion dissociation. For example, it has been used to determine magic numbers (i.e., preferred stoichiometries) in sodium fluoride clusters[94] and to gain detailed understanding about the energetics and dynamics of peptide fragmentation[95,96] with vital contributions toward developing the mobile proton model for peptide dissociation.[97,98] A complete knowledge of peptide fragmentation pathways will allow prediction of peptide MS/MS spectra, in terms of both product ion *m/z* values and relative abundances, which would greatly enhance MS/MS-based protein identification.

Similar to high-energy CID, SID can be advantageous for dissociating MALDI-generated singly charged peptide ions and for accessing higher-energy fragmentation pathways such as amino acid side-chain losses, which can differentiate between leucine and isoleucine.[70] Several MALDI-SID combinations have been reported in the literature.[99–102] SID has also been applied toward intact protein dissociation: For example, McLafferty and co-workers[3] applied this activation method to fragment the 29-kDa protein carbonic anhydrase. These authors stated that secondary fragmentation was minimized in SID compared to CID and IRMPD. This characteristic facilitates spectral interpretation. More recent work by Wysocki and co-workers[103] applied SID to the dissociation of protein complexes and compared the resulting fragmentation patterns to those from low-energy CID. Intriguingly, large differences were seen with more symmetric dissociation in SID versus the previously observed highly asymmetric[104,105] dissociation in CID. Figure 16.5 shows an example involving CID and SID of the 11+ charge state of the cytochrome *c* noncovalent dimer.[103] In CID (Figure 16.5a), the main dissociation pathway is formation of one highly charged (8+) monomer and one monomer with less charge (3+) (accounting for the total charge of 11). Previous work has explained this asymmetric behavior based on proton transfer between monomers and significant unfolding of the highly charged monomer.[104,106,107] In SID (Figure 16.5b), the main dissociation pathway of the same

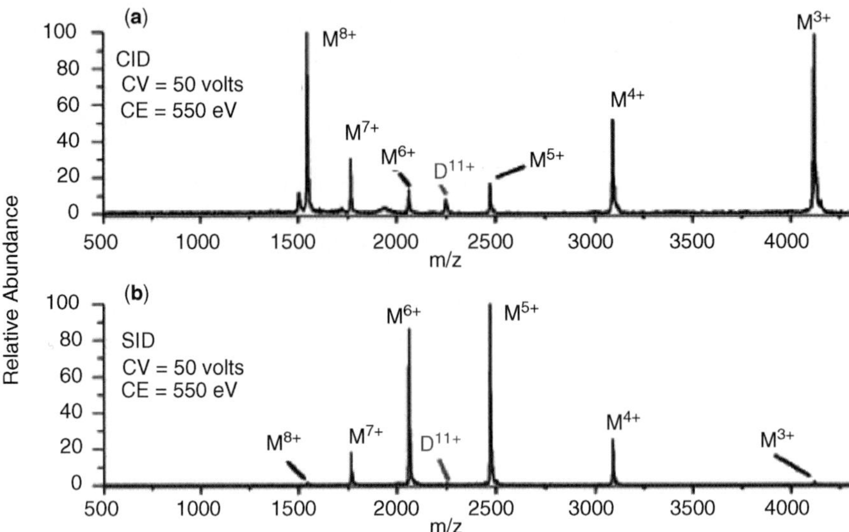

Figure 16.5. Comparison of tandem mass spectra for the 11+ charge state of the cytochrome *c* dimer by low-energy CID (**a**) and SID (**b**). The precursor ion is shown in gray. The voltage difference between the source hexapole and the collision cell (CID) or the surface (SID) is listed as ΔV. The collision energy (CE), the product of ΔV and the precursor ion charge state, is also given for each spectrum. (Adapted from Ref. 103 with permission from the American Chemical Society.)

11+ dimer is formation of monomers with 5 and 6 charges, respectively—that is, the most symmetric fragmentation possible for this charge state. The stark contrast observed between CID and SID is likely due to the very different timescales of these two processes (see Figure 16.1): Because sufficient energy for dissociation is imparted on a very short time scale ($<10^{-10}$ s) in SID, there is not enough time for unfolding of one monomer. From a practical standpoint, asymmetric dissociation can limit the amount of structural information that can be derived from MS/MS spectra because the *m/z* values of, for example, a quadruply charged dimer and a doubly charged monomer are the same if the molecular weight of monomers is the same. Thus, peak assignments can be ambiguous. SID may play an important future role in elucidating the stoichiometry of protein complexes because charge appears to distribute in proportion to molecular weight, thereby facilitating spectral interpretation.

16.3 PHOTON-INDUCED ACTIVATION METHODS

Although MS/MS activation is most commonly performed using CID, alternative activation methods are gaining importance. In MS-based photodissociation experiments, ion activation (and subsequent dissociation) occurs through the absorption of one or more photons (i.e., single or multiphoton dissociation). Photon-induced dissociation can be carried out over a wide range of photon wavelengths (energies)—the ultraviolet (UV), visible (VIS), and infrared (IR) regions of the electromagnetic spectrum. Notably, with the growing popularity of trapped ion MS instruments, there has been a dramatic increase in the use of IR photodissociation techniques.

16.3.1 Infrared Multiphoton Dissociation (IRMPD)

16.3.1.1 Mechanism

The frequency of IR radiation is comparable to those of the vibrational modes of molecules, and activation of isolated gaseous ions occurs by absorption through IR-active vibrational modes. The energy of a single IR photon is generally too small to overcome the fragmentation reaction energy barrier. For example, the energy per photon from a 10.6 μm CO_2 laser is only 0.117 eV. Consequently, the absorption of multiple IR photons (typically tens or hundreds) is required for dissociation (i.e., IR multiphoton dissociation or IRMPD). High-power (MW cm^{-2}) pulsed lasers are particularly useful for the activation of small polyatomic ions. Resonant or near-resonant absorption of multiple photons "forces" the ions up a ladder of discrete vibrational energy levels until the ion reaches the vibrational quasi-continuum, wherein the density of vibrational states becomes sufficiently large that there is at least one vibrational energy level that matches the energy of the laser. In this case, excitation readily occurs by incoherent absorption of multiple photons, and rapid intramolecular vibrational redistribution effectively transfers the energy out of the pumped mode and into other vibrational modes.[108]

Low-power (<100 W cm^{-2}) continuous-wave (cw) lasers, such as a cw CO_2 laser, combined with long irradiation times (> ms) can also be used to effect IRMPD.[108] This approach was first demonstrated by Beauchamp and co-workers[7,109] for the dissociation of ions of small organic molecules and related noncovalent complexes. Owing to their high internal energies, large ions at room temperature already exist in the quasi-continuum and can be readily activated, nonresonantly, using a cw laser. This type of IRMPD is generally implemented using FT-ICR or other trapping MS instruments. Because of the relatively long irradiation times required, relaxation processes can be significant. Collisional or radiative cooling can offset the increase in energy due to photon absorption, thereby slowing the rate of dissociation. The importance of the collisional cooling will depend on the background pressure. In FT-ICR MS, which is performed at ultrahigh vacuum (< ntorr), this mechanism of cooling is not significant, whereas in quadrupole ion traps, which normally operate at ~mtorr pressures, these effects can be significant. However, the effects of collisional relaxation can be mitigated using "thermally assisted" IRMPD, wherein the average internal energy of the ions is increased (leading to greater dissociation) by heating the trap bath gas.[110]

The general kinetic scheme for unimolecular dissociation reactions initiated by IR multiphoton absorption is given below for the hypothetical ion AB^+:

$$AB^+ + h\nu \underset{k_{-1}}{\overset{k_1}{\rightleftarrows}} AB^{+*} \xrightarrow{k_d} A^+ + B \tag{16.3}$$

where AB^{+*} is the activated species and A^+ and B are the reaction products, k_1 and k_{-1} are the energy-dependent rate constants for absorption and emission of radiation, respectively, and k_d is the energy-dependent rate constant for dissociation. The dissociation kinetics are characterized by an induction period followed by first-order decay of the irradiated ion. The induction period is the time required to increase the internal energy of the ions above the threshold energy for dissociation.[74] When the dissociation rate is slow compared to the rates of energy transfer, the rapid energy exchange limit (REX) is achieved and the internal energy distribution of the reactant ion will resemble that of a Boltzmann distribution.[74] In such cases, although ions are heated by IR radiation at a single frequency, the internal energy distribution resembles that achieved in BIRD experiments (as discussed in Section 16.3.2.2). The corresponding temperature will depend on the nature of the ion and the laser characteristics (frequency and intensity) and can be established by comparing the dissociation kinetics with those determined by BIRD.

The advantages of IRMPD over CID for MS/MS of gaseous ions are numerous. IRMPD does not require a collision gas or the translational excitation of ions. As a result, there are fewer ions lost to scattering. Another advantage is the ease of controlling the extent of reactant ion dissociation and secondary fragmentation by varying laser power and irradiation time. The ability to generate secondary product ions, which can provide important structural information, without the need for additional stages of MS is often advantageous. Additionally, IRMPD has been shown to be more efficient than CID at fragmenting large ions, a feature that is particularly valuable in the analysis of biomolecules.[5] A disadvantage of IRMPD is the preferential sampling of low-energy dissociation pathways. Furthermore, the IRMPD technique generally requires the use of trapping MS instruments.

16.3.1.2 IRMPD Application

IRMPD is used extensively for the structural characterization and identification of biopolymers (peptides, proteins, DNA, and oligosaccharides) and their noncovalent complexes, as well as organic molecules and pharmaceuticals. Additionally, IR dissociation experiments, in particular when performed with tunable IR light sources, can provide a wealth of information about (a) the structures and conformations of gaseous ions, (b) the nature of their reactions, and (c) the thermodynamics, kinetics, and dynamics of the chemical processes. In general, the fragmentation mass spectra of biopolymers obtained by IRMPD (using 10.6-μm light from a CO_2 laser) are similar to those observed with low-energy CID. For example, both IRMPD and CID of protonated peptides yield predominantly *y*- and *b*-type ions. These are low-energy product ions that are generated from the cleavage of the peptide backbone amide bonds. A side-by-side comparison of the fragmentation spectra of several model peptides (e.g., bradykinin and melittin) using IRMPD and low-energy CID, implemented with a quadrupole ion trap, revealed that the sequence coverage was similar in the two cases, but that there were more abundant low-*m*/*z* ions, which are potentially useful for analyzing post-translational modifications, observed in the case of IRMPD.[111] The advantages of IRMPD over CID are more pronounced for modified peptides, such as phosphorylated peptides. The phosphate group is an excellent chromophore for 10.6-μm radiation. As a result, phosphorylated peptide ions typically exhibit significantly greater fragmentation efficiencies than do nonphosphorylated peptides[111,112]; Figure 16.6 shows an example from work by Crowe and Brodbelt[111] involving doubly protonated angiotensin II. This feature can be exploited for the selective dissociation of phosphorylated peptides present in peptide mixtures. A recent example involves development of a phosphate-containing cross-linker to selectively detect chemically cross-linked peptides.[113] N-terminal sulfonation of peptides can also be used to aid in simplifying peptide tandem MS spectra by reducing the dissociation activation energies, which promotes greater fragmentation and improved sequence coverage.[114] For example, IRMPD of N-terminally sulfonated peptide ions in a quadrupole ion trap can yield complete *y*-type ion series, while the low-mass cutoff problem associated with ion traps prohibits the analysis of low mass product ions with CID (Figure 16.7).

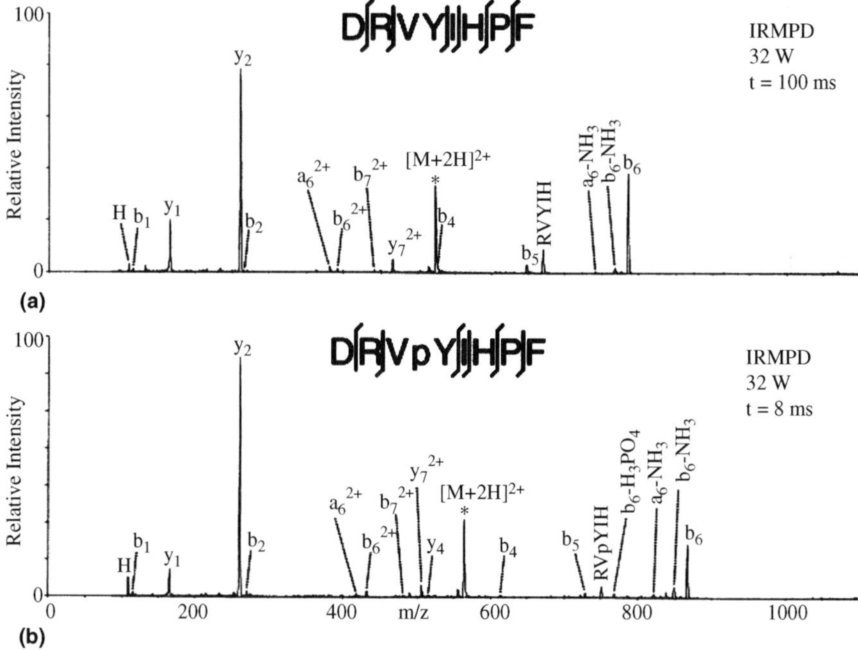

Figure 16.6. IRMPD mass spectra of (**a**) doubly protonated angiotensin II (32-W laser power, 100-ms irradiation time) and (**b**) doubly protonated phosphorylated angiotensin II (32-W laser power, 8-ms irradiation time). The precursor ions are indicated by asterisks. (Adapted from Ref. 111 with permission from Elsevier Science.)

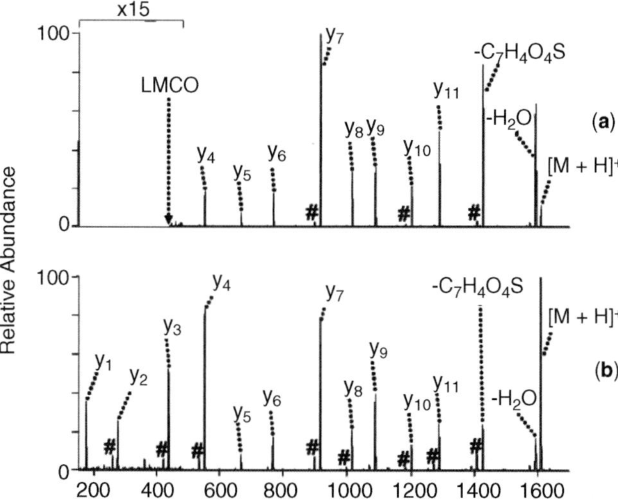

Figure 16.7. ESI-MS/MS spectra of the N-terminally sulfonated peptide HSDAVFTDNYTR from (**a**) low-energy CID and (**b**) IRMPD. LMCO, low-mass cutoff; #, satellite peaks due to loss of H_2O. (Adapted from Ref. 114 with permission from the American Chemical Society.)

IRMPD also represents a valuable tool for the structural analysis of oligosaccharides[115,116] and glycopeptides (N- and O-linked).[117,118] The fragmentation of oligosaccharides with IRMPD is similar to that observed with low-energy CID and results primarily in B-/Y- and C-/Z-type ions produced from cleavage at the glycosidic linkages, although cross-ring cleavage may also occur. However, with CID, multiple stages of activation are frequently required to obtain extensive sequence information, particularly for larger oligosaccharides. In contrast, with IRMPD it is possible to control the extent of secondary fragmentation by adjusting the duration and laser power of the irradiation step, thereby eliminating the need for additional stages of activation. The derivatization of oligosaccharides can provide additional control of the fragmentation process. For example, labeling of oligosaccharides with an IR-active boronic acid reagent (in which a phosphonate group is incorporated into boronic acid) has been shown to enhance the photon absorption process in IRMPD.[119] Also, the mass spectra exhibit a characteristic primary fragmentation pathway involving cleavage from only the nonreducing end, which simplifies interpretation of the mass spectra and, consequently, the determination of oligosaccharide sequences.

Recently, IRMPD experiments carried out with trapping instruments (usually FT-ICR mass analyzers) using pulsed, tunable light sources have successfully produced IR "action" spectra for a variety of gaseous ions, including those of biomolecules (amino acids and small peptides). Briefly, the IRMPD spectrum is generated by monitoring the change in the relative abundance of the reactant ion as a function of radiation frequency. Resonant photon absorption increases the internal energy of the ion, leading to dissociation. Such IRMPD action spectra, combined with electronic structure calculations, can provide direct insights into the structures and conformations of small gaseous ions. For example, IRMPD spectra of cationized arginine ions obtained at wavelengths of 800–1900 cm^{-1} were shown to clearly resolve the charge solvated and salt-bridge structures of these ions.[120] In another recent study, variable wavelength IRMPD performed on the protonated tripeptides, $GlyGlyGlyH^+$ and $AlaAlaAlaH^+$, provided the first direct evidence that the amide oxygens can serve as protonation sites in gaseous peptides.[121]

16.3.2 Blackbody Infrared Radiative Dissociation (BIRD)

16.3.2.1 Mechanism

Ions trapped within the essentially "collision-free" environment of an FT-ICR ion cell can become activated (leading to dissociation) through the absorption of blackbody radiation emitted from the walls of the ion cell. This dissociation technique, originally referred to as *zero-pressure thermal radiation ion dissociation* or ZTRID,[122] is commonly known as *blackbody infrared radiative dissociation* or BIRD.[9] The BIRD technique has emerged as a powerful method for measuring the dissociation kinetics of isolated gaseous ions, particularly those of large biological molecules and their noncovalent complexes. The corresponding kinetic parameters can provide insights into the structures and stabilities of the gaseous ions, as well as the nature of the dissociation mechanisms. The BIRD method produces structurally informative product ions that are similar and, in some cases, complementary to those observed with other activation techniques such as CID and IRMPD and therefore represents a useful addition to the arsenal of activation techniques available for the sequencing of biopolymers.

The distribution of radiation emitted from the walls of the ion cell, which is assumed to behave as an ideal blackbody radiator, is given by the Planck radiation law:

$$\rho(\nu) = \frac{8\pi}{c^3} \frac{h\nu^3}{\exp\left(\frac{h\nu}{k_B T}\right) - 1} \tag{16.4}$$

where $\rho(\nu)$ is the energy density of the radiation at a frequency ν, h is the Planck constant (6.626×10^{-34} J s), c is the speed of light in vacuum (2.998×10^{8} m s^{-1}), k_B is the Boltzmann constant (1.38×10^{-23} J K^{-1}), and T is temperature (in K). At typical ion cell temperatures, which range from room temperature up to ~250°C, the distribution of emitted radiation is dominated by IR photons (Figure 16.8). For example, at 200°C (473 K) the most intense radiation is emitted at 6.2 μm. Consequently, activation of the ions typically occurs by the excitation of vibrational modes through the absorption of multiple IR photons, analogous to the slow heating mechanism underlying IRMPD.

There are two basic requirements for the implementation of BIRD. First, a very low pressure is required such that collisional energy transfer to and from the trapped ions is negligible. Second, because BIRD is a slow heating method, long storage times (typically seconds to minutes) are necessary to achieve extensive fragmentation of the ions. These requirements are easily satisfied using an FT-ICR mass analyzer because ions can be trapped within the ion cell by crossed electric and magnetic fields for times ranging from milliseconds to hours. The ultra-high vacuum ($<10^{-9}$ torr) associated with FT-ICR MS ensures a near collision-free environment for the ions trapped within the cell. For carrying out quantitative kinetic measurements with BIRD there is a third requirement: The corresponding temperature of the blackbody radiation field must be accurately known.

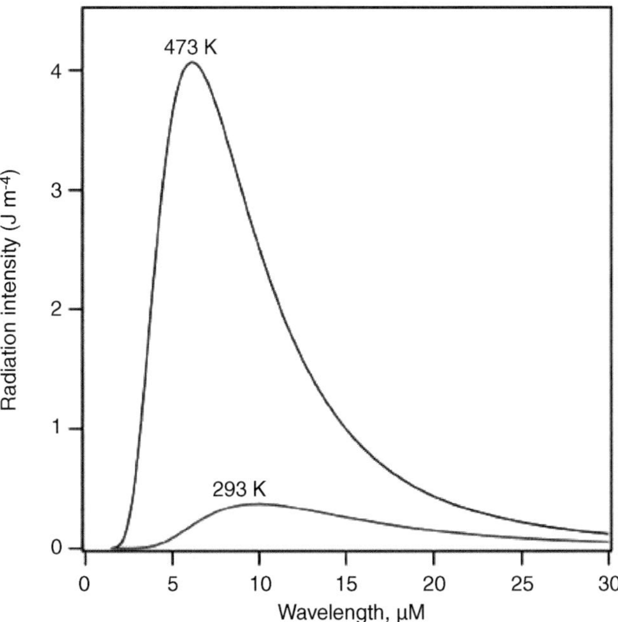

Figure 16.8. Planck blackbody energy distribution at 20°C (293) and 200°C (473 K).

The simplest way to implement the BIRD experiment involves heating the portion of the vacuum chamber that surrounds the ion cell.[9] This implementation is also inexpensive and readily achieved on commercial instruments. The heated vacuum chamber radiatively heats the walls of the ion cell; and, in turn, the temperature of the ion cell defines the reaction temperature in BIRD. The cell temperature can be established by placing thermocouples inside or near the ion cell[9] or by using ion thermometry reactions, in which temperature-dependent dissociation rate constants of thermometer ions establish the reaction temperature.[106] Limitations of this "indirect" approach to controlling the temperature of the ion cell include the modest range of accessible temperatures, typically 20–250°C (293–523 K), as well as the possibility of nonuniform temperatures within the cell.[106] When a broader temperature range or a more uniform cell temperature is required, ion cells employing "direct" temperature control may be used. Several different designs of temperature-controlled ion cells have been reported,[123–125] including one that employs both direct heating and cooling systems and can operate at temperatures ranging from −196 to 165°C (77 to 438 K).[125] Low cell temperatures are beneficial for studying kinetically labile complexes such as small solvated amino acid ions, which have short kinetic lifetimes at ambient temperature.[126] An alternative method for carrying out BIRD involves the use of heated filaments, placed in the proximity of the ion cell, to generate a near-blackbody IR radiation field that serves to heat the ions within the cell.[127,128] The principal advantage of the heated filament approach is that heating of the ions can be carried out in a pulsed fashion and much higher temperatures can be achieved, up to 2300°C. At these high temperatures, the emission of both visible and IR photons is significant and activation may involve both vibrational and electronic excitation. However, because of the small surface area of the filaments that are typically used, the radiation power is low. Also, the radiation field experienced by the trapped ions is a mixture of fields produced by the hot field and the cold walls of the ion cell. As a result, assigning the reaction temperature is not straightforward. The application of BIRD carried out with heated filaments has been described for the dissociation of small organic ions[127,129,130] and small bio-ions.[128]

Time-resolved BIRD measurements provide an opportunity to precisely measure the kinetics of dissociation for isolated gaseous ions. The BIRD technique can, in principle, be used to study the dissociation kinetics of gaseous ions of any size. However, interpretation of the kinetic data depends on the size of the ions (or more accurately, the relative rates of absorption, emission, and dissociation). The general kinetic scheme for unimolecular reactions initiated by photon absorption at zero pressure is analogous to that described for IRMPD [see Eq. (16.3) in Section 16.3.1.1). Application of the steady-state approximation (i.e., the assumption that $d[AB^{+*}]/dt \approx 0$) gives the apparent rate constant (k_{uni}) for unimolecular dissociation:

$$k_{uni} = k_1 k_d / (k_{-1} + k_d) \tag{16.5}$$

For small, reactive ions (typically less than 100 degrees of freedom),[131] the rate of dissociation can greatly exceed the rate of photon emission ($k_{-1} \ll k_d$). In such cases, ions that are activated above the threshold dissociation energy (E_0) rapidly dissociate and the ion population never reaches thermal equilibrium. The ion energy population under these conditions can be modeled as a Boltzmann distribution that is truncated above E_0 (referred to as truncated Boltzmann distribution[132]). For intermediate-size ions that fall between the small molecule and large molecule limits, the dissociation rates are comparable to those of photon emission ($k_{-1} \approx k_d$) and the "truncated Boltzmann" analysis is no longer valid. In this case, a greater fraction of the ion population has energy states above E_0, but a thermal distribution is not achieved. This situation is analogous to the low-pressure fall-off region observed for collisionally activated unimolecular reactions.[133] In these situations, extraction of meaningful kinetic parameters from time-dependent BIRD data requires the use of master equation modeling, as described elsewhere.[132] For large molecules, the rate of energy equilibration with the surroundings by photon exchange is much faster than the rate of unimolecular dissociation ($k_{-1} \gg k_d$), and the ion population achieves thermal equilibrium; that is, the REX limit is achieved, and the BIRD dissociation kinetics will be equivalent to those measured in the high-pressure limit at the same temperature. Whether the REX limit is achieved in a BIRD experiment depends on several factors: the size of the molecule (the number and frequency of the vibrational degrees of freedom), the reaction temperature, and the nature of the dissociation reaction(s). From theoretical modeling of hydrocarbons, a rough lower limit of 500 degrees of freedom has been suggested as being necessary to achieve the REX limit,[8] while for peptides (and presumably other polar molecules) the REX limit is likely achieved at molecular weights of >1.6 kDa.[134]

The temperature-dependence of the BIRD-derived rate constants is normally analyzed according to the Arrhenius equation:

$$k_{\text{uni}}(T) = Ae^{-Ea/RT} \tag{16.6}$$

The Arrhenius activation energy (E_a) and pre-exponential factor (A) are determined from the slope and y-intercept, respectively, of the Arrhenius plot ($\ln k$ versus T^{-1}). Shown in Figure 16.9 are illustrative kinetic and Arrhenius plot constructed from the BIRD-derived rate constants for the dissociation of protonated ions of a protein–trisaccharide complex composed of a 27-kDa single-chain fragment (scFv) of a monoclonal antibody and its native trisaccharide ligand, αGal[αAbe]αMan.[135] It is important to note that the kinetic data obtained for ions of any size can be subjected to Arrhenius analysis. However, the Arrhenius parameters are only meaningful for ions that are within the REX limit. Under conditions where thermal equilibrium is not achieved, the measured E_a and A-factors will be lower than the values that would be obtained for an ion population with a thermal (Boltzmann) energy distribution.

From a comparison of the Arrhenius equation and the thermodynamic formulation of transition state theory of unimolecular reactions, it is seen that the E_a is related to the enthalpy of activation ($\Delta H^{\ddagger}$) and the A-factor is related to the entropy of activation ($\Delta S^{\ddagger}$):[136]

$$E_a = \Delta H^{\ddagger} + RT \tag{16.7}$$

$$A = (ek_BT/h)\exp(\Delta S^{\ddagger}/R) \tag{16.8}$$

The $\Delta H^{\ddagger}$ and $\Delta S^{\ddagger}$ terms correspond to the difference in enthalpy and entropy, respectively, between the transition state and the reactant ion. As described below, the Arrhenius and corresponding activation parameters can provide insight into the structures and stabilities of gaseous ions, as well as the nature of the dissociation mechanisms. Thus, BIRD is a versatile activation method that has a number of valuable applications, in particular the determination of ion dissociation kinetic parameters. Its major drawbacks are

(a)

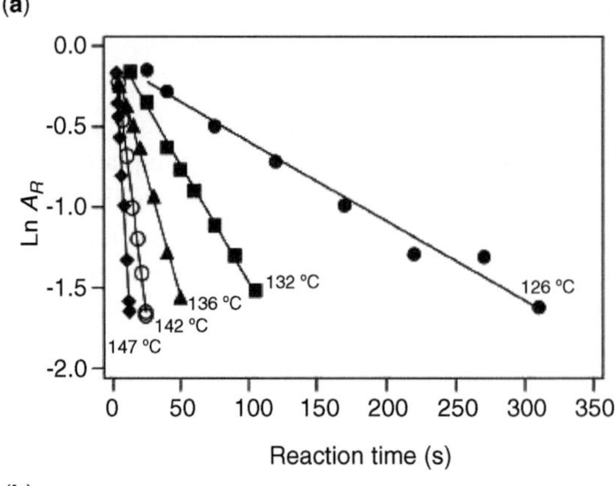

(b)

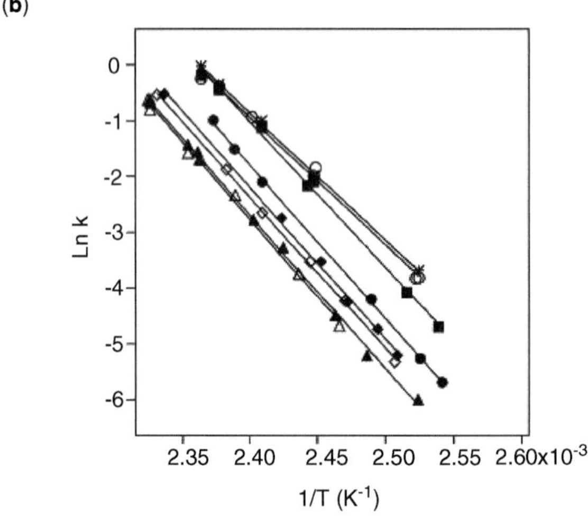

Figure 16.9. (**a**) Kinetic plots obtained by BIRD for the dissociation of the 7 + charge state of a 27-kDa single-chain monoclonal antibody fragment (scFv)-trisaccharide complex, (scFv + αGal[αAbe]αMan)$^{+7}$, at the reaction temperatures indicated (Gal, galactose; Abe, abequose; Man, mannose). (**b**) Arrhenius plots obtained for the loss of neutral trisaccharide ligand from the (scFv + αGal[αAbe]αMan)$^{n+}$ ions at $n = +6$ (◆), +7 (◇), +8 (Δ), +9 (▲), +10 (●), +11 (■), +12 (○), +13 (✱). (Adapted from Ref. 137 with permission from the American Chemical Society.)

(a) the often lengthy time (many hours) required for temperature equilibration of the ion cell and (b) the time-consuming nature of the measurements.

16.3.2.2 BIRD Application

The use of blackbody IR radiation to dissociate isolated gaseous ions within the cell of an FT-ICR mass spectrometer was first demonstrated by McMahon and co-workers[137] for the dissociation of small, weakly bound ion clusters of the type $H_3O^+(H_2O)_n$, $Cl^-(H_2O)_n$, and $H_3O^+((CH_3)_2O)(H_2O)_n$. These ions were found to readily lose neutral water molecules at ambient temperature. At the typical reaction temperatures achievable with BIRD, dissociation of the noncovalent interactions within ionic clusters (e.g., ion–solvent interactions) that have bond dissociation energies of 70–200 kJ mol^{-1}

(17–48 kcal mol^{-1}) readily occurs, and the BIRD technique has been used extensively to study the binding energies and structures of a wide variety of small ionic clusters, including hydrated metal ions[138] and metal cationized amino acids,[139] as well as proton-bound amino acids.[140] The dissociation kinetics generally fall in the intermediate size regime and master equation analysis of the unimolecular kinetics measured by BIRD is necessary. Where available, the BIRD-derived binding energies agree very well with values determined by other methods, including high-pressure mass spectrometry, indicating that BIRD can serve as a convenient and reliable method for evaluating the dissociation energies of ionic clusters.

The BIRD technique has also been applied to gaseous ions of peptides, oligodeoxynucleotides, and oligosaccharides. In general, the dissociation pathways are similar to those observed by other slow heating methods such as IRMPD. For example, BIRD of protonated peptides proceeds, in part, by the cleavage of amide bonds, leading to *a*-, *b*-, and *y*-type sequence ions.[141] However, the observation of low-energy pathways, such as the loss of H_2O or NH_3, is not uncommon. The majority of protein ions are kinetically stable at the highest accessible BIRD temperatures and they do not undergo dissociation. However, there are a few examples of the successful application of BIRD to gaseous protein ions as large as 42 kDa.[142] BIRD of protonated proteins produces some sequence-informative product ions, similar and in some cases complementary, to those generated by CID or IRMPD, although pathways leading to the loss of small neutrals are also significant. Deprotonated oligodeoxynucleotides and protonated oligosaccharides are also highly susceptible to BIRD and the dissociation pathways are similar to those observed with IRMPD and CID, as discussed in Section 16.2.1.2.[143,144]

BIRD studies have been carried out on a number of protein complexes, including antibody–antigen,[135] enzyme–substrate/inhibitor,[145] lectin–carbohydrate,[145] and bacterial toxin multiprotein complexes.[106] At the temperatures investigated, <200°C, protein complexes tend to undergo cleavage of their stabilizing noncovalent interactions. Notably, these BIRD measurements have revealed a new regime of unimolecular kinetics—that is, one that is characterized by large E_a values $>160\,\mathrm{kJ\,mol^{-1}}$ ($40\,\mathrm{kcal\,mol^{-1}}$) and large *A*-factors, $>10^{20}\,\mathrm{s^{-1}}$.[106] For example, Arrhenius parameters for the dissociation of the protonated (scFv + αGal[αAbe]αMan)$^{n+}$ ions at $n \leq 10$ are $\sim 55\,\mathrm{kcal\,mol^{-1}}$ and $\sim 10^{27}\,\mathrm{s^{-1}}$.[135] Typical *A*-factors measured for the dissociation of small gaseous ions rarely exceed $10^{17}\,\mathrm{s^{-1}}$.[136] The large E_a and *A*-factors commonly found for the dissociation of protein complexes are thought to reflect the multiple stabilizing intermolecular interactions that must be broken in the activated complex. These interactions contribute in an additive fashion to $\Delta H^{\ddagger}$ and $\Delta S^{\ddagger}$.[135]

On their own, the Arrhenius parameters for the dissociation of large noncovalent protein complexes are difficult to interpret in terms of the structure of the complex and the precise nature of the intermolecular interactions. Recently, BIRD combined with a functional group replacement strategy (FGR) has emerged as a powerful method to identify individual interactions within gaseous ions of protein–ligand complexes and to quantify the strength of the interactions.[146] To identify whether a particular functional group, either on the protein or ligand, is involved in binding, the group is modified (chemically or biochemically) in such a way that any preexisting interaction is lost. A decrease in E_a upon modification indicates that the particular functional group under scrutiny stabilized the complex. Furthermore, the difference in E_a (i.e., ΔE_a) provides a measure of the strength of the interaction:

$$\Delta E_a = E_a(\text{unmodified complex}) - E_a(\text{modified complex}) \tag{16.9}$$

The BIRD-FGR technique was recently applied to (scFv + αGal[αAbe]αMan) complex to obtain the first detailed and quantitative description of the intermolecular interactions within the gaseous ions of a protein–ligand complex.[135] BIRD was applied to protonated forms of the (scFv + αGal[αAbe]αMan) complex and modified complexes consisting of single-point scFv mutants and monodeoxy analogs of the ligand to generate gas-phase interaction maps over a range of charge states. Comparison of the gas-phase map and solved crystal structures revealed that specific intermolecular hydrogen (H) bonds, at least at certain charge states, were preserved upon transfer of the complex from solution to the gas phase by electrospray ionization. However, evidence for the formation of nonspecific interactions—that is, interactions not present in solution but which form in the gas phase—was also reported. Importantly, the average energies for the three H-bond donor/acceptor pairs identified in the (scFv + αGal[αAbe]αMan)$^{n+}$ ions are in good agreement with theoretical values calculated for model systems.

16.3.3 Ultraviolet Photodissociation (UVPD)

16.3.3.1 Mechanism

Another type of photodissociation technique employs photons in the ultraviolet region to fragment gaseous ions. This technique, commonly referred to as ultraviolet photodissociation (UVPD), has been used to study the structure and dissociation dynamics of a variety of small polyatomics. More recently, the technique has attracted attention as a method for the structural characterization of biopolymers—in particular, peptides. The key advantages of UVPD over other photodissociation techniques, such as IRMPD, are (a) the ability to generate highly endothermic product ions, similar to those observed in high-energy CID, which can provide greater structural information, and (b) the possibility of tuning the fragmentation patterns by varying the excitation wavelength. Additionally, as discussed below, UVPD is more readily implemented on a wider range of MS instruments, compared to other photodissociation techniques.

Excitation of gaseous ions by UV radiation relies on electronic transitions resulting from photon absorption. The transition from the ground state to an electronically excited state occurs by resonant photon absorption and can cause dissociation of the excited ion. The energy separation between electronic states is much larger than the separation between vibrational levels; therefore, the mechanism of UV excitation is significantly different from IR excitation considered for IRMPD and BIRD. Photon energies in the UV region are comparable or larger than those of typical covalent bonds. As a result, the absorption of a single photon can be sufficient to induce dissociation of isolated gaseous ions and, in some cases, can lead to secondary fragmentation.

Much of the current understanding of the mechanism of UVPD of isolated ions comes from the study of small gaseous ions. In small polyatomic ions with numerous available electronic states, a number of different types of photodissociation mechanisms are possible.[147] The simplest situation is photoexcitation to a repulsive electronic surface, which leads directly to dissociation in $\sim 10^{-13}$ s (Figure 16.10). Such a direct dissociation mechanism has been implicated in UVPD (at 366 nm) of CH_3Cl^+, which dissociates exclusively to CH_3^+ and Cl, instead of the lower-energy pathway that leads to CH_2Cl^+ and H.[148] In this case, excitation leads to a surface that is C–H bonding, but C–Cl antibonding. Photoexcitation to an electronically excited bound state, which is predissociated by a nearby dissociative state, may also occur (Figure 16.10). Alternatively,

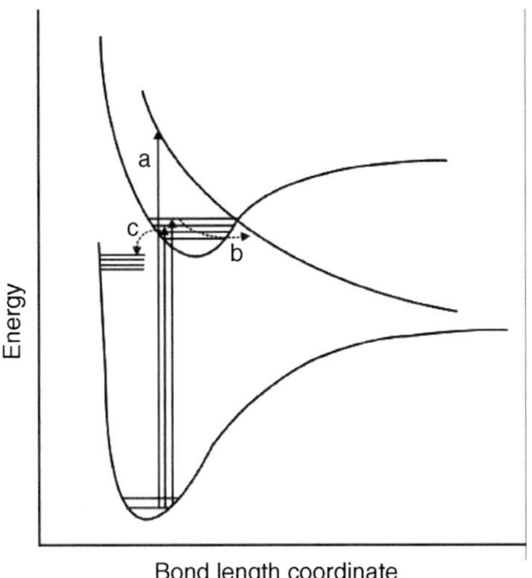

Figure 16.10. Schematic illustration of several possible types of photodissociation for a polyatomic ion: (**a**) Direct photodissociation on a repulsive surface; (**b**) electronic predissociation; (**c**) vibrational predissociation.

internal conversion (intersystem crossing) can lead to a vibrationally excited ion in its ground electronic state, which then undergoes fragmentation (Figure 16.10). Indeed, it is believed that this latter mechanism dominates for most small polyatomic ions. For example, the dissociation of $C_4H_8{}^+$ by irradiation with a Xe arc lamp results in two sets of products:

$$C_4H_8^+ \rightarrow C_4H_7^+ + H \tag{16.10a}$$

$$C_4H_8^+ \rightarrow C_3H_5^+ + CH_3 \tag{16.10b}$$

for which the branching ratios can be described using RRKM theory, a statistical theory of unimolecular decay based on the assumption of rapid intramolecular vibrational energy redistribution.[149] The observation of rearrangement reactions for some ions following electronic excitation is further evidence that UVPD of small polyatomics can proceed via vibrationally hot, electronically ground-state ions. The mechanisms underlying UVPD of large molecules, including biopolymers, are currently a matter of debate. Reilly and co-workers have argued in favor of a nonergodic mechanism of dissociation for protonated peptides, whereby initial cleavage involves electronically excited ions.[154] However, recent UVPD data suggest that dissociation occurs predominantly from ions in their ground electronic state.[150]

Laser UVPD requires a light source that matches an absorption band of the gaseous analyte ion. Amide-containing molecules, such as peptides, strongly absorb UV radiation at ~190 nm, with appreciable absorption centered at ~160 and ~140 nm.[150] As a result, UVPD of peptides and related molecules is readily achieved with ArF (wavelength 193 nm, energy 6.4 eV per photon) and F_2 (wavelength 157 nm, energy 7.9 eV per photon) excimer lasers. The Nd:YAG laser (fourth harmonic at 266 nm, 4.7 eV) has also been used for UVPD. Many biological molecules do not contain chromophores that overlap with common UV laser wavelengths, thereby limiting the application of UVPD. In such situations, appropriate chromophores may be introduced chemically to the molecule of interest.

The majority of UVPD data for gaseous ions has been obtained using trapping mass analyzers. The long observation times associated with trapping instruments enables high fragmentation efficiencies to be achieved, >80%. Ultra-high-vacuum conditions are not required for efficient UVPD because of the high energy deposited per photon and the speed of electronic excitation ($\sim 10^{-15}$ s), compared to collisional relaxation (~ms).[151] As a result, UVPD is readily implemented not only with FT-ICR MS, but also with ion traps that operate at much higher pressures. Because dissociation requires the absorption of only a single photon, UVPD can also be implemented on magnetic sector and time-of-flight instruments, although the fragmentation efficiency is generally much lower compared to trapping instruments. Importantly, the type of mass analyzer used for UVPD can influence the photofragmentation spectra. For example, comparative UVPD experiments performed on the peptide Glu-fibrinopeptide B revealed that more abundant low-energy (pathway) product ions were observed with a quadrupole ion trap (QIT), compared to a TOF instrument.[152] These spectral differences may reflect the longer experimental window associated with the QIT, which allows a greater contribution of slower or secondary reactions to the mass spectra.

16.3.3.2 UVPD Application

Much of the current interest in UVPD is related to its potential for sequencing biopolymers—in particular, peptides. The UVPD technique has a number of attractive features that include the ability to access both high-energy and low-energy product ions, the ability to manipulate the energy deposited based on the laser wavelength used (which provides some control of the fragmentation pathways), and the ease of implementation of the method on a wide variety of mass analyzers. The application of UVPD to gaseous peptide ions, di- and tripeptides exposed to 193-nm radiation, was first reported by Bowers et al.[153] Since then, UVPD of protonated peptides and small proteins has been performed at various excitation wavelengths and with a variety of mass analyzers. Typically, UVPD mass spectra of protonated peptides display features that are common to both high- and low-energy CID, as well as some features that are unique to UVPD. Depending on the sequence and the wavelength, UVPD can lead to abundant a- and x-type product ions. Notably, UVPD at 157 nm has been shown to lead to complete and exclusive series of x-type ions for singly protonated peptides with an arginine located near the C-terminus,[11] while a complete and exclusive series of a-type ions was observed when the arginine is located near N-terminus.[154] In some instances, the detected a-type ions are 1 Da heavier than are standard a ions[155]; these are referred to as $(a + 1)$ ions. It has been proposed that the observation of $(a + 1)$ and x-type ions is evidence that primary fragmentation in UVPD may proceed via homolytic radical cleavage between the α-carbon and carbonyl carbon to produce the complementary odd electron $(a + 1)$ and $(x + 1)$ ions (Scheme 16.1).[154] For singly charged peptides, wherein only one radical product can retain the charge, a series of either $(a + 1)$-type or $(x + 1)$-type ions is produced. Radical elimination from these ions results in the formation of a- and x-type ions or d-, v-, and w-type ions (Scheme 16.1).[154] The d-, v-, and w-ions, which involve cleavage of amino acid side chains, can also be produced by high-energy CID. Finally, b- and y-type ions, which are preferentially formed in low-energy CID, are also commonly observed in UVPD, particularly at lower photon energies. Such product ions, which require hydrogen (proton) migration to form, are presumably produced directly from the reactant ion in its ground electronic state.

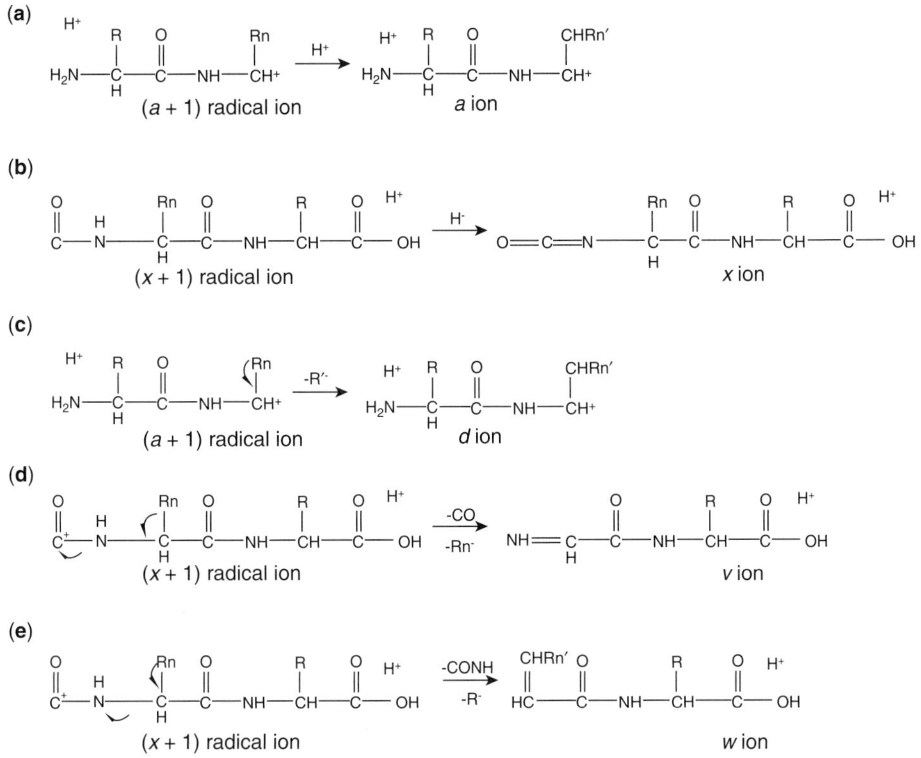

Scheme 16.1. Radical elimination to yield even-electron products from initially formed radical products in 157-nm UVPD (**a, b**). Generation of side-chain loss fragments through secondary fragmentation of initially formed radical products (**c–e**). (Reproduced from Ref. 154 with permission from Elsevier Science.)

The efficiency of UVPD peptide fragmentation, as well as the nature of the product ions, can be influenced by the covalent attachment of chromophores (e.g., Alexa Fluor 350 (AF350), 7-amino-4-methyl-coumarin-3-acetic acid, and dinitrophenyl) to the peptide.[156] UVPD of chromophore-labeled peptides proceeds with greater efficiency, not only for the reactant ion but also for the fragments that retain the chromophore functionality. Also, UVPD of doubly protonated chromophore-derivia-tized peptides at 355 nm leads to a significant reduction in spectral complexity compared to low-energy CID or UVPD at 157 or 193 nm. For example, the array of *b*- and *y*-type ions generated by CID of the doubly protonated N-terminally labeled peptide AF350-FSWGAEGQR was reduced to a single series of *y*-type ions with UVPD at 355 nm.[156]

UVPD also holds promise for the structural characterization of oligosaccharides. UVPD (at 157 nm) of singly charged oligosaccharides results in the cleavage of glycosidic bonds, as well as cross-ring fragmentation. Overall, UVPD of linear oligosaccharides leads to spectra that are less informative than those obtained by high-energy CID: Many cross-ring fragments obtained by CID are missing in UVPD spectra, and there is a deficiency of informative low-mass ions. However, UVPD of Girard's T (GT) reagent derivatized or cationized oligosaccharides yields more abundant high-energy cross-ring fragments and better sequence coverage than either low- or high-energy CID. For example, a nearly complete series of glycosidic Z- and Y-type and cross-ring X-type ions, with the charge remaining at the reducing end, were produced by UVPD (at 157 nm) of GT-derivatized maltoheptaose.[157]

The nucleic acid bases strongly absorb UV radiation in the 220- to 290-nm range and, consequently, DNA ions are highly susceptible to UVPD. Recent UVPD experiments performed on deprotonated DNA $[\text{6-mer}]^{3-}$ ions with 260-nm radiation revealed two competing pathways: (a) formation of low-energy (*a*-base) and *w*-type fragments and (b) electron detachment. [158] The yield of electron detachment was found to be base-dependent and increases with decreasing ionization potential of the nucleobase: $G > A > C > T$. According to these preliminary data, UVPD of DNA anions yields product ions that provide less sequence coverage than do other activation techniques, such as IRMPD and CID. However, CID of the radical ions generated from electron detachment (termed electron photodetachment dissociation, EPD) shows promise for characterizing higher-order structure.[159]

16.4 ELECTRON-MEDIATED ACTIVATION METHODS

The traditional manner of ionizing volatile compounds for mass spectrometric analysis involves irradiation with energetic (~70 eV) electrons. At this electron kinetic energy, sufficient energy transfer to neutral gaseous analyte molecules can occur to overcome their ionization thresholds; that is, one electron is removed, resulting in formation of analyte radical cations. Due to both the radical nature of the resulting analyte ions, and the energetic nature of this process, analyte ions generally undergo extensive fragmentation. This overall reaction is summarized in Eq. (16.11):

$$\mathrm{M} + e^{-}_{70\ eV} \rightarrow \mathrm{M}^{+\bullet *} + 2e^{-} \rightarrow \text{fragments} + 2e^{-} \qquad (16.11)$$

where M denotes the neutral analyte molecule and the asterisk denotes excitation, which can be either vibrational or electronic. This reaction has high analytical value due to its reproducible nature and extensive libraries have been generated for the identification of unknowns based on their electron ionization mass spectra. However, this reaction in itself is not an MS/MS method because ionization and fragmentation occurs concurrently and, without precursor ion selection, connectivity between precursor and product ions is lost in samples that are not highly pure. Also, it is only applicable to volatile analytes.

Cody and Freiser first applied electron irradiation as an activation method in 1979 to dissociate radical cations from substituted benzene.[12] They referred to this method as electron impact excitation of ions from organics (EIEIO). However, these authors found that fragmentation patterns were not that different from high-energy CID of the same species. McLafferty and co-workers later applied 70-eV electron irradiation to fragment sodium-cationized gramicidin S, generated by laser desorption ionization.[160] This experiment is quite different from the ones by Cody and Freiser because precursor ions are even-electron species rather than radical cations. In the McLafferty work, the types of product ions generated from EIEIO were the same as those observed in CID and 193-nm photodissociation, although the product ion molecular weights were smaller, on average, following electron irradiation. Because no tremendous advantages were evident from these experiments, EIEIO did not become a widespread MS/MS activation method. However, in 1998, Zubarev, Kelleher, and McLafferty introduced electron capture dissociation[13] (see below), and a surge in research on electron-mediated activation methods began due to the unique

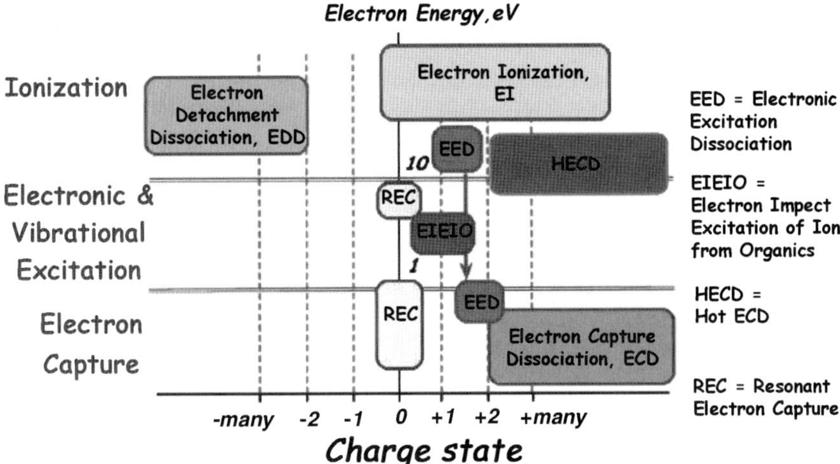

Figure 16.11. Energy ranges and applicable precursor ion polarity and charge state for various electron-mediated activation methods in tandem mass spectrometry.

fragmentation behavior of ECD and its analytical value. An excellent summary panel of electron-based activation methods, created by Zubarev and co-workers,[161] is provided in Figure 16.11. This panel shows how to differentiate these methods (all discussed below except for resonant electron capture, which applies to neutral species rather than ions) based on energy imparted into precursor ions and the ion polarity that is amenable to the particular technique.

16.4.1 Electron Capture Dissociation (ECD)

16.4.1.1 Mechanism

ECD[13,162–167] involves irradiation of multiply charged cationic even-electron species with low-energy (<1 eV) electrons. Negatively charged electrons are attracted by the cationic precursor ions and can be captured into a molecular orbital of the precursors to create a radical species with one charge less than the original ion. This charge-reduced radical intermediate typically undergoes rapid dissociation into unique product ions. For example, for peptides, cleavage of backbone N–C_α bonds occurs rather than the amide (C(O)–N) bond cleavages that are typical in most other activation methods. The overall ECD fragmentation pathway for multiply protonated species is shown in Eq. (16.12):

$$[\mathrm{M}+n\mathrm{H}]^{n+} + e^-_{<1\,eV} \rightarrow [\mathrm{M}+n\mathrm{H}]^{(n-1)+\bullet} \rightarrow \text{fragments} \qquad (16.12)$$

where M denotes the neutral analyte molecule, H denotes a proton, and n denotes the number of protons.

The original ECD implementation by Zubarev, Kelleher, and McLafferty involved an FT-ICR mass analyzer with an ion cell fitted with extra trapping electrodes to allow trapping of both electrons and precursor cations.[13] The FT-ICR mass analyzer is ideally suited for ECD because very light electrons can be stored at the same time as large biomolecular ions, thereby maximizing their interaction. Later FT-ICR implementations[168,169] showed that extra trapping electrodes are not necessary for efficient ECD. The main challenge with implementing ECD in quadrupole ion traps is their inability to store ions with a mass lower than ~20 amu; that is, electrons cannot adopt stable trajectories in such devices due to radio-frequency heating. However, successful ECD in quadruple ion traps has been demonstrated, either by adding a weak magnetic field to assist electron confinement[170,171] or by using a digital ion trap in which electrons can be injected while the electric field is constant.[172] Due to the requirement for multiply charged precursor ions, ESI is the preferred ionization method for ECD. The original ECD implementations involved standard heated filaments as electron sources. Improved (faster and more efficient) ECD has been demonstrated with indirectly heated dispenser cathodes, which provide higher electron currents with a significantly lower energy spread.[173] A more recent development is the use of a cold cathode, which eliminates the degassing and thermal activation issues that can be problematic at the high operating temperatures of dispenser cathodes.[174]

The mechanism of ECD is heavily debated in the literature.[16–19,175–178] Most of the discussion has focused on peptide fragmentation because the majority of ECD applications have involved peptide and protein dissociation, and several unique analytical advantages have been identified for this type of analysis, as discussed in the following section. The original ECD article states that ECD is a nonergodic process; that is, covalent bond cleavage following electron–proton recombination is rapid (faster than picosecond time scale) and occurs prior to energy redistribution in the molecule. Support for a nonergodic mechanism has been published by Breuker et al.,[16] including nearly no change in the fragmentation pattern of ubiquitin 13+ ions upon heating from 25°C to 125°C (recall from Section 16.2.1.1 that thermally raising precursor ion internal energy prior to activation increased both fragmentation efficiency and the number of fragmentation pathways in CID[29]), and the fact that activation of thermalized ubiquitin $[M + 12H]^{11+\bullet}$ ions remaining after ECD of $[M + 12H]^{12+}$ precursor ions only resulted in $H^\bullet$ ejection rather than the commonly observed extensive backbone cleavage. By contrast, Turecek showed that for small model systems, including aminoketyl and cation radicals from β-alanine, *N*-methylamide, *N*-acetyl-1,2-diaminoethane, N_α-acetyl lysine amide, and N_α-glycyl glycine amide, combined density functional theory and Møller–Plesset perturbational calculations indicate extremely facile unimolecular dissociation of thermalized ions, thereby rendering it unnecessary to invoke nonergodicity for the dissociation of ECD intermediates.[19] More recent work by Williams and co-workers[18] used extensively hydrated divalent metal ions as nanocalorimeters to measure (based on known/estimated binding energies of water molecules) the internal energy deposition resulting from electron capture. Their results suggest that, for these types of clusters, the vast majority of the recombination energy resulting from electron capture is converted into internal energy of the reduced precursor ions, thereby indicating that such cluster ions dissociate in an ergodic manner. However, in follow-up work, the same authors observed that loss of water molecules occurred exclusively for larger clusters with more than 30 water molecules whereas loss of a hydrogen atom and water occurred exclusively for clusters with less than 22 water molecules.[179] For a Mg cluster with 5 water molecules, significantly less energy than the estimated recombination energy is deposited. Thus, these authors suggest that dissociation of smaller clusters may be nonergodic.

In addition to the timescale controversy, another topic subject to debate is *where/how* the ECD electron is captured within the precursor ion. McLafferty and co-workers first proposed the so-called "hot hydrogen" or "Cornell" mechanism for peptide/protein ECD

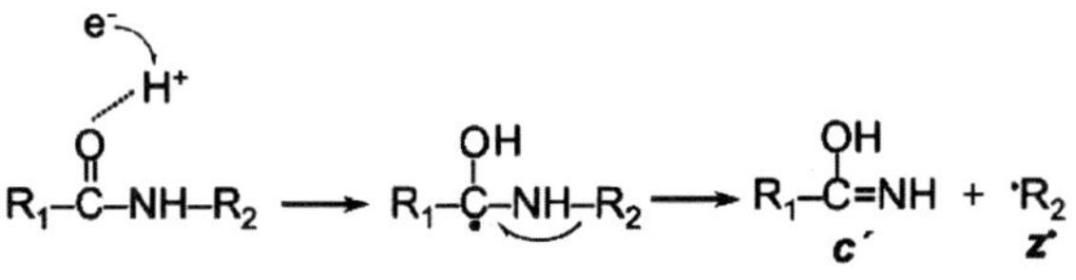

Scheme 16.2. The hot hydrogen mechanism for ECD of peptides. Electron capture at a charge center (solvated to carbonyl groups) is followed by hydrogen atom transfer to a backbone carbonyl, resulting in cleavage of N–C_α bonds. (Reproduced from Ref. 180 with permission from Elsevier Science.)

in 1999.[175] In this mechanism, neutralization via electron capture is believed to occur at a protonated site (e.g., lysine or arginine side chain), solvated onto one or several backbone carbonyls. This neutralization releases a hot (recombination energy ~5–6 eV)[16] hydrogen atom that is captured by a backbone carbonyl group to form an aminoketyl radical, which rapidly dissociates via N–C_α backbone bond cleavage to form N-terminal *c*- and C-terminal $z^\bullet$-type product ions, as illustrated in Scheme 16.2.[180] In an alternative mechanism (the "Utah–Washington" or "amide superbase" mechanism), proposed by Simons and co-workers[176,181] and by Syrstad and Turecek,[177] the electron is instead proposed to be directly captured into electronically excited states, such as the σ^* orbital of a disulfide bond (which undergoes facile cleavage in ECD[175]) or a backbone amide π^* orbital. Computational work has shown that such direct electron attachment can be rendered exothermic through Coulomb stabilization by protons or other positive charges located within a few angstroms of the particular chemical bond.[176,181] In the amide superbase mechanism, coined by Syrstad and Turecek,[177] the excited amide acts as a "superbase" that will abstract a proton from a nearby chemical group, thereby causing N–C_α bond cleavage, as shown in Scheme 16.3. Turecek and co-workers have suggested that the Cornell and Utah–Washington mechanisms may both be valid, depending on the nature of the charge carrier (e.g., lysine versus arginine).[177,182] Zubarev and co-workers[178] have also proposed that electron capture may occur at a neutral intramolecular hydrogen bond followed by hydrogen transfer. Finally, McLafferty and co-workers[175] suggested initial electron capture into an excited Rydberg state. Sobczyk and Simons[183] have shown that through-bond electron transfer from an excited Rydberg state to a disulfide bond σ^* orbital is prompt and associated with a small barrier. Furthermore, Turecek et al.[184] recently invoked the role of Rydberg-like orbitals in collisional electron transfer dissociation of extended conformations of dipeptide cations.

A third ECD mechanistic question is what kind of secondary reactions may occur following ECD but prior to product ion detection. It was noted early on that ECD product ion yields decreased with increasing protein molecular weight, although extensive

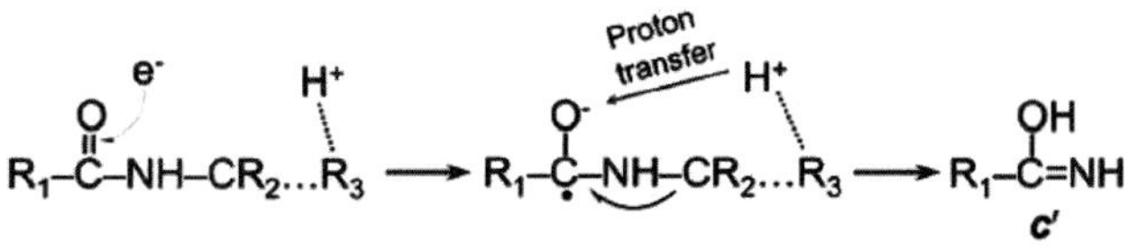

Scheme 16.3. The amide superbase mechanism for ECD of peptides. Electron capture in an excited state of a backbone amide group is followed by proton transfer for charge neutralization and subsequent N–C_α bond cleavage. (Reproduced from Ref. 180 with permission from Elsevier Science.)

electron capture to yield multiply charge-reduced species was observed. McLafferty and co-workers[185] proposed that this lack of product ions is due to retention of intramolecular noncovalent interactions such as hydrogen bonds and salt bridges; that is, covalent N–C_α bonds may be cleaved in precursor ions, but the two resulting fragments remain bound such that the overall mass does not change. This potential problem, which also exists for larger peptides, can be overcome via so-called "activated ion" (AI) ECD, in which ECD is combined with mild collisional or IR activation to disrupt noncovalent interactions.[186–191] Work by O'Connor and co-workers[192,193] has shown that relatively long-lived $c/z^\bullet$ product ion complexes exist also for smaller peptides and that hydrogen migration occurs within these complexes to generate $c^\bullet/z$-type product ions. Zubarev and co-workers[194] have performed an extensive characterization of such hydrogen rearrangements in a large data set of doubly protonated tryptic peptides and found that the occurrence of hydrogen migration can be as high as 47%. The role of precursor ion gas-phase conformation in ECD has also been investigated by, for example, Heeren and co-workers[195] and Williams and co-workers.[196] The former group characterized the ECD behaviors of the linear peptide substance P and the cyclic peptide gramicidin S as a function of ICR cell temperature. At 86 K, only two backbone product ions were observed for each peptide versus eight and five, respectively, at room temperature. The authors concluded that the more specific fragmentation at low temperature reflects the reduced conformational heterogeneity. The latter group used a high-field asymmetric waveform ion mobility spectrometry (FAIMS) device to separate different gas-phase conformations of ubiquitin prior to ECD. The electron capture efficiencies and product ion abundances of different conformers of the same charge state were drastically different. Finally, O'Connor and co-workers[197] have proposed that initial ECD covalent bond cleavage can be followed by a radical cascade, explaining the extensive fragmentation that is observed in cyclic peptides.

The main reason behind the excitement surrounding ECD is the unique fragmentation behavior observed for peptides and proteins and the resulting unique analytical information that can be gained from ECD spectra. These advantages are discussed in more detail in the following section. The main drawbacks of ECD are the requirement for multiply charged cationic precursors, which can preclude analysis of small and acidic molecules. In addition, the fragmentation efficiency is typically lower than that of, for example, CID and IRMPD. One explanation for this behavior is that product ions remain in the electron beam and can be neutralized at longer irradiation times.

16.4.1.2 ECD Application

As described above, ECD of peptides and proteins results in N–C_α rather than amide backbone bond cleavage, thereby producing c/z-type product ions rather than the b/y-type ions that are observed with other activation methods. This cleavage preference has several advantages, including the following three: (1) Combination of MS/MS data from ECD and slow heating techniques such as CID or IRMPD results in so-called "golden" product ion pairs with characteristic mass differences—for example, -17.03 amu for b/c-type ions and 16.02 amu for $y/z^\bullet$-type ions.[198] Such information allows a particular product ion in an MS/MS spectrum to be assigned as containing the N-terminus or C-terminus, respectively, which is highly valuable for *de novo* peptide sequencing and for peptide identification from genomic data. Recent work by Tsybin et al.[199] has shown that similar assignments can be made based on ECD data alone if ECD is performed with and without ion activation (see Section 16.4.1.1 for a description of activated ion ECD). Here, less radical *c*-type ions and less even-electron *z*-type ions are seen with increasing ion internal energy prior to ECD. Such product ions are due to hydrogen migration in long-lived $c/z^{\bullet}$ product ion complexes and mixtures of odd- and even-electron product ions of the same type are commonly observed. Thus, if the radical component (which is 1 amu lighter) decreases with increasing activation, a product ion can be assigned as N-terminal and vice versa. (2) Sequence information obtainable from ECD data is generally highly complementary to that obtained from slow heating methods due to the very different cleavage mechanism, which results in different cleavage preferences.[200] This complementarity has been utilized by Zubarev and co-workers in the development of a *de novo* sequencing algorithm, which uses the combination of ECD and CID fragmentation.[201] These authors demonstrated that this approach has the same level of efficiency and reliability as conventional database-identification strategies, but with alleviation of the problems associated with relying on only CID for sequencing, such as incomplete fragmentation patterns. (3) More extensive sequence information is generally obtained from ECD alone versus CID alone,[202,203] presumably because ECD does not show as strong cleavage preferences as does CID. For example, Creese and Cooper[204] recently showed that overall protein sequence coverage from bottom-up LC-ECD-MS/MS was lower than that from LC-CID/MS/MS. However, LC-ECD-MS/MS resulted in longer peptide sequence tags, which provided greater confidence in protein assignment.

The different energetics/timescale in ECD compared to slow-heating MS/MS techniques also yields great analytical advantages. First, ECD has developed into a key technology for characterizing protein post-translational modifications (PTMs), which tune protein biological function. Such modifications, including, for example, phosphorylation, glycosylation, and sulfation, are often labile and therefore easily lost in, for example, CID and IRMPD (see Section 16.2.1.1 for a brief discussion on CID of phosphorylated species). By contrast, numerous examples[186,205–213] have shown that labile PTMs are retained upon backbone amine bond cleavage in ECD, which allows their precise location within a peptide to be determined. Figure 16.12 shows a comparison,

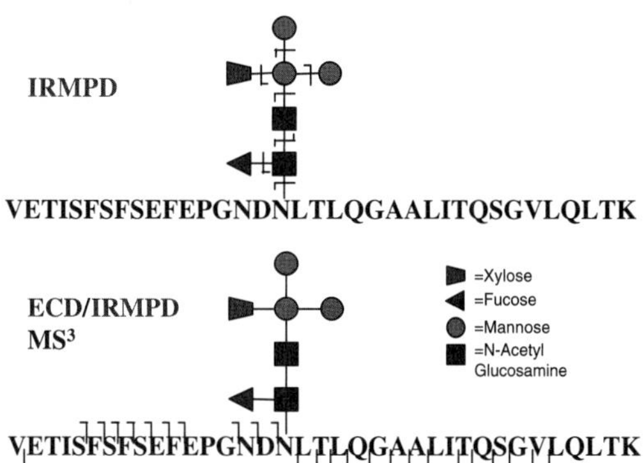

Figure 16.12. Summary of the structural information obtained from ECD and IRMPD MS/MS and MS^3 of a 5-kDa *N*-glycosylated tryptic peptide. IRMPD of the even-electron $[M + 3H]^{3+}$ ion resulted in cleavage of all glycosidic bonds, whereas ECD followed by IRMPD of the $[M + 3H]^{2+\bullet}$ radical ion formed upon electron irradiation resulted in cleavage of 25 of 35 backbone bonds. (Adapted from Ref. 189 with permission from the American Chemical Society.)

performed by Hakansson et al.,[189] between bond cleavages observed in IRMPD and ECD/IRMPD MS^3 (a version of activated ion ECD in which the charge-reduced, nonfragmented radical species is further isolated and subjected to IR photon irradiation) of a 5-kDa *N*-glycopeptide. In IRMPD, extensive cleavage of the glycan glycosidic bonds is observed, which is useful for determining glycan composition and partial structure. The ECD/IRMPD MS^3 data are strictly complementary in that only extensive peptide backbone fragmentation is observed (25 out of 36 amine bonds). ECD has also become a popular choice for analyzing histone modifications, typically acetylation, methylation, and phosphorylation.[214–218] Although the former two modifications are not labile in CID, ECD allows top-down rather than bottom-up analysis (due to the more extensive backbone bond cleavage described above), which provides a more holistic picture of the highly complex modification patterns. Another intriguing example of ECD for PTM analysis is work by O'Connor and co-workers,[219,220] who utilized this MS/MS method to differentiate the isomeric amino acids aspartate and isoaspartate based on unique diagnostic ions. Conversion of aspartate to isoaspartate is believed to contribute to the aging of proteins and to protein folding disorders such as Alzheimer's disease.

A second application of the ability of ECD to cleave covalent bonds while retaining more labile interactions is the probing of protein gas-phase structure and folding. McLafferty and co-workers have utilized the retention of noncovalent interactions in ECD to probe the gas-phase folding and unfolding of cytochrome *c*[221] and ubiquitin.[222] Protein unfolding was accomplished by thermal activation either through heating of the ion cell of an FT-ICR mass analyzer or by IR photon irradiation. More unfolded conformations result in more ECD product ions due to disruption of noncovalent interactions in *c*/*z* product ion complexes. The ubiquitin experiments even allowed determination of unfolding enthalpies, conformational melting temperatures, and kinetics of unfolding and refolding.[222] These data suggest that, under the conditions examined, gas-phase structures for these two proteins do not correlate well with solution-phase structures. In another example, Zubarev and co-workers[223] used circular dichroism, charge state distributions, and ECD to compare how D-amino acid substitutions influence the solution- and gas-phase structures of tryptophan cage protein. These authors found that some gas-phase structural features were similar to those in solution.

ECD has also been used to probe *inter*molecular noncovalent interactions. The idea here is that, because such interactions remain intact, biomolecular interaction regions should be possible to determine. Zubarev and co-workers[224] analyzed nonspecific peptide dimers and specific glycopeptide antibiotic–peptide complexes by ECD and showed that cleavage of backbone covalent bonds does occur without breaking the noncovalent interactions responsible for maintaining the complexes. More recently, Loo and co-workers[225] applied ECD to reveal ligand binding sites in the α-synuclein (AS)–spermine complex. AS binding to polycations such as spermine is believed to have a role in its aggregation, which, in turn, is believed to be linked to Parkinson's disease. The ECD experiments located spermine binding to the C-terminus of AS,[225] consistent with solution-phase studies. Heck, Heeren and co-workers[226] compared ECD and SORI-CID of an 84-kDa heptameric gp31 complex. In this example, unexpected asymmetric dissociation of the 21+ charge state of the complex into a monomer and a hexamer was seen in ECD. However, the charge separation over the two products was highly proportional to molecular weight, similar to the SID experiments by Wysocki and co-workers, discussed in Section 16.2.3.2. The authors speculate that the ejected monomer retains more of its original structure in ECD as compared to SORI-CID (which results in asymmetric charge partitioning).[226]

A third area in which rapid, potentially nonergodic, dissociation would be highly valuable is in the fragmentation of deuterated peptides from solution-phase amide hydrogen/deuterium exchange experiments. Such experiments are performed to determine protein solution-phase structures and interaction sites between, for example, proteins and their ligands.[227–231] Following exposure to D_2O, exchange is quenched by low pH and temperature; proteins are then digested with pepsin (which functions at low pH) and the resulting peptides are subjected to MS analysis. MS/MS of deuterated peptides is desired because fragmentation patterns may more precisely reveal the locations of nondeuterated amides, thereby providing higher structural resolution on areas protected from exchange (and thus involved in, for example, interacting with a ligand). However, numerous studies have shown that slow-heating techniques such as CID result in hydrogen/deuterium migration (scrambling) prior to backbone dissociation and, therefore, information on the location of nondeuterated amides is lost.[232–237] Some evidence for limited deuterium scrambling was presented in the very first ECD publication.[13] In follow-up work, Kweon and Hakansson[238] showed that limited scrambling appeared to occur when analyzing *c*-type ions from deuterated melittin. However, analysis based on *z*-type ions did not correlate with previous NMR studies, possibly due to the radical nature of *z* ions promoting hydrogen scrambling. In more recent work, Rand and Jorgensen[239] developed a peptide probe more suitable for characterizing the degree of scrambling in MS/MS. With this probe, they showed that very little scrambling occurs in ECD when instrument ion acceleration voltages are kept low and when precursor ion selection windows are kept broad.[240]

In addition to the extensive application of ECD to peptide and protein analysis, this MS/MS activation method has been applied to other types of macromolecules, including polymers,[241–243] nucleic acids,[244–246] lantibiotics,[247] and carbohydrates.[248,249] In all cases, complementary fragmentation behavior was observed compared to slow-heating MS/MS techniques. For polymer analysis, McLafferty and co-workers[242] showed that efficient sequencing of polyethylene glycol/polypropylene glycol co-polymers can be performed with ECD whereas CID results in significant rearrangements, rendering spectral interpretation challenging.[242] ECD of DNA[244,245] and RNA[246] oligonucleotides has been demonstrated by Hakansson and co-workers, who found that, for DNA, nucleobase and water loss (which dilute the product ion signal) was reduced compared to IRMPD. In addition, retention of intramolecular hydrogen bonds was seen for adenine-containing oligodeoxynucleotides, thereby suggesting that characterization of nucleic acid gas-phase structures may be possible, similar to the protein work described above. For RNA, ECD showed a strong nucleobase dependence, suggesting that initial electron capture occurs at the nucleobases. Heeren, Heck, and co-workers[247] utilized ECD to locate the position of intramolecular monosulfide bridges in lantibiotics, including nisin A, nisin Z, mersacidin, and lacticin 481. This work showed that $c^{\bullet}/z$ rather than $c/z^{\bullet}$ product ions are preferentially formed in the vicinity of monosulfide bridges. ECD of carbohydrates was first demonstrated by Zubarev and co-workers.[249] These authors investigated the fragmentation behavior of doubly protonated chitooligosaccharides (which contain free amines at each residue and thereby undergo facile multiple protonation). However, mostly glycosidic cleavages between monosaccharide residues were observed with only minor differences compared to CID of the same species. More recent work by Adamson and Hakansson[248] explored ECD of divalent metal-adducted oligosaccharides. Here, analysis is not limited to amino sugars because the divalent metal provides the two charges required for ECD. Furthermore, divalent metal-adduction resulted in extensive, sometimes *dominant*, cross-ring fragmentation, crucial for determining saccharide linkages. Cross-ring fragmentation observed from ECD is complementary to that obtained from metal-assisted IRMPD (see also Sections 16.2.1.2 and 16.2.2.2 for more details on cross-ring fragmentation).

The use of divalent metal adducts as charge carriers in ECD as a means to circumvent the requirement for multiple charging has also allowed application of ECD to acidic sulfopeptides[207] and to smaller molecules, including phosphate-containing metabolites[250] and phospholipids.[251] For sulfopeptides, metal adducts were observed to stabilize highly labile sulfate groups and thereby allow their localization (by contrast, complete sulfate loss was seen in ECD of the corresponding protonated species).[207] For phosphate-containing metabolites, ECD of calcium complexes provided complementary fragmentation behavior compared to SORI-CID, including ribose cross-ring cleavages and generation of hydrated product ions from hydrated precursor ions.[250] O'Hair and co-workers showed that ECD of divalent metal complexes of phosphocholines resulted in rich and complex chemistry, including charge reduction and fragmentation involving losses of a methyl radical, trimethylamine, and acyl chains.[251] Thus, ECD provides structurally useful information on the phospholipid, including the nature of the head group, the acyl chains, and the positions of the acyl chains.

16.4.2 Hot Electron Capture Dissociation (HECD)

16.4.2.1 Mechanism

Hot ECD is a variant of ECD that uses more energetic electrons than regular ECD. HECD was introduced by Zubarev and co-workers,[252] based on the realization that there is a second local maximum for electron capture around 10 eV (the first maximum occurs at less than 0.2 eV with a 2–3 orders of magnitude decrease in cross section at 1 eV[253]). The excess energy involved in HECD compared to ECD is believed to cause electronic excitation via inelastic ion-electron collisions prior to electron capture, thereby decelerating the hot electron and facilitating its capture.[252] The overall reaction scheme can be written as follows:

$$[\mathrm{M}+n\mathrm{H}]^{n+} + e^-_{\sim 10\ eV} \rightarrow [\mathrm{M}+n\mathrm{H}]^{(n-1)+\bullet *} \rightarrow \text{fragments} \qquad (16.13)$$

where M denotes the neutral analyte molecule, H denotes a proton, n denotes the number of protons, and the asterisk denotes excitation. The excess energy in HECD causes extensive secondary fragmentation. Of particular interest is amino acid side-chain fragmentation within the isomeric residues leucine and isoleucine (loss of $^\bullet CH(CH_3)_2$ versus loss of $^\bullet CH_2CH_3$ from $z^\bullet$-type ions to generate so-called *w*-type ions), thereby allowing their differentiation. Figure 16.13, from Zubarev and co-workers,[252] shows product ion abundances from a tryptic decapeptide as a function of electron energy. Two clearly separated maxima are observed for N–C_α backbone bond cleavages, one at ~0 eV and one around 7 eV, where the second maximum corresponds to the hot ECD regime. At higher electron energies, amide bond cleavage (*b*/*y*-type product ions) dominates, which is likely due to an EIEIO-type process,[12,160] as discussed above. The richness of HECD product ion spectra can be an advantage, but the complexity could also be a disadvantage when working with unknowns. An optimum workflow may involve regular ECD followed by HECD.

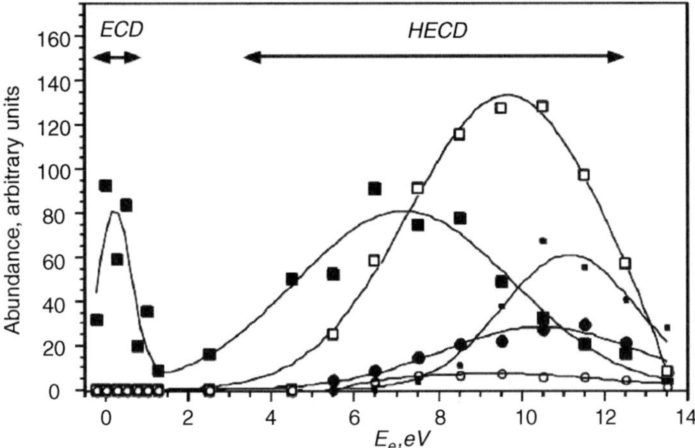

Figure 16.13. Product ion abundances versus electron energy, E_e, for 250-ms irradiation of a doubly protonated tryptic decapeptide from signal recognition particle of *Saccharomyces cerevisiae*. Large filled squares represent N–C_α bond cleavages, open squares represent C–N bond cleavages, open circles represent z_4^+ fragments, and filled circles represent w_4^+ fragments. Smaller filled squares represent product ions (C–N bond cleavages) from singly protonated precursor ions. (Adapted from Ref. 252 with permission from Elsevier Science.)

16.4.2.2 HECD Application

Zubarev and co-workers[254] have applied HECD to assign leucine/isoleucine residues in the 135-amino-acid bovine milk protein PP3. A bottom-up approach was used in which tryptic peptides were subjected to HECD. Twenty out of 25 Leu/Ile residues could be determined based on *w* ion formation. The same research group also subjected a 12-residue peptide to various MS/MS activation methods and showed that HECD yielded the highest number of inter-residue backbone cleavages—that is, at least two different kinds of product ions per position.[255] Another discovery from this group involves a new type of product ion from HECD, so called *u*-type ions.[256] Such ions are also produced by secondary fragmentation of $z^\bullet$-type ions but, instead of fragmentation within the side chain closest to the N-terminus of the $z^\bullet$ product ion, generation of a γ-lactam via bond formation between the N-terminal α carbon and the β carbon of the second amino acid residue results in side chain cleavage within the latter. *u*-type ions can be used in addition to *w*-type ions to increase the number of Leu/Ile assignments in proteins. Recent work by O'Connor and co-workers[257] showed the utility of HECD for analysis of permethylated sodiated linear and branched oligosaccharides. Permethylation is known to increase sugar cross-ring cleavage in CID (see Section 16.2.1.2).[65–67] However, HECD was shown to provide complementary fragmentation patterns for branched *N*-glycans as compared to CID, including X-type cross-ring fragments containing the reducing end of the glycan.

16.4.3 Electron Transfer Dissociation (ETD)

16.4.3.1 Mechanism

From the previous two sections, it is clear that ECD has tremendous analytical utility, both stand alone but, particularly, when combined with complementary slow-heating MS/MS activation methods. However, one drawback of ECD is the difficulties associated with implementing this technique on instruments other than FT-ICR mass spectrometers. The latter type of instrument is expensive and often requires highly skilled users. Hunt and co-workers[14] sought an alternative approach and came up with the clever solution to use an anion electron carrier rather than a free electron to trigger electron capture by multiply charged cationic peptides. Because the anion electron carrier can be heavier than the low-mass cutoff value, such an experiment is more straightforward to implement in quadrupole ion traps, which are less expensive and more widespread than FT-ICR instruments. The fragmentation scheme of this reaction, termed electron transfer dissociation, is shown in Eq. (16.14):

$$[M + nH]^{n+} + A^{-\bullet} \rightarrow [M + nH]^{(n-1)+\bullet} + A \rightarrow \text{fragments} \qquad (16.14)$$

where M denotes the neutral analyte molecule, H denotes a proton, n denotes the number of protons, and A denotes the electron carrier.

The main challenge for realizing ETD was to find an appropriate anion that would transfer an electron upon encountering a multiply protonated peptide rather than accepting a proton from the peptide. The first ETD publication (from 2004)[14] reported successful use of anthracene radical anions (generated by chemical ionization, CI). Follow-up work by the same group explored alternative anions.[258] Some of these anions react exclusively via proton transfer, others react by proton and electron transfer, while 9,10-diphenyl anthracene behaved predominantly as an electron transfer agent. However, compared to anthracene, charge reduction without dissociation of product ion pairs was more pronounced although the electron transfer capacity was higher for 9,10-diphenyl anthracene. The same article[258] also reported successful use of $SO_2^{-\bullet}$ in a linear quadrupole ion trap, results that were later reproduced in a three-dimensional quadrupole ion trap by McLuckey and co-workers.[259]

In order to enable encounters, and hence reactions, between multiply protonated peptides and anions, instrumentation geometries also need to be considered. In the original work by Hunt and co-workers,[14] a linear quadrupole ion trap was used with an ESI source for peptide ionization at the front and a CI source for anion generation at the back. The instrument control software had to be modified to allow injection of both peptide cations and reactant anions into the ion trap. McLuckey and co-workers[260] utilized a pulsed dual nano-ESI/atmospheric pressure chemical ionization source to implement ETD on a commercial quadrupole/time-of-flight (Q-Q-TOF) mass spectrometer. To enable mutual storage of oppositely charged ions from this dual source, an auxiliary radio frequency was superimposed on the end lenses of the second quadrupole during the reaction period to confine the reagent anions. The same group also introduced a strategy for forming reagent anions via ESI.[261] Most ESI-generated anions, however, react via proton rather than electron transfer, and an alternative approach in which precursor reagent anions are collisionally activated to form products that are ETD-reactive was realized. Several potential precursor anions were identified with 2-(fluoranthene-8-carbonyl) benzoic acid having the most attractive set of characteristics. This same strategy (using the reagent anion generated via CO_2 loss from 9-anthracenecarboxylic acid) was later adopted by Coon and co-workers[262] to implement ETD on a hybrid linear quadrupole ion trap/orbitrap mass spectrometer. This group in collaboration with Muddiman and co-workers[263] also reported on an alternative pulsed ESI source, which is physically separated in space rather than being separated in time. This configuration provided improved spray stability and decreased switching times. ETD has also been implemented in an external hexapole of a hybrid FT-ICR mass spectrometer utilizing a CI source mounted above the hexapole for reagent anion generation.[264]

The mechanism of ETD is believed to be very similar to that of ECD, supported by generation of the same peptide product ion types ($c/z^\bullet$-type ions rather than b/y-type ions), preferential cleavage of disulfide bonds,[265] and retention of higher-order structure. The latter characteristic can pose a problem in peptide analysis because charge-reduced, nonfragmented species can dominate product ion spectra, thereby yielding very little to no sequence information. This problem is particularly severe for doubly protonated peptides, as has been documented in detail by Coon and co-workers[266] for a set of ~4000 peptides. Several strategies have been implemented to circumvent this problem. McLuckey and co-workers[259] utilized CID of the charge-reduced species to increase sequence coverage of doubly protonated peptides in ETD with $SO_2^{-\bullet}$ reagent anions in a three-dimensional ion trap. The same group also performed ETD with nitrobenzene reagent anions at elevated bath gas temperatures and reported higher peptide sequence coverage compared to room temperature, presumably due to disruption of intramolecular noncovalent interactions that prevent product ion pairs from separating. Coon and co-workers[267] also used supplemental CID of charge-reduced species, which they termed ETcaD, to improve sequence coverage for doubly protonated peptides. Another solution involves use of *m*-nitrobenzyl alcohol in the ESI solvent to "supercharge"[268] peptides prior to ETD. This approach, presented by Jensen and co-workers,[269] showed that the predominant charge state for BSA tryptic peptides changed from 2+ to 3+.

ETD has also been reported in the negative ion mode with xenone radical cations as reagent, which mainly resulted in backbone C_α–C(O) bond cleavage to yield *a*- and *x*-type product ions.[270] Although phosphate loss was observed for phosphopeptides,[270] the ability to operate in both ion polarities is an advantage compared to ECD, which is only applicable to cations. The other main advantage of ETD over ECD is its more facile implementation in quadrupole ion traps. A potential disadvantage is that the electron energy cannot be easily controlled and, therefore, experiments analogous to hot ECD are not possible.

16.4.3.2 ETD Application

Because of the close analogy to ECD, the analytical advantages of ETD are similar and its application has therefore focused on similar biochemical analysis. Pandey and co-workers[271] recently performed a large-scale (~19,000 peptides) comparison between CID and ETD and showed that MS/MS data from the two techniques are complementary. Thus, a combination of these two techniques should be highly valuable for *de novo* sequencing. As described for ECD, it is also highly valuable, both for *de novo* sequencing and for peptide-based protein identification via database searching, to be able to assign whether a particular product ion in an MS/MS spectrum is N-terminal or C-terminal. In ETD, McLuckey and co-workers[272] have reported reactions between radical *z*-type ions and residual molecular oxygen in the ion trap, thereby producing $[z^\bullet + 32]$-type product ions, which allow assignment of *z* ions. In follow-up work, the same group showed that the ability to confidently determine which product ions are *z* ions in a database search is heavily dependent on the quality of ETD spectra.[273] Coon and co-workers[274] have recently reported on the distinct chemical nature of radical *z*-type ions from ECD and ETD compared to their complementary even-electron *c*-type ions [i.e., *c*-type ions have an odd number of atoms with an odd valence (e.g., N and H), whereas $z^\bullet$-type ions contain an even number of atoms with an odd valence]. By extension, the mass of a *c*-type ion cannot be the same as that of a $z^\bullet$-type ion. From a practical standpoint, however, very high mass accuracy is required to accomplish this distinction although high mass accuracy instrumentation is becoming more widespread. By contrast, *b* and *y*-type product ions from CID are not chemically distinct in this same sense.

ETD is emerging as a valuable tool for top-down proteomics. One concern with ETD in low-resolution quadrupole ion traps is that, because product ions from intact proteins are quite large and multiply charged, their charge states cannot be readily assigned because the isotopes are not resolved, thereby rendering product ion assignments ambiguous. A solution to this problem was presented by Hunt and co-workers, who utilized a second ion/ion reaction following ETD, namely proton transfer from highly charged product ions to the carboxylate anion of benzoic acid.[275,276] This proton transfer reaction (PTR) can be controlled to generate only singly or doubly protonated species, such that charge state assignment is no longer ambiguous; although, because quadrupole ion traps typically have an upper *m/z* limit of ~2000–4000, only product ions covering the ends of the protein will be observed. However, this approach provides simultaneous sequence information (~15–40 residues) from the protein N- and C-terminus, which together with the intact molecular weight of the protein can be used to identify the protein and potential PTMs.[276] The same group used this approach to identify 46 out of 55 known unique proteins from the *Escherchia coli* 70S ribosomal protein complex in a single, 90-min, on-line chromatography experiment.[277] This combined ETD/PTR[276] approach is illustrated in Figure 16.14 for ubiquitin 13 + ions. Panel A shows the product ion spectrum following 15-ms reaction with fluoranthene radical anions to yield extensive but unresolved highly charged *c*- and $z^\bullet$-type ions. Panels C–E show the spectral appearance after various time periods of proton transfer to benzoic acid anions to yield lower, resolvable product ion charge states. Panel E shows the sequence coverage from the N- and C-termina of the protein, considering only singly charged product ions.

As for ECD, one of the main reasons for the excitement about ETD is its ability to localize PTMs in peptides and proteins. The very first ETD publication showed extensive backbone fragmentation of a doubly phosphorylated synthetic peptide without phosphate loss.[14] The same group used ETD to identify 29 phosphorylation sites (19 novel and 10 previously reported) as well as a novel glycosylation site on the adhesion adapter protein, paxillin.[278] Large-scale phosphorylation analysis by ETD has been demonstrated by Hunt and co-workers,[279] who analyzed phosphorylation sites on proteins from *Saccharomyces cerevisiae*, and by Pandey and co-workers,[280] who performed global phosphoproteome analysis of human embryonic kidney 293T cells. In the latter example, 60% more phosphopeptides were identified with ETD compared to CID. A combination of ETD and CID identified 80% of the known phosphorylation sites in more than 1000 phosphopeptides. One potential problem with ETD of phosphopeptides is that, preferably, at least three positive charges are needed. McLuckey and co-workers[281] demonstrated an approach in which negative-ion mode (where phosphopeptides ionize well due to their acidity) is used for initial ionization followed by proton transfer from amino-terminated dendrimers to accomplish charge inversion into phosphopeptide cations suitable for ETD.

ETD has also been shown to provide similar data as ECD for *N*-glycopeptides in that extensive backbone fragmentation is seen rather than extensive glycan fragmentation, which is typical in CID and IRMPD (see Figure 16.12 for ECD data).[282] Other examples of

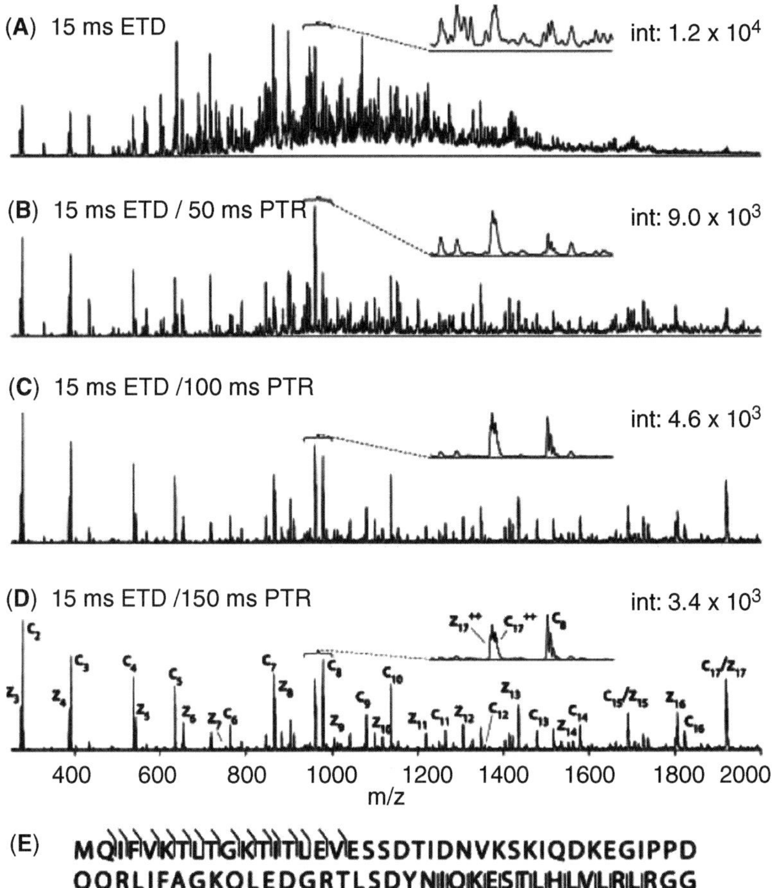

Figure 16.14. Tandem mass spectrum of ubiquitin generated by sequential ion/ion reactions. (**A**) Whole protein dissociation (ubiquitin 13 +, *m/z* 659) after a 15-ms reaction with the radical anion of fluoranthene. Note production of several hundred highly charged unresolved *c*- and *z*-type product ions. (**B–D**) The subsequent reaction of these products with even-electron anions of benzoic acid for 50 ms (**B**), 100 ms (**C**), and 150 ms. (**D**). Note the gradual degradation of multiply charged products, leaving predominately doubly and singly charged fragments after 150 ms. (**E**) The resulting sequence coverage considering only singly charged product ions. Each spectrum is the average of ~50 spectra (~30-s acquisition), and the *y* axis indicates the relative ion abundance. (Reproduced from Ref. 276 with permission from The National Academy of Sciences of the USA.)

PTM analysis with ETD include sulfation,[283] oxidation,[284] glycation,[285] and differentiation between aspartic and isoaspartic acid.[286] A recent study by Coon and co-workers[287] utilized ETD for mapping PTMs on the tail of histone H4 from human embryonic stem cells. These authors were able to decipher 74 discrete combinatorial PTM codes and quantify striking changes in methylation and acetylation patterns occurring as the cells underwent differentiation. So far, ETD has had very limited application to molecules other than peptides and proteins. One example, by McLuckey and co-workers,[288] involves doubly sodiated glycerophosphocholine lipids. Here, ETD product ions provided information on carbon number and degree of unsaturation.

16.4.4 Electron Detachment Dissociation (EDD)

16.4.4.1 Mechanism

Electron detachment dissociation was introduced in 2001 by Zubarev and co-workers[289] as a negative-ion mode alternative to ECD. EDD involves more energetic electrons than ECD, but not as energetic as in electron ionization: typically around 10–30 eV. In EDD, multiply charged anions are irradiated with electrons to impart sufficient energy to eject one electron from the precursor ion. This anion ionization generates a radical anionic intermediate different from that in ECD and ETD but which still undergoes facile dissociation. The overall EDD fragmentation scheme is shown in Eq. (16.15):

$$[\mathrm{M}-n\mathrm{H}]^{n-}+e^{-}_{>10\,\mathrm{eV}}\rightarrow[\mathrm{M}-n\mathrm{H}]^{(n-1)-\bullet *}+2e^{-}\rightarrow\text{fragments} \qquad (16.15)$$

where M denotes the neutral analyte molecule, H denotes a proton, n denotes the number of protons, and the asterisk denotes excitation, which can be either vibrational or electronic.

EDD has been implemented in both FT-ICR[289–292] and quadrupole ion trap[293] mass analyzers, mostly with indirectly heated dispenser cathodes as electron sources. Hakansson and co-workers[292] demonstrated successful EDD with a heated filament, although use of a dispenser cathode in the same instrument yielded both higher EDD efficiency and additional fragmentation channels. The first EDD publication,[289] based on FT-ICR EDD data, proposed a peptide dissociation mechanism involving electron–hole recombination. In this mechanism, the positive hole created by electron ejection is mobile and can recombine with an electron, thereby resulting in electronic excitation (recombination energy of ~5 eV) followed by bond cleavage. In the first publication, the authors focused mainly on observed N–C_α bond cleavage (to yield c- and z-type ions) in the sulfated peptide caerulein and noted that the recombination energy and type of bond cleavage was analogous to ECD. Furthermore, most product ions retained the labile sulfate group. In a later publication by the same group, based on quadrupole ion trap EDD data, backbone C_α–C(O) bond cleavage (to yield $a^\bullet$ and x-type product ions) was seen as the preferred fragmentation pathway in peptides.[293] The cleavage preference discrepancy compared to the earlier publication was explained based on the different ion temperatures in FT-ICR versus quadrupole ion traps. Intriguingly, peptide backbone C_α–C(O) bond cleavage is also preferred in negative ion mode ETD[270] and in 157-nm UV photodissociation,[11] which is believed to involve electronic excitation and initial homolytic bond cleavage.[154] Zubarev and co-workers[293] proposed a mechanism for $a^\bullet/x$-type formation in EDD, which involves electron detachment from a deprotonated backbone amide nitrogen followed by radical driven bond cleavage, as depicted in Scheme 16.4. *Ab initio* calculations showed that a unidirectional radical attack with bond cleavage N-terminal rather than C-terminal to the initial radical site is favored with about 74.2 kJ mol^{-1}, consistent with experimental data.[293] Further computational work by Simons and co-workers[294] confirmed this directional preference and also showed that C_α–C(O) backbone cleavage is favored over amino acid side chain loss, except for loss of a tyrosine side chain.

Kalli and Hakansson have shown that disulfide bond cleavage is favored in EDD,[295] similar to ECD[175] and negative ion mode low-energy CID.[296] These authors also showed that loss of a tryptophan side chain can compete with disulfide bond cleavage. Because tryptophan has the lowest vertical ionization energy of the amino acids, this behavior was proposed to be due to direct electron detachment from tryptophan. Recent work by Amster and co-workers[297] and by Yang and Hakansson[298] has measured the EDD fragmentation efficiency as functions of key experimental parameters, including electron

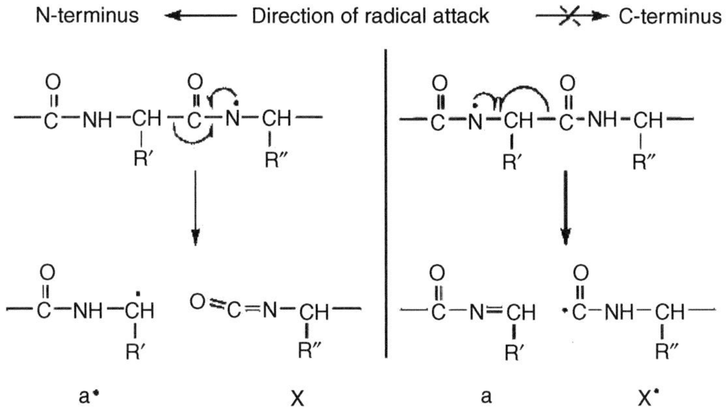

N-terminus ← Direction of radical attack ⇸ C-terminus

Scheme 16.4. Proposed mechanism for $a^{\bullet}/x$-type product ion formation in peptide EDD. This fragmentation pathway is favored over $a/x^{\bullet}$-type product ion formation by $74.2\,\mathrm{kJ\,mol^{-1}}$. (Rreproduced from Ref. 293 with permission from John Wiley & Sons.)

current, electron number, electron energy, dispenser cathode heating current, electron beam duration, charge state of the precursor ion, oligomer length, and precursor ion number. From these measurements, it was found that EDD fragmentation efficiency increases with increasing precursor ion charge state[297,298] and electron energy optimization appears to peak around 16–22 eV.[298] Amster and co-workers[299] have also performed a more detailed characterization of precursor ion charge state dependence in EDD of sulfated glycosaminoglycans (GAGs). These authors found that, when the degree of ionization exceeds the number of sulfate groups, sulfate loss is significantly reduced, suggesting that SO_3 loss is a consequence of electron detachment from negatively charged sulfate groups rather than negatively charged carboxylate groups. Preferential electron detachment from carboxylate versus sulfate is thermodynamically preferred, provided that carboxylate is in its ionized state. As described below, EDD has been shown to provide unique fragmentation characteristics for acidic molecules such as GAGs, nucleic acids, sialylated oligosaccharides, and sulfo- and phosphopeptides, all of which undergo facile ionization in negative-ion mode. The main drawback of the technique is the low fragmentation efficiency.

16.4.4.2 EDD Application

As discussed above, the first EDD publication reported retention of a labile sulfate group in peptide product ions. Later work demonstrated retention of phosphate in product ions from tyrosine-,[293] histidine,-[300] and serine-phosphorylated peptides.[301] This characteristic suggests that EDD should be a valuable tool for characterization of acidic PTMs. However, work by Kweon and Hakansson[301] showed that, for larger phosphopeptides with multiple phosphorylations, very limited fragmentation occurred in EDD; thus ECD appeared to be a better alternative. On the other hand, Jensen and co-workers[300] showed that, for a histidine-phosphorylated peptide, the phosphate group was more efficiently retained in EDD compared to both ECD and ETD although fragmentation efficiency in EDD was low. Hakansson and co-workers applied EDD to nucleic acid analysis and showed that rich fragmentation patterns are achieved for both DNA[292] and RNA[246] with minimal nucleobase loss. For DNA, fragmentation patterns were similar to those in ECD, but higher sensitivity was achieved with EDD due to the advantage of using the negative ion mode. For RNA, the nucleobase dependence seen in ECD (see Section 16.4.1.2) was not present and backbone fragmentation was similar to that of DNA. The latter behavior is different from that in low-energy CID, which yields different fragmentation pathways for DNA and RNA as described in Section 16.2.1.2. Furthermore, backbone cleavage in a DNA duplex was reported without rupture of the hydrogen bonds between nucleobases,[292] thereby suggesting that EDD should be applicable to higher-order structure characterization. This suggestion was followed up by Mo and Hakansson,[302] who applied EDD, IRMPD, activated ion (AI) EDD, and EDD/IRMPD MS^3 to the characterization of three isomeric 15-mer DNAs with different sequences and predicted solution-phase structures. These authors found that all three 15-mers had higher-order structures in the gas phase, although preferred structures were only predicted for two of them in solution. Nevertheless, EDD, AI EDD, and EDD/IRMPD MS^3 experiments yielded different cleavage patterns with less backbone fragmentation for the more stable solution-phase structure as compared to the other two 15-mers. By contrast, no major differences were observed in IRMPD, although the extent of backbone cleavage was higher with that technique for all three 15-mers.

Perhaps the most exciting EDD application area to date is the structural characterization of glycosaminoglycans and other carbohydrates. Amster and co-workers first applied EDD to doubly deprotonated GAG tetrasaccharides and compared EDD product ion spectra to those from low-energy CID and IRMPD.[291] These authors found that EDD produces information-rich spectra with both cross-ring and glycosidic cleavage product ions, whereas CID and IRMPD mainly result in glycosidic bond rupture. EDD seemed promising for locating sites of sulfation in GAGs, a longstanding mass spectrometric challenge due to the lability of sulfate groups. In follow-up work, the same group demonstrated that EDD can distinguish the GAG epimers iduronic acid and glucuronic acid based on diagnostic product ions.[303] These diagnostic products are not seen in CID or IRMPD and are believed to result from the EDD radical-driven chemistry involving hydrogen abstraction across a sugar ring, which is highly dependent on the precise stereochemistry—that is, whether the hydrogen is on the same side of the ring as the radical site or not. In more recent work, the Amster group extended their work to dermatan sulfate 10-mers and, once again, demonstrated that EDD provides more extensive fragmentation than IRMPD with far less loss of SO_3 from labile sulfate modifications and more cross-ring fragmentation.[304] Similar advantages of EDD have been reported by Adamson and Hakansson[305], who applied EDD to linear and branched neutral and sialylated oligosaccharides. The latter publication showed that EDD provides complementary structural information compared to both CID and IRMPD. In addition, EDD resulted in more extensive cross-ring fragmentation compared to the other two techniques. A comparison between EDD and low-energy CID fragmentation patterns for the branched sialylated oligosaccharide LSTb is shown in Figure 16.15. Finally, Marshall, Nilsson, and co-workers[290] applied EDD to the characterization of a ganglioside. These authors reported that EDD provided extensive fragmentation but that spectral interpretation was rendered difficult by frequent hydrogen rearrangements.

16.4.5 Electronic Excitation Dissociation (EED) and Electron-Induced Dissociation (EID)

16.4.5.1 Mechanism

A final category of gas-phase ion–electron reactions as activation methods for tandem mass spectrometry involves interactions between singly charged precursor ions and free electrons (note that ECD, hot ECD, ETD, and EDD require at least doubly charged precursor ions). The first reaction of this type, involving singly protonated peptides, was reported by Zubarev

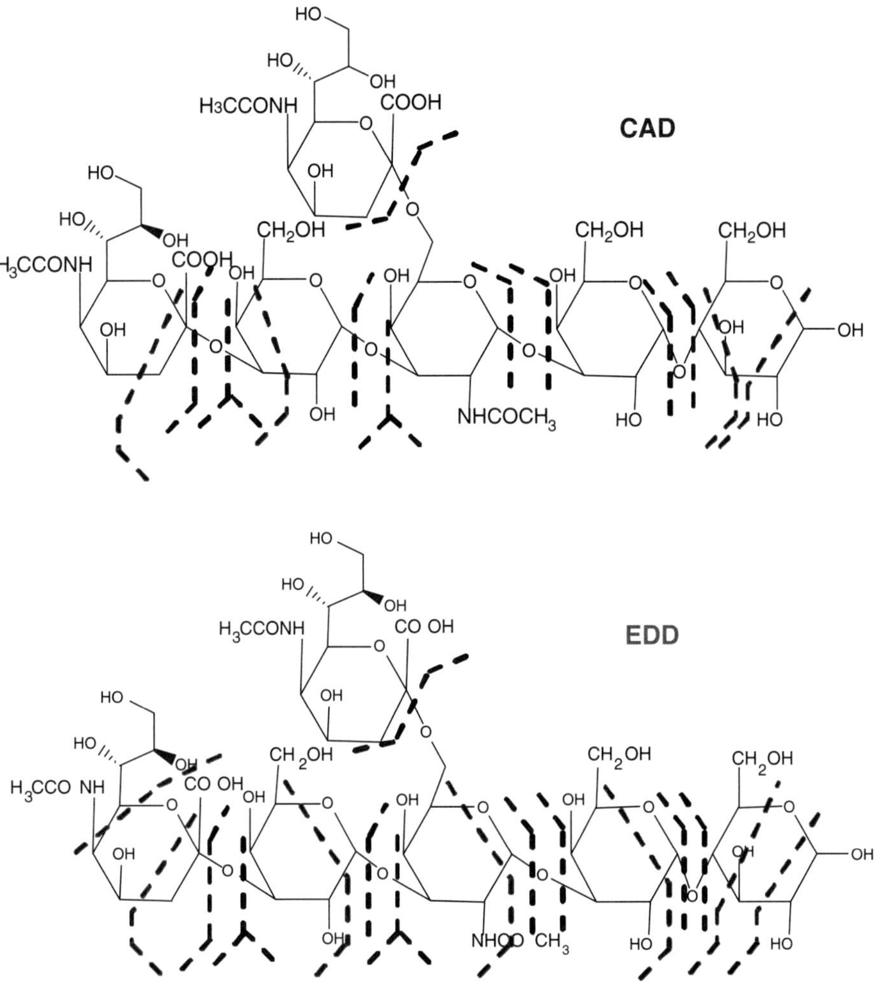

Figure 16.15. Fragmentation patterns observed following EDD and low-energy CID of a doubly deprotonated sialylated branched oligosaccharide. Cross-ring cleavages are highlighted in purple. (Adapted from Ref. 305 with permission from Elsevier Science.)

and co-workers[306] and termed electronic excitation dissociation. In EED, singly protonated precursors are first further ionized by >10-eV electrons (so-called tandem ionization mass spectrometry, TIMS)[253] to form doubly charged distonic cations and free electrons with one protonated and one radical site [see Eq. (16.15a)]. These doubly charged cations can subsequently recapture an electron to yield an electronically excited ion that is isoelectronic with the original precursor ion. This electronically excited ion may undergo facile dissociation due to the high exothermicity (10–12 eV) of electron capture. The overall EED fragmentation scheme is shown in Equations (16.15a) and (16.15b):

$$[\mathrm{M+H}]^{+} + e^{-}_{>10\,\mathrm{eV}} \rightarrow [\mathrm{M+H}]^{2+\bullet} + 2e^{-} \tag{16.15a}$$

$$[\mathrm{M+H}]^{2+\bullet} + e^{-}_{\mathrm{slow}} \rightarrow [\mathrm{M+H}]^{+*} \rightarrow \text{fragments} \tag{16.15b}$$

where M denotes the neutral analyte molecule, H denotes a proton, and the asterisk denotes electronic excitation.

Product ions observed from irradiation of singly charged cations with moderate energy (10–30 eV) electrons may also be a consequence of direct vibrational or electronic excitation—that is, an EIEIO process.[12] This process, depicted in Eq. (16.16a), may also be referred to as electron-induced dissociation, EID.[249,307] Recent work by Lioe and O'Hair[308] reported EID of singly protonated aromatic amino acids and cystine. These authors showed that fragmentation channels in EID were dramatically different from those in low-energy CID, including, for example, formation of benzyl cations in EID versus NH_3 and (H_2O + CO) losses in CID of aromatic amino acids. EID was also superior to CID in cleaving the disulfide bond in cystine. Due to similarities between EID and UV photodissociation and electron ionization, the former technique was proposed to proceed through both electronic and vibrational excitation. EID of singly deprotonated metabolite anions (see Eq. (16.16b) has been explored by Hakansson

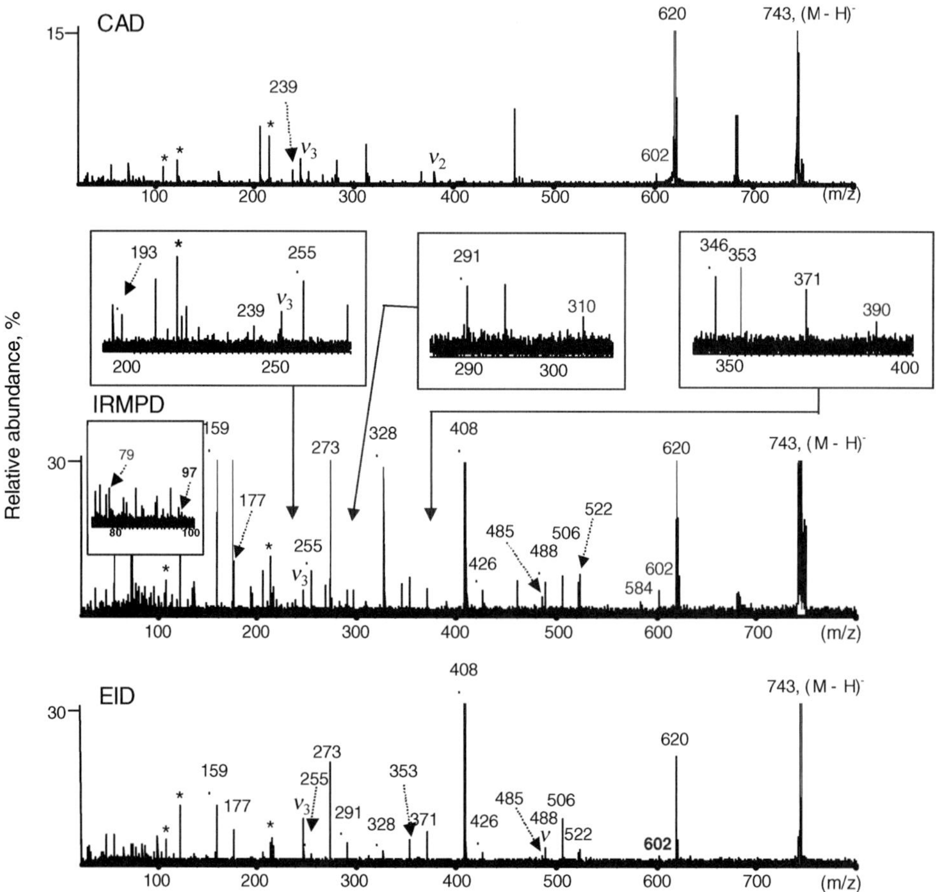

Figure 16.16. CID (7-V collision cell DC offset, 15 scans, **top**), IRMPD (10 W, 70 ms, 15 scans, middle), and EID (16 eV, 6 s, 15 scans, **bottom**) MS/MS spectra of nicotinic acid adenine dinucleotide phosphate. Product ions labeled with filled circles result from phosphoester and phosphoanhydride bond cleavages and are more abundant in IRMPD than in EID. ν_2 and ν_3 indicate harmonic peaks. Product ions with gray labels correspond to secondary H_2O losses. Electronic noise peaks are marked with asterisks. (Reproduced from Ref. 309 with permission from the American Chemical Society.)

and co-workers[309] as described below.

$$[M+H]^{+} + e^{-}_{>10\ eV} \rightarrow [M+H]^{+*} + e^{-} \rightarrow \text{fragments} \qquad (16.16a)$$

$$[M-H]^{-} + e^{-}_{>10\ eV} \rightarrow [M-H]^{-*} + e^{-} \rightarrow \text{fragments} \qquad (16.16b)$$

where M denotes the neutral analyte molecule, H denotes a proton, and the asterisk denotes vibrational or electronic excitation. The main advantage of EED and EID is their applicability to singly charged ions, thereby rendering them compatible with MALDI ionization. Drawbacks are limited fragmentation efficiency and, to date, a limited understanding of the underlying mechanisms.

16.4.5.2 EED and EID Application

In the original EED publication by Zubarev and co-workers,[306] fragmentation of a 10-residue peptide was compared to ECD of the corresponding doubly protonated species. Only one *c*-type ion was observed following EED, whereas seven $z^{\bullet}$ and two *c*-type ions were seen following ECD. For singly protonated substance P, three *c* ions and eight *a* ions were observed. Gas-phase ionization of peptides with acidic C-termini was accompanied by CO_2 loss, whereas peptides with amidated C-termini displayed a loss of 74 Da. In a later publication by the same group,[310] it was confirmed that even-electron *a* and *c* ions are the main products from peptide EED. The data presented in that work showed three *c* ions and nine *a* ions following EED of substance P, corresponding to cleavage between all but one amino acid residue. EED of singly protonated angiotensin II resulted in four even-electron *a* ions but no *c* ions. The second EED publication[310] also presented mechanisms for the 74-Da loss observed for C-terminally amidated peptides. In summary, EED provides peptide backbone cleavages complementary to those from ECD and CID. In particular, the observed *a* ion formation does not discriminate against the N-terminal side of proline, a cleavage site which is only rarely observed in ECD.[311]

Hakansson and co-workers[309] have applied negative ion mode EID to metabolite analysis. These authors found that, for phosphate-containing metabolites, EID provided complementary fragmentation compared to both low-energy CID and IRMPD, although EID fragmentation efficiency was lower compared to the two former techniques. A comparison of CID, IRMPD, and EID for nicotinic acid adenine dinucleotide phosphate is shown in Figure 16.16. O'Hair and co-workers[312] recently used EID to examine clusters of the zwitterionic amino acid betaine. Once again, complementary fragmentation behavior was observed compared to low-energy CID.

16.5 SUMMARY

There exists a wide range of ion activation techniques suitable for MS/MS applications. The different methods, which are distinguished by the timescale of activation, the distribution of energy deposited into precursor ions, and the molecular modes involved in excitation, can lead to very different fragmentation outcomes for a given precursor ion. CID, which is performed using either low- or high-energy collisions with neutral atoms or small molecules, is the most common form of ion activation for MS/MS. Low-energy CID, whereby ions are gradually heated through successive energetic collisions, is an exceptionally versatile activation method and can be implemented on all common MS instruments. Low-energy CID has been applied to virtually all classes of ionizable compounds and has proven to be a particularly valuable activation technique for MS/MS analysis of peptides and other biopolymers. High-energy CID, which is implemented using beam-type instruments, may involve direct electronic and vibrational excitation. The fragmentation spectra are generally quite different from those produced by low-energy CID, with the higher internal energies achieved in high-energy CID favoring higher energy dissociation pathways that often result in structurally informative product ions. High-energy CID is also well suited for MS/MS of large singly charged ions, such as those produced by MALDI. SID, which involves energetic collisions with surfaces, can be implemented on a variety of mass analyzers, including quadrupole, Q-TOF and FT-ICR instruments. The higher internal energies afforded by SID, compared to low-energy CID, are advantageous for the dissociation of large singly charged ions and for accessing higher energy fragmentation pathways. Although MS/MS is most commonly performed using collisional activation, alternative methods are gaining prominence. Photon induced dissociation, whereby gas-phase ions are activated by the absorption of a single or multiple photons, can be carried out at a wide range of photon wavelengths. With the growing popularity of trapping instruments, there has been a dramatic increase in the use of IR photodissociation techniques. Precursor ion activation in IRMPD involves the absorption of multiple IR photons, usually produced from a cw CO2 laser. IRMPD fragmentation spectra are generally similar to those observed with low-energy CID, although IRMPD has the advantage of not requiring a collision gas. Additionally, the extent of dissociation of precursor ions is conveniently controlled by the choice of laser power and irradiation time. IRMPD, when performed with a tunable IR source, allows for the determination of IR spectra, which can provide novel insights into structures of gas-phase ions. The BIRD technique, whereby ions trapped in the ion cell of an FT-ICR mass spectrometer are activated by the absorption of blackbody IR radiation emitted from the wall of the cell, has emerged as an important method for determining the kinetic parameters for dissociation of gas-phase ions. UVPD involves electronic excitation of ions through the absorption of UV radiation and is carried out using both beam and trapping instruments. UVPD has the advantage that it produces both low- and high-energy product ions and much of the current interest in the technique is related to peptide sequencing applications. Recently, there has been substantial growth in the use of electron mediated methods, particularly for proteomics applications. ECD (and HECD), which relies on the capture of free electrons by multiply protonated ions, has emerged as an important activation method for the characterization of proteins and their post-translational modifications. ETD involves an anion electron carrier to affect electron transfer to multiply protonated precursor ions. Fragmentation by ETD is similar to that observed with ECD, but the technique is more easily implemented on trapping instruments other than FT-ICR. In EDD, a negative ion mode alternative to ECD, irradiation of multiply charged anions with electrons causes ejection of an electron from precursor ions and subsequent fragmentation of the resulting radical anions. EDD provides unique fragmentation characteristic for acidic molecules and, as such, is well suited for the analysis of acidic biopolymers such as sialylated oligosaccharides, nucleic acids, and sulfo- and phosphopeptides. While electron mediated methods are normally applied to multiply charged precursor ions, they can also be used to induce dissociation of singly charged ions. Fragmentation in EED and EID involves the interaction between singly charged precursor ions and free electrons. This interaction can lead to the formation of doubly charged distonic cations, which can subsequently recapture an electron and undergo fragmentation (EED), in a manner analogous to ECD. Alternatively, electron irradiation can result in electronic and/or vibrational excitation of precursor ions and their subsequent fragmentation (EID).

REFERENCES

1. McLafferty, F. W. *Tandem Mass Spectrometry*, John Wiley & Sons, New York, 1983.
2. McLuckey, S. A. Principles of collisional activation in analytical mass spectrometry. *J. Am. Soc. Mass Spectrom.* 1992, *3*, 599–614.
3. Chorush, R. A.; Little, D. P.; Beu, S. C.; Wood, T. D.; McLafferty, F. W. Surface induced dissociation of multiply protonated proteins. *Anal. Chem.* **1995**, *67*, 1042–1046.
4. McCormack, A. L.; Jones, J. L.; Wysocki, V. H. Surface-induced dissociation of multiply protonated peptides. *J. Am. Soc. Mass Spectrom.* **1992**, *3*, 859–862.
5. Grill, V.; Shen, J.; Evans, C.; Cooks, R. G. Collisions of ions with surfaces at chemically relevant energies: Instrumentation and phenomena. *Rev. Sci. Instr.* **2001**, *72*, 3149–3179.
6. Little, D. P.; Speir, J. P.; Senko, M. W.; O'Connor, P. B.; McLafferty, F. W. Infrared multiphoton dissociation of large multiply-charged ions for biomolecule sequencing. *Anal. Chem.* **1994**, *66*, 2809–2815.
7. Woodin, R. L.; Bomse, D. S.; Beauchamp, J. L. Multiphoton dissociation of molecules with low power continuous wave infrared laser radiation. *J. Am. Chem. Soc.* **1978**, *100*, 3248–3250.
8. Dunbar, R. C.; McMahon, T. B. Activation of unimolecular reactions by ambient blackbody radiation. *Science* **1998**, *279*, 194–197.
9. Price, W. D.; Schnier, P. D.; Williams, E. R. Tandem mass spectrometry of large biomolecule ions by blackbody infrared radiative dissociation. *Anal. Chem.* **1996**, *68*, 859–866.
10. Guan, Z.; Kelleher, N. L.; O'Connor, P. B.; Aaserud, D. J.; Little, D. P.; McLafferty, F. W. 193 nm photodissociation of larger multiply-charged biomolecules. *Int. J. Mass Spectrom. Ion Processes* **1996**, *157/158*, 357–364.
11. Thompson, M. S.; Cui, W. D.; Reilly, J. P. Fragmentation of singly charged peptide ions by photodissociation at $\lambda = 157$ nm. *Angew. Chem. Int. Ed.* **2004**, *43*, 4791–4794.
12. Cody, R. B.; Freiser, B. S. Electron-impact excitation of ions from organics—Alternative to collision-induced dissociation. *Anal. Chem.* **1979**, *51*, 547–551.
13. Zubarev, R. A.; Kelleher, N. L.; McLafferty, F. W. Electron capture dissociation of multiply charged protein cations. A nonergodic process. *J. Am. Chem. Soc.* **1998**, *120*, 3265–3266.
14. Syka, J. E. P.; Coon, J. J.; Schroeder, M. J.; Shabanowitz, J.; Hunt, D. F. Peptide and protein sequence analysis by electron transfer dissociation mass spectrometry. *Proc. Natl. Acad. Sci. U.S.A* **2004**, *101*, 9528–9533.
15. McLuckey, S. A.; Goeringer, D. E. slow heating methods in tandem mass spectrometry. *J. Mass Spectrom.* **1997**, *35*, 461–474.
16. Breuker, K.; Oh, H. B.; Lin, C.; Carpenter, B. K.; McLafferty, F. W. Nonergodic and conformational control of the electron capture dissociation of protein cations. *Proc. Natl. Acad. Sci. U.S.A.* **2004**, *101*, 14011–14016.
17. Jones, J. W.; Sasaki, T.; Goodlett, D. R.; Turecek, F. Electron capture in spin-trap capped peptides. An experimental example of ergodic dissociation in peptide cation-radicals. *J. Am. Soc. Mass Spectrom.* **2007**, *18*, 432–444.
18. Leib, R. D.; Donald, W. A.; Bush, M. F.; O'Brien, J. T.; Williams, E. R. Internal energy deposition in electron capture dissociation measured using hydrated divalent metal ions as nanocalorimeters. *J. Am. Chem. Soc.* **2007**, *129*, 4894–4895.
19. Turecek, F. N–C-alpha bond dissociation energies and kinetics in amide and peptide radicals. Is the dissociation a non-ergodic process? *J. Am. Chem. Soc.* **2003**, *125*, 5954–5963.
20. Lifshitz, C.; Laskin, J. *Principles of Mass Spectrometry Applied to Biomolecules*, John Wiley & Sons, New York, 2006.
21. Vachet, R. W.; Glish, G. L. Effects of heavy gases on the tandem mass spectra of peptide ions in the quadrupole ion trap. *J. Am. Soc. Mass Spectrom.* **1996**, *7*, 1194–1202.
22. Gauthier, J. W.; Trautman, T. R.; Jacobson, D. B. Sustained off-resonance irradiation for CAD involving FTMS—CAD technique that emulates infrared multiphoton dissociation. *Anal. Chim. Acta* **1991**, *246*, 211–225.
23. Louris, J. N.; Cooks, R. G.; Syka, J. E. P.; Kelley, P. E.; Stafford, G. C.; Todd, J. F.; Instrumentation, applications, and energy deposition in quadrupole ion trap tandem mass spectrometry. *Anal. Chem.* **1987**, *59*, 1677–1685.
24. Xia, Y.; Liang, X.; McLuckey, S. A. Ion trap versus low-energy beam-type collision-induced dissociation of protonated ubiquitin ions. *Anal. Chem.* **2006**, *78*, 1218–1227.
25. Laskin, J.; Futrell, J. H. Collisional activation of peptide ions in FT-ICR mass spectrometry. *Mass Spectrom. Rev.* **2003**, *22*, 158–181.
26. Vekey, K. Multiply charged ions. *Mass Spectrom. Rev.* **1995**, *14*, 195–225.
27. Yu, W.; Vath, J. E.; Huberty, M. C.; Martin, S. A. Identification of the facile gas-phase cleavage of the Asp-Pro and Asp-Xxx peptide bonds in matrix-assisted laser desorption time-of-flight mass spectrometry. *Anal. Chem.* **1993**, *65*, 3015–3023.
28. Qin, J.; Chait, B. T. Preferential fragmentation of protonated gas-phase peptide ions adjacent to acidic amino acid residues. *J. Am. Chem. Soc.* **1995**, *117*, 5411–5412.
29. Racine, A. H.; Payne, A. H.; Remes, P. M.; Glish, G. L. Thermally assisted collision induced dissociation in a

quadrupole ion trap mass spectrometer. *Anal. Chem.* **2006**, *78*, 4609–4614.

30. Fenn, J. B.; Mann, M.; Meng, C. K.; Wong, S. F.; Whitehouse, C. M. Electrospray ionization for mass spectrometry of large biomolecules. *Science* **1989**, *246*, 64–71.
31. Karas, M.; Hillenkamp, F. Laser desorption ionization of proteins with molecular masses exceeding 10000 daltons. *Anal. Chem.* **1988**, *60*, 2299–2301.
32. Tanaka, K.; Waki, H.; Ido, Y.; Akita, S.; Yoshida, Y.; Yoshida, T. Protein and polymer analyses up to *m/z* 100,000 by laser ionization time-of-flight mass spectrometry. *Rapid Commun. Mass Spectrom.* **1988**, *2*, 151–153.
33. Bowie, J. H.; Brinkworth, C. S.; Dua, S. Collision-induced fragmentations of the (M – H)(–) parent anions of underivatized peptides: An aid to structure determination and some unusual negative ion cleavages. *Mass Spectrom. Rev.* **2002**, *21*, 87–107.
34. Paizs, B.; Suhai, S. Fragmentation pathways of protonated peptides. *Mass Spectrom. Rev.* **2005**, *24*, 508–548.
35. Wysocki, V. H.; Tsaprailis, G.; Smith, L. L.; Breci, L. A. Special feature: Commentary—mobile and localized protons: A framework for understanding peptide dissociation. *J. Mass Spectrom.* **2000**, *35*, 1399–1406.
36. Hopley, C.; Bristow, T.; Lubben, A.; Simpson, A.; Bul, E.; Klagkou, K.; Herniman, J.; Langley, J. Towards a universal product ion mass spectral library—Reproducibility of product ion spectra across eleven different mass spectrometers. *Rapid Commun. Mass Spectrom.* **2008**, *22*, 1779–1786.
37. Smith, C. A.; O'Maille, G.; Want, E. J.; Qin, C.; Trauger, S. A.; Brandon, T. R.; Custodio, D. E.; Abagyan, R.; Siuzdak, G. METLIN—A metabolite mass spectral database. *Ther. Drug Monit.* **2005**, *27*, 747–751.
38. Schlosser, A.; Pipkorn, R.; Bossemeyer, D.; Lehmann, W. D. Analysis of protein phosphorylation by a combination of elastase digestion and neutral loss tandem mass spectrometry. *Anal. Chem.* **2001**, *73*, 170–176.
39. Staack, R. F.; Hopfgartner, G. New analytical strategies in studying drug metabolism. *Anal. Bioanal. Chem.* **2007**, *388*, 1365–1380.
40. Naxing Xu, R.; Fan, L.; Rieser, M. J.; El-Shourbagy T. A. Recent advances in high-throughput quantitative bioanalysis by LC–MS/MS. *J. Pharm. Biomed Anal.* **2007**, *44*, 342–355.
41. Kassel, D. B. Applications of high-throughput ADME in drug discovery. *Curr. Opin. Chem. Biol.* **2004**, *8*, 339–345.
42. Hopfgartner, G.; Bourgogne, E. Quantitative high-throughput analysis of drugs in biological matrices by mass spectrometry. *Mass Spectrom. Rev.* **2003**, *22*, 195–214.
43. Zhang, H.; Henion, J. Comparison between liquid chromatography–time-of-flight mass spectrometry and selected reaction monitoring liquid chromatography—Mass Spectrometry for Quantitative Determination of Idoxifene in human plasma. *J. Chromatogr. B* **2001**, *757*, 151–159.
44. Rosen, R. Mass Spectrometry for monitoring micropollutants in water. *Curr. Opin. Biotechnol.* **2007**, *18*, 246–251.
45. Gros, M.; Petrovic, M.; Barcelo, D. Multi-residue analytical methods using LC-tandem MS for the determination of pharmaceuticals in environmental and wastewater samples: A review. *Anal. Bioanal. Chem.* **2006**, *386*, 941–952.
46. Pico, Y.; Blasco, C.; Font, G. Environmental and food applications of LC-tandem mass spectrometry in pesticide-residue analysis: An overview. *Mass Spectrom. Rev.* **2004**, *23*, 45–85.
47. Kraemer, T.; Paul, L. D. Bioanalytical procedures for determination of drugs of abuse in blood. *Anal. Bioanal. Chem.* **2007**, *388*, 1415–1435.
48. Maurer, H. H. Multi-analyte procedures for screening for and quantification of drugs in blood, plasma, or serum by liquid chromatography-single stage or tandem mass spectrometry (LC-MS or LC-MS/MS) relevant to clinical and forensic toxicology. *Clin. Biochem.* **2005**, *38*, 310–318.
49. Wood, M.; Laloup, M.; Samyn, N.; Fernandez, M. D. R.; de Bruijn, E. A.; Maes, R. A. A.; De Boeck, G. Recent applications of liquid chromatography-mass spectrometry in forensic science. *J. Chromatogr. A 1130*, 3–15.
50. Liu, H. B.; Lin, D. Y.; Yates, J. R. Multidimensional separations for protein/peptide analysis in the post-genomic era. *Biotechniques* **2002**, *32*, 898–911.
51. Fournier, M. L.; Gilmore, J. M.; Martin-Brown, S. A.; Washburn, M. P. Multidimensional separations-based shotgun proteomics. *Chem. Rev.* **2007**, *107*, 3654–3686.
52. Wells, J. M.; McLuckey, S. A. Collision-induced dissociation (CID) of peptides and proteins. *Biol. Mass Spectrom.* **2005**, *402*, 148–185.
53. Lubec, G.; Afjehi-Sadat, L. Limitations and pitfalls in protein identification by mass spectrometry. *Chem. Rev.* **2007**, *107*, 3568–3584.
54. Kelleher, N. L.; Lin, H. Y.; Valaskovic, G. A.; Aaserud, D. J.; Fridriksson, E. K.; McLafferty, F. W. Top down versus bottom up protein characterization by tandem high-resolution mass spectrometry. *J. Am. Chem. Soc.* **1999**, *121*, 806–812.
55. McLuckey, S. A.; Van Berkel, G. J.; Glish, G. L. Tandem mass spectrometry of small, multiply-charged oligonucleotides. *J. Am. Soc. Mass Spectrom.* **1992**, *3*, 60–70.
56. Kirpekar, F.; Krogh, T. N. RNA Fragmentation studied in a matrix-assisted laser desorption/ionization tandem quadrupole/orthogonal time-of-flight mass spectrometer. *Rapid Commun. Mass Spectrom.* **2001**, *15*, 8–14.

57. Schurch, S.; Bernal-Mendez, E.; Leumann, C. J. Electrospray tandem mass spectrometry of mixed-sequence RNA/DNA oligonucleotides. *J. Am. Soc. Mass Spectrom.* **2002**, *13*, 936–945.
58. Ni, J.; Pomerantz, S. C.; Rozenski, J.; Zhang, Y.; McCloskey, J. A. Interpretation of oligonucleotide mass spectra for determination of sequence using electrospray ionization and tandem mass spectrometry. *Anal. Chem.* **1996**, *68*, 1989–1999.
59. Null, A. P.; Muddiman, D. C. Perspectives on the use of electrospray ionization Fourier transform ion cyclotron resonance mass spectrometry for short tandem repeat genotyping in the post-genome Era. *J. Mass Spectrom.* **2001**, *36*, 589–606.
60. Monn, S. T. M.; Schurch, S. New aspects of the fragmentation mechanisms of unmodified and methylphosphonate-modified oligonucleotides. *J. Am. Soc. Mass Spectrom.* **2007**, *18*, 984–990.
61. Tromp, J. M.; Schurch, S. Gas-phase dissociation of oligoribonucleotides and their analogs studied by electrospray ionization tandem mass spectrometry. *J. Am. Soc. Mass Spectrom.* **2005**, *16*, 1262–1268.
62. Wan, K. X.; Gross, M. L. Fragmentation mechanisms of oligodeoxynucleotides: Effects of replacing phosphates with methylphosphonates and thymines with other bases in T-rich sequences. *J. Am. Soc. Mass Spectrom.* **2001**, *12*, 580–589.
63. Wu, J.; McLuckey, S. A. Gas-phase fragmentation of oligonucleotide ions. *Int. J. Mass Spectrom.* **2004**, *237*, 197–241.
64. Dias, N.; Stein, C. A. Antisense oligonucleotides: Basic concepts and mechanisms. *Mol. Cancer Ther.* **2002**, *1*, 347–355.
65. Viseux, N.; deHoffmann, E.; Domon, B. Structural analysis of permethylated oligosaccharides by electrospray tandem mass spectrometry. *Anal. Chem.* **1997**, *69*, 3193–3198.
66. Weiskopf, A. S.; Vouros, P.; Harvey, D. J. Electrospray ionization-ion trap mass spectrometry for structural analysis of complex N-linked glycoprotein oligosaccharides. *Anal. Chem.* **1998**, *70*, 4441–4447.
67. Sheeley, D. M.; Reinhold, V. N. Structural characterization of carbohydrate sequence, linkage, and branching in a quadrupole ion trap mass spectrometer: Neutral oligosaccharides and N-linked glycans. *Anal. Chem.* **1998**, *70*, 3053–3059.
68. Haddon, W. F.; McLafferty, F. W. Metastable ion characteristics. VII. Collision-induced metastables. *J. Am. Chem. Soc.* **1968**, *90*, 4745–4746.
69. Jennings, K. R. Collision-induced decompositions of aromatic molecular ions. *Int. J. Mass Spectrom. Ion. Phys.* **1968**, *1*, 227–235.
70. Papayannopoulos, I. A. The interpretation of collision induced dissociation tandem mass spectra of peptides. *Mass Spectrom. Rev.* **1995**, *14*, 49–73.
71. Cheng, C.; Gross, M. L. Applications and mechanisms of charge-remote fragmentation. *Mass Spectrom. Rev.* **2000**, *19*, 398–420.
72. Harvey, D. J.; Bateman, R. H.; Green, B. N. High-energy collision-induced fragmentation of complex oligosaccharides ionized by matrix-assisted laser desorption/ionization mass spectrometry. *J. Mass Spectrom.* **1997**, *32*, 167–187.
73. Mechref, Y.; Novotny, M. V.; Krishnan, C. Structural characterization of oligosaccharides using MALDI-TOF/TOF tandem mass spectrometry. *Anal. Chem.* **2003**, *75*, 4895–4903.
74. Laskin, J.; Futrell, J. H. Activation of large ions in FT-ICR mass spectrometry. *Mass Spectrom. Rev.* **2005**, *24*, 135–167.
75. Wysocki, V. H.; Joyce, K. E.; Jones, C. M.; Beardsley, R. L. Surface-induced dissociation of small molecules, peptides, and non-covalent protein complexes. *J. Am. Soc. Mass Spectrom.* **2008**, *19*, 190–208.
76. Cooks, R. G.; Terwilliger, D. T.; Ast, T.; Beynon, J. H.; Keough, T. Surface modified mass spectrometry. *J. Am. Chem. Soc.* **1975**, *97*, 1583–1585.
77. Wysocki, V. H.; Ding, J. M.; Jones, J. L.; Callahasn, J. H.; King, F. L. Surface-induced dissociation in tandem quadrupole mass spectrometers—A comparison of 3 designs. *J. Am. Soc. Mass Spectrom.* **1992**, *3*, 27–32.
78. Williams, E. R.; Fang, L. L.; Zare, R. N. Surface induced dissociation for tandem time-of-flight mass spectrometry. *Int. J. Mass Spectrom.* **1993**, *123*, 233–241.
79. Galhena, A. S.; Dagan, S.; Jones, C. M.; Beardsley, R. L.; Wysocki, V. H. Surface-induced dissociation of peptides and protein complexes in a quadrupole/time-of-flight mass spectrometer. *Anal. Chem.* **2008**, *80*, 1425–1436.
80. Morris, M. R.; Riederer, D. E.; Winger, B. E.; Cooks, R. G.; Ast, T.; Chidsey, C. E. D. Ion surface collisions at functionalized self-assembled monolayer surfaces. *Int. J. Mass Spectrom. Ion Processes* **1992**, *122*, 181–217.
81. Cooks, R. G.; Ast, T.; Pradeep, T.; Wysocki, V. H. Reactions of ions with organic surfaces. *Acc. Chem. Res.* **1994**, *27*, 316–323.
82. Dongre, A. R.; Somogyi, A.; Wysocki, V. H. Surface-induced dissociation: An effective tool to probe structure, energetics and fragmentation mechanisms of protonated peptides. *J. Mass Spectrom.* **1996**, *31*, 339–350.
83. Gu, C. G.; Wysocki, V. H.; Harada, A.; Takaya, H.; Kumadaki, I. Dissociative and reactive hyperthermal ion-surface collisions with Langmuir–Blodgett films terminated by $CF3(CH2)_n$, *n*-perfluoroalkyl, or *n*-alkyl groups. *J. Am. Chem. Soc.* **1999**, *121*, 10554–10562.
84. Beck, R. D.; Rockenberger, J.; Weis, P.; Kappes, M. M. Fragmentation of C-60(+) and higher fullerenes by surface impact. *J. Chem. Phys* **1996**, *104*, 3638–3650.
85. Laskin, J.; Futrell, J. H. Energy transfer in collisions of peptide ions with surfaces. *J. Chem. Phys* **2003**, *119*, 3413–3420.
86. Schultz, D. G.; Hanley, L. Shattering of $SiMe^{3+}$ during surface-induced dissociation. *J. Chem. Phys* **1998**, *109*, 10976–10983.

87. SCHULTZ, D. G.; WAINHAUS, S. B.; HANLEY, L.; DESAINTECLAIRE, P.; HASE, W. L. Classical dynamics simulations of $SiMe^{3+}$ ion-surface scattering. *J. Chem. Phys* **1997**, *106*, 10337–10348.
88. KUBISTA, J.; DOLEJSEK, Z.; HERMAN, Z. Energy partitioning in collisions of slow polyatomic ions with surfaces: Ethanol molecular ions on stainless steel surfaces. *Eur. Mass Spectrom.* **1998**, *4*, 311–319.
89. WIEGHAUS, A.; SCHMIDT, L.; POPOVA, A. M.; KOMAROV, V. V.; JUNGCLAS, H. Fragmentation of polyatomic molecules by grazing incidence surface-induced dissociation (GI-SID). *J. Mass Spectrom.* **1999**, *34*, 1178–1184.
90. WORGOTTER, R.; KUBISTA, J.; ZABKA, J.; DOLEJSEK, Z.; MARK, T. D.; HERMAN, Z. Surface-induced reactions and decomposition of the benzene molecular ion C_6H^{6+}: Product ion intensities, angular and translational energy distributions. *Int. J. Mass Spectrom.* **1998**, *174*, 53–62.
91. WINGER, B. E.; LAUE, H. J.; HORNING, S. R.; JULIAN, R. K.; LAMMERT, S. A.; RIEDERER, D. E.; COOKS, R. G. Hybrid BEEQ tandem mass spectrometer for the study of ion surface collision processes. *Rev. Sci. Instr.* **1992**, *63*, 5613–5625.
92. BECK, R. D.; WEIS, P.; BRAUCHLE, G.; ROCKENBERGER, J. Tandem time-of-flight mass spectrometer for cluster-surface scattering experiments. *Rev. Sci. Instr.* **1995**, *66*, 4188–4197.
93. GOLOGAN, B.; GREEN, J. R.; ALVAREZ, J.; LASKIN, J.; COOKS, R. G. Ion/surface reactions and ion soft-landing. *Phys. Chem. Chem. Phys.* **2005**, *7*, 1490–1500.
94. BECK, R. D.; ST. JOHN, P.; HOMER, M. L.; WHETTEN, R. L. Impact-induced cleaving and melting of alkali-halide nanocrystals. *Science* **1991**, *253*, 879–883.
95. BAILEY, T. H.; LASKIN, J.; FUTRELL, J. H. Energetics of selective cleavage at acidic residues studied by time-and energy-resolved surface-induced dissociation in FT-ICR MS. *Int. J. Mass Spectrom.* **2003**, *222*, 313–327.
96. LASKIN, J.; BAILEY, T. H.; FUTRELL, J. H. Shattering of peptide ions on self-assembled monolayer surfaces. *J. Am. Chem. Soc.* **2003**, *125*, 1625–1632.
97. DONGRE, A. R.; JONES, J. L.; SOMOGYI, A.; WYSOCKI, V. H. Influence of peptide composition, gas-phase basicity, and chemical modification on fragmentation efficiency: Evidence for the mobile proton model. *J. Am. Chem. Soc.* **1996**, *118*, 8365–8374.
98. TSAPRAILIS, G.; NAIR, H.; SOMOGYI, A.; WYSOCKI, V. H.; ZHONG, W.; FUTRELL, J. H.; SUMMERFIELD, S. G.; GASKELL, S. J. Influence of secondary structure on the fragmentation of protonated peptides. *J. Am. Chem. Soc.* **1999**, *121*, 5142–5154.
99. WIEGHAUS, A.; SCHMIDT, L.; POPOVA, A. M.; KOMAROV, V. V.; JUNGCLAS, H. Grazing incidence surface-induced dissociation of protonated peptides generated by matrix-assisted laser desorption/ionization. *Rapid Commun. Mass Spectrom.* **2000**, *14*, 1654–1661.
100. STONE, E.; GILLIG, K. J.; RUOTOLO, B.; FUHRER, K.; GONIN, M.; SCHULTZ, A.; RUSSELL, D. H. Surface-induced dissociation on a MALDI-ion mobility-orthogonal time-of-flight mass spectrometer: Sequencing peptides from an "in-solution" protein digest. *Anal. Chem.* **2001**, *73*, 2233–2238.
101. LASKIN, J.; BECK, K. M.; HACHE, J. J.; FUTRELL, J. H. Surface-induced dissociation of ions produced by matrix-assisted laser desorption/ionization in a Fourier transform ion cyclotron resonance mass spectrometer. *Anal. Chem.* **2004**, *76*, 351–356.
102. GAMAGE, C. M.; FERNANDEZ, F. M.; KUPPANNAN, K.; WYSOCKI, V. H. Submicrosecond surface-induced dissociation of peptide ions in a MALDI TOF MS. *Anal. Chem.* **2004**, *76*, 5080–5091.
103. JONES, C. M.; BEARDSLEY, R. L.; GALHENA, A. S.; DAGAN, S.; WYSOCKI, V. H. Symmetrical gas-phase dissociation of noncovalent protein complexes via surface collisions. *J. Am. Chem. Soc.* **2006**, *128*, 15044–15045.
104. JURCHEN, J. C.; GARCIA, D. E.; WILLIAMS, E. R. Further studies on the origins of asymmetric charge partitioning in protein homodimers. *J. Am. Soc. Mass Spectrom.* **2004**, *15*, 1408–1415.
105. SOBOTT, F.; MCCAMMON, M. G.; ROBINSON, C. V. Gas-phase dissociation pathways of a tetrameric protein complex. *Int. J. Mass Spectrom.* **2003**, *230*, 193–200.
106. FELITSYN, N.; KITOVA, E. N.; KLASSEN, J. S. Thermal decomposition of a gaseous multiprotein complex studied by blackbody infrared radiative dissociation. Investigating the origin of the asymmetric dissociation behavior. *Anal. Chem.* **2001**, *73*, 4647–4661.
107. BENESCH, J. L. P.; AQUILINA, J. A.; RUOTOLO, B. T.; SOBOTT, F.; ROBINSON, C. V. Tandem mass spectrometry reveals the quaternary organization of macromolecular assemblies. *Chem. Biol.* **2006**, *13*, 597–605.
108. THORNE, L. R.; BEAUCHAMP, J. L.In Bowers, M. T. (Ed.), *Gas-Phase Ion Chemistry*, Vol. *3*, Academic Press, New York, 1984, pp. 41–97.
109. BOMSE, D. S.; WOODIN, R. L.; BEAUCHAMP, J. L. Molecular activation with low-intensity CW infrared laser radiation—multiphoton dissociation of ions derived from diethyl Ether. *J. Am. Chem. Soc.* **1979**, *101*, 5503–5512.
110. PAYNE, A. H.; GLISH, G. L. Thermally assisted infrared multiphoton photodissociation in a quadrupole ion trap. *Anal. Chem.* **2001**, *73*, 3542–3548.
111. CROWE, M. C.; BRODBELT, J. S. Infrared multiphoton dissociation (IRMPD) and collisionally activated dissociation of peptides in a quadrupole ion trap with selective IRMPD of phosphopeptides. *J. Am. Soc. Mass Spectrom.* **2004**, *15*, 1581–1592.
112. FLORA, J. W.; MUDDIMAN, D. C. Gas-phase ion unimolecular dissociation for rapid phosphopeptide mapping by IRMPD in a Penning ion trap: An energetically favored process. *J. Am. Chem. Soc.* **2004**, *124*, 6546–6547.
113. GARDNER, M. W.; VASICEK, L. A.; SHABBIR, S.; ANSLYN, E. V.; BRODBELT, J. S. Chromogenic cross-linker for the characterization of protein structure by infrared

multiphoton dissociation mass spectrometry. *Anal. Chem.* **2008**, *80*, 4807–4819.
114. Wilson, J. J.; Brodbelt, J. S. Infrared multiphoton dissociation for enhanced *de novo* sequence interpretation of N-terminal sulfonated peptides in a quadrupole ion trap. *Anal. Chem.* **2006**, *78*, 6855–6862.
115. Shi, S. D.-H.; Hendrickson, C. L.; Marshall, A. G.; Seigel, M. M.; Kong, F.; Carter, G. T. structural validation of saccharomicins by high resolution and high mass accuracy FTICR MS and IRMPD mass spectrometry. *J. Am. Soc. Mass Spectrom.* **1999**, *10*, 1285–1290.
116. Xie, Y. M.; Lebrilla, C. B. Infrared multiphoton dissociation of alkali metal-coordinated oligosaccharides. *Anal. Chem.* **2003**, *75*, 1590–1598.
117. Lancaster, K. S.; An, H. J.; Li, B. S.; Lebrilla, C. B. Interrogation of N-linked oligosaccharides using infrared multiphoton dissociation in FT-ICR mass spectrometry. *Anal. Chem.* **2006**, *78*, 4990–4997.
118. Zhang, J. H.; Schubothe, K.; Li, B. S.; Russell, S.; Lebrilla, C. B. Infrared multiphoton dissociation of O-linked mucin-type oligosaccharides. *Anal. Chem.* **2005**, *77*, 208–214.
119. Pikulski, M.; Hargrove, A.; Shabbir, S. H.; Anslyn, E. V.; Brodbelt, J. S. Sequencing and characterization of oligosaccharides using infrared multiphoton dissociation and boronic acid derivatization in a quadrupole ion trap. *J. Am. Soc. Mass Spectrom.* **2007**, *18*, 2094–2106.
120. Forbes, M. W.; Bush, M. F.; Polfer, N. C.; Oomens, J.; Dunbar, R. C.; Williams, E. R.; Jockusch, R. A. Infrared spectroscopy of arginine cation complexes: Direct observation of gas-phase zwitterions. *J. Phys. Chem. A* **2007**, *111*, 11759–11770.
121. Wu, R. H.; McMahon, T. B. Infrared multiple photon dissociation spectroscopy as structural confirmation for GlyGlyGlyH(+) and AlaAlaAlaH (+) in the gas phase. Evidence for amide oxygen as the protonation site. *J. Am. Chem. Soc.* **2007**, *129*, 11312–11313.
122. Dunbar, R. C.; McMahon, T. B.; Tholmann, D.; Tonner, D. S.; Salahub, D. R.; Wei, D. Zero-pressure thermal-radiation-induced dissociation of gas-phase cluster ions: comparison of theory and experiment for $(H_2O)_2Cl$- and $(H_2O)_3Cl$-. *J. Am. Chem. Soc.* **1995**, *117*, 12819–12825.
123. Heeren, R. M. A.; Vekey, K. A Novel method to determine collisional energy transfer efficiency by fourier transform ion cyclotron resonance mass spectrometry. *Rapid Commun. Mass Spectrom.* **1998**, *12*, 1175–1181.
124. Jurchen, J. C.; Williams, E. R. Origin of asymmetric charge partitioning in the dissociation of gas-phase protein homodimers. *J. Am. Chem. Soc.* **2003**, *125*, 2817–2826.
125. Guo, X. H.; Duursma, M.; Al-Khalili, A.; McDonnell, L. A.; Heeren, R. M. A. Design and performance of a new FT-ICR cell operating at a temperature range of 77–438 K. *Int. J. Mass Spectrom.* **2004**, *231*, 37–45.
126. Jockusch, R. A.; Lemoff, A. S.; Williams, E. R. Hydration of valine–cation complexes in the gas phase: On the number of water molecules necessary to form a zwitterion. *J. Phys. Chem. A* **2001**, *105*, 10929–10942.
127. Sena, M.; Riveros, J. M. Thermal dissociation of acetophenone molecular ions activated by infrared radiation. *J. Phys. Chem. A* **1997**, *101*, 4384–4391.
128. Wong, R. L.; Robinson, E. W.; Williams, E. R. Activation of protonated peptides and molecular ions of small molecules using heated filaments in fourier transform ion cyclotron resonance mass spectrometry. *Int. J. Mass Spectrom.* **2004**, *234*, 1–9.
129. Sena, M.; Riveros, J. M. Ring-hydrogen participation in the keto–enol isomerization of the acetophenone radical cation. *Chem. Eur. J.* **2000**, *6*, 785–793.
130. Giroldo, T.; Riveros, J. M. Keto–enol isomerization of gas-phase 2′-methylacetophenone molecular ions probed by high-temperature near-blackbody-induced dissociation, ion-molecule reactions, and *ab initio* calculations. *J. Phys. Chem. A* **2002**, *106*, 9930–9938.
131. Dunbar, R. C. BIRD (blackbody infrared radiative dissociation): Evolution, principles, and applications. *Mass Spectrom. Rev.* **2004**, *23*, 127–158.
132. Dunbar, R. C. Kinetics of thermal unimolecular dissociation by ambient infrared radiation. *J. Phys. Chem.* **1994**, *98*, 8705–8712.
133. Steinfield, J. I. *Chemical Kinetics and Dynamics*, Prentice Hall, Upper Saddle River, NJ, 1999.
134. Price, W. D.; Williams, E. R. Activation of peptide ions by blackbody radiation: Factors that lead to dissociation kinetics in the rapid energy exchange limit. *J. Phys. Chem. A* **1997**, *101*, 8844–8852.
135. Kitova, E. N.; Seo, M.; Roy, P. N.; Klassen, J. S. Elucidating the intermolecular interactions within a desolvated protein–ligand complex. An experimental and computational study. *J. Am. Chem. Soc.* **2008**, *130*, 1214–1226.
136. Benson, S. W. *Thermochemical Kinetics*, 2nd ed. John Wiley & Sons, New York, 1976.
137. Tholmann, D.; Tonner, D. S.; McMahon, T. B. Spontaneous unimolecular dissociation of small cluster ions, $(H_3O^+)L(n)$ and Cl-$(H_2O)(n)$ ($n = 2$–4), under Fourier transform ion cyclotron resonance conditions. *J. Phys. Chem.* **1994**, *98*, 2002–2004.
138. Rodriguez-Cruz, S. E.; Jockusch, R. A.; Williams, E. R. Hydration energies of divalent metal ions, $Ca^{2+}(H_2O)(n)$ ($n = 5$–7) and $Ni^{2+}(H_2O)(n)$ ($n = 6$–8), obtained by blackbody infrared radiative dissociation. *J. Am. Chem. Soc.* **1998**, *120*, 5842–5843.
139. Jockusch, R. A.; Price, W. D.; Williams, E. R. Structure of cationized arginine (Arg-M + , M = H, Li, Na, K, Rb, and Cs) in the gas phase: Further evidence for zwitterionic arginine. *J. Phys. Chem. A* **1999**, *103*, 9266–9274.

140. Price, W. D.; Schnier, P. D.; Williams, E. R. Binding energies of the proton-bound amino acid dimers GlyGly, AlaAla, GlyAla, and LysLys measured by blackbody infrared radiative dissociation. *J. Phys. Chem. B* **1997**, *101*, 664–673.
141. Schnier, P. D.; Price, W. D.; Jockusch, R. A.; Williams, E. R. Blackbody infrared radiative dissociation of bradykinin and its analogues: energetics, dynamics, and evidence for salt-bridge structures in the gas phase. *J. Am. Chem. Soc.* **1996**, *118*, 7178–7189.
142. Ge, Y.; Horn, D. M.; McLafferty, F. W. Blackbody infrared radiative dissociation of larger (42 kDa) multiply charged proteins. *Int. J. Mass Spectrom.* **2001**, *210/211*, 203–214.
143. Klassen, J. S.; Schnier, P. D.; Williams, E. R. Blackbody infrared radiative dissociation of oligonucleotide anions. *J. Am. Soc. Mass Spectrom.* **1998**, *9*, 1117–1124.
144. Daneshfar, R.; Klassen, J. S. Arrhenius activation parameters for the loss of neutral nucleobases from deprotonated oligonucleotide anions in the gas phase. *J. Am. Soc. Mass Spectrom.* **2004**, *15*, 55–64.
145. Shoemaker, G. K.; Kitova, E. N.; Palcic, M. M.; Klassen, J. S. Equivalency of binding sites in protein–ligand complexes revealed by time-resolved tandem mass spectrometry. *J. Am. Chem. Soc.* **2007**, *129*, 8674–8675.
146. Kitova, E. N.; Bundle, D. R.; Klassen, J. S. Partitioning of solvent effects and intrinsic interactions in biological recognition. *Angew. Chem. Int. Ed.* **2004**, *43*, 4183–4186.
147. Dunbar, R. C.In Bowers, M. T. (Ed.), *Gas Phase Ion Chemistry*, Vol. 2, Academic Press, New York, 1979.
148. Orth, R. G.; Dunbar, R. C. Photofragmentation of CH_3Cl^+ at 366 nm. *J. Chem. Phys* **1978**, *68*, 3254–3259.
149. Riggin, M.; Orth, R.; Dunbar, R. C. Photodissociation of C_4H^{8+} ions—Comparison of experimental branching ratios with unimolecular decomposition theory. *J. Chem. Phys.* **1976**, *65*, 3365–3371.
150. Yoon, S. H.; Chung, Y. J.; Kim, M. S. Time-resolved photodissociation of singly protonated peptides with an arginine at the N-terminus: a statistical interpretation. *J. Am. Soc. Mass Spectrom.* **2008**, *19*, 645–655.
151. Clark, L. B. Polarization assignments in the vacuum UV spectra of the primary amide, carboxyl, and peptide groups. *J. Am. Chem. Soc.* **1995**, *117*, 7974–7986.
152. Thompson, M. S.; Cui, W. D.; Reilly, J. P. Factors that impact the vacuum ultraviolet photofragmentation of peptide ions. *J. Am. Soc. Mass Spectrom.* **2007**, *18*, 1439–1452.
153. Bowers, W. D.; Delbert, S.-S.; Hunter, R. L.; McIver, R. T. Jr. Fragmentation of oligopeptide ions using ultraviolet laser radiation and fourier transform mass spectrometry. *J. Am. Chem. Soc.* **1984**, *106*, 7288–7289.
154. Cui, W. D.; Thompson, M. S.; Reilly, J. P. Pathways of peptide ion fragmentation induced by vacuum ultraviolet light. *J. Am. Soc. Mass Spectrom.* **2005**, *16*, 1384–1398.
155. Zhang, L.; Cui, W.; Thompson, M. S.; Reilly, J. P. Structures of alpha-type ions formed in the 157 nm photodissociation of singly-charged peptide ions. *J. Am. Soc. Mass Spectrom.* **2006**, *17*, 1315–1321.
156. Wilson, J. J.; Brodbelt, J. S. MS/MS Simplification by 355 nm ultraviolet photodissociation of chromophore-derivatized peptides in a quadrupole ion trap. *Anal. Chem.* **2007**, *79*, 7883–7892.
157. Devakumar, A.; Thompson, M. S.; Reilly, J. P. Fragmentation of oligosaccharide ions with 157 nm vacuum ultraviolet light. *Rapid Commun. Mass Spectrom.* **2005**, *19*, 2313–2320.
158. Gabelica, V.; Rosu, F.; Tabarin, T.; Kinet, C.; Antoine, R.; Broyer, M.; De Pauw, E.; Dugourd, P. Base-dependent electron photodetachment from negatively charged DNA strands upon 260-nm laser irradiation. *J. Am. Chem. Soc.* **2007**, *129*, 4706–4713.
159. Gabelica, V.; Tabarin, T.; Antoine, R.; Rosu, F.; Compagnon, I.; Broyer, M.; De Pauw, E.; Dugourd, P. Electron photodetachment dissociation of DNA polyanions in a quadrupole ion trap mass spectrometer. *Anal. Chem.* **2006**, *78*, 6564–6572.
160. Wang, B.-H.; McLafferty, F. W. Electron impact excitation of ions from larger organic molecules. *Org. Mass Spectrom.* **1990**, *25*, 54–556.
161. Kjeldsen, F.; Silivra, O. A.; Ivonin, I. A.; Zubarev, R. A. Electron detachment dissociation in quadrupole ion trap highlights dominant C(alpha)-C backbone fragmentation in polypeptide radical anions. In *Proceedings of The 52nd ASMS Conference on Mass Spectometry and Allied Topics;* Nashville, TN, May 23–27, 2004 CD-ROM.
162. Cooper, H. J.; Hakansson, K.; Marshall, A. G. The role of electron capture dissociation in biomolecular analysis. *Mass Spectrom. Rev.* **2005**, *24*, 201–222.
163. McLafferty, F. W.; Horn, D. M.; Breuker, K.; Ge, Y.; Lewis, M. A.; Cerda, B.; Zubarev, R. A.; Carpenter, B. K. Electron capture dissociation of gaseous multiply charged ions by Fourier transform ion cyclotron resonance. *J. Am. Soc. Mass Spectrom.* **2001**, *12*, 245–249.
164. Zubarev, R. A. Reactions of polypeptide ions with electrons in the gas phase. *Mass Spectrom. Rev.* **2003**, *22*, 57–77.
165. Zubarev, R. A. Electron capture dissociation tandem mass spectrometry. *Curr. Opin. Biotechnol.* **2004**, *15*, 12–16.
166. Zubarev, R. A.; Haselmann, K. F.; Budnik, B.; Kjeldsen, F.; Jensen, F. Towards an understanding of the mechanism of electron capture dissociation: A historical perspective and modern ideas. *Eur. Mass Spectrom.* **2002**, *8*, 337–349.
167. Bakhtiar, R.; Guan, Z. Q. Electron capture dissociation mass spectrometry in characterization of

peptides and proteins. *Biotechnol. Lett.* **2006**, *28*, 1047–1059.

168. Axelsson, J.; Palmblad, M.; Hakansson, K.; Hakansson, P. Electron capture dissociation of substance P using a commercially available Fourier transform ion cyclotron resonance mass spectrometer. *Rapid Commun. Mass Spectrom.* **1999**, *13*, 474–477.

169. Hakansson, K.; Emmett, M. R.; Hendrickson, C. L.; Marshall, A. G. High sensitivity electron capture dissociation tandem FT-ICR mass spectrometry of microelectrosprayed peptides. *Anal. Chem.* **2001**, *73*, 3605–3610.

170. Baba, T.; Hashimoto, Y.; Hasegawa, H.; Hirabayashi, A.; Waki, I. Electron capture dissociation in a radio frequency ion trap. *Anal. Chem.* **2004**, *76*, 4263–4266.

171. Silivra, O. A.; Kjeldsen, F.; Ivonin, I. A.; Zubarev, R. A. Electron capture dissociation of polypeptides in a three-dimensional quadrupole ion trap: Implementation and first results. *J. Am. Soc. Mass Spectrom.* **2005**, *16*, 22–27.

172. Ding, L.; Brancia, F. L. Electron capture dissociation in a digital ion trap mass spectrometer. *Anal. Chem.* **2006**, *78*, 1995–2000.

173. Tsybin, Y. O.; Hakansson, P.; Budnik, B. A.; Haselmann, K. F.; Kjeldsen, F.; Gorshkov, M.; Zubarev, R. A. Improved low-energy electron injection systems for high rate electron capture dissociation in fourier transform ion cyclotron resonance mass spectrometry. *Rapid Commun. Mass Spectrom.* **2001**, *15*, 1849–1854.

174. Tsybin, Y. O.; Quinn, J. P.; Tsybin, O. Y.; Hendrickson, C. L.; Marshall, A. G. Electron capture dissociation implementation progress in fourier transform ion cyclotron resonance mass spectrometry. *J. Am. Soc. Mass Spectrom.* **2008**, *19*, 762–771.

175. Zubarev, R. A.; Kruger, N. A.; Fridriksson, E. K.; Lewis, M. A.; Horn, D. M.; Carpenter, B. K.; McLafferty, F. W. Electron capture dissociation of gaseous multiply-charged proteins is favored at disulfide bonds and other sites of high hydrogen atom affinity. *J. Am. Chem. Soc.* **1999**, *121*, 2857–2862.

176. Sobczyk, M.; Anusiewicz, W.; Berdys-Kochanska, J.; Sawicka, A.; Skurski, P.; Simons, J. Coulomb-assisted dissociative electron attachment: Application to a model peptide. *J. Phys. Chem. A* **2005**, *109*, 250–258.

177. Syrstad, E. A.; Turecek, F. Toward a general mechanism of electron capture dissociation. *J. Am. Soc. Mass Spectrom.* **2005**, *16*, 208–224.

178. Patriksson, A.; Adams, C.; Kjeldsen, F.; Raber, J.; van der Spoel, D.; Zubarev, R. A. Prediction of N–C-alpha bond cleavage frequencies in electron capture dissociation of Trp-cage dications by force-field molecular dynamics simulations. *Int. J. Mass Spectrom.* **2006**, *248*, 124–135.

179. Leib, R. D.; Donald, W. A.; Bush, M. F.; O'Brien, J. T.; Williams, E. R. Nonergodicity in electron capture dissociation investigated using hydrated ion nanocalorimetry. *J. Am. Soc. Mass Spectrom.* **2007**, *18*, 1217–1231.

180. Liu, H.; Hakansson, K. Divalent metal–peptide interactions probed by electron capture dissociation of trications. *J. Am. Soc. Mass Spectrom.* **2006**, *17*, 1731–1741.

181. Sawicka, A.; Skurski, P.; Hudgins, R. R.; Simons, J. Model calculations relevant to disulfide bond cleavage via electron capture influenced by positively charged groups. *J. Phys. Chem. B* **2003**, *107*, 13505–13511.

182. Chen, X. H.; Turecek, F. The arginine anomaly: Arginine radicals are poor hydrogen atom donors in electron transfer induced dissociations. *J. Am. Chem. Soc.* **2006**, *128*, 12520–12530.

183. Sobczyk, M.; Simons, J. The role of excited rydberg states in electron transfer dissociation. *J. Phys. Chem. B* **2006**, *110*, 7519–7527.

184. Turecek, F.; Chen, X.; Hao, C. Where does the electron go? Electron distribution and reactivity of peptide cation radicals formed by electron transfer in the gas phase. *J. Am. Chem. Soc.* **2008**, *130*, 8818–8833.

185. Horn, D. M.; Ge, Y.; McLafferty, F. W. Activated ion electron capture dissociation for mass spectral sequencing of larger (42 kDa) proteins. *Anal. Chem.* **2000**, *72*, 4778–4784.

186. Shi, S. D.-H.; Hemling, M. E.; Carr, S. A.; Horn, D. M.; Lindh, I.; McLafferty, F. W. Phosphopeptide/phosphoprotein mapping by electron capture dissociation mass spectrometry. *Anal. Chem.* **2001**, *73*, 19–22.

187. Ge, Y.; Lawhorn, B. G.; Einaggar, M.; Strauss, E.; Park, J. H.; Begley, T. P.; McLafferty, F. W. Top down characterization of larger proteins (45 kDa) by electron capture dissociation mass spectrometry. *J. Am. Chem. Soc.* **2002**, *124*, 672–678.

188. Cooper, H. J.; Case, M. A.; McLendon, G. L.; Marshall, A. G. ESI FT-ICR mass spectrometric analysis of metal-ion selected dynamic protein libraries. *J. Am. Chem. Soc.* **2003**, *125*, 5331–5339.

189. Hakansson, K.; Chalmers, M. J.; Quinn, J. P.; McFarland, M. A.; Hendrickson, C. L.; Marshall, A. G. Combined electron capture and infrared multiphoton dissociation for multistage MS/MS in an FT-ICR mass spectrometer. *Anal. Chem.* **2003**, *75*, 3256–3262.

190. Chalmers, M. J.; Hakansson, K.; Johnson, R.; Smith, R.; Shen, J.; Emmett, M. R.; Marshall, A. G. Protein kinase A phosphorylation characterized by tandem Fourier transform ion cyclotron resonance mass spectrometry. *Proteomics* **2004**, *4*, 970–981.

191. Sze, S. K.; Ge, Y.; Oh, H.; McLafferty, F. W. Plasma ECD for the characterization of large proteins by top-down mass spectrometry. *Anal. Chem.* **2003**, *75*, 1599–1603.

192. Lin, C.; O'Connor, P. B.; Cournoyer, J. J. Use of a double resonance electron capture dissociation experiment to probe fragment intermediate lifetimes. *J. Am. Soc. Mass Spectrom.* **2006**, *17*, 1605–1615.

193. O'Connor, P. B.; Lin, C.; Cournoyer, J. J.; Pittman, J. L.; Belyayev, M.; Budnik, B. A. Long-lived electron capture dissociation product ions experience radical migration via hydrogen abstraction. *J. Am. Soc. Mass Spectrom.* **2006**, *17*, 576–585.
194. Savitski, M. M.; Kjeldsen, F.; Nielsen, M. L.; Zubarev, R. A. Hydrogen rearrangement to and from radical z fragments in electron capture dissociation of peptides. *J. Am. Soc. Mass Spectrom.* **2007**, *18*, 113–120.
195. Mihalca, R.; Kleinnijenhuis, A. J.; McDonnell, L. A.; Heck, A. J. R.; Heeren, R. M. A. Electron capture dissociation at low temperatures reveals selective dissociations. *J. Am. Soc. Mass Spectrom.* **2004**, *15*, 1869–1873.
196. Robinson, E. W.; Leib, R. D.; Williams, E. R. The role of conformation on electron capture dissociation of ubiquitin. *J. Am. Soc. Mass Spectrom.* **2006**, *17*, 1469–1479.
197. Leymarie, N.; Costello, C. E.; O'Connor, P. B. Electron capture dissociation initiates a free radical reaction cascade. *J. Am. Chem. Soc.* **2003**, *125*, 8949–8958.
198. Horn, D. M.; Zubarev, R. A.; McLafferty, F. W. Automated *de novo* sequencing of proteins by tandem high-resolution mass spectrometry. *Proc. Natl. Acad. Sci. USA* **2000**, *97*, 10313–10317.
199. Tsybin, Y. O.; He, H.; Emmett, M. R.; Hendrickson, C. L.; Marshall, A. G. Ion activation in electron capture dissociation to distinguish between N-terminal and C-terminal product ions. *Anal. Chem.* **2007**, *79*, 7596–7602.
200. Savitski, M. M.; Kjeldsen, F.; Nielsen, M. L.; Zubarev, R. A. Complementary sequence preferences of electron capture dissociation and vibrational excitation in fragmentation of polypeptide polycations. *Angew. Chem. Int. Ed.* **2006**, *45*, 5301–5303.
201. Savitski, M. M.; Nielsen, M. L.; Kjeldsen, F.; Zubarev, R. A. Proteomics-grade *de novo* sequencing approach. *J. Proteome Res.* **2005**, *4*, 2348–2354.
202. Kruger, N. A.; Zubarev, R. A.; Horn, D. M.; McLafferty, F. W. Electron capture dissociation of multiply charged peptide cations. *Int. J. Mass Spectrom.* **1999**, *185/186/187*, 787–793.
203. Zubarev, R. A.; Horn, D. M.; Fridriksson, E. K.; Kelleher, N. L.; Kruger, N. A.; Lewis, M. A.; Carpenter, B. K.; McLafferty, F. W. Electron capture dissociation for structural characterization of multiply charged protein cations. *Anal. Chem.* **2000**, *72*, 563–573.
204. Creese, A. J.; Cooper, H. J. Liquid chromatography electron capture dissociation tandem mass spectrometry (LC-ECD-MS/MS) versus liquid chromatography collision-induced dissociation tandem mass spectrometry (LC-CID-MS/MS) for the identification of proteins. *J. Am. Soc. Mass Spectrom.* **2007**, *18*, 891–897.
205. Hakansson, K.; Cooper, H. J.; Emmett, M. R.; Costello, C. E.; Marshall, A. G.; Nilsson, C. L. Electron capture dissociation and infrared multiphoton dissociation MS/MS of an *N*-glycosylated tryptic peptide yield complementary sequence information. *Anal. Chem.* **2001**, *73*, 4530–4536.
206. Kelleher, N. L.; Zubarev, R. A.; Bush, K.; Furie, B.; Furie, B. C.; McLafferty, F. W.; Walsh, C. T. Localization of labile posttranslational modifications by electron capture dissociation: The case of gamma-carboxyglutamic acid. *Anal. Chem.* **1999**, *71*, 4250–4253.
207. Liu, H.; Hakansson, K. Electron capture dissociation of tyrosine *O*-sulfated peptides complexed with divalent metal cations. *Anal. Chem.* **2006**, *78*, 7570–7576.
208. Mirgorodskaya, E.; Roepstorff, P.; Zubarev, R. A. Localization of *O*-glycosylation sites in peptides by electron capture dissociation in a Fourier transform mass spectrometer. *Anal. Chem.* **1999**, *71*, 4431–4436.
209. Stensballe, A.; Norregaard-Jensen, O.; Olsen, J. V.; Haselmann, K. F.; Zubarev, R. A. Electron capture dissociation of singly and multiply phosphorylated peptides. *Rapid Commun. Mass Spectrom.* **2000**, *14*, 1793–1800.
210. Kjeldsen, F.; Haselmann, K. F.; Budnik, B. A.; Sorensen, E. S.; Zubarev, R. A. Complete characterization of posttranslational modification sites in the bovine milk protein PP3 by tandem mass spectrometry with electron capture dissociation at the last stage. *Anal. Chem.* **2003**, *75*, 2355–2361.
211. Mormann, M.; Macek, B.; de Peredo, A. G.; Hofsteenge, J.; Peter-Katalinic, J. Structural studies on protein *O*-fucosylation by electron capture dissociation. *Int. J. Mass Spectrom.* **2004**, *234*, 11–21.
212. Renfrow, M. B.; Cooper, H. J.; Tomana, M.; Kulhavy, R.; Hiki, Y.; Toma, K.; Emmett, M. R.; Mestecky, J.; Marshall, A. G.; Novak, J. Determination of aberrant *O*-glycosylation in the IgA1 hinge region by electron capture dissociation Fourier transform ion cyclotron resonance mass spectrometry. *J. Biol. Chem.* **2005**, *280*, 19136–19145.
213. Adamson, J. T.; Hakansson, K. Infrared multiphoton dissociation and electron capture dissociation of high-mannose type glycopeptides. *J. Proteome Res.* **2006**, *5*, 493–501.
214. Pesavento, J. J.; Kim, Y. B.; Taylor, G. K.; Kelleher, N. L. Shotgun annotation of histone modifications: A new approach for streamlined characterization of proteins by top down mass spectrometry. *J. Am. Chem. Soc.* **2004**, *126*, 3386–3387.
215. Zhang, L. W.; Freitas, M. A. Comparison of peptide mass mapping and electron capture dissociation as assays for histone posttranslational modifications. *Int. J. Mass Spectrom.* **2004**, *234*, 213–225.
216. Medzihradszky, K. F.; Zhang, X.; Chalkley, R. J.; Guan, S.; McFarland, M. A.; Chalmers, M. J.; Marshall, A. G.; Diaz, R. L.; Allis, C. D.; Burlingame, A. L. Characterization of tetrahymena histone H2B variants and posttranslational populations by electron capture dissociation (ECD) Fourier

transform ion cyclotron mass spectrometry (FT-ICR MS). *Mol. Cell. Proteomics* **2004**, *3*, 872–886.

217. Siuti, N.; Roth, M. J.; Mizzen, C. A.; Kelleher, N. L.; Pesavento, J. J. Gene-specific characterization of human histone H2B by electron capture dissociation. *J. Proteome Res.* **2006**, *5*, 233–239.
218. Garcia, B. A.; Siuti, N.; Thomas, C. E.; Mizzen, C. A.; Kelleher, N. L. Characterization of neurohistone variants and post-translational modifications by electron capture dissociation mass spectrometry. *Int. J. Mass Spectrom.* **2007**, *259*, 184–196.
219. Cournoyer, J. J.; Pittman, J. L.; Ivleva, V. B.; Fallows, E.; Waskell, L.; Costello, C. E.; O'Connor, P. B. Deamidation: Differentiation of aspartyl from isoaspartyl products in peptides by electron capture dissociation. *Protein Sci.* **2005**, *14*, 452–463.
220. Cournoyer, J. J.; Lin, C.; Bowman, M. J.; O'Connor, P. B. Quantitating the relative abundance of isoaspartyl residues in deamidated proteins by electron capture dissociation. *J. Am. Soc. Mass Spectrom.* **2007**, *18*, 48–56.
221. Horn, D. M.; Breuker, K.; Frank, A. J.; McLafferty, F. W. Kinetic intermediates in the folding of gaseous protein ions characterized by electron capture dissociation mass spectrometry. *J. Am. Chem. Soc.* **2001**, *123*, 9792–9799.
222. Breuker, K.; Oh, H.; Horn, D. M.; Cerda, B. A.; McLafferty, F. W. Detailed unfolding and folding of gaseous ubiquitin ions characterized by electron capture dissociation. *J. Am. Chem. Soc.* **2002**, *124*, 6407–6420.
223. Adams, C. M.; Kjeldsen, F.; Patriksson, A.; van der Spoel, D.; Graslund, A.; Papadopoulos, E.; Zubarev, R. A. Probing solution- and gas-phase structures of Trp-cage cations by chiral substitution and spectroscopic techniques. *Int. J. Mass Spectrom.* **2006**, *253*, 263–273.
224. Haselmann, K. F.; Jorgensen, T. J. D.; Budnik, B. A.; Jensen, F.; Zubarev, R. A. Electron capture dissociation of weakly bound polypeptide polycationic complexes. *Rapid Commun. Mass Spectrom.* **2002**, *16*, 2260–2265.
225. Xie, Y. M.; Zhang, J.; Yin, S.; Loo, J. A. Top-down ESI-ECD-FT-ICR mass spectrometry localizes noncovalent protein–ligand binding sites. *J. Am. Chem. Soc.* **2006**, *128*, 14432–14433.
226. Geels, R. B. J.; van der Vies, S. M.; Heck, A. J. R.; Heeren, R. M. A. Electron capture dissociation as structural probe for noncovalent gas-phase protein assemblies. *Anal. Chem.* **2006**, *78*, 7191–7196.
227. Hoofnagle, A. N.; Resing, K. A.; Ahn, N. G. Protein analysis by hydrogen exchange mass spectrometry. *Annu. Rev. Biophys. Biomol. Struct.* **2003**, *32*, 1–25.
228. Kaltashov, I. A.; Eyles, S. J. Crossing the boundary to study protein dynamics and function: Combination of amide hydrogen exchange in solution and ion fragmentation in the gas phase. *J. Mass Spectrom.* **2002**, *37*, 557–565.
229. Konermann, L.; Simmons, D. A. Protein folding kinetics and mechanisms studied by pulse labeling and mass spectrometry. *Mass Spectrom. Rev.* **2003**, *22*, 1–26.
230. Miranker, A.; Robinson, C. V.; Radford, S. E.; Aplin, R. T.; Dobson, C. M. Detection of transient protein folding populations by mass spectrometry. *Science* **1993**, *262*, 896–900.
231. Smith, D. L.; Deng, Y.; Zhang, Z. J. Probing the non-covalent structure of proteins by amide hydrogen exchange and mass spectrometry. *J. Mass Spectrom.* **1997**, *32*, 135–146.
232. Buijs, J.; Hagman, C.; Hakansson, K.; Richter, J. -H.; Hakansson, P.; Oscarsson, S. Inter- and intra-molecular migration of peptide amide hydrogens during electrospray ionization. *J. Am. Soc. Mass Spectrom.* **2001**, *12*, 410–419.
233. Demmers, J. A. A.; Rijkers, D. T. S.; Haverkamp, J.; Killian, J. A.; Heck, A. J. R. Factors affecting gas-phase deuterium scrambling in peptide ions and their implications for protein structure determination. *J. Am. Chem. Soc.* **2002**, *124*, 11191–11198.
234. Hagman, C.; Hakansson, P.; Buijs, J.; Hakansson, K. Inter-molecular migration during collisional activation monitored by hydrogen/deuterium exchange FT-ICR tandem mass spectrometry. *J. Am. Soc. Mass Spectrom.* **2004**, *15*, 640–647.
235. Hoerner, J. K.; Xiao, H.; Dobo, A.; Kaltashov, I. A. Is there hydrogen scrambling in the gas phase? Energetic and structural determinants of proton mobility within protein ions. *J. Am. Chem. Soc.* **2004**, *126*, 7709–7717.
236. Johnson, R. S.; Krylov, D.; Walsh, K. A. Proton mobility within electrosprayed peptide ions. *J. Mass Spectrom.* **1995**, *30*, 386–387.
237. Jorgensen, T. J. D.; Gardsvoll, H.; Ploug, M.; Roepstorff, P. Intramolecular migration of amide hydrogens in protonated peptides upon collisional activation. *J. Am. Chem. Soc.* **2005**, *127*, 2785–2793.
238. Kweon, H. K.; Hakansson, K. Site-specific amide hydrogen exchange in melittin probed by electron capture dissociation fourier transform ion cyclotron resonance mass spectrometry. *Analyst* **2006**, *131*, 275–280.
239. Rand, K. D.; Jorgensen, T. J. D. Development of a peptide probe for the occurrence of hydrogen (H-1/H-2) scrambling upon gas-phase fragmentation. *Anal. Chem.* **2007**, *79*, 8686–8693.
240. Rand, K. D.; Adams, C. M.; Zubarev, R. A.; Jorgensen, T. J. D. Electron capture dissociation proceeds with a low degree of intramolecular migration of peptide amide hydrogens. *J. Am. Chem. Soc.* **2008**, *130*, 1341–1349.
241. Cerda, B. A.; Horn, D. M.; Breuker, K.; Carpenter, B. K.; McLafferty, F. W. Electron capture dissociation of multiply-charged oxygenated cations. A nonergodic process. *Eur. Mass Spectrom.* **1999**, *5*, 335–338.
242. Cerda, B. A.; Horn, D. M.; Breuker, K.; McLafferty, F. W. Sequencing of specific copolymer oligomers by

electron-capture-dissociation mass spectrometry. *J. Am. Chem. Soc.* **2002**, *124*, 9287–9291.

243. Koster, S.; Duursma, M. C.; Boon, J. J.; Heeren, R. M. A.; Ingemann, S.; van Benthem, R. A. T. M.; de Koster, C. G. Electron capture and collisionally activated dissociation mass spectrometry of doubly charged hyperbranched polyesteramides. *J. Am. Soc. Mass Spectrom.* **2003**, *14*, 332–341.
244. Hakansson, K.; Hudgins, R. R.; Marshall, A. G.; O'Hair, R. A. J. Electron capture dissociation and infrared multiphoton dissociation of oligodeoxynucleotide dications. *J. Am. Soc. Mass Spectrom.* **2003**, *14*, 23–41.
245. Schultz, K. N.; Hakansson, K. Rapid Electron Capture dissociation of mass-selectively accumulated oligodeoxynucleotide dications. *Int. J. Mass Spectrom.* **2004**, *234*, 123–130.
246. Yang, J.; Hakansson, K. Fragmentation of oligoribonucleotides from gas-phase ion-electron reactions. *J. Am. Soc. Mass Spectrom.* **2006**, *17*, 1369–1375.
247. Kleinnijenhuis, A. J.; Duursma, M. C.; Breukink, E.; Heeren, R. M. A.; Heck, A. J. R. Localization of intramolecular monosulfide bridges in lantibiotics determined with electron capture dissociation. *Anal. Chem.* **2003**, *75*, 3219–3225.
248. Adamson, J. T.; Hakansson, K. Electron capture dissociation of oligosaccharides ionized with alkali, alkaline earth, and transition metals. *Anal. Chem.* **2007**, *79*, 2901–2910.
249. Budnik, B. A.; Haselmann, K. F.; Elkin, Y. N.; Gorbach, V. I.; Zubarev, R. A. Applications of electron-ion dissociation reactions for analysis of polycationic chitooligosaccharides in Fourier transform mass spectrometry. *Anal. Chem.* **2003**, *75*, 5994–6001.
250. Liu, H. C.; Yoo, H. J.; Hakansson, K. Characterization of phosphate-containing metabolites by calcium adduction and electron capture dissociation. *J. Am. Soc. Mass Spectrom.* **2008**, *19*, 799–808.
251. James, P. F.; Perugini, M. A.; O'Hair, R. A. J. Electron capture dissociation of complexes of diacylglycerophosphocholine and divalent metal ions: Competition between charge reduction and radical induced phospholipid fragmentation. *J. Am. Soc. Mass Spectrom.* **2008**, *19*, 978–986.
252. Kjeldsen, F.; Haselmann, K.; Budnik, B. A.; Jensen, F.; Zubarev, R. A. Dissociative capture of hot electrons by polypeptide polycations: An Efficient process accompanied by secondary fragmentation. *Chem. Phys. Lett.* **2002**, *356*, 201–206.
253. Zubarev, R. A.; Nielsen, M. L.; Budnik, B. A. Tandem ionization mass spectrometry of biomolecules. *Eur. J. Mass Spectrom.* **2000**, *6*, 235–240.
254. Kjeldsen, F.; Haselmann, K. F.; Sorensen, E.; Zubarev, R. A. Distinguishing of Ile/Leu amino acid residues in the PP3 Protein by (hot) electron capture dissociation in Fourier transform ion cyclotron resonance mass spectrometry. *Anal. Chem.* **2003**, *75*, 1267–1274.
255. Haselmann, K. F.; Budnik, B. A.; Kjeldsen, F.; Nielsen, M. L.; Olsen, J. V.; Zubarev, R. A. Electronic excitation gives informative fragmentation of polypeptide cations and anions. *Eur. J. Mass Spectrom.* **2002**, *8*, 117–121.
256. Kjeldsen, F.; Zubarev, R. A. Secondary losses via γ-lactam formation in hot electron capture dissociation: A missing link to complete *de novo* sequencing of proteins?. *J. Am. Chem. Soc.* **2003**, *125*, 6628–6629.
257. Zhao, C.; Xie, B.; Chan, S. Y.; Costello, C. E.; O'Connor, P. B. Collisionally activated dissociation and electron capture dissociation provide complementary structural information for branched Permethylated Oligosaccharides. *J. Am. Soc. Mass Spectrom.* **2008**, *19*, 138–150.
258. Coon, J. J.; Syka, J. E. P.; Schwartz, J. C.; Shabanowitz, J.; Hunt, D. F. Anion dependence in the partitioning between proton and electron transfer in ion/ion reactions. *Int. J. Mass Spectrom.* **2004**, *236*, 33–42.
259. Pitteri, S. J.; Chrisman, P. A.; Hogan, J. M.; McLuckey, S. A. Electron transfer ion/ion reactions in a three-dimensional quadrupole ion trap: Reactions of doubly and triply protonated peptides with SO_2 center dot-. *Anal. Chem.* **2005**, *77*, 1831–1839.
260. Xia, Y.; Chrisman, P. A.; Erickson, D. E.; Liu, J.; Liang, X. R.; Londry, F. A.; Yang, M. J.; McLuckey, S. A. Implementation of ion/ion reactions in a quadrupole/time-of-flight tandem mass spectrometer. *Anal. Chem.* **2006**, *78*, 4146–4154.
261. Huang, T. Y.; Emory, J. F.; O'Hair, R. A. J.; McLuckey, S. A. Electron-transfer reagent anion formation via electrospray ionization and collision-induced dissociation. *Anal. Chem.* **2006**, *78*, 7387–7391.
262. McAlister, G. C.; Phanstiel, D.; Good, D. M.; Berggren, W. T.; Coon, J. J. Implementation of electron transfer dissociation on a hybrid linear ion trap–orbitrap mass spectrometer. *Anal. Chem.* **2007**, *79*, 3525–3534.
263. Williams, D. K.; McAlister, G. C.; Good, D. M.; Coon, J. J.; Muddiman, D. C. Dual electrospray ion source for electron transfer dissociation on a hybrid linear ion trap–orbitrap mass spectrometer. *Anal. Chem.* **2007**, *79*, 7916–7919.
264. Kaplan, D. A.; Hartmer, R.; Speir, J. P.; Stoermer, C.; Gumerov, D.; Easterling, M. L.; Brekenfeld, A.; Kim, T.; Laukien, F.; Park, M. A. Electron transfer dissociation in the hexapole collision cell of a hybrid quadrupole–hexapole Fourier transform ion cyclotron resonance mass spectrometer. *Rapid Commun. Mass Spectrom.* **2008**, *22*, 271–278.
265. Chrisman, P. A.; Pitteri, S. J.; Hogan, J. M.; McLuckey, S. A. SO_2^- Electron transfer ion/ion reactions with disulfide linked polypeptide ions. *J. Am. Soc. Mass Spectrom.* **2005**, *16*, 1020–1030.
266. Good, D. M.; Wirtala, M.; McAlister, G. C.; Coon, J. J. Performance characteristics of electron transfer dissociation mass spectrometry. *Mol. Cell. Proteomics* **2007**, *6*, 1942–1951.

267. Swaney, D. L.; McAlister, G. C.; Wirtala, M.; Schwartz, J. C.; Syka, J. E. P.; Coon, J. J. Supplemental activation method for high-efficiency electron transfer dissociation of doubly protonated peptide precursors. *Anal. Chem.* **2007**, *79*, 477–485.
268. Iavarone, A. T.; Williams, E. R. Supercharging in electrospray ionization: Effects on signal and charge. *Int. J. Mass Spectrom.* **2002**, *219*, 63–72.
269. Kjeldsen, F.; Giessing, A. M. B.; Ingrell, C. R.; Jensen, O. N. Peptide sequencing and characterization of post-translational modifications by enhanced ion-charging and liquid chromatography electron-transfer dissociation tandem mass spectrometry. *Anal. Chem.* **2007**, *79*, 9243–9252.
270. Coon, J. J.; Shabanowitz, J.; Hunt, D. F.; Syka, J. E. P. Electron transfer dissociation of peptide anions. *J. Am. Soc. Mass Spectrom.* **2005**, *16*, 880–882.
271. Molina, H.; Matthiesen, R.; Kandasamy, K.; Pandey, A. Comprehensive comparison of collision induced dissociation and electron transfer dissociation. *Anal. Chem.* **2008**, *80*, 4825–4835.
272. Xia, Y.; Chrisman, P. A.; Pitteri, S. J.; Erickson, D. E.; McLuckey, S. A. Ion/molecule reactions of cation radicals formed from protonated polypeptides via gas-phase ion/ion electron transfer. *J. Am. Chem. Soc.* **2006**, *128*, 11792–11798.
273. Liu, J.; Liang, X. R.; McLuckey, S. A. On the Value of Knowing a z(center dot) Ion for What It Is. *J. Proteome Res.* **2008**, *7*, 130–137.
274. Hubler, S. L.; Jue, A.; Keith, J.; McAlister, G. C.; Craciun, G.; Coon, J. J. Valence parity renders z(center dot)-type ions chemically distinct. *J. Am. Chem. Soc.* **2008**, *130*, 6388–6394.
275. Stephenson, J. L.; McLuckey, S. A. Ion/ion reactions in the gas phase: Proton transfer reactions involving multiply-charged proteins. *J. Am. Chem. Soc.* **1996**, *118*, 7390–7397.
276. Coon, J. J.; Ueberheide, B.; Syka, J. E. P.; Dryhurst, D. D.; Ausio, J.; Shabanowitz, J.; Hunt, D. F. Protein identification using sequential ion/ion reactions and tandem mass spectrometry. *Proc. Natl. Acad. Sci. U.S. A.* **2005**, *102*, 9463–9468.
277. Chi, A.; Bai, D. L.; Geer, L. Y.; Shabanowitz, J.; Hunt, D. F. Analysis of intact proteins on a chromatographic time scale by electron transfer dissociation tandem mass spectrometry. *Int. J. Mass Spectrom.* **2007**, *259*, 197–203.
278. Schroeder, M. J.; Webb, D. J.; Shabanowitz, J.; Horwitz, A. F.; Hunt, D. F. Methods for the detection of paxillin post-translational modifications and interacting proteins by mass spectrometry. *J. Proteome Res.* **2005**, *4*, 1832–1841.
279. Chi, A.; Huttenhower, C.; Geer, L. Y.; Coon, J. J.; Syka, J. E. P.; Bai, D. L.; Shabanowitz, J.; Burke, D. J.; Troyanskaya, O. G.; Hunt, D. F. Analysis of phosphorylation sites on proteins from *Saccharomyces cerevisiae* by electron transfer dissociation (ETD) mass spectrometry. *Proc. Natl. Acad. Sci. USA* **2007**, *104*, 2193–2198.
280. Molina, H.; Horn, D. M.; Tang, N.; Mathivanan, S.; Pandey, A. Global proteomic profiling of phosphopeptides using electron transfer dissociation tandem mass spectrometry. *Proc. Natl. Acad. Sci. USA* **2007**, *104*, 2199–2204.
281. Gunawardena, H. P.; Emory, J. F.; McLuckey, S. A. Phosphopeptide anion characterization via sequential charge inversion and electron transfer dissociation. *Anal. Chem.* **2006**, *78*, 3788–3793.
282. Hogan, J. M.; Pitteri, S. J.; Chrisman, P. A.; McLuckey, S. A. Complementary structural information from a tryptic N-linked glycopeptide via electron transfer ion/ion reactions and collision-induced dissociation. *J. Proteome Res.* **2005**, *4*, 628–632.
283. Medzihradszky, K. F.; Guan, S.; Maltby, D. A.; Burlingame, A. L. Sulfopeptide fragmentation in electron capture and electron transfer dissociation. *J. Am. Soc. Mass Spectrom.* **2007**, *18*, 1617–1624.
284. Srikanth, R.; Wilson, J.; Bridgewater, J. D.; Numbers, J. R.; Lim, J.; Olbris, M. R.; Kettani, A.; Vachet, R. W. Improved sequencing of oxidized cysteine and methionine containing peptides using electron transfer dissociation. *J. Am. Soc. Mass Spectrom.* **2007**, *18*, 1499–1506.
285. Zhang, Q. B.; Frolov, A.; Tang, N.; Hoffmann, R.; van de Goor, T.; Metz, T. O.; Smith, R. D. Application of electron transfer dissociation mass spectrometry in analyses of non-enzymatically glycated peptides. *Rapid Commun. Mass Spectrom.* **2007**, *21*, 661–666.
286. O'Connor, P. B.; Cournoyer, J. J.; Pitteri, S. J.; Chrisman, P. A.; McLuckey, S. A. Differentiation of aspartic and isoaspartic acids using electron transfer dissociation. *J. Am. Soc. Mass Spectrom.* **2006**, *17*, 15–19.
287. Phanstiel, D.; Brumbaugh, J.; Berggren, W. T.; Conard, K.; Feng, X.; Levenstein, M. E.; McAlister, G. C.; Thomson, J. A.; Coon, J. J. Mass spectrometry identifies and quantifies 74 unique histone H4 isoforms in differentiating human embryonic stem cells. *Proc. Natl. Acad. Sci. USA* **2008**, *105*, 4093–4098.
288. Liang, X. L. R.; Liu, J.; Leblanc, Y.; Covey, T.; Ptak, A. C.; Brenna, J. T.; McLuckey, S. A. Electron transfer dissociation of doubly sodiated glycerophosphocholine lipids. *J. Am. Soc. Mass Spectrom.* **2007**, *18*, 1783–1788.
289. Budnik, B. A.; Haselmann, K. F.; Zubarev, R. A. Electron detachment dissociation of peptide di-anions: An electron–hole recombination phenomenon. *Chem. Phys. Lett.* **2001**, *342*, 299–302.
290. McFarland, M. A.; Marshall, A. G.; Hendrickson, C. L.; Nilsson, C. L.; Fredman, P.; Mansson, J. E. Structural characterization of the GM1 ganglioside by infrared multiphoton dissociation/electron capture dissociation, and electron detachment dissociation

electrospray ionization FT-ICR MS/MS. *J. Am. Soc. Mass Spectrom.* **2005**, *16*, 752–762.

291. Wolff, J. J.; Amster, I. J.; Chi, L.; Linhardt, R. J. Electron detachment dissociation of glycosaminoglycan tetrasaccharides. *J. Am. Soc. Mass Spectrom.* **2007**, *18*, 234–244.

292. Yang, J.; Mo, J.; Adamson, J. T.; Hakansson, K. Characterization of oligodeoxynucleotides by electron detachment dissociation Fourier transform ion cyclotron resonance mass spectrometry. *Anal. Chem.* **2005**, *77*, 1876–1882.

293. Kjeldsen, F.; Silivra, O. A.; Ivonin, I. A.; Haselmann, K. F.; Gorshkov, M.; Zubarev, R. A. C(alpha)-C backbone fragmentation dominates in electron detachment dissociation of gas-phase polypeptide polyanions. *Chem. Eur. J.* **2005**, *11*, 1803–1812.

294. Anusiewicz, I.; Jasionowski, M.; Skurski, P.; Simons, J. Backbone and side-chain cleavages in electron detachment dissociation (EDD). *J. Phys. Chem. A* **2005**, *109*, 11332–11337.

295. Kalli, A.; Hakansson, K. Preferential cleavage of S–S and C–S bonds in electron detachment dissociation and infrared multiphoton dissociation of disulfide-linked peptide anions. *Int. J. Mass Spectrom.* **2007**, *263*, 71–81.

296. Chrisman, P. A.; McLuckey, S. A. Dissociations of disulfide-linked gaseous polypeptide/protein anions: Ion chemistry with implications for protein identification and characterization. *J. Proteome Res.* **2002**, *1*, 549–557.

297. Leach, F. E.; Wolff, J. J.; Laremore, T. N.; Linhardt, R. J.; Amster, I. J. Evaluation of the experimental parameters which control electron detachment dissociation, and their effect on the fragmentation efficiency of glycosaminoglycan carbohydrates. *Int. J. Mass Spectrom.* **2008**, *276*, 110–115.

298. Yang, J.; Hakansson, K. Characterization and optimization of electron detachment dissociation Fourier transform ion cyclotron resonance mass spectrometry. *Int. J. Mass Spectrom.* **2008**, *276*, 144–148.

299. Wolff, J. J.; Laremore, T. N.; Busch, A. M.; Linhardt, R. J.; Amster, I. J. Influence of charge state and sodium cationization on the electron detachment dissociation and infrared multiphoton dissociation of glycosaminoglycan oligosaccharides. *J. Am. Soc. Mass Spectrom.* **2008**, *19*, 790–798.

300. Kleinnijenhuis, A. J.; Kjeldsen, F.; Kallipolitis, B.; Haselmann, K. F.; Jensen, O. N. Analysis of histidine phosphorylation using tandem MS and ion-electron reactions. *Anal. Chem.* **2007**, 7450–7456.

301. Kweon, H. K.; Hakansson, K. Metal oxide-based enrichment combined with gas-phase ion-electron reactions for improved mass spectrometric characterization of protein phosphorylation. *J. Proteome Res.* **2008**, *7*, 745–755.

302. Mo, J.; Hakansson, K. Characterization of nucleic acid higher order structure by high resolution tandem mass spectrometry. *Anal. Bioanal. Chem.* **2006**, *386*, 675–681.

303. Wolff, J. J.; Chi, L.; Linhardt, R. J.; Amster, I. J. Distinguishing glucuronic from iduronic acid in glycosaminoglycan tetrasaccharides by using electron detachment dissociation. *Anal. Chem.* **2007**, *79*, 2015–2022.

304. Wolff, J. J.; Laremore, T. N.; Busch, A. M.; Linhardt, R. J.; Amster, I. J. Electron detachment dissociation of dermatan sulfate oligosaccharides. *J. Am. Soc. Mass Spectrom.* **2008**, *19*, 294–304.

305. Adamson, J. T.; Hakansson, K. Electron detachment dissociation of neutral and sialylated oligosaccharides. *J. Am. Soc. Mass Spectrom.* **2007**, *18*, 2162–2172.

306. Nielsen, M. L.; Budnik, B. A.; Haselmann, K. F.; Olsen, J. V.; Zubarev, R. A. Intramolecular hydrogen atom transfer in hydrogen-deficient polypeptide radical cations. *Chem. Phys. Lett.* **2000**, *330*, 558–562.

307. Gord, J. R.; Horning, S. R.; Wood, J. M.; Cooks, R. G.; Freiser, B. S. Energy deposition during electron induced dissociation. *J. Am. Soc. Mass Spectrom.* **1993**, *4*, 145–151.

308. Lioe, H.; O'Hair, R. A. J. Comparison of collision-induced dissociation and electron-induced dissociation of singly protonated aromatic amino acids, cystine and related simple peptides using a hybrid linear ion trap-FT-ICR mass spectrometer. *Anal. Bioanal. Chem.* **2007**, *389*, 1429–1437.

309. Yoo, H. J.; Liu, H. C.; Hakansson, K. Infrared multiphoton dissociation and electron-induced dissociation as alternative MS/MS strategies for metabolite identification. *Anal. Chem.* **2007**, *79*, 7858–7866.

310. Nielsen, M. L.; Budnik, B. A.; Haselmann, K. F.; Zubarev, R. A. Tandem MALDI/EI ionization for tandem Fourier transform ion cyclotron resonance mass spectrometry of polypeptides. *Int. J. Mass Spectrom.* **2003**, *226*, 181–187.

311. Cooper, H. J.; Hudgins, R. R.; Hakansson, K.; Marshall, A. G. Secondary fragmentation of linear peptides in electron capture dissociation. *Int. J. Mass Spectrom.* **2003**, *228*, 723–728.

312. Feketeova, L.; Khairallaha, G. N.; O'Hair, R. A. J. Intercluster chemistry of protonated and sodiated betaine dimers upon collision induced dissociation and electron induced dissociation. *Eur. J. Mass Spectrom.* **2008**, *14*, 107–110.

Chapter 17

Dissociation of Even-Electron Ions

Carlos Afonso, [*] ***Richard B. Cole,*** [†] ***and Jean-Claude Tabet*** [*]

[*] *Laboratoire de Chimie Structurale Organique et Biologique, Université Pierre & Marie Curie, Paris, France*
[†] *Department of Chemistry, University of New Orleans, New Orleans, Louisiana*

Electrospray and MALDI Mass Spectrometry: Fundamentals, Instrumentation, Practicalities, and Biological Applications, Second Edition, Edited by Richard B. Cole

17.1 SOME GENERALITIES ON EVEN-ELECTRON ION ACTIVATION

17.1.1 Internal Energy Distributions and Consequences

17.1.1.1 Initial Internal Energy of the Survivor Ions

All of these ionization methods (except EI) are used for producing intact protonated (or deprotonated) molecules (as well as adduct ions) yielding, by the dearth of fragment ions, a simplification of mass spectra useful for analytical purposes. However, such even-electron molecular species are of varying stability in accordance with their chemical nature (positive versus negative species) and their internal energies (E_{int}) that orient their respective dissociations toward competitive and/or consecutive prompt dissociations displayed in mass spectra.[1] This energy described in terms of an average or distribution depends, in particular, on the pressure conditions used for formation of charged molecular species.

a. Vacuum Conditions. Under vacuum conditions (from 10^{-7} to 10^{-1} torr), only the ionization step influences the E_{int} magnitude and distribution:

i. *Under gas-phase ionization conditions* (e.g., in chemical ionization), the maximum E_{int} available to a positive MH^+ ion is related to the exothermicity [required for efficient gas-phase (GP) proton exchange from the reagent GH^+ ion] of ion–molecule reactions[2] corresponding to the difference between the basicities of sites of polyfunctional compounds and that of G. The sites of the former are characterized by higher proton affinities (PAs) than that of the reagent G ($PA_{(G)}$ is the enthalpy variation of the following endothermic reaction: $GH^+ \rightarrow G + H^+$). Different sites can be competitively protonated to give rise to formation of a mixture of isomeric MH^+ ions (convertible by possible proton migration or hindered by steric or stereochemical effects) and/or isomeric adduct ions.[3] These competitive processes result in a broadening of the E_{int} distribution, although collisional cooling regulates the extent. Concerning the formation of negatively charged even-electron $[M-H]^-$ ions, similar thermochemical rules must be applied—that is, based upon the ion–molecule reaction exothermicity that depends on the difference of acidities (i.e., ΔH^0_{acid}) between M and G (e.g., $\Delta H^0_{acid(G)}$ is the enthalpy variation of the following endothermic reaction: $G \rightarrow [G-H]^- + H^+$). However, the even-electron $[M-H]^-$ anions often present an enhanced GP stability compared to the respective protonated molecules in positive-ion mode, due to their chemical nature and their lower E_{int} values which are less dependent on the ion preparation exothermicity. Note that independent of the ion polarity, due to multiple collisional relaxations, this initially acquired internal energy ($E_{vib\ mode}$) of even-electron species is significantly lower than that acquired by odd-electron ions under electron ionization conditions where both electronic and vibrational excitation may occur.[4]

ii. *Under "vacuum" desorption conditions* (e.g., in MALDI[5]), charged molecular species are produced within a lower level of vibrational excitation. However, the situation is more complicated since, independent of the chosen matrix (e.g., from the cinnamic acid family), matrix mH^+ and $[m-H]^-$ ions promptly decompose mainly by major H_2O loss and minor CO_2 release, respectively. This contrasts to the observed stability of the analyte molecular species generally characterized by a much larger size and, thus, by more degrees of freedom, allowing internal energy dispersion. Furthermore, the chosen matrix in MALDI indirectly plays a major role in determining the dissociation rate constants of charged aggregates[6] to give rise to the formation of "naked" charged molecular species[*]. It should be pointed out that the size of the charged aggregates could depend on the thermochemical properties of the employed matrix (as described in the *lucky survivor ion model*[7]).

b. Atmospheric Pressure Conditions. Under atmospheric pressure conditions, the charged molecular species are mainly produced as charged solvated forms (comprised of both the solvent and the analyte within a certain distribution of stoichiometries related to both the solvent and analyte properties and temperature) from either droplets[8] (in ESI, nanospray, and sonic spray) or clusters (in DESI, AP-MALDI) or by multiple ion–molecule

[*]The charged aggregates are produced by intra-aggregate proton transfer after formation of electronically excited neutral aggregates (as initial intermediates in the ionization step) during plume formation after the laser ablation step.

reactions[9] (in APCI and APPI). An essential difference of ESI is that it provides multiply charged molecules in contrast to the production of noncationized and singly charged species created by ion–molecule reactions (e.g., in APCI).

In all cases, under AP conditions, size (or size distribution) of the solvated, charged molecular species is related to solvent properties, source temperature, and heated gas flow rate.[10] The respective E_{int} values of these charged species reflect the AP source conditions, and thus, are considered as a thermal-like distribution related to the source temperature.[11] However, this internal energy is not sufficient for producing completely naked ionic species, except for those produced via a direct desorption step. In order to produce such desolvated ionic species, the E_{int} of the surviving charged aggregates is increased through a collisional cascade occurring under reduced pressure conditions (i.e., involving a larger mean free path), after ion acceleration in a zone beyond a small-diameter skimmer. Therefore, the E_{int} carried by naked even-electron molecular species depends mainly on the desolvation conditions. This internal energy may be significantly higher than the residual energy reached during charged aggregate formation.[12] The latter promptly decompose by consecutive losses of noncovalently attached solvent molecules from the charged analyte, rather than by covalent bond cleavages of the analyte. However, after the desolvation cascade, such covalent bond cleavages may occur at higher skimmer voltages (which correspond to higher energy uptakes compared to processes occurring at AP) where a quasi thermal E_{int} distribution is in place.

Finally, the location of the charge on the molecular species is difficult to define because the protonated/deprotonated sites[*] are in fact initially solvated. Thus, during the desolvation step, a potential analyte charge site may or may not be protonated, depending upon the basicity (or acidity) difference[13] between that protonated/deprotonated site and the solvent. Charge (e.g., proton) location may also be affected by the acceleration at the skimmer, which directly influences E_{int} of the charged aggregates (in magnitude and distribution). Finally, with the internal energy of the desorbed/desolvated ions being considered as a quasi-thermal distribution, it can be described by a theoretical temperature called the "characteristic temperature" (T_{char})[14] that depends on the nature of the precursor ions and on the employed instrument. This parameter can be evaluated by studying the extent to which precursor ions, prepared from charged "thermometer molecules" that have well-known molecular and thermochemical properties,[15] survive to the detector.

Thus, under vacuum conditions, the internal energy (E_{int}) of even-electron molecular species (or naked closed-shell systems) is related to the ionization thermochemistry, which lacks precise control and is therefore not fine-tunable. On the other hand, the desolvation step of charged aggregates, formed under AP conditions, is adjustable externally by variation of the skimmer voltage. Therefore, the average internal energy (and the energy distribution) of solvent-free even-electron molecular species can be externally controlled. However, as the positive or negative charge can be located at different sites, it is expected that the charge location influences the competitive fragmentation pathways and thus, the nature, the structures and the internal energies of the observed product ions. Finally, it can be considered that the surviving naked even-electron molecular species still contain a residual internal energy whose magnitude is below the critical energy value of the lowest threshold decomposition pathway (as limited by the time-window required by the instrument for detection of the dissociations).

[*]These sites correspond to the basic and acidic functional groups that are conformationally accessible on the analyte.

c. Intact and Isomerized Charged Molecular Species Emerging from Desolvation Steps. The formed naked ions are generally maintained intact or can be accompanied by prompt fragmentations directed by the positive or negative charge. However, an intermediate situation can be observed which involves molecular species isomerization into ion–dipole (or ion–neutral) complexes.[16,17*] Such even-electron ion–neutral complexes may survive during the flight to the activation cell if they have sufficient lifetimes. Such ion–neutral complex formation is consistent with the behavior of protonated dialkyl ethers and deprotonated esters[18] [Eqs. (17.2) and (17.3))]:

$$[\mathbf{R'OR+H}]^{+} \rightarrow \{\mathbf{R'OH;\ R^{+}}\} \leftrightarrows \{\mathbf{R'OH;\ H^{+}(R\text{-}H)}\} \tag{17.2}$$

$$[\mathbf{R'CH\text{-}COOR}]^{-} \rightarrow \{\mathbf{R'CH{=}C{=}O;\ {}^{-}OR}\} \leftrightarrows \{\mathbf{R'C{\equiv}C\text{-}O^{-};HOR}\} \tag{17.3}$$

In order to be observed, naked precursor ions must have internal energies lower, or only slightly higher, than the threshold energy required for bond cleavage, in order to avoid in-flight dissociation. The fact that precursor ions with internal energies barely above the threshold energy can survive past the mass analyzer results in a so-called kinetic shift; that is, they are detected as intact ions only because not enough time was given for them to dissociate. One of the required conditions to promote such isomerization [Eq. (17.2) or (17.3)] is the thermochemical properties—for example, basicity (or acidity) of each partner that constituted the protonated (or deprotonated) heterodimers. The closer these reactions are to being thermoneutral, the longer will be the ion–neutral complex's stability and lifetime. However, to facilitate their detection, it is better to use an instrument that features a large time-window (long observation time) for ion dissociation, which means that observation of the weak kinetic shift processes is promoted. In addition, under such conditions, isomerization (or rearrangement) prior to dissociation are favored compared to simple cleavages. In all cases, although both the ion–molecule complex and the initial precursor ion have the same *m/z* ratio, part of the structural information is lost. For instance, if the broken bond(s) is (are) vicinal to the chiral sites, stereochemical effects can be lost, leading to ambiguous conclusions.

The following discussion will involve mainly dissociation processes occurring under low-energy activation conditions. This is chosen for enlightening especially the role played by the formation of ion–neutral complexes on decomposition behavior. They allow one to rationalize low-energy dissociations strictly as charge-driven processes rather than as charge-remote (charge-distant) pathways. Thus, the mechanisms proposed herein are based on the application of this "charge-driven" concept even though "proof" of their existence has not yet been provided by isotopic labeling or by theoretical calculations (the molecules are too large). More of the chosen examples of ion–neutral complexes that follow pertain to negative ions, however, it should be noted that positive-ion fragmentations are subject to analogous treatment.

*This concept was introduced by Williams et al.[16] from odd-electron ions prepared in EI from alkanes: $R'R^{+\bullet} \rightarrow [R'^{+},\ R^{\bullet}] \leftrightarrows [(R'-H)^{+\bullet},\ (R+H)]$ [Eq. (17.1)] yielding in competition odd-electron $[R'-H]^{+\bullet}$ or even-electron R'^{+} fragment ions, with the final distribution depending upon which is the less endothermic pathway (Stevenson–Audier rule). Longevialle[17] described a similar model from odd-electron 3,17-bifunctional steroidal systems in which long-distance group migration takes place, allowing for exothermic or weakly endothermic hydrogen transfer. Generally, they are "two-body" complexes, but three-body ion–neutrals have also been described, especially by the groups of Audier and Nibbering.

17.1.1.2 Internal Energy After Ion Excitation in MS/MS Experiments

After their selection, gas-phase excitation of singly charged ions can occur by using different activation techniques, and this step involves, for instance, kinetic energy interconversion and redistribution into internal energy modes. Such energy redistribution occurs much faster than the subsequent fragmentation processes; thus, the activation and dissociation steps must be discussed independently. An older but still most popular ion activation mode is collisional activation using either a target gas (collision-induced dissociation, CID)[19a,b] or a solid surface (surface-induced dissociation, SID)[19c]; but more recently, infrared multiphoton dissociation (IRMPD)[20] has emerged as an activation mode for charged ions yielding structural information complementary to that obtained by CID. For ion-beam experiments, two regimes of collision conditions (i.e., kinetic energy availablity as E_{lab}) must be considered for the selected precursor ion: (i) low-energy (e.g., <300 eV) and (ii) high-energy collision conditions (from several hundred electron volts to 10 keV)[4].

a. Low-Energy Collision Conditions[21]. In this activation regime (as well as in IRMPD), kinetic energy interconversion into internal modes takes place through vibrational coupling, and relatively high vibrational levels of the fundamental electronic state can be reached. On the other hand, the maximum kinetic energy (Q_{max}) that can be converted into internal energy is related to the center-of-mass kinetic energy E_{cm} of the target–ion complex [i.e., $(m_T + m_{ion})$] defined from E_{lab} by $E_{cm} = E_{lab}[m_T/(m_T + m_{ion})]$, where m_T and m_{ion} are the masses of the neutral and ionic reactants, respectively. This approximate relation is only relevant under low-energy conditions (see below for high-energy case). It can further be approximated that the energy interconversion efficiency and redistribution into internal modes is enhanced in a complex having a longer lifetime (especially at very low kinetic energy or for a large-size atomic target). This energy transferred to ions subjected to low-energy CID results in a slight internal energy increase, along with a corresponding modest widening of its energy distribution. Although relatively narrow, it is sufficient to allow fragmentation. The set of parameters, associated with the number and energy of collisions (i.e., collision cell offset and target gas pressure), explain why the exact "fingerprint" of the activation spectrum is strongly dependent on the experimental conditions.

b. High-Energy Collision Conditions[22]. Analogous rationalization cannot be applied to selected ions submitted to the high-energy CID regime because the resulting situation is more complex. Indeed, electronic excitation levels can be reached and their redistribution into vibrational modes of fundamental electronic state (internal conversion) cannot be completely achieved; thus isolated electronic excited levels may continue to exist, yielding particular dissociation mechanisms that cannot occur under low collision energies. Furthermore, the average internal energy is substantially raised, as is the width of the internal energy distribution which now will very likely include a high-energy "tail." The description of Massey's criterion reflects mainly the consequences of the ion–target interaction time that is very short, implying that essentially electronic excitations are involved directly.[23] On the other hand, ion–target complexes that have a very long interaction time directly favor vibrational energy mode excitations at the fundamental electronic level. Consequently, from the high-energy collision regime (in the kilo-electron-volt laboratory energy range with sectors or TOF/TOF instruments), the activation spectra recorded are not significantly modified by the instrument timing properties because of their narrow time-window characteristic and their broad E_{int} distribution. Thus, these CID spectra can be considered to be almost uniform—that is, independent of the mass spectrometer.

c. Mechanistic Differences Between Low-/High-Energy Collision Regimes: An Example. Under low-energy collisions, mainly positive- or negative-charge-induced cleavages (charge-promoted processes, CPP) due to excited vibrational modes within a narrow distribution are produced. These processes allow differentiation of charged isomers and stereoisomers when experimental conditions are maintained constant. This contrasts with the high-energy collision regime that involves electronic excitation and which can yield radical cleavages where the charge is merely a "spectator" to the cleavage (charge-remote processes, CRP).[24] These different trends reflect ion-beam MS/MS experiments in which a fast heating takes place rather than the slow heating process that occurs in quadrupole ion trap instruments and which favors mainly charge-promoted dissociations. Orientation of decomposition pathways toward the CPP/CRP mechanisms also depends on the kinetic shift related to the decomposition time-window of the employed instrument.[25] Indeed, the processes occurring through loose transition states having high-frequency factors (i.e., simple cleavages) are favored with instruments characterized by short time-windows.[26] In the case of CRP, the production of a radical ion is a particular case that merits some attention because it can appear under a high-energy collision regime, or it can be chemically promoted under low-energy collisions from cationized species—for example, copper complexes allowing internal Cu(II) → Cu(I) reduction concomitant with apparition of a radical site[27,28] — or it can be produced by electron capture dissociation (ECD) employed for dissociations of multiply charged ions.

d. Estimation of Vibrational Internal Energy Involved During Ion Decomposition Within Specific Decomposition Time-Window. The precursor ion internal energy and its distribution are determinant parameters for orientation of the fragmentations which depend on the thermokinetic properties of the competitive decomposition pathways (critical energy E^0, frequency factor ν, and degrees of freedom n). Rice–Ramsperger–Kassel–Marcus (RRKM) theory will be useful for the rationalization of the fragment ion abundances in relation to the previously discussed parameters including the instrument time-window.[29] This significant energy excess above the transition state (i.e., critical energy plus kinetic shift) is expressed by a theoretical parameter called effective temperature (T_{eff}) that depends on the ion and reaction properties as well as the instrumentation time-window for decomposition observation.[30] This theoretical term has found use in various assumptions, and approximations have been employed for calculation of dissociation rate constants from kinetic treatment. It is a useful term for comparing energies transferred to ionic species that are (i) desorbed and/or ionized in the gas phase and (ii) involved in competitive dissociations.

17.2 NATURE OF PRECURSOR IONS

17.2.1 Even-Electron Versus Odd-Electron Species and Intact Precursor Ions Versus Isomerized Molecular Species

Generally, produced under desorption conditions, the closed-shell molecular species are characterized by a larger gas-phase (GP) stability compared to odd-electron molecular species that present broad internal energy distributions when they are prepared under EI conditions. Indeed, the latter show significant gas-phase instabilities due to (i) their large E_{int} values and (ii) the presence of an unpaired electron located either at the charge site or at a different site (i.e., distonic* molecular species). Actually, a free-electron located at a specific site rapidly promotes prompt dissociations that are directed from the position of the unpaired electron. For such species, which can be considered to be excited radical ions, *charge* location is a secondary consideration in the dissociation processes. This behavior differs from the even-electron species where prompt dissociations can be compared to the organic chemistry reactions described in textbooks, where protonated (or deprotonated) sites play the role of acid/base initiators of reactions. Thus, the decomposition processes follow the charge-driven pathways occurring by:

i. *direct neighboring C–X bond cleavages* (X being C or heteroatom), at the α–β position, yielding product ions†;

ii. *skeleton rearrangement* involving one (or more) proton (or hydride) transfer(s) as isomerization step(s) followed by C–X cleavages considered either as stepwise processes, or more rarely, as concerted pathways from even-electron precursor ions.

17.2.2 Protonated/Deprotonated Versus Cationized/Anionized Species

A proton initially present at a basic site (of relatively high basicity) may endothermically migrate to a more labile functional group (i.e., monovalent group with a lower basicity), leading to formation of fragile ions (e.g., protonated alcohols or ethers) which promptly dissociate especially by release of a neutral of low proton affinity. A stronger stability is shown from deprotonated species. In addition to direct cleavages induced by negative charge, two possible migration pathways can occur: (i) A proton from an acidic carbon atom can migrate to induce bond cleavage, and (ii) a hydride transferred from a charged site (e.g., carbide or alkoxide) promotes ion dissociation.

Alkali-cationized species are generally more stable than the protonated species because of the reduced mobility of the charge away from the alkali metal; however, in the case of the presence of an acidic site, proton migration to a basic site may occur, producing an anion site able to solvate the alkali ion (i.e., salt).[31] On the other hand, this means that cationized zwitterionic forms are generated that are destabilized, owing to the protonated site that induces ion dissociation with alkali metal retention in the product ions.

From transition metal complexes, molecular reactivity can be enhanced because of metal cation coordination that yields a particular structural ligand arrangement and, thus, destabilization of certain bonds not involved in the metal coordination.[27,32] Conversely, the analyte converted to anionic form by attachment of a deprotonated mineral or organic acid reagent (X^-) reflects either the presence of a labile proton site (but insufficiently acid

*Odd-electron ions can take the form of distonic species where the charge site is away from the radical site. Distonic species form by isomerization prior to dissociation in a stepwise pathway. Prompt CRP are responsible for structural modifications either by homolytic bond C–C cleavage, yielding ring opening from alicyclic systems (or ion–neutral complex formation from aliphatic compounds) or by hydrogen atom migration. Alternatively, following formation of such distonic species, consecutive isomerization induced by the radical site can take place especially from polycyclic systems, whereas from noncyclic systems, additional H radical migration may occur but mainly, ion dissociation proceeds.

†In certain cases, this process can be the first step of parent ion isomerization. The latter occurs either *via* ring opening from cyclic compounds or via ion–dipole complex formation (*vide infra*) prior to dissociation; the isomeric forms progress down the lowest activation energy pathways to yield fragment ions.

relative to the acid XH reagent to be removed) or the absence of a labile proton, an electrophilic function.[33] The former gives rise to formation of an anionic $[M + X]^-$ complex stabilized by a noncovalent interaction via hydrogen bonding {i.e., $[(M - H)^- H^+ \cdots X^-]$} whereas the latter, associated with electrophilic attack, yields covalent bond creation via formation of a tetravalent intermediate.[34] The latter mainly decomposes by either anion X^- reagent restitution or, possibly, generation of odd-electron $M^{-\bullet}$ species (if the electron affinity of M is close to that of $X^\bullet$). Decomposition of the former noncovalent adduct ion leads to X^- and $[M - H]^-$ via competitive anion desolvation as occurring with carbohydrate solvating chloride anions.[35,36]

Interestingly, if formation of $[M - H]^-$ cannot occur directly from an ion source, but under adduct ion activation the deprotonated molecules can be produced, then consecutive dissociation can take place, leading to smaller-size fragment ions. In all cases, ion activation allows the transfer of sufficient vibrational energy to reach an adequate dissociation rate constant to allow fragmentation in the required time-window. Low-energy collision processes and IRMPD lead to more predictable decomposition pathways than do higher-energy routes.

17.3 DISSOCIATION OF SINGLY CHARGED SPECIES

Small-size compounds ionized under ESI conditions are mainly singly protonated or deprotonated molecules although doubly charged dicarboxylate compounds were described in the literature,[37] but generally this is not the case. For this reason, the dissociation of even-electron species are herein reported only for singly charged systems. The main relevant examples are provided from purified organic compounds as well as from other chemical samples such as food pollutants, herbicides, pesticides, drugs, and metabolites. This relatively short list is intended as a representative subset of studies reported in the literature concerning the decomposition of protonated or deprotonated low-molecular-weight compounds. Decompositions produced from their positive and negative even-electron species are reported together for discussion, taking into consideration the differences on the charge location according to the positive or negative polarity of the studied ions. One typical example can demonstrate such a situation, and it concerns the R′-CH(R)-CO-OR″ ester compounds (with R as hydrogen or alkyl branching). In this case, (i) protonation can take place on one of the oxygen atoms (with the carbonyl group being the most basic), and (ii) deprotonation is regioselectively oriented toward the α-position of the carboxylic group. Because of space limitations, it is not possible to cover every possible circumstance or eventuality that may be incurred in the vast domain of decompositions of even electron ions. Instead, this chapter will give just a few examples that we consider to be exemplary and pedagogic. Detail of nuances and exceptions can be culled from the extensive literature on particular subtopics of interest to the reader.

In positive mode, the proton is mainly located on a lone electron pair site of monovalent (–X), divalent (=X), or trivalent (≡X) functions and more rarely at a carbon–carbon unsaturation. Generally, its location is related to the difference of proton affinity between the samples and the employed ESI solvent (Table 17.1).

From the aggregate, solvent evaporates, and as the last solvent molecules depart, protons situate themselves on the most basic sites available (sites more basic than the solvent). Conversely, in the negative mode, deprotonation yielding negative ions takes place from sites that bear a mobile proton (Table 17.2), during the endothermic desolvation steps yielding neutral solvent releases, considering that the solvent/sample aggregates are smaller

Table 17.1. Relative Thermochemical State Functions of Usual Solvents[a]

Solvents	$PA_{(S)}$ (kJ/mol)	Solvents	$\Delta H^0_{acid(S)}$ (kJ/mol)
Water	691.	Water	1633.
Methanol	754.3	Methanol	1596.
Ethanol	776.4	Ethanol	1583.
Isopropanol	793.0	Isopropanol	1569.
Acetonitrile	779.2	Acetonitrile	1560.
Acetone	812.	Acetone	1543.
Acetic acid	783.7	Acetic acid	1540.
Formic acid	742.0	Formic acid	1445.
Triethylamine	981.8	Carbonic acid	1551.
Ammonia[b]	853.6		

[a] From NIST webbook.

[b] Produced from ammonium buffer.

Table 17.2. Relative Thermochemical State Functions of Monofunctional Compounds[a]

Monofunctional Compounds	$PA_{(M)}$ (kJ/mol)	$\Delta H^0_{acid(M)}$ (kJ/mol)	Monofunctional Compounds	$PA_{(M)}$ (kJ/mol)	$\Delta H^0_{acid(M)}$ (kJ/mol)
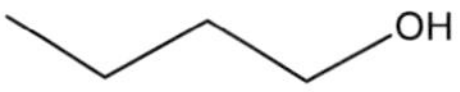 OH	789.2	1570.	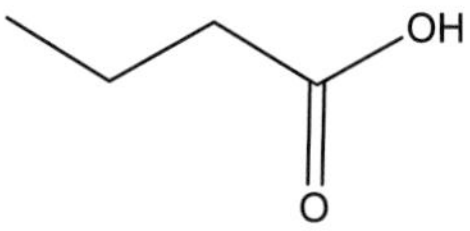 OH, O	—	1451.
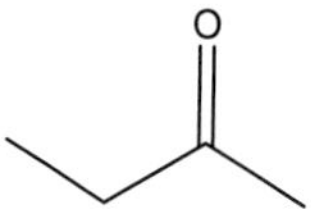 O	827.3	1536.	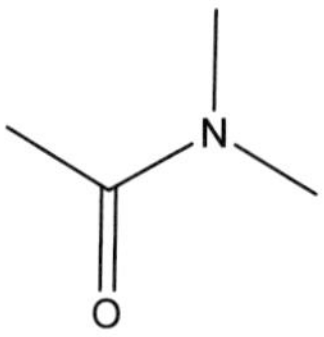 N, O	908.0	1568.
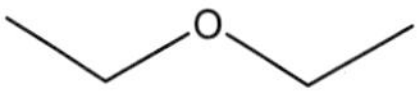 O	828.4	—	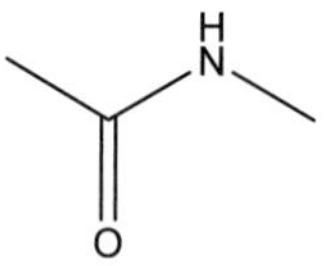 H N, O	888.5	1514
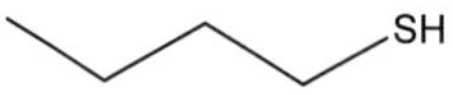 SH	801.7	1480.	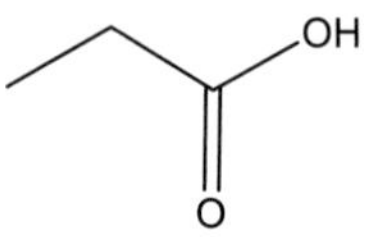 OH, O	797.2	1454.
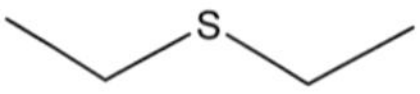 S	856.7	—	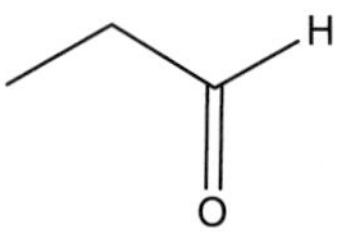 H, O	786.0	1528.
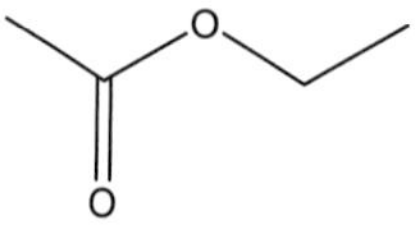 O, O	835.7	1555.			

[a] From NIST webbook.

than those formed in positive mode.[38*] Thus, based upon the gas-phase acidity, a proton can be removed at a proton-carrying heteroatom—for example, from alcohols, thiols (and to a lesser extent amines), carboxylic acids, and amides—or at the α-position of CO groups (e.g., ketones, aldehydes, and esters or substituted amides), α to CN groups (e.g., imines, oximes), and α to nitrogen atoms in nitriles and nitroalkanes. Furthermore, it is possible to produce carbide anions by proton removal from (i) allylic and benzylic positions, (ii) acetylenic or allenic position, and (iii) more rarely, phenylic CH sites. [39] This presentation cannot be exhaustive because of the existence of many possible dissociation pathways. Nevertheless, many typical mechanism examples will be treated to illustrate different decomposition processes.

17.3.1 Monofunctional and Bifunctional Even-Electron Ions

Dissociations of small-size even-electron species prepared under chemical ionization (or in APCI and more rarely in ESI) reported either from the past literature or from unpublished results from our own laboratory are herein discussed to illustrate mechanisms. In addition, for a better rationalization of the even-electron ion behavior toward cleavages, stereochemical effects will be discussed in order to bring to light the effects of distant or neighboring atoms (or groups) on the dissociation rate constant. A lot of examples are provided from (i) Mandelbaum's group for positive ions and (ii) Bowie's group for negative ions. The mechanism examples presented herein are fragmentation processes promoted by the positive or negative charge. This signifies that mainly low-energy excitation processes are discussed rather than the high-energy range for ion activation. Consequently, charge-remote processes will be discussed in the lipid part of this chapter.

17.3.1.1 Thermochemical Orientation of Direct Cleavage Dissociations and Neutral Release Involving Endothermic Internal Proton Transfer

The presence of the proton at a heteroatom (O or N atom) may yield direct cleavage to release a small neutral at some probability related to the heteroatom's proton affinity. To illustrate this assertion, Cooks and co-workers[40] showed that the ω-alkyl amino alcohols, according to the length of their flexible backbones, are able to favorably lose water rather than ammonia; initial protonation of the amine group is favored, but water is a better leaving group. Two more examples involve rigid substituted skeletons. In a steroid example, the dissociation of the protonated *trans*-3-dimethylamino-pregnan-2-ol epimer more favorably yields water loss over dimethyl amine release, as compared to what is observed from the *cis* epimer (where hydrogen bonding hinders water loss).[41] For decalins, a comparison[42] of the CID spectra of 3β-dimethylamino-5α-*trans* and 3β-methoxy-5α-*trans* decalins reveals that mainly two product ions, namely, $(CH_3)_2NH_2^+$ and $[MH - (CH_3)_2NH]^+$, result from decompositions of the former whereas the latter presents only $[MH - CH_3OH]^+$. Methanol loss yields a carbocation at $C_{(3)}$ of this latter decalin without production of $CH_3OH_2^+$ ($PA_{(CH_3OH)}$ being lower than that of unsaturated decalin) (Scheme 17.1). Conversely, from the former, a significant competition exists that favors $(CH_3)_2NH_2^+$ over $[MH - (CH_3)_2NH]^+$.

*The higher is the exothermicity of proton solvation, the less multisolvation of that proton is required. Thus, more neutrals typically solvate protons in positive clusters (i.e., $[M_nH]^+$, n higher than 2) than anions solvate protons in negative clusters (i.e., $[2(A - H)^-, H^+]^-$).

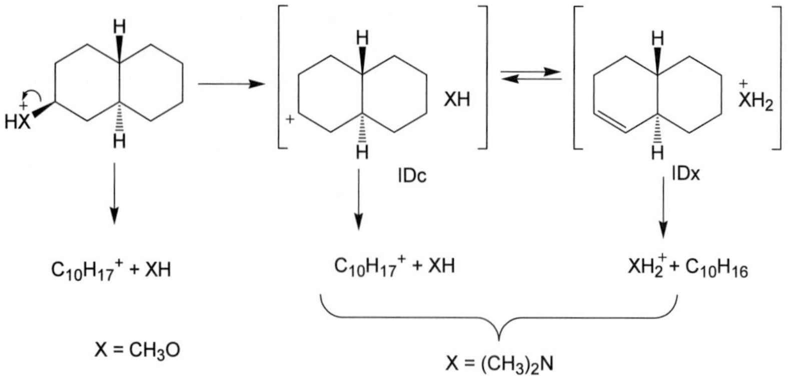

Scheme 17.1. Competitive direct cleavages and isomerization processes into ion–neutral complexes from functionalized *trans*-decaline. (Adapted with permission from Ref. 42.)

This change of behavior is related to the higher PA of $(CH_3)_2NH$ than that of the produced alkenes. This means that the previous direct cleavage does not take place, and formation of the ion–neutral ID_c complex able to isomerize into ID_x via exothermic internal proton transfer is preferred prior to cleavage (Scheme 17.1). Respectively, the ID_c and ID_x complexes competitively dissociate into $C_{10}H_{17}{}^+$ and $(CH_3)_2NH_2{}^+$, the latter ion being strongly favored by the higher proton affinity of $(CH_3)_2NH$ relative to $C_{10}H_{16}$ when both are solvating the proton.

In the case of bifunctional aliphatic compounds comprised of different functions, as illustrated by the Cooks example above,[40] the neutral elimination occurring with the larger rate constant is produced from the protonated group characterized by the lower basicity. This behavior is possible when intramolecular interaction occurs by conformational change yielding a better internal charge solvation. Generally, this results in a fast proton transfer from one site to the second by an endothermic process. From alicyclic systems with more rigid conformations, interaction between competitively protonated sites of distinct basicity depends on their relative stereochemistry. Consequently, the neutral release is related to the initial charge position as well as to its possible migration, and kinetic effects appear (e.g., due to steric effects and anchimeric assistance, *vide infra*).

For instance, as shown in Scheme 17.2, from the protonated *cis*-1-methoxy-1-butyl-4-dimethylaminocyclohexane, M_cH^+, CH_3OH loss occurs either directly from the $M_cH^+_{(O)}$ form or subsequent to endothermic proton transfer from the very basic Me_2N-group [$M_cH^+_{(N)}$ form] to the CH_3O site.[43] This transfer is possible only through a boat and/or a twist conformation. The *trans* isomer (M_t) cannot lose CH_3OH from the very stable $M_tH^+_{(N)}$ protonated form, and only the very minor $M_tH^+_{(O)}$ form may eliminate methanol.

Recently, a study of deprotonated 3-hydroxyflavone dissociation showed the possibility to explain heterocyclic cleavage via a quasi-concerted retro-Diels–Alder process by considering dissociation via diradical anion formation.[44] This was evidenced by calculations which show that the less endothermic pathway is the 1,3-cross-ring cleavage occurring through the lowest-energy transition state and leading to the formation of the deprotonated $C_6H_5–C{\equiv}C–O^-$ anion (ion ${}^{1,3}B^-$). The complementary cleavage (0,4-cross-ring cleavage) into $C_6H_5–CO–C(–CO)–O^-$ (ion ${}^{0,4}B^-$) by benzyne release was calculated to have a significantly higher transition state by more than $200\,kJ \cdot mol^{-1}$.

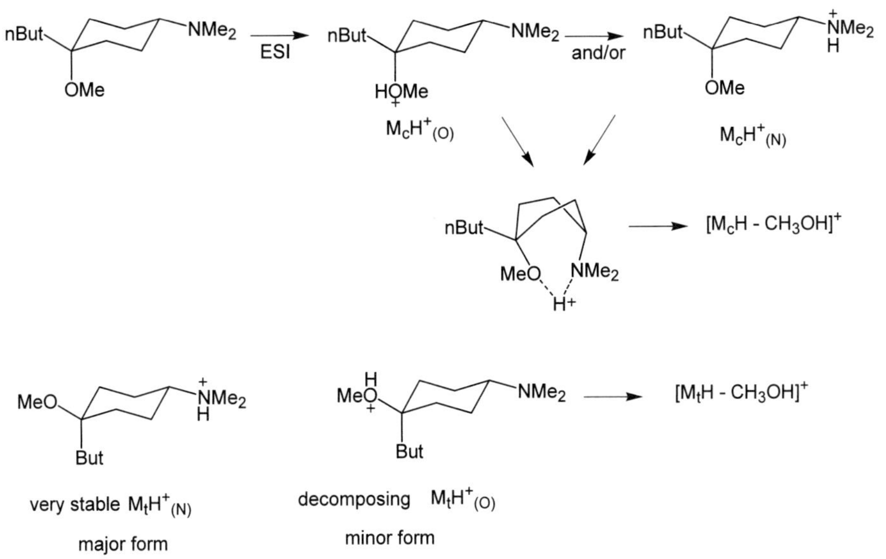

Scheme 17.2. Methanol release from the M_cH^+ within boat conformation and from the minor protonated $M_tH^+_{(O)}$ form. (Adapted with permission from Ref. 43.)

17.3.1.2 Kinetic Isomerization into Ion–Neutral Complexes Progressing Through Thermochemical Orientation

Interestingly, from protonated systems, in order to explain the observed nontotal regioselectivity observed for labeled hexene elimination (nonspecific H/D exchanges) from 2,2,4,4-D_4*sec*-hexyl *m*-dimethylamino ethers, the contribution of ion–dipole complex formation prior to dissociation is considered[45] without (or with) charge retention at the initial protonation site (i.e., Me_2N-site, Scheme 17.3). This ion isomerization requires as the first step an endothermic proton transfer from the protonated tertiary amine to the aromatic system (Scheme 17.3). This complex is able to undergo reversible proton (or deuteron)

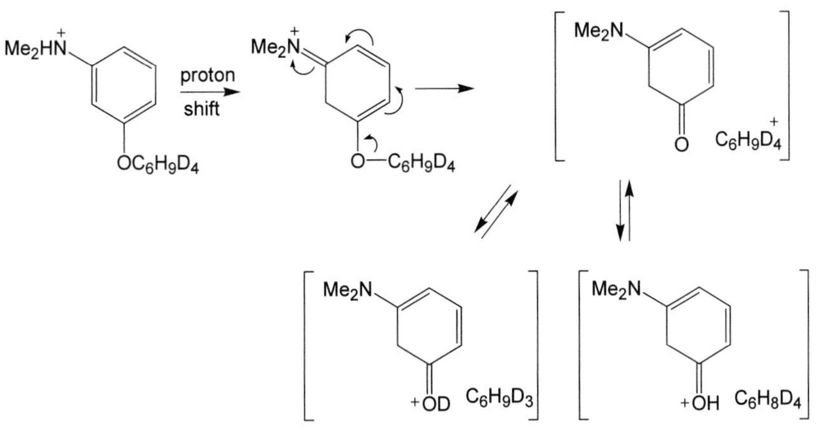

Scheme 17.3. Ion–dipole complex generates competitive proton and deuteron transfers by reversible stepwise processes under low-energy collision conditions. (Adapted with permission from Ref. 45.)

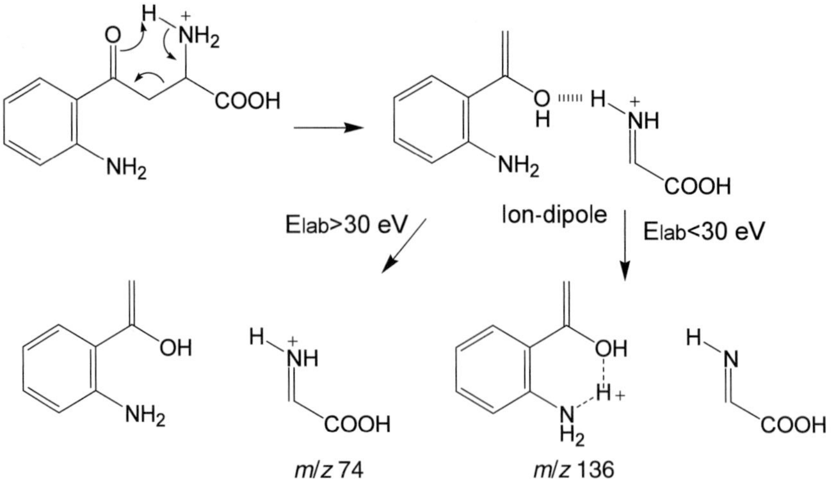

Scheme 17.4. Decomposition mechanism of competitive formation of protonated imine and ortho acetyl aniline. (Adapted with permission from Ref. 46.)

transfer(s) to finally give rise to formation of protonated *N*-dimethyl phenol. The proton may return to the nitrogen atom to restore the aromaticity that should offer improved stability.

Dissociation of protonated kynurenine[46] (*m/z* 209) can be representative of the role played by formation of ion–neutral complexes for production of complementary ions, allowing internal proton transfer such as HOOC-HC$=NH_2^+$ (*m/z* 74) and protonated *ortho*-aminoacetophenone (*m/z* 136) (Scheme 17.4). The former is favored at high activation energy, allowing a direct cleavage of the ion–neutral complex into iminium species. Conversely, under lower activation energy, from longer-lifetime complexes involving low-frequency factor pathways, proton transfer can take place, leading to the formation of the *m/z* 136 ion.

From more complex systems such as sulfometuron methyl (mol.Wt. 364, Scheme 17.5), which are sulfonylurea herbicides comprised of two aromatic sides (e.g., aromatic sulfamide and pyrimidine moieties) linked by a urea group, deprotonation may take place regioselectively on the NH groups of the urea, which is able to generate two isomeric deprotonated ions, namely, $[M_{pyr} - H]^-$ (at the pyrimidine side) or $[M_{sulf} - H]^-$ (at the sulfamide side). Subjected to low-energy collision processes, they dissociate into ion–dipole ID_{pyr} or ID_{sulf} complexes, respectively (Scheme 17.5).[47] Interestingly, the ID_{sulf} form leads to only one anion at *m/z* 122 as the major pathway (only one moiety may be deprotonated). This contrasts with ID_{pyr}, which may internally exchange a proton to give rise to the formation of a low-abundance complementary pair of product ions (*m/z* 214 and *m/z* 148). Note that the *m/z* 214 ion is able to release methanol yielding the *m/z* 182 ion.

From larger systems such as the estradiol-17β fatty acid,[48,49] unexpected long distance proton transfer may occur from two types of deprotonated molecules (Scheme 17.6). However, this example is exceptional since the stepwise process goes forward via exothermic (or thermoneutral) internal proton transfer before dissociation. In this case, a thermal-like process is assumed and this was confirmed by H/D exchange experiments. However, from the $[M - H]^-$ ion with the negative charge localized at the ester group α position, a charge directed process yielding loss of the fatty acid moiety can be proposed. Thus, two pairs of complementary product ions such as (*m/z* 255)/(*m/z* 253) or

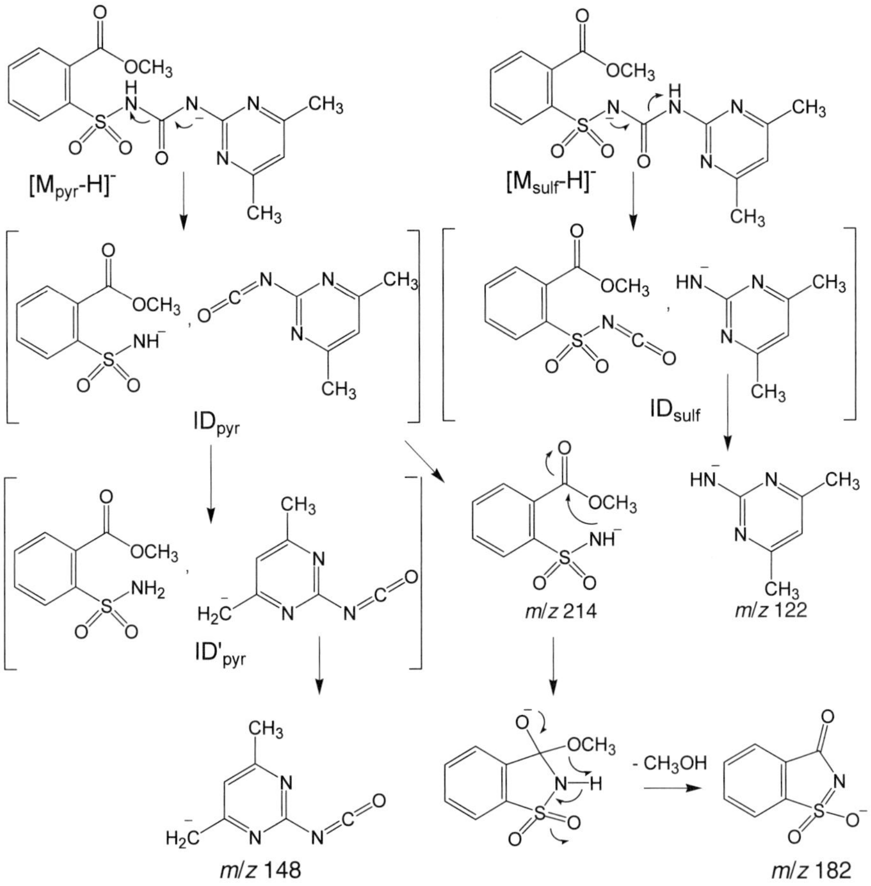

Scheme 17.5. Formation and dissociation of isomeric ion–neutral complexes according to the deprotonated site from deprotonated sulfometuron methyl. (Adapted with permission from Ref. 47.)

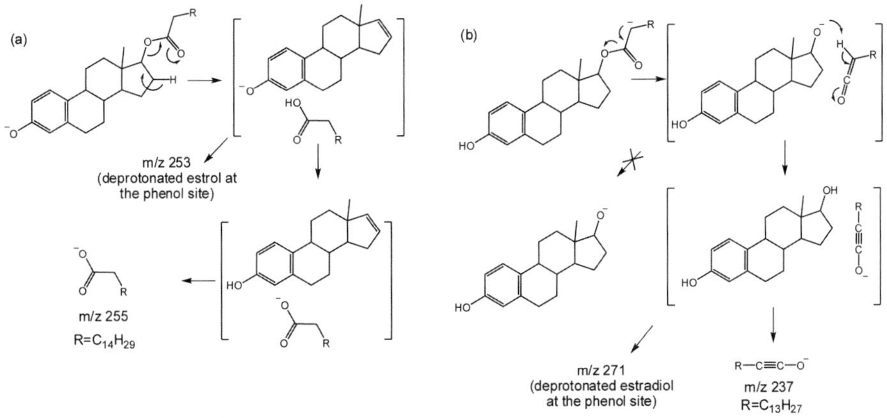

Scheme 17.6. Ion–neutral complexes that allow long distance proton transfer along the steroid skeleton to yield two pairs of complementary product ions: (a) thermal process, (b) charge driven process. (Adapted with permission from Ref. 48.)

(*m/z* 237)/(*m/z* 271) are generated by each ion–dipole complex dissociation; the first pair is related to the deprotonated intact fatty acid and deprotonated estrol-17-en, in contrast with the second pair, which is connected to the deprotonated ketenic form and deprotonated estra-3-17-diol. Such long-distance proton transfers imply that both of the implicated ion–dipole complexes have long lifetimes.

Ion–neutral complexes involving aromatic systems are known to be readily formed; and under low-energy CID conditions, relatively large complexes can be generated. They can involve an entire reorganization of the skeleton, thereby producing fragment ions generated through intramolecular electrophilic aromatic substitution processes known in organic chemistry. For instance, *cis/trans* protonated dibenzyl cyclohexyl-1,4-diesters (as well as the corresponding diether analogs) decompose into the $C_{14}H_{13}^{+}$ ion (*m/z* 181) as the main product ion (Scheme 17.7) in competition with the very minor $[MH - 180]^{+}$ (protonated 1,4-dicarboxylic acid cyclohexane) complementary ion.[50]

The hydrocarbon ion $C_{14}H_{13}^{+}$ corresponds to a dibenzyl ion that was unexpected from the known structure of the studied analyte. In fact, the protonated molecule isomerizes into an ion–neutral complex prior to dissociation. This occurs by benzyl ion release that remains associated with the benzyl site of the monoester of cyclohexyl-1,4-diacid neutral through formation of both noncovalent π and covalent σ complexes that evolve toward covalent attachment at *ortho* (and *para*) positions (Scheme 17.7). This last step is unusual since, from the ion–neutral complex, covalent bond formation takes place rather than dissociation of the ion–neutral complex. Rearrangement of such a covalent system evolves toward stabilization by re-aromatization via proton transfer from the σ complex to the neighboring $-COO^{-}$ group activating benzyl bond cleavage. By this route, 1,4-dicarboxylic cyclohexane is

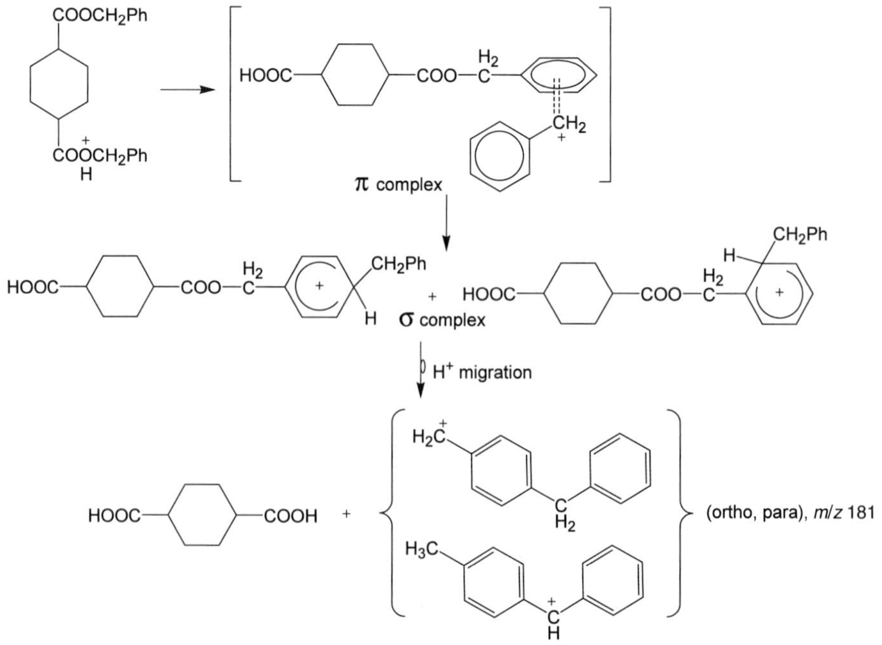

Scheme 17.7. Formation of ion–dipole complex allowing intramolecular electrophilic aromatic substitution via production of π and σ complexes as reaction intermediates. (Adapted with permission from Ref. 50.)

released and formation of the $C_{14}H_{13}^+$ ion (*m*/z 181) is accomplished (Scheme 17.7). Several examples of this type of reaction are found in literature from Grutzmacher's[51] and Kuck's[52] groups. From the latter, hydride transfer is considered to occur from such aromatic ion–dipole complexes.

17.3.1.3 Direct Cleavage and Stepwise Distant Proton Migration to Assist Group Release

Usually, the loss of water from trivalent functions such as carboxylic acids is considered to occur by CO–OH_2^+ bond cleavage. This structure is produced either directly during the ionization step or through a 1,3-proton transfer from the protonated –C(OH) = OH^+ carbonyl to its hydroxy moiety. The latter pathway involves a relatively high transition state compared to those required by 1,2-, 1,4-, 1,5- and 1,6-proton migrations. In the case of compounds constituted by a pair of such trivalent functions, long-distance interactions through space, involving less tight transition states, can allow such proton transfers. From typical examples such as fumaric and maleic acids (and derivatives), different distance proton migrations can be implicated,[53] explaining why the latter favorably yields loss of a water molecule, in contrast to the former. Protonated monomethyl fumaric ($M_{(E)}H^+$) and maleic ($M_{(Z)}H^+$) esters under low-energy CID conditions present especially two product ions: $[MH - H_2O]^+$ and $[MH - CH_3OH]^+$. The $[MH - H_2O]^+/[MH - CH_3OH]^+$ relative abundance ratio is significantly lower with the E compared to the Z geometry (Scheme 17.8).

A similar behavior is observed from competitive dissociations of protonated monoamides of maleic and fumaric acids which lead to the formation of $[MH - H_2O]^+$ and $[MH - NH_3]^+$, respectively. They are accompanied by the presence of NH_4^+, although the loss of water corresponds to the base peak from the Z stereochemistry but is of lower abundance from the E isomer. From fumarate monomethyl ester or monoamide, the major pathway for protonated molecule dissociation corresponds to the loss of XH as methanol or ammonia, respectively, which suggests that the modified carboxylic group is the preferred protonation site (Scheme 17.8). Consequently, the favorable loss of water from the Z isomers (not only for maleic acid, but also for the monoester and monoamide derivatives) indicates that the water loss rate constant, via 1,6-H^+ transfer, is much larger than that occurring from the E isomer which involves either 1,3-H^+ transfer (a symmetry forbidden process) or a multistep proton migration which is characterized by higher transition state level(s) (Scheme 17.8).

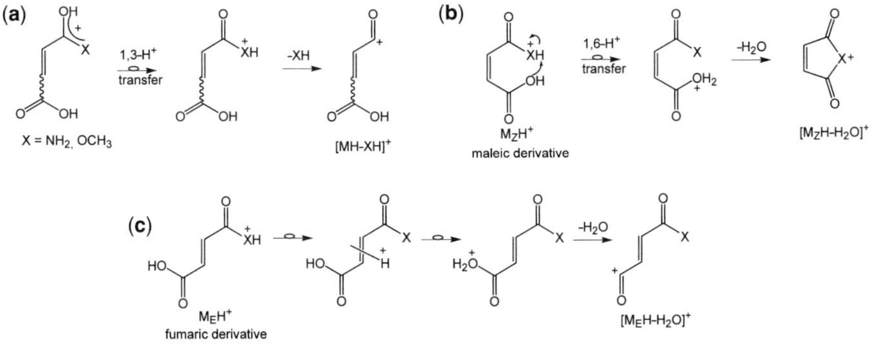

Scheme 17.8. (**a**) 1,3-H^+ transfer yielding loss of XH (NH_3 and CH_3OH from monoamide and monomethyl ester, respectively) independent of the Z/E geometry; (**b**) water loss by 1,6-H^+ migration from Z; and (**c**) stepwise 1,4- and 1,3-H^+ transfer from the E isomer. (Adapted with permission from Ref. 53.)

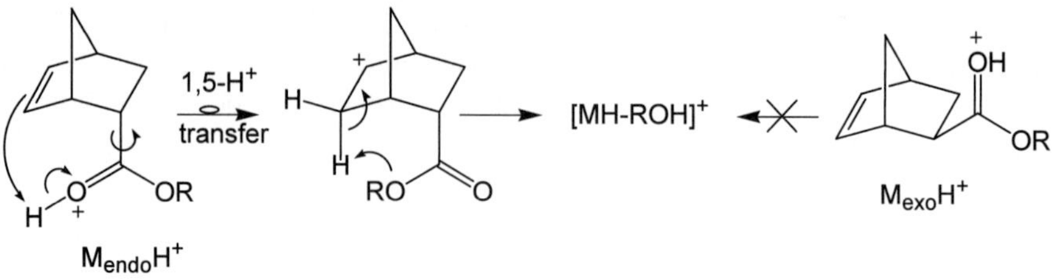

Scheme 17.9. Reversible 1,5-proton migration allowing methanol release from the *endo*-ester isomer. (Adapted with permission from Ref. 54.)

The symmetry forbidden character of the 1,3-proton transfer provides the explanation for the absence of methanol loss from the protonated methyl *exo*-bicyclo[2.2.2]oct-5-ene-2-carboxylate under low-energy collision conditions, although this elimination is favored from its protonated *endo* isomer.[54] Scheme 17.9 shows the origin of the observed stereospecificity. In this example, the double bond is used as proton "*acceptor*," allowing a reversible 1,5 proton transfer from the *endo* protonated ester, followed by methanol loss. From the *exo* ester such release does not occur because this 1,5 proton transfer is not possible and the 1,3-proton transfer pathway is forbidden. Interestingly, in the case of the *trans*-ethyl methyl esters of the bicyclo[2.2.2]oct-5-ene-2,3-dicarboxylic acid system, the alcohol loss is also regioselective for the *endo* ester, although a very minor alcohol release for the *exo* ester can be detected.

17.3.1.4 Direct Cleavage and Hydride Migration for Assisting Group Release

The study of rigid conformational systems is very informative concerning the influence of neighboring groups on dissociation rate constants. On the other hand, the ascending degree of substitution in primary, secondary, and tertiary ethers presents an increased propensity toward carboxy bond cleavage due to the improved carbocation stabilization with increasing branching degree.[55] It has been shown[56] that methoxy groups situated axially on mono- or bicyclic skeletons were more quickly eliminated as compared to equatorial groups. Furthermore, the rate constant is enhanced when (i) an alkyl group is vicinal to the functionalized ring carbon atom and (ii) both the alkyl and methoxy groups are on the same side of the ring (*cis* isomer). Likely, steric effects can destabilize the protonated system, which readily decomposes. More importantly, cleavage assistance can occur due to 1,2-hydride migration from the tertiary carbon atom to the carbon atom bearing the protonated methoxy group, thereby assisting methanol elimination and yielding a tertiary carbocation. It has been calculated that with such assistance, the transition state was lowered by approximately 10 kcal mol^{-1} compared to that of the direct cleavage.[57] From the *trans* isomer, the methanol elimination rate constant is significantly lowered. Furthermore, ring contraction according to a Meerwein-like transposition may also be considered by a less favored stepwise pathway.

A similar behavior characterizes dissociation of the acetate derivatives via a favorable elimination of acetic acid from the *cis*-2-methylcyclohexyl acetate,[58] which involves anchimeric assistance by a 1,2-hydride migration. This does not occur from the 3- or 4-methyl positional isomers, which would require 1,3- and 1,4-hydride migrations and therefore do not present large *cis/trans* stereochemical effects that are observed from the 2-methyl isomers. Concerning *cis*-1,2-bifunctional derivatives (e.g., diethers or diacetates),

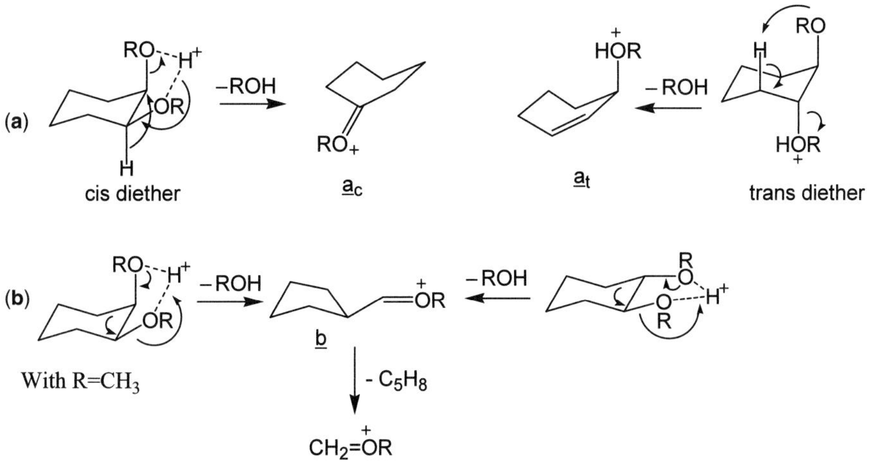

Scheme 17.10. (**a**) 1,2-Hydride migration to form $\underline{a}_c$ and 1,4-proton transfer producing $\underline{a}_t$ from the *cis*- and *trans*-1,2-dimethoxycyclohexane isomers, respectively; and (**b**) competitive ring contraction by a Meerwein-like transposition yielding cationized aldehyde able to dissociate into $CH_2{=}OR^{+}$. (Adapted with permission from Ref. 59.)

the first neutral loss (e.g., methanol or acetic acid) occurs via a similar 1,2-hydride migration. From the 1,2-dimethoxy derivative, this process leads to formation of the cationized cyclohexanone $\mathbf{a_c}$ (Scheme 17.10a) whereas from the *trans* isomer, hydride transfer being unavailable, methanol release proceeds through a 1,4-proton transfer to give rise to formation of protonated 3-methoxycyclohexene $\mathbf{a_t}$ (Scheme 17.10a). These processes occur from the *cis*- and *trans*-1,2-dimethoxy stereomers, respectively, in competition with ring contraction.[59] This last pathway requires a C–C bond transposition to assist the loss of the protonated methoxy group (Scheme 17.10b) yielding the product **b** form as a functional isomer of the $\mathbf{a_c}$ and $\mathbf{a_t}$ forms. The **b** form is evidenced by its consecutive cleavage into diagnostic $CH_2{=}OR^+$ ions.

The formation of ion–dipole complexes from bifunctional compounds can yield suppression (or attenuation) of stereochemical effects.[3] Indeed, from the *cis*- and *trans*-1-alkoxy-1-methyl-4-(2-alkoxy-2-propyl)cyclohexanes, the fact that regioselective methanol loss induced by a 1,2-hydride transfer is not stereospecific suggests a possible isomerization of the protonated precursor ion into a common ion–neutral complex (Scheme 17.11). From this complex, long-distance proton transfer can be considered to be independent of the initial *cis/trans* stereochemistry. Such isomerization is especially favored from long lifetime precursor ions such as those that exist during ion storage in an ion trap instrument.[60]

Stepwise processes are involved for dissociation of protonated tetracyclic alkaloids related to the Aspidosperma[61] skeleton containing a *cis* or *trans* stereochemistry that characterizes the B/D fused rings (Scheme 17.12). The main stereochemical effect concerns B/D ring cleavage yielding acetamide neutral release after opening of the D ring. This elimination is strongly favored from the *trans* epimer. This D ring opening is promoted by 1,5-hydride transfer inducing an antiparallel bond cleavage possible only from the *trans* epimer. The resulting open D ring then decomposes by an additional hydride transfer concomitant with B-ring enlargement and acetamide release (Scheme 17.12a). Furthermore, the release of the benzyl protecting group of the indolic moiety takes place, yielding an ion–neutral complex (Scheme 17.12b). From this adduct, an internal electron transfer from the indolic moiety occurs (characterized by an ionization energy close to that of the

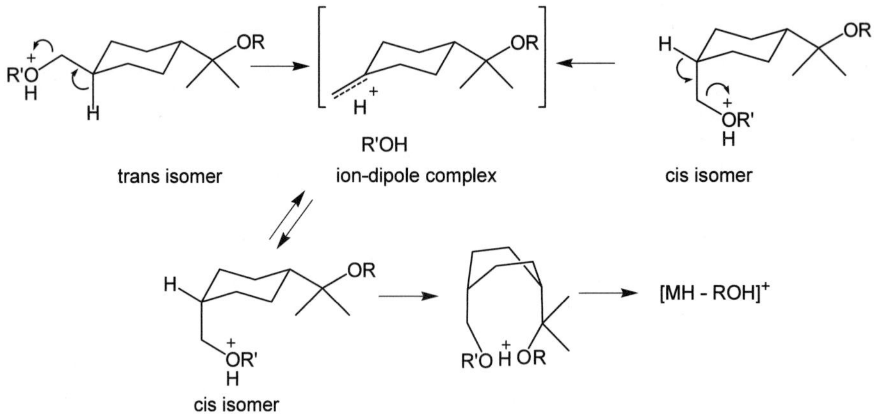

Scheme 17.11. Isomerization into a common ion–dipole complex by a 1,2-hydride migration yielding a proton-bridged structure that loses methanol independent of the initial configuration. (Adapted with permission from Ref. 60.)

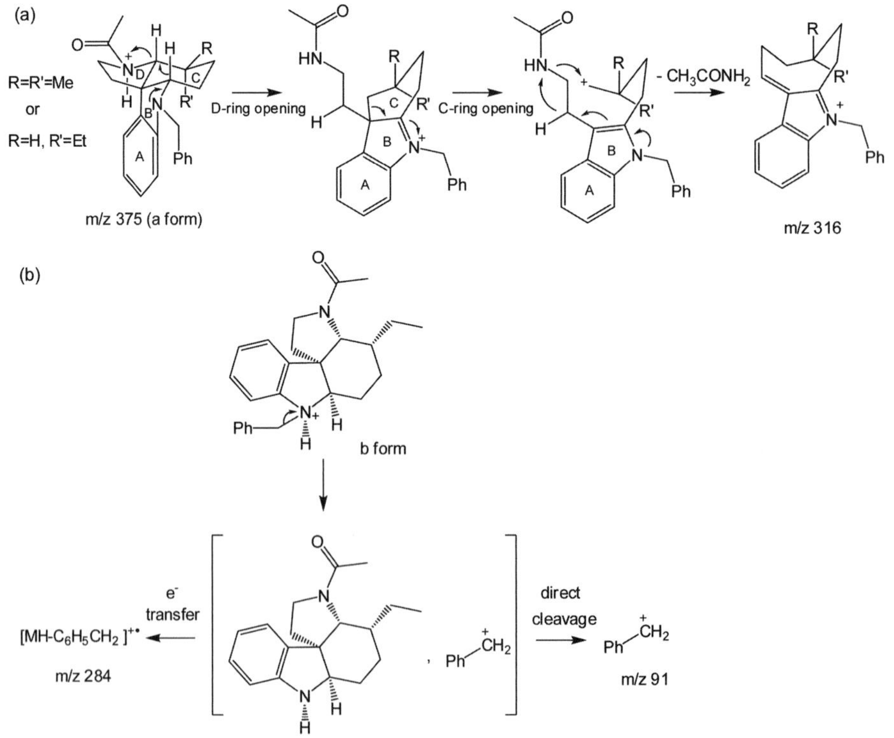

Scheme 17.12. Stepwise decompositions of protonated Aspidosperma alkaloid occurring by either (**a**) stereochemical D ring cleavage yielding loss of acetamide neutral or (**b**) ion–dipole complex formation by *N*-benzyl bond cleavage of the protecting group of the indolic system decomposing into odd-electron product ions. (Adapted with permission from Ref. 61.)

benzyl radical), leading to release of the benzyl radical (i.e., formation of the odd-electron indolic product ion at *m/z* 284) competitively with benzyl cation formation (*m/z* 91).

From deprotonated $[M-H]^-$ molecules, direct cleavage promoted by charge is usually observed with mono-, di-, and trivalent functions. Here, three typical examples [Eqs. (17.4)–(17.7)] are given as stepwise reactions involving isomerization into ion–neutral complexes prior to dissociation. In particular, Bowie's group contributed heavily to the development of this mechanistic approach for small molecules (alicyclic, aliphatic, or aromatic compounds),[62] amino acids, and small peptides. They are generated by hydride release (similar to the hyperconjugation phenomenon[63]) or carbide formation, followed by intramolecular ion–neutral competitive proton transfer(s), the more favored being those with less endothermic pathways:

$$\text{R-CH}_2\text{-(R}'\text{)CH-O}^- \rightarrow [\text{R-CH}_2\text{-(R}'\text{)C=O, H}^-] \rightarrow \begin{cases} \text{H}_2 + \text{RCH=(R}'\text{)C-O}^- & \text{(a)} \\ \text{or/and} & \\ \text{RH} + \text{CH}_2\text{=(R}'\text{)C-O}^- & \text{(b)} \end{cases} \quad (17.4)$$

(with R and R' = alkyl or hydrogen)

$$\text{R-CH}^-\text{-(R}'\text{)C=O} \rightarrow [\text{R-CH=C=O, R}'^-] \rightarrow \begin{cases} \text{R-CH=C=O} + \text{R}'^- & \text{(a)} \\ \text{or/and} & \\ \text{R-C}\equiv\text{C-O}^- + \text{R}'\text{H} & \text{(b)} \end{cases} \quad (17.5)$$

$$\text{RCH}_2\text{-CH}^-\text{-CO-CH}_2\text{R}' \rightarrow [\text{CH}_2\text{=CH-CO-CH}_2\text{R}',\text{R}^-] \rightarrow \begin{cases} \text{CH}_2\text{=CH-CO CH}_2\text{R}' + \text{R}^- & \text{(a)} \\ \text{or/and} & \\ \text{CH}_2\text{=CH-CO-CH}^-\text{R}' + \text{RH} & \text{(b)} \end{cases} \quad (17.6)$$

$$\text{R-CH}^-\text{-CO-OR}' \rightarrow [\text{R-CH=C=O, RO}^-] \rightarrow \begin{cases} \text{RCH=C=O} + \text{RO}^- & \text{(a)} \\ \text{or/and} & \\ \text{RC}\equiv\text{C-O}^- + \text{R-OH} & \text{(b)} \end{cases} \quad (17.7)$$

Under kilo-electron-volt energy range CID, hydride transfers are usually occurring during stereospecific dissociations of deprotonated molecules. These take place typically from deprotonated monovalent $-X^-$ functions [essentially, primary and secondary alcohols and thiols, Eqs. (17.4a, b)]. Early examples concern first a stereospecific hydride transfer from a secondary alkoxide site of *trans*-1,4-cyclohexanediol anion[64] to the opposite hydroxy group (Scheme 17.13) to eliminate H_2. From the *cis* isomer, the process is different

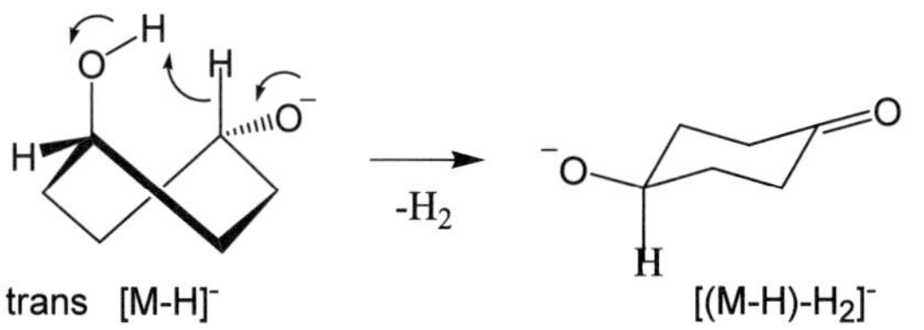

Scheme 17.13. Stereospecific H_2 elimination from *trans* deprotonated 1,4-cyclohexanediol. (Adapted with permission from Ref. 64.)

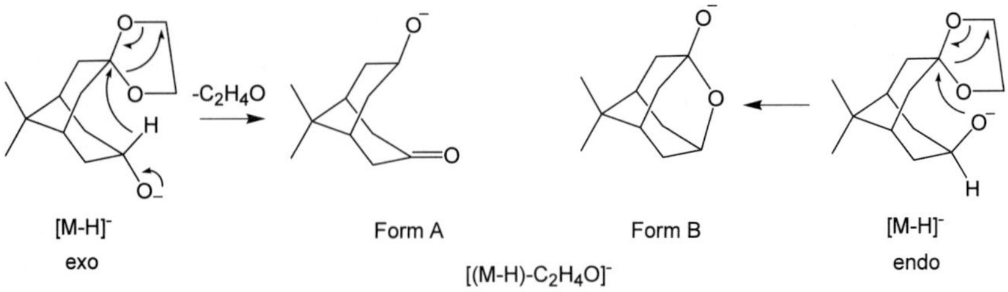

Scheme 17.14. Stereospecific ketal reduction by a distant hydride transfer promoted by the alkoxide site. (Adapted with permission from Ref. 65.)

since it involves a 1,2-H_2 elimination with a sharply lowered rate constant. A free hydride inside the ion–neutral complex may allow stereospecific distant C–O bond reduction[65] from the *exo*-3-hydroxy,7-ketal 9,9-dimethyl bicyclo[3.3.1] in contrast with the *endo* isomer, where the alkoxide site directly induces ethylene-ketal hydrolysis rather than its reduction (Scheme 17.14). Furthermore, under low-energy collision conditions, C–C bond reduction can also take place by hydride transfer as shown from the deprotonated 3,16α- and 3,16β-hydroxy-17-ketoandrostane epimers (Scheme 17.15)[66]. In this case, methane release is stereospecific to the 16α-hydroxy epimer. The deprotonated hydroxy activates hydride transfer, yielding the $C_{(13)}$–$C_{(18)}H_3$ bond reduction to release methane.

The following example shows that the stereospecific skeletal isomerization of deprotonated 3α,11β-diol 5β-androstan-17-one into an ion–neutral complex may require several steps (Scheme 17.16). This sequential process is initiated by a 1,2-hydride transfer promoted by the alkoxide of the A ring, causing its opening. This ring cleavage is followed by (i) a C–C bond cleavage concomitant with one proton transfer and (ii) a second proton migration to lead to an ion–neutral complex. Its direct dissociation yields a product alkoxide ion constituted of fused B/C/D rings, whereas exothermic internal proton transfer within the ion–neutral complex allows stereospecific release of the deprotonated methyl vinyl ketone that is not favored from the 3β epimer.[67]

Under high-energy collisions, a particular anionic behavior which yields production of unexpected odd-electron species can be observed from the dissociation of deprotonated aliphatic ketones as shown by Bowie and co-workers.[68] Open-shell ions may also form in low-energy collisions as shown by Harrison and co-workers.[69] The latter work studied the effect of collision energy on the evolution in abundances of the alkene/alkane losses [Eq. (17.5)] compared to the radical release that became increasingly favored at higher collision energies. From 2-pentanone, radical fragment ions can be formed via two possible mechanisms (Eq. 17.8a and Eq. 17.8b) depending upon the charge location.

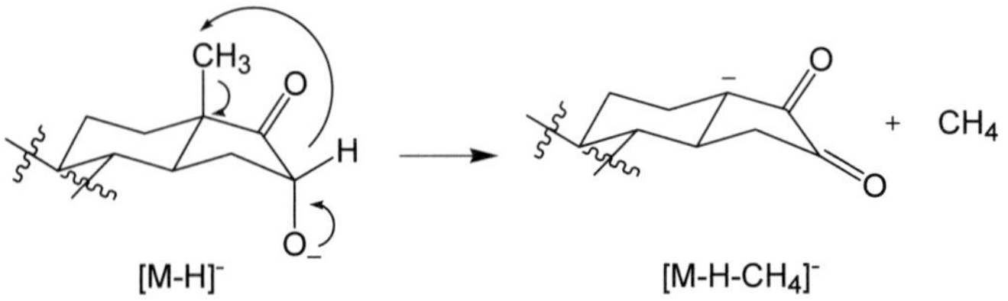

Scheme 17.15. Stereospecific C–CH_3 bond reduction by 1,4-hydride transfer stereospecifically promoted by the alkoxide site from 3,16α-hydroxy 17-ketoandrostane. (Adapted with permission from Ref. 66.)

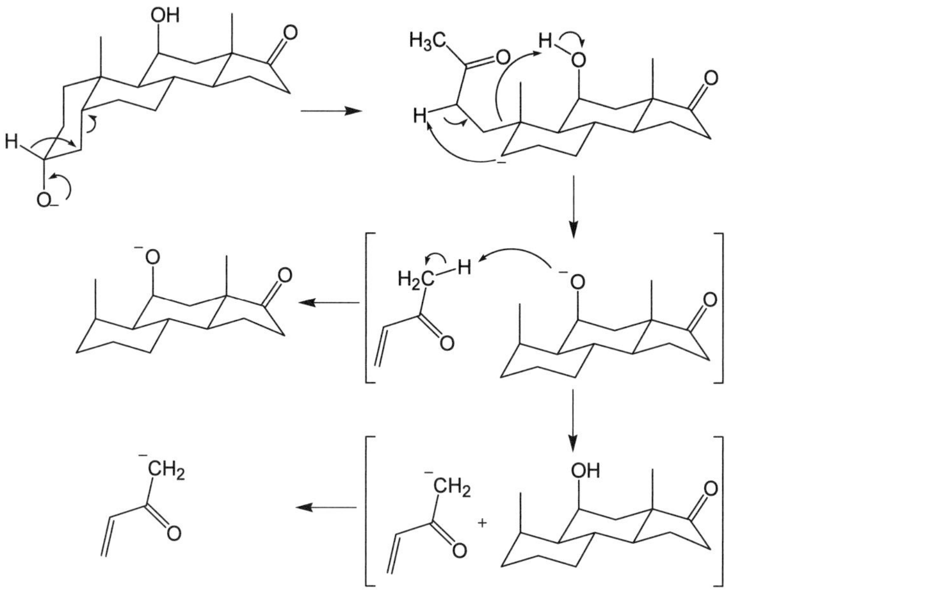

Scheme 17.16. Stepwise process from deprotonated 3α,11β-diol 5β-androstan-17-one initiated by a 1,2-hydride transfer to reach the required ion–dipole complex for providing stereospecific deprotonated methyl vinyl ketone. (Adapted with permission from Ref. 67.)

$$^{-}CH_2\text{-}CO\text{-}CH_2CH_2CH_3 \rightarrow {}^{-}CH_2\text{-}CO\text{-}CH_2^{\bullet} + {}^{\bullet}CH_2CH_3 \quad (17.8a)$$

$$CH_3\text{-}CO\text{-}CH^{-}CH_2CH_3 \rightarrow CH_3\text{-}CO\text{-}CH^{-\bullet} + {}^{\bullet}CH_2CH_3 \quad (17.8b)$$

17.3.1.5 Cyclization by Anchimeric Assistance Through Internal Nucleophilic Substitution

This concerns intramolecular backside nucleophilic attack generally occurring from bifunctional (or more highly substituted) aliphatic or alicyclic backbones. The substituents are monovalent functions (i.e., alcohols/thiols, ethers/thioethers, or primary, secondary, or tertiary amines). With an aliphatic backbone (e.g., 1,2-diether alkanes), free C–C bond rotations are occurring; because of this, an antiparallel conformation, which facilitates the required 1,2-hydride transfer, allows dissociation via alcohol release. In addition to 1,2-hydride migration, other pathways may also be favorable, such as intramolecular nucleophilic substitution. For example, from 1,4-diether alkanes, when the lone electron pair of one alkyl ether oxygen atom is conformationally able to approach and to attack the carbon atom bearing the second alkoxy group (the heavier one) which is activated by the ionizing proton, it is possible to produce a cationized tetrahydrofuran form via a stepwise process.[70] In general, elimination of the larger size alcohol is slightly favorable, but this is somewhat dependent on the presence of neighboring interacting alkyl groups that can result in steric hindrance impending obtainment of the transition state of the reaction. This is the case when the alkyl group is linked to the carbon atom at the β position of the more reactive leaving group. Indeed, alcohol loss is regioselective and controlled by the resulting steric effects:

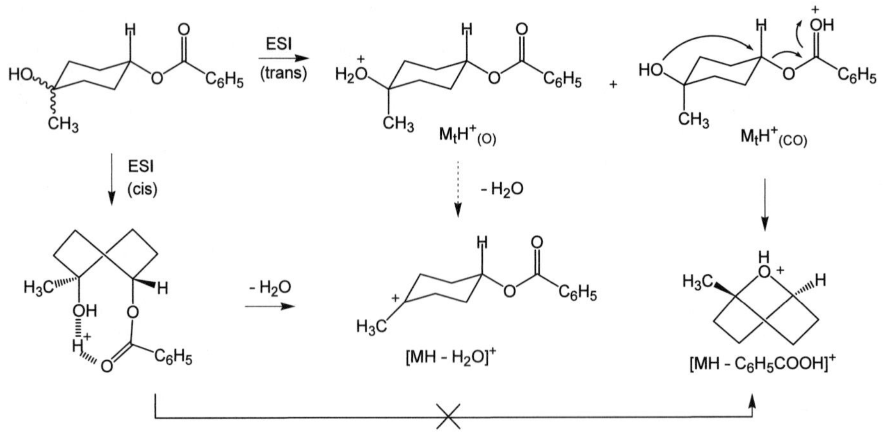

Scheme 17.17. Stereochemical cleavages of *cis*/*trans*-4-hydroxy-4-methylcyclohexyl benzoates (M_cH^+/ M_tH^+) and competitive protonated $M_cH^+_{(O)}$ and $M_cH^+_{(CO)}$ forms from the *cis* isomer. (Adapted with permission from Ref. 71.)

A larger neighboring alkyl substituent will lead to a lower rate constant for alcohol release. Actually, this rationalizes the elimination of alcohols from smaller ether groups distant from the alkyl substituent.

From bifunctional alicyclic compounds presenting more rigid conformations, similar anchimeric assistance is observed. For instance, the protonated *trans*-4-hydroxy-4-methylcyclohexyl benzoate prepared under CI/ammonia[71] and self-chemical ionization,[60] as well as in ESI conditions (Scheme 17.17), can lead stereospecifically to the protonated bicyclo[2,2,1]hexane–ether system by benzoic acid loss. In the case of the protonated *cis* isomer, water loss is the major pathway in contrast to the *trans* epimer from which this loss is hindered (Scheme 17.17). Indeed, protonation of the benzoate group, the latter having a relatively high PA value compared to the tertiary alcohol, is a favored process, but benzoic acid loss requires a long timescale. However, anchimeric assistance due to the opposite nucleophilic attack—that is, by the tertiary hydroxyl attacking the tertiary carbon holding the leaving group—strongly accelerates the benzoic acid elimination rate, resulting in stereospecific formation of the protonated ether $H_3C(HO^+)C_6H_9$ (Scheme 17.17).

Deprotonated molecules are typically able to produce direct internal nucleophilic substitutions and associated reactions involving skeletal transposition. Such is the case for the deprotonated monomethoxide of 2,3-dimethyl butane 2,3-diol.[72] By a 1,2-nucleophilic attack the negative charge of the alkoxide site yields the formation of an ion-neutral complex constituted of methoxide solvated by 2,3-dimethyl-2,3-oxobutane. The epoxide opening leads to methanol loss by proton transfer from a methyl group to the methoxide moiety. On the other hand, in this type of reaction involving the formation of ion–neutral complexes, a Meerwein-like process is promoted by alkoxide charge migration, which induces the neighboring C–C bond migration (Scheme 17.18a). It results in a ring contraction that assists the release of methoxide anion that is stabilized in the ion–neutral complex.[72] Methanol is then lost by internal deprotonation of the α-position enolizable aldehyde to give rise to formation of the enolic anion. A second example treats decompositions, analogous to the pinacol rearrangement, of the deprotonated 2-phenyl, 3-methylbutane 2,3-diol.[73] The negative charge induces competitive 1,2-methyl and phenyl migration yielding OH^-

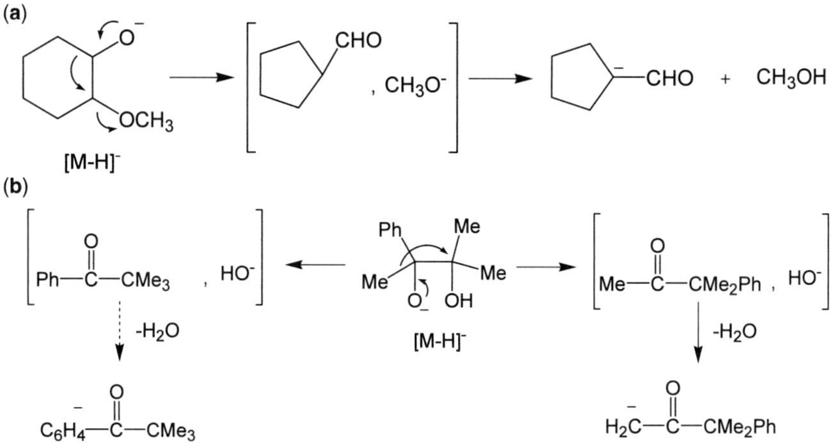

Scheme 17.18. Examples of (a) ring contraction by C–C bond migration and (b) competitive group migration from pinacol analogs through methanol and water losses, respectively. (Adapted with permission from Refs. 72 and 73.)

release, thereby forming an ion–neutral complex. This reagent is able to remove the proton α to the carbonyl group, yielding the enolate anion (Scheme 17.18b).

Transposition of substituents takes place from aromatic compounds based on *ortho* effects from hydroxyphenyl ketones[74] via assumed nucleophilic attack on the carbonyl carbon atom of the phenoxide site to give a tight tetravalent intermediate that promptly decomposes through benzyne neutral release and formation of a carboxylate anion (Scheme 17.19a). This reaction is hindered from *meta*- and *para*-substituted phenols. Alternatively, radical alkane loss is also observed that can be rationalized by considering the formation of an ion–neutral complex (Scheme 17.19b) comprised of quinone-like and alkylide groups. The relatively low ionization energy allows the generation of odd-electron quinone-like species and the elimination of the alkane radical (Scheme 17.19b).

Other nucleophilic substitution processes were described through anion–neutral complex formation—for instance, *para*-disubstituted aromatic compounds, such as

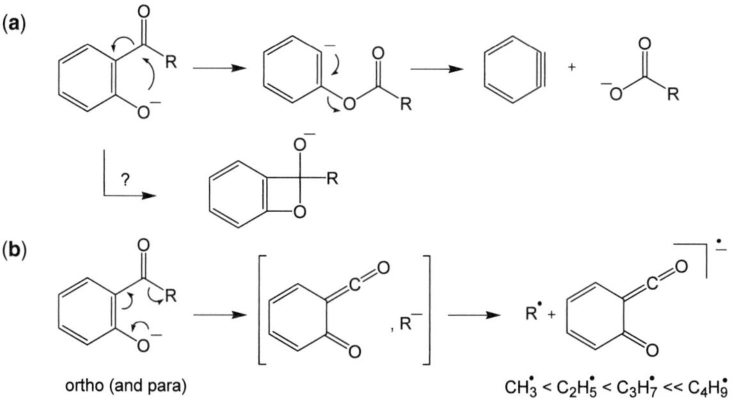

Scheme 17.19. Ortho effects from hydroxyphenyl ketones responsible for formation of (a) the carboxylate ion and (b) odd-electron homoquinone with radical alkyl release. (Adapted with permission from Ref. 74.)

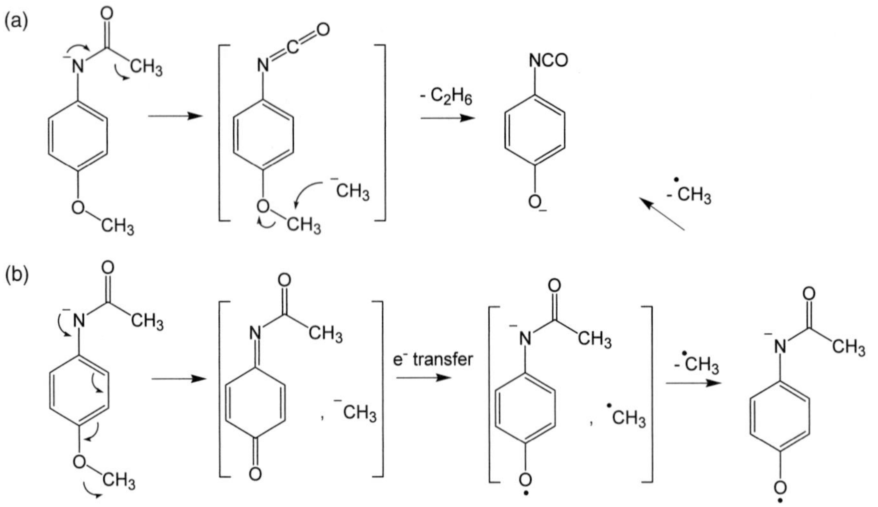

Scheme 17.20. Internal distant nucleophilic substitution from an ion–neutral complex prepared from *para* deprotonated methoxyacetanilide. (Adapted with permission from Ref. 75.)

methoxy acetanilide from which the negative charge at the nitrogen site induces C–C bond cleavage yielding methylide (CH_3^-) release with interaction from the neutral (Scheme 17.20).[75] Under high-energy collision, the vibrational mobility of the methylide reagent induces internal nucleophilic attack at the distant methoxy group leading to neutral ethane elimination and stable phenoxide formation (Scheme 17.20a). Furthermore, under these activation conditions, in competition with the previous process, radical CH_3• loss from the methoxy site leads to the major product ion being a distonic odd-electron anion able to yield a second CH_3• loss (Scheme 17.20b).

17.4 SMALL PEPTIDE *DE NOVO* SEQUENCING

Identification of proteins by tandem mass spectrometry is an important application of modern mass spectrometry. In the case of the analysis of polypeptides obtained from the proteolytic digestion of proteins from an organism of known genome, this identification work can be performed using database searching based on the predictability of the peptide mass fingerprint[76] and/or fragmentation pattern.[77,78] In the case of unsequenced genomes or when no specific enzyme is used, *de novo* sequencing is needed. Indeed the ability to perform reliable identifications based on existing algorithms is strongly reduced when no specific enzyme such as trypsin is used, and when the product ions are multiply charged.[79] In the same way, some peptides do not lead to extensive sequence information because of dominant specific cleavages, abundant noninformative side-chain losses, or internal fragmentations that limit the results obtainable by automated database searching. The following section aims to provide some basic knowledge needed for *de novo* sequencing. The discussion will involve mainly the behavior of protonated peptides under the widely used collisional and photoactivation techniques that involve excitation of vibrational modes—that is, CID and IRMPD, respectively.[80]

a. Nomenclature; Charge-Driven Cleavages: Conventional Cleavages Directed by the Charge; Model of the Mobile Proton; Small Releases Versus Backbone Cleavage. The mechanisms of peptide backbone cleavages have been widely studied and were the topic of several reviews.[81] Under low-energy CID, protonated peptides involve mainly amide bond cleavages leading to y_i and b_j ions according to the Roepstorff nomenclature[82] modified by Biemann[83] (Scheme 17.21). This behavior is initiated by proton migration to the amide bond nitrogen, which leads to weakening of the backbone. Such proton transfer involves, therefore, the existence of a "mobile" proton as proposed by Wysocki.[84,85] In this way, the carbon atom of the amide bond can undergo a nucleophilic attack to form a protonated five-membered ring oxazolone[86,87] and the subsequent release of the C-terminal moiety of the peptide as shown in Scheme 17.22. In the proposed model, the oxazolone and C-terminal peptide parts can remain attached by noncovalent association forming an ion–dipole complex[88] that undergoes a rearrangement leading to the formation of a proton-bound dimer. This proton-bound dimer can competitively lead to the formation of a y_i or b_j ion, depending on the relative proton affinity of each partner of the dimer.[89] This behavior is analogous to the competition in decomposition pathways that takes place during the dissociation of a proton-bound mixed dimer with the Cooks kinetic method.[90,91]

This now accepted mechanism, involving the formation a b_j oxazolone ion, was the result of numerous studies. Originally, it was considered that the b_j ions were acylium cations obtained through the heterolytic dissociation of a protonated amide bond. This was consistent with their ability to lose CO and lead to *a* ions. However, it is not possible from such a structure to explain why b_1 ions are generally not observed except when the N-terminal residue is acylated or contains a long side chain.[92]. Harrison and co-workers[93]

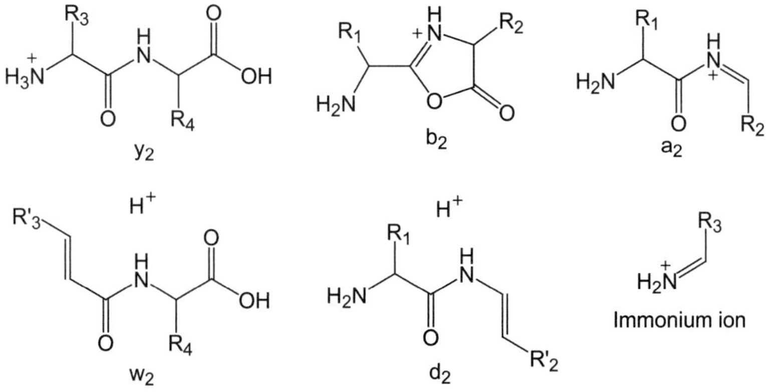

Scheme 17.21. Nomenclature for peptide fragmentations. (Adapted with permission from Ref. 82.)

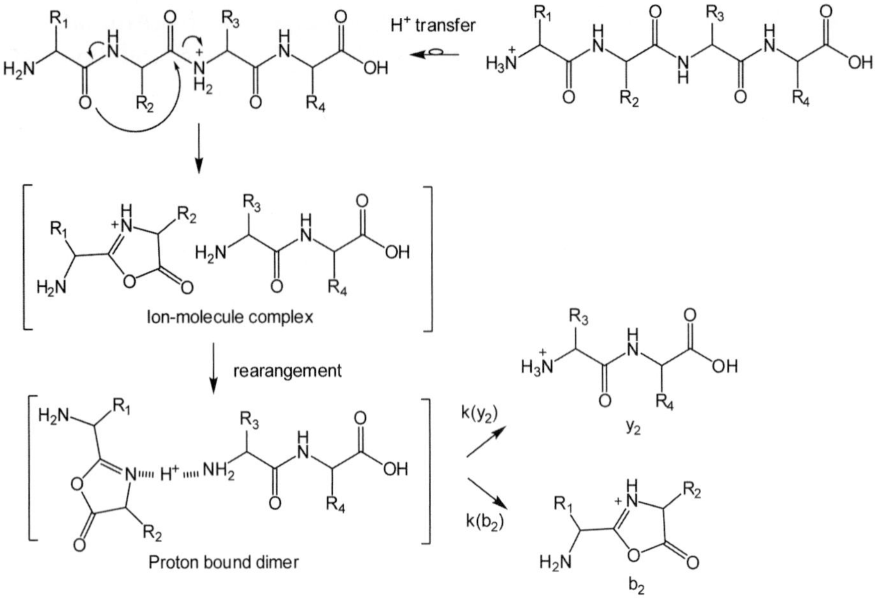

Scheme 17.22. General mechanism for the production of *b* and *y* ions.

were the first to consider the existence of this oxazolone form. This structure was further confirmed from experimental and theoretical results.

b. Structures of Protonated Peptides: Single-Charged Site Species, Zwitter-ionic and Salt-bridge; Role of the Polarizability and Solvated Charge by Peptide Folding. Under low-energy CID, fragmentations are generally considered to be charge directed. The localization of the charge is therefore an important point in the understanding of gas-phase behavior of protonated peptides.[94] Several amino acid proton affinity scales have been published by a few groups.[95,96] The evolution of the (gas-phase) proton affinity is R > K > H > W > E > P > Q > M > Y > N > F > T > I > L > V > D > S > A > C > G. Therefore, the most basic sites in peptides are (a) the side chains of Arg and Lys and (b) the N-terminal, which explains why *y* ions are generally favored from peptides originating from tryptic digestion.[97]

The ability of a proton to induce a backbone cleavage is also related to its mobility as proposed by Wysocki.[84,85] The "mobile proton" model coincides with the notion that under low-energy collision conditions the main cleavages are charge-directed. Peptides are relatively complex species that present numerous protonation sites. After ion activation, a proton can migrate endothermically to less stable protonation sites and thereby induce competitive fragmentations (Scheme 17.22). Therefore, vibrationally excited protonated peptides can exist in various forms that can lead to different cleavages. A given decomposition pathway is accessible only if the process is not overly endothermic. If the peptide contains an arginine (highly basic) residue, as is often the case with tryptic peptides, then movement of this proton will be less energetically favorable. The proton is thus considered as sequestered, and it is not directly involved in the fragmentation process. In this case, charge remote fragmentation can be produced. The "mobile proton" model allows one to understand why the amide bond nitrogen,

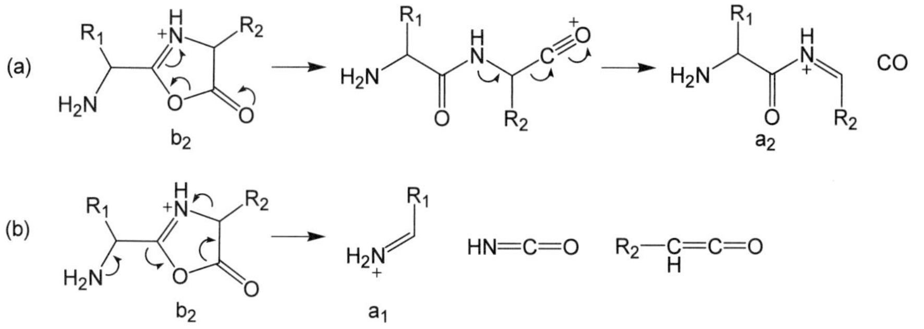

Scheme 17.23. Pathways for the formation of; (a) a_n and (b) a_{n-1} ions consecutively from b_n. (Adapted with permission from Ref. 105.)

which is not a thermodynamically favored protonation site, can be protonated after ion activation and induce amide bond weakening and cleavage. It also allows rationalization of the charge state dependence of peptide fragmentations,[98] the sequence dependence of the gas-phase stabilities of protonated peptides,[84] and the specific decompositions of certain peptides containing acidic amino acids[99] (*vide infra*). It should be pointed out that the "mobile proton" model is now validated by the use of H/D exchange experiments[100] as well as by using theoretical calculations.[101,102] It was observed, for instance, that the position of a deuterium is randomized among the exchangeable sites after ion activation.[103,104]

The production of *a* ions derived from consecutive decompositions of *b* ions can occur by two main pathways as shown by Harrison and co-workers[105] (Scheme 17.23). Other pathways from the protonated peptide leading directly to a_i/y_j ions have also been proposed.[106]

c. Orientation of Positive Precursor Fragmentations Induced by Basic and Acidic Residues; Linear Relationship Between Log y/b and Basicity. The production of y_i/b_j ions is not completely random along the peptide sequence. Substantial differences in product ion abundances depend on various parameters such as charge state and amino acid composition. In some cases, particular cleavages may be missing, which limits peptide sequence determination. It is well known that the presence of certain amino acids, such as proline or aspartic acid, will lead to specific cleavages with the so-called proline[107,108] and aspartic acid[109] effects. In particular, it has been shown that decomposition at the N-terminal side chain of proline is favored (Scheme 17.24). Proline is a cyclic amino acid, and its N-terminal part is a ternary amine with a significantly enhanced proton affinity. In addition, because of the presence of the ring, steric hindrance severely limits cleavage on the C-terminal part of the amino acid.[85] This point is consistent with the fact that the b_j ion is actually a protonated oxazolone.[110]

The cases of the acidic amino acids (i.e., aspartic acid and glutamic acid) are particularly interesting, and their examination also provides strong evidence in support of the "mobile proton" model. In practice, it appears that in the absence of a "mobile proton," fragmentation at the C-terminal side of the acidic residue is strongly favored.[109,111–113] Therefore, this selective cleavage occurs in the case of peptides that present an Asp or Glu residue and that have a number of added protons that is lower or equal to the number of arginine residues in the amino acid sequence. In some cases, only one product ion may produced,[114] especially when the acidic residue effect is coupled with the

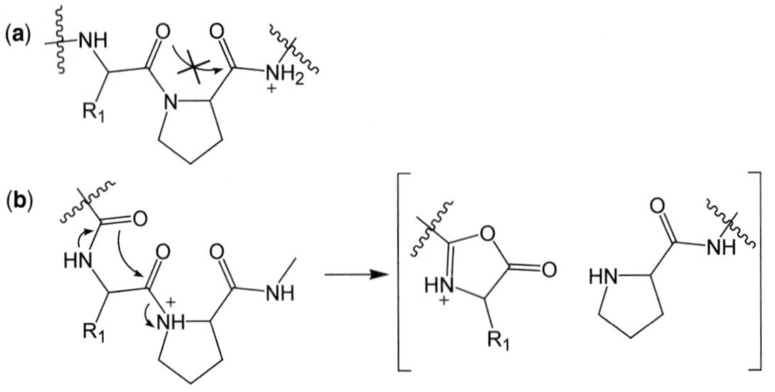

Scheme 17.24. (**a**) Steric hindrance due to proline ring and (**b**) role of the basic proline ternary amine that favors N-terminal cleavage. (Adapted with permission from Ref. 85.)

proline effect.[115,116] This acidic residue effect was further studied both experimentally and by theoretical calculations.[116,117]

Various mechanisms have been proposed to explain the role of the acidic residue in the selective C-terminal side (Scheme 17.25). The original mechanism prosposed by Yu et al.[115] suggested the existence of a salt-bridge intermediate (Scheme 17.25a). Others consider that this salt-bridge structure can be stabilized by intermolecular interactions with the protonated arginine.[118] But theoretical calculations seem to confirm that the concerted mechanism proposed in Scheme 17.25b is the most favorable energetically.[116,117] This fragmentation does not appear to be induced by the charge (i.e., it is charge remote), which explains why this process is generally observed in the case of peptides with no mobile protons.

A selective cleavage on the C-terminal side of histidine has also been demonstrated.[119] This behavior occurs when a mobile proton is available.[85] In this case, the transfer of one proton to the histidine side chain induces the fragmentation as shown in Scheme 17.26. Generally, the production of *b* ions is favored because of the basic character of the histidine side chain.

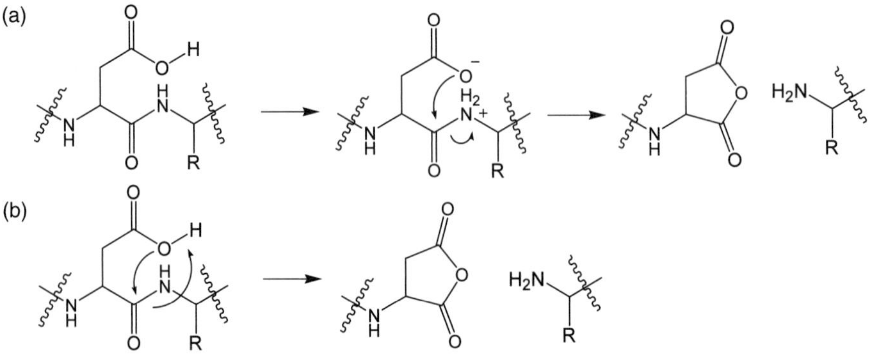

Scheme 17.25. Mechanisms for the Asp C-terminal cleavage with no mobile proton involving (**a**) an intermediate salt bridge (adapted with permission from Ref. 110) and (**b**) a concerted process (adapted with permission from Refs. 116 and 117.)

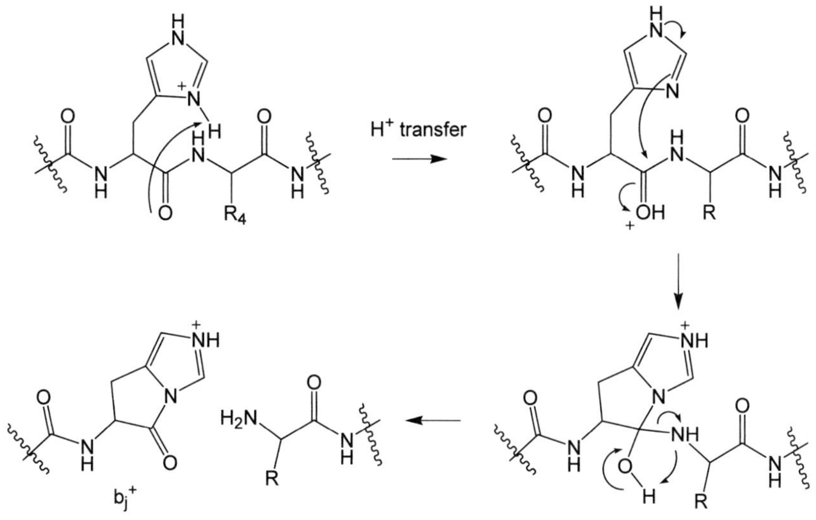

Scheme 17.26. Enhanced cleavage at C-terminal of histidine. (Adapted with permission from Ref. 85.)

d. Charge Remote Processes for Distinguishing the L/I Side Chains and Other Isomeric Residues; α/β Peptide Linkage Distinction. Amino acids are characterized on the peptide sequence by the mass difference between ions of the *y* or *b* series. Obviously, it is not possible to differentiate the isomeric L and I amino acids in this manner. However, it has been shown that under high-energy collision, side-chain cleavages of the w_n type can be produced, thereby allowing such differentiation. High-energy collision conditions (i.e., in the kilo-electron-volt range) are obtained in magnetic sector instruments[120,121] and also in TOF/TOF instruments that are becoming very popular.[122] The w_n ion mentioned above is obtained through the consecutive dissociation of the $z + 1$ as shown in Scheme 17.27, and this pathway leads to a loss of propyl or

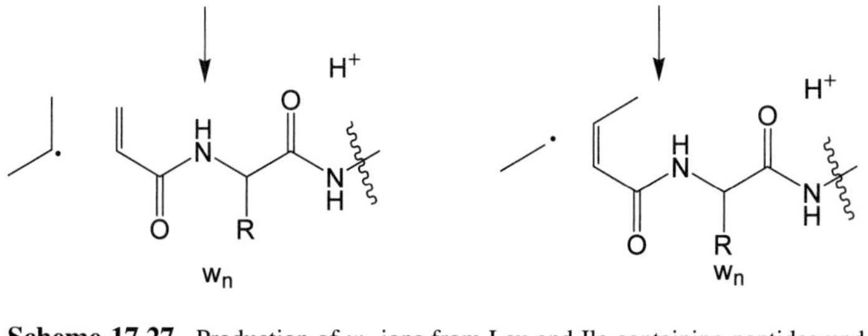

Scheme 17.27. Production of w_n ions from Leu and Ile containing peptides under high-energy collision conditions. (Adapted with permission from Ref. 120.)

ethyl radical for Leu or Ile, respectively. It should be noted that such w_n ions are observed mainly from peptides containing a basic residue on the C-terminus. Under low-energy collision conditions, such differentiation cannot be directly performed in this way, but it was shown that differentiation is possible through the consecutive decomposition of the Leu/Ile immonium ion at *m/z* 86. For instance, Hulst et al.[123] used in-source CID to produce the *m/z* 86 immonium that is activated in the collision cell of a triple quadrupole instrument. In the same way, Armirotti et al.[124] used MS^n experiments carried out on an ion trap instrument. In both cases, decomposition of the immonium ion of isoleucine at *m/z* 86 leads to a diagnostic fragment ion at *m/z* 69 by loss of ammonia that is produced to a lower extent by leucine.

Others used the capability of peptide–copper[125] complexes to produce radical species after ion activation for Leu/Ile differentiation. Wee et al.,[126] inspired by the work of Siu and co-workers,[127] used $[[Cu(II)(L_3)M]^{2+}$ complexes (where L_3 is a tridendate ligand) to produce $M^{+\bullet}$ ions leading consecutively to even-electron *y* ions together with side-chain radical losses of $\bullet CH_3$ and $CH_3CH_2\bullet$ for isoleucine and $(CH_3)CH_2^{\bullet}$ for leucine. In the same way, Tabet's group used $[M - 3H + Cu(II)]^-$ binary complexes to obtain similar behavior.[128,129] This was possible by the reduction of Cu^{II} into Cu^{I} after collisional activation that generates a radical, thereby inducing C–C bond cleavage. Such copper complexes are interesting because they allowed demonstration of the existence of zwitterionic forms of MCu^{II} ions[130] that led (under low-energy CID) to both positive (e.g., NH_3 loss) and negative charge-driven (e.g., tyrosine side chain) fragmentations.[131–133]

e. Decomposition of Multiply Charged Peptides Under CID Conditions. When a peptide has only one added proton, each cleavage will lead only to one product ion, generally of the *y* or *b* type, if the charge is retained on the N- or C-terminal fragment, respectively. As discussed previously, this will depend on the relative proton affinities of the cleaved moieties; the more basic moieties will retain the charge. In the case of multiply charged peptides, the situation is different. With two added protons on the precursor ion, doubly charged product ions can lead to the release of a neutral, but the major product ions are singly charged. In this latter case, complementary *y* and *b* ions are produced in a single cleavage. In addition, for tryptic peptides that contain an arginine residue in their C-terminus, the presence of two protons (or more) generally allows one to achieve a better sequence coverage than is obtainable from the singly charged species. Indeed, in this case, according to the "mobile proton" model, one proton is sequestered by the arginine residue, whereas the additional protons are available to activate amide bond cleavages. This explains why, during the analysis of tryptic digests, doubly charged peptides are generally preferred to obtain the maximum sequence coverage and easier protein identification using MS/MS in conjunction with database searching. Statistical studies on the influence of the charge state have been carried out by different groups.[134,135] Based on the "mobile proton" model, Zhang[136] proposed a kinetic model to predict low-energy CID spectra of multiprotonated peptides.

Another factor that can be considered is the folding of protonated peptides to enhance local charge self-solvation.[137–140] A nonmobile proton localized on an arginine side chain may assist cleavages by polarization of an amide bond as shown in Scheme 17.28. Such folding will also depend on the charge state. From such a conformation, even peptides with nonmobile protons can lead to fragmentation that can be considered as charge assisted rather than charge remote. Peptides with higher numbers of protons will be less folded because of coulombic repulsion.[141]

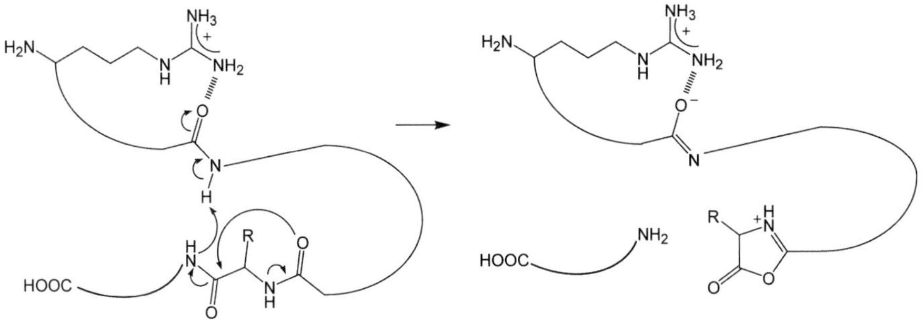

Scheme 17.28. Peptide folding can explain the role of the nonmobile proton in peptide backbone cleavage.

17.5 OLIGOSACCHARIDES

Carbohydrates (or saccharides) are highly abundant biological compounds composed of covalently linked monosaccharides that carry as a hallmark a tremendous diversity in linkage types including linkage position and stereochemical (anomeric) configuration. Due to this tremendous inherent diversity, the analytical characterization of saccharides is quite challenging and problematic. A general, systematic nomenclature to classify mass spectrometric cleavages of oligosaccharides was introduced by Domon and Costello[142] by analogy to that of peptide cleavage nomenclature (Scheme 17.29). According to this widely accepted nomenclature, the universe of cleavages that may occur at the glycosidic bond carries the designation of B, C, Y, and Z cleavages, whereas cross-ring cleavages are designated as A or X. The A, B, and C labels are differentiated from the X, Y, and Z labels in that the former all correspond to fragments that have lost the reducing end as a neutral, whereas the latter all correspond to fragmentations with charge retention on the reducing end. Double fragmentations can also occur, and the C/Z combination has been dubbed a "D-type" fragmentation.[143,144] The favored pathways for ion decompositions are heavily influenced by the nature of the precursor ion—that is, protonated or cationized molecule, or cationic or anionic adduct species. In addition, the particular structural features of a given

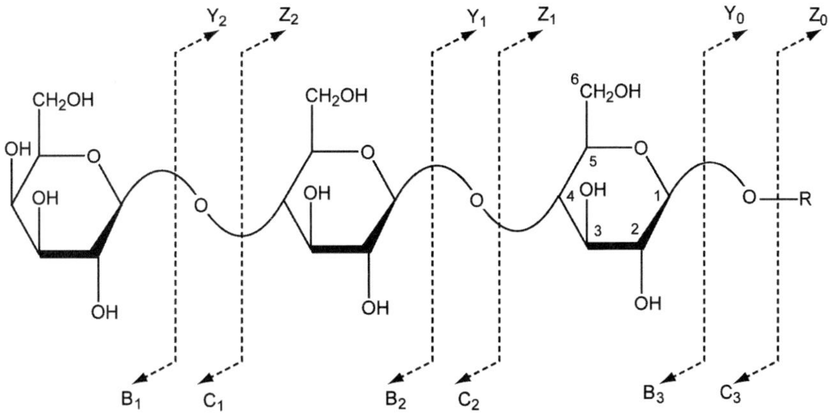

Scheme 17.29. Nomenclature for saccharide fragmentations. (Adapted with permission from Ref. 142.)

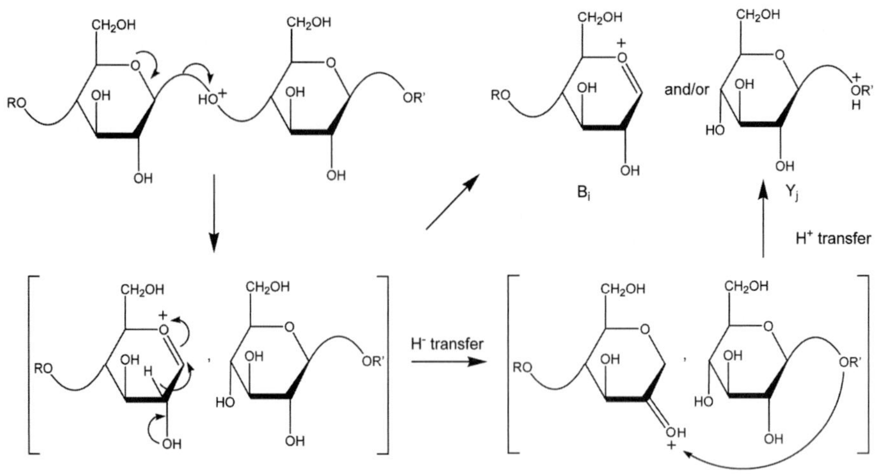

Scheme 17.30. Formation of B_i/Y_i product ion via ion–dipole complex formation.

saccharide, such as the presence of sialic acid and/or sites of sulfation, can significantly influence the preferred routes of decomposition.

17.5.1 Glycosidic Bond Cleavages

From protonated precursors, regardless of the method used for ionization, because of the ease with which a stabilized oxonium ion may be produced, an ionizing proton that is initially situated directly on a glycosidic oxygen has a high tendency to lead to a B-type cleavage[142,145] (Scheme 17.30). Alternatively, in order to explain the Y ions that are commonly observed in decompositions from protonated oligosaccharides, cleavage promoted by hydride transfer(s) can be implicated, including one from the nonreducing end departing neutral to the reducing end fragment ion, thereby creating a new site of unsaturation on the lost neutral (Scheme 17.30).

Likely, B/Y (or C/Z) cleavages involve the formation of ion–dipole complexes. This allows one to explain the competitive observation of both of the complementary product ions, as presented in Scheme 17.30. Direct cleavage of the ion–dipole complex yields the B_i ion, whereas a 1,2-hydride transfer followed by proton transfer yields the Y_j ion.

Note that the C_i/Z_j complementary ions can be produced through the formation of another ion–dipole complex involving C_4–O bond cleavage (see Scheme 17.29). This cleavage is assisted by H^- transfer from the C_3 to C_4 position yielding the protonated ketone (Scheme 17.31). The direct dissociation of this ion–dipole complex yields the Z_j product ions, whereas the C_i ion is produced after an intra-molecular proton transfer. The C_4–O bond cleavage can lead to sugar ring opening (*vide infra*).

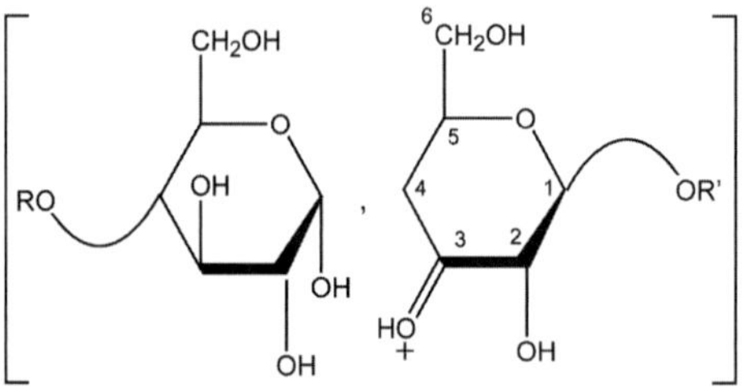

Scheme 17.31. Ion dipole complex resulting from intra-molecular proton transfer.

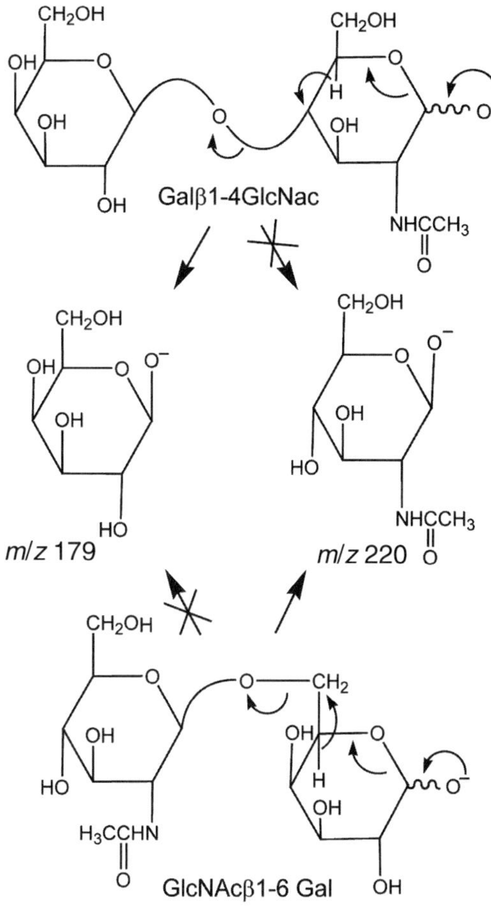

Scheme 17.32. Dominant C_1 decomposition pathways of isomeric Galβ1-4GlcNac and GlcNacβ1-6Gal. The Y_1 cleavages are not observed. (Adapted with permission from Ref. 146.)

An interesting and instructional comparison of the decomposition behavior of two isomeric disaccharides in the negative-ion mode was performed by Garozzo et al.[146] We use this example to shed light on some general tendencies for glycosidic cleavages from oligosaccharides. A key feature of these two sugars is that, although they are positional isomers, the mass of the respective reducing end units is different, which can simplify interpretation of the product ion spectra. The Galβ1-4GlcNac disaccharide yields a dominant peak at *m/z* 179, thus unequivocally indicating C_1-type cleavage with charge retention on the glycosidic oxygen (Scheme 17.32). This occurs by ring opening followed by 1–2 hydride transfer that allows C_4–O reduction and formation of $[Gal - H]^-$. Analogously, the isomeric GlcNacβ1-6Gal yields *m/z* 220 as a large product ion, again confirming a C_1–type cleavage with the glycosidic oxygen bearing the charge. These C_1 cleavages are the dominant routes of glycosidic bond cleavage for the two disaccharides independent of the N-acetyl group location. In the original paper,[146] the authors cite the appearance of the two ions at *m/z* 179 and 220 as evidence that, in each case, initial deprotonation occurred at the hydroxyl located at the reducing end anomeric carbon. Although no indication of mechanism was given in the original paper, we assign logical pathways for each in

Scheme 17.32. This example is particularly informative because in many disaccharides, the adjacent sugars are isomeric; when that is the case, C_1 and Y_1 cleavages yield ions of the same mass. Here, the difference in masses of the reducing ends gives a definitive indication that the dominant cleavage corresponds to loss of the reducing end monosaccharide as a neutral. This can be rationalized as occurring from precursor $[M - H]^-$ molecules that are initially deprotonated at the anomeric hydroxy group (the most acidic of the various OH groups). The fact that the reducing end is readily cleaved and lost also results in the fortuitous ability to sequentially decompose larger oligosaccharides by successive losses of the monosaccharide unit at the reducing end in multiple stages of negative ion MS/MS. Sequencing of small oligosaccharides has been achieved using this strategy.[147]

Anion attachment has been developed as an alternative negative ion general strategy for the simultaneous analysis of neutral and acidic oligosaccharides using both electrospray[148,149] and MALDI.[150,151] The use of chloride anions has been particularly successful for anion attachment to oligosaccharides due largely to the close gas-phase acidities of glucose and HCl. Moreover, in CID experiments, it has been demonstrated that the first step in decompositions of most chloride adducts of oligosaccharides, $[M + Cl]^-$, is loss of HCl to form the deprotonated molecule, $[M - H]^-$, and this occurs in competition with Cl^- regeneration. Chlorine-containing product ions were reported to appear only during decompositions of nonreducing disaccharides such as those that contain a terminal sucrose at the "downstream" end wherein the linkage is formed between two reducing hydroxyl groups on the glucose and the fructose units.[148,151]

17.5.2 Cross-Ring Cleavages

Besides the glycosidic bond cleavages (B, C, Y, and Z ions), cross-ring cleavages (i.e., $^{0,2}A_n$, $^{2,4}A_n$ and $^{2,5}A_n$) are particularly useful for determination of linkage and branching patterns (Scheme 17.33). In general, cross-ring cleavages occur competitively with glycosidic bond cleavages. The relative abundances of the product ions are related to the nature of the precursor ion and to the conditions of ion activation. For instance, glycosidic bond cleavages tend to dominate when the precursor ion is a large protonated oligosaccharide; thus, less detailed information (e.g., concerning branching structure) is obtained.[152] From deprotonated species, it has been shown that native (i.e., underivatized) oligosaccharides favorably yield glycosidic cleavages as well as A-type cross-ring cleavages,[153,154] whereas

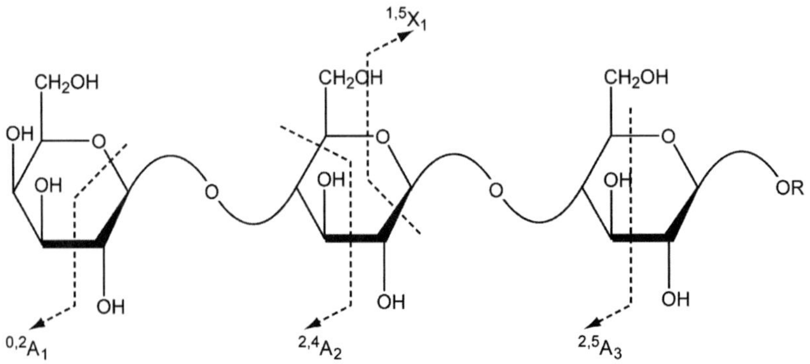

$^{1,5}X_1$

CH2OH CH2OH CH2OH
OH O O O
OH O OH O OH OR
OH OH OH

$^{0,2}A_1$ $^{2,4}A_2$ $^{2,5}A_3$

Scheme 17.33. Nomenclature for cross-ring saccharide fragmentations. (Adapted with permission from Ref. 142.)

permethylated species yield more abundant cross-ring cleavages. Metal ions present different properties allowing some control of fragmentation orientation. For instance, the binding energies of alkali metal ions decrease in the order $Li^+ > Na^+ > K^+ > Rb^+ > Cs^+$.[155] Small metal ions such as Li^+ appear to favor charge-induced processes and therefore the rupture of glycosidic bonds as it reduces the dissociation threshold. In contrast, K^+, Rb^+, and Cs^+ yield mainly the loss of the metal ion.[156,157] When cross-ring cleavages occur, fragmentation does not appear to be highly dependent upon the alkali metal ion, which is consistent with the consideration that cross-ring cleavages are charge remote. However, in the presence of a cation, salt formation may promote proton migration. In this case, charge-driven cross-ring cleavage can occur, but at a rather long distance from the metal ion.

In general, high-energy collision conditions (i.e., kilo-electron-volt range) tend to favor cross-ring cleavages.[158] For instance, under such activation conditions, only $^{2,4}A_n$ ion production is observed from the maltoheptaose $[M-H]^-$ ion.[159] Moreover, Lemoine et al. investigated the role of activation conditions on alkali metal cationized and permethylated oligosaccharides.[160] It was shown that from $[M+Na]^+$ ions under high-energy collision conditions, $^{1,5}X_n$ fragments dominate the spectra. By contrast, activation using slow-heating methods as occurs in a quadrupole ion trap tends to favor glycosidic bond cleavages. Similarly, the likelihood of cross-ring cleavages increases if the number of labile glycosidic bonds is diminished (i.e., when HexNAc residues are present). Scheme 17.34 presents proposed mechanisms for the production of $^{0,2}A$, $^{2,4}A$, and $^{2,5}A$ cross-ring cleavages from deprotonated precursor ions.

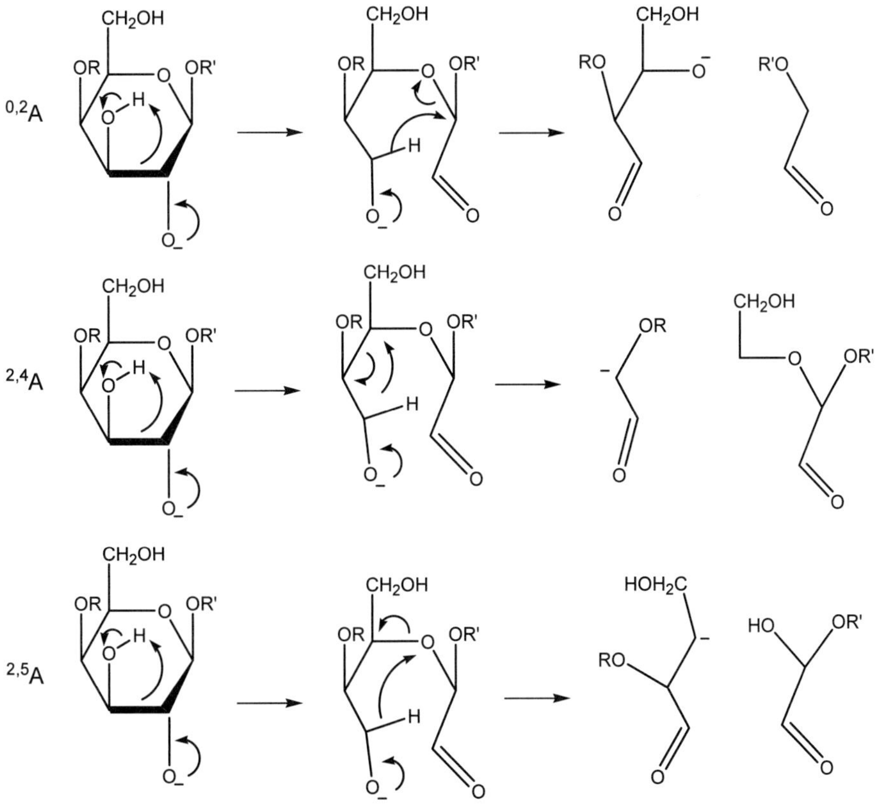

$^{0,2}A$

$^{2,4}A$

$^{2,5}A$

Scheme 17.34. Proposed mechanisms for cross-ring cleavages of deprotonated oligosaccharides.

17.6 LIPIDS

The synthesis and functioning of lipids involves the cooperative activity of enzymes, transporting proteins and receptor sites in biological organisms. Historically, the working definition of the term "lipids" referred to the hydrophobic components of a partitioning of a biological sample between water and an immiscible organic solvent. To add rigor to the definition, an attempt has been made recently to classify lipids into eight categories[161]: fatty acyls (FA), glycerolipids (GL), glycerophospholipids (GP), sphingolipids (SP), sterol lipids (ST), prenol lipids (PR), saccharolipids (SL), and polyketides (PK). Owing to the diversity of these compound types, a comprehensive treatment of lipid decomposition behavior in mass spectrometry is beyond the scope of this chapter; nonetheless, a few examples will be developed.

The first example is that of a 1,2-diglyceride (a glycerolipid) that is comprised of two fatty acyl chains that are of different chain length (1-stearin-2-palmitin). From the lithium adduct generated by electrospray, a competition exists between the losses of the two fatty acyl chains.[162] The predominant product ion peaks correspond to losses of these side chains as lithium fatty acetates (Scheme 17.35). In this case, the loss of lithium acetate is assisted by the nucleophilic carbonyl oxygen, which attacks the electrophilic methylene site. The 1-stearin chain is lost more readily than the 2-palmitin chain. Minor pathways leading to losses of the two corresponding fatty ketenes are also observed (Scheme 17.35); these latter ions, in turn, may lose LiOH. The final products shown in Scheme 17.35 are charge-stabilized by five-membered ring formation.[162]

Among the glycerophospholipids, phosphatidylcholine is a type of headgroup that contains a single nitrogen atom (ammonium group) and that, in the CID spectrum, can produce even-electron product ions containing one nitrogen atom that, according to the nitrogen rule, will appear at even *m*/*z* values. A pedagogic example is the MALDI post-source decay spectrum of protonated lysophosphatidylcholine (*m*/*z* 496, Scheme 17.36).[163] The ionizing proton may attach to a variety of oxygen atoms and, thus, lead to four main even-electron product ion structures, two of which retain the single nitrogen atom and thus yield even *m*/*z* value ions; *m*/*z* 184 is the base peak in the spectrum and is diagnostic of the

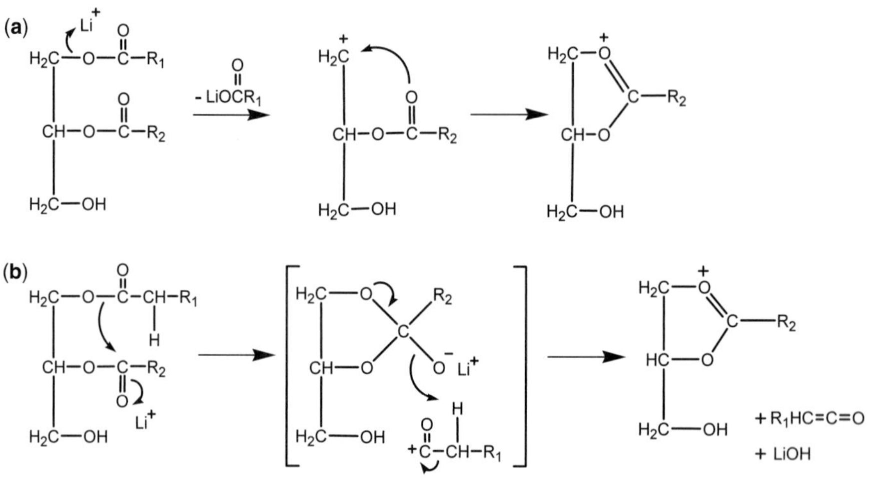

(a) (b) $- LiOCR_1$ $+ R_1HC{=}C{=}O$ $+ LiOH$

Scheme 17.35. Decomposition mechanisms for a lithium adduct of a 1,2-diglyceride. (Adapted with permission from Ref. 162.) (**a**) Loss of lithium carboxylate; (**b**) loss of long-chain ketene.

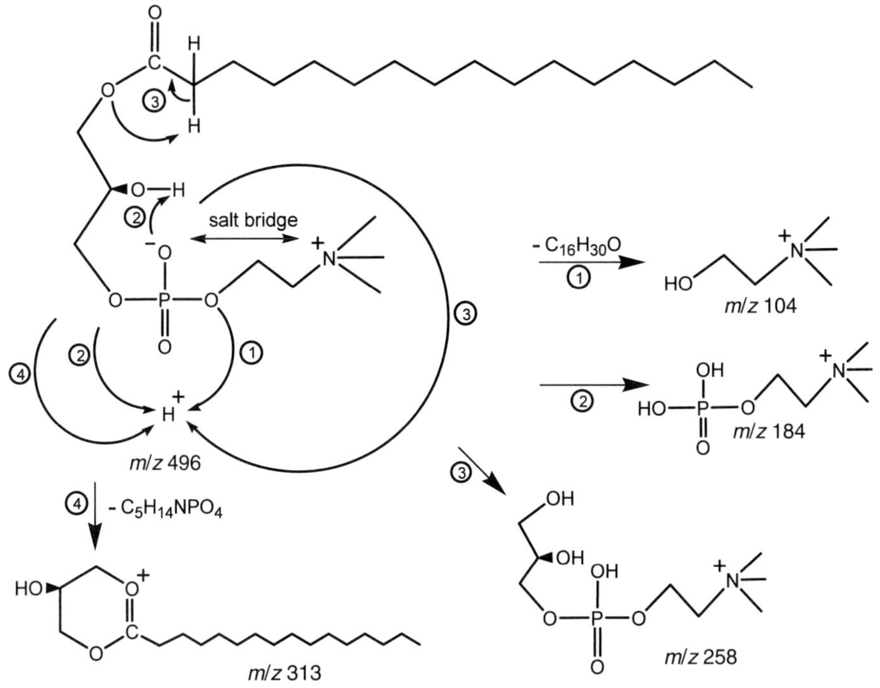

Scheme 17.36. (**Top**) Product ion mass spectrum of protontated 16:0 lysophophatidylcholine precursor. (Adapted with permission from Ref. 163.) (**Bottom**) Proposed mechanisms of formation of the four major product ions.

phosphatidylcholine headgroup (Figure 17.1 and Scheme 17.36). This fragmentation behavior involves a spectator quarternary amine and, by considering the likely existence of a salt bridge, the additional mobile proton can induce the observed decompositions.

In contrast to the previous example, fatty acyl lipid chains may undergo cleavages in a complex set of unimolecular reactions that are physically remote from the charge site. This class of reactions, which can apply to nonlipids as well, is referred to as "charge-remote" fragmentation.[164,165] When a reaction is considered to be charge-remote, it can be difficult to assess (a) whether the reaction was initiated by a homolytic bond cleavage to produce a radical site or (b) whether the reaction proceeded exclusively through even-electron

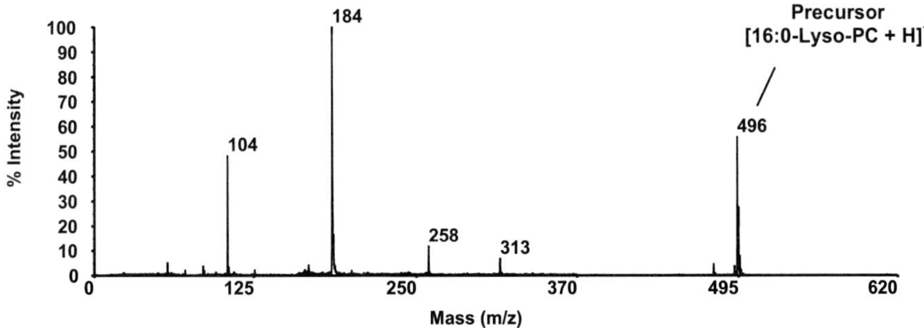

Figure 17.1. MALDI post-source decay spectrum of protonated lyso-phosphatidylcholine yielding m/z 184 diagnostic ion. (Adapted with permission from Ref. 163.)

reactants. This holds true whether one is considering anionic or cationic species. Originally, when considering the mechanism of carbon–carbon bond cleavage in long-chain hydrocarbons, Gross and co-workers[166] proposed a 1,4 H_2 elimination mechanism whereby terminal double bonds are formed on each side of the cleavage site on a hydrocarbon chain, and a proton and a hydride combine to give H_2 loss (Scheme 17.37A). A few years later, Wysocki and Ross[167] not only proposed radical cleavages to account for radical product ions, but also implicated them in the production of saturated and unsaturated even-electron product ions that were rationalized as being formed via an intermediate ion–molecule complex (Scheme 17.37B). The possibilities for radical involvement in hydrocarbon chain

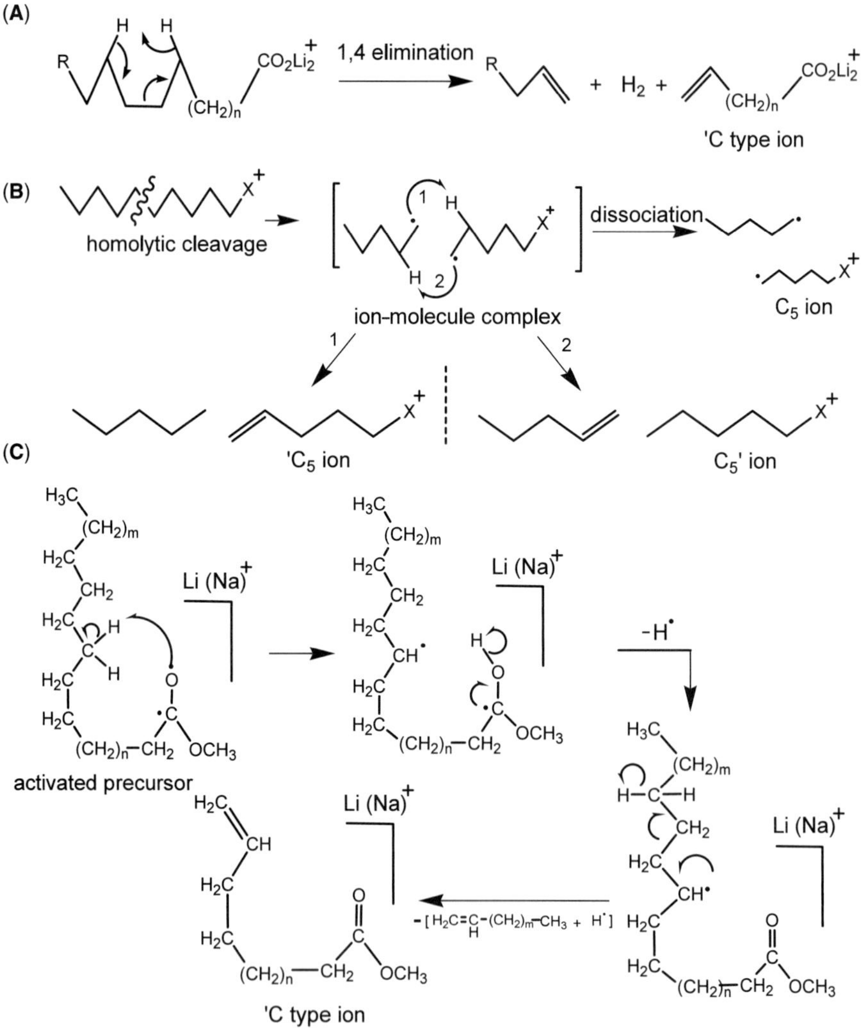

Scheme 17.37. Proposed mechanisms for decompositions of long-chain hydrocarbons and formation of C-type ions: (**A**) 1,4-Elimination involving even-electron hydrogen transfers. (Adapted with permission from Ref. 5.) (**B**) Homolytic cleavage and ion–molecule complex formation leading to two possibilities for transfer of hydrogen radical. (Adapted with permission from Ref. 167.) (**C**) Initial diradical formation upon high-energy CID of alkali-cationized fatty acid ester precursor leading to hydrogen radical departures. (Adapted with permission from Ref. 168.)

cleavages were elaborated upon further by Claeys et al.,[168] who proposed that a diradical carbonyl (formed upon high-energy CID) was interacting with the hydrocarbon chain. After an initial stepwise hydrogen radical abstraction by the second radical to yield a nonradical ester site, two double bonds were ultimately formed, each at the expense of a hydrogen radical departure (Scheme 17.37C). In general, it seems fair to state that radical intermediates can be rationalized as becoming more prominent in such cleavages under high-energy (kilo-electron volt) collisions as opposed to the low-energy (10–100 eV) regime.

According to the systematic nomenclature proposed by Griffiths et al.[169] for lipid decompositions, cleavages of carbon–carbon bonds such as those shown in Scheme 17.37 with charge retention on the carboxyl containing portion are designated by C_n, where the subscript n indicates the number of carbon atoms in the fragment ion. If charge retention is with the methyl terminus fragment, the ion is described by F_n. Placement of the prime symbol ($'$) to the left of the letter signifies that this product ion is deficient in one hydrogen atom relative to a fragment ion formed by homolytic cleavage at the same cleavage site of a hypothetical precursor molecule, $M^{-\bullet}$. Multiple hydrogen deficiencies are denoted by multiple prime symbols to the left of the letter. In the event of hydrogen surpluses in the product ion, the appropriate number of prime symbols is placed to the right of the letter and subscript. Because $(M-H)^-$ is deficient one hydrogen relative to $M^{-\bullet}$, it is designated as $'M$ and those even-electron C fragment ions that undergo double bond formation at the cleavage site are designated $''C_n$ fragments according to the original nomenclature of Griffiths et al.[169] But in order to make the nomenclature applicable to a wider variety of precursors (e.g., $[M+Li]^+$ or $[M-H+2Li]^+$), Claeys et al.[170] proposed a slight modification wherein only one prime was used instead of two to indicate that an ion is deficient in one hydrogen compared to the hypothetical product ion that would be formed by homolytic cleavage of any type of precursor ion at the same bond; thus the $''C_n$ fragments defined by Griffiths et al. became $'C_n$ fragments according to Claeys et al.[170]

Formally, fragmentations are considered to be charge-remote when there is no significant interaction between the charge and the reaction sites. In the pioneering work from Gross's group on the mechanisms of charge-remote fragmentation, it was recognized early on that a complete absence of interaction may not be possible to achieve due to the flexibility of hydrocarbon chains.[164] More recently, Harvey[171] has developed a perspective on how the charge may have an influence on the preferred sites of cleavage for fatty acyl chains even when the charge is located on a functional group many atoms away. Harvey accounts for some product ions by postulating an initial hydride (H^-) transfer from a –CH_2– group to the charge site as being responsible for initiating positive-ion-mode decompositions (Scheme 17.38). Conversely, in the negative-ion mode, a proton (H^+) transfer to the charge site initiates decompositions. As a result of this type of transfer, the charge site is moved to the center of the aliphatic chain. However, because charge retention in the product ions occurs preferentially on the side of the molecule that initially held the charge (leading to m/z 211 in the Scheme 17.38 example), and because the product ion at m/z 153 that would have resulted from the analogous cleavage with charge retention on the other side of the alkyl chain charge site was not observed, it was postulated[171] that (a) the charge remains partially associated with the initial (carboxy) site at the moments leading up to alkyl chain cleavage and (b) the losses leading to m/z 211 are concerted processes. Although not explicitly stated in the article, like the radical mechanism of Claeys et al.[170] (Scheme 17.37c), this charge-associated mechanism must be subject to conformational constraints; that is, m/z 211 production requires a close proximity of the departing hydride to the charge site. Moreover, potential hydride departures from neighboring carbon atoms on the fatty acyl chain will be competing transfers to the charge site that require somewhat

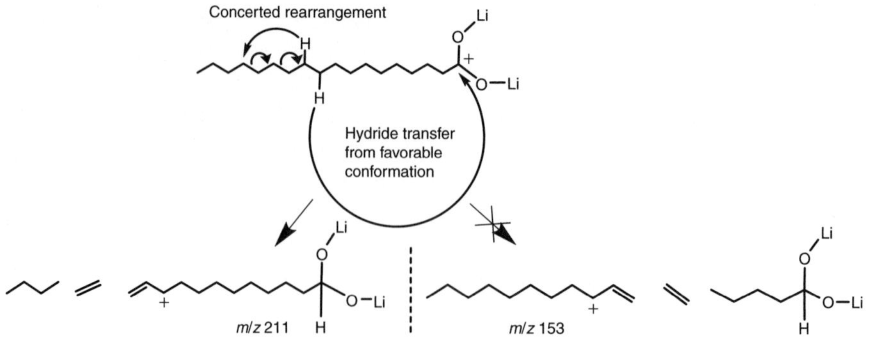

Scheme 17.38. Proposed mechanism for the formation of $[M - C_nH_{2n+2}]^+$ ions in high-energy positive ion CID of $[M - H + 2Li]^+$ (M = octadecanoic acid) precursor. Hydride transfer to the charge site occurs in concert with a second hydride transfer to yield the *m/z* 211 product ion plus ethylene and butane neutral fragments. The second pathway to the right was not observed. (Adapted with permission from Ref. 171.)

different conformations. A nonrandom distribution of product ions originating from hydride transfers (analogous to that shown in Scheme 17.38 except that the hydrogen atoms were initially held at different carbon atoms on the unsaturated hydrocarbon chain) will thus be observed in the high-energy CID mass spectrum.

Another class of compounds that is cited as being susceptible to charge-remote fragmentations is the sterol lipids and their closely related metabolites. A specific example is that of the bile acids that are formed upon oxidation of cholesterol. Bile acids typically consist of a steroid backbone that may exist as a free acid or may be conjugated with, for example, taurine or glycine (Scheme 17.39). These conjugates contain a single free acid group (e.g., sulfonate or carboxylate) that is deprotonated in negative-ion-mode mass spectrometry. Griffiths et al.[172] have developed an extension of the above-mentioned systematic nomenclature used for lipids to account for all charge-remote fragmentations of bile acids (Scheme 17.39A). The rings are labeled a, b, c, and d in moving from left to right in the Scheme, and the corresponding capitalized letter is used to indicate which ring is opened and broken in a cross-ring cleavage. The subscripts 1, 2, and 3 indicate the cross-ring cleavage sites on that ring. Again, the "prime" symbol (') is used to keep track of hydrogen movement during decompositions. As above, the placement of the prime to the left of the letter signifies that this product ion is deficient in one hydrogen atom relative to a fragment ion formed by homolytic cleavage(s) at the same cleavage site(s) of a hypothetical precursor molecule, $M^{-\bullet}$. Multiple hydrogen deficiencies or hydrogen surpluses in the product ion follow the same rules as above. Because $(M - H)^-$ is deficient one hydrogen relative to $M^{-\bullet}$, it is designated as 'M; and when no hydrogen transfers occur during decomposition, such as in every case shown in Scheme 17.39b, all fragments are designated with a single prime to the left.[173]

Charge-remote ring cleavage may take place at either the a, b, or c rings, resulting in ring opening, loss of neutral ethylene(s), and double-bond formation on the product ion (Scheme 17.39B). The three ring cleavages take place in an analogous fashion with no structural feature to make one route dominant over the rest. Conversely, when a site of unsaturation exists—for example, on the b-ring (Scheme 17.39, bottom right)—the double bond directs a specific fragmentation route—that is, a retro-Diels–Alder process that results in the B_3 product as the dominant fragment ion.[173]

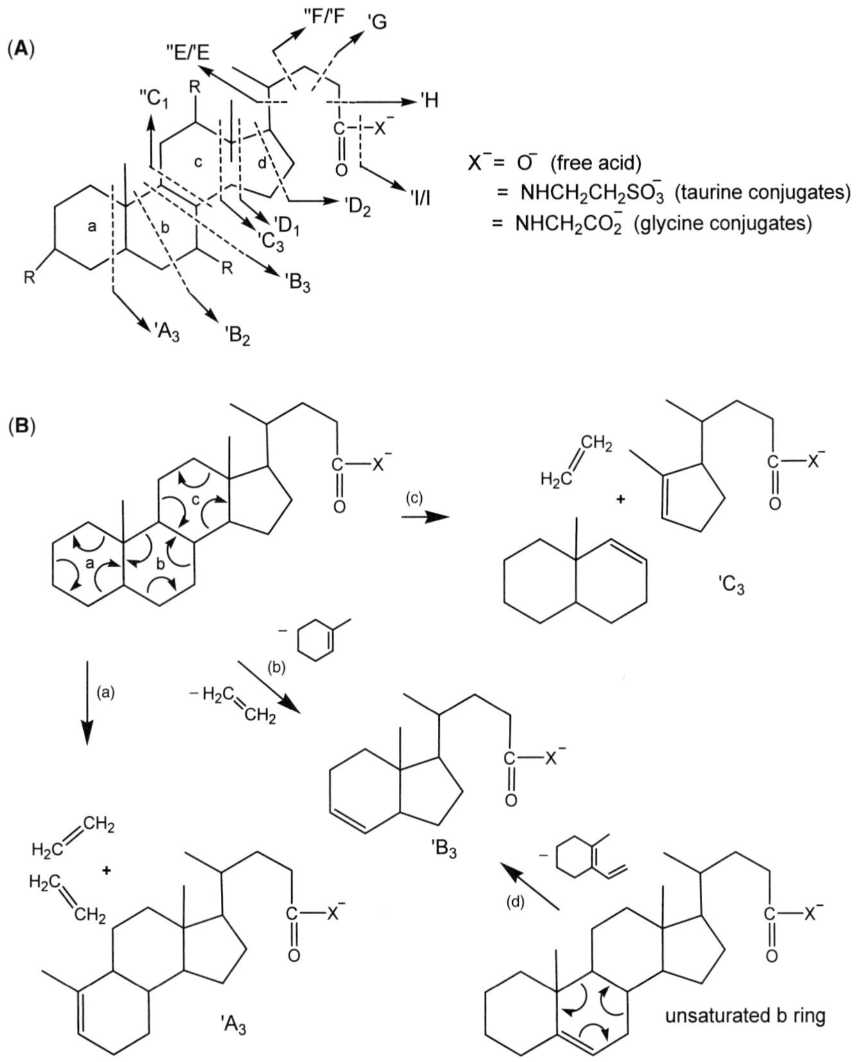

Scheme 17.39. (**A**) Nomenclature (see text) describing charge-remote fragmentation of bile acids (metabolites of cholesterol). (**B**) Mechanisms of ring cleavage leading to $'A_3$, $'B_3$, and $'C_3$ product ions from negatively charged precursors of bile acids. Lower right shows retro-Diels–Alder route to $'B_3$ ion from unsaturated b-ring precursor. (Adapted with permission from Ref. 173.)

17.7 CONCLUSION

This review on the dissociation of even electron systems (positively or negatively charged) demonstrates first the role played by the internal energy of the charged molecular species produced "in source" on the orientation of the fragmentation processes. The observed favored direct reaction involves covalent bond cleavage requiring a relatively low appearance energy (E_0 plus the kinetic shift of the process). Consequently, in the case of two competitive cleavages having similar transition state properties (i.e., frequency factors and looseness/tightness), under lower internal energy conditions the less endothermic pathway, involving a spatially accessible mobile proton, is the most favored. This pathway corresponds to the process yielding the loss of the neutral presenting the lowest basicity (positive mode) or acidity (negative mode). At low energies near the threshold for decomposition, when transition state properties are nearly identical for two competitive pathways, the kinetic factors dictating the preferred reaction will closely reflect the thermochemical orientation (that considers only initial and final states). When a large amount of excess energy is present, obtained by a high level of internal energy uptake by the precursor ions, the competitive decomposition processes have a tendency to reach similar rate constants. However, at energies well above the thresholds, decomposition pathways passing through loose transition states are heavily favored. The tight transition pathway will thus be disfavored even if it is less endothermic (thermodynamically favored). Tight transition states imply the existence of a non-negligible reverse activation barrier and a low state density. By operating instead in the low-energy collision regime, the tight transition pathway will become favored. Generally, even-electron precursor ions yield even-electron product ions, however, in the case of aromatic compounds comprised of several heteroatoms, especially nitrogen, it is possible to generate in competition, odd- and even-electron product ions.

In the low-energy range of ion activation, the observed direct cleavages can be assisted by either a proton or a hydride transfer, as well as by nucleophilic (or electrophilic) attack through a conformational folding, allowing charge solvation and/or ion dissociation by a lowering of the transition state energy level. Note that if functional group migration is less usual, ylide group transfer may take place, yielding (for example) (a) ring contraction and (b) alkane release from positive and negative ions, respectively. These processes can be analogized to those described in organic chemistry textbooks.

Precursor ion structures can undergo isomerization into ion–neutral complexes during or prior to the activation step, which generally involves proton solvation by two neutrals, that is, $A..H^{+}...B$ (positive ions), or by two anions, that is, $(A-H)^{-}..H^{+}...(B-H)^{-}$ (negative mode). In order to have a long lifetime, allowing a stepwise isomerization/dissociation process, the partners of the ion–neutral complexes must present a similar basicity (for positive ions) or acidity (for negative ions) that confers them higher gas-phase stabilities. The production of such complexes for larger systems can be useful to explain dissociations when either (a) the ionized group and the release site are too distant from one another or (b) an adequate conformation to allow charge transfer (proton, hydride, electron, etc) does not exist. Thus, charge transfer can take place by a free neutral (or charged group) migration (induced by vibration/rotation) without involving charge-remote processes that are favored under high-energy collision conditions. A typical example can be given by considering dissociation of deprotonated liposaccharides containing one or several phosphate groups. It was assumed some time ago that the release of fatty acid side chains was occurring via charge-remote-like processes[174] under low-energy collision. More recently, by considering ion–dipole complex formation, it was easier to rationalize the consecutive losses of fatty acid side chains by charge promoted dissociation.[175] Such mechanisms, involving ion– neutral complex intermediates, now allow the rationalization of peptide dissociations and, in particular, they take into account the position of the basic residues on the cleavage orientation.

REFERENCES

1. McLafferty, F. W. Unimolecular decompositions of even-electron ions. *Org. Mass Spectrom.* **1980**, *15*, 114–121.
2. Harrison, A. G. *Chemical Ionization Mass Spectrometry*, CRC Press, New York, 1983.
3. Splitter, J. S.; Turecek, F. (Eds.). *Applications of Mass Spectrometry to Organic Stereochemistry*, VCH Publishers, New York, 1994, p. 705.
4. Busch, K. L.; Glish, G. L.; McLuckey, S. A. *Mass Spectrometry/Mass Spectrometry. Techniques and Applications of Tandem Mass Spectrometry*, VCH Publishers, New York, 1988.
5. (a) Karas, M.; Bachmann, D.; Bahr, U.; Hillenkamp, F. Matrix-assisted ultraviolet laser desorption of non-volatile compounds. *Int. J. Mass Spectrom. Ion Proc.* **1987**, *78*, 53–68. (b) Gabelica, V.; Schulz, E.; Karas, M. Internal energy build-up in matrix-assisted laser desorption/ionization. *J. Mass Spectrom.* **2004**, *39*, 579–593.
6. (a) Fournier, I.; Brunot, A.; Tabet, J. C.; Bolbach, G. Delayed extraction experiments using a repulsing potential before ion extraction: Evidence of non-covalent clusters as ion precursor in UV MALDI. Part II—Dynamic effects with α-cyano-4-hydroxycinnamic acid matrix. *J. Mass Spectrom.* **2005**, *40*, 50–59. (b) Alves, S.; Fournier, F.; Afonso, C.; Wind, F.; Tabet, J.-C. Gas-phase ionization/desolvation. processes and their effect on protein charge state distribution under matrix-assisted laser desorption/ionization conditions. *Eur. J. Mass Spectrom.* **2006**, *12*, 369–383. (c) Livadaris, V.; Blais, J. C.; Tabet, J. C. Formation of non-specific protein cluster ions in matrix-assisted laser desorption/ionization: abundances and dynamical aspects. *Eur. J. Mass Spectrom.* **2000**, *6*, 409–413.
7. Karas, M.; Gluckmann, M.; Schafer, J. Ionization in matrix-assisted laser desorption/ionization: Singly charged molecular ions are the lucky survivors. *J. Mass Spectrom.* **2000**, *35*, 1–12.
8. Whitehouse, C. M.; Dreyer, R. N.; Yamashita, M.; Fenn, J. B. Electrospray interface for liquid chromatography and mass spectrometers. *Anal. Chem.* **1985**, *57*, 675–679.
9. (a) Caroll, D. I.; Dzidic, I.; Stillwell, R. N.; Haegele, K. D.; Horning, E. C. Atmospheric pressure ionization mass spectrometry. Corona dicharge ion source for use in a liquid chromatograph-mass-spectrometer-computer analytical system. *Anal.Chem.* **1975**, *47*, 2369. (b) Carroll, D. I.; Dzidic, I.; Horning, E. C.; Stillwell, R. N. Atmospheric-pressure ionization mass spectrometry. *Appl. Spec. Rev.* **1981**, *17*, 337–406. (c) Rondeau, D.; Vogel, R.; Tabet, J. C. Unusual atmospheric pressure chemical ionization conditions for detection of organic peroxides. *J. Mass Spectrom.* **2003**, *38*, 931–940. (d) Robb, D. B.; Covey, T. R.; Bruins, A. P. Atmospheric pressure photoionization: An ionization method for liquid chromatography–mass spectrometry. *Anal. Chem.* **2000**, *72*, 3653–3659.
10. Cole, R. B. (Ed.), *Electrospray Ionization Mass Spectrometry: Fundamentals, Instrumentation, and Applications*, Wiley & Sons, New York, 1997, p. 577.
11. Collette, C.; De Pauw, E. Calibration of the internal energy distribution of ions produced by electrospray. *Rapid Commun. Mass Spectrom.* **1998**, *12*, 165–170.
12. (a) Gabelica, V.; De Pauw, E.; Karas, M. Influence of the capillary temperature and the source pressure on the internal energy distribution of electrosprayed ions. *Int. J. Mass Spectrom.* **2004**, *231*, 189–195. (b) Naban-Maillet, J.; Lesage, D.; Bossée, A.; Gimbert, Y.; Sztaray, J.; Vekey, K.; Tabet, J. C. Internal energy distribution in electrospray ionization. *J. Mass Spectrom.* **2005**, *40*, pp. 1–8.
13. Kebarle, P. Equilibrium studies of the solvated proton by high pressure mass spectrometry. Thermodynamic determinations and implications for the electrospray ionization process. *J. Mass Spectrom.* **1997**, *32*, pp. 922–929.
14. Drahos, L.; Heeren, R. M. A.; Collette, C.; De Pauw, E.; Vekey, K. Thermal energy distribution observed in electrospray ionization. *J. Mass Spectrom.* **1999**, *34*, 1373–1379.
15. Derwa, F.; De Pauw, E.; Natalis, P. New basis for a method for the estimation of secondary ion internal energy distribution in soft ionization techniques. *Org. Mass Spectrom.* **1991**, *26*, 117–118.
16. (a) Bowen, R. D.; Williams, D. H. Non-concerted unimolecular reactions of ions in the gas-phase: The importance of ion–dipole interactions in carbonium ion isomerizations. *Int. J. Mass Spectrom. Ion Physics* **1979**, *29*, pp. 47–55. (b) Bowen, R. D.; Williams, D. H. Unimolecular reactions of isolated organic ions. The importance of ion–dipole interactions. *J. Am. Chem. Soc.* **1980**, *102*, pp. 2752–2756.
17. (a) Longevialle, P. Ion–neutral complexes in the unimolecular reactivity of organic cations in the gas phase. *Mass Spectrom. Reviews.* **1992**, *11*, 157–92. (b) Longevialle, P.; Botter, R. The interaction between ionic and neutral fragments from the same parent ion in the mass spectrometer. *Int. J. Mass Spectrom. Ion Phys.* **1983**, *47*, pp. 179–182.
18. Fournier, F.; Remaud, B.; Blasco, T.; Tabet, J. C. Ion–dipole complex formation from deprotonated phenol fatty acid esters evidenced by using gas-phase labeling combined with tandem mass spectrometry. *J. Am. Soc. Mass Spectrom.* **1993**, *4*, 343–51.
19. (a) Cooks, R. G.; Howe, I.; Williams, D. H. Structure and fragmentation mechanisms of organic ions in the mass spectrometer. *Org. Mass Spectrom.* **1969**, *2*, 137–156. (b) McLafferty, F. W.; Bente, P.

F.; KORNFELD, R.; TSAI, S. C.; HOWE, I. Metastable ion characteristics. XXII. Collisional activation spectra of organic ions. *J. Am. Chem. Soc.* **1973**, *95*, 2120–2129. (c) MABUD, M. A.; DEKREY, M. J.; COOKS, R. G. Surface-induced dissociation of molecular Ions. *Int. J. Mass Spectrom. Ion Processes.* **1985**, *67*(3), 285–294.
20. LITTLE, D. P.; SPEIR, J. P.; SENKO, M. W.; O'CONNOR, P. B.; MCLAFFERTY, F. W. Infrared multiphoton dissociation of large multiply charged ions for biomolecule sequencing. *Anal. Chem.* **1994**, *66*, 2809–2815.
21. NACSON, S.; HARRISON, A. G.; DAVIDSON, W. R. Effect of Method of Ion Preparation on Low-Energy Collision-Induced Dissociation Mass Spectra. *Org. Mass Spectrom.* **1986**, *21*, 317–319.
22. COOKS, R. G.; AST, T.; KRALJ, B.; KRAMER, V.; ZIGON, D. Internal energy distributions deposited in doubly and singly charged tungsten hexacarbonyl ions generated by charge stripping, electron impact, and charge exchange. *J. Am. Soc. Mass Spectrom.* **1990**, *1*, 16–27.
23. COOKS, R. G. *Collision Spectroscopy*, Plenum Press, New York, 1978, p. 383.
24. (a) GROSS, M. L. Charge-remote fragmentation: An account of research on mechanisms and applications. *Int. J. Mass Spectrom.* **2000**, *200*, 611–624. (b) BAMBAGIOTTI, A. M.; CORAN, S. A.; GIANNELLINI, V.; VINCIERI, F. F.; DAOLIO, S.; TRALDI, P. Hydroxyl ion negative chemical ionization and collisionally activated dissociation mass analysed ion kinetic energy spectra for an easy mass spectrometric characterization of fatty acid methyl esters. *Org. Mass Spectrom.* **1983**, *18*, 133–134.
25. (a) WYSOCKI, V. H.; ROSS, M. M.; HORNING, S. R.; COOKS, R. G. Remote-site (charge-remote) fragmentation. *Rapid Comm. Mass Spectrom.* **1988**, *2*, 214–217. (b) WYSOCKI, V. H.; BIER, M. E.; COOKS, R. G. Internal energy requirements for remote site fragmentation. *Org. Mass Spectrom.* **1988**, *23*, 627–633. (c) CHENG, C.; PITTENAUER, E.; GROSS, M. Charge-remote fragmentations are energy-dependent processes. *J. Am. Soc. Mass. Spectrom.* **1998**, *9*, pp. 840–844.
26. (a) WYSOCKI, V. H.; ROSS, M. M.; HORNING, S. R.; COOKS, R. G. Remote-site (charge-remote) fragmentation. *Rapid Commun. Mass Spectrom.* **1988**, *2*, pp. 214–217. (b) WYSOCKI, V. H.; BIER, M. E.; COOKS, R. G. Internal energy requirements for remote site fragmentation. *Org. Mass Spectrom.* **1988**, *23*, 627–633. (c) WHALEN, K.; GROSSERT, J.S.; BOYD, R. Ion dissociation reactions induced in a high-pressure quadrupole collision cell. *Rapid Commun Mass Spectrom.* **1995**, *9*, 1366–1375. (d) ADAMS, J.; GROSS, M. L. Energy requirement for remote charge site ion decompositions and structural information from collisional activation of alkali metal cationized fatty alcohols. *J. Am. Chem. Soc.* **1986**, *108*, 6915–6921. (e) CHENG, C.; PITTENAUER, E.; GROSS, M. L. Charge-Remote Fragmentations are energy-dependent processes. *J. Am. Soc. Mass Spectrom.* **1998**, *9*, 840–844.
27. AFONSO, C.; RIU, A.; XU, Y.; FOURNIER, F.; TABET, J.-C. Structural characterization of fatty acids cationized with copper by electrospray ionization mass spectrometry under low-energy collision-induced dissociation. *J. Mass Spectrom.* **2005**, *40*, 342–349.
28. (a) JI, H.; MORRE, J.; VOINOV, V. G.; DEINZER, M. L.; BAROFSKY, D. F. Distinguishing between cis/trans isomers of monounsaturated fatty acids by FAB MS. *Anal. Chem.* **2007**, *79*, 1519–1522. (b) JI, H.; VOINOV, V. G.; DEINZER, M. L.; BAROFSKY, D. F. Charge-remote metastable ion decomposition of free fatty acids under FAB MS: Evidence for biradical ion structures. *Anal. Chem.* **2007**, *79*, 2822–2826.
29. (a) BAER, T.; MAYER, P. M. Statistical Rice–Ramsperger–Kassel–Marcus quasi-equilibrium theory calculations in mass spectrometry. *J. Am. Soc. Mass Spectrom.* **1997**, *8*, 103–115. (b) DRAHOS, L.; VEKEY, K. Mass kinetics: A theoretical model of mass spectra incorporating physical processes, reaction kinetics and mathematical descriptions. *J. Mass Spectrom.* **2001**, *36*, 237–263.
30. (a) DRAHOS, L.; SZTARAY, J.; VEKEY, K. Theoretical calculation of isotope effects, kinetic energy release and effective temperatures for alkylamines. *Int. J. Mass Spectrom.* **2003**, *225*, 233–248. (b) WU, L.; DENAULT, J. W.; COOKS, R. G.; DRAHOS, L.; VEKEY, K. Alkali chloride cluster ion dissociation examined by the kinetic method: Heterolytic bond dissociation energies, effective temperatures, and entropic effects. *J. Am. Soc. Mass Spectrom.* **2002**, *13*, 1388–1395. (c) DRAHOS, L.; VEKEY, K. How closely related are the effective and the real temperature. *J. Mass Spectrom.* **1999**, *34*, 79–84.
31. COLE, R. B.; TABET, J. C.; BLAIS, J. C. Stabilities of fast atom bombardment desorbed alkali metal adducts of pyrimidine molecules. *Int. J. Mass Spectrom. Ion Proc.* **1990**, *98*, 269–283.
32. (a) GRUENE, P.; TRAGE, C.; SCHROEDER, D.; SCHWARZ, H. Solvent–ligand effects on structures of iron halide cations in the gas phase. *Eur. J. Inorg. Chem.* **2006**, *22*, 4546–4552. (b) O'HAIR, R. A. J. The 3D quadrupole ion trap mass spectrometer as a complete chemical laboratory for fundamental gas-phase studies of metal mediated chemistry. *Chem. Commun.* **2006**, *14*, 1469–1481.
33. ZHU, J.; COLE, R. B. Ranking of gas-phase acidities and chloride affinities of monosaccharides and linkage specificity in collision-induced decompositions of Negative Ion Electrospray-Generated chloride adducts of oligosaccharides. *J. Am. Soc. Mass Spectrom.* **2001**, *12*, 1193–1204.
34. MANCEL, V.; SELLIER, N.; LESAGE, D.; FOURNIER, F.; TABET, J. C. Gas phase enantiomeric distinction of (R)- and (S)-aromatic hydroxy esters by negative ion chemical ionization mass spectrometry using a chiral

reagent gas. *Int. J. Mass Spectrom.* **2004**, *237*, 185–195.

35. ZHU, J.; COLE, R. B. Formation and decompositions of chloride adduct ions, $[M + Cl]^-$, in negative ion electrospray ionization mass spectrometry. *J. Am. Soc. Mass Spectrom.* **2000**, *11*, 932–941.
36. CAI, Y.; JIANG, Y.; COLE, R. B. Anionic adducts of oligosaccharides by matrix-assisted laser desorption/ionization time-of-flight (MALDI-TOF) mass spectrometry. *Anal. Chem.* **2003**, *75*, 1638–1644.
37. KUMAR, M. R.; PRABHAKAR, S.; KUMAR, M. K.; REDDY, T. J.; VAIRAMANI, M. Negative ion electrospray ionization mass spectral study of dicarboxylic acids in the presence of halide ions. *Rapid Commun. Mass Spectrom.* **2004**, *18*, 1109–1115.
38. (a) VIERA, N. E.; GILLIGAN, J. J.; YERGEY, A. L.In *Gas Phase Energetics of the Interactions Between Water and Amino Acids.* Proceedings of the 51st ASMS Conference on Mass Spectrometry and Allied Topics, Montreal, Canada, June 8–12, 2003. (b) MEOT-NER (MAUTNER), M. The ionic hydrogen bond. *Chem. Rev.* **2005**, *105*, 213–284. (c) MEOT-NER (MAUTNER), M.; SPELLER, C. V. Multicomponent cluster ions. 2. Comparative stabilities of cationic and anionic hydrogen-bonded networks: Mixed clusters of water and hydrogen cyanide. *J. Phys. Chem.* **1989**, *93*, 3663–3666. (d) MEOT-NER (MAUTNER), M. Heats of hydration of organic ions: Predictive relations and analysis of solvation factors based on ion clustering. *J. Phys. Chem.* **1987**, *91*, 417–426. (e) MEOT-NER (MAUTNER), M.; HAMLET, P.; HUNTER, E. P.; FIELD, F. H. Internal and external solvation of polyfunctional ions. *J. Am. Chem. Soc.* **1980**, pp. *102*, pp. 6393–6399. (f) Works of Bowers as well as of Williams or Kebarle.
39. LEE, R. E.; SQUIRES, R. R. Anionic homoaromaticity: A gas-phase experimental study. *J. Am. Chem. Soc.* **1986**, *108*, 5078–5086.
40. (a) WYSOCKI, V. H.; BURINSKY, D. J.; COOKS, R. G. Competitive dehydration & deamination of α, ω-amino alcohols and α, ω-amino acids in the gas phase. *J. Org. Chem.* **1985**, *50*, 1287–1291. (b) DAVIS, D. V.; COOKS, R. G. Site of protonation and bifunctional group interactions in α, ω-hydroxyalkylamines. *Org. Mass Spectrom.* **1981**, *16*, 176–179.
41. (a) LONGEVIALLE, P.; MILNE, G. W. A.; FALES, H. M. Chemical ionization mass spectrometry of complex molecules. XI. Stereochemical and conformational effects in the isobutane chemical ionization mass spectra of some steroidal amino alcohols. *J. Am. Chem. Soc.* **1973**, *95*, 6666–6669. (b) LONGEVIALLE, P.; GIRARD, P. J.; ROSSI, J. P.; TICHY, M. Isobutane chemical ionization as a tool for evaluating the proportions of conformational isomers of alicyclic, amino-alcohols at equilibrium in gas phase. *Org. Mass Spectrom.* **1980**, *15*, 268.
42. Unpublished results from the Tabet' group.
43. VAIS, V.; ETINGER, A.; MANDELBAUM, A. Intramolecular proton transfers in stereoisomeric gas-phase ions and the kinetic nature of the protonation process upon chemical ionization. *J. Mass Spectrom.* **1999**, *34*, 755–760.
44. LEWARS, E. G.; MARCH, R. E. Fragmentation of 3-hydroxyflavone: A computational and mass spectrometric study. *Rapid Comm. Mass Spectrom.* **2007**, *21*, pp. 1669–1679.
45. ZHANG, K.; BOUCHONNET, S.; SERAFIN, S. V.; MORTON, T. H. Stereochemical analysis of deuterated alkyl chains by charge-remote fragmentations of protonated parent ions. *Int. J. Mass Spectrom.* **2003**, *227*, 175–189.
46. VAZQUEZ, S.; TRUSCOTT, R. J. W.; O'HAIR, R. A. J.; WEIMANN, A.; SHEIL, M. M. A study of kynurenine fragmentation using electrospray tandem mass spectrometry. *J. Am. Soc. Mass Spectrom.* **2001**, *12*, 786–794.
47. WINNIK, W.; BRUMLEY, W.; BETOWSKI, L. Negative-ion mass spectrometry of sulfonylurea herbicides. *J. Mass Spectrom.* **1995**, *30*, 1574–1580.
48. FOURNIER, F.; TABET, J. C.; DEBRAUWER, L.; RAO, D.; PARIS, A.; BORIES, G. Structural Dependence of anion/dipole complex upon the deprotonated site in $[M - H]^-$ ions of 17β-estradiol-17-stearate ester in ammonia negative-ion chemical ionization. *Rapid Commun. Mass Spectrom.* **1991**, *5*, 44–47.
49. DEBRAUWER, L.; PARIS, A.; RAO, D.; FOURNIER, F.; TABET, J. C. Mass spectrometric studies on 17β-estradiol-17-fatty acid esters: Evidence for the formation of Anion–dipole intermediates. *Org. Mass Spectrom.* **1992**, *27*(6), 709–719.
50. (a) EDELSON-AVERBUKH, M.; MANDELBAUM, A. Chemistry and stereochemistry of Benzyl–benzyl interactions in MH^+ ions of dibenzyl esters upon chemical ionization and collision-induced dissociation conditions. *J. Mass Spectrom.* **1997**, *32*, 515–524. (b) EDELSON-AVERBUKH, M.; MANDELBAUM, A. Intramolecular electrophilic aromatic substitution in gas-phase-protonated difunctional compounds containing one or two arylmethyl groups. *J. Mass Spectrom.* **2003**, *38*, 1169–1177.
51. WENIGER, K.; JOST, M.; GRUTZMACHER, H. F. Mass spectrometry of Di-*tert*-butylnaphthalenes and benzyl-tert-butylnaphthalenes: Hydrogen Migration and intermediate ion-molecule complexes in *tert*-butylated naphthalene ions. *Eur. Mass Spectrom.* **1999**, *5*, pp. 101–115.
52. MATTHIAS, C.; CARTONI, A.; KUCK, D. Ion/neutral complexes generated during Unimolecular fragmentation: Intra-complex hydride abstraction by *tert*-butyl cations from electron-rich and electron-Poor 1,3-diphenylpropanes. *Int. J. Mass Spectrom.* **2006**, *255–256*, 195–212.
53. TU, Y. P.; HARRISON, A. G. Stereochemical applications of mass spectrometry. Part 5—fragmentation of protonated and methylated maleic and fumaric acid and derivatives. *J. Mass Spectrom.* **1998**, *33*, 858–871.
54. KHASELEV, N.; MANDELBAUM, A. The role of the C:C double bond in alcohol elimination from MH^+ ions of

unsaturated bicyclic esters upon chemical ionization. *J. Mass Spectrom.* **1995**, *30*, 1533–1538.

55. Denekamp, C.; Mandelbaum, A. Configurational effects on internal proton transfers and ion–neutral complex formation in stereoisomeric 1,4-di(alkoxymethyl)cyclohexanes on chemical ionization. *J. Mass Spectrom.* **1995**, *30*, 1421–1428.
56. Morlender-Vais, N.; Mandelbaum, A. The role of hydrogen migration in the mechanism of alcohol elimination from MH^+ ions of ethers upon chemical ionization. *J. Mass Spectrom.* **1997**, *32*, 1124–1132.
57. Ben Ari, J.; Karni, M.; Apeloig, Y.; Mandelbaum, A. The role of hydride migration in the mechanism of alcohol elimination from protonated ethers upon chemical ionization experiment and theory. *Int. J. Mass Spectrom.* **2003**, *228*, 297–306.
58. Kuzmenkov, I.; Etinger, A.; Mandelbaum, A. Role of hydrogen migration in the mechanism of acetic acid elimination from MH^+ ions of acetates on chemical ionization and collision-induced dissociation. *J. Mass Spectrom.* **1999**, *34*, 797–803.
59. (a) Wolfschuetz, R.; Schwarz, H.; Blum, W.; Richter, W. J. Detection of a "Wagner–Meerwein" ring contraction via proton induced water elimination from 1,2-cyclohexanediol. *Org. Mass Spectrom.* **1979**, *14*, 462–475. (b) Gaumann, T.; Houriet, R.; Stahl, D.; Tabet, J. C.; Heinrich, N.; Schwarz, H. Further examples of skeletal rearrangements of the Wagner–Meerwein type in chemical ionization mass spectrometry: The case of $[C_6H_9]^+$ ions. *Org. Mass Spectrom.* **1983**, *18*, 215–218.
60. Rathahao, E.; Perlat, M. C.; Fournier, F.; Tabet, J. C. Stereochemical effects enhanced by using selective "self-ionization" under electron ionization conditions in a quadrupole ion trap mass spectrometer. *Int. J. Mass Spectrom.* **1999**, *193*, 161–179.
61. Gregoire, S.; Dugat, D.; Fournier, F.; Gramain, J.-C.; Tabet, J.-C. Stereospecific decompositions induced by low-energy collision-induced dissociation of MH^+ ions of epimeric alkaloids. *Rapid Comm. Mass Spectrom.* **1995**, *9*, 9–12.
62. (a) Bowie, J. H. Reactions of nucleophiles in the gas phase—The last decade: An overview. *Mass Spectrom. Rev.* **1984**, *3*, 1–37. (b) Bowie, J. H. Collisional processes of negative ions. *Analytical and mechanistic aspects. Adv. in Mass Spectrom.* **1986**, *10*, (Pt A), 553–565.
63. Noest, A. J.; Nibbering, N. M. Homoconjugation vs. charge dipole interaction effects in the carbanion stabilization in the gas phase. *J. Am. Chem. Soc.* **1980**, *102*, 6427–6429.
64. Gaeumann, T.; Stahl, D.; Tabet, J. C. The mechanism of the loss of hydrogen and water from 1,4-cyclohexanediol anion and related compounds. *Org. Mass Spectrom.* **1983**, *18*, 263–266.
65. Tabet, J. C.; Hanna, I.; Fetizon, M. Stereospecific hydride transfer under NCI/hydroxide ion conditions. 2—Origins of the C_2H_4O elimination from stereoisomeric alkoxyketal derivatives. *Org. Mass Spectrom.* **1985**, *20*, pp. 61–64.
66. Popot, M. A.; Garcia, P.; Fournier, F.; Bonnaire, Y.; Tabet, J. C. Different approaches to the identification of a cortisol isomer compound in horse urine. *The Analyst.* **1998**, *123*, 2649–2652.
67. Garcia, P.; Popot, M. A.; Fournier, F.; Tabet, J.-C. Long-distance stereochemical effects in deprotonated epimeric androstanediols in tandem mass spectrometry. *Rapid Commun. Mass Spectrom.* **1995**, *9*, pp. 23–26.
68. (a) Currie, G. J.; Stringer, M. B.; Bowie, J. H.; Holmes, J. L. Collision-induced carbanion rearrangements. The elimination of ethene from the 3,3-dimethylheptan-4-one enolate ion. *Aust. J. Chem.* **1987**, *40*, 1365. (b) Stringer, M. B.; Bowie, J. H.; Holmes, J. L. Carbanion rearrangements. Collision-induced dissociations of the enolate on of heptan-4-one. *J. Am. Chem. Soc.* **1986**, *108*, 3888.
69. Donnelly, A.; Chowdhury, S.; Harrison, A. G. Fragmentation reactions of the enolate Ions of 2-pentanone. *Org. Mass Spectrom.* **1989**, *24*, 89–93.
70. Ben Ari, J.; Navon, I.; Mandelbaum, A. The effect of steric hindrance on the relative rates of anchimerically assisted alcohol eliminations from MH^+ Ions of 2-substituted 1,4-dialkoxybutanes upon CI and CID. *Int. J. Mass Spectrom.* **2006**, *249/250*, 433–445.
71. Le Meillour, S.; Tabet, J. C. Ion–molecule reactions in the gas phase. XI(1); regioselective SN_2 induced by ammonia on the $[M + NH_4]^+$ adduct ions of the 4-hydroxy-4-methylcyclohexyl benzoate isomers. *Spectroscopy* **1988**, *5*, 135–148.
72. Dua, S.; Whait, R. B.; Alexander, M. J.; Hayes, R. N.; Lebedev, A. T.; Eichinger, P. C. H.; Bowie, J. H. The search for the gas-phase negative ion pinacol rearrangement. *J. Am. Chem. Soc.* **1993**, *115*, 5709–5715.
73. Dua, S.; Whait, R. B.; Alexander, M. J.; Bowie, J. H. The search for the gas-phase negative-ion Pinacol rearrangement. 2-aryl rearrangements. *Org. Mass Spectrom.* **1993**, *28*, 1155–1160.
74. Attygalle, A.; Ruzicka, J.; Varughese, D.; Sayed, J. An unprecedented ortho effect in mass spectrometric fragmentation of even-electron negative ions from hydroxyphenyl carbaldehydes and ketones. *Tetra. Lett.* **2006**, *47*, 4601–4603.
75. Blanksby, S. J.; Dua, S.; Bowie, J. H. Unusual cross-ring S_N2 reactions of $[M-H]^-$ Ions of methoxyacetanilides. *Rapid Comm. Mass Spectrom.* **1995**, *9*, pp. 177–1779.
76. Henzel, W. J.; Billeci, T. M.; Stults, J. T.; Wong, S. C.; Grimley, C.; Watanabe, C. Identifying proteins from two-dimensional gels by Molecular mass searching of peptide fragments in protein sequence databases. *Proc. Natl. Acad. Sc. USA* **1993**, *90*, 5011–5015.
77. Eng, J. K.; McCormack, A. L.; Yates, J. R., III. An approach to correlate tandem mass spectral data of peptides with amino acid sequences in a protein

database. *J. Am. Soc. Mass Spectrom.* **1994**, *5*, 976–989.
78. Perkins, D. N.; Pappin, D. J. C.; Creasy, D. M.; Cottrell, J. S. Probability-based protein Identification by searching sequence databases using mass spectrometry data. *Electrophoresis.* **1999**, *20*, 3551–3567.
79. Tabb, D. L.; Huang, Y.; Wysocki, V. H.; Yates, J. R., III. Influence of basic residue content on fragment ion peak intensities in low-energy collision-induced dissociation spectra of peptides. *Anal. Chem.* **2004**, *76*, pp. 1243–1248.
80. Sleno, L.; Dietrich, A. V. Ion activation methods for tandem mass spectrometry. *J. Mass Spectrom.* **2004**, *39*, 1091–1112.
81. Paizs, B.; Suhai, S. Fragmentation pathways of protonated peptides. *Mass Spectrom. Rev.* **2005**, *24*, 508–548.
82. Roepstorff, P. Proposal for a common nomenclature for sequence ions in mass spectra of peptides. *Biomed. Mass Spectrom.* **1984**, *11*, 601.
83. Biemann, K. Contributions of mass spectrometry to peptide and protein structure. *Biomed. Environ. Mass Spectrom.* **1988**, *16*, 99–111.
84. Dongré, A. R.; Jones, J. L.; Somogyi, A.; Wysocki, V. H. Influence of peptide composition, gas-phase basicity, and chemical modification on fragmentation efficiency: Evidence for the mobile proton model. *J. Am. Chem. Soc.* **1996**, *118*, 8365–8374.
85. Wysocki, V. H.; Tsaprailis, G.; Smith, L. L.; Breci, L. A. Mobile and localized protons: A framework for understanding peptide dissociation. *J. Mass Spectrom.* **2000**, *35*, 1399–1406.
86. Yalcin, T.; Csizmadia, I. G.; Peterson, M. R.; Harrison, A. G. The structure and fragmentation of B_n ($n > 3$) ions in peptide spectra. *J. Am. Soc. Mass Spectrom.* **1996**, *7*, 233–242.
87. Schlosser, A.; Lehmann, W. D. Five-membered ring formation in unimolecular reactions of peptides: A key structural element controlling low-energy collision-induced dissociation of peptides. *J. Mass Spectrom.* **2000**, *35*, 1382–1390.
88. Tu, Y.-P.; Harrison, A. G. Fragmentation of protonated amides through intermediate ion–neutral complexes: neighboring group participation. *J. Am. Soc. Mass Spectrom.* **1998**, *9*, 454–462.
89. Paizs, B.; Suhai, S. Towards understanding some ion intensity relationships for the tandem mass spectra of protonated peptides. *Rapid Commun. Mass Spectrom.* **2002**, *16*, 1699–1702.
90. Cooks, R. G.; Kruger, T. L. Intrinsic basicity determination using metastable ions. *J. Am. Chem. Soc.* **1977**, *99*, 1279.
91. Cooks, R. G.; Koskinen, J. T.; Thomas, P. D. The kinetic method of making thermochemical determinations. *J. Mass Spectrom.* **1999**, *34*, 85–92.
92. Tu, Y.-P.; Harrison, A. G. The B_1 ion derived from methionine is a stable species. *Rapid Commun. Mass Spectrom.* **1998**, *12*(13), 849–851.
93. Nold, M. J.; Wesdemiotis, C.; Yalcin, T.; Harrison, A. G. Amide bond dissociation in protonated peptides. Structures of the N-terminal ionic and neutral fragments. *Int. J. Mass Spectrom. Ion Proc.* **1997**, *164*, 137–153.
94. Hill, H. H.; Hill, C. H.; Asbury, G. R.; Wu, C.; Matz, L. M.; Ichiye, T. Charge location on gas phase peptides. *Int. J. Mass Spectrom.* **2002**, *219*, 23–37.
95. Afonso, C.; Modeste, F.; Breton, P.; Fournier, F.; Tabet, J.-C. Proton affinities of the commonly occurring L-amino acids by using electrospray ionization–ion trap mass spectrometry. *Eur. J. Mass Spectrom.* **2000**, *6*, 443–449.
96. Bleiholder, C.; Suhai, S.; Paizs, B. Revising the proton affinity scale of the naturally occurring α-amino acids. *J. Am. Soc. Mass Spectrom.* **2006**, *17*, 1275–1281.
97. Tabb, D. L.; Smith, L. L.; Breci, L. A.; Wysocki, V. H.; Lin, D.; Yates, J. R., III. Statistical characterization of ion trap tandem mass spectra from doubly charged tryptic peptides. *Anal. Chem.* **2003**, *75*, 1155–1163.
98. Engel, B. J.; Pan, P.; Reid, G. E.; Wells, J. M.; McLuckey, S. A. Charge state dependent fragmentation of gaseous protein ions in a quadrupole ion trap: Bovine ferri-, ferro-, and apo-cytochrome C. *Int. J. Mass Spectrom.* **2002**, *219*, 171–187.
99. Bailey, T. H.; Laskin, J.; Futrell, J. H. Energetics of selective cleavage at acidic residues studied by time- and energy-resolved surface-induced dissociation in FT-ICR MS. *Int. J. Mass Spectrom.* **2003**, *222*, 313–327.
100. Harrison, A. G.; Yalcin, T. Proton mobility in protonated amino acids and peptides. *Int. J. Mass Spectrom. Ion Processes.* **1997**, *165*, 339–347.
101. Csonka, I. P.; Paizs, B.; Lendvay, G.; Suhai, S. Proton mobility in protonated peptides: A joint molecular orbital and RRKM study. *Rapid Commun. Mass Spectrom.* **2000**, *14*, 417–431.
102. Paizs, B.; Suhai, S. Theoretical study of the main fragmentation pathways for protonated glycylglycine. *Rapid Commun. Mass Spectrom* **2001**, *15*, 651–663.
103. Johnson, R. S.; Krylov, D.; Walsh, K. A. Proton mobility within electrosprayed ions. *J Mass Spectrom.* **1995**, *30*, 386–387.
104. Hoerner, J. K.; Xiao, H.; Dobo, A.; Kaltashov, I. A. Is there hydrogen scrambling in the gas phase? Energetic and structural determinants of proton mobility within protein ions. *J. Am. Chem. Soc.* **2004**, *126*, 7709–7717.
105. Ambihapathy, K.; Yalcin, T.; Leung, H.-W.; Harrison, A. G. Pathways to immonium ions in the fragmentation of protonated peptides. *J. Mass Spectrom.* **1997**, *32*, 209–215.
106. Paizs, B.; Schnölzer, M.; Warnken, U.; Suhai, S.; Harrison, A. G. Cleavage of the amide bond of protonated dipeptides. *Phys. Chem. Chem. Phys.* **2004**, *6*, 2691–2699.
107. Loo, J. A.; Edmonds, C. G.; Smith, R. D. Tandem mass spectrometry of very large molecules. 2. Dissociation of multiply charged proline-containing proteins from

electrospray ionization. *Anal. Chem.* **1993**, *65*, 425–438.

108. Vaisar, T.; Urban, J. Probing the proline effects in CID of protonated peptides. *J. Mass Spectrom.* **1996**, *31*, 1185–1187.
109. Qin, J.; Chait, B. T. Preferential fragmentation of protonated gas-phase peptide ions adjacent to acidic amino acid residues. *J. Am. Chem. Soc.* **1995**, *117*, 5411–5412.
110. Grewal, R. N.; El Aribi, H.; Harrison, A. G.; Siu, K. M. W.; Hopkinson, A. C. Fragmentation of protonated tripeptides: The proline effect revisited. *J. Phys. Chem. B* **2004**, *108*, 4899–4908.
111. Burlet, O.; Orkiszewski, R. S.; Ballard, K. D.; Gaskell, S. J. Charge promotion of low-energy fragmentations of peptide ions. *Rapid Commun. Mass Spectrom.* **1992**, *6*, 658–662.
112. Bailey, T. H.; Laskin, J.; Futrell, J. H. Energetics of selective cleavage at acidic residues studied by time- and energy-resolved surface-induced dissociation in FT-ICR MS. *Int. J. Mass Spectrom.* **2003**, *222*, 313–327.
113. Gu, C.; Tsaprailis, G.; Breci, L.; Wysocki, V. H. Selective gas-phase cleavage at the peptide bond C-terminal to aspartic acid in fixed-charge derivatives of Asp-containing peptides. *Anal. Chem.* **2000**, *72*, 5804–5813.
114. Yao, Z.-P.; Afonso, C.; Fenselau, C. Rapid microorganism identification with on-slide proteolytic digestion followed by matrix-assited laser desorption/ ionization tandem mass Spectrometry. *Rapid Commun. Mass Spectrom.* **2002**, *16*, 1953–1956.
115. Yu, W.; Vath, J. E.; Huberty, M. C.; Martin, S. A. Identification of the facile gas-phase cleavage of the Asp-Pro and Asp-Xxx peptide bonds in matrix-assisted laser desorption time-of-flight mass spectrometry. *Anal.Chem.* **1993**, *65*, 3015–3023.
116. Paizs, B.; Suhai, S.; Hargittai, B.; Hruby, V. J.; Somogyi, A. *Ab initio* and MS/MS studies on protonated peptides containing basic and acidic amino acid residues: I. solvated proton versus salt-bridged structures and the cleavage of the terminal amide bond of protonated RD-NH_2. *Int. J. Mass Spectrom.* **2002**, *219*, 203–232.
117. Rožman, M. Aspartic Acid Side Chain Effect - Experimental and Theoretical Insight. *J. Am. Soc. Mass Spectrom.* **2007**, *18*, pp. 121–127.
118. Price, W. D.; Schnier, P. D.; Jockush, R. A.; Williams, E. R. Unimolecular Reaction Kinetics in the High-Pressure Limit Without Collisions. *J. Am. Chem. Soc.* **1996**, *118*, pp. 10640–10644.
119. Tsaprailis, G.; Nair, H.; Zhong, W.; Kuppannan, K.; Futrell, J. H.; Wysocki, V. H. A mechanistic investigation of the enhanced cleavage at histidine in the gas-phase dissociation of protonated peptides. *Anal. Chem.* **2004**, *76*, 2083–2094.
120. Johnson, R. S.; Martin, S. A.; Biemann, K.; Stults, J. T.; Watson, J. T. Novel fragmentation process of peptides by collision-induced decomposition in a tandem mass spectrometer: Differentiation of leucine and isoleucine. *Anal.Chem.* **1987**, *59*, 2621–2625.
121. Martin, S. A.; Biemann, K. A comparison of keV atom bombardment mass spectra of Peptides Obtained with a Two-Sector Mass Spectrometer with those from a four-sector tandem mass spectrometer. *Int. J. Mass Spectrom. Ion Proc.* **1987**, *78*, 213–228.
122. Medzihradszky, K. F.; Campbell, J. M.; Baldwin, M. A.; Falick, A. M.; Juhasz, P.; Vestal, M. L.; Burlingame, A. L. The characteristics of peptide collision-induced dissociation using a high-performance MALDI-TOF/TOF tandem mass spectrometer. *Anal. Chem.* **2000**, *72*, 552–558.
123. Hulst, A. G.; Kientz, C. E. Differentiation between the isomeric amino acids leucine and isoleucine using low-energy collision-induced dissociation tandem mass spectroscopy. *J. Mass Spectrom.* **1996**, *31*, 1188–1190.
124. Armirotti, A.; Millo, E.; Damonte, G. How to discriminate between leucine and isoleucine by low energy ESI-TRAP MS^n. *J. Am. Soc. Mass Spectrom.* **2007**, *18*, 57–63.
125. Tureček, F. Copper–biomolecule complexes in the gas phase. The ternary way. *Mass Spectrom. Rev.* **2007**, *26*, 563–582.
126. Wee, S.; O'Hair, R. A. J.; McFadyen, W. D. Side-chain radical losses from radical cations allows distinction of leucine and isoleucine residues in the isomeric peptides Gly-XXX-Arg. S. *Rapid Commun. Mass Spectrom.* **2002**, *16*, 884–890.
127. Chu, I. K.; Rodriquez, C. F.; Lau, T.-C.; Hopkinson, A. C.; Siu, K. W. M. Molecular radical cations of oligopeptides. *J. Phys. Chem. B.* **2000**, *104*, 3393–3397.
128. Bossée, A.; Fournier, F.; Tasseau, O.; Bellier, B.; Tabet, J. C. Electrospray mass spectrometric study of anionic complexes of enkephalins with Cu(II): Regioselective distinction of Leu/Ile at the C-terminus induced by metal reduction. *Rapid Commun. Mass Spectrom.* **2003**, *17*, 1229.
129. Bossée, A.; Afonso, C.; Fournier, F.; Tasseau, O.; Pepe, C.; Bellier, B.; Tabet, J.-C. Anionic copper complex fragmentations from enkephalins under low-energy collision-induced dissociation in an ion trap mass spectrometer. *J. Mass Spectrom.* **2004**, *39*, 903–912.
130. Boutin, M.; Bich, C.; Afonso, C.; Fournier, F.; Tabet, J.-C. Negative-charge driven fragmentations for evidencing zwitterionic forms from doubly charged coppered peptides. *J. Mass Spectrom.* **2007**, *42*, 25–35.
131. Waugh, R. L.; Bowie, J. H. Collision induced dissociations of deprotonated peptides: Dipeptides containing phenylalanine, tyrosine, histidine, and tryptophan. *Int. J. Mass Spectrom. Ion Process.* **1991**, *107*, 333.
132. Bowie, J. H.; Brinkworth, C. S.; Dua, S. Collision-induced fragmentations of the $[M - H]^-$ parent anions of underivatized peptides: An aid to structure

determination and some unusual negative ion cleavages. *Mass Spectrom. Rev.* **2002**, *21*, 87.

133. Marzluff, E. M.; Campbell, S.; Rodgers, M. T.; Beauchamp, J. L. Low-energy dissociation pathways of small deprotonated peptides in the gas phase. *J. Am. Chem. Soc.* **1994**, *116*, 7787–7796.
134. Tabb, D. L.; Smith, L. L.; Breci, L. A.; Wysocki, V. H.; Lin, D.; Yates, J. R. Statistical characterization of ion trap tandem mass spectra from doubly charged tryptic peptides. *Anal. Chem.* **2003**, *75*, 1155–1163.
135. Huang, Y.; Triscari, J. M.; Tseng, G. C.; Pasa-Tolic, L.; Lipton, M. S.; Smith, R. D.; Wysocki, V. H. Statistical characterization of the charge state and residue dependence of low-energy CID peptide dissociation patterns. *Anal. Chem.* **2005**, *77*, 5800–5813.
136. Zhang, Z. Prediction of low-energy collision-induced dissociation spectra of peptides with three or more charges. *Anal. Chem.* **2005**, *77*, 6364–6373.
137. Wyttenbach, T.; von Helden, G.; Bowers, M. T. Gas-phase conformation of biological molecules: Bradykinin. *J. Am. Chem. Soc.* **1996**, *118*, 8355–8364.
138. Gross, D. S.; Schnier, P. D.; Rodriguez-Cruz, S. E.; Fagerquist, C. K.; Williams, E. R. Conformations and folding of lysozyme ions *in vacuo*. *Proc. Natl. Acad. Sci. USA* **1996**, *93*, 3143–3148.
139. Vachet, R. W.; Asam, M. R.; Glish, G. L. Secondary interactions affecting the dissociation patterns of arginine-containing peptide ions. *J. Am. Chem. Soc.* **1996**, *118*, 6252–6256.
140. Tsaprailis, G.; Nair, H.; Somogyi, A.; Wysocki, V. H.; Zhong, W.; Futrell, J. H.; Summerfield, S. G.; Gaskell, S. J. Influence of secondary structure on the fragmentation of protonated peptides. *J. Am. Chem. Soc.* **1999**, *121*, 5142–5154.
141. Collings, B. A.; Douglas, D. J. Conformation of gas-phase myoglobin ions. *J. Am. Chem. Soc.* **1996**, *118*, 4488–4489.
142. Domon, B.; Costello, C. E. A systematic nomenclature for carbohydrate fragmentations in FAB-MS/MS spectra of glycoconjugates. *Glyco. J.* **1988**, *5*(4), 397–409.
143. Chai, W.; Piskarev, V.; Lawson, A. M. Negative-ion electrospray mass spectrometry of neutral underivatized oligosaccharides. *Anal. Chem.* **2001**, *73*, 631–657.
144. Zaia, J.; Miller, M. J. C.; Seymour, J. L.; Costello, C. E. The Role of Mobile Protons in Negative Ion CID of Oligosaccharides. *J. Am. Soc. Mass Spec.* **2007**, *18*(5), 952–960.
145. Harvey, D. J. Collision-induced fragmentation of underivatised N-linked carbohydrates ionized by electrospray. *J. Mass Spec.* **2000**, *35*, 1178–1190.
146. Garozzo, D.; Giuffrida, M.; Impallomeni, G.; Ballistreri, A.; Montaudo, G. Determination of linkage position and identification of the reducing end in linear oligosaccharides by negative-ion fast atom bombardment mass-spectrometry. *Anal. Chem.* **1990**, *62* (3), 279–286.
147. Ashline, D.; Singh, S.; Hanneman, A.; Reinhold, V. Congruent strategies for carbohydrate sequencing. 1. Mining structural details by $MS^{n.}$ *Anal. Chem.* **2005**, *77*, 6250–6262.
148. Zhu, J.; Cole, R. B. Ranking of gas-phase acidities and chloride affinities of monosaccharides and linkage specifity in collision-induced decompositions of negative ion electrospray-generated chloride adducts of oligosaccharides. *J. Am. Soc. Mass Spectrom.* **2001**, *12*, 1193–1204.
149. Cai, Y.; Concha, M. C.; Murray, J. S.; Cole, R. B. Evaluation of the role of multiple hydrogen bonding in offering stability to negative ion adducts in electrospray mass spectrometry. *J. Am. Soc. Mass Spectrom.* **2002**, *13*, 1360–1369.
150. Cai, Y.; Jiang, Y.; Cole, R. B. Anionic adducts of oligosaccharides by matrix-assisted laser desorption/ionization time-of-flight (MALDI-TOF) mass spectrometry. *Anal. Chem.* **2003**, *75*, 1638–1644.
151. Guan, B.; Cole, R. B. MALDI linear-field reflectron TOF post-source decay analysis of underivatized oligosaccharides: Determination of glycosidic linkages and anomeric configurations using anion attachment. *J. Am. Soc. Mass Spectrom.* **2008**, *19*, 1119–1131.
152. Harvey, D. J. Collision-induced fragmentation of underivatized N-linked carbohydrates ionized by electrospray. *J. Mass Spectrom.* **2000**, *35*(10), 1178–1190.
153. Chai, W.; Piskarev, V.; Lawson, A. M. Negative-ion electrospray mass spectrometry of neutral underivatized oligosaccharides. *Anal. Chem.* **2001**, *73*(3), 651–657.
154. Pfenninger, A.; Karas, M.; Finke, B.; Stahl, B. Structural analysis of underivatized neutral human milk oligosaccharides in the negative ion mode by nano-electrospray MSn (Part 1: Methodology). *J. Am. Soc. Mass Spectrom.* **2002**, *13*(11), 1331–1340.
155. Cancilla, M. T.; Penn, S. G.; Carroll, J. A.; Lebrilla, C. B. Coordination of alkali metals to oligosaccharides dictates fragmentation behavior in matrix assisted laser desorption/ionization Fourier transform mass spectrometry. *J. Am. Chem. Soc.* **1996**, *118*, 6736–6745.
156. Park, Y.; Lebrilla, C. B. Application of fourier transform ion cyclotron resonance mass spectrometry to oligosaccharides. *Mass Spec. Rev.* **2005**, *24*, 232–264.
157. Cancilla, M. T.; Wong, A. W.; Voss, L. R.; Lebrilla, C. B. Fragmentation reactions in the mass spectrometry analysis of neutral oligosaccharides. *Anal. Chem.* **1999**, *71*, 3206–3218.
158. Harvey, D. J.; Bateman, R. H.; Green, M. R. High-energy collision-induced fragmentation of complex oligosaccharides ionized by matrix-assisted laser desorption/ionization mass spectrometry. *J Mass Spectrom.* **1997**, *32*, 167–187.

159. Gillece-Castro, B. L.; Burlingame, A. L. Oligosaccharide Characterization with High-Energy Collision-Induced Dissociation Mass Spectrometry. *Met. Enzy.* **1990**, *193*, 689–712.
160. Lemoine, J.; Fournet, B.; Despeyroux, D.; Jennings, K. R.; Rosenberg, R.; de Hoffmann, E. Collision-induced dissociation of alkali metal cationized and permethylated oligosaccharides: Influence of the collision energy and of the collision gas for the assignment of linkage position. *J. Am. Soc. Mass Spec.* **1993**, *4*(3), 197–203.
161. Fahy, E.; Subramaniam, S.; Brown, H. A.; Glass, C. K.; Merrill, A. H. Jr.; Murphy, R. C.; Raetz, C. R. H.; Russell, D. W.; Seyama, Y.; Shaw, W.; Shimizu, T.; Spener, F.; van Meer, G.; VanNieuwenhze, M. S.; White, S. H.; Witztum, J. L.; Dennis, E. A. A comprehensive classification system for lipids. *J. Lipid Res.* **2005**, *46*, 839–861.
162. Ham, B. M.; Jacob, J. T.; Keese, M. M.; Cole, R. B. Identification, quantification and comparison of major nonpolar lipids in normal and dry eye tear lipidomes by electrospray tandem mass spectrometry. *J. Mass Spectrom.* **2004**, *39*, 1321–1336.
163. Ham, B. M.; Jacob, J. T.; Cole, R. B. MALDI-TOF MS of phosphorylated lipids in biological fluids using immobilized metal affinity chromatography and a solid ionic crystal matrix. *Anal. Chem.* **2005**, *77*, 4439–4447.
164. Cheng, C.; Gross, M. L. Applications and mechanisms of charge-remote fragmentation. *Mass Spec Rev.* **2000**, *19*, 398–420.
165. Gross, M. L. Charge-remote fragmentation: An account of research on mechanisms and applications. *Int. J.Mass Spectrom.* **2000**, *200*, 611–624.
166. Jensen, N. J.; Tomer, K. B.; Gross, M. L. Gas-phase ion decomposition occurring remote to a charge site. *J. Am. Chem. Soc.* **1985**, *107*, 1863–1868.
167. Wysocki, V. H.; Ross, M. M. Charge-remote fragmentation of gas-phase ions: mechanistic and energetic considerations in the dissociation of long-chain functionalized alkanes and alkenes. *Int. J. Mass Spec. Ion Proc.* **1991**, *104*, 179–211.
168. Claeys, M.; Nizigiyimana, L.; Van Den Heuvel, H.; Vedernikova, I.; Haemers, A. Charge-remote and charge-proximate fragmentation processes in alkali-cationized fatty acid esters upon high-energy collisional activation. A new mechanistic proposal. *J. Mass Spec.* **1998**, *33*, 631–643.
169. Griffiths, W. J.; Yang, Y.; Lindgren, J. A.; Sjoevall, J. Charge remote fragmentation of fatty acid anions in 400 eV collisions with xenon atoms. *Rapid Commun. Mass Spec.* **1996**, *10*, 21–28.
170. Claeys, M.; Nizigiyimana, L.; Van den Heuvel, H.; Derrick, P. J. Mechanistic aspects of charge-remote fragmentation in saturated and mono-unsaturated fatty acid derivatives. Evidence for homolytic cleavage. *Rapid Commun. Mass Spec.* **1996**, *10*, 770–774.
171. Harvey, D. J. A new charge-associated mechanism to account for the production of fragment ions in the high-energy CID spectra of fatty acids. *J. Am. Soc. Mass Spec.* **2005**, *16*, 280–290.
172. Griffiths, W. J.; Brown, A.; Reimendal, R.; Yang, Y.; Zhang, J.; Sjoevall, J. A comparison of fast-atom bombardment and electrospray as methods of ionization in the study of sulfated- and sulfonated lipids by tandem mass spectrometry. *Rapid Commun. Mass Spec.* **1996**, *10*, 1169–1174.
173. Griffiths, W. J. Tandem mass spectrometry in the study of fatty acids, bile acids, and steroids. *Mass Spec. Rev.* **2003**, *22*(2), 81–152.
174. (a) Boué, S. M.; Cole, R. B. Confirmation of the structure of lipid A from *Enterobacter agglomerans* by electrospray ionization tandem mass spectrometry. *J. Mass Spec.* **2000**, *35*, 361–368. (b) Hsu, F. F.; Turk, J. J. Charge-remote and charge-driven fragmentation processes in diacyl glycerophosphoethanolamine upon low-energy collisional activation: A mechanistic proposal. *J. Am. Soc. Mass Spec.* **2000**, *11*, 892. (c) Kussak, A.; Weintraub, A. Quadrupole ion-trap mass spectrometry to locate fatty acids on lipid a from gram-negative bacteria. *Anal. Biochem.* **2002**, *307*, 131–137.
175. Madalinski, G.; Fournier, F.; Wind, F.-L.; Afonso, C.; Tabet, J.-C. Gram-negative bacterial lipid A analysis by negative electrospray ion trap mass spectrometry: stepwise dissociations of deprotonated species under low energy CID conditions. *Int. J. Mass Spectrom.* **2006**, *249/250*, 77–92.

Part V

Biological Applications of ES and MALDI

Chapter 18

The Role of Mass Spectrometry for Peptide, Protein, and Proteome Characterization

A. Jimmy Ytterberg,[*] Jason Dunsmore,[*] Shirley H. Lomeli,[*] Mario Thevis,[*] Yongming Xie,[*] Rachel R. Ogorzalek Loo,[†] and Joseph A. Loo[*,†]

[*]*Department of Chemistry and Biochemistry, University of California—Los Angeles, Los Angeles, California*
[†]*Department of Biological Chemistry, David Geffen School of Medicine, University of California—Los Angeles, Los Angeles, California*

Electrospray and MALDI Mass Spectrometry: Fundamentals, Instrumentation, Practicalities, and Biological Applications, Second Edition, Edited by Richard B. Cole

18.1 INTRODUCTION

The measurement of large polypeptides has benefited tremendously from the development of electrospray ionization (ESI) and matrix-assisted laser desorption/ionization (MALDI) coupled to gas-phase analyzers such as mass spectrometry (MS) and ion mobility spectrometry (IMS). It was only 20 years or so ago when the measurement of a linear peptide of mass 1000 Da was a challenging task. Frank Field, for whom the American Chemical Society's Frank H. Field and Joe L. Franklin Award for Outstanding Achievement in Mass Spectrometry is co-named, remarked in 1986 at the 3rd Texas Symposium on Mass Spectrometry, "Success and Failure at *m/z* 10,000 and Beyond," that "the mass region of real interest for proteins is *m/z* 40,000–100,000, and one can only speculate as to whether such monster gaseous ions can be produced."[1] But John Fenn's work demonstrated that even a simple quadrupole mass analyzer with an upper limit of mass-to-charge ratio (*m/z*) 1500 could "weigh" molecules in excess of 100,000 Da because, on average, a protein will pick up one proton per 10 amino acid residues by ESI, thereby shifting the *m/z* of all proteins into a very similar and easily measurable range.[2] Koichi Tanaka[3] and Franz Hillenkamp[4] demonstrated that MALDI coupled with time-of-flight (TOF) analyzers can desorb, ionize, and measure large singly charged proteins beyond *m/z* 100,000. For the developments to measure these "flying elephants"[2] by ESI and MALDI, the 2002 Nobel Prize in Chemistry was awarded, and it is now obvious how very large molecules can be measured reliably and sensitively.

In a scenario that occurs often in a protein biochemistry lab setting, a researcher has isolated a protein of unknown function, or he/she has overexpressed a protein in *Escherichia coli* and wishes to characterize it. A common practice today is to submit a small amount of the protein to a core mass spectrometry laboratory for a molecular weight measurement. Using either ESI or MALDI, a molecular weight with a precision and accuracy of $\pm 0.05\%$ or better can be measured. This, of course, depends heavily on the purity of the protein sample, the relative size of the protein, the presence or absence of post-translational modifications (PTM) (e.g., phosphorylation, glycosylation, etc.), the resolution of the mass analyzer, and so on. Primary structure information (i.e., amino acid sequence) can be determined by either tandem mass spectrometry (MS/MS) sequencing of enzymatically derived peptide fragments of the original intact protein or, in favorable cases, direct MS/MS of the intact protein. In addition, the presence of protein post-translational modifications can be elucidated by these "bottom-up" and "top-down" mass spectrometry methods. Over the past few years, the sensitivity and specificity of mass spectrometry, coupled with separation techniques such as liquid chromatography, have improved to a large degree, such that protein identification, PTM detection, and even protein quantification can be derived from very complex mixtures (e.g., tissues, biofluids, cell lysates, etc.).

Protein quaternary structure can be addressed primarily by ESI-MS, but in some examples by either ESI-MS or MALDI-MS (e.g., covalent crosslinked protein complexes). Proteins are involved in molecular recognition, including the recognition of signals by receptors, of substrates and their transition states by enzymes, and of ligands by proteins. Quaternary structure refers to the arrangement of subunits in multi-subunit proteins. Intermolecular noncovalent interactions are responsible for aggregation of folded polypeptide chains into multimers, which determines a protein system's quaternary structure. The ability of ESI to ionize macromolecules without disrupting covalent bonds and maintain weak noncovalent interactions is key for the MS-based study of biological complexes.[5–9] A molecular mass measurement provides a direct determination of the stoichiometry of the binding partners in the complex, even for multi-ligand hetero-complexes. ESI-MS can measure proteins and their complexes in aqueous solution at near neutral pH. The protein interactions are often sufficiently retained upon the transition to the gas phase that the size and binding stoichiometry can be measured.

18.2 PROTEIN IDENTIFICATION, QUANTIFICATION, AND STRUCTURE CHARACTERIZATION BY MASS-SPECTROMETRY-BASED METHODS

18.2.1 Bottom-Up Protein Identification

Current approaches to protein identification are dominated by two primary philosophies. A more traditional bottom-up approach typically "sorts" proteins by size and isoelectric point using two-dimensional polyacrylamide gel electrophoresis (2D-PAGE), excises the protein spots of interest (e.g., unique to a disease state or treatment), and "breaks" the proteins by enzymatically digesting them within the gel matrix. Eluted protein fragments are analyzed by mass spectrometry to measure their sizes and to break them further, to obtain sequence information (e.g., MS/MS with typically, collisionally activated dissociation, or CAD). Computer programs can readily calculate how proteins predicted from genome sequences would be cleaved by enzymes and subsequently by MS/MS methods; by comparing these predicted fragments to the measured MS and MS/MS data, proteins can be identified uniquely.[10] Rarely are all of the protein pieces recovered; fortunately, only a few are required to posit an identity. For example, sufficiently accurate mass measurements (low parts-per-million range) performed on one or two tryptic peptides, each with corresponding MS/MS product ion spectra, are sufficient to identify faint silver-stained protein spots from polyacrylamide gels.[11]

In contrast, an alternative but complementary bottom-up philosophy advocates first enzymatically or chemically cleaving ("breaking") a complex mixture of cellular proteins and then "sorting" the pieces by one or more steps of chromatography. MS analyzes the recovered fragments as in the previous approach, and computers match the fragments to the proteins from which they are derived. Embodiments of this latter philosophy include multidimensional protein identification technology (MudPIT).[12,13] Complex protein mixtures are cleaved with endoproteinase Lys-C and/or trypsin; or, alternatively, less soluble components are cleaved with CNBr in formic acid. Following cleavage, peptide mixtures are separated by reversed-phase high-performance liquid chromatography (HPLC), or a first dimension of strong cation exchange chromatography (SCX) followed by a second dimension of reversed-phase HPLC. Peptides eluted from HPLC are analyzed on-line by MS/MS. This strategy has demonstrated impressive results for large-scale protein identification for a variety of biochemical systems, including yeast and *E. coli.*

The major advantage of this latter "shotgun" approach is that it circumvents the challenges inherent in protein separation and recovery by transforming the problem into one of peptide analysis. (A) Peptides (especially tryptic peptides) behave much more predictably in reversed-phase HPLC and are much less likely to bind irreversibly. (B) Rather than attempting to perform microchemistry and analysis on minute quantities of purified protein, chemistry is performed on a larger quantity of material, present as a complex mixture of peptides. Losses from irreversible binding are less critical. (C) Sequence information from as few as one or two tryptic peptides is sufficient to link a protein to its gene. (D) Shotgun proteomic methods project the digest products from all protein isoforms onto the single set of peptides corresponding to cleavage of the full-length gene product, clearly valuable when attempting to identify heterogeneous proteins. A concern relating to MudPIT is that the vast number of peptides and their dramatic range of abundances (6–7 orders of magnitude) could overload the capacity of the column chromatography and of the data-dependent MS/MS acquisition, ultimately skewing detection toward more abundant species. In a data-dependent experiment, as full *m/z*-range mass spectra are acquired continuously in LC-MS mode, any ion detected with a signal intensity above a predefined threshold will trigger the mass spectrometer to switch over to an MS/MS operation. Thus, the mass spectrometer switches back and forth between MS (molecular mass information) and MS/MS modes (sequence information) in a single LC run. The data-dependent scanning capability can dramatically increase the capacity and throughput for protein identification, but its efficiency is highly dependent on the MS-to-MS/MS cycle time required to acquire sufficient signal-to-noise mass spectra relative to the chromatographic profile. Coeluting or closely eluting peptides may go unmeasured if the duty cycle is relatively long. Moreover, more time can be spent analyzing peptides from the more abundant proteins present in the mixture. Therefore, less abundant proteins may be missed or analyzed less frequently in such a data-dependent strategy.

Both approaches excel at linking expression products to genes, but they assume that prior to the analysis, the protein existed as predicted by the genome sequence—that is, full-length and unmodified. In exchange for sensitivity and speed, shotgun methods sever links to the fragments' heritage, erasing much of the means to characterize splicing variants and post-translational modifications. The former approach, when pursued with 2D-PAGE, does expose large deviations from predictions—for example, migration in the SDS (sodium dodecyl sulfate) dimension inconsistent with a given molecular weight, focusing far from the predicted isoelectric point, or nonbinding of appropriate antibodies. Unfortunately, few tools are available for follow-up efforts—that is, for characterizing the proteins. A higher-resolution recording of the intact proteins' characteristics before they were cleaved would be desirable.

18.2.2 Protein Identification Algorithms for Spectral Processing

Scientists have been using algorithms to interpret MS spectra since the 1960s. An algorithm is a specific set of rules used to solve a problem. In the past, algorithms used for interpreting MS/MS spectra were performed manually, but today they are most often implemented in the form of a computer program. Search algorithms have been used to identify compounds from mass spectra for several decades. An early mass spectral database search program was written in 1970 by Heller and Feldmann.[14] As mass spectrometers became more sophisticated and were able to analyze large biomolecules, this general approach was adapted to peptide and protein identification. As of the end of 2009, the UniProtKB/Swiss-Prot protein database contained over 500,000 sequence entries, up from only 76,000 entries in 2000, and is growing at a nearly exponential rate.[15]

High-throughput protein identification is a goal of many proteomics studies. ESI-MS/MS is a popular tool for these studies because mass analysis of complex mixtures can be readily automated when coupled with LC, and protein identifications have dramatically higher confidence levels as compared to peptide mass fingerprinting. In bottom-up studies using ESI–MS/MS, peptide identification is the first step in protein identification. In this section, we will examine several computer algorithms used to identify peptide sequences from MS/MS data, including those used in *de novo* sequencing and database searches.

18.2.2.1 De Novo Sequencing

De novo peptide sequencing is performed without any prior knowledge of the amino acid sequence from, for example, a protein sequence database. Thus, *de novo* sequencing can be used to determine the sequence of a peptide when little or no sequence information is available for the organism being studied. This can be performed also for identifying novel proteins and post-translationally or chemically modified proteins. *De novo* sequencing is used also when peptide sequence information for higher eukaryotes is not available. Because the DNA of higher eukaryotes has many intron/exon boundaries that must be correctly interpreted, the use of software to encode a higher eukaryotic nucleotide sequence to a peptide sequence is prone to error because of the increased chance of mistaking noncoding DNA for coding DNA. Genomes of bacteria and lower eukaryotes have relatively few introns and can ordinarily be sequenced using a database search if the genomic sequence is available—that is, unless the proteins undergo extensive post-translational modification, or a covalent processing event that changes the properties of a protein.

PAAS 3 (*p*robable *a*mino *a*cid *s*equence), one of the first *de novo* sequencing algorithms, calculates every possible amino acid sequence for the peptide mass in question, and then it uses the experimental MS/MS spectrum to determine probable sequences.[16] Because all permutations and combinations of the peptide sequence are considered, this approach is extremely computationally expensive for peptides greater than 800 Da. To correct the shortcoming of PAAS 3, a more efficient *de novo* sequencing algorithm was developed that uses graph theory, a branch of mathematics that uses graphs, or mathematical structures to model pairwise relations between objects from a specific collection, to convert *m/z* values to ions of a single type; for example, a spectrum consisting of a, b, and y ions would be converted to a spectrum of only b ions.[17] In theory, the resulting spectrum of b ions would contain a "mass ladder," from which the amino acid sequence could be deduced by calculating the mass differences between peaks (see Figure 18.1).

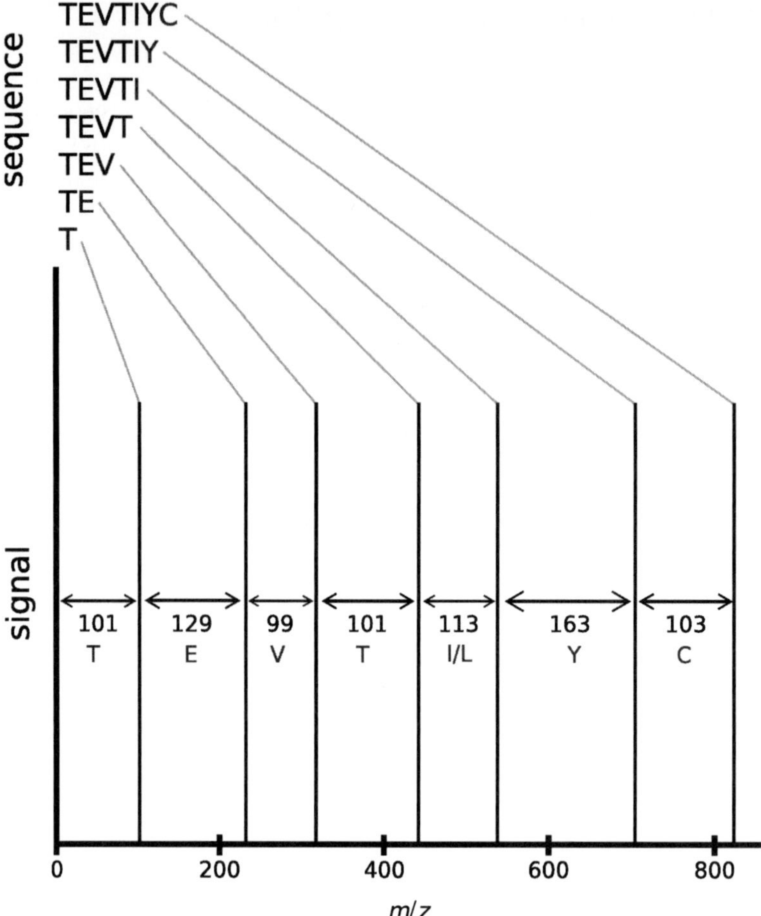

Figure 18.1. A mass spectrum containing a peptide mass ladder. The peptide sequence is determined from the mass differences between adjacent peaks.

Many *de novo* sequencing algorithms have been developed, but none are currently in widespread use due to fundamental limitations with the *de novo* approach to peptide identification. In order to obtain meaningful results from a *de novo* sequencing experiment, the MS/MS data must be of high quality, having low noise and complete peptide fragmentation. Only a fraction of the MS/MS spectra collected in a high-throughput, shotgun proteomics experiment are typically of sufficient quality to use for *de novo* sequencing.

Another weakness of the *de novo* sequencing method is that it often produces multiple sequence possibilities; and without further information, it is difficult to discern the correct sequence. However, if the sequences identified by the *de novo* algorithm are used as input for a database search, the sequences can be scored and ranked based on homology to protein sequences in the database. Lutefisk, a *de novo* sequencing program, and CIDentify, a variant of the FASTA (stands for "Fast All," reflecting that is it applicable for both fast protein and nucleotide comparisons) program for homology-based database searching, use this approach to *de novo* sequencing.[18,19] The MS BLAST (Basic Local Alignment Search Tool) search routine can be used as an alternative to FASTA or CIDentify, providing advantages in throughput from its use of the faster BLAST algorithm.[20]

18.2.2.2 Searching Sequence Databases

Searching MS/MS data against a protein or genomic sequence database is the workhorse for high-throughput peptide identification. (If a genomic sequence database is used, then it must be encoded to a peptide sequence in six reading frames, three forward and three in reverse, prior to searching.) The method consists of (1) performing a virtual enzymatic digest on the protein sequences in the database, (2) filtering the peptide fragments based on mass to find candidate peptides, (3) predicting the fragmentation pattern of each of the candidate peptides, and (4) assigning scores to each of the candidate peptides based on how well they match the observed fragmentation pattern. Algorithms that search sequence databases vary most in the way they assign scores to the candidate peptides.

Two protein databases, NCBI (National Center for Biotechnology Information) and SwissProt, are commonly used for peptide sequencing and protein identification. The NCBI database is a non-curated compilation of public databases and is used for identifying proteins of organisms for which little is known. If the protein from the organism being studied is not in the database, then one hopes that a related homolog from another species can be found. SwissProt is a curated, highly annotated, nonredundant database better suited for identifying proteins of well-studied or model organisms.

Database searching programs require several user-defined parameters, including which database and taxonomy to search, which fixed or variable modifications to consider, and mass-tolerance, to name a few. It is generally better to start with narrower parameters (e.g., taxonomy: mouse) and then work toward broader parameters (e.g., taxonomy: mammals) to eliminate the bias that can occur if results from the broader search are realized before adjusting parameters.

The MS-Tag program, implemented in Protein Prospector, was the first to identify proteins from MS/MS spectra by searching sequence databases.[21] Unlike the other database-searching programs, MS-Tag does not calculate a "score." Instead, it simply ranks peptides by the number of fragment ions that match to a sequence in the database. Oftentimes, this method of ranking peptides is too low-resolution to differentiate between the top-ranking peptides and is supplemented with a method to score the peptides.[22]

SEQUEST is a program that identifies peptides, and subsequently proteins, based on correlation of the MS/MS product ion data to sequence databases.[23] It was originally designed for low-resolution, low-mass accuracy data acquired with ion-trap mass spectrometers. Two values are used to score peptide matches, X_{corr} (cross-correlation) and ΔCn. X_{corr} is a measure of similarity between observed and predicted spectra and is a robust method for estimating accuracy of the matches. It is found by summing overlapping peaks from an overlay of the observed and predicted spectra. As a result, signal intensity is an important factor in the X_{corr} values. ΔCn is the difference between the X_{corr} values of the first- and second-ranked sequences and is used as a crude assessment of the relative quality of the match. The value of ΔCn is dependent on database size, search parameters, and sequence homologies. Because of this, ΔCn is a weak relative score, making SEQUEST results prone to false-positives and thereby requiring that the results be subjected to further downstream statistical validation.

Mascot is another popular database-searching program that emphasizes the quality of the match using a probability-based approach to peptide identification.[24] In contrast to

SEQUEST, Mascot was designed to handle higher-resolution, higher-mass accuracy data. Mascot's scoring system is based on the probability that the match between the experimental data and each candidate sequence in the database is a chance event. A statistical significance cutoff, usually a 5% false-positive rate, is calculated for each set of search results based on a number of factors that determine the size of the search space—for example, sequence database size, mass tolerance, number of variable modifications, and so on.[24] A limitation of probability-based scoring is that small peptides always have low statistical significance because the increased number of candidate sequences increases the probability that the sequence would be in the database by random chance. Peptides containing extended repeats also have low statistical significance, due to the assumption Mascot makes that database sequences are random.

A number of other programs have been developed that complement the scoring of Mascot and SEQUEST. Peptide Prophet calculates a discriminant score and estimates the probability of peptide assignment made by SEQUEST or Mascot of being correct.[25] X!Tandem is another database-searching program with strong relative scoring, a good complement to SEQUEST's X_{corr}.[26] The Scaffold software probabilistically combines search results from SEQUEST, Mascot, and X!Tandem to identify greater than 40% more proteins than any one software alone.[27]

Database searching does not ordinarily identify extensively post-translationally modified proteins or sequence variants. When this problem occurs, high-quality MS/MS spectra can return low-confidence matches. Those spectra can then be searched using an error-tolerant database search routine, which considers a large number of unsuspected modifications and primary sequence variations.[28] However, this approach is very computationally expensive. InsPecT is a program that implements an efficient algorithm to identify post-translational modifications by using database filters.[29]

Traditionally, the accuracy of database search results was estimated based on how good the highest-matching peptide score was relative to the next-highest-matching peptide score. The problem with this approach is that it provides little information about the probability that the assignment is correct. For this reason, it is essential to incorporate statistical validation into proteomics experiments. Furthermore, for results from different database-searching software to be related to each other, a single method for statistical validation must be used. One method that has been recently gaining in popularity is the use of decoy databases.[30] A decoy database is either a randomized or reversed version of the protein database being searched. Following the search of the original, unmodified protein database, the decoy database is searched using the same parameters and the false-error rate is estimated based on the number of significant matches to the decoy database. For an in-depth review of statistical validation of proteomics data, see Nesvizhskii et al.[31]

18.2.3 Experimental Methods for *De Novo* Protein Identification and Sequencing

Mass spectrometry-based protein identification protocols have propelled proteomics to the forefront of biomedical research. Provided with available genomic and/or protein sequence information in databases, protein identification by MALDI-based peptide mass fingerprint mapping and LC-MS/MS and MALDI-TOF/TOF peptide sequencing with CAD fragmentation is efficient. However, in situations in which sufficient genome and protein sequence data are unavailable, mass spectrometric methods can be used for *de novo* protein sequencing.[32] The mass measurement accuracy of MS provides a unique capability to generate peptide sequence. *De novo* sequencing is especially augmented through incorporation of isotopic labeling to deconvolute peptide fragmentation patterns and to identify C-terminal sequences. For example, trypsin digestion in the presence of ^{18}O-labeled water allows y- and b-product ions to be differentiated (Figure 18.2).[33] When proteins are cleaved into peptides by trypsin, H_2O is incorporated into each cleavage site (OH on the C-terminal side of the site and H on the N-terminal). By performing the proteolysis in $H_2{}^{18}O$, stable isotopes can be incorporated into the peptides. Thus, trypsinolysis allows the C-terminal end of the Arg/Lys-peptide to be labeled with -^{18}OH (-^{16}OH from normal water). Subsequent CAD–MS/MS spectra of the ^{18}O-labeled tryptic peptides generates y products that incorporate the ^{18}O-label and are shifted by 2-Da compared to y products from the corresponding unlabeled peptide. (The b-ion fragments remain the same for both the labeled and unlabeled peptides.) The amino acid sequence can be immediately determined from knowledge of which product ions are y fragments. This also allows the C-terminal tryptic peptide of the protein to be identified, because this tryptic peptide remains unlabeled in the presence of $H_2{}^{18}O$.

The method for *de novo* protein sequencing is demonstrated by using a 23-kDa protein expressed in Sf9 insect (*Spodoptera frugiperda*, fall armyworm) cells. The protein isolated from Sf9 cells was separated by SDS-PAGE, and the band of interest was subjected to in-gel trypsin digestion, MALDI-TOF-MS peptide mapping, and LC-MS/MS peptide sequencing using both a quadrupole time-of-flight (QTOF) mass spectrometer and a quadrupole

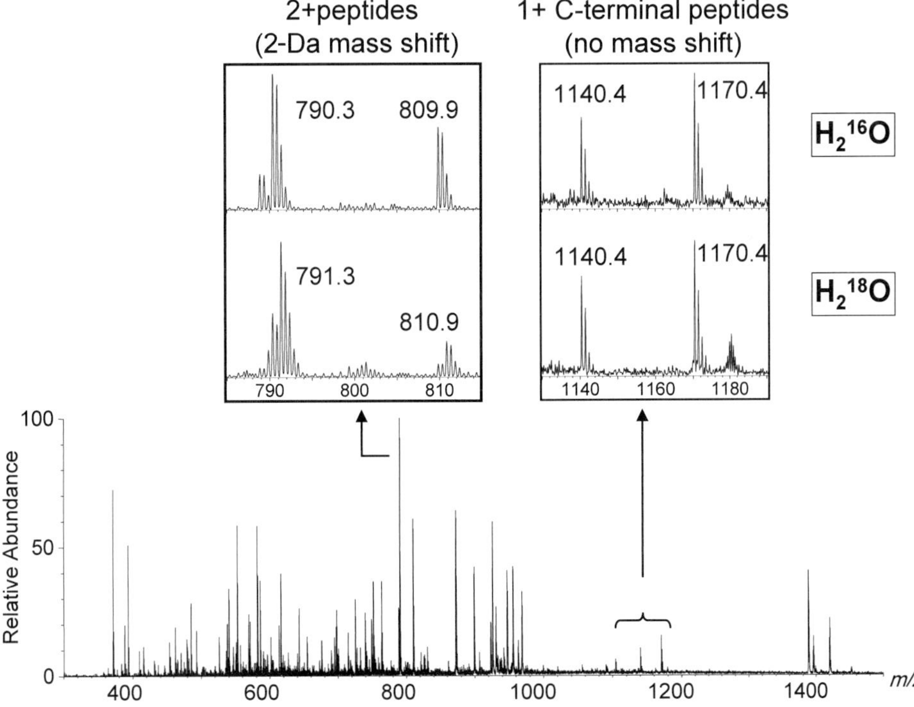

Figure 18.2. ESI-QTOF-MS of a trypsin digestion of a mixture of 23-kDa glutathione *S*-transferase proteins expressed in Sf9 insect (*Spodoptera frugiperda*, fall armyworm). The trypsin digests were generated from normal water (**insets, top**) and ^{18}O-labeled water (**insets, bottom**). Most of the tryptic peptides should be shifted by 2 Da (1 *m/z* units for 2+-charged peptides) with $H_2{}^{18}O$/trypsin digestions, with the exception of the C-terminal peptides (**right inset**).

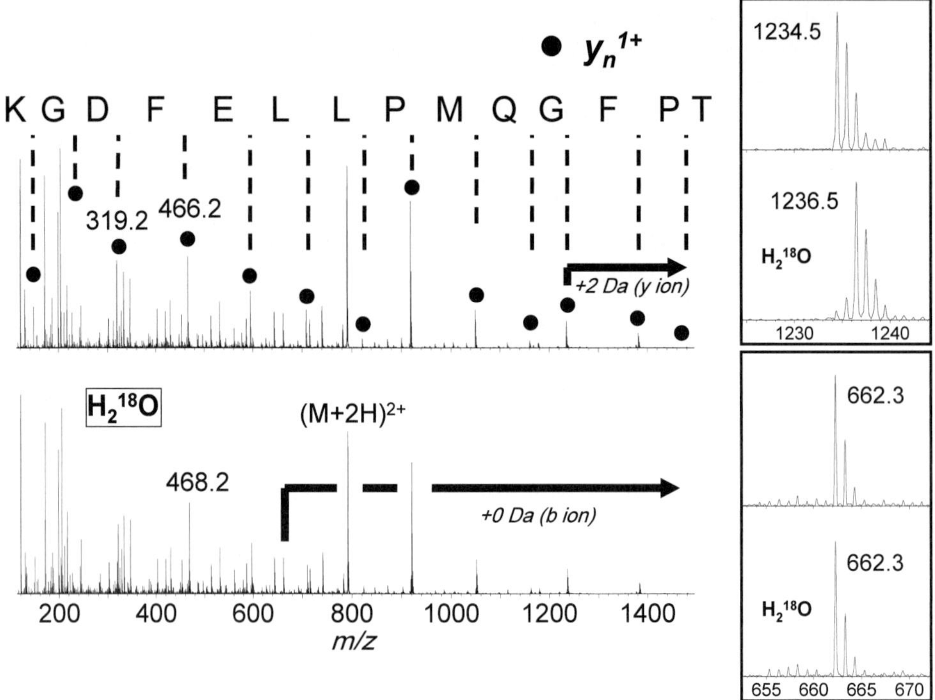

Figure 18.3. ESI-QTOF-MS/MS spectra of tryptic peptide generated from normal water (**top**) and ^{18}O-labeled water (**bottom**). Product ions shifted by 2 Da are y-type ions, whereas b ions show no increase in mass. Product y ions are indicated by the dot symbols.

ion-trap mass spectrometer. No clear identifications were obtained from database sequencing searching using both SEQUEST and Mascot search routines. Previous Edman sequencing of the N-termini and BLAST searches of the measured N-terminal sequence generated a potential identification from similarity with protein glutathione *S*-transferase (GST) from *Manduca sexta* (tobacco hawkmoth, GTS2_MANSE). Trypsin digestion with ^{18}O-labeled water allows y- and b-product ions to be differentiated and the C-terminal peptides to be identified (Figure 18.2). In combination with sub-5-ppm mass accuracy from the ESI-QTOF mass spectrometer, several tryptic peptides could be sequenced in their entirety (Figure 18.3). Using the *Manduca sexta* GST sequence as a template, partial protein sequences could be pieced together. Furthermore, additional sequencing data generated by in-source decay (ISD) MALDI-TOF of the intact proteins aided the alignment of the peptide fragments within the framework of the overall sequence. In-source dissociation of MALDI-generated ions shows promise for sequencing intact proteins because of its insensitivity to N- or C-terminal blockage, the long lengths of contiguous sequence obtained, and its relatively simple interpretation. Previously, we demonstrated a c-product ion series spanning 43 residues from regions near the N-terminus, obtained from 84.6-kDa cobalamin-independent methionine synthase (*metE*), from an *E. coli* cell lysate.[34] MALDI-ISD of Sf9-GST generated product ions beginning near residue 19 from the N-terminus and extending greater than 20 residues in length (Figure 18.4).

Endogenous protein contaminants from insect cell lines have been described previously.[35] However, the identification and sequence of these proteins were not established.

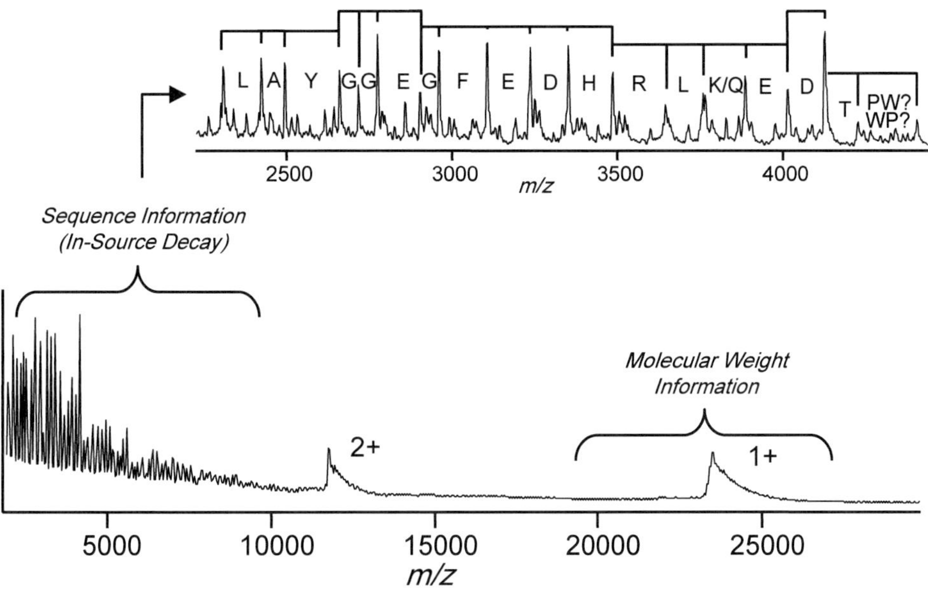

Figure 18.4. MALDI-TOF (Applied Biosystems Voyager Elite) in-source decay mass spectrum of one of the 23-kDa GST proteins from *Spodoptera frugiperda*. From the low-*m/z* region of the spectrum showing ISD product ions, considerable sequence information centered around residue 19 can be derived for this protein.

Very few protein sequences in Insecta are entered into the SWISS-PROT database, and even fewer protein sequences from genus *Spodoptera* are found in the database (78 in UniProtKB/SWISS-PROT Release 55.6, 01-July-2008). Moreover, very few GST proteins from the Lepidoptera family are available. Because none of the tryptic peptides from *Spodoptera frugiperda* GST matched identically to the template (*Manduca sexta* GST), mass-spectrometry-based methods for *de novo* sequencing is a powerful approach for identification of unknown proteins that may be homologous to proteins from other species.

MALDI–MS is especially favored for combining micro-scale chemical and/or enzymatic reactions with MS analysis, and it provides an alternative method for *de novo* sequencing. On-probe reduction of disulfide bonds and enzymatic digestion are commonly performed with microliter or even sub-microliter sample volumes. In contrast to many proteomic applications that focus on identifying proteins, the task of comprehensively characterizing a protein (e.g., a recombinant protein therapeutic) often requires more than simple trypsin digestion and subsequent MS analysis. In attempting to confirm protein termini by mapping tryptic peptides, frequently one or both terminal peptides are not recovered or are not observed. Obviously, access to both MALDI and LC-ESI techniques increases the chance that an important peptide will be revealed, but some peptides require a dogged pursuit if their mass and sequence are to be exposed.

Many structural biology and biochemistry studies rely on the precise identification of the N-/C-terminal sequence of a protein (e.g., signal peptide identification, determining the cleavage site and specificity of a novel protease, etc.). Digestion of proteins in arrays of carboxypeptidase Y at different concentrations with MALDI-MS-based readout can be employed to derive C-terminal sequence information for proteins.[36] The amino-terminus of a protein can be derived by wet chemistries similar to Edman degradation coupled to MS-readout. This method of "protein ladder sequencing," reported first by Kent and

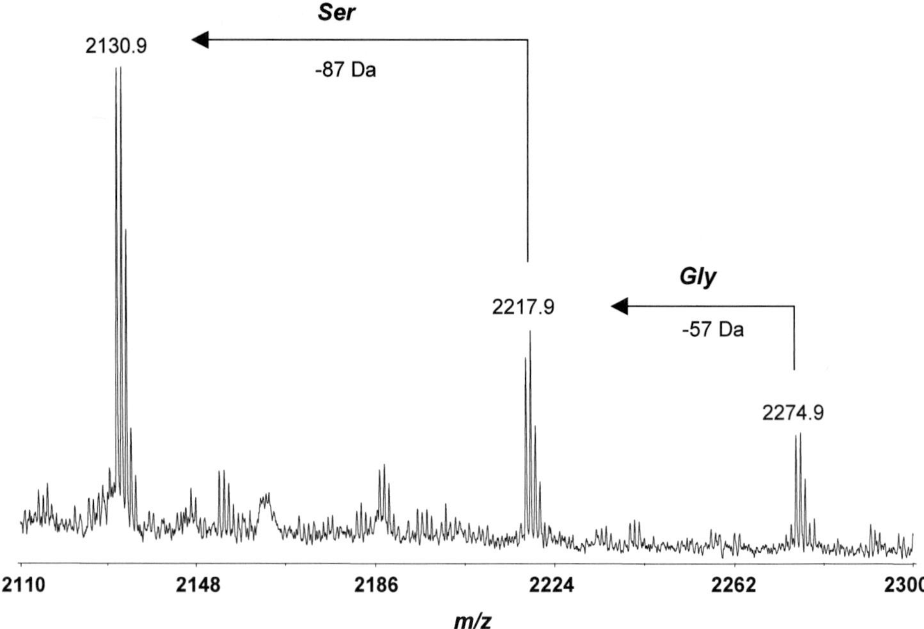

Figure 18.5. MALDI-TOF mass spectrum from protein ladder sequencing of the N-terminal peptide of cofilin (G-S-R-S-G-V-A-...). A stepwise degradation was carried out in the presence of 5% phenylisocyanate/95% phenylisothiocyanate. The peptide was subjected to two cycles of ladder generating chemistry.[38]

co-workers,[37] generates a series of sequence-defining concatenated sets of peptide fragments, each differing from the next by a single residue through stepwise protein degradation in the presence of a small amount of terminating agent. Figure 18.5 depicts an example of peptide ladder sequencing with a MALDI-MS readout. Cofilin is a member of the actin-depolymerizing factor (ADF)/cofilin family of proteins and is a key regulator of actin dynamics.[38] A 2-kDa peptide was generated from Lys-C digestion of cofilin (2 nmol) and isolated by HPLC with fraction collection. To confirm the first three residues as Gly-Ser-Arg, the lyophilized peptide was ladder sequenced. Peptide ladder sequencing consisted of two steps, as shown in Figure 18.5. After two cycles, the mass spectra of the cleavage products were recorded by MALDI–MS.

18.2.4 Non-Gel-Based Methods for Protein Quantification

Methods for protein detection using staining or incorporation of radioactive isotopes can be very powerful if combined with appropriate separation techniques (e.g., 2D-PAGE). One of the disadvantages with this type of approach is that if more than one protein is present in a band or a spot on a gel, it can be difficult, if not impossible, to determine which of the proteins constitute the major component. If the identification is part of a comparative experiment, it would be difficult to determine what protein(s) is responsible for the change in staining intensity. Comparisons between two states are usually performed in parallel experiments (e.g., 2D-PAGE), which require a large number of replicates to control for gel-to-gel variation.

An alternative approach that is becoming increasingly popular is to base protein quantification on the proteolytic peptides themselves using the information in the mass spectra, because the intensity of peptide ions can be correlated to their concentration. The conversion factor will depend on their respective ionization efficiencies, which will depend not only on the amino acid sequence of the peptide, but also on the conditions during ionization (e.g., what other ions are present at the same time competing for the ionizing protons). Other factors that affect the ion abundances are instrument-related effects such as the transfer of ions through the instrument and the detector response.

This is best compensated for by using an internal standard—that is, a peptide or protein that would have the same behavior and response by mass spectrometry. The ideal internal standard is the same protein or peptide. By generating peptide or protein with the same amino acid composition, but enriched for heavier isotopes (i.e., ^{18}O, ^{2}H, ^{15}N, ^{13}C), the chemical properties would be the same (though note that the retention in C18-reversed-phase chromatography can be different for peptides containing ^{2}H and ^{15}N than for those containing only ^{1}H and ^{14}N,[39] but they would have different masses that can be detected in the mass spectrometer). The larger the number of heavier isotopes incorporated into the peptide, the greater the mass difference between the heavy and light species. Either the isotope-labeled peptides can be synthesized, as in the case of AQUA (*vide infra*), or whole proteomes can be labeled and compared (i.e., so-called comparative proteomics). Labeling proteomes can also be achieved *in vivo* using metabolic labeling or by chemical or proteolytic labeling.

Recently a large number of promising studies have been reported that rely on label-free quantification. The studies depend on highly reproducible sample introduction, chromatography, and mass spectrometry to minimize technical variation and the normalization is extraordinarily important. In the following sections, the different non-gel-based quantification schemes will be discussed with illustrating examples from the literature. In some cases, quantification is based on MALDI-MS, but in general the quantification approach would have been similar regardless of the type of mass spectrometer available.

18.2.4.1 Metabolic Labeling

Metabolic labeling has been used to study intracellular processes for more than half a century, and it has been indispensable for the elucidation of many biochemical pathways and kinetic studies of protein translation, targeting, and processing.[40] Most of these studies relied on metabolic precursors or amino acids labeled with radioactive isotopes that are relatively easy to detect at very low levels. Metabolic labeling using stable isotopes is also routinely used in structural studies by nuclear magnetic resonance (NMR) spectroscopy. In contrast, comparative proteomics using metabolic incorporation of stable isotopes was not demonstrated until the late 1990s by Oda et al.[41] While radioactive isotopes can be detected at low levels of incorporation, stable isotopes need to be incorporated at high levels for detection by mass spectrometry. On the other hand, stable isotopes do not have the same safety requirements.

There are two main approaches for metabolic labeling: through ^{15}N-enriched media or through labeled amino acids. In the first approach, the cells are labeled by growing them on a media where the nitrogen source is enriched in ^{15}N. For each cell division the number of ^{15}N atoms incorporated increases, until the incorporation reaches the level of enrichment of the media. The two samples that are to be compared and that are enriched for either ^{14}N or ^{15}N are then mixed together. The proteins are extracted, separated, and enzymatically digested. The resulting peptides can then be quantified from the mass spectra or the LC response, and protein ratios can be calculated. It is important to note that the shape of the isotope envelope is different when the peptides are enriched for ^{15}N, which needs to be taken into consideration when the quantification is performed. It is important also to realize that the number of nitrogen atoms in the peptide depends on the amino acid sequences, which makes it difficult, if not impossible, to identify pairs of heavy and light labeled peptides prior to identification. On the other hand, the identification of the two different forms is relatively simple.

In the study by Oda et al.,[41] differences between wild-type *Saccharomyces cerevisiae* (yeast) and a strain lacking the G1 cyclin *CLN2* could be quantified. It was shown also that it was possible to quantify differences in the phosphorylation states with the same methodology. Since that report, several multicellular organisms have also been labeled: *Caenorhabditis elegans* (nematode),[42] *Drosophila melanogaster* (fruit fly),[42] and *Rattus norvegicus* (rat).[43,44] In the first study involving metabolic labeling of rats, Wu et al.[43] showed that the incorporation of ^{15}N varied between different types of tissue. In tissue with high protein turnover (e.g., liver and plasma), they achieved higher than 90% mean enrichment of ^{15}N. In other tissues with slower turnover, the enrichment was much lower (e.g., brain showed only 74% enrichment). The enrichment also varied between the amino acids within the tissues (e.g., in the liver, isoleucine showed 92.85% $\pm$ 0.03 enrichment, while glycine only showed 86.53% $\pm$ 0.17). To compensate for the variable incorporation, they developed software to generate extracted mass chromatograms that contained the whole isotope envelope for the ^{15}N-enriched peptides. The labeled tissue was then used as internal standard when comparing liver from untreated and cycloheximide-treated rats. Cycloheximide is a compound known to inhibit protein elongation. In the study,[43] 310 proteins were quantified, in which 65 proteins were found to have greater than two-fold change between the treatments. In a later study,[44] the authors improved the protocol for breeding the rats and showed that by labeling the mice for two generations they could achieve close to 95% ^{15}N-enrichment in both liver and brain tissue.

In the second approach, the cells are labeled by feeding them amino acids containing a set number of stable isotopes.[40] Only essential amino acids can be used (i.e., amino acids that the cells cannot synthesize themselves), and it is crucial that the media do not contain any unlabeled forms of the amino acids. It is equally important that the essential amino acid is not metabolized into other amino acids inside the cells—for example, the metabolic conversion of arginine to proline in eukaryotes.[45] Amino acids that are not essential for the organism can be essential in cell culture. One such example is arginine, which is not essential for adult vertebrae but cannot be synthesized by the cell in cell culture. Normally, only one or two different amino acids are used at a time. Some of the advantages using this method are as follows: The cell culture techniques for the labeling are the same as for normal cell cultures; it is easy to achieve close to 100% incorporation into all proteins relatively quickly (i.e., after six cell divisions); normal protocols can be used for protein extraction, digestion, and peptide separation; the mass spacings between the heavy and light peptides are consistent and can be calculated, making it possible to identify pairs without knowing their sequence; the shape of the isotope envelope does not change significantly. A disadvantage is that the cells need to be labeled in cell culture, which means that only unicellular organisms or cell lines derived from higher organisms can be used.

Three groups, headed by Chen, English, and Mann, published independently studies using this technique in 2002,[46–48] and all three used deuterated (^{2}H) leucine to label the cells. The term "SILAC" (*S*table *I*sotope *L*abeling of *A*mino acids in *C*ell *c*ulture), coined by Mann and colleagues, has become commonly used for the method. Shortly thereafter, Mann and colleagues improved the methodology[49] using arginine and lysine, and by switching to amino acids enriched for ^{13}C, which in contrast to ^{2}H does not have differences in retention time compared to the lighter isotope.[39] Using arginine and lysine for the labeling ensures that all tryptic peptides are labeled, which maximizes the number of peptides that can be quantified. The group has since used the methodology in a number of ground-breaking papers: in the determination of protein–protein interactions (to distinguish specific from nonspecific co-purification),[50] the determination of the effect of three different metabolic inhibitors that each affect the morphology of the human nucleolus (i.e., isolated organelles),[51] the determination of the changes in the yeast phosphoproteome in response to a pheromone factor,[52] the use of multiplexing in SILAC (i.e., the direct comparison of more than two states),[53] and the determination of temporal responses to stimulation of the epidermal growth factor receptor.[53,54] Examples from other groups using SILAC are: a study by Grønborg et al. to identify markers for pancreatic cancer,[55] a study to quantify the changes in phosphorylation state of the Kv2.1 potassium channel in response to calcineurin,[56] the quantification of changes in the proteome during human B-cell differentiation,[57] a study by Gruhler et al. to determine the changes in *Arabidopsis thaliana* suspension cells in response to abiotic stress,[58] the determination of the changes in the lipid raft proteome in response to viral infection,[59] and a study by Naumann et al. quantifying the changes in the thylakoid membrane from *Chlamydomonas reinhardtii* (a unicellular algae) in response to iron deficiency.[60]

Metabolic labeling is well-suited when subcellular fractionation is necessary or complexes are to be compared, and it is likely to be an important tool for comparative proteomics in the future. However, it should be pointed out that cell lines, derived from multicellular organism, will have significant differences from the intact organism. An excellent review by Beynon and Pratt[40] discusses the experimental issues with metabolic stable isotope labeling and describes a large number of studies published.

18.2.4.2 Chemical Labeling

In contrast to metabolic labeling, chemical labeling is not limited to samples from simple organisms, organisms from which cell cultures can be derived, or tissue types. Instead the main issues are the types of amino acid that is to be labeled, the yield of reaction, the yield of by-products, and stability of the label. It is also easier to vary the difference in mass between the two forms of isotope labels, or even multiplex the procedure. The choice of amino acid to label is important not only for the type of protein labeled (e.g., integral membrane or soluble proteins), but also for the number of quantitative measurements that can be obtained. Some amino acids (e.g., tryptophan) do not occur very frequently in proteins; and if used as a target for labeling, they would not result in a high degree of incorporation of stable isotopes into the proteins/peptides. The presence of certain amino acids can also vary between organisms, but also between organelles within the cells. For example, 40% of the proteins in the thylakoid membrane of the plant model organism *Arabidopsis thaliana* lack cysteines.[61] To date, the two functional groups that have received the most attention are the sulfhydryl groups on cysteines (e.g., isotope-coded affinity tag, or ICAT) and the free amines on lysines and N-termini (e.g., isobaric tags for relative and absolute quantification, or iTRAQ).

Any chemistry that does not yield full incorporation of the label would obviously not be useful, but it is equally important that the reaction produce a minimum of side products. Even when the reaction conditions are closely monitored, side products can sometimes not be avoided (e.g., partial labeling of tyrosines when using the iTRAQ reagent). The side products do not necessarily cause major problems because the quantification can be corrected for computationally. When reversible labels are used, it is crucial to minimize the loss of the label. Reversible labels have been used successfully (e.g., isotope labeling through esterification[62]), but the vast majority of the studies published to date using chemical labeling have used stable labels.

A few "nonstandard" chemical labels that have been used successfully include dimethylation of free amines (i.e., lysines and N-termini) using reductive amination,[63] guanidination of lysines,[64] and alkylation of cysteines by acrylamide.[65,66] In practice, the commercially available labels (e.g., ICAT and iTRAQ) have received the most widespread use. Ones that are based on chemical reactions require a deeper understanding of chemistry (in contrast to the kits) and also computational skills, because there are very few generally available programs that can be used for automated quantification of the data sets.

a. ICAT. The first chemical tag used for quantitative proteomics was the so-called ICAT developed by Gygi et al.[67] The tag was composed of three parts: a biotin moiety for affinity purification, a linker region containing the stable isotopes, and a thiol-specific reactive group. The stable isotopes consisted of eight ^{1}H or ^{2}H, resulting in a mass difference of 8-Da for each tag added to the peptide between the light and the heavy forms. Reducing the complexity of the sample to only peptides containing cysteines, the authors showed that they could dig deeper into the yeast proteome and identify low-abundance proteins. The restriction of peptides also made the search space smaller, thereby simplifying protein identification. Quantification using the tags showed a relatively small variation and error (within 12% for a marker mixture). Even though the ICAT approach proved to be very powerful, a number of disadvantages were reported in the literature. Firstly, incorporation of ^{2}H has been shown to have slightly different retention times in reversed-phase chromatography; the more of the isotope incorporated, the larger the difference.[39] If the retention time is sufficiently different, isotope-labeled peptides might experience different suppression effects increasing the error in quantification. Secondly, the retention time is greatly influenced by the biotin tag so that tagged peptides eluted in a relatively small window.[68] Thirdly, the tags were relative large (close to 600 Da). The increased size significantly reduced the effective size of the precursor peptides,[68] with the result that the identified peptides were shorter and that many peaks in the MS/MS spectrum originated from fragmentation of the tag itself.[69]

A second generation of the tag, cICAT, was developed where ^{13}C was used instead of ^{2}H and an acid-cleavable site was introduced into the linker.[68,70] In contrast to ^{2}H and ^{15}N, ^{13}C-labeled peptides have identical retention times compared to the light isotope version, and the removal of the biotin moiety increased the spread in reversed-phase LC by a factor of three.[68] The tagging method has been used in a large number of studies in a range of organisms, including bacteria,[71] yeast,[70] maize,[72] mouse,[68] and humans.[73]

b. Isobaric Tags. When using isobaric tags for stable isotope labeling, as the name suggests, there are no mass differences between the labeled peptides. Instead, reporter ions are generated when the peptides are fragmented. This is done by using a reagent consisting of three parts: a reporter group, a mass balancing group, and a reactive group (e.g., amine-specific, as in the case of iTRAQ). The peak area of the reporter ions are compared between the different states to determine the relative amounts, while the fragment ions derived from the peptide have identical mass regardless of the labeling tag and overlap, thereby improving overall sensitivity. In contrast to other isotope labeling techniques, isobaric tags do not increase the complexity of the sample [e.g., tags that generate two different forms double the number of peptides in the sample, and tags that generate three forms (i.e., multiplexing) generate three times the number of peptides]; and in principle, all peptides that are identified are also quantified, which is not the case for other labeling techniques. In practice, some spectra still need to be discarded—for example, due to fragmentation of more than one peptide at the same time. Sensitivity is improved, because peptides that by themselves are too low to be selected for fragmentation now contain the contribution of two or more states. In the case of iTRAQ, the tags have been shown to not interfere with affinity purification.[74,75]

Isobaric tags were introduced by Thompson et al.[76] under the name "tandem mass tags." The tags had a relatively large mass (close to 540 Da) and generated reporter ions at 287.2 and 290.2. About one year later, Ross et al.[77] reported tags with significant improvements. Their iTRAQ method consisted of four different tags of relatively small mass (144 Da added to the modified peptides) that generated reporter ions in a region with a minimum of interfering peptide fragment ions (*m/z* 113–117). The four different tags made multiplexing possible, significantly reducing the number of replicate measurements needed. More recently, the iTRAQ strategy has been expanded by the availability of an 8-plex version of the isobaric reagent for the quantification of proteins in complex mixtures, with reporter ions measured at *m/z* 113–119 and 121. The iTRAQ method also made it possible to introduce internal standard peptides (e.g., AQUA, or absolute quantification[78]) into the quantification strategy.

Using a clever combination of biochemical separation and quantification, Lilley and co-workers[79] devised an assay to determine subcellular localization. Organelles and large complexes were fractionated using density gradient centrifugation, and the relative amounts of individual proteins were determined using stable isotope labeling. By comparing the abundance profiles of protein markers from organelles and profiles of proteins with unknown localization, they verified the markers by their co-migration in the density gradient and were able to assign location to proteins of previously unknown location. Initially the comparisons were based on ICAT and SDS-PAGE,[79] which was later compared to iTRAQ and 2D-LC.[80] By using multiplexing (i.e., comparing four fractions at the same time instead of two), the confidence of localization was greatly enhanced. Even with partial overlap in the density gradient, "true" localization could be deciphered for close to 527 proteins.

Jensen and co-workers[75] showed that isobaric tags can be used in combination with 2D-PAGE and affinity purification using Fe(III)-IMAC (immobilized metal ion affinity chromatography) to quantify differences in phosphorylation states. They were able to quantify the relative amounts of a peptide with one phosphate at different amino acid positions by HPLC, because the three isoforms have different retention times.

Zhang et al.[74] used iTRAQ in combination with immunopurification of phosphorylated tyrosines and IMAC to study the response to stimulation of the epidermal growth factor receptor. By comparing samples collected at different time points (prior to stimulation; and 5, 10, and 30 minutes post-stimulation), the authors showed temporal differences between phosphorylation sites. In total they identified 78 phosphorylated sites from 58 different proteins and were able to group the sites according to quantitative profiles (i.e., the timing of the phosphorylation) to gain insight into the signaling cascade.

Gan et al.[81] used iTRAQ to perform repeated experiments to estimate the variation caused by technical variation (by repeated measurement of the same sample), experimental variation (repeated measurement of several samples), and biological variation (measurement of independent starting material). The study showed that for iTRAQ, the biological variation was larger than both the technical and experimental variation, which was expected.

18.2.4.3 Proteolytic ^{18}O Labeling

As discussed for *de novo* peptide sequencing, by performing the proteolysis in $H_2{}^{18}O$, stable isotopes can be incorporated into the peptides. However, the proteases interact not only with the substrates (proteins), but also with the products (peptides) and each time there is a chance of a second ^{18}O being incorporated. While the first incorporation is fast with high yield, the second incorporation of ^{18}O is slow. To achieve more than 98% incorporation of two oxygens, five rounds of binding and release of the peptide are required. The rate of the incorporation is dependent, on peptide sequence, and therefore it varies from peptide to peptide. The maximum yield of the incorporation is also dependent on the purity of the $H_2{}^{18}O$. This can result in incomplete incorporation and with a resulting mixture of peptides with one or two ^{18}O incorporated. Because the difference between labeled and unlabeled is only 2 Da, the isotope envelopes overlap, and this needs to be taken into account.

Mirgorodskaya et al.[82] was one of the first reports to demonstrate that the method could be used for relative protein quantification. While the study demonstrated quantification of single proteins, a study was published shortly thereafter using a similar method for the comparison of two viral proteomes.[33] Several groups have attempted to compensate for the limitations of the technique by (a) improving the reaction conditions (see review[83]) uncoupling the digestion and labeling steps and optimizing them separately[84] and (b) using immobilized trypsin to increase the enzyme concentration. The presence of protease also can cause some back exchange, which can hinder quantification. Labeling needs to be performed during the digestion or immediately after, which can also be problematic.

Labeling using ^{18}O has been used successfully in the quantification of smaller proteomes and protein mixtures, but there are still issues (such as the variable incorporation level, ease of use, software for quantification, etc.) that need to be improved before the method will get widespread use in large-scale quantification of highly complex proteomes.

18.2.4.4 Absolute Quantification

Most of the techniques used in comparative proteomics determine the *relative* amounts between two proteomes. In contrast, AQUA determines the absolute concentration of a protein in a sample. Already in 1996, Barr et al.[85] demonstrated that synthesized peptides containing stable isotopes can be used as internal standards to determine absolute concentrations. The method did not get widespread attention until after Gerber et al.[78] showed that the technique can be used in a proteomic setting compatible with SDS-PAGE and is able to measure post-translational modifications. Peptides matching the proteins of interest were synthesized with one amino acid labeled with stable isotopes. The peptides were added to the digestion mixture, and the ratio between the unlabeled and labeled peptides were quantified using LC-MS by selected reaction monitoring with two transitions for both the native and the labeled peptides. The method was demonstrated by the quantification of two low-abundance yeast proteins and quantification of the phosphorylation sites on a human phosphoprotein. When selecting a peptide sequence as the internal standard, sequences containing amino acids that are frequently modified (e.g., methionine) should be avoided. When quantifying a sequence containing a ragged end (i.e., RR, KK, KR, or RK), which normally generates both the fully cleaved and the miscleaved peptide, the labeled peptide should be extended so that it contains the cleavage site. Even after careful selection of the sequence to avoid variable modifications, there are some pitfalls. Not all peptides will give good signal intensity or they might not be detected at all (owing to issues with digestion, absorption to surfaces, or poor ionization). Synthesizing peptides with stable isotopes are expensive, and each peptide can only be used for one peptide sequence. During the past few years, several groups have published solutions to some of the problems faced with AQUA-based quantification.

Beynan et al.[86] addressed the costs for peptides syntheses by generating an artificial gene containing 25 tryptic peptides from 25 different proteins of interest, together with a C-terminal histidine tag for purification. A plasmid containing the gene was overexpressed in *E. coli* grown on ^{15}N-enriched media. Following cell lysis and affinity purification, the protein concentration was determined and the metabolically labeled protein was used for AQUA experiments. Because plasmids can be maintained infinitely by amplification (though it would be important to constantly make sure that no mutations are introduced), they provide an unlimited supply of labeled peptides. The peptides used for the quantification were called "Q peptides," and the artificial gene was called "QCAT" (from concatenation of Q peptides).

As peptides are chosen for synthesis to use for quantification, one can never be sure that the peptide will be detected by MS (unless it has already been observed in previous experiments). Mallik et al.[87] has tried to improve the chances of finding suitable peptides by creating an algorithm that can predict "proteotypic" peptides. Their definition of a proteotypic peptide is a peptide that is detected at least 50% of the time that the protein is identified. To train the predictor, more than 600,000 mass spectra matching more than 16,000 peptides from approximately 4030 yeast proteins were used. The predictor was then validated on a human test set. However, experimental confirmation (i.e., to what degree peptides suggested by the algorithm are detected in a quantification experiment) is still lacking. Not only did the introduction of iTRAQ and multiplexing with SILAC open up the possibility of simultaneously quantifying relative amounts from different samples, but also the inclusion of AQUA peptides allowed the accurate determination of the absolute amount of specific proteins.[77]

18.2.4.5 Label-Free Quantification

Label-free quantification is a very attractive method, because there are no extra costs involved. Instead, there are much higher demands on the reproducibility of technical parameters, such as sample loading, chromatography, and mass accuracy, and more experimental replicates are needed. In the following sections, different approaches to label-free quantification will be discussed: the extracted ion chromatogram (XIC), spectral counting (SC), the protein abundance index (PAI), and the LCMS^E method.

The first approach, XIC, is based on extracting the chromatograms for identified peptide ions and calculating the areas under each of the LC traces. Bondarenko et al.[88] showed that by adding all peptide areas matching a protein and using an internal standard, they could accurately determine the relative amounts when spiking 200 fmol and 400 fmol of myoglobin to human sera. The standard could be either (a) an endogenous protein that does not change between the samples or (b) an exogenous protein added prior to digestion. Using a similar approach, Wang et al.[89] created software to identify proteins within a sample that do not change and then used the protein(s) as internal standard(s). One of the files (i.e., chromatograms) was then chosen arbitrarily as a reference file, and all the other files were normalized to it. They then verified the method by adding variable amounts of myoglobin and cytochrome C to human sera. The method was improved further,[90] and it was used to compare a human colon cancer cell line and a cell line derived from the first but lacking p53. Out of 118 identified proteins, many showed a statistically significant difference between the two samples.

The second approach, spectral counting (SC), is based on the assumption that the more abundant a protein is, the more frequently its peptides will be selected for fragmentation (MS/MS). By using protein markers added at different concentrations to a yeast lysate, Liu et al.[91] showed that the number of scans (or MS/MS spectra) matching a protein in two-dimensional LC–MS/MS experiments had a linear relationship to the protein concentration covering two orders of magnitude. Similarly, Zybailov et al.[92] compared spectral counting with metabolic (^{15}N) labeling and showed that there is a positive correlation between the two, verifying that spectral counting generates meaningful quantitative data. They suggest that the deviation from a perfect correlation is due to the limited dynamic range of metabolic labeling and to outliers. A later study that also compared metabolic labeling with SC found that the primary difference between SC and metabolic labeling was in sensitivity to changes in expression when the peptide ions either have a low signal-to-noise ratio or only have few spectral counts.[93] Metabolic labeling was more sensitive and better at reliably distinguishing small differences. At a twofold change, metabolic labeling was able to detect 50% more significant changes in protein abundance. At high signal-to-noise ratio (and SC), within the dynamic range of metabolic labeling, the two methods generated similar results. The SC method was improved further by normalizing the spectral count to the protein length and in addition, a new normalized abundance factor was defined: NSAF (or the number of spectral counts matching a protein divided by the length of the protein and the sum of all spectral counts divided by their protein length).[94] The natural logarithm has a Gaussian-shaped distribution and standard statistical analyses can be performed. By analyzing enriched membrane fractions from yeast, statistically significant changes as small as 1.5-fold could be detected.

Old et al.[95] compared XIC and SC quantification of treated and untreated human cell lines. It was determined that SC is more sensitive to detect changes, but that XIC is more accurate in determining the ratios. They also found that SC has a large dynamic range and that the ratios from SC matched the expected within a factor of 2.3. The errors were significantly higher with SC than with XIC. On the other hand, SC was able to quantify three times more protein than XIC and all of the proteins quantified by XIC were also quantified by SC. SC showed good reproducibility between 2D-LC experiments and had the advantage that two proteins do not need to have identical peptides in order to be compared.

The third method of label-free quantification is based on the fact that the higher the protein concentration, the larger the number of matching peptides identified. Two groups independently used this phenomenon to define a score for relative abundance. Sanders et al.[96] defined the score as the average number of peptides matching to one protein in a set of fractions divided by the molecular mass and multiplied by 10,000. They used the score to distinguish specific from nonspecific interactions from immunopurified complexes involved in transcription in yeast. Rappsilber et al.[97] named the score "protein abundance index" (PAI) and defined it as the number of observed peptides divided by the number of theoretically observable peptides. Similar to the Sanders et al.[97] report, they used the score to identify members of the human spliceosome. Later they improved the score by showing that the value, $10^{PAI} - 1$, is proportional to the protein concentration in a mixture and called the new index the "exponentially modified protein abundance index" (emPAI).[98] They showed that in a mixture, the protein content (molar %) equals the emPAI divided by the sum of the emPAI for all proteins present multiplied by 100. To validate the method, they determined the absolute amount of 46 proteins in a lysate from a metabolically labeled cell line (SILAC) using synthetic peptides (AQUA). The protein abundances of the 46 proteins determined by SILAC were then compared with the emPAI, and the deviations were on average within 63%.

The last method of label-free quantification is the so-called LCMS^E method presented by Silva et al.[99] The methods discussed previously involve data-dependent acquisition (i.e., an MS scan followed by a number of MS/MS scans of selected peptide precursors). In contrast, LCMS^E is a data-independent method where the instrument is measuring in MS mode the entire time (i.e., no precursor ion selection) but cycles alternately through high and low collision energies. At low energy (10 eV) the peptide masses are detected, and then at higher energy (by raising the collision energy in the collision cell of a quadrupole time-of-flight analyzer to 28–35 eV) the product ions are detected. Each cycle takes approximately 4 seconds. Precursor and product ions are matched by chromatographic co-migration, and the peptides are identified using reconstructed MS/MS spectra from the co-migrating fragments and by their accurate mass measured, within 5 parts-per-million (ppm) using a lock mass. The samples are analyzed three times and similar total ion signals are required. Only peptides that are present in all three runs are used for subsequent analyses. The three runs are normalized to one protein that statistically does not vary between the runs (i.e., consistent signal and large sequence coverage). The peptide signal intensities are then used to calculate peptide ratios. The methodology relies heavily on high mass accuracy and reproducible chromatography.

18.2.5 Top-Down Protein Analysis

18.2.5.1 Defining "Top-Down" Mass Spectrometry Methodologies

The molecular mass of an intact protein defines the native covalent state of a gene's product including the effects of post-transcriptional/translational modifications, and associated heterogeneity, that are modulated by the actions of other gene products. According to Kelleher, "the most rigorous definition of a top-down experiment involves high-resolution measurement of an intact molecular weight value M_r and direct fragmentation of protein ions in the gas phase. The accuracy of an M_r value can vary on different instruments, but it is most useful if the value is <2 Da in order to better match the accuracy obtained for smaller protein fragments (often in the low parts-per-million range).[100]" Moreover, the bottom-up strategy "refers to the dominant experiment in proteomics today wherein digestion to small peptides occurs without determining an accurate M_r value of the intact protein."

Certainly, one can dispute the specificities of these definitions. Are measurements with resolving power sufficient for resolution of the isotopic envelope of the intact protein required for the top-down methodology? In most cases, high resolving power translates directly to higher-accuracy measurements. Because dissociation of multiply charged molecules usually leads to the formation of multiply charged product ions, resolution of the isotopic ion distributions of the products allows for determination of product ion charge states. Although lower-resolution top-down measurements with ESI-MS/MS were demonstrated in the early 1990s using triple quadrupole instruments,[101,102] and later with QTOF analyzers,[103] higher-resolution measurements with Fourier transform ion cyclotron resonance (FT-ICR)[104–106] and most recently demonstrated with OrbiTrap analyzers[107] reduce uncertainties that allow completely unknown protein sequences and PTMs to be addressed.

Although the most practiced method for MS-based protein sequencing combines proteolytic cleavage of the intact protein with subsequent sequencing of the resulting peptide fragments by tandem mass spectrometry, especially with LC–MS/MS and MALDI–MS/MS, there is value in the direct measurement and fragmentation of intact proteins, or top-down sequencing.[100,108] The 2002 Nobel Prize in Chemistry was awarded to John Fenn and Koichi Tanaka for the development of two different soft ionization techniques that allowed MS of large intact biological macromolecules; implicit in this is the fact that such measurements are useful. The fragmentation pattern from CAD of large proteins can generate sufficient information for identification from sequence databases, particularly when combined with accurate mass measurements of both the intact molecule and its product ions.[109] Enhanced efficiency for CAD of multiply charged molecules[110] generates multiply charged sequence-informative product ions from ESI-produced gas-phase ions for biomolecules as large as 66-kDa serum albumin proteins[102] and more recently from proteins greater than 200 kDa.[111] CAD-MS of multiply charged proteins often yield multiply charged products. The ESI-MS/MS products from proteins can be interpreted as resulting from either the N- or C-termini, or from cleavage near proline residues, even with data from low-resolving-power analyzers. However, the determination of precursor and product charge state is amenable to higher-resolving-power instruments such as FT-ICR mass spectrometers, yielding data that are interpretable unambiguously. This top-down sequencing approach, as coined by McLafferty,[108] can yield substantial sequence information by fragmenting intact proteins directly, especially from FT-ICR-MS armed with infrared multiphoton dissociation (IRMPD)[112] and more recently with electron capture dissociation (ECD)[105,113] and electron transfer dissociation (ETD).[114]

Technologies that offer the opportunity to use MS for intact protein mass determination allow rapid assessment of signal peptide cleavage sites and post-translational heterogeneity. A well-resolved mass spectrum of an intact protein defines its native covalent profile as well as its associated heterogeneity. A comprehensive mass spectrometric approach that integrates both intact protein molecular mass measurement (top-down) and proteolytic fragment identification (bottom-up) would represent a powerful approach for proteome/protein characterization.

18.2.5.2 Electron Capture Dissociation and Electron Transfer Dissociation for Top-Down Mass Spectrometry

New modes of ion activation/dissociation have enhanced the prospects for more complete protein sequence characterization. Specifically, ECD, a mode in which low-energy electrons are captured by multiply charged molecules generated by ESI to effect nonergodic dissociation of the biomolecule, has been shown to be more efficient for top-down protein sequencing compared to CAD. McLafferty has demonstrated that the combination of ECD followed by heating (e.g., IRMPD) increases sequencing yields even further.[105] The advantages of ECD for top-down sequencing are illustrated by the following example. The pathological hallmark of Parkinson's disease (PD) is the presence of intracellular inclusions, called Lewy bodies (LB), in the dopaminergic neurons of the substantia nigra and several other brain regions. Filamentous α-synuclein protein is the major component of these deposits, and its aggregation is believed to play an important role in PD and several other neurodegenerative diseases. Using a hybrid linear ion-trap–ion cyclotron resonance Fourier transform mass spectrometer (Thermo Fisher LTQ-FT) and the ProSight PTM computational tool developed by Kelleher and co-workers[115] to interpret the fragmentation pattern, conventional CAD of the 14+-charged α-synuclein protein produces b- and y-type products that cover approximately 50% of the sequence. The high resolving power of the FT-ICR is indispensable for charge state determination of the product ions (Figure 18.6). ECD of the same 14+-charged α-synuclein protein produces c- and z-type products that cover approximately 50% of the sequence as well. However, ECD generates fragments from

14459.3
1033.8849
15+
1113.3384
965.0270
1205.9480
904.7134
851.5537
14420 14440 14460 14480 14500 14520
mass
1315.4905
10+
1447.0382
1607.5974
1808.4288
400 600 800 1000 1200 1400 1600 1800
m/z

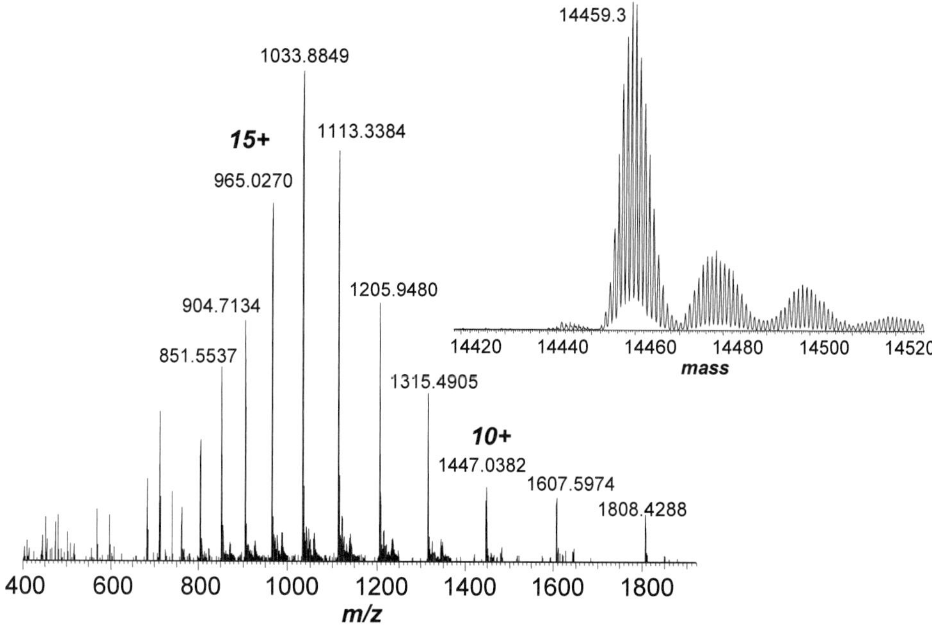

Figure 18.6. High-resolution nanoESI-FT-ICR MS (7 Tesla Thermo Fisher LTQ-FT) of human α-synuclein in water/acetonitrile with 0.1% formic acid. **Inset:** Spectrum deconvoluted to the mass domain.

regions not covered by CAD. Combining the information from CAD and ECD yields over 60% of the complete sequence from experiments that spanned only a few minutes (Figure 18.7).[113]

Especially promising is the capability of ECD for determining the sites of protein post-translational modifications. Glycosylation and phosphorylation moieties are often labile under standard activation modes. The resulting MS/MS experiments yield little information on modification sites. ECD studies suggest that bonds remote to the modification sites are more labile, and the modification remains intact for MS/MS experiments. This is especially true for modifications such as phosphorylation,[116–120] acetylation,[121–123] deamidation,[124]

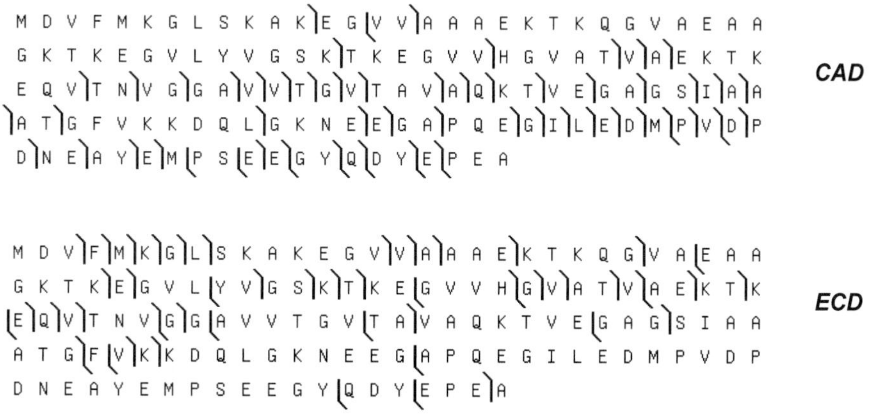

M D V F M K G L S K A K E G V V A A A E K T K Q G V A E A A
G K T K E G V L Y V G S K T K E G V V H G V A T V A E K T K
E Q V T N V G G A V V T G V T A V A Q K T V E G A G S I A A
A T G F V K K D Q L G K N E E G A P Q E G I L E D M P V D P
D N E A Y E M P S E E G Y Q D Y E P E A
CAD
M D V F M K G L S K A K E G V V A A A E K T K Q G V A E A A
G K T K E G V L Y V G S K T K E G V V H G V A T V A E K T K
E Q V T N V G G A V V T G V T A V A Q K T V E G A G S I A A
A T G F V K K D Q L G K N E E G A P Q E G I L E D M P V D P
D N E A Y E M P S E E G Y Q D Y E P E A
ECD

Figure 18.7. Sequence coverage from CAD and ECD of the 14+ charged precursor ion of α-synuclein protein.

trimethylation,[122] N- and O-linked glycosylation,[118,125] and disulfide bond formation.[126] It has been reported also that isobaric Ile and Leu can be distinguished by ECD.[127] Work by Kruppa and co-workers demonstrate the advantages of ECD for elucidating the sites of protein cross-linking.[128–131] McLafferty's work has promoted the capabilities of ECD for studying the tertiary structure and conformers of proteins.[132–134]

Similar to ECD, electron transfer dissociation generates fragmentation of biomolecules by transferring an electron to the multiply charged protein cation, producing an odd electron species which then undergoes cleavage in the gas phase.[114] The electron for ETD is produced by a reagent gas, such as fluoranthene, that is subjected to negative-ion chemical ionization to produce a radical anion and subsequently transfers an electron to the polypeptide ion species. Also similar to ECD, ETD generates primarily c- and z-type products with higher sequence coverage than conventional CAD processes.[114] The analysis of proteins on a chromatographic timescale has been achieved via a combination of ETD for dissociation and proton-transfer reaction (PTR) for charge state reduction.[135] Charge state reduction is a useful means for simplifying the ETD-MS/MS spectra, because multiply charged product ions are charge-reduced to the 1 + state by ion–ion reactions with a suitable base, such as benzoic acid. The ETD-PTR top-down feature performed with lower resolving power linear ion-trap analyzers on a chromatographic timescale has been demonstrated for small proteins, such as 8.6-kDa ubiquitin and ribosomal proteins,[135] and proteins from a soluble extract of *E. coli*.[136] With higher-resolving-power analyzers, such as the OrbiTrap, an option that does not require PTR charge reduction will be available, as product ion charge states can be measured directly from the resolved isotope peaks. Preliminary top-down ETD-OrbiTrap[137] and ETD-FT-ICR[138] analysis of ubiquitin and 17-kDa myoglobin demonstrate the potential of this methodology.

18.3 NONCOVALENT PROTEIN COMPLEXES

18.3.1 ESI-MS and Protein Stoichiometry

The role of protein assemblies in normal cellular processes and diseases warrants a practical and sensitive method for the study of large macromolecular complexes. X-ray crystallography and nuclear magnetic resonance (NMR) spectroscopy provide unrivaled high-resolution structural information. However, protein crystallization is traditionally time-consuming, NMR is limited by the size of the protein target, and both methods require large quantities of purified analyte compared to MS. Without the requirement of high-resolution structures, ESI-MS can provide crucial functional information about proteins and their noncovalent complexes, such as assembly states, stoichiometry, conformational changes, and binding affinity constants.[5,6] These details are comparable to that offered by modern analytical ultracentrifugation techniques, chromatographic methods (with light scattering and fluorescence detection), surface plasmon resonance (SPR), and differential scanning and isothermal calorimetry. However, ESI-MS sensitivity is generally superior. Furthermore, ESI-MS techniques may be used in isolation or be coupled with on-line and off-line separation methods to characterize protein components in heterogeneous mixtures. MS is more generally suited for these measurements because it measures an inherent fundamental property of proteins and protein complexes (i.e., molecular mass) with an accuracy that substantially surpasses other analytical techniques.

ESI mass spectra of proteins are usually obtained with solution conditions that unfold the regular, ordered structure of proteins, such as 50 : 50 water:acetonitrile (v/v) mixtures

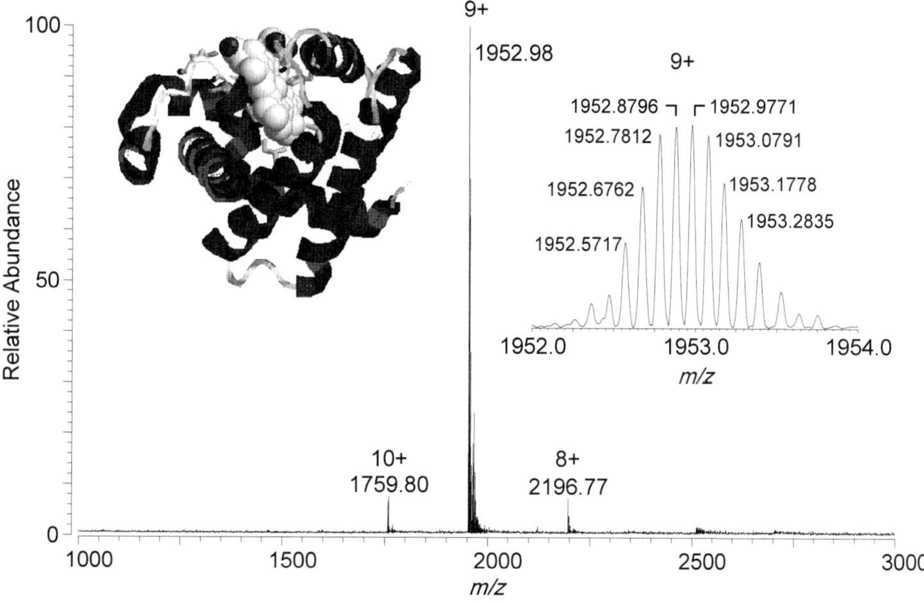

Figure 18.8. High-resolution nanoESI mass spectrum of equine myoglobin (0.2 mg mL^{-1}, pH 7.5, 10 mM NH_4HCO_3) with a 7-tesla FT-ICR mass spectrometer (IonSpec Ultima).

with organic acids (e.g., acetic acid, trifluoroacetic acid, formic acid) to lower the pH. Although these solutions conditions can yield intense signals for the protein, information is lost regarding the possible presence of interacting ligands and other proteins. Moreover, the high-order structure of the protein, dictated by weak interactions, can be disrupted by varying the environmental conditions, such as pH and solvent composition, and by the addition of other denaturants (e.g., detergents). This is demonstrated by the high-resolution FT-ICR mass spectrum of equine myoglobin (Figure 18.8) and ecotin (Figure 18.9). Myoglobin binds noncovalently the protoporphyrin IX (heme) molecule at near neutral pH; acidifying the solution yields the 16.9-kDa apoprotein from the loss of the 616-Da heme group. Ecotin is a serine protease inhibitor from *E. coli.*[139] The ESI mass spectrum of ecotin in an acidic pH solution shows a mass of 16,090 Da, which is consistent with the molecular weight of the monomer (Figure 18.9, top). Most protein complexes are denatured at acid pH. The mass spectrum of the protein obtained from a near-neutral physiological pH solution shows a molecular weight of 32,180 Da, or two times the mass of the monomer value (Figure 18.9, bottom). This is consistent with ecotin functioning as a homodimeric complex. Thus, as with most cases investigated by ESI-MS, the stoichiometry of the complex observed in the gas phase is consistent with that expected for the solution phase complex.

Such ESI-MS measurements of the noncovalent complexes can be helpful to confirm protein function by defining the quaternary structure. The putative *Spodoptera frugiperda* glutathione *S*-transferase described in Figures 18.2–18.4 was measured by nanoelectrospray MS using near-neutral pH solvents to help retain their native structures. The unknown protein shows a protein dimer as the most stable complex, consistent with the function of GST proteins from other species. From the three distinct protein monomers with molecular masses approximately 23 kDa measured by ESI-MS at pH 2.5, at least six hetero- and homodimers of 46 kDa are measured from a pH 7.5 solution (Figure 18.10). Nearly all GST

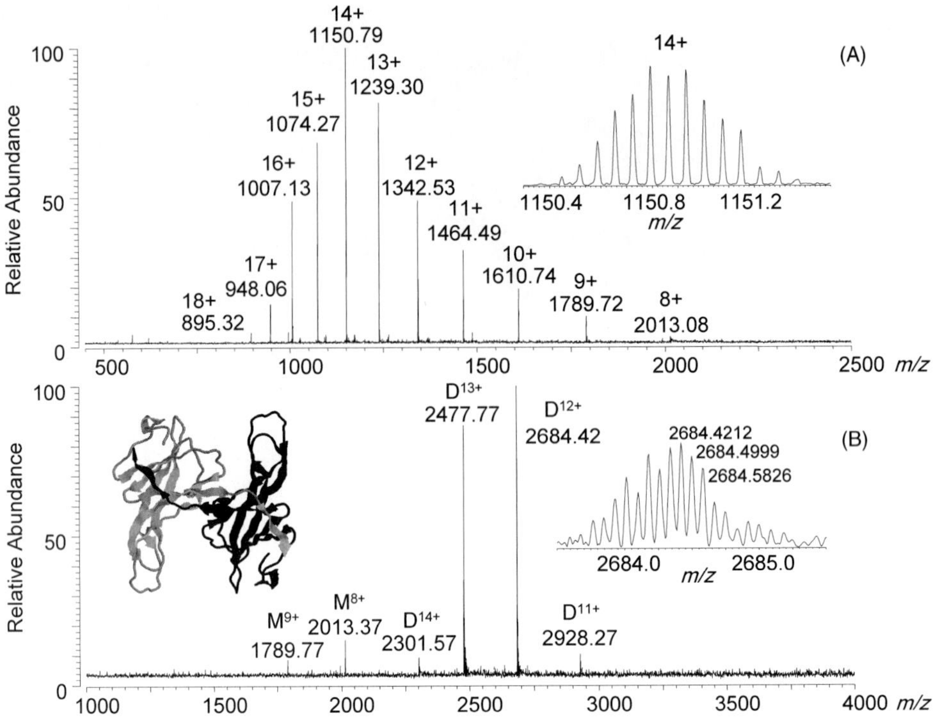

Figure 18.9. High-resolution nanoESI-FT-ICR (7-tesla IonSpec Ultima) mass spectra of *E. coli* ecotin in (**A**) acidic pH solution (0.01 mg mL^{-1}, water/acetonitrile, 1% acetic acid) showing ions for the 16-kDa protein denatured monomer and (**B**) pH 7.5 solution (0.1 mg mL^{-1}, 10 mM NH_4HCO_3) showing ions for the native 32-kDa dimer complex.

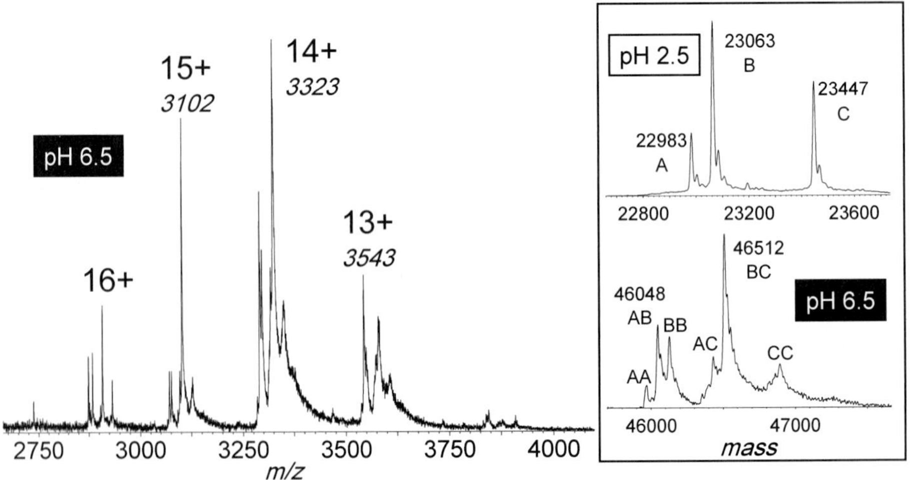

Figure 18.10. NanoESI-QTOF mass spectrum (**left**) of glutathione *S*-transferase proteins from *Spodoptera frugiperda*. ESI-MS of the protein sample from water/acetonitrile, 4% acetic acid solution indicate a mixture of three proteins (A, B, and C) with monomeric molecular mass of ~23 kDa (**right top**, deconvoluted to the mass domain). ESI-MS of the mixture from pH 7.5 solution (10 mM ammonium bicarbonate) show six protein homo- and heterodimers from all pairwise combinations of the three proteins (**right bottom**).

enzymes function as dimeric proteins; the measurements for the putative GST from *Spodoptera frugiperda* are consistent with this structure.

18.3.2 Tandem Mass Spectrometry of Noncovalent Complexes

Mass spectrometry studies have suggested that elements of the solution phase complex are preserved in the dehydrated complex.[140] In general, a high correlation between the ESI-MS data and expectations from the solution state world has been found.[141] The precise three-dimensional structure of the gas-phase molecule or complex may or may not be the same as the solvated species; however, there may be some structural elements that are preserved upon lifting a biomolecule into the gaseous state. In some examples reported, other physical characteristics of the gas-phase complex, such as their dissociation behavior and chemistries designed to probe topographical features, are consistent with the complex formed in solution. Collision cross-section measurements by ESI–ion mobility and ion scattering measurements,[142–144] and blackbody infrared dissociation (BIRD)[145] have suggested that different conformers can be measured for gas-phase proteins. The conserved bioactivity of a virus sample collected post-ESI mass spectrometry demonstrated that the ESI and desolvation processes are not destructive, but do not indicate directly the structure of the gas-phase (or solution phase) complex.[146] The CAD behavior of a small, cyclic naturally derived polypeptide was found to correlate to its solution structure.[141] The naturally isolated peptide is composed of a peptide "tail" region that inserts into a seven-residue "loop" section, and the tail is held inside by noncovalent forces. A synthetic version of the same sequence peptide was found to have the tail region external to the loop section. The MS/MS spectra for the two peptide forms show dramatically different fragmentation patterns, which are consistent with their respective solution structures. This system represents an example in which the peptide solution structure is consistent with the gas phase, and the results suggest that probing the gas-phase structure of biomolecules can be used to elucidate their solution phase geometries.

A report by Zubarev has suggested that ECD possesses capabilities to dissociate uniquely covalent bonds and leave noncovalent bonds intact.[147] An ESI-ECD-MS/MS strategy of protein noncovalent complexes may reveal the contact interface for protein–protein and protein–ligand interactions. This top-down sequencing approach for noncovalent protein complexes may represent a unique technique for determining the sites of ligand interactions. In general, CAD of noncovalently bound protein–ligand complexes yields the apo-protein and the liberated ligand and little information on the sites of binding. However, Xie et al. demonstrate that the sites of ligand binding of a noncovalent protein–ligand complex between the 14.5-kDa protein, α-synuclein, and a small molecule ligand may be determined by an ECD-based approach.[113]

A previous NMR study suggested that spermine (202 Da), a natural polycation that accelerates aggregation of α-synuclein, binds to the C-terminal acidic region of the protein with a solution binding affinity (equilibrium solution dissociation constant, K_D) of 0.6 mM.[148] The stability for such weakly bound ligands is enhanced in the gas phase because of charge–charge electrostatic interactions. ECD FT-ICR-MS of the AS-spermine complex yields information on the ligand binding sites of the protein. For the 14 + -charged AS-spermine complex, nondissociative charge reduction to the 13 + charge is the predominant channel observed. No products for the free spermine ligand were observed. However, c- and z-type products that retain binding of the spermine ligand localized near the C-terminus were observed for the 14+-charged holo-protein. Product ion yields were enhanced slightly by infrared (IR) laser heating of the ions after ECD, consistent with the proposed mechanism that hydrogen bonds are retained upon ECD but are dissociated by further heating. In total, for the 14+-charged AS-spermine complex, 82 unique product ions were measured. Nearly 50% of the products retain spermine binding. From the ECD product ions generated, ligand binding is localized between residues G106 and P138. The previous NMR study measured AS backbone chemical shift changes in ^{1}H-^{15}N HSCQ (heteronuclear single quantum correlation) spectra upon addition of spermine in residues located in the highly acidic Q109-A140 C-terminal region.[148] Thus, the ECD-MS data are consistent with the NMR study. These experiments demonstrate that spermine binding is localized to the C-terminus of α-synuclein, not only in solution but also in the gas-phase complex. The transition from solution to the gas phase apparently does not alter significantly the positional information of intermolecular interactions.

18.3.3 Ion Mobility–Mass Spectrometry of Proteins

Complementing ESI mass spectrometry techniques, ESI ion mobility spectrometry (IMS) methods have been used to probe the structural details of proteins and protein assemblies.[7,8,149,150] Highlighting the practical application of IMS in biological chemistry research, ESI–MS/IMS data on amyloid beta proteins support proposed mechanisms of protein aggregation involved in Alzheimer's disease.[151] An alternative technique of ion mobility analysis of charged–reduced ESI molecules is the gas-phase electrophoretic mobility molecular analyzer (GEMMA), or macroIMS (macroion mobility spectrometer). Such GEMMA measurements of biological macromolecules have been extended to the MDa range.[7,150,152,153]

The GEMMA is composed of an electrospray unit, a differential mobility analyzer (DMA), and a condensation particle counter (CPC).[152] The protein solution is sprayed into a mist of highly charged droplets that dry and are then charge reduced by a bipolar neutralizer composed of the α-radiation emitter, ^{210}Po. Most of the resulting particles will be neutral, a size-dependent percentage of the particles will possess one charge, and a lesser fraction of particles will display multiple charges. The DMA separates the particles based on their electrophoretic mobility in air. Only negatively or only positively charged particles within a narrow range of electrophoretic mobilities are selected for detection by the CPC.

The electrophoretic mobility diameter (EMD) is defined as the diameter of a singly charged sphere with the same electrophoretic mobility as the particle. The volume (V_{EMD}) of the particle is calculated by utilizing the geometric formula for the volume of the sphere. The effective density of the particles can then be calculated as the molecular weight (MW) divided by this calculated volume, V_{EMD}, and Avogadro's number. For a series of spherical particles displaying similar compactness, there will be a linear relationship between molecular weight and volume. For reported proteins and protein complexes measured by ESI-GEMMA, there is a general correlation between EMD and molecular weight.[5,7,152,154] Displaying an average effective density of 0.6 g/mL, this data set included proteins as small as oxytocin (3 kDa) and protein complexes as large as ribonucleoprotein vault complexes (>10 MDa).

The complementary nature of ESI-MS and ESI-IMS is illustrated in Figure 18.11. The proteasome is an intracellular protease complex that is responsible for degradation of most unfolded proteins in the cytosol and in the nucleus. The 20S proteasome is composed of four heptameric rings ($\alpha_7\beta_7\beta_7\alpha_7$) stacked in a barrel fashion, forming an internal cavity where the protease activity can be found. ESI-MS of the α-ring from the archaeon *M. thermophila*

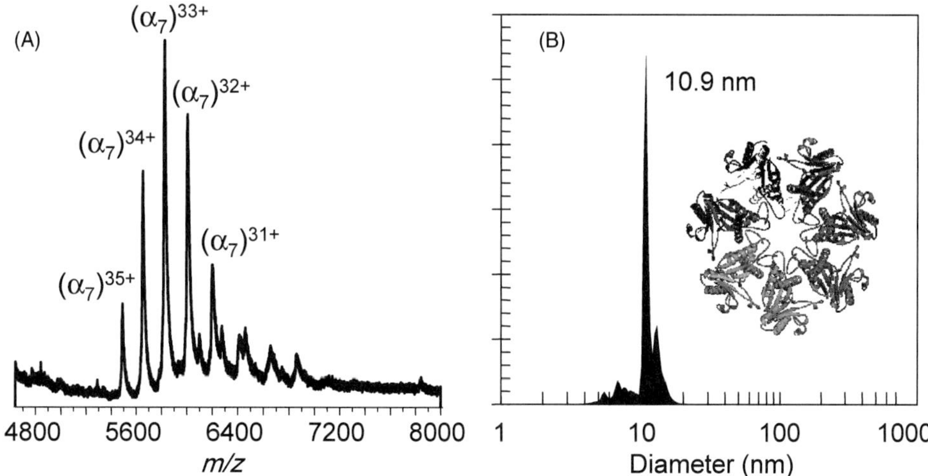

Figure 18.11. (**A**) ESI-QTOF mass spectrum of the α_7-complex from the *Methanosarcina thermophila* 20S proteasome (pH 7.5). The measured molecular mass of the heptameric complex is 192,056 ± 19 (191,865, theory). (**B**) ESI-GEMMA (TSI, Inc.) of the α_7-complex shows an electrophoretic mobility diameter of 10.9 nm.

20S proteasome under native solution conditions shows a total molecular mass of 191.9 kDa (Figure 18.11A), consistent with the expected value (192.1 kDa). (ESI-MS of the full 20S proteasome yields a mass of 690.5 kDa (theoretical value = 689.3 kDa) with multiply charged molecules found at *m/z* 11,000.[7]) ESI-GEMMA experiments of the α-ring yields an electrophoretic mobility diameter of 10.9 nm (Figure 18.11B). The crystal structure of the 20S proteasome barrel showed a width of 11.3 nm,[155] relatively consistent with the diameter measurement for the α_7 complex.

Figure 18.12 shows the ESI-GEMMA profiles for six protein complexes ranging in size from 65 kDa to 815 kDa. Vaults are the largest known ribonucleoproteins (up to 13 MDa).

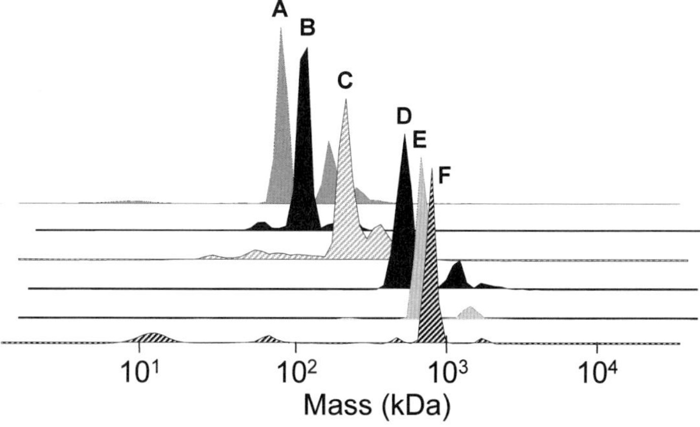

Figure 18.12. ESI-GEMMA of protein complexes (pH 7.5, 1 nM concentration each, analyte flowrate 50 nL min^{-1}). (**A**) 65-kDa streptavidin tetramer; (**B**) 93-kDa yeast enolase dimer; (**C**) 192-kDa *M. thermophila* 20S proteasome α_7 ring; (**D**) 483-kDa equine ferritin 24-mer; (**E**) 693-kDa *M. thermophila* 20S proteasome $\alpha_7\beta_7\ \beta_7\ \alpha_7$; (**F**) 815-kDa 14-mer GroEL complex.

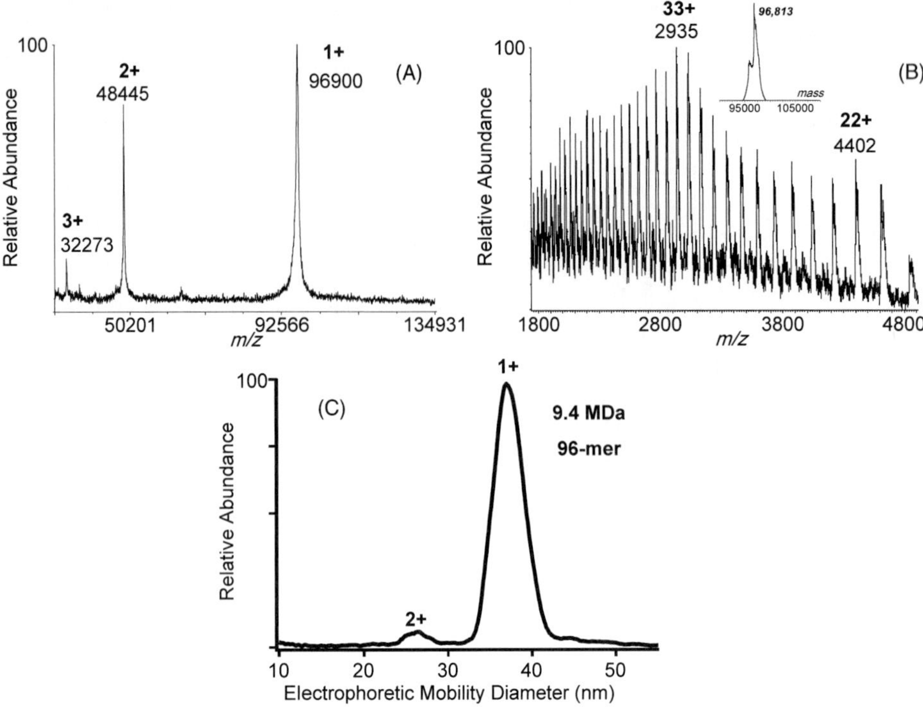

Figure 18.13. Molecular weight measurements of the human major vault protein (MVP). MALDI-TOF-MS (**A**) and ESI-QTOF-MS (**B**) of the denatured MVP protein (pH < 3) show the molecular weight of the MVP chain to be 96.8 kDa. (**C**) ESI-GEMMA of native MVP (pH 6.8) demonstrates the assembly of the MVP vault to a 96-mer, 9.4-MDa complex.

They are ubiquitously expressed in the cytoplasm and are highly conserved in eukaryotes. The most abundant protein, major vault protein (MVP, 96 kDa), forms the shell of the hollow ellipsoid-like vault structure as a 96-mer protein complex (9 MDa).[5,154,156] Both ESI-QTOF-MS and MALDI-TOF-MS confirm the 96-kDa molecular weight of the denatured MVP (Figures 18.13A and 18.13B). ESI ion mobility under native solution conditions (20 mM ammonium acetate, pH 6.8) of the MVP-composed vault shows full assembly of the 96-mer, 9-MDa complex (Figure 18.13C). For most protein complexes, sample acquisitions can be in the range of 3–10 minutes and sample consumption is very low. Although the resolving power of GEMMA is relatively low, comparable to that of size exclusion chromatography with a resolution of 7–18, its excellent sensitivity makes it a practical device for many applications in protein analysis.

18.4 CONCLUDING REMARKS

The human race has a longstanding fascination for making large objects "fly." Witness the Hindenburg—the dirigible built by the Zeppelin Company in the mid-1930s that measured over 880 feet long and still holds the record as the largest flying aircraft. Or consider the "Spruce Goose," the Hercules aircraft built by tycoon Howard Hughes and Hughes Aircraft completed after World War II. Its 320-ft wingspan was the largest for any aircraft ever built. Both the Hindenburg and the Spruce Goose failed to reach any sustained longevity for varied (explosive, in the former's case) reasons. However, it is safe to say that the applications resulting from the use of ESI and MALDI will continue for some time to come, and an important biomedical application will be to propel large proteins and their complexes into mass spectrometers and ion mobility devices for measuring their sizes. Field's "monster gaseous ions" and Fenn's "flying elephants" generated by ionization sources such as ESI and MALDI are destined for longevity within the scientific community.

ACKNOWLEDGMENTS

The UCLA Functional Proteomics Center was established and equipped by a grant to UCLA from the W. M. Keck Foundation. We are grateful to Drs. Emil Reilser and Sabrina Benchaar (UCLA Department of Chemistry and Biochemistry), Drs. Leonard Rome and Valerie Kickhoefer (UCLA Department of Biological Chemistry), and Dr. Tom Mueller (Pfizer Global Research and Development, Ann Arbor, MI) for the collaborative projects involving the proteins cofilin, major vault protein, and glutathione-*S*-transferase, respectively. JAL acknowledges support from the UCLA Jonsson Comprehensive Cancer Center, The UCLA Molecular Biology Institute, the U.S. Department of Energy for funding of the UCLA-DOE Institute for Genomics and Proteomics, and the National Institutes of Health (RR20004, AI067769, GM085402). SHL acknowledges support from the NIH/NIGMS Ruth L. Kirschstein Individual NRSA Fellowship Program (GM075384).

REFERENCES

1. Field, F. In McNeal, C. J. (Ed.), *Mass Spectrometry in the Analysis of Large Molecules*, John Wiley & Sons, Chichester, **1986**, pp. 213–214.
2. Fenn, J. B. Electrospray wings for molecular elephants (Nobel lecture). *Angew. Chem. Int. Ed.* **2003**, *42*, 3871–3894.
3. Tanaka, K. Gentle mass spectrometry—The origin of macromolecule ionization by laser irradiation (Nobel lecture). *Angew. Chem. Int. Ed.* **2003**, *42*, 3861–3870.
4. Karas, M.; Hillenkamp, F. Laser desorption ionization of proteins with molecular masses exceeding 10000 daltons. *Anal. Chem.* **1988**, *60*, 2299–2301.
5. Kaddis, C. S.; Loo, J. A. Native protein mass spectrometry and ion mobility: ESI and large flying proteins. *Anal. Chem.* **2007**, *79*, 1779–1784.
6. Loo, J. A. Studying noncovalent protein complexes by electrospray ionization mass spectrometry. *Mass Spectrom. Rev.* **1997**, *16*, 1–23.
7. Loo, J. A.; Berhane, B.; Kaddis, C. S.; Wooding, K. M.; Xie, Y.; Kaufman, S. L.; Chernushevich, I. V. Electrospray ionization mass spectrometry and ion mobility analysis of the 20S proteasome complex. *J. Am. Soc. Mass Spectrom.* **2005**, *16*, 998–1008.
8. Ruotolo, B. T.; Giles, K.; Campuzano, I.; Sandercock, A. M.; Bateman, R. H.; Robinson, C. V. Evidence for macromolecular protein rings in the absence of bulk water. *Science* **2005**, *310*, 1658–1661.
9. van den Heuvel, R. H. H.; Heck, A. J. R. Native protein mass spectrometry: From intact oligomers to functional machineries. *Curr. Opin. Chem. Biol.* **2004**, *8*, 519–526.
10. Patterson, S. D.; Aebersold, R. Mass spectrometric approaches for the identification of gel-separated proteins. *Electrophoresis* **1995**, *16*, 1791–1814.
11. Nielsen, M. L.; Bennett, K. L.; Larsen, B.; Moniatte, M.; Mann, M. Peptide end sequencing by orthogonal MALDI tandem mass spectrometry. *J. Proteome Res.* **2002**, *1*, 63–71.
12. Washburn, M. P.; Wolters, D.; Yates, J. R. Large-scale analysis of the yeast proteome by multidimensional protein identification technology. *Nature Biotechnol.* **2001**, *19*, 242–247.
13. Wolters, D. A.; Washburn, M. P.; Yates, J. R., III An automated multidimensional protein identification technology for shotgun proteomics. *Anal. Chem.* **2001**, *73*, 5683–5690.
14. Heller, S. R. The history of the NIST/EPA/NIH mass spectral database. *Today's Chemist at Work* **1999**, *8*, 49–50.

15. UniProtKB/Swiss-Prot protein knowledgebase release 52.1 statistics. Http://expasy.Org/sprot/relnotes/relstat. Html.
16. Sakurai, T.; Matsuo, T.; Matsuda, H.; Katakuse, I. PAAS 3, a computer program to determine probable sequence of peptides from mass spectrometric data. *Biomed. Mass Spectrom.* **1984**, *11*, 396–399.
17. Bartels, C. Fast algorithm for peptide sequencing by mass spectroscopy. *Biol. Mass Spectrom.* **1990**, *19*, 363–368.
18. Taylor, J. A.; Johnson, R. S. Sequence database searches via *de novo* peptide sequencing by tandem mass spectrometry. *Rapid Commun. Mass Spectrom.* **1997**, *11*, 1067–1075.
19. Taylor, J. A.; Johnson, R. S. Implementation and uses of automated *de novo* peptide sequencing by tandem mass spectrometry. *Anal. Chem.* **2001**, *73*, 2594–2604.
20. Shsevchenko, A.; Sunyaev, S.; Loboda, A.; Shevchenko, A.; Bork, P.; Ens, W.; Standing, K. G. Charting the proteomes of organisms with unsequenced genomes by MALDI–quadrupole time-of-flight mass spectrometry and BLAST homology searching. *Anal. Chem.* **2001**, *73*, 1917–1926.
21. Clauser, K. R.; Baker, P.; Burlingame, A. L. Role of accurate mass measurement (+/−10 ppm) in protein identification strategies employing MS or MS/MS and database searching. *Anal. Chem.* **1999**, *71*, 2871–2882.
22. Fernandez, F. M.; Smith, L. L.; Kuppannan, K.; Yang, X.; Wysocki, V. H. Peptide sequencing using a patchwork approach and surface-induced dissociation in sector-TOF and dual quadrupole mass spectrometers. *J. Am. Soc. Mass Spectrom.* **2003**, *14*, 1387–1401.
23. Eng, J. K.; McCormack, A. L.; Yates, J. R., III An approach to correlate tandem mass spectral data of peptides with amino acid sequences in a protein database. *J. Am. Soc. Mass Spectrom.* **1994**, *5*, 976–989.
24. Perkins, D. N.; Pappin, D. J. C.; Creasy, D. M.; Cottrell, J. S. Probability-based protein identification by searching sequence databases using mass spectrometry data. *Electrophoresis* **1999**, *20*, 3551–3567.
25. Keller, A.; Nesvizhskii, A. I.; Kolker, E.; Aebersold, R. Empirical statistical model to estimate the accuracy of peptide identifications made by MS/MS and database search. *Anal. Chem.* **2002**, *74*, 5383–5392.
26. Craig, R.; Beavis, R. C. A method for reducing the time required to match protein sequences with tandem mass spectra. *Rapid Commun. Mass Spectrom.* **2003**, *17*, 2310–2316.
27. Searle, B. C.; Turner, M. *Identifying peptides the way experts do: A scientist-driven scoring system for improving ms/ms peptide identification confidence*, Presented at the Association of Biomedical Research Facilities (ABRF) Meeting, Long Beach, CA, 2006.
28. Creasy, D. M.; Cottrell, J. S. Error tolerant searching of uninterpreted tandem mass spectrometry data. *Proteomics* **2002**, *2*, 1426–1434.
29. Tanner, S.; Shu, H.; Frank, L.; Wang, A.; Zandi, M.; Mumby, E.; Pevzner, P. A.; Bafna, V. Inspect: Identification of posttranslationally modified peptides from tandem mass spectra. *Anal. Chem.* **2005**, *77*, 4626–4639.
30. Elias, J. E.; Haas, W.; Faherty, B. K.; Gygi, S. P. Comparative evaluation of mass spectrometry platforms used in large-scale proteomics investigations. *Nature Methods* **2005**, *2*, 667–675.
31. Nesvizhskii, A. I.; Aebersold, R. Analysis, statistical validation and dissemination of large-scale proteomics datasets generated by tandem MS. *Drug Discovery Today* **2004**, *9*, 173–181.
32. Shevchenko, A.; Chernushevich, I.; Ens, W.; Standing, K. G.; Thomson, B.; Wilm, M.; Mann, M. Rapid "*de novo*" peptide sequencing by a combination of nanoelectrospray, isotopic labeling and a quadrupole/time-of-flight mass spectrometer. *Rapid Commun. Mass Spectrom.* **1997**, *11*, 1015–1024.
33. Yao, X.; Afonso, C.; Fenselau, C. Dissection of proteolytic ^{18}O labeling: Endoprotease-catalyzed ^{16}O-to-^{18}O exchange of truncated peptide substrates. *J. Proteome Res.* **2003**, *2*, 147–152.
34. Ogorzalek Loo, R. R.; Cavalcoli, J. D.; VanBogelen, R. A.; Mitchell, C.; Loo, J. A.; Moldover, B.; Andrews, P. C. Virtual 2-D gel electrophoresis: Visualization and analysis of the *E. coli* proteome by mass spectrometry. *Anal. Chem.* **2001**, *73*, 4063–4070.
35. Bichet, P.; Mollat, P.; Capdevila, C.; Sarubbi, E. Endogenous glutathione-binding proteins of insect cell lines: Characterization and removal from glutathione s-transferase (GST) fusion proteins. *Protein Express. Purif.* **2000**, *19*, 197–201.
36. Patterson, D. H.; Tarr, G. E.; Regnier, F. E.; Martin, S. A. C-terminal ladder sequencing via matrix-assisted laser desorption mass spectrometry coupled with carboxypeptidase Y time-dependent and concentration-dependent digestions. *Anal. Chem.* **1995**, *67*, 3971–3978.
37. Chait, B. T.; Wang, R.; Beavis, R. C.; Kent, S. B. H. Protein ladder sequencing. *Science* **1993**, *262*, 89–92.
38. Benchaar, S. A.; Xie, Y.; Phillips, M.; Loo, R. R. O.; Galkin, V. E.; Orlova, A.; Thevis, M.; Muhlrad, A.; Almo, S. C.; Loo, J. A.; Egelman, E. H.; Reisler, E. Mapping the interaction of cofilin with subdomain 2 on actin. *Biochemistry* **2007**, *46*, 225–233.
39. Regnier, F. E.; Riggs, L.; Zhang, R.; Xiong, L.; Liu, P.; Chakraborty, A.; Seeley, E.; Sioma, C.; Robert, A. Thompson Comparative proteomics based on stable isotope labeling and affinity selection. *J. Mass Spectrom.* **2002**, *37*, 133–145.
40. Beynon, R. J.; Pratt, J. M. Metabolic labeling of proteins for proteomics. *Mol. Cell. Proteomics* **2005**, *4*, 857–872.

41. Oda, Y.; Huang, K.; Cross, F. R.; Cowburn, D.; Chait, B. T. Accurate quantitation of protein expression and site-specific phosphorylation. *Proc. Natl. Acad. Sci. USA* **1999**, *96*, 6591–6596.
42. Krijgsveld, J.; Ketting, R. F.; Mahmoudi, T.; Johansen, J.; Artal-Sanz, M.; Verrijzer, C. P.; Plasterk, R. H. A.; Heck, A. J. R. Metabolic labeling of *C. elegans* and *D. melanogaster* for quantitative proteomics. *Nature Biotechnol.* **2003**, *21*, 927–931.
43. Wu, C. C.; MacCoss, M. J.; Howell, K. E.; Matthews, D. E.; Yates, J. R. Metabolic labeling of mammalian organisms with stable isotopes for quantitative proteomic analysis. *Anal. Chem.* **2004**, *76*, 4951–4959.
44. McClatchy, D. B.; Dong, M.-Q.; Wu, C. C.; Venable, J. D.; Yates, J. R. ^{15}N metabolic labeling of mammalian tissue with slow protein turnover. *J. Proteome Res.* **2007**, *6*, 2005–2010.
45. Van Hoof, D.; Pinkse, M. W. H.; Ward-Van Oostwaard, D.; Mummery, C. L.; Heck, A. J. R.; Krijgsveld, J. An experimental correction for arginine-to-proline conversion artifacts in SILAC-based quantitative proteomics. *Nature Methods* **2007**, *4*, 677–678.
46. Zhu, H.; Pan, S.; Gu, S.; Bradbury, E. M.; Chen, X. Amino acid residue specific stable isotope labeling for quantitative proteomics. *Rapid Commun. Mass Spectrom.* **2002**, *16*, 2115–2123.
47. Jiang, H.; English, A. M. Quantitative analysis of the yeast proteome by incorporation of isotopically labeled leucine. *J. Proteome Res.* **2002**, *1*, 345–350.
48. Ong, S.-E.; Blagoev, B.; Kratchmarova, I.; Kristensen, D. B.; Steen, H.; Pandey, A.; Mann, M. Stable isotope labeling by amino acids in cell culture, SILAC, as a simple and accurate approach to expression proteomics. *Mol. Cell. Proteomics* **2002**, *1*, 376–386.
49. Ong, S.-E.; Kratchmarova, I.; Mann, M. Properties of ^{13}C-substituted arginine in stable isotope labeling by amino acids in cell culture (SILAC). *J. Proteome Res.* **2003**, *2*, 173–181.
50. Blagoev, B.; Kratchmarova, I.; Ong, S.-E.; Nielsen, M.; Foster, L. J.; Mann, M. A proteomics strategy to elucidate functional protein-protein interactions applied to EGF signaling. *Nature Biotechnol.* **2003**, *21*, 315–318.
51. Andersen, J. S.; Lam, Y. W.; Leung, A. K. L.; Ong, S.-E.; Lyon, C. E.; Lamond, A. I.; Mann, M. Nucleolar proteome dynamics. *Nature* **2005**, *433*, 77–83.
52. Gruhler, A.; Olsen, J. V.; Mohammed, S.; Mortensen, P.; Faergeman, N. J.; Mann, M.; Jensen, O. N. Quantitative phosphoproteomics applied to the yeast pheromone signaling pathway. *Mol. Cell. Proteomics* **2005**, *4*, 310–327.
53. Blagoev, B.; Ong, S.-E.; Kratchmarova, I.; Mann, M. Temporal analysis of phosphotyrosine-dependent signaling networks by quantitative proteomics. *Nature Biotechnol.* **2004**, *22*, 1139–1145.
54. Dengjel, J.; Akimov, V.; Olsen, J. V.; Bunkenborg, J.; Mann, M.; Blagoev, B.; Andersen, J. S. Quantitative proteomic assessment of very early cellular signaling events. *Nature Biotechnol.* **2007**, *25*, 566–568.
55. Gronborg, M.; Kristiansen, T. Z.; Iwahori, A.; Chang, R.; Reddy, R.; Sato, N.; Molina, H.; Jensen, O. N.; Hruban, R. H.; Goggins, M. G.; Maitra, A.; Pandey, A. Biomarker discovery from pancreatic cancer secretome using a differential proteomic approach. *Mol. Cell. Proteomics* **2006**, *5*, 157–171.
56. Park, K.-S.; Mohapatra, D. P.; Misonou, H.; Trimmer, J. S. Graded regulation of the kv2.1 potassium channel by variable phosphorylation. *Science* **2006**, *313*, 976–979.
57. Romijn, E. P.; Christis, C.; Wieffer, M.; Gouw, J. W.; Fullaondo, A.; van der Sluijs, P.; Braakman, I.; Heck, A. J. R. Expression clustering reveals detailed co-expression patterns of functionally related proteins during B cell differentiation: A proteomic study using a combination of one-dimensional gel electrophoresis, LC-MS/MS, and stable isotope labeling by amino acids in cell culture (SILAC). *Mol. Cell. Proteomics* **2005**, *4*, 1297–1310.
58. Gruhler, A.; Schulze, W. X.; Matthiesen, R.; Mann, M.; Jensen, O. N. Stable isotope labeling of *Arabidopsis thaliana* cells and quantitative proteomics by mass spectrometry. *Mol. Cell. Proteomics* **2005**, *4*, 1697–1709.
59. Mannova, P.; Fang, R.; Wang, H.; Deng, B.; McIntosh, M. W.; Hanash, S. M.; Beretta, L. Modification of host lipid raft proteome upon hepatitis C virus replication. *Mol. Cell. Proteomics* **2006**, *5*, 2319–2325.
60. Naumann, B.; Stauber, E. J.; Busch, A.; Sommer, F.; Hippler, M. N-terminal processing of LHCA3 is a key step in remodeling of the photosystem I-light-harvesting complex under iron deficiency in *Chlamydomonas reinhardtii*. *J. Biol. Chem.* **2005**, *280*, 20431–20441.
61. Sun, Q.; Emanuelsson, O.; van Wijk, K. J. Analysis of curated and predicted plastid subproteomes of arabidopsis. Subcellular compartmentalization leads to distinctive proteome properties. *Plant Physiol.* **2004**, *135*, 723–734.
62. Goodlett, D. R.; Keller, A.; Watts, J. D.; Newitt, R.; Yi, E. C.; Purvine, S.; Eng, J. K.; Haller, P. v.; Aebersold, R.; Kolker, E. Differential stable isotope labeling of peptides for quantitation and *de novo* sequence derivation. *Rapid Commun. Mass Spectrom.* **2001**, *15*, 1214–1221.
63. Hsu, J. L.; Huang, S. Y.; Chow, N. H.; Chen, S. H. Stable-isotope dimethyl labeling for quantitative proteomics. *Anal. Chem.* **2003**, *75*, 6843–6852.
64. Warwood, S.; Mohammed, S.; Cristea, I. M.; Evans, C.; Whetton, A. D.; Gaskell, S. J. Guanidination chemistry for qualitative and quantitative proteomics. *Rapid Commun. Mass Spectrom.* **2006**, *20*, 3245–3256.

65. Sechi, S. A method to identify and simultaneously determine the relative quantities of proteins isolated by gel electrophoresis. *Rapid Commun. Mass Spectrom.* **2002**, *16*, 1416–1424.
66. Gehanne, S.; Cecconi, D.; Carboni, L.; Righetti, P. G.; Domenici, E.; Hamdan, M. Quantitative analysis of two-dimensional gel-separated proteins using isotopically marked alkylating agents and matrix-assisted laser desorption/ionization mass spectrometry. *Rapid Commun. Mass Spectrom.* **2002**, *16*, 1692–1698.
67. Gygi, S. P.; Rist, B.; Gerber, S. A.; Turecek, F.; Gelb, M. H.; Aebersold, R. Quantitative analysis of complex protein mixtures using isotope-coded affinity tags. *Nat. Biotechnol.* **1999**, *17*, 994–999.
68. Hansen, K. C.; Schmitt-Ulms, G.; Chalkley, R. J.; Hirsch, J.; Baldwin, M. A.; Burlingame, A. L. Mass spectrometric analysis of protein mixtures at low levels using cleavable ^{13}C-isotope-coded affinity tag and multidimensional chromatography. *Mol. Cell. Proteomics* **2003**, *2*, 299–314.
69. Borisov, O. V.; Goshe, M. B.; Conrads, T. P.; Rakov, V. S.; Veenstra, T. D.; Smith, R. D. Low-energy collision-induced dissociation fragmentation analysis of cysteinyl-modified peptides. *Anal. Chem.* **2002**, *74*, 2284–2292.
70. Li, J.; Steen, H.; Gygi, S. P. Protein profiling with cleavable isotope-coded affinity tag (cICAT) reagents: The yeast salinity stress response. *Mol. Cell. Proteomics* **2003**, *2*, 1198–1204.
71. Molloy, M. P.; Donohoe, S.; Brzezinski, E. E.; Kilby, G. W.; Stevenson, T. I.; Baker, J. D.; Goodlett, D. R.; Gage, D. A. Large-scale evaluation of quantitative reproducibility and proteome coverage using acid cleavable isotope coded affinity tag mass spectrometry for proteomic profiling. *Proteomics* **2005**, *5*, 1204–1208.
72. Majeran, W.; Cai, Y.; Sun, Q.; van Wijk, K. J. Functional differentiation of bundle sheath and mesophyll maize chloroplasts determined by comparative proteomics. *Plant Cell* **2005**, *17*, 3111–3140.
73. Sethuraman, M.; Clavreul, N.; Huang, H.; McComb, M. E.; Costello, C. E.; Cohen, R. A. Quantification of oxidative posttranslational modifications of cysteine thiols of p21ras associated with redox modulation of activity using isotope-coded affinity tags and mass spectrometry. *Free Rad. Biol. Med.* **2007**, *42*, 823–829.
74. Zhang, Y.; Wolf-Yadlin, A.; Ross, P. L.; Pappin, D. J.; Rush, J.; Lauffenburger, D. A.; White, F. M. Time-resolved mass spectrometry of tyrosine phosphorylation sites in the epidermal growth factor receptor signaling network reveals dynamic modules. *Mol. Cell. Proteomics* **2005**, *4*, 1240–1250.
75. Sachon, E.; Mohammed, S.; Bache, N.; Jensen, O. N. Phosphopeptide quantitation using amine-reactive isobaric tagging reagents and tandem mass spectrometry: Application to proteins isolated by gel electrophoresis. *Rapid Commun. Mass Spectrom.* **2006**, *20*, 1127–1134.
76. Thompson, A.; Schafer, J.; Kuhn, K.; Kienle, S.; Schwarz, J.; Schmidt, G.; Neumann, T.; Johnstone, R.; Mohammed, A. K.; Hamon, C. Tandem mass tags: A novel quantification strategy for comparative analysis of complex protein mixtures by MS/MS. *Anal. Chem.* **2003**, *75*, 1895–1904.
77. Ross, P. L.; Huang, Y. N.; Marchese, J. N.; Williamson, B.; Parker, K.; Hattan, S.; Khainovski, N.; Pillai, S.; Dey, S.; Daniels, S.; Purkayastha, S.; Juhasz, P.; Martin, S.; Bartlet-Jones, M.; He, F.; Jacobson, A.; Pappin, D. J. Multiplexed protein quantitation in *Saccharomyces cerevisiae* using amine-reactive isobaric tagging reagents. *Mol. Cell. Proteomics* **2004**, *3*, 1154–1169.
78. Gerber, S. A.; Rush, J.; Stemman, O.; Kirschner, M. W.; Gygi, S. P. Absolute quantification of proteins and phosphoproteins from cell lysates by tandem MS. *Proc. Natl. Acad. Sci. USA* **2003**, *100*, 6940–6945.
79. Dunkley, T. P. J.; Watson, R.; Griffin, J. L.; Dupree, P.; Lilley, K. S. Localization of organelle proteins by isotope tagging (LOPIT). *Mol. Cell. Proteomics* **2004**, *3*, 1128–1134.
80. Dunkley, T. P. J.; Hester, S.; Shadforth, I. P.; Runions, J.; Weimar, T.; Hanton, S. L.; Griffin, J. L.; Bessant, C.; Brandizzi, F.; Hawes, C.; Watson, R. B.; Dupree, P.; Lilley, K. S. Mapping the arabidopsis organelle proteome. *Proc. Natl. Acad. Sci. USA* **2006**, *103*, 6518–6523.
81. Gan, C. S.; Chong, P. K.; Pham, T. K.; Wright, P. C.; Technical, experimental, and biological variations in isobaric tags for relative and absolute quantitation (iTRAQ). *J. Proteome Res.* **2007**, *6*, 821–827.
82. Mirgorodskaya, O. A.; Kozmin, Y. P.; Titov, M. I.; Körner, R.; Sönksen, C. P.; Roepstorff, P. Quantitation of peptides and proteins by matrix-assisted laser desorption/ionization mass spectrometry using ^{18}O-labeled internal standards. *Rapid Commun. Mass Spectrom.* **2000**, *14*, 1226–1232.
83. Miyagi, M.; Rao, K. C. S. Proteolytic ^{18}O-labeling strategies for quantitative proteomics. *Mass Spectrom. Rev.* **2007**, *26*, 121–136.
84. Heller, M.; Mattou, H.; Menzel, C.; Yao, X. Trypsin catalyzed ^{16}O-to-^{18}O exchange for comparative proteomics: Tandem mass spectrometry comparison using MALDI–TOF, ESI–QTOF, and ESI–ion trap mass spectrometers. *J. Am. Soc. Mass Spectrom.* **2003**, *14*, 704–718.
85. Barr, J.; Maggio, V.; Patterson, D., Jr.; Cooper, G.; Henderson, L.; Turner, W.; Smith, S.; Hannon, W.; Needham, L.; Sampson, E. Isotope dilution—Mass spectrometric quantification of specific proteins: Model application with apolipoprotein A-I. *Clin. Chem.* **1996**, *42*, 1676–1682.
86. Beynon, R. J.; Doherty, M. K.; Pratt, J. M.; Gaskell, S. J. Multiplexed absolute quantification in proteomics

using artificial QCAT proteins of concatenated signature peptides. *Nature Methods* **2005**, *2*, 587–589.

87. Mallick, P.; Schirle, M.; Chen, S. S.; Flory, M. R.; Lee, H.; Martin, D.; Ranish, J.; Raught, B.; Schmitt, R.; Werner, T.; Kuster, B.; Aebersold, R. Computational prediction of proteotypic peptides for quantitative proteomics. *Nature Biotechnol.* **2007**, *25*, 125–131.
88. Bondarenko, P. V.; Chelius, D.; Shaler, T. A. Identification and relative quantitation of protein mixtures by enzymatic digestion followed by capillary reversed-phase liquid chromatography–tandem mass spectrometry. *Anal. Chem.* **2002**, *74*, 4741–4749.
89. Wang, W.; Zhou, H.; Lin, H.; Roy, S.; Shaler, T. A.; Hill, L. R.; Norton, S.; Kumar, P.; Anderle, M.; Becker, C. H. Quantification of proteins and metabolites by mass spectrometry without isotopic labeling or spiked standards *Anal. Chem.* **2003**, *75*, 4818–4826.
90. Wang, G.; Wu, W. W.; Zeng, W.; Chou, C.-L.; Shen, R.-F. Label-free protein quantification using LC-coupled ion trap or FT mass spectrometry: Reproducibility, linearity, and application with complex proteomes. *J. Proteome Res.* **2006**, *5*, 1214–1223.
91. Liu, H.; Sadygov, R. G.; Yates, J. R. A model for random sampling and estimation of relative protein abundance in shotgun proteomics. *Anal. Chem.* **2004**, *76*, 4193–4201.
92. Zybailov, B.; Coleman, M. K.; Florens, L.; Washburn, M. P. Correlation of relative abundance ratios derived from peptide ion chromatograms and spectrum counting for quantitative proteomic analysis using stable isotope labeling. *Anal. Chem.* **2005**, *77*, 6218–6224.
93. Hendrickson, E. L.; Xia, Q.; Wang, T.; Leigh, J. A.; Hackett, M. Comparison of spectral counting and metabolic stable isotope labeling for use with quantitative microbial proteomics. *Analyst* **2006**, *131*, 1335–1341.
94. Zybailov, B.; Mosley, A. L.; Sardiu, M. E.; Coleman, M. K.; Florens, L.; Washburn, M. P. Statistical analysis of membrane proteome expression changes in *Saccharomyces cerevisiae*. *J. Proteome Res.* **2006**, *5*, 2339–2347.
95. Old, W. M.; Meyer-Arendt, K.; Aveline-Wolf, L.; Pierce, K. G.; Mendoza, A.; Sevinsky, J. R.; Resing, K. A.; Ahn, N. G. Comparison of label-free methods for quantifying human proteins by shotgun proteomics. *Mol. Cell. Proteomics* **2005**, *4*, 1487–1502.
96. Sanders, S. L.; Jennings, J.; Canutescu, A.; Link, A. J.; Weil, P. A. Proteomics of the eukaryotic transcription machinery: Identification of proteins associated with components of yeast tfiid by multidimensional mass spectrometry. *Mol. Cell. Biol.* **2002**, *22*, 4723–4738.
97. Rappsilber, J.; Ryder, U.; Lamond, A. I.; Mann, M. Large-scale proteomic analysis of the human spliceosome. *Genome Res.* **2002**, *12*, 1231–1245.
98. Ishihama, Y.; Oda, Y.; Tabata, T.; Sato, T.; Nagasu, T.; Rappsilber, J.; Mann, M. Exponentially modified protein abundance index (emPAI) for estimation of absolute protein amount in proteomics by the number of sequenced peptides per protein. *Mol. Cell. Proteomics* **2005**, *4*, 1265–1272.
99. Silva, J. C.; Denny, R.; Dorschel, C.; Gorenstein, M. V.; Li, G.-Z.; Richardson, K.; Wall, D.; Geromanos, S. J. Simultaneous qualitative and quantitative analysis of the *Escherichia coli* proteome: A sweet tale. *Mol. Cell. Proteomics* **2006**, *5*, 589–607.
100. Kelleher, N. L. Top-down proteomics. *Anal. Chem.* **2004**, *76*, 197A–203A.
101. Loo, J. A.; Edmonds, C. G.; Smith, R. D. Primary sequence information from intact proteins by electrospray ionization tandem mass spectrometry. *Science* **1990**, *248*, 201–204.
102. Loo, J. A.; Edmonds, C. G.; Smith, R. D. Tandem mass spectrometry of very large molecules: Serum albumin sequence information from multiply charged ions formed by electrospray ionization. *Anal. Chem.* **1991**, *63*, 2488–2499.
103. Thevis, M.; Ogorzalek Loo, R. R.; Loo, J. A. Mass spectrometric characterization of transferrins and their fragments derived by reduction of disulfide bonds. *J. Am. Soc. Mass Spectrom.* **2003**, *14*, 635–647.
104. Horn, D. M.; Zubarev, R. A.; McLafferty, F. W. Automated *de novo* sequencing of proteins by tandem high-resolution mass spectrometry. *Proc. Natl. Acad. Sci. USA* **2000**, *97*, 10313–10317.
105. Ge, Y.; Lawhorn, B. G.; ElNaggar, M.; Strauss, E.; Park, J. H.; Begley, T. P.; McLafferty, F. W. Top down characterization of larger proteins (45 kDa) by electron capture dissociation mass spectrometry. *J. Am. Chem. Soc.* **2002**, *124*, 672–678.
106. Meng, F.; Cargile, B. J.; Miller, L. M.; Forbes, A. J.; Johnson, J. R.; Kelleher, N. L. Informatics and multiplexing of intact protein identification in bacteria and the archaea. *Nat. Biotechnol.* **2001**, *19*, 952–957.
107. Macek, B.; Waanders, L. F.; Olsen, J. V.; Mann, M. Top-down protein sequencing and MS^3 on a hybrid linear quadrupole ion trap-orbitrap mass spectrometer. *Mol. Cell. Proteomics* **2006**, *5*, 949–958.
108. Kelleher, N. L.; Lin, H. Y.; Valaskovic, G. A.; Aaserud, D. J.; Fridriksson, E. K.; McLafferty, F. W. Top down versus bottom up protein characterization by tandem high-resolution mass spectrometry. *J. Am. Chem. Soc.* **1999**, *121*, 806–812.
109. Mortz, E.; O'Connor, P. B.; Roepstorff, P.; Kelleher, N. L.; Wood, T. D.; McLafferty, F. W.; Mann, M. Sequence tag identification of intact proteins by matching tanden mass spectral data against sequence data bases. *Proc. Natl. Acad. Sci. USA* **1996**, *93*, 8264–8267.
110. Smith, R. D.; Loo, J. A.; Barinaga, C. J.; Edmonds, C. G.; Udseth, H. R. Collisional activation and collision-activated dissociation of large multiply

charged polypeptides and proteins produced by electrospray ionization. *J. Am. Soc. Mass Spectrom.* **1990**, *1*, 53–65.
111. Han, X.; Jin, M.; Breuker, K.; McLafferty, F. W. Extending top-down mass spectrometry to proteins with masses greater than 200 kiloDaltons. *Science* **2006**, *314*, 109–112.
112. Hofstadler, S. A.; Sannes-Lowery, K. A.; Griffey, R. H. Infrared multiphoton dissociation in an external ion reservoir. *Anal. Chem.* **1999**, *71*, 2067–2070.
113. Xie, Y.; Zhang, J.; Yin, S.; Loo, J. A. Top-down ESI-ECD-FT-ICR mass spectrometry localizes noncovalent protein–ligand binding sites. *J. Am. Chem. Soc.* **2006**, *128*, 14432–14433.
114. Syka, J. E. P.; Coon, J. J.; Schroeder, M. J.; Shabanowitz, J.; Hunt, D. F. Peptide and protein sequence analysis by electron transfer dissociation mass spectrometry. *Proc. Natl. Acad. Sci. USA* **2004**, *101*, 9528–9533.
115. Taylor, G. K.; Kim, Y.-B.; Forbes, A. J.; Meng, F.; McCarthy, R.; Kelleher, N. L. Web and database software for identification of intact proteins using "top down" mass spectrometry. *Anal. Chem.* **2003**, *75*, 4081–4086.
116. Cooper, H. J.; Hakansson, K.; Marshall, A. G. The role of electron capture dissociation in biomolecular analysis. *Mass Spectrom. Rev.* **2005**, *24*, 201–222.
117. Meng, F.; Forbes, A. J.; Miller, L. M.; Kelleher, N. L. Detection and localization of protein modifications by high resolution tandem mass spectrometry. *Mass Spectrom. Rev.* **2005**, *24*, 126–134.
118. Kjeldsen, F.; Haselmann, K. F.; Budnik, B. A.; Sorensen, E. S.; Zubarev, R. A. Complete characterization of posttranslational modification sites in the bovine milk protein PP3 by tandem mass spectrometry with electron capture dissociation as the last stage. *Anal. Chem.* **2003**, *75*, 2355–2361.
119. Shi, S. D.-H.; Hemling, M. E.; Carr, S. A.; Horn, D. M.; Lindh, I.; McLafferty, F. W. Phosphopeptide/phosphoprotein mapping by electron capture dissociation mass spectrometry. *Anal. Chem.* **2001**, *73*, 19–22.
120. Stensballe, A.; Jensen, O. N.; Olsen, J. V.; Haselmann, K. F.; Zubarev, R. A. Electron capture dissociation of singly and multiply phosphorylated peptides. *Rapid Commun. Mass Spectrom.* **2000**, *14*, 1793–1800.
121. Pesavento, J. J.; Kim, Y.-B.; Taylor, G. K.; Kelleher, N. L. Shotgun annotation of histone modifications: A new approach for streamlined characterization of proteins by top down mass spectrometry. *J. Am. Chem. Soc.* **2004**, *126*, 3386–3387.
122. Medzihradszky, K. F.; Zhang, X.; Chalkley, R. J.; Guan, S.; McFarland, M. A.; Chalmers, M. J.; Marshall, A. G.; Diaz, R. L.; Allis, C. D.; Burlingame, A. L. Characterization of tetrahymena histone H2b variants and posttranslational populations by electron capture dissociation (ECD) Fourier transform ion cyclotron mass spectrometry (FT-ICR MS). *Mol. Cell. Proteomics* **2004**, *3*, 872–886.
123. Zhang, L.; Freitas, M. A. Comparison of peptide mass mapping and electron capture dissociation as assays for histone posttranslational modifications. *Int. J. Mass Spectrom.* **2004**, *234*, 213–225.
124. Cournoyer, J. J.; Pittman, J. L.; Ivleva, V. B.; Fallows, E.; Waskell, L.; Costello, C. E.; O'Connor, P. B. Deamidation: Differentiation of aspartyl from isoaspartyl products in peptides by electron capture dissociation. *Protein Sci.* **2005**, *14*, 452–463.
125. Mirgorodskaya, E.; Roepstorff, P.; Zubarev, R. A. Localization of *O*-glycosylation sites in peptides by electron capture dissociation in a fourier transform mass spectrometer. *Anal. Chem.* **1999**, *71*, 4431–4436.
126. Zubarev, R. A.; Kruger, N. A.; Fridriksson, E. K.; Lewis, M. A.; Horn, D. M.; Carpenter, B. K.; McLafferty, F. W. Electron capture dissociation of gaseous multiply-charged proteins is favored at disulfide bonds and other sites of high hydrogen atom affinity. *J. Am. Chem. Soc.* **1999**, *121*, 2857–2862.
127. Kjeldsen, F.; Haselmann, K. F.; Sorensen, E. S.; Zubarev, R. A. Distinguishing of Ile/Leu amino acid residues in the PP3 protein by (hot) electron capture dissociation in Fourier transform ion cyclotron resonance mass spectrometry. *Anal. Chem.* **2003**, *75*, 1267–1274.
128. Kruppa, G. H.; Schoeniger, J.; Young, M. M. A top down approach to protein structural studies using chemical cross-linking and Fourier transform mass spectrometry. *Rapid Commun. Mass Spectrom.* **2003**, *17*, 155–162.
129. Leavell, M. D.; Novak, P.; Behrens, C. R.; Schoeniger, J. S.; Kruppa, G. H. Strategy for selective chemical cross-linking of tyrosine and lysine residues. *J. Am. Soc. Mass Spectrom.* **2004**, *15*, 1604–1611.
130. Novak, P.; Young, M. M.; Schoeniger, J. S.; Kruppa, G. H. A top-down approach to protein structure studies using chemical cross-linking and Fourier transform mass spectrometry. *Eur. J. Mass Spectrom.* **2003**, *9*, 623–631.
131. Novak, P.; Haskins, W. E.; Ayson, M. J.; Jacobsen, R. B.; Schoeniger, J. S.; Leavell, M. D.; Young, M. M.; Kruppa, G. H. Unambiguous assignment of intramolecular chemical cross-links in modified mammalian membrane proteins by Fourier transform–tandem mass spectrometry. *Anal. Chem.* **2005**, *77*, 5101–5106.
132. Oh, H.; Breuker, K.; Sze, S. K.; Ge, Y.; Carpenter, B. K.; McLafferty, F. W. Secondary and tertiary structures of gaseous protein ions characterized by electron capture dissociation mass spectrometry and photofragment spectroscopy. *Proc. Natl. Acad. Sci. USA* **2002**, *99*, 15863–15868.
133. Breuker, K.; Oh, H.; Lin, C.; Carpenter, B. K.; McLafferty, F. W. Nonergodic and conformational control of the electron capture dissociation of protein

cations. *Proc. Natl. Acad. Sci. USA* **2004**, *101*, 14011–14016.

134. Breuker, K.; McLafferty, F. W. Native electron capture dissociation for the structural characterization of noncovalent interactions in native cytochrome c. *Angew. Chem. Int. Ed.* **2003**, *42*, 4900–4904.

135. Udeshi, N. D.; Shabanowitz, J.; Hunt, D. F.; Rose, K. L. Analysis of proteins and peptides on a chromatographic timescale by electron-transfer dissociation MS. *FEBS J.* **2007**, *274*, 6269–6276.

136. Bunger, M. K.; Cargile, B. J.; Ngunjiri, A.; Bundy, J. L.; Stephenson, J. L., Jr. Automated proteomics of *E. coli* via top-down electron-transfer dissociation mass spectrometry. *Anal. Chem.* **2008**, *80*, 1459–1467.

137. McAlister, G. C.; Berggren, W. T.; Griep-Raming, J.; Horning, S.; Makarov, A.; Phanstiel, D.; Stafford, G.; Swaney, D. L.; Syka, J. E. P.; Zabrouskov, V.; Coon, J. J. A proteomics grade electron transfer dissociation-enabled hybrid linear ion trap–orbitrap mass spectrometer. *J. Proteome Res.* **2008**, *7*, 3127–3136.

138. Kaplan, D. A.; Hartmer, R.; Speir, J. P.; Stoermer, C.; Gumerov, D.; Easterling, M. L.; Brekenfeld, A.; Kim, T.; Laukien, F.; Park, M. A. Electron transfer dissociation in the hexapole collision cell of a hybrid quadrupole-hexapole Fourier transform ion cyclotron resonance mass spectrometer. *Rapid Commun. Mass Spectrom.* **2008**, *22*, 271–278.

139. Gillmor, S. A.; Takeuchi, T.; Yang, S. Q.; Craik, C. S.; Fletterick, R. J. Compromise and accommodation in ecotin, a dimeric macromolecular inhibitor of serine proteases. *J. Mol. Biol.* **2000**, *299*, 993–1003.

140. Gross, D. S.; Schnier, P. D.; Rodriguez-Cruz, S. E.; Fagerquist, C. K.; Williams, E. R. Conformations and folding of lysozyme ions *in vacuo*. *Proc. Natl. Acad. Sci. USA* **1996**, *93*, 3143–3148.

141. Loo, J. A.; He, J.-X.; Cody, W. L. Higher order structure in the gas phase reflects solution structure. *J. Am. Chem. Soc.* **1998**, *120*, 4542–4543.

142. Collings, B. A.; Douglas, D. J. Conformation of gas-phase myoglobin ions. *J. Am. Chem. Soc.* **1996**, *118*, 4488–4489.

143. Covey, T.; Douglas, D. J. Collision cross sections for protein ions. *J. Am. Soc. Mass Spectrom.* **1993**, *4*, 616–623.

144. Hoaglund-Hyzer, C. S.; Counterman, A. E.; Clemmer, D. E. Anhydrous protein ions. *Chem. Rev.* **1999**, *99*, 3037–3079.

145. Gross, D. S.; Zhao, Y.; Williams, E. R. Dissociation of heme-globin complexes by blackbody infrared radiative dissociation: Molecular specificity in the gas phase? *J. Am. Soc. Mass Spectrom.* **1997**, *8*, 519–524.

146. Siuzdak, G.; Bothner, B.; Yeager, M.; Brugidou, C.; Fauquet, C. M.; Hoey, K.; Chang, C.-M. Mass spectrometry and viral analysis. *Chem. Biol.* **1996**, *3*, 45–48.

147. Haselmann, K. F.; Jorgensen, T. J. D.; Budnik, B. A.; Jensen, F.; Zubarev, R. A. Electron capture dissociation of weakly bound polypeptide polycationic complexes. *Rapid Commun. Mass Spectrom.* **2002**, *16*, 2260–2265.

148. Fernández, C. O.; Hoyer, W.; Zweckstetter, M.; Jares-Erijman, E. A.; Subramaniam, V.; Griesinger, C.; Jovin, T. M. NMR of α-synuclein-polyamine complexes elucidates the mechanism and kinetics of induced aggregation. *EMBO J.* **2004**, *23*, 2039–2046.

149. Clemmer, D. E.; Hudgins, R. R.; Jarrold, M. F. Naked protein conformations: Cytochrome c in the gas phase. *J. Am. Chem. Soc.* **1995**, *117*, 10141–10142.

150. Thomas, J. J.; Bothner, B.; Traina, J.; Benner, W. H.; Siuzdak, G. Electrospray ion mobility spectrometry of intact viruses. *Spectroscopy* **2004**, *18*, 31–36.

151. Bernstein, S. L.; Wyttenbach, T.; Baumketner, A.; Shea, J. E.; Bitan, G.; Teplow, D. B.; Bowers, M. T. Amyloid beta-protein: Monomer structure and early aggregation states of Abeta42 and itsPpro19 alloform. *J. Am. Chem. Soc.* **2005**, *127*, 2075–2084.

152. Bacher, G.; Szymanski, W. W.; Kaufman, S. L.; Zollner, P.; Blaas, D.; Allmaier, G. Charge-reduced nano electrospray ionization combined with differential mobility analysis of peptides, proteins, glycoproteins, noncovalent protein complexes and viruses. *J. Mass Spectrom.* **2001**, *36*, 1038–1052.

153. Kaufman, S. L.; Kuchumov, A. R.; Kazakevich, M.; Vinogradov, S. N. Analysis of a 3.6-MDa hexagonal bilayer hemoglobin from *Lumbricus terrestris* using a gas-phase electrophoretic mobility molecular analyzer. *Anal. Biochem.* **1998**, *259*, 195–202.

154. Kaddis, C. S.; Lomeli, S. H.; Yin, S.; Berhane, B.; Apostol, M. I.; Kickhoefer, V. A.; Rome, L. H.; Loo, J. A. Sizing large proteins and protein complexes by electrospray ionization mass spectrometry and ion mobility. *J. Am. Soc. Mass Spectrom.* **2007**, *18*, 1206–1216.

155. Lowe, J.; Stock, D.; Jap, B.; Zwickl, P.; Baumeister, W.; Huber, R. Crystal structure of the 20S proteasome from the archaeon *T. acidophilum* at 3.4 Å resolution. *Science* **1995**, *268*, 533–539.

156. Poderycki, M. J.; Kickhoefer, V. A.; Kaddis, C. S.; Raval-Fernandes, S.; Johansson, E.; Zink, J. I.; Loo, J. A.; Rome, L. H. The vault exterior shell is a dynamic structure that allows incorporation of vault-associated proteins into its interior. *Biochemistry* **2006**, *45*, 12184–12193.

Chapter 19

Carbohydrate Analysis by ESI and MALDI

David J. Harvey

Glycobiology Institute, Department of Biochemistry, University of Oxford, Oxford, United Kingdom

Electrospray and MALDI Mass Spectrometry: Fundamentals, Instrumentation, Practicalities, and Biological Applications, Second Edition, Edited by Richard B. Cole

19.1 INTRODUCTION

19.1.1 Sources and General Structures of Carbohydrates

Carbohydrates are the most structurally diverse and abundant compounds found in nature. They range from monosaccharides to polymeric compounds such as cellulose (D-Glcβ (1→4)- linear repeat) and chitin (D-GlcNAcβ(1→4)-linear), with molecular weights that can exceed 1 mDa. Unlike the linear polymers such as nucleic acids and proteins, oligo- and polymeric carbohydrates usually exist as branched structures because linkage of the constituent monosaccharides can occur at any of the hydroxyl groups of the adjacent monosaccharide. Consequently, very large numbers of isomers are possible; it has been calculated, for example, that for a simple hexasaccharide, there are more than 1.05×10^{12} possible isomeric structures,[1] presenting an "isomer barrier" to the analyst. It is doubtful, therefore, that any single-method analytical technique will ever provide all of the structural information necessary to characterize large oligosaccharides. Fortunately, however, this hypothetical situation does not occur in nature because the biosynthetic pathways creating most complex carbohydrates are specific and limited by the available glycosyltransferases. Consequently, only a very few of the theoretically possible isomers are ever encountered, a property that can be utilized structurally if the source of the glycan is known.

The basic building blocks of oligosaccharides are monosaccharides with the general formula $C_nH_{2n}O_n$, where $n=3$ (trioses) to 9 with $n=6$ (hexoses) being the most abundant. The monosaccharides contain $n-1$ hydroxy groups and one aldehyde (reducing sugars) or keto group. Many modified hexoses exist such as those with missing hydroxyl groups or hydroxy groups replaced by, for example, a primary amine or acylamino group. Hydroxy groups can also be methylated, acetylated, or oxidized. Most of these monosaccharides exist in a number of isomeric forms as the result of equilibration between linear and cyclic forms (typically six-membered (pyranose) or five-membered (furanose) rings) and the presence of chiral carbon atoms. Rings can also exist in different conformations—for example, chair or boat (see, for example, the book by Kennedy[2] for more information). Oligo- (2–10 units) and polysaccharides (>10 units) are usually formed by a condensation reaction between the anomeric hydroxyl group (the group formed upon cyclization of the linear form of the monosaccharide) of one monosaccharide and any of the other hydroxy groups of a second residue. Formation of a bond to the anomeric carbon prevents ring opening and fixes the ring size and conformation of that monosaccharide. Branching arises when two or more sugars are linked to the adjacent monosaccharides.

In addition to existence in the free state, oligosaccharides are frequently found conjugated to other molecules such as proteins, lipids (e.g., ceramides), and steroids. Mammalian, plant, and insect glycoproteins contain glycans attached via GlcNAc to asparagine (N-linked) or either serine or threonine (O-linked), whereas Gram-positive bacteria and archaea can have linkages such as Glc→Asn, Gal→Thr, GalNAc→Asn (*Halobacterium halobium*), Rha→Asn (*Mathanothrix soehngenii*), and Glc→Tyr, Gal→Tyr (*Thermoanaerobacter thermohydrosulfuricus*). Carbohydrate also forms part of the glycosylphosphatidylinositol (GPI) anchors that attach some glycoproteins to cell membranes. Glycosphingolipids contain acyl sphingosine (ceramide, Cer) linked to a (usually) short carbohydrate chain of which seven series are recognized, depending on the structure of the sugar chain. Sphingolipids of the ganglio series are the most common, and they contain a D-Galβ(1→3)-D-GalNAcβ(1→4)-D-Galβ(1→4)-D-Glcβ(1→) chain that bears one or more Neu5Ac (sialic acid) residues (gangliosides). Bacterial lipooligosaccharides from Gram-negative bacteria usually consist of multiple repeats of a short oligosaccharide chain attached to a core oligosaccharide which, in turn, is linked to an anchoring moiety, known as lipid A, via the acidic sugar, 2-keto-3-deoxyoctulosonic acid (Kdo). These polysaccharides frequently contain monosaccharide residues such as rhamnose and quinovosamine, which are not found in higher organisms.[3] Lipid A is a glycolipid that usually consists of two 1,6-linked glucosamines substituted with two phosphate groups and up to six N- and O-linked fatty acids and hydroxy fatty acids, whose function, like that of the GPI anchors, is to anchor the molecule into the cell membrane. Many naturally occurring plant metabolites such as steroids and triterpenes are heavily glycosylated (glycosides), and carbohydrate is also a constituent of the nucleic acids.

This chapter will discuss methods for the analysis of many of these carbohydrate types using the newer methods of ionization, namely, matrix-assisted laser desorption/ionization (MALDI) and electrospray ionization (ESI). Recent reviews include those by Harvey,[4–6] Zaia,[7] Jang-Lee et al.,[8] Haslam et al.,[9] and Campa et al.[10] [capillary electrophoresis/mass spectrometry (CE/MS)].

19.2 IONIZATION

19.2.1 Electron Impact (EI)

Although beyond the scope of this chapter, it is worth remembering that EI has enjoyed considerable success in the analysis of small carbohydrates, particularly monosaccharides, and is still valuable as the ionization method in combined gas chromatography/mass spectrometry (GC/MS) systems for the determination of composition and linkage. "Methylation analysis"[11–14] whereby polysaccharides are permethylated, hydrolyzed and then further acetylated is still a valuable technique for structural investigation and should not be discarded in the light of the more recently developed techniques such as ESI and MALDI. Methods for permethylation of carbohydrates have recently been reviewed.[15]

19.2.2 Fast-Atom Bombardment (FAB)

FAB, introduced in 1981,[16] was the major ionization method for oligosaccharides for the next decade. Although it is still used to monitor products of chemical synthesis, it is now regarded as mainly of historical interest with the advent of ESI and MALDI, which have much lower backgrounds. Nevertheless, much excellent work with carbohydrates has been reported using this method (e.g., see Refs. 17–20), which was one of the first techniques that allowed large intact carbohydrates to be ionized efficiently.

19.2.3 Matrix-Assisted Laser Desorption/Ionization (MALDI)

The use of MALDI to ionize carbohydrates was first reported by Karas and Hillenkamp in one of their early papers on the technique[21] and applied to N-linked glycans by Mock et al.[22] in 1991. It has since become one of the most popular techniques for glycan analysis (for reviews see Refs. 4, 23, and 24). Many matrices have been developed, but one of the earliest, 2,5-dihydroxybenzoic acid (DHB),[25] is still the most popular. Although the dried droplet method of sample preparation works reasonably well, more homogeneous targets and, consequently, better performance have been achieved by techniques such as mixing the DHB with various additives including 2-hydroxy-5-methoxy-benzoic acid[26] (the mixture is commonly known as super-DHB), 1-amino-*iso*-quinoline (HIQ),[27] fucose,[28] or spermine.[29] Recrystallization of the initially dried sample spot from ethanol[30] is also popular in producing a more homogeneous target surface and better mixing of sample and matrix. Studies by Kussmann et al.[31] have suggested that there is considerable fractionation in targets prepared by the standard dried droplet technique. 2,4,6-Trihydroxyacetophenone (THAP) has been reported to be more appropriate for sialylated glycans, particularly when mixed with ammonium citrate[32] because it results in reduced loss of sialic acid, a problem with the "hotter" matrices such as DHB. 6-Aza-2-thiothymine (ATT) has also been reported to cause less sialic acid loss.[32] Among newer matrices are immobilized carbon nanotubes[33] and pencil "lead"[34]; these matrices do not produce abundant matrix ions and are, thus, suitable for small molecules.

MALDI produces mainly $[M + Na]^+$ ions from neutral glycans[22,23,35] (Figure 19.1a), although other cations can be introduced by doping the matrix with an appropriate salt. Negative ions of neutral carbohydrates have only been reported occasionally from specific

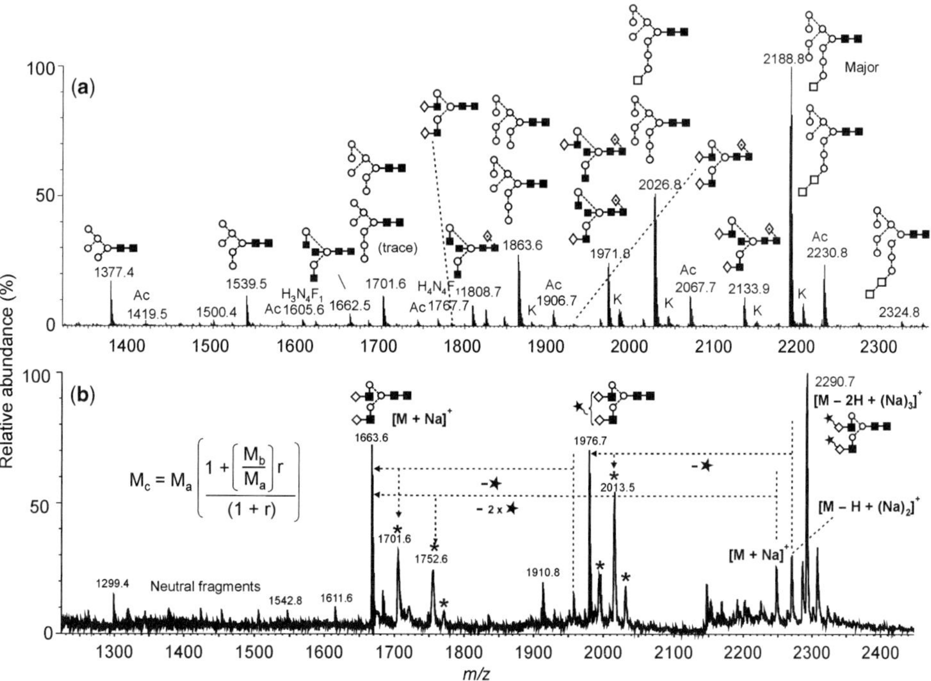

Figure 19.1. (**a**) Positive-ion, reflectron MALDI–TOF spectrum of N-linked glycans released with hydrazine from chicken immunoglobulin Y. Major ions are $[M + Na]^+$. These are accompanied by $[M + K]^+$ ions, 16 mass units higher. The ions marked OAc are *O*-acetylated by-products of the reacetylation reaction that is necessary after hydrazine release. (**b**) Positive-ion, reflectron MALDI–TOF spectrum of a disialylated biantennary N-linked glycan showing loss of sialic acid residues to give focused ions and PSD ions (marked with an asterisk). The formula links the masses of the parent, fragment, and metastable ions. Key to symbols used for the constituent monosaccharides in this and subsequent figures: □ glucose, ◇ galactose, ■ GlcNAc, ○ mannose; ◇ fucose, ★ *N*-acetylneuraminic (sialic) acid. Full line represents β-linkage; broken line represents α-linkage. The bond angle shows the linkage position.

matrices such as β-carbolines.[36,37] Sialylated glycans produce a mixture of $[M-H]^-$, $[M+Na]^+$ and $[M-nH+(n+1)Na]^+$ ions, but there is considerable loss of sialic acid by both in-source and post-source fragmentation (Figure 19.1b). In order to minimize the effects of fragmentation, many investigators examine sialylated glycans with time-of-flight (TOF) instruments operated in linear mode.[38] However, this only removes the post-source component of the fragmentation. Migration of the labile acidic proton is the cause of the instability of these compounds which can, thus, be prevented by derivatization such as methyl ester formation,[39] permethylation,[40] conversion to amides[41] or by ion pairing with quaternary ammonium or phosphonium salts.[42] This technique not only stabilizes the sialic acids but also prevents negative-ion and salt formation, thus allowing quantitative glycan profiling to be made in the positive-ion mode.

Atmospheric-pressure MALDI ion sources[43] are capable of producing less energetic ions than vacuum MALDI ion sources and enable sialylated glycans to be observed with reduced loss of sialic acid.[44] The matrix has a pronounced effect on fragmentation; with a standard nitrogen laser, "softness" of the matrix is roughly in the order alpha-cyano-4-hydroxycinnamic acid (CHCA) $\gg$ DHB > sinapinic acid (SA) ~THAP > ATT > hydroxypicolinic acid (HPA).[45] Infrared lasers generally produce less in-source fragmentation; glycerol or 3-nitrobenzyl alcohol matrices provide the low energy ionization necessary for preventing fragmentation of sialylated glycans.[46–48] Use of a Peltier-cooled sample stage has enabled water (ice) to be used as the matrix to provide conditions approaching those found physiologically.[49]

Sulfated glycans are also unstable and eliminate sulfate easily.[50] Their sodium salts, however, are more stable and can survive positive-mode ionization by MALDI.[51,52] It has been found that whereas phosphated glycans, which have the same nominal mass as their sulfated counterparts, are stable enough to survive the MALDI process as free acids, sulfated glycans are invariably seen in the positive-ion spectra only as sodium salts, allowing ready differentiation.[52] Ion pairing with the tripeptide Lys-Lys-Lys has also been used for sulfate stabilization on *N*-glycans,[53] and its use also enables isobaric phosphates and sulfates to be differentiated in their fragmentation spectra.[54] Whereas the sulfates showed preferential cleavage of the oxygen–sulfur bond, the phosphates preferred to eliminate the ligand. The effect of the position of phosphate substitution on high-mannose glycans has also been studied; whereas 6-linked phosphates are stable, 1-linked phosphates undergo cleavage in their PSD spectra.[55] Larger sulfated molecules such as heparin fragments have been stabilized for MALDI analysis by ion pairing with small peptides. Initially, the synthetic peptide IRRERNKMAAAKSRNRRRELTDTL was used,[56] but later, synthetic peptides such as $(R)_6$PYRL, $(RG)_{10}$, and $(RG)_{15}$ proved effective.[57] The best results were obtained when the number of basic arginines of the peptide equaled the number of sulfates and these arginines were spaced with glycine residues. The best matrix was 3-hydroxypicolinic acid. The molecular weights revealed the number of sulfate and acetate groups for the smaller heparin fragments, but the larger fragments still showed a tendency to eliminate sulfate (as SO_3). Larger fragments could be stabilized with the peptide, angiotensin. Several investigators have since used the $(RG)_{10}$ and $(RG)_{15}$ peptides for stabilizing sulfated glycans. Thus, Ueoka et al.[58] examined a condroitin-H-derived hexasaccharide from hagfish (*Eptatretus burgeri*) and Shriver et al.[59] have studied heparin decasaccharides. In both cases, caffeic acid was used as the matrix, resulting in only small amounts of sulfate loss. Use of this method has been summarized.[60] Sulfated carbohydrates are more stable under ESI conditions, allowing them to be examined by liquid chromatography–mass spectrometry (LC/MS) without ion pairing.[61,62] More information on sulfated carbohydrates can be found in the review by Zaia.[7]

Although not generally regarded as a quantitative technique, MALDI can provide accurate measurements of glycan compositions, provided that target inhomogenieties are overcome by acquiring spectra with many laser shots directed to several target positions.[30] Unlike peptides where ionization by protonation reflects the proton affinity of the constituent amino acids, the abundance of the $[M + Na]^+$ ions normally formed by carbohydrates appear to show little variation with structure.[35,63] Several investigators have shown excellent linear correlations between sample size and signal strength.[30,64] Ionic liquid matrices prepared from a standard MALDI matrix and an organic base such as butylamine[65] can overcome the problems associated with target inhomogeneity.[66] These matrices can also be used to analyze the very labile sulfated carbohydrates as illustrated by the use of 1-methyl-imidazolium α-cyano-4-hydroxycinnamate and butylammonium 2,5-dihydroxybenzoate ionic liquid matrices for the detection of picomole amounts of the sodium salts of sucrose octasulfate and an octasulfated pentasaccharide.[67]

19.2.4 Electrospray Ionization (ESI)

Ionization of carbohydrates by electrospray can result in several types of ions, depending on the ion source conditions and additives to the solvent. Multiple charging is common, particularly in negative-ion mode with acidic glycans, giving a spectrum that does not accurately reflect the glycan profile. Furthermore, signal suppression can be a problem, particularly when compounds are present that ionize in preference to the carbohydrates. Consequently, ESI is inferior to MALDI for profiling glycan mixtures. $[M + Na]^+$ ions, as in MALDI, can be obtained by use of high cone voltages,[68] but these ions are usually accompanied by a considerable amount of in-source fragmentation. Sensitivity generally falls as a function of increasing mass such that it is difficult to record $[M + Na]^+$ ions from the larger glycans.[68–74] Both $[M + H]^+$ and $[M + 2H]^{2+}$ ions can be formed under milder ion-source conditions, particularly from derivatized glycans that contain an amine group. In the negative-ion mode, carbohydrates form $[M - H]^-$ ions or adducts with various anions.[75–81] Anion addition stabilizes the ions and increases sensitivity. Nano-electrospray appears to be particularly appropriate for glycan analysis because it has been reported not to suffer from loss of signal at high mass to the same extent as classical electrospray.[82] A further advantage of electrospray over MALDI is that the milder ionization conditions cause little, if any, sialic acid loss.

The compatibility of electrospray with liquid separation techniques has resulted in LC/MS becoming extensively used for carbohydrate analysis in recent years. LC/MS[83] and micro-separation methods suitable for glycan analysis, particularly when coupled to mass spectrometry, have recently been reviewed.[84,85] Normal-phase separations provide good correlation between structure, molecular weight, and retention time,[86] allowing some structural information to be obtained directly from the elution profile, although the buffers often result in doubly charged ions in the mass spectra. Wuhrer et al.[87,88] have described a nanoscale LC/MS technique using normal-phase (NP) LC/MS that is capable of producing fragmentation data from small N-linked glycans at the low-femtomole level. Reversed-phase HPLC systems have also been used as illustrated by a study of the 2-aminobenzamide (2-AB) derivatives of small oligosaccharides and N-linked glycans.[89] $[M + Na]^+$ ions were the major products, and these were successively fragmented in an ion-trap instrument to provide detailed structural information. Porous graphitized carbon is also popular for carbohydrate separations,[90,91] and graphatized carbon nanoflow columns (0.6 μL/min) have been reported to increase sensitivity by 10-fold over conventional columns with a detection limit in the low-femtomole range for N-linked glycans[92]; cyclodextrin-based columns have also been used.[93]

The somewhat limited resolving power of most conventional high-performance liquid chromatography (HPLC) columns has resulted in other separation techniques such as electrochromatography[94] and capillary electrophoresis[95–97] being investigated, but there is still much need for improvement. Ion mobility separations have also been reported[98]; but separations, so far, have been restricted to small molecules. "Chip-based" methods employing arrays of small sprayers built into silicon chips enable the ESI techniques to be automated and have been exploited in the carbohydrate field.[99–101]

19.3 SAMPLE HANDLING

Although MALDI and, to a lesser extent, ESI are comparatively tolerant to the presence of contaminants, it is important to desalt samples before analysis in order to produce acceptable spectra. Suitable methods include the use of resins,[102] aminopropyl silica in a HILAC micro-elution plate,[103] or graphitized carbon.[104] Resins are usually packed into microcolumns or added directly to the MALDI target.[105] Alternatively, membranes such as low-molecular cutoff dialysis membranes or Nafion 117[106] can be used. The latter technique is particularly effective and involves floating the membrane on the surface of water and leaving a droplet, typically about 1 μL, of the aqueous glycan solution on the membrane surface for about 15 min. The solution can then be examined directly by either MALDI or ESI. Peptidic material can be removed from glycan samples with C18 such as that incorporated into ZipTips. However, some of this peptidic material is too hydrophilic to stick to the resin effectively, but has been made more hydrophobic to improve adsorption to C18 or graphatized carbon by reacting amino groups with sodium 2,4,6-trinitro-benzene-1-sulfonate.[107]

Among other clean-up methods is a technique to capture glycans with reducing termini, of the type released from glycoproteins, by allowing these termini to react with oxylamino-containing polymers followed by release with mild acid.[108]

19.3.1 Derivatization

Derivatization of the reducing terminus, usually by reductive amination[109–111] (Figure 19.2), is necessary for detection of carbohydrates by optical methods and is frequently used for mass spectral analysis for purposes such as enhancement of signals or modification of fragmentation patterns. The reductive amination reaction with 2-aminopyridine (2-AP) can be reversed to recover the carbohydrate.[112] Conversion to aminobenzoic acid alkyl esters increase hydrophobicity and can improve sensitivity,[113,114] particularly for FAB ionization. Small aromatic amines such as 2-aminobenzamide (2-AB),[115] 2-aminoacridone (AMAC),[116,117] benzylamine,[118] or 2-AP[119] are used extensively for fluorescent detection and increase the proton affinity of the molecules such that $[M + H]^+$ ions are formed sometimes in preference to the more usual $[M + Na]^+$, particularly with electrospray ionization where they have been reported to enhance signal strength.[120] Derivatization with an amine-containing derivative,[72] or more particularly one that has a constitutive cationic charge,[120–122] can lead to pronounced increases in sensitivity. Derivatization with cationic,[123] amino, or hydrophobic derivatives has also been used to increase the relative abundances of carbohydrate peaks in the presence of contaminants such as peptides.

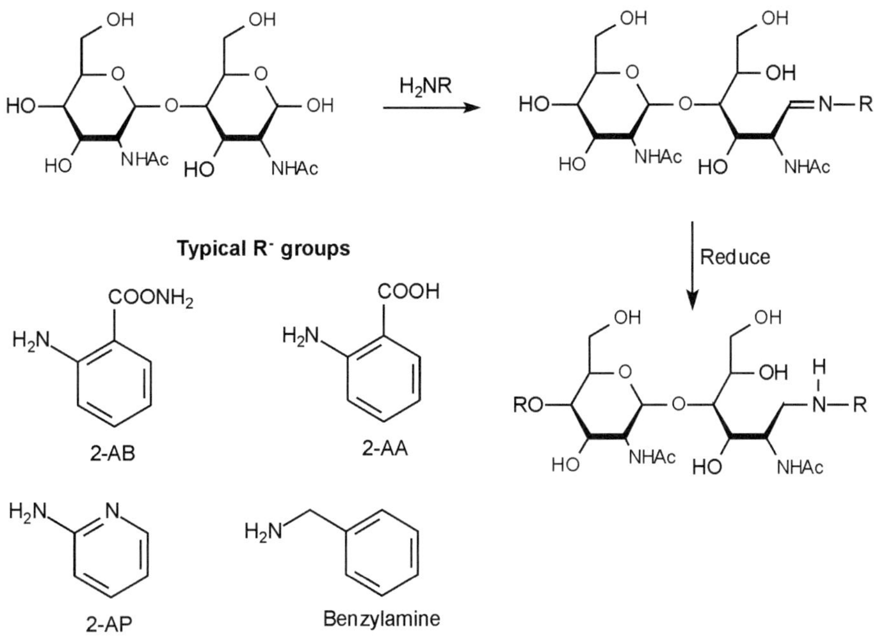

Figure 19.2. Reductive amination reaction for derivatization of the reducing terminus of carbohydrates (shown for the chitobiose core of *N*-glycans) and some of the typical amines used for the reaction.

Changes in fragmentation patterns are governed to a large extent by the type of precursor ion that is formed.[124] Protonation of the derivative localises the charge resulting in a spectrum dominated by Y-type cleavages (see below). Similarly, in the negative ion spectra of 2-aminobenzoic acid (2-AA) derivatives, the charge is localized on the acid group and diagnostic fragmentation produced by proton abstraction from hydroxy groups is suppressed.[125] Küster et al.[126] have studied the effect of a number of derivatives on the high-energy fragmentation of *N*-glycans; more cross-ring fragments were seen than with underivatized compounds. An ^{0,2}A-ion from the reducing terminus (loss of the derivative) was often the most abundant ion in the spectrum. Derivatives containing a bromine atom have been used as labeling reagents because of the distinctive bromine isotope pattern that allows all fragment ions containing the derivatized end of the molecule to be identified.[127,128] Alternative derivatization strategies that avoid some of the clean-up methods necessitated by reductive amination, such as the need to remove salts, include reactions with phenylhydrazine[129–131] and with the Fmoc reagent, 9-fluorenylmethyl chloroformate.[132] The latter derivatives can be decomposed to recover the native glycans.

19.4 FRAGMENTATION

Fragment ions can be formed within the ion source of the mass spectrometer [in-source decay (ISD) or prompt fragmentation], within the flight tube of time-of-flight instruments [post-source decay (PSD)],[133] or by collisional activation (CID) in a collision cell. ISD, which is common with FAB spectra, can be a problem because Y-type ions are isobaric with native glycans, thus distorting the glycan profile. PSD ions are sometimes difficult to observe, are often poorly resolved, and generally require relatively large amounts of glycan. CID-produced ions, on the other hand, suffer from of none of these disadvantages. High-quality fragmentation spectra can now be obtained from both electrosprayed and MALDI-generated ions[71,134–136] by use of Q-TOF-type instruments. The recent review by Zaia[7] provides a good overview of carbohydrate fragmentation. Fragmentation of carbohydrates occurs by two main pathways, glycosidic cleavage involving the breaking of a bond between sugar rings and cleavage of the bonds comprising the rings (cross-ring cleavage). Glycosidic cleavages from even-electron ions of the type $[M + Na]^+$ result in the loss of neutral molecules and are accompanied by hydrogen migrations. A third type of fragmentation that was proposed to involve a six-membered transition state and transfer of a carbon-attached hydrogen atom has recently been suggested[137]; the fragment ions effectively eliminate two oxygen functions to leave the expelled neutral particle with a carbonyl group.

19.4.1 Nomenclature of Fragment Ions

The accepted nomenclature for describing the fragmentation of carbohydrates is that introduced by Domon and Costello[138] (Figure 19.3) in 1988. Ions that retain the charge at the reducing terminus are designated X (cross-ring), Y, and Z (glycosidic), whereas ions with the charge at the nonreducing terminus are A (cross ring), B, and C (glycosidic) (Figure 19.3). Sugar rings are numbered from the nonreducing end for A, B, and C ions and from the reducing end for the others. Ions are designated by a subscript number that follows the letter to show the bond that is broken. However, to avoid this number changing for glycans that have different length chains, we have introduced a modification in which ions numbering from the reducing end have the subscript R, R – 1, etc., and those from the nonreducing end, NR, NR – 1, etc.[139] Fragments from branched-chain glycans are distinguished by Greek letters, with α representing the largest chain. Cross-ring fragments are given superscript numbers showing the bonds cleaved.

19.4.2 Fragmentation Modes of Different Ion Types

The type of precursor ion produced from the carbohydrate has a significant effect on the type of fragmentation that is produced, but the resulting spectrum appears to be relatively independent of the way in which the precursor ion is formed.

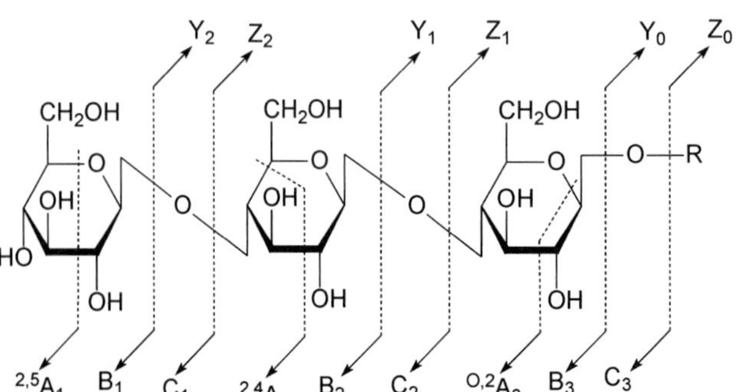

Figure 19.3. Nomenclature introduced by Domon and Costello[138] for describing the fragmentation of carbohydrates.

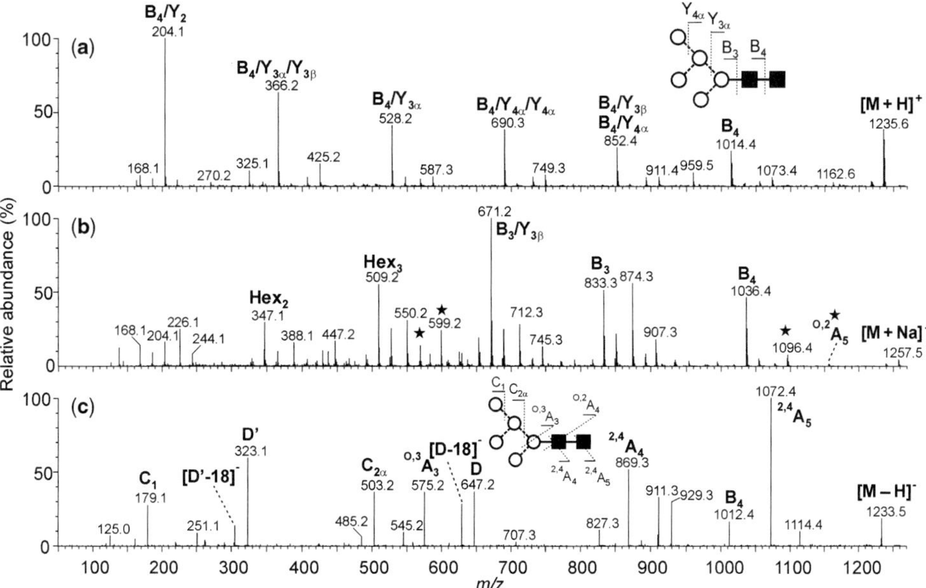

Figure 19.4. (**a**) CID Spectrum of the $[M + H]^+$ ion from the high-mannose N-linked glycan, $Man_5GlcNAc_2$. (**b**) CID spectrum of the $[M + Na]^+$ ion from $Man_5GlcNAc_2$. Ions marked with a star are cross-ring products. (**c**) CID Spectrum of the $[M - H]^-$ ion from $Man_5GlcNAc_2$.

19.4.2.1 $[M + H]^+$ Ions

Protonated molecules decompose much more readily than metal-cationized species and yield mainly B- and Y-type glycosidic cleavage ions with very little or no cross-ring fragments[68] (Figure 19.4a). Thus, they are of limited use for determining the detailed structure of unknown carbohydrates. Another problem is the tendency for rearrangement reactions to occur,[140–144] particularly when the carbohydrate has been derivatized at the reducing terminus.[145,146] Carbohydrates that have been derivatized by reductive amination contain a secondary amine that readily attracts a proton. This localizes the charge and produces a spectrum that contains mainly Y-type glycosidic fragments as illustrated by the PSD spectrum of 2-AP derivatives of maltoheptaose[120] (recorded using CHCA as the matrix). When examined by LC/MS using normal-phase HPLC columns and an ammonium formate buffer, a range of $[M + 2X]^{2+}$ ions where $X = H^+$, NH_4^+, or Na^+ can be produced. If one of the X atoms is hydrogen, fragmentation is normally similar to that of the $[M + H]^+$ ion with the production of abundant singly charged Y ions providing sequence information.

19.4.2.2 $[M + \text{Alkali Metal}]^+$ Ions

In contrast to the fragmentation of $[M + H]^+$ ions, fragmentation of $[M + \text{alkali metal}]^+$ ions is more difficult. The ease with which various ions decompose follows the order $H = Li^+ > Na^+ > K^+ > Cs^+$. $[M + Cs]^+$ ions, in fact, do not fragment other than to give Cs^+.[68,147,148] Internal fragments (losses from two or more terminal sites) are common and, in addition, A-type cross-ring fragments are present,[68,149] although generally in fairly low abundances from the low-energy spectra that are typically produced by Q-TOF instruments[68,69,71] (Figure 19.4b). More abundant cross-ring fragment ions, particularly X-type, can be produced at higher energies of the type found in TOF/TOF-type mass spectrometers[150–154] or formed by high-energy photodissociation.[155] The $[M + Na]^+$ ions of permethylated carbohydrates have also been examined with TOF/TOF instruments. Their spectra also contained many A- and X-type cross-ring fragments, but, in general, B- and Y-ions formed by cleavage adjacent to GlcNAc residues dominated the spectra.[156–158] The cross-ring cleavages have allowed the linkage of methylated sialic acids to be determined.[40] An additional advantage of the fragmentation of the $[M + metal]^+$ ions is that the rearrangements reported from the $[M + H]^+$ ions are absent.[159]

Infrared multiphoton dissociation (IRMPD) of $[M + Li]^+$ and $[M + Na]^+$ ions appears to give very similar fragmentation to that observed by CID[160] and enables fragmentation of larger N-linked glycans to be observed in Fourier transform ion cyclotron resonance (FTICR) mass spectrometers when it could not be obtained by CID.[161] IRMPD of $[M + nH]^{n+}$ ions of large glycopeptides, on the other hand, show preferential fragmentation of the peptide chain.[162]

19.4.2.3 [M + Other Metals]⁺

$[M + Ag]^+$ ions have been used to produce characteristically different spectra relative to adducts of group I metals from some monosaccharides, allowing them to be differentiated whereas $[M + Cu]^+$ ions did not.[163] With N-linked glycans, silver appears to be particularly good at cleaving glycosidic bonds but does not offer any particular advantage for structural determination.[164] Among doubly charged cations, calcium appears to be particularly effective both for increasing sensitivity and for fragmenting carbohydrates. Doubly charged ions are the major ions formed with these metals, although copper has a tendency to form singly charged ions.[165]

19.4.2.4 [M − H]⁻ and [M + Anion]⁻ Ions

Until recently, fragmentation of negative ions has received less attention than that of positive ions, even though it has been known for many years that many carbohydrates can be induced to ionize by removal of a proton from a hydroxyl group. Acidic (anionic) carbohydrates that contain groups such as sulfate, phosphate, or sialic acid form $[M - H]^-$ or $[M - nH]^{n-}$ ions very readily. Fragmentation of $[M - H]^-$ ions from neutral carbohydrates (Figure 19.4c) tends to be more specific than that of $[M + H]^+$ or $[M + metal]^+$ ions because of specific reactions initiated by abstraction of protons from individual hydroxyl groups. The resulting spectra often contain less ambiguous structural information than positive-ion spectra as examples below will demonstrate. Doubly charged negative ions from neutral carbohydrates behave similarly.[166] Fragmentation of $[M + anion]^-$ ions where the anion is typically chloride, nitrate, or phosphate[76,80,81,167–169] yield similar fragment ions to the $[M - H]^-$ ions because the first stage of fragmentation involves proton abstraction by the anion to give the $[M - H]^-$ ion. Unlike positive-ion spectra that contain mainly B- and Y-type fragments, the negative-ion spectra tend to be dominated by C and cross-ring cleavage fragments.

19.4.3 Source of Fragment Ions

Fragment ions from carbohydrates can arise in various parts of the mass spectrometer; some are detrimental to an analysis, but many provide vital structural information.

19.4.3.1 In-Source Decay (ISD) Ions

Fragment ions from very fast decomposition of the molecular ions that occur within the ion source are focused with the molecular ions. Such ions can be the result of Y-type glycosidic cleavages that give rise to compositions indistinguishable from those of native glycans and which will distort a glycan profile. Frequently, however, B-type and cross-ring fragments will also be present to alert the user of the possible presence of the Y-ions. In-source fragmentation is particularly prominent with carbohydrates containing sialic acid.

19.4.3.2 Post-Source Decay (PSD) Ions

Ions that decompose more slowly can give rise to PSD fragments if decomposition occurs between the ion source and the detector. Huberty et al.[170] first noted that, for linear TOF instruments, much of the signal produced by glycopeptides that contained sialylated glycans consisted of fragment ions formed by loss of sialic acid, as could be seen by switching on the reflectron. By stepping the reflectron voltage, so that fragment ions were focused over the entire mass range, Huberty et al.[170] also observed strong glycosidic PSD fragment ions from a neutral biantennary glycan. These ions were found to be produced from cleavage between the constituent monosaccharides, leading the way for general fragmentation studies following MALDI ion production.[133] As the result of ion kinetics, the abundance of these PSD ions depends on the relative time that the ions spend in the ion source compared with that involved in traversing the flight tube. Thus, long in-source delay times, can have an adverse effect on the relative abundance of PSD ions. Kaufmann et al.[171] have proposed that an additional loss of PSD fragments can be due to a reduced collisional activation in delayed-extraction sources. Cross-ring fragmentation appears to occur in a shorter time frame than glycosidic cleavages, and it has been observed to produce fragments in ISD spectra.[172]

PSD ions manifest themselves in reflectron–TOF spectra as broad metastable ions whose positions, unlike those ions produced in magnetic sector instruments, are dependent on the make of instrument and on the focusing conditions. However, a relationship between the masses of the precursor and fragment ions and the apparent mass of the metastable (PSD) ion has been reported.[173] The PSD ions can be focused by varying the reflectron potential or by the use of a curved-field reflectron, but, frequently, the ion peaks are much broader than those of ISD ions and, without a curved-field reflectron, the spectrum has to be acquired in stages and each subspectrum stitched together to give the final result. Thus, the abundance relationship of ions in different sections of the spectra can be distorted. Structural analysis using PSD ions suffers from a number of other disadvantages. Ion production is difficult to control and generally needs relatively large amounts (e.g., up to 100 pmol) of carbohydrate. To some extent, judicious matrix selection can be used to control fragmentation. This dependence can be illustrated by the PSD spectra of 2-amino-pyridine (2-AP) derivatives of maltoheptaose recorded by Okamoto et al.[120] CHCA efficiently protonated the nitrogen atom of the derivative, and the resulting $[M + H]^+$ ion subsequently fragmented to give only a Y-series of fragments as the result of the charge localization on the reducing terminus. 2,5-DHB, on the other hand, produced only a $[M + Na]^+$ ion with the charge localized on the sodium atom rather than on the carbohydrate. This ion fragmented to give a complex mixture of B and Y ions together with some cross-ring fragments.

19.4.3.3 Collisional-Induced Decomposition (CID)

The most efficient method of carbohydrate fragmentation is CID because fragmentation is isolated from ion production and can be controlled by application of varying voltages on the collision cell. The conditions for fragmentation obtained with Q-TOF-type instruments are relatively low energy, producing mainly glycosidic cleavages with low amounts of mainly A-type cross-ring fragments. Both ESI[68] and MALDI[71,134,174–176] ion sources can be used. Higher-energy collisions that produce X-type cleavages can be obtained with TOF/TOF instruments,[154] and magnetic sector instruments can be fitted with an orthogonal TOF analyzer.[70,177]

19.4.4 MS^n

Sequential fragmentation or “disassembly” of carbohydrates is emerging as a powerful method for carbohydrate analysis. Several successive stages of fragmentation can be observed in an ion-trap instrument, particularly where cleavages can occur adjacent to GlcNAc residues.[178,179] The technique is particularly valuable when interfaced with HPLC.[89] MS^n spectra have shown that many of the fragment ions in the MS^2 spectra from complex glycans arise from several pathways, a property that can present problems for the interpretation of “unknown” spectra.[139] Deguchi et al.[180] have chosen both peptide and carbohydrate-derived ions from the MS^2 spectrum of glycopeptides and fragmented each at the MS^3 stage in both positive- and negative-ion modes to obtain data on both halves of the molecule. In another approach for the total analysis of glycoproteins, the compounds were digested in-gel with trypsin followed by successive stages of fragmentation in an ion-trap–TOF instrument. Information on the carbohydrate was obtained in the early stages of fragmentation until finally all of the sugar had been removed, allowing the peptide to fragment to reveal its sequence by a series of y ions.[181] Ashline et al.[182] have studied the fragmentation of small permethylated oligosaccharides as alkali metal adducts and identified ions that are characteristic of, for example, the nature of the terminal monosaccharide residue. Larger, N-linked glycans have been studied by Ojima et al.,[183] who suggest that some isomer information is available in the form of peak intensity in the positive ion MS^3 spectra. Further information on anomeric configuration can be obtained with energy-resolved spectra at various stages of MS^n. It has been noted[184] that for the $[M + Na]^+$ ions from linear oligosaccharides, α-linkages fragment at a lower energy than β-ones. However, the authors were not sure if the relationship would be apparent with all types of carbohydrate.

Although MALDI–ion-trap–TOF instruments produce well-resolved fragment ions from carbohydrates, it has been noted that considerable metastable ion formation can occur during the ion-trapping process[139,185]; thus, they are less useful than MALDI-TOF instruments for profiling sialylated glycans. Although most work on successive (MS^n) fragmentation requires trapping instruments, it is possible to make use of in-source fragmentation to provide the MS^2 spectrum and then to fragment the product ions in a collision cell in the conventional manner to provide the MS^3 spectrum.[186] However, the method is only satisfactory for single compounds when there is no ambiguity as to the source of the MS^2 fragments.

19.5 COMPUTER INTERPRETATION OF MS DATA

Although most spectra interpretation has, until recently, been performed manually, several investigators[187–192] have experimented with computer-aided interpretation and library-search techniques. The program by Mizuno et al.,[187] which is claimed to be able to interpret spectra in only a few seconds, assigns compositions to fragment ions in PSD spectra and constructs predicted spectra from known structures to compare with the experimental data. The approach taken by Gaucher et al.[188] is more empirical but is limited to carbohydrates with less than 10 residues. Only N-linked glycans are addressed by the software developed by Ethier et al.,[189,190] whereas the program developed by Joshi (GlycosidIQ™)[192] for negative-ion spectra is much more comprehensive and is available at the GlycoSuite web site from Proteome Systems (https://tmat.proteomesystems.com/glycosuite/). The site also contains a carbohydrate database, GlycoSuiteDB, and Proteome Systems additionally has a tool, GlycoMod[193] (http://www.expasy.org/tools/glycomod/), for obtaining glycan composition from glycan masses. GlycoX is an algorithm that takes the accurate mass of a glycopeptide as measured with an FTICR mass spectrometer and the glycan spectra and computes not only the glycosylation site but also the composition of the glycans at each site.[194] The program can cope with several glycosylation sites per molecule. The web tool Glycofragment compares peaks in the experimental MS^2 mass spectrum with the masses calculated from structures in the SweetDB database.[191,195] Kameyama et al.[196] have described a method using MS^3 data obtained from oligosaccharides prepared using glycosyltransferases that matches these data with the experimental spectra, whereas MS^n data are used in the OSCAR algorithm[197] for interrogating data in the FragLib library.[198] STAT is another web-based algorithm that works with MS^n data[188] after calculating all possible structures based on the molecular weight. The program requires input from the user as to, for example, what monosaccharides are likely to be present and which of a set of possible compositions is most likely, based on the monosaccharides present. Use of the program was illustrated by applications to *N*-glycans and glycans from bacteria. A program, CartoonistTwo, that claims to be an improvement because it takes more account of minor fragments, has also been developed and applied to O-linked glycans.[199] URLs to available databases can be found in the reviews by Pérez and Mulloy[200] and von der Lieth et al.[201]

A program, STEP (statistical test of equivalent pathways), has been developed to determine whether ions in MS/MS experiments originate directly from precursor ions or whether they are secondary products. STEP ratios are calculated by comparing the relative abundances of product ions in two MS/MS experiments. Product ions arising directly from the precursor ion always have STEP ratios that are less than or equal to 1 whereas ions that result from secondary fragmentation pathways have STEP ratios that are significantly larger than the primary ions.[202]

Following the demise of the Complex Carbohydrate Database, a number of initiatives have been launched to aid glycomics. Foremost among these is KEGG (Kyoto Encyclopedia of Genes and Genomics[203,204]http://www.genome.jp/kegg/glycan which has its own database of carbohydrate structures, including those from the CarbBank database and a database of known biosynthetic pathways. Further biosynthetic pathways can be found at the CAZY web site (http://afmb.cnrs-mrs.fr/CAZY).[205] The Consortium for Functional Glycomics (http://www.functionalglycomics.org/) is attempting to link structural databases with information on biosynthesis and genomics in collaboration with KEGG and CAZY[206] and the Glycosciences.de portal (http://www.glycosciences.de)[207] provides access to the former CarbBank database.

An algorithm, named "cartoonist," has been developed to automatically annotate peaks in MALDI spectra. Symbolic structures are selected from a library and attached to the peaks with a confidence score but without any attempt to verify the structure by techniques such as fragmentation analysis.[208]

19.6 APPLICATION TO SPECIFIC CARBOHYDRATE TYPES

Space does not permit a discussion of all carbohydrate types, so only some selected examples will be given.

19.6.1 Unconjugated Oligosaccharides and Polysaccharides

These compounds include (a) the smaller polysaccharides such as those found in milk and (b) very large compounds such as starch, dextran, and constituents of cell walls.

19.6.1.1 Milk Sugars

Early work with neutral and acidic fucosylated lactose polymers with molecular weights of up to 8000 from human milk indicated that DHB and super-DHB were most appropriate MALDI matrices for neutral sugars, whereas 3-aminoquinoline was for acidic glycans.[209] However, Finke et al.[210,211] preferred a mixed matrix of ATT with diammonium hydrogen citrate for the acidic glycans. These investigators have shown by high-performance[210] and preparative anion-exchange chromatography (AEC)[211] that many of the peaks detected by MALDI contained several isomers emphasizing the need to combine mass spectrometric techniques with chromatographic techniques for detailed structural analysis of these compounds. Such chromatographic techniques include micellar electrokinetic chromatography (MEKC) on a C-18 stationary phase for separation of the human milk sugars as methyl-, ethyl-, or butyl-benzoate or 4-*n*-heptyloxyaniline derivatives[212] and HPLC of fluorescent 2-AMAC derivatives.[213] In both studies, MALDI was used in parallel with ESI. In the study by Charlwood et al.[213] the MALDI spectra from DHB/HIC gave mixtures of $[M + H]^+$, $[M + Na]^+$ and $[M + K]^+$ ions; the $[M + H]^+$ ions presumably arose from protonation of the derivative. The MS/MS spectra, performed with a Q-TOF mass spectrometer on $[M + 2H]^{2+}$ ions, provided information on the location of the fucose residues. These sugars have also been separated by thin-layer chromatography (TLC) and analyzed directly from the plate by both IR- and UV-MALDI.[214] For IR-MALDI, the matrix was glycerol, whereas for UV-MALDI a mixed matrix of glycerol and CHCA was used. The highest sensitivity was found with IR-MALDI. Negative-ion MS^2 and MS^3 spectra yield abundant C and cross-ring fragments[215] providing extensive structural information enabling several new structures to be identified from human milk.[216] Because of the high fucose content of some milk sugars, some ambiguity can arise from deductions of composition from a molecular weight measurement because the mass difference between five fucose residues and two *N*-acetyl-lactosamine (Gal-β-(1→4)-GlcNAc) units is only 0.042 Da.[217] In addition, milk from humans as well as milk sugars from several other species such as elephant,[214] Japanese black bear (*Ursus thibetanus japonicus*),[218] and polar bear (*Ursus maritimus*)[219] have also been examined.

19.6.1.2 Large Polysaccharides

Large polymers of the type found in plant cell walls are difficult to analyze directly. Therefore, most studies use prefractionation by size-exclusion chromatography (SEC) or reduction to smaller units by partial acidic hydrolysis or enzymolysis. For mixtures of large polymers such as dextrans ionized by MALDI (Figure 19.5), there is a fall in signal intensity with increasing mass[220,221] that has been attributed to signal suppression by the smaller,

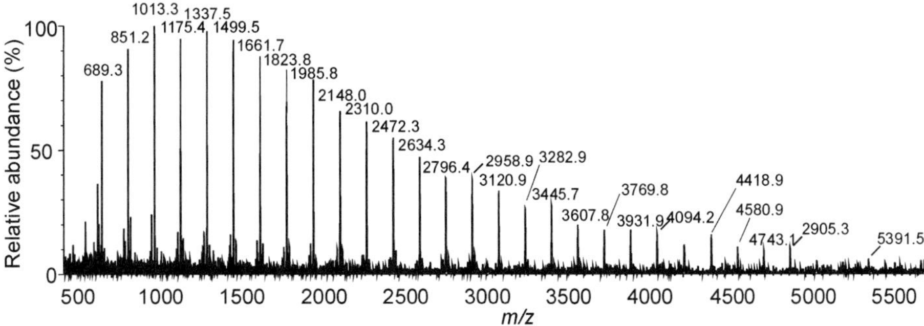

Figure 19.5. Positive-ion MALDI–TOF mass spectrum of dextran from *Leuconostoc mesenteroides*.

but more abundant, components. After fractionation of the mixtures, compounds with higher molecular weights were seen. However, this suppression effect was attributed by Mohr et al.[27] to the aggregation of the larger molecules as seen with synthetic polymer samples. Signal suppression by MALDI matrix ions can also affect ion abundance in the low mass region, probably by transient desensitization of the TOF detector. In general, however, there appears to be a good correlation between profiles obtained by MALDI, HPLC, high-performance anion exchange chromatography (HPAEC), and FAB MS.

Fractionation by SEC can be used to reveal larger constituents when polymer mixtures contain a disproportionately large amount of lower homologues.[222] SEC has, for example, shown promise for the analysis of dextran oligomers (dextran 5000 and dextran 12,000)[223] and SEC-MALDI, combined with partial acid hydrolysis has been used to examine hemicelluloses from several tree species.[224] Spectra of cleaved fragments, following digestion with enzymes such as endopolygalacturonase or endoxylanase A, can be used to determine the repeating unit for simple polymers, but they often fail for more complicated structures because of features such as random acetylation or the presence of sodium salts. These polymers can, however, be examined after chemical treatment such as permethylation or deacetylation. Hemicelluloses, or xylans, are often difficult to extract and hydrolyze, although prior delignification[225] or microwave irradiation[226] considerably improves the yield.

19.6.2 Protein-Linked Glycans

Proteins can contain three types of attached glycans: O-linked glycans that are attached to either serine or threonine but with no directing consensus sequence; N-linked glycans attached to asparagine in an Asn-Xxx-Ser(or Thr) motif, where Xxx is any amino acid except for proline; and lastly, glycans forming part of a glycosylphosphatidylinositol lipid anchor. Recently, a few glycoproteins have been reported in which cysteine replaces serine or threonine in the N-linked consensus sequence.[227] About 50% of all proteins appear to be glycosylated with the glycan portion responsible for many of the biophysical properties of the molecules.[228,229] O-Linked glycans from mammalian systems tend to be relatively diverse structures, whereas the *N*-glycans, although often larger, contain a trimannosyl chitobiose pentasaccharide core that is attached to the protein by a GlcNAc link and have well-defined overall structures. Their biosynthesis[230] is outlined in Figure 19.6 and involves the attachment of the glycan $Glc_3Man_9GlcNAc_2$ (I, Figure 19.6) to the protein during transcription followed by the action of a series of glycosidases and glycosyltransferases.[230]

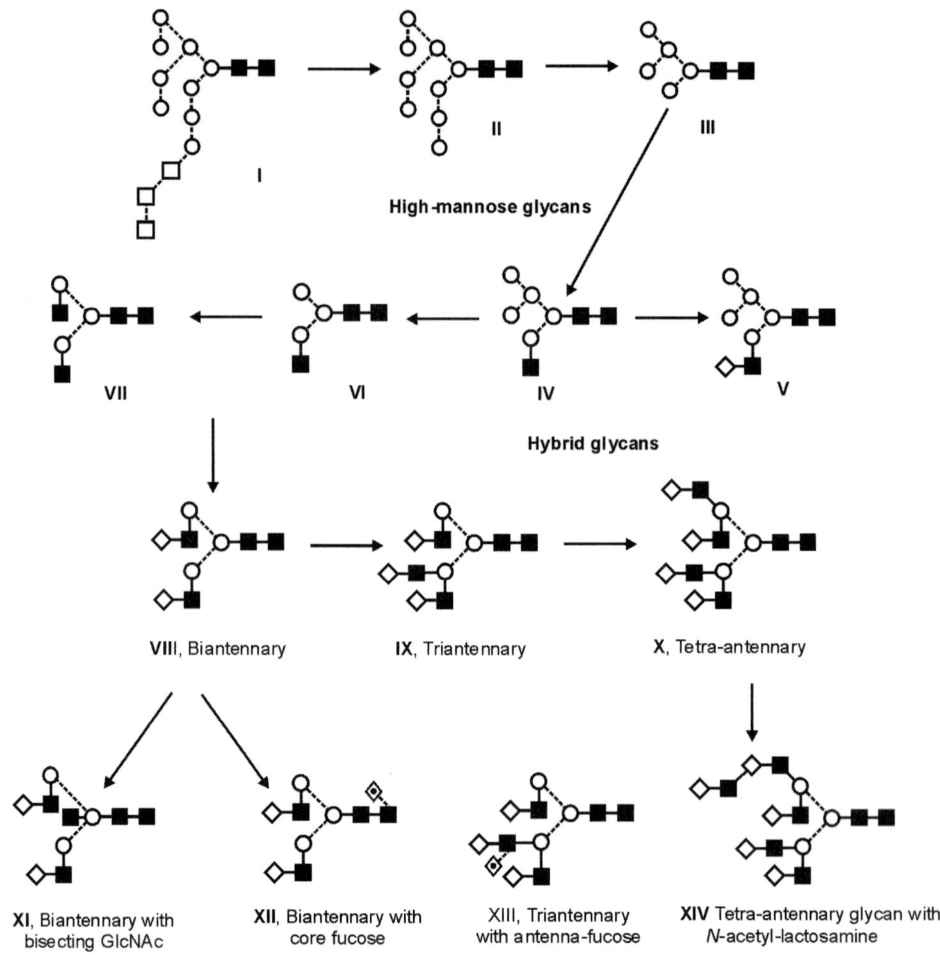

Figure 19.6. Biosynthesis of N-linked glycans. The structures that are shown are only a few of those possible. Arrows are illustrative and do not necessarily depict intact biochemical pathways. The antennae of the complex glycans usually terminate in *N*-acetyl-neuraminic acid in either α2-3 or α2-6 linkage.

Sites of attachment may be fully occupied with one or many glycans, or they may be partially occupied or vacant. Occupancy of several sites by many glycans can result in large numbers of individual glycoproteins (known as glycoforms) for each protein chain. It has been estimated, for example, that *Desmodus rotundus* salivary plasminogen activator, with its two N- and four O-linked sites, contains in excess of 330,000 individual molecular species if the glycans were randomly distributed.[231,232]

19.6.2.1 Mass Spectrometry of Intact Glycoproteins

Although both MALDI and ESI are capable of recording spectra of intact glycoproteins, individual glycoforms can only be resolved from small proteins (up to about 20–40 kDa) containing a limited number of glycans, preferably attached to a single site. An example is

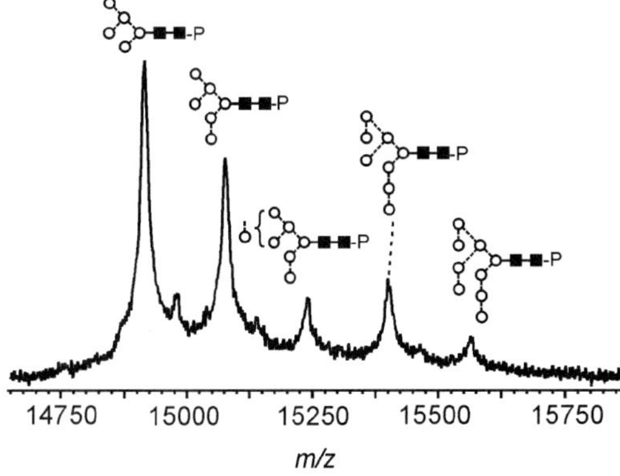

Figure 19.7. Positive-ion MALDI mass spectrum of glycoforms of ribonuclease B. P represents a Protein.

shown in Figure 19.7. If the mass of the protein is known, the glycan mass can be obtained by taking the difference; and the glycan composition can be derived in terms of the constituent monosaccharides. N- and O-linked glycans usually contain only a limited number of monosaccharide types (hexose, *N*-acetylhexose, deoxy-hexose, etc.), and combinations of the masses of these monosaccharides (Table 19.1) frequently match the experimental glycan mass for common N-and O-linked glycans. Larger glycoproteins, and those with several glycosylation sites, usually give unresolved mass peaks, particularly by MALDI, and further analysis requires either proteolysis or removal of the intact carbohydrates for subsequent analysis. A recently reported method for recovering glycoproteins from mixtures used

Table 19.1. Residue Masses of Common Monosaccharides

Monosaccharide	Residue Formula	Residue Mass[a,b]
Pentose	$C_5H_8O_4$	132.042 132.116
Deoxyhexose	$C_6H_{10}O_4$	146.078 146.143
Hexose	$C_6H_{10}O_5$	162.053 162.142
Hexosamine	$C_6H_{11}NO_4$	161.069 161.157
HexNAc	$C_8H_{13}NO_5$	203.079 203.179
Hexuronic acid	$C_6H_8O_6$	176.032 176.126
N-Acetylneuraminic acid	$C_{11}H_{17}NO_8$	291.095 291.258
N-Glycolylneuraminic acid	$C_{11}H_{17}NO_9$	307.090 307.257

[a]Top figure represents monoisotopic mass (based on C = 12.000000, H = 1.007825, N = 14.003074, O = 15.994915), bottom figure represents average mass (based on C = 12.011, H = 1.00794, N = 14.0067, O = 15.9994.

[b]The masses of the intact glycans can be obtained by addition of the residue masses given above, plus the mass of the terminal group (H_2O for an unmodified glycan) 18.011 (monoisotopic) and 18.152 (average) and the mass of any reducing terminal or other derivative.

magnetic beads containing immobilized aminophenylboronic acid. The glycoproteins were analysed by MALDI directly from the beads with sinapinic acid as the matrix.[233]

19.6.2.2 N-Linked Glycans

a. Site Analysis. Site analysis involves the determination of the glycan occupancy at each consensus sequence site; several methods are available. Commonly, the protein is cleaved with a protease such as trypsin to isolate each site to a different glycopeptide with analysis by either HPLC or LC/MS. Some peptides prove difficult to digest, but the use of a new surfactant "RapiGest SF™" greatly aids digestion as shown for the glycoprotein α1-acid glycoprotein.[234] Problems arise when it is not possible to produce peptides containing only one glycosylation site and when the peptides are too large for efficient mass spectral analysis. In addition, mass spectrometric detection of glycopeptides in the presence of peptides is often difficult because of suppression effects.[235] Carr et al.[236] have developed a method for glycopeptide identification in which small protonated glycan fragment ions such as *m/z* 163 (hexose), 204 (HexNAc), or 366 (Hex-HexNAc) are monitored throughout a LC/MS analysis. Comparison of the traces from these ions with the total ion chromatogram identifies those peptides containing attached glycans. Similar ion monitoring can be achieved from $[M + H]^+$ ions generated by MALDI with the advantage that spectra are not complicated by the production of multiply charged ions that can occur during ESI.[237,238] Lochnit and Geyer[239] have collected HPLC effluent on a MALDI plate with the aid of a robot (Probot) MALDI spotter and report improved spectral quality of both peptides and glycopeptides.

Removal of glycans from either the intact glycoprotein or a glycopeptide with the enzyme peptide *N*-glycosidase F (PNGase-F) leaves aspartic acid in place of the asparagine at the N-linked site of the protein. Consequently, if the peptide sequence is known, the occupancy at the site can be deduced. In cases where the sequence is not known, the aspartic acid can be identified by partial ^{18}O incorporation if the digestion is performed in 40% ^{18}O-enriched water.[240]

Lattová et al.[241] have developed a method for detection of the released glycans by MALDI in the presence of the peptide. Normally, the glycans ionize inefficiently as $[M + Na]^+$ ions such that their signals are swamped by the stronger $[M + H]^+$ ions from the peptide. However, after derivatization with phenylhydrazine, the glycans formed $[M + H]^+$ ions with comparable efficiency. Thus the glycan profile and site occupancy (Asn to Asp conversion) could be monitored in a single spectrum. Basic and quaternary ammonium-containing derivatives linked via hydrazine chemistry have also been used for this purpose.[123]

Lee et al.[242] have heated glycopeptides with trifluoroacetic acid (TFA) in a microwave oven and degraded the glycans (from horseradish peroxidase) such that only GlcNAc was left at the glycosylation site. This technique marks the glycosylation site but removes information about the attached glycan. An alternative technique uses a nonspecific protease, such as pronase to cleave the protein, leaving glycans containing asparagine or a very short peptide chain.[243–245] Although reducing the molecular weight considerably, this method can produce several peptide fragments from each site, thus complicating the analysis.

b. Glycan Release. Glycan structures can be examined with the glycopeptides derived as above; or, more commonly, the glycans are removed from the glycoproteins either chemically or enzymatically and analysed separately.[246] Chemical release, usually with hydrazine, has the advantage of being nonselective but frequently introduces artifacts. It has been calculated[247] that as much as 25% of the total glycans are modified at the reducing terminus and that these compounds can never be converted into the parent sugar. Anhydrous hydrazine releases both N-[248–250] and O-linked[249] glycans. O-Linked glycans are specifically released at 60°C, whereas 95°C is needed to release the N-linked sugars. Hydrazine preserves the reducing terminus of the glycan; but, because it cleaves all peptide bonds, it removes the acetyl groups from the *N*-acetylamino sugars, potentially resulting in a loss of information. Consequently, reacetylation is usually performed after hydrazinolysis; a recent method employs a carbon column for simultaneous reacetylation and hydrazine removal.[251]

Several enzymes are available for releasing *N*-glycans. The most popular is PNGase-F,[252] an amidase that releases the intact glycan as the glycosylamine. The released glycosylamine readily hydrolyzes to the glycan or can be encouraged to do so by incubation with dilute acetic acid. However, without this treatment, some glycosylamine can be retained, particularly if the glycan release has been performed in the presence of ammonium-containing buffers.[253] PNGase-F releases most *N*-glycans except those containing fucose α1-3 linked to the reducing-terminal GlcNAc.[254] In these situations, PNGase-A is effective. Endoglycosidase-H (endo-H), another popular enzyme, cleaves the chitobiose core of high-mannose and hybrid glycans, leaving the core GlcNAc with any attached fucose bound to the protein. Thus, information on the presence of core fucosylation is not available from the spectra of the resulting glycans. Reducing termini are left intact with all of these release methods, allowing the glycans to be derivatized with fluorescent or other reagents to aid detection if non-mass-spectrometric methods are to be used for subsequent analysis.

Because PNGase F releases the glycans as the glycosylamines, it is possible to prepare derivatives directly by reaction with carbonyl reagents. Chen and Novotny[255] have prepared such derivatives from 2-methyl-3-oxo-4-phenyl-2,3-dihydrofuran-2-yl acetate. To avoid hydrolysis of the glycosylamine, glycans were tagged immediately after release. MALDI analysis was from arabinosazone and it was noted that, when excess salt was present, washing the dried matrix:sample spot with cold, deionized water was beneficial.

Several investigators have developed methods for releasing glycans from glycoproteins within SDS-PAGE[102,256–259] or isoelectric focussing gels.[260] Usually, a low-density gel is used to allow PNGase to diffuse in readily. Glycans are usually recovered from the surrounding solution, although dissolution of the gel has also been used.[261] PNGase A is larger than PNGase-F (75.5 as compared to 35 KDa), does not readily penetrate gels, and only releases glycans from smaller peptides.[262] Consequently, the glycoprotein can be digested with, for example, trypsin prior to incubation.[263]

c. Profiling of Released Glycans. Released glycans can be profiled by normal-phase HPLC[264] after derivatization with a fluorescent dye such as 2-AB[86,115,265] or 2-AP,[266–268] and some structural information can be obtained because retention time is directly proportional to glycan size.[86] The resolutions that can be obtained with standard columns, however, are generally too low to deconvolute complex mixtures. In general, mass-spectrometry-based methods are superior. A recent comparison of three techniques, FAB, MALDI, and ESI, for glycan profiling of 2-AB-derivatized glycans[269] found that MALDI gave the simplest spectra because the profiles were not complicated by the presence of doubly charged ions (common with ESI) or fragments and did not require permethylation as was necessary for FAB analysis. Furthermore, MALDI appears to give quantitative profiles of the glycans.[30,63]

Mass measurements yield composition information in terms of the isobaric monosaccharide content as described above (Table 19.1). Some investigators use such masses to search carbohydrate databases for suitable structures, but this approach can lead to incorrect assignments unless further structural details are determined. Traditionally, structural details are obtained by sequential exoglycosidase digestions (see Dwek et al.[270] for suitable enzymes) with profiling by HPLC or MALDI-MS.[271,272] Several investigators have developed methods for glycan release[273] and sequencing directly on a MALDI target. The method developed by Küster et al.[274] involves sequential incubation of the glycans with the relevant enzyme, spectral recording, and removal of the matrix (DHB) prior to analysis of the next digest. Starting with 100 pmol of glycan, it was possible to conduct three successive enzyme digestions before the amount of sample became insufficient to give a MALDI signal. Other investigators[273,275] have developed similar methods using mixtures of exoglycosidase (exoglycosidase arrays) on different target spots in order to avoid removal of the matrix at each stage. The neutral matrix, ATT, rather than the acidic DHB, has been used so that enzyme digests could be performed in the presence of a matrix that did not affect the enzyme activities[276,277]. Even when mass spectral fragmentation, particularly in negative-ion mode, is the major analytical technique, the use of exoglycosidases is still important for determination of the nature (mannose, glucose, etc.) of constituent monosaccharides.

d. Fragmentation of Negative ions from N-Linked Glycans. Fragmentation of negative ions of the type $[M - H]^-$ or $[M + \text{anion}]^-$, in Q-TOF-type instruments, yields both glycosidic and cross-ring fragments, as in positive ion spectra but with major differences. The glycosidic ions tend to be mainly of the C rather than the B or Y type; and cross-ring fragments, particularly the A-type fragments often dominate[278–281] (Figures 19.4c and 19.8). However, more B- and Y-type cleavages appear to be produced in TOF/TOF instruments.[152] Fragmentation mechanisms tend to be much more specific than those seen in positive-ion mode because they are initiated by loss of individual protons from hydroxyl groups. Consequently, fragment ions tend to be very diagnostic of specific structural features[282,283] as shown in Figure 19.8 for the formation of the prominent $^{2,4}A_R$ ion from N-linked glycans. They can often reveal structural features, such as the presence of a bisecting GlcNAc residue (Figure 19.8c), that are difficult to determine by techniques such as exoglycosidase digestion.

One of the most prominent fragments is a $^{2,4}A_R$ cleavage of the reducing-terminal GlcNAc residue (Figures 19.4c and 19.8) which confirms the $\beta 1 \rightarrow 4$ linkage between the GlcNAc residues and indicates the presence or absence of 6-linked fucose on the core GlcNAc (fucose is eliminated in the neutral fragment, Figures 19.8a,c). Although this $^{2,4}A_R$ ion is not present in the spectra of glycans derivatized at the reducing terminus by reductive amination, because of the open nature of the GlcNAc ring, B_R and $^{2,4}A_{R-1}$ (cleavage of the penultimate GlcNAc residue) ions carry similar information with respect to core fucosylation. Another very diagnostic ion (named ion D) contains the 6-antenna and the branching mannose residue from the core and, consequently, allows the composition of both the 6- and 3- (by difference) antennae to be determined. The mass of this ion allows isomers of, for example, the high-mannose glycans to be identified[282] (*m/z* 647, Figure 19.4c). Ion D is accompanied by another ion, 18 mass units lower, formed by loss of H_2O. When a bisecting GlcNAc residue is present, the D ion, which will contain the bisecting GlcNAc residue, eliminates this GlcNAc as a neutral molecule (221 u) to give what is usually a very abundant ion (Figure 19.8c). Antenna composition is revealed by a cross-ring cleavage of the mannose residues to give an ion consisting of the antenna plus

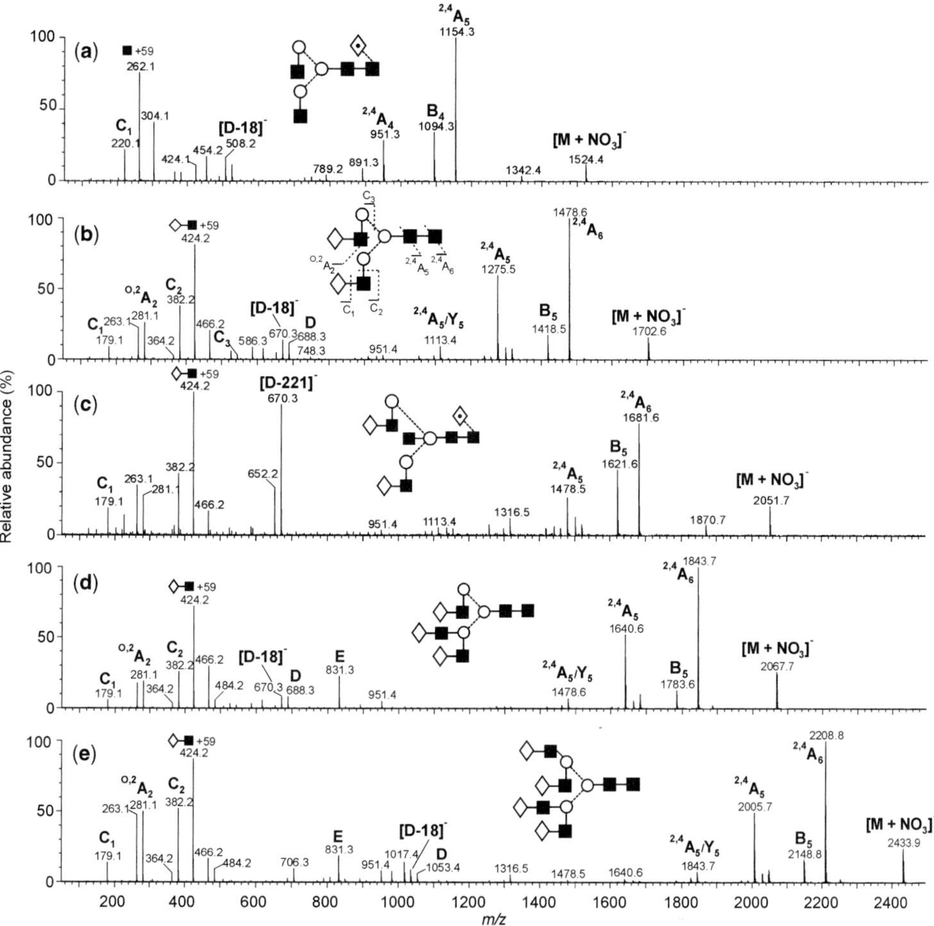

Figure 19.8. Negative-ion MS/MS spectra of *N*-glycans. (**a**) The position of the core fucose is defined by the masses of the $^{2,4}A_4$ and $^{2,4}A_5$ ions. The C_1 ion at *m/z* 220 indicates GlcNAc at the ends of the antennae and the D and $[D-18]^-$ ions at *m/z* 526 and 508, respectively, shows that the 6-antenna contains only mannose and GlcNAc. (**b**) The C_1 ion at *m/z* 179 in this and spectra c–e indicates galactose at the ends of the antennae; and the D and $[D-18]^-$ ions at *m/z* 688 and 670, respectively, shows that the 6-antenna contains Gal-GlcNAc. (**c**) The presence of the bisecting GlcNAc residue produces the abundant $[M-221]^-$ ion at *m/z* 670. (**d**) The branched 3-antenna gives rise to the abundant E ion at *m/z* 831. The D and $[D-18]^-$ ions at remain at *m/z* 688 and 670, respectively. (**e**) The branching pattern is revealed by the E ion at *m/z* 831 and the D and $[D\text{-}18]^-$ ions at *m/z* 1053 and 1035. A third diagnostic ion is present at *m/z* 1017 ($[D\text{-}36]^-$).

H and -O-CH=CH-O$^-$ (59 u)—for example, Gal-GlcNAc-O-CH=CH-O$^-$ (*m/z* 424)—and, consequently, the presence of substituents such as fucose or α-galactose can easily be spotted. An $^{O,4}A$ cleavage of the mannose residue from each antenna also gives a prominent fragment allowing the two possible triantennary complex glycans to be differentiated because this ion contains the substituent at the 4- but not the 6-position. Table 19.2 lists the most important of these diagnostic ions. An additional advantage of negative-ion fragmentation is that isomeric compounds yield ions with differing *m/z* values that are much more useful than the differences in abundance of the same mass ion that frequently characterizes isomers in positive-ion spectra.[133]

Table 19.2. Ions Defining Structural Features in the Negative-Ion Spectra of N-Linked Glycans

Structural feature	Ion	Ionic composition	*m/z*
Composition	Molecular	$[M + X]^-$	—
Antenna sequence	C	Gal, Man, Glc	179
		GlcNAc, GalNAc	220
		[Fuc]Gal	325
		Gal-[Fuc]GlcNAc	528
		αGal-Gal, Man-Man	341
		Man-[Man]Man	503
		GalNAc-GlcNAc	423
Antenna composition	F	Man	262
		GlcNAc	303
		Gal-GlcNAc	424
		Gal-[Fuc]GlcNAc	570
		αGal-Gal-GlcNAc	586
		GalNAc-GlcNAc	465
		$(\text{Gal-GlcNAc})_2$	789
		$(\text{Gal-GlcNAc})_2\text{Fuc}$	935
		$(\text{Gal-GlcNAc})_3$	1154
Fucose at 6-position of reducing terminus	$^{2,4}A_R$	$[M - Cl - 307]^-$	$[M - 342]^-$
		$[M - NO_3 - 307]^-$	$[M - 369]^-$
		$[M - H_2PO_4 - 307]^-$	$[M - 405]^-$
Absence of fucose at 6-position of reducing terminus	$^{2,4}A_R$	$[M - Cl - 161]^-$	$[M - 196]^-$
		$[M - NO_3 - 161]^-$	$[M - 223]^-$
		$[M - H_2PO_4 - 161]^-$	$[M - 259]^-$
Composition of 6-antenna	D and $[D - 18]^-$ ($[D - 36]^-$)	GlcNAc	526, 508
		Gal-GlcNAc	688, 670
		Gal-[Fuc]GlcNAc	834, 816
		$(\text{Gal-GlcNAc})_2$	1053, 1035 (1017)
		$(\text{GalGlcNAc})_2\text{Fuc}$	1199, 1181 (1163)
		Man_1	647, 629
		Man_2	809, 791
		Man_3	971, 953
	$^{0,3}A_{R-2}$ and $^{0,4}A_{R-2}$	GlcNAc	292
		GalGlcNAc	454
		Gal-[Fuc]GlcNAc	600
		$(\text{Gal-GlcNAc})_2$	819
		$(\text{GalGlcNAc})_2\text{Fuc}$	965
		Man_3	251
		Man_4	413
		Man_5	575
Composition of 3-antenna	$^{0,4}A_{R-3}$ (E) ion	Gal-GlcNAc	466
		GlcNAc_2	507
		Gal-GlcNAc_2	669
		$(\text{Gal-GlcNAc})_2$	831
		Gal-[Fuc]GlcNAc	977

Table 19.2. *(Continued)*

Structural feature	Ion	Ionic composition	*m/z*
Presence of bisect	Abundant $[D-221]^-$ ion	$GlcNAc_2$	508
		Gal-$GlcNAc_2$	670
		Gal-[Fuc]GlcNAc	816
		$(Gal\text{-}GlcNAc)_2$	1035
		$(GalGlcNAc)_2Fuc$	1181
		Man_3	629
Presence of sialic acid	B_1	Neu5Ac	290
		Neu5Gc	306
Presence of $\alpha2\rightarrow6$-linked sialic acid	$^{0,4}A_2$-CO_2	Neu5Ac	306
		Neu5Gc	322

Although the $[M-H]^-$ ions formed from neutral carbohydrates tend to be rather unstable and can show extensive fragmentation in the ion source of the mass spectrometer, stabilization can be achieved by anion adduction with anions such as Cl^-, Br^-, NO_3^-, or $H_2PO_4^-$. Fortunately, these ions produce similar fragmentation spectra to the $[M-H]^-$ ions after the initial loss of the corresponding acids (HCl, etc.) by hydrogen abstraction.[81] Acidic carbohydrates, such as those containing sialic acid or carboxyl-containing derivatives such as 2-AA,[125] ionize by loss of one or more protons from acidic groups to give $[M-nH]^{n-}$ ions with localized charges and, consequently, restricted fragmentation.

19.6.2.3 O-Linked Glycans

O-Linked glycans are generally smaller and more varied in structure than their N-linked counterparts.[284] There is no single 'core' structure or a structural motif for locating the glycans on the protein chain other than that the glycans are attached mainly to serine or threonine. A database of *O*-glycan structures is available.[285] It is common to find several O-linked glycosylation sites in close proximity such that proteolysis rarely produces fragments with single sites as with the *N*-glycans. Consequently, peptides containing several glycans are often examined either as intact molecules or after stripping the glycans to their attached GalNAc residue. These depleted molecules provide data on the glycan attachment sites but little structural information of the individual glycan compositions. Several other methods have been developed; for example, Mirgorodskaya et al.[286] have used partial vapor-phase hydrolysis to generate N- and C-terminal peptide ladders from which the positions of the glycans could be determined from their MALDI spectra.

a. Glycan Release. Because there is no universal enzyme for release of O-linked glycans, these compounds are usually released chemically; reductive β-elimination with sodium hydroxide and borohydride is the most common method, although hydrazinolysis, as mentioned above, can also be used. Reduction of the sugar is necessary with β-elimination to avoid degradation in the basic medium, but, unfortunately, reduction removes the reducing terminus of the glycan, thus preventing the attachment of chromophores or fluorescent tags by reductive amination. Ammonia in the presence of ammonium carbonate has recently been used as an alternative release method.[287,288] The milder conditions of this reaction do not produce the "peeling" reaction that necessitates the reductive stage when sodium hydroxide is used. Consequently, this reductive stage could be dispensed with. Clean-up of the product was minimal as all reagents were volatile. The protein was left intact, albeit with a dehydro-serine or threonine marking the original *O*-glycosylation site by mass difference. In a modification to this procedure, glycans were released with ammonium hydroxide and ammonium carbonate as above and reacted with boric acid to give the reducing sugars. These were cleaned with a C18 Sep-Pak cartridge and a graphitized carbon column. The intact reducing terminus allowed the glycans to be derivatized with AMAC or 8-aminonaphthalene-1,3,6-trisulfonic acid (ANTS) by reductive amination. The labeled glycans were examined by fluorophore-assisted carbohydrate electrophoresis (FACE) and eluted from the resulting gel bands for MALDI MS.[289] Methylamine or ethylamine at 50°C has also been used to release *O*-glycans,[290] leaving the alkylamine attached to the protein as a label for the position of the original glycans. An in-gel β-elimination release method using the conventional reagents has been described; recovery of the released glycans was achieved after peracetylation or permethylation.[291] Reduction in the classic β-elimination procedure can, however, have its advantages; by use of deuterated reagents, a deuterium atom can be introduced into the sugar allowing comparative quantitative experiments to be performed. Thus, for example, Xie et al.[292] have used this reaction to compare glycosylation in two eggs from *Xenopus laevis*.

b. Location of O-Linked Glycosylation Sites. A method for locating *O*-glycosylation sites described by Müller et al.[293] involves partial deglycosylation with trifluoromethanesulfonic acid to the level of core-GlcNAc residues. The glycoprotein was then cleaved with the Arg-C-specific endopeptidase, clostripain, to yield tandem repeat icosapeptides that were analyzed by MALDI/PSD from CHCA matrix. Similar experiments on synthetic glycopeptides have also been reported.[294,295] Attempts to locate *O*-glycans on peptide chains by fragmentation sometimes fail[296] because of preferential loss of the glycans catalyzed by proton migration from the normal $[M + nH]^{n+}$ ions that are usually generated. Elimination of the proton has been proposed as a way of overcoming this reaction and has been achieved by electron-capture dissociation in an ICR instrument.[297] It has been argued that ionization by charge localization could achieve the same result by eliminating the ionizing protons. Consequently, Czeszak et al.[298] derivatized glycopeptides at their amino terminus with a phosphonium group and showed that the molecules, when studied by MALDI/PSD, undergo predictable A-type fragmentation of the peptide chain without loss of the attached glycans. In contrast, CID on the doubly charged protonated phosphonium cation resulted in the predominant loss of the sugar moiety. Experiments were conducted with only GalNAc attached to the peptide, but the method may be applicable to peptides carrying larger glycans. Alternatively, the *O*-glycans could be degraded to GalNAc as in the method described by Müller et al.[293]

c. Fragmentation of O-Linked Glycans. Positive-ion MS/MS spectra have been used extensively for structural determination of released *O*-glycans, particularly those from egg jelly-coat mucins.[299–304] Spectral interpretation has been aided by a procedure termed the "catalog-library approach" whereby fragment ions from small oligosaccharides are matched in the spectra of larger glycans.[305] These approaches provide information on sequence and linkage of the oligosaccharides, but exoglycosidase digestion needs to be applied in order to obtain information on the nature of the constituent monosaccharides.[306,307] Negative-ion fragmentation spectra of O-linked glycans appear to offer similar specificity to that seen in the spectra of the N-linked glycans. Karlsson et al.[308] have investigated negative-ion spectra of O-linked alditols from salivary mucin MUC5B and found major Z- and Y-type fragments. C- and A-type cleavages provided information on the structure of the reducing termini. However, cross-ring fragments were not as abundant as in the spectra of the N-linked glycans. Positive-ion IRMPD spectra of these alditols, as with the *N*-glycans, are very similar to low-energy CID spectra.[309]

19.6.3 Glycans and Glycolipids from Bacteria

MALDI and electrospray ionization, usually in negative-ion mode because of the common occurrence of phosphate groups, have been used extensively for examination of glycolipids from bacteria.[310] Many of these compounds contain unusual monosaccharides, not found in mammalian glycans, that can present problems and which require additional techniques such as GC/MS in order to identify them. The intact molecules are typically extracted using hot phenol and water, although hot ammonium isobutyrate has recently been found to be effective for extraction of lipid A.[311] A few of the intact molecules are small enough to be examined directly, but most are too large, have a high hydrocarbon content and problematic phosphate groups, and must be degraded into their constituent parts (lipid A, O-chain, etc.). Mild acid hydrolysis is used to separate the lipid A portion, and hydrolysis with TFA can be used to isolate the carbohydrate repeat unit, when present.

MALDI spectra of intact molecules have been reported for several bacteria. Sample preparation is critical for obtaining good spectra. A recent method[312] involved initial dissolution with brief sonication of a small amount of the intact R-type LPS in a mixture of methanol/water (1:1) containing 5 mM ethylenediaminetetraacetic acid (EDTA). A few microliters of this solution was desalted on a small piece of Parafilm with some grains of Dowex 50WX8-200 cation-exchange beads in their ammonium form. 0.3 μL of this solution and the same volume of 20 mM dibasic ammonium citrate was deposited on a thin layer of homogeneous matrix composed of THAP and nitrocellulose. The spectra from bacteria such as *Shewanella pacifica* and *Xanthomonas campestris* contained, in addition to molecular ions in the region of 3 kDa, ions defining the lipid A portion and fragments from the glycan. Other intact molecules whose spectra have been recorded include those from *Erwinia carotovora*[313] and *Bdellovibrio bacteriovorus*,[314] the latter compound having α-D-mannose in place of the usual phosphate, making it more amenable to MALDI analysis. The spectra of these small lipopolysaccharide (LPS) molecules display a range of ions representing heterogeniety in the lipid portion, whereas the spectrum from a larger LPS from *Escherichia coli* O164 (28 kDa) showed only a broad peak.[315]

Deacylation is frequently used to aid the production of MALDI spectra. For example, LPS from *Haemophilus influenzae*, *H. ducreyi*, and *Salmonella typhimurium* has been ionized from 2,5-DHB containing 1-HIQ following removal of the O-linked fatty acids with hydrazine,[316] and LPS from *Chlamydia trachomatis* has produced a spectrum from THAP following deacylation.[317] Both deacylation and dephosphorylation were employed by Olsthoorn et al.[318] to obtain a PSD spectrum from the LPS from a rough strain (lacking the O-specific chain) of *Klebsiella pneumoniae*. Many other examples can be found in the literature.

19.6.3.1 Lipid A

Much of the early mass spectrometry on the analysis of lipid A was performed with laser ionization without a matrix.[319–322] Problems with the phosphate groups, such as phosphate loss,[323] can be overcome by methylation with, for example, diazomethane[324]; deacylation of the esterified hydroxy fatty acids also appears to improve signal quality in many cases (for example see Refs. 325–327). Although DHB and, to a lesser extent, THAP and ATT appear to be the most commonly used MALDI matrices for lipid A, *nor*-harmane has recently been shown to give good results.[328] ESI and LC/MS are being increasingly used in this field, recent studies include the use of an ESI quadrupole time-of-flight tandem mass spectrometer (Q-TOF) for the structural determination of Lipid A from *Pseudomonas corrugate*[329] and EI-ion-trap MS^n for studies of lipid A from *Pseudomonas aeruginosa* strains[330] and *Escherichia coli*.[331] The latter methods were able to reveal details about the acyl substitution.

19.6.3.2 O-Chains

The O-specific chains of these compounds are frequently too long to be examined directly by mass spectrometry, but, because they are usually composed of multiple repeats of small oligosaccharides, structures are usually determined by sequencing these oligosaccharides following mild acid hydrolysis. The literature is extensive; and because of the complex nature of these molecules and the occurrence of unusual monosaccharides, many techniques such as NMR and GC/MS are usually used in conjunction with ESI and MALDI MS to determine the detailed structures. Recent examples are O-chains from *Plesiomonas shigelloides*,[332] *Bordetella avium* ATCC 5086,[333] *Agrobacterium radiobacter*,[334] and *Bordetella trematum*.[335]

19.6.4 Glycosphingolipids

19.6.4.1 Analysis of Intact Glycosphingolipids

Early work on the analysis of glycosphingolipids by FAB has been reviewed.[336] Today, much of the work is performed by ESI, particularly in negative-ion mode. LC/MS has been used by several investigators and reviewed by Merrill et al.[337] As many glycosphingolipids are sialylated, they suffer the same loss of sialic acid as seen in the spectra of the glycoproteins, particularly when ionized by MALDI[338]; stabilization can be achieved by methylation.[39,338] It has also been found that MALDI at elevated pressures (1–10 mbar) is a softer technique than true vacuum MALDI, enabling the sialylated molecules to be observed intact[339,340] (Figure 19.9). Automated chip technology interfaced to a Q-TOF mass spectrometer has recently been applied to ganglioside analysis and has been claimed to be superior to capillary-based ESI in terms of both resolution and sensitivity.[100,341]

Recently, there has been some interest in the use of FTICR instruments for ganglioside analysis. McFarland et al.[342] have compared IRMPD, electron capture dissociation (ECD), and electron detachment dissociation (EDD) to obtain fragmentation of electrosprayed ions. All three ionization techniques provided extensive fragmentation of both protonated and deprotonated molecules exemplified for the GM1 ganglioside. ECD gave extensive structural information, including identification of both halves of the ceramide. IRMPD provided similar glycan fragmentation but no cleavage of the *N*-acetyl moiety as observed in the ECD spectra. Cleavage between the fatty acid and the long-chain base of the ceramide

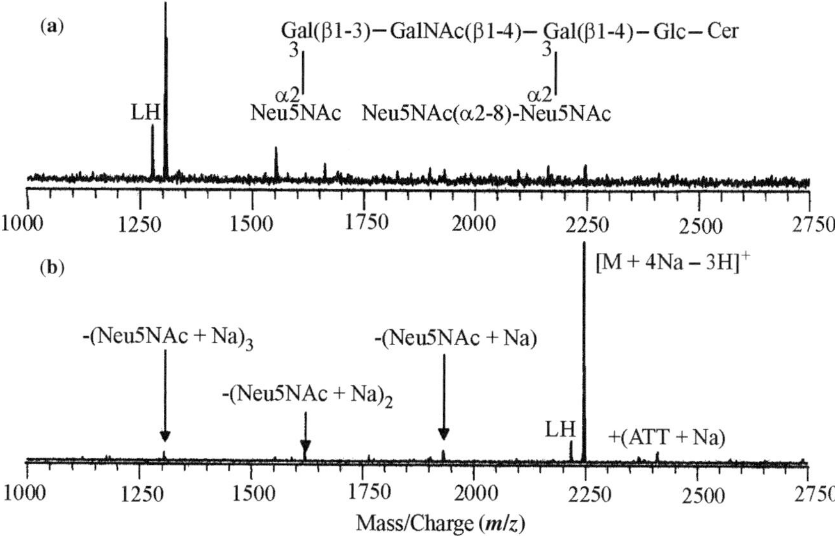

Figure 19.9. MALDI spectrum of the ganglioside GT1B desorbed from ATT (**a**) without collision gas and (**b**) with collision gas. (From Ref. 339 with permission from John Wiley and Sons Ltd.)

moiety was seen by IRMPD in the negative, but not in the positive, ion spectra. Sustained off-resonance irradiation (SORI) CID was used by Vukelic et al.[343] in 2005 for the first structural elucidation of polysialylated gangliosides, and informative fragmentation patterns of the sialic acid groups were obtained.

Ganglioside spectra have been obtained directly from the TLC plates often used to separate these compounds. Use of DHB, or a mixture of DHB and 2-amino-5-nitropyridine, as the MALDI matrix has provided an increase in sensitivity of one to two orders of magnitude over that obtained by liquid secondary ion mass spectrometry (LSIMS). However, the method was still further improved by heat transfer of the glycolipids to a PVDF membrane attached to a thin film of adhesive to the MALDI target.[344] In a related method, the plate was sprayed with sinapinic acid and introduced into the ion source of the mass spectrometer that was operated at elevated pressure to minimize loss of sialic acid.[345] Desorption has also been achieved with an IR laser using glycerol as the matrix. In this case, better results were obtained with a Q-TOF[346] than with an FTICR instrument,[347] probably because of the vapor pressure of the glycerol. Nakamura et al.[348] have ionized glycolipids from TLC plates to which the matrix DHB in acetonitrile was added with a microsyringe. Spectra were acquired with an ion-trap–TOF mass spectrometer, allowing MS^3 spectra to be acquired for detailed structural studies.

19.6.4.2 Glycan Release and Analysis of Delipidated Glycans

The analysis of glycosphingolipids is complicated by the presence of homologous acyl groups and/or sphingosine chains that result in multiple peaks in the spectra. For the analysis of the carbohydrate moiety, the ceramides can be removed enzymatically with an endoglycoceramidase to reduce the heterogeniety. Endoglycoceramidase produces glycans with an intact reducing terminus, allowing them to be derivatized by reductive amination. 2-AP appears to have been the most used of the common derivatization reagents although Wing et al.[349] have used 2-AB labeling of the glycan following removal of the ceramide with ceramide glycanase. By employing this method, several investigators have probed the structure of the glycan portion of glycosphingolipids with measurements by both MS and HPLC. As an example of the technique, highly fucosylated glycans in the eggs of the human parasite *Schistosoma mansoni*[350] have been identified with the major antigenic motif being α-Fuc-(1 → 3)-GalNAc →.[351] However, it has been reported that fucose residues could be lost from compounds of this type when ionization was performed by MALDI.[352]

19.6.5 Synthetic Carbohydrates

Mass spectrometry (FAB, MALDI, and ESI modes of ionization) are used extensively to monitor and confirm products of chemical synthesis. Unfortunately, important details such as the type of MALDI or FAB matrix or the type of instrument used to record the spectra are often missing from these publications as the techniques tend to be regarded as routine. There are a few areas—for example, analysis of glycodendrimers and carbohydrate–protein complexes—where the high-mass capability of MALDI is particularly useful.

19.6.5.1 Glycodendrimers and Glycoclusters

Synthetic glycodendrimers contain arrays of carbohydrate expressed on the surface of large molecules and can have molecular weights in excess of 100 kDa. Thus, for example, the spectrum of a fourth-generation 6-suberoyl maltose poly-lysine dendrimer (Figure 19.10) contained broad peaks attributed to dendrimer oligomers connected by dicarboxyester bridges at *m/z* 11,821, 25,770, 39,188, 52,739, 75,625, and 79,043.[353] A mannose-functionalized sixth generation dendrimer with a mass of around 100 kDa (Figure 19.11) has given a broad MALDI–TOF spectrum, although the observed mass was somewhat lower than the expected mass (106,204 rather than the calculated 109,268 Da). The discrepancy, which also occurred to a lesser extent with the fourth- and fifth-generation dendrimers, was attributed to the absence of 44 terminal amino groups. It was proposed that MALDI MS could account for deletions in the dendrimers more accurately than NMR because of the repetitive nature of the dendrimer structure.[354,355] However, it has been pointed out that MALDI, but not electrospray ionization, sometimes causes some decomposition of these compounds and may, therefore, suggest defects in the molecules where none exist.[356]

19.6.5.2 Carbohydrate–Protein Complexes

The ability of MALDI to measure increases in molecular weight of proteins, when they are conjugated to other molecules, has been utilized in a number of laboratories for studies on protein–protein interactions and of drug binding to proteins. For example, MALDI–TOF MS has been used to monitor lactose (molecular weight 342) binding to human serum albumin (mass 66,480). Mass shifts of 7030 and 7880 in two experiments showed a loading value of 21–24 molecules of lactose.[357] Similarly, Tanaka et al.[358] have shown that up to 15 steroidal alkaloid glycosides become attached to bovine serum albumin in an immunostaining technique for solamargine.

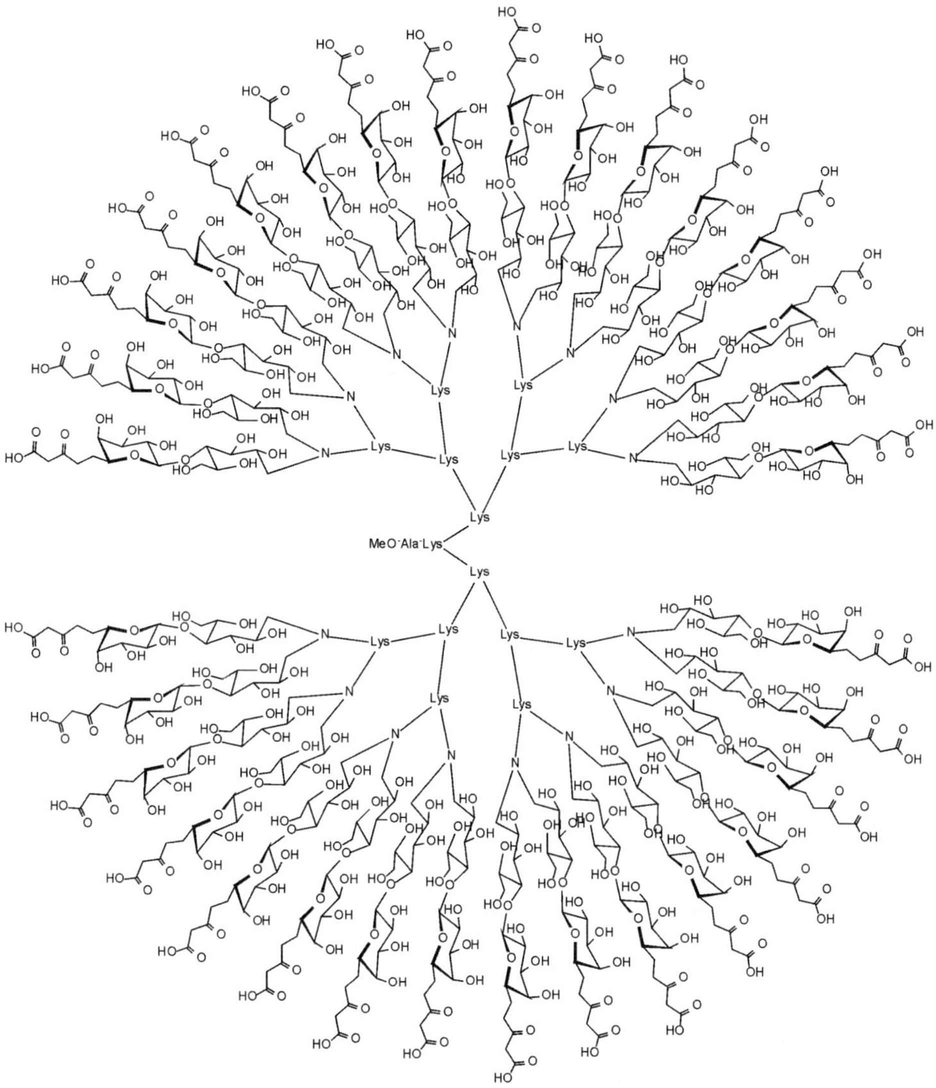

Figure 19.10. Structure of a fourth-generation 6-suberoyl-maltose poly-lysine dendrimer.

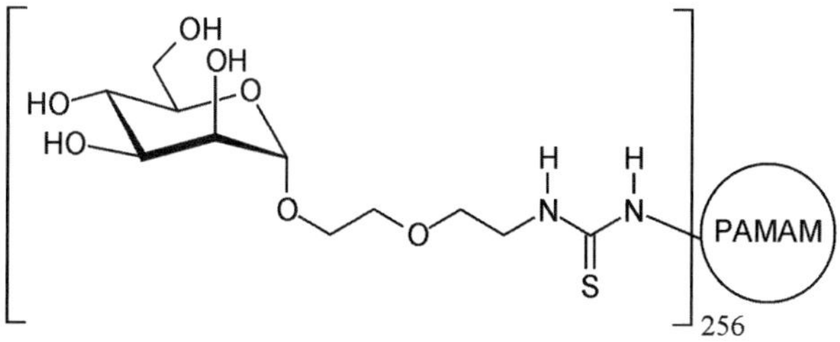

Figure 19.11. Structure of a mannose-functionalized glycodendrimer (PAMAM stands for polyamidoamine).

19.7 CONCLUSIONS

The modern ionization techniques of MALDI and ESI have enabled most types of carbohydrate to be amenable to analysis by mass spectrometry and progress in the field shows no sign of slowing. New instrumentation such as the Orbitrap™ mass spectrometer and instruments that are capable of ion-mobility separations provide the opportunity for new method development and could possibly provide answers to questions that are not currently answerable by existing techniques.

ABBREVIATIONS

2-AA, 2-aminobenzoic acid; 2-AB, 2-aminobenzamide; AMAC, 2-aminoacridone; ANTS, 8-aminonaphthalene-1,3,6-trisulfonic acid; 2-AP, 2-aminopyridine; AEC, anion-exchange chromatography; ATT, 6-azo-2-thiothymine; CE, capillary electrophoresis; Cer, ceramide; CHCA, alpha-cyano-4-hydroxycinnamic acid; CID, collisional-induced decomposition; DHB, dihydroxybenzoic acid; ECD, electron capture dissociation; EDD, electron detachment dissociation; EDTA, ethylenediaminetetraacetic acid; EI, electron impact; endo-H, endoglycosidase H; ESI, electrospray ionization; FAB, fast-atom bombardment; FACE, fluorophore-assisted carbohydrate electrophoresis; FT, Fourier transform; Fuc, fucose; Gal, galactose; GalNAc, *N*-acetylhexosamine; GC/MS, gas-chromatography/mass spectrometry; Glc, glucose; GlcNAc, *N*-acetylglucosamine; GPI, glycosylphosphatidylinositol; Hex, hexose; HexNAc, *N*-acetylaminohexose; HIQ, 1-amino-*iso*-quinoline; HPA, hydroxypicolinic acid; HPLC, high-performance liquid chromatography; ICR, ion cyclotron resonance; IR, infrared; IRMPD, infrared multiphoton dissociation; ISD, in-source decay; KEGG, *Kyoto Encyclopedia of Genes and Genomics*; Kdo, 2-keto-3-deoxyoctulosonic acid; LC/MS, liquid chromatography–mass spectrometry; LPS, lipopolysaccharide; LSIMS, liquid secondary ion mass spectrometry; MALDI, matrix-assisted laser desorption/ionization; MEKC, micellar electrokinetic chromatography; MS, mass spectrometry; Neu5Ac, *N*-acetylneuraminic acid (sialic acid); NMR, nuclear magnetic resonance; NP, normal phase; PAGE, polyacrylamide gel electrophoresis; PAMAM, polyamidoamine; PNGase, protein *N*-glycosidase; PSD, post-source decay; PVDF, polyvinylidine difluoride; Q, quadrupole; SDS, sodium dodecylsulfate; SEC, size-exclusion chromatography; SORI, sustained off-resonance irradiation; STEP, statistical test of equivalent pathways; TFA, trifluoroacetic acid; THAP, 2,4,6-trihydroxyacetophenone; TLC, thin-layer chromatography; TOF, time-of-flight; UV, ultraviolet.

REFERENCES

1. Laine, R. A. A calculation of all possible oligosaccharide isomers both branched and linear yields 105×10^{12} structures for a reducing hexasaccharide: the Isomer Barrier to development of single-method saccharide sequencing or synthetic systems. *Glycobiology* **1994**, *4*, 759–767.
2. Kennedy, J. F. *Carbohydrate Chemistry*, Clarendon Press; Oxford, **1988**.
3. Messner, P.; Allmaier, G.; Schäffer, C.; Wugeditsch, T.; Lortal, S.; König, H.; Neimetz, R.; Dorner, M. Biochemistry of S-layers. *FEMS Microbiol. Rev.* **1997**, *20*, 25–46.
4. Harvey, D. J. Matrix-assisted laser desorption/ionization mass spectrometry of carbohydrates and glycoconjugates. *Int. J. Mass Spectrom.* **2003**, *226*, 1–35.
5. Harvey, D. J. Proteomic analysis of glycosylation: structural determination of *N*- and *O*-linked glycans by mass spectrometry. *Expert Rev. Proteomics* **2005**, *2*, 87–101.
6. Harvey, D. J. Structural determination of *N*-linked glycans by matrix-assisted laser desorption/ionization and electrospray ionization mass spectrometry. *Proteomics* **2005**, *5*, 1774–1786.

7. Zaia, J. Mass spectrometry of oligosaccharides. *Mass Spectrom. Rev.* **2004**, *23*, 161–227.
8. Jang-Lee, J.; North, S. J.; Sutton-Smith, M.; Goldberg, D.; Panico, M.; Morris, H.; Haslam, S.; Dell, A. Glycomic profiling of cells and tissues by mass spectrometry: Fingerprinting and sequencing methodologies. *Methods Enzymol.* **2006**, *415*, 59–86.
9. Haslam, S. M.; North, S. J.; Dell, A. Mass spectrometric analysis of *N*- and *O*-glycosylation of tissues and cells. *Curr. Opin. Struct. Biol.* **2006**, *16*, 584–591.
10. Campa, C.; Coslovi, A.; Flamigni, A.; Rossi, M. Overview on advances in capillary electrophoresis–mass spectrometry of carbohydrates: A tabulated review. *Electrophoresis* **2006**, *27*, 2027–2050.
11. Lindberg, B. Methylation analysis of polysaccharides. *Methods Enzymol.* **1972**, *28*, 178–195.
12. Lindberg, B.; Lönngren, J. Methylation analysis of complex carbohydrates: General procedure and application for sequence analysis. *Methods Enzymol.* **1978**, *50*, 3–33.
13. Hellerqvist, C. G. Linkage analysis using Lindberg method. *Methods Enzymol.* **1990**, *193*, 554–573.
14. Levery, S. B. In Large, D.G.; Warren, C. D. (Eds.), *Glycopeptides and Related Compounds: Synthesis, Analysis and Applications*, Marcel Dekker, New York, **1997**, pp. 541–592.
15. Ciucanu, I. Per-*O*-methylation reaction for structural analysis of carbohydrates by mass spectrometry. *Anal. Chim. Acta* **2006**, *576*, 147–155.
16. Barber, M.; Bordoli, R. S.; Sedgwick, R. D.; Tyler, A. N. Fast atom bombardment of solids (FAB): A new ion source for mass spectrometry. *J. Chem. Soc. Chem. Commun.* **1981**, 325–327.
17. Dell, A. FAB Mass spectrometry of carbohydrates. *Adv.Carbohydrate Chem. Biochem.* **1987**, *45*, 19–72.
18. Dell, A.; Carman, N. H.; Tiller, P. R.; Thomas-Oates, J. E. Fast atom bombardment mass spectrometric strategies for characterising carbohydrate-containing biopolymers. *Biomed. Environ. Mass Spectrom.* **1987**, *16*, 19–24.
19. Dell, A.; Thomas-Oates, J. E. In Biermann, C. J., McGinnis, G. D. (Eds.), *Analysis of Carbohydrates by GLC and MS*; CRC Press, Boca Raton, FL, **1989**, pp. 217–235.
20. Dell, A.; Morris, H. R. Glycoprotein structure determination by mass spectrometry. *Science* **2001**, *291*, 2351–2356.
21. Karas, M.; Bachmann, D.; Bahr, U.; Hillenkamp, F. Matrix-assisted ultraviolet laser desorption of non-volatile compounds. *Int. J. Mass Spectrom. Ion Processes* **1987**, *78*, 53–68.
22. Mock, K. K.; Davy, M.; Cottrell, J. S. The analysis of underivatised oligosaccharides by matrix-assisted laser desorption mass spectrometry. *Biochem. Biophys. Res. Commun.* **1991**, *177*, 644–651.
23. Harvey, D. J. Matrix-assisted laser desorption/ionization mass spectrometry of carbohydrates. *Mass Spectrom. Rev.* **1999**, *18*, 349–451.
24. Harvey, D. J. Analysis of carbohydrates and glycoconjugates by matrix-assisted laser desorption/ionization mass spectrometry: An update covering the period 1999–2000. *Mass Spectrom. Rev.* **2006**, *25*, 595–662.
25. Strupat, K.; Karas, M.; Hillenkamp, F. 2,5-Dihydroxybenzoic acid: A new matrix for laser desorption-ionization mass spectrometry. *Int. J. Mass Spectrom. Ion Processes* **1991**, *111*, 89–102.
26. Karas, M.; Ehring, H.; Nordhoff, E.; Stahl, B.; Strupat, K.; Hillenkamp, F.; Grehl, M.; Krebs, B. Matrix-assisted laser desorption/ionization mass spectrometry with additives to 2,5-dihydroxybenzoic acid. *Org. Mass Spectrom.* **1993**, *28*, 1476–1481.
27. Mohr, M. D.; Börnsen, K. O.; Widmer, H. M. Matrix-assisted laser desorption/ionization mass spectrometry: Improved matrix for oligosaccharides. *Rapid Commun. Mass Spectrom.* **1995**, *9*, 809–814.
28. Gusev, A. I.; Wilkinson, W. R.; Proctor, A.; Hercules, D. M. Improvement of signal reproducibility and matrix/comatrix effects in MALDI analysis. *Anal. Chem.* **1995**, *67*, 1034–1041.
29. Mechref, Y.; Novotny, M. V. Matrix-assisted laser desorption/ionization mass spectrometry of acidic glycoconjugates facilitated by the use of spermine as a comatrix. *J. Am. Soc. Mass Spectrom.* **1998**, *9*, 1292–1302.
30. Harvey, D. J. Quantitative aspects of the matrix-assisted laser desorption mass spectrometry of complex oligosaccharides. *Rapid Commun. Mass Spectrom.* **1993**, *7*, 614–619.
31. Kussmann, M.; Nordhoff, E.; Rehbek-Nielsen, H.; Haebel, S.; Rossel-Larsen, M.; Jakobsen, L.; Gobom, J.; Mirgorodskaya, E.; Kroll-Kristensen, A.; Palm, L.; Roepstorff, P. Matrix-assisted laser desorption/ionization mass spectrometry sample preparation techniques designed for various peptide and protein analytes. *J Mass Spectrom.* **1997**, *32*, 593–601.
32. Papac, D. I.; Wong, A.; Jones, A. J. S. Analysis of acidic oligosaccharides and glycopeptides by matrix assisted laser desorption/ionization time-of-flight mass spectrometry. *Anal. Chem.* **1996**, *68*, 3215–3223.
33. Ren, S.-F.; Zhang, L.; Cheng, Z.-H.; Guo, Y. L. Immobilized carbon nanotubes as matrix for MALDI-TOF-MS analysis: Applications to neutral small carbohydrates. *J. Am. Soc. Mass Spectrom.* **2005**, *16*, 333–339.
34. Black, C.; Poile, C.; Langley, J.; Herniman, J. The use of pencil lead as a matrix and calibrant for matrix-assisted laser desorption/ionisation. *Rapid Commun. Mass Spectrom.* **2006**, *20*, 1053–1060.
35. Stahl, B.; Steup, M.; Karas, M.; Hillenkamp, F. Analysis of neutral oligosaccharides by matrix-assisted laser desorption/ionization mass spectrometry. *Anal. Chem.* **1991**, *63*, 1463–1466.

36. Nonami, H.; Tanaka, K.; Fukuyama, Y.; Erra-Balsells, R. β-Carboline alkaloids as matrices for UV-matrix-assisted laser desorption/ionization time-of-flight mass spectrometry in positive and negative ion modes. Analysis of proteins of high molecular mass, and of cyclic and acyclic oligosaccharides. *Rapid Commun. Mass Spectrom.* **1998**, *12*, 285–296.
37. Yamagaki, T.; Suzuki, H.; Tachibana, K. In-source and postsource decay in negative-ion matrix-assisted laser desorption/ionization time-of-flight mass spectrometry of neutral oligosaccharides. *Anal. Chem.* **2005**, *77*, 1701–1707.
38. Tsarbopoulos, A.; Bahr, U.; Pramanik, B. N.; Karas, M. Glycoprotein analysis by delayed extraction and post-source decay MALDI-TOF-MS. *Int. J. Mass Spectrom. Ion Processes* **1997**, *169/170*, 251–261.
39. Powell, A. K.; Harvey, D. J. Stabilisation of sialic acids in *N*-linked oligosaccharides and gangliosides for analysis by positive ion matrix-assisted laser desorption–ionization mass spectrometry. *Rapid Commun. Mass Spectrom.* **1996**, *10*, 1027–1032.
40. Mechref, Y.; Kang, P.; Novotny, M. V. Differentiating structural isomers of sialylated glycans by matrix-assisted laser desorption/ionization time-of-flight/time-of-flight tandem mass spectrometry. *Rapid Commun. Mass Spectrom.* **2006**, *20*, 1381–1389.
41. Sekiya, S.; Wada, Y.; Tanaka, K. Derivatization for stabilizing sialic acids in MALDI-MS. *Anal. Chem.* **2005**, *77*, 4962–4968.
42. Ueki, M.; Yamaguchi, M. Analysis of acidic carbohydrates as their quaternary ammonium or phosphonium salts by matrix-assisted laser desorption/ionization mass spectrometry. *Carbohydrate Res.* **2005**, *340*, 1722–1731.
43. Moyer, S. C.; Marzilli, L. A.; Woods, A. S.; Laiko, V. V.; Doroshenko, V. M.; Cotter, R. J. Atmospheric pressure matrix-assisted laser desorption/ionization (AP MALDI) on a quadrupole ion trap mass spectrometer. *Int. J. Mass Spectrom.* **2003**, *226*, 133–150.
44. Zhang, J.; LaMotte, L.; Dodds, E. D.; Lebrilla, C. B. Atmospheric pressure MALDI Fourier transform mass spectrometry of labile oligosaccharides. *Anal. Chem.* **2005**, *77*, 4429–4438.
45. Schulz, E.; Karas, M.; Rosu, F.; Gabelica, V. Influence of the matrix on analyte fragmentation in atmospheric pressure MALDI. *J. Am. Soc. Mass Spectrom.* **2006**, *17*, 1005–1013.
46. Von Seggern, C. E.; Moyer, S. C.; Cotter, R. J. Liquid infrared atmospheric pressure matrix-assisted laser desorption/ionization ion trap mass spectrometry of sialylated carbohydrates. *Anal. Chem.* **2003**, *75*, 3212–3218.
47. Von Seggern, C. E.; Zarek, P. E.; Cotter, R. J. Fragmentation of sialylated carbohydrates using infrared atmospheric pressure MALDI ion trap mass spectrometry from cation-doped liquid matrixes. *Anal. Chem.* **2003**, *75*, 6523–6530.
48. Tan, P. V.; Taranenko, N. I.; Laiko, V. V.; Yakshin, M. A.; Prasad, C. R.; Doroshenko, V. M. Mass spectrometry of *N*-linked oligosaccharides using atmospheric pressure infrared laser ionization from solution. *J. Mass Spectrom.* **2004**, *39*, 913–921.
49. Von Seggern, C. E.; Gardner, B. D.; Cotter, R. J. Infrared atmospheric pressure MALDI ion trap mass spectrometry of frozen samples using a Peltier-cooled sample stage. *Anal. Chem.* **2004**, *76*, 5887–5893.
50. Dell, A.; Morris, H. R.; Greer, F.; Redfern, J. M.; Rogers, M. E.; Wisshaar, G.; Hiyama, J.; Renwick, A. G. C. Fast-atom-bombardment mass spectrometry of sulphated oligosaccharides from ovine lutropin, *Carbohydrate Res.* **1991**, *209*, 33–50.
51. Wheeler, S. F.; Harvey, D. J. Extension of the in-gel release method for structural analysis of neutral and sialylated *N*-linked glycans to the analysis of sulphated glycans. *Anal. Biochem.* **2001**, *296*, 92–100.
52. Harvey, D. J.; Bousfield, G. R. Differentiation between sulphated and phosphated carbohydrates in low-resolution matrix-assisted laser desorption/ionization mass spectra. *Rapid Commun. Mass Spectrom.* **2005**, *19*, 287–288.
53. Irungu, J.; Dalpathado, D. S.; Go, E. P.; Jiang, H.; Ha, H.-V.; Bousfield, G. R.; Desaire, H. Method for characterizing sulfated glycoproteins in a glycosylation site-specific fashion, using ion pairing and tandem mass spectrometry. *Anal. Chem.* **2006**, *78*, 1181–1190.
54. Zhang, Y.; Jiang, H.; Go, E. P.; Desaire, H. Distinguishing phosphorylation and sulfation in carbohydrates and glycoproteins using ion-pairing and mass spectrometry. *J. Am. Soc. Mass Spectrom.* **2006**, *17*, 1282–1288.
55. Takashiba, M.; Chiba, Y.; Jigami, Y. Identification of phosphorylation sites in *N*-linked glycans by matrix-assisted laser desorption/ionization time-of-flight mass spectrometry. *Anal. Chem.* **2006**, *78*, 5208–5213.
56. Juhasz, P.; Biemann, K. Mass spectrometric molecular-weight determination of highly acidic compounds of biological significance via their complexes with basic polypeptides. *Proc. Natl. Acad. Sci. USA* **1994**, *91*, 4333–4337.
57. Juhasz, P.; Biemann, K. Utility of non-covalent complexes in the matrix-assisted laser desorption ionization mass spectrometry of heparin-derived oligosaccharides. *Carbohydrate Res.* **1995**, *270*, 131–147.
58. Ueoka, C.; Nadanaka, S.; Seno, N.; Khoo, K.-H.; Sugahara, K. Structural determination of novel tetra- and hexasaccharide sequences isolated from chondroitin sulfate H (oversulfated dermatan sulphate) of hagfish notochord. *Glycoconjugate J.* **1999**, *16*, 291–305.
59. Shriver, Z.; Raman, R.; Venkataraman, G.; Drummond, K.; Turnbull, J.; Toida, T.; Linhardt, R.;

BIEMANN, K.; SASISEKHARAN, R. Sequencing of 3-O sulfate containing heparin decasaccharides with a partial antithrombin III binding site. *Proc. Natl. Acad. Sci. USA* **2000**, *97*, 10359–10364.

60. BIEMANN, K. Four decades of structure determination by mass spectrometry: from alkaloids to heparin. *J. Am. Soc. Mass Spectrom.* **2002**, *13*, 1254–1272.
61. KARLSSON, N. G.; SCHULZ, B. L.; PACKER, N. H.; WHITELOCK, J. M. Use of graphitised carbon negative ion LC-MS to analyse enzymatically digested glycosaminoglycans. *J. Chromatogr. B* **2005**, *824*, 139–147.
62. YAGI, H.; TAKAHASHI, N.; YAMAGUCHI, Y.; KIMURA, N.; UCHIMURA, K.; KANNAGI, R.; KATO, K. Development of structural analysis of sulfated *N*-glycans by multidimensional high performance liquid chromatography mapping methods. *Glycobiology* **2005**, *15*, 1051–1060.
63. NAVEN, T. J. P.; HARVEY, D. J. Effect of structure on the signal strength of oligosaccharides in matrix-assisted laser desorption/ionization mass spectrometry on time-of-flight and magnetic sector instruments. *Rapid Commun. Mass Spectrom.* **1996**, *10*, 1361–1366.
64. SIEMIATKOSKI, J.; LYUBARSKAYA, Y.; HOUDE, D.; TEP, S.; MHATRE, R. A comparison of three techniques for quantitative carbohydrate analysis used in characterization of therapeutic antibodies. *Carbohydrate Res.* **2006**, *341*, 410–419.
65. MANK, M.; STAHL, B.; BOEHM, G. 2,5-Dihydroxybenzoic acid butylamine and other ionic liquid matrixes for enhanced MALDI-MS analysis of biomolecules. *Anal. Chem.* **2004**, *76*, 2938–2950.
66. THOLEY, A.; HEINZLE, E.; IONIC, (liquid) matrices for matrix-assisted laser desorption/ionization mass spectrometry—Applications and perspectives. *Anal. Bioanal. Chem.* **2006**, *386*, 24–37.
67. LAREMORE, T. N.; MURUGESAN, S.; PARK, T.-J.; AVCI, F. Y.; ZAGOREVSKI, D. V.; LINHARDT, R. J. Matrix-assisted laser desorption/ionization mass spectrometric analysis of uncomplexed highly sulfated oligosaccharides using ionic liquid matrices. *Anal. Chem.* **2006**, *78*, 1774–1779.
68. HARVEY, D. J. Collision-induced fragmentation of underivatised N-linked carbohydrates ionized by electrospray. *J. Mass Spectrom.* **2000**, *35*, 1178–1190.
69. REINHOLD, V. N.; REINHOLD, B. B.; COSTELLO, C. E. Carbohydrate molecular weight profiling, sequence, linkage and branching data: ES-MS and CID. *Anal. Chem.* **1995**, *67*, 1772–1784.
70. HARVEY, D. J.; BATEMAN, R. H.; GREEN, M. R. High-energy collision-induced fragmentation of complex oligosaccharides ionized by matrix-assisted laser desorption/ionization mass spectrometry. *J. Mass Spectrom.* **1997**, *32*, 167–187.
71. HARVEY, D. J.; BATEMAN, R. H.; BORDOLI, R. S.; TYLDESLEY, R. Ionization and fragmentation of complex glycans with a Q-TOF mass spectrometer fitted with a MALDI ion source. *Rapid Commun. Mass Spectrom.* **2000**, *14*, 2135–2142.
72. HARVEY, D. J. *N*-[2-Diethylamino)ethyl-4-aminobenzamide derivatives for high sensitivity mass spectrometric detection and structure determination of N-linked carbohydrates. *Rapid Commun. Mass Spectrom.* **2000**, *14*, 862–871.
73. HARVEY, D. J. Electrospray mass spectrometry and collision-induced fragmentation of 2-aminobenzamide-labelled neutral N-linked glycans. *The Analyst* **2000**, *125*, 609–617.
74. HARVEY, D. J. Electrospray mass spectrometry and fragmentation of N-linked carbohydrates derivatised at the reducing terminus. *J. Am. Soc. Mass Spectrom.* **2000**, *11*, 900–915.
75. WONG, A. W.; CANCILLA, M. T.; VOSS, L. R.; LEBRILLA, C. B. Anion dopant for oligosaccharides in matrix-assisted laser desorption/ionization mass spectrometry. *Anal. Chem.* **1999**, *71*, 205–211.
76. COLE, R. B.; ZHU, J. Chloride ion attachment in negative ion electrospray ionization mass spectrometry. *Rapid Commun. Mass Spectrom.* **1999**, *13*, 607–611.
77. WONG, A. W.; WANG, H.; LEBRILLA, C. B. Selection of anionic dopant for quantifying desialylation reactions with MALDI-FTMS. *Anal. Chem.* **2000**, *72*, 1419–1425.
78. ZHU, J.; COLE, R. B. Formation and decomposition of chloride adduct ions, $[M + Cl]^-$, in negative ion electrospray ionization mass spectrometry. *J. Am. Soc. Mass Spectrom.* **2000**, *11*, 932–941.
79. CAI, Y.; CONCHA, M. C.; MURRAY, J. S.; COLE, R. B. Evaluation of the role of multiple hydrogen bonding in offering stability to negative ion adducts in electrospray mass spectrometry. *J. Am. Soc. Mass Spectrom.* **2002**, *13*, 1360–1369.
80. CAI, Y.; JIANG, Y.; COLE, R. B. Anionic adducts of oligosaccharides by matrix-assisted laser desorption/ionization time-of-flight mass spectrometry. *Anal. Chem.* **2003**, *75*, 1638–1644.
81. HARVEY, D. J. Fragmentation of negative ions from carbohydrates: Part 1; Use of nitrate and other anionic adducts for the production of negative ion electrospray spectra from *N*-linked carbohydrates. *J. Am. Soc. Mass Spectrom.* **2005**, *16*, 622–630.
82. BAHR, U.; PFENNINGER, A.; KARAS, M.; STAHL, B. High sensitivity analysis of neutral underivatized oligosaccharides by nanoelectrospray mass spectrometry. *Anal. Chem.* **1997**, *69*, 4530–4535.
83. WUHRER, M.; DEELDER, A. M.; HOKKE, C. H. Protein glycosylation analysis by liquid chromatography–mass spectrometry. *J. Chromatogr. B* **2005**, *825*, 124–133.
84. NOVOTNY, M. V.; MECHREF, Y. New hyphenated methodologies in high-sensitivity glycoprotein analysis. *J. Separation Sci.* **2005**, *28*, 1956–1968.
85. MECHREF, Y.; NOVOTNY, M. V. Miniaturized separation techniques in glycomic investigations. *J. Chromatogr. B* **2006**, *841*, 65–78.

86. Royle, L.; Mattu, T. S.; Hart, E.; Langridge, J. I.; Merry, A. H.; Murphy, N.; Harvey, D. J.; Dwek, R. A.; Rudd, P. M. An analytical and structural database provides a strategy for sequencing *O*-glycans from microgram quantities of glycoproteins. *Anal. Biochem.* **2002**, *304*, 70–90.
87. Wuhrer, M.; Koeleman, C. A. M.; Deelder, A. M.; Hokke, C. H. Normal-phase nanoscale liquid chromatography–mass spectrometry of underivatized oligosaccharides at low-femtomole sensitivity. *Anal. Chem.* **2004**, *76*, 833–838.
88. Wuhrer, M.; Koeleman, C. A. M.; Hokke, C. H.; Deelder, A. M. Nano-scale liquid chromatography–mass spectrometry of 2-aminobenzamide-labeled oligosaccharides at low femtomole sensitivity. *Int. J. Mass Spectrom.* **2004**, *232*, 51–57.
89. Morelle, W.; Page, A.; Michalski, J.-C. Electrospray ionization ion trap mass spectrometry for structural characterization of oligosaccharides derivatized with 2-aminobenzamide. *Rapid Commun. Mass Spectrom.* **2005**, *19*, 1145–1158.
90. Karlsson, J.; Momcilovic, D.; Wittgren, B.; Schülein, M.; Tjerneld, F.; Brinkmalm, G. Enzymatic degradation of carboxymethyl cellulose hydrolyzed by the endoglucanases Cel5A, Cel7B, and Cel45A from *Humicola insolens* and Cel7B, Cel12A and Cel45Acore from *Trichoderma reesei. Biopolymers* **2002**, *63*, 32–40.
91. Friedl, C. H.; Lochnit, G.; Zähringer, U.; Bahr, U.; Geyer, R. Structural elucidation of zwitterionic carbohydrates derived from glycosphingolipids of the porcine parasitic nematode *Ascaris suum. Biochem. J.* **2003**, *369*, 89–102.
92. Karlsson, N. G.; Wilson, N. L.; Wirth, H.-J.; Dawes, P.; Joshi, H.; Packer, N. H. Negative ion graphitised carbon nano-liquid chromatography/mass spectrometry increases sensitivity for glycoprotein oligosaccharide analysis. *Rapid Commun. Mass Spectrom.* **2004**, *18*, 2282–2292.
93. Liu, Y.; Urgaonkar, S.; Verkade, J. G.; Armstrong, D. W. Separation and characterization of underivatized oligosaccharides using liquid chromatography and liquid chromatography–electrospray ionization mass spectrometry. *J. Chromatogr. A* **2005**, *1079*, 146–152.
94. Que, A. H.; Mechref, Y.; Huang, Y.; Taraszka, J. A.; Clemmer, D. E.; Novotny, M. V. Coupling capillary electrochromatography with electrospray Fourier transform mass spectrometry for characterizing complex oligosaccharide pools. *Anal. Chem.* **2003**, *75*, 1684–1690.
95. Zamfir, A.; Konig, S.; Althoff, J.; Peter-Katalinc, J. Capillary electrophoresis and off-line capillary electrophoresis-electrospray ionization quadrupole time-of-flight tandem mass spectrometry of carbohydrates. *J. Chromatogr., A* **2000**, *895*, 291–299.
96. Demelbauer, U. M.; Plematl, A.; Kremser, L.; Allmaier, G.; Josic, D.; Rizzi, A. Characterization of glyco isoforms in plasma derived human antithrombin by on-line capillary zone electrophoresis–electrospray ionisation–quadrupole ion trap-mass spectrometry of the intact glycoproteins. *Electrophoresis* **2004**, *25*, 2026–2032.
97. Zamfir, A.; Peter-Katalinic, J. Capillary electrophoresis-mass spectrometry for glycoscreening in biomedical research. *Electrophoresis* **2004**, *25*, 1949–1963.
98. Clowers, B. H.; Dwivedi, P.; Steiner, W. E.; Hill, H. H. J.; Bendiak, B. Separation of sodiated isobaric disaccharides and trisaccharides using electrospray ionisation–atmospheric pressure ion mobility–time of flight mass spectrometry. *J. Am. Soc.Mass Spectrom.* **2005**, *16*, 660–669.
99. Froesch, M.; Bindila, L. M.; Baykut, G.; Allen, M.; Peter-Katalinic, J.; Zamfir, A. D. Coupling of fully automated chip electrospray to Fourier transform ion cyclotron resonance mass spectrometry for high-performance glycoscreening and sequencing. *Rapid Commun. Mass Spectrom.* **2004**, *18*, 3084–3092.
100. Zamfir, A.; Vakhrushev, S.; Sterling, A.; Niebel, H. J.; Allen, M.; Peter-Katalinic, J. Fully automated chip-based mass spectrometry for complex carbohydrate system analysis. *Anal. Chem.* **2004**, *76*, 2046–2054.
101. Zhang, S.; Chelius, D. Characterization of protein glycosylation using chip-based infusion nanoelectrospray linear ion trap. *J. Biomol. Techniques* **2004**, *15*, 120–133.
102. Küster, B.; Wheeler, S. F.; Hunter, A. P.; Dwek, R. A.; Harvey, D. J. Sequencing of N-linked oligosaccharides directly from protein gels: In-gel deglycosylation followed by matrix-assisted laser desorption/ionization mass spectrometry and normal-phase high performance liquid chromatography. *Anal. Biochem.* **1997**, *250*, 82–101.
103. Yu, Y. Q.; Gilar, M.; Kaska, J.; Gebler, J. C. A rapid sample preparation method for mass spectrometric characterization of N-linked glycans. *Rapid Commun. Mass Spectrom.* **2005**, *19*, 2331–2336.
104. Packer, N. H.; Lawson, M. A.; Jardine, D. R.; Redmond, J. W. A general approach to desalting oligosaccharides released from glycoproteins. *Glycoconjugate J* **1998**, *15*, 737–747.
105. Rouse, J. C.; Vath, J. E. On-the-probe sample cleanup strategies for glycoprotein-released carbohydrates prior to matrix-assisted laser desorption–ionization time-of-flight mass spectrometry. *Anal. Biochem.* **1996**, *238*, 82–92.
106. Börnsen, K. O.; Mohr, M. D.; Widmer, H. M. Ion exchange and purification of carbohydrates on a Nafion (R) membrane as a new sample pretreatment for matrix-assisted laser desorption–ionization mass spectrometry. *Rapid Commun. Mass Spectrom.* **1995**, *9*, 1031–1034.
107. Nakano, M.; Kakehi, K.; Lee, Y. C. Sample clean-up method for analysis of complex *N*-glycans released from glycopeptides. *J. Chromatogr. A* **2003**, *1005*, 13–21.

108. Nishimura, S.-I.; Niikura, K.; Kurogochi, M.; Matsushita, T.; Fumoto, M.; Hinou, H.; Kamitani, R.; Nakagawa, H.; Deguchi, K.; Miura, N.; Monde, K.; Kondo, H. High-throughput protein glycomics: Combined use of chemoselective glycoblotting and MALDI-TOF/TOF mass spectrometry. *Angew. Chem. Int. Ed. English* **2005**, *44*, 91–96.
109. Hase, S. Precolumn derivatization for chromatographic and electrophoretic analysis of carbohydrates. *J. Chromatogr. A* **1996**, *720*, 173–182.
110. Gao, X. B.; Yang, J. H.; Huang, F.; Wu, X.; Li, L.; Sun, C. X. Progresses of derivatization techniques for analyses of carbohydrates. *Anal. Lett.* **2003**, *36*, 1281–1310.
111. Lamari, F. N.; Kuhn, R.; Karamanos, N. K. Derivatization of carbohydrates for chromatographic, electrophoretic and mass spectrometric structure analysis. *J. Chromatogr. B* **2003**, *793*, 15–36.
112. Suzuki, S.; Fujimori, T.; Yodoshi, M. Recovery of free oligosaccharides from derivatives labeled by reductive amination. *Anal. Biochem.* **2006**, *354*, 94–103.
113. Poulter, L.; Burlingame, A. L. Desorption mass spectrometry of oligosaccharides coupled with hydrophobic chromophores. *Methods Enzymol.* **1990**, *193*, 661–689.
114. Takao, T.; Tambara, Y.; Nakamura, A.; Yoshino, K.-I.; Fukuda, H.; Fukuda, M.; Shimonishi, Y. Sensitive analysis of oligosaccharides derivatised with 4-aminobenzoic acid 2-(diethylamino)ethyl ester by matrix-assisted laser desorption/ionization mass spectrometry. *Rapid Commun. Mass Spectrom.* **1996**, *10*, 637–640.
115. Bigge, J. C.; Patel, T. P.; Bruce, J. A.; Goulding, P. N.; Charles, S. M.; Parekh, R. B. Nonselective and efficient fluorescent labeling of glycans using 2-aminobenzamide and anthranilic acid. *Anal. Biochem.* **1995**, *230*, 229–238.
116. Okafo, G.; Burrow, L.; Carr, S. A.; Roberts, G. D.; Johnson, W.; Camilleri, P. A coordinated high-performance liquid chromatographic, capillary electrophoretic, and mass spectrometric approach for the analysis of oligosaccharide mixtures with 2-aminoacridone. *Anal. Chem.* **1996**, *68*, 4424–4430.
117. Okafo, G.; Langridge, J.; North, S.; Organ, A.; West, A.; Morris, M.; Camilleri, P. High-performance liquid chromatographic analysis of complex N-linked glycans derivatized with 2-aminoacridone. *Anal. Chem.* **1997**, *69*, 4985–4993.
118. Morelle, W.; Michalski, J.-C. Sequencing of oligosaccharides derivatized with benzylamine using electrospray ionisation–quadrupole time of flight-tandem mass spectrometry. *Electrophoresis* **2004**, *25*, 2144–2155.
119. Hase, S.; Ibuki, T.; Ikenaka, T. Reexamination of the pyridylamination used for fluorescence labelling of oligosaccharides and its application to glycoproteins. *J. Biochemistry (Tokyo)* **1984**, *95*, 197–203.
120. Okamoto, M.; Takahashi, K.; Doi, T.; Takimoto, Y. High-sensitivity detection and postsource decay of 2-aminopyridine-derivatized oligosaccharides with matrix-assisted laser desorption-ionization mass spectrometry. *Anal. Chem.* **1997**, *69*, 2919–2926.
121. Naven, T. J. P.; Harvey, D. J. Cationic derivatization of oligosaccharides with Girard's T reagent for improved performance in matrix-assisted laser desorption/ionization and electrospray mass spectrometry. *Rapid Commun. Mass Spectrom.* **1996**, *10*, 829–834.
122. Gouw, J. W.; Burgers, P. C.; Trikoupis, M. A.; Terlouw, J. K. Derivatization of small oligosaccharides prior to analysis by matrix-assisted laser desorption/ionization using glycidyltrimethylammonium chloride and Girard's reagent T. *Rapid Commun. Mass Spectrom.* **2002**, *16*, 905–912.
123. Shinohara, Y.; Furukawa, J.; Niikura, K.; Miura, N.; Nishimura, S.-I. Direct *N*-glycan profiling in the presence of tryptic peptides on MALDI-TOF by controlled ion enhancement and suppression upon glycan-selective derivatization. *Anal. Chem.* **2004**, *76*, 6989–6997.
124. Lattová, E.; Snovida, S.; Perreault, H.; Krokhin, O. Influence of the labeling group on ionization and fragmentation of carbohydrates in mass spectrometry. *J. Am. Soc. Mass Spectrom.* **2005**, *16*, 683–696.
125. Harvey, D. J. Collision-induced fragmentation of negative ions from *N*-linked glycans derivatized with 2-aminobenzoic acid. *J. Mass Spectrom.* **2005**, *40*, 642–653.
126. Küster, B.; Naven, T. J. P.; Harvey, D. J. Effect of the reducing-terminal substituents on the high energy collision-induced dissociation matrix-assisted laser desorption/ionization mass spectra of oligosaccharides. *Rapid Commun. Mass Spectrom.* **1996**, *10*, 1645–1651.
127. Li, M.; Kinzer, J. A. Structural analysis of oligosaccharides by a combination of electrospray mass spectrometry and bromine isotope tagging of reducing-end sugars with 2-amino-5-bromopyridine. *Rapid Commun. Mass Spectrom.* **2003**, *17*, 1462–1466.
128. Harvey, D. J. Halogeno-substituted 2-aminobenzoic acid derivatives for negative ion fragmentation studies of N-linked carbohydrates, *Rapid Commun. Mass Spectrometry* **2005**, *19*, 397–400.
129. Lattova, E.; Perreault, H. Labelling saccharides with phenylhydrazine for electrospray and matrix-assisted laser desorption–ionization mass spectrometry. *J. Chromatogr. B* **2003**, *793*, 167–179.
130. Lattova, E.; Perreault, H. Profiling of *N*-linked oligosaccharides using phenylhydrazine derivatization and mass spectrometry. *J. Chromatogr. A* **2003**, *1016*, 71–87.
131. Lattova, E.; Perreault, H.; Krokhin, O. Matrix-assisted laser desorption/ionization tandem mass spectrometry and post-source decay fragmentation study of phenylhydrazones of *N*-linked oligosaccharides from ovalbumin. *J. Am. Soc. Mass Spectrom.* **2004**, *15*, 725–735.

132. Kamoda, S.; Nakano, M.; Ishikawa, R.; Suzuki, S.; Kakehi, K. Rapid and sensitive screening of *N*-glycans as 9-fluorenylmethyl derivatives by high-performance liquid chromatography: A method which can recover free oligosaccharides after analysis. *J. Proteome Res.* **2005**, *4*, 146–152.
133. Spengler, B.; Kirsch, D.; Kaufmann, R.; Lemoine, J. Structure analysis of branched oligosaccharides using post-source decay in matrix-assisted laser desorption/ionization mass spectrometry. *J. Mass Spectrom.* **1995**, *30*, 782–787.
134. Shevchenko, A.; Loboda, A.; Shevchenko, A.; Ens, W.; Standing, K. G. MALDI Quadrupole time-of-flight mass spectrometry: a powerful tool for proteomic research. *Anal. Chem.* **2000**, *72*, 2132–2141.
135. Verhaert, P.; Uttenweiler-Joseph, S. de Vries, M. Loboda, A. Ens, W. Standing, K. G. Matrix-assisted laser desorption/ionization quadrupole time-of-flight mass spectrometry: An elegant tool for peptidomics. *Proteomics* **2001**, *1*, 118–131.
136. Loboda, A. V.; Krutchinsky, A. N.; Bromirski, M.; Ens, W.; Standing, K. G. A quadrupole/time-of-flight mass spectrometer with a matrix-assisted laser desorption/ionization source: Design and performance. *Rapid Commun. Mass Spectrom.* **2000**, *14*, 1047–1057.
137. Spina, E.; Sturiale, L.; Romeo, D.; Impallomeni, G.; Garozzo, D.; Waidelich, D.; Glueckmann, M. New fragmentation mechanisms in matrix-assisted laser desorption/ionization time-of-flight/time-of-flight tandem mass spectrometry of carbohydrates. *Rapid Commun. Mass Spectrom.* **2004**, *18*, 392–398.
138. Domon, B.; Costello, C. E. A systematic nomenclature for carbohydrate fragmentations in FAB-MS/MS spectra of glycoconjugates. *Glycoconjugate J.* **1988**, *5*, 397–409.
139. Harvey, D. J.; Martin, R. L.; Jackson, K. A.; Sutton, C. W. Fragmentation of *N*-linked glycans with a MALDI-ion trap time-of-flight mass spectrometer. *Rapid Commun. Mass Spectrom.* **2004**, *18*, 2997–3007.
140. Kovácik, V.; Hirsch, J.; Kovác, P.; Heerma, W.; Thomas-Oates, J.; Haverkamp, J. Oligosaccharide characterization using collision-induced dissociation fast atom bombardment mass spectrometry: Evidence for internal monosaccharide residue loss. *J. Mass Spectrom.* **1995**, *30*, 949–958.
141. Brüll, L. P.; Heerma, W.; Thomas-Oates, J.; Haverkamp, J.; Kovácik, V.; Kovác, P. Loss of internal 1-6 substituted monosaccharide residues from underivatized and per-*O*-methylated trisaccharides. *J. Am. Soc. Mass Spectrom.* **1997**, *8*, 43–49.
142. Warrack, B. M.; Hail, M. E.; Triolo, A.; Animati, F.; Seraglia, R.; Traldi, P. Observation of internal monosaccharide losses in the collisionally activated dissociation mass spectra of anthracycline aminodisaccharides. *J. Am. Soc. Mass Spectrom.* **1998**, *9*, 710–715.
143. Mattu, T. S.; Royle, L.; Langridge, J.; Wormald, M. R. Van den Steen, P. E. Van Damme, J. Opdenakker, G. Harvey, D. J. Dwek, R. A. Rudd, P. M. O-Glycan analysis of natural human neutrophil gelatinase B using a combination of normal-phase HPLC and online tandem mass spectrometry: Implications for the domain organization of the enzyme. *Biochemistry* **2000**, *39*, 15695–15704.
144. Wuhrer, M.; Koeleman, C. A.; Hokke, C. H.; Deelder, A. M. Mass spectrometry of proton adducts of fucosylated *N*-glycans: Fucose transfer between antennae gives rise to misleading fragments. *Rapid Commun. Mass Spectrom.* **2006**, *20*, 1747–1754.
145. Franz, A. H.; Lebrilla, C. B. Evidence for long-range glycosyl transfer reactions in the gas phase. *J. Am. Soc. Mass Spectrom.* **2002**, *13*, 325–337.
146. Harvey, D. J.; Mattu, T. S.; Wormald, M. R.; Royle, L.; Dwek, R. A.; Rudd, P. M. "Internal residue loss": Rearrangements occurring during the fragmentation of carbohydrates derivatized at the reducing terminus. *Anal. Chem.* **2002**, *74*, 734–740.
147. Ngoka, L. C.; Gal, J.-F.; Lebrilla, C. B. Effects of cations and charge types on the metastable decay rates of oligosaccharides. *Anal. Chem.* **1994**, *66*, 692–698.
148. Cancilla, M. T.; Penn, S. G.; Carroll, J. A.; Lebrilla, C. B. Coordination of alkali metals to oligosaccharides dictates fragmentation behavior in matrix assisted laser desorption ionization/Fourier transform mass spectrometry. *J. Am. Chem. Soc.* **1996**, *118*, 6736–6745.
149. Orlando, R.; Bush, C. A.; Fenselau, C. Structural analysis of oligosaccharides by tandem mass spectrometry: Collisional activation of sodium adduct ions. *Biomed. Environ. Mass Spectrom.* **1990**, *19*, 747–754.
150. Mechref, Y.; Novotny, M. V.; Krishnan, C. Structural characterization of oligosaccharides using MALDI-TOF/TOF tandem mass spectrometry. *Anal. Chem.* **2003**, *75*, 4895–4903.
151. Stephens, E.; Maslen, S. L.; Green, L. G.; Williams, D. H. Fragmentation characteristics of neutral *N*-linked glycans using a MALDI-TOF/TOF tandem mass spectrometer. *Anal. Chem.* **2004**, *76*, 2343–2354.
152. Wuhrer, M.; Deelder, A. M. Negative-mode MALDI-TOF/TOF-MS of oligosaccharides labeled with 2-aminobenzamide. *Anal. Chem.* **2005**, *77*, 6954–6959.
153. Lewandrowski, U.; Resemann, A.; Sickmann, A. Laser-induced dissociation/high-energy collision-induced dissociation fragmentation using MALDI-TOF/TOF-MS instrumentation for the analysis of neutral and acidic oligosaccharides. *Anal. Chem.* **2005**, *77*, 3274–3283.
154. Morelle, W.; Slomianny, M.-C.; Diemer, H.; Schaeffer, C.; van Dorsselaer, A.; Michalski, J.-C. Structural characterization of 2-aminobenzamide-derivatized oligosaccharides using a matrix-assisted laser desorption/ionization two-stage time-of-flight tandem mass spectrometer. *Rapid Commun. Mass Spectrom.* **2005**, *19*, 2075–2084.

155. DEVAKUMAR, A.; THOMPSON, M. S.; REILLY, J. P. Fragmentation of oligosaccharide ions with 157 nm vacuum ultraviolet light. *Rapid Commun. Mass Spectrom.* **2005**, *19*, 2313–2320.
156. MORELLE, W.; SLOMIANNY, M. C.; DIEMER, H.; SCHAEFFER, C.; DORSSELAER, A. V.; MICHALSKI, J. C. Fragmentation characteristics of permethylated oligosaccharides using a matrix-assisted laser desorption/ionization two-stage time-of-flight (TOF/TOF) tandem mass spectrometer. *Rapid Commun. Mass Spectrom.* **2004**, *18*, 2637–2649.
157. YU, S. Y.; WU, S. W.; KHOO, K. H. Distinctive characteristics of MALDI-Q/TOF and TOF/TOF tandem mass spectrometry for sequencing of permethylated complex type *N*-glycans. *Glycoconjugate J.* **2006**, *23*, 355–369.
158. KUROGOCHI, M.; NISHIMURA, S.-I. Structural characterization of *N*-glycopeptides by matrix-dependent selective fragmentation of MALDI-TOF/TOF tandem mass spectrometry. *Anal. Chem.* **2004**, *76*, 6097–6101.
159. BRÜLL, L. P.; KOVÁCIK, V.; THOMAS-OATES, J. E.; HEERMA, W.; HAVERKAMP, J. Sodium-cationized oligosaccharides do not appear to undergo "internal residue loss" rearrangement processes on tandem mass spectrometry. *Rapid Commun. Mass Spectrom.* **1998**, *12*, 1520–1532.
160. XIE, Y.; LEBRILLA, C. B. Infrared multiphoton dissociation of alkali metal-coordinated oligosaccharides. *Anal. Chem.* **2003**, *75*, 1590–1598.
161. LANCASTER, K. S.; AN, H. J.; LI, B.; LEBRILLA, C. B. Interrogation of *N*-linked oligosaccharides using infrared multiphoton dissociation in FT-ICR mass spectrometry. *Anal. Chem.* **2006**, *78*, 4990–4997.
162. ADAMSON, J. T.; HÅKANSSON, K. Infrared multiphoton dissociation and electron capture dissociation of high-mannose type glycopeptides. *J. Proteome Res.* **2006**, *5*, 493–501.
163. BOUTREAU, L.; LÉON, E.; SALPIN, J.-Y.; AMEKRAZ, B.; MOULIN, C.; TORTAJADA, J. Gas-phase reactivity of silver and copper coordinated monosaccharide cations studied by electrospray ionization and tandem mass spectrometry. *Eur. J. Mass Spectrom.* **2003**, *9*, 377–390.
164. HARVEY, D. J. Ionization and fragmentation of *N*-linked glycans as silver adducts by electrospray mass spectrometry. *Rapid Commun. Mass Spectrom.* **2005**, *19*, 484–492.
165. HARVEY, D. J. Ionization and collision-induced fragmentation of *N*-linked and related carbohydrates using divalent cations. *J. Am. Soc. Mass Spectrom.* **2001**, *12*, 926–937.
166. TAKEGAWA, Y.; DEGUCHI, K.; ITO, S.; YOSHIOKA, S.; NAKAGAWA, H.; NISHIMURA, S.-I. Structural assignment of isomeric 2-aminopyridine-derivatized oligosaccharides using negative-ion MS^n spectral matching. *Rapid Commun. Mass Spectrom.* **2005**, *19*, 937–946.
167. CAI, Y.; COLE, R. B. Stabilization of anionic adducts in negative ion electrospray mass spectrometry. *Anal. Chem.* **2002**, *74*, 985–991.
168. JIANG, Y.; COLE, R. B. Oligosaccharide analysis using anion attachment in negative mode electrospray mass spectrometry. *J. Am. Soc. Mass Spectrom.* **2005**, *16*, 60–70.
169. YAMAGAKI, T.; SUZUKI, H.; TACHIBANA, K. Semiquantitative analysis of isomeric oligosaccharides by negative-ion mode UV-MALDI TOF postsource decay mass spectrometry and their fragmentation mechanism study at *N*-acetyl hexosamine moiety. *J. Mass Spectrom.* **2006**, *41*, 454–462.
170. HUBERTY, M. C.; VATH, J. E.; YU, W.; MARTIN, S. A. Site-specific carbohydrate identification in recombinant proteins using MALD-TOF MS. *Anal. Chem.* **1993**, *65*, 2791–2800.
171. KAUFMANN, R.; CHAURAND, P.; KIRSCH, D.; SPENGLER, B. Post-source decay and delayed extraction in matrix-assisted laser desorption/ionization-reflectron time-of-flight mass spectrometry. Are there trade-offs? *Rapid Commun. Mass Spectrom.* **1996**, *10*, 1199–1208.
172. NAVEN, T. J. P.; HARVEY, D. J.; BROWN, J.; CRITCHLEY, G. Fragmentation of complex carbohydrates following ionization by matrix-assisted laser desorption with an instrument fitted with time-lag focusing. *Rapid Commun. Mass Spectrom.* **1997**, *11*, 1681–1686.
173. HARVEY, D. J.; HUNTER, A. P.; BATEMAN, R. H.; BROWN, J.; CRITCHLEY, G. The relationship between in-source and post-source fragment ions in the MALDI mass spectra of carbohydrates recorded with reflectron–TOF mass spectrometers. *Int. J. Mass Spectrom. Ion Processes* **1999**, *188*, 131–146.
174. CHURNUSHEVICH, I. V.; ENS, W.; STANDING, K. J. Orthogonal injection TOFMS for analysing biomolecules. *Anal. Chem.* **1999**, *71*, 452A–461A.
175. KRUTCHINSKY, A. N.; LOBODA, A. V.; SPICER, V. L.; DWORSCHAK, R.; ENS, W.; STANDING, K. G. Orthogonal injection of matrix-assisted laser desorption/ionization ions into a time-of-flight spectrometer through a collisional damping interface. *Rapid Commun. Mass Spectrom.* **1998**, *12*, 508–518.
176. HUNNAM, V.; HARVEY, D. J.; PRIESTMAN, D. A.; BATEMAN, R. H.; BORDOLI, R. S.; TYLDESLEY, R. Ionization and fragmentation of neutral and acidic glycosphingolipids with a Q-TOF mass spectrometer fitted with a MALDI ion source. *J. Am. Soc. Mass Spectrom.* **2001**, *12*, 1220–1225.
177. CLAYTON, E.; BATEMAN, R. H. Time-of-flight mass analysis of high-energy collision-induced dissociation fragment ions. *Rapid Commun. Mass Spectrom.* **1992**, *6*, 719–720.
178. WEISKOPF, A. S.; VOUROS, P.; HARVEY, D. J. Electrospray ionization-ion trap mass spectrometry for structural analysis of complex N-linked glycoprotein oligosaccharides. *Anal. Chem.* **1998**, *70*, 4441–4447.
179. WEISKOPF, A. S.; VOUROS, P.; HARVEY, D. J. Characterization of oligosaccharide composition and

structure by quadrupole ion trap mass spectrometry. *Rapid Commun. Mass Spectrom.* **1997**, *11*, 1493–1504.
180. Deguchi, K.; Ito, H.; Takegawa, Y.; Shinji, N.; Nakagawa, H.; Nishimura, S. I. Complementary structural information of positive- and negative-ion MS^n spectra of glycopeptides with neutral and sialylated *N*-glycans. *Rapid Commun. Mass Spectrom.* **2006**, *20*, 741–746.
181. Takemori, N.; Komori, N.; Matsumoto, H. Highly sensitive multistage mass spectrometry enables small-scale analysis of protein glycosylation from two-dimensional polyacrylamide gels. *Electrophoresis* **2006**, *27*, 1394–1406.
182. Ashline, D.; Singh, S.; Hanneman, A.; Reinhold, V. Congruent strategies for carbohydrate sequencing. 1. Mining structural details by MS^n. *Anal. Chem.* **2005**, *77*, 6250–6262.
183. Ojima, N.; Masuda, K.; Tanaka, K.; Nishimura, O. Analysis of neutral oligosaccharides for structural characterization by matrix-assisted laser desorption/ionization quadrupole ion trap time-of-flight mass spectrometry. *J. Mass Spectrom.* **2005**, *40*, 380–388.
184. Kurimoto, A.; Daikoku, S.; Mutsuga, S.; Kanie, O. Analysis of energy-resolved mass spectra at MS^n in a pursuit to characterize structural isomers of oligosaccharides. *Anal. Chem.* **2006**, *78*, 3461–3466.
185. Demelbauer, U. M.; Zehl, M.; Plematl, A.; Allmaier, G.; Rizzi, A. Determination of glycopeptide structures by multistage mass spectrometry with low-energy collision-induced dissociation: Comparison of electrospray ionization quadrupole ion trap and matrix-assisted laser desorption/ionization quadrupole ion trap reflectron time-of-flight approaches. *Rapid Commun. Mass Spectrom.* **2004**, *18*, 1575–1582.
186. Wuhrer, M.; Deelder, A. M. Matrix-assisted laser desorption/ionization in-source decay combined with tandem time-of-flight mass spectrometry of permethylated oligosaccharides: targeted characterization of specific parts of the glycan structure. *Rapid Commun. Mass Spectrom.* **2006**, *20*, 943–951.
187. Mizuno, Y.; Sasagawa, T.; Dohmae, N.; Takio, K. An automated interpretation of MALDI/TOF postsource decay spectra of oligosaccharides. 1. Automated peak assignment. *Anal. Chem.* **1999**, *71*, 4764–4771.
188. Gaucher, S. P.; Morrow, J.; Leary, J. A. STAT: A saccharide topology analysis tool used in combination with tandem mass spectrometry. *Anal. Chem.* **2000**, *72*, 2331–2336.
189. Ethier, M.; Saba, J. A.; Ens, W.; Standing, K. G.; Perreault, H. Automated structural assignment of derivatized complex *N*-linked oligosaccharides from tandem mass spectra. *Rapid Commun. Mass Spectrometry* **2002**, *16*, 1743–1754.
190. Ethier, M.; Saba, J. A.; Spearman, M.; Krokhin, O.; Butler, M.; Ens, W.; Standing, K. G.; Perreault, H. Application of the StrOligo algorithm for the automated structure assignment of complex N-linked glycans from glycoproteins using tandem mass spectrometry. *Rapid Commun. Mass Spectrom.* **2003**, *17*, 2713–2720.
191. Lohmann, K. K. von der Lieth, C.-W. GLYCO-FRAGMENT: A web tool to support the interpretation of mass spectra of complex carbohydrates. *Proteomics* **2003**, *3*, 2028–2035.
192. Joshi, H. J.; Harrison, M. J.; Schulz, B. L.; Cooper, C. A.; Packer, N. H.; Karlsson, N. G. Development of a mass fingerprinting tool for automated interpretation of oligosaccharide fragmentation data. *Proteomics* **2004**, *4*, 1650–1664.
193. Cooper, C. A.; Gasteiger, E.; Packer, N. H. GlycoMod—A software tool for determining glycosylation compositions from mass spectrometric data. *Proteomics* **2001**, *1*, 340–349.
194. An, H. J.; Tillinghast, J. S.; Woodruff, D. L.; Rocke, D. M.; Lebrilla, C. B. A new computer program (GlycoX) to determine simultaneously the glycosylation sites and oligosaccharide heterogeneity of glycoproteins. *J. Proteome Res.* **2006**, *5*, 2800–2808.
195. Lohmann, K. K.; von der Lieth, C.-W. GlycoFragment and GlycoSearchMS: web tools to support the interpretation of mass spectra of complex carbohydrates. *Nucleic Acids Res.* **2004**, *32*, W261–W266.
196. Kameyama, A.; Kikuchi, N.; Nakaya, S.; Ito, H.; Sato, T.; Shikanai, T.; Takahashi, Y.; Takahashi, K.; Narimatsu, H. A Strategy for identification of oligosaccharide structures using observational multistage mass spectral library. *Anal. Chem.* **2005**, *77*, 4719–4725.
197. Zhang, H.; Singh, S.; Reinhold, V. N. Congruent strategies for carbohydrate sequencing. 2. FragLib: An MS^n spectral library. *Anal. Chem.* **2005**, *77*, 6263–6270.
198. Lapadula, A. J.; Hatcher, P. J.; Hanneman, A. J.; Ashline, D. J.; Zhang, H.; Reinhold, V. N. Congruent strategies for carbohydrate sequencing. 3. OSCAR: an algorithm for assigning oligosaccharide topology from MS^n data. *Anal. Chem.* **2005**, *77*, 6271–6279.
199. Goldberg, D.; Bern, M.; Li, B.; Lebrilla, C. B. Automatic determination of *O*-glycan structure from fragmentation spectra. *J. Proteome Res.* **2006**, *5*, 1429–1434.
200. Pérez, S.; Mulloy, B. Prospects for glycoinformatics. *Curr. Opinion Struct. Biol.* **2005**, *15*, 517–524.
201. von der Lieth, C.-W.; Lütteke, T.; Frank, M. The role of informatics in glycobiology research with special emphasis on automatic interpretation of MS spectra. *Biochim. Biophys. Acta* **2006**, *1760*, 568–577.
202. Bandu, M. L.; Wilson, J.; Vachet, R. W.; Dalpathado, D. S.; Desaire, H. STEP (statistical test of equivalent pathways) analysis: A mass spectrometric method for carbohydrates and peptides. *Anal. Chem.* **2005**, *77*, 5886–5893.
203. Aoki, K.; Yamaguchi, A.; Ueda, N.; Akutsu, T.; Mamitsuka, H.; Goto, S.; Kanehisa, M. KCaM (KEGG

Carbohydrate Matcher): a software tool for analyzing the structures of carbohydrate sugar chains. *Nucleic Acids Res.* **2004**, *32*, W267–W272.

204. Hashimoto, K.; Goto, S.; Kawano, S.; Aoki-Kinoshita, K. F.; Ueda, N.; Hamajima, M.; Kawasaki, T.; Kanehisa, M. KEGG as a glycome informatics resource. *Glycobiology* **2006**, *16*, 63R–70R.

205. Coutinho, P. M.; Henrissat, B. In Gilbert, H. J.; Davies, G., Henrissat, B.; Svensson, B. (Eds.), *Recent Advances in Carbohydrate Bioengineering*, The Royal Society of Chemistry, Cambridge, **1999**, pp. 3–12.

206. Raman, R.; Venkataraman, M.; Ramakrishnan, S.; Lang, W.; Raguram, S.; Sasisekharan, R. Advancing glycomics: Implementation strategies at the consortium for functional glycomics. *Glycobiology* **2006**, *16*, 82R–90R.

207. Lütteke, T.; Bohne-Lang, A.; Loss, A.; Goetz, T.; Frank, M.; von der Lieth, C.-W. GLYCOSCIENCES.de: An Internet portal to support glycomics and glycobiology research. *Glycobiology* **2006**, *16*, 71R–81R.

208. Goldberg, D.; Sutton-Smith, M.; Paulson, J.; Dell, A. Automatic annotation of matrix-assisted laser desorption/ionization *N*-glycan spectra. *Proteomics* **2005**, *5*, 865–875.

209. Stahl, B.; Thurl, S.; Zeng, J.; Karas, M.; Hillenkamp, F.; Steup, M.; Sawatzki, G. Oligosaccharides from human milk as revealed by matrix-assisted laser desorption/ionization mass spectrometry. *Anal. Biochem.* **1994**, *223*, 218–226.

210. Finke, B.; Stahl, B.; Pfenninger, A.; Karas, M.; Daniel, H.; Sawatzki, G. Analysis of high-molecular-weight oligosaccharides from human milk by liquid chromatography and MALDI-MS. *Anal. Chem.* **1999**, *71*, 3755–3762.

211. Finke, B.; Mank, M.; Daniel, H.; Stahl, B. Offline coupling of low-pressure anion-exchange chromatography with MALDI-MS to determine the elution order of human milk oligosaccharides. *Anal. Biochem.* **2000**, *284*, 256–265.

212. Schmid, D.; Behnke, B.; Metzger, J.; Kuhn, R. Nano-HPLC–mass spectrometry and MEKC for the analysis of oligosaccharides from human milk. *Biomed. Chromatogr.* **2002**, *16*, 151–156.

213. Charlwood, J.; Tolson, D.; Dwek, M.; Camilleri, P. A detailed analysis of neutral and acidic carbohydrates in human milk. *Anal. Biochem.* **1999**, *273*, 261–277.

214. Dreisewerd, K.; Kölbl, S.; Peter-Katalinic, J.; Berkenkamp, S.; Pohlentz, G. Analysis of native milk oligosaccharides directly from thin-layer chromatography plates by matrix-assisted laser desorption/ionization orthogonal-time-of-flight mass spectrometry with a glycerol matrix. *J. Am. Soc. Mass Spectrom.* **2006**, *17*, 139–150.

215. Pfenninger, A.; Karas, M.; Finke, B.; Stahl, B. Structural analysis of underivatized neutral human milk oligosaccharides in the negative ion mode by nano-electrospray MS^n (Part 1: Methodology). *J. Am. Soc. Mass Spectrom.* **2002**, *13*, 1331–1340.

216. Pfenninger, A.; Karas, M.; Finke, B.; Stahl, B. Structural analysis of underivatized neutral human milk oligosaccharides in the negative ion mode by nano-electrospray MS^n (Part 2: Application to isomeric mixtures). *J. Am. Soc. Mass Spectrom.* **2002**, *13*, 1341–1348.

217. Pfenninger, A.; Karas, M.; Finke, B.; Stahl, B.; Sawatzki, G. Mass spectrometric investigations of human milk oligosaccharides. *Adv. Exp. Med. Biol.* **2001**, *501*, 279–284.

218. Urashima, T.; Sumiyoshi, W.; Nakamura, T.; Arai, I.; Saito, T.; Komatsu, T.; Tsubota, T. Chemical characterization of milk oligosaccharides of the Japanese black bear, *Ursus thibetanus japonicus*. *Biochim. Biophys. Acta* **1999**, *1472*, 290–306.

219. Urashima, T.; Yamashita, T.; Nakamura, T.; Arai, I.; Saito, T.; Derocher, A. E.; Wiig, O. Chemical characterization of milk oligosaccharides of the polar bear, *Ursus maritimus*. *Biochim. Biophys. Acta* **2000**, *1475*, 395–408.

220. Garozzo, D.; Impallomeni, G.; Spina, E.; Sturiale, L.; Zanetti, F. Matrix assisted laser desorption/ionization mass spectrometry of polysaccharides. *Rapid Commun. Mass Spectrom.* **1995**, *9*, 937–941.

221. Hao, C.; Ma, X.; Fang, S.; Liu, Z.; Liu, S.; Song, F.; Liu, J. Positive- and negative-ion matrix-assisted laser desorption/ionization mass spectrometry of saccharides. *Rapid Commun. Mass Spectrom.* **1998**, *12*, 345–348.

222. Garozzo, D.; Spina, E.; Cozzolino, R.; Cescutti, P.; Fett, W. F. Studies on the primary structure of short polysaccharides using SEC MALDI mass spectrometry. *Carbohydrate Res.* **1999**, *323*, 139–146.

223. Deery, M. J.; Stimson, E.; Chappell, C. G. Size-exclusion chromatography/mass spectrometry applied to the analysis of polysaccharides. *Rapid Commun. Mass Spectrom.* **2001**, *15*, 2273–2283.

224. Jacobs, A.; Dahlman, O. Characterization of the molar masses of hemicelluloses from wood and pulps employing size exclusion chromatography and matrix-assisted laser desorption ionization time-of-flight mass spectrometry. *Biomacromolecules* **2001**, *2*, 894–905.

225. Dahlman, O.; Jacobs, A.; Liljenberg, A.; Olsson, A. I. Analysis of carbohydrates in wood and pulps employing enzymatic hydrolysis and subsequent capillary zone electrophoresis. *J. Chromatogr. A* **2000**, *891*, 157–174.

226. Jacobs, A.; Lundqvist, J.; Stålbrand, H.; Tjerneld, F.; Dahlman, O. Characterization of water-soluble hemicelluloses from spruce and aspen employing SEC/MALDI mass spectrometry. *Carbohydrate Res.* **2002**, *337*, 711–717.

227. Satomi, Y.; Shimonishi, Y.; Takao, T. *N*-Glycosylation at Asn in the Asn-Xaa-Cys motif of human transferrin. *FEBS Lett.* **2004**, *576*, 51–56.

228. Varki, A. Biological roles of oligosaccharides: All of the theories are correct. *Glycobiology* **1993**, *3*, 97–130.

229. Dwek, R. A. Glycobiology: Towards understanding the function of sugars. *Chemical Rev.* **1996**, *96*, 683–720.

230. Kornfeld, R.; Kornfeld, S. Assembly of asparagine-linked oligosaccharides. *Annu. Rev. Biochem.* **1985**, *54*, 631–664.
231. Chakel, J. A.; Pungor, E. Jr.; Hancock, W. S.; Swedberg, S. A. Analysis of recombinant DNA-derived glycoproteins via high-performance capillary electrophoresis coupled with off-line matrix-assisted laser desorption ionization time-of-flight mass spectrometry. *J. Chromatogr. B* **1997**, *689*, 215–220.
232. Apffel, A.; Chakel, J. A.; Hancock, W. S.; Souders, C.; M'Timkulu, T.; Pungor, E. Jr. Application of high-performance liquid chromatography-electrospray ionization mass spectrometry and matrix-assisted laser desorption ionization time-of-flight mass spectrometry in combination with selective enzymatic modifications in the characterization of glycosylation patterns in single-chain plasminogen activator. *J. Chromatogr. A* **1996**, *732*, 27–42.
233. Lee, J. H.; Kim, Y.; Ha, M. Y.; Lee, E. K.; Choo, J. Immobilization of aminophenylboronic acid on magnetic beads for the direct determination of glycoproteins by matrix assisted laser desorption Ionization mass spectrometry. *J. Am. Soc. Mass Spectrom.* **2005**, *16*, 1456–1460.
234. Imre, T.; Schlosser, G.; Pocsfalvi, G.; Siciliano, R.; Molnár-Szöllsi, É.; Kremmer, T.; Malorni, A.; Vékey, K. Glycosylation site analysis of human alpha-1-acid glycoprotein (AGP) by capillary liquid chromatography—electrospray mass spectrometry. *J. Mass Spectrom.* **2005**, *40*, 1472–1483.
235. Annesley, T. M. Ion suppression in mass spectrometry. *Clin. Chem.* **2003**, *49*, 1041–1044.
236. Carr, S. A.; Huddleston, M. J.; Bean, M. F. Selective identification and differentiation of *N*- and *O*-linked oligosaccharides in glycoproteins by liquid chromatography–mass spectrometry. *Protein Sci.* **1993**, *2*, 183–196.
237. Krokhin, O.; Ens, W.; Standing, K. G.; Wilkins, J.; Perreault, H.; Site-specific, *N*-glycosylation analysis: Matrix-assisted laser desorption/ionization quadrupole–quadrupole time-of-flight tandem mass spectral signatures for recognition and identification of glycopeptides. *Rapid Commun. Mass Spectrom.* **2004**, *18*, 2020–2030.
238. Wada, Y.; Tajiri, M.; Yoshida, S. Hydrophilic affinity isolation and MALDI multiple-stage tandem mass spectrometry of glycopeptides for glycoproteomics. *Anal. Chem.* **2004**, *76*, 6560–6565.
239. Lochnit, G.; Geyer, R. An optimized protocol for nano-LC-MALDI-TOF-MS coupling for the analysis of proteolytic digests of glycoproteins. *Biomed. Chromatogr.* **2005**, *18*, 841–848.
240. Gonzalez, J.; Takao, T.; Hori, H.; Besada, V.; Rodriguez, R.; Padron, G.; Shimonishi, Y. A method for determination of *N*-glycosylation sites in glycoproteins by collision-induced dissociation analysis in fast atom bombardment mass spectrometry: Identification of the positions of carbohydrate-linked asparagine in recombinant α-amylase by treatment with peptide-*N*-glycosidase F in ^{18}O-labelled water. *Anal. Biochem.* **1992**, *205*, 151–158.
241. Lattová, E.; Kapková, P.; Krokhin, O.; Perreault, H. Method for investigation of oligosaccharides from glycopeptides: Direct determination of glycosylation sites in proteins. *Anal. Chem.* **2006**, *78*, 2977–2984.
242. Lee, B.-S.; Krishnanchettiar, S.; Lateef, S. S.; Gupta, S. Characterization of oligosaccharide moieties of glycopeptides by microwave-assisted partial acid hydrolysis and mass spectrometry. *Rapid Commun. Mass Spectrom.* **2005**, *19*, 1545–1550.
243. Juhasz, P.; Martin, S. A. The utility of nonspecific proteases in the characterization of glycoproteins by high-resolution time-of-flight mass spectrometry. *Int. J. Mass Spectrom. Ion Processes* **1997**, *169/170*, 217–230.
244. Coddeville, B.; Girardet, J.-M.; Plancke, Y.; Campagna, S.; Linden, G.; Spik, G. Structure of the *O*-glycopeptides isolated from bovine milk component PP3. *Glycoconjugate J.* **1998**, *15*, 371–378.
245. An, H. J.; Peavy, T. R.; Hedrick, J. L.; Lebrilla, C. B. Determination of *N*-glycosylation sites and site heterogeneity in glycoproteins. *Anal. Chem.* **2003**, *75*, 5628–5637.
246. Davies, M. J.; Hounsell, E. F. HPLC and HPAEC of oligosaccharides and glycopeptides. *Methods Mol. Biol.* **1998**, *76*, 79–100.
247. Bendiac, B.; Cumming, D. A. Hydrazinolysis-*N*-reacetylation of glycopeptides and glycoproteins. Model studies using 2-acetamido-1-*N*-(L-aspart-4-oyl)-2-deoxy-α-D-glucopyranosylamine. *Carbohydrate Res.* **1985**, *144*, 1–12.
248. Takasaki, S.; Misuochi, T.; Kobata, A. Hydrazinolysis of asparagine-linked sugar chains to produce free oligosaccharides. *Methods Enzymol.* **1982**, *83*, 263–268.
249. Patel, T.; Bruce, J.; Merry, A.; Bigge, C.; Wormald, M.; Jaques, A.; Parekh, R. Use of hydrazine to release in intact and unreduced form both N- and O-linked oligosaccharides from glycoproteins. *Biochemistry* **1993**, *32*, 679–693.
250. Merry, A. H.; Neville, D. C. A.; Royle, L.; Matthews, B.; Harvey, D. J.; Dwek, R. A.; Rudd, P. M. Recovery of intact 2-aminobenzamide-labeled *O*-glycans released from glycoproteins by hydrazinolysis. *Anal. Biochem.* **2002**, *304*, 91–99.
251. Tanabe, K.; Ikenaka, K. In-column removal of hydrazine and *N*-acetylation of oligosaccharides released by hydrazinolysis. *Anal. Biochem.* **2006**, *348*, 324–326.
252. Tarentino, A. L.; Gómez, C. M.; Plummer, T. H. Jr. Deglycosylation of asparagine-linked glycans by peptide:*N*-glycosidase F. *Biochemistry* **1985**, *24*, 4665–5671.
253. Küster, B.; Harvey, D. J. Ammonium-containing buffers should be avoided during enzymatic release of

glycans from glycoproteins when followed by reducing terminal derivatization. *Glycobiology* **1997**, *7*, vii–ix.

254. Tretter, V.; Altmann, F.; März, L. Peptide-N_4-(*N*-acetyl-glucosaminyl)asparagine amidase F cannot release glycans with fucose attached α-(1→3) to the asparagine-linked *N*-acetylglucosamine residue. *Eur. J. Biochem.* **1991**, *199*, 647–652.
255. Chen, P.; Novotny, M. V. 2-Methyl-3-oxo-4-phenyl-2,3-dihydrofuran-2-yl acetate: A fluorogenic reagent for detection and analysis of primary amines. *Anal. Chem.* **1997**, *69*, 2806–2811.
256. Charlwood, J.; Skehel, J. M.; Camilleri, P. Immobilisation of antibodies in gels allows the improved release and identification of glycans. *Proteomics* **2001**, *1*, 275–284.
257. Mills, P. B.; Mills, K.; Johnson, A. W.; Clayton, P. T.; Winchester, B. G. Analysis by matrix assisted laser desorption/ionisation–time of flight mass spectrometry of the post-translational modifications of α_1-antitrypsin isoforms separated by two-dimensional polyacrylamide gel electrophoresis. *Proteomics* **2001**, *1*, 778–786.
258. Charlwood, J.; Skehel, J. M.; Camilleri, P. Analysis of N-linked oligosaccharides released from glycoproteins separated by two-dimensional gel electrophoresis. *Anal. Biochem.* **2000**, *284*, 49–59.
259. Charlwood, J.; Bryant, D.; Skehel, J. M.; Camilleri, P. Analysis of *N*-linked oligosaccharides: Progress towards the characterization of glycoprotein-linked carbohydrates. *Biomol. Eng.* **2001**, *18*, 229–240.
260. Zhou, Q.; Park, S.-H.; Boucher, S.; Higgins, E.; Lee, K.; Edmunds, T. *N*-Linked oligosaccharide analysis of glycoprotein bands from isoelectric focusing gels. *Anal. Biochem.* **2004**, *335*, 10–16.
261. Callewaert, N.; Vervecken, W.; Van Hecke, A.; Contreras, R. Use of a meltable polyacrylamide matrix for sodium dodecyl sulfate–polyacrylamide gel electrophoresis in a procedure for *N*-glycan analysis on picomole amounts of glycoproteins. *Anal. Biochem.* **2002**, *303*, 93–95.
262. Kolarich, D.; Altmann, F. *N*-Glycan analysis by matrix-assisted laser desorption/ionization mass spectrometry of electrophoretically separated nonmammalian proteins: Application to peanut allergen Ara h 1 and olive pollen allergen Ole e 1. *Anal. Biochem.* **2000**, *285*, 64–75.
263. Altmann, F.; Paschinger, K.; Dalik, T.; Vorauer, K. Characterisation of peptide-N^4-(*N*-acetyl-β-glucosaminyl)asparagine amidase A and its *N*-glycans. *Eur. J. Biochem.* **1998**, *252*, 118–123.
264. Guile, G. R.; Rudd, P. M.; Wing, D. R.; Prime, S. B.; Dwek, R. A. A rapid high-resolution high-performance liquid chromatographic method for separating glycan mixtures and analyzing oligosaccharide profiles. *Anal. Biochem.* **1996**, *240*, 210–226.
265. Rudd, P. M.; Guile, G. R.; Küster, B.; Harvey, D. J.; Opdenakker, G.; Dwek, R. A. Oligosaccharide sequencing technology. *Nature* **1997**, *388*, 205–207.
266. Hase, S. Analysis of sugar chains by pyridylamination. *Methods Mol. Biol.* **1993**, *14*, 69–80.
267. Kuraya, N.; Hase, S. Analysis of pyridylaminated *O*-linked sugar chains by two-dimensional sugar mapping. *Anal. Biochem.* **1996**, *233*, 205–211.
268. Otake, Y.; Fujimoto, I.; Tanaka, F.; Nakagawa, T.; Ikeda, T.; Menon, K. K.; Hase, S.; Wada, H.; Ikenaka, K. Isolation and characterization of an *N*-linked oligosaccharide that is significantly increased in sera from patients with non-small cell lung cancer. *J. Biochem. (Tokyo)* **2001**, *129*, 537–542.
269. Viseux, N.; Hironowski, X.; Delaney, J.; Domon, B. Qualitative and quantitative analysis of the glycosylation pattern of recombinant proteins. *Anal. Chem.* **2001**, *73*, 4755–4762.
270. Dwek, R. A.; Edge, C. J.; Harvey, D. J.; Wormald, M. R.; Parekh, R. B. Analysis of glycoprotein-associated oligosaccharides. *Annu. Rev. Biochem.* **1993**, *62*, 65–100.
271. Sutton, C. W.; O'Neill, J. A.; Cottrell, J. S. Site-specific characterization of glycoprotein carbohydrates by exoglycosidase digestion and laser desorption mass spectrometry. *Anal. Biochem.* **1994**, *218*, 34–46.
272. Harvey, D. J.; Rudd, P. M.; Bateman, R. H.; Bordoli, R. S.; Howes, K.; Hoyes, J. B.; Vickers, R. G. Examination of complex oligosaccharides by matrix-assisted laser desorption/ionization mass spectrometry on time-of-flight and magnetic sector instruments. *Org. Mass Spectrom.* **1994**, *29*, 753–765.
273. Mechref, Y.; Novotny, M. V. Mass spectrometric mapping and sequencing of N-linked oligosaccharides derived from submicrogram amounts of glycoproteins. *Anal. Chem.* **1998**, *70*, 455–463.
274. Küster, B.; Naven, T. J. P.; Harvey, D. J. Rapid approach for sequencing neutral oligosaccharides by exoglycosidase digestion and matrix-assisted laser desorption/ionization time-of-flight mass spectrometry. *J. Mass Spectrom.* **1996**, *31*, 1131–1140.
275. Colangelo, J.; Orlando, R. On-target exoglycosidase digestions, MALDI-MS for determining the primary structures of carbohydrate chains. *Anal. Chem.* **1999**, *71*, 1479–1482.
276. Schmitt, S.; Glebe, D.; Alving, K.; Tolle, T. K.; Linder, M.; Geyer, H.; Linder, D.; Peter-Katalinic, J.; Gerlich, W. H.; Geyer, R. Analysis of the pre-S2 N- and O-linked glycans of the M surface protein from human hepatitis B virus. *J. Biol. Chem.* **1999**, *274*, 11945–11957.
277. Geyer, H.; Schmitt, S.; Wuhrer, M.; Geyer, R. Structural analysis of glycoconjugates by on-target enzymatic digestion and MALDI-TOF-MS. *Anal. Chem.* **1999**, *71*, 476–482.
278. Chai, W.; Piskarev, V.; Lawson, A. M. Negative-ion electrospray mass spectrometry of neutral underivatized oligosaccharides. *Anal. Chem.* **2001**, *73*, 651–657.
279. Chai, W.; Piskarev, V.; Lawson, A. M. Branching pattern and sequence analysis of underivatized

oligosaccharides by combined MS/MS of singly and doubly charged molecular ions in negative-ion electrospray mass spectrometry. *J. Am. Soc. Mass Spectrom.* **2002**, *13*, 670–679.

280. SAGI, D.; PETER-KATALINIC, J.; CONRADT, H. S.; NIMTZ, M. Sequencing of tri- and tetraantennary *N*-glycans containing sialic acid by negative mode ESI QTOF tandem MS. *J. Am. Soc. Mass Spectrom.* **2002**, *13*, 1138–1148.

281. WHEELER, S. F.; HARVEY, D. J. Negative ion mass spectrometry of sialylated carbohydrates: Discrimination of *N*-acetylneuraminic acid linkages by matrix-assisted laser desorption/ionization-time-of-flight and electrospray-time-of-flight mass spectrometry. *Anal. Chem.* **2000**, *72*, 5027–5039.

282. HARVEY, D. J. Fragmentation of negative ions from carbohydrates: Part 2, Fragmentation of high-mannose N-linked glycans. *J. Am. Soc. Mass Spectrom.* **2005**, *16*, 631–646.

283. HARVEY, D. J. Fragmentation of negative ions from carbohydrates: Part 3, Fragmentation of hybrid and complex N-linked glycans. *J. Am. Soc. Mass Spectrom.* **2005**, *16*, 647–659.

284. HOUNSELL, E. F.; DAVIES, M. J.; RENOUF, D. V. O-Linked protein glycosylation structure and function. *Glycoconjugate J.* **1996**, *13*, 19–26.

285. COOPER, C. A.; WILKINS, M. R.; WILLIAMS, K. L.; PACKER, N. H. BOLD—A biological O-linked glycan database. *Electrophoresis* **1999**, *20*, 3589–3598.

286. MIRGORODSKAYA, E.; HASSAN, H.; WANDALL, H. H.; CLAUSEN, H.; ROEPSTORFF, P. Partial vapor-phase hydrolysis of peptide bonds: A method for mass spectrometric determination of *O*-glycosylated sites in glycopeptides. *Anal. Biochem.* **1999**, *269*, 54–65.

287. HUANG, Y.; MECHREF, Y.; NOVOTNY, M. V. Microscale nonreductive release of O-linked glycans for subsequent analysis through MALDI mass spectrometry and capillary electrophoresis. *Anal. Chem.* **2001**, *73*, 6063–6069.

288. HUANG, Y.; KONSE, T.; MECHREF, Y.; NOVOTNY, M. V. Matrix-assisted laser desorption/ionization mass spectrometry compatible β-elimination of O-linked oligosaccharides. *Rapid Commun. Mass Spectrom.* **2002**, *16*, 1199–1204.

289. ROBBE, C.; CAPON, C.; FLAHAUT, C.; MICHALSKI, J.-C. Microscale analysis of mucin-type *O*-glycans by a coordinated fluorophore-assisted carbohydrate electrophoresis and mass spectrometry approach. *Electrophoresis* **2003**, *24*, 611–621.

290. HANISCH, F. G.; JOVANOVIC, M.; PETER-KATALINIC, J. Glycoprotein identification and localization of *O*-glycosylation sites by mass spectrometric analysis of deglycosylated/alkylaminylated peptide fragments. *Anal. Biochem.* **2001**, *290*, 47–59.

291. TAYLOR, A. M.; HOLST, O.; THOMAS-OATES, J. Mass spectrometric profiling of O-linked glycans released directly from glycoproteins in gels using in-gel reductive β-elimination. *Proteomics* **2006**, *6*, 2936–2946.

292. XIE, Y.; LIU, J.; ZHANG, J.; HEDRICK, J. L.; LEBRILLA, C. B. Method for the comparative glycomic analyses of O-linked, mucin-type oligosaccharides. *Anal. Chem.* **2004**, *76*, 5186–5197.

293. MÜLLER, S.; GOLETZ, S.; PACKER, N.; GOOLEY, A.; LAWSON, A. M.; HANISCH, F.-G. Localization of *O*-glycosylation sites on glycopeptide fragments from lactation-associated MUC1. All putative sites within the tandem repeat are glycosylation targets in vivo. *J. Biol. Chem.* **1997**, *272*, 24780–24793.

294. GOLETZ, S.; THIEDE, B.; HANISCH, F.-G.; SCHULTZ, M.; PETER-KATALINIC, J.; MÜLLER, S.; SEITZ, O.; KARSTEN, U. A sequencing strategy for the localization of *O*-glycosylation sites of MUC1 tandem repeats by PSD-MALDI mass spectrometry. *Glycobiology* **1997**, *7*, 881–896.

295. GOLETZ, S.; LEUCK, M.; FRANKE, P.; KARSTEN, U. Structure analysis of acetylated and non-acetylated O-linked MUC1-glycopeptides by post-source decay matrix-assisted laser desorption/ionization mass spectrometry. *Rapid Commun. Mass Spectrom.* **1997**, *11*, 1387–1398.

296. ALVING, K.; PAULSEN, H.; PETER-KATALINIC, J. Characterization of *O*-glycosylation sites in MUC2 glycopeptides by nanoelectrospray QTOF mass spectrometry. *J. Mass Spectrom.* **1999**, *34*, 395–407.

297. MIRGORODSKAYA, E.; ROEPSTORFF, P.; ZUBAREV, R. A. Localization of *O*-glycosylation sites in peptides by electron capture dissociation in a Fourier transform mass spectrometer. *Anal. Chem.* **1999**, *71*, 4431–4436.

298. CZESZAK, X.; MORELLE, W.; RICART, G.; TÉTAERT, D.; LEMOINE, J. Localization of the *O*-glycosylated sites in peptides by fixed-charge derivatization with a phosphonium group. *Anal. Chem.* **2004**, *76*, 4320–4324.

299. MORELLE, W.; GUYETANT, R.; STRECKER, G. Structural analysis of oligosaccharide-alditols released by reductive β-elimination from the oviductal mucins of *Rana dalmatia. Carbohydrate Res.* **1998**, *306*, 435–443.

300. MORELLE, W.; STRECKER, G. Structural analysis of hexa to dodecaoligosaccharide-alditols released by reductive β-elimination from oviducal mucins of *Bufo bufo. Glycobiology* **1997**, *7*, 1129–1151.

301. MORELLE, W.; STRECKER, G. Structural analysis of oligosaccharide-alditols released by reductive β-elimination from oviducal mucins of *Bufo bufo*: Characterization of the carbohydrate sequence Gal(α1-3)GalNAc(α1-3)[Fuc(α1-2)]Gal. *Glycobiology* **1997**, *7*, 777–790.

302. MORELLE, W.; STRECKER, G. Structural analysis of a new series of oligosaccharide-alditols released by reductive β-elimination from oviducal mucins of *Rana utricularia. Biochem. J.* **1998**, *330*, 469–478.

303. MORELLE, W.; CABADA, M. O.; STRECKER, G. Structural analysis of oligosaccharide-alditols released by reductive β-elimination from the jelly coats of the

anuran *Bufo arenarum*. *Eur. J. Biochem.* **1998**, *252*, 253–260.

304. Tseng, K.; Xie, Y.; Seeley, J.; Hedrick, J. L.; Lebrilla, C. B. Profiling with structural elucidation of the neutral and anionic O-linked oligosaccharides in the egg jelly coat of *Xenopus laevis* by Fourier transform mass spectrometry. *Glycoconjugate J.* **2001**, *18*, 309–320.
305. Tseng, K.; Hedrick, J. L.; Lebrilla, C. B. Catalog-library approach for the rapid and sensitive structural elucidation of oligosaccharides. *Anal. Chem.* **1999**, *71*, 3747–3754.
306. Xie, Y.; Tseng, K.; Lebrilla, C. B.; Hedrick, J. L. Targeted use of exoglycosidase digestion for the structural elucidation of neutral O-linked oligosaccharides. *J. Am. Soc. Mass Spectrom.* **2001**, *12*, 877–884.
307. Zhang, J.; Lindsay, L. L.; Hedrick, J. L.; Lebrilla, C. B. Strategy for profiling and structure elucidation of mucin-type oligosaccharides by mass spectrometry. *Anal. Chem.* **2004**, *76*, 5990–6001.
308. Karlsson, N. G.; Schulz, B. L.; Packer, N. H. Structural determination of neutral O-linked oligosaccharide alditols by negative ion LC-electrospray-MS^n. *J. Am. Soc. Mass Spectrom.* **2004**, *15*, 659–672.
309. Zhang, J.; Schubothe, K.; Li, B.; Russell, S.; Lebrilla, C. B. Infrared multiphoton dissociation of O-linked mucin-type oligosaccharides. *Anal. Chem.* **2005**, *77*, 208–214.
310. van Baar, B. L. M. Characterisation of bacteria by matrix-assisted laser desorption/ionization and electrospray mass spectrometry. *FEMS Microbiol. Rev.* **2000**, *24*, 193–219.
311. El Hamidi, A.; Tirsoaga, A.; Novikov, A.; Hussein, A.; Caroff, M. Microextraction of bacterial lipid A: easy and rapid method for mass spectrometric characterization. *J. Lipid Res.* **2005**, *46*, 1773–1778.
312. Sturiale, L.; Garozzo, D.; Silipo, A.; Lanzetta, R.; Parrilli, M.; Molinaro, A. New conditions for matrix-assisted laser desorption/ionization mass spectrometry of native bacterial R-type lipopolysaccharides. *Rapid Commun. Mass Spectrom.* **2005**, *19*, 1829–1834.
313. Fukuoka, S.; Knirel, Y. A.; Lindner, B.; Moll, H.; Seydel, U.; Zähringer, U. Elucidation of the structure of the core region and the complete structure of the R-type lipopolysaccharide of *Erwinia carotovora* FERM P-7576. *Eur. J. Biochem.* **1997**, *250*, 55–62.
314. Schwudke, D.; Linscheid, M.; Strauch, E.; Appel, B.; Zähringer, U.; Moll, H.; Müller, M.; Brecker, L.; Gronow, S.; Lindner, B. The obligate predatory *Bdellovibrio bacteriovorus* possesses a neutral lipid A containing α-D-mannoses that replace phosphate residues: Similarities and differences between the lipid As and the lipopolysaccharides of the wild type strain *B. bacteriovorus* HD100 and its host-independent derivative HI100. *J. Biol. Chem.* **2003**, *278*, 27502–27512.
315. Linnerborg, M.; Weintraub, A.; Widmalm, G. Structural studies of the O-antigen polysaccharide from the enteroinvasive *Escherichia coli* O164 cross-reacting with *Shigella dysenteriae* type 3. *Eur. J. Biochem.* **1999**, *266*, 460–466.
316. Gibson, B. W.; Engstrom, J. J.; John, C. M.; Hines, W.; Falick, A. M. Characterization of bacterial lipooligosaccharides by delayed extraction matrix-assisted laser desorption ionization time-of-flight mass spectrometry. *J. Am. Soc. Mass Spectrom.* **1997**, *8*, 645–658.
317. Rund, S.; Lindner, B.; Brade, H.; Holst, O. Structural analysis of the lipopolysaccharide from *Chlamydia trachomatis* serotype L2. *J. Biol. Chem.* **1999**, *274*, 16819–16824.
318. Olsthoorn, M. M. A.; Haverkamp, J.; Thomas-Oates, J. E. Mass spectrometric analysis of *Klebsiella pneumoniae* ssp. *pneumoniae* rough strain R20 (O1 (−): K20(−)) lipopolysaccharide preparations: Identification of novel core oligosaccharide components and three 3-deoxy-D-*manno*-oct-2-ulopyranosonic artefacts. *J. Mass Spectrom.* **1999**, *34*, 622–636.
319. Seydel, U.; Lindner, B.; Wollenweber, H.-W.; Rietschel, E. T. Structural studies on the lipid component of enterobacterial lipopolysaccharides by laser desorption mass spectrometry. Location of acyl groups on the lipid A backbone. *Eur. J. Biochem.* **1984**, *145*, 505–509.
320. Allmaier, G.; Schmid, E. R.; Hagspiel, K.; Kubicek, C. P.; Karas, M.; Hillenkamp, F. Strategy for the characterization of the glycoprotein endoglucanase I isolated from *Trichoderma reesei*: Combination of plasma desorption and UV laser desorption time-of-flight mass spectrometry. *Anal. Chim. Acta* **1990**, *241*, 321–327.
321. Kulshin, V. A.; Zähringer, U.; Lindner, B.; Frasch, C. E.; Tsai, C.-M.; Dmitriev, B. A.; Rietschel, E. T. Structural characterization of the lipid A component of pathogenic *Neisseria meningitides*. *J. Bacteriol.* **1992**, *174*, 1793–1800.
322. Moran, A. P.; Lindner, B.; Walsh, E. J. Structural characterization of the lipid A component of *Helicobacter pylori* rough- and smooth-form lipopolysaccharides. *J. Bacteriol.* **1997**, *179*, 6453–6463.
323. Aussel, L.; Brisson, J.-R.; Perry, M. B.; Caroff, M. Structure of the lipid A of *Bordetella hinzii* ATCC 51730. *Rapid Commun. Mass Spectrom.* **2000**, *14*, 595–599.
324. Kaltashov, I. A.; Doroshenko, V.; Cotter, R. J.; Takayama, K.; Qureshi, N. Confirmation of the structure of lipid A derived from the lipopolysaccharide of *Rhodobacter sphaeroides* by a combination of MALDI, LSIMS and tandem mass spectrometry. *Anal. Chem.* **1997**, *69*, 2317–2322.
325. White, K. A.; Kaltashov, I. A.; Cotter, R. J.; Raetz, C. R. H. A mono-functional 3-deoxy-D-manno-octulosonic acid (Kdo) transferase and a Kdo kinase in extracts of *Haemophilus influenzae*. *Journal of Biological Chemistry* **1997**, *272*, 16555–16563.

326. Brabetz, W.; Lindner, B.; Brade, H. Comparative analysis of secondary gene products of 3-deoxy-D-*manno*-oct-2-ulosonic acid transferases from *Chlamydiaceae* in *Escherichia coli* K-12. *Eur. J. Biochem.* **2000**, *267*, 5458–5465.
327. Rund, S.; Lindner, B.; Brade, H.; Holst, O. Structural analysis of the lipopolysaccharide from *Chlamydophila psittaci* strain 6BC. *Eur. J. Biochem.* **2000**, *267*, 5717–5726.
328. Casabuono, A. C.; D'Antuono, A.; Sato, Y.; Nonami, H.; Ugalde, R.; Lepek, V.; Erra-Balsells, R.; Couto, A. S. A matrix-assisted laser desorption/ionization mass spectrometry approach to the lipid A from *Mesorhizobium loti. Rapid Commun. Mass Spectrom.* **2006**, *20*, 2175–2182.
329. Corsaro, M. M.; Dal Piaz, F.; Lanzetta, R.; Naldi, T.; Parrilli, M. Structure of lipid A from *Pseudomonas corrugata* by electrospray ionization quadrupole time-of-flight tandem mass spectrometry. *Rapid Commun. Mass Spectrom.* **2004**, *18*, 853–858.
330. Bedoux, G.; Vallée-Réhel, K.; Kooistra, O.; Zähringer, U.; Haras, D. Lipid A components from *Pseudomonas aeruginosa* PAO1 (serotype O5) and mutant strains investigated by electrospray ionization ion-trap mass spectrometry. *J. Mass Spectrom.* **2004**, *39*, 505–513.
331. Lee, C.-S.; Kim, Y.-G.; Joo, H.-S.; Kim, B.-G. Structural analysis of lipid A from *Escherichia coli* O157:H7:K-using thin-layer chromatography and ion-trap mass spectrometry. *J. Mass Spectrom.* **2004**, *39*, 514–525.
332. Czaja, J.; Jachymek, W.; Niedziela, T.; Lugowski, C.; Aldova, E.; Kenne, L. Structural studies of the O-specific polysaccharide from *Plesiomonas shigelloides* strain CNCTC 113/92. *Eur. J. Biochem.* **2000**, *267*, 1672–1679.
333. Larocque, S.; Brisson, J.-R.; Thérisod, H.; Perry, M. B.; Caroff, M. Structural characterization of the O-chain polysaccharide isolated from *Bordetella avium* ATCC 5086: variation on a theme. *FEBS Lett.* **2003**, *535*, 11–16.
334. De Castro, C.; Bedini, E.; Garozzo, D.; Sturiale, L.; Parrilli, M. Structural determination of the O-chain moieties of the lipopolysaccharide fraction from *Agrobacterium radiobacter* DSM. *Eur. J. Org. Chem.* **2004**, 3842–3849.
335. Vinogradov, E.; Caroff, M. Structure of the *Bordetella trematum* LPS O-chain subunit. *FEBS Lett.* **2005**, *579*, 18–24.
336. Egge, H.; Peter-Katalinic, J. Fast atom bombardment mass spectrometry for structural elucidation of glycoconjugates. *Mass Spectrom. Rev.* **1987**, *6*, 331–393.
337. Merrill, A. H. J.; Sullards, M. C.; Allegood, J. C.; Kelly, S.; Wang, E. Sphingolipidomics: High-throughput, structure-specific, and quantitative analysis of sphingolipids by liquid chromatography tandem mass spectrometry. *Methods* **2005**, *36*, 207–224.
338. Juhasz, P.; Costello, C. E. Matrix-assisted laser desorption ionization time-of-flight mass spectrometry of underivatized and permethylated gangliosides. *J. Am. Soc. Mass Spectrom.* **1992**, *3*, 785–796.
339. O'Connor, P. B.; Costello, C. E. A high pressure matrix-assisted laser desorption/ionization Fourier transform mass spectrometry ion source for thermal stabilization of labile biomolecules. *Rapid Commun. Mass Spectrom.* **2001**, *15*, 1862–1868.
340. O'Connor, P. B.; Mirgorodskaya, E.; Costello, C. E. High pressure matrix-assisted laser desorption/ionizaton Fourier transform mass spectrometry for minimization of ganglioside fragmentation. *J. Am. Soc. Mass Spectrom.* **2002**, *13*, 402–407.
341. Vukelic, Žarei M.; Peter-Katalinic, J.; Zamfir, A. D. Analysis of human hippocampus gangliosides by fully-automated chip-based nanoelectrospray tandem mass spectrometry. *J. Chromatogr. A* **2006**, *1130*, 238–245.
342. McFarland, M. A.; Marshall, A. G.; Hendrickson, C. L.; Nilsson, C. L.; Fredman, P.; Månsson, J.-E. Structural characterization of the GM1 ganglioside by infrared multiphoton dissociation, electron capture dissociation, and electron detachment dissociation electrospray ionization FT-ICR MS/MS. *J. Am. Soc. Mass Spectrom.* **2005**, *16*, 752–762.
343. Vukelic, Žamfir A. D.; Bindila, L.; Froesch, M.; Peter-Katalinic, J.; Usuki, S.; Yu, R. K. Screening and sequencing of complex sialylated and sulfated glycosphingolipid mixtures by negative ion electrospray Fourier transform ion cyclotron resonance mass spectrometry. *J. Am. Soc. Mass Spectrom.* **2005**, *16*, 571–580.
344. Guittard, J.; Hronowski, X.; Costello, C. E. Direct matrix-assisted laser desorption/ionization mass spectrometric analysis of glycosphingolipids on thin layer chromatographic plates and transfer membranes. *Rapid Commun. Mass Spectrom.* **1999**, *13*, 1838–1849.
345. Ivleva, V. B.; Sapp, L. M.; O'Connor, P. B.; Costello, C. E. Ganglioside analysis by thin-layer chromatography matrix-assisted laser desorption/Ionization orthogonal time-of-flight mass spectrometry. *J. Am. Soc. Mass Spectrom.* **2005**, *16*, 1552–1560.
346. Dreisewerd, K.; Müthing, J.; Rohlfing, A.; Meisen, I.; Vukelic, Ž.; Peter-Katalinic, J.; Hillenkamp, F.; Berkenkamp, S. Analysis of gangliosides directly from thin-layer chromatography plates by infrared matrix-assisted laser desorption/ionization orthogonal time-of-flight mass spectrometry with a glycerol matrix. *Anal. Chem.* **2005**, *77*, 4098–4107.
347. Ivleva, V. B.; Elkin, Y. N.; Budnik, B. A.; Moyer, S. C.; O'Connor, P. B.; Costello, C. E. Coupling thin-layer chromatography with vibrational cooling matrix-assisted laser desorption/ionization Fourier transform mass spectrometry for the analysis of ganglioside mixtures. *Anal. Chem.* **2004**, *76*, 6484–6491.

348. Nakamura, K.; Suzuki, Y.; Goto-Inoue, N.; Yoshida-Noro, C.; Suzuki, A. Structural characterization of neutral glycosphingolipids by thin-layer chromatography coupled to matrix-assisted laser desorption/ionization quadrupole ion trap time-of-flight MS/MS. *Anal. Chem.* **2006**, *78*, 5736–5743.
349. Wing, D. R.; Garner, B.; Hunnam, V.; Reinkensmeier, G.; Andersson, U.; Harvey, D. J.; Dwek, R. A.; Platt, F. M.; Butters, T. D. High performance liquid chromatography analysis of ganglioside carbohydrates at the picomol level after ceramide glycanase digestion and fluorescent labelling with 2-aminobenzamide. *Anal. Biochem.* **2001**, *298*, 207–217.
350. Wuhrer, M.; Kantelhardt, S. R.; Dennis, R. D.; Doenhoff, M. J.; Lochnit, G.; Geyer, R. Characterization of glycosphingolipids from *Schistosoma mansoni* eggs carrying Fuc(α1-3) GalNAc, GalNAc(β1-4)[Fuc(α1-3)]GlcNAc- and Gal (β1-4)[Fucα1-3)]GlcNAc- (Lewis X) terminal structures. *Eur. J. Biochem.* **2002**, *269*, 481–493.
351. Kantelhardt, S. R.; Wuhrer, M.; Dennis, R. D.; Doenhoff, M. J.; Bickle, Q.; Geyer, R. Fuc(α1→3) GalNAc-: The major antigenic motif of *Schistosoma mansoni* glycolipids implicated in infection sera and keyhole-limpet haemocyanin cross-reactivity. *Biochem. J.* **2002**, *366*, 217–223.
352. Suzuki, M.; Suzuki, A. Structural characterization of fucose-containing oligosaccharides by high-performance liquid chromatography and matrix-assisted laser desorption/ionization time-of-flight mass spectrometry. *Biol. Chem.* **2001**, *382*, 251–257.
353. Baigude, H.; Katsuraya, K.; Okuyama, K.; Yachi, Y.; Sato, S.; Uryu, T. Synthesis of dicarboxylate oligosaccharide multilayer terminal functionality upon poly(lysine) dendrimer scaffolding. *J. Polymer Sci. Part A* **2002**, *40*, 3622–3633.
354. Woller, E. K.; Cloninger, M. J. Mannose functionalization of a sixth generation dendrimer. *Biomacromolecules* **2001**, *2*, 1052–1054.
355. Woller, E. K.; Cloninger, M. J. The lectin-binding properties of six generations of mannose-functionalized dendrimers. *Org. Lett.* **2002**, *4*, 7–10.
356. Baytekin, B.; Werner, N.; Luppertz, F.; Engeser, M.; Bruggemann, J.; Bitter, S.; Henkel, R.; Felder, T.; Schalley, C. A. How useful is mass spectrometry for the characterization of dendrimers? *Int J. Mass Spectrom.* **2006**, *249–250*, 138–148.
357. Siegel, M. M.; Tsou, H.-R.; Lin, B.; Hollander, I. J.; Wissner, A.; Karas, M.; Ingendoh, A.; Hillenkamp, F. Determination of the loading values for high levels of drugs and sugars conjugated to proteins by matrix assisted ultraviolet laser desorption/ionization mass spectrometry. *Biol. Mass Spectrom.* **1993**, *22*, 369–376.
358. Tanaka, H.; Putalun, W.; Tsuzaki, C.; Shoyama, Y. A simple determination of steroidal alkaloid glycosides by thin-layer chromatography immunostaining using monoclonal antibody against solamargine. *FEBS Lett.* **1997**, *404*, 279–282.

Chapter 20

Applications of ESI and MALDI to Lipid Analysis

Xianlin Han

Division of Bioorganic Chemistry and Molecular Pharmacology, Department of Medicine, Washington University School of Medicine, St. Louis, Missouri

Electrospray and MALDI Mass Spectrometry: Fundamentals, Instrumentation, Practicalities, and Biological Applications, Second Edition, Edited by Richard B. Cole

20.1 INTRODUCTION

20.1.1 Definition and Classification of Lipids

Lipids can be broadly defined as a group of organic compounds in living organisms, most of which are insoluble in water, but soluble in nonpolar solvents. Therefore, based on this definition, petroleum products from fossil materials or synthetic organic compounds must be excluded from this category. Lipids are the main constituents of plant and animal cell membranes and are major components of lipoproteins in serum. Some lipids are conjugated with carbohydrates, which are known as lipopolysaccharides. Most lipids are comprised of two domains. One domain is largely hydrophobic, meaning that it is not generally solvated in polar solvents (e.g., water) while the other is often polar or hydrophilic, which is readily soluble in polar solvents. In biological cells, most lipids are amphiphilic molecules (i.e., having both hydrophobic and hydrophilic portions). Prominent exceptions to this include (a) cholesterol, which is predominantly nonpolar except for its hydroxyl group, and (b) triacylglycerols, which are largely comprised of nonpolar aliphatic groups of varying chain lengths and degrees of unsaturation.

Lipids play many essential roles in living organisms, including (1) the constitution of cellular membranes which provide hydrophobic barriers to separate cellular compartments, (2) serving as an optimal matrix to facilitate transmembrane protein function, (3) a source for the precursors of lipid secondary messengers during signal transduction, and (4) a reservoir of fuel for biological processes. Increasingly, experimental results support a rationale that lipids are associated with many human diseases (e.g., diabetes and obesity, atherosclerosis and stroke, cancer, psychiatric disorders, neurodegenerative diseases and neurological disorders, as well as infectious diseases). Therefore, a new discipline in lipid research, called "lipidomics," is rapidly expanding owing to the recognition of the importance of lipids in multiple epidemic diseases and the necessity for a systems biology approach to study lipids.

Tens of thousands of possible lipid molecular species have been predicted to exist in the cellular lipidome at the level of attomole to nanomole lipid per milligram of protein. Fortunately, this large array of potential lipids has been categorized into a relatively small number of groups based upon similarities in their chemical structures. For example, individual lipid molecular species that possess an identical polar head group (e.g., phosphocholine, phosphoethanolamine, or phosphoserine attached to the *sn*-3 position of a glycerol backbone) are each categorized into a separate lipid class (e.g., choline glycerophospholipid (GPCho), ethanolamine glycerophospholipid (GPEtn), or serine glycerophospholipid (GPSer), respectively). A comprehensive classification system of lipid classes has recently been made available.[1]

Alternatively, lipids are also commonly divided into nonpolar lipids and polar lipids based on their overall hydrophobicity. The nonpolar lipids include fatty acids and their derivatives (e.g., long-chain alcohols and waxes), glycerol-derived acyl lipids (e.g., monoacylglycerols (MG), diacylglycerols (DG), triacylglycerols (i.e., fats or oils) (TG)), and steroids, while the polar lipids include glycerophospholipids (e.g., GPCho, GPEtn, and GPSer), sphingoid base-derived lipids (i.e., sphingolipids) (e.g., ceramides (Cer), cerebrosides (e.g., galactosylceramides, (GalCer), gangliosides, sphingomyelins (SM), and sulfatides (ST)).

In general, there are tens of thousands of potential lipid molecular species present in the cellular lipidome. Most cellular lipid molecular species can be simply expressed by a minimal number of general structures that are comprised of varying "building block" components. For example, all molecular species of all glycerol-based lipid classes (including MG, DG, TG, GPCho, GPEtn, GPEtn, and GPSer) are multiple discrete covalent assemblies of a backbone (i.e., glycerol) with combinations of various aliphatic chains (containing typically 14–22 carbons with variable degrees of unsaturation) and a wide variety of polar head groups (which can also be absent). Therefore, all molecular species of all glycerol-based lipid classes can be expressed by a general structure with three building blocks linked to a glycerol backbone as shown in Figure 20.1. In this structure, building blocks I and II can be either a hydrogen or an acyl chain or an aliphatic chain linked by an ester, ether, or vinyl ether bond. Building block III can vary from a hydrogen or acyl (or aliphatic) chain in MG and DG, to an acyl (aliphatic) chain in TG, to various sugar ring(s) and their derivatives in glycolipids, and to phosphoesters (e.g., phosphate, phosphocholine, phosphoethanolamine, phosphoglycerol, phosphoserine, and phosphoinositol) in phospholipids and lysophospholipids.

All molecular species containing a sphingoid base have now been collectively designated as the sphingolipidome.[2] A comprehensive classification and nomenclature of the sphingolipidome can be found at the website of www.sphingomap.com. Similar to glycerol-based molecular species, all sphingoid base molecular species can also be represented by a general structure with three building blocks (Figure 20.2). Building block I represents a different polar moiety (linked to the oxygen at the C1 position of the sphingoid base) including hydrogen, phosphoethanolamine, phosphocholine, galactose, glucose, lactose, sulfated galactose, and other complex sugar groups (corresponding to Cer, ceramide phosphoethanolamine, SM, GalCer, GluCer, lactosyl Cer, ST, and other glycosphingolipids such as gangliosides, respectively) (Figure 20.2). Building block II represents a fatty acyl moiety, which is acylated to the primary amine at the C2 position of the sphingoid base. A variety of fatty acyl chains including those which contain a hydroxyl group (usually located at the alpha or omega position) (Figure 20.2) can occupy this position. Building block III represents the aliphatic chain present in all sphingoid bases, which is linked through a carbon–carbon bond to the C3 position. This aliphatic chain varies in length, degree of unsaturation, presence of a branch in the chain, and the presence of an additional hydroxyl group (Figure 20.2). Thousands of possible sphingolipid molecular species can be readily constructed from the combination of these three building blocks.

A third important category of lipid compounds are the sterols, all of which possess a four-ringed structure (Figure 20.3A). The most abundant sterols in mammals are cholesterol and cholesterol esters (Figures 20.3B and 20.3C, respectively). These three general structures (glycerol based, sphingoids-based, and sterol), in addition to some building block components themselves such as free fatty acids and their derivatives (e.g., eicosanoids), cover most (if not all) of the lipid classes in a cellular lipidome. The contents of these lipid classes are estimated to account for over 95% of the total cellular lipid mass in most cases. Accordingly,

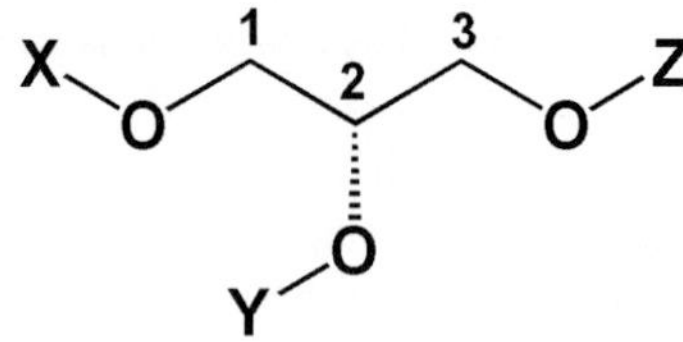

−X = **−H**

$-CH_2$-CH_2-R_1 (Defined as plasmanyl in phospholipids)

−CH=CH-R_1 (Defined as plasmenyl in phospholipids)

$-\overset{O}{\overset{\|}{C}}$-$CH_2$-$R_1$ (Defined as phosphatidyl in phospholipids)

R_1 is usually an unbranched saturated or unsatuarted aliphatic chain containing 12 to 20 carbons

....

−Y = **−H**

$-\overset{O}{\overset{\|}{C}}$-$R_2$

$-CH_2 R_2$

R_2 is usually an unbranched saturated or unsatuarted aliphatic chain containing 13 to 21 carbons

....

−Z = **−H**

$-\overset{O}{\overset{\|}{C}}$-$R_3$

$-CH_2 R_3$

R_3 is usually an unbranched saturated or unsatuarted aliphatic chain containing 13 to 21 carbons

HO, OH, O, HO, OH (inositol ring)

$-\overset{O}{\overset{\|}{P}}$-OH, O^-

$-\overset{O}{\overset{\|}{P}}$-$OCH_2CH_2NH_3^+$, O^-

$-\overset{O}{\overset{\|}{P}}$-$OCH_2CH_2NMe_3^+$, O^-

$-\overset{O}{\overset{\|}{P}}$-O–CH₂CH(OH)CH₂OH, O^-

$-\overset{O}{\overset{\|}{P}}$-O–CH₂CH($CO_2^-$)$NH_3^+$, O^-

....

Figure 20.1. General structure of glycerol-based lipids. Three building blocks (X, Y, and Z) are linked to the hydroxyl groups of a glycerol backbone. Potential candidates of the building blocks are listed.

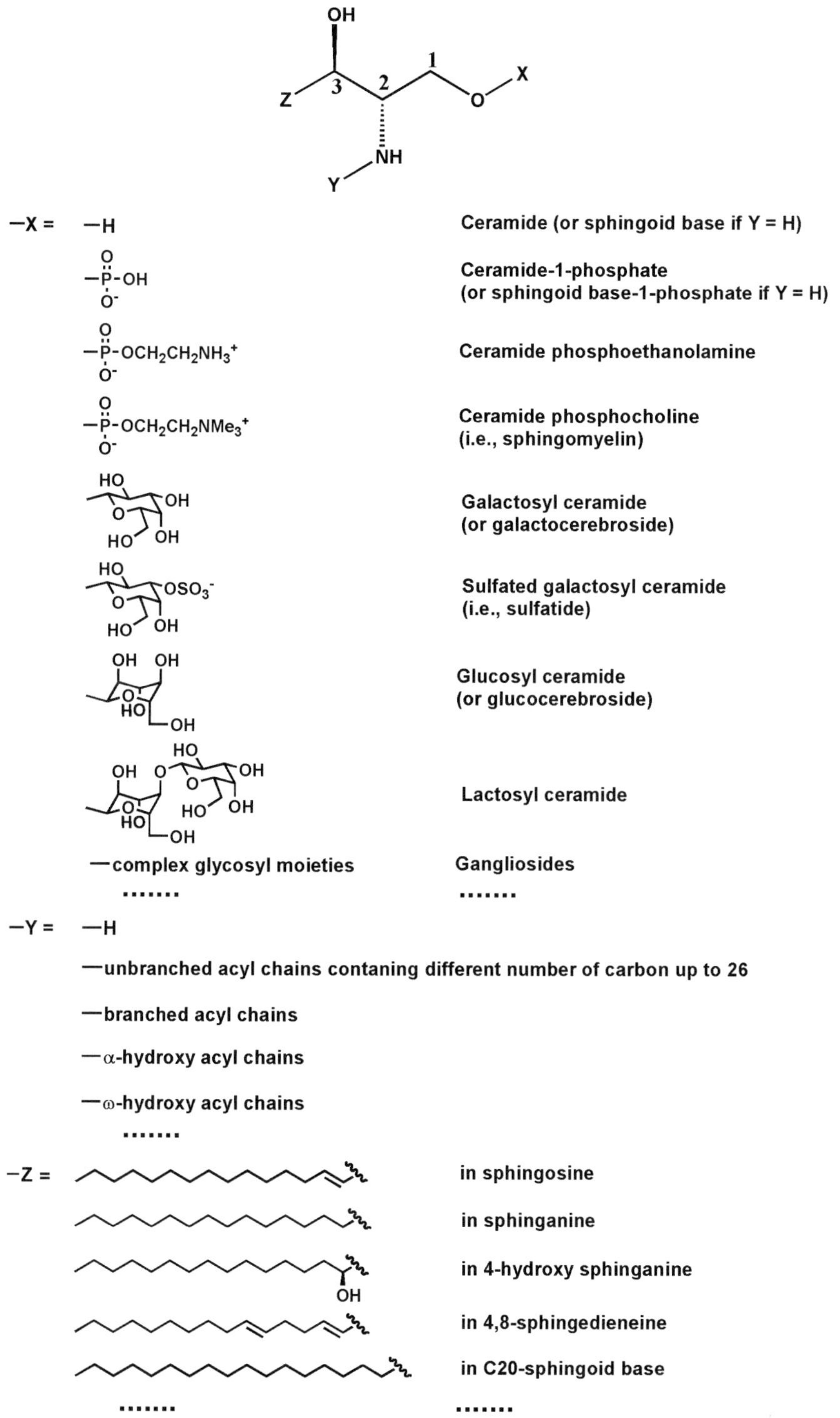

OH

Z 3 2 1 O X

NH

Y

−X = −H — Ceramide (or sphingoid base if Y = H)

$-P(=O)(O^-)-OH$ — Ceramide-1-phosphate (or sphingoid base-1-phosphate if Y = H)

$-P(=O)(O^-)-OCH_2CH_2NH_3^+$ — Ceramide phosphoethanolamine

$-P(=O)(O^-)-OCH_2CH_2NMe_3^+$ — Ceramide phosphocholine (i.e., sphingomyelin)

Galactosyl ceramide (or galactocerebroside)

Sulfated galactosyl ceramide (i.e., sulfatide)

Glucosyl ceramide (or glucocerebroside)

Lactosyl ceramide

—complex glycosyl moieties — Gangliosides

....

−Y = —H

—unbranched acyl chains contaning different number of carbon up to 26

—branched acyl chains

—α-hydroxy acyl chains

—ω-hydroxy acyl chains

....

−Z = in sphingosine

in sphinganine

in 4-hydroxy sphinganine

in 4,8-sphingedieneine

in C20-sphingoid base

....

Figure 20.2. General structure of sphingoid-based lipids. The building block I (X) represents a different polar moiety (linked to the oxygen at the C1 position of sphingoid base). The building block II (Y) represents fatty acyl chains (acylated to the primary amine at the C2 position of sphingoid base) with or without the presence of a hydroxyl group that is usually located at the alpha or omega position. The building block III (Z) represents the aliphatic chains in all of possible sphingoid bases, which are carbon–carbon-linked to the C3 position of sphingoid bases and vary with the aliphatic chain length, degree of unsaturation, the presence of branch, and the presence of an additional hydroxyl group.

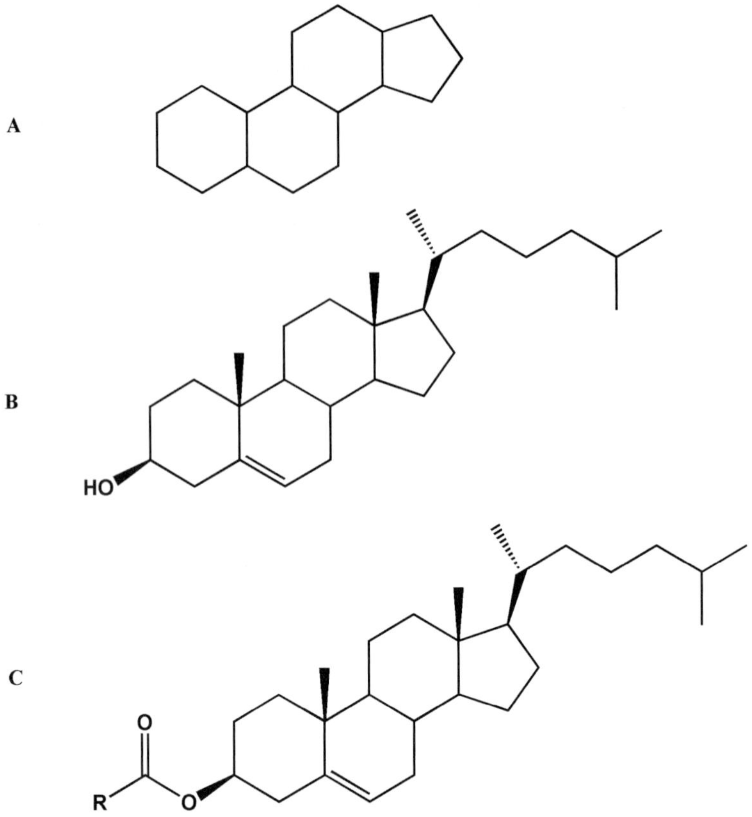

A

B

HO

C

O

R O

Figure 20.3. The structures of sterol skeleton (**A**), cholesterol (**B**), and cholesterol esters (**C**).

if one is able to identify and quantify the molecular species covered by these three general structures, the analysis of a given cellular lipidome (by mass) is largely achieved.

20.1.2 Brief History in Mass Spectrometry (MS) for Lipid Analysis

Electron ionization (EI) was used to study a variety of lipids since it was introduced in the late 1940s.[3] Chemical ionization (CI) made a major impact in the analysis of lipids by enhancing molecular weight information and facilitating specific structural details since it was introduced in the late 1960s.[4] A variety of other ionization techniques (e.g., thermospray ionization, field ionization, field desorption, plasma desorption, laser desorption, and fast atom bombardment (FAB)) have been applied in the past for the mass spectrometric (MS) analyses of lipids when they were introduced in the history of mass spectrometry. Each of these techniques, particularly FAB/MS, possessed unique advantages for lipid analysis at the time of their development and have contributed to the understanding of the ion chemistry of lipids. However, many of these ionization techniques also have limitations in their applications. Low sensitivity, high background noise from matrix, significant in-source fragmentation, and difficulty in operation are some primary drawbacks. Therefore, they have seldom been applied for lipid analyses in the recent literature and have been replaced by newly developed techniques.

It is noteworthy that FAB/MS has historically played an important role in the development of MS-based intact lipid analyses. Many fragmentation patterns of different polar lipid classes were characterized either by FAB/MS through source fragmentation[5,6] or by FAB/MS/MS.[7–9] These structural determinations established the basic principles of lipid ion chemistry.[10] Since FAB ionization is quite soft, molecular ions of lipids are predominant in FAB mass spectra in most cases by using a proper matrix (e.g., glycerol). Therefore, FAB/MS has been used to profile individual molecular species of a phospholipid class after that lipid class is separated from a crude lipid extract, although the profile may not reflect the quantitative composition of the lipid class since differential source fragmentation of different molecular species may alter the profile.[11,12] Without question, these studies were the prototypes of the current large-scale analyses of lipids.

As a technique, neutral loss scanning was introduced for detection of the loss of a common neutral fragment after collision-induced dissociation (CID) from its corresponding quasi-molecular ions, so that all individual molecular species of an entire lipid class or group containing the lost neutral fragment can be spectrometrically "isolated" if the lost neutral fragment carries the information of a lipid class or group.[13,14] This technique, along with other MS/MS techniques that have since been developed, provides a foundation for the large-scale analyses of lipids which has now become an important part of a newly emerging research field in lipids (i.e., lipidomics).

Atmospheric-pressure chemical ionization (APCI) has been used for analysis of relatively polar lipids in comparison to those analyzed by EI and CI. APCI/MS spectra are relatively simple, with a protonated molecular ion as the most common base peak. Many classes of lipids, including free fatty acids, phospholipids, sterols, and TG, are ionizable by APCI.[15,16] Although APCI is a relatively soft ionization technique, the comparatively harsh conditions employed relative to electrospray ionization (ESI) still induce some degree of fragmentation. After they were made commercially available in the late 1980s and early 1990s, both ESI and MALDI mass spectrometers were extensively used for analysis of almost all nonvolatile lipids in their intact forms.[17–21] Summaries for the majority of these studies can be found in multiple valuable review articles.[22–24]

20.1.3 Impacts of New MS Developments on Lipid Analysis

Multiple new developments related to the different components of mass spectrometers (e.g., ion source and analyzer) have been made in the last several years. For example, atmospheric-pressure photoionization (APPI) has become the newest family member of atmospheric-pressure ionization (API) techniques in addition to ESI and APCI. APPI has provided a new approach to analyze compounds not readily ionized by ESI with improved ionization efficiency in comparison to APCI. Therefore, APPI serves as a complementary alternative to ESI for lipid analysis. Since APPI requires less heat for desolvation than APCI, thermally labile compounds can be analyzed with fewer concerns for thermal chemical alterations or degradation in the ionization source. In addition, APPI offers lower detection limits, higher signal intensities, and higher signal to noise ratios in comparison to APCI.[25] Therefore, APPI has been well-appreciated for its ability to analyze many neutral lipid classes, including free fatty acids (and their esters), MG, DG, TG, sterols, fat-soluble vitamins, and even polar phospholipids.[25–28]

Off-axis electrospray devices employing a pneumatic sheath gas have recently been widely employed in commercially available API ion sources. This improvement dramatically increases the ionization efficiency through the separation of ions from neutral molecules and establishes an environment of complete desolvation. This improvement has had a significant impact on the global analysis of individual lipid molecular species from lipid extracts of biological samples without pre-chromatographic separation.[29] Although mass spectrometers containing a highly accurate mass analyzer (e.g., Fourier transform (FT) MS) have previously been used for lipid analysis,[21,30–33] it can be anticipated that the commercial introduction of the Orbitrap[34] will further accelerate the elemental analysis of the composition of lipids.

Time-of-flight secondary ion mass spectrometry (TOF-SIMS) that employs gold or silver ions as primary ions has been used to directly detect cellular lipids in intact tissue in both positive- and negative-ion modes.[35–37] Deprotonated cholesterol, ST, and GPIns molecular ions are prominent in negative-ion TOF-SIMS analyses of brain tissue, whereas various silver-cholesterol adducts and protonated phosphocholine ions from GPCho are predominantly detected in the positive-ion mode. Mapping of these ions shows the spatial distributions of various lipid classes in brain sections. Although these studies represent the initial applications of this technique, mapping of subcellular membrane lipids and determination of spatial and dynamic relationships of lipids at the subcellular level will be possible in the near future.

20.1.4 The Scope of the Chapter

Due to the development of new types of instruments and new techniques, the power of MS has recently been greatly appreciated for the analysis of lipids. The dramatic increase in the application of MS to the field of lipid research can be demonstrated by the yearly histograph for literature published over the last 10 years (Figure 20.4) as obtained by SciFinder Scholar

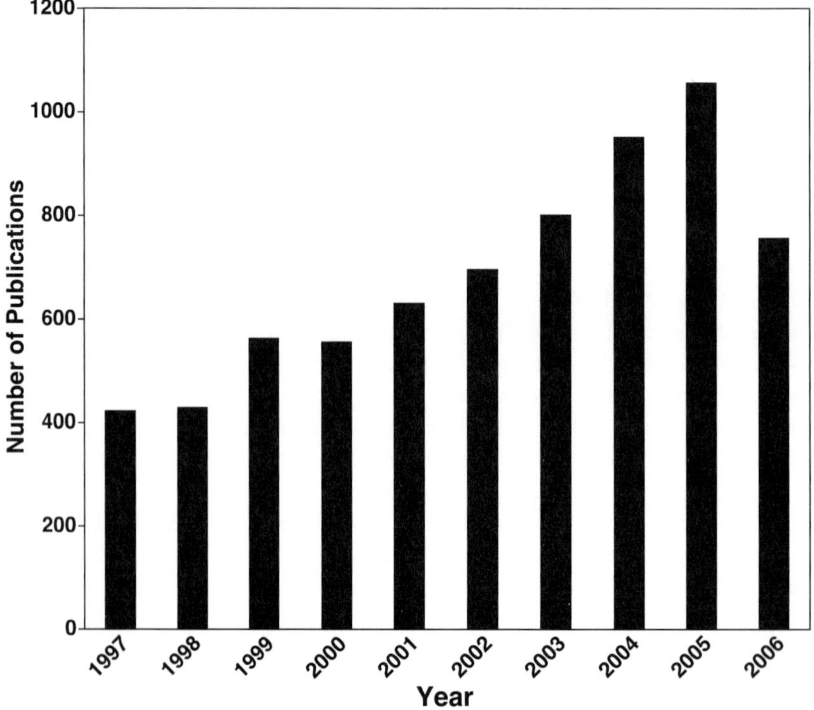

Figure 20.4. Yearly histogram of mass spectrometry for lipid analysis obtained by SciFinder Scholar using "lipids" and "mass spectrometry" as keywords. The year 2006 contains only the first 10 months.

using "lipids" and "mass spectrometry" as keywords. These studies have been mainly focused on two areas of active research: (1) structural characterization of known lipid classes and identification of novel lipid classes and molecular species through studies of their fragmentation patterns and chemical and physical properties and (2) quantitation of individual lipid molecular species in the range of femtomole to low picomole levels (per milligram of protein) in a cell, tissue, or body fluid sample. Many outstanding studies have been published recently in these two areas of research, although, because of space limitations, it will be impossible to discuss all of them in detail here. A particular focus of this review will be the ionization features of lipids by MALDI/MS and ESI/MS and the methodologies used for quantitation of individual lipid molecular species.

20.2 CHARACTERIZATION OF LIPIDS BY ESI/MS AND MALDI/MS

20.2.1 Characterization of Lipids by ESI/MS

The ESI ion source is amongst the softest ionization devices in which only quasi-molecular ions are generated and in-source CID can be minimized in most of cases. The electrospray ionization process is so soft that even solvent adducts, dimers, and other complexes that normally only exist as weakly bound complexes in solution are observable as distinct ion peaks.[18,38,39]

In the positive-ion mode, lipids tend to form small cation adducts. Formation of these positively charged lipid adducts depends on the concentration and affinity of the small cations for each lipid molecular species. Sodium ions are the most common in crude lipid extracts of biological samples, and they possess high affinity for polar lipids [even nonpolar lipids (e.g., TG)]. Therefore, if no other modifier(s) is (are) added to the infused solution, sodiated lipid molecular species (i.e., $[M+Na]^+$), particularly for zwitterionic lipids (e.g., GPCho and SM) are usually the most abundant ions displayed in positive-ion ESI mass spectra (Figure 20.5).[18,40] However, when a modifier is added to the sprayed solution, the quasi-molecular lipid ions will be the adducts of the cation that is from the modifier. For example, organic acids (e.g., formic or acetic acid) are commonly employed as modifiers, particularly when ESI/MS is coupled with LC, resulting in protonated molecular species of the majority of lipid classes being displayed in positive-ion ESI mass spectra. Similarly, protonated phospholipid molecular species (in most cases) or ammonium adducts of TG are also detected when an ammonium salt (e.g., ammonium formate or acetate) is used as a modifier in the sprayed lipid solution. Lithium adducts of lipid molecular species, however, are detected when LiOH or another lithium salt (e.g., LiCl, lithium formate, or lithium acetate) is added to an infused lipid solution.[41,42]

The detected lipid molecular ions are either deprotonated molecules or anionic adducts of lipids in the negative-ion mode, depending on whether the molecular species of lipids carry a separable ionic bond. Thus, all the molecular species of acidic lipid classes (e.g., GPEtn, GPIns, GPSer, GPA, GPGro, CL, ST, free fatty acid, eicosanoids, acyl-CoA, and sphingosine-1-phosphate) form as deprotonated ions,[18,29,43] whereas molecular species of polar neutral lipid classes (e.g., cerebroside, glycolipids) or strong zwitterionic lipid classes (e.g., GPCho and SM) are detected as their anionic adducts (e.g., $[M + Cl]^-$, $[M + CH_3COO]^-$, and $[M + HCOO]^-$).[19,40,44]

Since ESI/MS facilitates quasi-molecular ions of lipids in most cases, MS/MS analysis with low-energy CID after ESI is commonly used for structural characterization of individual molecular species of lipid classes. Most early studies in this area can be found

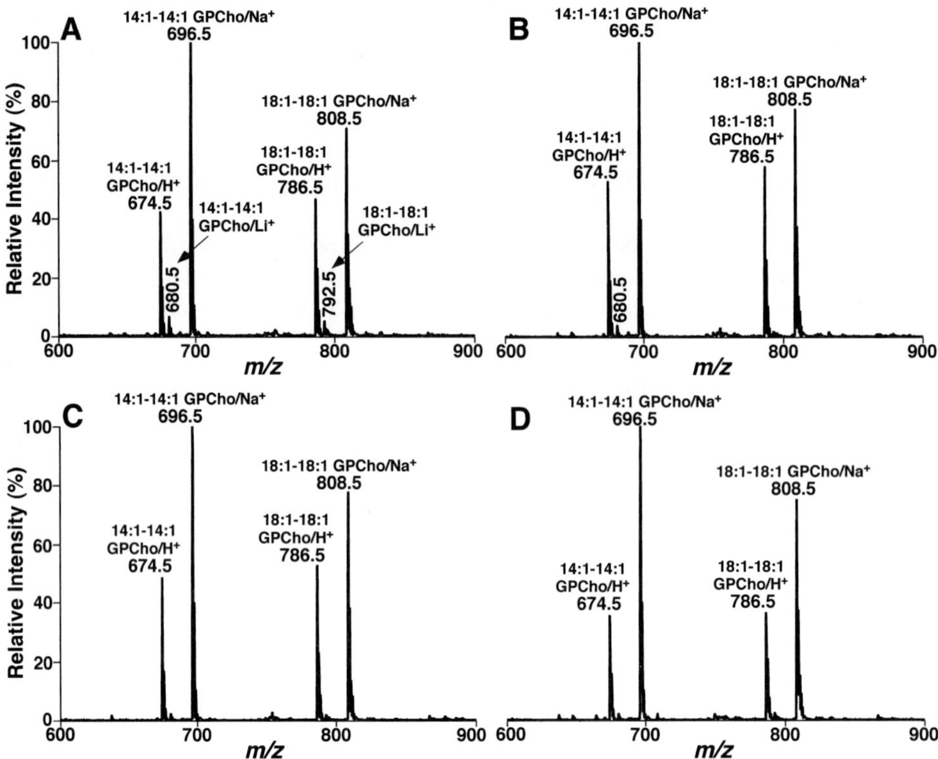

Figure 20.5. The effects of sample preparation on adduct ions of phosphatidylcholine mixture. The lipid mixture comprised of 2 pmol/μL each of 14:1–14:1 and 18:1–18:1 GPCho molecular species was extracted against 50 mM LiCl, (**A**) 50 mM ammonium acetate, (**B**) 50 mM NH_4Cl, (**C**) and tenfold diluted PBS (**D**) by a modified Bligh and Dyer method. Positive-ion ESI mass spectra were acquired using a TSQ Quantum Ultra ESI mass spectrometer (Thermo, San Jose, CA) at a flow rate of 4 μL/min.

in multiple review articles.[22–24] In recent years, Hsu and Turk have performed a series of studies on the structural characterization and fragmentation mechanisms of different lipid classes in unprecedented detail. Their work has recently been reviewed.[45–48] Therefore, these articles should be consulted for a more detailed description of the structural characterization of a specific lipid class. Herein, a brief discussion of the fragmentation properties and structural features of phospholipids is given.

In the positive-ion mode, product-ion spectra of the $[M + H]^+$ ions of phospholipids are often simple, and the structurally informative ions are minimal. For example, protonated phosphocholine (*m/z* 184) is always a predominant fragment from protonated GPCho and SM molecular species. However, product-ion spectra of the alkaline adducts of phospholipids in the forms of $[M + alk]^+$ and/or $[M + 2alk - H]^+$ ions (alk = Li, Na, K, ...) give robust fragment ions that can be readily analyzed for structural identification. In the positive-ion mode, charge-remote fragmentation processes play the predominant role in the fragmentation, and loss of the fatty acid substituent at *sn*-1 is a more favorable pathway than the analogous loss of the fatty acid substituent at *sn*-2. Therefore, the regiospecificities of *sn*-1 and *sn*-2 lysoGPCho and lysoGPEtn can be readily identified from either $[M + Na]^+$ or $[M + Li]^+$ ions.[49,50]

In the negative-ion mode, charge-driven fragmentations are the major processes, and the gas-phase basicity of the precursor ions is the determinant that leads to the loss of the fatty acid substituent as an acid or as a ketene. The distinction of the fatty acid substituents at *sn*-1 from those at *sn*-2 is based on the observation that loss of the fatty acyl group at *sn*-2 as an acid or as a ketene is more favorable than that at *sn*-1 (i.e., $[M-H-R_2CO_2H]^-$ is more abundant than $[M-H-R_1CO_2H]^-$; $[M-H-R'_2CH{=}CO]^-$ is more abundant than $[M-H-R'_1CH{=}CO]^-$), due to steric constraints. The gas-phase basicity of the $[M-H]^-$ ions of phospholipids is governed by the various polar head groups, which are readily distinguishable by tandem mass spectrometry. The fragmentation processes are also influenced by the modification and the degree of unsaturation of the fatty acid substituents.[40]

20.2.2 Characterization of Lipids by MALDI/MS

Positive-ion MALDI TOF/MS analysis of GPCho and SM molecular species usually shows two quasi-molecular ions corresponding to protonated and sodiated adducts, respectively. The peak intensity ratio of these ions depends on availability of sodium in matrix. The protonated molecular ion of these phospholipids yields an exclusive fragment at *m/z* 184 corresponding to the phosphocholine head group. The Na^+ adduct gives a more informative fragmentation pattern as previously described.[51] Although phosphocholine-containing lipid molecular species can give intense positive ion signals, desorption/ionization of these molecules in the negative-ion mode is very poor.[52]

The positive-ion MALDI-TOF mass spectrum of GPEtn is characterized by the presence of a specific fragment ion corresponding to the loss of the phosphoethanolamine head group.[51] GPEtn molecular species can be readily ionized in the negative-ion mode with a lower sensitivity in comparison to that in the positive-ion mode. Negative-ion MALDI mass spectra of GPEtn are usually dominated by the matrix adducts of GPEtn.

Positive-ion MALDI mass spectra of GPSer molecular species usually exhibit an additional ion peak corresponding to $[M-H+2Na]^+$ in addition to protonated and sodiated molecular ions. Furthermore, a fragment ion (similar to that of GPEtn) corresponding to the loss of the phosphoserine head group is also present in the positive-ion MALDI mass spectrum of GPSer, indicating that the cleavage of the polar head group is the most pronounced fragmentation characteristic of phospholipids.[51] The negative-ion MALDI mass spectrum of GPSer shows a base peak corresponding to $[M-H]^-$ and other intense ion peaks corresponding to $[M+Na-2H]^-$ and related matrix adducts.

Many classes of polar lipids have been characterized by MALDI/MS.[53,54] MALDI/MS has also been employed for characterization of nonpolar lipid classes such as cholesterol and TG. Positive-ion MALDI mass spectra of TG exclusively display the sodium adduct as its quasi-molecular ion. In the fragment region, MALDI mass spectra of TG molecular species are featured by the presence of ions corresponding to the loss of sodium fatty acyl carboxylate(s).[51] A post-source decay technique has been commonly employed for characterization of polar lipids as previously described.[51] Since MALDI QqTOF or TOF/TOF mass spectrometers have been commercially available, true tandem MS analyses of lipids through MALDI have been reported.[55]

20.3 APPLICATIONS OF ESI/MS FOR LIPID ANALYSIS

In this section, we will summarize the application of ESI/MS for lipid analysis. Again, instead of attempting to list all of the applications in the literature which have been done in recent years, this section will mainly highlight the principles of the methodologies related to these applications. The following discussions will cover the uniqueness of the ESI ion source as a separation device for lipid analysis, outline the development of multidimensional MS after ESI/MS, and describe the methods and potential concerns of using ESI/MS for lipid quantitation.

20.3.1 Intrasource Separation and Selective Ionization of Lipids by ESI

An essential feature of ESI is that charge separation occurs in molecules containing an ionic bond in the ion source under high electrical potential (typically ~4 kV).[56,57] Specifically, if there are both positively and negatively charged moieties present in the infused solution, then in the positive-ion mode, the electrospray ion source selectively generates cations in the gas phase, whereas the anions are selectively retained at the end of the spray capillary or dispersed as electrically neutral molecules in the gas phase and are eventually disposed of as waste after oxidation/reduction reactions.[58] Similarly, in the negative-ion mode, the ion source selectively generates anions in the gas phase and removes the cationic moieties to waste after redox chemistry has occurred. Therefore, an electrospray ion source is functionally analogous to an electrophoretic separation device conducted with a continuously equilibrating mobile phase.[56–58]

In fact, an electrospray ion source is more broadly capable than an electrophoretic cell to resolve multiple types of analytes since the ion source can generate gas phase ions from electrically neutral but polar compounds. Thus, when the analytes in the infused solution do not carry separable charge(s) but possess intrinsic dipoles, these compounds can be induced to interact with small cation(s) (e.g., H^+, Li^+, Na^+, $NH_4{}^+$, K^+) or anion(s) (e.g., OH^-, Cl^-, formate, acetate) to yield adduct ions in the positive- or negative-ion mode, respectively, in a high electric field. The ionization efficiencies of these electrically neutral compounds depend on their inherent dipoles, the concentration of the small matrix ions, the affinity of the small ions for the analytes, and the electrochemical properties of the resultant adducts.[29]

The physical properties of the ESI ion source can lead to separate and selective ionization of different lipid classes (i.e., intrasource separation and selective ionization), due to their different electrical properties which largely depend on the nature of the polar head groups. Such separation of different lipid classes in the ion source is analogous to the use of an ion-exchange column to separate individual lipid classes.[59] However, in contrast to ion-exchange chromatography, intrasource separation is rapid, direct, and reproducible and avoids artifacts inherent to chromatography-based systems.[60]

Based on the electrical propensities, all the lipid classes can be reclassified into three main categories.[29,61] The lipid classes in the first category are those carrying at least one net negative charge under weakly acidic pH conditions (i.e., near pH 5) and are therefore called anionic lipids. Lipid classes in this category include cardiolipin, GPGro, GPIns and its polyphosphate derivatives, GPSer, GPA, ST, acyl-CoA, and anionic lysophospholipids. The lipid classes in the second category are those that are electrically neutral under weakly acidic pH conditions, but become negatively charged under alkaline pH conditions. Therefore, they are referred to as weakly anionic lipids. Lipid classes in this category include GPEtn, lysoGPEtn, nonesterified fatty acids and their derivatives, bile acids, and Cer. The remaining lipid classes belong to the third category which includes GPCho, lysoGPCho, SM, mono-hexosylceramide, acylcarnitine, DG, TG, cholesterol and its esters, and so on. This category of lipid classes are referred to as electrically neutral but polar or polarizable lipids.

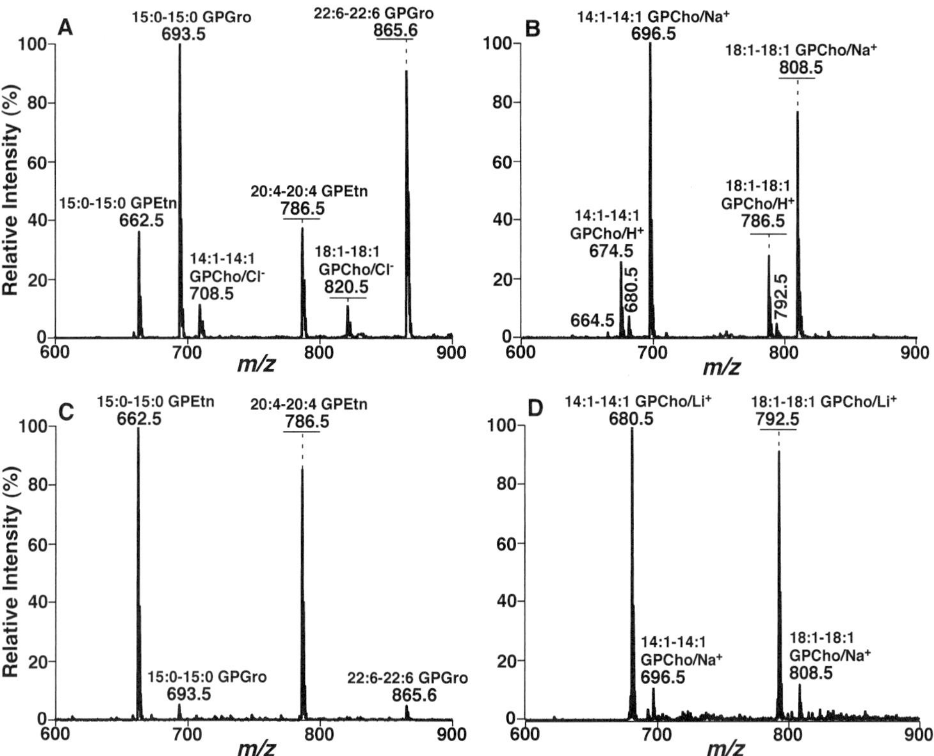

Figure 20.6. Intrasource separation of a mixture of phospholipids. The phospholipid mixture is comprised of 15:0–15:0 and 22:6–22:6 GPGro (1 pmol/μL each), 14:1–14:1 and 18:1–18:1 GPCho (10 pmol/μL each), and 15:0–15:0 and 20:4–20:4 GPEtn (15 pmol/μL each) in 1:1 $CHCl_3$/MeOH. Panels **A** and **C** show mass spectra acquired in the negative-ion mode and Panels **B** and **D** show mass spectra acquired in the positive-ion mode in the absence (Panels **A** and **B**) or presence (Panels **C** and **D**) of LiOH. The horizontal bars indicate the ion peak intensities after ^{13}C de-isotoping and normalization of molecular species in each class to the one with lower molecular weight.

A practical strategy for the separation of these different types of lipids based on their differential intrinsic electrical properties has been discussed in detail[29,61–63] and has been demonstrated through a model mixture of phospholipids that represent the three categories of lipids (Figure 20.6).[64] Specifically, the first category of lipids (i.e., anionic lipids) can be directly and selectively ionized in the negative-ion mode and can be analyzed from diluted lipid extracts by negative-ion ESI/MS. Next, we make the diluted lipid extract solution mildly basic by addition of a small amount of LiOH (or other suitable base) prior to injection into the mass spectrometer, and then the second category of lipids (i.e., weakly anionic lipids) can be analyzed by negative-ion ESI/MS. Most of the other lipid classes (except categories 1 and 2) belong to a third category that can be analyzed directly from the mildly alkalinized diluted lipid extract (i.e., the solution used in last step) in positive-ion ESI/MS because lipids in the first and second categories are now anionic under these conditions. Through this approach, a comprehensive series of mass spectra with respect to each of the aforementioned conditions can be obtained for each category of lipids from any biological samples. For example, Figure 20.7 displays three totally different mass spectra that profile the three categories of lipid classes from a lipid extract of mouse cervical dorsal root ganglia (~2 mg of wet weight tissue).

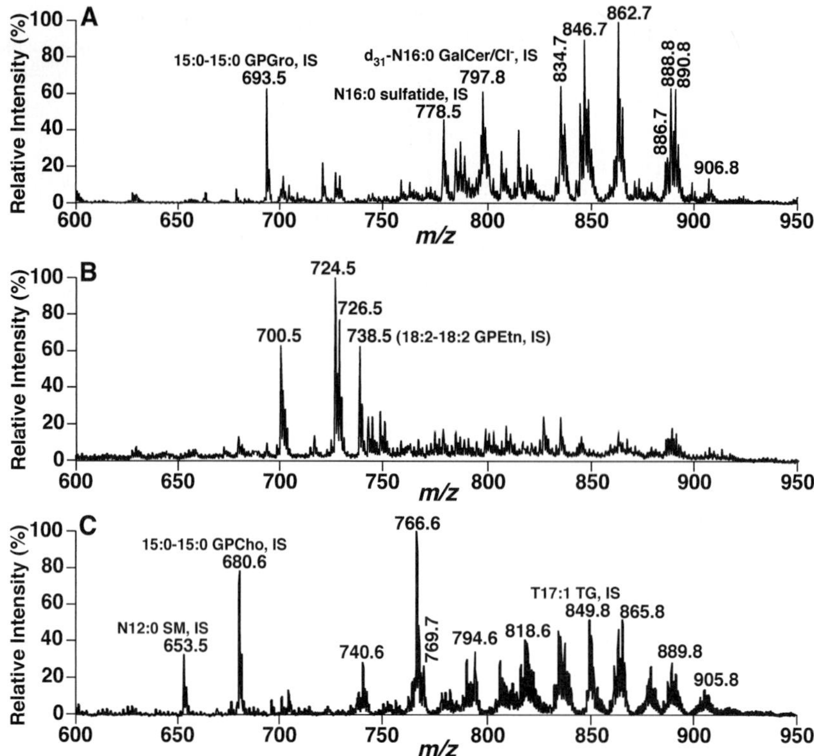

Figure 20.7. ESI/MS analyses of the lipidome of mouse cervical dorsal root ganglia after intrasource separation. The lipid extract of mouse cervical dorsal root ganglia was prepared by a modified Bligh and Dyer procedure as previously described.[38] Spectrum **A** was acquired in the negative-ion mode directly from the lipid extract that was diluted to a concentration of less than 50 pmol of total lipids/μl. Spectrum **B** was acquired again in the negative-ion mode from the diluted lipid solution as used for spectrum **A** after addition of approximately 25 pmol LiOH/μL to the lipid solution. Spectrum **C** was acquired in the positive-ion mode from the identical diluted lipid solution as used in spectrum **B** after direct infusion. "IS" stands for internal standard. All mass spectral traces are displayed after being normalized to the base peak in each individual spectrum.

20.3.2 Multidimensional Mass Spectrometry

We recognized that each peak in the mass spectra of lipid extracts of biological samples potentially represents at least one and very often more than one lipid molecular species, particularly in those mass spectra acquired by instruments with low mass accuracy. At this stage, one could perform product ion analyses of individual ion peaks to identify the molecular species underneath each ion peak. However, we recognized that most of these biological lipid species are linear combinations of aliphatic chains, glycerol backbones, and/or polar head groups, each of which represents a building block of the molecular species under consideration as discussed previously in the chapter.[63] Identification of these building blocks can be accomplished by two other powerful tandem mass spectrometric techniques (i.e., neutral loss scanning and precursor ion scanning) by monitoring the specific loss of a neutral fragment or the yield of a fragmental ion in the tandem MS modes. Therefore, all of these building blocks of the different lipid classes constitute an additional dimension to the molecular ions present in the original mass spectrum obtained from the first dimension. The

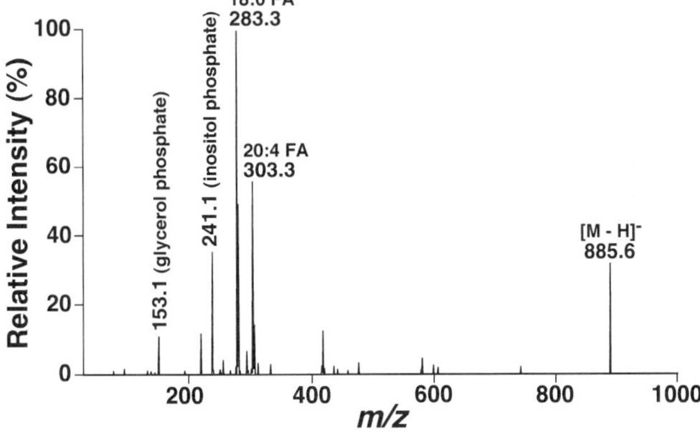

Figure 20.8. Product-ion mass spectrum of 18:0–20:4 GPIns acquired by using a QqQ mass spectrometer.

"crossed" peaks of a given primary molecular ion in the first dimension with the building blocks in the second dimension represent the fragments of this given molecular ions under the conditions employed. Analysis of these crossed peaks (i.e., the individual fragments) thereby determines the structure of the given molecular ion as well as its isobaric constituents.[63]

For example, from CID analysis of synthetic phosphatidylinositol (GPIns), we have found that the major fragments of GPIns molecular species are those corresponding to fatty acyl carboxylate(s), inositol phosphate (*m/z* 241.1), and glycerophosphate (*m/z* 153.1) derivatives (Figure 20.8). Therefore, GPIns molecular species of a lipid extract of a biological sample can be readily identified through a 2D-MS analysis of these building blocks [i.e., precursor ion (PI) scanning of all potential fatty acyl carboxylates, inositol phosphate, and glycerophosphate derivatives]. Figure 20.9 shows such a 2D-MS analysis of GPIns molecular species in the lipid extract of mouse myocardium. For example, the peak at *m/z* 885.6 in the first dimension is crossed with the building blocks of PI 153.1, PI 241.2, PI 283.3, and PI 303.3 (the right broken line in Figure 20.9) in addition to the higher ion intensity of *m/z* 885.6 in PI 303.3 than in PI 283.3, thereby identifying this species as 18:0–20:4 GPIns. Similarly, the peak at *m/z* 857.6 can be identified as 16:0–20:4 GPIns (the left broken line in Figure 20.9).

This basic two-dimensional mass spectrum can be changed in terms of peak intensities and profiles by varying instrumental conditions such as ionization parameters (e.g., source temperature and spray voltage) and fragmentation conditions (e.g., collision gas pressure, collision energy, and collision gas), among others.[63] Therefore, besides the basic 2D mass spectral profile in which the second dimension is constructed with specific lipid building blocks, each series of ramped variation in a particular instrumental condition facilitates the generation of an additional dimension and can potentially be useful for lipid analyses.[63] Collectively, all of these dimensions constitute multidimensional mass spectrometry.

20.3.3 Quantitation of Lipids by ESI/MS and Lipidomics

Lipidomics is a new and rapidly spreading field.[61,65,66] Lipidomics is defined as the analysis and characterization of lipids and their spatial and dynamic interactions with other lipids and other cellular components. It is implemented in a systems approach by integrating many different modern techniques including MS and the separation sciences.[61] Large

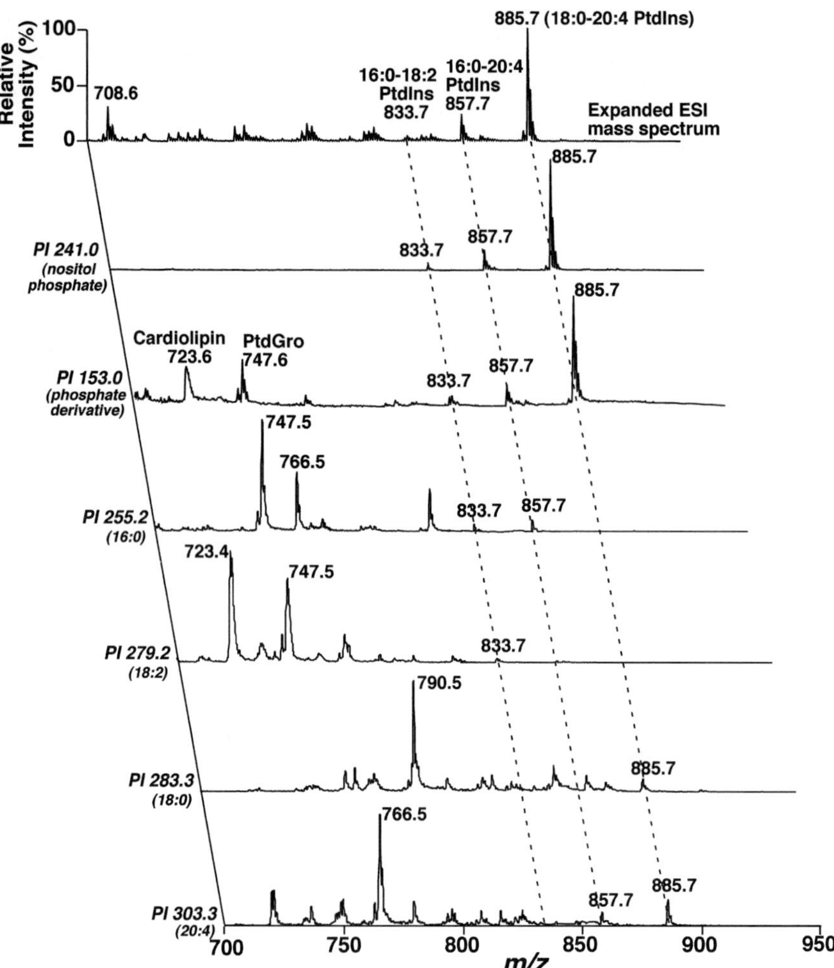

Figure 20.9. Two-dimensional mass spectrometric analysis of GPIns molecular species in a lipid extract of mouse myocardium. The lipid extract of mouse myocardium was prepared by a modified Bligh and Dyer procedure as previously described.[95] Each MS or MS/MS trace of the 2D ESI mass spectrum was acquired by sequentially programmed, customized scans operating under Xcalibur software. For negative-ion tandem mass spectrometry in the precursor-ion (PI) mode, the first quadrupole was scanned in the selected mass range and the second quadrupole was used as a collision cell while the third quadrupole was fixed to monitor the ion of interest (i.e., either inositol phosphate, glycerophosphate, or a fatty acyl carboxylate fragmented from GPIns molecular species). All mass spectral traces were displayed after being normalized to the base peak in each individual trace.

scale-identification of cellular lipids and quantification of alterations in lipid classes, subclasses, and individual molecular species (i.e., the lipidome) induced by a disease, a genetic defect, a drug treatment, or cellular signaling are the essential tasks of lipidomics at its current stage.

The minimal in-source CID inherent in the ion source of ESI not only dramatically enhances the ionization sensitivity of ESI/MS, but also makes ESI/MS a powerful tool for quantitative applications since collision-induced fragmentation is a process that depends on the molecular structure of the analytes. Accordingly, ESI has currently become the method of choice for the quantitative analyses of individual lipid molecular species, and ESI/MS greatly facilitates the development of lipidomics.

There are three principal, independently developed ESI/MS-based approaches currently utilized for lipidomics analyses, each of which possesses different advantages and limitations which have been extensively discussed previously.[29] The three main ESI/MS-based approaches are (1) HPLC-coupled ESI/MS, (2) tandem mass spectrometry-based shotgun lipidomics, and (3) multidimensional MS-based shotgun lipidomics after intra-source separation. An alternative MS-based analytical technique has been developed recently for profiling relative changes between different cellular states and should be of interest in the future.[67,68] Both of the "shotgun lipidomics" approaches are implemented after direct infusion, and potentially both can be used for global analyses of individual lipid molecular species directly from a lipid extract of a biological sample without pre-chromatographic separation. In the LC-coupled ESI/MS method, both normal-phase and reversed-phase columns have been employed for lipid profiling. In the following part of this section, the principles, advantages, and limitations of each approach will be briefly discussed and followed by a description of general concerns that are related to the accurate quantitation of individual lipid molecular species.

20.3.3.1 HPLC-MS Approach for Lipidomics

In this approach, ESI/MS is essentially used as a mass detector (i.e., similar to a UV detector or other detectors, which is connected to an HPLC system) in most cases. The determined total ion current (TIC) of each individual molecular ion can be extracted and a TIC chromatograph can be reconstructed. The reconstructed ion peak area of each molecular species can be compared either to a standard curve of the molecular species or to the reconstructed ion peak area of an internal standard obtained under identical experimental conditions. The combination of ESI/MS detection with HPLC separation and the specificity of the TIC compared to other detection modalities make this approach a potential choice for lipid profiling and quantitation in lipidomics if either one of the following conditions is met. First, an appropriate standard curve of the lipid molecular species of interest is carefully established under identical experimental conditions.[69] Second, a stable isotopically labeled internal standard of each lipid molecular species must be made.[70] Third, it is validated that ionization efficiencies of each individual molecular species of a class are identical under experimental conditions after considering some correction factors (see below).[71–73] In practice, such a combination has been employed in many applications for the identification of lipid molecular species. However, lipid quantitation using this methodology on a large scale is quite limited,[71] although analysis of a small number of lipids whose standard curves can be generated is quite common.[74]

Although column separation can enrich low-abundance molecular species and eliminate the interactions of different lipid molecular species or so-called ion suppression, the difficulties employing this methodology on a large scale for quantitative analysis are also apparent and should be recognized. First, it is either very difficult to establish a standard curve for each individual lipid molecular species from a particular biological preparation or it is unpractical to make stable isotopically labeled compounds for each individual lipid molecular species of a particular biological sample. Second, even if it were practical to have a standard curve for each lipid compound, the determined standard curve is externally measured and the involvement of multiple steps of sample preparation, separation, and analysis can introduce experimental errors in each step which are propagated for the entire process. Third, when a normal-phase HPLC column is employed for separation of different lipid classes, different lipid molecular species in a class are not uniformly distributed in the eluted peak (i.e., each individual molecular species of a class may possess its own distinct retention time and peak shape due to differential interactions with the stationary phase). If a reversed-phase HPLC column is used to resolve individual lipid molecular species, then the relatively polar mobile phase that is commonly employed will cause difficulties with solubility in a molecular-species-dependent manner, leading to different apparent ionization efficiencies of each individual lipid molecular species. If a solvent gradient is employed to resolve individual molecular species by reversed-phase HPLC, changes in the components of the mobile phase may also cause an ionization stability problem. Fourth, differential loss of lipids on the column is also not unusual.[60] Finally, while reversed-phase chromatography eliminates certain lipid–lipid interactions, there is a large (up to a 1000-fold) increase in the amount of specific lipid–lipid interactions between the identical types of lipid molecular species (i.e., homodimer formation), which often contributes to lipid aggregation during reversed-phase HPLC. These are only a few examples of potentially problematic issues among many others regarding quantification of lipid species by ESI/MS coupled with HPLC.

20.3.3.2 Tandem Mass-Spectrometry-Based Shotgun Lipidomics

In this approach, quantitative analyses of lipids after ESI is entirely based on tandem mass spectrometry.[75–77] At least two molecular species of a lipid class of interest are added as internal standards to the lipid extract of a biological sample. The choice of these standards should involve at least two criteria. One is that these compounds should be absent, or present in very minimal amounts, in the original lipid extract. The other is that these compounds should be representative of the physical properties (e.g., the length of acyl chains and the degree of unsaturation) of the entire lipid class of interest as closely as possible. Then, a unique MS/MS analysis in either neutral loss or precursor-ion mode relative to the lipid class is performed. The concentration of each individual molecular species of the class of interest is then calculated from its ion peak intensity determined by tandem mass spectrometric comparison with the internal standards. Using these standards, many experimental factors that may introduce experimental error are essentially eliminated. The advantages of this methodology are apparent and the method is straightforward. Therefore, it has become increasingly popular for analyses of many selected lipid classes. However, some limitations of this method exist. For example, selection of the standards to meet the two criteria is not easy; the differential fragmentation kinetics of different molecular species of a class may affect accurate quantitation; there exists a limited dynamic range for some lipid classes for each of which a sensitive tandem MS method is lacking; the type II effect of ^{13}C isotopomer on quantitation of the neighboring peaks has so far been neglected in all applications up to date (among others see discussions in Ref. 29).

20.3.3.3 Multidimensional MS-Based Shotgun Lipidomics after Intrasource Separation

After separation of different lipid classes in the ESI ion source (i.e., intrasource separation) and identification of individual molecular species by multidimensional MS, quantitation of these lipid molecular species is conducted using a two-step procedure.[44,63,78] First, quantitation of the abundant and non-overlapping molecular species of a class is achieved by ratiometric comparisons with a preselected internal standard of the class after ^{13}C de-isotoping. This is done because response factors of different individual molecular species of a polar lipid class are essentially identical in the low concentration region after ^{13}C

de-isotoping.[63,78–80] It should be specifically pointed out that the response factors of different nonpolar lipid molecular species to ESI/MS are quite different and must be predetermined for the quantitation of these classes of lipids.[42] Next, all the determined molecular species of the class of interest plus the preselected internal standard are used as standards to determine the content of other low-abundance or overlapping molecular species using one or multiple tandem mass traces (each of which represents a specific building block of the class of interest) by two-dimensional MS.

This second step is equivalent to the second method (i.e., tandem mass spectrometry) as described above. However, the standards used in this step are endogenous, which can generally provide a more comprehensive representation of the physical properties of structurally similar, but less abundant, lipid molecular species within a selected class in comparison to externally added standards. This second step in the quantitation process provides an extended linear dynamic range of quantitation by eliminating background noise and by filtering the overlapping molecular species through a multidimensional approach.[29] The limitation of this methodology is that the experimental error for the molecular species measured in the second step of quantitation is propagated and is larger than those for the first step. To minimize this error propagation, it is critical to reduce the experimental error in the first step. For example, the error introduced by an elevated baseline is inevitable, but can be minimized through background subtraction. In addition, the baseline-induced experimental error can also be reduced through utilizing an appropriate concentration of the selected internal standard so that its ion peak intensity is comparable to the base peak intensity of the lipid class of interest.

Figure 20.10 shows a typical example of quantitation of GPCho molecular species in a lipid extract of mouse dorsal root ganglia. Lithiated GPCho molecular species at *m/z* 740.6, 766.6, 794.6, and 812.6 in the one-dimensional mass spectrum (Figure 20.10A) were quantitated by comparison of their ion abundances with that of the GPCho internal standard at *m/z* 680.6 after ^{13}C de-isotoping. The contents of lithiated SM molecular ions at *m/z* 709.5, 737.5, and 819.6 (Figure 20.10A) were similarly determined by comparison with the SM internal standard at *m/z* 653.5. Then, other minor or overlapping molecular species of GPCho and SM were repeatedly assessed from the trace of NL 183.1 (Figure 20.10B) by using all quantified GPCho and SM molecular species as standards, respectively.

20.3.4 Conclusion

ESI/MS is the most developed, prominent, and successful mass spectrometric methodology in lipidomics to date (see Refs. 2, 29, 63, and 66 for recent reviews and a recently published special issue entitled "Lipidomics: Developments and Applications" in the journal *Frontiers in Bioscience*). The advantages of ESI/MS are severalfold. First, a complete and highly efficient quantitative analysis of lipid classes, subclasses, and individual molecular species without prior chromatographic separation is feasible by employing intrasource separation. Second, a higher signal-to-noise ratio is present in ESI/MS in comparison to other traditional MS approaches (e.g., FAB/MS). Third, the ionization efficiency or instrument response factor of individual molecular species in a polar lipid class primarily depends on the electrical property of the lipid class, and the differences of the response factors between molecular species in a polar lipid class are within experimental error when the experiment is performed at low lipid concentration. Fourth, there is a near-linear relationship between the relative responses of molecular ions and the masses of individual lipids over a wide dynamic range in the low concentration regime for lipids with large dipole moments. In lipids without

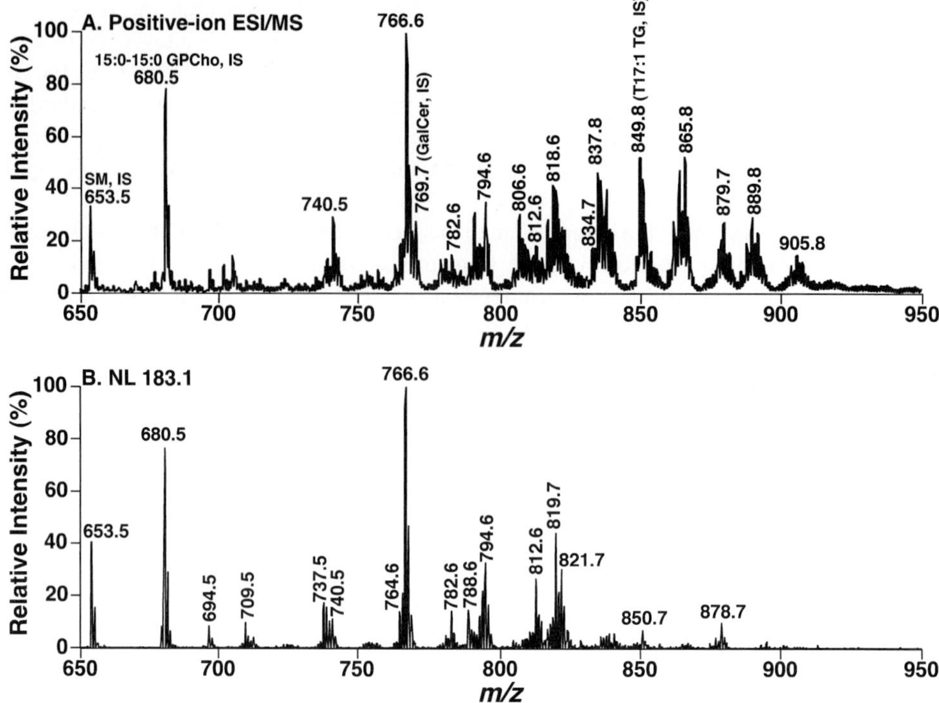

Figure 20.10. A typical example of quantitation of GPCho and SM molecular species in a lipid extract of mouse cervical dorsal root ganglia. First, the presence of lithiated GPCho and SM molecular species in survey scan (**A**) was determined through the characterization of a building block corresponding to GPCho and SM head group via neutral loss of 183.1 u (i.e., phosphocholine) (**B**). Identification of the fatty acyl chains of GPCho molecular species were repeatedly performed through either neutral loss analysis of all potential fatty acids in the positive-ion mode or the 2D MS array analysis of chlorinated GPCho species in the negative-ion mode. Such an analysis is remarkable for the identification of the presence of low-abundance GPCho molecular species containing long fatty acyl chains such as 22:1–22:6 pGPCho at *m/z* 878.6. After identification, lithiated GPCho molecular species at *m/z* 740.6, 766.6, 794.6, and 812.6 in the survey scan (**A**) were quantitated by comparison of these ion intensities with that of the GPCho internal standard at *m/z* 680.6 after de-isotoping. The contents of lithiated SM molecular species at *m/z* 709.5, 737.5, and 819.6 (**A**) were similarly determined by comparison with the SM internal standard at *m/z* 653.5 in the spectrum. Then, other minor or overlapping molecular species of GPCho and SM were assessed from NL 183.1 (**B**) by using all quantified GPCho and SM molecular species as standards, respectively. "IS" stands for internal standards.

large dipoles, correction factors should be predetermined. Finally, the reproducibility of sample measurements is excellent ($< 5\%$ of experimental error if a standard for each lipid class is included and used to normalize these measurements). Thus, it is evident that ESI/MS and ESI/MS/MS have become essential tools for the advancement of the lipidomics field.

20.4 APPLICATIONS OF MALDI/MS FOR LIPID ANALYSIS

20.4.1 MALDI/MS Lipid Analysis of Biological Samples

MALDI/MS has been employed to study a considerable number of biologically relevant lipid mixtures (see Refs. 53 and 81 for reviews). Owing to the limited space of this chapter,

Table 20.1. Selected Applications of MALDI-TOF MS to Biological Sample Analysis

Application	Interesting Features
Liver, spleen, cerebrum and cerebellum[90]	First application of "delayed extraction" MALDI-TOF MS to the analysis of biological lipid extracts. The main focus of the study was on sphingolipids.
Human neutrophils[91]	Examination of the effect of drugs on the phospholipid composition.
Human spermatozoa[92]	Assessment of the effect of cryo-storage on the phospholipid composition of human spermatozoa through measurement of the GPCho/lysoGPCho content.
Human lipoproteins[93]	The effects of HOCl-treatment and phospholipase A_2 activity on lipoprotein lipid composition were studied.
Bull spermatozoa[94]	The high content of ether phospholipids in animal spermatozoa.
Porcine lens[82]	A quantitative analysis of the major phospholipid classes (e.g., GPGro, GPEtn, and GPSer) in the lenses was performed.
Brain tissue[55]	A comprehensive analysis of brain GPCho in the presence of Li^+.
Tear films[54]	A comprehensive study of various phospholipid classes using immobilized metal affinity chromatography and solid *p*-nitroaniline-butyric acid ionic crystal matrix.

only general concerns are discussed while some biological applications employing MALDI/MS are only selectively listed in Table 20.1 with brief comments. It should be noted that most studies have been focused on the analysis of the GPCho class, largely due to its higher sensitivity (resulting from the stable positive charge on the quaternary ammonium moiety) in comparison to other lipid classes in lipid extracts from biological samples.[52] This selective ionization of GPCho molecular species makes analysis of individual molecular species from other lipid classes difficult in the positive-ion mode. Unlike in ESI/MS, severe source fragmentation results from the use of most matrices in the negative-ion mode of MALDI/MS. In the negative-ion mode, fatty acyl carboxylates are predominant, but the high propensity toward fragmentation makes the analysis of acidic phospholipids (e.g., GPEtn) difficult. Therefore, pre-chromatographic separation of different lipid classes by HPLC or TLC appears necessary, followed by analysis of these individual lipid classes in the positive-ion mode or in the negative-ion mode by using a acidic matrix such as *para*-nitroaniline (PNA) as previously described.[82]

MALDI/MS possesses multiple advantages for the analysis of lipid mixtures from biological sources, including speed, convenience, high sensitivity (pmol), reproducibility (samples can often be re-analyzed at a later time), and high throughput. However, there are several potential complicating factors of MALDI/MS for this application. First, matrix compound(s) are ionized under most experimental conditions, which may complicate the lipid analysis in the low mass-to-charge region. Second, the post-source decay present in MALDI/MS is a double-edged sword; that is, it is useful for structural elucidation, but problematic for quantitation due to differential analyte fragmentation kinetics. Third, the presence of multiple adducts and/or ion forms of each lipid molecular species is quite common in both positive- and negative-ion modes, which further complicates analysis of individual molecular species of a lipid mixture for both identification and quantitation. Fourth, the presence of lipid–lipid interactions and lipid aggregation effects during crystallization make the distribution of analytes in the sample spot heterogeneous. Fifth, accurate quantitative results of lipid analyses by current MALDI/MS instruments are thus far less than ideal, and little progress has been made toward the direct quantitation of lipids by MALDI/MS, particularly in complex mixtures. Therefore, at its present stage of development, MALDI/MS can be used more as a qualitative tool to rapidly screen lipid profiles of a biological sample than as a quantitative instrument ready for lipidomics.

20.4.2 Lipid Profiling Directly from Tissue Samples by MALDI/MS

MALDI/MS has already been successfully used for direct mapping and imaging of peptides and proteins in mammalian tissues[83] and even single cells[84] for many years. Recently, this technique has been extended to lipid profiling in individual zooplankton,[85] muscle fibers,[86] and brain tissues.[55,87] For direct lipid analysis of brain tissue, frozen brain tissue should be sectioned as thinly as possible using a cryostat. Different matrices and matrix concentrations have been tested and optimized for direct lipid profiling.[87] Similar to the MALDI/MS analysis of lipid extracts of biological samples as discussed above, ion peaks corresponding to GPCho molecular species are also prominent in MALDI/MS spectra of intact tissue in the positive-ion mode. Either protonated molecular species[87] or sodium adducts[85] of GPCho are abundant in these MALDI/MS spectra, depending on the matrix and other reagents employed. Intriguingly, very abundant sodiated TG molecular species can also be readily detected from biological samples under some experimental conditions.[85] These preliminary studies should lead to lipid maps of tissues[37] and circumvent current problems through the development of new strategies.

However, it should be noted that lipid profiling of tissues by MALDI/MS in the negative-ion mode has not yet been reported, although it has been found that deprotonated cholesterol, sulfatide, and GPIns molecular species can be readily detected in negative-ion mapping of brain tissue by time-of-flight secondary ion MS.[35,36] Moreover, quantitation of individual lipid molecular species by MALDI/MS is still lagging in comparison to ESI/MS. Therefore it will likely remain a significant challenge in the near future to directly profile lipids in a semiquantitative manner from intact tissues.

20.4.3 Developments and Applications of Ionic-Liquid Matrices for Lipid Analysis

The presence of lipid–lipid interactions and lipid aggregation during crystallization renders the lipid distribution in the sample spot heterogeneous, which generally leads to a reduced detection limit, decreased reproducibility, and poor quantitative results. To improve the spot homogeneity for lipid analysis, one of the advances that has recently been made is the introduction of ionic-liquid (or ionic-solid) matrices.[54,88,89] In general, studies applying ionic-liquid matrices have demonstrated multiple advantages for lipid analysis in comparison to previously used matrices. For example, ionic-liquid matrices possess strong UV absorbances, thereby reducing post-source decay and increasing signal intensity.[88] Moreover, ionic-liquid matrices have shown promise in improving the homogeneity of the lipid distribution in MALDI target spots. Sample homogeneity is a key to improving analysis reproducibility, thereby potentially facilitating the quantitative analysis of lipid molecular species by MALDI/MS.

20.5 SUMMARY

This chapter has discussed the applications of MALDI and ESI techniques for lipid analyses for both structural characterization and quantitation. In particular, the methodologies that are currently available for quantitation of individual lipid molecular species based on both ESI/MS and MALDI/MS are discussed in some detail. The limitations associated with each method and the potential concerns related to accurate quantitation of lipids have also been addressed. Overall, ESI/MS is the principal modality for global lipid identification and quantitation at the present time; MALDI/MS can be selected as a primary screening technique before conducting large-scale lipid analyses and future technological developments may contribute to its enhanced usage. Therefore, a combination of both techniques will dramatically accelerate the development of lipidomics.

ACKNOWLEDGMENTS

This work was supported by National Institute on Aging Grants R01 AG23168 and R01AG031675, National Institute of Health grant P01 HL57278, and the Neurosciences Education and Research Foundation. The author is grateful to Dr. Christopher M. Jenkins for his comments and to Dr. Kui Yang and Ms Hua Cheng for their technical help.

REFERENCES

1. Fahy, E.; Subramaniam, S.; Brown, H. A.; Glass, C. K.; Merrill, A. H. Jr.; Murphy, R. C.; Raetz, C. R.; Russell, D. W.; Seyama, Y.; Shaw, W.; Shimizu, T.; Spener, F.; van Meer, G.; VanNieuwenhze, M. S.; White, S. H.; Witztum, J. L.; Dennis, E. A. A comprehensive classification system for lipids. *J. Lipid Res.* **2005**, *46*, 839–861.
2. Merrill, A. H. Jr.; Sullards, M. C.; Allegood, J. C.; Kelly, S.; Wang, E. Sphingolipidomics: High-throughput, structure-specific, and quantitative analysis of sphingolipids by liquid chromatography tandem mass spectrometry. *Methods* **2005**, *36*, 207–224.
3. Nier, A. O. A mass spectrometer for isotope and gas analysis. *Rev. Sci. Inst.* **1947**, *18*, 398–411.
4. Munson, M. S. B.; Field, F. H. Chemical ionization mass spectrometry. I. General introduction, *J. Am. Chem. Soc.* **1966**, *88*, 2621–2630.
5. Fenwick, G. R.; Eagles, J.; Self, R. Fast atom bombardment mass spectrometry of intact phospholipids and related compounds. *Biomed. Mass Spectrom.* **1983**, *10*, 382–386.
6. Ayanoglu, E.; Wegmann, A.; Pilet, O.; Marbury, G. D.; Hass, J. R.; Djerassi, C. Mass spectrometry of phospholipids. Some applications of desorption chemical ionization and fast atom bombardment. *J. Am. Chem. Soc.* **1984**, *106*, 5246–5251.
7. Jensen, N. J.; Tomer, K. B.; Gross, M. L. Fast atom bombardment and tandem mass spectrometry of phosphatidylserine and phosphatidylcholine. *Lipids* **1986**, *21*, 580–588.
8. Jensen, N. J.; Tomer, K. B.; Gross, M. L. FAB MS/MS for phosphatidylinositol, -glycerol, -ethanolamine and other complex phospholipids. *Lipids* **1987**, *22*, 480–489.
9. Costello, C. E.; Vath, J. E. Tandem mass spectrometry of glycolipids. *Methods Enzymol.* **1990**, *193*, 738–768.
10. Matsubara, T.; Hayashi, A. FAB/mass spectrometry of lipids. *Prog. Lipid Res.* **1991**, *30*, 301–322.
11. Lehmann, W. D.; Kessler, M. Fatty acid profiling of phospholipids by field desorption and fast atom bombardment mass spectrometry. *Chem. Phys. Lipids* **1983**, *32*, 123–135.
12. Gross, R. W. High plasmalogen and arachidonic acid content of canine myocardial sarcolemma: A fast atom bombardment mass spectroscopic and gas chromatography–mass spectroscopic characterization. *Biochemistry* **1984**, *23*, 158–165.
13. Heller, D. N.; Murphy, C. M.; Cotter, R. J.; Fenselau, C.; Uy, O. M. Constant neutral loss scanning for the characterization of bacterial phospholipids desorbed by fast atom bombardment. *Anal. Chem.* **1988**, *60*, 2787–2791.
14. Cole, M. J.; Enke, C. G. Direct determination of phospholipid structures in microorganisms by fast atom bombardment triple quadrupole mass spectrometry. *Anal. Chem.* **1991**, *63*, 1032–1038.
15. Byrdwell, W. C. Atmospheric pressure chemical ionization Mass spectrometry for analysis of lipids. *Lipids.* **2001**, *36*, 327–346.

16. Byrdwell, W. C. APCI-MS lipid analysis. *Oily Press Lipid Library* **2003**, *16*, 171–253.
17. Weintraub, S. T.; Pinckard, R. N.; Hail, M. Electrospray ionization for analysis of platelet-activating factor. *Rapid Commun. Mass Spectrom.* **1991**, *5*, 309–311.
18. Han, X.; Gross, R. W. Electrospray ionization mass spectroscopic analysis of human erythrocyte plasma membrane phospholipids. *Proc. Natl. Acad. Sci. USA* **1994**, *91*, 10635–10639.
19. Kerwin, J. L.; Tuininga, A. R.; Ericsson, L. H. Identification of molecular species of glycerophospholipids and sphingomyelin using electrospray mass spectrometry. *J. Lipid Res.* **1994**, *35*, 1102–1114.
20. Kim, H. Y.; Wang, T. C.; Ma, Y. C. Liquid chromatography/mass spectrometry of phospholipids using electrospray ionization. *Anal. Chem.* **1994**, *66*, 3977–3982.
21. Marto, J. A.; White, F. M.; Seldomridge, S.; Marshall, A. G. Structural characterization of phospholipids by matrix-assisted laser desorption/ionization Fourier transform ion cyclotron resonance mass spectrometry. *Anal. Chem.* **1995**, *67*, 3979–3984.
22. Murphy, R. C.; Fiedler, J.; Hevko, J. Analysis of nonvolatile lipids by mass spectrometry. *Chem. Rev.* **2001**, *101*, 479–526.
23. Pulfer, M.; Murphy, R. C. Electrospray mass spectrometry of phospholipids. *Mass Spectrom. Rev.* **2003**, *22*, 332–364.
24. Griffiths, W. J. Tandem mass spectrometry in the study of fatty acids, bile acids, and steroids. *Mass Spectrom. Rev.* **2003**, *22*, 81–152.
25. Cai, S. S.; Syage, J. A. Comparison of atmospheric pressure photoionization, atmospheric pressure chemical ionization, and electrospray ionization mass spectrometry for analysis of lipids. *Anal. Chem.* **2006**, *78*, 1191–1199.
26. Raffaelli, A.; Saba, A. Atmospheric pressure photoionization mass spectrometry. *Mass Spectrom. Rev.* **2003**, *22*, 318–331.
27. Kostiainen, R.; Kauppila, T. J. Analysis of steroids by liquid chromatography—Atmospheric pressure photoionization mass spectrometry. In Byrdwell, W. C. (Ed.), *Modern Methods For Lipid Analysis By Liquid Chromatography/Mass Spectrometry And Related Techniques*, AOCS Press; Champaign, IL, 2005, pp. 472–487.
28. Delobel, A.; Touboul, D.; Laprevote, O. Structural characterization of phosphatidylcholines by atmospheric pressure photoionization mass spectrometry. *Eur. J. Mass Spectrom.* **2005**, *11*, 409–417.
29. Han, X.; Gross, R. W. Shotgun lipidomics: Electrospray ionization mass spectrometric analysis and quantitation of the cellular lipidomes directly from crude extracts of biological samples. *Mass Spectrom. Rev.* **2005**, *24*, 367–412.
30. Fridriksson, E. K.; Shipkova, P. A.; Sheets, E. D.; Holowka, D.; Baird, B.; McLafferty, F. W. Quantitative analysis of phospholipids in functionally important membrane domains from RBL-2H3 mast cells using tandem high-resolution mass spectrometry. *Biochemistry* **1999**, *38*, 8056–8063.
31. Ivanova, P. T.; Cerda, B. A.; Horn, D. M.; Cohen, J. S.; McLafferty, F. W.; Brown, H. A. Electrospray ionization mass spectrometry analysis of changes in phospholipids in RBL-2H3 mastocytoma cells during degranulation. *Proc. Natl. Acad. Sci. USA* **2001**, *98*, 7152–7157.
32. Jones, J. J.; Batoy, S. M.; Wilkins, C. L. A comprehensive and comparative analysis for MALDI FTMS lipid and phospholipid profiles from biological samples. *Comput. Biol. Chem.* **2005**, *29*, 294–302.
33. Ishida, M.; Yamazaki, T.; Houjou, T.; Imagawa, M.; Harada, A.; Inoue, K.; Taguchi, R. High-resolution analysis by nano-electrospray ionization Fourier transform ion cyclotron resonance mass spectrometry for the identification of molecular species of phospholipids and their oxidized metabolites. *Rapid Commun. Mass Spectrom.* **2004**, *18*, 2486–2494.
34. Hardman, M.; Makarov, A. A. Interfacing the Orbitrap mass analyzer to an electrospray ion source. *Anal. Chem.* **2003**, *75*, 1699–1705.
35. Sjovall, P.; Lausmaa, J.; Nygren, H.; Carlsson, L.; Malmberg, P. Imaging of membrane lipids in single cells by imprint-imaging time-of-flight secondary ion mass spectrometry. *Anal. Chem.* **2003**, *75*, 3429–3434.
36. Sjovall, P.; Lausmaa, J.; Johansson, B. Mass spectrometric imaging of lipids in brain tissue. *Anal. Chem.* **2004**, *76*, 4271–4278.
37. Altelaar, A. F.; Klinkert, I.; Jalink, K.; de Lange, R. P.; Adan, R. A.; Heeren, R. M.; Piersma, S. R. Gold-enhanced biomolecular surface imaging of cells and tissue by SIMS and MALDI mass spectrometry. *Anal. Chem.* **2006**, *78*, 734–742.
38. Cheng, H.; Jiang, X.; Han, X. Alterations in lipid homeostasis of mouse dorsal root ganglia induced by apolipoprotein E deficiency: A shotgun lipidomics study. *J. Neurochem.* **2007**, *101*, 57–76.
39. Thomas, M. C.; Mitchell, T. W.; Blanksby, S. J. A comparison of the gas phase acidities of phospholipid headgroups: Experimental and computational studies. *J. Am. Soc. Mass Spectrom.* **2005**, *16*, 926–939.
40. Han, X.; Gross, R. W. Structural determination of picomole amounts of phospholipids via electrospray ionization tandem mass spectrometry. *J. Am. Soc. Mass Spectrom.* **1995**, *6*, 1202–1210.
41. Hsu, F.-F.; Bohrer, A.; Turk, J. Formation of lithiated adducts of glycerophosphocholine lipids facilitates their identification by electrospray ionization tandem mass spectrometry. *J. Am. Soc. Mass Spectrom.* **1998**, *9*, 516–526.
42. Han, X.; Gross, R. W. Quantitative analysis and molecular species fingerprinting of triacylglyceride molecular species directly from lipid extracts of

biological samples by electrospray ionization tandem mass spectrometry. *Anal. Biochem.* **2001**, *295*, 88–100.

43. Jiang, X.; Han, X. Characterization and direct quantitation of sphingoid base-1-phosphates from lipid extracts: A shotgun lipidomics approach. *J. Lipid Res.* **2006**, *47*, 1865–1873.

44. Han, X.; Cheng, H. Characterization and direct quantitation of cerebroside molecular species from lipid extracts by shotgun lipidomics. *J. Lipid Res.* **2005**, *46*, 163–175.

45. Hsu, F.-F.; Turk, J. Electrospray ionization with low-energy collisionally activated dissociation tandem mass spectrometry of complex lipids: Structural characterization and mechanism of fragmentation. In Byrdwell, W. C. (Ed.), *Modern Methods for Lipid Analysis by Liquid Chromatography/Mass Spectrometry and Related Techniques*, AOCS Press, Champaign, IL, 2005, pp. 61–178.

46. Hsu, F. F.; Turk, J. Analysis of sphingomyelins. In Caprioli, R. M. (Ed.), *The Encyclopedia of Mass Spectrometry*; Vol. 3, Elsevier: New York, 2005, pp. 430–447.

47. Hsu, F. F.; Turk, J. Analysis of sulfatides. In Caprioli, R. M. (Ed.), *The Encyclopedia of Mass Spectrometry*, Elsevier New York, 2005, Vol. 3, pp. 473–492.

48. Hsu, F. F.; Turk, J. Electrospray ionization with low-energy collisionally activated dissociation tandem mass spectrometry of glycerophospholipids: Mechanisms of fragmentation and structural characterization. *J. Chromatogr. B*, **2009,**, *877*, 2673–2695.

49. Han, X.; Gross, R. W. Structural determination of lysophospholipid regioisomers by electrospray ionization tandem mass spectrometry. *J. Am. Chem. Soc.* **1996**, *118*, 451–457.

50. Hsu, F.-F.; Turk, J.; Thukkani, A. K.; Messner, M. C.; Wildsmith, K. R.; Ford, D. A. Characterization of alkylacyl, alk-1-enylacyl and lyso subclasses of glycerophosphocholine by tandem quadrupole mass spectrometry with electrospray ionization. *J. Mass Spectrom.* **2003**, *38*, 752–763.

51. Al-Saad, K. A.; Zabrouskov, V.; Siems, W. F.; Knowles, N. R.; Hannan, R. M.; Hill, H. H. Jr. Matrix-assisted laser desorption/ionization time-of-flight mass spectrometry of lipids: Ionization and prompt fragmentation patterns. *Rapid Commun. Mass Spectrom.* **2003**, *17*, 87–96.

52. Petkovic, M.; Schiller, J.; Muller, M.; Benard, S.; Reichl, S.; Arnold, K.; Arnhold, J. Detection of individual phospholipids in lipid mixtures by matrix-assisted laser desorption/Ionization time-of-flight mass spectrometry: Phosphatidylcholine prevents the detection of further species. *Anal. Biochem.* **2001**, *289*, 202–216.

53. Schiller, J.; Suss, R.; Arnhold, J.; Fuchs, B.; Lessig, J.; Muller, M.; Petkovic, M.; Spalteholz, H.; Zschornig, O.; Arnold, K. Matrix-assisted laser desorption and ionization time-of-flight (MALDI-TOF) mass spectrometry in lipid and phospholipid research. *Prog. Lipid Res.* **2004**, *43*, 449–488.

54. Ham, B. M.; Jacob, J. T.; Cole, R. B. MALDI-TOF MS of phosphorylated lipids in biological fluids using immobilized metal affinity chromatography and a solid ionic crystal matrix. *Anal. Chem.* **2005**, *77*, 4439–4447.

55. Jackson, S. N.; Wang, H. Y.; Woods, A. S. *In situ* structural characterization of phosphatidylcholines in brain tissue using MALDI-MS/MS. *J. Am. Soc. Mass Spectrom.* **2005**, *16*, 2052–2056.

56. Ikonomou, M. G.; Blades, A. T.; Kebarle, P. Electrospray-ion spray: A comparison of mechanisms and performance. *Anal. Chem.* **1991**, *63*, 1989–1998.

57. Tang, L.; Kebarle, P. Effect of the conductivity of the electrosprayed solution on the electrospray current. Factors determining analyte sensitivity in electrospray mass spectrometry. *Anal. Chem.* **1991**, *63*, 2709–2715.

58. Gaskell, S. J. Electrospray: Principles and practice. *J. Mass Spectrom.* **1997**, *32*, 677–688.

59. Gross, R. W.; Sobel, B. E. Isocratic high-performance liquid chromatography separation of phosphoglycerides and lysophosphoglycerides. *J. Chromatogr.* **1980**, *197*, 79–85.

60. DeLong, C. J.; Baker, P. R. S.; Samuel, M.; Cui, Z.; Thomas, M. J. Molecular species composition of rat liver phospholipids by ESI-MS/MS: The effect of chromatography. *J. Lipid Res.* **2001**, *42*, 1959–1968.

61. Han, X.; Gross, R. W. Global analyses of cellular lipidomes directly from crude extracts of biological samples by ESI mass spectrometry: A bridge to lipidomics. *J. Lipid Res.* **2003**, *44*, 1071–1079.

62. Han, X.; Yang, J.; Cheng, H.; Ye, H.; Gross, R. W. Towards fingerprinting cellular lipidomes directly from biological samples by two-dimensional electrospray ionization mass spectrometry. *Anal. Biochem.* **2004**, *330*, 317–331.

63. Han, X.; Gross, R. W. Shotgun lipidomics: Multi-dimensional mass spectrometric analysis of cellular lipidomes. *Expert Rev. Proteomics.* **2005**, *2*, 253–264.

64. Han, X.; Yang, K.; Yang, J.; Fikes, K. N.; Cheng, H.; Gross, R. W. Factors influencing the electrospray intrasource separation and selective ionization of glycerophospholipids. *J. Am. Soc. Mass Spectrom.* **2006**, *17*, 264–274.

65. Lagarde, M.; Geloen, A.; Record, M.; Vance, D.; Spener, F. Lipidomics is emerging. *Biochim. Biophys. Acta* **2003**, *1634*, 61.

66. Wenk, M. R. The emerging field of lipidomics. *Nat. Rev. Drug Discovery* **2005**, *4*, 594–610.

67. Guan, X. L.; He, X.; Ong, W. Y.; Yeo, W. K.; Shui, G.; Wenk, M. R. Non-targeted profiling of lipids during kainate-induced neuronal injury. *FASEB J.* **2006**, *20*, 1152–1161.

68. Guan, X. L.; Wenk, M. R. Mass spectrometry-based profiling of phospholipids and sphingolipids in extracts

from *Saccharomyces cerevisiae. Yeast* **2006**, *23*, 465–477.

69. Lieser, B.; Liebisch, G.; Drobnik, W.; Schmitz, G. Quantification of sphingosine and sphinganine from crude lipid extracts by HPLC electrospray ionization tandem mass spectrometry. *J. Lipid Res.* **2003**, *44*, 2209–2216.
70. Ekroos, K.; Chernushevich, I. V.; Simons, K.; Shevchenko, A. Quantitative profiling of phospholipids by multiple precursor ion scanning on a hybrid quadrupole time-of-flight mass spectrometer. *Anal. Chem.* **2002**, *74*, 941–949.
71. Hermansson, M.; Uphoff, A.; Kakela, R.; Somerharju, P. Automated quantitative analysis of complex lipidomes by liquid chromatography/mass spectrometry. *Anal. Chem.* **2005**, *77*, 2166–2175.
72. Sparagna, G. C.; Johnson, C. A.; McCune, S. A.; Moore, R. L.; Murphy, R. C. Quantitation of cardiolipin molecular species in spontaneously hypertensive heart failure rats using electrospray ionization mass spectrometry. *J. Lipid Res.* **2005**, *46*, 1196–1204.
73. Sommer, U.; Herscovitz, H.; Welty, F. K.; Costello, C. E. LC-MS-based method for the qualitative and quantitative analysis of complex lipid mixtures. *J. Lipid Res.* **2006**, *47*, 804–814.
74. Liebisch, G.; Drobnik, W.; Reil, M.; Trumbach, B.; Arnecke, R.; Olgemoller, B.; Roscher, A.; Schmitz, G. Quantitative measurement of different ceramide species from crude cellular extracts by electrospray ionization tandem mass spectrometry (ESI-MS/MS). *J. Lipid Res.* **1999**, *40*, 1539–1546.
75. Brugger, B.; Erben, G.; Sandhoff, R.; Wieland, F. T.; Lehmann, W. D. Quantitative analysis of biological membrane lipids at the low picomole level by nano-electrospray ionization tandem mass spectrometry. *Proc. Natl. Acad. Sci. USA* **1997**, *94*, 2339–2344.
76. Ekroos, K.; Shevchenko, A. Simple two-point calibration of hybrid quadrupole time-of-flight instruments using a synthetic lipid standard. *Rapid Commun. Mass Spectrom.* **2002**, *16*, 1254–1255.
77. Welti, R.; Wang, X. Lipid Species profiling: A high-throughput approach to identify lipid compositional changes and determine the function of genes involved in lipid metabolism and signaling. *Curr. Opin. Plant Biol.* **2004**, *7*, 337–344.
78. Han, X.; Yang, K.; Yang, J.; Cheng, H.; Gross, R. W. Shotgun lipidomics of cardiolipin molecular species in lipid extracts of biological samples. *J. Lipid Res.* **2006**, *47*, 864–879.
79. Han, X.; Gubitosi-Klug, R. A.; Collins, B. J.; Gross, R. W. Alterations in individual molecular species of human platelet phospholipids during thrombin stimulation: Electrospray ionization mass spectrometry-facilitated identification of the boundary conditions for the magnitude and selectivity of thrombin-induced platelet phospholipid hydrolysis. *Biochemistry* **1996**, *35*, 5822–5832.
80. Koivusalo, M.; Haimi, P.; Heikinheimo, L.; Kostiainen, R.; Somerharju, P. Quantitative determination of phospholipid compositions by ESI-MS: Effects of acyl chain length, unsaturation, and lipid concentration on instrument response. *J. Lipid Res.* **2001**, *42*, 663–672.
81. Schiller, J.; Sub, R.; Fuchs, B.; Muller, M.; Zschornig, O.; Arnold, K. MALDI-TOF MS in lipidomics. *Front. Biosci.* **2006**, *12*, 2568–2579.
82. Estrada, R.; Yappert, M. C. Regional phospholipid analysis of porcine lens membranes by matrix-assisted laser desorption/ionization time-of-flight mass spectrometry. *J Mass spectrom.* **2004**, *39*, 1531–1540.
83. Stoeckli, M.; Chaurand, P.; Hallahan, D. E.; Caprioli, R. M. Imaging mass spectrometry: A new technology for the analysis of protein expression in mammalian tissues. *Nat. Med.* **2001**, *7*, 493–496.
84. Li, L.; Garden, R. W.; Sweedler, J. V. Single-cell MALDI: A new tool for direct peptide profiling. *Trends Biotechnol.* **2000**, *18*, 151–160.
85. Ishida, Y.; Nakanishi, O.; Hirao, S.; Tsuge, S.; Urabe, J.; Sekino, T.; Nakanishi, M.; Kimoto, T.; Ohtani, H. Direct analysis of lipids in single zooplankter individuals by matrix-assisted laser desorption/ionization mass spectrometry. *Anal. Chem.* **2003**, *75*, 4514–4518.
86. Touboul, D.; Piednoel, H.; Voisin, V.; De La Porte, S.; Brunelle, A.; Halgand, F.; Laprevote, O. Changes in phospholipid composition within the dystrophic muscle by matrix-assisted laser desorption/ionization mass spectrometry and mass spectrometry imaging. *Eur. J. Mass Spectrom.* **2004**, *10*, 657–664.
87. Jackson, S. N.; Wang, H. Y.; Woods, A. S. Direct profiling of lipid distribution in brain tissue using MALDI-TOFMS. *Anal. Chem.* **2005**, *77*, 4523–4527.
88. Li, Y. L.; Gross, M. L.; Hsu, F.-F. Ionic-liquid matrices for improved analysis of phospholipids by MALDI-TOF mass spectrometry. *J. Am. Soc. Mass Spectrom.* **2005**, *16*, 679–682.
89. Jones, J. J.; Batoy, S. M.; Wilkins, C. L.; Liyanage, R.; Lay, J. O. Jr. Ionic liquid matrix-induced metastable decay of peptides and oligonucleotides and stabilization of phospholipids in MALDI FTMS analyses. *J. Am. Soc. Mass Spectrom.* **2005**, *16*, 2000–2008.
90. Fujiwaki, T.; Yamaguchi, S.; Sukegawa, K.; Taketomi, T. Application of delayed extraction matrix-assisted laser desorption ionization time-of-flight mass spectrometry for analysis of sphingolipids in tissues from sphingolipidosis patients. *J. Chromatogr. B Biomed. Sci. Appl.* **1999**, *731*, 45–52.
91. Schiller, J.; Arnhold, J.; Benard, S.; Muller, M.; Reichl, S.; Arnold, K. Lipid analysis by matrix-assisted laser desorption and ionization mass spectrometry: A methodological approach. *Anal. Biochem.* **1999**, *267*, 46–56.

92. Schiller, J.; Arnhold, J.; Glander, H. J.; Arnold, K. Lipid analysis of human spermatozoa and seminal plasma by MALDI-TOF mass spectrometry and NMR spectroscopy—Effects of freezing and thawing. *Chem. Phys. Lipids* **2000**, *106*, 145–156.
93. Schiller, J.; Zschornig, O.; Petkovic, M.; Muller, M.; Arnhold, J.; Arnold, K. Lipid analysis of human HDL and LDL by MALDI-TOF mass spectrometry and (31) P-NMR. *J. Lipid Res.* **2001**, *42*, 1501–1508.
94. Schiller, J.; Muller, K.; Suss, R.; Arnhold, J.; Gey, C.; Herrmann, A.; Lessig, J.; Arnold, K.; Muller, P. Analysis of the lipid composition of bull spermatozoa by MALDI-TOF mass spectrometry—A cautionary note. *Chem. Phys. Lipids* **2003**, *126*, 85–94.
95. Han, X.; Yang, J.; Cheng, H.; Yang, K.; Abendschein, D. R.; Gross, R. W. Shotgun lipidomics identifies cardiolipin depletion in diabetic myocardium linking altered substrate utilization with mitochondrial dysfunction. *Biochemistry.* **2005**, *44*, 16684–16694.

Chapter 21

In Vitro ADME Profiling and Pharmacokinetic Screening in Drug Discovery

Daniel B. Kassel

Takeda San Diego, Inc., San Diego, California

Electrospray and MALDI Mass Spectrometry: Fundamentals, Instrumentation, Practicalities, and Biological Applications, Second Edition, Edited by Richard B. Cole

21.1 INTRODUCTION

Focus within the pharmaceutical industry has been to increase the likelihood of successfully developing clinical candidates by optimizing the components of the discovery process [target identification → chemical design → synthesis → compound analysis → compound purification → registration → biological and ADME (absorption, distribution, metabolism, and excretion) screening]. By optimizing each step in the iterative discovery process, it is expected that compound attrition rate will be reduced dramatically as compounds advance into preclinical development. A myriad of new technologies has been introduced to streamline this process. Notably, advances in genomics, high-throughput screening, combinatorial chemistry, parallel synthesis, automation, miniaturization, and analytical chemistry have enabled large numbers of potent (active) and selective compounds to be identified at early stages of drug discovery. The field of analytical chemistry, and specifically the technique of liquid chromatography/mass spectrometry (LC/MS), enjoys important roles throughout the discovery process. Once considered primarily an enabling tool for medicinal chemists, LC/MS is now a key technology incorporated at just about every stage of the drug discovery process.

Drug discovery programs typically initiate as follows. Once a therapeutic area has been selected, the next step is to identify a biological target. Target selection is one of the most challenging aspects of drug discovery. A key factor influencing target selection is an assessment of the validation state of the target. The validation state spans (1) clinically validated targets (i.e., the mechanism of action established in humans), (2) semivalidated targets (animal disease models, genetic knockouts, and SiRNA and/or RNAi) that directly link the biological target with disease state in animal models, and (3) unvalidated targets (candidate proteins/genes identified as being up- or down-regulated in disease relative to healthy tissues/cells). Technological advances in the field of analytical chemistry have been made since the human genome initative was launched that have greatly facilitated biological target identification, too numerous to review here.[1] Once a target has been selected, the next step is to establish a high-throughput screen (HTS) to support screening of compound collections and synthetic libraries, a primary source of initial hits/actives. Almost without exception, this requires an assay that is amenable to microtiter plate format and an assay readout that is also plate-based (e.g., fluorescence-based). Historically, mass spectrometry has been considered too slow to compete with HTS methods, primarily because the technique is serial-based, as opposed to the HTS screening methods that are parallel-based (96-well, 384-well, and 1536-well formats). Recent advances in high-throughput mass-spectrometry-based systems are making this technique more attractive for HTS, especially in instances where a direct measurement of the endogenous substrate turnover is desirable (e.g., 11b-HSD1 enzymatically converts cortisol to cortisone).[2] In combination with

Table 21.1. Characteristics of a Developable Drug

- Good aqueous solubility
- Good pharmacokinetic profile for the intended route/frequency of dosing
- Balanced clearance
- Metabolized by several cytochrome P450s (as opposed to a single isoform)
- No chemically reactive metabolites
- Minimal P450 or P-glycoprotein (P_{gp}) inhibition
- Minimal P450 induction
- Not highly plasma protein bound (< 99%)
- Good safety margin

structure–activity relationship (SAR) data generated from HTS, chemists incorporate knowledge of protein three-dimensional structures and utilize computational tools (i.e., *in silico* methods that measure diversity and "drug-likeness" as well as two-dimensional and three-dimensional pharmacophore models [descriptors] that predict biological activity) to support iterative compound design, synthesis, and biological testing. Once the hits or actives have been identified, the process of hit refinement and lead optimization is initiated. Typically, this involves creation of a dedicated chemistry team that utilizes high-throughput organic synthesis and conventional medicinal chemistry strategies to rapidly converge on qualified leads [so-called hit-to-lead (H2L) stage]. LC/MS plays an extremely important role during the H2L stage, providing key enabling analysis and purification capabilities to the medicinal chemist.

The aspect that a compound is active and selective does not necessarily make it an attractive drug development candidate. Rodrigues described in a recent review the "desirable ADME properties" of a developable drug (see Table 21.1) and suggests that early access to ADME information enhances development success.[3] It is generally agreed that the "ideal" development candidate should contain the majority if not all of these biopharmaceutical properties shown in Table 21.1 prior to candidate selection for development. Traditionally, these developability criteria were not addressed until very late in the drug discovery process, often at or near the time of preclinical candidate selection. Not surprisingly, as a result, poor pharmacokinetics was identified as a key reason for compound attrition during clinical development, as described in a landmark paper by Kola and Landis.[4]

21.2 ADME SCREENING IN DRUG DISCOVERY

In order to identify chemotypes and lead compounds that contain these desirable developability properties, it is recognized that studies that assess ADME/PK should be initiated as early as possible in the discovery process.[5–7] By doing so, it is anticipated that the likelihood of development success will be maximized and development time reduced. This shift from late stage optimization of ADME/PK properties to a strategy of identifying potential liabilities early in the discovery process has taken hold within the pharmaceutical community. Traditionally, the discovery phase focused primarily on generating SARs. The new paradigm of early ADME/PK screening adds the dimension of structure-ADME

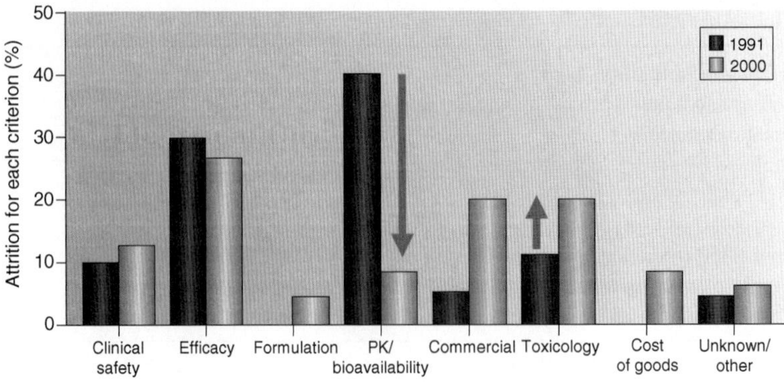

Figure 21.1. Attrition during drug development. Compounds failing to make it through drug development due to pharmacokinetic-related issues reduced from approximately 40% in 1990 to less than 10% in 2000. In part, this can be explained by pulling forward the evaluation of metabolic stability, CYP inhibition, plasma protein binding, and pharmacokinetics into drug discovery. (Reprinted with permission from *Nat. Rev. Drug Discovery* **2004**, *3*, 711–715.)

relationships in parallel to structure-activity relationships as an integral part of the iterative drug discovery process.[8–11] In the paper by Kola and Landis[4] and shown in Figure 21.1, failures in clinical development today are now no longer primarily attributed to pharmacokinetic issues but have shifted to issues such as toxicity, efficacy, and market factors.

Because of the large number of hits that are now routinely identified from screening compound collections and gene family compound libraries, the industry has recognized the need for high-throughput ADME assays. Most ADME assays can be run in a high-throughput fashion, principally because of the widespread incorporation of liquid chromatography/mass spectrometry (LC/MS) and liquid chromatography/tandem mass spectrometry (LC/MS/MS).[12,13] LC/MS and LC/MS/MS have become the preferred techniques for *in vitro* ADME analyses, principally as a result of enhanced sensitivity, selectivity, and ease of automation relative to traditional analytical methods. The selectivity advantages of LC/MS have made possible the ability to analyze endogenous and nonfluorescent probe substrates in cytochrome inhibition assays,[14] enabled rapid permeability assessment (e.g., Caco-2 assay),[15] provided faster methods for assessing lipophilicity and solubility of drug leads, and provided much more facile assessment of liver metabolism,[16,17] for which many examples are highlighted below.

To achieve high throughput, it is important that these assays be brought forward into the discovery process as early as possible. One approach is to initiate ADME assays at the time of biological screening. Once compounds are registered, they are generally plated and arrayed in 96-well microtiter plates at a concentration of 10 mM in DMSO. Many *in vitro* ADME screens may be performed in microtiter plate format, and it is at the time of biological screening that a number of daughter plates may also be generated for high-throughput ADME, as shown in Figure 21.2. In addition to the metabolic stability assays, plasma protein binding can be performed in microtiter plate format using both the ultrafiltration method[18] and equilibirium dialysis method.[19] In addition, both solubility and log *P*/log *D* screens have been performed in microtiter plate format.[20,21]

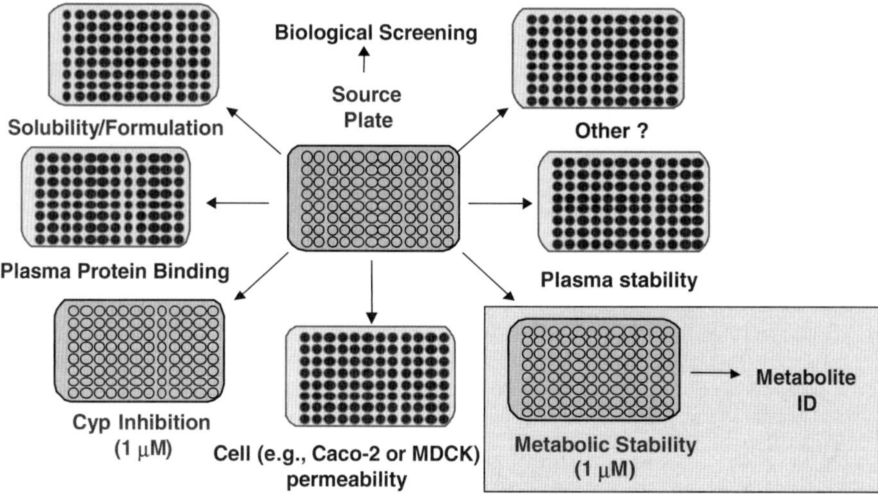

Figure 21.2. Streamlined approach to evaluating ADME/PK properties of molecules at the time of biological screening. Typically, organizations apply a tiered ADME approach to the evaluation of lead compounds, first assessing metabolic liabilities and/or potential for drug–drug interactions prior to *in vivo* pharmacokinetic studies. Generally, during lead generation (hits with confirmed activities of $\leq$10 uM), compounds are evaluated for their "druggability" using *in silico* methods that calculate clogP, polar surface area, and so on. During hit expansion and lead optimization, significant time-savings can be made if compounds are evaluated for biological activity concurrent with *in vitro* ADME screening. This enables promising chemotypes to be identified early, thereby increasing their chance to be considered for candidate selection.

21.3 *IN VITRO* ABSORPTION AND PERMEABILITY SCREENING

The property of absorption is of great interest because oral dosing is by far the preferred route of administration to treat most diseases (with the obvious exception of life-threatening diseases such as cancer). For compounds to reach systemic circulation and hence their site of action following an oral dose, they must pass through several key hurdles (i.e., survive first-pass effects), as described by Kerns and Li[7] and shown in Figure 21.3. The first barrier an orally administered compound encounters is disintegration/dissolution. Factors affecting disintegration and dissolution include, but are not limited to: size of particle, form (amorphous versus crystalline), solubility, log *P*, and so on. Once in solution, a fraction of the compound may be absorbed (primarily in the lumen of the small intestine). The primary factors affecting absorption include permeability and efflux. The fraction of the drug that is absorbed is transferred to the portal vein, where it is presented to the liver (the primary organ of xenobiotic transformation) prior to reaching systemic circulation. The compound may be either a substrate for or an inhibitor of the cytochrome P_{450} metabolizing enzymes (i.e., the monooxidases primarily responsible for metabolizing xenobiotics). The extent to which a compound is metabolized by the P_{450} metabolizing enzymes as well as the extent to which a compound is a substrate for the efflux transporter, P-glycoprotein (P_{gp}) directly affects the bioavailability of the compound. Also, knowledge of the inhibitory effect of a drug molecule on any one of the key cytochrome P_{450} metabolizing enzymes is crucial to preventing undesirable drug–drug interactions when drugs are co-administered (a more and more common occurrence in the clinic).

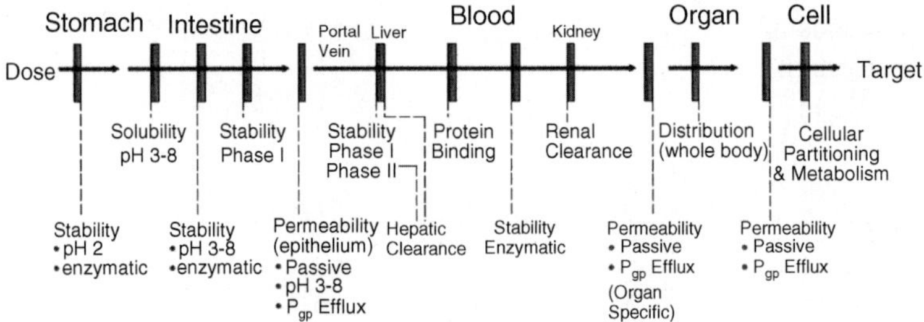

Figure 21.3. The various hurdles a compound must overcome before reaching systemic circulation and its target of action. For an orally administered drug, the compound must survive the acidic environment of the stomach in tact, it must be capable of undergoing dissolution, and a fraction of the drug must be absorbed into the intestine. The fraction absorbed is then subject to oxidative and conjugative metabolism in the liver prior to reaching systemic circulation. These factors, along with plasma protein binding and volume of distribution, ultimately determine the amount of drug that is available to reach its site of action. (Reprinted with permission from Kerns et al., *Drug Discovery Today* **2003**, *8*(7) 316–323.)

Orally administered drugs are absorbed primarily by the intestines via one of three primary processes: (a) passive diffusion, (b) paracellular transport, or (c) active transport: Most lipophilic drugs (approximately 80%) diffuse through the lipid bilayer cell membrane by the process of passive diffusion. Paracellular transport, on the other hand, involves the transport of hydrophilic molecules through gaps between cells in the intestinal epithelium. This process is regulated by tight junctions between cells. Transporter-mediated absorption occurs for small, hydrophilic molecules that are substrates for protein transporters. Drugs that are substrates of the uptake transporters can enter the systemic circulation after oral administration via this pathway. Efflux transporters (most notably P_{gp}) pump xenobiotics that have entered the intestinal epithelium back into the lumen.

21.3.1 Caco-2

The Caco-2 system is perhaps the most widely used cell culture model to assess intestinal absorption. The Caco-2 cell line is derived from human colon carcinoma tissue and is used to mimic intestinal mucosal epithelium and therefore mimic human absorption. Caco-2 cells, upon differentiation, develop tight cell junctions as well as uptake and efflux transporters, which are expressed in intestinal mucosa. Caco-2 cells are widely used for screening compounds for their permeability as a surrogate for human absorption as well as for assessing efflux (P_{gp}-mediated). Caco-2 cells are generally plated on the surfaces of 24-well microtiter plates whereby they form monolayers with tight liquid junctions after 21-day cell growth. In microtiter plate format, these differentiated cells are well-suited to both automated liquid handling transfer steps (dosing) and direct autosampling for LC/MS and LC/MS/MS analysis. For passive diffusion measurement, compounds are typically dosed into the apical compartment at a concentration of 1–10 μM. The flux of compound from the apical to basalateral membrane is measured after a period of time (typically 3 h). The apparent permeability (P_{app}) is calculated based on this flux. Compounds exhibiting low permeability are classified with $P_{app} < 1 \times 10^{-6}$ cm/s, moderate permeability 1×10^{-6} cm/s $\leq P_{app} \leq 10 \times 10^{-6}$ cm/s, and high permeability $P_{app} > 10 \times 10^{-6}$ cm/s. Often, bi-directional permeability studies (apical to basolateral and basolateral to apical) are performed in Caco-2 cells to evaluate both passive diffusion and efflux potential. In this case, compounds are dosed either on the apical (*A*) or basalateral (*B*) sides, and their apparent permeabilities calculated. Compounds exhibiting *B*/*A* ratios of ≥ 3 are considered to be transported out of the intestinal mucosa via active transport (P_{gp}-mediated). For measurements of flux of drug from apical to basolateral and from basolateral to apical is generally carried out by LC/MS and LC/MS/MS analysis. Standard LC/MS methods include direct autosampling from the Caco-2 chamber. High-throughput LC/MS/MS methods have been developed, including cassette dosing,[22] dual-column LC/MS/MS.[23] Recently, on-line extraction by turbulent flow chromatography coupled to tandem mass spectrometric detection was used to analyze multiple compounds in a single analytical run. The authors combined cassette with multiple reaction monitoring mass spectrometry to increase analysis throughput.

21.3.2 Madin–Darby Canine Kidney (MDCK) Cell Model

The Madin–Darby canine kidney (MDCK) cell model is another commonly used cell monolayer system for the assessment of human intestinal absorption. MDCK cell lines offer a couple of key advantages over Caco-2 cells, including the ability to reach full differentiation in 3–7 days (as opposed to 21 days for Caco-2) as well as the ability to be co-transfected with P_{gp}, the primary intestinal efflux transporter. Good correlation has been observed between MDCK permeability and human absorption.[24] The primary disadvantage of the MDCK cell line is that it is derived from canine, rather than human, cells; and, as such, expression of transporters and metabolizing enzymes is not identical.

21.3.3 Parallel Artifical Membrane Permeability Assay (PAMPA)

Parallel artificial membrane permeability assay (PAMPA), offers a fast and robust tool for screening permeability of candidate molecules in early discovery. The method involves coating a membrane surface of a 96-well plate with a lecithin solution containing dodecane and measuring the passive diffusion of drug molecules from the donor well to receiver well. The PAMPA system is an artificial lipophilic membrane surface that mimics the intestinal lumen. The amount of transfer of drug from the donor to receiver compartment is typically determined either using a UV plate reader or by LC/MS (for higher sensitivity experiments or for situations in which no substantial UV chromophore exists for the compound class). The technique is used to estimate permeability for compounds with a passive transcellular diffusion mechanism. Preferably it may serve as a prescreening tool for permeability ranking when multiple *in vitro* models are introduced to address GI permeability. Typically, sample is prepared by diluting directly from a 10 mM working stock solution of test compound to 1 μM or 10 μM in physiological pH buffer. The test compound is dosed into the donor well, and after a period of time (typically 1–2 h) the transfer of compound to the receiver compartment is measured and reported as a permeability flux value (similar to the Caco-2 assay). The technique has been shown by Kansy et al.[25] and by Adveef[26] to correlate very well with human absorption. One of the principal drawbacks to the method is that when run under physiological conditions, extensive nonspecific binding to the membrane surface can occur, complicating the interpretation of results.

Recently, the double-sink PAMPA assay was introduced to circumvent this problem and to further improve the predictiveness of the method for estimating absorption.[27] The double-sink method involves dosing the test compound into the donor compartment at pH 5 with the receiver compartment at physiological pH. The reason for creating a pH gradient in this *in vitro* model is to mimic the *in vivo* situation. As a compound passes through the gastrointestinal tract (GIT), it is exposed to different pH microenvironments (pH 1–8). A pH gradient is formed, and it is this gradient that drives the transfer of compounds across membrane barriers. This creates a pH gradient between the acceptor and donor wells and a chemical scavenger system present in the acceptor wells (the second sink). This model has proved useful for high-throughput screening because it operates on a faster timescale than traditional PAMPA and has greatly reduced membrane retention, thereby facilitating quantitation of highly lipophilic compounds.

21.4 CYTOCHROME P450 METABOLISM STUDIES

The cytochrome P450 family of enzymes plays one of the most important roles in xenobiotic metabolism. The CYP isozymes are a group of heme-containing enzymes held within the lipid bilayers of the endoplasmic reticulum of hepatocytes (liver cells). The enzymes are responsible primarily for the biotransformation (oxidative metabolism) of endogenous substrates and xenobiotics, mainly in the liver, and to a lesser extent in intestine, kidney, lungs, and brain. Of all the CYP enzymes reported in literature, the major metabolizing enzymes are CYP1A2, CYP2C8, CYP2C9, CYP2C19, CYP2D6, and CYP3A4, and as many as 50% of drugs are metabolized by CYP3A4 alone.[28] All compounds, independent of route of administration, are subject to biotransformation in the liver, and the extent to which a compound is metabolized in the liver is driven by a number of factors, including liver blood flow, protein binding, and, importantly, whether or not the compound is a substrate for a metabolizing enzyme. Compounds delivered via an oral route of administration are subject to first-pass metabolism in the liver upon intestinal absorption. Once absorbed, the drug passes through the portal vein and reaches the liver, where it is subjected to Phase I oxidative metabolism (e.g., oxidation, reduction, or hydrolysis), Phase II conjugation (e.g., glucuronidation, sulfation, or glutathionylation) may be Phase I-dependent or -independent. Phase 2 enzymes include uridine diphosphate (UDP)-dependent glucuronosyl transferase (UGT), phenol sulfotransferase (PST), estrogen sulfotransferase (EST), and glutathione-*S*-transferase (GST), all of which exist in multiple isoforms.[29]

21.4.1 Metabolic Stability Screening

The metabolic stability of a drug candidate (sometimes referred to as metabolic rate) reflects the amount of drug metabolized by liver enzymes per unit time and is one of the first barriers a compound needs to overcome before it can undergo systemic circulation and reach its target of action. The *in vitro* microsomal (often referred to as metabolic) stability assay has been employed in drug discovery as a precursor to *in vivo* pharmacokinetic profiling for the specific purpose of ranking compounds based on their intrinsic clearance (calculated) values and to predict *in vivo* clearance.[30,31] Microsomes are endoplasmic reticulumn membrane vesicles prepared by homogenization of the liver, followed by a centrifugation at high speed (9000*g*) to remove whole cells, free nuclei, plasma membranes, and mitochondria. The resulting supernatant is then centrifuged at even higher speed, typically 100,000*g*, to pellet the endoplasmic reticulum, which contains the P450 isoforms and one of the phase 2 conjugating enzymes, UGT. Microsomes, although primarily used to assess Phase I metabolism, can also be used to study glucuronidation by addition of UDP-glucuronic acid (UDPGA) and the pore opener, alemethicin.[32] Microsomes are readily available, and the assay is relatively cheap to employ and easy to run. The ability to predict *in vivo* clearance from *in vitro* metabolic stability data in the absence of knowledge of Phase II metabolism is challenging. Alternatives to microsomes that enable assessment of both Phase I and Phase II metabolism (e.g., glucuronidation, sulfonylation, etc.) include S9 fractions and cryopreserved and fresh hepatocytes, although the ability to predict *in vivo* clearance from these *in vitro* systems has proved challenging as well. This is principally because these *in vitro* assays do not take into account other processes that affect clearance, such as plasma protein binding. Importantly though, *in vitro* metabolic stability assays have proved extremely useful in the following ways: (a) to aid in the selection of the most attractive scaffold (starting points) to minimize the number of iterations required to produce a suitable drug candidate, (b) to prioritize/rank compounds for *in vivo* profiling, and (c) to weed out compounds that are metabolized rapidly *in vitro*.

Mass spectrometry has been used primarily to assess xenobiotic metabolism by the CYP enzymes.[33–35] In many of these studies, compounds are evaluated for metabolic stability over a time course or at a single endpoint determination. Compounds are incubated with microsomes + NADPH at a substrate concentration presumed to be at or near their K_m values (typically 1 μM). At a predetermined endpoint, the incubations are quenched by (a) addition of either acetonitrile or trichloroacetic acid and (b) the protein precipitate pelleted by centrifugation. An aliquot of the supernatant is then analyzed by LC/MS/MS, using either selected ion monitoring or multiple reaction monitoring. The $T_{1/2}$ and intrinsic clearance (Clint) values are determined and reported to project teams.

The analytical community is constantly being challenged to identify methods for enhanced throughput. The simplest way to enhance analysis throughput is to incorporate fast HPLC run times. Samples can be analyzed one at a time or as mixtures in as little as 30 s to 1 min per sample by applying fast gradients compatible with mass spectrometric detection to assess ADME properties.[34,35] Figure 21.4 shows an example of how fast chromatography is applied to characterizing probe substrates for cytochrome P450 metabolizing enzymes. These fast analyses are achieved using short chromatographic columns (e.g., 2.1 × 20 mm; 3- or 5-μm particle size). Generic gradients are typically employed

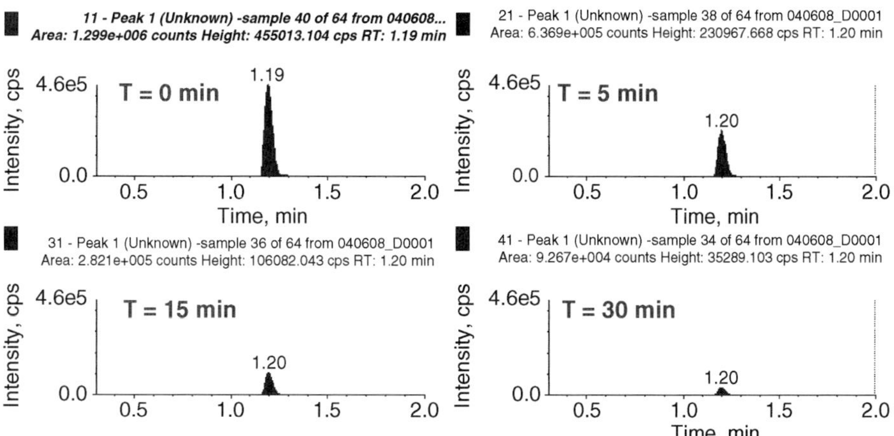

Figure 21.4. A typical time-course metabolic stability analysis. Compound is incubated with human liver microsomes in the presence of NADPH. At predetermined times, the reactions are quenched (using acetonitrile or trichloroacetic acid) and analyzed by LC/MS/MS (MRM mode). Parent drug signal is monitored over this time course (peak area determination).

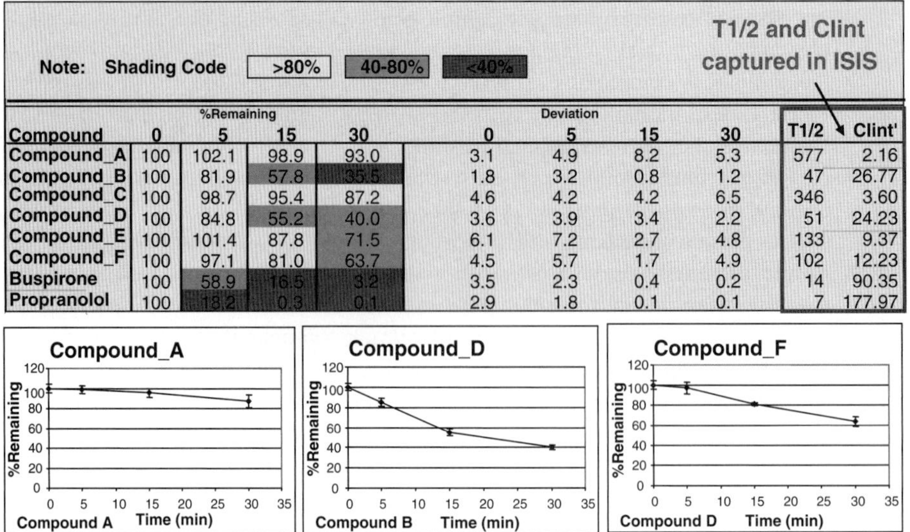

Compound	%Remaining 0	5	15	30	Deviation 0	5	15	30	T1/2	Clint'
Compound_A	100	102.1	98.9	93.0	3.1	4.9	8.2	5.3	577	2.16
Compound_B	100	81.9	57.8	35.5	1.8	3.2	0.8	1.2	47	26.77
Compound_C	100	98.7	95.4	87.2	4.6	4.2	4.2	6.5	346	3.60
Compound_D	100	84.8	55.2	40.0	3.6	3.9	3.4	2.2	51	24.23
Compound_E	100	101.4	87.8	71.5	6.1	7.2	2.7	4.8	133	9.37
Compound_F	100	97.1	81.0	63.7	4.5	5.7	1.7	4.9	102	12.23
Buspirone	100	58.9	16.5	3.2	3.5	2.3	0.4	0.2	14	90.35
Propranolol	100	18.2	0.3	0.1	2.9	1.8	0.1	0.1	7	177.97

Figure 21.5. $T_{1/2}$ and intrinsic clearance measurements are made taking into account milligrams of microsomal protein used in the metabolic stability assay and the species used (e.g.; human, rat). Compounds are generally "binned" according to their metabolic stability, as either stable, moderately stable, moderately unstable, or unstable light, medium, and dark shading, respectively. This information is used, in conjunction with other information (e.g.; PAMPA, biological data, selectivity data, etc.) to prioritize compounds for pharmacokinetic screening.

(e.g., 5% to 95% organic in 1.5 min with a 30-s column reequilibration time). Compounds generally afford peaks that are symmetrical with peak widths less than 0.05 min at baseline. The time-course metabolic stability of the drug is shown ($t = 0$, 5, 15, and 30 min). Time-course measurement of the disappearance of parent compound allows for a relatively accurate estimation of $T_{1/2}$ and intrinsic clearance (Clint). Project teams are also interested in both graphical and tabular presentation of data following in vitro ADME profiling. Shown in Figure 21.5 is a partial summary report of a plate of eight reference compounds and 88 test compounds received from a drug discovery project. The values of the remaining percentages of parent compounds are color-coded for easy visualization: green, >80%; orange; 80–40%; red, <40%. The project chemist receives both (a) the summary report which helps "bin" the compounds into distinct classes of microsomal stability and (b) the time-course stability plots for all compounds submitted for HT microsomal stability analysis. This information helps to prioritize compounds for further consideration as potential drug candidates.

Some groups choose to perform a single time-point determination ($t = 0$ and 20 min) as a mechanism for further accelerating analysis throughput. Monolithic columns have been evaluated more recently to reduce analysis times further. Monolithic columns are attractive in that they tend to operate at very low back-pressures and are hence capable of being operated at very high flow rates. Chromatographic integrity is generally not compromised because peak capacity and resolution on monoliths is ostensibly independent of flow rate. Typically, a 4.6×50-mm monolith column operates at a flow rate of 4–6 mL/min. Performance is similar to that obtained using 3-μm packed columns, but with flow rates much greater than those normally employed. Thus, the theoretical advantage is that chromatographic run times can be reduced by a factor of 3–5 times without loss in performance. Van de Merbel et al.[36] described the use of monoliths in quantitative bioanalysis, showing the advantage of these columns over conventional C18 supports for the determination of estradiol in plasma.

Ultra-performance liquid chromatography (UPLC) has gained substantial interest from the analytical community recently, owing to the fact that separations can be achieved in substantially shorter times than by standard fast HPLC or monolith methods, without compromising chromarographic integrity. This is explained by the use of small particles ($<2\ \mu m$) and by operating the HPLC at high flow rates, both possible due to advances in HPLC technology, allowing separations to be carried out at ultra-high pressures (up to 15,000 psi). Members of the author's laboratory adopted a post-incubation pooling strategy based on clogD combined with UPLC to accelerate metabolic stability screening of candidate drugs.[37] Following incubation and termination of reactions by addition of trichloroacetic acid/acetontrile, four-component pools were prepared (based on clogD and MW) and analyzed by UPLC/MS/MS. The results are shown in Figure 21.6. By analyzing in pools of four compounds, the effective run time per compound can be reduced to less than 30 s per sample.

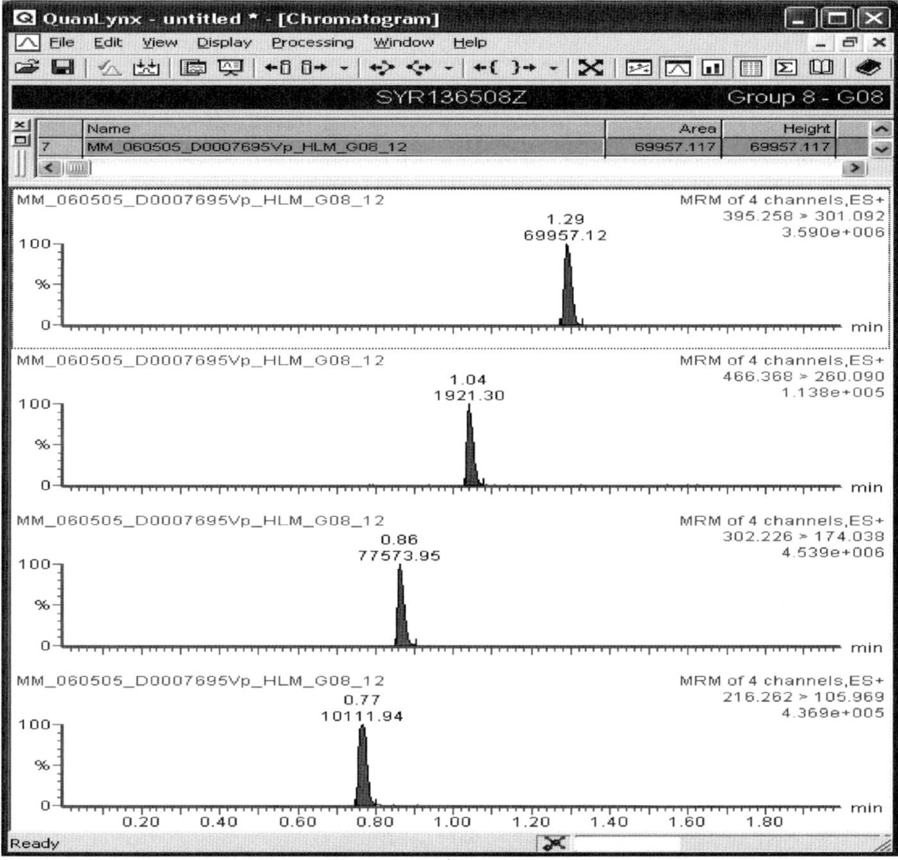

Figure 21.6. Ultra-performance liquid chromatography/MS/MS chromatograms for four compounds pooled, post-microsomal incubations, based on clogD. The four compounds are baseline separated using a 1.7-μm particles and a fast generic gradient.

21.4.2 Cytochrome P450 Inhibition and Induction Screening

Compounds that are potent inhibitors or inducers of one or more cytochrome P450 (CYP450) enzymes have a potential for drug–drug interactions. More and more frequently, drugs are being prescribed in combination with other medications to enhance their therapeutic potential (through synergistic or complimentary mechanisms of action). The risk of such a strategy, should the compounds not be evaluated for drug–drug interaction potential prior to candidate selection, is that one drug may inhibit the metabolism of a co-administered drug and the affected drug may, as a consequence, be toxic. Alternatively, if the compound causes CYP450 enzyme induction, the net result may be altered plasma concentrations of the co-administered drug, rendering it ineffective (i.e., sufficient plasma levels not reached) or less effective (*in vivo* half-life reduced) due to excessive metabolism by the induced enzyme. To reduce the risk of adverse *in vivo* drug–drug interactions, CYP450 inhibition assays are performed to evaluate whether candidate molecules in lead generation and lead optimization are inhibitors of the drug-metabolizing CYP450 enzymes. The drug candidates are incubated with recombinant enzymes (baculovirus, human lymphoblasts) or with liver microsomes or hepatocytes in the presence of substrates specific for the principal drug-metabolizing enzymes. Reduction in the rate of substrate conversion to product in the presence and absence of the candidate molecules allows one to determine IC_{50} (concentration that decreases substrate turnover by 50%) and K_i (inhibitory constant) values against the isolated enzyme and hence measure its inhibitory potential. IC_{50} or K_i values as compared with the intended plasma concentration would suggest a high potential of the drug to cause drug interactions with coadministered drugs, which are substrates of the affected enzyme.

There have been numerous studies conducted using both endogenous substrates and fluorescent probe substrates for the purpose of evaluating CYP inhibition potential of test compounds. Cohen et al.[38] demonstrated a lack of correlation between CYP inhibitors *in vitro* when using fluorescent versus conventional P450 probe substrates, arguing for LC/MS-based approaches.

21.4.3 Post-Incubation Pooling Analysis by LC/MS/MS

Most groups employ fast liquid chromatography with tandem mass spectrometry (LC/MS/MS) for routine screening of drug candidates for inhibition of the major human cytochrome P450 (CYP) isozymes, CYP1A2. CYP2C8, CYP2C9, CYP2C19, CYP2D6, and CYP3A4.[39–41] Figure 21.7 shows the LC/MS/MS analysis of the products of the reactions of the six CYP isozymes with their probe substrates (CYP1A2, phenacetin, CYP2C8, amodiaquine; CYP2C9, diclofenac; CYP2C19, *S*-mephenytoin; CYP2D6, bufuralol; CYP3A4, midazolam). The total analysis time for the baseline separation of the six products of the CYP reactions is 2 min, equating to an effective run time of 20 s/CYP). Measurement of the inhibitory potential of a drug candidate against the CYP enzymes is generally done either at a single concentration (typically 10 μM) or over a concentration range, so that an IC_{50} determination can be obtained. The generally agreed rule of thumb is that compounds exhibiting an IC_{50} value of $\geq$10 μM are unlikely to exhibit significant drug–drug interactions. Shown in Figure 21.8 are the dose response curves and calculated IC_{50} values for the standard probe inhibitors of the various CYP enzymes (CYP1A2, furafylline; CYP2C8, montelukast; CYP2C9, sulfaphenazole; CYP2C19, tranylcypromine; CYP2D6, quinidine; CYP3A4, ketoconazole) evaluated by LC/MS/MS. Replicate

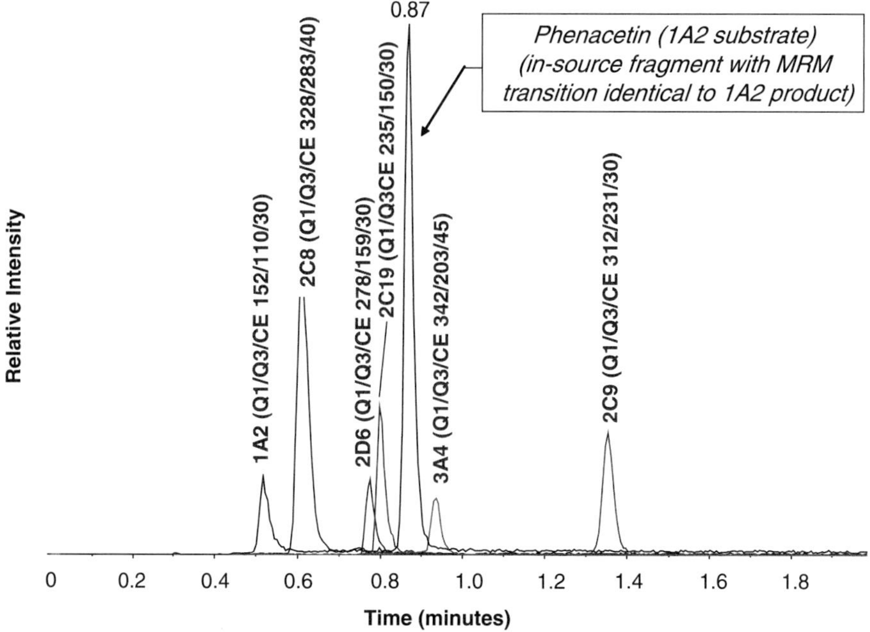

Figure 21.7. Cassette analysis of six CYP probe substrates (CYP1A2, CYP2C8, CYP2C9, CYP2C19, CYP2D6, CYP3A4) separated on a 5-μm phenylcolumn (1 mm × 50 mm) operated at 700 μL/min. The samples were detected in the MRM mode on an API4000 triple quadrupole mass spectrometer.

determinations were made, and the results between the two separate analyses compared favorably. Equally, the results of this LC/MS/MS method compared favorably with literature values.[42–44]

Peng et al.[45] incorporated a monolithic silica rod column to allow fast flow rates to significantly reduce chromatographic run times. In their study, major metabolites of six CYP-specific probe substrates for the five P450 isoforms were monitored and quantified to determine IC_{50} values of five drug compounds against each CYP450 isozyme. Human liver microsomal incubation samples at each test compound concentration were combined and analyzed simultaneously by the LC/MS/MS method. Each pooled sample containing six substrates and an internal standard was separated and detected in only 24 s. The combination of ultrafast chromatography and sample pooling techniques was shown to significantly increase sample throughput and shortened assay turnaround time, allowing a large number of compounds to be screened rapidly for potential P450 inhibitory activity, to aid in compound selection and optimization in drug discovery.

CYP inhibition screening is an extremely valuable *in vitro* screen for prioritizing compounds for preclinical development. However, it is important to understand the relationship between *in vitro* inhibitory potential, plasma protein binding, and effective plasma concentration. Kajosaari et al.[46] showed that pioglitazone, a widely prescribed drug for the treatment of Type II diabetes, is found *in vitro* to be an inhibitor of CYP2C8 and CYP3A4. *In* vivo, however, the compound does not increase the plasma concentrations of co-administered drugs that are known to be substrates for CYP2C8 and CYP3A4 (repaglinide, in their particular study). The authors concluded that the inhibitory effect of pioglitazone on CYP2C8 and CYP3A4 is very weak *in vivo*, owing to its extensive plasma protein binding.[46]

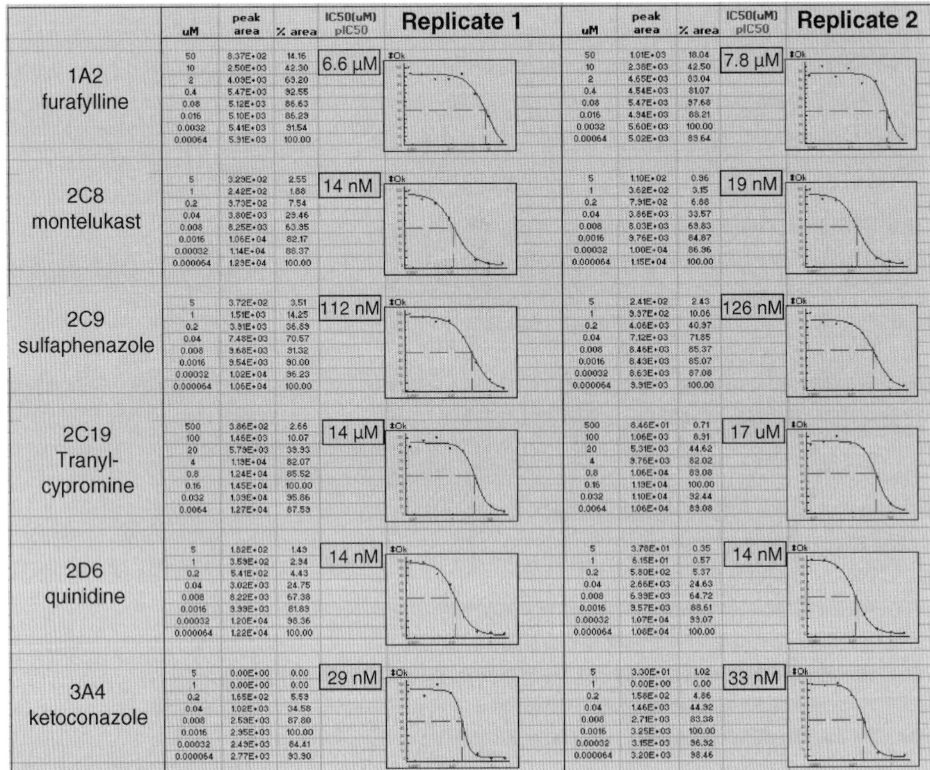

	uM	peak area	% area	IC50(uM) pIC50 (Replicate 1)	uM	peak area	% area	IC50(uM) pIC50 (Replicate 2)
1A2 furafylline	50	8.37E+02	14.16	6.6 μM	50	1.01E+03	18.04	7.8 μM
	10	2.50E+03	42.30		10	2.38E+03	42.50	
	2	4.09E+03	69.20		2	4.65E+03	83.04	
	0.4	5.47E+03	92.55		0.4	4.54E+03	81.07	
	0.08	5.12E+03	86.63		0.08	5.47E+03	97.68	
	0.016	5.10E+03	86.29		0.016	4.94E+03	88.21	
	0.0032	5.41E+03	91.54		0.0032	5.60E+03	100.00	
	0.00064	5.91E+03	100.00		0.00064	5.02E+03	89.64	
2C8 montelukast	5	3.29E+02	2.55	14 nM	5	1.10E+02	0.96	19 nM
	1	2.42E+02	1.88		1	3.62E+02	3.15	
	0.2	9.73E+02	7.54		0.2	7.91E+02	6.88	
	0.04	3.80E+03	29.46		0.04	3.86E+03	33.57	
	0.008	8.25E+03	63.95		0.008	8.03E+03	69.83	
	0.0016	1.06E+04	82.17		0.0016	9.76E+03	84.87	
	0.00032	1.14E+04	88.37		0.00032	1.00E+04	86.96	
	0.000064	1.29E+04	100.00		0.000064	1.15E+04	100.00	
2C9 sulfaphenazole	5	3.72E+02	3.51	112 nM	5	2.41E+02	2.43	126 nM
	1	1.51E+03	14.25		1	9.97E+02	10.06	
	0.2	3.91E+03	36.89		0.2	4.06E+03	40.97	
	0.04	7.48E+03	70.57		0.04	7.12E+03	71.85	
	0.008	9.68E+03	91.32		0.008	8.46E+03	85.37	
	0.0016	9.54E+03	90.00		0.0016	8.43E+03	85.07	
	0.00032	1.02E+04	96.23		0.00032	8.63E+03	87.08	
	0.000064	1.06E+04	100.00		0.000064	9.91E+03	100.00	
2C19 Tranyl-cypromine	500	3.86E+02	2.66	14 μM	500	8.46E+01	0.71	17 uM
	100	1.46E+03	10.07		100	1.06E+03	8.91	
	20	5.79E+03	39.93		20	5.31E+03	44.62	
	4	1.19E+04	82.07		4	9.76E+03	82.02	
	0.8	1.24E+04	85.52		0.8	1.06E+04	89.08	
	0.16	1.45E+04	100.00		0.16	1.19E+04	100.00	
	0.032	1.39E+04	95.86		0.032	1.10E+04	92.44	
	0.0064	1.27E+04	87.59		0.0064	1.06E+04	89.08	
2D6 quinidine	5	1.82E+02	1.49	14 nM	5	3.78E+01	0.35	14 nM
	1	3.59E+02	2.94		1	6.15E+01	0.57	
	0.2	5.41E+02	4.43		0.2	5.80E+02	5.37	
	0.04	3.02E+03	24.75		0.04	2.66E+03	24.63	
	0.008	8.22E+03	67.38		0.008	6.99E+03	64.72	
	0.0016	9.99E+03	81.89		0.0016	9.57E+03	88.61	
	0.00032	1.20E+04	98.36		0.00032	1.07E+04	99.07	
	0.000064	1.22E+04	100.00		0.000064	1.08E+04	100.00	
3A4 ketoconazole	5	0.00E+00	0.00	29 nM	5	3.30E+01	1.02	33 nM
	1	0.00E+00	0.00		1	0.00E+00	0.00	
	0.2	1.65E+02	5.59		0.2	1.58E+02	4.86	
	0.04	1.02E+03	34.58		0.04	1.46E+03	44.92	
	0.008	2.59E+03	87.80		0.008	2.71E+03	83.38	
	0.0016	2.95E+03	100.00		0.0016	3.25E+03	100.00	
	0.00032	2.49E+03	84.41		0.00032	3.15E+03	96.92	
	0.000064	2.77E+03	93.90		0.000064	3.20E+03	98.46	

Figure 21.8. Standard probe inhibitiors for CYP1A2, CYP 2C8, CYP 2C9, CYP 2C19, CYP 2D6, and CYP 3A4 were incubated at eight inhibitor concentrations and analyzed by LC/MS/MS using cassette analysis, as described in the previous figure. Experimentally derived IC_{50} values for each of the probe inhibitors compared favorably with literature values.

21.5 HIGHER-THROUGHPUT ANALYTICAL METHODS TO SUPPORT ADME EVALUATION

21.5.1 Parallel Chromatography for Accelerated CYP Metabolism and Inhibition Screening

Parallel HPLC has shown promise for high-throughput ADME screening. The parallel LC-MS methods allow multiple samples to be analyzed in parallel by injecting discrete compounds onto multiple columns and detecting them simultaneously in a single mass spectrometer ion source. Numerous groups have developed and implemented parallel LC-MS methods to support HT ADME studies.[47–50] Existing LC-MS systems may be converted into parallel LC-MS systems with minimal modification, as shown in Figure 21.9. Successful parallel analytical ADME analyses have been achieved using a simple valve manifold to split the flow from a binary HPLC system evenly between eight analytical columns. A generic high-throughput parallel LC-MS system consists of a high-pressure binary solvent delivery pumping system, a multiple probe autosampler, a switching valve, and a single quadrupole mass spectrometer equipped with a electrospray ionization (with or without indexed sprayer capabilities). Eight samples are injected into the eight injection ports simultaneously and onto eight separate microbore columns. The pump flow rate is split into eight equivalent streams

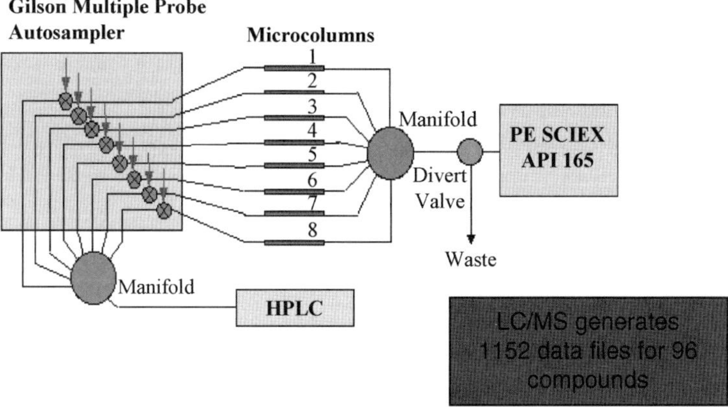

Figure 21.9. Eight-channel parallel LC-MS system consisting of gradient HPLC pumps, a multiple probe autosampler, a single quadrupole mass spectrometer equipped with an electrospray ion source, eight microbore columns (10 mm × 1 mm i.d.; 3 µm, HQ-C18), and a switching valve. The volume of the sample loops is 20 µL and full loop injections were used for all experiments. Total mobile phase flow rate is 2.0 mL/min (0.25 mL/min for each column). The eight-channel LC-MS system allows up to four plates of compounds (or 4 × 1152 samples) to be run in a single day.

using a Valco manifold before entering the multiple probe autosampler. Eight microbore columns are connected to eight injection valves of the autosampler. For non-indexed mass spectrometric detection, the outlets of the columns are recombined using a second valve manifold just prior to the ion source. The flow is then passed through a flow divert valve before entering the mass spectrometer. The in-line flow divert valve is used to ensure that undesirable materials eluting at the solvent front are diverted to waste to keep the ion source from becoming contaminated. When the valve is in the sampling position, the mobile phase is passed directly into the ion source without splitting. Xu et al.[35] reported that over 100,000 samples were analyzed for early ADME assessment by successful incorporation of this parallel configuration. The system was used for assessing the time-course metabolic stability (4 time-points in triplicate) for hits identified from screening of lead generation libraries. For each compound, a total of 12 samples are generated (4 time-points in triplicate) and a single plate of compounds therefore yields a total of 1152 samples requiring analysis. Single-column systems operating in the sequential sampling mode are capable of analyzing roughly one plate in a single day (assuming a 1-min cycle time). On the other hand, a parallel array of eight columns enables up to 8 plates to be analyzed in a single day on a single instrument (theoretical maximum throughput).

The commercially available multiplexed (MUX) electrospray interface, which introduces multiple LC flows directly into an "indexed" electrospray ion source, has also been applied successfully for high-throughput ADME applications. Using a triple quadrupole mass spectrometer with MUX interface, Yang[49] identified the main advantage of parallel LC-MS/MS as being four times the throughput relative to single-column systems. However, disadvantages were reported as (a) cross-talk between the sprayers (negligible at concentrations <100 ng/mL but as high as 0.08% at 1000 ng/mL), (b) sensitivity less than that of a single sprayer interface (about three times lower than the single sprayer interface), and (c) total cycle time longer than that of a single sprayer interface (hence not compatible with ultra-fast chromatography). The MUX technique was validated for rabbit, rat, mouse, and dog plasma, and the authors concluded that the technique is well-suited for simultaneous method validations and early discovery studies.

Recently, "massively" parallel nano-LC systems have been commercialized and offer the potential for yet higher throughput ADME applications. Recently, a 24-channel microfluidic HPLC system was introduced for high throughput analysis.[51] Fractions are collected off-line and then analyzed by high throughput flow injection mass spectrometry or fast HPLC. In principle, these microfluidic systems should be capable of performing 48 or even 96 chromatographic separations in parallel and may be further coupled with emerging high-throughput techniques, such as matrix-assisted laser desorption ionization (MALDI) (1 s per sample).

21.5.2 Quantitative MALDI

MALDI coupled with triple quadrupole mass spectrometry has been shown recently to be a potentially ultra-high-throughput technique for small molecule quantification. MALDI analyses can be achieved in 1 s per sample with adequate sensitivity, accuracy and precision for *in vitro* ADME quantification, as presented at the *55th ASMS Conference on Mass Spectrometry and Allied Topics*.[52,53] For quantitation by MALDI, microsomal incubations are performed and the reactions quenched by acetonitrile precipitation. Immediately afterwards, the samples are applied to 96- or 384-well SPE clean-up and the eluate applied to the MALDI target. Figure 21.10 shows an example of microsomal stability analysis by the

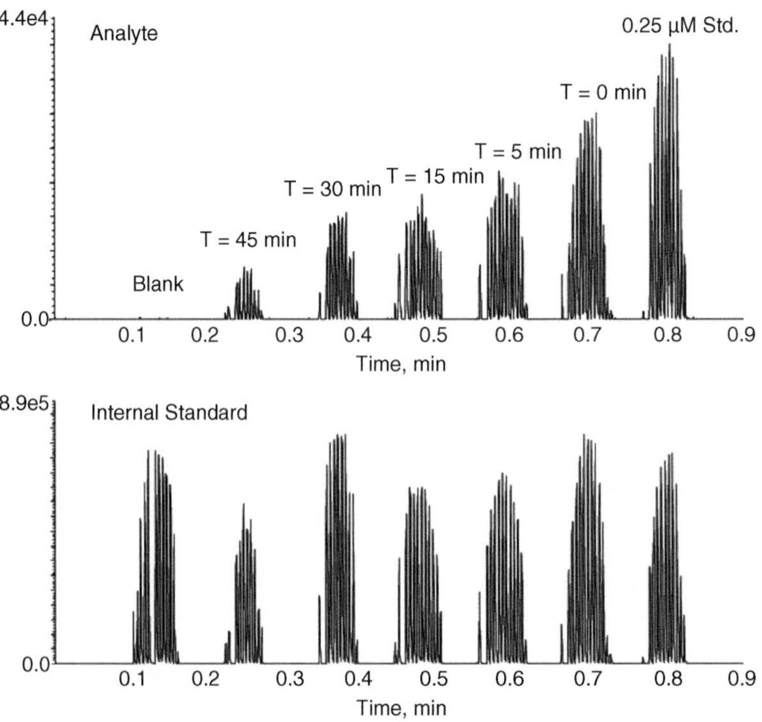

Figure 21.10. Example of a microsomal incubation time course evaluated by MALDI using a triple quadrupole mass spectrometer operated in the MRM mode. Following incubation with liver microsomes, the reactions were quenched, samples were cleaned-up by off-line SPE, and then the eluate spotted onto a MALDI target where it was mixed with matrix. Analyte signal was measured across the entire cystal surface, resulting in a chromatographic-like signal that is amenable to peak integration. (Reprinted with permission from *Anal. Chem.* **2005**, *77*, 5643–5654.)

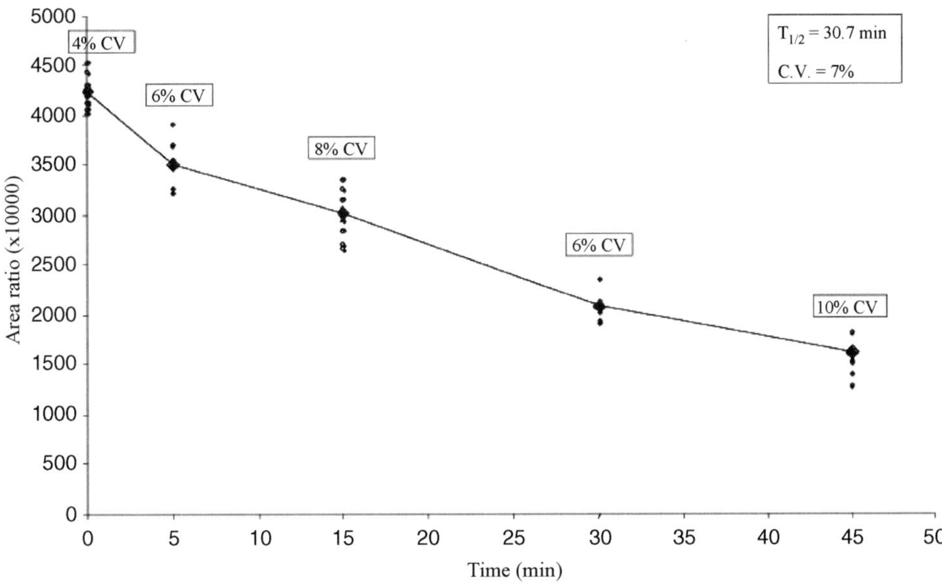

Figure 21.11. Time-course metabolic stability plot for an analyte spotted onto a MALDI target and quantified by MRM on a triple quadrupole mass spectrometer. (Reprinted with permission from *Anal. Chem.* **2005**, *77*, 5643–5654.)

MALDI MRM method. Microsomal incubations are terminated by addition of acetonitrile at predefined time-points—in this case, 5 min, 15 min, 30 min, and 45 min. Following off-line SPE clean-up, each of the samples is combined with MALDI matrix and is dried, and resultant crystals form. The laser rasters across each of the crystals at approximately 1-s intervals, thereby generating a chromatographic-like peak, where the analyte is found in lowest abundance at the periphery of the crystal and in highest abundance at or around the center of the spot. An internal standard, typically added at the time of reaction quenching, shows good sample-to-sample response. Peak integration is performed and the results reproduced as an analyte response versus time profile, as shown in Figure 21.11. The coefficients of variation are reported, showing that the method can be applied successfully to semiquantitation of analytes derived from complex matrices.

The primary drawback to quantitation by MALDI is that off-line sample preparation (SPE) is required. Salts and matrix interferences need to be removed prior to analyte analysis due to the potential for ion suppression effects, as well as possible deleterious effects on crystal formation, a prerequisite for successful MALDI analysis. Owing to the fact that off-line clean-up is required, it is unclear whether MALDI then offers a substantial advantage over high-throughput flow injection analysis whereby samples can be analyzed on a similar timescalc. An alternative, potentially significantly less expensive approach to MALDI, multiple reaction monitoring (MRM) quantitation, is to perform parallel flow injection analysis (FIA)-MS/MS. Previously, Kassel and co-workers[54] and, later, Morand et al.[55] showed that multiple injectors could be placed in series and HT FIA MS analysis achieved on a timescale comparable with this newly introduced MALDI methodology (2–3 s per sample with a total analysis time of 5–10 min per microtiter plate). Owing to the acceptable sample clean-up afforded by off-line, plate-based SPE systems, it should be possible to apply a high-throughput off-line SPE clean-up step followed by parallel flow injection analysis for evaluating metabolic stability and CYP inhibition.

21.5.3 Ultra-High Speed On-Line SPE/MS

Recently, a high-throughput on-line SPE/MS system (so-called RapidFire™), developed by BioTrove, was introduced as an alternative to the MALDI and parallel FIA-MS systems, both of which require off-line SPE clean-up. Samples in a microtiter plate are loaded one at a time onto a SPE cartridge (e.g., C18) at high flow rates (typically 1 mL/min). Sample loading, washing, elution, and SPE cartridge reequilibration are achieved in 6 s. The flow stream is continuously monitored using a triple quadrupole mass spectrometer operated in the MRM mode. The speed of the technique results from a novel sampling device coupled with high-speed switching valves to allow for rapid on-line clean-up and elution of target analyte into the mass spectrometer. The RapidFire™ technique was recently evaluated for CYP inhibition screening.[56] In this method, test compounds were incubated as discrete samples with the major CYP metabolizing enzymes at eight inhibitor concentrations; reactions were quenched by addition of acetonitrile and samples were pooled for cassette analysis. The effective run time per sample reported was less than 2 s, and analyses were carried out in less than one-tenth the time required by fast LC/MS/MS. Figure 21.12 shows the partial results from the RapidFire™ analysis of the pooled CYP inhibition screening samples. A total of eight replicate injections were made for each pool of six CYP reaction

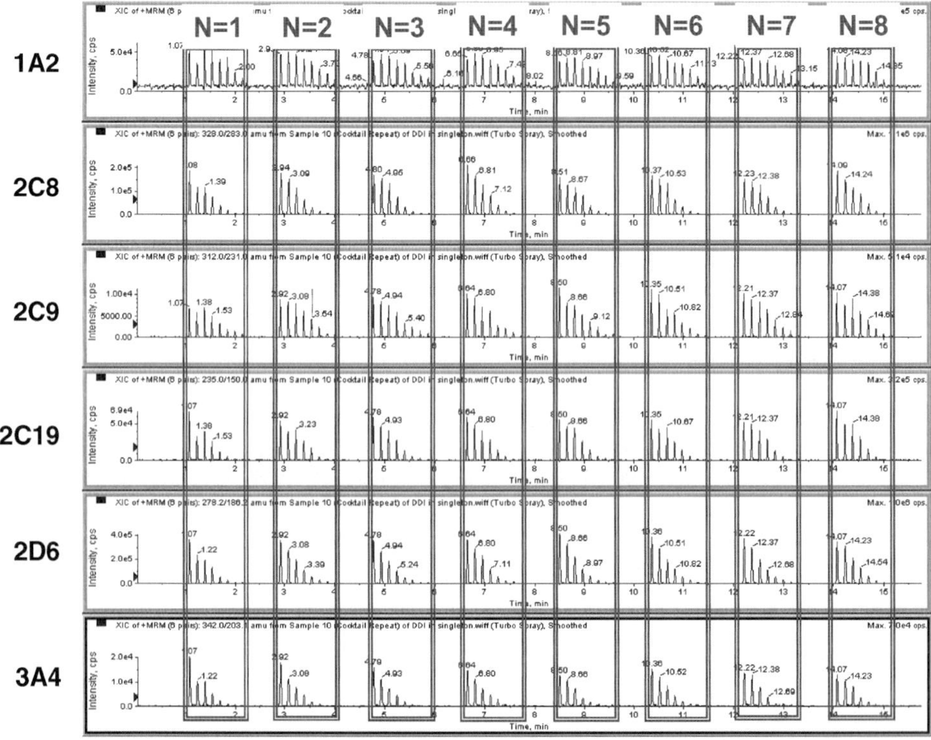

Figure 21.12. High-throughput on-line SPE/MS/MS of a cocktail of six CYP enzymes, probe substrates, and inibitors evaluated at eight different inhibitor concentrations. CYP enzymes were combined post-incubation at each inhibitor concentration and analyzed one at a time using ultra-fast on-line SPE/MS/MS. Each group of six samples (combining CYP1A2, CYP 2C8, CYP 2C9, CYP 2C19, CYP 2D6, CYP 3A4 incubations) was injected every 9 s, resulting in an effective run time of 1 s per sample.

products evaluated against test compound at eight inhibitor concentrations. Eight samples were analyzed in approximately 1 min and when evaluated in "pools" of six reaction mixtures, 64 independent measurements were made in the same amount of time, leading to an effective run time of <2 s per incubation.

The RapidFire™ method is well-suited to studies such as CYP inhibition screening and for monitoring any enzymatic reaction, because the same substrate/product pair is monitored for each incubation. Thus, it is possible to optimize the on-line SPE/MS/MS method for each specific product formed from an enzymatic assay. For studies requiring evaluation of the appearance/disappearance of the test compound (e.g., metabolic stability screening), an optimized SPE/MS/MS method for each compound must be developed, which may present challenges if the analytes being evaluated have substantially different physicochemical properties.

21.6 AUTOMATED DATA PROCESSING IS INSTRUMENTAL IN ACHIEVING HIGH-THROUGHPUT ADME

Although great strides have been made in reducing analysis times as described above, the true bottlenecks to providing rapid turnaround of *in vitro* ADME information to project teams are: liquid handling steps involved with sample incubations, data processing, data analysis and data reporting. The incorporation of robotic liquid handlers and automated sample prepartation is absolutely required to achieve high throughput. Equally, the generation of analytical data using automated instrumentation has produced a bottleneck since data can be generated faster than it can be analyzed. Automated data acquisition software and hardware has fueled the proliferation of mass spectrometry (MS)-based computer software applications to facilitate capture and analysis of mass spectral data and provide the information necessary for decision-making. The automated post-data acquisition analysis strategy is to extract the most appropriate information required for decision-making in as streamlined a manner as possible. For example, a time-course assessment of metabolic stability (e.g., four time-points and analyses in triplicate) generates 1152 samples for every plate of compounds submitted. Manual processing of so many samples would clearly render the data processing and data reporting rate-limiting. To address this, numerous groups have combined the power of vendor software programs that automate peak area determinations with visual basic programming to provide simple, yet elegant, methods for data processing and data reporting.[57]

21.7 "ALL-IN-ONE" METABOLIC STABILITY AND METABOLITE DETECTION STUDIES

In vitro metabolic stability assays are a valued first-pass assessment of potential metabolic liabilities. However, detailed information about the location of metabolic "soft-spots" is particularly useful in understanding whether the observed liabilities are specific to a molecule's core or introduced as part of a side chain in the lead optimization process. Whether the metabolic liability is associated with the core or side chain has clear implications for the degree to which the liability can be engineered out of a chemical series. Up until recently, metabolic stability screening and metabolite detection have been decoupled processes, that is, the metabolic stability assays are typically performed at a physiologically relevant concentration (e.g., 0.5–1 μM), and follow-up metabolite ID studies involve a second incubation at a higher substrate concentration to ensure generation of sufficiently high-quality MS/MS information to support structure elucidation. Sensitivity enhancements in mass spectrometry instrumentation have enabled metabolite ID information to be obtained from the same incubates used to assess metabolic stability; the first of such reports was by Kantharaj et al.[58] In particular, the authors showed for a number of cytochrome P450 substrates (verapamil, propanolol, cisapride, and flunariazine) that they were not only able to estimate metabolic turnover from a single run, but also able to identify major metabolites as well.

To perform detailed metabolism studies on candidate compounds has been a laborious, time-consuming task. Tuning, method setup, and assessing rates of metabolism by either selected ion monitoring (SIM), selected reaction monitoring (SRM), or precursor information-dependent acquisition (IDA) for metabolite ID typically have been distinct, independent processes requiring significant time and manual data interpretation by the investigator. In an effort to automate this process, prototype software was recently conceived and evaluated to automate metabolic stability assays by both Xu[59] and Detlefsen.[60] These software programs were developed to provide full automation of the following: (a) on-line quantitation to determine the rate of parent loss as a function of incubation time, (b) "intelligent" selection (i.e., qualitative trigger) of compounds for detailed metabolite ID (based on percent loss of parent at a fixed time point), (c) selection of a suitable product ion for metabolite determination using precursor ion scanning, (d) creation of a custom optimized MS1 and precursor IDA, (e) analysis of both sample and control, and (f) metabolite data analysis by Metabolite ID software. The software determines metabolically labile compounds by comparing the percent parent remaining at a specific time-point with user-defined criteria for triggering follow-up IDA acquisitions. The IDA data acquired allow the user to extract useful metabolite information using associated Metabolite ID software. Base peak chromatograms extracted from the precursor ion survey IDA experiment show a number of metabolites detected for the target compound. The analyst is then well-positioned to use this information to narrow down the search of peaks and retrieve IDA MSMS data for structural information.

Recently, with high-resolution MS/MS systems offering <5 ppm mass accuracies in both the MS and MS/MS mode, it has been shown that it is possible to extract and detect metabolites from very complex total ion current chromatograms (TIC) using the technique of mass defect filtering. Scientists from Bristol Myers Squibb showed that the majority of metabolites will add (or subtract) a mass to the precursor that results in a mass defect shift of less than 50 mDa relative to the mass defect of the precursor compound.[61] The total ion current chromatogram can be mass defect filtered, resulting in a highly enriched TIC, as shown in Figure 21.13. From the filtered TIC, it is possible to rapidly identify candidate metabolites (hydroxylated metabolites of buspirone shown in the lower chromatographic trace).

21.8 HIGH-THROUGHPUT PHARMACOKINETIC SCREENING

As stated earlier, combinatorial chemistry, parallel synthesis, and high-throughput screening have led to unprecendeted rates of hits and leads being generated, requiring "developability" assessment. First-tier *in vitro* ADME studies, such as metabolic stability screening, CYP inhibition screening, and so on, are the first steps toward optimizing the "drugability" of these hits and leads. The information from these *in vitro* screens allows for

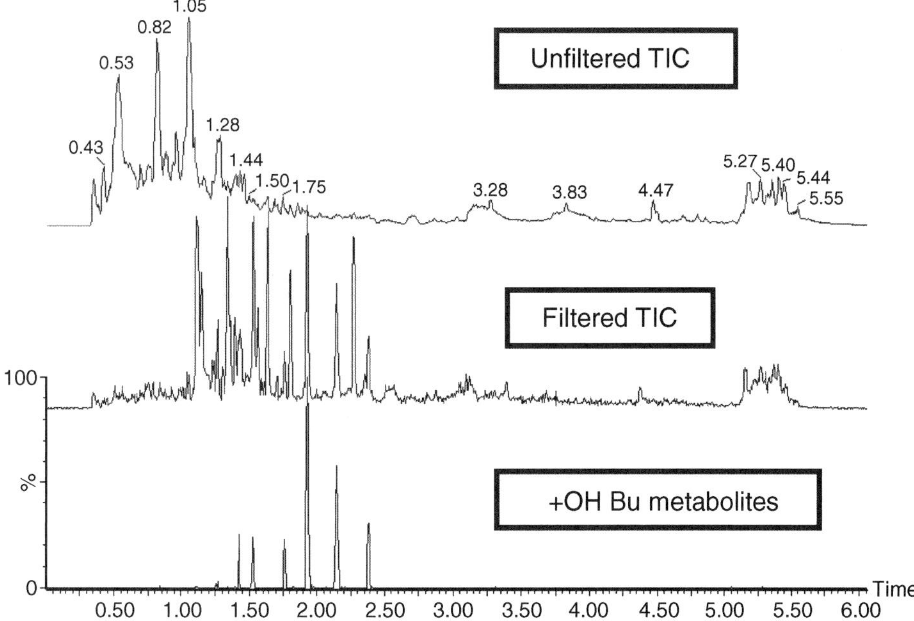

Figure 21.13. **Top**: Total ion current (TIC) chromatogram for a buspirone microsomal incubation. **Middle**: Filtererd TIC using mass-defect filtering. **Bottom**: Extracted ion chromatograms for hydroxylated metabolites of buspirone. (Reprinted with permisiion, *Drug Metab. Disp.***2006**, *34*(10), 1722–1733.)

prioritization of compounds for secondary ADME/Tox studies as well as *in vivo* pharmacokinetic screening. *In vivo* studies in animals are critical to advancing hits to leads, and leads to clinical candidates, because ultimately the *in vivo* animal models are still the best surrogates for predicting pharnacokinetics and pharmacodynamics in humans. Conventional DMPK methods are not high-throughput and have often impeded the fast development of new drug candidates. Mice and rats are classically used for *in vivo* PK studies. Interindividual variation is often high; for example, plasma concentrations could range twofold between animals, resulting in high variation in the data and difficulties in interpretation due to the differences in the expression of drug metabolism enzymes and genetic polymorphism. Thus, typical pharmacokinetic studies require dosing of compounds into multiple animals and taking multiple time-points, either from within the same animal (rat) or from multiple animals (mouse). During discovery, a typical and comprehensive study design is to evaluate the fate of a test compound administered to three animals via intravenous route of administration and three animals via the oral route of administration over a 24-h time course (typically 11 blood samples per animal plus a pre-dose blood draw to serve as a control). The result is 72 blood samples collected for each compound evaluated for *in vivo* pharmacokinetics. Many bioanalytical groups in drug discovery organizations (pharmaceutical and biotechnology) evaluate between 10 and 20 compounds per week (unpublished report, *55th ASMS Conference on Mass Spectrometry and Allied Topics, Drug Discovery Short Course*) with average turnaround times of 2–5 days upon receipt of plasma samples from the in-life groups. In addition to the numerous plasma samples requiring analysis, for each compound, an acceptable (optimized) bioanalytical method must be developed.

Table 21.2. Pooling Strategies for Pharmacokinetic Evaluation of Drug Candidates

Pooling Strategies	Disadvantages
• Cassette (*N*-in-1) dosing	• Drug–drug interaction potential
• Multi-analysis (pool after individual dosing)	• Reduced sensitivity and increased complexity
• Single sample screens (pooled or single C_{max} sample collected)	• Loss of information on shape of plasma concentration–time curve

21.8.1 Cassette Dosing

An approach to enhancing pharmacokinetic analysis throughput is to profile multiple compounds simultaneously, known as cassette (or N-in-1) dosing. In essence, cassette dosing is a compound "pooling" strategy whereby compounds are profiled as mixtures so as to increase throughput, reduce the total number of samples to be analyzed and hence reduce overall analysis times. Cassette dosing strategies have been used principally for rapid pharmacokinetic profiling of drug leads. Halm and Berman et al.[62] were the first to describe the application of cassette ("N-in-one") dosing to facilitate rapid pharmacokinetic screening. Olah et al.[63] evaluated and published a method for an $N = 22$ in a dog PK study. The number of compounds to pool is governed typically by sensitivity and solubility limits. As the number of compounds included in the "pool" increases, the concentration of each individual component is lowered and the greater the potential for synergistic and/or antagonistic effects. Additional methods for streamlining *in vivo* PK analysis are summarized in Table 21.2. Although analysis times may be reduced by these methods, it comes at a price.

The one principal drawback with the cassette dosing strategy is that the risk for drug–drug interactions is exacerbated, which can lead to both false positives and negatives.[64] Korfmacher et al.[65] effectively addressed this issue by implementing a a "cassette planning" strategy and devised the technique "Cassette Accelerated Rapid Rat Screen" ("CARRS") as a means for increasing *in vivo* pharmacokinetic throughput. Ostensibly, this approach can be described as one in which drug candidates are dosed individually ($n = 2$ rats per compound) in batches of six compounds per set, and then samples are pooled across the two rats to provide a smaller number (six per compound) of test samples for analysis. This method streamlines the process, from compound selection for PK screening to reporting of pharmacokinetic data. Kormacher reported that the method is being applied to evaluation of more than 70 compounds per week in Schering Plough's discovery programs (unpublished communication, *55th ASMS Conference on Mass Spectrometry and Allied Topics, Drug Metabolism Workshop*).

21.8.2 Faster and Higher-Resolution Pharmacokinetic Screening

The constant challenge is to strike the right balance between speed and quality. Short columns (2-mm-i.d. × 30-mm length) operated at high flow rates (1 mL/min) running generic (steep) gradients (5% to 95% acetonitrile in 1 min) are typical for pharmacokinetic analysis, with average run times of 2 min per sample (including sample loading and column reequilibration). The risk of using high-speed LC/MS/MS methods is coelution of either metabolites or matrix with the analyte of interest. Glucuronide metabolites are particularly problematic because they are susceptible to in-source fragmentation of the glucuronide moiety back to parent compound, resulting in an overestimate of analyte concentration. To reduce the risk of interferences due to plasma matrix, vehicle matrix, or metabolites, groups have begun to adopt UPLC/MS/MS as the method of choice for plasma bioanalysis. Again, by utilizing smaller dimension chromatographic particle sizes (below 2 μm) and by operating the columns at high pressure (approximately 15,000 psi), chromatographic resolution is greatly enhanced relative to the conventional high-speed LC/MS/MS methods, leading to better separation between analyte and interferent.

21.9 CONCLUSIONS

One of the greatest challenges of analytical chemistry in an era of high-throughput drug discovery has been to balance the need for high throughput while maintaining an analytical standard of high quality. Dramatic advances in sample analysis and purification throughput have been achieved incorporating fast chromatography coupled with mass spectrometry and by performing these analyses both in series and in parallel. Throughput has also been impacted dramatically by the ability to seamlessly link all of the processes downstream of library synthesis, including automating the uploading of sample lists for automated data acquisition, automating the assessment of compound purity, and automating post-purification analysis and associated sample handling. Fast emerging as the next bottleneck in drug discovery is to design and synthesize more "drug-like" molecules so as to increase the likelihood of their developmental success. Innovations in high-throughput *in vitro* and *in vivo* ADME analysis are beginning to be realized, too, with the introduction of fast serial and fast parallel chromatography coupled with mass spectrometry. As the amount of analytical data increases for each and every molecule synthesized, the key will be how to glean from this data the information content that ultimately accelerates the drug discovery process.

Early ADME assessment of compounds is occurring at nearly every stage of the discovery process, from lead generation through lead optimization. This has occurred for two principal reasons, the first being an enlightened view as to the importance of optimizing on drug-like properties in addition to potency and selectivity. Secondly, and perhaps more importantly, early ADME assessment is occurring due to innovations in analytical chemistry and the widespread proliferation of LC/MS/MS technology. Automated sample preparation, data acquisition, and data processing have enabled ADME profiling studies to move into the high-throughput realm. It was only a few years ago that it was suggested that high-throughput ADME would be difficult to achieve. Continued innovations by the analytical community have successfully challenged this view.

REFERENCES

1. Aebersold, R.; Mann, M. Mass spectrometry-based proteomics. *Nature* **2003**, *422*(6928), 198–207.
2. Xu, R.; Sang, B.-C.; Navre, M.; Kassel, D. B. Cell-based assay for screening 11b-hydroxysteroid inhibitors using liquid chromatography/tandem mass spectrometry detection. *Rapid Commun. Mass Spectrom.* **2006**, *20*, 1–5.
3. Rodrigues, A. D. New technologies and approaches for increasing drug candidate survivability: Lead identification to lead optimization. In Lee, M. (Ed.), CPSA Digest, Princeton, NJ, 2001, CPSA.
4. Kola, I.; Landis, J. Can the pharmaceutical industry reduce attrition rates? *Nat. Rev. Drug Discovery* **2004**, *3*, 711–715.

5. Thompson, T. N. Optimization of metabolic stability as a goal of modern drug design. *Med. Res. Rev.* 2001, *21,* 412–449.
6. Smith, D. A.; van de Waterbeemd, H. Pharmacokinetics and metabolism in early drug discovery. *Curr. Opin. Chem. Biol.* **1999**, *3*, 373–378.
7. Kerns, E. H.; Li, D. Pharmaceutical profiling in drug discovery. *Drug Discovery Today* **2003**, *8*, 316–323.
8. Frenette, R.; Blouin, M.; Brideau, C.; Chauret, N.; Ducharme, Y.; Friesen, R. W.; Hamel, P.; Jones, T. R.; Laliberte, F.; Li, C.; Masson, P.; McAuliffe, M.; Girard, Y. Substituted 4-(2,2-diphenylethyl)pyridine-*N*-oxides as phosphodiesterase-4 inhibitors: SAR study directed toward the improvement of pharmacokinetic parameters. *Bioorg. Med. Chem. Lett.* **2002**, *12*, 3009–3013.
9. Wyatt, P. G.; Allen, M. J.; Chilcott, J.; Jickin, G.; Miller, N. D.; Woollard, P. M. Structure–activity relationship investigations of a potent and selective benzodiazepine oxytocin antagonist. *Bioorg. Med. Chem. Lett.* **2001**, *11*, 1301–1305.
10. Caldwell, G. W. Compound optimization in early- and late-phase drug discovery: Acceptable pharmacokinetic properties utilizing combined physicochemical, *in vitro* and *in vivo* screens. *Curr. Opin. Drug Discovery* **2000**, *3*, 30–41.
11. Sinko, P. J. Drug selection in early drug development: screening for acceptable pharmacokinetic properties using combined *in vitro* and computational approaches. *Curr. Opin. Drug Discovery Dev.* **1999**, *2*, 42–48.
12. Rossi, D. T.; Sinz, M. *Mass Spectrometry in Drug Discovery*, Marcel Dekker, New York, 2002.
13. Ackermann, B. L.; Berna, M. J.; Murphy, A. T. Recent advances in use of LC/MS/MS for quantitative high-throughput bioanalytical support of drug discovery. *Curr. Top. Med. Chem.* **2002**, *2*, 53–66.
14. Rodrigues, A. D.; Lin, J. H. Screening of drug candidates for their drug–drug interaction potential. *Curr. Opin. Chem. Biol.* **2001**, *5*, 396–401.
15. Li, Y.; Shin, Y. G.; Kosmeder, J. W.; Hirschelman, W. H.; Pezzutto, J. M.; van Breemen, R. B. Increasing the throughput and productivity of Caco-2 cell permeability assays using liquid chromatography–mass spectrometry: Application to resveratrol absorption and metabolism. *Comb. Chem. High Throughput Screen* **2003**, *6*, 757–767.
16. Korfmacher, W. A.; Cox, K. A.; Bryant, M. S.; Veals, J.; Ng, K.; Watkins, R.; Lin, C. HPLC-API/MS/MS: A powerful tool for integrating drug metabolism into the drug discovery process. *Drug Discovery Today* **1997**, *2*, 532–537.
17. Eddershaw, P. J.; Dickins, M. Advances in drug metabolism screening. *Pharm. Sci. Technol. Today* **1999**, *2*, 13–19.
18. Fung, E. N.; Chen, Y.; Lau, H.; Y. Y. Semi-automatic high throughput determination of plasma protein binding using a 96-well plate filtrate assembly and fast liquid chromatography–tandem mass spectrometry. *J. Chromaotgr. B Anal. Technol. Biomed. Life Sci.* **2003**, *795*, 187–194.
19. Kariv, I.; Cao, H.; Oldenburg, K. R. Development of a high throughput equilibrium dialysis method. *J. Pharm. Sci.* **2001**, *90*, 580–587.
20. Qian, M. J. A fast screening method to measure equilibrium solubility in early drug discovery process. In *AAPS Annual Meeting and Exposition, Denver, CO, 2001. AAPS.*
21. Wilson, D. M.; Wang, X.; Walsh, E.; Rourick, R. A. High throughput log D determination using liquid chromatography-mass spectrometry. *Comb. Chem. High Throughput Screen.* **2001**, *4*, 511–519.
22. Stevenson, C. L.; Augustijns, P. F.; Hendren, R. W. Use of Caco-2 cells and LC/MS/MS to screen a peptide combinatorial library for permeable structures. *Int. J. Pharm.* **1999**, *177*, 103–115.
23. Janiszewski, J. S.; Rogers, K. J.; Whalen, K. M.; Cole, M. J.; Liston, T. E.; Duchoslav, E.; Fouda, H. G. A high-capacity LC/MS system for the bioanalysis of samples generated from plate-based metabolic screening. *Anal. Chem.* **2001**, *73*, 1495–1501.
24. Balimane, P. V.; Chong, S. Cell culture-based models for intestinal permeability: A critique. *Drug Discovery Today* **2005**, *10*, 335–343.
25. Kansy, M.; Avdeef, A.; Fischer, H. Advances in screening for membrane permeability: High-resolution PAMPA for medicinal chemists. *Drug Discovery Today: Technologies* **2005**, *1*, 349–355.
26. Avdeef, A. The Rise of PAMPA. *Expert Opin. Drug Metab. Toxicol.* **2005**, *1*, 325–342.
27. Avdeef, A.; Artursson, P.; Neuhoff, S.; Lazorova, L.; Grasjo, J.; Tavelin, S. Caco-2 permeability of weakly basic drugs predicted with the double-sink PAMPA pKa(flux) method. *Eur. J. Pharm. Sci.* **2005**, *24*, 333–349.
28. Danielson, P. B. The cytochrome P450 superfamily: biochemistry, evoluition and drug metabolism in humans. *Curr. Drug Metab.* **2002**, *3*, 561–597.
29. McCarver, D. G.; Hines, R. N. The ontogeny of human drug-metabolizing enzymes: Phase II conjugation enzymes and regulatory mechanisms. *J. Pharmacol. Exp. Ther.* **2002**, *300*, 361–366.
30. Obach, R. S. Prediction of human clearance of twenty-nine drugs from hepatic microsomal intrinsic clearance data: An examination of *in vitro* half-life approach and nonspecific binding to microsomes. *Drug Metab. Dispos.* **1999**, *27*, 1350–1359.
31. Thompson, T. N. Early ADME in support of drug discovery: The role of metabolic stability studies. *Curr. Drug Metab.* **2000**, *3*, 215–241.
32. Patten, C. J. New technologies for assessing UDP-glucuronosyltransferase (UGT) metabolism in drug discovery and development. *Drug Discovery Today* **2006**, *3*, 73–78.
33. Di, L.; Kerns, E. H.; Hong, Y.; Kleintop, T. A.; McConnel, O. J.; Huryn, D. M. Optimization of a higher throughput microsomal stability screening assay

for profiling drug discovery candidates. *J. Biomolec. Screen.* **2003**, *8*, 453–462.

34. Korfmacher, W. A.; Palmer, C. A.; Nardo, C.; Meynell, K. D.; Grotz, D.; Cox, K.; Lin, C. C.; Elicone, C.; Liu, C.; Duchoslav, E. Development of an automated mass spectrometry system for the quantitative analysis of liver microsomal incubation samples: A tool for rapid screening of new compounds for metabolic stability. *Rapid Commun. Mass Spectrom.* **1999**, *13*, 901–907.
35. Xu, R.; Nemes, C.; Jenkins, K. M.; Rourick, R. A.; Kassel, D. B. Application of parallel liquid chromatography/mass spectrometry for high throughput microsomal stability screening of compound libraries. *J. Am. Soc. Mass Spectrom.* **2002**, *13*, 155–165.
36. van de Merbel, N. C.; Poelman, H. Experiences with monolithic LC phases in quantitative bioanalysis. *J. Pharm. Biomed. Anal.* **2003**, *33*, 495–504.
37. Lim, K.; Manuel, M.; Pang, S.; Cramlett, J.; Hascal, D.; Kassel, D. B. Pooling strategy to increase throughput and reduce interference for microsomal stability assay by high-resolution chromatography/tandem mass spectrometry. In *54th ASMS Conference on Mass Spectrometry and Allied Topics*, Seattle, WA, 2006. ASMS.
38. Cohen, L. H.; Remley, M. J.; Raunig, D.; Vaz, A. D. N. In vitro drug interactions of cytochrome P450: An evaluation of fluorogenic to conventional substrates. *Drug Metab. Dispos.* **2003**, *31*, 1005–1015.
39. Bu, H. Z.; Magis, L.; Knuth, K.; Teitelbaum, P. High-throughput cytochrome P450 (CYP) inhibition screening via cassette probe-dosing strategy. I. Development of direct injection/on-line guard cartridge extraction/tandem mass spectrometry for the simultaneous detection of CYP probe substrates and their metabolites. *Rapid Commun. Mass Spectrom.* **2000**, *14*, 1619–1624.
40. Dierks, E. A.; Stams, K. R.; Lim, H.-K.; Cornelius, G.; Zhang, H.; Ball, S. E. A method for the simultaneous evaluation of the activities of seven major human drug-metabolizing cytochrome P450s using an *in vitro* cocktail of probe substrates and fast gradient liquid chromatography tandem mass spectrometry. *Drug Metab. Dispos.* *29*, **2001**, 23–29.
41. Scott, R. J.; Palmer, J.; Lewis, I. A.; Pleasance, S. Determination of a "GW cocktail" of cytochrome P450 probe substrates and their metabolites in plasma and urine using automated solid phase extraction and fast gradient liquid chromatography tandem mass spectrometry. *Rapid Commun. Mass Spectrom.* **1999**, *13*, 2305–2319.
42. Walsky, R. L.; Obach, R. S. Validated assays for human cytochrome P450 activities. *Drug Metab. Dispos.* **2004**, *32*, 647–660.
43. Weaver, R.; Graham, K. S.; Beattie, I. G.; Riley, R. J. CYTOCHROME P450 inhibition using recombinant proteins and mass spectrometry/multiple reaction monitoring technology in a cassette incubation. *Drug Metab. Dispos.* **2003**, *31*, 955–966.
44. Walsky, R. L.; Gaman, E. A.; Obach, R. S. Examination of 209 drugs for inhibition of cytochrome P450 2C8. *J. Clin. Pharmacol.* **2005**, *45*, 68–78.
45. Peng, S. X.; Barbone, A.; Ritchie, D. M. High-throughput cytochrome P450 inhibition assays by ultrafast gradient liquid chromatography with tandem mass spectrometry using monolithic columns. *Rapid Commun. Mass Spectrom.* **2003**, *17*, 509–518.
46. Kajosaari, L. I.; Jaakkola, T.; Neuvonen, P. J.; Backman, J. T. Pioglitazone, an *in vitro* inhibitor of CYP2C8 and CYP3A4 does not increase the plasma concentrations of the CYP2C8 and CYP3A4 substrate repaglinide. *Eur. J. Clin. Pharm.* **2006**, *62*, 217–223.
47. Hiller, D. L. High throughput quantitation using indexed multiprobe electrospray technology in support of drug discovery. In *48th ASMS Conference on Mass Spectrometry and Allied Topics*, Long Beach, CA, **2000**.
48. Bayliss, M. K,; Little, D.; Mallett, D. N.; Plumb, R. S. Parallel ultra-high flow rate liquid chromatography with mass spectrometric detection using a multiplex electrospray source for direct, sensitive determination of pharmaceuticals in plasma at extremely high throughput. *Rapid Commun. Mass Spectrom.* **2000**, *14*, 2039–2045.
49. Yang, L.; Mann, T. D.; Little, D.; Wu, , N.; Clement, R. P.; Rudewicz, P. J. Evaluation of a four-channel multiplexed electrospray triple quadrupole mass spectrometer for the simultaneous validation of LC/MS/MS methods in four different preclinical matrixes. *Anal. Chem.* **2001**, *73*, 1740–1747.
50. Korfmacher, W. A.; Veals, J.; Dunn-Meynell, K.; Zhang, X.; Tucker, G.; Cox, K. A.; Lin, C.-C. Demonstration of the capabilities of a parallel high performance liquid chromatography tandem mass spectrometry system for use in the analysis of drug discovery plasma samples. *Rapid Commun Mass Spectrom.* **1999**, *13*, 1991–1998.
51. Lloyd, T. L.; Dean, D.; Guazzotti, S.; Barbero, R. Not so fast: A novel way to employ longer HPLC separations without slowing down your ms. In *55th ASMS Conference on Mass Spectrometry and Allied Topics*, Indianapolis, IN, June 2007, book of abstracts.
52. Gobey, J.; Cole, M.; Janiszewski, J.; Covey, T.; Chau, T.; Kovarik, P.; Corr, J. Characterization and performance of MALDI on a triple quadrupole mass spectrometer for analysis and quantification of small molecules. *Anal. Chem.* *77*, **2005**, 5643–5654.
53. Zhong, F.; Scott, G.; Labre, D.; Ghobarah, H.; Corr, J. Evaluation of MALDI triple quadrupole mass spectrometry for high throughput drug–drug interaction screening. In *55th ASMS Conference on Mass Spectrometry and Allied Topics*, Indianapolis, IN, June 2007, book of abstracts.
54. Wang, T.; Zeng, L.; Strader, T.; Burton, L.; Kassel, D. B. A new ultra-high throughput method for characterizing combinatorial libraries incorporating a multiple probe autosampler coupled with flow injection mass spectrometry analysis. *Rapid Commun. Mass Spectrom.* **1998**, *12*(16), 1123–1129.

55. Morand, K. L.; Burt, T. M.; Regg, B. T.; Chester, T. Techniques for increasing the throughput of flow injection mass spectrometry. *Anal. Chem.* **2001**, *73*(2), 247–252.
56. Kassel, D. B.; Lim, K.; Ozbal, C. C.; LaMarr, W. A.; Development of an ultra high throughput MS/MS CYP inhibition assay, In *55th ASMS Conference on Mass Spectrometry and Allied Topics*, Indianapolis, IN, June 2007, book of abstracts.
57. Williams, A. Applications of computer software for the interpretation and management of mass spectrometry data in pharmaceutical science. *Curr. Top. Med. Chem.* **2002**, *2*, 99–107.
58. Kantharaj, E.; Tuytelaars, A.; Proost Pascale, E. A.; Ongel, Z.; Van Assouw, H. P.; Gilssen, R. A. H. J. Simultaneous measurement of drug metabolic stability and identification of metabolites using ion-trap mass spectrometry. *Rapid Commun. Mass Spectrom.* **2003**, *17*, 2661–2668.
59. Xu, R.; Duchoslav, E.; Aparicio, A.; Jones, E. B.; Tita, S.; Kassel, D. B. Streamlined approaches to metabolic stability assessment and metabolite profiling in drug discovery. In *51st ASMS Conference on Mass Spectrometry and Allied Topics*, Montreal, Canada, 2003. ASMS.
60. Detlefsen, D. J.; Whitney, J. L.; Hail, M. E.; Joseph, J. L.; Sanders, M.; Nugent, K. D. A total analysis solution for metabolic stability and detailed metabolite profiling. In *51st ASMS Conference on Mass Spectrometry and Allied Topics*, Montreal, Canada, 2003. ASMS.
61. Zhu, M.; Ma, L.; Zhang, D.; Ray, K.; Zhao, W.; Humphreys, W. G.; Skiles, G.; Sanders, M.; Zhang, H. Detection and characterization of metabolites in biological matrices using mass defect filtering of liquid chromatography/high resolution mass spectrometry data. *Drug Metab. Dispos.* **2006**, *34*, 1722–1733.
62. Berman, J.; Halm, K.; Adkinson, K.; Shaffer, J. E. Simultaneous pharmacokinetic screening of a mixture of compounds in the dog using API LC/MS/MS analysis for increased throughput. *J. Med. Chem.* **1997**, *40*, 827–829.
63. Olah, T. V.; McLoughlin, D. A.; Gilbert, J. D. The simultaneous determination of mixtures of drug candidates by liquid chromatography/atmospheric pressure chemical ionization mass spectrometry as an *in vivo* drug screening procedure. *Rapid Commun. Mass Spectrom.* **1997**, *11*, 17–23.
64. White, R. E.; Manitpisitkul, P. Pharmacokinetic theory of cassette dosing in drug discovery screening. *Drug Metab Dispos.* **2001**, *29*, 957–966.
65. Korfmacher, W. A.; Cox, K. A.; Ng, K. J.; Veals, J.; Hsieh, Y.; Wainhaus, S.; Broske, L.; Prelusky, D.; Nomeir, A.; White, R. E. Cassette-accelerated rapid rat screen: A systematic procedure for the dosing and liquid chromatography/atmospheric pressure ionization tandem mass spectrometric analysis of new chemical entities as part of new drug discovery. *Rapid Commun. Mass Spectrom.* **2001**, *15*(5), 335–340.

Index

Electrospray and MALDI Mass Spectrometry: Fundamentals, Instrumentation, Practicalities, and Biological Applications, Second Edition, Edited by Richard B. Cole